DRUG DISCOVERY TOXICOLOGY

DRUG DISCOVERY TOXICOLOGY

From Target Assessment to Translational Biomarkers

Edited by

YVONNE WILL
J. ERIC McDUFFIE
ANDREW J. OLAHARSKI
BRANDON D. JEFFY

Published by John Wiley & Sons, Inc., Hoboken, New Jersey
Published simultaneously in Canada

For general information on our other products and services or for technical support, please contact our Customer Care Department within the United States at (800) 762-2974, outside the United States at (317) 572-3993 or fax (317) 572-4002.

Wiley also publishes its books in a variety of electronic formats. Some content that appears in print may not be available in electronic formats. For more information about Wiley products, visit our web site at www.wiley.com.

Library of Congress Cataloging-in-Publication Data:

Names: Will, Yvonne, editor.
Title: Drug discovery toxicology : from target assessment to translational biomarkers / edited by Yvonne Will [and three others].
Description: Hoboken, New Jersey : John Wiley & Sons, Inc., [2016] | Includes bibliographical references and index.
Identifiers: LCCN 2015039627 (print) | LCCN 2015050089 (ebook) | ISBN 9781119053330 (cloth) | ISBN 9781119053323 (Adobe PDF) | ISBN 9781119053392 (ePub)
Subjects: LCSH: Drugs–Toxicology. | Drugs–Testing. | Toxicity testing. | High throughput screening (Drug development)
Classification: LCC RA1238 .D75 2016 (print) | LCC RA1238 (ebook) | DDC 615.9/02–dc23
LC record available at http://lccn.loc.gov/2015039627

Printed in the United States of America

10   9   8   7   6   5   4   3   2

CONTENTS

LIST OF CONTRIBUTORS xxi

FOREWORD xxv

PART I INTRODUCTION 1

1 Emerging Technologies and their Role in Regulatory Review 3
  Thomas J. Colatsky

 1.1 Introduction 3
 1.2 Safety Assessment in Drug Development and Review 4
  1.2.1 Drug Discovery 4
  1.2.2 Preclinical Development 5
 1.3 The Role of New Technologies in Regulatory Safety Assessment 6
  1.3.1 *In Silico* Models for Toxicity Prediction 6
  1.3.2 Cell-Based Assays for Toxicity Prediction 7
 1.4 Conclusions 8
 References 8

PART II SAFETY LEAD OPTIMIZATION STRATEGIES 13

2 Small-Molecule Safety Lead Optimization 15
  Donna M. Dambach

 2.1 Background and Objectives of Safety Lead Optimization Approaches 15
 2.2 Target Safety Assessments: Evaluation of Undesired Pharmacology and
  Therapeutic Area Considerations 16
 2.3 Implementing Lead Optimization Strategies for Small Molecules 16
  2.3.1 Strategic Approach 17
  2.3.2 Application of Prospective Models 17
  2.3.3 Application of Retrospective Models 22
 2.4 Conclusions 23
 References 23

**3   Safety Assessment Strategies and Predictive Safety of
Biopharmaceuticals and Antibody Drug Conjugates**   **27**

Michelle J. Horner, Mary Jane Hinrichs and Nicholas Buss

3.1   Background and Objectives   27
3.2   Target Safety Assessments: Strategies to Understand Target Biology
      and Associated Liabilities   28
      3.2.1   Target Safety Assessment for Biopharmaceuticals Targeting
              the Immune System   28
3.3   Strategic Approaches for Biopharmaceuticals and ADCs   29
      3.3.1   Modality-Associated Risks   29
      3.3.2   mAbs   29
      3.3.3   ADCs   30
      3.3.4   On-Target Toxicity   30
      3.3.5   Off-Target Toxicity   32
      3.3.6   Evaluation of Novel Warheads   32
      3.3.7   Evaluation of New ADC Technologies   33
3.4   Predictive Safety Tools for Large Molecules   33
      3.4.1   Immunogenicity   33
      3.4.2   Specialized Assays for Detection of ADCC, CDC, and ADCP   33
      3.4.3   Immunotoxicity Testing   34
      3.4.4   Predicting and Assessing Unintended Adverse Consequences   34
3.5   Strategies for Species Selection   34
3.6   Strategy for Dose-Ranging Studies for Safety Evaluation of Biopharmaceuticals   35
3.7   Conclusions   35
      References   36

4   Discovery and Development Strategies for Small Interfering RNAs   **39**

Scott A. Barros and Gregory Hinkle

4.1   Background   39
      4.1.1   RNAi Molecular Mechanism   39
      4.1.2   Conjugate siRNAs for Hepatic Targets   39
4.2   Target Assessments   40
      4.2.1   Large Gene Families   40
      4.2.2   Short Transcripts   40
      4.2.3   Genes with Rapid mRNA Turnover   40
      4.2.4   Selecting among Alternate Transcript Variants   41
4.3   siRNA Design and Screening Strategies   41
      4.3.1   siRNA Design   41
      4.3.2   Chemical Modification of siRNA   42
      4.3.3   Screening of siRNA Therapeutics   42
4.4   Safety Lead Optimization of siRNA   45
      4.4.1   Immunostimulation Screening   45
      4.4.2   Toxicology Screening in Rodents   46
      4.4.3   Points to Consider for Chemically Modified Nucleotides   47
4.5   Integration of Lead Optimization Data for Candidate Selection and Development   48
4.6   Conclusions   49
      References   49

PART III   BASIS FOR *IN VITRO–IN VIVO* PK TRANSLATION   **53**

5   Physicochemistry and the Off-Target Effects of Drug Molecules   **55**

Dennis A. Smith

5.1   Lipohilicity, Polar Surface Area, and Lipoidal Permeability   55
5.2   Physicochemistry and Basic ADME Properties for High Lipoidal
      Permeability Drugs   56

5.3   Relationship between Volume of Distribution (V_d) and Target Access
        for Passively Distributed Drugs   58
5.4   Basicity, Lipophilicity, and Volume of Distribution as a Predictor
        of Toxicity (T): Adding The T to ADMET   59
5.5   Basicity and Lipophilicity as a Predictor of Toxicity (T):
        Separating the D from T in ADMET   60
5.6   Lipophilicity and PSA as a Predictor of Toxicity (T): Adding the T to ADMET   60
5.7   Metabolism and Physicochemical Properties   61
5.8   Concentration of Compounds by Transporters   61
5.9   Inhibition of Excretion Pumps   63
5.10   Conclusions   64
        References   65

**6   The Need for Human Exposure Projection in the Interpretation
      of Preclinical *In Vitro* and *In Vivo* ADME Tox Data   67**
      Patrick Poulin

6.1   Introduction   67
6.2   Methodology Used for Human PK Projection in Drug Discovery   67
        6.2.1   Prediction of Plasma Concentration–Time Profile by
                  Using the Wajima Allometric Method   68
        6.2.2   Prediction of Plasma and Tissue Concentration–Time
                  Profiles by Using the PBPK Modeling Approach   68
        6.2.3   Integrative Approaches of Toxicity Prediction Based on the
                  Extent of Target Tissue Distribution   70
6.3   Summary of the Take-Home Messages from the Pharmaceutical
        Research and Manufacturers of America CPCDC Initiative
        on Predictive Models of Human PK from 2011   72
        6.3.1   PhRMA Initiative on the Prediction of CL   75
        6.3.2   PhRMA Initiative on the Prediction of Volume of Distribution 75
        6.3.3   PhRMA Initiative on the Prediction of Concentration–Time Profile   75
        6.3.4   Lead Commentaries on the PhRMA Initiative   76
        References   77

7   ADME Properties Leading to Toxicity   82
      Katya Tsaioun

7.1   Introduction   82
7.2   The Science of ADME   83
7.3   The ADME Optimization Strategy   83
7.4   Conclusions and Future Directions   89
        References   90

PART IV   PREDICTING ORGAN TOXICITY   93

8   Liver   95
      *J. Gerry Kenna, Mikael Persson, Scott Q. Siler, Ke Yu, Chuchu Hu, Minjun Chen,
      Joshua Xu, Weida Tong, Yvonne Will and Michael D. Aleo*

8.1   Introduction   95
8.2   DILI Mechanisms and Susceptibility   96
8.3   Common Mechanisms that Contribute to DILI   98
        8.3.1   Mitochondrial Injury   98
        8.3.2   Reactive Metabolite-Mediated Toxicity   100
        8.3.3   BSEP Inhibition   102
        8.3.4   Complicity between Dual Inhibitors of BSEP
                  and Mitochondrial Function   105
8.4   Models Systems Used to Study DILI   108

	8.4.1	High Content Image Analysis	108
	8.4.2	Complex Cell Models	110
	8.4.3	Zebrafish	111
8.5		*In Silico* Models	114
8.6		Systems Pharmacology and DILI	118
8.7		Summary	119
		References	121

9 Cardiac   **130**

David J. Gallacher, Gary Gintant, Najah Abi-Gerges, Mark R. Davies, Hua Rong Lu,
Kimberley M. Hoagland, Georg Rast, Brian D. Guth, Hugo M. Vargas and Robert L. Hamlin

9.1		General Introduction	130
9.2		Classical *In Vitro/Ex Vivo* Assessment of Cardiac Electrophysiologic Effects	133
	9.2.1	Introduction	133
	9.2.2	Subcellular Techniques	134
	9.2.3	Ionic Currents	134
	9.2.4	AP/Repolarization Assays	135
	9.2.5	Proarrhythmia Assays	136
	9.2.6	Future Directions: Stem Cell-Derived CMs	136
	9.2.7	Conclusions	136
9.3		Cardiac Ion Channels and *In Silico* Prediction	137
	9.3.1	Introduction	137
	9.3.2	High-Throughput Cardiac Ion Channel Data	137
	9.3.2	*In Silico* Approaches	137
9.4		From Animal *Ex Vivo/In Vitro* Models to Human Stem Cell-Derived CMs for Cardiac Safety Testing	140
	9.4.1	Introduction	140
	9.4.2	Currently Available Technologies	140
	9.4.3	Conclusions	141
9.5		*In Vivo* Telemetry Capabilities and Preclinical Drug Development	141
	9.5.1	Introduction	141
	9.5.2	CV SP Evaluations Using Telemetry	142
	9.5.3	Evaluation of Respiratory Function Using Telemetry	143
	9.5.4	Evaluation of CNS Using Telemetry	143
	9.5.5	Evaluation of Other Systems Using Telemetry	143
9.6		Assessment of Myocardial Contractility in Preclinical Models	144
	9.6.1	Introduction	144
	9.6.2	Gold Standard Approaches	144
	9.6.3	*In Vitro* and *Ex Vivo* Assays	145
	9.6.4	*In Vivo* Assays	145
	9.6.5	Translation to Clinic	146
9.7		Assessment of Large Versus Small Molecules in CV SP	147
	9.7.1	Introduction	147
	9.7.2	CV SP Evaluation	147
9.8		Patients do not Necessarily Respond to Drugs and Devices as do Genetically Identical, Young Mature, Healthy Mice!	148
	9.8.1	Conclusions	152
		References	152

10 Predictive *In Vitro* Models for Assessment of Nephrotoxicity and Drug–Drug Interactions *In Vitro*   **160**

Lawrence H. Lash

10.1		Introduction	160
	10.1.1	Considerations for Studying the Kidneys as a Target Organ for Drugs and Toxic Chemicals	160

10.1.2 Advantages and Limitations of *In Vitro* Models
in General for Mechanistic Toxicology and Screening
of Potential Adverse Effects 161
10.1.3 Types of *In Vitro* Models Available for Studying Human Kidneys 162
10.2 Biological Processes and Toxic Responses of the Kidneys that are
Normally Measured in Toxicology Research and Drug
Development Studies 163
10.3 Primary Cultures of hPT Cells 164
10.3.1 Methods for hPT Cell Isolation 164
10.3.2 Validation of hPT Primary Cell Cultures 165
10.3.3 Advantages and Limitations of hPT Primary Cell Cultures 165
10.3.4 Genetic Polymorphisms and Interindividual Susceptibility 166
10.4 Toxicology Studies in hPT Primary Cell Cultures 166
10.5 Critical Studies for Drug Discovery in hPT Primary Cell Cultures 168
10.5.1 Phase I and Phase II Drug Metabolism 168
10.5.2 Membrane Transport 168
10.6 Summary and Conclusions 168
10.6.1 Advantages and Limitations of Performing Studies
in hPT Primary Cell Cultures 168
10.6.2 Future Directions 169
References 170

11 Predicting Organ Toxicity *In Vitro*: Bone Marrow **172**
Ivan Rich and Andrew J. Olaharski

11.1 Introduction 172
11.2 Biology of the Hematopoietic System 172
11.3 Hemotoxicity 173
11.4 Measuring Hemotoxicity 173
11.4.1 Uses of the CFC Assay 173
11.4.2 *In Vitro/In Vivo* Concordance 175
11.4.3 Limitations of the CFC Assay 175
11.5 The Next Generation of Assays 175
11.6 Proliferation or Differentiation? 175
11.7 Measuring and Predicting Hemotoxicity *In Vitro* 176
11.8 Detecting Stem and Progenitor Cell Downstream Events 177
11.9 Bone Marrow Toxicity Testing During Drug Development 177
11.10 Paradigm for *In Vitro* Hemotoxicity Testing 178
11.11 Predicting Starting Doses for Animal and Human Clinical Trials 179
11.12 Future Trends 179
11.13 Conclusions 180
References 180

12 Predicting Organ Toxicity *In Vitro*: Dermal Toxicity **182**
Patrick J. Hayden, Michael Bachelor, Mitchell Klausner and Helena Kandárová

12.1 Introduction 182
12.2 Overview of Drug-Induced Adverse Cutaneous Reactions 182
12.3 Overview of *In Vitro* Skin Models with Relevance to
Preclinical Drug Development 183
12.4 Specific Applications of *In Vitro* Skin Models and Predictive
In Vitro Assays Relevant to Pharmaceutical Development 184
12.4.1 Skin Sensitization 184
12.4.2 Phototoxicity 185
12.4.3 Skin Irritation 187
12.5 Mechanism-Based Cutaneous Adverse Effects 187
12.5.1 Percutaneous Absorption 187

		12.5.2	Genotoxicity	188
		12.5.3	Skin Lightening/Melanogenesis	188
	12.6	Summary		188
	References			189

13   *In Vitro* Methods in Immunotoxicity Assessment   193

Xu Zhu and Ellen Evans

	13.1	Introduction and Perspectives on *In Vitro* Immunotoxicity Screening	193
	13.2	Overview of the Immune System	194
	13.3	Examples of *In Vitro* Approaches	196
		13.3.1 Acquired Immune Responses	196
		13.3.2 Fcγ Receptor and Complement Binding	196
		13.3.3 Assessment of Hypersensitivity	196
		13.3.4 Immunogenicity of Biologics	198
		13.3.5 Immunotoxicity Due to Myelotoxicity	198
	13.4	Conclusions	198
	References		199

14   Strategies and Assays for Minimizing Risk of Ocular Toxicity during Early Development of Systemically Administered Drugs   201

Chris J. Somps, Paul Butler, Jay H. Fortner, Keri E. Cannon and Wenhu Huang

	14.1	Introduction	201
	14.2	*In Silico* and *In Vitro* Tools and Strategies	201
	14.3	Higher-Throughput *In Vivo* Tools and Strategies	202
		14.3.1 Ocular Reflexes and Associated Behaviors	202
		14.3.2 Noninvasive Ophthalmic Examinations	206
	14.4	Strategies, Gaps, and Emerging Technologies	208
		14.4.1 Strategic Deployment of *In Silico, In Vitro,* and *In Vivo* Tools	208
		14.4.2 Emerging Biomarkers of Retinal Toxicity	210
	14.5	Summary	210
	References		210

15   Predicting Organ Toxicity *In Vivo*—Central Nervous System   214

Greet Teuns and Alison Easter

	15.1	Introduction	214
	15.2	Models for Assessment of CNS ADRs	214
		15.2.1 *In Vivo* Behavioral Batteries	214
		15.2.2 *In Vitro* Models	215
	15.3	Seizure Liability Testing	216
		15.3.1 Introduction	216
		15.3.2 Medium/High Throughput Approaches to Assess Seizure Liability of Drug Candidates	216
		15.3.3 *In Vivo* Approaches to Assess Seizure Liability of Drug Candidates	217
	15.4	Drug Abuse Liability Testing	218
		15.4.1 Introduction	218
		15.4.2 Preclinical Models to Test Abuse Potential of CNS-Active Drug Candidates	219
	15.5	General Conclusions	222
		15.5.1 *In Vitro*	222
		15.5.2 *In Vivo*	223
	References		223

16 Biomarkers, Cell Models, and *In Vitro* Assays for Gastrointestinal Toxicology 227

Allison Vitsky and Gina M. Yanochko

16.1 Introduction 227
16.2 Anatomic and Physiologic Considerations 228
    16.2.1 Oral Cavity 228
    16.2.2 Esophagus 228
    16.2.3 Stomach 228
    16.2.4 Small and Large Intestine 229
16.3 GI Biomarkers 229
    16.3.1 Biomarkers of Epithelial Mass, Intestinal Function,
            or Cellular Damage 229
    16.3.2 Biomarkers of Inflammation 230
16.4 Cell Models of the GI Tract 231
    16.4.1 Cell Lines and Primary Cells 231
    16.4.2 Induced Pluripotent Stem Cells 232
    16.4.3 Coculture Systems 232
    16.4.4 3D Organoid Models 233
    16.4.5 Organs-on-a-Chip 235
16.5 Cell-Based *In Vitro* Assays for Screening and Mechanistic
     Investigations to GI Toxicity 235
    16.5.1 Cell Viability 236
    16.5.2 Cell Migration 236
    16.5.3 Barrier Integrity 236
16.6 Summary/Conclusions/Challenges 236
References 236

**17 Preclinical Safety Assessment of Drug Candidate-Induced Pancreatic
Toxicity: From an Applied Perspective** **242**

Karrie A. Brenneman, Shashi K. Ramaiah and Lauren M. Gauthier

17.1 Drug-Induced Pancreatic Toxicity 242
    17.1.1 Introduction 242
    17.1.2 Drug-Induced Pancreatic Exocrine Toxicity in Humans:
            Pancreatitis 243
    17.1.3 Mechanisms of Drug-Induced Pancreatic Toxicity 244
17.2 Preclinical Evaluation of Pancreatic Toxicity 245
    17.2.1 Introduction 245
    17.2.2 Risk Management and Understanding the Potential
            for Clinical Translation 245
    17.2.3 Interspecies and Interstrain Differences in Susceptibility
            to Pancreatic Toxicity 246
17.3 Preclinical Pancreatic Toxicity Assessment: *In Vivo* 247
    17.3.1 Routine Assessment 247
    17.3.2 Specialized Techniques 248
17.4 Pancreatic Biomarkers 249
    17.4.1 Introduction 249
    17.4.2 Exocrine Injury Biomarkers in Humans and Preclinical Species 250
    17.4.3 Endocrine/Islet Functional Biomarkers for Humans and
            Preclinical Species 252
    17.4.4 A Note on Biomarkers of Vascular Injury Relevant
            to the Pancreas 253
    17.4.5 Author's Opinion on the Strategy for Investments to Address
            Pancreatic Biomarker Gaps 253

17.5		Preclinical Pancreatic Toxicity Assessment: *In Vitro*	253
	17.5.1	Introduction to Pancreatic Cell Culture	253
	17.5.2	Modeling *In Vitro* Toxicity *In Vitro*, Testing Translatability, and *In Vitro* Screening Tools	254
	17.5.3	Case Study 1: Drug Candidate-Induced Direct Acinar Cell Toxicity *In Vivo* with Confirmation of Toxicity and Drug Candidate Screening *In Vitro*	255
	17.5.4	Case Study 2: Drug Candidate-Induced Microvascular Injury at the Exocrine–Endocrine Interface in the Rat with Unsuccessful Confirmation of Toxicity *In Vitro* and No Pancreas-Specific Monitorable Biomarkers Identified	256
	17.5.5	Emerging Technologies/Gaps: Organotypic Models	256
17.6		Summary and Conclusions	257
		Acknowledgments	258
		References	258

PART V   ADDRESSING THE FALSE NEGATIVE SPACE—INCREASING PREDICTIVITY   **261**

18   Animal Models of Disease for Future Toxicity Predictions   **263**
Sherry J. Morgan and Chandikumar S. Elangbam

18.1		Introduction	263
18.2		Hepatic Disease Models	264
	18.2.1	Hepatic Toxicity: Relevance to Drug Attrition	264
	18.2.2	Hepatic Toxicity: Reasons for Poor Translation from Animal to Human	264
	18.2.3	Available Hepatic Models to Predict Hepatic Toxicity or Understand Molecular Mechanisms of Toxicity: Advantages and Limitations	264
18.3		Cardiovascular Disease Models	268
	18.3.1	Cardiac Toxicity: Relevance to Drug Attrition	268
	18.3.2	Cardiac Toxicity: Reasons for Poor Translation from Animal to Human	268
	18.3.3	Available CV Models to Predict Cardiac Toxicity or Understand Molecular Mechanisms of Toxicity: Advantages and Limitations	269
18.4		Nervous System Disease Models	270
	18.4.1	Nervous System Toxicity: Relevance to Drug Attrition	270
	18.4.2	Nervous System Toxicity: Reasons for Poor Translation from Animal to Human	270
	18.4.3	Available Nervous System Models to Predict Nervous System Toxicity or Understand Molecular Mechanisms of Toxicity: Advantages and Limitations	270
18.5		Gastrointestinal Injury Models	273
	18.5.1	Gastrointestinal (GI) Toxicity: Relevance to Drug Attrition	273
	18.5.2	Gastrointestinal Toxicity: Reasons for Poor Translation from Animal to Human	273
	18.5.3	Available Gastrointestinal Animal Models to Predict Gastrointestinal Toxicity or Understand Molecular Mechanisms of Toxicity: Advantages and Limitations	274
18.6		Renal Injury Models	279
	18.6.1	Renal Toxicity: Relevance to Drug Attrition	279
	18.6.2	Renal Toxicity: Reasons for Poor Translation from Animal to Human	279

18.6.3 Available Renal Models to Predict Renal Toxicity or
Understand Molecular Mechanisms of Toxicity: Advantages
and Limitations   280
18.7 Respiratory Disease Models   282
18.7.1 Respiratory Toxicity: Relevance to Drug Attrition   282
18.7.2 Respiratory Toxicity: Reasons for Adequate Translation
from Animal to Human   282
18.7.3 Available Respiratory Models to Predict Respiratory Toxicity
or Understand Molecular Mechanisms of Toxicity:
Advantages and Limitations   282
18.8 Conclusion   285
References   287

19 The Use of Genetically Modified Animals in Discovery Toxicology   298

Dolores Diaz and Jonathan M. Maher

19.1 Introduction   298
19.2 Large-Scale Gene Targeting and Phenotyping Efforts   299
19.3 Use of Genetically Modified Animal Models in Discovery Toxicology   300
19.4 The Use of Genetically Modified Animals in Pharmacokinetic and
Metabolism Studies   303
19.4.1 Drug Metabolism   303
19.4.2 Drug Transporters   306
19.4.3 Nuclear Receptors and Coordinate Induction   307
19.4.4 Humanized Liver Models   308
19.5 Conclusions   309
References   309

**20 Mouse Population-Based Toxicology for Personalized Medicine
and Improved Safety Prediction   314**

Alison H. Harrill

20.1 Introduction   314
20.2 Pharmacogenetics and Population Variability   314
20.3 Rodent Populations Enable a Population-Based Approaches
to Toxicology   316
20.3.1 Mouse Diversity Panel   317
20.3.2 CC Mice   318
20.3.3 DO Mice   319
20.4 Applications for Pharmaceutical Safety Science   320
20.4.1 Personalized Medicine: Development of Companion
Diagnostics   320
20.4.2 Biomarkers of Sensitivity   320
20.4.3 Mode of Action   322
20.5 Study Design Considerations for Genomic Mapping   322
20.5.1 Dose Selection   322
20.5.2 Model Selection   322
20.5.3 Sample Size   323
20.5.4 Phenotyping   324
20.5.5 Genome-Wide Association Analysis   324
20.5.6 Candidate Gene Analysis   324
20.5.7 Cost Considerations   325
20.5.8 Health Status   325
20.6 Summary   326
References   326

PART VI   STEM CELLS IN TOXICOLOGY   **331**

21   Application of Pluripotent Stem Cells in Drug-Induced Liver Injury Safety Assessment   **333**

Christopher S. Pridgeon, Fang Zhang, James A. Heslop, Charlotte M.L. Nugues, Neil R. Kitteringham, B. Kevin Park and Christopher E.P. Goldring

21.1	The Liver, Hepatocytes, and Drug-Induced Liver Injury	333
21.2	Current Models of DILI	334
	21.2.1 Primary Human Hepatocytes	334
	21.2.2 Murine Models	336
	21.2.3 Cell Lines	336
	21.2.4 Stem Cell Models	337
21.3	Uses of iPSC HLCs	338
21.4	Challenges of Using iPSCs and New Directions for Improvement	339
	21.4.1 Complex Culture Systems	340
	21.4.2 Coculture	340
	21.4.3 3D Culture	340
	21.4.4 Perfusion Bioreactors	341
21.5	Alternate Uses of HLCs in Toxicity Assessment	341
	References	342

22   Human Pluripotent Stem Cell-Derived Cardiomyocytes: A New Paradigm in Predictive Pharmacology and Toxicology   **346**

Praveen Shukla, Priyanka Garg and Joseph C. Wu

22.1	Introduction	346
22.2	Advent of hPSCs: Reprogramming and Cardiac Differentiation	347
	22.2.1 Reprogramming	347
	22.2.2 Cardiac Differentiation	347
22.3	iPSC-Based Disease Modeling and Drug Testing	349
22.4	Traditional Target-Centric Drug Discovery Paradigm	354
22.5	iPSC-Based Drug Discovery Paradigm	354
	22.5.1 Target Identification and Validation: "Clinical Trial in a Dish"	356
	22.5.2 Safety Pharmacology and Toxicological Testing	356
22.6	Limitations and Challenges	358
22.7	Conclusions and Future Perspective	359
	Acknowledgments	360
	References	360

23   Stem Cell-Derived Renal Cells and Predictive Renal *In Vitro* Models   **365**

Jacqueline Kai Chin Chuah, Yue Ning Lam, Peng Huang and Daniele Zink

23.1	Introduction	365
23.2	Protocols for the Differentiation of Pluripotent Stem Cells into Cells of the Renal Lineage	367
	23.2.1 Earlier Protocols and the Recent Race	367
	23.2.2 Protocols Designed to Mimic Embryonic Kidney Development	369
	23.2.3 Rapid and Efficient Methods for the Generation of Proximal Tubular-Like Cells	372
23.3	Renal *In Vitro* Models for Drug Safety Screening	376
	23.3.1 Microfluidic and 3D Models and Other Models that have been Tested with Lower Numbers of Compounds	376
	23.3.2 *In Vitro* Models that have been Tested with Higher Numbers of Compounds and the First Predictive Renal *In Vitro* Model	376
	23.3.3 Stem Cell-Based Predictive Models	377

23.4 Achievements and Future Directions   378
Acknowledgments   379
Notes   379
References   379

**PART VII   CURRENT STATUS OF PRECLINICAL *IN VIVO*
               TOXICITY BIOMARKERS   385**

**24 Predictive Cardiac Hypertrophy Biomarkers in Nonclinical
     Studies   387**
Steven K. Engle

24.1 Introduction to Biomarkers   387
24.2 Cardiovascular Toxicity   387
24.3 Cardiac Hypertrophy   388
24.4 Diagnosis of Cardiac Hypertrophy   389
24.5 Biomarkers of Cardiac Hypertrophy   389
24.6 Case Studies   392
24.7 Conclusion   392
References   393

25 Vascular Injury Biomarkers   397
Tanja S. Zabka and Kaïdre Bendjama

25.1 Historical Context of Drug-Induced Vascular Injury
      and Drug Development   397
25.2 Current State of DIVI Biomarkers   398
25.3 Current Status and Future of *In Vitro* Systems to
      Investigate DIVI   402
25.4 Incorporation of *In Vitro* and *In Vivo* Tools in Preclinical
      Drug Development   403
25.5 DIVI Case Study   403
References   403

26 Novel Translational Biomarkers of Skeletal Muscle Injury   407
Peter M. Burch and Warren E. Glaab

26.1 Introduction   407
26.2 Overview of Drug-Induced Skeletal Muscle Injury   407
26.3 Novel Biomarkers of Drug-Induced Skeletal Muscle
      Injury   409
      26.3.1 Skeletal Troponin I (sTnI)   409
      26.3.2 Creatine Kinase M (CKM)   409
      26.3.3 Myosin Light Chain 3 (Myl3)   409
      26.3.4 Fatty Acid-Binding Protein 3   410
      26.3.5 Parvalbumin   410
      26.3.6 Myoglobin   410
      26.3.7 MicroRNAs   410
26.4 Regulatory Endorsement   411
26.5 Gaps and Future Directions   411
26.6 Conclusions   412
References   412

27   Translational Mechanistic Biomarkers and Models for Predicting Drug-Induced Liver Injury: Clinical to *In Vitro* Perspectives   **416**

Daniel J. Antoine

27.1   Introduction   416
27.2   Drug-Induced Toxicity and the Liver   417
27.3   Current Status of Biomarkers for the Assessment of DILI   418
27.4   Novel Investigational Biomarkers for DILI   419
        27.4.1   Glutamate Dehydrogenase   419
        27.4.2   Acylcarnitines   420
        27.4.3   High-Mobility Group Box-1 (HMGB1)   420
        27.4.4   Keratin-18 (K18)   421
        27.4.5   MicroRNA-122 (miR-122)   421
27.5   *In Vitro* Models and the Prediction of Human DILI   422
27.6   Conclusions and Future Perspectives   423
References   424

PART VIII   KIDNEY INJURY BIOMARKERS   **429**

28   Assessing and Predicting Drug-Induced Kidney Injury, Functional Change, and Safety in Preclinical Studies in Rats   **431**

Yafei Chen

28.1   Introduction   431
28.2   Kidney Functional Biomarkers (Glomerular Filtration and Tubular
        Reabsorption)   433
        28.2.1   Traditional Functional Biomarkers   433
        28.2.2   Novel Functional Biomarkers   434
28.3   Novel Kidney Tissue Injury Biomarkers   435
        28.3.1   Urinary *N*-Acetyl-β-D-Glucosaminidase (NAG)   435
        28.3.2   Urinary Glutathione *S*-Transferase α (α-GST)   435
        28.3.3   Urinary Renal Papillary Antigen 1 (RPA-1)   435
        28.3.4   Urinary Calbindin D28   435
28.4   Novel Biomarkers of Kidney Tissue Stress Response   436
        28.4.1   Urinary Kidney Injury Molecule-1 (KIM-1)   436
        28.4.2   Urinary Clusterin   436
        28.4.3   Urinary Neutrophil Gelatinase-Associated Lipocalin (NGAL) 436
        28.4.4   Urinary Osteopontin (OPN)   437
        28.4.5   Urinary L-Type Fatty Acid-Binding Protein (L-FABP)   437
        28.4.6   Urinary Interleukin-18 (IL-18)   437
28.5   Application of an Integrated Rat Platform (Automated Blood
        Sampling and Telemetry, ABST) for Kidney Function and
        Injury Assessment   437
References   439

29   Canine Kidney Safety Protein Biomarkers   **443**

Manisha Sonee

29.1   Introduction   443
29.2   Novel Canine Renal Protein Biomarkers   443
29.3   Evaluations of Novel Canine Renal Protein Biomarker Performance   444
29.4   Conclusion   444
References   445

30   Traditional Kidney Safety Protein Biomarkers and Next-Generation Drug-Induced Kidney Injury Biomarkers in Nonhuman Primates   **446**

Jean-Charles Gautier and Xiaobing Zhou

30.1   Introduction   446
30.2   Evaluations of Novel NHP Renal Protein Biomarker Performance   447
30.3   New Horizons: Urinary MicroRNAs and Nephrotoxicity in NHPs   447
References   447

31   Rat Kidney MicroRNA Atlas   **448**

Aaron T. Smith

31.1   Introduction   448
31.2   Key Findings   448
References   449

32   MicroRNAs as Next-Generation Kidney Tubular Injury Biomarkers in Rats **450**

Heidrun Ellinger-Ziegelbauer and Rounak Nassirpour

32.1   Introduction   450
32.2   Rat Tubular miRNAs   450
32.3   Conclusions   451
References   451

33   MicroRNAs as Novel Glomerular Injury Biomarkers in Rats   **452**

Rachel Church

33.1   Introduction   452
33.2   Rat Glomerular miRNAs   452
References   453

34   Integrating Novel Imaging Technologies to Investigate Drug-Induced Kidney Toxicity   **454**

Bettina Wilm and Neal C. Burton

34.1   Introduction   454
34.2   Overviews   455
34.3   Summary   456
References   456

35   *In Vitro* to *In Vivo* Relationships with Respect to Kidney Safety Biomarkers   **458**

Paul Jennings

35.1   Renal Cell Lines as Tools for Toxicological Investigations   458
35.2   Mechanistic Approaches and *In Vitro* to *In Vivo* Translation   459
35.3   Closing Remarks   460
References   460

36   Case Study: Fully Automated Image Analysis of Podocyte Injury Biomarker Expression in Rats   **462**

Jing Ying Ma

36.1   Introduction   462
36.2   Material and Methods   462
36.3   Results   463
36.4   Conclusions   465
References   465

37   Case Study: Novel Renal Biomarkers Translation to Humans   **466**
Deborah A. Burt

37.1   Introduction   466
37.2   Implementation of Translational Renal Biomarkers
in Drug Development   466
37.3   Conclusion   467
References   467

38   Case Study: MicroRNAs as Novel Kidney Injury Biomarkers in Canines   **468**
Craig Fisher, Erik Koenig and Patrick Kirby

38.1   Introduction   468
38.2   Material and Methods   468
38.3   Results   468
38.4   Conclusions   470
References   470

39   Novel Testicular Injury Biomarkers   **471**
Hank Lin

39.1   Introduction   471
39.2   The Testis   471
39.3   Potential Biomarkers for Testicular Toxicity   472
39.3.1   Inhibin B   472
39.3.2   Androgen-Binding Protein   472
39.3.3   SP22   472
39.3.4   Emerging Novel Approaches   472
39.4   Conclusions   473
References   473

PART IX   BEST PRACTICES IN BIOMARKER EVALUATIONS   **475**

40   Best Practices in Preclinical Biomarker Sample Collections   **477**
Jaqueline Tarrant

40.1   Considerations for Reducing Preanalytical Variability in Biomarker Testing   477
40.2   Biological Sample Matrix Variables   477
40.3   Collection Variables   480
40.4   Sample Processing and Storage Variables   480
References   480

41   Best Practices in Novel Biomarker Assay Fit-for-Purpose Testing   **481**
Karen M. Lynch

41.1   Introduction   481
41.2   Why Use a Fit-for-Purpose Assay?   481
41.3   Overview of Fit-for-Purpose Assay Method Validations   482
41.4   Assay Method Suitability in Preclinical Studies   482
41.5   Best Practices for Analytical Methods Validation   482
41.5.1   Assay Precision   482
41.5.2   Accuracy/Recovery   484
41.5.3   Precision and Accuracy of the Calibration Curve   484
41.5.4   Lower Limit of Quantification   484
41.5.5   Upper Limit of Quantification   484
41.5.6   Limit of Detection   485

41.5.7   Precision Assessment for Biological Samples   485
41.5.8   Dilutional Linearity and Parallelism   485
41.5.9   Quality Control   486
41.6   Species- and Gender-Specific Reference Ranges   486
41.7   Analyte Stability   487
41.8   Additional Method Performance Evaluations   487
References   487

42   Best Practices in Evaluating Novel Biomarker Fit for Purpose and Translatability   489

Amanda F. Baker

42.1   Introduction   489
42.2   Protocol Development   489
42.3   Assembling an Operations Team   489
42.4   Translatable Biomarker Use   490
42.5   Assay Selection   490
42.6   Biological Matrix Selection   490
42.7   Documentation of Patient Factors   491
42.8   Human Sample Collection Procedures   491
          42.8.1   Biomarkers in Human Tissue Biopsy and Biofluid Samples   491
42.9   Choice of Collection Device   491
          42.9.1   Tissue Collection Device   491
          42.9.2   Plasma Collection Device   492
          42.9.3   Serum Collection Device   492
          42.9.4   Urine Collection Device   492
42.10   Schedule of Collections   492
42.11   Human Sample Quality Assurance   492
          42.11.1   Monitoring Compliance to Sample Collection Procedures   492
          42.11.2   Documenting Time and Temperature from Sample Collection to Processing   492
          42.11.3   Optimal Handling and Preservation Methods   492
          42.11.4   Choice of Sample Storage Tubes   493
          42.11.5   Choice of Sample Labeling   493
          42.11.6   Optimal Sample Storage Conditions   493
42.12   Logistics Plan   493
42.13   Database Considerations   493
42.14   Conclusive Remarks   493
References   493

43   Best Practices in Translational Biomarker Data Analysis   495

Robin Mogg and Daniel Holder

43.1   Introduction   495
43.2   Statistical Considerations for Preclinical Studies of Safety Biomarkers   496
43.3   Statistical Considerations for Exploratory Clinical Studies of Translational Safety Biomarkers   497
43.4   Statistical Considerations for Confirmatory Clinical Studies of Translational Safety Biomarkers   498
43.5   Summary   498
References   498

44   Translatable Biomarkers in Drug Development: Regulatory Acceptance and Qualification   **500**

John-Michael Sauer, Elizabeth G. Walker and Amy C. Porter

44.1   Safety Biomarkers   500
44.2   Qualification of Safety Biomarkers   501
44.3   Letter of Support for Safety Biomarkers   502
44.4   Critical Path Institute's Predictive Safety Testing Consortium   502
44.5   Predictive Safety Testing Consortium and its Key Collaborations   504
44.6   Advancing the Qualification Process and Defining Evidentiary Standards   505
References   506

PART X   CONCLUSIONS   **509**

45   Toxicogenomics in Drug Discovery Toxicology: History, Methods, Case Studies, and Future Directions   **511**

Brandon D. Jeffy, Joseph Milano and Richard J. Brennan

45.1   A Brief History of Toxicogenomics   511
45.2   Tools and Strategies for Analyzing Toxicogenomics Data   513
45.3   Drug Discovery Toxicology Case Studies   519
        45.3.1   Case Studies: Diagnostic Toxicogenomics   520
        45.3.2   Case Studies: Predictive Toxicogenomics   521
        45.3.3   Case Studies: Mechanistic/Investigative Toxicogenomics   523
        45.3.4   Future Directions in Drug Discovery Toxicogenomics   524
References   525

46   Issue Investigation and Practices in Discovery Toxicology   **530**

Dolores Diaz, Dylan P. Hartley and Raymond Kemper

46.1   Introduction   530
46.2   Overview of Issue Investigation in the Discovery Space   530
46.3   Strategies to Address Toxicities in the Discovery Space   532
46.4   Cross-Functional Collaborative Model   533
46.5   Case-Studies of Issue Resolution in The Discovery Space   536
46.6   Data Inclusion in Regulatory Filings   538
References   538

ABBREVIATIONS   **540**

CONCLUDING REMARKS   **542**

INDEX   **543**

LIST OF CONTRIBUTORS

Najah Abi-Gerges, AnaBios Corporation, San Diego, CA, USA

Michael D. Aleo, Investigative Toxicology, Drug Safety Research and Development, Pfizer Inc., Groton, CT, USA

Daniel J. Antoine, MRC Centre for Drug Safety Science and Department of Molecular and Clinical Pharmacology, The Institute of Translational Medicine, University of Liverpool, Liverpool, UK

Michael Bachelor, MatTek Corporation, Ashland, MA, USA

Amanda F. Baker, Arizona Health Sciences Center, University of Arizona, Tucson, AZ, USA

Scott A. Barros, Investigative Toxicology, Alnylam Pharmaceuticals Inc., Cambridge, MA, USA

Kaïdre Bendjama, Transgene, Illkirch-Graffenstaden, France

Eric A.G. Blomme, AbbVie, Pharmaceutical Research & Development, North Chicago, IL, USA

Richard J. Brennan, Preclinical Safety, Sanofi SA, Waltham, MA, USA

Karrie A. Brenneman, Toxicologic Pathology, Drug Safety Research and Development, Pfizer Inc., Andover, MA, USA

Peter M. Burch, Investigative Pathology, Drug Safety Research and Development, Pfizer Inc., Groton, CT, USA

Deborah A. Burt, Biomarker Development and Translation, Drug Safety Research and Development, Pfizer Inc., Groton, CT, USA

Neal C. Burton, iThera Medical, GmbH, Munich, Germany

Nicholas Buss, Biologics Safety Assessment, MedImmune, Gaithersburg, MD, USA

Paul Butler, Global Safety Pharmacology, Drug Safety Research and Development, Pfizer Inc., La Jolla, CA, USA

Keri E. Cannon, Toxicology, Halozyme Therapeutics, Inc., San Diego, CA, USA

Minjun Chen, Division of Bioinformatics and Biostatistics, National Center for Toxicological Research, US Food and Drug Administration (NCTR/FDA), Jefferson, AZ, USA

Yafei Chen, Mechanistic & Investigative Toxicology, Discovery Sciences, Janssen Research & Development, L.L.C., San Diego, CA, USA

Jacqueline Kai Chin Chuah, Institute of Bioengineering and Nanotechnology, The Nanos, Singapore

Rachel Church, University of North Carolina Institute for Drug Safety Sciences, Chapel Hill, NC, USA

Thomas J. Colatsky, Division of Applied Regulatory Science, Office of Clinical Pharmacology, Office of Translational Sciences, Center for Drug Evaluation and Research, US Food and Drug Administration, Silver Spring, MD, USA

Donna M. Dambach, Safety Assessment, Genentech Inc., South San Francisco, CA, USA

Mark R. Davies, QT-Informatics Limited, Macclesfield, England

Dolores Diaz, Discovery Toxicology, Safety Assessment, Genentech, Inc., South San Francisco, CA, USA

Alison Easter, Biogen Inc., Cambridge, MA, USA

Heidrun Ellinger-Ziegelbauer, Investigational Toxicology, GDD-GED-Toxicology, Bayer Pharma AG, Wuppertal, Germany

Chandikumar S. Elangbam, Pathophysiology, Safety Assessment, GlaxoSmithKline, Research Triangle Park, NC, USA

Steven K. Engle, Lilly Research Laboratories, Division of Eli Lilly and Company, Lilly Corporate Center, Indianapolis, IN, USA

Ellen Evans, Immunotoxicology Center of Emphasis, Drug Safety Research and Development, Pfizer Inc., Groton, CT, USA

Craig Fisher, Drug Safety Evaluation, Takeda California Inc., San Diego, CA, USA

Jay H. Fortner, Veterinary Science & Technology, Comparative Medicine, Pfizer Inc., Groton, CT, USA

David J. Gallacher, Global Safety Pharmacology, Janssen Research & Development, a division of Janssen Pharmaceutica N.V., Beerse, Belgium

Priyanka Garg, Stanford Cardiovascular Institute, Institute for Stem Cell Biology and Regenerative Medicine, Department of Medicine, Division of Cardiology, Stanford University School of Medicine, Stanford, CA, USA

Lauren M. Gauthier, Investigative Toxicology, Drug Safety Research and Development, Pfizer Inc., Andover, MA, USA

Jean-Charles Gautier, Preclinical Safety, Sanofi, Vitry-sur-Seine, France

Gary Gintant, Integrative Pharmacology, Integrated Science & Technology, AbbVie, North Chicago, IL, USA

Christopher E.P. Goldring, MRC Centre for Drug Safety Science, Department of Molecular and Clinical Pharmacology, The Institute of Translational Medicine, University of Liverpool, Liverpool, UK

Warren E. Glaab, Systems Toxicology, Investigative Laboratory Sciences, Safety Assessment, Merck Research Laboratories, West Point, PA, USA

Brian D. Guth, Drug Discovery Support, Boehringer Ingelheim Pharma GmbH & Co. KG, Biberach (Riss), Germany; DST/NWU, Preclinical Drug Development Platform, Faculty of Health Sciences, North-West University, Potchefstroom, South Africa

Robert L. Hamlin, Department of Veterinary Medicine and School of Biomedical Engineering, The Ohio State University, Columbus, OH, USA

Alison H. Harrill, Department of Environmental and Occupational Health, Regulatory Sciences Program, The University of Arkansas for Medical Sciences, Little Rock, AR, USA

Dylan P. Hartley, Drug Metabolism and Pharmacokinetics, Array BioPharma Inc., Boulder, CO, USA

Patrick J. Hayden, MatTek Corporation, Ashland, MA, USA

James A. Heslop, MRC Centre for Drug Safety Science, Department of Molecular and Clinical Pharmacology, The Institute of Translational Medicine, University of Liverpool, Liverpool, UK

Gregory Hinkle, Bioinformatics, Alnylam Pharmaceuticals Inc., Cambridge, MA, USA

Mary Jane Hinrichs, Biologics Safety Assessment, MedImmune, Gaithersburg, MD, USA

Kimberly M. Hoagland, Integrated Discovery and Safety Pharmacology, Department of Comparative Biology and Safety Sciences, Amgen Inc., Thousand Oaks, CA, USA

Daniel Holder, Biometrics Research, Merck Research Laboratories, West Point, PA, USA

Michelle J. Horner, Comparative Biology and Safety Sciences (CBSS) – Toxicology Sciences, Amgen Inc., Thousand Oaks, CA, USA

Chuchu Hu, Division of Bioinformatics and Biostatistics, National Center for Toxicological Research, US Food and Drug Administration (NCTR/FDA), Jefferson, AZ, USA; Zhejiang Institute of Food and Drug Control, Hangzhou, China

Peng Huang, Institute of Bioengineering and Nanotechnology, The Nanos, Singapore

Wenhu Huang, General Toxicology, Drug Safety Research and Development, Pfizer Inc., La Jolla, CA, USA

Brandon D. Jeffy, Exploratory Toxicology, Celgene Corporation, San Diego, CA, USA

Paul Jennings, Division of Physiology, Department of Physiology and Medical Physics, Medical University of Innsbruck, Innsbruck, Austria

Raymond Kemper, Discovery and Investigative Toxicology, Drug Safety Evaluation, Vertex Pharmaceuticals, Boston, MA, USA

Helena Kandárová, MatTek *In Vitro* Life Science Laboratories, Bratislava, Slovak Republic

J. Gerry Kenna, Fund for the Replacement of Animals in Medical Experiments (FRAME), Nottingham, UK

Patrick Kirby, Drug Safety and Research Evaluation, Takeda Boston, Takeda Pharmaceuticals International Co., Cambridge, MA, USA

Neil R. Kitteringham, MRC Centre for Drug Safety Science, Department of Molecular and Clinical Pharmacology, The Institute of Translational Medicine, University of Liverpool, Liverpool, UK

Mitchell Klausner, MatTek Corporation, Ashland, MA, USA

Erik Koenig, Molecular Pathology, Takeda Boston, Takeda Pharmaceuticals International Co., Cambridge, MA, USA

Yue Ning Lam, Institute of Bioengineering and Nanotechnology, The Nanos, Singapore

Lawrence H. Lash, Department of Pharmacology, School of Medicine, Wayne State University, Detroit, MI, USA

Hank Lin, Drug Safety Research and Development, Pfizer Inc., Cambridge, MA, USA

Hua Rong Lu, Global Safety Pharmacology, Janssen Research & Development, a division of Janssen Pharmaceutica N.V., Beerse, Belgium

Karen M. Lynch, Safety Assessment, GlaxoSmithKline, King of Prussia, PA, USA

Jing Ying Ma, Molecular Pathology, Discovery Sciences, Janssen Research & Development, L.L.C., San Diego, CA, USA

Jonathan M. Maher, Discovery Toxicology, Safety Assessment, Genentech, Inc., South San Francisco, CA, USA

Sherry J. Morgan, Preclinical Safety, AbbVie, Inc., North Chicago, IL, USA

J. Eric McDuffie, Mechanistic & Investigative Toxicology, Discovery Sciences, Janssen Research & Development, San Diego, CA, USA

Joseph Milano, Milano Toxicology Consulting, L.L.C., Wilmington, DE, USA

Robin Mogg, Early Clinical Development Statistics, Merck Research Laboratories, Upper Gwynedd, PA, USA

Rounak Nassirpour, Biomarkers, Drug Safety Research and Development, Pfizer Inc., Andover, MA, USA

Charlotte M.L. Nugues, MRC Centre for Drug Safety Science, Department of Molecular and Clinical Pharmacology, The Institute of Translational Medicine, University of Liverpool, Liverpool, UK

Andrew J. Olaharski, Toxicology, Agios Pharmaceuticals, Cambridge, MA, USA

B. Kevin Park, MRC Centre for Drug Safety Science, Department of Molecular and Clinical Pharmacology, The Institute of Translational Medicine, University of Liverpool, Liverpool, UK

Mikael Persson, Lundbeck, Valby, Denmark; Currently at AstraZeneca, Molndal, Sweden

Amy C. Porter, Predictive Safety Testing Consortium (PSTC), Critical Path Institute (C-Path), Tucson, AZ, USA

Patrick Poulin, Associate Professor, Department of Occupational and Environmental Health, School of Public Health, IRSPUM, Université de Montréal, Montréal, Québec, Canada and Consultant, Québec city, Québec, Canada

Christopher S. Pridgeon, MRC Centre for Drug Safety Science, Department of Molecular and Clinical Pharmacology, The Institute of Translational Medicine, University of Liverpool, Liverpool, UK

Shashi K. Ramaiah, Biomarkers, Drug Safety Research and Development, Pfizer Inc., Cambridge, MA, USA

Georg Rast, Drug Discovery Support, Boehringer Ingelheim Pharma GmbH & Co. KG, Biberach (Riss), Germany

Ivan Rich, Hemogenix, Inc., Colorado Springs, CO, USA

John-Michael Sauer, Predictive Safety Testing Consortium (PSTC), Critical Path Institute (C-Path), Tucson, AZ, USA

Praveen Shukla, Stanford Cardiovascular Institute, Institute for Stem Cell Biology and Regenerative Medicine, Department of Medicine, Division of Cardiology, Stanford University School of Medicine, Stanford, CA, USA

Scott Q. Siler, The Hamner Institute, Research Triangle Park, NC, USA

Aaron T. Smith, Investigative Toxicology, Eli Lilly and Company, Indianapolis, IN, USA

Dennis A. Smith, Independent Consultant, Canterbury, UK

Chris J. Somps, Investigative Toxicology, Drug Safety Research and Development, Pfizer Inc., Groton, CT, USA

Manisha Sonee, Mechanistic & Investigative Toxicology, Discovery Sciences, Janssen Research & Development, L.L.C., Spring House, PA, USA

Jaqueline Tarrant, Development Sciences-Safety Assessment, Genentech Inc., South San Francisco, CA, USA

Greet Teuns, Janssen Research & Development, Janssen Pharmaceutica N.V., Beerse, Belgium

Weida Tong, Division of Bioinformatics and Biostatistics, National Center for Toxicological Research, US Food and Drug Administration (NCTR/FDA), Jefferson, AZ, USA

Katya Tsaioun, Safer Medicine Trust, Cambridge, MA, USA

Hugo M. Vargas, Integrated Discovery and Safety Pharmacology, Department of Comparative Biology and Safety Sciences, Amgen Inc., Thousand Oaks, CA, USA

Allison Vitsky, Biomarkers, Drug Safety Research and Development, Pfizer Inc., La Jolla, CA, USA

Elizabeth G. Walker, Predictive Safety Testing Consortium (PSTC), Critical Path Institute (C-Path), Tucson, AZ, USA

Yvonne Will, Investigative Toxicology, Drug Safety Research and Development, Pfizer Inc., Groton, CT, USA

Bettina Wilm, Department of Cellular and Molecular Physiology, The Institute of Translational Medicine, The University of Liverpool, Liverpool, UK

Joseph C. Wu, Stanford Cardiovascular Institute, Institute for Stem Cell Biology and Regenerative Medicine, Department of Medicine, Division of Cardiology, Stanford University School of Medicine, Stanford, CA, USA

Joshua Xu, Division of Bioinformatics and Biostatistics, National Center for Toxicological Research, US Food and Drug Administration (NCTR/FDA), Jefferson, AZ, USA

Xu Zhu, Immunotoxicology Center of Emphasis, Drug Safety Research and Development, Pfizer Inc., Groton, CT, USA

Gina M. Yanochko, Investigative Toxicology, Drug Safety Research and Development, Pfizer Inc., La Jolla, CA, USA

Ke Yu, Division of Bioinformatics and Biostatistics, National Center for Toxicological Research, US Food and Drug Administration (NCTR/FDA), Jefferson, AZ, USA

Tanja S. Zabka, Development Sciences-Safety Assessment, Genentech Inc., South San Francisco, CA, USA

Fang Zhang, MRC Centre for Drug Safety Science, Department of Molecular and Clinical Pharmacology, The Institute of Translational Medicine, University of Liverpool, Liverpool, UK

Xiaobing Zhou, National Center for Safety Evaluation of Drugs, Beijing, China

Daniele Zink, Institute of Bioengineering and Nanotechnology, The Nanos, Singapore

FOREWORD

Discovering drugs with good efficacy and safety profiles is a very complex and difficult task. The magnitude of the challenge is best illustrated by the size of the research and development (R&D) investments needed for driving a new molecular entity (NME) to approval. Multiple factors contribute to this level of difficulty, let alone the fact that biology and diseases are, by themselves, extremely complex. There is good consensus that safety and efficacy represent the two most important aspects for success and are, not surprisingly, considered the two major causes for failure in development. Trying to predict safety and toxicity in humans is not a recent area of interest but has been emphasized much earlier in the drug discovery process over the past decade. This makes a lot of business sense, given that even minor improvements in toxicity-related attrition at the development stage translate in significant overall increases in R&D productivity and meaningful benefit to patients.

Toxicologists, in their effort to predict toxicity, have always tried to develop new models or technologies. In particular, a large volume of scientific literature covers characterization of *in vitro* models for toxicology applications. In spite of experimental inconsistencies among users and across published studies, there is no doubt that progress has been made in understanding the characteristics of those models. Some have clear and often insurmountable limitations, but others have sufficiently robust characteristics to be useful for small-molecule lead optimization or for mechanistic investigations of toxic effects. However, practices and implementations across companies are quite different, and any opportunity for scientists to share their experience and recommendations can only help move the field forward. One common theme across companies, however, is the effort to move safety assessment earlier in the drug discovery and development process, at least at the lead optimization stage but preferentially as early as target selection.

In the pharmaceutical industry, toxicology support at the discovery stage is a different approach from toxicology activities at the development stage. The role of the discovery toxicologist is to participate in collaboration with other functions in the selection of molecules with optimal properties (e.g., physicochemical, pharmacokinetic, pharmacological, safety) but also in the prioritization of therapeutic targets with a reasonable probability of success. The latter requires scientists to develop a fundamental understanding of the biology of the target not only in terms of potential therapeutic benefits but also in terms of potential safety liabilities. In the past, this aspect was a relatively low priority in most pharmaceutical companies with most efforts focused on pharmacology and medicinal chemistry. However, recent experience in most companies indicates that target-related safety issues are more frequent than previously thought and can be development limiting. This becomes even more relevant given the improved ability of medicinal chemists and toxicologists to rapidly and reliably eliminate molecules with intrinsic reactive properties.

Beyond target biology, various tools are currently used for compound optimization for absorption, distribution, metabolism, and excretion (ADME), pharmacokinetics, and toxicology properties, as reviewed in the first part of this comprehensive book. These tools include, among others, *in silico* models; high-throughput binding assays; cell-based assays with biochemical, impedance, or high-content imaging endpoints; or lower-throughput specialized assays, such as the Langendorff assay or three-dimensional *in vitro* models. Irrespective of their level of complexity and sophistication, all these assays must be interpreted in the context of all other relevant data to properly influence compound selection and optimization. Hence, the main challenge for toxicologists supporting discovery projects is usually not data generation but mostly interpretation and communication of these data in a timely manner. This implies that data need to be generated at the appropriate time to be useful and interpreted in the context of large numbers of other data points. To address these issues, a robust discovery toxicology organization needs to have access to the appropriate logistical support, as well as informatics and computational tools, an aspect that is currently often not emphasized enough. In contrast, models focused on predicting toxicity for specific tissues are difficult to use in a prospective manner but can be extremely useful for optimization against a target organ toxicity already identified in animals with lead molecules.

Animal models do not predict all possible toxic events in humans, but it is important to keep in mind that their negative predictive value is extremely high. As such, they fulfill their main objective very well. In other words, they allow drug developers to test novel molecules in humans without undue safety risks. This is best illustrated by the extremely rare major safety issues encountered in first-in-human studies. Therefore, to further improve toxicity prediction, one valuable approach is to identify the gaps in the current nonclinical models used for toxicity prediction and try to fill these. Solutions include, for instance, the use of nontraditional animal models, such as genetically engineered or diseased rodent models, the rapidly evolving stem cell field with the development of human induced pluripotent stem cell (iPSC)-based systems, the development of safety biomarkers with better performance characteristics compared to current biomarkers, or the use of information-rich technologies that help bring mechanistic clarity.

The past decade has witnessed an increased number of precompetitive consortia, such as the Predictive Safety Testing Consortium and the Innovative Medicine Initiative, which have fueled the pace of research progress in predictive toxicology. These precompetitive collaborations represent ideal forums to share ideas and experience but also to test in an efficient and systematic way new methods for toxicity prediction. These collaborative efforts will undeniably accelerate the development of novel models or biomarkers that will ultimately benefit patients and support animal welfare efforts. Companies and scientists should be encouraged to be actively involved in those forums.

The book edited by my colleagues Drs. Yvonne Will, J. Eric McDuffie, Andrew J. Olaharski, and Brandon D. Jeffy provides a very comprehensive view of the current state of the art of discovery toxicology in the pharmaceutical industry. The various components of discovery toxicology are presented in a coherent and logical manner through a series of parts and chapters authored by renowned contributors combining impressive cumulative years of experience in the field. These chapters accurately reflect the current thinking and toolbox available to the toxicologist working in the pharmaceutical industry and also reflect on future possibilities. The authors and editors should be applauded for their efforts to comprehensively and didactically share this knowledge. This book will undoubtedly become a reference for all of us involved in the toxicological assessment of pharmaceutical experimental compounds.

Eric A.G. Blomme, DVM, PhD, Diplomate of the American College of Veterinary Pathologists
Senior Research Fellow, Vice-President of Global Preclinical Safety
AbbVie Inc.
North Chicago, IL, USA
E-mail address: eric.blomme@abbvie.com

PART I

INTRODUCTION

1

EMERGING TECHNOLOGIES AND THEIR ROLE IN REGULATORY REVIEW

THOMAS J. COLATSKY

Division of Applied Regulatory Science, Office of Clinical Pharmacology, Office of Translational Sciences, Center for Drug Evaluation and Research, US Food and Drug Administration, Silver Spring, MD, USA

1.1 INTRODUCTION

The sequencing of the human genome and the emergence of other omics-based technologies have provided drug discoverers with powerful new tools that can be used as a framework for understanding disease mechanisms and predicting patient outcomes (Venter, 2000; Venter et al., 2001; Castle et al., 2002; Kennedy, 2002; Goodsaid, 2003; Guerreiro et al., 2003; Witzmann and Grant, 2003; Walgren and Thompson, 2004; Robertson, 2005; Kell, 2006; Lindon et al., 2007; Clarke and Haselden, 2008). Since the turn of the century, pharmaceutical scientists have been able to incorporate these approaches into their work: to identify specific molecular targets involved in disease initiation and progression; to establish links between animal models and clinical activity at the level of genes, proteins, and pathways; and to devise new ways of measuring and monitoring drug response. In contrast to finding drugs that act at proven drug targets and behaved

"correctly" in established preclinical tests, discovery efforts were directed toward screening against sets of novel and sometimes closely related molecular targets that had not yet been thoroughly validated in medical practice, using new preclinical models and assays to confirm therapeutic benefits and define potential toxicities, and streamlined development strategies to obtain early proof of concept in clinical trials (Food Drug Administration, 2006a; Sarapa, 2007; Butz and Morelli, 2008; Takimoto, 2008). Importantly, the vast multidimensional data sets generated by genomics, proteomics, metabolomics, and other reductionist approaches were accompanied by the development of new computational methods needed to cut through the noise and variability associated with in these complex measurements and to assign therapeutic significance to the data. The emergence of systems biology provided an organizational framework that attempted to address the need to reconstitute these data sets into a functioning organic whole (Butcher et al., 2004; Hood and Perlmutter, 2004; Fischer, 2005; Edwards and Preston, 2008).

Not surprisingly, as more innovation and opportunity entered the drug discovery process, the risk of clinical failure did not always go down, except perhaps in cases where disease or toxicity was found to have a relatively straightforward etiology involving a single gene or a well-characterized and understood biochemical process. Despite impressive technological advances, late-stage attrition remained a problem in drug development, and serious and sometimes rare or unexpected adverse events continued to be seen during clinical investigations or postapproval (Arrowsmith, 2011a, b; Arrowsmith and Miller, 2013). Regulatory agencies interpreted this unexpected attrition to indicate that critical

Drug Discovery Toxicology: From Target Assessment to Translational Biomarkers, First Edition. Edited by Yvonne Will, J. Eric McDuffie, Andrew J. Olaharski, and Brandon D. Jeffy.
© 2016 John Wiley & Sons, Inc. Published 2016 by John Wiley & Sons, Inc.

gaps still existed in the preclinical testing pathway and the translation of preclinical toxicology findings to clinical outcomes of interest. Some of these critical gaps can be traced to how regulatory toxicology studies are currently conducted. These studies tend to use healthy animals and are designed to identify robust toxicities that depend on dose and exposure rather than conditional effects triggered by individual susceptibilities or interactions with disease and disease comorbidities. Toxicology studies are also designed to characterize the possibility and type of toxicity and to suggest an initial "safe" human dose range rather than to determine the expected clinical prevalence and magnitude of the effect. In some cases, species differences in basic physiology and how a drug may be transported or biotransformed will confound the translation of preclinical findings to human patients. As a result, while preclinical safety data can reasonably predict clinical risk under appropriate testing conditions (Ewart et al., 2014; Holzgrefe et al., 2014), a lack of concordance can sometimes be found between preclinical and clinical findings, including the observation of toxicities in animal models that have no observed correlate in clinical experience (Olson et al., 1998, 2000; Alden et al., 2011; Wang and Gray, 2014).

To help address these issues and promote the advancement of new technologies, the FDA has issued several documents that define key regulatory science priorities as well as a process for introducing new tools into drug development. Beginning with the publication of the FDA's Critical Path Initiative and Opportunities List in 2004, these documents highlight the need for new methods in toxicology, including the evaluation and development of more predictive models and assays; the identification and performance characterization of more reliable biomarkers; and the application of *in silico* approaches and large data sets to organize and interpret diverse safety data (Food Drug Administration, 2004a, b, 2006b, 2011; Woodcock, 2007). In parallel and in response, the pace of scientific innovation has accelerated, with numerous emerging technologies being positioned as transformative new drug development tools with the potential to improve safety assessment and reduce the possibility of late-stage attrition. Recent attempts to "humanize" animal models (Cheung and Gonzalez, 2008; Zhang et al., 2009; Shultz et al., 2012) and to replicate human response *in vitro* using organotypic cultures (Schmeichel and Bissell, 2003; Huh et al., 2011; Mathur et al., 2013; Sung et al., 2013; Abaci and Shuler, 2015) and induced pluripotent stem cells (iPSCs) (Sirenko et al., 2013, 2014a, b; Kolaja, 2014; Doherty et al., 2015) have opened additional avenues for assessing human drug safety and efficacy. New *in silico* and *in vitro* approaches are being proposed to assess the risk of drug-induced proarrhythmia (Mirams et al., 2011, 2012; Johannesen et al., 2014; Sager et al., 2014) and to strengthen safety signals detected during postmarket pharmacovigilance (Szarfman et al., 2004; Harpaz et al., 2013; Liu et al., 2013; White et al., 2013).

In some cases, new regulatory pathways have been developed to improve the prediction of clinical risk based on fresh

insights into toxicity mechanisms. One example is using assays based on the human ether-a-go-go-related gene (hERG) channel, which is believed to encode the native cardiac potassium channel responsible for generating the rapid delayed rectifier potassium current (IKr) in the human heart (Kiehn et al., 1995; Sanguinetti et al., 1995). The recognition that some drugs can trigger torsade de pointes (TdP), a serious and usually fatal cardiac arrhythmia, by excessively prolonging ventricular repolarization through block of IKr led to the development of a new approach for assessing cardiac safety, currently embodied in the International Council on Harmonisation (ICH) S7B and E14 guidelines (FDA, 2005a, b; ICH, 2005). This new pathway involves testing drug effects on the hERG channel in a clonal cell line expression system (Hammond and Pollard, 2005), with confirmation of any notable findings in the clinical Thorough QT (TQT) study, which measures changes in the electrocardiographic QT interval (Darpo et al., 2006).

The purpose of this chapter is to identify specific questions that may arise when evaluating the potential regulatory impact of a new technology as well as the type of criteria that can be used to determine whether a new tool has general applicability as a basis for regulatory decision-making in drug development.

1.2 SAFETY ASSESSMENT IN DRUG DEVELOPMENT AND REVIEW

1.2.1 Drug Discovery

The likelihood that a new chemical will become a safe and effective therapeutic product is typically assessed at multiple stages in the drug development process. In the discovery phase, potential drug candidates are screened broadly for toxicity issues to eliminate those with obvious liabilities, using a variety of methods including computational analyses based on chemical structures or the evaluation of possible off-target effects in comprehensive panels of *in vitro* assays covering a wide range of pharmacological targets and activity endpoints. It is important to recognize that there are no specific regulatory recommendations governing how early assessments of drug safety should be made. It is up to the sponsor to determine the specific technologies and acceptance criteria needed to support advancing a candidate to the next decision point. The scope and thoroughness of the testing done at this stage of development are intended to provide comfort to the sponsor that the candidate drug warrants further investment. Early adopters of emerging technologies may use novel data sets to complement and support the results obtained in more traditional studies, but the weight given these additional data will depend on the level of comfort that management has in the credibility of the assay and the degree to which the technology has been validated. In all cases, the decisions made during the discovery phase will be company specific and shaped by current knowledge about the molecular target and concerns about the pharmacologic class or therapeutic indication, some of which may be known publicly but much of which may be proprietary to the company and contained in its base of institutional knowledge. For example, structural alerts generated by quantitative structure–activity relationship ((Q)SAR) models are commonly used during lead optimization to flag potential drug candidates based on their predicted safety profiles (Kruhlak et al., 2012). Measuring the transcriptional changes generated by a drug candidate and comparing them to a reference database of standard known toxicants is another example of exploratory research that can be conducted on to assess and reduce risk in candidate selection (Ganter et al., 2006; Judson et al., 2012; Bouhifd et al., 2015). These types of early evaluation typically combine the use of commercial assay kits, models, and analytical tools integrated with unique methods and data sets developed internally by each company.

1.2.2 Preclinical Development

As a drug candidate advance from lead selection into preclinical development, the safety studies conducted take on increasing importance in shaping the downstream development program and its probability of success. Rather than supporting the feasibility of a particular lead candidate within a company's larger research and development portfolio, study results now become the basis for a series of regulatory decisions that will inform the design, cost, and duration of the clinical development program. The appearance of organ toxicities in animal studies will define the dose ranges expected to be safely tolerated in humans and the drug concentrations that can be targeted to explore compound efficacy as fully as possible. While some toxicology studies are typically done later in development (e.g., carcinogenicity, reproductive toxicology), the earliest toxicology studies are intended to select a safe starting dose for humans and address the following specific questions: (i) Is there one or more target organ toxicities and are these toxicities reversible? (ii) What is the margin of safety between a clinical and a toxic dose? (iii) Can the relationship between critical pharmacodynamic–toxicodynamic endpoints and pharmacokinetic parameters be predicted?

Regulatory guidelines currently exist for the conduct of the toxicology and safety pharmacology studies intended to characterize the toxicities that might be expected to occur under the conditions of the proposed clinical trials (International Council on Harmonization, 2001, 2010; Food Drug Administration, 2005a). Safety pharmacology studies evaluate the functional effects of a candidate drug on a core battery of key organ systems (cardiovascular, central nervous system, respiratory) using therapeutic plasma concentrations and above. In designing a safety pharmacology program to support a new regulatory submission, the ICH S7A Tripartite Guideline encourages the use of new technologies and methodologies, as long as they are relevant, sound, and

scientifically valid (International Council on Harmonization, 2010). Sponsors may select from a wide range of *in vivo* and *in vitro* test systems to identify adverse pharmacodynamic and/or pathophysiological effects and the mechanism(s) by which these effects are produced. Supplemental safety data can also be generated as needed for other organ systems, including renal/urinary, the autonomic nervous system, the gastrointestinal system, and others, when there may be reasons for concern. Compliance with the principles of good laboratory practices (GLP) is generally required in the conduct of these studies, to ensure the reliability and quality of the data obtained, with justification for any safety pharmacology and follow-up studies not conducted under GLP. However, studies intended to characterize the primary and secondary pharmacologic effects of a new drug candidate can be conducted under non-GLP conditions.

In conjunction with the series of core battery and supplemental safety pharmacology studies, and prior to the initiation of clinical trials, sponsors must also characterize the concentrations of drug achieved over a range of doses considered to be therapeutic and toxicological. In addition, information on how a drug is metabolized is important, to allow for a comparison of human and animal metabolites and their associated risk of producing toxicity. These data will be used to support the selection of the most appropriate species and dose regimen for the nonclinical toxicology studies and ultimately to relate exposure levels to toxicity findings.

Information on acute toxicity is used to predict human tolerability and the possible consequences of drug overdose. Typically, acute toxicity is assessed in a single-dose toxicology study conducted in two mammalian species (rodent and nonrodent) using the intended clinical route of administration as well as parenteral dosing, but other approaches can also be considered (e.g., dose escalation studies, short-duration dose-ranging studies, or studies that achieve large or maximal exposures). The need for repeat dose toxicology studies is determined by the expected duration of treatment, the therapeutic indication, and the nature of the clinical trials described in the clinical development plan. As a general rule, repeat dose studies are also conducted in two species with durations that are equal to or exceed the duration of the human clinical trials up to a maximum of 6 months (rodent species) and 9 months (nonrodent species), with a minimum of 2 weeks.

These preclinical toxicology studies provide an estimate of the first dose that can be used in human trials. The no observed adverse effect level (NOAEL) is defined by the FDA as "the highest dose tested in an animal species that does not produce a significant increase in adverse effects in comparison to the control group" (Food Drug Administration, 2005c). It is important to note that any observed adverse event that can be considered biologically significant will determine the NOAEL; there is no need to demonstrate that the observation is statistical significance. The findings that determine the NOAEL may include the observation of overt toxicity (e.g., clinical signs, gross and histopathology lesions), changes in the levels of toxicity biomarkers (e.g., hepatic enzyme levels as surrogates for liver injury), and exaggerated pharmacodynamic effects. Once a NOAEL is determined, it is converted to a human equivalent dose (HED) using scaling techniques based on differences in body surface area between animals and humans. The lowest HED is obtained in the most sensitive animal species and usually informs the decision on initial clinical dosing, but in some cases, sponsors can justify using data from a less sensitive species and a higher HED if it can be argued as being more relevant in the assessment of human risk.

1.3 THE ROLE OF NEW TECHNOLOGIES IN REGULATORY SAFETY ASSESSMENT

Regulatory agencies have made a long-standing commitment to identify and promote the application of new technologies to drug, with the goal of reducing or replacing the need for animal studies and improving the prediction of clinical risk. However, before any advanced scientific method can be adopted as a basis for regulatory decision-making, it must be considered scientifically valid and be available to sponsors as a viable option for generating reliable and reproducible data. To assist researchers in gaining regulatory acceptance for new drug development tools, the FDA has established a formal qualification process that considers the requirements for establishing an assay as technically valid, as well as the process for generating the supporting data needed to define the specific utility of the measurement and the type of regulatory decisions it will be able to support (Food Drug Administration, 2014). Currently, the FDA's drug development tool process has centered on the qualification of three different types of tools: (i) new biomarkers intended for use in assessing drug safety and efficacy, (ii) patient reported outcome (PRO) rating instruments intended for use in clinical trials, and (iii) animal models intended to support product approval under the Animal Rule. However, the FDA's drug development tool process can also support other approaches as they become available. For example, *in vitro* assays may be determined to fall within the scope of the current process if they generate biomarkers used to predict drug safety. So far, it has been reported that five drug development tools have been qualified with ~80 applications being considered in the three qualification program areas noted earlier (Parekh et al., 2015), including biomarkers for monitoring renal and cardiac toxicity with better performance characteristics than conventional surrogates (Dieterle et al., 2010; Harpur et al., 2011; Hausner et al., 2013; Ennulat and Adler, 2015). Research within the FDA has focused on collaborating in the collection of the qualification data sets and on evaluating and setting standards for data quality and the analytical methods used to anchor biomarker performance to the endpoints of interest (Rouse et al., 2011, 2012,

2014; Goodwin et al., 2014; Shea et al., 2014; Amur et al., 2015; Rouse, 2015).

While the FDA's formal drug development tool qualification process currently does not extend beyond biomarkers, PROs, and animals models, the agency is considering other ways of recognizing the regulatory utility of an emerging technology and expressing confidence in its regulatory use. This includes issuing a "Letter of Acceptance" that deems a new tool "fit for purpose," such as was done to support the use of a simulation tool developed by the Critical Path Institute's Coalition Against Major Diseases (CAMD) as an aid in the design and interpretation of clinical trials for drugs intended to treat mild to moderate Alzheimer's disease (Rogers et al., 2012; Ito et al., 2013; Panegyres et al., 2014; Romero et al., 2015). By using this clinical trial simulation tool, which has been made available as a public resource, researchers can explore different outcomes in "virtual" Alzheimer's disease trials that build on knowledge about anonymized placebo responses extracted from prior clinical studies.

A key concept in the drug development tool qualification process is that of "context of use." The context of use is a clear and concise statement that specifies how and when the tool will be used in drug development and the conditional boundaries for its use as justified by the data submitted to support its qualification. The context of use is described in terms of its general area of use (e.g., nonclinical or clinical, pharmacodynamics, disease, or toxicology), its specific area of use (e.g., in clinical trial design, disease monitoring, dose or patient selection, assessment of drug effects including efficacy and toxicity), the critical parameters governing its use (e.g., drug or drug class specific, prognostic or diagnostic, type of assay platform), and the specific regulatory decision it is intended to inform. For the qualification of animal models, the context of use statement must include those details needed to replicate the model, including a description of the animals and challenge agent to be used, treatment information, descriptions of the primary and secondary endpoints, and the value ranges for the quality criteria determining successful implementation of the model in other labs.

1.3.1 *In Silico* Models for Toxicity Prediction

Drug developers and regulatory agencies already rely heavily on the use of modeling and simulation technologies to guide decision-making and to predict clinical outcomes. *In silico* models are used throughout drug development, early on in discovery to help identify and validate new drug targets, later in development to select appropriate doses for first-in-human trials and to estimate doses in special populations, and in all phases to set boundaries on the types of drug product manufacturing changes permitted under quality by design. However, unlike the assays and biomarkers considered under the FDA's drug development tool guidance, computational models are viewed as dynamic and in need of revision as soon as new knowledge becomes available about the chemical and biological process they are intended to represent. Consequently, modeling and simulation in drug development are seen as "fit for purpose" and tightly constrained by the specific data sets used to calibrate and validate model performance.

While the current drug development tool qualification process does not extend to the use of *in silico* models, the recent ICH M7 guidelines issued for the use of (Q)SAR models to assess the genotoxicity of drugs, metabolites, and product contaminants/impurities refer to a set of principles for model validation developed by the Organisation for Economic Co-operation and Development (OECD) (International Council on Harmonization, 2014). The OECD principles state that, to be considered valid for regulatory use, a (Q)SAR model should be associated with the following information: a defined endpoint; an unambiguous algorithm; a defined domain of applicability (i.e., context of use); appropriate measures of goodness of fit, robustness, and predictivity; and a mechanistic interpretation, if possible. There are clear parallels between these requirements and those applied to the technical validation and qualification of new drug development tools as currently implemented by the FDA. This may be useful to consider as a framework for evaluating the general regulatory utility of an *in silico* model.

1.3.2 Cell-Based Assays for Toxicity Prediction

As noted previously, the purpose of preclinical toxicology testing is to identify potential organ toxicities and the drug levels at which they occur so that a safe starting dose in human trials can be determined. New technologies intended to replace or supplement existing safety assessment pathways should have this as their ultimate goal. In cases where *in vitro* assays using human cells or cell lines are used, including iPSC-derived organotypic cells, initial questions to be asked include: (i) How closely does the assay replicate or predict the human outcome of interest? (ii) Can the assay provide knowledge about the drug concentration ranges producing the effect? (iii) Are the results sufficiently robust and reproducible across laboratories and studies to support a regulatory (vs. company internal) decision on product safety? In addition, it will be important to demonstrate that the relevant drug effects on the specific endpoints of interest can be distinguished from changes seen solely due to experimental constraints and conditions. Finally, concordance should be demonstrated with current approaches before new technologies are adopted for regulatory use.

One example of a cell-based assay that has been successfully incorporated into the safety assessment pathway is the assessment of drug-induced proarrhythmia risk based on block of the cardiac repolarization current IKr and the clinical assessment of the electrocardiographic QT interval, as discussed in the ICH S7B and ICH E14 harmonized guidelines. While the regulatory recommendations for assessing IKr pharmacology are quite broad and allow for the use of either native or expressed channels as systems for the study of IKr pharmacology, heterologous expression of the hERG channel in a clonal cell line is widely used as a readily accessible human test system that meets the basic requirements for accepted regulatory use: it is scientifically valid and robust, assay protocols can be standardized, the results are reasonably reproducible, and the measured endpoint is considered relevant for assessing human risk. The assay is also attractive for drug developers because it can be performed using either manual or high-throughput automated patch clamp methods, making it possible to screen larger compound libraries in the drug discovery phase prior to candidate selection. The hERG assay is most often conducted at room temperature using a hERG channel assembled as 1a subunits due to improved expression and ease of measurement (see, e.g., Chen et al., 2007), even though in the adult human heart the IKr channel appears to exist as the combination of 1a/1b subunits (London et al., 1997; Jones et al., 2004, 2014). Studies have shown that heteromeric hERG 1a/1b currents are much larger in magnitude and exhibit faster gating kinetics than channels composed of hERG 1a subunits only (Sale et al., 2008), and also exhibit different drug sensitivities (Abi-Gerges et al., 2011), potentially confounding the assessment of clinical risk. The use of room temperature in the hERG assay represents an additional factor to consider when evaluating predictivity of the assay, as raising the temperature increases current magnitude and also speeds the kinetics of channel gating (Milnes et al., 2010). Finally, drug effects have been typically measured in terms of IC_{50} values, despite the recognition that channel block is dynamic with a marked dependence on transmembrane voltage, channel state, and the frequency of stimulation. A final challenge is in relating the concentration used *in vitro* to the drug concentrations predicted for efficacy, taking into account protein binding, to provide a window between therapeutic and toxic levels.

Despite these apparent limitations, the hERG assay and the subsequent clinical TQT study have been able to identify potentially torsadogenic drugs early on and prevent their entry into the market. However, some clinically important drugs have been found to block IKr and prolong the QT interval at therapeutic plasma concentrations, but not to be proarrhythmic. The almost decade of experience with the regulatory pathways outlined in ICH S7B and ICH E14 has indicated that while the hERG–TQT paradigm may be highly sensitive to potentially torsadogenic drugs, it is not very accurate in predicting actual clinical risk. Consequently, there is a concern that a number of new drugs with interesting and therapeutically important profiles may have been terminated early in development due to a positive hERG result. Ventricular repolarization in the heart is a complex process that depends on the time- and voltage-dependent interactions of a variety of ion channels and membrane transport mechanisms. In many cases, drugs that block hERG also have activity at other ion channels that can exacerbate or mask the effect of a reduction of IKr on QT prolongation and the appearance of ventricular arrhythmia.

To address this limitation, the FDA's Center for Drug Evaluation and Research is collaborating with a wide range of scientists representing industry, academic, and nonprofit groups, including the Cardiovascular Safety Research Consortium, the Safety Pharmacology Society, and ILSI-HESI, to develop and characterize a new way of approaching the prediction of drug-induced proarrhythmia. The Comprehensive *In Vitro* Proarrhythmia Assay (CiPA) initiative is proposing to integrate measurements of drug effects on multiple cardiac ion channels with *in silico* models of the human ventricular myocyte and the results from studies using iPSC-derived cardiomyocytes to create a mechanism-based ranking of torsadogenic risk for investigational drugs while eliminating the need for the clinical TQT study and concerns about its potential false positives (Sager et al., 2014; Fermini et al., 2015).

1.4 CONCLUSIONS

The effort needed to advance a drug from discovery through development to approval remains time and resource intensive, and despite best efforts, unanticipated adverse events leading to late-stage attrition or market withdrawal can still occur. As scientific advances continue to yield with new tools and technologies with better performance characteristics and predictive power than the traditional assays and biomarkers used in drug development, it will become increasingly important to see that these approaches are thoroughly tested and rigorously validated and find their way into regulatory decision-making.

REFERENCES

Abaci HE and Shuler ML (2015) Human-on-a-chip design strategies and principles for physiologically based pharmacokinetics/pharmacodynamics modeling. *Integr Biol (Camb).* 7(4):383–91.

Abi-Gerges N, Holkham H, Jones EM, Pollard CE, Valentin JP and Robertson GA (2011) hERG subunit composition determines differential drug sensitivity. *Br J Pharmacol.* 164(2b):419–32.

Alden CL, Lynn A, Bourdeau A, Morton D, Sistare FD, Kadambi VJ and Silverman L (2011) A critical review of the effectiveness of rodent pharmaceutical carcinogenesis testing in predicting for human risk. *Vet Pathol.* 48(3):772–84.

Amur S, LaVange L, Zineh I, Buckman-Garner S and Woodcock J (2015) Biomarker Qualification: toward a multi-stakeholder framework for biomarker development, regulatory acceptance, and utilization. *Clin Pharmacol Ther.* 98(1):34–46.

Arrowsmith J (2011a) Trial watch: phase II failures: 2008–2010. *Nat Rev Drug Discov.* 10(5):328–9.

Arrowsmith J (2011b) Trial watch: phase III and submission failures: 2007–2010. *Nat Rev Drug Discov.* 10(2):87.

Arrowsmith J and Miller P (2013) Trial watch: phase II and phase III attrition rates 2011–2012. *Nat Rev Drug Discov.* 12(8):569.

Bouhifd M, Andersen ME, Baghdikian C, Boekelheide K, Crofton KM, Fornace AJ Jr, Kleensang A, Li H, Livi C, Maertens A, McMullen PD, Rosenberg M, Thomas R, Vantangoli M, Yager JD, Zhao L and Hartung T (2015) The human toxome project. *ALTEX.* 32(2):112–24.

Butcher EC, Berg EL and Kunkel EJ (2004) Systems biology in drug discovery. *Nat Biotechnol.* 22(10):1253–9.

Butz RF and Morelli G (2008) Innovative strategies for early clinical R&D. *IDrugs.* 11(1):36–41.

Castle AL, Carver MP and Mendrick DL (2002) Toxicogenomics: a new revolution in drug safety. *Drug Discov Today.* 7(13):728–36.

Chen MX, Sandow SL, Doceul V, Chen YH, Harper H, Hamilton B, Meadows HJ, Trezise DJ and Clare JJ (2007) Improved functional expression of recombinant human ether-a-go-go (hERG) K+ channels by cultivation at reduced temperature. *BMC Biotechnol.* 7:93.

Cheung C and Gonzalez FJ (2008) Humanized mouse lines and their application for prediction of human drug metabolism and toxicological risk assessment. *J Pharmacol Exp Ther.* 327(2):288–99.

Clarke CJ and Haselden JN (2008) Metabolic profiling as a tool for understanding mechanisms of toxicity. *Toxicol Pathol.* 36(1):140–7.

Darpo B, Nebout T and Sager PT (2006) Clinical evaluation of QT/QTc prolongation and proarrhythmic potential for nonantiarrhythmic drugs: the International Conference on Harmonization of Technical Requirements for Registration of Pharmaceuticals for Human Use E14 guideline. *J Clin Pharmacol.* 46(5):498–507.

Dieterle F, Sistare F, Goodsaid F, Papaluca M, Ozer JS, Webb CP, Baer W, Senagore A, Schipper MJ, Vonderscher J, Sultana S, Gerhold DL, Phillips JA, Maurer G, Carl K, Laurie D, Harpur E, Sonee M, Ennulat D, Holder D, Andrews-Cleavenger D, Gu YZ, Thompson KL, Goering PL, Vidal JM, Abadie E, Maciulaitis R, Jacobson-Kram D, Defelice AF, Hausner EA, Blank M, Thompson A, Harlow P, Throckmorton D, Xiao S, Xu N, Taylor W, Vamvakas S, Flamion B, Lima BS, Kasper P, Pasanen M, Prasad K, Troth S, Bounous D, Robinson-Gravatt D, Betton G, Davis MA, Akunda J, McDuffie JE, Suter L, Obert L, Guffroy M, Pinches M, Jayadev S, Blomme EA, Beushausen SA, Barlow VG, Collins N, Waring J, Honor D, Snook S, Lee J, Rossi P, Walker E, Mattes W (2010) Renal biomarker qualification submission: a dialog between the FDA-EMEA and Predictive Safety Testing Consortium. *Nat Biotechnol.* 28(5):455–62.

Doherty KR, Talbert DR, Trusk PB, Moran DM, Shell SA and Bacus S (2015) Structural and functional screening in human induced-pluripotent stem cell-derived cardiomyocytes accurately identifies cardiotoxicity of multiple drug types. *Toxicol Appl Pharmacol.* 285(1):51–60.

Edwards SW and Preston RJ (2008) Systems biology and mode of action based risk assessment. *Toxicol Sci.* 106(2):312–8.

Ennulat D and Adler S (2015) Recent successes in the identification, development, and qualification of translational biomarkers: the next generation of kidney injury biomarkers. *Toxicol Pathol.* 43(1):62–9.

Ewart L, Aylott M, Deurinck M, Engwall M, Gallacher DJ, Geys H, Jarvis P, Ju H, Leishman D, Leong L, McMahon N, Mead A, Milliken P, Suter W, Teisman A, Van Ammel K, Vargas HM, Wallis R and Valentin JP (2014) The concordance between nonclinical and phase I clinical cardiovascular assessment from a cross-company data sharing initiative. *Toxicol Sci.* 142(2):427–35.

Fermini B, Hancox JC, Abi-Gerges N, Bridgland-Taylor M, Chaudhary KW, Colatsky T, Correll K, Crumb W, Damiano B, Erdemli G, Gintant G, Imredy J, Koerner J, Kramer J, Levesque P, Li Z, Lindqvist A, Obejero-Paz CA, Rampe D, Sawada K, Strauss DG, Vandenberg JI (2015) A new perspective in the field of cardiac safety testing through the comprehensive in vitro proarrhythmia assay paradigm. *J Biomol Screen.* pii:1087057115594589. [Epub ahead of print]

Fischer HP (2005) Towards quantitative biology: integration of biological information to elucidate disease pathways and to guide drug discovery. *Biotechnol Annu Rev.* 11:1–68.

Food Drug Administration (2004) *Challenges and Opportunity on the Critical Path to New Medical Products* (FDA Maryland).

Food Drug Administration (2004) *Critical Path Opportunity List* (FDA Maryland).

Food Drug Administration (2005a) *International Conference on Harmonisation Guidance on S7B Nonclinical Evaluation of the Potential for Delayed Ventricular Repolarization (QT Interval Prolongation) by Human Pharmaceuticals* (FDA Maryland).

Food Drug Administration (2005b) *Guidance for Industry: Clinical Evaluation of QT/QTc Interval Prolongation and Proarrhythmic Potential for Non-Antiarrhythmic Drugs* (FDA Maryland).

Food Drug Administration (2005c) *Estimating the Maximum Safe Starting Dose in Initial Clinical Trials for Therapeutics in Adult Healthy Volunteers* (FDA Maryland).

Food Drug Administration (2006a) *Guidance for Industry, Investigators, and Reviewers: Exploratory IND Studies* (FDA Maryland).

Food Drug Administration (2006b) *Critical Path Opportunity Report* (FDA Maryland).

Food Drug Administration (2011) *Advancing Regulatory Science at FDA: A Strategic Plan* (FDA Maryland).

Food Drug Administration (2014) *Guidance for Industry and FDA Staff: Qualification Process for Drug Development Tools* (FDA Maryland).

Ganter B, Snyder RD, Halbert DN and Lee MD (2006) Toxicogenomics in drug discovery and development: mechanistic analysis of compound/class-dependent effects using the DrugMatrix database. *Pharmacogenomics.* 7(7):1025–44.

Goodsaid FM (2003) Genomic biomarkers of toxicity. *Curr Opin Drug Discov Devel.* 6(1):41–9.

Goodwin D, Rosenzweig B, Zhang J, Xu L, Stewart S, Thompson K and Rouse R (2014) Evaluation of miR-216a and miR-217 as potential biomarkers of acute pancreatic injury in rats and mice. *Biomarkers.* 19(6):517–29.

Guerreiro N, Staedtler F, Grenet O, Kehren J and Chibout SD (2003) Toxicogenomics in drug development. *Toxicol Pathol.* 31(5):471–9.

Hammond TG, Pollard CE (2005) Use of *in vitro* methods to predict QT prolongation. *Toxicol Appl Pharmacol.* 207(2 Suppl):446–50.

Harpaz R, DuMouchel W, LePendu P, Bauer-Mehren A, Ryan P and Shah NH (2013) Performance of pharmacovigilance signal detection algorithms for the FDA adverse event reporting system. *Clin Pharmacol Ther.* 93:539–46.

Harpur E, Ennulat D, Hoffman D, Betton G, Gautier JC, Riefke B, Bounous D, Schuster K, Beushausen S, Guffroy M, Shaw M, Lock E, Pettit S and HESI Committee on Biomarkers of Nephrotoxicity (2011) Biological qualification of biomarkers of chemical-induced renal toxicity in two strains of male rat. *Toxicol Sci.* 122(2):235–52.

Hausner EA, Hicks KA, Leighton JK, Szarfman A, Thompson AM and Harlow P (2013) Qualification of cardiac troponins for nonclinical use: a regulatory perspective. *Regul Toxicol Pharmacol.* 67(1):108–14.

Holzgrefe H, Ferber G, Champeroux P, Gill M, Honda M, Greiter-Wilke A, Baird T, Meyer O and Saulnier M (2014) Preclinical QT safety assessment: cross-species comparisons and human translation from an industry consortium. *J Pharmacol Toxicol Methods.* 69(1):61–101.

Hood L and Perlmutter RM (2004) The impact of systems approaches on biological problems in drug discovery. *Nat Biotechnol.* 22(10):1215–7.

Huh D, Hamilton GA and Ingber DE (2011) From 3D cell culture to organs-on-chips. *Trends Cell Biol.* 21:745–54.

International Council on Harmonization (2001) S7A: Safety pharmacology studies for human pharmaceuticals (Geneva, Switzerland).

International Council on Harmonization (2005) S7B: Safety pharmacology studies for human pharmaceuticals (Geneva, Switzerland).

International Council on Harmonization (2010) M2(R2) Guidance on nonclinical safety studies for the conduct of human clinical trials and marketing authorization for pharmaceuticals (Geneva, Switzerland).

International Council on Harmonization (2014) M7: Assessment and control of DNA reactive (mutagenic) impurities in pharmaceuticals to limit potential carcinogenic risk (Geneva, Switzerland).

Ito K, Corrigan B, Romero K, Anziano R, Neville J, Stephenson D and Lalonde R (2013) Understanding placebo responses in Alzheimer's disease clinical trials from the literature meta-data and CAMD database. *J Alzheimers Dis.* 37(1):173–83.

Johannesen L, Vicente J, Gray RA, Galeotti L, Loring Z, Garnett CE, Florian J, Ugander M, Stockbridge N and Strauss DG (2014) Improving the assessment of heart toxicity for all new drugs through translational regulatory science. *Clin Pharmacol Ther.* 95(5):501–8.

Jones EM, Roti EC, Wang J, Delfosse SA and Robertson GA (2004) Cardiac IKr channels minimally comprise hERG 1a and 1b subunits. *J Biol Chem.* 279(43):44690–4.

Jones DK, Liu F, Vaidyanathan R, Eckhardt LL, Trudeau MC and Robertson GA (2014) hERG 1b is critical for human cardiac repolarization. *Proc Natl Acad Sci U S A.* 111(50):18073–7.

Judson RS, Martin, MT, Egeghy P, Gangwal S, Reif DM, Kothiya P, Wolf M, Cathey T, Transue T, Smith D, Vail J, Frame A, Mosher S, Cohen Hubal EA and Richard AM (2012) Aggregating data for computational toxicology applications: the U.S. Environmental Protection Agency (EPA) Aggregated Computational Toxicology Resource (ACToR) System. *Int J Mol Sci.* 13(2): 1805–31.

Kell DB (2006) Systems biology, metabolic modelling and metabolomics in drug discovery and development. *Drug Discov Today.* 11(23–24):1085–92.

Kennedy S (2002) The role of proteomics in toxicology: identification of biomarkers of toxicity by protein expression analysis. *Biomarkers.* 7(4):269–90.

Kiehn J, Wible B, Ficker E, Taglialatela M and Brown AM (1995) Cloned human inward rectifier K+ channel as a target for class III methanesulfonanilides. *Circ Res.* 77(6):1151–5.

Kolaja K (2014) Stem cells and stem cell-derived tissues and their use in safety assessment. *J Biol Chem.* 289(8):4555–61.

Kruhlak NL, Benz RD, Zhou H and Colatsky TJ (2012) (Q)SAR modeling and safety assessment in regulatory review. *Clin Pharmacol Ther.* 91(3):529–34.

Lindon JC, Holmes E and Nicholson JK (2007) Metabonomics in pharmaceutical R&D. *FEBS J.* 274(5):1140–51.

Liu M, Hu Y and Tang B (2013) Role of text mining in early identification of potential drug safety issues. *Methods Mol Biol.* 1159:227–51.

London B, Trudeau MC, Newton KP, Beyer AK, Copeland NG, Gilbert DJ, Jenkins NA, Satler CA and Robertson GA (1997) Two isoforms of the mouse ether-a-go-go-related gene coassemble to form channels with properties similar to the rapidly activating component of the cardiac delayed rectifier K+ current. *Circ Res.* 81(5):870–8.

Mathur A, Loskill P, Hong S, Lee J, Marcus SG, Dumont L, Conklin BR, Willenbring H, Lee LP and Healy KE (2013) Human induced pluripotent stem cell-based microphysiological tissue models of myocardium and liver for drug development. *Stem Cell Res Ther.* 4(Suppl 1):S14.

Milnes JT, Witchel HJ, Leaney JL, Leishman DJ, Hancox JC (2010) Investigating dynamic-protocol dependence of hERG potassium channel inhibition at 37°C: cisparide versus dofetilide. *J Pharmacol Toxicol Methods.* 61:178–91.

Mirams GR, Cui Y, Sher A, Fink M, Cooper J, Heath BM, McMahon NC, Gavaghan DJ and Noble D (2011) Simulation of multiple ion channel block provides improved early prediction of compounds' clinical torsadogenic risk. *Cardiovasc Res.* 91(1):53–61.

Mirams GR, Davies MR, Cui Y, Kohl P and Noble D (2012) Application of cardiac electrophysiology simulations to proarrhythmic safety testing. *Br J Pharmacol.* 167(5):932–45.

Olson H, Betton G, Stritar J and Robinson D (1998) The predictivity of the toxicity of pharmaceuticals in humans from animal data—an interim assessment. *Toxicol Lett.* 102–103:535–8.

Olson H, Betton G, Robinson D, Thomas K, Monro A, Kolaja G, Lilly P, Sanders J, Sipes G, Bracken W, Dorato M, Van Deun K,

Smith P, Berger B and Heller A (2000) Concordance of the toxicity of pharmaceuticals in humans and in animals. *Regul Toxicol Pharmacol.* 32(1):56–67.

Panegyres PK, Chen HY and the Coalition against Major Diseases (CAMD) (2014) Early-onset Alzheimer's disease: a global cross-sectional analysis. *Eur J Neurol.* 21(9):1149–54.

Parekh A, Buckman-Garner S, McCune S, O'Neill R, Geanacopoulos M, Amur S, Clingman C, Barratt R, Rocca M, Hills I and Woodcock J (2015) Catalyzing the Critical Path Initiative: FDA's progress in drug development activities. *Clin Pharmacol Ther.* 97(3):221–33.

Robertson DG (2005) Metabonomics in toxicology: a review. *Toxicol Sci.* 85(2):809–22.

Rogers JA, Polhamus D, Gillespie WR, Ito K, Romero K, Qiu R, Stephenson D, Gastonguay MR and Corrigan B (2012) Combining patient-level and summary-level data for Alzheimer's disease modeling and simulation: a β regression meta-analysis. *J Pharmacokinet Pharmacodyn.* 39(5):479–98.

Romero K, Ito K, Rogers JA, Polhamus D, Qiu R, Stephenson D, Mohs R, Lalonde R, Sinha V, Wang Y, Brown D, Isaac M, Vamvakas S, Hemmings R, Pani L, Bain LJ, Corrigan B and the Alzheimer's Disease Neuroimaging Initiative; Coalition Against Major Diseases (2015) The future is now: model-based clinical trial design for Alzheimer's disease. *Clin Pharmacol Ther.* 297(3):210–4.

Rouse R (2015) Regulatory forum opinion piece*: blinding and binning in histopathology methods in the biomarker qualification process. *Toxicol Pathol.* 43(6):757–9.

Rouse R, Zhang J, Stewart SR, Rosenzweig BA, Espandiari P and Sadrieh NK (2011) Comparative profile of commercially available urinary biomarkers in preclinical drug-induced kidney injury and recovery in rats. *Kidney Int.* 79(11):1186–97.

Rouse R, Siwy J, Mullen W, Mischak H, Metzger J and Hanig J (2012) Proteomic candidate biomarkers of drug-induced nephrotoxicity in the rat. *PLoS One.* 7(4):e34606.

Rouse R, Min M, Francke S, Mog S, Zhang J, Shea K, Stewart S and Colatsky T (2014) Impact of pathologists and evaluation methods on performance assessment of the kidney injury biomarker, kim-1. *Toxicol Pathol.* 43(5):662–74.

Sager PT, Gintant G, Turner JR, Pettit S and Stockbridge N (2014) Rechanneling the cardiac proarrhythmia safety paradigm: a meeting report from the Cardiac Safety Research Consortium. *Am Heart J.* 167(3):292–300.

Sale H, Wang J, O'Hara TJ, Tester DJ, Phartiyal P, He JQ, Rudy Y, Ackerman MJ and Robertson GA (2008) Physiological properties of hERG 1a/1b heteromeric currents and a hERG 1b-specific mutation associated with Long-QT syndrome. *Circ Res.* 103(7):e81–e95.

Sanguinetti MC, Jiang C, Curran ME and Keating MT (1995) A mechanistic link between an inherited and an acquired cardiac arrhythmia: HERG encodes the IKr potassium channel. *Cell.* 81(2):299–307.

Sarapa N (2007) Exploratory IND: a new regulatory strategy for early clinical drug development in the United States. *Ernst Schering Res Found Workshop.* 59:151–63.

Schmeichel KL and Bissell MJ (2003) Modeling tissue-specific signaling and organ function in three dimensions. *J Cell Sci.* 116:2377–88.

Shea K, Stewart S and Rouse R (2014) Assessment standards: comparing histopathology, digital image analysis, and stereology for early detection of experimental cisplatin-induced kidney injury in rats. *Toxicol Pathol.* 42(6):1004–15.

Shultz LD, Brehm MA, Garcia-Martinez JV and Greiner DL (2012) Humanized mice for immune system investigation: progress, promise and challenges. *Nat Rev Immunol.* 12(11):786–98.

Sirenko O, Cromwell EF, Crittenden C, Wignall JA, Wright FA and Rusyn I (2013) Assessment of beating parameters in human induced pluripotent stem cells enables quantitative in vitro screening for cardiotoxicity. *Toxicol Appl Pharmacol.* 273(3):500–7.

Sirenko O, Hesley J, Rusyn I and Cromwell EF (2014a) High-content assays for hepatotoxicity using induced pluripotent stem cell-derived cells. *Assay Drug Dev Technol.* 12(1):43–54.

Sirenko O, Hesley J, Rusyn I and Cromwell EF (2014b) High-content high-throughput assays for characterizing the viability and morphology of human iPSC-derived neuronal cultures. *Assay Drug Dev Technol.* 12(9–10):536–47.

Sung JH, Esch MB, Prot JM, Long CJ, Smith A, Hickman JJ and Shuler ML (2013) Microfabricated mammalian organ systems and their integration into models of whole animals and humans. *Lab Chip.* 13:1201–12.

Szarfman A, Tonning JM and Doraiswamy PM (2004) Pharmacovigilance in the 21st century: new systematic tools for an old problem. *Pharmacotherapy.* 24(9):1099–104.

Takimoto CH (2008) Phase 0 clinical trials in oncology: a paradigm shift for early drug development? *Cancer Chemother Pharmacol.* 63(4):703–9.

Venter JC (2000) Genomic impact on pharmaceutical development. *Novartis Found Symp.* 229:14–5; discussion 15–8.

Venter JC, Adams MD, Myers EW, Li PW, Mural RJ, Sutton GG, Smith HO, Yandell M, Evans CA, Holt RA, Gocayne JD, Amanatides P, Ballew RM, Huson DH, Wortman JR, Zhang Q, Kodira CD, Zheng XH, Chen L, Skupski M, Subramanian G, Thomas PD, Zhang J, Gabor Miklos GL, Nelson C, Broder S, Clark AG, Nadeau J, McKusick VA, Zinder N, Levine AJ, Roberts RJ, Simon M, Slayman C, Hunkapiller M, Bolanos R, Delcher A, Dew I, Fasulo D, Flanigan M, Florea L, Halpern A, Hannenhalli S, Kravitz S, Levy S, Mobarry C, Reinert K, Remington K, Abu-Threideh J, Beasley E, Biddick K, Bonazzi V, Brandon R, Cargill M, Chandramouliswaran I, Charlab R, Chaturvedi K, Deng Z, Di Francesco V, Dunn P, Eilbeck K, Evangelista C, Gabrielian AE, Gan W, Ge W, Gong F, Gu Z, Guan P, Heiman TJ, Higgins ME, Ji RR, Ke Z, Ketchum KA, Lai Z, Lei Y, Li Z, Li J, Liang Y, Lin X, Lu F, Merkulov GV,

Milshina N, Moore HM, Naik AK, Narayan VA, Neelam B, Nusskern D, Rusch DB, Salzberg S, Shao W, Shue B, Sun J, Wang Z, Wang A, Wang X, Wang J, Wei M, Wides R, Xiao C, Yan C, Yao A, Ye J, Zhan M, Zhang W, Zhang H, Zhao Q, Zheng L, Zhong F, Zhong W, Zhu S, Zhao S, Gilbert D, Baumhueter S, Spier G, Carter C, Cravchik A, Woodage T, Ali F, An H, Awe A, Baldwin D, Baden H, Barnstead M, Barrow I, Beeson K, Busam D, Carver A, Center A, Cheng ML, Curry L, Danaher S, Davenport L, Desilets R, Dietz S, Dodson K, Doup L, Ferriera S, Garg N, Glueksmann A, Hart B, Haynes J, Haynes C, Heiner C, Hladun S, Hostin D, Houck J, Howland T, Ibegwam C, Johnson J, Kalush F, Kline L, Koduru S, Love A, Mann F, May D, McCawley S, McIntosh T, McMullen I, Moy M, Moy L, Murphy B, Nelson K, Pfannkoch C, Pratts E, Puri V, Qureshi H, Reardon M, Rodriguez R, Rogers YH, Romblad D, Ruhfel B, Scott R, Sitter C, Smallwood M, Stewart E, Strong R, Suh E, Thomas R, Tint NN, Tse S, Vech C, Wang G, Wetter J, Williams S, Williams M, Windsor S, Winn-Deen E, Wolfe K, Zaveri J, Zaveri K, Abril JF, Guigó R, Campbell MJ, Sjolander KV, Karlak B, Kejariwal A, Mi H, Lazareva B, Hatton T, Narechania A, Diemer K, Muruganujan A, Guo N, Sato S, Bafna V, Istrail S, Lippert R, Schwartz R, Walenz B, Yooseph S, Allen D, Basu A, Baxendale J, Blick L, Caminha M, Carnes-Stine J, Caulk P, Chiang YH, Coyne M, Dahlke C, Mays A, Dombroski M, Donnelly M, Ely D, Esparham S, Fosler C, Gire H, Glanowski S, Glasser K, Glodek A, Gorokhov M, Graham K, Gropman B, Harris M, Heil J, Henderson S, Hoover J, Jennings D, Jordan C, Jordan J, Kasha J, Kagan L, Kraft C, Levitsky A, Lewis M, Liu X, Lopez J, Ma D, Majoros W, McDaniel J, Murphy S, Newman M, Nguyen T, Nguyen N, Nodell M, Pan S, Peck J, Peterson M, Rowe W, Sanders R, Scott J, Simpson M, Smith T, Sprague A, Stockwell T, Turner R, Venter E, Wang M, Wen M, Wu D, Wu M, Xia A, Zandieh A and Zhu X (2001) The sequence of the human genome. *Science.* 291(5507):1304–51.

Walgren JL and Thompson DC (2004) Application of proteomic technologies in the drug development process. *Toxicol Lett.* 149(1–3):377–85.

Wang B and Gray G (2014) Concordance of noncarcinogenic endpoints in rodent chemical bioassays. *Risk Anal.* 35(6):1154–66.

White RW, Tatonetti NP, Shah NH, Altman RB and Horvitz E (2013) Web-scale pharmacovigilance: listening to signals from the crowd. *J Am Med Inform Assoc.* 20(3):404–8.

Witzmann FA and Grant RA (2003) Pharmacoproteomics in drug development. *Pharmacogenomics J.* 3(2):69–76.

Woodcock J (2007) FDA's critical path initiative. *Drug Discov Today Technol.* 4(2):33.

Zhang B, Duan Z and Zhao Y (2009) Mouse models with human immunity and their application in biomedical research. *J Cell Mol Med.* 13(6):1043–58.

PART II

SAFETY LEAD OPTIMIZATION STRATEGIES

2

SMALL-MOLECULE SAFETY LEAD OPTIMIZATION

DONNA M. DAMBACH

Safety Assessment, Genentech Inc., South San Francisco, CA, USA

2.1 BACKGROUND AND OBJECTIVES OF SAFETY LEAD OPTIMIZATION APPROACHES

While the strategy of embedding nonclinical safety scientists on teams in the discovery space is now well established, this approach, often called "discovery" or "exploratory" toxicology, is <15 years old (Sasseville et al., 2004; Dambach and Gautier, 2006; Kramer et al., 2007; Hornberg et al., 2014a, b). Discovery toxicology was initially instituted for small molecules and was borne out of the shift to structure-based, combinatorial chemistry design strategies, that is, rational drug design, and high-throughput screening (HTS) approaches that delivered hundreds of molecules, coupled with the success of applying a "candidate lead optimization" approach for the identification of ADME/PK parameters that resulted in reducing attrition due to PK and bioavailability (Kerns and Di, 2003; Kola and Landis, 2004; Kramer et al., 2007).

With regard to small-molecule therapeutics, the discovery phase is the period when there is intense medicinal chemistry design activity to identify a drug candidate with the optimal characteristics related to pharmacology, pharmaceutics, ADME/PK, and safety. At this early stage, target candidate criteria are established to guide the "lead optimization" strategy for the significant characteristics desired of a drug candidate (the "lead"). From a nonclinical safety perspective, the lead optimization strategy is customized for each target in the context of risk tolerance for the intended therapeutic area (e.g., life-threatening/unmet medical need versus non-life-threatening) and includes activities to evaluate possible pharmacology-mediated ("on-target") activity and chemical structure-mediated ("off-target") activity that may result in undesired effects. The strategy is implemented through the use of a combination of computational (*in silico*), *in vitro*, and *in vivo* models, similar to the approach taken to evaluate efficacy, pharmaceutics, and ADME/PK properties. In particular, the most typical paradigm used for lead candidate selection utilizes a tiered approach in which computational and/or *in vitro* assays are used as first- and/or second-tier assessments of a scaffold series or compounds of interest to "flag" potential issues, followed by confirmation of findings in additional, more physiologically relevant *in vitro* models, *in vivo* assays, or a standard regulatory assay to build a weight of evidence for a potential translatable relevance and characterization of a liability (Fig. 2.1).

Front-loading safety assessment activities in parallel with those for efficacy, ADME/PK, and pharmaceutics properties enables chemists to incorporate the best overall features of a molecule during the period of chemical design, removes the worst molecules, and allows time to investigate and characterize underlying mechanisms of toxicity and the identified liabilities as to their clinical significance to inform the clinical development plans and minimize clinical attrition due to toxicity. The inclusion of discovery safety activities also has value for biotherapeutics, although the approaches utilized will differ based on the molecular platform (e.g., peptide, RNA, monoclonal antibody, antibody–drug conjugate).

Drug Discovery Toxicology: From Target Assessment to Translational Biomarkers, First Edition. Edited by Yvonne Will, J. Eric McDuffie, Andrew J. Olaharski, and Brandon D. Jeffy.
© 2016 John Wiley & Sons, Inc. Published 2016 by John Wiley & Sons, Inc.

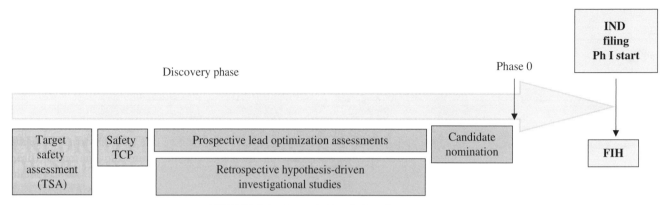

FIGURE 2.1   Key activities during discovery phase.

Regardless of the molecular platform, the overall goals of safety lead optimization are as follows:

- To identify the best candidate for advancement with a thorough understanding of the identified risks and their translation to humans
- To make most effective use of resources and minimize animal use that results in decisions to remove the worst candidates early (prior to clinical studies)

This chapter will describe safety lead optimization strategic approaches for small molecules, including commonly used tools and data integration approaches to inform decision-making.

2.2   TARGET SAFETY ASSESSMENTS: EVALUATION OF UNDESIRED PHARMACOLOGY AND THERAPEUTIC AREA CONSIDERATIONS

The purpose of target safety assessment (TSA) is to gain insight into the potential liabilities that may result from engaging a target and to understand the risk tolerance for liabilities based on the intended therapeutic area. The information derived from this analysis is used by nonclinical safety scientists to (i) educate project teams of the safety considerations that are used to formulate the lead optimization and overall safety strategy for the targeted therapeutic, (ii) generate a target candidate profile that defines the key characteristics of a drug-like molecule, and (iii) formulate the safety lead optimization activities customized to the target, including investigation of theoretical risks. The TSA is performed early in drug discovery, typically at initiation of the project when there is no chemical matter (for small molecules). An example of the information gathered for a TSA is shown in Table 2.1.

A key component of the TSA is an evaluation of the biology of the target and the effects of engaging that target, either inhibition or activation, through review of the literature and internal company expertise. Information gathered includes the expression distribution (cell and organ), isoforms, species similarities or differences in activity or homology, and any mechanistic data available, including effects in genetically engineered mouse or rat models. The outcome of this evaluation will inform theoretical/possible target organ toxicities, and identification of species similarities or differences that may impact the choice of nonclinical species or provide awareness of differential target organ toxicity. This information may trigger prospective investigation of the significance of a potential or theoretical liability identified and inclusion of specific counter-assays or endpoints for decision-making around the feasibility or "druggability" of the target or identifying a lead candidate. Two recent examples of investigational and counter-assay approaches are described by Zabka et al. (2015) and Tarrant et al. (2015).

A second component of the TSA is the examination of the tolerance of risk or liability for the intended therapeutic area. In particular, the risk tolerance is often greater for life-threatening indications or indications of unmet medical need, for example, ICH S9-designated indications, versus indications where there are numerous treatment modalities or indications that require chronic dosing, that is, non-life-threatening or ICH M3-designated indications (ICH, 2009a, b). The assessment of the therapeutic area tolerance involves collaboration with clinical safety scientists and clinicians, review of competitive landscape, an understanding of combination or cotherapies that might be administered, and common comorbidities found in the patient population. An additional component of the TSA would include any specific considerations of the target that would impact the success of the drug candidate, for example, distribution considerations such as brain penetration.

2.3   IMPLEMENTING LEAD OPTIMIZATION STRATEGIES FOR SMALL MOLECULES

As mentioned earlier, discovery safety activities have value for small molecules and biotherapeutics. However, the approaches and assays utilized will differ based on the molecular platform.

TABLE 2.1   Target Safety Assessment (TSA) Elements

Assessment Component	Items for Consideration	Example Evaluation Resources
Target pharmacology	• Potential undesired on-target pharmacology • Distribution of the target across tissues • Species homology of the target • Target isoforms • Transgenic (GERM) strategy	• Literature and available databases • Internal knowledge • Genetic animal models (KO, cKO, Tg mice) • Genetic mutations in humans
Therapeutic area and patient population	• Level of safety required • Unacceptable/undesirable clinical AEs • Potential interactions with comeds	• Engage clinical and drug safety representative • Literature and public databases
Competitive landscape	• Safety of competitors (same target) • Safety of competitors (same/related pathway)	• Literature and public databases • Competitive intelligence • Internal assessment of competitors (biologics or small molecules)
Core screening strategy components (small molecule)	• Promiscuity and selectivity • Intrinsic cytotoxicity • Genetic toxicology (non-life-threatening) • Specific target organ toxicity • Target organ assessment: cardiovascular, hepatic, hematopoietic • ADME/metabolic risk assessment	• Physicochemical risk profile; secondary pharmacology binding assays; target selectivity assays (e.g., kinase) • Primary hepatocyte viability • Ames II; *in vitro* micronucleus; GreenScreen, etc. • Tissue-specific models (e.g., CV, hematopoietic, hepatic) • Toxicophore assessment; covalent binding/burden; metabolic stability; metabolite profile (reactive metabolite formation); metabolic phenotyping; transporter assessment; blood–brain barrier assessment; tissue accumulation assessment

GERM, genetically modified rodent models.

2.3.1   Strategic Approach

Safety lead optimization strategies for small molecules encompass assessments for both undesired pharmacologically mediated ("on-target") and chemical structure-mediated ("off-target") toxicities. These safety activities are integrated concurrently with other lead optimization activities performed by the discovery teams including those for efficacy, ADME/PK, and pharmaceutics. This approach is most successful because it is occurring at the time to inform chemical design and build in the most useful characteristics of a candidate molecule. Because the tools and approaches used for safety lead optimization are identical to those used for the optimization of other molecule characteristics, namely, computational (*in silico*), in vitro, ex vivo, and *in vivo* models, discovery teams are well versed with the type of data generated by these types of approaches. However, to be most impactful, the context regarding how these approaches will be used for decision-making should be communicated to project teams, that is, to advance the most promising candidate, and safety lead optimization is based on a tiered, weight-of-evidence approach that requires robust investigation of issues, appropriate application of qualified models/assays, and rational interpretation of the data with an understanding of the limitations of each model platform. It is by no means a "checkbox" or "unsupervised" approach.

Safety lead optimization strategies employ what Kramer et al. (2007) described as "prospective" and "retrospective" approaches (Table 2.2). "Prospective" approaches are meant to "flag" or "predict" a potential undesired outcome of a known or established mechanism of toxicity, for example, genetic toxicity or engagement of endogenous ligands of clinical relevance, for which there is a level of confidence in the translational relevance. "Retrospective" approaches are traditionally applied to characterize a liability that has been identified, usually the result of a finding *in vivo* either in nonclinical species or humans or a theoretical risk identified by the TSA; this approach is based in hypothesis-driven investigative work.

2.3.2   Application of Prospective Models

For small molecules, most prospective (aka "predictive") models are used to identify potential chemical structure-mediated, off-target effects so that chemists can design away from unwanted features and chose the candidate with the overall best profile. The application of prospective models largely encompasses the evaluation of (i) selectivity/promiscuity, (ii) secondary and safety pharmacology, (iii) intrinsic cytotoxicity, (iv) ADME-based drivers of toxicity, and (v) genotoxicity. This core profiling battery is based on drivers

TABLE 2.2 Lead Optimization Assessment Components

Prospective Lead Optimization Core Components	Candidate Target Profile
Selectivity	• High selectivity • Low promiscuity • No/acceptable functional effects (agonist/antagonist)
Intrinsic cytotoxicity	• Low
ADME/PK risk	• Low intrinsic clearance • No/minimal reactive metabolite formation/covalent binding (burden) • Acceptable distribution profile (tissue/cellular)
Genetic toxicity (non-life-threatening indications)	• Eliminate mutagens and clastogens • Minimize aneugens
Cardiovascular toxicity	• Acceptable electrophysiological and hemodynamic profiles
Hepatic toxicity	• Minimize risk factors for iDILI
Target-specific counter-screens (identified through TSA)	• Acceptable profiles
Retrospective assessments	**Activity**
Hypothesis-driven investigation	• Follow up to TSA assessment of possible on-target risks and develop customized counter-screens • Follow up to observed toxicities during *in vivo* studies to determine drivers (on-target versus off-target) and characterize translation risk and monitoring

iDILI, idiosyncratic drug-induced liver injury.

of attrition that are well established ("what is known"). The goal is to identify highly selective molecules with minimal translatable *in vivo* effects, minimal intrinsic cytotoxicity, and acceptable ADME characteristics (minimal reactivity and accumulation). For non-life-threatening indications, candidate molecules should not be genotoxic (mutagenic, clastogenic, or aneugenic). In addition to this core profiling battery, it is common practice to employ *in vitro* target organ assays to assess the three major historical causes of clinical attrition: cardiovascular, hepatic, and hematopoietic toxicities (Stevens and Baker, 2009; Laverty et al., 2011). Finally, a company may have committed to specific technological approaches, for example, transcriptional profiling, that may be applied as part of the lead optimization strategy. Important components of lead optimization paradigms are that standardized assays screen against known risks; there is little (<10 mg) or no compound (in the case for *in silico* models) required; assays have rather quick turnaround (<2 weeks); and the positive or negative predictive values, as well as limitations, of the assays to predict clinical or nonclinical toxicities are generally well understood. Together these components enable facile review of data for decision-making.

2.3.2.1 Selectivity and Secondary Pharmacology Assessments

The purpose of minimizing promiscuity and enhancing selectivity is to reduce the potential for off-target effects, which are also known as secondary pharmacology effects. Both promiscuity and selectivity should be considered, as there is an important distinction between these traits. Promiscuity is a measure of the propensity of a molecule to bind other targets,

whereas selectivity takes promiscuity into account; it is often used to indicate a biochemical safety margin, that is, a molecule is indicated to be X-fold selective between the intended target ligand and an undesired target ligand. These are important distinctions because a molecule may be considered highly selective over an intended therapeutic range, but during toxicology studies, the achieved exposures may reach the range of off-target engagement and result in unintended effects. An understanding of these potential effects is important for the interpretation of findings in toxicology studies. Promiscuity and selectivity are assessed through integrated evaluation of the physicochemical structures of a molecule, known structure–activity relationship alerts (usually via computational assessments), and measures of secondary pharmacology, which include *in vitro* (and *in silico*) ligand binding and cell-based function assays and *in vivo* studies to establish translational exposure–effect relationships.

Highly promiscuous drugs have a higher failure rate when compared with successfully marketed drugs (Whitebread et al., 2005; Azzaoui et al., 2007; Bowes et al., 2012). The physicochemical parameters of high lipophilicity ($cLogP > 3$), ionization state ($pK_a > 6$), and molecular size (>500 Da) have been correlated with increased promiscuity and *in vivo* toxicity (Leeson and Springthorpe, 2007; Waring, 2010; Diaz et al., 2013). Compounds, like basic amines, with high pK_a values, for example, >6, are highly ionized at physiological pH and tend to interact with membrane phospholipids and become trapped in acidic organelle compartments, that is, mitochondrial intermembrane space and lysosomes, where they can cause dysfunction (Yokogawa et al., 2002; Diaz et al., 2013; Poulin et al., 2013). Additionally, compounds with low topological polar surface area (TPSA) can more readily cross membranes and distribute into tissues. Hughes et al. (2008) identified a TPSA of **<75** (low charge) associated with increased risk of adverse events *in vivo*. These physicochemical parameters should be part of the chemistry design considerations for safety optimization.

Promiscuity and selectivity are further evaluated using secondary pharmacology screening approaches. *In vitro* ligand binding screens are commonly used and are composed of endogenous receptors, ion channels, and enzymes with known physiological roles in clinical liabilities. These screening platforms are available through commercial research organizations or are often present as part of internal assay repertoires. Furthermore, assay platforms are often available that can be customized for a specific molecular target indication (e.g., kinase targets) or to evaluate specific organ effects, for example, CNS targets for potential to suicidality/abuse liability (Muller et al., 2015).

These screens are used to obtain a general assessment of promiscuity and selectivity based on the overall binding ("hit") rates across the assay panel, as well as insight into potential issues related to specific endogenous ligands that are bound (Bowes et al., 2012). Although thresholds for promiscuity are somewhat arbitrarily defined and thus may vary, Azzaoui et al. (2007) defined a compound's target hit rate (THR) as the ratio of the number of targets bound (at >50% inhibition at $10\,\mu M$) to the total number of targets tested: "highly promiscuous" compounds have a THR of ≥20% and "selective" compounds have a THR of ≤5%. Databases containing ligand–drug interaction information and computational models (e.g., CEREP BioPrint™) have been developed based on structural similarity to predict off-target binding of drugs, and although potentially biased toward a particular chemical space (i.e., G-protein-coupled receptor (GPCR) targets), they may be useful for informing chemical design (Whitebread et al., 2005). An additional, recently described computational approach that expands the biological target space for potential off-target ligands, which are unrelated to conventional off-target assessments, may also prove useful (Lounkine et al., 2012).

A typical approach to the integration and decision-making with regard to promiscuity and selectivity is as follows: chemists work with safety scientists to understand the physicochemical properties that trend with a higher propensity for toxicity as they design their molecules to identify the balance across all drug-like parameters. Representative molecules of various chemical structural series are "screened" using an *in vitro* ligand binding and/or *in silico* binding model to assess the number of overall ligand interactions (promiscuity) and evaluate the specific ligands with regard to their potential functional impact. As ligand binding screens cannot assess agonistic versus antagonistic activity, for molecules of interest, translational risk assessment is enabled with follow-up functional evaluation of ligands identified in the screen using cell-based assays, *ex vivo* organ assays (e.g., aortic ring contractility or hanging heart models), or *in vivo* studies if species translatability or relevance of the target is known. Furthermore, these follow-up assessments allow for the calculation of "safety margins" as related to *in vitro* biochemical potency or *in vivo* efficacy. For situations in which off-target engagement results in important, translatable functional effects, the *in vitro* models can then be incorporated into the tiered screening paradigm to select molecules that lack the identified liability.

2.3.2.2 Intrinsic Cytotoxicity Assessments Determining whether a compound is intrinsically cytotoxic is a key parameter for assessing risk of toxicity *in vivo* (Cockerell et al., 2002; Benbow et al., 2010; Greene et al., 2010; Sutherland et al., 2012). The determination of general cell viability using *in vitro* models can be used as a first-tier assessment for comparing chemical scaffolds or candidate compounds. The cell type chosen for these first-tier assessments is dependent on both scientific rational and practical logistics. The use of primary cells (e.g., cells derived directly from normal organs) provides information thought to most closely resemble "normal" *in vivo* cell functional states, for example, metabolism. The application of cryopreserved primary cells has made the use of these primary cells amenable to HTS (Li, 2007). Immortalized cells typically are derived from abnormal tissues (e.g., cancerous) and thus may not represent a "normal, adult" cell but are more readily available for screening applications than primary cells; however, these cells often lack or have altered cell functions, such as metabolic capabilities or overexpression of receptors. Additionally, potential antiproliferative (cytostatic) effects observed with immortalized cells that are highly proliferative need to be differentiated from direct cytotoxicity. Nonetheless, it is common for an immortalized cell line to be used for the first-tier, higher-throughput assessment of potential cytotoxicity followed by a more directed approach, which may include evaluation in other cell types, to confirm and translate risk. Furthermore, molecules of interest having some degree of cytotoxicity can be further characterized using specific mechanistic endpoints, for example, mitochondrial effects, in various cell-based models (McKim, 2010; Will and Dykens, 2014). This tiered approach enables scaffold or compound selection, investigation of underlying drivers of cytotoxicity, and translational risk assessment. Likewise, with regard to retrospective or hypothesis-driven assessments, cell-based models can be used to investigate toxicity observed in animal studies and to evaluate the *in vitro* to *in vivo* translation and cross species sensitivity to an identified toxicant (Diaz et al., 2013) or to build specific computational models (Pai et al., 2013).

The *in vitro* cytotoxicity assessment of the selected candidate can also inform the design of animal studies or the choice of *in vivo* models to improve characterization of risk. For example, if a compound is selected that has evidence of cytotoxicity associated with mitochondrial dysfunction, a transgenic mouse model that has enhanced susceptibility to mitochondrial toxicants (i.e., superoxide dismutase (SOD) +/−) may be considered to evaluate translatable risk to humans (Ong et al., 2006, 2007; Kashimshetty et al., 2009; Ramachandran et al., 2011).

One important historical limitation of *in vitro* cytotoxicity assays has been the extrapolation of *in vitro* outcomes, for example, IC_{50} (50% inhibitor concentration) values, to *in vivo* exposure–effect relationships that can inform either animal studies or clinical trials. Currently there are no commonly agreed-upon methods to accomplish this *in vitro–in vivo* extrapolation. Typically, the output of these *in vitro* models is binary, that is, toxic or not toxic, and is most commonly used for ranking compounds or compound series. However, approaches have been proposed to calculate *in vitro–in vivo* exposure–effect extrapolations that may be useful to predict *in vivo* risk (Cockerell et al., 2002; Benbow et al., 2010; Greene et al., 2010). For example, an exposure relationship comparing the ratio of a projected *in vivo* concentration to an *in vitro* LC_{50} (50% lethal concentration) could be used to predict a high-risk potential (i.e., maximum serum concentration $(C_{max})/LC_{50} > 1$) for an adverse outcome versus a low-risk potential (i.e., $C_{max}/LC_{50} < 0.01$) (Greene et al., 2010). Using an alternative method, drugs with an $LC_{50} \leq 50\,\mu M$ were shown to have fivefold increase of demonstrating toxicity *in vivo* if a C_{max} of $10\,\mu M$ (total drug) was achieved when compared with a drug with an $LC_{50} > 50\,\mu M$ (Greene et al., 2010). Using these approaches, the LC_{50} can be used to predict at what C_{avg} (average serum concentration for AUC0-24) or C_{max} toxicity may be observed compared with projected efficacious C_{max} values, that is, projected safety margin. This analysis can be used either to prioritize compounds, to set dose ranges in toxicology studies, or, in some cases, to provide insights into translation and enable decision-making.

2.3.2.3 Focused Target Organ Assessments

It is not uncommon to have *in vitro* models to assess potential target organ toxicity coupled with *in vivo* assessments during lead optimization. This is particularly the case with regard to the organs associated with the most common cause of attrition due to toxicity: cardiovascular (electrophysiology and hemodynamic), hepatic (idiosyncratic drug-induced liver injury), and hematopoietic (Stevens and Baker, 2009; Laverty et al., 2011). Furthermore, depending on the historical causes of attrition within a company or a program, additional target organ systems may be employed. The *in vitro* models take into consideration specific physiological conditions of the particular organ, for example, electrophysiology, metabolism, proliferation, and unique homeostatic systems. Some of the most mature and useful integrated lead optimization strategies involve the use of *in vitro* modeling platforms that evaluate drug effects on cardiovascular electrophysiology and function. For electrophysiology assessments, these include well-established cell-based, high-throughput ion channel patch clamp assays (e.g., hERG, hNav 1.5, hCa 1.2) and hanging heart models, as well as more recent uses of human and rodent stem cell-derived cardiomyocytes to evaluate integrated effects that are coupled with confirmatory *in vivo* assessments (Clements and Thomas, 2014; Nakamura et al., 2014; Sager et al., 2014). With regard to the evaluation of cardiovascular functional outputs, there are established *ex vivo* models (e.g., hanging heart and aortic ring), as well as more recently developed functional outputs using impedance to measure beat rate as a surrogate for cardiac contractility in stem cell- or pluripotent cell-based models (Scott et al., 2014; Doherty et al., 2015; Peters et al., 2015).

The use of *in vitro* and *in vivo* assessments of hematopoietic toxicity is also a well-established paradigm for lead optimization screening, as well as issue investigation, that has demonstrated good *in vitro–in vivo* concordance (Pereira et al., 2007; Olaharski et al., 2009). Furthermore, recently developed *in vitro* models offer higher throughput useful for screening and the ability to recapitulate lineage development that can be used to interrogate underlying mechanisms of toxicity (Tarrant et al., 2015; Uppal et al., 2015). Finally, hepatocytes (primary or immortalized) are commonly used to "predict" possible hepatotoxicity (DILI). However, this approach when used alone has limited value aside from assessing intrinsic cytotoxicity, as the pathogenesis of DILI is multifactorial, that is, the result of many risk factors. Therefore, more recently strategies to assess DILI rely on assessment of multiple endpoints in determining potential risk and evaluating underlying mechanisms of toxicity (Aleo et al. 2014; Dambach, 2014).

Taken together, one prudent use of target organ-specific models is in the context of hypothesis-driven investigation or when such an approach has been demonstrated to have some translational value for a particular program, in which case the model may be used in a screening mode to rank or identify lead compounds. Included in this approach is the ability to assess potential species differences *in vitro* as a means to understand translation to humans (Diaz et al., 2013). Lastly, although beyond the scope of this chapter, there are extensive efforts to develop "organ(s)-on-chip" or microphysiological systems that will hopefully be able to predict outcomes either as single organ systems or as part of an integrated, multi"organ" system, but these platforms have not been fully evaluated and their utility has not been fully realized (Sutherland et al., 2013; Wikswo, 2014).

2.3.2.4 ADME Assessments Related to Toxicity

The fundamental basis for toxicity is exposure to an offending molecule over some period of time. Key ADME-associated drivers of toxicity include distribution (subcellular, cellular, organ), which is related to absorption, clearance and elimination, and metabolism, which may generate reactive metabolites (RMs). The well-established concept of a "C_{max}

effect" relates to rapid onset of toxicity often at a high concentration. Commonly this effect is due to secondary pharmacology, that is, engagement of an off-target, endogenous receptor, transporter, or enzyme, whereas the concept of a serum "concentration versus time curve (AUC) effect" relates to a longer time over threshold to onset, which is likely due to complex mechanisms that take time to manifest, such as effects on downstream cellular processes over time, or time for subcellular accumulation of the molecule. These two exposure–effect concepts are helpful in establishing toxicity in the context of "time over threshold" but do not contribute to the evaluation of the actual potential mechanisms of ADME-related drivers of toxicity, which are multifactorial. As such, several key parameters can be evaluated with regard to their contributions to toxicity. These include intrinsic clearance, RM formation/covalent binding, metabolic phenotyping and metabolite prediction, distribution and clearance, and transporter effects.

In vitro intrinsic clearance estimated by primary hepatocytes or liver microsomes can indicate potential species differences in the rate of drug clearance and metabolic profiles that may be important in investigating the translatable relevance of toxicity findings in animals to humans. Furthermore, disconnects between intrinsic clearance estimated from liver microsomes versus hepatocytes and *in vitro–in vivo* clearance disconnects in nonclinical species may indicate other non-CYP450 enzyme involvement in drug metabolism. Determining which non-CYP450 enzymes are involved may result in the identification of metabolizing enzymes with known polymorphisms or with species differences (e.g., epoxide hydrolase, aldehyde oxidase, *N*-acetyltransferase, or flavin-containing monooxygenase (FMO)) that could impact outcomes of toxicity studies or clinical trials (Beedham et al., 1987; Janmohamed et al., 2004; O'Brien et al., 2004).

The formation of RMs is well recognized as an important risk factor for the development of toxicity especially DILI (Park et al., 2011). Recent studies evaluating the potential contribution of covalent binding and dose to hepatotoxicity outcome provide evidence for these risk factors (Nakayama et al., 2009; Thompson et al., 2011; Sakatis et al., 2012). For example, based on the work of Sakatis et al. (2012), drugs with a dose of ≥100 mg/day that also demonstrated either glutathione adduct formation (medium or high intensity), metabolism-dependent cytochrome P450 (CYP) inhibition ($>5 \times$ decrease in IC_{50}), or covalent binding (>200 pmol eq/protein) predicted $>80\%$ of clinical hepatotoxicants. Furthermore, assessments that integrated measures of RM formation (covalent binding burden), cellular drug accumulation (transporter inhibition), and cytotoxicity (mitochondrial dysfunction) distinguished hepatotoxicants from nonhepatotoxicants (Thompson et al., 2011). Thus, the evaluation of the potential to form RMs and covalent binding potential coupled with projected dose and cytotoxicity endpoints shows utility as part of an integrated risk assessment.

Compound structure, metabolic profile, and metabolic pathway prediction may provide clues for potential reactive substructures (toxicophores) and for assessing the potential impact of altering a major metabolic pathway, which may result in accumulation of a metabolite or shunting to another metabolic pathway that may be predicted to be toxic. Metabolic phenotyping to determine which CYP450 enzymes are important for the metabolism of a drug and, importantly, whether a particular CYP450 may be responsible for the major metabolic route will help determine the risk impact of inhibiting or inducing that particular enzyme or whether there is a polymorphism that should be considered as a possible risk factor for toxicity. Additionally, effects on phase II enzymes should also be considered. For example, inhibitors of uridine diphosphate glucuronyl transferase (UGT) can potentially cause hyperbilirubinemia, as this is the conjugating enzyme for the endogenous processing of bilirubin (Zucker et al., 2001; Fevery, 2008).

Furthermore, it is important to gain an understanding of the potential distribution of a molecule at the cellular, organ, and systemic levels and how this may impact toxicity with regard to the other parameters described earlier (i.e., potential for off-target interactions, level of intrinsic cytotoxicity, and subcellular target), as well as the potential for metabolic drivers of toxicity. Physicochemical characteristics of a molecule, like lipophilicity and ionization state, are important determinants of distribution and potential accumulation. These characteristics are fixed and unique to each molecule and are key components for consideration in chemical design. Importantly, physicochemical characteristics can impact cellular and subcellular distribution and accumulation (e.g., high pK_a and accumulation in cell membranes and acidic compartments). Likewise, they impact systemic distribution, for example, entry into the CNS via the blood–brain barrier (Chico et al., 2009). Finally, examination of parameters such a volume of distribution (V_d) and clearance (Cl) can identify risk for the accumulation of drugs with other risk factors for toxicity, for example, intrinsic cytotoxicity, that may result in toxicity due to long exposures over a toxic threshold. For example, Poulin et al. (2013) found a relationship between high V_d and low clearance and the increased potential for toxicity. Furthermore, Sutherland et al. (2012) identified V_{ss} (steady state) and intrinsic cytotoxity in rat hepatocytes as key indicators of potential *in vivo* toxicity; in particular, the combination of high V_{ss} and low IC_{50} for intrinsic cytotoxicity resulted in a fivefold lower lowest observable adverse effect level (LOAEL).

Chemical structure also determines engagement of various cellular transporters that may affect cellular concentrations of a drug, as well as elimination, that might impact target organ toxicity. Inhibition of xenobiotic efflux transporters may result in the accumulation of a drug within cells, resulting in toxicity for an intrinsically cytotoxic drug that would not have been predicted based on plasma drug concentrations. The evaluation of potential impact of transporter inhibition on toxicity has largely focused on hepatic and renal transporters. For example, when screening for cholestasis potential, compounds can be assessed for the inhibition of various hepatic transporters such as bile salt export pump (BSEP), multidrug resistance-associated protein 2 (MRP2), Na+–taurocholate cotransporting polypeptide (NTCP), and organic acid-transporting polypeptide (OATP) (Ansede et al., 2010; International Transporter Consortium, 2010). Using an *in vitro* model, Morgan et al. (2010) determined that potent BSEP inhibition (IC_{50} of $\leq 25\,\mu M$) associated with known hepatotoxicants could be distinguished from nonhepatotoxicants ($IC_{50} > 100\,\mu M$). The same group found improved predictivity when they related the IC_{50} to clinical exposures for several transporters (Morgan et al., 2013).

Thus, the discovery safety scientist can partner closely with their DMPK scientist to evaluate various ADME/PK parameters with regard to informing potential toxicity risks or in the context of investigating an identified toxicity in animals or in humans.

2.3.2.5 *Genotoxicity Assessments* The purpose of genotoxicity lead optimization screening is to eliminate or minimize the advancement of molecules with direct effects on DNA or chromosomes, as these have been linked to the development of carcinogenic effects. This strategy is particularly applicable for drugs intended to treat non-life-threatening medical conditions, where genetic toxicity is not an acceptable adverse effect. The models utilized are also applicable for evaluating potential safety risks of chemical process intermediates for nonpatients exposed to the drugs related to environmental and occupational exposure, as well as assessment of potential genotoxic impurities (ICH M7, 2014). With regard to lead optimization activities to assess genetic toxicity risk, the most commonly applied assays are those meant to predict the outcome of the regulatory assays for mutagenicity, clastogenicity, and aneugenicity (bacterial and mammalian) (ICH S2, 2011). This approach has been highly successful in reducing or eliminating the advancement of drugs with genotoxic risk, particularly with regard to drugs intended for non-life-threatening therapeutic indications (Kirkland et al., 2011). As such, inclusion of this screening during the lead optimization phase is widely applied across the pharmaceutical industry.

The genetic toxicity safety lead optimization strategy is principally made up of both computational (*in silico*) models and *in vitro* assays for mutagenicity and aneugenicity/clastogenicity that have minimal compound requirements and allow for HTS. There are numerous *in silico* and *in vitro* assays available commercially that can be readily incorporated into a lead optimization strategy. Each approach should be evaluated for the best strategic fit; in addition to the literature, contract research organization scientists who are experts in

this area are a valuable resource to help facilitate this evaluation. Typically *in silico* models are used as the first tier to evaluate a compound series or individual compounds to assess mutagenic risk. Computational models are commonly applied early in the drug design phase to "flag" potential toxicophores or physicochemical characteristics that are associated with DNA interaction (Hillebrecht et al., 2011; Ford, 2013). Confirmation of an *in silico* prediction is typically made using an *in vitro* mutagenicity screening assay. *In vitro* screening models are used to assess both mutagenicity and potential aneugenicity or clastogenicity (e.g., see Gee et al., 1998; Diehl et al., 2000; Diaz et al., 2007). With regard to *in vitro* mutagenicity and aneugenicity/clastogenicity assays, a positive result in a screening assay that is not related to a known mechanism of action of the drug often leads to investigation of the cause, for example, parent or metabolite, impurity, or technical (Kirkland et al., 2007; Thybaud et al., 2007). Follow-up confirmatory assessments to determine risk are performed often using *in vitro* investigational (e.g., centromere labeling) or standard regulatory assays and/or *in vivo* mammalian assessments (Kirkland et al., 2011).

For non-life-threatening therapeutic indications, genetic toxicity assessments are an important early component of the lead optimization strategy, and the focus of decision-making is to remove any genotoxic molecules/chemical series or understand the translatable risk to humans. However, given the significant impact of advancing a potential mutagen or clastogen, it is a very common practice to simply remove the offending chemical or series. For life-threatening therapeutic indications, non-good laboratory practice (GLP) screens are often performed in the discovery space to identify potential risk, in particular related to occupational exposures, versus being used to identify an acceptable compound for advancement.

2.3.3 Application of Retrospective Models

Investigation or characterization of identified or potential liabilities is an important part of the lead optimization strategy to enable informed, rational decision-making (Table 2.2). The goals of investigative work are twofold: (i) identify the cause of the toxicity (i.e., on-target versus off-target) to inform chemical design, a backup strategy, or continued "druggability" of the target and (ii) inform and characterize translatable risk to humans, which includes the identification of potential safety biomarkers and exposure–effect relationships. Retrospective or hypothesis-driven models are applied to evaluate identified theoretical risks or target organ toxicity that has been identified (usually *in vivo*) or to confirm an *in silico* or *in vitro* screening outcome. In all cases, applied models need to be qualified as to their ability to accurately predict the translatable outcome (either in animals or humans). These model platforms may include computational, *in vitro*, *ex vivo*, and *in vivo* models that are often used in combination to build a weight of evidence for internal decision-making. The models are almost exclusively research based (i.e., non-GLP).

There are numerous examples in the literature demonstrating the application of these models to elucidate mechanisms of toxicity. Theoretic risks typically are identified during the TSA as either potential undesired pharmacology or based on competitive intelligence regarding adverse effects identified either nonclinically or clinically with a competitor compound or target/drug class. Alternatively, target organ toxicity is identified during efficacy, pharmacokinetic, or pilot toxicity studies. In any case, it is paramount that a hypothesis-driven research plan be developed that defines essential questions to be addressed and answered. Common themes in investigative approaches include determination of on-target versus off-target effects and mechanisms, cross species translation, and reversibility assessments. With regard to on-target versus off-target effects, "tool" molecules that are structurally distinct and/or structurally similar molecules that lack pharmacological activity are often compared to the offending molecule; likewise, the evaluation of genetically engineered rodent models is often evaluated. When target organ toxicity is identified *in vivo,* gaining an understanding of the potential translation to humans is a major focus but is a significant challenge. Such an investigation may rely upon the presence of the target organ toxicity in two species, as well as a cross species susceptibility assessment for the target organ *in vitro* to assess the potential translation to humans. Often *in vitro* studies utilize a cell type identified histologically in the target organ to assess species translation and to investigate underlying cellular drivers of toxicity. This knowledge can be used to determine a possible structure–activity relationship or to identify risk factors for patients based on the cellular mechanisms, for example, mitochondrial toxin.

The works of Diaz et al. (2013), Tarrant et al. (2015), Uppal et al. (2015), and Zabka et al. (2015) are recent examples of utilizing cell-based models to investigate toxicity observed in animal or clinical studies and to evaluate the *in vitro to in vivo* translation and cross species sensitivity to an identified toxicant.

2.4 CONCLUSIONS

Safety lead optimization efforts are used to maximize resources and early decision-making for the purposes of reducing costs and animal use and removing the least favorable compounds early during the discovery phase, when there is most flexibility in optimizing the candidate molecule. The ultimate goal is to identify a lead development candidate that has the most superior safety characteristics and, for those safety issues identified, a well-characterized risk assessment to reduce clinical attrition due to toxicity and inform the clinical development plan.

To that end, the components of the safety lead optimization strategy should include (i) a TSA to evaluate potential target-related issues and therapeutic area considerations, (ii) a defined candidate safety characteristic profile, and (iii) an active investigation of issues to facilitate rational drug design and decision-making around molecule advancement and druggability. The candidate safety characteristic profile is made up of "core characteristics" plus those customized based on the TSA. The core characteristics are those related to known chemical-based causes of attrition and can be summarized as a candidate that is highly selective and has low promiscuity with minimal/acceptable functional translation of identified off-target engagement, low intrinsic cytotoxicity, and low ADME/PK risk, and for a drug candidate targeting a non-life threatening condition, no genetic toxicity, in particular, mutagenicity or clastogenicity.

Finally, for such a strategy to work, there needs to be a cross functional culture that embraces the inclusion of safety lead optimization strategies and approaches as part of the holistic assessment and identification of candidate molecules during the drug discovery phase; this includes engagement of chemists, pharmacologists, DMPK, and pharmaceutics scientists, as well as an understanding of how safety scientists utilize the data to build a weight-of-evidence approach for safety assessment.

REFERENCES

Aleo, M., Luo, Y., Swiss, R., Bonin, P.D., Potter, D.M., Will, Y. (2014). Human drug-induced liver injury severity is highly associated with dual inhibition of mitochondrial function and bile salt export pump. *Hepatology*, 60, 1015–1022.

Ansede, J.H., Smith, W.R., Perry, C.H., St. Claire III, R.L., Brouwer, K.R. (2010). An in vitro assay to assess transporter-based cholestatic hepatotoxicity using sandwich-cultured rat hepatocytes. *Drug Metabolism and Disposition*, 38, 276–280.

Azzaoui, K., Hamon, J., Faller, B., Whitebread, S., Jacoby, E., Bender, A., Jenkins, J.L., Urban, L. (2007). Modeling promiscuity based on in vitro safety pharmacology profiling data. *ChemMedChem*, 2, 874–888.

Beedham, C., Bruce, S.E., Critchley, D.J., Al-Tayib, Y., Rance, D.J. (1987). Species variation in hepatic aldehyde oxidase activity. *European Journal of Drug Metabolism and Pharmacokinetics*, 12, 307–310.

Benbow, J.W., Aubrecht, J., Banker, M.J., Nettleton, D., Aleo, M.D. (2010). Predicting safety toleration of pharmaceutical chemical leads: cytotoxicity correlations to exploratory toxicity studies. *Toxicology Letters*, 197, 175–182.

Bowes, J., Brown, A.J., Hamon, J., Jarolimek, W., Sridhar, A., Waldron, G., Whitebread, S. (2012). Reducing safety-related drug attrition: the use of in vitro pharmacological profiling. *Nature Reviews in Drug Discovery*, 11, 909–922.

Chico, L.K., Van Eldik, L.J., Watterson, D.M. (2009). Targeting protein kinases in central nervous system disorders. *Nature Reviews Drug Discovery*, 8, 892–909.

Clements, M., Thomas, N. (2014). High-throughput multi-parameter profiling of electrophysiological drug effects in human embryonic stem cell derived cardiomyocytes using multi-electrode arrays. *Toxicological Sciences*, 140, 445–461.

Cockerell, G.L., McKim, J.M.,Vonderrecht, S.L. (2002). Strategic importance of research support through pathology. *Toxicological Pathology*, 30, 4–7.

Dambach, D.M. (2014). *Predictive ADMET: Integrated Approaches in Drug Discovery and Development*, John Wiley & Sons, Inc., Hoboken, NJ, pp. 433–465.

Dambach, D.M., Gautier, J.C. (2006). *Toxicologic Biomarkers*, Taylor and Francis, New York, pp. 143–164.

Diaz, D., Scott, A., Carmichael, P., Shi, W., Cosatles, C. (2007). Evaluation of an automated in vitro micronucleus assay in CHO-K1 cells. *Mutation Research*, 630, 1–13.

Diaz, D., Ford, K.A., Hartley, D.P., Harstad, E.B., Cain, G.R., Achilles-Poon, K., Nguyen, T., Peng, J., Zheng, Z., Merchant, M., Sutherlin, D.P., Gaudino, J.J., Kaus, R., Lewin-Koh, S.C., Choo, E.F., Liederer, B.M., Dambach, D.M. (2013). Pharmacokinetic drivers of toxicity for basic molecules: strategy to lower pKa results in decreased tissue exposure and toxicity for a small molecule Met inhibitor. *Toxicology and Applied Pharmacology*, 266, 86–94.

Diehl, M.S., Willaby, S.L., Synder, R.D. (2000). Comparison of the results of a modified miniscreen and the standard bacterial reverse mutation assays. *Environmental and Molecular Mutagenesis*, 35, 72–77.

Doherty, K.R., Talbert, D.R., Trusk, P.B., Moran, D.M., Shell, S.A., Bacus, B. (2015). Structural and functional screening in human induced-pluripotent stem cell-derived cardiomyocytes accurately identifies cardiotoxicity of multiple drug types. *Toxicology and Applied Pharmacology*, 285, 51–60.

Fevery, J. (2008). Bilirubin in clinical practice: a review. *Liver International*, 28, 592–603.

Ford, K. (2013). Role of electrostatic potential in the in silico prediction of molecular bioactivation and mutagenesis. *Molecular Pharmaceutics*, 10, 1171–1182.

Gee, P., Sommers, C.H., Melick, A.S., Gidrol, X.M., Todd, M.D., Burris, R.B., Nelson, M.E., Klemm, R.C., Zeiger, E. (1998). Comparison of responses of base-specific Salmonella tester strains with the traditional strains for identifying mutagens: the results of a validation study. *Mutation Research*, 412, 115–130.

Greene, N., Aleo, M.D., Louise-May, S., Price, D.A., Will, Y. (2010). Using in vitro cytotoxicity assay to aid in compound selection for in vivo safety studies. *Bioorganic & Medicinal Chemistry Letters*, 20, 5308–5312.

Hillebrecht, A., Muster, W., Brigo, A., Kansy, M., Weiser, T., Singer, T. (2011). Comparative evaluation of in silico systems for ames test mutagenicity prediction: scope and limitations. *Chemical Research in Toxicology*, 24, 843–854.

Hornberg, J.J., Laursen, M., Brenden, N., Persson, M., Thougaard, A.V., Toft, D.B., Mow, T. (2014a). Exploratory toxicology as an integrated part of drug discovery. Part I: why and how. *Drug Discovery Today*, 19, 1131–1136.

Hornberg,J.J.,Laursen,M.,Brenden,N.,Persson,M.,Thougaard,A.V., Toft, D.B., Mow, T. (2014b). Exploratory toxicology as an integrated part of drug discovery. Part II: screening strategies. *Drug Discovery Today*, 19, 1137–1144.

Hughes, J.D., Blagg, J., Price, D.A., Bailey, S., DeCrescenzo, G.A., Devraj, R.V., Ellsworth, E., Fobian, Y.M., Gibbs, M.E., Gilles, R.W., Greene, N., Huang, E., Krieger-Burke, T., Loesel, J., Wager, T., Whiteley, L., Zhang, Y. (2008). Physicochemical drug properties associated with in vivo toxicological outcomes. *Bioorganic & Medicinal Chemistry Letters*, 18, 4872–4875.

International Conference on Harmonization. (2009a). Topic S9: Nonclinical evaluation for anticancer pharmaceuticals. http://www.ich.org/fileadmin/Public_Web_Site/ICH_Products/Guidelines/Safety/S9/Step4/S9_Step4_Guideline.pdf (accessed October 12, 2015).

International Conference on Harmonization. (2009b). Guidance on nonclinical safety studies for the conduct of human clinical trials and marketing authorization for pharmaceuticals M3(R2). http://www.ich.org/fileadmin/Public_Web_Site/ICH_Products/Guidelines/Multidisciplinary/M3_R2/Step4/M3_R2__Guideline.pdf (accessed October 12, 2015).

International Conference of Harmonization. (2011). Guidance on genotoxicity testing and data interpretation for pharmaceuticals intended for human use S2(R1). http://www.ich.org/fileadmin/Public_Web_Site/ICH_Products/Guidelines/Safety/S2_R1/Step4/S2R1_Step4.pdf (accessed October 12, 2015).

International Conference on Harmonization. (2014). Assessment and control of DNA reactive (mutagenic) impurities in pharmaceuticals to limit potential carcinogenic risk M7. http://www.ich.org/fileadmin/Public_Web_Site/ICH_Products/Guidelines/Multidisciplinary/M7/M7_Step_4.pdf (accessed October 12, 2015).

International Transporter Consortium. (2010). Membrane transporters in drug development. *Nature Reviews Drug Discovery*, 9, 215–236.

Janmohamed, A., Hernandez, D., Phillips, I.R., Shephand, E.A. (2004). Cell, tissue, sex, and developmental stage-specific expression of mouse flavin-containing monooxygenases (FMOs). *Biochemical Pharmacology*, 68, 73–83.

Kashimshetty, R., Desai, V.G., Kale, V.M., Lee, T., Moland, C.L., Branham,W.S.,New,L.S.,Chan,E.C.,Younis, H.,Boelsterli, U.A. (2009). Underlying mitochondrial dysfunction triggers flutamide-induced oxidative liver injury in a mouse model of idiosyncratic drug toxicity. *Toxicology and Applied Pharmacology*, 238, 150–159.

Kerns, E.H., Di, L. (2003). Pharmaceutical profiling in drug discovery. *Drug Discovery Today*, 8, 316–323.

Kirkland, D.J., Aardema, M., Banduhn, N., Carmichael, P., Fautz, R., Meunier, J.-R., Pfuhler, S. (2007). In vitro approaches to develop weight of evidence (WoE) and mode of action (MoA) discussions with positive in vitro genotoxicity results. *Mutagenesis*, 22, 161–175.

Kirkland, D., Reeve, L., Gatehouse, D., Vanparys, P. (2011). A core in vitro genotoxicity battery comprising the Ames test plus the in vitro micronucleus test is sufficient to detect rodent carcinogens and in vivo genotoxins. *Mutation Research*, 721, 27–73.

Kola, I., Landis, J. (2004). Can the pharmaceutical industry reduce attrition rates? *Nature Reviews Drug Discovery*, 3, 711–715.

Kramer, J.A., Sagartz, J.E., Morris, D.L. (2007). The application of discovery toxicology and pathology towards the design of safety pharmaceutical lead candidates. *Nature Reviews Drug Discovery*, 6, 636–649.

Laverty, H., Benson, C., Cartwright, E., Cross, M., Garland, C., Hammond, T., Holloway, C., McMahon, N., Milligan, J., Park, B., Pirmohamed, M., Pollard, C., Radford, J., Roome, N., Sager, P., Singh, S., Suter, T., Suter, W., Trafford, A., Volders, P., Wallis, R., Weaver, R., York, M., Valentin, J. (2011). How can we improve our understanding of cardiovascular safety liabilities to develop safety medicines? *British Journal of Pharmacology*, 163, 675–693.

Leeson, P.D., Springthorpe, B. (2007). The influence of drug-like concepts on decision making in medicinal chemistry. *Nature Reviews Drug Discovery*, 6, 881–890.

Li, A.P. (2007). Human hepatocytes: isolation, cryopreservation and applications in drug development. *Chemico-Biological Interactions*, 168, 16–29.

Lounkine, E., Keiser, M.J., Whitebread, S., Mikhailov, D., Hamon, J., Jenkins, J.L., Lavan, P., Weber, E., Doak, A.K., Cote, S., Shoichet, B.K., Urban, L. (2012). Large-scale prediction and testing of drug activity on side-effect targets. *Nature*, 486, 361–367.

McKim, J.M. (2010). Building a tiered approach to in vitro predictive toxicity screening: a focus on assays with in vivo relevance. *Combinatorial Chemistry & High Throughput Screening*, 13, 188–206.

Morgan, R.E., Trauner, M., van Staden, C.J., Lee, P.H., Ramachandran, B., Eschenberg, M., Afshari, C.A., Qualls, C.W., Lightfoot-Dunn, R., Hamadeh, H.K. (2010). Interference with bile salt export pump function is a susceptibility factor for human liver injury in drug development. *Toxicological Sciences*, 118, 485–500.

Morgan, R.E., van Staden, C.J., Chen, Y., Kaylanaraman, N., Kalanzi, J., Dunn II, R.T., Afshari, C.A., Hamadeh, H.K. (2013). A multifactorial approach to hepatobiliary transporter assessment enables improved therapeutic compound development. *Toxicological Sciences*, 136, 216–241.

Muller, P.Y., Dambach, D., Gemzik, B., Hartmann, A., Ratcliffe, S., Trendelenburg, C., Urban, L. (2015). Integrated risk assessment of suicidal ideation and behavior in drug development. *Drug Discovery Today*, 20, 1135–1142.

Nakamura, Y., Matsuo, J., Miyamoto, N., Ojima, A., Ando, K., Kanda, Y., Sawada, K., Sugiyama, A., Sekino, Y. (2014). Assessment of testing methods for drug-induced repolarization delay and arrhythmias in an ips cell-derived cardiomyocyte sheet: multi-site validation study. *Journal of Pharmacological Science*, 124, 494–501.

Nakayama, S., Atsumi, R., Takakusa, H., Kobayashi, Y., Kurihara, A., Nagai, Y., Naiai, D., Okazaki, O. (2009). A zone classification system for risk assessment of idiosyncratic drug toxicity using daily dose and covalent binding. *Drug Metabolism and Disposition*, 37, 1970–1977.

O'Brien, P.J., Chan, K., Silber, P.M. (2004). Human and animal hepatocytes in vitro with extrapolation in vivo. *Chemico-Biological Interactions*, 150, 97–114.

Olaharski, A.J., Uppal, H., Cooper, M., Platz, S., Zabka, T.S., Kolaja, K.L. (2009). In vitro to in vivo concordance of a high throughput assay of bone marrow toxicity across a diverse set of drug candidates. *Toxicology Letters*, 188, 98–103.

Ong, M.M.K., Wang, A.S., Loew, K.Y., Khoo, Y.M., Boelsterli, U.A. (2006). Nimesulide-induced hepatic mitochondrial injury in heterozygous SOD+/− mice. *Free Radical Biology & Medicine*, 40, 420–429.

Ong, M.M.K., Latchoumycandane, C., Boelsterli, U.A. (2007). Troglitazone-induced hepatic necrosis in an animal model of silent genetic mitochondrial abnormalities. *Toxicological Sciences*, 97, 205–213.

Pai, R., Wei, B.Q., Chang, J., Crawford, J., Young, W., Ortwine, D., Misner, D., Dambach, D. (2013). Application of an in silico approach to predict intrinsic in vitro cytotoxicity for compounds in primary human hepatocytes during preclinical development. *The Toxicologist*, 132, 189.

Park, B.K., Boobis, A., Clarke, S., Goldring, C.E.P., Jones, D., Kenna, J.G., Lambert, C., Laverty, H.G., Naisbitt, D.J., Nelson, S., Nicoll-Griffith, D.A., Obach, R.S., Routledge, P., Smith, D.A., Tweedie, D.J., Vermeulen, N., Williams, D.P., Wilson, I.D., Baillie, T.A. (2011). Managing the challenge of chemically reactive metabolites in drug development. *Nature Reviews Drug Discovery*, 10, 292–306.

Pereira, C., Clarke, E., Damen, J. (2007). *Stem Cell Assays*. Methods in Molecular Biology, Humana Press, Totowa, NJ, 407, pp. 177–208.

Peters, M.F., Lamore, S.D., Guo, L., Scott, C.W., Kolaja, K.L. (2015). Human stem cell-derived cardiomyocytes in cellular impedance assays: bringing cardiotoxicity screening to the front line. *Cardiovascular Toxicology*, 15, 127–139.

Poulin, P., Dambach, D.M., Hartley, D.H., Ford, K., Theil, P., Harstad, E., Halladay, J., Choo, E., Boggs, J., Liederer, B., Dean, B., Diaz, D. (2013). An algorithm for evaluating potential tissue drug distribution in toxicology studies from readily available pharmacokinetic parameters. *Journal of Pharmaceutical Science*, 102, 3816–3829.

Ramachandran, A., Lebofsky, M., Weinman, S.A., Jaeschke, H. (2011). The impact of partial manganese superoxide dismutase (SOD2)-deficiency on mitochondrial oxidant stress, DNA fragmentation and liver injury during acetaminophen hepatotoxicity. *Toxicology and Applied Pharmacology*, 251, 226–233.

Sager, P.T., Gintant, G., Turner, J.R., Pettit, S., Stockbridge, N. (2014). Rechanneling the cardiac proarrhythmia safety paradigm: a meeting report from the Cardiac Safety Research Consortium. *American Heart Journal*, 167, 292–300.

Sakatis, M.Z., Reese, M.J., Harrell, A.W., Taylor, M.A., Baines, I.A., Chen, L., Bloomer, J.C., Yang, E.Y., Ellens, H.M., Ambroso, J.L., Lovatt, C.A., Ayrton, A.D. Clarke, S.E. (2012). Preclinical strategy to reduce clinical hepatotoxicity using in vitro bioactivation data for >200 compounds. *Chemical Research in Toxicology*, 25, 2067–2082.

Sasseville, V.G., Lane, J.H., Kadambi, V.J., Bouchard, P., Lee, F.W., Balani, S., Miwa, G.T., Smith, P.F., Alden, C.L. (2004). Testing paradigm for prediction of development-limiting barriers and human drug toxicity. *Chemico-Biological Interactions*, 150, 9–25.

Scott, C.W., Zhang, X., Abi-Gerges, N., Lamore, S.D., Abassi, Y.A., Peters, M.F. (2014). An impedance-based cellular assay using human iPSC-derived cardiomyocytes to quantify modulators of cardiac contractility. *Toxicological Sciences*, 142, 331–338.

Stevens, J.L., Baker, T.K. (2009). The future of drug safety testing: expanding the view and narrowing the focus. *Drug Discovery Today*, 14, 162–167.

Sutherland J.J., Raymond, J.W., Stevens, J.L., Baker, T.K., Watson, D.E. (2012). Relating molecular properties and in vitro assay results to in vivo drug disposition and toxicity outcomes. *Journal of Medicinal Chemistry*, 55, 6455–6466.

Sutherland, M.L., Fabre, K.M., Tagle, D.A. (2013). The national institutes of health microphysiological systems program focuses on a critical challenge in the drug discovery pipeline. *Stem Cell Research and Therapy*, 4, 1–5.

Tarrant, J.M., Dhawan, P., Singh, J., Zabka, T.S., Clarke, E., DosSantos, E., Dragovich, P.S., Sampath, D., Lin, T., McCray, B., La, N., Nguyen, T., Kauss, A., Dambach, D., Misner, D.L., Diaz, D., Uppal, H. (2015). Preclinical models of nicotinamide phosphoribosyltransferase inhibitor-mediated hematotoxicity and mitigation by co-treatment with nicotinic acid. *Toxicology Mechanisms and Methods*, 25, 201–211.

Thompson, R.A., Isin, E.M., Li, Y., Weaver, R., Weidolf, L., Wilson, I., Claesson, A., Page, K., Dolgos, H., Kenna, J.G. (2011). Risk assessment and mitigation strategies for reactive metabolites in drug discovery and development. *Chemico-Biological Interactions*, 192, 65–71.

Thybaud, V., Aardeama, M., Clements, J., Dearfield, K., Galloway, S., Hayashi, M., Jacobson-Kram, D., Kirkland, D., MacGregor, J.T., Marzin, D., Ohyama, W., Schuler, M., Suzuki, H., Zeiger, E. (2007). Strategy for genotoxicity testing: hazard identification and risk assessment in relation to in vitro testing. *Mutation Research*, 627, 41–58.

Uppal, H., Doudement E., Mahapatra, K., Darbonne, W.C., Bumbaca, D., Shen, B.Q., Du, X., Saad, O., Bowles, K., Olsen, S., Lewis Phillips, G.D., Hartley, D., Sliwkowski, M.X., Girish, S., Dambach, D., Ramakrishnan, V. (2015). Potential mechanisms for thrombocytopenia development with trastuzumab emtansine (T-DM1). *Clinical Cancer Research*, 21, 123–133.

Waring, M.J. (2010). Lipophilicity and drug discovery. *Expert Opinion in Drug Discovery*, 5, 235–248.

Whitebread, S., Hamon, J., Bojanic, D., Urban, L. (2005). In vitro safety pharmacology profiling: an essential tool for successful drug development. *Drug Discovery Today*, 10, 1421–1433.

Wikswo, J.P. (2014). The relevance and potential roles of microphysiological systems in biology and medicine. *Experimental Biology and Medicine*, 239, 1061–1072.

Will, Y., Dykens, J. (2014). Mitochondrial toxicity assessment in industry—a decade of technology development and insight. *Expert Opinion in Drug Metabolism and Toxicology*, 10, 1061–1067.

Yokogawa, K., Ishizaki, J., Ohkuma, S., Miyamoto, K. (2002). Influence of lipophilicity and lysosomal accumulation on tissue distribution kinetics of basic drugs: a physiologically based pharmacokinetic model. *Methods and Finding in Experimental Clinical Pharmacology*, 24, 81–93.

Zabka, T., Singh, J., Dhawan, P., Liederer, B.M., Oeh, J., Kauss, M.A., Xiao, Y., Zak, M., Lin, T., McCray, B., La, N., Nguyen, T., Beyer, J., Farman, C., Uppal, H., Dragovich, P.S., O'Brien, T., Sampath, D., Misner, D.L. (2015). Retinal toxicity, in vivo and in vitro, associated with inhibition of nicotinamide phosphoribosyltransferase. *Toxicological Sciences*, 144, 163–172.

Zucker, S.D., Qin, X., Rouster, S.D., Yu, F., Green, R.M., Keshavan, P., Feinberg, J., Sherman, K.E. (2001). Mechanism of indinavir-induced hyperbilirubinemia. *Proceedings of the National Academy of Sciences of the United States of America*, 98, 12671–12676.

3

SAFETY ASSESSMENT STRATEGIES AND PREDICTIVE SAFETY OF BIOPHARMACEUTICALS AND ANTIBODY DRUG CONJUGATES

MICHELLE J. HORNER[1], MARY JANE HINRICHS[2] AND NICHOLAS BUSS[2]

[1]Comparative Biology and Safety Sciences (CBSS) – Toxicology Sciences, Amgen Inc., Thousand Oaks, CA, USA

[2]Biologics Safety Assessment, MedImmune, Gaithersburg, MD, USA

3.1 BACKGROUND AND OBJECTIVES

Discovery of new targets and improved knowledge of cellular mechanisms and pathophysiology have resulted in the rapid expansion of novel technologies for developing new medicines, including biopharmaceuticals with high degrees of selectivity for specific targets. However, uncertainty remains regarding many cellular pathways and the concordance of physiological and pathological effects between humans and animals. In 1982, insulin (Humulin) received approval from the Division of Metabolic and Endocrine Drugs (CDER; Leader et al., 2008). In the late 1980s, murine monoclonal antibodies (mAbs) were in clinical development but were associated with allergic reactions, antidrug antibodies (ADAs), and reduced effector function. Thereafter, antibody engineering attempted to improve effector function and reduce immunogenicity through the development of chimeric mouse–human antibodies and humanized and fully human mAbs. Increasing innovation has resulted in expanded diversity in therapeutic modalities such as bispecific T cell engagers and antibody drug conjugates (ADCs); each of these has its own uncertainties, which may lead to a higher risk of unintentional, unrecognized, or poorly understood effects in humans. From a nonclinical perspective, it is important to have a clear understanding of how these properties influence early risk assessment and nonclinical safety studies.

It is generally assumed that, unlike new chemical entities, biopharmaceuticals have a better safety profile due to high binding specificity to the intended targets (and low off-target promiscuity), low susceptibility for biotransformation via drug metabolizing enzymes, and high similarity to endogenous human proteins (especially through more recent use of molecular engineering techniques to "humanize" the biopharmaceutical and minimize immunogenicity). Despite these general properties, clinical experience with biopharmaceutical drugs has revealed profound, and sometimes unexpected, adverse drug reactions (e.g., TGN1412; Farzaneh et al., 2007; Attarwala, 2010). Predicting adverse reactions of biopharmaceuticals is made even more difficult today with a rapidly expanding array of platforms that include traditional constructs (recombinant human proteins and monovalent antibodies), antibody fragments (e.g., variable fragment (Fab), scFv), peptibodies, ADC, and bispecific molecules. Considering the unique and complex nature of biopharmaceuticals, the goals of predictive safety for biopharmaceuticals is to institute derisking strategies that will predict liabilities and adverse events that may impede progression of novel biopharmaceuticals during clinical development. Predictive safety assessment of a biopharmaceutical begins with the target safety assessment, which identifies putative risks based on the nature of the target and the intended mechanism of action (MOA). Information obtained from the target

Drug Discovery Toxicology: From Target Assessment to Translational Biomarkers, First Edition. Edited by Yvonne Will, J. Eric McDuffie, Andrew J. Olaharski, and Brandon D. Jeffy.

safety assessment shapes subsequent data collection efforts focused on the potential risks associated with the modality, target characteristics, intended MOA, and unforeseen/unintended effects to enable appropriate risk management. This chapter provides guidance for adopting strategies to maintain and improve nonclinical predictive safety approaches, *in vitro* and/or *in vivo*, to support the clinical development of biopharmaceuticals including ADCs. The points covered are intended to guide decision-making and planning of derisking strategies in the best manner possible, based on scientific and drug development needs.

3.2   TARGET SAFETY ASSESSMENTS: STRATEGIES TO UNDERSTAND TARGET BIOLOGY AND ASSOCIATED LIABILITIES

Evaluation of target biology from a safety perspective can elucidate definite and/or putative liabilities of a biopharmaceutical agent in organs that express those targets. For biopharmaceuticals, many of the toxicities can be attributed to exaggerated pharmacology, and these can often be identified during a target safety assessment (TSA; for a discussion on small-molecule TSA, please see Chapter 2). During this exercise, potential target organs and toxicities are identified based on target expression and function, genetic or disease models (both human and animal), homology across species, and competitive intelligence. Special consideration is also warranted when the MOA for a biopharmaceutical is immunomodulatory, especially when there is a potential for cytokine release syndrome (CRS).

Prediction of putative liabilities can generally be deduced based on knowledge of the specific cells and/or mediators affected by the novel therapeutic. The evidence base for understanding the putative liabilities of modulating a specific target of interest is primarily acquired through the public domain (e.g., PubMed, Summary Basis of Approvals for New Drug Applications and Biologic License Applications). The data need to be critically evaluated to establish a weight of evidence for the potential liability to occur and to define a hierarchy of concern that takes into account the indication and patient population being pursued. Conducting subsequent *in vitro* or *in vivo* studies (or including additional endpoints on standard toxicology studies) may be needed to fill critical data gaps in order to better understand the likelihood of a liability occurring.

When evaluating the data for a target safety assessment, it is important to keep in mind whether the goal of the target modulation is to normalize a condition through inhibition or enhancement of a pathway or if the target is being used to identify cells that are to be killed (e.g., by immune cells or a conjugated toxin). Likewise, an understanding of the differences in target expression between the disease state in the human clinical population and the healthy nonclinical species is important for translation of predicted risks and observed toxicities (e.g., Kyoto Encyclopedia of Genes and Genomes (KEGG) disease, pathway, and drug databases).

Understanding the biological pathway of the target is a critical aspect of the target safety assessment. A review of the literature for effects of modulating the target signaling pathway should be conducted, and the ramifications of any physiological changes should be considered in the context of both the nonclinical species and the human disease population. Pathway network mapping can be used to identify downstream impact due to modulation of the target (e.g., Ingenuity Pathway Analysis, KEGG, GeneGo, etc.). A general understanding and awareness of upstream and downstream modulators will provide additional perspective on potential liabilities. For example, if a receptor–ligand interaction is disrupted, all other ligands or receptors that may be affected should be considered. Evaluating knockout or transgenic phenotypes and/or known human mutations and polymorphisms can be useful in interrogating the biology of a target. It is important to consider how the target will be modulated to decide whether the mutation data is relevant for understanding potential toxicities.

Another important aspect to determine is the likelihood of the biopharmaceutical binding to the target protein in the animal species being examined for relevance. This is generally accomplished by exploring nonclinical target-related databases for tissue expression and homology across species. Sequence information can be searched through various databases, including the NCBI-UniGene database (for nucleotide sequences) and the Swiss Prot/ExPASy database (for protein sequence and structure motifs), and using sequence comparison tools. It is worth mentioning that these *in silico* tools mostly contain human, rat, and mouse information, while nonhuman primate (NHP) and dog information is limited. Information on transcript abundance can be generated using various platforms (NCBI-UniGene, GeneCards, the gene expression database GEO, and the Novartis Research Foundation (GNF) transcript expression database). In addition, generation of microarray, *in situ* hybridization, or immunohistochemistry (IHC) expression data can be useful to fill in any gaps. The expression pattern in the potential species for nonclinical studies can then be compared to the expression pattern in human tissues. However, it is important to remember that making quantitative comparisons across species using these approaches is very challenging and that sequence similarities do not imply identical biological responses.

3.2.1   Target Safety Assessment for Biopharmaceuticals Targeting the Immune System

Distinct strategic approaches for assessing potential effects on the immune system should be considered in the predictive safety for biopharmaceuticals, whether the molecule is anticipated to have a targeted effect on the immune system or not. There is no specific regulatory guidance addressing the nonclinical safety assessment of immunomodulators, but specific comments relating to immunomodulators are made in several regulatory guidance documents. Under ICH S6, the potential for adverse events of biologically derived therapeutics is acknowledged, including the potential for injection site reactions, impacts on both cellular and humoral immunity, and alterations in the expression of surface markers. Distinct clinical concerns exist for immunosuppressive and immunostimulating agents, and the nonclinical assessment should be tailored to address these specific concerns. Biopharmaceuticals for treatment of inflammatory and autoimmune diseases or to prevent organ transplant rejection are often designed to target peripheral and central immune cell populations and mediators (cytokines, growth factors, complement components) in order to deplete them or suppress their function. Pronounced suppression or depletion of on-target immune cells and mediators on a chronic basis can lead to an increased risk of opportunistic bacterial, fungal, or parasitic infection, chronic viral infections or virally induced cancers, and/ or reactivation of JC-1 virus (e.g., Bartt, 2006). Biopharmaceuticals intended for cancer therapy may also impact normal lymphocytes/myeloid cells and can lead to cytopenia, immunotoxicity, and tissue injury (e.g., Aster, 2007). Therapeutic agonistic antibodies (directed to antigens on the cell surface) have the potential to trigger systemic CRS in man by cross-linking and clustering of the antigen target on immune cells by the Fab arms, by interaction with constant fragment (Fc) receptors on NK cells and neutrophils, or by a combination of the two. Systemic and local presence of these molecules and the associated inflammation and hemodynamic effects damage tissues and can result in disseminated intravascular coagulation, organ failure, and death if left untreated (Meng et al., 2015). These severe clinical sequelae of CRS typically do not occur in animal models, in spite of generating high systemic levels of cytokines. *In vitro* studies with human cells may be of more value in trying to assess the risk of CRS prior to FIH studies. Conducting *in vitro* cytokine release assays for lead candidate molecules during the early antibody selection process shows due diligence in the assessment of human risk, can provide useful comparative data against known positives, and can be a useful complement to *in vivo* animal studies (Finco et al., 2014). It may also reflect what was already expected based on the immunopharmacology and structure of molecules evaluated. There are a number of molecular characteristics (e.g., agonistic or stimulatory actions) that increase the potential to stimulate cytokine release (EMEA/CHMP/ SWP/28367/2007), some of which are highlighted in the final report of the Expert Scientific Group on Phase 1 clinical trials (2006).

3.3 STRATEGIC APPROACHES FOR BIOPHARMACEUTICALS AND ADCs

Given the broad array of biopharmaceutical modalities and targets, a standard approach to predictive safety is not realistic. A successful nonclinical biopharmaceutical safety program, however, will identify potential safety risks based on the intended target, MOA, and previous experience with similar products. In addition, it is necessary to demonstrate that a pharmacologically relevant species (ICH S6R1, 2011) is used in the safety studies and that the model is a suitable surrogate for prediction of human safety (e.g., Bussiere et al., 2009), which is an important differentiator from the nonclinical safety assessment of small molecules. The following objectives should be considered to aid in safety assessments of biopharmaceuticals prior to first-in-human (FIH) dosing and also during clinical development:

- Identify known or predicted risks associated with a given modality.
- Identify known or predicted risks based on the nature of the target, the intended MOA of the target, and relevance of animal species to predict such risks associated with a given biopharmaceutical (discussed in Section 3.2).
- Determine any off-target activity and associated targets that may require development of monitoring strategies in the clinic.

Core predictive safety approaches for derisking biopharmaceuticals are outlined in Fig. 3.1.

3.3.1 Modality-Associated Risks

Approaches for conventional biopharmaceuticals (e.g., antibody, protein modalities) have been established with a number of variations within each of these categories. From a safety perspective, each of these modalities offer unique advantages and challenges related to the mode of pathway interrogated, pharmacokinetics/biodistribution, opportunities to shape target specificity or cell–cell interactions, and safety. Thus, nonclinical scientists should understand the inherent risks associated with a given modality as well as the target-associated liabilities in their evaluations of a given agent.

3.3.2 mAbs

In addition to target-associated toxicities, intrinsic toxicities associated with mAb modalities may be predicted based on the specific construct. As a class, mAb-based therapies include whole antibodies, antibody fragments, Fc-modified peptides and proteins, bi-/multispecific constructs, ADCs, effector-modified antibodies, and many other engineered

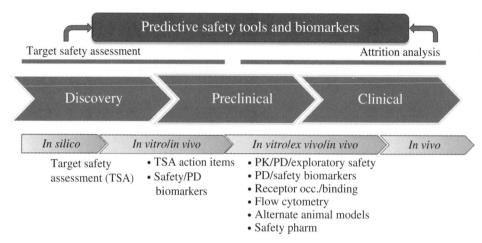

FIGURE 3.1   Predictive safety strategy for biopharmaceuticals.

constructs that incorporate Fab or Fc components. The engineered mAb constructs may modify or eliminate the Fab and/or Fc arms of an antibody to modulate exposure (half-life and biodistribution) and effector activity/immune stimulation and carry warheads to target tissues (e.g., Holliger and Hudson, 2005; Baeuerle et al., 2009; Alley et al., 2010; Beck et al., 2010). Fc-mediated effector function if strong and polyclonal (and persistent due to the long half-life of mAbs) could lead not only to the desired activation of cancer-specific immune cells but also to the undesirable activation of autopathogenic cells and development of autoimmunity.

Interaction between the Fc region of human IgG (particularly IgG1 and IgG3) and FcγRs on effector cells can cause antibody-dependent cell-mediated cytotoxicity (ADCC) or antibody-dependent cell-mediated phagocytosis (ADCP), whereas interaction with the complement component C1q can drive complement-mediated cytotoxicity (CDC; Chan and Carter, 2010; Moore et al., 2010). ADCC and CDC can be beneficial when the target antigen is present on a tumor cell. However, when the target antigen is present on lymphocytes, Fc/FcγR interactions may result in adverse reactions. Potential liabilities associated with Fc/FcγR interactions include infusion reactions/cytokine release, thrombocytopenia, and PK changes. Therefore, antibody isotopes are chosen based on indication and desired pharmacological activity. The selection of the IgG isotype (1, 2, or 4) and the design of the Fc portion of the antibody to minimize or enhance Fc-mediated ADCC, ADCP, and CDC can have a major influence on the toxicity to target and nontarget tissues.

3.3.3   ADCs

The approach behind the development of ADCs as oncology therapeutics is to combine the selectivity of an antibody toward tumor antigens with the potency of a cytotoxic agent in an effort to increase target specificity and reduce systemic toxicities associated with unconjugated chemotherapies (Chari, 2008). This facilitates accumulation of high concentrations of the warhead within the tumor environment while maintaining low systemic concentrations and reducing the toxic effects associated with chemotherapeutics. ADCs consist of a potent cytotoxic small molecule—that is, warhead—covalently linked to a tumor-targeting mAb via a cleavable or noncleavable linker. Changes to any of these variables—that is, antibody target, warhead MOA, linker type, and conjugation chemistry—can significantly impact safety and PK and need to be considered during lead candidate selection (Hinrichs and Dixit, 2015). The conjugation process produces a complex molecule with unique properties that are distinct from those of the individual components (Shen et al., 2012). As such, the nonclinical safety assessment of an ADC must consider the hybrid nature of the conjugate, in addition to the principles of large and small-molecule toxicity testing (Table 3.1; Roberts et al., 2013). One of the most important aspects of ADC safety assessment is the concept of on-versus off-target toxicity. On-target toxicity refers to specific killing of target-expressing normal tissues, while off-target toxicity refers to toxicity related to inappropriate release of the warhead outside target-expressing cells. A thorough understanding of this concept is necessary to guide lead candidate selection of ADC as clinical development has been limited by a combination of on- and off-target toxicities in patients (Hinrichs and Dixit, 2015).

3.3.4   On-Target Toxicity

The ideal ADC target is highly expressed on tumor cells and absent on normal cells; however, in reality, nearly all target proteins have some degree of normal tissue expression (Kessler et al., 2006). This is problematic as any target-expressing normal cell is potentially susceptible to ADC-mediated killing. The most well-known clinical example of on-target toxicity is bivatuzumab mertansine, an antihuman

TABLE 3.1   General Differences between Biopharmaceuticals, Small-Molecule Pharmaceuticals, and ADCs

Attribute	Small-Molecule Pharmaceutical	Biopharmaceutical	ADC
Process	Chemically derived	Biologically derived	Chemically and biologically derived
Size	<1 kDa	>10 kDa	>10 kDa
Absorption, distribution, metabolism, and excretion	Distribution to many organs/tissues, metabolized to active and nonactive metabolites	Limited distribution, catabolized to amino acids	Catabolism/target-mediated clearance of mAb; hepatic and/or renal elimination of warhead
Species specificity	Typically species independent	High species specificity	High species specificity for "on target"; species independent for "off target"
Toxicological effects	May be related to metabolite and unrelated to pharmacology	Generally associated with pharmacology or immunogenicity	On-/off-target toxicity related to MOA of cytotoxic warhead
Route of administration	Oral, inhaled, or parenteral	Parenteral	Intravenous
Dosing frequency	Daily	Intermittent	Intermittent
Immunogenicity	Rare	Common	Can occur

CD44v6 antibody conjugated to DM1 (Tijink et al., 2006). The target, CD44v6, is expressed on normal keratinocytes and epithelial cells in the cornea and tonsils (Mackay et al., 1994). Unexpected findings of serious dose-dependent skin toxicity, including toxic epidermal necrolysis (TEN), in patients resulted in early termination of the Phase 1 trial. This potential for target-mediated killing of normal cells highlights the need to have a thorough understanding of target expression prior to lead candidate selection (Tijink et al., 2006).

As stated earlier for the safety assessment of a pharmacological target, an initial evaluation of normal tissue expression should be conducted by performing a comprehensive literature search that includes publically available databases. However, it is advisable to follow up this information with internal confirmatory studies as soon as possible due to reproducibility issues associated with publically generated data (Persson et al., 2006). Despite recent technological advances, IHC remains the gold standard technique to evaluate target expression in normal tissues. IHC offers the advantage of visualizing target expression at the cellular level, thus enabling determination of cell type, anatomical location, and/or membrane expression (Riva et al., 2014). On the other hand, IHC is both time-consuming and resource-intensive due to issues with assay optimization, reagent availability, and the need for a trained pathologist to interpret the slides. While other techniques, such as quantification of mRNA levels—for example, RT-PCR and *in situ* hybridization—offer increased specificity and quantification over IHC, they lack the ability to distinguish target expression at the protein and cellular level. Therefore, IHC remains the best way to assess normal tissue expression at this time.

While critical to early safety evaluation, target expression alone is not sufficient to predict the potential for on-target toxicity of ADCs. This is due to the fact that there are other factors—such as anatomical location (i.e., membrane vs.

cytoplasm, apical vs. basal), vital nature (e.g., cardiomyocytes vs. epithelial cells), and proliferative/regenerative potential of the target cell—that also impact the potential for on-target toxicity (Carter and Senter, 2008). As a result, target expression data should not be evaluated in isolation; rather it should be evaluated as part of a bigger picture. This type of assessment can be difficult to conduct in the absence of *in vivo* safety data; however, there are some general principles that can be applied. For instance, the general properties of antibody biodistribution can be useful to inform risk assessment (Alley et al., 2009). Antibodies are generally restricted to plasma space and do not easily cross cell membranes (Shah and Betts, 2012). As a result, it is possible that target expression in antibody-restricted areas might not pose a significant risk for on-target toxicity. A good example to highlight this concept is prostate-specific membrane antigen (PSMA-) ADC, which targets PSMA overexpressed in prostate cancers (Tagawa et al., 2010). PSMA is also expressed in normal prostate, small intestine, proximal renal tubules, and salivary glands (Troyer et al., 1995); however, the major dose-limiting toxicities (DLT) associated with PSMA-ADC treatment are neutropenia and peripheral neuropathy. The current hypothesis is that localization of PSMA to the luminal/brush border side of normal cells, which are not typically exposed to circulating antibodies, is protective against anti-PSMA-ADC toxicity (Tagawa et al., 2010). In addition, it is also very important to understand the MOA of the cytotoxic warhead during target safety evaluation. To date, the most commonly used warheads are microtubule inhibitors—for example, DM1 and DM4—that block mitosis of rapidly dividing cells (Lodish et al., 2000). The reversible nature of this binding generally minimizes cytotoxicity to slowly dividing cells. Therefore, the proliferative index and/or the microtubule activity of the cell should be considered when assessing the potential for on-target toxicity. However, this relationship does not hold true for other classes of warheads—for example, DNA inhibitors—that target both slowly and rapidly dividing cells. In this case, more stringent criteria should be applied to target selection due to the increased risk for on-target toxicity. Comparison of relative target expression in tumor cells versus normal cells can also be used to inform the target safety assessment. Significant overexpression in tumors versus low-level expression in normal cells could provide a therapeutic window for target-mediated cytotoxicity (Momburg et al., 1987). Relative expression levels can first be confirmed by flow cytometry or IHC but then followed up with *in vitro* cytotoxicity assays to determine whether or not the difference in target expression translates to differences in susceptibility to target-mediated killing. While the limitations of cell culture must be considered, these data can be useful to build confidence during the target evaluation process.

While the above approaches can significantly derisk a proposed target, the remaining safety liabilities can only be resolved by *in vivo* safety assessment. Due to the antibody component, species selection for the ADC needs to follow the same principles used for unconjugated biopharmaceuticals (discussed later in this chapter). Lack of cross-reactivity to the target in commonly used toxicology species generally precludes use of rodent models to assess potential on-target toxicity. As a result, on-target safety evaluation is usually limited to NHP toxicology studies. While use of the NHP involves ethical and practical considerations that usually preclude early toxicology studies, it could be advantageous to run a small NHP study to evaluate the potential for on-target toxicity for targets with particular liabilities. Another approach would be to generate a rodent surrogate; however, the costs/benefits of these types of evaluations would have to be carefully considered.

3.3.5 Off-Target Toxicity

Despite the promise of targeted delivery of highly potent cytotoxic agents to tumor cells, off-target toxicity continues to pose a problem for ADC development (Carter and Senter, 2008). The maximum tolerated dose (MTD) of the vast majority of ADC in clinical development ranges from 2 to 5 mg/kg, despite significant differences in normal target expression, antibody subtype, and linker and/or warhead selection (Keizer et al., 2010). In addition, a recent analysis of DLT from Phase 1 trials of ADC showed that the majority of the toxicities—that is, neutropenia, thrombocytopenia, peripheral neuropathy, elevated liver enzymes, and gastrointestinal toxicity—were related to the known effects of microtubule-inhibiting agents (Hinrichs and Dixit, 2015). While the exact mechanisms are not known, they are thought to include conjugate instability, nonspecific uptake of ADC by FcRn or FcγR, and/or pinocytosis (Roberts et al., 2013). A well-known example highlighting the importance of off-target toxicity is Kadcyla® (trastuzumab-DM1), where the most common adverse effects are thrombocytopenia and elevated liver enzymes despite significant lack of target expression on platelets and liver cells. Due to the potential clinical liabilities, it is important to identify the potential off-target toxicities of novel ADC constructs—for example, novel warheads, conjugation technologies, and/or linkers—prior to lead candidate selection. Since these toxicity studies are largely designed to evaluate off-target toxicity, non-clinical safety assessment is not limited by species cross-reactivity and can be conducted in rodents.

3.3.6 Evaluation of Novel Warheads

Understanding the toxicity profile of the warhead is essential to understanding the potential off-target toxicities of a lead ADC candidate. While the safety of commonly used warheads such as DM1, DM4, MMAE, MMAF, SN38, and calicheamicin has been thoroughly investigated, recent clinical data has led to renewed focus on generating novel warheads/linkers with greater potency and/or potential for greater therapeutic index (Carter and Senter, 2008). Due to the high potential for off-target toxicity, the safety profile of novel warheads should be carefully evaluated prior to lead candidate selection (Roberts et al., 2013). However, it is important to note that conjugation significantly alters the safety and pharmacokinetics profile of the warhead to the extent that it is generally not useful to conduct a full panel of *in vitro* screening studies as for small molecules (Hinrichs and Dixit, 2015). For example, conjugation significantly alters exposure to the warhead. In its unconjugated form, warheads typically reach rapid maximal concentration (C_{max}) within minutes and undergo renal and/or hepatic clearance. As conjugates, small molecules undergo slow release from the ADC before being excreted (Han and Zhao, 2014). As a result, detectable levels of warhead in the plasma are generally close to the lowest level of quantification (LLOQ) following administration of an ADC (Han and Zhao, 2014). In the case of T-DM1, which is dosed at 3.6 mg/kg, the mean C_{max} of free DM1 is only 5.5 ng/ml in patients (Girish et al., 2012). Therefore, the decision to conduct *in vitro* safety screens should be evaluated on a case-by-case approach. For example, it might be appropriate to limit the *in vitro* assessment of cardiac ion channel inhibition (e.g., hERG) to a non-GLP screening study if preliminary data demonstrate limited inhibition at anticipated dose levels. In contrast, genotoxicity studies should generally be performed as these molecules are anticipated to be genotoxic at very low-dose levels due to their intended MOA.

While *in vitro* screens for warheads might not be as valuable as for other small molecules, there is a still a need to understand the nonclinical toxicity profile of the free warhead. As described previously, these studies can be conducted in rodents, as species cross-reactivity is not an issue for cytotoxic small molecules. In general, a single-dose rat toxicology study should be sufficient to identify the organs of toxicity for cytotoxic agents targeting rapidly dividing cells as ADCs are typically dosed on an every 3-week schedule (Hinrichs and Dixit, 2015).

3.3.7 Evaluation of New ADC Technologies

ADC technologies are rapidly advancing to enable development of next-generation molecules with improved efficacy and/or safety. These technologies include advances in linker types, conjugation chemistry, warhead MOA, and antibody engineering (Carter and Senter, 2008). Given the fact that changes to any of these parameters can significantly impact the safety and pharmacokinetic profile of an ADC, it is important to evaluate the nonclinical safety of novel technologies prior to lead selection. For the most part, new technologies are being designed with the intention to decrease off-target toxicity. One area of considerable interest is site-specific technology, which uses site-directed mutations to conjugate the warhead to the antibody. Recent data have demonstrated that switching from random conjugation to site-directed conjugation can significantly improve the tolerability of ADC (Junutula et al., 2008). At present, *in vitro* studies are not very useful to evaluate novel ADC technologies, largely due to inability of cell-based systems to assess the impact of changes to the biodistribution and clearance of ADC. For example, Strop et al. recently published data demonstrating that ADC engineered with two different conjugation sites behaved similarly in analytical and *in vitro* experiments but had dramatically different pharmacokinetic profiles *in vivo* (Strop et al., 2013). Therefore, at present, rodent toxicology studies remain the gold standard for evaluating novel ADC technologies.

3.4 PREDICTIVE SAFETY TOOLS FOR LARGE MOLECULES

3.4.1 Immunogenicity

In the context of nonclinical safety assessment, immunogenicity is defined as an antigen-stimulated immune response leading to the production of ADAs via T-dependent or T-independent B cell mechanisms (Baker et al., 2010). This is a common occurrence in toxicology studies conducted with protein therapeutics, as these are nonnative proteins that are recognized as foreign by the host species. While not inherently toxic, ADA can have a significant impact on the nonclinical assessment of biopharmaceuticals. In particular, ADA binding to its target induces the formation of immune complexes, which can clear the drug from systemic circulation (decreasing exposure) or induce immune complex-mediated toxicities (Rojko et al., 2014). This is an important consideration for nonclinical safety

assessment as it can confound interpretation of the toxicology studies. While immunogenicity in animal studies is not predictive of immunogenicity in man, it poses a significant problem for nonclinical safety assessment (Swanson and Bussiere, 2012). Ideally, immunogenic potential should be included in the criteria used for lead candidate selection. Currently, there are *in silico* tools available that use the amino acid sequence of the biopharmaceutical to identify potential T cell epitopes or peptides that bind to MHC molecules (De Groot et al., 2008). There are also *in vitro* T cell assays that show the potential of the biopharmaceutical to activate CD4+ T cells. Additionally, human immune system xenograft mouse models are also being considered as a way to evaluate potential human immunogenicity (Brinks et al., 2013). However, the predictive value of *in silico, in vitro,* and *in vivo* testing is still unclear, thereby limiting their applicability to lead candidate selection.

3.4.2 Specialized Assays for Detection of ADCC, CDC, and ADCP

ADCC assays measure complementarity determining region (CDR) and Fc-mediated activities that result in cytotoxicity. These assays, based on cell viability, assess killing of target cells by measuring the release of specific probes from prelabeled target cells using fluorescent dyes such as calcein-AM (Gómez-Román et al., 2005; Bracher et al., 2007). Cells such as NK92/FcγRIIIA (V) and FcγRIIIA-genotyped primary NK cells are used as effector cells for Fc effector function-enhanced and canonical Fc-bearing IgG1 molecules, respectively. Additional ADCC models can be established with primary NK cells from the nonclinical toxicology species. Complement activation by mAbs can cause direct tumor cell lysis or enhance ADCC. CDC is dependent on antibody isotype; IgG3 followed by IgG1 are the most effective isotypes for stimulating the classic complement cascade. Both isotypes bind to C1q, leading to formation of C3b on the surface of antibody-coated tumor cells near the site of complement activation (Gelderman et al., 2004). Specialized assays are performed with nonheat-treated serum or commercially available complement fractions. To date, however, the complement-dependent hemolytic assay (CH50) has been the most utilized to measure the functional lytic capacity of complement. Target cell killing can be assessed by several cell viability reagents such as alamarBlue, CellTiter-Glo, LDH release, or calcein-AM release. ADCP measures destruction of target cells via monocyte- or macrophage-mediated phagocytosis. ADCP assays use PBMC-derived cells or U937 cells differentiated to the mononuclear type. The phagocytosis readout requires tracking fluorescently tagged target cells by either flow cytometry or confocal microscopy. The ADCP assay can also be conducted in cells from the cynomolgus monkey (Li et al., 2012), which can help in the interpretation of *in vivo* studies. This technique uses fluorescein-labeled platelets or fluorescent beads as target cells for phagocytosis by monocytes/macrophages.

3.4.3   Immunotoxicity Testing

For many biopharmaceuticals, the desired pharmacology involves modulation of the immune system (Green and Black, 2000). While there are no specific regulatory guidelines for immunotoxicity evaluation of biopharmaceuticals, similar approaches can be considered as for small molecules. This involves a weight of evidence approach using data generated in the toxicology studies, as well as additional immunotoxicology studies (or inclusion of endpoints). Data from the toxicology studies can include endpoints such as blood cell immune-phenotyping and histopathological evaluation of lymphoid tissues, including thymus, spleen, lymph nodes, gut-associated lymphoid tissue, and bone marrow (ICH S6R1, 2011). Immune function can also be assessed *in vivo* by including a T cell-dependent antibody response (TDAR) assay in the repeat-dose toxicology study. This assay, which evaluates both primary and secondary antibody responses to keyhole limpet hemocyanin (KLH) or tetanus toxoid (TT), evaluates many aspects of the immune response including antigen processing and presentation, differentiation, upregulation of cell surface receptors, secretion of cytokines, somatic hypermutation, and immunoglobulin class (isotype) switching (Lebrec et al., 2014).

3.4.4   Predicting and Assessing Unintended Adverse Consequences

A common strategy for early prediction of unintended effects with biopharmaceuticals is to conduct pilot and/or dose-ranging studies to identify significant safety signals and to evaluate a dose response. Some of the safety signals revealed in these studies could include exaggerated pharmacology, target organ toxicity, reversibility of toxicities, which could affect development of the biopharmaceutical. Evaluation of a dose response can inform submaximum pharmacology, maximum pharmacology, and oversaturation of the target. Pharmacodynamic (PD) endpoints, safety biomarkers, and/or target coverage assays should be considered for inclusion in exploratory *in vivo* studies with biopharmaceuticals. In addition, these studies should be conducted in the appropriate, relevant species.

3.5   STRATEGIES FOR SPECIES SELECTION

For both small molecules and biopharmaceuticals, regulatory guidelines state that nonclinical safety testing should be conducted in both a rodent and a nonrodent species (ICH M3R2, 2009; ICH S6R1, 2011). For some

types of biopharmaceuticals—depending on the target—it is possible to generate lead candidates with multispecies cross-reactivity (Bussiere et al., 2009). However, for most biopharmaceuticals, the high target specificity and species selectivity prevents the molecule from engaging with the target in most species commonly used for toxicity testing (i.e., mouse, rat, rabbit, and dog) due to lack of homology. In this case, the guidelines state that use of one relevant species, typically NHP, may be appropriate and that a single species toxicology program is justified (ICH S6R1, 2011). However, it is still important for the nonclinical scientist to provide robust justification for using NHPs and consider all available alternatives (e.g., dog, minipig) to adhere to the principles of 3R—that is, replacement, reduction, and refinement of animal use (Pfuhler et al., 2009).

From a practical perspective, species cross-reactivity should be considered an important criterion for lead candidate selection. In the absence of cross-reactivity, there are limited options for assessing the potential safety risks during nonclinical development, which in turn impacts the conduct of the clinical trials (Muller et al., 2009). Ideally, a lead candidate with both rodent and nonrodent cross-reactivity would be identified, as this enables the molecule to be evaluated in pharmacology studies and if needed early rodent toxicology studies. However, if not possible, all available protein engineering efforts should be undertaken to identify a lead with NHP cross-reactivity (Doria-Rose and Joyce, 2015).

Similar to the target assessment, sequence homology is equally important in the selection of the toxicology species and involves an initial analysis of identity to assess potential activity/binding to the orthologue (De Haan et al., 2011). High sequence homology builds confidence, while low sequence homology suggests that the molecule might not engage with the target in a pharmacologically meaningful way. As lead candidates are identified, they are screened for species cross-reactivity by assessing relative binding affinities (K_d) to the human target and nonclinical orthologues, followed by *in vitro* cell-based potency assays. Taken together, these data support the justification for species selection in subsequent toxicology studies. However, it is valuable to include a PD endpoint in the toxicology study to demonstrate pharmacological activity in the species of choice (De Haan et al., 2011). If a relevant species cannot be identified, use of transgenic mice that express the human target or a rodent surrogate could be considered for toxicity testing (ICH S6R1, 2011). However, these approaches require significant development that will impact timelines and should be considered carefully. For small-molecule drugs, the predictive toxicology toolbox of *in vitro* testing provides early alternatives to animal testing to screen out suboptimal candidate molecules due to concerns regarding potential for target organ toxicity. However, many of these tests are not relevant for biopharmaceuticals (of high MW) and are therefore not considered useful. Data from *in vitro* binding affinity and functionality studies for biopharmaceuticals alongside an appropriate PD marker will ensure use of the most appropriate species, generate good quality *in vivo* data, and likely reduce the number of animals used. Therefore, it is critical to understand species cross-reactivity early on in discovery so that an appropriate nonclinical testing strategy can be put into place.

The strongest line of evidence for selecting a pharmacologically relevant species is the demonstration of a similar PD effect across species, utilizing molecular markers reflecting appropriate responses to the biopharmaceutical. In the case where there is a known molecular marker of the test article effect in humans, illustration of a similar change in that marker in a nonclinical species under consideration lends confidence to the MOA. For novel molecules, markers of effect on the pathway could serve to demonstrate the desired effect (or lack thereof) of the molecule in the species under evaluation for relevance. Ultimately, the acceptance criteria for the selection of relevant animal species are based on a combination of sequence homology, binding data, *in vitro* functional assays, and *in vivo* PD responses.

3.6 STRATEGY FOR DOSE-RANGING STUDIES FOR SAFETY EVALUATION OF BIOPHARMACEUTICALS

Similar to small-molecule drugs, safety assessment of biopharmaceuticals should be conducted in two relevant species, as discussed. However, a single species can be used if no other relevant species can be identified, and for many biopharmaceuticals, toxicology studies are performed in NHP. Justification of conducting toxicology studies in only one species includes providing data that indicates a lack of pharmacology in the other common nonclinical species.

Primary tests such as in-life observations, body weight measurements, and standard clinical pathology (hematology and serum chemistry) are often included in dose range-finding studies. Since these studies are exploratory in nature, gross pathology, organ weights, and extended histopathology are conducted on a case-by-case basis, typically to explore specific safety questions. Toxicokinetics (TK) and PD are critical endpoints to include on these studies as these data feed into dose-level selection for subsequent pivotal studies.

The route and frequency of administration should be as close as possible to that proposed for clinical use. Consideration should be given to TK and bioavailability of the product in a relevant species being used and the volume that can be administered to the test animals. Dose levels should be selected to provide suitable multiples over the proposed clinical dose and information on a dose–response relationship. For some classes of products with little to no toxicity, it may not be possible to define a specific maximum dose based on safety. In these cases, a scientific justification of the rationale for the dose selection and projected multiples of human exposure should be provided. To justify high-dose selection, consideration should be given to the expected pharmacological/physiological effects and the intended clinical use. Where a product has a lower affinity to or potency in the cells of the selected species than in human cells, testing of higher doses may be required. For high-risk medicinal products, the minimum anticipated biological effect level (MABEL) approach has been recommended by the EMEA (2007) for calculation of the first dose in man, and this should be taken into consideration when selecting doses for toxicology studies to support these products.

3.7 CONCLUSIONS

Designing safety evaluation programs for biopharmaceuticals is unique compared to small-molecule drugs due to fundamental differences in the biophysical attributes of the molecules. Two of the most important differences are the high-affinity target binding and the large molecular weight of biopharmaceuticals. These attributes of large molecules impact nonclinical safety assessment as species selection is often limited due to lack of cross-reactivity to the target in animals and most of the *in vitro* screens used for small molecules are not applicable to biopharmaceuticals. Approaches to large-molecule safety assessment have previously been guilty of unnecessary/inappropriate testing of these products, perhaps to meet precedents set by others, to meet perceived regulatory expectations (i.e., to "check the box"), or because such tests are part of small-molecule testing programs. As drug developers and regulators become more experienced in developing biopharmaceuticals, coupled with advances in scientific understanding of these products

TABLE 3.2 Specialized Approaches to Derisk Biopharmaceutical Modalities

	Proteins	Fc-Containing Antibodies	Antibody Drug Conjugates
Specialized assays	• Immunomodulation • Cytokine release • Immunogenicity	• Fc effector function • Cytokine release	• Cytotoxicity assay with ADC
Purpose	• Risk mitigation in the clinic for cytokine release syndrome	• Potency is an indication of activity • Cross-verify effector function in nonclinical species to understand safety-related findings	• Potency for risk mitigation

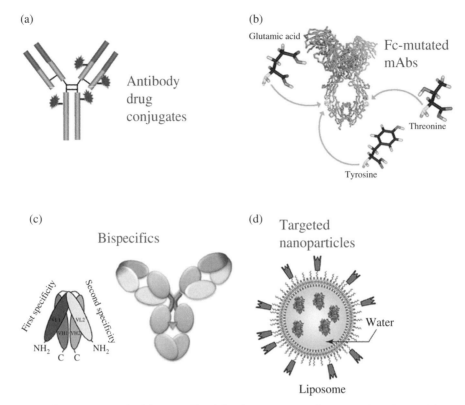

FIGURE 3.2   Complexity of biopharmaceutical formats. The following represent a few examples of novel biopharmaceutical platforms currently in clinical development. (a) Antibody drug conjugates consisting of cytotoxic small molecules chemically conjugated to a monoclonal antibody, (b) Fc-mutated monoclonal antibodies with mutations in Fc region to modify effector function, (c) bispecific molecules that bind to two different targets, and (d) targeted nanoparticles that mediate targeted delivery of lipid nanoparticles. Adapted from Chames et al. (2009), Buss et al. (2012), Trail (2013), Solaro et al. (2010), and Brien et al. (2013).

through clinical and laboratory experience, scientific rationale will prevail in the design of nonclinical testing programs, and more relevant test systems will be used. Finally, possession of a predictive safety toolkit for large molecules such as that outlined in Table 3.2 will address target biology, have effective tools that can be used in customized "flow schemes" that are modality specific, and will allow consideration of specialized questions regarding cross-species relevance and translation to the clinic (Fig. 3.2).

REFERENCES

Alley SC, Zhang X, Okeley NM, Anderson M, Law C-L, Senter PD, et al. The pharmacologic basis for antibody-auristatin conjugate activity. *Journal of Pharmacology and Experimental Therapeutics*. 2009;330(3):932–8.

Alley SC, Okeley NM, Senter PD. Antibody-drug conjugates: targeted drug delivery for cancer. *Current Opinion in Chemical Biology*. 2010;14(4):529–37.

Aster RH. Drug induced thrombocytopenia. In *Platelets* (A.D. Michelson, ed.), 2nd ed., pp. 887–902. Burlington, MA: Elsevier; 2007.

Attarwala H. TGN1412: from discovery to disaster. *Journal of Young Pharmacists*. 2010;2(3):332–6.

Baeuerle PA, Kufer P, Bargou R. BiTE: teaching antibodies to engage T-cells for cancer therapy. *Current Opinion in Molecular Therapeutics*. 2009;11(1):22–30.

Baker MP, Reynolds HM, Lumicisi B, Bryson CJ. Immunogenicity of protein therapeutics: the key causes, consequences and challenges. *Self/Nonself—Immune Recognition and Signaling*. 2010;1(4):314–22.

Bartt RE. Multiple sclerosis, natalizumab therapy, and progressive multifocal leukoencephalopathy. *Current Opinion in Neurology*. 2006;19(4):341–9.

Beck A, Wurch T, Bailly C., Corvaia N. Strategies and challenges for the next generation of therapeutic antibodies. *Nature Reviews Immunology*. 2010;10(5):345–52.

Bracher M, Gould HJ, Sutton BJ, Dombrowicz D, Karagiannis SN. Three-colour flow cytometric method to measure antibody-dependent tumor cell killing by cytotoxicity and phagocytosis. *Journal of Immunological Methods*. 2007;323(2):160–71.

Brien JD, Sukupolvi-Petty S, Williams KL, Lam CY, Schmid MA, Johnson S, et al. Protection by immunoglobulin dual-affinity retargeting antibodies against dengue virus. *Journal of Virology*. 2013;87(13):7747–53.

Brinks V, Weinbuch D, Baker M, Dean Y, Stas P, Kostense S, et al. Preclinical models used for immunogenicity prediction of therapeutic proteins. *Pharmaceutical Research*. 2013;30(7): 1719–28.

Buss NA, Henderson SJ, McFarlane M, Shenton JM, de Haan L. Monoclonal antibody therapeutics: history and future. *Current Opinion in Pharmacology*. 2012;12(5):615–22.

Bussiere JL, Martin P, Horner M, Couch J, Flaherty M, Andrews L, Beyer J, Horvath C. Alternative strategies for toxicity testing of species-specific biopharmaceuticals. *International Journal of Toxicology*. 2009;28(3):230–53.

Carter PJ, Senter PD. Antibody-drug conjugates for cancer therapy. *Cancer Journal (Sudbury, Massachusetts)*. 2008;14(3):154–69.

Chames P, Van Regenmortel M, Weiss E, Baty D. Therapeutic antibodies: successes, limitations and hopes for the future. *British Journal of Pharmacology*. 2009;157(2):220–33.

Chan AC, Carter PJ. Therapeutic antibodies for autoimmunity and inflammation. *Nature Reviews Immunology*. 2010;10(5):301–16.

Chari RV. Targeted cancer therapy: conferring specificity to cytotoxic drugs. *Accounts of Chemical Research*. 2008;41(1):98–107.

De Groot AS, McMurry J, Moise L. Prediction of immunogenicity: in silico paradigms, ex vivo and in vivo correlates. *Current Opinion in Pharmacology*. 2008;8(5):620–6.

De Haan L, Henderson S, McFarlane M, Scott A, Shenton J. Species selection. *European Biopharmaceutical Review*. 2011; 12(155).

Doria-Rose NA, Joyce MG. Strategies to guide the antibody affinity maturation process. *Current Opinion in Virology*. 2015;11: 137–47.

EMEA CPMP, Guideline on strategies to identify and mitigate risks for first-in-human clinical trials with investigational medicinal products (EMEA/CHMP/SWP/28367), 2007.

Expert Scientific Group on Phase 1 Clinical Trials. Final Report, 30 November 2006.

Farzaneh L, Kasahara N, Farzaneh F. The strange case of TGN1412. *Cancer Immunology, Immunotherapy*. 2007;56(2):129–34.

Finco D, Grimaldi C, Fort M, Walker M, Kiessling A, Wolf B, et al. Cytokine release assays: current practices and future directions. *Cytokine*. 2014;66(2):143–55.

Gelderman KA, Tomlinson S, Ross GD, Gorter A. Complement function in mAb-mediated cancer immunotherapy. *Trends in Immunology*. 2004;25(3):158–64.

Girish S, Gupta M, Wang B, Lu D, Krop IE, Vogel CL, et al. Clinical pharmacology of trastuzumab emtansine (T-DM1): an antibody-drug conjugate in development for the treatment of HER2-positive cancer. *Cancer Chemotherapy and Pharmacology*. 2012;69(5):1229–40.

Gómez-Román VR, Patterson LJ, Venzon D, Liewehr D, Aldrich K, Florese R, Robert-Guroff M. Vaccine-elicited antibodies mediate antibody-dependent cellular cytotoxicity correlated with significantly reduced acute viremia in rhesus macaques challenged with SIVmac251. *Journal of Immunology*. 2005;74(4):2185–9.

Guideline for Requirements for First in Man Clinical Trials for Potential High Risk Medicinal Products. Committee for Medicinal Products for Human Use (CHMP). EMEA/CHMP/SWP/28367/2007. Corr.23 Regulatory Focus 41. May 2007.

Green J, Black L. Status of preclinical safety assessment for immunomodulatory biopharmaceuticals. *Human & Experimental Toxicology*. 2000;19(4):208–12.

Han TH, Zhao B. ADME considerations for the development of antibody-drug conjugates. *Drug Metabolism and Disposition: The Biological Fate of Chemicals*. 2014;42:1914–20.

Hinrichs MJ, Dixit R. Antibody drug conjugates: nonclinical safety considerations. *The AAPS Journal*. 2015;17:1055–64.

Holliger P, Hudson PJ. Engineered antibody fragments and the rise of single domains. *Nature Biotechnology*. 2005;23(9):1126–36.

ICH Harmonized Tripartite Guideline M3(R2). Guidance on nonclinical safety studies for the conduct of human clinical trials and marketing authorization for pharmaceuticals, 2009.

ICH Harmonized Tripartite Guideline S6(R1). Note for guidance on preclinical safety evaluation of biotechnology-derived pharmaceuticals, 2011 (CPMP/ICH/302/95).

Junutula JR, Raab H, Clark S, Bhakta S, Leipold DD, Weir S, et al. Site-specific conjugation of a cytotoxic drug to an antibody improves the therapeutic index. *Nature Biotechnology*. 2008; 26(8):925–32.

Keizer RJ, Huitema AD, Schellens JH, Beijnen JH. Clinical pharmacokinetics of therapeutic monoclonal antibodies. *Clinical Pharmacokinetics*. 2010;49(8):493–507.

Kessler M, Goldsmith D, Schellekens H. Immunogenicity of biopharmaceuticals. *Nephrology, Dialysis, Transplantation*. 2006;21 Suppl 5:v9–v12.

Leader B, Baca QJ, Golan DE. Protein therapeutics: a summary and pharmacological classification. *Nat Rev Drug Discov*. 2008;7(1):21–39.

Lebrec H, Molinier B, Boverhof D, Collinge M, Freebern W, Henson K, et al. The T-cell-dependent antibody response assay in nonclinical studies of pharmaceuticals and chemicals: study design, data analysis, interpretation. *Regulatory Toxicology and Pharmacology*. 2014;69(1):7–21.

Li N, Ludmann S, He C, Anest L, Narayanan P. Generation of macrophages from cynomolgus bone marrow as a model to estimate effects of drugs on innate immunity. *Society for Leukocyte Biology Annual Meeting*. 2012;161:S64.

Lodish H, Berk A, Zipursky SL, Matsudaira P, Baltimore D, Darnell J. *Microtubule Dynamics and Associated Proteins. Molecular Cell Biology*. 4th ed. New York: W.H. Freeman; 2000.

Mackay C, Terpe H, Stauder R, Marston W, Stark H, Günthert U. Expression and modulation of CD44 variant isoforms in humans. *The Journal of Cell Biology*. 1994;124(1):71–82.

Meng L, Romano A, Smith E, Macik G, Grosh WW. Disseminated intravascular coagulation and immune hemolytic anemia, possibly evans syndrome, after oxaliplatin and bevacizumab infusion for metastatic colon adenocarcinoma: a case report and literature review. *Clinical Colorectal Cancer*. 2015;14(1):e1–e3.

Momburg F, Moldenhauer G, Hammerling GJ, Moller P. Immunohistochemical study of the expression of a Mr 34,000 human epithelium-specific surface glycoprotein in normal and malignant tissues. *Cancer Research*. 1987;47(11):2883–91.

Moore GL, Chen H, Karki S, Lazar GA. Engineered Fc variant antibodies with enhanced ability to recruit complement and mediate effector functions. *mAbs*. 2010;2(2):181–9.

Muller PY, Milton M, Lloyd P, Sims J, Brennan FR. The minimum anticipated biological effect level (MABEL) for selection of first human dose in clinical trials with monoclonal antibodies. *Current Opinion in Biotechnology*. 2009;20(6):722–9.

Persson A, Hober S, Uhlen M. A human protein atlas based on antibody proteomics. *Current Opinion in Molecular Therapeutics*. 2006;8(3):185–90.

Pfuhler S, Kirkland D, Kasper P, Hayashi M, Vanparys P, Carmichael P, et al. Reduction of use of animals in regulatory genotoxicity testing: Identification and implementation opportunities-Report from an ECVAM workshop. *Mutation Research*. 2009;680 (1–2):31–42.

Riva MA, Manzoni M, Isimbaldi G, Cesana G, Pagni F. Histochemistry: historical development and current use in pathology. *Biotechnic & Histochemistry*. 2014;89(2):81–90.

Roberts SA, Andrews PA, Blanset D, Flagella KM, Gorovits B, Lynch CM, et al. Considerations for the nonclinical safety evaluation of antibody drug conjugates for oncology. *Regulatory Toxicology and Pharmacology*. 2013;67(3):382–91.

Rojko JL, Evans MG, Price SA, Han B, Waine G, DeWitte M, et al. Formation, clearance, deposition, pathogenicity, and identification of biopharmaceutical-related immune complexes: review and case studies. *Toxicologic Pathology*. 2014;42(4): 725–64.

Shah DK, Betts AM. Towards a platform PBPK model to characterize the plasma and tissue disposition of monoclonal antibodies in preclinical species and human. *Journal of Pharmacokinetics and Pharmacodynamics*. 2012;39(1):67–86.

Shen BQ, Xu K, Liu L, Raab H, Bhakta S, Kenrick M, et al. Conjugation site modulates the in vivo stability and therapeutic activity of antibody-drug conjugates. *Nature Biotechnology*. 2012;30(2):184–9.

Solaro R, Chiellini F, Battisti A. Targeted delivery of protein drugs by nanocarriers. *Materials*. 2010;3:1929–80.

Strop P, Liu SH, Dorywalska M, Delaria K, Dushin RG, Tran TT, et al. Location matters: site of conjugation modulates stability and pharmacokinetics of antibody drug conjugates. *Chemistry & Biology*. 2013;20(2):161–7.

Swanson SJ, Bussiere JL. Immunogenicity assessment in nonclinical studies. *Current Opinion in Microbiology* 2012;15:1–11.

Tagawa ST, Beltran H, Vallabhajosula S, Goldsmith SJ, Osborne J, Matulich D, et al. Anti-prostate-Specific membrane antigen-based radioimmunotherapy for prostate cancer. *Cancer*. 2010; 116(S4):1075–83.

Tijink BM, Buter J, de Bree R, Giaccone G, Lang MS, Staab A, et al. A phase I dose escalation study with anti-CD44v6 bivatuzumab mertansine in patients with incurable squamous cell carcinoma of the head and neck or esophagus. *Clinical Cancer Research*. 2006;12(20 Pt 1):6064–72.

Trail PA. Antibody drug conjugates as cancer therapeutics. *Antibodies*. 2013;2(1):113–29.

Troyer JK, Beckett ML, Wright GL, Jr. Detection and characterization of the prostate-specific membrane antigen (PSMA) in tissue extracts and body fluids. *International Journal of Cancer Journal International du Cancer*. 1995;62(5):552–8.

4

DISCOVERY AND DEVELOPMENT STRATEGIES FOR SMALL INTERFERING RNAs

SCOTT A. BARROS[1] AND GREGORY HINKLE[2]

[1]*Investigative Toxicology, Alnylam Pharmaceuticals Inc., Cambridge, MA, USA*

[2]*Bioinformatics, Alnylam Pharmaceuticals Inc., Cambridge, MA, USA*

4.1 BACKGROUND

4.1.1 RNAi Molecular Mechanism

RNAi is a naturally occurring cellular mechanism for regulating gene expression mediated by "small interfering RNAs" (siRNAs). Synthetic siRNAs are typically 19–25 base pair double–stranded oligonucleotides in a staggered duplex with a 2–4 nucleotide overhang at one or both of the 3′ ends and can be designed to target any endogenous messenger RNA (mRNA) transcript. When introduced into cells, the guide (or antisense) strand of the siRNA loads into an enzyme complex called the RNA-induced silencing complex (RISC), which subsequently binds to its complementary mRNA sequence, cleaves the target mRNA and thereby suppresses expression of the target protein encoded by the mRNA (Elbashir et al., 2001a, b). The ability to selectively degrade the mRNA encoding a disease-related protein offers an attractive approach as a novel therapeutic modality.

4.1.2 Conjugate siRNAs for Hepatic Targets

Since unmodified siRNAs are rapidly eliminated and do not achieve significant tissue distribution upon systemic administration (Soutschek et al., 2004), various targeted delivery strategies target distribution to tissues and facilitate uptake of siRNAs into a relevant cell type. One approach used successfully in animal models (including rodents and nonhuman primates [NHP]) and humans employs intravenous (IV) delivery of siRNA in lipid nanoparticle (LNP) formulations (Soutschek et al., 2004; Morrissey et al., 2005;

Geisbert et al., 2006; Judge et al., 2006; Zimmermann et al., 2006; Coelho et al., 2013; Tabernero et al., 2013). Another approach delivers siRNA to target tissues via attachment or conjugation of a small-molecule ligand to the siRNA that facilitates tissue distribution, receptor-mediated cellular uptake, and cytosolic release of the siRNA (Soutschek et al., 2004). The most thoroughly validated method uses a conjugate carbohydrate ligand covalently linked to the siRNA that enables the direct hepatocyte-specific uptake via the asialoglycoprotein receptor (ASGPR). The ASGPR receptor is a member of the C-type lectin family and reversibly binds glycoproteins with terminal, *N*-acetylgalactosamine (GalNAc) residues (Ashwell and Morell, 1974). ASGPR is expressed on the cell surface of hepatocytes at a high copy number (0.5–1 million per cell) (Baenziger and Fiete, 1980; Schwartz et al., 1980) and facilitates clearance of desialylated glycoproteins from the blood (Geffen and Spiess, 1992). Binding of the carbohydrate ligand to the ASGPR leads to receptor-mediated endocytosis of the ligand–receptor complex followed by release of its cargo and rapid recycling of the receptor to the cell surface for successive rounds of uptake. Due to its high level and specificity of expression on hepatocytes, ability to mediate multiple rounds of uptake, and its ligand specificity, the ASGPR receptor has been a widely used target in animals for liver-specific drug and gene delivery (Wu et al., 2002).

Attachment of a triantennary GalNAc ligand to an siRNA ("conjugate siRNA") enables binding and subsequent hepatocyte uptake via the ASGPR, resulting in engagement of the RNAi pathway and downregulation of the target

Drug Discovery Toxicology: From Target Assessment to Translational Biomarkers, First Edition. Edited by Yvonne Will, J. Eric McDuffie, Andrew J. Olaharski, and Brandon D. Jeffy.

protein. Potent, dose dependent, and reversible RNAi-mediated reduction in protein levels has been demonstrated in multiple programs in the clinic via both systemic LNP- and SC-based approaches (Wittrup and Lieberman, 2015).

This chapter broadly describes the lead optimization activities relevant to the development of novel conjugate siRNA therapeutics for hepatic targets via subcutaneous administration. For specific information regarding other therapeutic oligonucleotide technologies (e.g., antisense, aptamer, CpG, anti-miR), specific nanoparticle-based delivery systems, or non-hepatic targeting, readers are directed elsewhere (Bennett and Swayze, 2010; Kole et al., 2012).

4.2 TARGET ASSESSMENTS

Selecting the most appropriate gene or genes to target for a particular disease is well beyond the scope of this chapter, though in principle every gene in the genome is a potential target for therapeutic intervention via siRNA. In practice however, hepatic targets dominate the current clinical development landscape due to the relative maturity of liver-based drug delivery. Therapeutics targeting other tissues, including nerve and kidney, are actively being pursued. The appropriateness of a target gene as a candidate for an RNAi therapy is the very first consideration in the design of siRNA therapeutics. Discovering appropriate siRNA therapeutic designs for some genes is more challenging than others in part due to the greater likelihood of unwanted side effects resulting in potential safety implications in downstream development.

4.2.1 Large Gene Families

Designing siRNAs for genes with multiple paralogs can be difficult due to the sequence divergence among transcripts. In considering the plausibility of a target that is a member of a multigene family, two strategies naturally arise: if the paralogs have significantly overlapping function, then knockdown of all the paralogs may be necessary to ablate the cellular function of the proteins. Alternatively the goal can be the specific knockdown of a single paralog. Both cases provide significant design challenges. In the case requiring the simultaneous knockdown of multiple gene transcripts, the principle challenge is discovering potent siRNAs with perfect or near perfect identity across all the paralogs. Establishing which genes and transcripts to include is the first step. The absence of consistent and reliable functional annotation can make this task particularly difficult. Unless the gene duplication events are recent, the sequence divergence among the members of the gene family, due primarily to third position synonymous mutations, can frustrate efforts to find even a single potential siRNA sequence candidate. Design alternatives are less than ideal. One strategy is to use

multiple siRNAs to cover several genes. Prima facie this strategy is appealing because of the opportunity for discovering highly potent siRNAs for each gene. In practice however the multiplicity of designs is not pragmatic. The screening effort for genes is additive, the number of unintended gene knock downs (i.e., off-targets) grows larger (due to the increased number of siRNAs), and perhaps most importantly for drug development, the manufacturing and quality control analytics for multiple siRNA mixes are greatly complicated. In the case where the aim is to specifically knockdown a single member of a gene family and leave the expression of other members unchanged, the challenge can be finding siRNA designs that are both potent and sufficiently specific for the intended target. For recently divergent genes, the degree of sequence similarity between paralogs may leave an inadequate number of potential distinct target sequences for discovery of a sufficiently potent siRNA. The most plausible strategy in this scenario is to identify regions for potential siRNA designs that maximize the sequence differences among the paralogs.

4.2.2 Short Transcripts

In deciding whether a gene is an appropriate therapeutic siRNA target, the length of the transcript is a crucial factor in the identification of a potential drug candidate. Short transcripts pose particular challenges because the sequence search space for discovery is necessarily limited. As a consequence, the number of potent sequences with identity to both the human target transcript and to nonclinical, toxicology species will likely be small. An inability to discover potent, cross-reactive siRNAs can be overcome by using toxicology species-specific siRNAs (e.g., rat or monkey), also known as "surrogate siRNAs," but the interpretation of toxicology findings is made more complex due to the difficulty of distinguishing chemistry or sequence chemistry-based interactions of the lead siRNA drug candidate.

4.2.3 Genes with Rapid mRNA Turnover

The systematic gene-by-gene analysis using RNAi technologies lead to the discovery that a small subset of gene transcripts appeared impervious to effective knockdown in that either screens of large numbers of siRNAs were required to find potent molecules, or in more extreme cases, no single siRNA significantly (>80%) knocked down the transcript levels *in vitro*. Analysis of turnover rates suggests that mRNAs with particularly faster turnover rates are more difficult to knockdown (Larsson et al., 2010). An open question remains as to whether phenotypes derived in part from genes with fast mRNA turnover rates are more sensitive to knockdown; the threshold for meaningful therapeutic impact by siRNA molecules may be lower for these genes. Difficult to knock down genes may require higher doses to achieve a therapeutically meaningful change in phenotype, but the higher doses of course risk increased toxicity.

4.2.4 Selecting among Alternate Transcript Variants

Many human genes produce two or more distinct transcript sequences through alternate or differential splicing of exons. More than 80% of human genes have alternate splice forms (Wang et al., 2008), many of which produce proteins with alternate amino acid sequences. Limiting the selection of siRNA candidates to the exon sequences conserved in all the transcript variants helps ensure that the final drug candidate will effectively target the intended gene. Some genes are annotated with highly truncated versions, so care must be exercised in reviewing the annotation for each transcript. The extremely truncated transcripts seen in the annotation of many genes are often rarely expressed and can leave insufficient sequence space to discover potent siRNAs.

4.3 siRNA DESIGN AND SCREENING STRATEGIES

4.3.1 siRNA Design

Once a target is chosen and the appropriate transcripts identified for inclusion in the design of siRNAs, the search begins for one or more lead siRNA designs. Several strategies are available for identifying the best siRNA for a given target in a given disease. The outline described herein presumes the candidate molecules will go through a primary *in vitro* screen in several cell lines followed by *in vivo* screening, typically in a rodent species.

Several distinct siRNA design criteria are independently evaluated in the selection of a lead RNAi therapeutic. The primary criteria are potency of the siRNA, specificity to the target gene, and cross-reactivity of the siRNA with respect to humans and all the attendant toxicology species.

4.3.1.1 Efficacy Prediction
The paramount criteria in the search for a potential therapeutic siRNA is potency, which is defined here as the concentration of the siRNA producing a 50% knockdown or half maximal inhibitory concentration (IC_{50}) of the target transcript. A brute force approach that determines the IC_{50} for every possible siRNA will always identify the most potent molecule, but the scale of such a synthesis and screening effort will necessarily be substantial, especially if the transcripts are long or varied. Bioinformatic predictions of potency are routinely applied to define a substantially smaller initial screening set than the brute force, "walk the gene" approach. Though the rules for the prediction of efficacy are not completely understood, the collective knowledge from the enormous number of siRNAs that have been synthesized and tested is sufficiently strong to confidently predict the subset of siRNA designs most likely to have good on-target potency.

4.3.1.2 Off-Target Prediction
The discovery of RNA-mediated transcript regulation naturally led to searches for the mechanisms of action as well the ability to accurately predict the regulatory targets of microRNAs. Sequence analysis of both the miRNA as well as in the cognate target mRNA quickly led to the discovery that the nucleotides of the miRNA in positions 2–9 have conserved complementarity to their target mRNAs and are the region of the miRNA that has the greatest impact on specificity. This region is known as the "seed" (Lewis et al., 2005). Substantial literature has since arisen around miRNA:mRNA seed pairing, and a great number of algorithms and applications built to better predict the on-target transcripts for each miRNA exist. The myriad rules for miRNA specificity by and large apply to siRNA design, though there are some exceptions, like RNA editing that accompanies some miRNA biosynthesis, which is not thought to play a role in siRNA-mediated knockdown (Kume et al., 2014). The creation of commercial and academic genome-wide RNAi libraries enabled the screening of model organisms as well as several mammalian species (Moffat et al., 2006). Understanding the consequences of applying siRNAs by measuring transcriptome-wide changes in gene expression, typically with microarrays and more recently with deep RNA-seq, very quickly led to the creation of very large datasets of siRNA-mediated both on- and off-target knockdown. Knowledge gleaned from these analyses confirmed the primary features of siRNAs that impact specificity and enabled the creation of bioinformatics tools to aid in the prediction of likely on- and off-targets (Echeverri et al., 2006).

4.3.1.3 Sequence-Based Off-Targets
The selection of siRNAs generally begins with the bioinformatics search for putative off-targets by comparing the siRNA sequence of both the sense and antisense strands with the entire transcriptome from humans as well as each of the toxicology species. A large and growing number of public and commercial applications exist to predict the likely off-targets for any given siRNA: most applications use similar criteria in their predictions. In assessing a list of candidate siRNA designs, the aim of the bioinformatics step is to determine both the number of potential off-targets as well as the degree of similarity between the siRNA sequence and the mRNA. The alignment and counting of seed matches is programmatically straight forward, but the interpretation of the predictions in a clinical application can be vexingly complex. Since specificity is largely driven through the complementarity of the seed region of the siRNA strand and the mRNA, preferential weight is typically given to the number of "seed matches" between the candidate siRNA and the transcriptional complement of a cell. Off-target prediction that includes the entire complement of possible

mRNAs is the most prudent course, though searches through tissue-specific expressed subsets of mRNAs can be employed. Example transcriptome collections include NCBI's REFSEQ (www.ncbi.nlm.nih.gov) and Ensembl (www.ensembl.org) databases.

We use a unique and proprietary algorithm designed to rank molecules for potential off-target effects. The algorithm aligns each sense and antisense strand to every possible complementary position on the entire transcriptome (all relevant species) and employs a position-specific scoring matrix that takes into consideration both the number and position of mismatches between the oligomer and the mRNA. Preferential scoring weights are assigned to mismatches based on the location within the oligomer (e.g., seed region vs. flanking regions). The aim of the exercise is to rank siRNAs against defined thresholds or "acceptance criteria" such that the most specific sequences are selected. Following algorithmic selection, the list of candidate off-target transcripts annotations can be evaluated with respect to expression in the target tissue and likelihood of causing unwanted phenotypic changes to assess the potential that downmodulation would cause toxicity. For example, a significant match to an olfactory gene is likely less significant than a more modest match to an RNA polymerase. The search is conducted for human as well as the toxicology species. For genes with short transcripts, or significant regions of low complexity sequence, or where the aim is to target a specific variant of a gene family, the threshold may by necessity be lower than for a long, highly conserved gene. In this case the off-target scores must be relaxed until a sufficient number of candidates remain for testing for efficacy and potency or the potential for toxicity must be evaluated *in vivo* in model systems.

4.3.1.4 Measuring Off-Targets with RNA-seq or Arrays

The measure of mRNA knockdown of both the targeted transcript as well as other messages in the transcriptome can be accomplished in a targeted manner using individual gene measures with techniques such as quantitative PCR or more broadly using expression arrays or through deep sequencing. The advantage of using single gene measures is speed and cost: assays for most human and toxicology species are available from commercial vendors, are easy to set up, and the results interpreted without sophisticated bioinformatics analyses. The number of genes easily measured using individual gene assays is quite limited yet the lists of potential off-targets generated through transcriptome-wide sequence searches can be quite large. Transcriptome-wide measures using microarrays or deep sequencing offer the opportunity to query the majority of transcripts in one experiment.

4.3.1.5 Cross-Species siRNAs

The number of potential siRNA candidate sequences depends on the length of the transcripts, the number of transcript variants, the number of

intended toxicology species, and of course the degree of conservation. In a typical three-way intersection of human, mouse or rat, and NHP, one can expect to find 0–30% of potential 19 mers are conserved across all three species. The demand for cross-species identity most typically results in the greatest constriction in numbers of siRNA candidate designs. Two sets of siRNAs with taxon-specific sequences, one rodent and one human–NHP, can be synthesized and screened when the pool of cross-reactive candidate siRNAs becomes so small as to preclude the likely discovery of a potent molecule. Toxicology studies using a surrogate siRNA (i.e., rodent specific), while not uncommon, come with the caveat that the off- and on-target effects for the analog molecule may not be representative of the drug candidate siRNA.

4.3.2 Chemical Modification of siRNA

Natural (unmodified) siRNAs have inherently poor drug-like properties. They are highly susceptible to exo- and endonuclease digestion in serum and tissues, do not readily cross cell membranes (size and polyanionic charge), nor bind to plasma proteins, the result of which is their rapid degradation and elimination following parenteral administration (Watts et al., 2008). In addition, some unmodified siRNAs may be recognized by host pattern recognition receptors (Toll-like receptors and retinoic acid inducible gene-I-like RNA helicases) of the innate immune system and stimulate a type-1 interferon response and/or a pro-inflammatory response (Hornung et al., 2005; Judge et al., 2005; Thompson and Locarnini, 2007).

A variety of chemical modifications have been implemented to overcome these obstacles (Watts et al., 2008). The most common modifications in oligonucleotide therapeutics (OTs) are depicted in Figure 4.1 and include phosphorothioate (PS) linkages (rather than phosphodiester), 2'-O-methyl (2'-O-Me), 2'-O-methoxyethyl (2'-MOE), 2'-fluoro (2'-F), bridged nucleic acid (BNA), or phosphorodiamidate morpholino oligomers (PMOs). The extent to which any base, sugar, or backbone modification is used as well as the position within the sequence is dependent on the specific type of oligonucleotide modality (single vs. double strand, RNAi vs. antisense mechanisms, etc.). For double-stranded siRNA, chemical modification of nucleotides affects duplex stability, resistance to nuclease digestion, protein binding, RNAi-mediated cleavage, and evasion of immuno-surveillance mechanisms, all of which serve to fine-tune the collective pharmacokinetic, pharmacodynamic, and toxicological properties of the molecule.

4.3.3 Screening of siRNA Therapeutics

Because siRNA therapeutics act through a common mechanism of action, robust screening pipelines for a broad variety of targets and indications can be established with a minimum

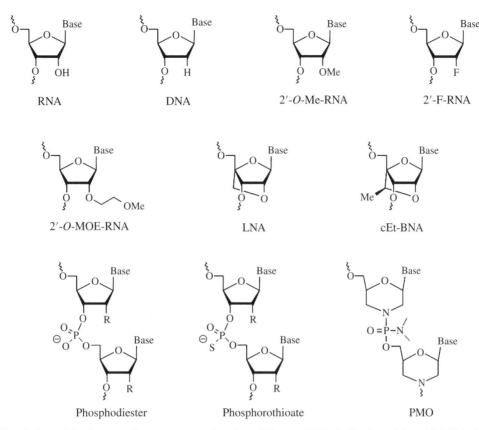

FIGURE 4.1 Chemical modifications to ribose sugar or phosphate linkages. BNA, bridged nucleic acid; LNA, locked nucleic acid; PMO, phosphorodiamidate morpholino oligomer.

of assay development. Since the ultimate goal is to affect the expression level of the target of interest, an endpoint measurement of target RNA is the ultimate assay for ranking siRNAs. In contrast, screens for small molecules or biologics often require a great deal of knowledge about the biology and the structure of the target in order for an appropriate assay to be developed. Another distinction between siRNAs and other classes of molecules is that siRNAs can be rationally designed based on an evolving set of characteristics that define potent and efficacious duplexes. Rational design serves to limit the number of molecules that must be screened to define a potent and specific lead candidate. Taken together, these properties of siRNAs serve to decrease the amount of time between target and lead identification to as little as several weeks.

High-throughput screens for siRNA therapeutics are distinct in several ways. The majority of published high-throughput screens are concerned with target discovery and use phenotypic selection as an endpoint; they generally employ siRNA libraries containing small numbers of siRNAs or shRNAs against each target (for review see Berns et al., 2004; Boutros and Ahringer, 2008). Genome-wide RNAi screens were first implemented in *Caenorhabditis elegans* with low-resolution "dissecting scope" assays. High-level functions were assigned to a large fraction of known and predicted genes. Libraries covering all the genes in mammals have been developed and applied to associate genes and phenotype (Kamath et al., 2003; Cullen and Arndt, 2005; Boutros and Ahringer, 2008). In these screens, the goal is typically the identification of an individual or small pool of siRNAs capable of target knockdown such that they phenocopy a loss-of-function mutation. Screens with genome-wide libraries are an efficient means to assign function to genes but unlikely to identify therapeutic candidates due to the low numbers of siRNAs per gene. In contrast, in screens for potential RNAi therapeutics, the goal is to identify the most potent, specific, and least toxic siRNA by screening a larger set of siRNA designs with diverse sequence targets as well as chemistry that target the disease gene. Unlike a strictly phenotypic screen, the goal is to identify an siRNA that results in target knockdown and an associated phenotype but is also capable of knockdown *in vivo* with a sufficient therapeutic index. Though a brute force gene walk, wherein every siRNA is synthesized and screened, could be applied, many factors serve to limit the number of duplexes screened. In practice screens are limited to siRNAs predicted through bioinformatics analyses to have significant activity. In addition, a major consideration for therapeutic development is cross-species conservation, which may be required for *in vivo* disease models as well as toxicology studies, as the drug transitions from research through development. Together these factors tend to limit the number of siRNAs screened from thousands to as few as dozens depending on how stringently the rational design rules are applied.

4.3.3.1 Selection of Cell Lines

The most important criterion in the selection of a cell line is sufficient expression of the target gene and transcript to enable accurate measures of knockdown via quantitative PCR. Generally cell lines with same tissue origin as the disease gene are employed. Other factors such as transfectability and ease of culture should also be considered. Though all cell-based screens are to some degree non-physiological, they serve as a good platform to rank siRNAs for their ability to knockdown a transcript and define an IC_{50} potency value in order to guide appropriate dosing regimens in subsequent *in vivo* experiments. We routinely use fresh or cryopreserved primary cells to further refine IC_{50} values. The disadvantage of primary cells is expense or time for isolation. In addition, primary cells are difficult to maintain for extended periods of time so siRNA activity must be measured within 24 h, which thereby limits their usefulness for studying the duration or cytotoxicity of siRNA knockdown.

4.3.3.2 Evaluation of Target Silencing

Though genome-wide measures of the change in expression levels such as RNA-seq or microarrays are amenable for primary, early stage siRNA screening, the number of replicates for sound statistics, and the relatively high expense make them impractical. Direct measures of the target gene in isolation are easily incorporated into modestly high-throughput screening operations using real-time, quantitative PCR. An alternative, less equipment-intensive alternative for measuring changes in expression level in a moderately high-throughput method is the use of branched DNA assays (Panomics), which only requires a plate reader and other common laboratory equipment including incubators and shakers. Genes not expressed endogenously in relevant cell lines can be screened using luminescence-based gene fusions, typically in a dual-luciferase system (dual-luc). Luminescent, fusion-based systems have the disadvantage of requiring the synthesis and cloning of a suitable construct, but the advantage is that any DNA fragment can be screened.

4.3.3.3 Lead Optimization of siRNA

Lead identification can be visualized as an upside down pyramid (Fig. 4.2) where an initial set of siRNA are filtered through a set of assays in order to hone in on one or more lead designs. Lead optimization, in contrast, best fits the visual metaphor of a diamond, whereby each potent lead is diversified through chemical modification of the nucleotides (Elbashir et al., 2001a; Czauderna et al., 2003) and subsequently refiltered. This process is very much iterative. In each round the

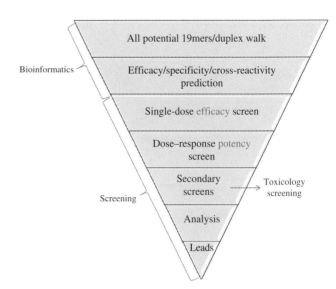

FIGURE 4.2 Screening funnel. As a screen progresses the number of potential leads is reduced as additional assays with increasing powers of discrimination are applied. Though some assays in a screening funnel, such as efficacy and potency screens, might be the same for all targets, target-specific secondary assays will often need to be implemented to select individual or small numbers of leads.

compounds are initially screened for efficacy, typically at a single low dose *in vitro* and benchmarked relative to the initial lead. Dose–response screens are then carried out on the few best compounds to identify duplexes with improved potency. The number of potential modifications at each position is great, so the combinatorial number of distinct siRNAs designs is extraordinarily large. Since each siRNA is typically synthesized in isolation, the fraction of the sequence space that can be effectively screened is vanishingly small. Nonetheless, nucleoside modifications are absolutely crucial for the development of siRNA therapeutics in order to improve duplex stability in serum, evade immune recognition, or improve potency (Elbashir et al., 2001a; Czauderna et al., 2003). Though in principal any base can be modified regardless of position, some modifications are known to reduce potency, while others might be beneficial. For example, placement of 2′-O-Me-modified bases in specific motifs allows for resistance to endonuclease-mediated degradation and off-target silencing (Czauderna et al., 2003; Jackson et al., 2006). The effect of these modifications on potency can be sequence dependent and therefore difficult to generalize. Since potency is only one of the important properties of a therapeutic siRNA, some sacrifice might be made in this regard to endow the lead with other beneficial characteristics, such as improved stability.

4.3.3.4 Off-Target Analysis

siRNAs can have a variety of off-target effects, including immune stimulation, interference with microRNA regulation, or off-target silencing due to partial sequence matching to a transcript other than the intended target (see Off-target Prediction above; also Sledz et al., 2003; Judge et al., 2005; Echeverri et al., 2006; Larsson et al., 2010). Immune effects as a class can be reduced through chemical modification of the siRNAs, as discussed in the following text. Direct inhibition of endogenous miRNA pathways is unlikely if the siRNA therapeutic potency is high and therefore the intracellular concentration low (Judge et al., 2006; John et al., 2007). The degree of inhibition of an off-target target will be impacted by the relevant tissue level expression, location of the target site within the mRNA, and mRNA turnover. We make direct measures of off-target transcript knockdown in order to evaluate their degree of inhibition. Direct screening by qPCR from the same RNA samples used for either single dose or dose–response screens is the most straightforward method to directly screen potential off-targets but is not always possible. Predicted off-targets might not be expressed in the cell line used for screening; these transcripts can either be screened in other cell lines, or the off-target sequence can be cloned into the dual-luc vector and screened exogenously as described previously. For off-targets, the entire cDNA or just the region that binds to the siRNA can be cloned into the dual-luc vector. Much like a screen, off-target silencing can be compared to on-target silencing first in a single high-dose experiment and then, if need be, in a dose response to demonstrate significant (i.e., orders of magnitude) difference in the on- and off-target potency.

4.4 SAFETY LEAD OPTIMIZATION OF siRNA

For each individual oligonucleotide modality (siRNA, antisense, anti-miR, aptamer, etc.), common features are typically observed in nonclinical toxicology studies that may be considered "class based," given the structural similarity of molecules within any given subclass. As the nonclinical safety database expands across sequence space (preclinical development programs), in vitro assays and in vivo study designs used in lead optimization become more uniform, and the screening paradigm becomes progressively more "plug and play." Test article-related effects due to (exaggerated) on-target pharmacology are readily identified when different siRNAs targeting different sequences of the same mRNA yield similar findings. Likewise, potential hybridization-based off-target effects may also be ruled out by different siRNAs binding to different sequences along the same mRNA transcript. For safety screening in discovery stages, the goal is primarily to identify any unexpected acute pro-inflammatory effects or potential nonhybridization-based, sequence-specific toxicities. Screening for acute immunostimulatory effects of siRNAs (as described in the following text) is conducted in an in vitro human whole blood cytokine assay and in vivo in an outbred mouse.

Toxicity screening for lead candidate selection is conducted in the rat and/or mouse. For candidates selected for development, the rat and the NHP (cynomolgus monkey) are used for subsequent nonclinical toxicology studies. The NHP is commonly used as the nonrodent species for nonclinical pharmacology and safety evaluations of OTs based on mRNA sequence homology to humans. Based on the historically minimal effects of conjugate siRNAs in subacute studies, the NHP is not used as a screening toxicology tool during lead selection.

4.4.1 Immunostimulation Screening

Once a small set (<10) of potent candidate siRNAs have been identified, the molecules are screened for potential acute pro-inflammatory effects in an in vitro human whole blood cytokine assay and in vivo in the outbred CD-1 mouse.

Previous in vitro studies evaluating immunostimulation by siRNAs have demonstrated the potential for unmodified siRNAs to induce robust cytokine responses. The immunostimulatory potential of siRNA can be minimized through various chemical modification of the 2' position of the sugar moiety of nucleosides (e.g., 2'-O-Methyl (2'-O-Me) and 2'fluoro (2'-F)) (Watts et al., 2008). Although the choice of in vitro culture systems is variable and user-dependent, we have found the most utility with whole blood preparations. Briefly, test articles are incubated for 24 h with freshly isolated human whole blood from multiple (3–4) individual donors. A transfection reagent may be employed to maximize intracellular delivery, as appropriate (i.e., the ligand/receptor used for hepatocellular delivery in vivo may not be ideal for free uptake incubations with whole blood cells). Immunostimulation in plasma is measured using a multiplexed cytokine/chemokine panel (Luminex; IL-1β, IL-RA, IL-6, IL-8, G-CSF, IP-10, MIP-1α, and TNF-α). Unmodified siRNA and mock transfection controls are included for comparison. This assay is intended to confirm that the chemical modification pattern used in any given siRNA duplex completely abrogates all acute stimulatory effects, as compared to unmodified siRNA.

Potential acute pro-inflammatory effects of candidate siRNAs are also evaluated in a rodent model—the outbred mouse being a good choice for early screening, requiring minimal test material availability and having the most commercially available reagents for immunoscreening. Animals are administered a single dose of test article at 10- to 100-fold multiples of efficacious exposures, and plasma cytokines/chemokines are evaluated 4 h postdose via similar multiplexed analysis (Luminex). As unmodified siRNAs do not have adequate plasma stability in vivo, other oligonucleotide agonists with canonical immunostimulatory motifs are used as comparator positive controls. Similar to the hWBA, this assay is intended to confirm that the chemical modification pattern used in any given siRNA duplex does not provoke any acute pro-inflammatory responses *in vivo*. It should be noted that plasma cytokine evaluations (and complement activation) are also typically included as part of subsequent NHP studies later in development.

The *in vitro* human whole blood assay and the outbred mouse are considered binary (yes/no) hazard identification tools in lead optimization screening for compound selection. In our experience to date, chemically modified conjugate siRNAs do not provoke acute pro-inflammatory effects in subsequent rodent or nonrodent studies or in clinical trials if they test negative in these two assays. Lacking positive controls, it is not yet feasible to assess whether the immunostimulation screening during lead optimization is predictive or relevant for human outcomes.

4.4.2 Toxicology Screening in Rodents

Following immunostimulation screening, a smaller subset (<5) of candidate siRNAs are typically advanced to repeat-dose subacute (14–28 days) general toxicology studies in outbred rodent species (Sprague Dawley rat, CD-1 mouse) for candidate selection. For conjugate siRNAs, single-dose tolerability studies are not informative or discriminatory for lead selection as most are well tolerated up to ICH limit doses. Test articles are administered via repeat subcutaneous injections and include modified Draize scoring for local tolerability, in addition to standard clinical and anatomic pathology endpoints and limited toxicokinetics (plasma and select tissues).

In our experience with conjugate siRNAs, the rat is the more sensitive rodent species for safety evaluations. All GalNAc–siRNAs share common rapid biodistribution from the blood to (primarily) the liver via ASGPR-mediated uptake and clearance primarily through the kidney. By design, the GalNAc-targeting ligand enables rapid distribution to the liver, resulting in short plasma half-lives. Given the intracellular trafficking via the endolysosomal system to cytoplasm to the RISC complex (Section 4.1.2), tissue half-lives may be greater than plasma depending on the metabolic stability conferred by the chemical modifications to the duplex. For lead candidate selections, tissue pharmacokinetics and/or toxicokinetics may be used for rank ordering, depending on the desired properties of the molecule. The appearance of test article-related findings in the rat are dose dependent and are commonly observed in the liver and kidney, the primary organs of distribution and elimination. Common findings for GalNAc–siRNAs in rat include hepatocellular vacuolation, "degeneration" (e.g., minimal single-cell necrosis), and "regeneration" (increased mitoses, hepatocellular karyomegaly). The degenerative changes generally occur at the higher dose levels tested. The criteria used for defining "adversity" across studies is dependent on observing degenerative/regenerative changes in association with functional deficits (e.g., elevations to serum

transaminases or total bilirubin). The common findings in the rat liver (across multiple targets/sequences) have not been dose limiting and are identified as reversible in all studies following a dose-free recovery period. This is particularly true of necrotic or degenerative lesions that were considered adverse, where these changes are typically fully reversible following recovery, whereas the hepatocellular vacuolization may persist in high-dose groups in the rat through the recovery period but always in the absence of causing functional deficits. Finally, findings in rat liver generally do not increase in severity with longer-term dosing schedules up to chronic dosing regimens of 6 months (or worsen during recovery) due to the tissue clearance of drug product. In the kidney, microscopic findings are typically limited to non-adverse basophilic granules in the proximal tubules (presumably the presence of test article) with no associated degenerative or functional changes.

Although the general pharmacokinetic and pharmacodynamic properties of conjugate siRNAs are similar in the mouse, toxicologic effects in the liver and kidney tend to be of lesser incidence/severity as compared to rat and occur at higher (>3×) dose levels on a milligram per kilogram basis. In our experience, the use of the outbred mouse is on an *ad hoc* basis for investigative toxicology studies as it does not represent the most sensitive species. This is in contrast to many single-strand antisense/antagomir oligonucleotide sponsors who employ the mouse as the rodent species for nonclinical development studies. For certain single-strand antisense molecules, the mouse may be a better model for characterizing associated nephrotoxicities (Frazier et al., 2014; Frazier, 2015).

It should be noted that for conjugate siRNAs, the hepatocellular findings commonly observed in rat toxicology studies (vacuolation, degeneration/regeneration) do not tend to occur in companion NHP toxicology studies. Common findings in NHP studies are related to the presence/accumulation of test article in target tissues following repeat subcutaneous administration of exaggerated dose multiples (e.g., basophilic granules in hepatocytes, pigmentation in phagocytic cells in liver, lymph node, and injection sites). These findings are non-adverse, not dose limiting, and have been identified to be reversible during recovery periods. Taken together, the implication is that the hepatocellular effects of conjugate siRNAs may be rat (rodent) specific, so appropriate caution and interpretation is warranted during the lead selection process in terms of setting "pass/fail" criteria.

As noted above, the more challenging dimension of early rodent screening is identifying potential non-hybridization-based, sequence-specific toxicities—the overall incidence of which tends to be relatively rare for modified oligonucleotides (and not necessarily identified more or less frequently in mouse vs. rat). There are recent published reports from single-stranded oligonucleotides that discuss the association of specific short sequence motifs with pronounced hepatotoxicity in rodent models (Hagedorn et al., 2013; Burdick et al., 2014). The underlying hypothesis is that there are short sequence motifs that may promote binding to intracellular proteins and/or pattern recognition receptors involved in activating acute antiviral-type responses that may ultimately be involved in the pathophysiology of liver injury. It follows then that given a large enough "gene walk" of oligonucleotides, one would encounter rare "bad actors" by chance inclusion of some offending sequence motif. This general principal has been demonstrated for single-stranded antisense molecules containing locked nucleic acid (LNA) chemistry (Burdick et al., 2014). The fact that this has been demonstrated for single-stranded antisense molecules is not to imply that this class is more or less subject to rare sequence-specific effects. Rather, the fact that the majority of single-strand antisense oligonucleotides employ fixed chemical modification patterns simply makes them more amenable to this type of retrospective analysis. For example, the antisense "gapmer" modification pattern employs a central core of ~8 DNA nucleotides flanked on both sides by chemically modified nucleotides such as LNAs or 2'-MOEs, with a full phosphorothioate backbone. Since the chemical modification pattern is fixed (e.g., 3LNA-8DNA-3LNA), the general physical–chemical properties of each oligonucleotide along a base-by-base gene walk is comparably similar, allowing for more robust analyses of sequence-specific differences. Across large screening sets, these types of libraries are more amenable to the identification of structure–activity relationships, such as short sequence pattern recognition.

By contrast, the chemical modification patterns used for double-stranded siRNAs are not fixed across molecules; the designs are guided by general rules but are ultimately empirically determined on a case-by-case basis during lead optimization. Therefore, the potential for inclusion of an offending sequence motif is theoretically the same as for single-stranded oligonucleotides; it is just that the motif is unlikely to be identified or predicted through retrospective analyses within siRNA screening sets. An additional confounding variable with siRNA is that the potential for protein binding by short specific sequence motifs may change based on modification pattern or strand location (sense/antisense) within the duplex. Given these complexities, the subacute rodent toxicity screening (with standard endpoints) becomes an effective hazard identification screen, albeit not a particularly effective mechanistic tool.

4.4.3 Points to Consider for Chemically Modified Nucleotides

In principle, all therapeutic oligonucleotides that enter cells will eventually be metabolized to individual monomeric forms and be subject to endogenous nucleotide salvage pathways and join endogenous nucleotide pools within any given cell. The actual extent and rate at which this occurs will rely on numerous factors including cell type, subcellular localization, nuclease activity, chemical modifications, and positioning and would need to be determined experimentally during traditional metabolite profiling studies. Nonetheless, the use of non-naturally occurring chemical modifications and subsequent liberation of these monomers may warrant considerations similar to antiviral or anticancer nucleoside analog drugs.

Although not specifically applicable to oligonucleotide product development, sponsors may consider the following basic tenets during lead optimization, as outlined in the Food and Drug Administration's Guidance for Antiviral Product Development (www.fda.gov). Since the same modification chemistry may be used in modular fashion across different sequences, any *in vitro* evaluation of individual modified nucleotides need only be conducted once in support of the oligonucleotide platform at large.

4.4.3.1 Exposure to Nonnaturally Occurring Monomeric Metabolites
Characterization of exposure to test article is fundamental to safety evaluations and risk assessment. For chemically modified nucleotides, developing semiquantitative or quantitative bioanalytical methods for monomeric forms in biological matrices is technically challenging, as monomers are very small polar molecules with masses likely to be nearly identical to the relatively abundant endogenous bases present in the background. In addition, nucleotides are normally in a dynamic state of flux between mono-, di-, and triphosphate forms, as governed by intracellular kinases and phosphatases, and the kinetics for each nucleotide (natural or nonnatural) will likely be different. Despite these challenges, determining the abundance of modified monomeric metabolites relative to the endogenous nucleotide form provides a context for relative exposures. For example, if the oligonucleotide drug product contained four 2'-F-U bases in the sequence, one could experimentally determine the relative abundance of the 2'-F-UMP versus endogenous UMP in liver samples from routine rodent PK/tox studies. Determination of this relative ratio helps to contextualize other *ex vivo/in vitro* evaluations.

4.4.3.2 Effects on Endogenous Polymerases
For nonnatural monomeric metabolites, the potential for interactions with endogenous polymerases (nuclear and/or mitochondrial) should be considered. Similar to nucleoside analog drugs, the potential for direct inhibitory effects on key nuclear and/or mitochondrial polymerases could be determined using routine *in vitro* enzyme kinetic assays. In addition, one may consider whether the nonnatural monomer is a potential substrate for incorporation by DNA or RNA polymerases. Chemical modifications at the 2'-hydroxyl position, for example, may alter steric conformations and change affinities for polymerases (incorporation, chain elongation) and/or exonucleases (repair processes). These types

of potential interactions are readily addressed in standard *in vitro* primer extension-type assays (Richardson et al., 2000; Lim and Copeland, 2001). The most common chemical modifications used in OTs, as monomeric metabolite forms, are not chain terminators and tend to be poor substrates for endogenous polymerases (very low potential for incorporation into nascent RNA or DNA) (Wright and Brown, 1990; Richardson et al., 2000). This is particularly evident in the presence of endogenous nucleotides (preferential substrates). To mimic theoretical *in vivo* exposures, competition assays (metabolite vs. endogenous nucleotide) may be conducted at the relative exposure ratios determined bioanalytically in rodent PK/tox studies.

4.4.3.3 *Potential Effects on Transcription or Translation*

To complement direct polymerase interaction assays, the potential effects of nonnatural monomeric metabolites on transcription and translation processes should also be considered. Cell-free recombinant transcription/translation kits are commercially available (bacterial or mammalian origin) to enable the characterization of individual monomeric metabolites. In principle, any "bad actor" chemically modified nucleotide could be identified via alterations to overall RNA yield/integrity. Again, these assays could be conducted in "competition-type" formats (modified NTP spiked into endogenous NTP pool) mimicking relative exposure ratios *in vivo*. For translation, synthetic RNA templates could be synthesized that contain the chemical modification in question to see whether the presence of that modification in an RNA template alters overall protein yield. As an early hazard ID-type screen, any nucleotide modifications that behave as chain terminators or have some other inhibitory activity should be readily identified.

4.4.3.4 *In Vitro Safety Evaluation*

Nonnatural monomeric metabolites should be evaluated in traditional cytotoxicity assays in appropriate cell-based models (tested as nucleosides, not parent oligonucleotide form). Consistent with the development of nucleoside analog drugs, the cell types should be representative of tissues of parent drug distribution (likely liver based for oligonucleotides). In addition, as mitochondrial toxicity is often associated with nucleoside analog drugs, mitochondrial-based toxicity endpoints could be included as part of the *in vitro* evaluations.

4.5 INTEGRATION OF LEAD OPTIMIZATION DATA FOR CANDIDATE SELECTION AND DEVELOPMENT

Final candidate selection for the development of an RNAi therapeutic integrates safety-based information from *in silico* target assessments, siRNA design features, and traditional screening in cell-based systems and rodent models. The mRNA target would ideally be well characterized and conserved across nonclinical species and humans. The siRNA sequence/design would have the desired pharmacokinetic and pharmacodynamics properties while minimizing the risk of potential off-target hybridization-based effects. The chemical modifications (backbone, sugar) used in the oligo design would have minimal potential for interaction with endogenous polymerases and minimal potential to cause mitochondrial toxicity in monomeric forms. The siRNA would have no apparent pro-inflammatory effects *in vitro/in vivo* and no significant (or novel) toxicity findings in subacute rodent toxicology studies. Because of the modular nature of siRNA design and chemistry, learnings from any individual molecule are often broadly applicable across other sequences and siRNA programs.

A detailed description of the nonclinical and regulatory strategies for the advancement of lead candidate conjugate siRNAs into clinical development is beyond the scope of this chapter. To date, there is no dedicated regulatory guidance for the nonclinical safety evaluation of OTs, which has resulted in considerable uncertainty with regards to harmonization across both investigators and regulators. OTs are considered small molecules by health and regulatory authorities worldwide and therefore follow the tenets of ICH M3(R2). However, there are some physical–chemical features of OTs that appear more in line with biotechnology-derived products (biologics), and hence some tenets described in ICH S6 may be applicable. Some of the comparative features of small molecules, OTs, and biologics are summarized in Table 4.1.

The Oligonucleotide Safety Working Group (OSWG) has recently published several white papers on the evaluation of exaggerated pharmacology of OTs (Kornbrust et al., 2013), reproductive and developmental toxicity testing of OTs (Cavagnaro et al., 2014), the potential lung toxicity of inhaled OTs (Alton et al., 2012), and recommendations for genotoxicity testing of OTs (Berman et al., submitted). Additional planned topics may also include approaches to safety pharmacology testing and immunomodulatory effects of OTs. The overarching purpose of these OSWG efforts is to share the collective experience of OT drug developers and to provide recommendations for best practices for nonclinical safety evaluations.

Our nonclinical IND-enabling paradigm for conjugate siRNAs follows a somewhat standard small-molecule approach. General toxicology studies are initiated with exploratory dose range-finding studies in rat and NHP (typically 1-month studies), followed by pivotal subchronic to chronic studies, as appropriate, which include full toxicokinetic evaluations and dose-free recovery periods. As noted earlier, conjugate siRNAs are generally well tolerated in nonclinical species, and traditional maximum tolerated dose (MTD) levels or early mortalities are rarely achieved even at ICH limit doses. More commonly, high dose levels are

TABLE 4.1 A Comparison of Drug Properties among Small Molecules, Oligonucleotides, and Biologics

Property[a]	Small Molecules	Oligonucleotides	Biologics
Molecular weight	Low (<1 kDa)	6–17 kDa	High
Manufacture	Chemical synthesis	Chemical synthesis	Biotechnology driven
Structure (drug substance)	Single entity of high purity; homogeneity	Single entity of high purity; homogeneity	Complex entity; heterogeneity
Target distribution	Intra- and extracellular	Intra- and extracellular	Extracellular
Structure–activity relationships in LO	Database often established	No database but subclass effects independent of sequence	No database
Metabolism	Species-specific metabolites	Catabolized to endogenous nucleotides	Catabolized to endogenous amino acids
Immunogenicity	+/−	+/− (variable depending on subclass)	+++
Species specificity	+/−	+/− (variable depending on subclass)	+++
Off-target toxicities	+++	+/− (variable)	+/−

[a]Adapted from Cavagnaro (2002).

determined based on appropriate multiples of efficacious dose levels and/or toxicokinetic parameters (i.e., high dose levels where exposures are no longer dose proportional). The dosing regimen and route of administration (SC) follow the intended clinical application.

Genetic toxicity testing follows the recommended test battery as outlined in Option 1 of ICH S2(R1), employing two *in vitro* assays (bacterial reversion mutation and chromosomal aberrations in hPBLs, +/− S9 metabolic activation) and the *in vivo* rat bone marrow micronucleus assay. Intracellular exposure to the full-length parent drug product has been demonstrated in all three assay systems (thereby applicable across different sequences).

Safety pharmacology studies (cardiovascular, respiratory, CNS) are conducted in the telemetered NHP and also employ single- or repeat-dose parallel study designs typical of other small-molecule drugs. Respiratory and/or CNS endpoints are often combined as part of the subchronic general toxicology studies. The *in vitro* hERG assay is typically not conducted because conjugate siRNAs, similar to a therapeutic protein, is a relatively large molecule (~16 kDa) that specifically targets ASGPRs predominantly expressed on hepatocytes. Tissue distribution to cardiac tissue is exceedingly low *in vivo*, and the drug product is unable to access, interact with, or block the inner pore of the hERG channel.

Taken together, this nonclinical approach to enable first-in-human clinical studies has been widely accepted by health and regulatory authorities worldwide for novel RNAi therapeutics and other oligonucleotide modalities.

4.6 CONCLUSIONS

RNAi is a naturally occurring cellular mechanism for regulating gene expression mediated by siRNAs. The ability to selectively degrade the mRNA encoding a disease-related protein has provided the opportunity to create a novel therapeutic modality. The process of candidate selection for the development of an RNAi therapeutic integrates safety-based information from *in silico* target assessments, siRNA design features, and traditional screening in cell-based systems and rodent models. The current clinical development landscape is dominated by hepatic targets, but therapies targeting other tissues are actively being pursued and hold promise to bring differentiated medicines to patients in need.

REFERENCES

Alton, E., Boushey, H., Garn, H., Green, F., Hodges, M., Martin, R., et al. (2012). Clinical expert panel on monitoring potential lung toxicity of inhaled oligonucleotides: consensus points and recommendations. *Nucleic Acid Ther*, 22(4), 246–254.

Ashwell, G., & Morell, A. (1974). The role of surface carbohydrates in the hepatic recognition and transport of circulating glycoproteins. *Adv Enzymol Relat Areas Mol Biol*, 41, 99–128.

Baenziger, J., & Fiete, D. (1980). Galactose and N-acetylgalactosamine-specific endocytosis of glycopeptides by isolated rat hepatocytes. *Cell*, 22, 611–620.

Bennett, C., & Swayze, E. (2010). RNA targeting therapeutics: molecular mechanisms of antisense oligonucleotides as a therapeutic platform. *Annu Rev Pharmacol Toxicol*, 50, 259–293.

Berman, C., Barros, S., Galloway, S., Kasper, P., Oleson, F., Priestly, C., et al. (2015). OSWG recommendations for genotoxicity testing of novel oligonucleotide-based therapeutics. (submitted, under review)

Berns, K., Hijmans, E., Mullenders, J., Brummelkamp, T., Velds, A., Heimerikx, M., et al. (2004). A large-scale RNAi screen in human cells identifies new components of the p53 pathway. *Nature*, 428, 431–437.

Boutros, M., & Ahringer, J. (2008). The art and design of genetic screens: RNA interference. *Nat Rev Genet*, 9 (7), 554–566.

Burdick, A., Sciabola, S., Mantena, S., Hollingshead, B., Stanton, R., Warneke, J., et al. (2014). Sequence motifs associated with hepatotoxicity of locked nucleic acid-modified antisense oligonucleotides. *Nucleic Acids Res*, 42 (8), 4882–4891.

Cavagnaro, J. (2002). Preclinical safety evaluation of biotechnology-derived pharmaceuticals. *Nat Rev Drug Discov*, 1 (6), 469–475.

Cavagnaro, J., Berman, C., Kornbrust, D., White, T., Campion, S., & Henry, S. (2014). Considerations for assessment of reproductive and developmental toxicity of oligonucleotide-based therapeutics. *Nucleic Acid Ther*, 24(5), 313–325.

Coelho, T., Adams, D., Silva, A., Lozeron, P., Hawkins, P., Mant, T., et al. (2013). Safety and efficacy of RNAi therapy for transthyretin amyloidosis. *N Engl J Med*, 369 (9), 819–829.

Cullen, L., & Arndt, G. (2005). Genome-wide screening for gene function using RNAi in mammalian cells. *Immunol Cell Biol*, 83 (3), 217–223.

Czauderna, F., Fechtner, M., Dames, S., Ayqun, H., Klippel, A., Pronk, G., et al. (2003). Structural variations and stabilising modifications of synthetic siRNAs in mammalian cells. *Nucleic Acids Res*, 31 (11), 2705–2716.

Echeverri, C., Beachy, P., Baum, B., Boultros, M., Buchholz, F., Chanda, S., et al. (2006). Minimizing the risk of reporting false positives in large-scale RNAi screens. *Nat Methods*, 3 (10), 777–779.

Elbashir, S. M., Harborth, J., Lendeckel, W., Yalcin, A., Weber, K., & Tuschl, T. (2001a). Duplexes of 21-nucleotide RNAs mediate RNA interference in cultured mammalian cells. *Nature*, 411, 494–498.

Elbashir, S., Lendeckel, W., & Tuschl, T. (2001b). RNA interference is mediated by 21- and 22-nucleotide RNAs. *Genes Dev*, 15 (2), 188–200.

Frazier, K. (2015). Antisense oligonucleotide therapies: the promise and the challenges from a toxicologic pathologist's perspective. *Toxicol Pathol*, 43 (1), 78–89.

Frazier, K., Sobry, C., Derr, V., Adams, M., Besten, C., De Kimpe, S., et al. (2014). Species-specific inflammatory responses as a primary component for the development of glomerular lesions in mice and monkeys following chronic administration of a second-generation antisense oligonucleotide. *Toxicol Pathol*, 42 (5), 923–935.

Geffen, I., & Spiess, M. (1992). Asialoglycoprotein receptor. *Int Rev Cytol*, 137B, 181–219.

Geisbert, T., Hensley, L., Kagan, E., Yu, E., Geisbert, J., Daddario-DiCaprio, K., et al. (2006). Postexposure protection of guinea pigs against a lethal ebola virus challenge is conferred by RNA interference. *J Infect Dis*, 193 (12), 1650–1657.

Hagedorn, P., Yakimov, V., Ottosen, S., Kammler, S., Nielsen, N., Hog, A., et al. (2013). Hepatotoxic potential of therapeutic oligonucleotides can be predicted from their sequence and modification pattern. *Nucleic Acid Ther*, 23 (5), 302–310.

Hornung, V., Guenthner-Biller, M., Bourquin, C., Ablasser, A., Schlee, M., Uematsu, S., et al. (2005). Sequence-specific potent induction of IFN-alpha by short interfering RNA in plasmacytoid dendritic cells through TLR7. *Nat Med*, 11 (3), 263–270.

Jackson, A., Burchard, J., Leake, D., Reynolds, A., Schelter, J., Guo, J., et al. (2006). Position-specific chemical modification of siRNAs reduces "off-target" transcript silencing. *RNA*, 12 (7), 1197–1205.

John, M., Constien, R., Akinc, A., Goldberg, M., Moon, Y., Spranger, M., et al. (2007). Effective RNAi-mediated gene silencing without interruption of the endogenous microRNA pathway. *Nature*, 449, 745–747.

Judge, A., Sood, V., Shaw, J., Fang, D., McClintock, K., & MacLachlan, I. (2005). Sequence-dependent stimulation of the mammalian innate immune response by synthetic siRNA. *Nat Biotechnol*, 23 (4), 457–462.

Judge, A., Bola, G., Lee, A., & MacLachlan, I. (2006). Design of noninflammatory synthetic siRNA mediating potent gene silencing in vivo. *Mol Ther*, 13 (3), 494–505.

Kamath, R., Fraser, A., Dong, Y., Poulin, G., Durbin, R., Gotta, M., et al. (2003). Systematic functional analysis of the *Caenorhabditis elegans* genome using RNAi. *Nature*, 421, 231–237.

Kole, R., Krainer, A., & Altman, S. (2012). RNA therapeutics: beyond RNA interference and antisense oligonucleotides. *Nat Rev Drug Discov*, 11 (2), 125–140.

Kornbrust, D., Cavagnaro, J., Levin, A., Foy, J., Pavco, P., Gamba-Vitalo, C., et al. (2013). Oligo safety working group exaggerated pharmacology subcommittee consensus document. *Nucleic Acid Ther*, 23(1), 21–28.

Kume, H., Hino, K., Galipon, J., & Ui-Tei, K. (2014). A-to-I editing in the miRNA seed region regulates target mRNA selection and silencing efficiency. *Nucleic Acids Res*, 42 (15), 10050–10060.

Larsson, E., Sander, C., & Marks, D. (2010). mRNA turnover rate limits siRNA and microRNA efficacy. *Mol Syst Biol*, 16 (6), 433.

Lewis, B., Burge, C., & Bartel, D. (2005). Conserved seed pairing, often flanked by adenosines, indicates that thousands of human genes are microRNA targets. *Cell*, 120 (1), 15–20.

Lim, S. E., & Copeland, W. C. (2001). Differential incorporation and removal of anitviral deoxynucleotides by human DNA polymerase gamma. *J Biol Chem*, 276(26), 23616–23623.

Moffat, J., Grueneberg, D., Yang, X., Kim, S., Kloepfer, A., Hinkle, G., et al. (2006). A lentiviral RNA library for human and mouse genes applied to an arrayed viral high-content screen. *Cell*, 124 (6), 1283–1298.

Morrissey, D., Lockridge, J., Shaw, L., Blanchard, K., Jensen, K., Breen, W., et al. (2005). Potent and persistent in vivo anti-HBV activity of chemically modified siRNAs. *Nat Biotechnol*, 23 (8), 1002–1007.

Richardson, F., Kuchta, R., Mazurkiewicz, A., & Richardson, K. (2000). Polymerization of 2′-fluoro- and 2′-O-methyl-dNTPs by human DNA polymerase alpha, polymerase gamma, and primase. *Biochem Pharmacol*, 59 (9), 1045–1052.

Schwartz, A., Rup, D., & Lodish, H. (1980). Difficulties in the quantitation of asialoglycoprotein receptors on the rat hepatocyte. *J Biol Chem*, 255 (19), 9033–9036.

Sledz, C., Holko, M., de Veer, M., Silverman, R., & Williams, B. (2003). Activation of the interferon system by short-interfering RNAs. *Nat Cell Biol*, 5 (9), 834–839.

Soutschek, J., Akinc, A., Bramlage, B., Charisse, K., Constien, R., Donoghue, M., et al. (2004). Therapeutic silencing of an endogenous gene by systemic administration of modified siRNAs. *Nature*, 11, 173–178.

Tabernero, J., Shapiro, G., LoRusso, P., Cervantes, A., Schwartz, G., Weiss, G., et al. (2013). First-in-humans trial of an RNA interference therapeutic targeting VEGF and KSP in cancer patients with liver involvement. *Cancer Discov*, 3 (4), 406–417.

Thompson, A., & Locarnini, S. (2007). Toll-like receptors, RIG-I-like RNA helicases and the antiviral immune response. *Immunol Cell Biol*, 85 (6), 435–445.

Wang, E., Sandberg, R., Luo, S., Khrebtukova, I., Zhang, L., Mayr, C., et al. (2008). Alternative isoform regulation in human tissue transcriptomes. *Nature*, 456, 470–476.

Watts, J., Deleavey, G., & Damha, M. (2008). Chemically modified siRNA: tools and applications. *Drug Discov Today*, 13 (19–20), 842–855.

Wittrup, A., & Lieberman, J. (2015). Knocking down disease: a progress report on siRNA therapeutics. *Nat Rev Genet*, 16, 543–551.

Wright, G., & Brown, N. (1990). Deoxyribonucleotide analogs as inhibitors and substrates of DNA polymerases. *Pharmacol Ther*, 47 (3), 447–497.

Wu, J., Nantz, M., & Zern, M. (2002). Targeting hepatocytes for drug and gene delivery: emerging novel approaches and applications. *Front Biosci*, 7, 17–25.

Zimmermann, T., Lee, A., Akinc, A., Bramlage, B., Bumcrot, D., Fedoruk, M., et al. (2006). RNAi-mediated gene silencing in non-human primates. *Nature*, 4, 111–114.

PART III

BASIS FOR *IN VITRO–IN VIVO* PK TRANSLATION

5

PHYSICOCHEMISTRY AND THE OFF-TARGET EFFECTS OF DRUG MOLECULES

DENNIS A. SMITH

Independent Consultant, Canterbury, UK

5.1 LIPOHILICITY, POLAR SURFACE AREA, AND LIPOIDAL PERMEABILITY

Cells form tissues and their exterior consists of a phospholipid/lipid bilayer (cell membrane). Proteins are associated with the cell membrane to provide signaling to the cell interior, maintain the intracellular homeostasis, or transport nutrients and other substances. Cells are linked to each other by junction proteins that have varying degrees of tightness. Small molecules can diffuse through these junctions that resemble an aqueous pore. This diffusion is termed paracellular. Paracellular diffusion is important for small, nonlipophilic (see following text) water-soluble drug molecules in the upper gastrointestinal tract and for most molecules to move from the vasculature to the extravascular space (water). Restriction of the junctions in the vasculature of the central nervous system (CNS) creates the blood–brain barrier (BBB) and limits diffusion to the lipoidal pathway by this manner (see following text).

Most drugs move through the cellular phospholipid/lipid bilayers in a process termed lipoidal diffusion. This passive bidirectional flux is correlated positively with lipophilicity and negatively with the hydrogen bonding potential of a molecule. This correlation should not be seen as a linear relationship but one in which positive and negative lipophilicities define probabilities of a particular mechanism of transport. The relationship between lipoidal permeability and lipophilicity is due to the alkyl chain interior of the bilayer. Compounds have to have positive lipophilicity. Hydrogen bonding groups have to undergo desolvation when they move from an aqueous extracellular environment to the lipophilic interior of membranes. Hydrogen bonding and the removal of solvent thus represent a large part of the energy cost involved in membrane permeability.

Lipophilicity can be quantified. The most used measure of lipophilicity is provided by octanol/buffer partitioning. Octanol is chosen to represent the best compromise of solvent to represent a biological membrane. The ratio of partitioning is normally expressed on a log scale due to the wide range of ratios for drugs or other compounds. The partitioning of the unionized molecule is referred to as the log P. This can be considered the intrinsic lipophilicity. In general the unionized molecule is that which is considered to cross the bilayer. Many drugs have groupings that can be ionized at physiological pH. For such compounds the degree of ionization at a particular pH is related to the pK_a (a measure of the strength of the ionizable moiety). The degree of ionization and the intrinsic lipophilicity can be expressed as a single term so that the portioning of a molecule at a specified pH is referred to as the log D. This is normally at physiological pH of 7.4. For charged molecules log D and log P can be related by the pK_a of the molecule. For monoprotic organic acids the relationship is log $([P/D]-1)=pH-pK_a$, and for monoprotic organic bases (BH^+ dissociating to B), the corresponding relationships is log $([P/D]-1)=pK_a-pH$.

Measurement of log P and log D can be automated, but there are many *in silico* methods available. Every fragment of an organic compound has a defined lipophilicity. Calculation of log P can be made computationally by summing the fragments to give a total lipophilicity (Leo et al., 1971; Rekker and Mannhold, 1992) A variety of methods

Drug Discovery Toxicology: From Target Assessment to Translational Biomarkers, First Edition. Edited by Yvonne Will, J. Eric McDuffie, Andrew J. Olaharski, and Brandon D. Jeffy.

are available, but the accuracy of the prediction can be masked by novel arrangement of fragments or atoms in a molecule. When calculated rather than measured, the value is referred to as clog P.

Calculation of pK_a is also routinely achieved by computational methods with several programs giving relatively low variance from measured values and identifying the site of ionization correctly (Shields and Seybold, 2013).

One limitation of octanol as a solvent is the hydroxyl group and in addition its relatively high (4%) aqueous content (Smith et al., 2012). Thus octanol supports hydrogen bonding and by itself will not completely characterize the behavior of a molecule. Alternative portioning systems such as cyclohexane are experimentally difficult, so the characterization of hydrogen bonding is normally carried out by simple calculations.

Hydrogen bonding potential is estimated by summing of hydrogen bond acceptors and donors or by more sophisticated techniques such as polar surface area (PSA). The PSA of a molecule is defined as the area of its van der Waals surface that arises from all oxygen and nitrogen atoms plus the hydrogen atoms attached to them. PSA is a therefore a measure of the ability of a compound to form hydrogen bonds and accounts for three-dimensional (3D) structural features such as shielding or burial of polar groups by other parts of the molecule. To calculate PSA requires a 3D structure of the molecule concerned and complex software. A simplified 2D computational system termed topological polar surface area (TPSA) has been published and uses fragment-based contributions (Ertl et al., 2000). This technique brings TPSA to a similar footing as cLog P, both being fragmental methods. The two descriptors can be readily calculated for all molecules, even for large data sets such as combinatorial or virtual libraries.

The disadvantage of not using a 3D representation of the molecule is that masked and internal hydrogen bonds are included in the overall area. Internal hydrogen bonding effectively removes one donor and acceptor for each internal hydrogen bond, thus effectively reducing the true PSA. Compounds such as the immunosuppressant cyclosporine and the HIV protease inhibitor atazanavir show this behavior (Alex et al., 2011). Both compounds have five H-bond donor functions (OH, NH), which reduce effectively to one in the biological environment. A further consideration is that hydrogen bond acceptors and donors are considered equal and the strength of the hydrogen bond is not considered (Desai et al., 2012).

Lipoidal or lipid permeability has been used to define the disposition characteristics of molecules (Wu and Benet, 2005). This permeability can be related to physicochemical properties as outlined earlier. The relationship can be extended to also the likely interaction sites for a molecule. Small, nonlipophilic molecules will rely on the paracellular pathway. Such molecules will be absorbed from the

gastrointestinal tract to a reasonable degree, but this absorption will decrease as molecular size (weight) increases. The limit for good aqueous pore absorption is around MW 250 Da in human, monkey, and rat and around 350 in dog. Once in the circulation, these nonlipophilic compounds will diffuse to the surface of the cells due to leaky vasculature. They will not diffuse freely across the BBB due to the restricted tight junctions. Only lipophilic drugs that can cross the cell membrane will be able to cross into the cell interior.

If a drug molecule has positive log D and low PSA ($<75 A^2$), passive lipoidal permeability is high and is a large fraction of the total permeability, owing to the large surface area of the cell membrane. Such lipophilic drugs are termed highly lipid permeable and readily cross the gastrointestinal tract, the vasculature, and the BBB and into cells.

Not all lipophilic drugs (as defined by log P or D in octanol) can be classed as highly lipid permeable. As hydrogen bonding increases (PSA $> 75 A^2$) then lipoidal diffusion rate decreases. These drugs show lipoidal diffusion, but now the influence of transporter proteins becomes important. With drugs of high lipid permeability, the bidirectional flux is so rapid that any transporter effects are of minor significance in overall flux. As the bidirectional passive flux decreases with lower lipid permeability, the effects of transporters become more important and can be the dominant force. This is explained in Figure 5.1.

As stated earlier, the physicochemical properties define the absorption, distribution, metabolism, and excretion (ADME) properties of a drug. To illustrate this in greater detail, Table 5.1 shows the ADME properties of drugs that are either hydrophilic (low lipid permeability), lipophilic but with high hydrogen bonding potential (medium lipid permeability), or lipophilic (high lipid permeability) with low hydrogen bonding potential.

5.2 PHYSICOCHEMISTRY AND BASIC ADME PROPERTIES FOR HIGH LIPOIDAL PERMEABILITY DRUGS

Clearance is the removal of drug and metabolite molecules from the body by metabolism, renal clearance, and "hepatobiliary excretion." Drugs of high lipoidal permeability show minimal or low transporter effects due to the high rates of bidirectional flux across membranes. Thus a drug of high lipid diffusion will be reabsorbed from the kidney tubule across the membranes of the cells even if filtered by the glomerulus or transported by uptake and efflux transporters in the kidney. As a result such drugs show negligible renal clearance. The concentrations in liver cells will also reflect those in the plasma due to same phenomena. Clearance of high lipid permeability drugs is therefore by metabolism. Relative concentrations in cells and the

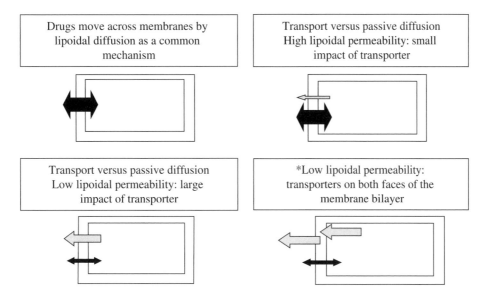

FIGURE 5.1 Simplified schematic showing the dependency on the significance of transporter flux on lipoidal permeability in determining overall flux of a compound across cell membranes These transporters can efflux compounds from the cell interior or transport them into cells (note only efflux and a single transporter shown). Thus drugs can be excluded from organs such as the brain or concentrated in some cases in organs such as the liver. Transporters play a major role in the excretion of drugs in kidney and liver cells, moving drugs from the circulation into the cell interior and then exporting them into the urine and bile. Transporters are actually present on both sides of the bilayer (apical and basolateral) as depicted in *bottom right.

TABLE 5.1   Physicochemistry and Lipid Permeability Classification of Drug Absorption, Bioavailability, Target Access, and Clearance

Lipid Permeability	Low	Medium	High
Physicochemistry	Low log D or P (<0)	log D or $P > 0$ (PSA $> 75 A^2$)	log D or $P > 0$ (PSA $< 75 A^2$)
Absorption	Low (*aliskiren*) unless MWt less than 250 Da and absorbed by paracellular route (*atenolol*)	Variable. Influenced by permeability and transporters (*nelfinavir*)	High via transcellular route (*propranolol*)
Bioavailability	As for absorption	As for absorption and metabolism	Variable. Influenced by metabolism
Access to target proteins	Extracellular including cell surface equilibrium with unbound drug in plasma (usually low ppb)	TBW but may be concentrated or effluxed from intracellular cytosol	TBW—drug will show identical unbound concentrations to that present in plasma
Clearance	Renal or biliary (possible transporter involvement)	Transporters and metabolism	Metabolism

circulation however have to take account of plasma protein binding, since the concentration gradients are driven by the unbound drug.

A common prtoblem in defining the disposition of a drug molecule in the body is the distinction between total and unbound drug. A reasonable assumption is that, unless a drug forms reactive species or is itself reactive, a drug will be in equilibrium with the proteins and membranes of the body and the actual "occupancy" of those proteins will be driven by the affinity of the drug for the protein (or membrane) and the concentration of unbound drug.

The disposition of a drug is normally described from the measurement of the total drug in plasma, which may be corrected for the degree of binding to yield the unbound drug in the circulation. In addition organs and tissues may be sampled, with the amounts or concentrations usually determined as total. This presents a dilemma as to how we conceptualize a drug molecule's disposition. One pharmacokinetic parameter that causes confusion is volume of distribution (V_d). This reflects a theoretical volume derived from the circulating concentrations, which relates to how much of the drug is outside the central circulation. It is normally derived from total drug in the circulation compared to dose. It is tempting to relate these values to on- and off-target interactions for a drug. This relationship is not correct as explained later.

5.3 RELATIONSHIP BETWEEN VOLUME OF DISTRIBUTION (V_d) AND TARGET ACCESS FOR PASSIVELY DISTRIBUTED DRUGS

The volume of distribution is a pharmacokinetic concept relating the concentrations of drug in the circulation to the total amount present in the body. To achieve this the total drug administered is divided by the plasma concentration to calculate a virtual volume. There are a number of different ways the volume term can be calculated, which have great significance to the interpretation. Normally volume terms are best obtained after intravenous administration. After a bolus dose of drug and allowing initial mixing of the drug in the circulation (occurs before the first sample point), the drug is distributed according to the volume of initial dilution or V_1. This volume will usually approximate to a value between blood or plasma volume and extracellular water volume. When a classical biphasic decline of log concentration with time is observed, distribution to tissues is assumed to be in equilibrium with efflux from the tissues when the linear decline (terminal half-life) is attained. Depending on the method of calculation, this is defined as $V_{d\beta}$ or V_{darea}. The traditional view of a distribution phase followed by an elimination phase is true only for low clearance drugs. Substantial clearance can occur during the distribution phase, meaning that the relationship between circulation concentration and tissue concentration is biased, since a substantial amount of the dose has been eliminated by the time tissue equilibrium is reached. For this reason a third value termed V_{dss} (volume of distribution at steady state) is calculated. This uses noncompartmental analysis, so is less dependent on defining the linear decline. It relates clearance to mean residence time for the drug in the body. It may also be a more accurate measure for relating the amount of drug in the circulation with the total drug in the body for drugs with rapid clearance compared to their distribution.

For a passively distributed drug, volume of distribution is a reflection largely of bulk low affinity but high-capacity binding to either the plasma (mainly albumin or α-1 acid glycoprotein) or the tissue membranes. It does not give an indication (except where the binding constituents are implicated in the toxicity) of whether the drug will be active at remote sites. To act on and off target, the drug must be free in aqueous solution. The actual distribution that is measured largely reflects the physicochemistry of a drug and its selectivity for various tissue constituents and plasma proteins (Smith et al., 2012):

1. A basic drug (ionized at physiological pH, $pK_a < 7.4$) will have affinity for phospholipid membranes due to interactions with acidic head groups. Bases bind to α-1 acid glycoprotein and albumin with moderate to strong affinity related to lipophilicity. Bases can also accumulate by pH effects within the cell as outlined in the following text. Basic drugs will be generally drawn from the circulation by these processes and will have volumes of distribution around 1–25 l/kg. Table 5.2 illustrates the interplay of binding affinity and concentration of the binding moiety in the human body for propranolol, a drug with a free fraction of 0.27 (73% bound) and V_d of 4.2 l/kg. It illustrates numerically high-capacity, nonspecific low affinity binding and compares it to typical drug target low-capacity high affinity binding.

2. Neutral compounds interact with the lipid portion of membranes and albumin. Both relationships are related to the lipophilicity of the drug. Volumes of distribution will generally be of the order of 0.7–4 l/kg.

3. Acids have low membrane affinity and very high affinity for albumin. Again lipophilicity is important, but very high plasma protein binding (due to the high concentration of albumin in plasma) occurs at lower lipophilicities than neutrals or bases. Extravascular water has considerable amounts of albumin present; thus acids may have larger volumes than plasma or blood volume. Acids will generally have volumes of between 0.1 and 0.4 l/kg.

Table 5.3 provides the volume of distributions for a series of drugs. All the drugs are of positive lipophilicity and of low PSA (or TPSA) indicative of high passive lipid permeability. The volumes of distribution range from low to high reflecting the above rules. Regardless of the apparent differences in "tissue distribution" implied by the volume term, the drugs show effectively identical unbound (free) drug concentrations throughout total body water (TBW), covering therefore plasma, extracellular water, intracellular water, synovial fluid, and in the CNS cerebrospinal fluid. Thus all will interact with their target regardless of location, in a manner governed solely by their affinity and unbound aqueous concentration (determined by dose and intrinsic clearance). This is exemplified by chlorpheniramine, which has a volume of distribution much greater than TBW (3 l/kg). Chlorpheniramine has a K_d for the H1 receptors in the brain, *in vivo*, calculated from CNS receptor occupancy (7 nM) using PET scanning, which is very similar to its unbound plasma concentration at therapeutic

TABLE 5.2   Binding Affinity of Racemic Propranolol and Concentration of the Binding Moiety in Plasma* and Whole Body[$a]

	Concentration in Human (μM)	K_d (μM)
AAG*	9	3.1
Albumin*	470	290
Phosphatidylserine[$]	2500	120
β-Adrenoceptor	0.007	0.013

[a]Propranolol is a drug with a free fraction of 0.14 (86% bound) and V_d of 4.2 l/kg. Phospholipid binding data from Small et al. (2011). β-adrenoceptor data is from dog heart (Merlet et al., 1993).

TABLE 5.3 Acidic, Neutral, and Basic Drugs with High Lipid Permeability That Achieve Effective Unity in Unbound (Free) Aqueous Drug Concentrations throughout the Body (Systemic and Tissue) despite Large Differences in Distribution Volume (See Text)

Drug	pK_a	$\log P$	$\log D$	PSA	Fraction Unbound	V_d (l/kg)	V_d(u) (l/kg)
Indomethacin	Acid 3.9	4.2	0.7	68.5	0.004	0.29	72.5
Ketoprofen	Acid 4.2	2.9	0.2	54.4	0.008	0.15	18.75
Fluconazole	—	0.5	0.5	71.8	0.87	0.7	0.8
Diazepam	—	2.8	2.8	32.7	0.013	1.1	84.6
Chlorpheniramine	Basic 9.1	3.4	1.5	16.2	0.28	3.2	11.4
Propranolol	Basic 9.4	2.9	0.6	41.0	0.14	4.2	60.6
Fluoxetine	Basic 10.5	3.9	1.4	21.3	0.06	35	583

The volumes reflect binding to proteins and membranes. Only fluconazole, with its moderate lipophilicity and neutral character (low nonspecific binding), has a volume of distribution related to physiology (total body water).

TABLE 5.4 Physicochemistry Properties and Volume of Distribution of Amiodarone and Azithromycin

Drug	pK_a	$\log P$	$\log D$	PSA	Fraction Unbound	V_d (l/kg)	V_d(u) (l/kg)
Amiodarone	Base 8.7	7.5	5.2	43	0.001	60	60,000
Azithromycin	Base (diprotic) 8.7 8.9	4.0	1.2	180	0.5	31	62

The extreme lipophilicities and resultant volume of distribution should be compared to the drugs in Table 5.1.

doses (Yanai et al., 1995). This is also very close to the K_d calculated from *in vitro* receptor experiments (3–4 nM). Table 5.1 also includes unbound volumes calculated by V_d/fraction unbound (fu). Overlap between the drug classes now is observed. This term now includes vascular and extravascular binding rather than total that only measures extravascular binding. The use of unbound values allows the significance of changes in intrinsic clearance and volume of distribution in a compound series and their influence on half-life to be easily understood without further calculation.

Of the acidic phospholipids, phosphatidylcholine is the major influence in the binding of basic drugs (Rodgers et al., 2005). Concentrations of phosphatidylserine in tissues (lung > kidney > live > muscle and heart > brain) reflect closely the ranking of the tissue partition coefficients (K_p) of many basic drugs, which are ionized at physiological pH. The actual concentrations are in addition governed by the lipophilicity of the compound reflecting the combination of hydrophobic and ion pair interactions with the lipid.

5.4 BASICITY, LIPOPHILICITY, AND VOLUME OF DISTRIBUTION AS A PREDICTOR OF TOXICITY (T): ADDING THE T TO ADMET

As described earlier, volume of distribution is not an indicator of unbound tissue concentration and largely reflects binding to plasma proteins or cell membranes and/

or lysosomal trapping. On some occasions the accumulation of drug in phospholipid or lysosomes can be associated with toxicity. Phospholipidosis is a lysosomal storage disease accompanied by excess accumulation of phospholipid in tissues. The volume of distribution of basic drugs relates to the concentrations or amount bound to phospholipid or trapped in lysosomes, and therefore in this specific case volume of distribution does reflect the amount of drug interacting at the nontarget site. Analysis of V_d and drug-induced phospholipidosis (Hanumegowda et al., 2010) indicates that for drugs causing this toxicity, V_d exceeds 5.4 l/kg. Of the drugs examined, 36% produced the toxicity. Considering the basic drugs in Table 5.2, fluoxetine is associated with phospholipidosis. Experimental evidence demonstrates that the distribution of this compound (Daniel and Wójcikowski, 1999) is mainly phospholipid binding rather than lysosomal trapping.

Further examples of drugs associated with phospholipidosis are shown in Table 5.4. Amiodarone with very high lipophilicity shows affinity for acidic phospholipids and neutral lipid and has a significant incidence of phospholipidosis and other toxicities both in animals and human (Roy et al., 2000). Azithromycin, due to its dibasic nature, undergoes a very high degree of lysosomal accumulation. The drug showed evidence of phospholipidosis in preclinical studies, but little evidence has been seen in man (Hopkins, 1991).

A case of linking volume of distribution, tissue distribution, and toxicities other than phospholipid has been reported.

The Met inhibitor GEN-203 caused liver and bone marrow toxicity in preclinical species. GEN-203 is a lipophilic base (containing N-ethyl-3-fluoro-4-aminopiperidine substituent) with a high log D (4.3) and a pK_a of 7.45 (Diaz et al., 2013). The compound had a moderately high volume of distribution ($V_d > 3$ l/kg) in mouse. An analog GEN-890 had lower basicity due to the addition of a second fluorine in the 3-position of the aminopiperidine and a lower volume of distribution ($V_d = 1$ l/kg). This analog had a much better toxicity profile, suggesting to the authors that the lower volume (and decreased overall distribution to the tissues) was responsible.

5.5 BASICITY AND LIPOPHILICITY AS A PREDICTOR OF TOXICITY (T): SEPARATING THE D FROM T IN ADMET

Ocular melanin is found in the uveal tract and in the pigmented epithelial layer of the retina. Many structurally and pharmacologically unrelated drugs from different therapeutic classes bind to melanin. Binding and persistence of compounds on the retina have been mostly observed in rodent tissue distribution studies, particularly whole-body autoradiography. Analysis of the type of compounds showing this relationship shows that they are exclusively lipophilic bases that are ionized at physiological pH. For instance, Zane et al. (1990) examined the relationship between the physicochemical characteristics of 27 new drug candidates and their distribution into the melanin-containing structure of the rat eye, the uveal tract. The drugs most likely to be distributed and ultimately retained at high concentrations were those containing strongly basic functionalities, such as piperidine or piperazine moieties and other amines. Furthermore, the more lipophilic, the greater the likelihood that binding occurred. Typical of such drugs are chloroquine and hydroxychloroquine, both of which not only bind to melanin but also cause retinal toxicity. Chloroquine is a strong base with a pK_a of 10.3 and lipophilic (log P 5.3) and has low TPSA (28 A^2). Electrostatic, hydrophobic, and van der Waals forces participate in the binding of chloroquine to melanin (Stępien and Wilczok, 1982). The highest affinity binding had affinities around 10 μM and was between the protonated ring system of chloroquine and the ortho-semiquinone groups of melanin. Other binding with affinities around 100 μM was due to ionic bonds between the protonated aliphatic nitrogen of chloroquine molecule and carboxyl groups of melanin. Van der Waals forces also occur between the aromatic rings of the drug and the aromatic indole nuclei of melanin.

Considerable inference has been made from melanin binding to ocular toxicity. Analysis of the data by Leblanc et al. (1998) indicates that in all cases, there are no direct consequences of drug–melanin binding. Any drug-related toxicity of the retina in humans and animals was described as unrelated to melanin binding with melanin binding and retinal toxicity viewed as two separate entities with the latter being related to the intrinsic toxicity of the compound rather than its ability to bind.

Retinal toxicity may show a relationship to the same physicochemical properties as melanin binding even though the mechanism is different (Khoh-Reiter et al., 2015). The retina is formed by layers containing seven different cell types. The photoreceptor cell is a specialized neuron in the retina that is capable of absorbing and converting light into electrophysiological signals. There is a constant renewal process for photoreceptors by retinal pigmented epithelial (RPE) cells. Drugs with a basic moiety have the potential to accumulate in the lysosome of RPE cells and impair the phagocytosis process, which clears the outer segments. Such an inhibition would lead to retinopathy. *In vitro* evaluation of known ocular toxic drugs showed the compounds that induced retinopathy clustered in the basic and lipophilic region. These drugs were shown to accumulate in lysosomes and inhibit phagocytosis, causing a blockage of the membrane trafficking process.

5.6 LIPOPHILICITY AND PSA AS A PREDICTOR OF TOXICITY (T): ADDING THE T TO ADMET

The same properties that govern the ADME of a drug in the body also define its interactions with on- and off-target proteins. Lipophilic interactions with a protein are always highly energy efficient since there is the addition of the hydrophobic binding energies and the energetically favorable elimination of water from the binding site. Hydrogen bonding interactions are less energy efficient in many cases due to the energy needed to eliminate water from the hydrogen bonding groups. Thus lipophilicity will tend to add affinity for a drug molecule regardless of the binding site. In contrast, hydrogen bonding interactions need to be more or highly specific for the particular binding site. Price et al. (2009) examined the influence of physicochemical properties in late-stage discovery compounds on *in vitro* selectivity panel assays (determining pharmacological selectivity) and short-term rodent toxicology studies. This analysis showed that compounds with higher lipophilicities and lower PSA had a much greater chance of interacting with multiple receptors. The authors concluded that a PSA of 75 A^2 and a cLog P value of 3 delineated the boundaries. Lipophilicity will always add positive binding energy, but hydrogen bonding groups can be favorable but have to overcome the negative energy of desolvation and this cost can only be repaid by specific interaction with a target protein. Note that the value of 75 A^2 is also the boundary for lipoidal diffusion discussed previously. The authors also reviewed short-term rodent toxicology studies of 2 weeks' duration. These were conducted during the discovery phase of the molecule.

TABLE 5.5   Observed Odds for Promiscuity and Toxicity

In Vitro Promiscuity	TPSA > 75 A^2	TPSA < 75 A^2
cLog P < 3	0.2	0.8
cLog P > 3	0.4	6.2
In Vivo Toxicology		
cLog P < 3	0.4	0.5
cLog P > 3	0.8	2.6

These were defined as multiple receptor interactions above 10 µM and toxicity effects above 1 µM unbound drug.

Toxicity was defined as any adverse event questioning compound progression. These were clustered at 1 µM and above plasma-free drug concentration or 10 µM and above total drug concentration. The incidence of toxicity was remarkably consistent with the promiscuity data (see Table 5.5), with similar boundaries of greater than PSA 75 A^2 and cLog P > 3 associated with the greatest incidence of toxicity.

Other groups (Azzaoi et al., 2007; Peters et al., 2009) have conducted similar *in vitro* assessments of promiscuity and selectivity and reached similar conclusions to the Price study: that increased lipophilicity and lower PSA decrease selectivity and increase promiscuity.

Trifluoperazine is a tricyclic drug fitting the profile of a lipophilic (log P 5), low TPSA (9.7 A^2) drug expected to show multiple interactions (promiscuity). It has a basic center with a pK_a of 8.4. The CNS actions of the drug are attributed to its inhibition of dopamine D2 receptors. It has a range of off-target pharmacology shown in Table 5.6. The most non-specific effect of compounds could be simple detergent effects on membranes causing disruption. Lipophilic bases and acids possess detergent-like properties, and this is shown in the table for effects on red blood cells (RBCs).

Relating this promiscuity to actual binding sites is possible. Compounds that block the K_v11.1 hERG channel prolong QT interval and can induce cardiac arrhythmia and sudden death. Many of these compounds are lipophilic bases. These physicochemical properties (Fernandez et al., 2004) can be rationalized to the importance of two aromatic residues, Tyr-652 and Phe-656, located in the S6 domain of hERG. The potency of drugs relates to their lipophilicity, which is explained by interactions with Phe-656. The importance of basicity is explained by a cation–π interaction between Tyr-652 and the basic tertiary nitrogen of these drugs.

5.7   METABOLISM AND PHYSICOCHEMICAL PROPERTIES

Metabolites may be responsible for or may contribute to the toxicity of a drug molecule. Metabolites may be unstable and chemically reactive. In these instances toxicity may be

TABLE 5.6   On- and Off-Target Interactions for a Promiscuous Drug Trifluoperazine (Log P 5, TPSA 9.7 A^2) as Defined by Price et al. (2009)

Target	IC$_{50}$ or Ki (µM)
Dopamine D2	0.002
Off-Target	
Dopamine D4	0.019
5HT4$_2$A	0.004
Adrenoceptor α1-A	0.022
K_v4.3 potassium channel	2
K_v11.1 potassium channel (hERG)	8
Na$_v$1.7 sodium channel	4
Na$_v$1.4 sodium channel	5
Calmodulin	6
Troponin	16
Protein S100	20
Protein kinase C	44
Membrane disruption of RBCs (threshold for observed changes and lysis)	60
50% lysis of RBCs	360

observed due to covalent modification of important proteins. Stable metabolites are formed for most lipophilic drugs (see Table 5.1) and can circulate and also are excreted in the urine and bile. Generally more interest is shown in the possible toxicity of phase 1 metabolites, normally formed by oxidative processes (e.g., cytochrome P450s), rather than phase 2 conjugation reactions (e.g., addition of glucuronic acid by glucuronyl transferases). The effects of metabolism are shown in Table 5.7 and should be considered against the promiscuity and toxicity data shown in Table 5.4 (Manners et al., 1988; Smith and Obach, 2010). It is likely that metabolites will be less likely to interact with proteins than the parent molecule. Where metabolites are equiactive or more potent against the target protein of the drug, this has been clearly related to the target structure–activity relationships (Smith and Obach, 2010). Although examples of toxicity produced by the interaction of a stable metabolite are known, they are actually remarkably rare in the literature probably testifying to the changes produced in the key physicochemical properties. An example of stable metabolite toxicity is provided later in the sulfate metabolite of troglitazone.

5.8   CONCENTRATION OF COMPOUNDS BY TRANSPORTERS

High local concentrations of free drug at the on- or off target proteins may occur due to the transport of a drug molecule (or metabolite). In almost all cases such drugs have defined physicochemical characteristics. In contrast to the compounds described in the section on basicity, lipophilicity, and volume of distribution as predictors of toxicity, the likelihood

TABLE 5.7 Changes in Physicochemical Properties Produced by Metabolism

Metabolic Step	Change in TPSA (A^2)	Change in Log P	pK_a and log $D_{7.4}$
Aliphatic hydroxylation	+20	−2	
Aromatic hydroxylation	+20	−0.7	
Dealkylation of tertiary amine	+9	−0.6 for methyl as leaving group; value increases with size of leaving fragment	Increase in basicity of 1 unit, reduction in log $D_{7.4}$ of 1 unit
Dealkylation of secondary amine	+14	−0.6 for methyl as leaving group; value increases with size of leaving fragment	
Oxidation of alcohol to carboxylic acid	+17	Little change	Introduction of acidic function. pK_a 3–5, reduction of log $D_{7.4}$ of 3–5 units
Glucuronidation of a phenol or alcohol	+127	−2.6	Introduction of acidic function. pK_a 3.2, reduction of log $D_{7.4}$ of 6.8 units[a]
Glucuronidation of a carboxylic acid	+127	−2.6	Usually introduces a stronger acid function and reduction of log $D_{7.4}$ of 3 units[b]
Sulfation of a phenol	+52	Little change	Introduction of acidic function. pK_a 1.7, reduction of log $D_{7.4}$ of 5.7 units[a]

[a] Applies to a neutral molecule. For a basic compound a zwitterion would be formed.
[b] Calculated for a nonsteroidal anti-inflammatory such as naproxen.

is that most of the compounds causing toxicity following concentration by transporters will be of low lipid permeability possibly due to high PSA (as described previously and detailed in Fig. 5.1). These drugs cross lipid membranes slowly so a transporter can create a high local free drug concentration in the tissue or cell. Lipid-permeable drugs, even if transported, will diffuse back across the membrane neutralizing any concentration effects.

Aminoglycosides are associated with renal toxicity and ototoxicity. An example of this class of drug is gentamicin, which has a log P of −2.88, five basic centers, and a TPSA of 214 A^2. In keeping with the low lipophilicity, the compound is renally cleared and has a volume of distribution of 0.25–0.3 l/kg. Gentamicin binds to megalin (pp 330, LRP2), a low-density lipoprotein receptor present in the cochlea of the inner ear and the proximal tubule of the kidney (Dagil et al., 2013). The binding is of low affinity (4 mM) but sufficient to transport gentamicin into the cell interior by endocytosis. The compound accumulates in the cells to high concentrations particularly in lysosomes and the endoplasmic reticulum.

Kidney toxicity is also seen when other carriers are implicated. For instance, uptake into proximal tubular cells includes human organic anion (HOAT) and cation (HOCT) transporters. Adefovir is an acyclic nucleotide phosphonate drug (a nucleoside reverse transcriptase inhibitor used in HIV); the prodrug (dipivoxil) form has a log P of 1.8 and a TPSA of 167 A^2. Adefovir and similar drugs such as cidofovir and tenofovir are transported into the proximal cells via HOAT (Fernandez-Fernandez et al., 2011). The transport gives rise to high concentrations that are believed to inhibit mitochondrial DNA polymerase-γ and give rise to mitochondrial dysfunction and resultant toxicity.

A converse situation arises where a compound selectivity for the target organ is provided by transporters. Cerivastatin is an inhibitor of hydroxymethylglutaryl-coenzyme A (HMG-CoA). The enzyme is important in cholesterol synthesis, which occurs predominantly in the liver. There are several drugs inhibiting HMG-CoA termed statins, and these represent the major therapy to reduce low-density lipoprotein. The major side effect is muscle toxicity (myopathy), which can occur in its severe form, rhabdomyolysis. Cerivastatin in comparison with other statins had an incidence of rhabdomyolysis up to 10-fold greater, resulting in its withdrawal from the marketplace. Among the likely mechanisms is the depletion of secondary metabolic intermediates (mevalonic acid and its metabolites). Statins are known to be actively transported into the hepatocyte, the site of the primary pharmacology by organic anion transporter protein B1 (OATP1B1). OATPB1 transport provides selectivity for the liver over other tissues such as the muscle. Cerivastatin was transported, but the intracellular to extracellular steady-state free drug concentration ratios of cerivastatin (8:1) are considerably lower (Paine et al., 2008) than a drug like atorvastatin (18:1), which has a lower incidence of muscle effects. The liver concentrations may reflect rate of transport into the hepatocyte, metabolism, and diffusion or transport from the hepatocyte. Both drugs are moderately to highly lipophilic with high TPSAs (cerivastatin, log P 3.4 and TPSA 99.9 A^2; atorvastatin, log P 4.4 and TPSA 114 A^2). Passive diffusion appeared the most important factor with cerivastatin having a threefold greater rate than atorvastatin (Paine et al., 2008). This difference in uptake was further confirmed in series of experiments comparing whole-cell and enzyme assays (microsomes) with calculated uptake values. The ratio of IC_{50} values for rosuvastatin, atorvastatin, pravastatin, fluvastatin, and cerivastatin were

75, 13, 8, 5 and 4, respectively, while the corresponding uptake ratios were 10–58, 63–490, 17–64, 15 and 6–14, respectively (Shitara et al., 2013). The conclusion was that statins except cerivastatin and fluvastatin were highly concentrated in hepatocytes.

A side effect of drugs is sometimes taste disturbance as evidenced by the antidiabetic biguanide drug metformin. Lipid-permeable drugs have salivary concentrations equal to unbound plasma concentrations due to passive diffusion. This is exemplified by drugs such as phenytoin (Ibarra et al., 2010) with a cLog P of 2.3 and a TPSA of 58 A^2. Metformin has low passive lipoidal permeability with physicochemical properties of clog P of −1.8, pK_a of 12.4, and TPSA of 89 A^2. Salivary glands express high levels of organic cation transporter-3 (OCT3). The transporter is present at both basolateral (blood-facing) and apical (saliva-facing) membranes of salivary gland acinar cells. Metformin was shown to be transported by OCT3/Oct3 *in vitro*. Metformin was also shown to be actively transported with a high level of accumulation in the salivary glands of wild-type mice. In contrast, active uptake and accumulation of metformin in salivary glands were not observed in *Oct3−/−* mice (Lee et al., 2014). Very high concentrations in the saliva of metformin explain the taste disturbance effects.

Transporters such as P-glycoprotein (Pgp) can act to exclude compounds from certain organs such as the brain. As discussed previously exclusion is most likely with drugs of moderate to low lipoidal permeability. In a study (Liu et al., 2004) looking at total brain uptake for a series of compounds, three common structure descriptors—log D, TPSA, and vsa_base (defined later)—were identified as the most important descriptors that correlate with BBB permeability. Log D, a measure of lipophilicity, has positive contribution to BBB permeability. TPSA measures a compound's polarity and hydrogen bonding potential and has negative contribution to BBB permeability. A larger TPSA value usually deters a compound from entering the brain. This is due to limited permeability and the resultant possible effect of efflux (see Fig. 5.1). Descriptor vsa_base represents van der Waals surface area of basic atoms and relates to the basicity of a compound. The positive contribution to BBB permeability indicates the role of phospholipid binding in total drug concentration. Its effect on net lipophilicity is incorporated in the log D term. As expected from earlier discussion in this chapter, the apparent total brain uptake for basic compounds was greater than neutral and acidic compounds. If unbound concentration in brain had been the subject of study rather than brain uptake, the descriptor vsa_base would have had little effect.

TPSA and lipoidal permeability are not absolute rules in terms of transporter involvement. Actual transport rate is obviously important, and it is likely that some compounds are substrates for multiple transporters, which could give very high rates. In addition as described earlier, hydrogen bond acceptors and donors are considered equal, and the strength of the hydrogen bond is not considered. Hydrogen bond donors are considered to be of more influence in membrane transit (Desai et al., 2012). These considerations are illustrated by consideration of the CNS effects of opioid agonists in mouse illustrated in Figure 5.2. All the compounds except morphine would be expected to have high lipoidal permeability, uniform free drug concentrations in TBW, and a close relationship between *in vitro* and *in vivo* pharmacodynamic measures. This is illustrated by comparison of *in vitro* Ki and *in vivo* EC_{50} values in Figure 5.2. Morphine does not correlate with being weaker *in vivo* due to its lowered penetration of the BBB due to its negative lipophilicity (log D).

Surprisingly loperamide also is much weaker *in vivo* due to poor penetration of the BBB, an effect attributed to transporter (Pgp) exclusion. The only outstanding difference in physicochemical properties to the high lipoidal permeability compounds is the sole hydrogen bond donor (Fig. 5.2). Since flux across MDCK cells in the presence of a Pgp inhibitor is equivalent to a high lipoidal permeability compound like propranolol, it can be concluded that it is likely that certain compounds have very high rates of transporter flux exceeding lipoidal permeability. Supporting the supposition of multiple transporters, detailed kinetic studies have identified an additional basolateral transporter, which assists Pgp-mediated efflux of digoxin and loperamide in MDCKII–hMDR1 cells (Acharya et al., 2008).

5.9 INHIBITION OF EXCRETION PUMPS

The bile salt export pump (BSEP) is located on the canalicular plasma membrane. Inhibition of the pump has been suggested to correlate with cholestatic drug-induced liver injury (DILI) in humans (Dawson et al., 2012). Physicochemistry properties can define drugs capable of inhibiting the pump, but the definitions are very broad, mainly suggesting that Log P values above 2.3 and molecular weights above 309 are needed for significant pump inhibition (Dawson et al., 2012). The relationship between BSEP inhibition and molecular physicochemical properties was further investigated (Warner et al., 2012) to increase the data set results and continued to show that lipophilicity and molecular size were significantly correlated with BSEP inhibition. To probe the effect of charge, compounds were divided into five ion class categories: neutral compounds, bases, acids, cations, and zwitterions. Across the full data set, the distribution between BSEP inhibition potency above and below the threshold value was almost completely even, and this was mirrored across the acids, bases, and neutral compounds. However, it is notable that cations formed the least likely class of BSEP inhibitors.

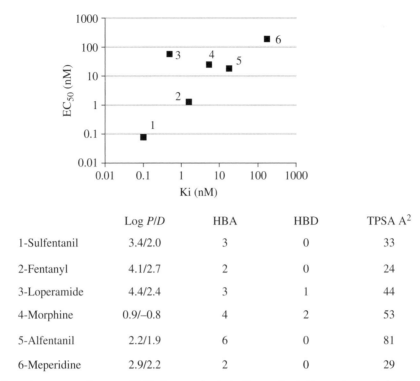

	Log P/D	HBA	HBD	TPSA A^2
1-Sulfentanil	3.4/2.0	3	0	33
2-Fentanyl	4.1/2.7	2	0	24
3-Loperamide	4.4/2.4	3	1	44
4-Morphine	0.9/–0.8	4	2	53
5-Alfentanil	2.2/1.9	6	0	81
6-Meperidine	2.9/2.2	2	0	29

FIGURE 5.2 Correlation of *in vitro* and *in vivo* CNS pharmacological effects of opioid agonists and their physicochemical properties. The weaker *in vivo* effects of morphine are explainable in contrast to those of loperamide.

In other studies by Pedersen et al. (2013), the major molecular properties related to BSEP interactions were charge, lipophilicity, hydrophobicity, and size. Transported BSEP substrates were preferably monovalent, negatively charged bile acids. Of the few nonbile acid substrates identified, all carried a negative net charge at physiological pH. However, unlike substrate interactions with BSEP, the majority of the BSEP inhibitors (58%) were unionized, highlighting an important difference between inhibitor and substrate molecular interactions. Positive charge was strongly associated with a lack of inhibition.

Studies of this type often use an IC_{50} value that does not provide mechanistic data. Detailed studies of troglitazone and its more potent sulfate metabolite do provide evidence, at least for this molecule that the inhibition is competitive with the bile salt substrate (Funk et al., 2001). A factor missing in simple calculation of IC_{50} or Ki against BSEP in predicting *in vivo* or whole-cell effects is the possibility of concentration of the compound in the liver as outlined for statins. This possibility is illustrated by the cholestatic sulfate metabolite of troglitazone, which not only circulates in high concentration but is also concentrated in the liver by OATP-C and possibly OATP8 transport (Nozawa et al., 2004). Troglitazone has a TPSA of $85\,A^2$ increased to $137\,A^2$ by sulfation.

Why drug-induced liver disease should be of low frequency when caused by secondary pharmacology can be partly explained by variation in the BSEP transporter. Ulzurrun et al. (2013) analyzed the combined effect on DILI development of the *ABCB11* 1331T>C polymorphism. The CC genotype was significantly associated with hepatocellular damage particularly in NSAID DILI cases and significantly linked to DILI development from drugs causing <50% BSEP inhibition.

5.10 CONCLUSIONS

Consideration of ADMET without understanding the physicochemical properties of a molecule is problematic. All the important physicochemical properties can be calculated with a reasonable degree of accuracy. Simple rules around log P and TPSA and charge at physiological pK (anion or cation) can allow division of molecules into likely effects of transporters such as exclusion or accumulation and understanding of total and unbound drug tissue concentrations and act as an indicator of likelihood of promiscuity. The value of these calculations is increased when a series (or even a pair) of molecules is compared. Clearly many aspects of the pharmacodynamics of on- and off-target effects of molecules will still need investigation, but having a framework to classify molecules is a great aid in predicting or characterizing these effects.

REFERENCES

Acharya, P., O'Connor, M.P., Polli, J.W., Ayrton, A., Ellens, H., and Bentz, J. (2008): Kinetic identification of membrane transporters that assist P-glycoprotein-mediated transport of digoxin and loperamide through a confluent monolayer of MDCKII-hMDR1 cells. *Drug Metab. Dispos.*, 36, 452–460.

Alex, A., Milan, D.S., Perez, M., Wakenhut, F., and Whitlock, G. (2011): Intramolecular hydrogen bonding to improve membrane permeability and absorption in beyond rule of five chemical space. *Med. Chem. Commun.*, 2, 669–674.

Azzaoi, K., Hamon, J., Faller, B., Jacoby, E., Bender, A., Jenkins, J.L., and Urban, L. (2007): Modeling promiscuity based on in vitro safety pharmacology profiling data. *ChemMedChem*, 2, 874–880.

Dagil, R., O'Shea, C., Nykjaer, A., Bonvin, A.M.J.J., and Kragelund, B.B. (2013): Gentamicin binds to the megalin receptor as a competitive inhibitor using the common ligand binding motif of complement type repeats: insight from the nmr structure of the 10th complement type repeat domain alone and in complex with gentamicin. *J. Biol. Chem.*, 288, 4424–4435.

Daniel, W.A. and Wójcikowski, J. (1999): Lysosomal trapping as an important mechanism involved in the cellular distribution of perazine and in pharmacokinetic interaction with antidepressants. *Eur. Neuropsychopharmacol.*, 9, 483–491.

Dawson, S., Stahl, S., Paul, N., Barber, J., and Kenna, J.G. (2012): In vitro inhibition of the bile salt export pump correlates with risk of cholestatic drug-induced liver injury in humans. *Drug Metab. Dispos.*, 40, 130–138.

Desai, P.V., Raub, T.J., and Blanco, M.-J. (2012): How hydrogen bonds impact P-glycoprotein transport and permeability. *Bioorg. Med. Chem. Lett.*, 22, 6540–6548.

Diaz, D., Ford, K.A., Hartley, D.P., Harstad, E.B., Cain, G.R., Achilles-Poon, K., Nguyen, T., Peng, J., Zheng, Z., Merchant, M., Sutherlin, D.P., Gaudino, J.J., Kaus, R., Lewin-Koh, S.C., Choo, E.F., Liederer, B.M., and Dambach, D.M. (2013): Pharmacokinetic drivers of toxicity for basic molecules: strategy to lower pK_a results in decreased tissue exposure and toxicity for a small molecule Met inhibitor. *Toxicol. Appl. Pharmacol.*, 266, 86–94.

Ertl, P., Rohde, B., and Selzer, P. (2000): Fast calculation of molecular polar surface area as a sum of fragment-based contributions and its application to the prediction of drug transport properties. *J. Med. Chem.*, 43, 3714–3717.

Fernandez, D., Ghanta, A., Kaufmann, G.W., and Sanguinetti M.C. (2004): Physicochemical features of the hERG channel drug binding site. *J. Biol. Chem.*, 279, 10120–10127.

Fernandez-Fernandez, B., Montoya-Ferrer, A., Sanz, A.B., Sanchez-Niño, M.D., Izquierdo, M.C., Poveda, J., Sainz-Prestel, V., Ortiz-Martin, N., Parra-Rodriguez, A., Selgas, R., Ruiz-Ortega, M., Egido, J., and Ortiz, A., Tenofovir nephrotoxicity: 2011 update (2011): *AIDS Research and Treatment*, 2011, Article ID 354908, 11 pages.

Funk, C., Ponelle, C., Scheuermann, G., and Pantze, M. (2001) Cholestatic potential of troglitazone as a possible factor contributing to troglitazone-induced hepatotoxicity: in vivo and in vitro interaction at the canalicular bile salt export pump (Bsep) in the rat. *Mol. Pharm.*, 59, 627–635.

Hanumegowda, U.M., Wenke, G., Regueiro-Ren, A., Yordanova, R., Corradi, J.P., and Adams, S.P.(2010): Phospholipidosis as a function of basicity, lipophilicity, and volume of distribution of compounds. *Chem. Res. Toxicol.*, 23, 749–755.

Hopkins, S. (1991): Clinical toleration and safety of azithromycin. *Am. J. Med.*, 91, S40–S45.

Ibarra, M., Vázquez, M., Fagiolino, P., Mutilva, F., and Canale, A. (2010): Total, unbound plasma and salivary phenytoin levels in critically ill patients. *J. Epilepsy Clin. Neurophysiol.*, 16 (2), 69–73.

Khoh-Reiter, S., Sokolowski, S.A., Jessen, B., Evans, M., Dalvie, D., and Lu, S. (2015): Contribution of membrane trafficking perturbation to retinal toxicity. *Toxicol. Sci.*, 145, 383–395.

Leblanc, B., Jezequel, S., Davies, T., Hanton, G., and Taradach, C. (1998): Binding of drugs to eye melanin is not predictive of ocular toxicity. *Regul. Toxicol. Pharmacol.*, 28, 124–132.

Lee, N., Duan, H., Hebert, M.F., Liang, C.J., Rice, K.M., and Wang, J. (2014): Taste of a pill: organic cation transporter-3 (OCT3) mediates metformin accumulation and secretion in salivary glands. *J. Biol. Chem.*, 289 (39), 27055–27064.

Leo, A., Hansch, C., and Elkins, D. (1971): Partition coefficients and their uses. *Chem. Rev.*, 71, 525–616.

Liu, X., Tu, M., Kelly, R.S., Chen, C., and Smith, B.J. (2004): Development of a computational approach to predict blood-brain barrier permeability. *Drug Metab. Dispos.*, 32, 132–139.

Manners, C.N., Payling, D.W., and Smith, D.A. (1988): Distribution coefficient, a convenient term for the relation of predictable physico-chemical properties to metabolic processes. *Xenobiotica*, 18, 331–350.

Merlet, P., Delforge, J., Syrota, A., Angevin, E., Crouzel, M., Valette, H., Loisance, D., Castaigne, A., and Randé, J.L.(1993): Positron emission tomography with 11C CGP-12177 to assess beta-adrenergic receptor concentration in idiopathic dilated cardiomyopathy. *Circulation*, 87, 1169–1178.

Nozawa, T., Sugiura, S., Nakajima, M., Goto, A., Yokoi, T., Nezu, J.I., Tsuji, A. Tamai, I., (2004): Involvement of organic anion transporting polypeptides in the transport of troglitazone sulfate: implications for understanding troglitazone hepatotoxicity. *Drug Metab. Dispos.*, 32, 291–294.

Paine, S.W., Parker, A.J., Gardiner, P., Webborn, P.J.H., Riley, R.J (2008): Prediction of the pharmacokinetics of atorvastatin, cerivastatin, and indomethacin using kinetic models applied to isolated rat hepatocytes. *Drug Metab. Dispos.*, 36 (7), 1365–1374.

Pedersen, J.M., Matsson, P., Bergström, C.A., Hoogstraate, J., Norén, A., LeCluyse, E.L., and Artursson, P. (2013). Early identification of clinically relevant drug interactions with the human bile salt export pump (BSEP/ABCB11). *Toxicol. Sci.*, 136, 328–343.

Peters, J.U., Schneides, P., Mattei, P., and Kansy, M. (2009): Pharmacological promiscuity: dependence on compound properties and target specificity in a set of recent Roche compounds. *ChemMedChem*, 4, 680–686.

Price, D., Blagg, J., Jones, L., Greene, N., and Wager, T. (2009): Physicochemical drug properties associated with in vivo toxicological outcomes: a review. *Expert Opin. Drug Metab. Toxicol.*, 5, 921–931.

Rekker, R.F. and Mannhold, R. (1992): *Calculation of Drug Lipophilicity: The Hydrophobic Fragmental Constant Approach*, VCH, Weinheim.

Rodgers, T., Leahy, D., and Rowland, M.(2005): Tissue distribution of basic drugs: accounting for enantiomeric, compound and regional differences amongst beta-blocking drugs in rat. *J. Pharm. Sci.*, 94, 1237–1248.

Roy, D., Talajic, M., Nattel, S., Dubuc, M., Thibault, B., and Guerra, P. (2000): Amiodarone oral. *Card. Electrophysiol. Rev.*, 4, 262–269.

Shields, G.C. and Seybold, P.G. (2013): *Computational Approaches for the Prediction of pK$_a$ Values*, CRC Press, London.

Shitara, Y., Maeda, K., Ikejiri, K., Yoshida, K., Horie, T., and Sugiyama, Y. (2013): Clinical significance of organic anion transporting polypeptides (OATPs) in drug disposition: their roles in hepatic clearance and intestinal absorption. *Biopharm. Drug Dispos.*, 34, 45–78.

Small, H., Gardner, I., Jones, H.M., Davis, J., and Rowland, M. (2011): Measurement of binding of basic drugs to acidic phospholipids using surface plasmon resonance and incorporation of the data into mechanistic tissue composition equations to predict steady-state volume of distribution. *Drug Metab. Dispos.*, 39, 1789–1793.

Smith, D.A. and Obach, R.S. (2010): Metabolites: have we MIST out the importance of structure and physicochemistry? *Bioanalysis*, 2, 1223–1233.

Smith, D.A., Allerton, C., Kalgutkar, A.S., Van De Waterbeemd, H., and Walker, D.K. (2012): *Pharmacokinetics and Metabolism in Drug Design*, Wiley-VCH, Verlag, Weinheim.

Stpęien, K.B. and Wilczok, T. (1982): Studies of the mechanism of chloroquine binding to synthetic dopa-melanin. *Biochem. Pharmacol.*, 31, 3359–3365.

Ulzurrun, E., Stephens, C., Crespo, E., Ruiz-Cabello, F., Ruiz-Nuñez, J., Saenz-López, P., Moreno-Herrera, I., Robles-Díaz, M., Hallal, H., Moreno-Planas, J.M., Maria, R., Cabello, M.R., Lucena, M.I., and Andrade, R.J. (2013): Role of chemical structures and the 1331T>C bile salt export pump polymorphism in idiosyncratic drug-induced liver injury. *Liver Int.*, 33, 1378–1385.

Warner, D.J., Chen, H., Cantin, L.-D., Kenna, J.G., Stahl, S., Walker, C.L., and Noeske, T. (2012): Mitigating the inhibition of human bile salt export pump by drugs: opportunities provided by physicochemical property modulation, in silico modeling, and structural modification. *Drug Metab. Dispos.*, 40, 2332–2341.

Wu, C.Y. and Benet, L.Z. (2005): Predicting drug disposition via application of BCS: transport/absorption/ elimination interplay and development of a biopharmaceutics drug disposition classification system. *Pharm. Res.*, 22, 11–23.

Yanai, K., Ryu, J.H., Watanabe, T., Iwata, R., Ido, T., Sawai, Y., Ito, K., and Itoh, M. (1995): Histamine H1 receptor occupancy in human brains after single oral doses of histamine H1 antagonists measured by positron emission tomography. *Br. J. Pharmacol.*, 116, 1649–1655.

Zane, P.A., Brindle, S.D., Gause, D.O., O'Buck, A.J., Raghavan, P.R., and Tripp, S.L. (1990): Physicochemical factors associated with binding and retention of compounds in ocular melanin of rats: correlations using data from whole-body autoradiography and molecular modeling for multiple linear regression analyses. *Pharm. Res.*, 7, 935–941.

6

THE NEED FOR HUMAN EXPOSURE PROJECTION IN THE INTERPRETATION OF PRECLINICAL *IN VITRO* AND *IN VIVO* ADME TOX DATA

PATRICK POULIN

Associate Professor, Department of Occupational and Environmental Health, School of Public Health, IRSPUM, Université de Montréal, Montréal, Québec, Canada and Consultant, Québec city, Québec, Canada

6.1 INTRODUCTION

The field of toxicology is currently undergoing a global paradigm shift to use of *in vitro* approaches for assessing the risks of chemicals and drugs in a more mechanistic and high-throughput manner than current approaches relying primarily on *in vivo* testing. In particular, allometric and physiological modeling methods can be used to predict the *in vivo* exposure conditions that would produce chemical concentrations in plasma and/or the target tissue equivalent to the concentrations at which effects were observed with *in vitro* assays of tissue/organ toxicity (Yoon et al., 2012; Poulin et al., 2015b). Therefore, this book chapter reviews the different modeling methods used for human pharmacokinetic (PK) projection in drug discovery with an emphasis on the prediction of tissue distribution in toxicology studies.

6.2 METHODOLOGY USED FOR HUMAN PK PROJECTION IN DRUG DISCOVERY

The prediction of human PK from preclinical *in vitro* and *in vivo* data remains an important goal in the drug discovery and development process to obtain an early estimate of efficacious human dose, dosing regimen, and safety window prior to first-in-human trials. From a clinical study perspective, under-prediction of human exposure would lead to safety concerns because actual exposures would be greater than expected, and inversely. The primary focus in the literature has been on the estimation of basic PK parameters such

as clearance (CL), volume of distribution at steady state (V_{ss}), and area under the curve (AUC) using different methodologies with less emphasis placed on the projection of human concentration–time profiles (Jones et al., 2011b; Khojasteh et al., 2011; Ring et al., 2011; Grime et al., 2013; Stepensky, 2013). However, the projection of the concentration–time profiles in humans from preclinical data is also necessary in discovery and development programs in which the maximum (C_{max}) or minimum (C_{min}) plasma concentration is thought to also be critical for toxicity and efficacy, respectively, as opposed to an overall exposure (AUC) estimate (Wajima et al., 2003; Fura et al., 2008; Poulin et al., 2011b; Vuppugalla et al., 2011, Yoon et al., 2012; Jones et al., 2013). In other words, the projection of concentration–time profiles in humans can be used to derive any associated PK parameters (e.g., C_{min}, C_{max}, AUC, $T_{1/2}$), which can be related to any toxicity or efficacy effect. An approach that is commonly used for predicting the concentration–time profiles in humans includes the use of standard equations derived from a one-compartmental kinetic model, which is likely not suitable for compounds that follow multicompartment kinetics. For example, the shortcomings include underestimation of the plasma half-life ($T_{1/2}$) and C_{min} or overestimation of C_{max} (Fura et al., 2008). The other methods reported in the literature for the prediction of human PK profiles (or exposure) from preclinical data include modified forms of the original allometric scaling technique proposed by Dedrick et al. (formally named the Wajima method) and physiologically based pharmacokinetic (PBPK) models (Wajima et al., 2003; Jones

Drug Discovery Toxicology: From Target Assessment to Translational Biomarkers, First Edition. Edited by Yvonne Will, J. Eric McDuffie, Andrew J. Olaharski, and Brandon D. Jeffy.

et al., 2006, 2011b, 2013; De Buck et al., 2007; Peters, 2008; Poulin et al., 2011b; Rowland et al., 2011; Vuppugalla et al., 2011; Thiel et al., 2014) (Fig. 6.1 and Table 6.1). Therefore, it is crucial to fully understand the assumptions in any scaling methodology being applied and the limitations inherent in the assumptions with respect to compounds they can be applied to.

6.2.1 Prediction of Plasma Concentration–Time Profile by Using the Wajima Allometric Method

The original Dedrick's method is relatively simple and is based on the concept of physiological time, wherein the concentration–time profile of preclinical species is transformed by correction factors to generate a human profile (Wajima et al., 2003; Vuppugalla et al., 2011). This method, however, is restricted to the use of CL and V_{ss} values projected only by simple allometry. As a consequence, the success of this method is largely dependent on the predictive power of interspecies allometric scaling. As there are several approaches for predicting the PK parameters other than allometric scaling, such as *in vitro–in vivo* correlation approaches, physiologically based models, *in silico* approaches based on molecular descriptors, and sophisticated model equations coupled with regression analysis methods, it becomes important to develop a general method for predicting concentration–time profile that reflects the PK parameters predicted independent of the interspecies scaling methodologies. Toward this end, another method that offers the flexibility to incorporate a broad spectrum of CL and V_{ss} values predicted by alternate techniques was proposed by Wajima et al. (2003) for the prediction of intravenous (IV) PK profiles.

The Wajima allometric method is used to extrapolate to human the plasma concentration–time profiles observed *in vivo* in each preclinical species (Wajima et al., 2003; Vuppugalla et al., 2011). Dose normalized curves are obtained by dividing the time axis with mean residence time (MRT set equals to V_{ss}/CL) and the concentration axis with dose/V_{ss} (C_{ss}) of the respective preclinical species. The plasma time course in humans after IV administration can be simulated using the normalized curve of the preclinical species and the predicted values of V_{ss} and CL in humans. In other words, this step is accomplished by multiplying the concentration and time scales of the normalized curve of each preclinical species by the predicted human C_{ss} and MRT, respectively (Wajima et al., 2003; Vuppugalla et al., 2011). To simulate the plasma concentration–time profiles following oral administration in humans, the additional parameters of average of preclinical species of fraction absorbed (F_{abs}) and rate constant of absorption (K_a) were used in combination with microconstants of the predicted IV profiles from compartmental analyses (Vuppugalla et al., 2011) (Fig. 6.1). In this context, due to the multitude of factors affecting bioavailability (F%)

in humans (e.g., F_{abs} and K_a), a conventional approach of the Wajima allometry is to assume that the value in humans is the average of all preclinical species (Wajima et al., 2003; Vuppugalla et al., 2011). In general, simple solutions/suspensions are used for per os studies in preclinical species, whereas more optimized formulations are used in the clinic. Therefore, it is likely that both F% and K_a will be significantly altered by formulation effects versus physicochemical properties of the drugs. Supporting this view, human F% could be under-predicted based on preclinical species data particularly for low-solubility compounds (Vuppugalla et al., 2011). Therefore, AUC and maximal time of absorption (T_{max}) would also be under-predicted. This may suggest a more rapid release from the less optimized formulations and likely amorphous material generally used in preclinical studies, which would contribute to a systematic over-prediction of C_{max} and, hence, under-prediction of T_{max} and AUC in humans after per os administrations (Vuppugalla et al., 2011). In other words, while the formulation used in preclinical and clinical studies are not similar, this may not be well served by the model. Nevertheless, the Wajima allometric method can be used to challenge any formulation effect based upon *in vivo* preclinical data.

Since the plasma PK profiles previously determined *in vivo* in each preclinical species can be extrapolated to human, more than one predictions to human can be provided (i.e., one prediction per species). While the PK profiles predicted to human from each preclinical species are superimposed, this probably gives an indication that the allometric principle is conserved in each species (i.e., the drug PK is governed by almost similar mechanisms in each species, and hence, their relation with body weight is similar) (Wajima et al., 2003; Vuppugalla et al., 2011). Inversely, while the PK profiles are not superimposed, this gives information that one or more species govern the PK differently, and hence, this could reduce the confidence in the prediction made to human. In this case, it may become difficult to identify which species is the closest to human in the absence of mechanistic studies. Overall, the Wajima allometric method principally relies on measured *in vivo* preclinical data on plasma concentration, and while the interspecies differences are covered, this allometric method could be more predictive (or as predictive) than *in vitro* data. Wajima allometry has been applied to small molecules only (Vuppugalla et al., 2011), and hence, application for biologics still needs to be investigated.

6.2.2 Prediction of Plasma and Tissue Concentration–Time Profiles by Using the PBPK Modeling Approach

By contrast to Wajima allometry, which is dedicated to the prediction of plasma concentration–time profile in humans only, a PBPK structural model can be used to simulate the concentration–time profile in both the plasma and target organs in any mammalian species (Wajima et al., 2003;

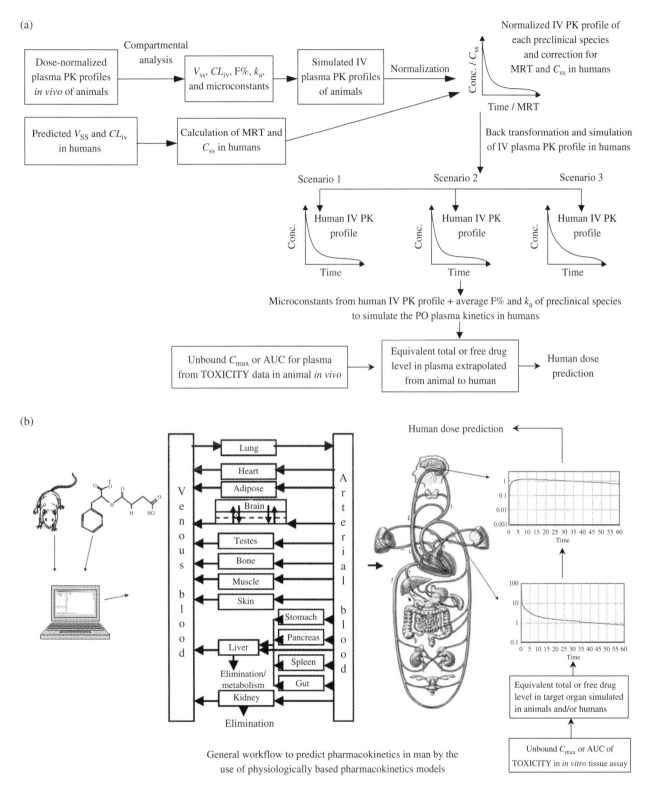

FIGURE 6.1 Illustration of the principle of the methods of Wajima allometry (a) and PBPK model (b) used to predict concentration–time profiles in a standard human. (a) Adapted from Vuppugalla et al. (2011, pp. 4111–4126). (b) Adapted from Yoon et al. (2012, pp. 633–652) and Poulin et al. (2015b).

TABLE 6.1   Principal Studies Related to the Wajima Allometry and PBPK Modeling Methods

| Methods | Matrices | | References | |
	Plasma PK	Tissue PK	Review Articles	Comparative Assessments
Wajima allometry	Yes	No	Wajima et al. (2003)	Vuppugalla et al. (2011)
PBPK modeling	Yes	Yes	Jones et al. (2013) and Rowland et al. (2011)	Poulin et al. (2011a), Parrott et al. (2005), Thiel et al. (2014), Peters (2008), Jones et al. (2006, 2011a), De Buck et al. (2007), and Chen et al. (2012)

Poulin et al., 2011b; Vuppugalla et al., 2011; Grime et al., 2013; Jones et al., 2013). The application of PBPK modeling is coming of age in drug development and regulation, reflecting significant advances over the past 15 years in the predictability of key PK parameters from *in vitro* data and in the availability of dedicated software platforms and associated databases (Jones et al., 2006, 2011b, 2013; De Buck et al., 2007; Peters, 2008; Poulin et al., 2011b; Rowland et al., 2011; Thiel et al., 2014) (Table 6.2). PBPK models comprise three major components: system-specific properties, drug properties, and the structural model (Rowland et al., 2011; Jones et al., 2013). System-specific properties include organ mass or volume, blood flow, and tissue composition. Drug properties include tissue affinity, plasma protein binding affinity, membrane permeability, enzymatic stability, and/or transporter activities. The structural model comprises the anatomical arrangement of the tissues and organs of the body, linked by blood perfusion. Unlike empirical models, the structures of which are dictated by the observed drug data, a PBPK structural model is independent of the drug and is the same for all mammalian species, although the degree of complexity often varies with the intended application. In other words, the PBPK model can provide simulations of PK to human based only upon the integration of diverse input data on absorption, distribution, metabolism, and excretion (ADME). Accordingly, any input data can be integrated in this model to predict the related ADME parameters (i.e., *in silico*, *in vitro*, physicochemical, and/or preclinical *in vivo* data); therefore, the PBPK model is the only tool that can provide prediction of plasma and tissue concentrations over time of drugs prior to any *in vivo* studies in mammals (Jones et al., 2006, 2011a, 2013; De Buck et al., 2007; Peters, 2008, Poulin et al., 2011b; Rowland et al., 2011; Thiel et al., 2014) (Fig. 6.1). The PBPK method has successfully been applied to both small and large molecules in several species (Jones et al., 2013). In addition, some authors linked the PBPK and allometric approaches to develop new rules for the interspecies scaling (Hall et al., 2012).

Because the PBPK model is built on a mechanistic framework, several aspects can fully be considered: (i) low-to-high dose and route-to-route extrapolations, (ii) interspecies differences, (iii) variability and uncertainty effects, (iv) impact of pharmaceutical formulation, (v) transporter effects and drug–drug interactions, and (vi) total versus free drug

TABLE 6.2   Commercially Available Softwares Used to Project Human Exposure of Drugs and Chemicals

| Trade Names | Methods Available | | References |
	Wajima Allometry	PBPK Modeling	
ADME WorkBench	Yes	Yes	www.admewb.com
Cloe PK	No	Yes	www.cyprotex.com
GastroPlus	No	Yes	www.simulation-plus.com
PK-Sim	No	Yes	www.systems-biology.com
SimBiology	No	Yes	www.mathworks.com
Simcyp	No	Yes	www.simcyp.com

level in the target organ. Furthermore, the PBPK models can simulate any mode of drug administration (e.g., IV, per os, dermal, and intraperitoneal). Yet in practice these issues are very important to adequately predict the human exposure to drugs in plasma and/or tissues. That is, some of the apparent lack of prediction of PBPK may be viewed differently had these additional factors been incorporated into the analysis (e.g., variability, formulation, transporters, DDI, permeation limitation) (Fenneteau et al., 2009; Hall et al., 2012; Jones et al., 2012a, b; Wagner et al., 2012; Hudachek and Gustafson, 2013; Chetty et al., 2014) (Table 6.2). In addition, the commercial softwares of PBPK models constantly improve the prediction of drug PK (Table 6.2). Therefore, the degree of confidence for the simulations in humans would increase when the PBPK model simulations in preclinical species are successful, and the main mechanisms of PK are known and well described for the tested drugs (Poulin et al., 2011b; Jones et al., 2013).

6.2.3   Integrative Approaches of Toxicity Prediction Based on the Extent of Target Tissue Distribution

Evaluating target tissue drug exposure in toxicology studies from readily available PK parameters would speed up the assessment of drug safety. In other words, the distribution of drugs into the tissues, that is, tissue accumulation, is a recognized contributor to drug toxicity; however, evaluation of tissue drug partitioning under *in vivo* condition is not routinely performed in toxicology studies, partly because it is resource-intensive. The endpoint is the prediction of values of tissue–plasma partition coefficients (K_p) from readily available input parameters, which are the parameters commonly used to estimate the degree to which a drug accumulates into the tissues. This issue can be approached with PBPK modeling methods, where the impacts of dosing regimen on tissue distribution can be predicted from the description of its controlling processes, although this could be a data-intensive approach. However, PBPK modeling provides an effective framework for conducting quantitative *in vitro* to *in vivo* extrapolation based on input K_p values. As said, the physiological structure of the PBPK model facilitates the incorporation of *in silico*- and *in vitro*-derived chemical-specific parameters in order to predict *in vivo* ADME (Kamgang et al., 2008; Chen et al., 2012; Jones et al., 2012b). In particular, the combination of *in silico* and *in vitro* parameter estimation with PBPK modeling can be used to predict the *in vivo* exposure conditions that would produce chemical concentrations in the plasma and/or target tissue equivalent to the concentrations at which effects were observed with *in vitro* assays of tissue/organ toxicity (Yoon et al., 2012; Poulin et al., 2015b). Related to this, a study focused on the use of PK modeling, and in particular PBPK modeling, to support the interpretation of human biomonitoring data from the perspective of exposure reconstruction and risk characterization (Clewell et al., 2008). Another study also linked a PBPK model to a toxicokinetic model for acetaminophen, and hence, the simulated liver concentration was associated to the *in vitro* cell viability due to hepatotoxicity (Péry et al., 2013).

In preclinical and clinical studies, total drug concentrations in the plasma or tissue are often correlated with pharmacodynamics (PD). However, the use of total tissue levels (e.g., tissue concentrations derived from homogenates) or biopsies to draw direct conclusions on drug activity are unwarranted and/or unreliable (Smith et al., 2010; Diaz et al., 2013; Mariappan et al., 2013). This is in contrast with the unbound (free) drug concentration at the target site, which should be more pharmacologically relevant. Therefore, the association of PBPK model with PD model has also been challenged (Smith et al., 2010; Timchalk et al., 2010; Diaz et al., 2013; Mariappan et al., 2013; Chetty et al., 2014; Cristofoletti and Dressman, 2014; Poulin, 2015a, b, c). Examples were published on the PBPK/PD modeling of specific toxicity effect (e.g., cholinesterase inhibition) and to assess the clinical relevance of current bioequivalence criteria as well as to predict the impact of differences in target binding capacity and target site drug concentrations on drug responses and variability (Timchalk et al., 2010; Chetty et al., 2014; Cristofoletti and Dressman, 2014). The basis of PK/PD research relies on the assumption that the unbound drug levels at the site of action is the relevant measure of drug effect; therefore, the toxicity and efficacy should be driven by free drug concentration in the target tissue (at the

site of action) (Smith et al., 2010; Diaz et al., 2013; Mariappan et al., 2013; Poulin, 2015a, b, c; Poulin et al., 2015a, b). Hence, it has recently been demonstrated that free drug concentrations in human tissues and plasma are not equal for passively permeable drugs that are ionized at the physiological pH (Berry et al., 2010; Poulin, 2015a, b). In effect, the free drug concentration in the tissue and plasma differed for the ionizable drugs due to the pH partition gradient between cells and plasma, but there was no difference for the neutral drugs. For a large number of compounds that are substrates for uptake and/or efflux transporters expressed at the tissue level, significant discrepancies between the free drug concentrations in plasma and tissue cells would also be expected due to the transport effect. Related to this, the impact of albumin concentration on the uptake of drugs in cells might involve mechanisms going beyond the free drug concentration hypothesis. Proceeding from the assumption that both the unbound and protein-bound drug fractions can be available for uptake, several authors have argued that the uptake of highly bound drugs in cells might be driven mainly by the albumin-facilitated uptake mechanism(s). Hence, a novel approach quantifying the additional contribution of the protein-bound drug complex and pH gradient effect in diverse *in vitro*-to-*in vivo* extrapolation procedures of drug uptake and clearance has been proposed and extensively validated by Poulin (2015b) and Poulin et al. (2015b). All of these observations have been tested both under *in vitro* static environments and *in vivo* dynamic environments (Berry et al., 2010; Poulin, 2015a, b, c, Poulin et al., 2015b). In addition, in a dynamic *in vivo* environment, the higher the drug CL and lipophilicity, the lower will be the free drug concentration in tissue, and inversely. Therefore, only considering the free drug concentration in plasma derived from a fu_p value determined in a static *in vitro* environment might be biased to guide drug design (the old paradigm), and hence, it is recommended to use a PBPK model to reproduce more accurately the *in vivo* condition in tissue (the new paradigm) (Poulin, 2015a, b, c). In other words, the importance of drug tissue partitioning compared to plasma can be recognized to guide drug design by using more accurate PBPK/PD modeling approaches (Poulin et al., 1999; Smith et al., 2010; Diaz et al., 2013; Mariappan et al., 2013; Poulin, 2015a, b). This newly developed paradigm can be used to predict free drug concentration in diverse tissue compartments for small molecules in toxicology and pharmacology studies, which can be leveraged to optimize the PK drivers of tissue distribution based upon physicochemical and physiological input parameters in an attempt to optimize free drug level in tissue. This is a clear advantage compared to the Wajima allometric method, which is only based on plasma data. Therefore, the PBPK model would gain in accuracy when the prediction of total and free drug concentration in the target organ can be simulated separately. In this context, some prediction models of tissue distribution, which are based on physiological and *in vitro* input data, have been used within a PBPK model to simulate the total and/or free drug concentration at different tissue levels (cellular, subcellular, and/or interstitial) (Poulin and Theil, 2000; Halifax and Houston, 2006; Kleist and Huisinga, 2007; Turner et al. 2007; Gertz et al., 2008; Schmitt, 2008; Berry et al., 2010; Haritova and Fink-Gremmels, 2010; Peyret et al., 2010; Poulin and Haddad, 2011, 2013; Ruark et al., 2014; Poulin, 2015a, b; Poulin et al., 2015b). These later approaches deserve further exploration for the PBPK/PD modeling of chemicals. The role of microdialysis should also be acknowledged; however, the microdialysis can only be used to estimate free drug concentration in the interstitial space based on *in vivo* data by contrast to the PBPK model where the free drug concentration could virtually be predicted in any compartment of each organ on the basis of *in vitro* or *in silico* data, which can be helpful in early drug discovery (Poulin, 2015a, b).

Alternatively, it can also be of interest to verify whether tissues can be treated as well-stirred compartments, and the influences of altered delivery through changes in blood flow and dosing regimen can be neglected. This premise assumes that the same mechanisms are present in each tissue, and hence, relationships among drug tissue partitioning can be developed, comparably to a physiologically based approach. The challenge is that, being potentially more empirical, such relationships may not totally account for the processes that determine the links of simple PK observations to tissue-specific and regimen-specific drug concentrations: processes such as nonlinear metabolism, saturation of binding, and non-steady-state distributions that depend on the dosing regimens and the interaction among uptake, redistribution, and tissue-specific CL processes. Despite these limitations, consistent patterns of relative partitioning into various tissues can be identified that can allow reasonable estimate with less intensive data collection, for which the input parameters are readily available in drug discovery, and hence, diverse prediction methods of tissue distribution were published in the literature as depicted in Table 6.3. Accordingly, the prediction of K_p values from readily available input parameters has been successfully demonstrated in several examples in PK studies either for steady-state or non-steady-state conditions for various routes of administration. The prediction models, however, focus on healthy tissues and do not incorporate data from tumors, which are also critical target organs for toxicity and efficacy. Recently, the prediction of K_p values has been successfully extended to further examples in oncology studies (Poulin et al., 2013a, 2015a; Yoshida et al., 2000). Overall, these prediction models of K_p values should be useful in selecting compounds based on their abilities to penetrate tissue and/or tumors in preclinical species and humans, thereby increasing the therapeutic index.

General rules for the quantitative evaluation of organ toxicity based on the extent of tissue distribution prediction have also been developed. In other words, the PK drivers of toxicity have been investigated. First, Diaz et al. (2013) studied the PK drivers of toxicity for basic molecules, and hence, these authors demonstrated that lowering the ionization constant (pK_a) values resulted in a decrease of tissue exposure (i.e., K_p values) and, hence, of toxicity. Accordingly, in a companion paper, potential tissue drug accumulation under non-steady-state condition in toxicology studies was predicted based on the input values of V_{ss} and CL respective of the pK_a value of each drug (Poulin et al., 2013b). Second, Hanumegowda et al. (2010) showed that phospholipidosis in tissues is a function of basicity, lipophilicity, and volume of distribution by correlating the *in vivo* V_{ss} values to the degree of phospholipidosis determined *in vitro*. The use of this later approach can be facilitated by predicting the input V_{ss} values from the diverse methods presented in Table 6.3. Third, anti-tumor activity of targeted and cytotoxic agents in murine subcutaneous tumor models was correlated with clinical response (Harvey et al., 2012). Finally, *in vitro*-to-*in vivo* extrapolation and species response comparisons for drug-induced liver injury have also been developed (Howell et al., 2012). Overall, the later prediction approaches are all based on total tissue concentration, but as said previously, these approaches might probably be improved by considering the free drug concentration in tissue instead (Smith et al., 2010; Diaz et al., 2013; Mariappan et al., 2013; Poulin, 2015a, b).

6.3 SUMMARY OF THE TAKE-HOME MESSAGES FROM THE PHARMACEUTICAL RESEARCH AND MANUFACTURERS OF AMERICA CPCDC INITIATIVE ON PREDICTIVE MODELS OF HUMAN PK FROM 2011

Pharmaceutical Research and Manufacturers of America (PhRMA) formed a pharmaceutical innovation steering committee, with the aim of reducing clinical lead attrition rates throughout the drug development cycle. PhRMA assembled three tactical groups, which focused on safety, efficacy, and human PK. The fundamental scientific benefit of the task force that focused on human PK was to have greater certainty in predictions of human PK based on preclinical *in vitro* and *in vivo* data. Therefore, the overall goal of this PhRMA initiative was to assess the predictability of human PK from preclinical data and to provide comparisons of available prediction methods from the literature, as appropriate, using a representative blinded data set of 108 drug candidates exhibiting a wide array of physicochemical and structural properties (i.e., low molecular weight leads clinical candidate compounds where 108 compounds were obtained following oral administration in humans, of which 19 had also been given intravenously). Broadly, the prediction methodologies have been divided into three areas: allometry, *in vitro–in vivo* extrapolation (IVIVE), and PBPK. Five manuscripts have been published in the literature that review

TABLE 6.3 Prediction Methods of Tissue Distribution in Humans

Methods		Parameters				References	
		Volume V_{ss}	Individual Tissue K_p Partitioning			Original Articles	Comparative Assessments
			Healthy Tissue	Tumor Tissue	Cellular or Subcellular		
Allometric models	Standard allometry	Yes	No	No	No	Jones et al. (2011b), Boxenbaum (1982), Sawada et al. (1984), Obach et al. (1997), Ward and Smith (2004), Jolivette and Ward (2005), and Mahmood (2010)	Jones et al. (2006, 2011b) and Poulin et al. (2011a)
	Øie–Tozer model	Yes	No	No	No	Khojasteh et al. (2011), Jones et al. (2011b), and Øie and Tozer (1979)	Jones et al. (2011b)
Correlation models	Species equivalency	Yes	Yes	No	No	Berry et al. (2011) and Bachmann et al. (1996)	Jones et al. (2011b)
	Healthy tissues	No	Yes	No	No	Poulin and Theil (2009), Poulin et al. (2011b, 2013a), Richter et al. (2006), Björkman (2002), Jansson et al. (2008), and Edginton and Yun (2013)	Poulin and Theil (2009), Poulin et al. (2011c, 2013), and Jansson et al. (2008)
Tissue composition-based models	Tumor tissues	No	No	Yes	No	Poulin et al. (2013a), and Yoshida et al. (2000)	
	Original models	Yes	Yes	No	No	Poulin and Theil (2000, 2002), Rodgers and Rowland (2006, 2007), Rodgers et al. (2012), Poulin and Haddad (2012), Haddad et al. (2000), Paterson and Mackay (1989), Berezhkovskiy (2004), and Small et al. (2001)	Jones et al. (2011b), Poulin et al. (2011b), Parrott et al. (2005), Thiel et al. (2014), Peters (2008), Jones et al. (2011a), and de Buck et al. (2007)
	Unified algorithm for healthy tissues	Yes	Yes	No	Yes	Poulin (2015a,b,c), Kleist and Huisinga (2007), Schmitt (2008), and Peyret et al. (2010)	
	Unified algorithm for tumor tissues	Yes	No	Yes	Yes	Poulin et al. (2015a)	
	Unified algorithm for microsomes	No	Yes	No	Yes	Poulin and Haddad (2011)	
	Unified algorithm for hepatocytes	No	Yes	No	Yes	Poulin and Haddad (2013)	
	Inclusion of specific binding	No	Yes	No	Yes	Mandula et al. (2006), Terasaki et al. (1984), Guohua and Morrie (2012), and Poulin and Krishnan (1996a)	
Kinetic model including transporter effect		Yes	Yes	No	No	Jones et al. (2012a), Hudachek and Gustafson (2013), Ruark et al. (2014), Haddad et al. (2010), Paine et al. (2008), and Poulin et al. (2013, 2015b), and Poulin et al. (2015b)	—
Arundel model		No	Yes	No	No	Arundel (1997), and Luttringer et al. (2003)	Jones et al. (2011b) and Luttringer et al. (2003)
In vitro–in vivo extrapolations (IVIVE)		Yes	Yes	No	No	Berry et al. (2010), Sui et al. (2009), Harashima et al. (1984), Valkó et al. (2001), Lin et al. (1982), Khalafallah and Jusko (1984), Schuhmann et al. (1987), Clausen and Bickel (1993), Ballard et al. (2000), Ballard and Rowland (2003), Ritschel and Hammer. (1980), Poulin (2015b), Poulin et al. (2015b), and Murphy et al. (1995)	—

(Continued)

TABLE 6.3 (*Continued*)

Methods	Parameters				References	
	Volume V_{ss}	Individual Tissue K_p Partitioning			Original Articles	Comparative Assessments
		Healthy Tissue	Tumor Tissue	Cellular or Subcellular		
QSAR/*in silico* models	Yes	Yes	No	No	Okezaki et al. (1988), Yokogawa et al. (1990), Abraham et al. (1994, 2014), Kaliszan and Markuszewski (1996), Lombardo et al. (1996, 2004, 2013a), Blakey et al. (1997), Nestorov et al. (1998), Luco (1999), Zhang and Zhang (2006), Berellini et al. (2009), Abraham (2014), Poulin and Krishnan (1996b, 1999), , Fauchécourt et al. (2001), Price and Krishnan (2011), de Jongh et al. (1997), Katritzky et al. (2005), Sprunger et al. (2008), and Zhivkova and Doytchinova (2011)	Stepensky (2013), Grime et al. (2013), Khojasteh et al. (2011), Jones et al. (2011b), Fagerholm (2007), Graham et al. (2012), Zou et al. (2012), Payne and Kenny (2002), and Obach (2007)
Disease state models	Yes	No	No	No	Ritschel and Kaul (1986), Berezhkovskiy (2011), and Eleveld et al. (2011)	—

the outcomes of this initiative (Jones et al., 2011a, 2013; Poulin et al., 2011a; Ring et al., 2011; Vuppugalla et al., 2011; Poulin et al., 2011c).

6.3.1   PhRMA Initiative on the Prediction of CL

From a clinical study perspective, under-prediction of CL would not lead to safety concerns because actual exposures would be less than expected, and inversely. A limited data set of 19 drugs had IV clinical PK data and were used in the analysis (Ring et al., 2011). *In vivo* and *in vitro* preclinical data were used to predict IV CL in humans by 29 different methods. In addition, 66 methods of predicting oral AUC were evaluated for 107 compounds using rational combinations of IV CL and bioavailability (F%) and direct scaling of observed oral CL from preclinical species. Across methods, the maximum success rate in predicting human IV CL for the 19 drugs was 100, 94, and 78% of the compounds with predictions falling within 10-fold, 3-fold, and 2-fold error, respectively, of the observed CL. In general, *in vivo* prediction methods performed slightly better than *in vitro*-based methods (at least in terms of measures of correlation and global concordance). The prediction accuracy of oral AUC was much lower; therefore, the greatest challenge to successful prediction of human oral CL is the estimate of F% in human. In this case, the oral AUC (CL) was particularly under-estimated. However, for many compounds, *in vivo* data from only two species (generally rat and dog) were available and/or the required *in vitro* data were missing, which meant some methods could not be properly evaluated. Another comprehensive assessment of human PK prediction challenged the use of primate data, and, hence, the methods employing monkey CL data yielded the best overall predictions compared to other single-species scaling methods (Lombardo et al., 2013a, b). It is recommended to initially use the IVIVE methods to predict CL in preclinical species and humans, putting the assumptions and compound properties into context. As *in vivo* data become available, these predictions should be reassessed and rationalized to indicate the level of confidence (uncertainty) in the human CL prediction. Nevertheless, recent developments in CL predictions particularly based upon *in vitro* data should improve the prediction accuracy in future (Paine et al., 2008; Haddad et al., 2010; Poulin, 2013). In addition, the commercial softwares of PBPK models constantly improve the prediction of CL for drugs (Table 6.2). For more information, additional comparative assessments of CL prediction methods were published in the literature (Table 6.1).

6.3.2   PhRMA Initiative on the Prediction of Volume of Distribution

From a clinical study perspective, under-prediction of V_{ss} (tissue distribution) would lead to safety concerns because actual tissue exposures would be higher than expected, and inversely. The same data set of 19 drugs was used in the analysis for V_{ss} (Jones et al., 2011b). *In vivo* and *in vitro* preclinical data were used to predict IV V_{ss} in humans by 24 different methods. Across methods, the maximum success rate in predicting human V_{ss} was 100, 94, and 78% of the compounds with predictions falling within 10-fold, 3-fold, and 2-fold error, respectively, of the observed V_{ss}. The human V_{ss} was either under- or over-predicted. Generally, the methods that made use of *in vivo* preclinical data were more predictive than those methods that relied solely on *in vitro* data. However, as mentioned for the prediction of CL, for many compounds, *in vivo* data from only two species (generally rat and dog) were available and/or the required *in vitro* data were missing, which meant some methods could not be properly evaluated. Again, it is recommended to initially use the *in vitro* tissue composition-based equations to predict V_{ss} in preclinical species and humans, putting the assumptions and compound properties into context. As *in vivo* data become available, these predictions should be reassessed and rationalized to indicate the level of confidence (uncertainty) in the human V_{ss} prediction. In addition, recent developments in the prediction of V_{ss} should improve the prediction accuracy in future (Table 6.3), and the commercial softwares of PBPK models constantly improve the prediction of V_{ss} for drugs (Table 6.2). Again, for more information, additional comparative assessments of V_{ss} prediction methods were published in the literature (Tables 6.1 and 6.3).

6.3.3   PhRMA Initiative on the Prediction of Concentration–Time Profile

A direct comparison of the results between PBPK model and the Wajima allometric-based method has been made for the same data sets and simulation scenarios (Poulin et al., 2011b; Vuppugalla et al., 2011). PBPK modeling fared no better than allometry. The results are indicative of a general under-prediction of drug exposure (AUC, C_{max}) over the time course simulated in human using the Wajima and PBPK approaches, which is more relevant after oral than IV administration. In other words, both approaches achieved similar levels of accuracy for the PhRMA IV data set, with respect to their effectiveness at simulating plasma concentration–time profiles in humans. Adding the absorption parameters in these two approaches, the prediction of oral PK in human resulted to a significant decrease in the prediction performance compared with IV PK, which again confirmed the greater difficulty of predicting drug absorption with each approach primarily due to an under-prediction of oral bioavailability (and, hence, of AUC). From a clinical study perspective, under-prediction of human exposure (AUC) would lead to safety concerns because actual exposures would be greater than expected, and inversely. There are likely to be many reasons for this failure, including the complex physiology of the gastrointestinal tract coupled with the complex processes occurring during absorption, especially following administration of sparingly soluble compounds, with markedly different formulations often used in preclinical development than tested in humans. Thus, it would be expected that predictions should be better particularly for class 1 compounds, highly soluble–highly permeable compounds, as was found and reported in a companion paper (Poulin et al., 2011b). In other words, attempts to predict the concentration–time profile were generally poor probably because of the physicochemical properties of the PhRMA data sets (lot of poorly soluble and poorly permeable compounds) (Wagner et al., 2012; Poulin et al., 2011a, c) (Poulin et al., 2011a, b). Nonetheless, recent PBPK models considering the impact of drug solubility effect would improve the prediction accuracy for such compounds. Again, commercial softwares of PBPK models continually improve the prediction drug PK, and for more information, additional comparative assessments of PBPK models were published in the literature (Table 6.1).

6.3.4 Lead Commentaries on the PhRMA Initiative

So where are things likely to go from here? Despite its limitations, allometry is likely to remain a maintained approach by industry for prediction of human PK in candidate selection for some years yet, although the benefit of using nonhuman primates is questionable. However, as our understanding of processes controlling PK improves, and the *in vitro* human biologic systems become increasingly more predictive of *in vivo* events, we will see the increasing adoption of the more mechanistic PBPK as the first line approach (Rowland and Benet, 2011).

Wajima allometry offers a genuine alternative to PBPK and in particular offers a tool to simulate plasma concentration–time profiles early in the drug discovery process when *in vivo* PK profile data in one or more preclinical species are available (Vuppugalla et al., 2011). However, mechanistic questions cannot be answered with the empirical Wajima allometry, which is a main disadvantage compared to the more physiological framework of PBPK model. That said, attempts to predict the concentration–time profile allometrically were generally poor interestingly; when predicting human concentration–time profiles, no one preclinical species was not found to be superior to the others nor were multiple species data shown to be superior to data from one species available (Vuppugalla et al., 2011). However, considering the poor predictability of bioavailability, one must recognize that, especially for the many poorly soluble drug candidates, the later-stage candidate selection studies in large animals almost always involve formulations that are closer to that used in first-in-human studies than used in the early-stage discovery rodent studies. In this specific case, the Wajima allometry method showed some advantage compared to the PBPK

model for the PhRMA data set of drugs available (Poulin et al., 2011b; Vuppugalla et al., 2011).

Physiologically based PK was evaluated using a generic model. Perhaps surprisingly to some, on average, this methodology fared no better than allometry in predicting human PK and sometimes proved inferior (Poulin et al., 2011b; Vuppugalla et al., 2011).

However, this needs to be put in perspective. Unlike allometry, which fundamentally can progress no further, PBPK, which is still in its infancy, is highly mechanistic, and current failures help highlight a lack of understanding of particular processes. Other advantages of this bottom-up approach are that it explains the observed *in vivo* behavior of drugs; requires minimal resources, including being animal sparing; and can be employed upstream, by linking processes to physicochemical and structure properties to help guide the drug discovery teams in the better design of compounds. In addition, unlike allometry (with the possible exception of its use in pediatrics), PBPK models extend beyond dose selection in first-in-human studies, by helping to guide various aspects of clinical development, such as prediction of the impact of drug–drug interactions, age, disease, and so forth on PK, design of experiments, as well as subsequent therapeutic use of compounds (Parrott et al., 2005; Fenneteau et al., 2009; Poulin et al., 2011b; Rowland et al., 2011; Jones et al., 2012a, 2013; Wagner et al., 2012; Hudachek and Gustafson, 2013; Chetty et al., 2014; Thiel et al., 2014). In the area of PBPK, a generic model was employed (Poulin et al., 2011a), which while very helpful in gaining some insights does not incorporate all the new features found in commercial PBPK software and the literature (Paine et al., 2008; Fenneteau et al., 2009; Haddad et al., 2010; Jones et al., 2012a; Wagner et al., 2012; Hudachek and Gustafson, 2013; Poulin, 2013; Chetty et al., 2014; Thiel et al., 2014) (see also Tables 6.1 and 6.3), which if employed may affect the success rate of this methodology. The quality of input data is also essential.

In another lead commentary, the influence of the compound selection process on the performance of human CL prediction methods used in the PhRMA initiative has also been discussed (Wong et al., 2012b). In this commentary, the influence of the compound selection process was examined by performing a probability analysis and examining the CL properties of compounds that are selected using an idealized drug discovery screening process focused on PK optimization. The results of the analysis suggest that the selection of screening species can influence the performance of various predictive models of human CL. The compounds whose CL properties are discordant between preclinical species used in the selection process and humans are more likely to be "screened out" by the compound selection process. This presents a problem since compounds that have low CL in humans but have high CL in the preclinical species would be eliminated. Elimination of these types of compounds from clinical evaluation will inadvertently improve the performance of the prediction methods (Wong et al., 2012a).

REFERENCES

Abraham MH. 2014. A simple method for estimating in vitro air-tissue and in vivo blood-tissue partition coefficients. *Chemosphere* 120C:188–191.

Abraham MH, Chadha HS, Mitchell TC. 1994. Hydrogen bonding. 3. Factors that influence the distribution of solutes between blood and brain. *J Pharm Sci* 83:1257–1268.

Abraham MH, Gola JMR, Ibrahim A, Acree WE, Liu X. 2014. The prediction of blood–tissue partitions, water–skin partitions and skin permeation for agrochemicals. *Pest Manag Sci* 70:1130–1137.

Arundel P. 1997. A multi-compartment model generally applicable to physiologically-based pharmacokinetics. 3rd IFAC Symposium: Modelling and control in biomedical systems, Warwick, UK, March 23–26, 1997.

Bachmann K, Pardoe D, White D. 1996. Scaling basic toxicokinetic parameters from rat to man. *Environ Health Perspect* 104:400–407.

Ballard P, Rowland M. 2003. Prediction of in vivo tissue distribution from in vitro data. 2. Influence of albumin diffusion from tissue pieces during an in vitro incubation on estimated tissue-to-unbound plasma partition coefficients (Kpu). *Pharm Res* 20:857–861.

Ballard P, Leahy DE, Rowland M. 2000. Prediction of in vivo tissue distribution from in vitro data 1. Experiments with markers of aqueous spaces. *Pharm Res* 17:660–663.

Berellini G, Springer C, Waters NJ, Lombardo F. 2009. In silico prediction of volume of distribution in human using linear and non linear models on a 669 compound dataset. *J Med Chem* 52:4488–4495.

Berezhkovskiy LM. 2004. Volume of distribution at steady state for a linear pharmacokinetic system with peripheral elimination. *J Pharm Sci* 93:1628–1640.

Berezhkovskiy LM. 2011. On the accuracy of estimation of basic pharmacokinetic parameters by the traditional noncompartmental equations and the prediction of the steady-state volume of distribution in obese patients based upon data derived from normal subjects. *J Pharm Sci* 100:2482–2497.

Berry LM, Roberts J, Be X, Zhao Z, Lin MH. 2010. Prediction of V_{ss} from in vitro tissue-binding studies. *Drug Metab Dispos* 38:115–121.

Berry LM, Li C, Zhao Z. 2011. Species differences in distribution and prediction of human V_{ss} from preclinical data. *Drug Metab Dispos* 39:2103–2116.

Björkman S. 2002. Prediction of the volume of distribution of a drug: which tissue: plasma partition coefficients are needed? *J Pharm Pharmacol* 54:1237–1245.

Blakey GE, Nestorv IA, Arundel PA, Aarons LJ, Rowland M. 1997. Quantitative structure-pharmacokinetics relationships: I. Development of a whole-body physiologically based model to characterize changes in pharmacokinetics across a homologous series of barbiturates in the rat. *J Pharmacokinet Biopharm* 25:277–312.

Boxenbaum H. 1982. Interspecies scaling, allometry, physiological time, and the ground plan of pharmacokinetics. *J Pharmacokinet Biopharm* 10:201–227.

Chen Y, Jin JY, Mukadam S, Malhi V, Kenny JR. 2012. Application of IVIVE and PBPK modeling in prospective prediction of clinical pharmacokinetics: strategy and approach during the drug discovery phase with four case studies. *Biopharm Drug Dispos* 33:85–98.

Chetty M, Rose RH, Abduljalil K, Patel N, Lu G, Cain T, Jamei M, Rostami-Hodjegan A. 2014. Applications of linking PBPK and PD models to predict the impact of genotypic variability, formulation differences, differences in target binding capacity and target site drug concentrations on drug responses and variability. *Front Pharmacol* 5:258.

Clausen J, Bickel MH. 1993. Prediction of drug distribution in distribution dialysis and in vivo from binding to tissues and blood. *J Pharm Sci* 82:345–349.

Clewell HJ, Tan YM, Campbell JL, Andersen ME. 2008. Quantitative interpretation of human biomonitoring data. *Toxicol Appl Pharmacol* 231:122–133.

Cristofoletti R, Dressman J. 2014. Use of physiologically based pharmacokinetic models coupled with pharmacodynamic models to assess the clinical relevance of current bioequivalence criteria for generic drug products containing ibuprofen. *J Pharm Sci* 103:3263–3275.

De Buck SS, Sinha VK, Fenu LA, Nijsen MJ, Mackie CE, Gilissen RA. 2007. Prediction of human pharmacokinetics using physiologically based modeling: a retrospective analysis of 26 clinically tested drugs. *Drug Metab Dispos* 35:1766–1780.

De Jongh J, Verhaar HJM, Hermes JLM. 1997. A quantitative property relationship (QSPR) approach to estimate in vivo tissue: blood partition coefficients of organic chemicals in rats and humans. *Arch Toxicol* 72:17–25.

Diaz D, Ford KA, Hartley DP, Harstad EB, Cain GR, Achiles-Poon K, Nguyen T, Peng J, Zheng Z, Merchant M, Sutherlin DP, Gaudino JJ, Kaus R, Lewin-Koh SC, Choo EF, Liederer BM, Dambach DM. 2013. Pharmacokinetic drivers of toxicity for basic molecules: strategy to lower pK_a results in decreased tissue exposure and toxicity for a small molecule Met inhibitor. *Toxicol Appl Pharmacol* 266:86–94.

Edginton AN, Yun YE. 2013. Correlation-based prediction of tissue-to-plasma partition coefficients using readily available input parameters. *Xenobiotica* 43:839–852.

Eleveld DJ, Proost JH, Absalom AR, Struys MRF. 2011. Obesity and allometric scaling of pharmacokinetics. *Clin Pharmacokinet* 50:751–753.

Fagerholm U. 2007. Prediction of human pharmacokinetics: evaluation of methods for prediction of volume of distribution. *J Pharm Pharmacol* 59:1181–1190.

Fauchécourt MO, Béliveau M, Krishnan K. 2001. Quantitative structure-pharmacokinetic relationship modelling. *Sci Total Environ* 274:125–135.

Fenneteau F, Poulin P, Nekka F. 2009. Physiologically based predictions of the impact of inhibition of intestinal and hepatic metabolism on human pharmacokinetics of CYP3A substrates. *J Pharm Sci* 99:486–514.

Fura A, Vyas V, Humphreys W, Chimalokonda A, Rodrigues D. 2008. Prediction of human oral pharmacokinetics using non-clinical data: examples involving four proprietary compounds. *Biopharm Drug Dispos* 29:455–468.

Gertz M, Kilford PJ, Houston B, Galetin A. 2008. Drug lipophilicity and microsomal protein concentration as determinants in the prediction of the fraction unbound in microsomals incubations. *Drug Metab Dispos* 36:535–542.

Graham H, Walker M, Jones O, Yated J, Galetin A, Aarons L. 2012. Comparison of in vivo and in silico methods used for prediction of tissue: plasma partition coefficients in rat. *J Pharm Pharmacol* 64:383–396.

Grime KH, Barton P, McGinnity DF. 2013. Application of in silico, in vitro and preclinical pharmacokinetic data for the effective and efficient prediction of human pharmacokinetics. *Mol Pharm* 10:1191–1206.

Guohua A, Morrie ME. 2012. A physiologically based pharmacokinetic model of mitoxantrone in mice and scale-up to humans: a semi-mechanistic model on corporating DNA and protein binding. *AAPS J* 14:352–354.

Haddad S, Poulin P, Krishnan K. 2000. Relative lipid content as the sole mechanistic determinant of the adipose tissue: blood partition coefficients of highly lipophilic organic chemicals. *Chemosphere* 40:839–843.

Haddad S, Poulin P, Funk C. 2010. Extrapolating in vitro metabolic interactions to isolated perfused liver: prediction of metabolic interactions between bufarolol, bunitrolol and debrisoquine. *J Pharm Sci* 99:4406–4426.

Halifax D, Houston BJ. 2006. Binding of drugs to hepatic microsomes: comments and assessment of current prediction methodology with recommendation for improvement. *Drug Metab Rev* 34:724–726.

Hall C, Lueshen E, Mošat A, Linninger AA. 2012. Interspecies scaling in pharmacokinetics: a novel whole-body physiologically based modeling framework to discover drug biodistribution mechanisms in vivo. *J Pharm Sci* 101:1221–1241.

Hanumegowda UM, Wenke G, Regueiro-Ren A, Yordanova R, Corradi JP, Adams SP. 2010. Phospholipidosis as a function of basicity, lipophilicity, and volume of distribution. *Chem Res Toxicol* 23:749–755.

Harashima H, Sugiyama Y, Sawada Y, Iga T, Hanano M. 1984. Comparison between in-vivo and in-vitro tissue-to-plasma unbound concentration ratios (Kp,f) of quinidine in rats. *J Pharm Pharmacol* 36:340–342.

Haritova AM, Fink-Gremmels J. 2010. A simulation model for the prediction of tissue: plasma partition coefficients for drug residues in natural casings. *Vet J* 185:278–284.

Harvey W, Choo EF, Alicke B, et al. 2012. Antitumor activity of targeted and cytotoxic agents in murine subcutaneous tumor models correlates with clinical response. *Clin Cancer Res* 18:3846–3852.

Howell BA, Yang Y, Kumar R, Woodhead JL, Harrill AH, Clewell HJ III, Andersen ME, Siler SQ, Watkins PB. 2012. In vitro to in vivo extrapolation and species response comparisons for drug-induced liver injury (DILI) using DILIsym™: a mechanistic, mathematical model of DILI. *J Pharmacokinet Pharmacodyn* 39:527–541.

Hudachek SF, Gustafson DL. 2013. Incorporation of ABCB1-mediated transport into a physiologically-based pharmacokinetic model of docetaxel in mice. *J Pharmacokinet Pharmacodyn* 40:437–449.

Jansson R, Bredberg U, Ashton M. 2008. Prediction of drug tissue to plasma concentration ratios using a measured volume of distribution in combination with lipophilicity. *J Pharm Sci* 97:2324–2339.

Jolivette LJ, Ward KW. 2005. Extrapolation of human pharmacokinetic parameters from rat, dog and monkey data: molecular properties associated with extrapolative success or failure. *J Pharm Sci* 94:1467–1483.

Jones HM, Parrott N, Jorga K, Lave T. 2006. A novel strategy for physiologically based predictions of human pharmacokinetics. *Clin Pharmacokinet* 45:511–542.

Jones RDO, Jones HM, Rowland M, Gibson CR, Yates JWT, Chien JY, Ring BJ, Adkison KK, Ku MS, He H, Vuppugalla R, Marathe P, Fischer V, Dutta S, Sinha VK, Björnsson T, Lavé T, Poulin P. 2011a. PhRMA CPCDC initiative on predictive models of human pharmacokinetics. 2. Comparative assessment of prediction methods of human volume of distribution. *J Pharm Sci* 100:4074–4089.

Jones HM, Gardner IB, Collard WT, Stanley PJ, Oxley P, Hosea NA, Plowchalk D, Gernhardt S, Lin J, Dickins M, Rahavendran SR, Jones BC, Watson KJ, Pertinez H, Kumar V, Cole S. 2011b. Simulation of human intravenous and oral pharmacokinetics of 21 diverse compounds using physiologically based pharmacokinetic modelling. *Clin Pharmacokinet* 50:331–347.

Jones HM, Barton HA, Lai Y, Bi YA, Kimoto E, Kempshall S, Tate SC, El-Kattan A, Houston JB, Galetin A, Fenner KS. 2012a. Mechanistic pharmacokinetic modeling for the prediction of transporter-mediated disposition in humans from sandwich culture human hepatocyte data. *Drug Metab Dispos* 40:1007–1017.

Jones HM, Dickins M, Youdim K, Gosset JR, Attkins NJ, Hay TL, Gurrell IK, Logan YR, Bungay PJ, Jones BC, Gardner IB. 2012b. Application of PBPK modelling in drug discovery and development at Pfizer. *Xenobiotica* 42:94–106.

Jones HM, Mayawala K, Poulin P. 2013. Dose selection based on physiologically-based pharmacokinetic (PBPK) modeling approaches. *AAPS J* 15:377–387.

Kaliszan R, Markuszewski M. 1996. Brain/blood distribution described by a combination of partition coefficient and molecular mass. *Int J Pharm* 145:9–16.

Kamgang E, Peyret T, Krishnan K. 2008. An integrated QSPR-PBPK modelling approach for in vitro-in vivo extrapolation of pharmacokinetics in rats. *SAR QSAR Environ Res* 19:669–680.

Katritzky AR, Minati K, Fara DC, Karelson M, William AE, Vitaly P, Solov E, Varnek A. 2005. QSAR modeling of blood:air and tissue:air partition coefficients using theoretical descriptors. *Bioorg Med Chem* 13:6450–6463.

Khalafallah N, Jusko WJ. 1984. Determination and prediction of tissue binding of prednisolone in the rabbit. *J Pharm Sci* 73:362–366.

Khojasteh CS, Wong H, Hop ECA. 2011. ADME properties and their dependence on physicochemical properties. In: *Drug metabolism and pharmacokinetics quick guide.* Springer, New York, pp. 165–181.

Kleist MV, Huisinga H. 2007. Physiologically-based pharmacokinetic modeling: a sub compartmentalized model of tissue distribution. *J Pharmacokinet Pharmacodyn* 34:789–906.

Lin JH, Sugiyama Y, Awazu S, Hanano M. 1982. In vitro and in vivo evaluation of the tissue-to-blood partition coefficient for physiological pharmacokinetic models. *J Pharmacokinet Biopharm* 10:637–647.

Lombardo F, Blake JF, Curatolo WJ. 1996. Computation of brain-blood partitioning of organic solutes via free energy calculations. *J Med Chem* 39:4750–4755.

Lombardo F, Obach RS, Shalaeva MY, Gao F. 2004. Prediction of human volume of distribution values for neutral and basic drugs. 2. Extended data set and leave-class-out statistics. *J Med Chem* 47:1242–1250.

Lombardo F, Waters NJ, Argikar UA, Dennehy MK, Gunduz M, Harriman SP, Berellini G, Rajlic IR, Obach RS. 2013a. Comprehensive assessment of human pharmacokinetic prediction based on in vivo animal pharmacokinetic data, Part 1: volume of distribution at steady state. *J Clin Pharmacol* 53:166–177.

Lombardo F, Waters NJ, Argikar UA, Dennehy MK, Zhan J, Gunduz M, Harriman SP, Berellini G, Liric Rajlic I, Obach RS. 2013b. Comprehensive assessment of human pharmacokinetic prediction based on in vivo animal pharmacokinetic data, part 2: clearance. *J Clin Pharmacol* 53:178–191.

Luco JM. 1999. Prediction of the brain-blood distribution of a large set of drugs from structurally derived descriptors using partial least-squares (PLS) modeling. *J Chem Inf Comput Sci* 39:396–404.

Luttringer O, Theil FP, Poulin P, Schmitt-Hoffmann AH, Guentert TW, Lave T. 2003. Physiologically based pharmacokinetic (PBPK) modeling of disposition of epiroprim in humans. *J Pharm Sci* 92:1990–2007.

Mahmood I. 2010. Theoretical versus empirical allometry: facts behind theories and application to pharmacokinetics. *J Pharm Sci* 99:2927–2933.

Mandula H, Parepally JMR, Feng R, Smith QR. 2006. Role of site specific binding to plasma albumin in drug availability to brain. *J Pharmacol Exp Ther* 317:667–675.

Mariappan TT, Mandlekar S, Marathe P. 2013. Insight into tissue unbound concentration: utility in drug discovery and development. *Curr Drug Metab* 14:324–340.

Murphy JE, Janszen DB, Gargas ML. 1995. An in vitro method for determination of tissue partition coefficients of non-volatile chemicals such as 2,3,7,8-tetrachlorodibenzo-p-dioxin and estradiol. *J Appl Toxicol* 15:147–152.

Nestorov I, Aarons l, Rowland M. 1998. Quantitative structure-pharmacokinetics relationships: II. A mechanistically based model to evaluate the relationship between tissue distribution parameters and compound lipophilicity. *J Pharmacokinet Biopharm* 26:521–545.

Obach RS. 2007. Prediction of human volume of distribution using in vivo, in vitro, and in silico approaches. *Annu Rep Med Chem* 42:469–488.

Obach RS, Baxter JG, Liston TE, Silber BM, Jones BC, MacIntyre F, Rance DJ, Wastall P. 1997. The prediction of human pharmacokinetic parameters from preclinical and *in vitro* metabolism data. *J Pharmacol Exp Ther* 283:46–58.

Øie S, Tozer TN. 1979. Effect of altered plasma protein binding on apparent volume of distribution. *J Pharm Sci* 66:1203–1205.

Okezaki E, Terasaki T, Nakamura M, Nagata O, Kato H, Tsuji A. 1988. Structure-tissue distribution relationship based on physiological pharmacokinetics for NY-198, a new antimicrobial agent, and the related pyridonecarboxylic acids. *Drug Metab Dispos* 16:865–874.

Paine SW, Parker AJ, Gardiner P, Webborn PJH, Riley RJ. 2008. Prediction of the pharmacokinetics of Atorvastatin, Cerivastatin, and Indomethacin using kinetic models applied to isolated rat hepatocytes. *Drug Metab Dispos* 36:1365–1374.

Parrott N, Paquereau N, Coassolo P, Lavé T. 2005. An evaluation of the utility of physiologically based models of pharmacokinetics in early drug discovery. *J Pharm Sci* 94:2327–2343.

Paterson S, Mackay D. 1989. Correlation of tissues, blood, and air partition coefficients of volatile organic chemicals. *Br J Ind Med* 46:321–328.

Payne PP, Kenny LC. 2002. Comparison of models for the estimation of biological partition coefficients. *J Toxic Environ Health A* 65:897–931.

Péry ARR, Brochot C, Zeman FA, Mombelli E, Desmots S, Pavan M, Fioravanzo E, Zaldívar JM. 2013. Prediction of dose-hepatotoxic response in humans based on toxicokinetic/toxicodynamic modeling with or without in vivo data: A case study with acetaminophen. *Toxicol Lett* 220:26–34.

Peters SA. 2008. Evaluation of a generic physiologically based pharmacokinetic model for lineshape analysis. *Clin Phamacokinet* 47:261–275.

Peyret T, Poulin P, Krishnan K. 2010. A unified algorithm for predicting partition coefficients for PBPK modeling of drugs and environmental chemicals. *Toxicol Appl Pharmacol* 249:197–207.

Poulin P. 2013. Prediction of total hepatic clearance by combining metabolism, transport, and permeability data in the in vitro–in vivo extrapolation methods: emphasis on an apparent fraction unbound in liver for drugs. *J Pharm Sci* 102:2085–2095.

Poulin P. 2015a. Drug distribution to human tissues: prediction and examination of the basic assumption in in vivo pharmacokinetics-pharmacodynamics (PK/PD) research. *J Pharm Sci* 104:2110–2118.

Poulin P. 2015b. Albumin and uptake of drugs in cells: additional validation exercises of a recently published equation that quantifies the albumin-facilitated uptake mechanism(s) in physiologically based pharmacokinetic and pharmacodynamic modeling research. *J Pharm Sci* 104. doi:10.1002/jps.24676.

Poulin P. 2015c. A paradigm shift in pharmacokinetics-pharmacodynamics (PK/PD) modeling: rule of thumb for an estimation of the free drug level in tissue compared to plasma to guide drug design. *J Pharm Sci* 104:2359–2368.

Poulin P, Haddad S. 2011. Microsome composition-based model as a mechanistic tool to predict non-specific binding in liver microsomes. *J Pharm Sci* 100:4501–4517.

Poulin P, Haddad S. 2012. Advancing prediction of tissue distribution and volume of distribution of highly lipophilic compounds from a simplified tissue composition-based model as a mechanistic animal alternative method. *J Pharm Sci* 101:2250–2261.

Poulin P, Haddad S. 2013. Hepatocyte composition-based model as a mechanistic tool for predicting the cell: medium partition coefficients of drugs in incubation mediums. *J Pharm Sci* 102:2806–2818.

Poulin P, Krishnan K. 1996a. A mechanistic algorithm for predicting blood: air partition coefficients of organic chemicals with the consideration of reversible binding in hemoglobin. *Toxicol Appl Pharmacol* 136:131–137.

Poulin P, Krishnan K. 1996b. Molecular structure-based prediction of the partition coefficients of organic chemicals for physiologically-based pharmacokinetic models. *Toxicol Method* 6:117–137.

Poulin P, Krishnan K. 1999. Molecular structure-based prediction of the toxicokinetics of inhaled vapors in humans. *Int J Toxicol* 18:7–18.

Poulin P, Theil FP. 2000. A priori prediction of tissue: plasma partition coefficients of drugs to facilitate the use of physiologically based pharmacokinetic models in drug discovery. *J Pharm Sci* 89:16–35.

Poulin P, Theil FP. 2002. Prediction of pharmacokinetics prior to *in vivo* studies. 1. Mechanism-based prediction of volume of distribution. *J Pharm Sci* 91:129–156.

Poulin, P, Theil FP. 2009. Development of a novel method for predicting human volume of distribution at steady-state of basic drugs and comparative assessment with existing methods. *J Pharm Sci* 98:4941–4961.

Poulin P, Béliveau M, Krishnan K. 1999. Mechanistic animal-replacement approaches for predicting pharmacokinetics of organic chemicals. In: *Toxicity assessments alternatives: methods, issues, opportunities*. Eds. H. Salem and S.A. Katz. Humana Press, Inc., Totowa, NJ, pp. 115–139.

Poulin P, Jones RDO, Jones HM, Gibson CR, Rowland M, Chien JY, Ring BJ, Adkison KK, Ku MS, He H, Vuppugalla R, Marathe P, Fischer V, Dutta S, Sinha VK, Bjornsson T, Lavé T, Yates JWT. 2011a. PhRMA CPCDC initiative on predictive models of human pharmacokinetics. 5. Prediction of plasma concentration-time profiles in human by using the physiologically-based pharmacokinetic (PBPK) approach. *J Pharm Sci* 100:4127–4157.

Poulin P, Ekin, S, Theil FP. 2011b. A hybrid approach to advancing quantitative prediction of tissue distribution of basic drugs in human. *Toxicol Appl Pharmacol* 250:194–212.

Poulin P, Jones RDO, Jones HM, Gibson CR, Rowland M, Chien JY, Ring BJ, Adkison KK, Ku MS, He H, Vuppugalla R, Marathe P, Fischer V, Dutta S, Sinha VK, Bjornsson T, Lavé T, Yates JWT, Ku MS. 2011c. PhRMA initiative on predictive models of human pharmacokinetics. 1. Goals, properties of the datasets and comparison with literature datasets. *J Pharm Sci* 100:4050–4073.

Poulin P, Hop ECA, Salphalti L, Liederer B. 2013a. Correlation of tissue: plasma partition coefficient between healthy tissues and subcutaneous xenografts of human tumor cell lines in mouse as a prediction tool of drug penetration in tumors. *J Pharm Sci* 102:1355–1369.

Poulin P, Dambach DM, Hartley DH, Ford K, Theil FP, Harstad E, Halladay J, Choo E, Boggs J, Liederer BM, Dean B, Diaz D. 2013b. An algorithm for evaluating potential tissue drug distribution in toxicology studies from readily available pharmacokinetic parameters. *J Pharm Sci* 102:3816–3829.

Poulin P, Chen YH, Ding X, Gould SE, Hop ECJ, Messick K, Oeh J, Liederer B. 2015a. Prediction of drug distribution in subcutaneous xenografts of human tumor cell lines and healthy tissues in mouse: application of the tissue composition-based model to antineoplastic agents. *J Pharm Sci* 104:1508–1521.

Poulin P, Burczynski FJ, Haddad S. 2015b. The role of extracellular binding proteins in the cellular uptake of drugs: impact on quantitative in vitro-to-in vivo extrapolations of toxicity and efficacy in physiologically based pharmacokinetic-pharmacodynamic research. *J. Pharm. Sci.* 104. doi: 10.1002/jps.24571.

Price K, Krishnan K. 2011. An integrated QSAR-PBPK modelling approach for predicting the inhalation toxicokinetics of mixtures of volatile organic chemicals in the rat. *SAR QSAR Environ Res* 22:107–128.

Richter W, Starke V, Whitby B. 2006. The distribution pattern of radioactivity across different tissues in quantitative whole-body autoradiography (QWBA) studies. *Eur J Pharm Sci* 28:155–165.

Ring BJ, Jones HM, Rowland M, Gibson CR, Yates JWT, Chien JY, Adkison KK, Ku MS, He H, Vuppugalla R, Jones RDO, Marathe P, Fischer V, Dutta S, Sinha VK, Björnsson T, Lavé T, Poulin P. 2011. PhRMA initiative on predictive models of human pharmacokinetics. 3. Comparative assessment of prediction methods of clearance. *J Pharm Sci* 100:4090–4110.

Ritschel WA, Hammer GV. 1980. Prediction of the volume of distribution from in vitro data and use for estimating the absolute extent of absorption. *Int J Clin Pharmacol Ther Toxicol* 18:298–316.

Ritschel WA, Kaul S. 1986. Prediction of apparent volume of distribution in obesity. *Methods Find Exp Clin Pharmacol* 8:239–247.

Rodgers T, Rowland M. 2006. Physiologically-based pharmacokinetics modeling 2: predicting the tissue distribution of acids, very weak bases, neutrals and zwitterions. *J Pharm Sci* 95: 1238–1252.

Rodgers T, Rowland M. 2007. Mechanistic approaches to volume of distribution predictions: understanding the processes. *Pharm Res* 24:918–933.

Rodgers T, Jones HM, Rowland M. 2012. Tissue lipids and drug distribution: dog versus rat. *J Pharm Sci* 101:4615–4625.

Rowland M, Benet LZ. 2011. Lead PK commentary: predicting human pharmacokinetics. *J Pharm Sci* 100:4047–4049.

Rowland M, Peck C, Tucker G. 2011. Physiologically based pharmacokinetic in drug development and regulatory science. *Annu Rev Pharmacol Toxicol* 51:45–73.

Ruark CD, Hack CE, Robinson PJ, Mahle DA, Gearhart JM. 2014. Predicting passive and active tissue: plasma partition coefficients: interindividual and interspecies variability. *J Pharm Sci* 103:2189–2198.

Sawada Y, Hanano M, Sugiyama Y, Harashima H, Iga T. 1984. Prediction of the volumes of distribution of basic drugs in humans based on data from animals. *J Pharmacokinet Biopharm* 12:587–596.

Schmitt, W. 2008. General approach for the calculation of tissue to plasma partition coefficients. *Toxicol In Vitro* 22:457–467.

Schuhmann G, Fichtl B, Kurz H. 1987. Prediction of drug distribution in vivo on the basis of in vitro binding data. *Biopharm Drug Dispos* 8:73–86.

Small H, Gardner I, Jones HM, Davis J, Rowland M. 2001. Measurement of binding of basic drugs to acidic phospholipids using surface plasmon resonance and incorporation of the data into mechanistic tissue composition equations to predict steady-state volume of distribution. *Drug Metab Dispos* 39:1789–1793.

Smith DA, Di L, Kerns EH. 2010. The effect of plasma protein binding on in vivo efficacy: misconceptions in drug discovery. *Nat Rev* 9:929–939.

Sprunger LM, Gibbs J, Acree WE, Abraham MH. 2008. Correlation of human and animal air-to-blood partition coefficients with a single linear free energy relationship model. *Mol Inf* 27:1130–1139.

Stepensky D. 2013. Prediction of drug disposition on the basis of its chemical structure. *Clin Pharmacokinet* 52:415–431.

Sui X, Sun J, Li H, Wang Y, Liu J, Liu X, Zhang W, Chen L, He Z. 2009. Prediction of volume of distribution values in human using immobilized artificial membrane partitioning coefficients, the fraction of compound ionized and plasma protein binding. *Eur J Med Chem* 44:4455–4460.

Terasaki T, Iga T, Sugiyama Y, Hanano M. 1984. Pharmacokinetic study on the mechanism of tissue distribution of doxorubicin: inter-organ and interspecies variation of tissue-to-plasma partition coefficients in rats, rabbits, and guinea pigs. *J Pharm Sci* 73:1359–1563.

Thiel C, Schneckener S, Krauss M, Ghallab A, Hofmann U, Kanacher T, Zellmer S, Gebhardt R, Hengstler JG, Kuepfer L. 2014. A systematic evaluation of the use of physiologically based pharmacokinetic modeling for cross-species extrapolation. *J Pharm Sci* 104:191–206.

Timchalk C, Hinderliter PM, Poet TS. 2010. Modeling cholinesterase inhibition. In: *Quantitative modeling in toxicology*. Eds. K. Krishnan and ME Andersen. John Wiley & Sons, Ltd, Chichester, pp. 122–149.

Turner DB, Rostami-Hodjegan A, Tucker GT, Yeo KR. 2007. Prediction of nonspecific hepatic microsomal binding from readily available physicochemical properties. *Drug Metab Rev* 38(S1):162.

Valkó KL, Nunhuck SB, Hill AP. 2001. Estimating unbound volume of distribution and tissue binding by in vitro HPLC-based human serum albumin and immobilised artificial membrane-binding measurements. *J Pharm Sci* 100:849–862.

Vuppugalla R, Jones RDO, Jones HM, Gibson CR, Rowland M, Chien JY, Ring BJ, Adkison KK, Ku MS, He H, Vuppugalla R, Marathe P, Fischer V, Dutta S, Sinha VK, Bjornsson T, Lavé T, Poulin P. 2011. PhRMA initiative on predictive models of human pharmacokinetics. 4. Prediction of plasma-concentration-time profiles from in vivo preclinical data by using the Wajima modeling approach. *J Pharm Sci* 100:4111–4126.

Wagner C, Jantratid E, Kesisoglou F, Vertzoni M, Reppas C, Dressman JB, 2012. Predicting the oral absorption of a poorly soluble, poorly permeable weak base using biorelevant dissolution and transfer model tests coupled with a physiologically based pharmacokinetic model. *Eur J Pharm Biopharm* 82:127–138.

Wajima T, Fukumura K, Yano Y, Oguma T. 2003. Prediction of human pharmacokinetics from animal data and molecular structural parameters using multivariate regression analysis: volume of distribution at steady state. *J Pharm Pharmacol* 55:939–949.

Ward KW, Smith BR. 2004. A comprehensive quantitative and qualitative evaluation of extrapolation of intravenous pharmacokinetic parameters from rat, dog, and monkey to humans. II. Volume of distribution and mean residence time. *Drug Metab Dispos* 32:612–619.

Wong H, Choo EF, Alicke B, Ding X, La H, McNamara E, Theil FP, Tibbitts J, Friedman LS, Hop CE, Gould SE. 2012a. Antitumor activity of targeted and cytotoxic agents in murine subcutaneous tumor models correlates with clinical response. *Clin Cancer Res* 18:3846–3852.

Wong H, Lewin-Koh SC, Theil FP, Hop CE. 2012b. Influence of the compound selection process on the performance of human clearance prediction methods. *J Pharm Sci* 101:509–515.

Yokogawa K, Nakashima E, Ishizaki J, Maeda H, Nagano T, Ichimura F. 1990. Relationships in the structure-tissue distribution of basic drugs in the rabbit. *Pharm Res* 7:691–696.

Yoon M, Campbell JL, Andersen ME, Clewell HJ. 2012. Quantitative in vitro to in vivo extrapolation of cell-based toxicity assay results. *Crit Rev Toxicol* 42:633–652.

Yoshida M, Kobunai T, Aoyagi K, Saito H, Utsugi T, Wierzba K, Yamada Y. 2000. Specific distribution of TOP-53 to the lung and lung-localized tumor is determined by its interaction with phospholipids. *Clin Cancer Res* 6:4396–4401.

Zhang H, Zhang Y. 2006. Convenient nonlinear model for predicting the tissue/blood partition coefficients of seven human tissues of neutral, acidic, and basic structurally diverse compounds. *J Med Chem* 49:5815–5829.

Zhivkova Z, Doytchinova I. 2011. Prediction of steady-state volume of distribution of acidic drugs by quantitative structure–pharmacokinetics relationships. *J Pharm Sci* 103:1253–1266.

Zou P, Zheng N, Yang Y, Yu LX, Sun D. 2012. Prediction of volume of distribution at steady state in humans: comparison of different approaches. *Expert Opin Drug Metab Toxicol* 8:855–872.

7

ADME PROPERTIES LEADING TO TOXICITY

Katya Tsaioun

Safer Medicine Trust, Cambridge, MA, USA

7.1 INTRODUCTION

Drug attrition that occurs in late clinical development or during postmarketing is a serious economic problem in the pharmaceutical industry (Kaitin and DiMasi, 2011). The cost for drug approvals has crossed the US$1 billion mark, and the cost of advancing a compound to phase 1 trials can reach up to US$100 million according to the Tufts Center for the Study of Drug Development (Dimasi et al., 2003). Given these huge expenditures, substantial savings can accrue from early recognition of problems that would demonstrate a compound's potential to succeed in development (Kola and Landis, 2004; Caldwell et al., 2009).

The costs associated with withdrawing a drug from the market are even greater. For example, terfenadine is both a potent hERG cardiac channel ligand and is metabolized by the liver enzyme CYP3A4. It is important to mention that if terfenadine is administered as a monotherapy, it is metabolized by CYP3A4 into fexofenadine, which is not a hERG inhibitor. However, in real clinical situations, terfenadine was frequently coadministered with CYP3A4 inhibitors ketoconazole and erythromycin (Honig et al., 1993). The consequent overexposure of patients to terfenadine resulted in increases in plasma terfenadine to levels that caused cardiac toxicity (Honig et al., 1992) resulting in the drug to be withdrawn from the market (FDA, 2009) at an estimated cost of US$6 billion. Another example is the broad-spectrum antibiotic trovafloxacin, which was introduced in 1997 and soon became Pfizer's top seller. The drug was metabolically activated *in vivo* and formed a highly reactive metabolite causing severe drug-induced hepatotoxicity (Ball et al., 1999). Trovafloxacin received a black box warning from the FDA in 1998 (Mandell and Tillotson, 2002) costing Pfizer US$8.5 billion in lawsuits. With the improvement in technologies to measure the impact of new molecules on cardiac ion channels such as hERG and other important absorption, distribution, metabolism, and excretion (ADME) parameters early in the drug discovery process, such liabilities are now recognized earlier allowing for safer analogs to be advanced to more resource-consuming formal preclinical and clinical studies.

The purpose of preclinical ADME, also referred to as early drug metabolism and pharmacokinetics (DMPK), is to reduce the risks similar to ones described earlier and avoid spending valuable resources on molecules that have poor pharmacokinetic (PK) properties. This strategy allows drug discovery resources to be focused on fewer but higher-quality drug candidates. In 1993, 40% of drugs failed in clinical trials because of PK and bioavailability problems in phase 1 clinical trials (Kubinyi, 2003). Since then, major technological advances have occurred in molecular biology and screening, which allow major aspects of ADME to be assessed earlier during the lead optimization stage. By the late 1990s, the pharmaceutical industry recognized the value of early ADME assessment and began routinely employing it with noticeable results. ADME and DMPK problems decreased from 40 to 11% (Kola and Landis, 2004). Presently, a lack of efficacy and human toxicity are the primary reasons for failure (Arrowsmith and Miller, 2013).

Drug Discovery Toxicology: From Target Assessment to Translational Biomarkers, First Edition. Edited by Yvonne Will, J. Eric McDuffie, Andrew J. Olaharski, and Brandon D. Jeffy.

The terms "druggability" and "drug-likeness" were first introduced by Dr. Christopher Lipinski, who proposed "Lipinski's rule of 5" due to the frequent appearance of a number "5" in the rules (Lipinski, 2000). The rule of 5 has come to be a compass for the drug discovery industry (Lipinski et al., 2001). It stipulates that small-molecule drug candidates should possess:

- A molecular weight <500 g/mol
- A partition coefficient (log *P*—a measure of hydrophobicity) <5
- No more than five hydrogen bond donors
- No more than 10 hydrogen bond acceptors

A compound with fewer than three of these properties is unlikely to become a successful orally bioavailable drug. But like every rule, it has notable exceptions. Such exceptions to Lipinski's rule of 5 that have become marketed drugs are molecules taken up by active transport mechanisms, natural compounds, oligonucleotides, and proteins.

The drug discovery industry is experiencing dramatic structural change and is no longer just the domain of traditional large pharmaceutical companies. Now venture-capital-funded start-ups, governments, venture philanthropy, and other nonprofit and academic organizations are important participants in the search for new drug targets, pathways, and molecules. Importantly, these organizations frequently form partnerships, sharing resources, capabilities, risks, and rewards of drug discovery. Thus, it is becoming increasingly important to ensure that investors', donors', and taxpayers' money is efficiently used so that new safe drugs for unmet medical needs may be delivered to the public. ADME profiling has been proven to play a crucial role in just that and has been demonstrated to be effective in preventing poor drug candidates from entering clinical development and acceleration of the discovery process.

7.2 THE SCIENCE OF ADME

Regulatory authorities have relied upon *in vivo* testing to predict the behavior of new molecules in the human body since the 1950s. Bioavailability, tissue distribution, pharmacokinetics, metabolism, and toxicity are assessed typically in one rodent and one nonrodent species prior to administering a drug to a human to evaluate drug pharmacokinetics and exposure in a clinical trial (phase 1). Biodistribution is assessed using radioactively labeled compounds later in development because it is expensive both in terms of synthesizing sufficient amounts of radioactively labeled compound and for performing the animal experiments (Oldendorf, 1970).

Pharmacodynamic (PD) effectiveness of test compounds is now routinely assessed initially using a battery *in vitro* models such as target binding or phenotypic screening, followed by confirmation through *in vivo* efficacy models in an appropriate animal model. The predictive abilities of these tests depend largely on the therapeutic area and the animal model (Tan et al., 2013; Veazey, 2013). Understanding the relationship between drug plasma and tissue concentrations (pharmacokinetics) and PD (termed PK/PD relationship) is crucial in supporting efficacy results and defining the therapeutic and safety windows. *In vivo* PK studies in a variety of animal models are routinely used for lead optimization to assess drug metabolism and absorption. It is important to note that there are significant differences in absorption and metabolism among species used in animal studies, which may cause conflicting predictions of degradation pathways of new chemical entities (NCEs). It is a standard practice now in pharmaceutical industry to use primary hepatocytes (Shih et al., 1999) and other models (Khetani et al., 2013) from different species in order to understand such species differences in metabolism, identify metabolites that are uniquely human, and select species for preclinical development that are most relevant to human.

7.3 THE ADME OPTIMIZATION STRATEGY

Historically, ADME studies were focused on *in vivo* assays. These are time and resource intensive and generally low-throughput assays resulting in their implementation later in the development process, when more resources are released to study the few molecules that have advanced to this stage. With the advances in cell and molecular biology, high-throughput screening, and miniaturization technologies in the 1990s and stem cell-derived models at the beginning of this century, early ADME studies have been developed to predict *in vivo* animal and human results, at a level of speed and cost-effectiveness appropriate for the discovery stage. This progress in the science of ADME has created a new paradigm that drug discovery programs follow in advancing compounds from hit to lead, from lead to advanced lead and to nominated clinical candidates. Now, early in the discovery phase, using human enzymes and human-origin cells, drug discovery programs are able to obtain highly actionable information about the drug-likeness of new molecules, the potential to reach target organ, and early indications of known human mechanisms of toxicities. Additionally, in discovery, an emphasis of DMPK is on nonclinical species PK prediction including the use of rodent and nonrodent primary hepatocytes. ADME assessment of varying complexity is currently routinely performed on compounds that have shown *in vitro* efficacy and in conjunction with or just prior to demonstrating early proof of principle *in vivo* (Tsaioun et al., 2009).

While ADME assays and principles are the same, the application of these principles is unique to each drug discovery program. The development path from discovery to IND is not a straight line and is dependent on the therapeutic area, route of administration, chemical series, commercial factors, and other parameters. Correspondingly, the importance of the various ADME assays is based upon the specifics of the drug discovery program. ADME assays can also be categorized into those that are routine and those reserved for more advanced profiling. This division is also a function of cost-effectiveness and the need for the specific information. For instance, data regarding induction of human liver enzymes and transporters are not relevant during the hit-to-lead phase and are normally obtained for fewer more advanced candidates. Table 7.1 lists major types of assays that are used in pharmaceutical discovery teams for prediction of ADME properties that underlie safety evaluation.

In some cases, the US FDA requests data from *in vitro* ADME assays. For example, *in vitro* drug–drug interaction (DDI) studies may now be conducted under the guidance from FDA in 2006 and EMA Guideline on the Investigation of Drug Interactions CPMP/EWP/560/95/Rev. 1 Corr. 2. The guidance documents precisely outline methods to conduct CYP450 inhibition and induction and P-gp interaction studies. It is now a standard practice for pharmaceutical companies to include the DDI information in IND submissions, which helps regulatory agencies to evaluate human metabolism and potential safety of drug candidates.

The way discovery and development teams use ADME tools is dependent on many parameters of the drug discovery program and is never formulaic and simple. It is useful to start from the ultimate goal, which is usually a therapeutic drug for a specific patient population and from there work backwards towards discovery. The project team first defines the target product profile (TPP), which includes indication, intended patient population, route of administration, and acceptable toxicities and ultimately will define the drug label. The TPP invariably will evolve during the life of the project, but having major parameters of TPP established initially maintains a collaboration and focus between disciplines such as biology and chemistry, discovery and development, and preclinical and clinical

groups. Once the TPP is finalized, then major design elements of the phase 2 and 3 clinical trials can be outlined, leading to questions about the tolerability, toxicity, and safety of the molecule. These parameters will then define the GLP toxicity studies in preclinical models, which will guide the team to the discovery and preclinical development data to be addressed in an early ADME program (Tsaioun et al., 2009).

How is this information implemented in the discovery phase? If a compound has high target receptor binding and biological activity in cells and in relevant *in vivo* animal models, what are the chances of it becoming a successful drug? A molecule needs to cross many barriers to reach its biological target. In order to reach it, a molecule must be in solution, and thus the first step is typically to assess the solubility of a compound. A solubility screen provides information about the NCE's solubility in fluids compatible with administration to humans. Chemical and metabolic stability is a further extension of the intrinsic properties of a molecule. Chemical stability in buffers, simulated gastric and intestinal fluids, and metabolic stability in plasma, hepatocytes, or liver microsomes of different species can be measured to predict the rate of breakdown of a compound in the different environments encountered in the human body on the way to its target.

The second step is to define the **absorption properties** and the bioavailability of a molecule if the TPP includes the development of orally bioavailable drug. Measurement of permeability across a cell monolayer such as Caco-2 model is a good predictor of high human oral bioavailability. For drugs directed at the brain, assessment of blood–brain barrier (BBB) penetration would be performed at this stage and is usually a key component of lead optimization campaigns. Passive BBB permeability may be assessed using BBB PAMPA, which use phospholipid layers of different compositions, whereas potential for active uptake or efflux may be determined using *in vivo* models or cell lines naturally expressing endogenous human intestinal or BBB transporters (such as Caco-2 cell line) or cell lines overexpressing specific transporters (such as MDCK-MDR1).

TABLE 7.1 ADME Assays That Inform Toxicity at Different Stages of Drug Discovery and Development

Stage of the Program	Absorption	Distribution	Metabolism
Hit to lead	Physicochemical properties (solubility, Log P) PAMPA	Plasma protein binding	Efficacy species and human liver microsomal stability, CAR and PXR transactivation assay
Lead optimization	Caco-2 MDCK-MDR1 *In vivo* PK (rodent) Major transporters (MDR-1, BCEP)	*In vivo* tissue distribution (rodent)	Efficacy species, toxicology species, and human primary hepatocytes stability and metabolite identification, CYP inhibition, CYP induction
IND-enabling studies	*In vivo* PK (rodent and nonrodent) Comprehensive transporter panel (Table 7.5) (FDA Guidance 2006 and FDA/EMA 2012)	*In vivo* tissue distribution (rodent and nonrodent)	CYP inhibition (Tables 7.2 and 7.3), CYP induction (Table 7.4) (FDA Guidance 2006 and FDA/EMA 2012)

Measurement of **binding to plasma proteins** indicates the degree of availability of the free compound in the blood circulation. This is critical as only unbound drugs are able to reach the target and exert their desirable pharmacologic and undesirable (toxic) effects.

Metabolism and drug–drug interaction issues are discovered by screening for inhibition and induction of **cytochrome P450 liver enzymes** (CYP450). FDA guidelines have specific recommendation for substrates (Table 7.2) and inhibitors (Table 7.3) to be used in these assays.

Cytochrome P450 (CYP) enzymes can be induced by many xenobiotics including some drugs through the mediation of nuclear receptors. Understanding CYP induction potential of an NCE is important to minimize the drug–drug interaction liability.

Traditionally, the CYP induction is determined by measuring the desired enzyme activity. However, because of enzyme inhibition, cytotoxicity, or other effects of some compounds, negative results generated from measuring enzyme activity could be false negative. New research showed that assays that measure the mRNA level of CYP enzymes are more sensitive and reliable. The mRNA measurement also generates much less false negative than the traditional enzyme activity assays in detecting CYP induction.

The constitutive androstane receptor (CAR) and pregnane X receptor (PXR) are important nuclear receptors involved in the regulation of cellular responses from exposure to many xenobiotics and various physiological processes. In recent years, CAR and PXR transactivation assays have frequently been used during the discovery stage as a screening tool for CYP induction is PXR reporter assays (Sinz et al., 2008).

Recently, in 2012, both FDA and EMA recommended that changes in the mRNA level of the target gene should be used as the endpoint to evaluate the cytochrome P450 (CYP1A2, CYP2B6, and CYP3A4) induction potential of a compound. High-quality plateable cryopreserved human hepatocytes should be used to generate results, which are acceptable for regulatory filing. The regulatory agencies also recommended that the assay should use hepatocyte preparations from at least three donors.

All these assays allow chemists, toxicologists, and biologists to obtain actionable information and provide a link between structure–activity relationship (SAR) and structure–property relationship (SPR) that drive decisions on the selection of chemical series and molecules.

The effect of a compound on CYP450 metabolism can be identified by determining the 50% inhibitory concentration (IC_{50}) for each CYP450. The relationships between the NCE

TABLE 7.2   Chemical Inhibitors for *In Vitro* Experiments (FDA Guidance for Industry, September 25, 2006)

CYP	Inhibitor Preferred	Ki (μM)	Inhibitor Acceptable	Ki (μM)
1A2	Furafylline	0.6–0.73	α-Naphthoflavone	0.01
2A6	Tranylcypromine	0.02–0.2	Pilocarpine	4
	Methoxsalen	0.01–0.2	Tryptamine	1.7
2B6			3-Isopropenyl-3-methyl diamantane	2.2
			2-Isopropenyl-2-methyl adamantane	5.3
			Sertraline	3.2
			Phencyclidine	10
			Triethylenethiophosphoramide (thiotepa)	4.8
			Clopidogrel	0.5
			Ticlopidine	0.2
2C8	Montelukast		Trimethoprim	32
	Quercetin	1.1	Gemfibrozil	69–75
			Rosiglitazone	5.6
			Pioglitazone	1.7
2C9	Sulfaphenazole	0.3	Fluconazole	7
			Fluvoxamine	6.4–19
			Fluoxetine	18–41
2C19			Ticlopidine	1.2
			Nootkatone	0.5
2D6	Quinidine	0.027–0.4		
2E1			Diethyldithiocarbamate	9.8–34
			Clomethiazole	12
			Diallyldisulfide	150
3A4/5	Ketoconazole	0.0037–0.18	Azamulin	
	Itraconazole	0.27, 2.3	Troleandomycin	17
			Verapamil	10, 24

TABLE 7.3 Preferred and Acceptable Chemical Substrates for *In Vitro* Experiments (September 25, 2006)

CYP	Substrate Preferred	Km (μM)	Substrate Acceptable	Km (μM)
1A2	Phenacetin-O-deethylation	1.7–152	7-Ethoxyresorufin-O-deethylation	0.18–0.21
			Theophylline-N-demethylation	280–1230
			Caffeine-3-N-demethylation	220–1565
			Tacrine 1-hydroxylation	2.8, 16
2A6	Coumarin-7-hydroxylation	0.30–2.3		
	Nicotine C-oxidation	13–162		
2B6	Efavirenz hydroxylase	17–23	Propofol hydroxylation	3.7–94
	Bupropion hydroxylation	67–168	S-Mephenytoin-N-demethylation	1910
2C8	Taxol 6 hydroxylation	5.4–19	Amodiaquine N-deethylation	2.4,
			Rosiglitazone para-hydroxylation	4.3–7.7
2C9	Tolbutamide methyl-hydroxylation	67–838	Flurbiprofen 4′-hydroxylation	6–42
	S-Warfarin 7-hydroxylation	1.5–4.5	Phenytoin-4-hydroxylation	11.5–117
	Diclofenac 4′-hydroxylation	3.4–52		
2C19	S-Mephenytoin 4′-hydroxylation	13–35	Omeprazole 5-hydroxylation	17–26
			Fluoxetine O-dealkylation	3.7–104
2D6	(±)-Bufuralol 1′-hydroxylation	9–15	Debrisoquine 4-hydroxylation	5.6
	Dextromethorphan O-demethylation	0.44–8.5		
2E1	Chlorzoxazone 6-hydroxylation	39–157	p-Nitrophenol 3-hydroxylation	3.3
			Lauric acid 11-hydroxylation	130
			Aniline 4-hydroxylation	6.3–24
3A4/5	Midazolam 1-hydroxylation	1–14	Erythromycin N-demethylation	33–88
			Dextromethorphan N-demethylation	133–710
			Triazolam 4-hydroxylation	234
	Testosterone 6 β-hydroxylation	52–94	Terfenadine C-hydroxylation	15
			Nifedipine oxidation	5.1–47

and metabolizing enzymes need to be evaluated in the context of the human effective exposure (C_{eff}) and maximum plasma concentrations (C_{max}). These human data are not normally available at early stages of discovery, but could be predicted by physiologically based pharmacokinetic (PBPK) models such as commercial PBPK models Simcyp and Simulations Plus. It is important to understand these transporter and CYP450 relationships for the following reasons:

1. The compound may affect the effective plasma concentrations of other concomitantly administered drugs if metabolized by the same CYPs (i.e., terfenadine).

2. If a molecule is a CYP inducer, it may increase the clearance rate of concomitantly administered drugs, which are metabolized by these CYPs. This may result in a decrease in these drugs' effective plasma concentrations, thus decreasing their pharmacologic effect.

3. Metabolites formed *via* CYP metabolism may be responsible for undesirable side effects such as organ toxicity.

4. The metabolite of a compound may actually be responsible for the compound's efficacy and not the parent compound. The metabolite may even have a better efficacy, safety, and PK profile than its parent. As a result, metabolism can be exploited to produce a better drug, which will impact the medicinal chemistry strategy.

5. The identification of drug-metabolizing enzymes involved in the major metabolic pathways of a compound assists to predict the probable drug–drug interactions in humans. This information also may be used to design human clinical trials to detect unnecessary drug–drug interaction.

6. CYP induction may alter the PK of the molecule. If a molecule induces CYPs that are responsible for its own metabolism, induction may have the effect of decreasing exposure on later days of a study versus the first day of the study and may prevent efficacy/toxicity.

FDA Guidance for Industry (2006) has listed the primary CYPs, substrates, and inhibitors that are used by the industry as a guidance to develop and run these assays (Tables 7.3 and 7.4).

Another family of molecules that influence drug exposure and subsequently potential adverse events are **drug transporters**. Over the past 25 years, a number of important human drug transporters have been identified that are expressed at the apical or basal side of the epithelial cells in various tissues (Ambudkar et al., 2003; Teodori et al., 2006; Ferreira et al., 2015). Most drug transporters belong to two superfamilies, ATP-binding cassette (ABC) and solute-linked carrier (SLC), including both cellular uptake and

TABLE 7.4 *In Vitro* **CYP Inducers (FDA Guidance for Industry 7/28/2011)**

CYP	*In Vitro* Inducer as Positive Controls	Recommended Concentration (μM) of the Positive Controls	Reported Fold Induction In Enzyme Activities
1A2	Omeprazole	25–100	14–24
	Lansoprazole	10	10
2B6	Phenobarbital	500–1000	5–10
2C8	Rifampin	10	2–4
2C9	Rifampin	10	3.7
2C19	Rifampin	10	20
2D6	None identified		
3A4	Rifampin	10–50	4–31

efflux transporters (Teodori et al., 2006). Major human transporters and examples of drugs reported to be substrates, inhibitors, or inducers of these transporters are listed in Table 7.5. The data shown in Table 7.5 and reports in the literature indicate the lack of specificity for many of the transporters, substrates, and inhibitors studied. The effect of drug transporters on permeability and the effect of drugs on transporter activity can be measured in cells that have been demonstrated to express such transporters endogenously (e.g., some subclones of Caco-2) or were transfected with specific transporters (e.g., MDCK-MDR1).

MDR1 (P-gp) is a transporter most studied and whose interactions are particularly important for CNS drugs due to high expression of these efflux transporters in the human BBB. BSEP/MRP2 interactions are important for the assessment of biliary clearance and also serve as a potential warning sign for hepatotoxicity. There are BSEP inhibition screening assays in sandwich culture hepatocytes that are routinely used in pharmaceutical industry at the stages of advanced lead profiling (Aleo et al., 2014).

Early knowledge about these interactions is instrumental to the medicinal chemistry strategy and helps drive lead optimization.

It is important to mention **human polymorphisms** that affect some of the CYPs and drug transporters. These are CYP3A4 (van der Weide and van der Weide, 2015), CYP2D6 (Zhou, 2009), CYP2C19 (Stingl and Viviani, 2015), and MDR1 (Ambudkar et al., 2003). CYP2D6 is the most extensively characterized polymorphic drug-metabolizing enzyme. A deficiency of the CYP2D6 enzyme is inherited as an autosomal recessive trait; these subjects (7% of Caucasians, about 1% of Orientals) are classified as poor metabolizers. Among the rest (extensive metabolizers), enzyme activity is highly variable, from extremely high in ultrarapid metabolizers to markedly reduced in intermediate metabolizers. The *CYP2D6* gene is highly polymorphic, with more than 70 allelic variants described so far. Of these, more than 15 encode an inactive or no enzyme at all. Others encode enzyme with reduced, "normal," or increased enzyme activity. The *CYP2D6* gene shows marked interethnic variability, with interpopulation differences in allele frequency

and existence of "population-specific" allelic variants, for instance, among Orientals and Black Africans. The CYP2D6 enzyme catalyses the metabolism of a large number of clinically important drugs including antidepressants, neuroleptics, some antiarrhythmics, lipophilic β-adrenoceptor blockers, and opioids. CYP2D6 has a particular clinical significance for treating of the patients with CNS disorders who frequently are on multiple medications (Dorado et al., 2007).

Unlike polymorphisms observed for drug-metabolizing enzymes, MDR1 polymorphisms are much less studied and their effects on drug disposition are less well established. There are published reports on the importance of genetic variations in the MDR1 transporter (Brambila-Tapia, 2013); however, in many cases, the reports have been inconsistent and in some cases conflicting (Kim et al., 2001). Many challenges remain in our understanding of the clinical relevance of genetic variations in transporters to drug disposition and drug–drug interactions.

As already mentioned, the ability of a drug to penetrate biological barriers such as cell membranes, intestinal walls, or the BBB is a significant determinant of its ability to get to the target organ and to produce undesirable adverse effects. For drugs that target the central nervous system (CNS) for indications such as stroke, depression, and schizophrenia, *in vitro* efficacy combined with the inability to penetrate the BBB typically results in poor *in vivo* efficacy in patients. Another obstacle caused by BBB permeability is that many drugs not intended as CNS therapeutics cause neurotoxicity. Early knowledge of this property is essential not only for CNS programs but also for other therapeutic indications.

The delivery of systemically administered drugs to the brain of mammals is limited by the BBB as it effectively isolates the brain from the blood because of the presence of tight junctions connecting the endothelial cells of the brain vessels. In addition, specific metabolizing enzymes and efflux pumps such as P-glycoprotein (P-gp) and the multidrug resistance protein (MDR-1), located within the endothelial cells of the BBB, actively pump exogenous molecules out of the brain (Schinkel, 1999). This is one of the reasons for CNS drugs having a notoriously high failure rate (Pangalos et al., 2007). In recent years, 9% of compounds

TABLE 7.5 Major Human Transporters (FDA Guidance for Industry, March 1, 2006)

Gene	Aliases	Tissue	Substrate	Inhibitor	Inducer
ABCB1	P-gp, MDR1	Intestine, liver, kidney, brain, placenta, adrenal, testes	Digoxin, fexofenadine, indinavir, vincristine, colchicine. topotecan, paclitaxel	Ritonavir, cyclosporine, verapamil, erythromycin, ketoconazole, itraconazole, quinidine, elacridar (GF120918) LY335979, valspodar (PSC 833)	Rifampin, St. John's Wort
ABCB4	MDR3	Liver	Digoxin, paclitaxel, vinblastine		
ABCB11	BSEP	Liver	Vinblastine		
ABCC1	MRP1	Intestine, liver, kidney, brain	Adefovir, indinavir		
ABCC2	MRP2, CMOAT	Intestine, liver, kidney, brain	Indinavir, cisplatin,	Cyclosporine	
ABCC3	MRP3, CMOAT2	Intestine, liver, kidney, placenta, adrenal	Etoposide, methotrexate, teniposide		
ABCC4	MRP4				
ABCC5	MRP5				
ABCC6	MRP6	Liver, kidney	Cisplatin, daunorubicin		
ABCG2	BCRP	Intestine, liver, breast, placenta	Daunorubicin, doxorubicin, topotecan, rosuvastatin, sulfasalazine	Elacridar (GF120918)	
SLCO1B1	OATP1B1, OATP-C OATP2	Liver	Rifampin, rosuvastatin, methotrexate, pravastatin, thyroxine	Cyclosporine rifampin	
SLCO1B3	OATP1B3, OATP8,	Liver	Digoxin, methotrexate, rifampin		
SLCO2B1	SLC21A9, OATP-B	Intestine, liver, kidney, brain	Pravastatin		
SLC10A1	NTCP	Liver, pancreas	Rosuvastatin		
SLC10A2	ASBT	Ileum, kidney, biliary tract			
SLC15A1	PEPT1	Intestine, kidney	Ampicillin, amoxicillin, captopril, valacyclovir		
SLC15A2	PEPT2	Kidney	Ampicillin, amoxicillin, captopril, valacyclovir		
SLC22A1	OCT-1	Liver	Acyclovir, amantadine, desipramine, ganciclovir metformin	Disopyramide, midazolam, phenformin, phenoxybenzamine quinidine, quinine, ritonavir, verapamil	
SLC22A2	OCT2	Kidney, brain	Amantadine, cimetidine, memantine	Desipramine, phenoxybenzamine quinine	
SLC22A3	OCT3	Skeletal muscle, liver, placenta, kidney, heart	Cimetidine	Desipramine, prazosin, phenoxybenzamine	
SLC22A4	OCTN1	Kidney, skeletal muscle, placenta, prostate, heart	Quinidine, verapamil		
SLC22A5	OCTN2	Kidney, skeletal muscle, prostate, lung, pancreas, heart, small intestine, liver	Quinidine, verapamil		
SLC22A6	OAT1	Kidney, brain	Acyclovir, adefovir, methotrexate, zidovudine	Probenecid, cefadroxil, cefamandole, cefazolin	
SLC22A7	OAT2	Liver, kidney	Zidovudine	Probenecid, cefadroxil, cefamandole, cefazolin	
SLC22A8	OAT3	Kidney, brain	Cimetidine, methotrexate, zidovudine	Probenecid, cefadroxil, cefamandole, cefazolin,	

ABC, ATP-binding cassette transporter superfamily; ASBT, apical sodium-dependent bile salt transporter; BCRP, breast cancer resistance protein; BSEP, bile salt export pump; MDR1, multidrug resistance; MRP, multidrug resistance-related protein; NTCP, sodium taurocholate cotransporting polypeptide; OAT, organic anion transporter; OCT, organic cation transporter; SLC, solute-linked carrier transporter family; SLCO, solute-linked carrier organic anion transporter family.

that entered phase 1 survived to launch, and only 3–5% of CNS drugs were commercialized (Pangalos et al., 2007). Greater than 50% of this attrition resulted from failure to demonstrate efficacy in phase 2 studies. Over the last decade, phase 2 failures have increased by 15%. Compounds with demonstrated efficacy against a target *in vitro* and in animal models frequently proved to be ineffective in humans. Many of these failures occur due to the inability to reach the CNS targets such as in stroke due to the lack of BBB permeability. For drugs targeted to reduce damage from a stroke, the delivery method, BBB permeability, and drug metabolism and clearance can provide life or death to a patient if the drug is not delivered to the target tissue in its active form in a matter of hours from the event.

Due to the extraordinary cost of drug development, it is highly desirable to have effective, cost-efficient, and high-throughput tools to measure BBB permeability before proceeding to expensive and time-consuming animal BBB permeability studies or human clinical trials. With *in vitro* tools available, promising drug candidates with ineffective BBB penetration may be improved by removing structural components that mediate interaction(s) with efflux proteins and/or lowering binding to brain tissue at earlier stages of development to increase intrinsic permeability (Liu and Chen, 2005).

The development of drugs targeting CNS requires precise knowledge of the drug's brain penetration. Ideally, this information would be obtained as early as possible to focus resources on compounds most likely to reach the target organ. The physical transport and metabolic composition of the BBB are highly complex. Numerous *in vitro* models have been designed to study kinetic parameters in the CNS, including noncerebral peripheral endothelial cell lines, immortalized rat brain endothelial cells, primary cultured bovine, porcine or rat brain capillary endothelial cells, and cocultures of primary brain capillary cells with astrocytes (Begley et al., 1999; Megard et al., 2002; Deli et al., 2005). *In vitro* BBB models must be carefully assessed for their capacity to reflect accurately the passage of drugs into the CNS *in vivo*.

Alternatively, several *in vivo* techniques have been used to estimate BBB passage of drugs directly in laboratory animals. *In vivo* transport across the BBB was first studied in the 1960s using the early indicator diffusion method (IDM) of Crone (1963). Other *in vivo* techniques were later proposed including brain uptake index (BUI) measurement (Oldendorf, 1970), *in situ* brain perfusion method (Takasato et al., 1984; Kakee et al., 1996), autoradiography, and intracerebral microdialysis (Elmquist and Sawchuk, 1997). Unfortunately these methods have limitations, including requiring sophisticated equipment, technical expertise, mathematical modeling, species differences, invasiveness, and low throughput—which render them unsuitable for use during early stages of drug discovery.

Artificial membrane permeability assays such as PAMPA (Chen et al., 2008) and BBB PAMPA offer a cost-effective and high-throughput method of screening for passively absorbed compounds but do not predict active transport in or out of the brain.

The development of a new coculture-based model of human BBB that enables the prediction of passive and active transport of molecules into the CNS has recently been reported (Josserand et al., 2006). This new model consists of primary cultures of human brain capillary endothelial cells cocultured with primary human glial cells (Megard et al., 2002; Josserand et al., 2006). The advantage of this system includes the use of human primary cells, avoiding species, age, and interindividual differences since the two cell types are removed from the same human donor and because of the demonstrated expression of functional efflux transporters such as P-gp, MRP-1, MRP-4, MRP-5, and BCRP. Such models have potential for the assessment of permeability of drug and specific transport mechanisms, which is not possible in artificial membrane assays (e.g., PAMPA) or other cell models due to incomplete expression of active transporters.

One important step in the development of any *in vitro* model is to correlate *in vitro* and *in vivo* data in order to validate experimental models and to assess the predictive power of the techniques (Pardridge et al., 1990). Reproducibility and predictive ability of new models are of paramount importance, and uniform detailed validation guidance is necessary for the industry to accelerate the adoption of valid modern science in drug development.

7.4 CONCLUSIONS AND FUTURE DIRECTIONS

A large amount of progress in the field of ADME profiling has occurred in the last quarter century. This progress has decreased the proportion of drug candidates failing in clinical trials for ADME reasons and providing important early input into safety and toxicity prediction of drug candidates.

Cell-based assays using established cell lines and cocultures have been used to determine toxicity to various organs, but many of these cell lines have lost some of the physiological activities present in normal cells. HepG2 cells, for instance, have greatly reduced levels of metabolic enzymes. Primary human hepatocytes can be used but are expensive, suffer from high donor-to-donor variability, and maintain their characteristics for only a short time. Three-dimensional models have been developed for cell-based therapies including micropatterned cocultures of human liver cells that maintain the phenotypic functions of the human liver for several weeks (Khetani et al., 2013). This development should provide more accurate information about toxicity when used in ADME screening and could be extended to other organ-specific cells leading to integrated tissue models in the "human on a chip" (Sung et al., 2014). The potential of stem cells to differentiate into cell lines of many different

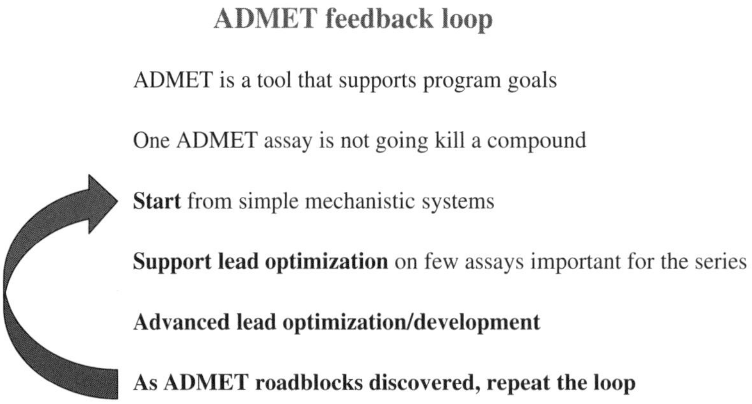

ADMET feedback loop

ADMET is a tool that supports program goals

One ADMET assay is not going kill a compound

Start from simple mechanistic systems

Support lead optimization on few assays important for the series

Advanced lead optimization/development

As ADMET roadblocks discovered, repeat the loop

FIGURE 7.1   ADMET feedback loop.

lineages may be exploited to develop human and animal stem cell-derived systems for major organ systems.

High-content screening (HCS) has been used for early cytotoxicity measurement since 2006 and provides great optimism (Persson et al., 2014). This method has been optimized for stem cell-derived models for the prediction of other organ toxicity.

Molecular profiling is another alternative and is defined as any combination or individual application of mRNA expression, proteomic, toxicogenomic, or metabolomic measurements that characterize the state of a tissue. This approach has been applied in an attempt to develop profiles or signatures of certain toxicities (Stoughton and Friend, 2005). Molecular profiles, in conjunction with agents that specifically perturb cellular systems, have been used to identify patterns of changes in gene expression and other parameters at subtoxic drug concentrations that might be predictive of hepatotoxicity including idiosyncratic hepatotoxicity (Ekins et al., 2006). In the future, larger data sets, high-throughput gene disruptions, and more-diverse profiling data will lead to more-detailed knowledge of disease pathways and will facilitate in target selection and the construction of detailed models of cellular systems for use in ADME screening to identify toxic compounds early in the discovery process. The combination of *in silico*, *in vitro*, and *in vivo* methods and models into multiple content data bases, data mining, human-on-a-chip systems, and predictive modeling algorithms, visualization tools, and high-throughput data-analysis solutions can be integrated to predict systems' ADME properties. Such models are starting to be built and should be widely available within the next decade. The use of these tools will lead to a greater understanding of the interactions of drugs with their targets and predict their toxicities.

Thus, while we see great advances in the prediction of human-specific toxicity mechanisms, it is clear that accurate prediction of human exposure is needed in order to put the *in vitro* data in the context of systemic and organ-specific

exposure to the compound of interest. This aspect is addressed in another chapter in this volume.

To conclude, the future drug development paradigm that is built on human biology-based *in vitro* models, *in silico* tools, and accurate prediction of human ADME and exposure is expected to provide a decrease in late-stage development failures and withdrawals of marketed drugs, faster timelines from discovery to market, and reduced development costs and larger number of patient-centric, safer therapies (Fig. 7.1).

REFERENCES

Aleo, M.D. et al., 2014. Human drug-induced liver injury severity is highly associated to dual inhibition of liver mitochondrial function and bile salt export pump. *Hepatology (Baltimore, Md.)*, 60, pp. 1015–1022.

Ambudkar, S.V. et al., 2003. P-glycoprotein: from genomics to mechanism. *Oncogene*, 22(47), pp. 7468–7485.

Arrowsmith, J. & Miller, P., 2013. Trial watch: phase II and phase III attrition rates 2011–2012. *Nature Reviews. Drug Discovery*, 12(8), p. 569.

Ball, P. et al., 1999. Comparative tolerability of the newer fluoroquinolone antibacterials. *Drug Safety*, 21(5), pp. 407–421.

Begley, J.G. et al., 1999. Altered calcium homeostasis and mitochondrial dysfunction in cortical synaptic compartments of presenilin-1 mutant mice. *Journal of Neurochemistry*, 72(3), pp. 1030–1039.

Brambila-Tapia, A.J.-L., 2013. MDR1 (ABCB1) polymorphisms: functional effects and clinical implications. *Revista de Investigación Clínica*, 65(5), pp. 445–454.

Caldwell, G.W. et al., 2009. ADME optimization and toxicity assessment in early- and late-phase drug discovery. *Current Topics in Medicinal Chemistry*, 9(11), pp. 965–980.

Chen, X. et al., 2008. A novel design of artificial membrane for improving the PAMPA model. *Pharmaceutical Research*, 25(7), pp. 1511–1520.

Crone, C., 1963. The permeability of capillaries in various organs as determined by use of the "indicator diffusion" method. *Acta Physiologica Scandinavica*, 58, pp. 292–305.

Deli, M.A. et al., 2005. Permeability studies on in vitro blood-brain barrier models: physiology, pathology, and pharmacology. *Cellular and Molecular Neurobiology*, 25(1), pp. 59–127.

Dimasi, J.A., Hansen, R.W., & Grabowski, H.G., 2003. The price of innovation: new estimates of drug development costs. *Journal of Health Economics*, 22(2), pp. 151–185.

Dorado, P., Peñas-Lledó, E.M., & Llerena, A., 2007. CYP2D6 polymorphism: implications for antipsychotic drug response, schizophrenia and personality traits. *Pharmacogenomics*, 8(11), pp. 1597–1608.

Ekins, S. et al., 2006. Algorithms for network analysis in systems-ADME/Tox using the MetaCore and MetaDrug platforms. *Xenobiotica*, 36(10–11), pp. 877–901.

Elmquist, W.F. & Sawchuk, R.J., 1997. Application of microdialysis in pharmacokinetic studies. *Pharmaceutical Research*, 14(3), pp. 267–288.

FDA, 2009. FDA terfenadine withdrawal. Available at: http://www.fda.gov/OHRMS/DOCKETS/98fr/06d-0344-gdl0001.pdf.

Ferreira, R.J., Dos Santos, D.J., & Ferreira, M.-J.U., 2015. P-glyco-protein and membrane roles in multidrug resistance. *Future Medicinal Chemistry*, 7(7), pp. 929–946.

Honig, P.K. et al., 1992. Changes in the pharmacokinetics and electrocardiographic pharmacodynamics of terfenadine with concomitant administration of erythromycin. *Clinical Pharmacology and Therapeutics*, 52(3), pp. 231–238.

Honig, P.K. et al., 1993. Effect of concomitant administration of cimetidine and ranitidine on the pharmacokinetics and electro-cardiographic effects of terfenadine. *European Journal of Clinical Pharmacology*, 45(1), pp. 41–46.

Josserand, V. et al., 2006. Evaluation of drug penetration into the brain: a double study by in vivo imaging with positron emission tomography and using an in vitro model of the human blood-brain barrier. *The Journal of Pharmacology and Experimental Therapeutics*, 316(1), pp. 79–86.

Kaitin, K.I. & DiMasi, J.A., 2011. Pharmaceutical innovation in the 21st century: new drug approvals in the first decade, 2000–2009. *Clinical Pharmacology and Therapeutics*, 89(2), pp. 183–188.

Kakee, A., Terasaki, T., & Sugiyama, Y., 1996. Brain efflux index as a novel method of analyzing efflux transport at the blood-brain barrier. *The Journal of Pharmacology and Experimental Therapeutics*, 277(3), pp. 1550–1559.

Khetani, S.R. et al., 2013. Use of micropatterned cocultures to detect compounds that cause drug-induced liver injury in humans. *Toxicological Sciences*, 132(1), pp. 107–117.

Kim, R.B. et al., 2001. Identification of functionally variant MDR1 alleles among European Americans and African Americans. *Clinical Pharmacology and Therapeutics*, 70(2), pp. 189–199.

Kola, I. & Landis, J., 2004. Can the pharmaceutical industry reduce attrition rates? *Nature Reviews. Drug Discovery*, 3(8), pp. 711–715.

Kubinyi, H., 2003. Drug research: myths, hype and reality. *Nature Reviews. Drug Discovery*, 2(8), pp. 665–668.

Lipinski, C.A., 2000. Drug-like properties and the causes of poor solubility and poor permeability. *Journal of Pharmacological and Toxicological Methods*, 44(1), pp. 235–249.

Lipinski, C.A. et al., 2001. Experimental and computational approaches to estimate solubility and permeability in drug discovery and development settings. *Advanced Drug Delivery Reviews*, 46(1–3), pp. 3–26.

Liu, X. & Chen, C., 2005. Strategies to optimize brain penetration in drug discovery. *Current Opinion in Drug Discovery & Development*, 8(4), pp. 505–512.

Mandell, L. & Tillotson, G., 2002. Safety of fluoroquinolones: An update. *The Canadian Journal of Infectious Diseases*, 13(1), pp. 54–61.

Megard, I. et al., 2002. A co-culture-based model of human blood-brain barrier: application to active transport of indinavir and in vivo-in vitro correlation. *Brain Research*, 927(2), pp. 153–167.

Oldendorf, W.H., 1970. Measurement of brain uptake of radiola-beled substances using a tritiated water internal standard. *Brain Research*, 24(2), pp. 372–376.

Pangalos, M.N., Schechter, L.E., & Hurko, O., 2007. Drug development for CNS disorders: strategies for balancing risk and reducing attrition. *Nature Reviews. Drug Discovery*, 6(7), pp. 521–532.

Pardridge, W.M. et al., 1990. Comparison of in vitro and in vivo models of drug transcytosis through the blood-brain barrier. *The Journal of Pharmacology and Experimental Therapeutics*, 253(2), pp. 884–891.

Persson, M. et al., 2014. High-content analysis/screening for predictive toxicology: application to hepatotoxicity and geno-toxicity. *Basic & Clinical Pharmacology & Toxicology*, 115(1), pp. 18–23.

Schinkel, A.H., 1999. P-Glycoprotein, a gatekeeper in the blood-brain barrier. *Advanced Drug Delivery Reviews*, 36(2–3), pp. 179–194.

Shih, H. et al., 1999. Species differences in hepatocyte induction of CYP1A1 and CYP1A2 by omeprazole. *Human & Experimental Toxicology*, 18(2), pp. 95–105.

Sinz, M., Wallace, G., & Sahi, J., 2008. Current industrial practices in assessing CYP450 enzyme induction: preclinical and clinical. *The AAPS Journal*, 10(2), pp. 391–400.

Stingl, J. & Viviani, R., 2015. Polymorphism in CYP2D6 and CYP2C19, members of the cytochrome P450 mixed-function oxidase system, in the metabolism of psychotropic drugs. *Journal of Internal Medicine*, 277(2), pp. 167–177.

Stoughton, R.B. & Friend, S.H., 2005. How molecular profiling could revolutionize drug discovery. *Nature Reviews. Drug Discovery*, 4(4), pp. 345–350.

Sung, J.H. et al., 2014. Using physiologically-based pharmacoki-netic-guided "body-on-a-chip" systems to predict mammalian response to drug and chemical exposure. *Experimental Biology and Medicine (Maywood, N.J.)*, 239, pp. 1225–1239.

Takasato, Y., Rapoport, S.I., & Smith, Q.R., 1984. An in situ brain perfusion technique to study cerebrovascular transport in the rat. *The American Journal of Physiology*, 247(3 Pt 2), pp. H484–H493.

Tan, S.M. et al., 2013. The modified selenenyl amide, M-hydroxy ebselen, attenuates diabetic nephropathy and diabetes-associated atherosclerosis in ApoE/GPx1 double knockout mice. *PloS one*, 8(7), p. e69193.

Teodori, E. et al., 2006. The functions and structure of ABC transporters: implications for the design of new inhibitors of Pgp and MRP1 to control multidrug resistance (MDR). *Current Drug Targets*, 7(7), pp. 893–909.

Tsaioun, K. et al., 2009. ADDME—Avoiding Drug Development Mistakes Early: central nervous system drug discovery perspective. *BMC Neurology*, 9 Suppl 1, p. S1.

Veazey, R.S., 2013. Animal models for microbicide safety and efficacy testing. *Current Opinion in HIV and AIDS*, 8(4), pp. 295–303.

van der Weide, K. & van der Weide, J., 2015. The influence of the CYP3A4*22 polymorphism and CYP2D6 polymorphisms on serum concentrations of aripiprazole, haloperidol, pimozide, and risperidone in psychiatric patients. *Journal of Clinical Psychopharmacology*, 35(3), pp. 228–236.

Zhou, S.-F., 2009. Polymorphism of human cytochrome P450 2D6 and its clinical significance: Part I. *Clinical Pharmacokinetics*, 48(11), pp. 689–723.

PART IV

PREDICTING ORGAN TOXICITY

8

LIVER

J. Gerry Kenna[1], Mikael Persson[2,*], Scott Q. Siler[3], Ke Yu[4], Chuchu Hu[4,5], Minjun Chen[4], Joshua Xu[4], Weida Tong[4], Yvonne Will[6] and Michael D. Aleo[6]

[1]*Fund for the Replacement of Animals in Medical Experiments (FRAME), Nottingham, UK*

[2]*Lundbeck, Valby, Denmark*

[3]*The Hamner Institute, Research Triangle Park, NC, USA*

[4]*Division of Bioinformatics and Biostatistics, National Center for Toxicological Research, US Food and Drug Administration (NCTR/FDA), Jefferson, AZ, USA*

[5]*Zhejiang Institute of Food and Drug Control, Hangzhou, China*

[6]*Investigative Toxicology, Drug Safety Research and Development, Pfizer Inc., Groton, CT, USA*

8.1 INTRODUCTION

In this chapter, we briefly summarize the current status of the development of *in silico* and *in vitro* models for prediction of drug-induced liver injury (DILI). Such approaches are needed during drug discovery to enable selection of drugs that have low propensity to cause both intrinsic DILI (which often can be detected in preclinical safety studies in animals) and idiosyncratic human DILI (which occurs infrequently and may not be observed until late in drug development or after regulatory approval). The currently available approaches focus on cell-based mechanisms by which drugs can cause injury to cells that may ultimately result in DILI *in vivo*, in particular reactive metabolite formation, mitochondrial dysfunction, and inhibition of the bile salt export pump (BSEP). Many drugs that cause idiosyncratic DILI exhibit more than one of these adverse liabilities. A multiassay panel is required to achieve good DILI sensitivity and specificity, while multiparametric evaluation of several toxicity endpoints is also advantageous. Data interpretation is greatly improved when account is taken of drug dose and/or *in vivo* drug exposure and use of physiologically based exposure modeling offers great additional promise. However the currently available *in vitro* assays do not explore the factors that may explain why only some humans develop idiosyncratic

DILI and hence do not provide insight into individual susceptibility. Studies undertaken using zebrafish suggest that these could provide a valuable *in vivo* model for DILI assessment, which is amenable to medium- to high-throughput compound testing. Knowledge-based expert systems and statistics-based quantitative structure–activity relationship (QSAR) models are two major types of modeling approaches. Intrinsic hepatotoxicity is believed to be associated with the molecular structure at a large degree and thus is likely predictable in theory using knowledge-based expert systems. However, most *in silico* models to date have been developed based solely on structural properties using various modeling algorithms, and their successes have been limited. This is likely to be due to inadequate understanding of DILI mechanisms and a relative lack of high-quality *in vivo* data on DILI mechanisms that can be modeled. Use of QSAR models to predict molecular initiating events (MIEs) related to DILI has been more successful and is useful for prioritizing chemicals during drug screening. Future efforts to improve modeling performance should focus on integrating the enhanced understanding of diverse DILI mechanisms into the modeling processes. Importantly, data from chemical properties, host factors, and environmental factors should be integrated and incorporated into the *in silico* models to improve DILI prediction.

*Currently at AstraZeneca, Molndal, Sweden.

Drug Discovery Toxicology: From Target Assessment to Translational Biomarkers, First Edition. Edited by Yvonne Will, J. Eric McDuffie, Andrew J. Olaharski, and Brandon D. Jeffy.

8.2 DILI MECHANISMS AND SUSCEPTIBILITY

The liver is a complex organ that contains many different cell types, all of which may be affected adversely following acute or chronic exposure to xenobiotics. Toxicity to the liver is caused by many hundreds of chemicals present in foods, agrochemicals, the environment, household products, herbal remedies, substances of abuse, and also numerous small-molecule prescription pharmaceuticals (Zimmerman, 1999). Furthermore, numerous different types of tissue pathology may result. These include cell degeneration, apoptosis, necrosis, steatosis, fibrosis, altered mitochondrial size or morphology, cellular hypertrophy, biliary hyperplasia, induction of apoptosis, and/or inflammatory cell infiltration. Many chemicals cause toxicity that is targeted to hepatocytes, which are the most abundant liver cell type, and hence is termed hepatocellular liver injury. The other most frequent types of chemically induced liver toxicity impair excretion of bile, which is a key physiological function of the liver, and thereby cause cholestatic or mixed hepatocellular/cholestatic liver injury (Zimmerman, 1999). The functional consequences of DILI range from mild and asymptomatic elevated plasma levels of alanine aminotransferase (ALT) and other enzymes released from damaged liver cells (often termed "liver enzymes") to symptomatic acute or chronic liver dysfunction. The most severe and concerning outcome is acute liver failure (ALF), which occurs following loss of a substantial proportion of functional liver cells and may result in fatality (Yuan and Kaplowitz, 2013).

Many compounds cause dose-dependent "intrinsic" liver injury, which occurs in many different animal species and in humans (e.g., that caused by the industrial chemical bromobenzene or the pharmaceutical acetaminophen/paracetamol) (Zimmerman, 1999). In addition, a large number of pharmaceuticals cause infrequent liver toxicity in some susceptible humans that is not evident in animals or in the vast majority of treated humans (Zimmerman, 1999; Kaplowitz and DeLeve, 2013). This type of DILI is termed "idiosyncratic DILI." In common with intrinsic DILI, the consequences of idiosyncratic DILI may range from mild and asymptomatic plasma liver enzyme elevations to severe symptomatic liver injury or even fatal liver failure. Although symptomatic liver injury caused by individual drugs that cause idiosyncratic DILI occurs very infrequently (typically in <1 in several thousand patients) and therefore is difficult to detect and diagnose, many hundreds of widely prescribed drugs may cause idiosyncratic DILI. Consequently, this adverse drug reaction is an important cause of serious human ill health and is the single most frequent cause of refusal to approve new drugs, cautionary labeling, and withdrawal from use of licensed drugs (Zimmerman, 1999; Kaplowitz and DeLeve, 2013). It is also a frequent reason for discontinuation of a new drug from nonclinical and clinical development (Laverty et al.,

2011). The mechanisms that underlie intrinsic DILI and idiosyncratic DILI are complex and involve multiple steps, which are outlined schematically in Fig. 8.1.

Both intrinsic DILI and idiosyncratic DILI are initiated by chemical insult to cells of the liver, which may arise in many different ways. Examples include pharmacological interactions with certain cell surface receptors (e.g., TRAIL and other cell death receptors) or intracellular receptors (e.g., the nuclear hormone receptor PPAR), generation of chemically reactive species (metabolites of the compounds themselves, reactive oxygen species (ROS), reactive lipid hydroperoxides, etc.), mitochondrial inhibition and/or uncoupling, inhibition of key hepatic enzymes regulating intermediary metabolism, and inhibition of biliary efflux transporters (Peraza et al., 2006; Zheng et al., 2004; Kaplowitz and DeLeve, 2013; Kenna et al., 2015). As discussed elsewhere in this chapter, the type of chemical insult can play an important role in determining the ultimate lesion (e.g., hepatocellular injury due to formation of chemically reactive species in hepatocytes, cholestatic liver injury due to inhibition of bile excretion). However, numerous compounds elicit multiple chemical insults. The currently available evidence indicates that essentially the same chemical insults are exhibited by drugs and chemicals that cause both intrinsic liver toxicity and idiosyncratic DILI (Thompson et al., 2011, 2012; Kaplowitz and DeLeve, 2013; Kenna et al., 2015).

The chemical insults trigger biological responses, which may be protective or may result in toxicity. Protective responses limit and control cellular damage and enable tissue repair and resolution. These may act within cells, where they involve the glutathione system and various stress response processes (which include heat shock protein and glucose-regulated protein systems, the antioxidant response, and the nrf-2 transcriptional cascade), or between cells (most notably via activation of the innate immune system and tissue repair) (Hayes and McLellan, 1999; Laskin, 2009; Fulda et al., 2010). Amplification and progression of cellular damage occur as toxicity develops and involves processes that occur within cells and between cells. For example, liver injury in animals and man following treatment with high doses of acetaminophen is initiated by a relatively simple chemical insult within hepatocytes (generation of a chemically reactive metabolite that reduces intracellular glutathione content) but then involves a concerted cascade of intracellular and intercellular events that involve covalent modification of hepatocyte proteins (Xie et al., 2014), followed by cytokine release and inflammatory cell infiltration and activation (i.e., activation of the innate immune system (Hinson et al., 2010; Williams and Jaeschke, 2012)). For many compounds activation of the innate immune system has been shown to act as a deleterious amplification cascade that markedly exacerbates liver toxicity. In addition, some drugs have the ability to cause liver injury

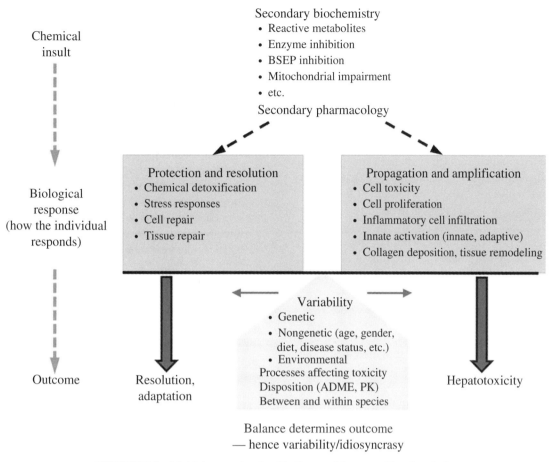

FIGURE 8.1   Multiple steps involved in liver toxicity caused by chemicals.

via activation of the adaptive immune system (Adams et al., 2010; Ju and Reilly, 2012). For these drugs (e.g., halothane or tienilic acid), covalent modification of liver proteins by chemically reactive metabolites produces neoantigens, which in susceptible patients trigger potent T cell and B cell responses (Pohl et al., 1988).

For chemicals that cause intrinsic liver toxicity, it is clear that the dose administered is an important factorial determinant of the manifestation liver toxicity. This is because generally there is a good correlation between the dose of chemical and the magnitude of chemical insult. Another important consideration is the duration of exposure, which can be rationalized as due to the extent of exposure to chemical insult. For numerous compounds that are hepatotoxic due to reactive metabolite formation (e.g., acetaminophen, bromobenzene), a further important susceptibility factor is the balance between metabolic bioactivation and detoxification of reactive intermediates via their conjugation to reduced glutathione (Park et al., 2011). Correlations between drug dose, the extent of chemical insult arising via metabolic bioactivation, and risk of liver injury in the human population have also been observed for many drugs that cause idiosyncratic DILI in humans

(Walgren et al., 2005; Park et al., 2011). The susceptibility factors that determine why some humans sustain idiosyncratic DILI, whereas the vast majority of exposed patients do not manifest DILI, remain poorly defined. HLA status can be an important inherited susceptibility factor for some drugs (e.g., flucloxacillin), which provides additional support for the inferred role of adaptive immune mechanisms in idiosyncratic DILI caused by at least some drugs (Monshi et al., 2011; Cargnin et al., 2014). However, attempts to identify genetically determined and nongenetically determined DILI susceptibility factors for many other drugs have been largely unsuccessful. This may be because multiple contributory factors are involved, where each in and of themselves are necessary but insufficient to enable the toxicity to arise (Ullrich, 2007).

Previous efforts to develop *in vitro* methods and *in silico* tools, which enable production of liver toxicity caused by drugs and chemicals, have been focused primarily upon evaluation of initiating direct chemical insults, rather than amplification processes or individual susceptibility factors. Approaches that appear to be particularly promising are outlined in the following, and their value for prediction of DILI is discussed.

8.3 COMMON MECHANISMS THAT CONTRIBUTE TO DILI

8.3.1 Mitochondrial Injury

Whereas the role of mitochondrial injury in liver toxicity caused by chemicals and environmental components such as insecticides and herbicides has been studied in academic settings for nearly half a century, drug-induced mitochondrial toxicity was not investigated until much later. Pharmaceutical companies did not pay much attention to this issue until the withdrawal of troglitazone due to liver injury and cerivastatin due to rhabdomyolysis. Both compounds were postulated to have caused their toxicity at least in part through mitochondrial impairment (Tirmenstein et al., 2002; Westwood et al., 2005). With the benefit of hindsight, it is surprising that the withdrawal of phenformin in 1968 due to severe lactic acidosis, a hallmark of mitochondrial toxicity, did not raise more widespread concern at the time about safety risks posed by damage to this organelle. Today, more than 200 drugs that cause severe liver injury (that either caused their removed from the market, resulted in a black box warning on the drug label, or has restricted their usage) have been shown to impair mitochondrial function (Pessayre et al., 2012; Porceddu et al., 2012). Furthermore, it has become increasingly apparent that mitochondrial toxicity seems to play a particularly important role in initiating idiosyncratic toxicity (Boelsterli and Lim, 2007).

Mitochondria generate most of the energy used by cells in the form of ATP, which is required to execute all essential cellular functions. When mitochondrial function is impaired, cellular function is impaired, which can lead to organ failure and in extreme circumstances to death. Mitochondria have an outer and an inner membrane. The latter harbors the electron transport chain, where ATP is produced. Mitochondria also contain their own DNA, which is small and encodes for 13 proteins and some transfer RNAs.

In the liver, mitochondria are also the main organelle involved in fatty acid oxidation (FAO). Long-chain fatty acid acyl-CoAs are transported into the mitochondria via the carnitine shuttle. Once inside the mitochondria, the fatty acyl-CoA is regenerated from acylcarnitine. The fatty acyl-CoA is split by successive beta-oxidation cycles into acetyl-CoA, which is degraded to CO_2 in the tricarboxylic acid cycle. Glycolysis (conversion of glucose to pyruvate) occurs in the cytoplasm of the hepatocyte. Pyruvate enters the mitochondria with the help of pyruvate dehydrogenase, which converts pyruvate, NAD^+, and CoA into NADH and acetyl-CoA. As in FAO, the acetyl-CoA formed via glycolysis is oxidized in the tricarboxylic acid cycle.

The processes of FAO, pyruvate oxidation, and acetyl-CoA oxidation generate the reducing equivalents NADH and FADH2. These are reoxidized by donating electrons to the mitochondrial electron chain, which consists of four complexes (I–IV) and the ATPase. The electrons pass along the chain to complex IV, the cytochrome oxidase, where they combine with oxygen to form water. The transfer of electrons along the electron transport chain is coupled to the transport of protons across the inner mitochondrial membrane, which creates an electrochemical gradient. If the cell needs energy (as indicated by a high cytosolic ADP/ATP ratio), ADP enters the mitochondria via the adenine nucleotide translocator in exchange for ATP. The entry of ADP causes the reentry of protons into the F_0 portion of the ATPase, which then causes a rotation in the F_1 portion of the ATPase and the conversion of ADP to ATP.

Most of the electrons donated to the mitochondrial electron transfer chain will react with oxygen to form water. However, a small percentage can leak and cause the formation of superoxide anion radicals, which, if not quickly detoxified, can lead to formation of other radicals and cause oxidative stress. Indeed the mitochondria are thought to be the main producer of ROS, which can damage proteins and nucleic acids and cause cellular damage. Mitochondria contain protective mechanisms that detoxify ROS (and so does the cell), such as glutathione, superoxide dismutase, peroxiredoxin, and thioredoxins, but they can be overwhelmed as an insult persists.

Mitochondrial function can be inhibited in numerous ways. For many hepatotoxic drugs, direct inhibition of one or more of the electron transport chain proteins (cerivastatin, chloroquine, fenofibrate, nefazodone) and inhibition of FAO have been demonstrated (salicylic acid, ibuprofen, pirprofen). In some cases the offending agent inhibits both processes (e.g., benzbromarone, perhexiline, troglitazone, valproic acid) (Pessayre et al., 2012). This would decrease ATP production and in the liver would manifest itself as liver cell necrosis and in some instances cholestasis and fibrosis. Whereas the lack of ATP is detrimental enough, the subsequent increase in ROS because of diminished electron flow will also contribute to injury. In addition, NADH and FADH2 cannot be reoxidized, which then leads to reduced oxidation of pyruvate, which instead is reduced to lactate causing lactic acidosis. In addition, the lack of NAD^+ and FAD^+ decreases beta-oxidation, which leads to accumulation of fatty acids in the form of triglycerides in cytoplasmic vacuoles (steatosis).

Drugs can also cause mitochondrial toxicity through uncoupling, a mechanism by which electron transport is separated from ATP synthesis as the drugs cause inner membrane damage that makes the hydrogens pass back and forth in a futile cycle, producing heat. Examples are weak acids and lipophilic cationic drugs. Tolcapone and nimesulide are examples of drugs that exhibit mitochondrial toxicity through uncoupling (Haasio et al., 2002; Nadanaciva et al., 2013).

Other mechanisms of drug-induced mitochondrial injury, which have been demonstrated for antivirals and antibiotics, are the inhibition of mtDNA transcription and translation, respectively (Nadanaciva et al., 2009). The deleterious consequences of these effects are not of acute nature, but will manifest over time (Selvaraj et al., 2014).

Pessayre et al. (2012) and Mehta et al. (2008) describe the effects on mitochondria of more than a hundred compounds that cause liver injury, which are categorized by mechanism of action/target and provide an excellent data resource.

In our mitochondrial studies over the past decade, we noted that many drugs inhibited more than one mitochondrial electron transport chain complex or mitochondrial target. For example, troglitazone was shown to most potently inhibit complex IV but also demonstrated inhibition of complexes II/III and V (Nadanaciva et al., 2007). We observed similar results for many other drugs as well, suggesting promiscuity of many pharmaceuticals. Our in-house analysis of more than 500 marketed drugs exhibiting mitochondrial toxicity revealed that the majority also had unfavorable physicochemical properties, such as high lipophilicity, which can be expected to result in tissue accumulation (clog$P$ > 3) (Will and Dykens, 2014).

Since mitochondrial dysfunction may result in significant adverse functional consequences, it is desirable to select candidate compounds during drug discovery that do not exhibit this adverse property. This requires screening assays that ideally can be run in high-throughput screening (HTS) format and can be used to define structure–activity relationships (SAR). When Pfizer started mitochondrial toxicity assessment in 2004, no suitable tools were available for this purpose. Mitochondrial function (as measured by oxygen consumption) was traditionally measured one compound at a time in a small chamber connected to a Clark electrode, and inhibition of mitochondrial electron transport chain complexes was performed using spectrophotometric cuvette assays. At that time it was impossible to generate dose–response curves on even a handful of compounds. Today, a number of HTS-applicable assays have been developed.

One HTS assay that now is used by many pharmaceutical companies is the "glucose–galactose" assay. This assay measures the ATP content of cells cultured with test compounds for 24 h in media containing high concentrations of either glucose or galactose as an energy source (Marroquin et al., 2007; Swiss et al., 2013). Cells cultured in media containing high glucose concentrations are resistant to ATP depletion caused by mitochondrial toxicants since they can compensate for loss of mitochondrial function by glycolysis, whereas cells cultured in media containing galactose cannot do so. This leads to a marked (>threefold) shift between the potencies of cellular ATP depletion caused by mitochondrial toxins in glucose versus galactose media. However, this shift will only occur with drugs/compounds where mitochondrial toxicity is the dominant mechanism of toxicity. For drugs that cause multifactorial toxicity, such as due to inhibition of BSEP, ROS formation, endoplasmic reticulum (ER) stress,

etc. in addition to mitochondrial toxicity, no difference will be detected in cell lines cultured in glucose and galactose media (Marroquin et al., 2007). In fact, Hynes et al. (2013) showed that only about 2–5% of mitochondrial toxicants can be detected using the glucose–galactose assay. Nevertheless, the assay is useful since it measures cell injury (as indicated by ATP depletion) caused by mitochondrial toxicity and has been commercialized (Swiss et al., 2013). Recently this assay was selected for evaluation by MIP-DILI (www.mip-dili.eu/), which is an IMI/EFPIA-funded consortium of 26 participants from the pharmaceutical industry, SMEs, and academic institutions tasked with building predictive models for DILI.

A more precise HTS assay of mitochondrial toxicity that can be deployed in early drug discovery is respiratory screening technology (RST), which uses soluble sensor molecules to measure oxygen consumption in isolated mitochondria in real time (Hynes et al., 2012). Oxygen consumption measurements can be considered a surrogate readout for mitochondrial bioenergetic function. By modulating the substrate that fuels the electron transport chain, for example, by selecting glutamate/malate or succinate, insight into the site of mitochondrial impairment can also be obtained if so desired. It is also possible to distinguish between inhibition of the electron transport chain and inhibition of fatty oxidation, both of which decrease oxygen consumption, and uncouplers which increase oxygen consumption. The RST assay can be also used to develop SAR and was used to derive an *in silico* tool for prediction of uncouplers (Naven et al., 2013). However, no SAR models have yet been built for inhibitors of mitochondrial function, even though more than 200 different drug classes are known to inhibit complex I alone. Mehta et al. (2008) have published a valuable review of drug classes and their mitochondrial targets.

Methods that assess effects of drugs on mitochondrial transcription and translation have also been described. However, these are labor-intensive multiday assays utilizing antibody approaches and sophisticated imaging techniques and so may be unsuitable for HTS use, although they still offer great insight into possible mechanisms of toxicity (Nadanaciva et al., 2011a).

When chemists are unable to use SAR to select away from mitochondrial liabilities, more complex assays that measure mitochondrial function in cells can be deployed to gain more insight into the risk of toxicity that may arise. For example, both mitochondrial and glycolytic activity can be measured simultaneously using soluble (Hynes et al., 2013) or solid sensor technology (Nadanaciva et al., 2012). In addition, cells can be permeabilized (Divakaruni et al., 2014), which allows testing of different substrates and provides insight into mechanisms of inhibition of bioenergetics and inhibition of FAO (Rogers et al., 2014). High content imaging has also been utilized to visualize effects of drugs that cause liver injury on mitochondrial function in liver-derived immortalized cell lines such as HepG2 (Garside et al., 2014) cultured primary rat (Nadanaciva et al., 2013) or human hepatocytes (Xu et al., 2012; Persson et al., 2014; Trask et al., 2014) or stem cell-derived hepatocyte-like cells (Sirenko et al., 2014). In these various studies, mitochondrial membrane potential was assessed using membrane potential-sensitive fluorescent dyes. However, it is important to note that this parameter is not generally considered to be a specific indicator of direct mitochondrial toxicity, since it can also be affected by cell injury caused by other mechanisms (e.g., following oxidative stress). Nevertheless, high content cell imaging is of value since it is a convenient approach that enables additional mechanistic parameters (ROS, cell death, apoptosis, DNA damage, etc.) to be measured alongside mitochondrial bioenergetic status (see Section 8.4.1 for detection of liver toxicity).

For many years mitochondrial toxicity was studied in isolation as an initiating mechanism that could lead to liver injury, and for a variety of drug classes excellent correlations were observed between the severity of human DILI observed in the clinic and potency of mitochondrial toxicity. Good illustrative examples are the thiazolidinediones (Nadanaciva et al., 2007), biguanides (Dykens et al., 2008a), nonsteroidal anti-inflammatory drugs (NSAIDs) (Nadanaciva et al., 2013), and antidepressants (Dykens et al., 2008b). It was also found that *in vitro* data interpretation become more meaningful when systemic exposure was taken into consideration (Xu et al., 2008; Nadanaciva et al., 2013). In fact, Porceddu et al. (2012) showed using >200 drugs a correlation with human DILI of >80% for drugs in which their IC_{50} values in the *in vitro* assay were <100-fold of the human maximum human plasma drug exposure.

The majority of drugs that inhibit mitochondrial function and cause human DILI did not exhibit evidence of overt liver toxicity in preclinical species that prevented them from entering the market. Furthermore, the incidence of human DILI is low and there is no evidence of overt dose dependence, that is, they cause idiosyncratic human DILI. In fact it is recognized that liver injury has the poorest translatability from preclinical species to human of all clinically concerning organ toxicities caused by drugs (Olson et al., 2000). Why animals and the vast majority of humans treated with drugs that cause mitochondrial toxicity do not exhibit evidence of DILI is unclear. This could be because healthy individuals typically exhibit a substantial reserve of mitochondrial capacity and animals used in preclinical drug safety studies are young, drug naïve, healthy, and nonobese and also lack genetic divergence. Similar explanations can be applied to humans as well, as described elegantly by Ullrich (2007).

Some failed drug candidates cause dose-dependent microvesicular hepatic steatosis and/or single-cell liver necrosis in animals, which could plausibly be due to mitochondrial toxicity, and retrospective investigative studies could confirm this. However, such compounds typically are not progressed into the clinic. In the absence of concerning overt histopathology finding, information on mitochondrial impairment is not provided by the plasma clinical chemistry parameters assessed routinely during preclinical safety testing of drugs in animals or routine monitoring of human clinical trials. For example, potentially informative parameters such as plasma lactate concentration, or activities of the enzymes glutamate dehydrogenase (GLDH) or ornithine carbamoyltransferase (OCT), are not generally evaluated in preclinical or clinical studies. Both enzymes are located in the mitochondria, and OCT is enriched in the liver so if it appears in the serum, it could be indicative of mitochondrial injury in the liver (Labbe et al., 2008). In addition, the mitochondrial function of hepatocytes isolated from animals exposed to drugs *in vivo* could be assessed directly, using the methods described earlier.

Boelsterli's group have described the use of the heterozygous mitochondrial superoxide dismutase (+/−) mouse as a model to gain further insight into the association between mitochondrial toxicity and idiosyncratic DILI and were able to demonstrate overt histopathological changes in the livers following administration of troglitazone (Lee et al., 2008). In addition, when these mice were treated with trovafloxacin (Hsiao et al., 2010) or flutamide (Kashimshetty et al., 2009), which also inhibit mitochondrial function and cause human DILI, biochemical changes indicative of mitochondrial toxicity were observed although there was no histopathological evidence of liver injury. This emphasizes that susceptibility to liver toxicity is a multifactorial process.

Work undertaken by the Pfizer group on the toxicity of nefazodone (Dykens et al., 2008b) provided the first hint that compounds might cause DILI by more than just mitochondrial toxicity. In the case of nefazodone, which exhibits clear mitochondrially mediated cell cytotoxicity (Fig. 8.2), inhibition of the BSEP also seemed to be an additional driver for toxicity (Kostrubsky et al., 2006). Years later, Aleo et al. (2014) confirmed that notion by showing that drugs that inhibit mitochondrial function also for the most part inhibit BSEP and inhibition of both processes correlates with severe DILI (necrosis, death). Moreover, many of the drugs investigated in the latter study have also have been shown to form chemically reactive metabolites, which are another main contributor to DILI. Now that it is known that most drugs cause DILI by multiple mechanisms, advancements are being made in the derivation of cumulative scores from multiple *in vitro* assays that can better predict human DILI (Thompson et al., 2012; Kenna et al., 2015; Shah et al., 2015).

8.3.2 Reactive Metabolite-Mediated Toxicity

Most foreign compounds are metabolized within the liver, and in general this process aids their elimination by converting relatively nonpolar compounds to more polar

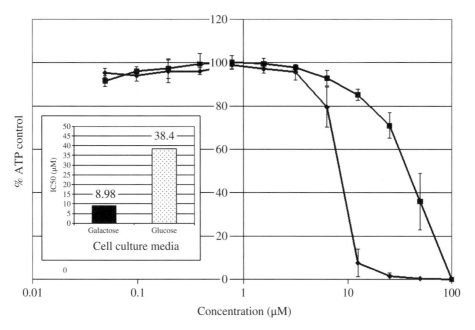

FIGURE 8.2 HepG2 cells grown in either glucose (closed square) or galactose (closed diamond) media were incubated with the indicated concentrations of nefazodone. As reflected by the left shift in the curve, galactose-grown cells were significantly more susceptible to nefazodone treatment (IC_{50} = 8.98 μM) compared with cells grown in glucose media (IC_{50} = 38.40 μM). Data are mean ± SD (n = 3) (Dykens et al. (2008b). Reproduced with permission of Oxford University Press).

metabolites that can be excreted directly into the bile or via the kidneys into the urine. Although the metabolites of most compounds are nontoxic, in some instances biotransformation results in the formation of chemically unstable species that react with cellular macromolecules. The interactions may be noncovalent (e.g., redox processes) or covalent and can result in organ toxicity (liver being the most common organ affected), various immune-mediated hypersensitivity reactions, mutagenesis, and tumor formation (Park et al., 2011). Mutagenesis and carcinogenesis arise as a consequence of DNA damage, while other adverse events (AE) are linked to chemical modification of proteins and in some instances lipids (Park et al., 2011).

The relationship between reactive metabolite formation and toxicity is best defined for DNA-reactive carcinogens and chemicals that cause dose-dependent organ toxicity that can be observed in animals as well as in humans. An excellent illustrative example is provided by the analgesic acetaminophen, which is safe in humans when ingested at therapeutic doses of up to 4 g/day. A therapeutic dose of acetaminophen is metabolized within the liver, primarily via sulfation and glucuronidation, and small amounts of a chemically reactive metabolite (*N*-acetyl-*p*-benzoquinone imine (NAPQI)) also are formed. The NAPQI is detoxified by conjugation to reduced glutathione. When acetaminophen is ingested in overdose, high levels of the NAPQI are formed that deplete hepatic glutathione stores and cause oxidative stress, bind covalently to hepatic proteins, activate the innate immune system, and ultimately result in hepatocellular

necrosis, which in the most severe cases leads to fatal liver failure (Hinson et al., 2010).

Many compounds other than acetaminophen cause dose-dependent reactive metabolite-mediated toxicity to liver or other target organs (e.g., the industrial chemical bromobenzene). Since such toxicity is evident from toxicity studies undertaken in animals, it is readily detected during nonclinical safety testing of candidate pharmaceuticals and so is not an important cause of human adverse drug reactions (with the very notable exception being acetaminophen). Although conventional *in vitro* and *in vivo* toxicological approaches cannot be used to assess the relevance of reactive metabolite formation to idiosyncratic DILI, a substantial weight of evidence approach supports a likely causal relationship. In particular, it has been observed that many drugs that cause idiosyncratic DILI are metabolized to reactive metabolites (e.g., troglitazone), and for several series of structurally and pharmacologically similar drugs, a good correlation has been observed between the extent of bioactivation and the frequency with which symptomatic liver injury can be observed in humans. A good example is the volatile anesthetics halothane, enflurane, isoflurane, and desflurane (Kenna, 2013). Patients who develop liver injury caused by these drugs have been shown to have circulating antibodies that recognize neoantigens formed by covalent binding of reactive metabolites to liver proteins. Such antibodies are not present in drug-exposed patients who do not sustain liver damage, and the inferred mechanism of liver injury is immunological. Similar mechanisms are considered to underlie idiosyncratic DILI caused by numerous other drugs that elicit drug-specific adaptive immune responses, for example, the diuretic ticrynafen and the antihypertensive dihydralazine (Pohl et al., 1988).

Managing reactive metabolite-associated issues during clinical development of drugs is very challenging, since at this stage there is likely to be little if any opportunity to alter the chemical structure of the molecule. Therefore it is preferable to use methods during drug discovery, while there is chemical choice, to proactively assess and minimize reactive metabolite formation. A variety of in silico tools can be used to aid chemical design. Databases can be used to identify, and ideally avoid, substructures known to be especially amenable to bioactivation, while a variety of computational tools enable prediction of especially reactive sites and can underpin design of alternative strategies by which reactivity can be reduced (e.g., http://www.moldiscovery.com/software/metasite/). These approaches should be used in conjunction with metabolic profiling studies, in which test compounds are incubated in metabolically competent systems (most commonly liver microsomes or isolated hepatocytes) and then analyzed by liquid chromatography–mass spectrometry (LC–MS). Ideally, metabolic profiling studies should be performed in the presence of glutathione or other small-molecule nucleophiles in order to trap any highly reactive and hence chemically unstable intermediates (Park et al., 2011).

However, it is important to be aware that qualitative avoidance of reactive metabolite formation is not a desirable or realistic goal during drug discovery. This is because chemical reactivity is a key feature of the mechanism of pharmacological action of numerous valuable drugs (Park et al., 2011). Furthermore, many compounds are metabolized via multiple pathways, and a compound for which bioactivation is a minor route of clearance is likely to raise much greater concern than a molecule where this is a major route. An additional important consideration is the drug dose, since bioactivation of a high-dose drug (given at hundreds of mg/day) can be expected to raise greater concern than that of a low-dose drug (given at <50 mg/day). Therefore quantitative criteria are needed to enable determination of whether or not to progress a candidate drug into development. This highlights an important limitation of in vitro LC–MS-based chemical trapping studies. They are unable to provide quantitative data unless the MS detector can be calibrated using authentic metabolite standards, which in practice are rarely available.

Quantitative determination of reactive metabolite formation can be achieved by determining irreversible covalent binding of radiolabeled test compound to proteins. The preferred test system for such studies is isolated human hepatocytes, which contain a broad variety of relevant enzymes and can be obtained commercially. Covalent binding studies provide invaluable information that can aid selection between potential lead series, or between short-listed compounds prior to clinical development, especially when the data are interpreted in conjunction with information on in vivo drug dose and the fraction of metabolic turnover that results in covalent binding (Nakayama et al., 2009; Thompson et al., 2012). Data generated by AstraZeneca have also indicated that when exposure- and turnover-adjusted in vitro hepatocyte covalent binding data are analyzed alongside in vitro data that quantifies other human DILI initiating mechanisms, highly specific and sensitive discrimination between hepatotoxic and nonhepatotoxic drugs can be achieved (Thompson et al., 2012; Kenna et al., 2015). However, they are unsuited to screening of large numbers of compounds since they can only be undertaken using radiolabeled test compounds.

8.3.3   BSEP Inhibition

BSEP is a liver-specific and ATP-dependent transport protein that mediates the excretion of bile salts into bile and is expressed on the apical plasma membrane domain (canalicular surface) of hepatocytes (Lam et al., 2010). Bile formation and excretion is an essential biological process in higher vertebrates and is an important route of xenobiotic elimination, which also plays a key role in intestinal dissolution and absorption of lipids, vitamins, and fat-soluble food components. Bile salts are synthesized within hepatocytes by cytochrome P450-mediated metabolism of cholesterol and are key components of bile (Hofmann and Hagey, 2008). Bile formation and bile flow are regulated physiologically by complex mechanisms, which in hepatocytes include nuclear hormone receptor-mediated transcriptional pathways and a variety of posttranscriptional processes (Kullak-Ublick et al., 2004; Gonzalez, 2012).

Impaired bile flow results in the accumulation of bile salts and other biliary constituents within the liver and is termed cholestasis. In humans, mutations or single nucleotide polymorphisms in the gene that encodes BSEP (ABCB11) cause reduced levels of transporter expression and impaired BSEP function. This results in the accumulation of high concentrations of bile salts within hepatocytes, where they act as detergents and cause cell toxicity (Strautnieks et al., 1998; Stieger et al., 2007). The most severe consequence is progressive familial intrahepatic cholestasis type 2. This is characterized by early-onset cholestasis soon after birth and subsequent progressive degenerative liver cell injury, which unless treated by liver transplantation is fatal. Less functionally severe ABCB11 gene mutations result in nonprogressive cholestasis, termed benign recurrent intrahepatic cholestasis type 2, and in intrahepatic cholestasis of pregnancy. Degenerative cholestatic liver injury has also been documented in homozygous Bsep−/− knockout mice (Zhang et al., 2012).

A variety of drugs that cause cholestatic or mixed cholestatic/hepatocellular DILI have been found to inhibit the activity of BSEP *in vitro* and to cause elevated concentrations of plasma bile acids *in vivo*, indicating functional impairment of bile flow and intrahepatic accumulation of cytotoxic bile acids (Fattinger et al., 2001; Funk et al., 2001). Evaluation of >250 drugs further demonstrated that the frequency and potency of BSEP inhibition by drugs that caused human DILI were markedly greater than those of drugs that did not cause DILI (Morgan et al., 2010, 2013; Dawson et al. 2012). Nevertheless, a variety of drugs that exhibited potent BSEP inhibition but did not cause human DILI were also identified. The majority of the "false-positive" drugs are administered to humans at low oral doses, or via nonoral routes that result in low plasma exposure, or are also part of very short courses of treatment. Therefore when drug dose and plasma exposure were also taken into account, an excellent correlation between BSEP inhibition and DILI was evident, and the ratio between plasma drug concentration and *in vitro* BSEP IC_{50} was a more informative indicator of DILI propensity than BSEP IC_{50} alone. In view of this, it was inferred that BSEP inhibition is a risk factor that, in susceptible humans, may initiate cholestatic DILI. Furthermore, it has been proposed that proactive evaluation of BSEP inhibition during drug discovery may aid in the identification and deselection of compounds that can cause DILI (Kenna, 2014).

Evaluation of the physicochemical properties of drugs that inhibit BSEP and drugs that do not revealed an excellent correlation between BSEP inhibitors (*in vitro* $IC_{50} < 300\,\mu M$), molecular weight >250 Da, Clog$P$ > 1.5, and nonpolar surface area >180 Å (Dawson et al., 2012; Warner et al., 2012). However, many drugs that exhibit these properties were found not to inhibit BSEP; therefore they cannot be used in isolation as a predictive *in silico* tool. Rather, these molecular descriptors can be used to discriminate between compounds that should be tested using *in vitro* BSEP inhibition assays and compounds where such testing is not required because BSEP inhibition can be considered highly unlikely (Warner et al., 2012).

The most commonly used *in vitro* method for evaluation of BSEP inhibition utilizes membrane vesicles prepared from insect cells (*Spodoptera frugiperda* Sf9 or Sf2) transiently transfected with a plasmid that encodes BSEP cDNA. The effects of test compounds on ATP-dependent accumulation of probe substrate into inverted (i.e., inside-out) membrane vesicles is readily evaluated following immobilization of the membranes on filter disks (Dawson et al., 2012; Warner et al., 2012). Exhaustive washing of the filter disks enables nonspecific adherence of untransported substrate to be minimized and high signal/noise to be obtained. Suitable membrane vesicles are available commercially, and medium- to high-throughput data generation can be achieved by the use of 96-well plate format. Most published studies have been undertaken using the radiolabeled and commercially available bile acid [^{3}H]-taurocholic acid. More physiologically relevant data can be obtained by

undertaking similar assays using purified apical plasma membrane vesicles obtained from liver tissue, which express BSEP in a native membrane environment that includes other biliary efflux transporters and apical membrane proteins (Stieger et al., 2000). However, generation of such membrane vesicles is technically challenging and labor intensive and requires availability of freshly isolated liver tissue, while thorough characterization of the purity of the vesicles is also needed. In addition, the amounts of material that can be obtained are relatively low, and these vesicles are not available commercially and therefore are unsuitable for use in high-volume compound screening.

A substantial number of drugs and other compounds have been evaluated using both apical liver plasma membrane vesicles and vesicles from transfected insect cells, and a good overall concordance between data obtained using the two assay systems has been observed. One notable exception is estradiol-17β, D-glucuronide, which inhibited BSEP activity in apical plasma membrane vesicles prepared from rat liver but not in rat BSEP-transfected Sf9 insect cells (Stieger et al., 2000). This compound is an endogenous metabolite that plays an important role in human intrahepatic cholestasis of pregnancy and is a substrate of the biliary efflux transporter multidrug resistance-associated protein 2 (MRP2) (Gerk and Vore, 2002). Further transport studies undertaken using membrane vesicles from Sf9 cells cotransfected with both BSEP and MRP2, plus apical plasma membrane vesicles prepared from livers of MRP2-deficient rats, revealed that BSEP inhibition by estradiol-17β, D-glucuronide could be observed only using membrane vesicles that coexpressed MRP2. The presumed explanation is that the compound *trans*-inhibits BSEP following its MRP2-mediated secretion into bile canaliculi (Stieger et al., 2000). Such cooperativity between BSEP and MRP2 appears to be an unusual feature of estradiol-17β,D-glucuronide and to date to has not been observed with other compounds or indeed with any pharmaceutical drugs.

An additional *in vitro* approach that can provide valuable insight into the possible influence on BSEP inhibition of cooperative interactions with other transporters is by studying vectorial transport of probe substrates using transfected cells cultured as confluent monolayers in transwell devices. This approach enables quantification of both basal-to-apical transport through the cells and apical efflux. In these studies it is important to use polarized epithelial cells that express distinct apical and basolateral sinusoidal plasma membrane domains and that also exhibit a low level of endogenous expression of transporters other than BSEP that may mediate apical probe substrate efflux. Appropriate cells include dog kidney-derived MDCK cell lines and porcine kidney-derived LLC-PK1 cells. It is also necessary to ensure that basolateral solute carriers are present in the cells that mediate probe substrate uptake. This may require coexpression of BSEP and relevant uptake transporter(s). MDCK cell lines coexpressing the human hepatic uptake transporter OATP1B3 (SLC21A8) and human MRP2 have been described (Cui et al., 2001), as have MDCK cells and LLC-PK1 cells that coexpress human BSEP and the hepatic basolateral uptake carrier NTCP, or their rat orthologues (Mita et al., 2006a, b; Fahrmayr et al., 2012). In view of the technical complexity and limited throughput of studies undertaken using transfected cells cultured in transwell devices, they are unsuitable for use as a high-volume screen and their most appropriate use is as investigational tools.

Neither membrane vesicle assays nor MDCK or LLC-PK1 cells exhibit significant drug biotransformation activities. Therefore these assay systems are unable to provide information on possible BSEP inhibition caused by drug metabolites. Since some drug metabolites exhibit markedly more potent BSEP inhibition than the corresponding parent drugs (e.g., troglitazone sulfate; Funk et al., 2001), this is an important limitation. Currently the most promising approach for investigation of possible metabolite-mediated BSEP inhibition involves the use of isolated hepatocytes maintained in culture on collagen-coated tissue culture plates and overlaid with Matrigel™ or gelled collagen. Under these conditions, "sandwich-cultured hepatocytes" (SCH) reestablish intracellular connections and regain plasma membrane polarity over the course of several days (Swift et al., 2010; De Bruyn et al., 2013). BSEP and other biliary efflux transporters become located on apical plasma membrane domains between adjacent cells and these apical domains are surrounded by impermeable tight junctions, which form sealed canalicular pockets, while basolateral plasma membrane domains are expressed at the interface between hepatocytes and the cell overlay. BSEP activity can be quantified by preloading cells with [^{3}H]-taurocholic acid, which is taken up from the culture medium via basolateral carriers. After washing to remove unaccumulated substrate, the cells are incubated either in buffer that contains Ca^{2+}/Mg^{2+} (which ensures that tight junctions are sealed) or in Ca^{2+}/Mg^{2+}-free buffer (which disrupts protein–protein interactions between connexins, leading to the release of material that otherwise remains enclosed within sealed canalicular pockets). Subtraction of substrate efflux into medium that contains Ca^{2+}/Mg^{2+} (i.e., transport mediated by basolateral transporters) from efflux into Ca^{2+}/Mg^{2+}-free medium (i.e., transport mediated by both basolateral transporters and apical BSEP) enables indirect estimation of BSEP-mediated apical transport. For a variety of test compounds, biliary excretion determined using this approach has been shown to correlate closely with biliary excretion observed in bile duct-cannulated rats in vivo (Liu et al., 1999). The method has also been used to demonstrate inhibition of BSEP-mediated apical excretion of [^{3}H]-taurocholate from rat and human SCH by numerous drugs found to inhibit BSEP activity when tested in membrane vesicle assays (Kostrubsky et al., 2006; Chu et al., 2013). When interpreting effects of test compounds on apical efflux from SCH, possible effects of

the compounds on basolateral probe substrate uptake also need to be considered (since the cells express both uptake and efflux transporters). This requires assessment of whether the extent of probe substrate uptake is affected by the compound. Interestingly, a number of drugs have found to inhibit both sinusoidal uptake transport activity and BSEP-mediated apical efflux (Chu et al., 2013).

Studies with SCH require substantial resources and expertise and are dependent upon access to plateable primary hepatocytes of high quality. In addition, a minimum of several days in culture is needed to enable SCH to repolarize. Therefore this model is best suited for low- to medium-throughput characterization of small numbers of test compounds and not high-volume compound screening.

Profiling of elevated plasma concentrations of bile acids provides evidence of functional inhibition of bile salt clearance in vivo. This has been observed in rats exposed to a variety of drugs that inhibit BSEP activity in vitro, and in some instances also in humans (Fattinger et al., 2001), which suggests that evaluation of plasma (or serum) bile acids could provide a useful noninvasive biomarker of in vivo BSEP inhibition. However, since uptake carriers on the basolateral plasma membrane domain mediate hepatic uptake of bile acids and therefore also play important roles in bile acid clearance (Kullak-Ublick et al., 2004), potential interactions involving these transporters need to be considered when interpreting effects of compounds on plasma bile acid concentrations. In addition, elevated bile acid concentrations have been observed in plasma and urine from rats treated in vivo with a variety of hepatotoxic drugs, some of which do not inhibit BSEP (e.g., acetaminophen, carbamazepine; Yamazaki et al., 2013). Therefore while elevated total plasma bile acid levels can provide a useful indirect index of impaired in vivo BSEP-mediated bile salt clearance, they may also arise due to other mechanisms and so cannot be considered a specific in vivo BSEP inhibition biomarker.

When interpreting in vitro BSEP inhibition data, it is necessary to take account of in vivo drug exposure (in addition to route of administration and treatment regimen). This can be achieved by estimating the ratios between total steady-state plasma drug concentrations (Css) and BSEP IC_{50} values. Morgan et al. (2013) have reported data obtained with >600 drugs and observed that 109 drugs exhibited Css/BSEP IC_{50} ratios >0.1, 95% of which cause DILI. Interestingly, this approach suggests that even relatively nonpotent in vitro BSEP inhibition (IC_{50} of several hundred micromolar) may be relevant for drugs that are administered at high doses and achieve high in vivo exposures (e.g., where total plasma concentrations >100 μM), which is consistent with a proposal that the overall in vitro BSEP IC_{50} "threshold of concern" may be as high of 300 μM (Dawson et al., 2012).

The emerging data linking BSEP inhibition with risk of human cholestatic DILI has been acknowledged recently by the International Transporter Consortium (ITC), which also has highlighted that there remain important gaps in our scientific understanding, and in the Guideline on Investigation of Drug Interactions issued by the European Medicines Agency (EMA) in 2012 (European Medicine Agency, 2012). The current recommendations focus on the need to investigate and understand possible BSEP interactions during drug development. However, identification of concerning levels of BSEP inhibition in drug development provides little or no opportunity to mitigate the problem by selecting an alternative compounds with reduced (or ideally no) BSEP liability. An alternative is to undertake prospective BSEP screening during drug discovery, when chemical choice is available, to select a compound with least possible BSEP inhibition potency. This is the approach now being undertaken by some pharmaceutical companies. If it is necessary to progress a compound into development that exhibits *in vitro* BSEP inhibition, it has been proposed that evaluation of total plasma bile acid concentrations alongside conventional markers of liver injury may be used to assess whether BSEP inhibition also occurs *in vivo* and to define safety margins (Kenna et al., 2015). In addition, a multiparametric approach that takes account of additional chemical insult has been recommended when assessing the potential human DILI risk posed by a drug that inhibits BSEP (Thompson et al., 2012; Kenna et al., 2015; Shah et al., 2015).

8.3.4 Complicity between Dual Inhibitors of BSEP and Mitochondrial Function

Beyond individual considerations of the involvement of certain hepatic transporters, like BSEP, and mitochondrial dysfunction on DILI, it is important to consider why these two risk factors can contribute so much to the manifestation of severe DILI in humans (Aleo et al., 2014). Bile acids, in general, appear to represent a more reliable marker of hepatic dysfunction regarding excretory abnormalities than serum bilirubin. Elevations in total bile acids can precede evidence of liver damage (as diagnosed by serum transaminase and bilirubin elevations) caused by hepatotoxic drugs by upward of 1 week (Fattinger et al., 2001). In order to explain this association, it is important to understand the intricate interactions between mitochondrial bioenergetics and transporter function within the liver.

It has been found that 90% of hepatic oxygen consumption is cyanide sensitive (Seifter and England, 1988) and that the major canalicular efflux transporters that mediate bile flow from the liver are ATP dependent (e.g., bile acids through BSEP and conjugated bilirubin through MRP2 transport) (Trauner and Boyer, 2003). Although one might think that hepatocellular ATP production is in excess of need, it has been shown that reductions in hepatic ATP levels are highly associated and tightly regulated with reduced biliary excretion. In rats treated with the hepatotoxic agent ethionine, even a relatively small (30%) reduction in liver ATP content is associated with a similar 20% reduction in bile flow (Slater and Delaney,

1970). Subsequently, others (Kamiike et al., 1985) showed an exquisite time-course and functional dynamic link between liver ATP content and bile flow in the rat under ischemic conditions and after ethionine and cyanide administration (see Fig. 8.3).

This functional link also has been observed in humans. In elegant experiments where bile samples were obtained from humans, relieved of obstructive jaundice by percutaneous transhepatic biliary drainage, and samples of liver were obtained during surgery (and then were measured for liver ATP content), hepatic ATP levels were found to be highly correlated with reductions in excretion into bile of bile acids (as BSEP substrates) and indocyanine green (as an MRP2 substrate) (Chijiiwa et al., 2002) (see Fig. 8.4). A 50% reduction in human liver ATP levels was associated with a 75% decline in bile acid output and a 50% decline in indocyanine green output, indicating that modest reductions in liver ATP content can compromise bile acid output and indocyanine green clearance by the liver. Because of this tight association, these authors suggested that bile acid output and indocyanine green clearance could be used as indirect indices of liver ATP content (Chijiiwa et al., 2002), even in hepatectomized patients (Kurumiya et al., 2003). This further suggests that these two endpoints could be useful surrogate biomarkers of mitochondrial dysfunction that leads to ATP depletion within the liver.

An association between pharmaceuticals that cause severe forms of DILI and dual inhibition of both BSEP and mitochondrial respiration was reported recently (Aleo et al., 2014).

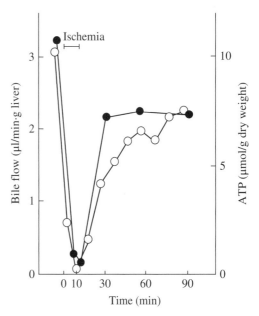

FIGURE 8.3 Suppression of bile excretion and the ATP level in liver ischemia and their recoveries during recirculation. (○) = bile flow rate; (●) cellular level of ATP in the rat liver. Original work reproduced from Kamiike et al. (1985) with kind permission from Wolters Kluwer Health.

(a)

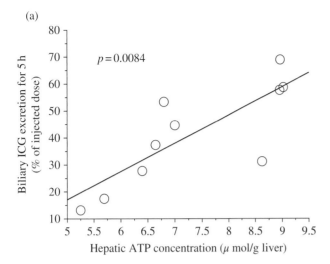

(b)

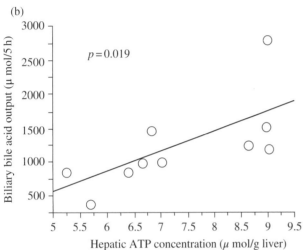

FIGURE 8.4 Indocyanine green excretion rate (a) and Biliary acid output (b) as a function of liver ATP content (Chijiiwa et al. (2002), figures 1 and 2. Reproduced with permission of Springer).

In this study, the majority of drugs that were potent inhibitors of both BSEP and mitochondria were classified as causing MOST-DILI concern, rather than LESS-DILI concern, in humans. This is based on a classification system in which drugs that were associated with severe liver injury leading to death or liver transplantation, or had serious warnings for liver injury, were categorized as MOST-DILI concern (Chen et al., 2011). When a similar analysis was undertaken using drug labeling information assigned by the US Food and Drug Administration (FDA), the majority of drugs that were dual and potent inhibitors of BSEP and mitochondrial bioenergetics were found to have been withdrawn from the marketplace or assigned a black box warning for liver injury. This association was also present, albeit to a lesser extent, in drugs that had liver injury in humans categorized as "warnings and precautions," "adverse reactions," or "no mention" of liver injury in their label (see Fig. 8.5).

High concentrations of bile acids are directly cytotoxic to hepatocytes through their detergent effects and hydrophobicity (Armstrong and Carey, 1982; Attili et al., 1986). Thus, if a drug is a dual and potent inhibitor both of BSEP and of mitochondrial function, the effects it may have on impairing these two fundamental systems can easily multiply itself due to the accumulation of bile acids within the liver, which leads to even more mitochondrial dysfunction and cell death.

In addition, there are several negative feedback processes that provide another plausible rationale for the association of inhibition of these two processes with severe forms of DILI in humans. The accumulation of hydrophobic bile acids within hepatocytes, either through direct inhibition of BSEP or a reduction in its transporter expression/function, has long been known to disrupt mitochondrial function (Rolo et al., 2000) due to the accumulation of bile acids within the liver. Primary bile acids reduce mitochondrial membrane potential to a greater extent than secondary bile acid species, which is in line with their hydrophobicity. Furthermore they also impair state 3 respiration (ADP dependent) and increase state 4 respiration (similar to classical uncouplers). This is believed to be caused by mitochondrial swelling induced by increased mitochondrial membrane permeability to protons. The impact of certain bile acids on mitochondrial energy production is stark and has important ramifications. For example, glycochenodeoxycholate can quickly and extensively decrease hepatocellular ATP content through inhibition of state 3 respiration. This, in turn, is followed by a sustained rise in cytosolic free calcium (Spivey et al., 1993), which then triggers calcium-dependent proteolysis and cell death. Most importantly, the depletion of ATP and consequent rise in intracellular calcium levels under these circumstances can be ameliorated through the addition of fructose, a glycolytic substrate in the liver. The protective effects of fructose have been shown when cyanide (an inhibitor of mitochondrial respiration), oligomycin (an inhibitor of the mitochondrial ATP synthase), or carbonyl cyanide m-cholorophenylhydrazone (CCCP, an uncoupler) was administered as single agent (Nieminen et al., 1994). However, the hepatoprotective effects of fructose were eliminated when cyanide and CCCP were added together. This is an extremely important observation since it suggests that drugs that have both mitochondrial inhibitory and uncoupling activities may abolish a very important compensatory and cytoprotective pathway of generating additional ATP through glycolysis. Thus, bile acids that accumulate within the liver due to the presence of a drug that can themselves both inhibit BSEP and mitochondrial function can further contribute to the disruption of mitochondrial function. These changes, along with oxidative stress (Rodrigues et al., 1998), also promote the induction of the mitochondrial permeability transition that in turn triggers cell death (Palmeira and Rolo, 2004) through proapoptotic signaling and caspase activation (Rodrigues et al., 1999; Sodeman et al., 2000). This specific

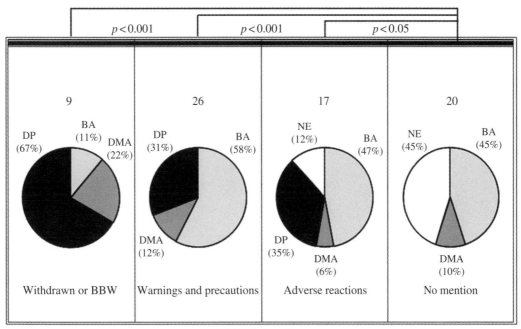

FIGURE 8.5   Breakdown of intrinsic bile salt export pump (BSEP) and mitochondrial inhibition liabilities as a function of FDA label categories. (Aleo et al. (2014), figure 4, pp. 1015–1022. Reproduced with permission from John Wiley and Sons). NE (white shading), no effect; BA (light gray shading), BSEP activity alone; DMA (dark gray shading), dual moderate activity; DP (black shading), dual potency (potent mitochondrial effect with BSEP inhibition). Overall significance was determined using Cochran-Mantel-Haenszel test (p <0.001) with values of pairwise comparisons noted within the figure or nonsignificant (NS) (p > 0.05).

event can be inhibited by administration of antioxidants (Yerushalmi et al., 2001).

The synergistic adverse effects arising from dual inhibition of mitochondrial function and BSEP by drugs can be further confounded by disease states in humans in which significant mitochondrial deficits are already present. Patients with type 1 diabetes diagnosed with hepatic glycogenosis (Sayuk et al., 2007) and type 2 diabetes (Dutu et al., 1985; Rabol et al., 2010) and people with nonalcoholic steatohepatitis (Pérez-Carreras et al., 2003) have low mitochondrial reserve function (Kurbatova et al., 2014), which may predispose them to DILI through these mechanisms (Massart et al., 2013). Background inflammation and oxidative stress are involved with the pathogenesis of diabetes (Akash et al., 2013), as is reduced antioxidant defense (Bonnefont-Rousselot et al., 2000). These processes would further impair the ability of the liver to cope with oxidative stress arising via mitochondrial inhibition/uncoupling caused by the drug itself or following the accumulation of cytotoxic bile acids following BSEP inhibition.

Furthermore, reduced expression or function of BSEP has been associated with an increased risk of DILI in humans (Lang et al., 2007).

There are many complex mechanistic and regulatory elements of BSEP expression that appear to be involved in modulating the severity of liver injury. These include

interactions with other hepatic uptake and efflux transporters and recently have been discussed in detail by other investigators (Rodrigues et al., 2014). This paragraph will only briefly highlight the complexity of the interactions (see Fig. 8.6). The first element to acknowledge is the dynamic interplay of specific bile acid pools within the liver that undergo enterohepatic recirculation and subsequent metabolism. Approximately 90% of bile acids entering the gut are reabsorbed, and as many as 20 cycles of gut reabsorption, liver uptake and metabolism, and liver efflux back into bile may be needed to eliminate a given bile acid completely from the host (Gonzalez, 2012). Bile acid metabolism and the relative abundance of individual bile acids can shape the manifestation of DILI, given the cytotoxic nature of certain highly hydrophobic bile acids and their poor affinity for efflux out of the liver through BSEP (e.g., lithocholic acid). The presence of certain bile acids can also serve as signaling substrates leading to changes in multiple transporter expression (BSEP, MRP4, NTCP, and OATP) via the nuclear hormone receptors FXR, PXR, and CAR. Drugs that can alter these natural response mechanisms for dealing with increased levels of hepatic bile acids (by inhibiting nuclear hormone receptor expression or function) can further aggravate the cholestatic and/or hepatocellular toxic response. Drug metabolites can have actions on the

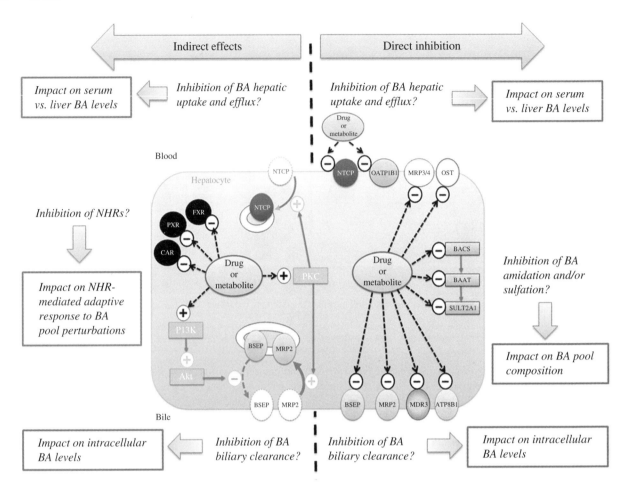

FIGURE 8.6   Summary of the potential mechanism(s) by which a drug or metabolite can impact the hepatobiliary disposition of BAs (Rodrigues et al. (2014), figure 3, pp. 566–574. Reproduced with permission of ASPET). (*See insert for color representation of the figure.*)

above pathways as well and in many cases have been shown to either inhibit (*cis-* or *trans-*inhibition) efflux transporters.

The key efflux transporters that appear to modulate the manifestation and severity of DILI, in addition to BSEP, are MRP2, which is present in the apical membrane, and MRP3/ MRP4, which are present in the basolateral membrane (Susukida et al., 2015). In this context, Morgan et al. (2013) showed the value of including multiple efflux transporters in the manifestation of DILI with several drugs showing inhibition of one or more MRP proteins in addition to BSEP. A follow-up publication showed that the dual interaction of BSEP with MRP4, more so than MRP3, was more important in increasing the odds of causing cholestatic DILI based on an analysis of 88 drugs (Köck et al., 2014). Although this line of research has led to some computational approaches for identifying potential interactions with drugs, it also showed that many similar chemical attributes (hydrophobic and hydrogen bond acceptor features) lead to the inhibitory effect, suggesting transporter inhibitory properties may easily overlap one another (Welch et al., 2015). With additional

time the ability to integrate or simulate these multifactorial elements will be needed to better approximate risks of liver injury in humans.

Overall, in view of the deleterious consequences of dual inhibition of mitochondrial function and BSEP inhibition, it is highly desirable to avoid these two synergistic safety liabilities as much as possible during the drug development process. This is especially important for drugs intended for patient populations in which BSEP expression, mitochondrial reserve capacity and function, and antioxidant status are compromised.

8.4   MODELS SYSTEMS USED TO STUDY DILI

8.4.1   High Content Image Analysis

During the last decade, high content imaging and analysis (HCA) has emerged as a powerful tool for assessing molecular, cellular, and tissue-based toxicity (O'Brien, 2014; Persson et al., 2014). Since HCA is based on automated imaging of cells or tissues, typically making use of fluorescent dyes, antibodies, or reporter systems, it offers some key advantages over the biochemical assays traditionally used for *in vitro* toxicology assessments. The main advantages are that the endpoints can be assessed on the single-cell level and that multiple endpoints can be assessed simultaneously. This enables the user to customize and multiplex endpoints, and if needed gate for particular cells of interest, which provides an integrated assessment of cellular or molecular toxicity and its mode of action. The multiplexing of different endpoints is only limited by the specificity of the employed dyes and antibodies and their spectral separation. This allows for highly customized assays where endpoints can be selected to detect specific mechanisms of action or for more generalized toxicities.

Whereas HCA and cell toxicity can be employed to predict various types of animal and human toxicities, even showing high concordance with systemic tolerance and the severity of general organ toxicity (Benbow et al., 2010), most interest and validations have been toward human clinical DILI. In the pharmaceutical industry, the HCA assays are generally geared toward general mechanisms, such as various states of cell health, in order to be a "one-shoe-fits-all" solution offering some, but limited, understanding of mechanism. This is a necessity when screening hundreds or thousands of compounds in a typical small-molecule drug discovery project and is most often enough to identify safety flaws in the molecules. This can be used to offer chemists active guidance on safer chemical series and direct their efforts toward safer drugs. In fact, most HCA assays for prediction of human DILI gain their predictivity primarily from mitochondrial-related and cytotoxicity-related parameters (O'Brien et al., 2006; Xu et al., 2008; Tolosa et al., 2012; Persson et al., 2014). Additional parameters, such as oxidative stress, lysosomal activity (Nadanaciva et al., 2011b; Lu et al., 2012), steatosis (Donato et al., 2012), imaging of canalicular bile salt transport (Qiu et al., 2015), glutathione levels, etc. can provide additional predictivity depending on the type of compounds assessed (see Fig. 8.7). Typically, the HCA assays for predicting human liver toxicity show sensitivities in the range of 40–90% and, most importantly, <10% false positives independent of whether the assay employs primary hepatocytes or cell lines such as HepG2 (O'Brien et al., 2006; Xu et al., 2008; Tolosa et al., 2012; Persson et al., 2013). The high predictivity seems to hold true for real-life applications in the pharmaceutical industry as well. In the MIP-DILI consortium it has been reported from individual pharmaceutical company partners (such as Janssen, GlaxoSmithKline, Orion Pharma, and Lundbeck) that their in-house HCA assays typically generate a sensitivity of 40–60% with <10% false positives when screening hundreds of reference compounds. Not surprisingly, relating the cellular toxicity to exposure or dose is key for reducing the false-positive rates and driving the predictivity (Persson et al., 2013; O'Brien, 2014). The safety margin needed for separating *in vitro* toxicity and human exposures causing risk for DILI is likely assay and model dependent, but once known, it can be used for guidance on safe doses or exposures (Persson et al., 2014).

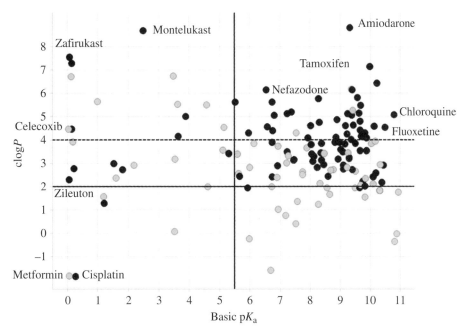

FIGURE 8.7 Clustering of compounds positive for the lysosomal mass endpoint according to their pK_a and clog$P$. The majority of compounds were associated with cell loss (black circle), especially compounds with clog$P$>4. In contrast almost all compounds with clog$P \leqslant 2$ showed no cell loss (gray circle) (Lu et al. (2012), pp. 613–620. Reproduced with permission of Elsevier).

It is important to note that HCA assays are often considered as state of the art for first-line toxicity screening at many larger pharmaceutical companies. This is because, once the required equipment is acquired, the assays are very cost effective when employing fluorescent dyes, have medium- to high-throughput capacity, are easy to handle, and most importantly provide high predictivity for human DILI with few false positives. By necessity, the early screening assays need to be simple and with high reproducibility. This has so far restricted most of the HCA assays to relatively simple two-dimensional cell cultures. These simple assays and models can later be followed in the screening cascade, when the number of acceptable compounds has been reduced, with more complex and sophisticated model systems.

8.4.2 Complex Cell Models

While simple model systems that measure multiple parameters, for example, by use of HCA, are ideally suited to screen out more than half of the compounds with safety liabilities in a closed automated fashion, there is still room for improvement as they do not enable detection of many drugs that cause DILI. The lack of "complete" sensitivity toward DILI is most often attributed to lack of full physiological metabolism capacity, low tissue-like morphology, low intercell communication, lack of transporters, lack of immune cells and other types of cells, limited possibilities for repeat-dose testing, and of course, the simple fact that not all mechanisms of toxicity are covered in the typical cell models used in these studies. Several more complex cell models have been generated in order to attempt to correct or circumvent these issues. None of the systems are ideal and able to address all the shortcomings of the simple models, but each one has its own strengths and weaknesses. The choice will come down to the mechanism of toxicity or issue one is trying to address or predict.

Micropatterned coculture of hepatocytes with other cell types, such as Kupffer cells or stromal cells, has been shown to address the decline of hepatic function and lack of prolonged cell survival often seen with hepatic cell lines or the common sandwich-cultured primary hepatocytes. Cocultivation allows exchange of soluble factors between cell types, which increases hepatocyte differentiation, maintains some drug-metabolizing activity, and prolongs time in culture to enable repeat-dose testing. Such systems typically allow several days of dosing in a more physiological manner than can be achieved during more conventional cell toxicity testing. Toxicity data obtained using human micropatterned cocultures of hepatocytes and stromal fibroblasts have been shown to predict human DILI with 60% sensitivity and <10% false positives, following 9 days of dosing with four repeat drug administrations and assessment of traditional biochemical markers of cell toxicity (Khetani et al., 2013).

A more advanced cell model is provided by three-dimensional (3D) spheroids, also known as microtissues. Like the micropatterned cocultures, the spheroids allow culture longevity for repeat-dose testing as well as coculturing of several cell types but have the advantage of naturally forming multicell microtissues with correct cell polarity. The 3D environment allows for more physiological cell-to-cell contact, exchange of soluble factors, and complex interactions (Messner et al., 2013). Although the composition of the spheroids is advanced, they can be formed relatively easily by self-assembly using gravity in hanging drops, ultralow attachment plates, or the InSphero™ technique without the use of scaffold material. Studies have shown that spheroid cultures of primary hepatocytes, stem cell-derived hepatocytes, or cell lines such as HepG2 induce or maintain hepatic functions, such as phase I and phase II xenobiotic-metabolizing enzymes, albumin secretion, bile transporters, and other relevant functions, in "long-term" culture (Ramaiahgari et al., 2014; Roth and Singer, 2014).

The spheroids are suitable for traditional biochemical toxicity assays, histopathology, and confocal imaging, both for acute and repeat-dose testing. However, while the spheroid systems hold large promise for the future of human DILI predictions, due to their relative complexity, there are currently no large validation studies that have assessed their DILI predictivity. Nonetheless, the spheroids have shown their superiority when compared with simpler models for detection of toxicity caused by drugs such as trovafloxacin under inflammatory conditions (Messner et al., 2013); this drug was withdrawn from the US market because it caused severe idiosyncratic DILI that was evident only following its administration to large numbers of humans.

Similar to spheroids, bioreactors and fluidic chambers typically maintain many hepatic functions such as high xenobiotic metabolism but have the advantage of being perfused in a more physiological manner (Bale et al., 2014). This allows for precise control of media flow for supplying fresh nutrients, removal of waste products such as drug metabolites, and buildup of gradients to provide oxygen zonation (Bhushan et al., 2013). These biomimetic livers, sometimes called liver-on-a-chip, show large potential, but further studies are needed to explore and realize their full potential as well as understand their limitations.

The complex cell models hold large promise for the future prediction of human DILI as they represent more physiologically relevant systems with cell-to-cell interactions; maintain much of the hepatic function, such as relatively high drug metabolism; and allow repeat-dose testing. Due to their relatively complex nature and cost, and being more labor intensive, they are mostly suitable as second-tier screening assays when the primary assays, such as HCA in two-dimensional cell cultures, have reduced the number of compounds to be tested. Currently, more extensive validations that extend beyond the positive proof-of-principle studies already undertaken using some drugs are needed to realize the potential of complex models and accurately position them in the screening cascades of the pharmaceutical industry.

8.4.3 Zebrafish

In the spirit of the reduction, refinement, and replacement (3Rs) of animal experimentation, the use of embryonic or larval zebrafish as a model for pharmaceutical screening of adverse effects of drug has appeal on a number of levels. Due to their diminutive status, embryonic and larval zebrafish also offer enormous potential to provide integrated safety information at a fraction of the bulk material costs typically associated with conducting a short-term rodent study (milligrams vs. grams). The many practical advantages of testing in this species have been reported often:

1. Live vertebrate animal with high degree of genetic and physiologic similarity to mammals.
2. Small embryos/larvae can be arrayed in multiwell plates, and compounds added to culture plate wells are absorbed through exposed surfaces and gills (this unfortunately can make compound uptake variable but is no different than any other culture systems).
3. Transparency enables visualization of multiple phenotypes and morphological changes by inverted phase microscopy either through expert technical assessment or automation.
4. High fecundity and rapid external development allow for quick generation of embryonic and larval zebrafish for experimentation, especially since most major organs are established by 5-day postfertilization (dpf).
5. Development of toxic responses to a number of drugs/small molecules comparable to mammals (endocrine disruption, reproductive and developmental toxicity, behavioral defects, carcinogenesis, cardiotoxicity, hepatotoxicity, neurotoxicity, and ototoxicity, to name a few).

Because of these features, zebrafish have been used as part of phenotypic screening for pharmacologic activity, organ system function, and toxicity. Since they are whole animals, they have the potential to provide insight into complex biological adverse effects that may not be adequately represented in isolated cell types or even complex 3D cell cultures. However, it is important to note that a complete understanding of the physiological and pharmacological similarities and differences between zebrafish and humans is not yet fully understood.

The anatomical structures of embryonic, larval, and adult zebrafish are well known (Menke et al., 2011), and the University of Oregon maintains an extensive website that contains anatomical references, developmental and staging atlases, and ontologies for researchers to use as an additional resource. This organization's website called ZFIN (http://zfin.org/) also contains a database of genomic information (ZFIN: The Zebrafish Model Organism Database) and is coordinated with the Zebrafish International Resource Center (ZIRC) to offer zebrafish transgenic lines and other important information for researchers in this area.

For pharmaceutical research and development, the use of zebrafish as an alternative vertebrate model is becoming more common for both pharmacology and toxicology work (Vogt et al., 2010; Peterson and Macrae, 2012), and elaborate robotic systems that are fully automated for injecting zebrafish embryos have been developed for HTS of embryonic zebrafish (Wang et al., 2007). Certainly as a general screening assay for acute lethality, there is a good correlation to rodent lethality, especially when differences in chemical classes are taken into account (Ali et al., 2011). The strength of correlation between zebrafish embryo LC_{50} values and data extracted from the literature on rodent LD_{50} values for 60 compounds from diverse pharmacological and chemical classes (alcohols, alkaloids, amides, carboxylic acids, glycosides, and others) has been examined (see Fig. 8.8).

Although the slopes of the relationship changes between chemical classes, there is a very good correlation within chemical class. This suggests that acute chemical toxicity in zebrafish can be used as a good surrogate for acute *in vivo* LD_{50} values (Parng et al., 2002). Even more exciting is the opportunity to automate reporter assay screening in a high-throughput manner using the whole organism. This concept was demonstrated by Walker et al. who developed a microplate reader with specific detection functions that could display endpoint information from transgenic zebrafish lines that incorporated global and regional expression of fluorescent-based reporters (Walker et al., 2012). Their particular methodology appeared extremely sensitive and reproducible and would offer researchers a robust and high-throughput format to discern mechanistic information at the chemical and genomic level that can complement phenotypic screening.

Many of these approaches can and have been utilized in some shape and form for screening pharmaceutical compounds for hepatotoxicity potential using zebrafish as a model organism (Vliegenthart et al., 2014). Zebrafish appear to offer many physiological similarities to higher vertebrates in terms of liver function. The liver itself is easily visualized in larval zebrafish, and by 72 hpf the liver is perfused with blood and is fully functional (Pack et al., 1996; Isogai et al., 2001). In terms of xenobiotic biotransformation capacities, larval zebrafish are metabolically competent, expressing numerous cytochrome P450 genes (Goldstone et al., 2010; Jones et al., 2010) and functional metabolism (Alderton et al., 2010; Jones et al., 2010). Although present in early

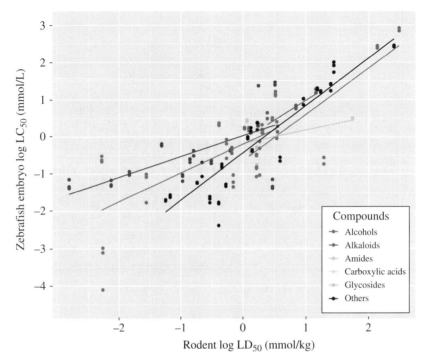

FIGURE 8.8 Linear regression model: rodent log LD_{50} and zebrafish embryo log LC_{50}. (Ali et al. (2011), figure 4, pp. e21076. Reproduced with permission of the Creative Commons Attribution License, which permits unrestricted use, distribution, and reproduction in any medium).

development at 24 hpf, cytochrome expression becomes highest in the larval stage after 96 hpf. Larval zebrafish at 96 hpf shows a wide range of phase I (oxidation, N-demethylation, O-deethylation, and N-dealkylation) and phase II (sulfation and glucuronidation) reactions using model substrates, suggesting that larval zebrafish could generate reactive or toxic metabolites (Chng et al., 2012). It is fair to say though that assumptions regarding absolute similarities in metabolic transformation of compounds in zebrafish to humans should not be readily made (Chng et al., 2012).

Hepatotoxicity assessments in larval zebrafish have focused on the following phenotypic endpoints: changes in liver size, liver abnormality (coloration), and yolk retention, which are easily visualized by phase contrast microscopy (Hill, 2011). The browning of the liver visualized by microscopy, along with liver size and yolk sac retention, correctly identified eight out of eight compounds as hepatotoxic and nonhepatotoxic drugs (He et al., 2013). In this small assessment, hepatotoxic drugs (acetaminophen, aspirin, tetracycline, sodium valproate, cyclophosphamide, and erythromycin) induced liver degeneration, reduced liver size, and delayed yolk sac absorption in larval zebrafish. A somewhat larger validation study, which used 33 drugs, was conducted using the phenotypic assessment alone. Of 33 pharmaceuticals classified by the US FDA as causing MOST- and LESS-DILI in humans, 9 of 13 of MOST-DILI compounds were identified, while only 5 of 20 LESS-DILI compounds were identified as hepatotoxic (Table 8.1). Although the sensitivity for detecting pharmaceuticals that

caused MOST-DILI was greater than that for LESS-DILI (69 vs. 25%, respectively), a positive finding in this assay alone provided no context to understand the degree of liver injury severity that a human may experience. Combining the results of this assay with those of other assays based on cell-based imaging of hepatotoxicity in cultured HepG2 cells (O'brien et al., 2006; Abraham et al., 2008) or human hepatocytes (Xu et al., 2008) held better promise at distinguishing compounds that had MOST-DILI classification (Hill et al., 2012). Using this approach, it appears that the severity of the injury potential can be captured by similar toxic findings in multiple models.

This approach has been used to segregate the toxic potential of several NSAIDs (Nadanaciva et al., 2013). Eleven NSAIDs representing different chemical classes were tested: three fenamic acids (flufenamic, tolfenamic, and mefenamic acid), three oxicams (meloxicam, sudoxicam, and piroxicam), two salicylates (aspirin and diflunisal), two acetic acid derivatives (diclofenac and sulindac plus its two metabolites), and a sulfonanilide (nimesulide). Comparison of data from the zebrafish assay with those from the two *in vitro* cell toxicity approaches showed that some of the NSAIDs that were potent in mitochondrial inhibition or uncoupling assays and cell-based toxicity assays (nimesulide, sulindac sulfide, and the three fenamic acids) were also acutely toxic in zebrafish. Evaluation of zebrafish morphology offered additional advantages to the other approaches in specifically highlighting GI toxicity of meloxicam, sulindac, and acetylsalicylic acid and liver toxicity with piroxicam. The

TABLE 8.1 Comparison of Outcomes between High Content Screening (HCS) Using HepG2 (Johnson & Johnson), Human Hepatocytes (Pfizer), and Zebrafish Phenotypic Hepatotoxicity Screening (Evotec) against FDA Liver Toxicity Knowledge Base (LTKB) DILI Classification

FDA LTKB ID	Drug Name	FDA LTKB DILI Potential	FDA LTKB DILI Label Section	J&J HCS Outcome	Zebrafish	Pfizer HCS Outcome
168	Troglitazone	MOST-DILI concern	Withdrawn	POS	POS	POS
645	Trovafloxacin	MOST-DILI concern	Withdrawn	NT	NEG	POS
46	Amiodarone	MOST-DILI concern	Boxed warning	POS	POS	POS
539	Danazol	MOST-DILI concern	Boxed warning	POS	POS	NEG
101	Flutamide	MOST-DILI concern	Boxed warning	NEG	POS	NEG
124	Nefazodone	MOST-DILI concern	Boxed warning	POS	POS	POS
144	Stavudine	MOST-DILI concern	Boxed warning	NEG	NEG	NEG
148	Sulindac	MOST-DILI concern	Warnings and precautions	NEG	FLAG	POS
84	Diclofenac	MOST-DILI concern	Warnings and precautions	NEG	POS	POS
150	Tacrine	MOST-DILI concern	Warnings and precautions	NEG	POS	NEG
152	Tamoxifen	MOST-DILI concern	Warnings and precautions	POS	POS	NEG
92	Erythromycin	MOST-DILI concern	Warnings and precautions	NT	FLAG	NEG
649	Zileuton	MOST-DILI concern	Warnings and precautions	NT	POS	POS
22	Indomethacin	LESS-DILI concern	Warnings and precautions	NEG	POS	NEG
204	Ketorolac	LESS-DILI concern	Warnings and precautions	NEG	NEG	NEG
134	Pioglitazone	LESS-DILI concern	Warnings and precautions	NT	NEG	NEG
1679	Piroxicam	LESS-DILI concern	Warnings and precautions	NT	NEG	NEG
1706	Propranolol	LESS-DILI concern	Warnings and precautions	NEG	NEG	NEG
287	Clozapine	LESS-DILI concern	Warnings and precautions	NT	POS	POS
380	Cyclosporine	LESS-DILI concern	Warnings and precautions	POS	NEG	NEG
36	Tetracycline	LESS-DILI concern	Warnings and precautions	NEG	FLAG	POS
59	Captopril	LESS-DILI concern	Adverse reactions	NEG	POS	NEG
58	Bupropion	LESS-DILI concern	Adverse reactions	NEG	NEG	NEG
13	Cyclophosphamide	LESS-DILI concern	Adverse reactions	NT	NEG	NEG
138	Ranitidine	LESS-DILI concern	Adverse reactions	NT	FLAG	NEG
1830	Trazodone	LESS-DILI concern	Adverse reactions	NT	POS	POS
647	Warfarin	LESS-DILI concern	Adverse reactions	NEG	POS	NEG
1148	Buspirone	LESS-DILI concern	Adverse reactions	NEG	NEG	NEG
95	Famotidine	LESS-DILI concern	Adverse reactions	NEG	NEG	NEG
108	Imipramine	LESS-DILI concern	Adverse reactions	POS	FLAG	NEG
1218	Cimetidine	LESS-DILI concern	Adverse reactions	NT	NEG	NEG
299	Furosemide	LESS-DILI concern	Adverse reactions	NEG	FLAG	NEG
87	Diphenhydramine	NO-DILI concern	No mention	NEG	NEG	NEG

Original work reproduced from Hill et al. (2012) with kind permission from Informa Healthcare.
FDA classification list according to Chen et al. (2011).
FLAG, borderline for hepatotoxic phenotype; NEG, negative; NT, not tested; POS, positive.

rank-order toxicity of the NSAIDs within their chemical class using the three experimental approaches was in accordance with the safety profile of these drugs in humans, where the reporting odds ratio (ROR) signal for each NSAID was calculated using the FDA's Adverse Event Reporting System (FAERS) database for adverse events reported for the liver and GI system. Therefore triangulating the results with several different mechanistic assays for hepatotoxicity can add additional resolution for detecting compounds with a potential to cause more severe forms of liver injury via multiple mechanisms. One difficulty with the zebrafish model was that some compounds were so acutely toxic to zebrafish larva that the liver phenotype was not evident. In these cases,

it was difficult to know what caused the toxic effect to the body and whether there was any indication of liver dysfunction (Nadanaciva et al., 2013).

Going beyond basic phenotypic screening and gene expression markers of liver injury (Driessen et al., 2014), Mesens et al. (2015) combined histopathologic and whole-mount *in situ* hybridization of liver-specific fatty acid binding protein 10a (*lfabp 10a*). Concentration–response relationships were found in that abnormal elevations in *lfabp 10a* expression were well correlated with histopathologic changes in the liver. Endpoints investigating the effect of drugs on heart rate, pericardial edema and hepatic glycogen content, and hyperplasia were of additional value in the assessment. These changes were more reflective of distinguishing between compounds that were or were not hepatotoxic versus trying to distinguish those that can cause severe liver injury in humans. While there are many techniques to explore the hepatotoxic potential of drugs in pharmaceutical development, it has yet to be shown that these assessments can be used in a prospective rather than a retrospective manner. Data obtained by HCA of cell imaging analysis of HepG2 cells (O'Brien et al., 2006) and human hepatocytes (Xu et al., 2008) have suggested that plasma exposure margins of 30-fold or 100-fold, respectively, when compared with *in vitro* IC_{50} values, can enable useful discrimination between drugs that can cause human DILI and nonhepatotoxic drugs. For zebrafish, a safety margin <10-fold segregated hepatotoxic from nonhepatotoxic compounds based on human exposure information (Mesens et al., 2015). The primary difficulty in applying these safety cutoffs in a prospective manner is that a true assessment of human exposure is absent during drug discovery, when this information can be most informative to compound design and selection. The estimated human efficacious exposure in drug discovery may be markedly higher or lower than that determined from human clinical trials, so a particular perceived safety margin can either increase or decrease with compound progression in the clinic.

8.5 *IN SILICO* MODELS

Early identification of potential of severe toxicity for the investigated compounds is a critical step in drug discovery. The US FDA's Advancing Regulatory Science Initiative embraces "Modernizing Toxicology" as one of eight focused areas (http://www.fda.gov/downloads/Science Research/SpecialTopics/RegulatoryScience/UCM228444. pdf). Developing *in silico* models to predict the toxicity of compounds provides an opportunity to reduce the time and cost to improve drug safety (Valerio, 2009), which has gained considerable impact on pharmaceutical industry and regulatory authorities (Durham and Pearl, 2001; Snyder, 2009).

In recent years, efforts to build *in silico* models aiming to predict hepatotoxicity have been reported, such as knowledge-based expert systems and statistical models (Matthews et al. 2009a, b). Expert systems are based on human knowledge with an in-depth understanding of hepatotoxicity mechanisms (Marchant et al., 2009). Statistical models are developed with a training data set composed of chemicals and the associated toxicity data in either quantitative or qualitative approaches using on various machine learning methods, including discriminant analysis (Pachkoria et al., 2007), Bayesian models (Ekins et al., 2010), artificial neural networks (ANN) (Cruz-Monteagudo

et al., 2008), k-nearest neighbor (k-NN) (Rodgers et al., 2010), and random forest (Chen et al., 2013b). Meanwhile, some advances have also been reported in developing knowledge-based expert systems, notably the Vertex cheminformatics platform (VERDI), ToxAlerts, Online Chemical Modeling Environment (OCHEM), and Derek for Windows (DfW) (Egan et al., 2004; Greene et al., 2010; Sushko et al., 2011, 2012).

While structural alerts for endpoints such as carcinogenicity, mutagenicity, and skin sensitization are well established, structural alerts for hepatotoxicity endpoints are poorly represented. For example, there are only two alerts for hepatotoxicity in version 8 of DfW, one of which is focused on aliphatic nitro-containing compounds and the other focused on pyrroline and pyrole esters (Greene et al., 2010). Recently, the relationship between the combination of daily dose and lipophilicity and risk for DILI was explored by using two independent data sets (Chen et al., 2013b), which demonstrated that apart from daily dose, lipophilicity contributes significantly to risk of DILI. In addition, it was revealed that drugs being substrates of cytochrome P450 enzymes have a higher likelihood of causing DILI, whereas drugs being inhibitors of P450 enzymes are prone to generating DILI when they are given at high daily dose (Yu et al., 2014). However, it was also reported that there is no obvious and straightforward association between idiosyncratic toxicity and structural alerts, and the major differentiating factor appears to be the daily dose (Stepan et al., 2011). Up to date, there is still a scarcity of *in silico* models for identifying organ-level endpoints such as hepatotoxicity (Li, 2004). The poor understanding of the hepatotoxicity mechanisms and lack of high-quality data sets (e.g., consensus in DILI annotation) are hindering the model development (Pearl et al., 2001; Simon-Hettich et al., 2006).

Completely accurate annotation of drugs with DILI potential is unrealistic currently due to the diverse clinical DILI manifestations, ambiguous causality between drug uptake and DILI outcomes, and low frequency of DILI occurrence in humans (Chen et al., 2011). Recently, some efforts have been made to develop plausible annotation schemes based on the DILI information from various data sources, including PubMed (Fourches et al., 2010; Greene et al., 2010; Gustafsson et al., 2014), the FDA FAERS (Ursem et al., 2009; Rodgers et al., 2010; Zhu and Kruhlak, 2014), the FDA-approved drug labeling (Chen et al., 2011), pharmaceutical databases (Lammert et al., 2008, 2010), and manually curated clinical hepatotoxicity databases (Xu et al., 2008; Ekins et al., 2010). Consequently, several hepatotoxicity data sets have been created (Chen et al., 2013a, b; Zhu and Kruhlak, 2014).

The US FDA Human Liver Adverse Effects Database (HLAED) is one of many useful sources for model development (Rodgers et al., 2010). This data set is composed of 490 compounds, and the hepatotoxic drugs are determined by the elevated liver enzymes with a composite module. Another example is an annotated data set composed of more than 300 drugs and chemicals assessed by the clinical hepatotoxicity data (Xu et al., 2008), which is later expanded with additional 237 compounds (Ekins et al., 2010). In that annotation, DILI-positive drugs include those withdrawn from the market mainly due to hepatotoxicity, those not marketed in the United States due to hepatotoxicity, those receiving black box warnings from the US FDA due to hepatotoxicity or having hepatotoxicity warnings on the labels, and those having well-established DILI causality and a significant number (>10) of independent clinical reports of serious hepatotoxicity that met the Hy's law criteria (Ekins et al., 2010).

Some discrepancy between the annotated data sets has been observed. Comparison of the data set created by the retrieval of DILI information from FAERS with that from the FDA-approved drug labels yielded 73% sensitivity and 73% specificity (Zhu and Kruhlak 2014), whereas the overall prediction accuracy of the *in silico* model to assess the drugs with DILI potential may be <70% (Chen et al., 2013b). The ambiguous annotation of drugs may substantially contribute to the limited model predictability.

Databases from regulatory agencies are also useful to annotate the drugs with DILI potential. For example, Suzuki et al. used reporting frequency of two customized terms ("overall liver injury" and "ALF") to screen DILI-positive drugs from WHO VigiBase and identified a total of 385 individual DILI-positive drugs (Suzuki et al., 2010). Zhu et al. used 37 hepatotoxicity-related preferred terms (PTs) from Medical Dictionary for Regulatory Activities (MedDRA; www.meddra.org/) to screen FAERS with a scalable approach and extracted a set of 2,029 unique and modelable drug entities with 13,555 drug AE combinations (Zhu and Kruhlak, 2014). Compared to other databases, the FDA-approved drug labeling is regulated by law (http://www.accessdata.fda.gov/scripts/cdrh/cfdocs/cfCFR/CFRSearch.cfm?fr=201.57) and authoritatively provides drug safety information with comprehensive assessment on the data obtained from clinical trials, literature reports, and postmarketing surveillance (Chen et al., 2011). By developing an annotation scheme based on DILI information from the FDA-approved drug labels, Chen et al. generated a benchmark data set providing different levels of DILI concern for 287 drugs (Chen et al., 2011), which was later expanded to 387 drugs (Chen et al., 2013b). The drugs are classified into three groups (e.g., MOST-DILI concern, LESS-DILI concern, and NO-DILI concern) based on the DILI severity described in the different labeling sections (e.g., boxed warning, warnings and precautions, and adverse reactions). However, manual reviewing of drug labels is required although limited.

In the future, drug annotations integrating multiple data sources including FDA drug labeling, WHO VigiBase,

PharmaPendium (http://www.elsevier.com/online-tools/pharma pendium), FAERS, PubMed, and other pharmaceutical databases will be a promising way to assess the potential of a drug to cause liver injury. Meanwhile, the viewpoints for a drug's hepatotoxicity from drug developers, regulatory agencies, academia, and clinicians can be collected, analyzed, and balanced to minimize the bias in determining the potential of drug hepatotoxicity.

Although *in silico* predictive models have been long investigated for various toxicological endpoints, most of the development of models to predict hepatotoxicity only occurred in the last 10 years after 2005 (Cheng and Dixon, 2003; Clark et al., 2004). The details about methodologies, data, and prediction performance for the majority of those models are summarized in a review paper (Przybylak and Cronin, 2012). Their prediction ability is varied, with the observed concordance ranging from 56 to 88.9%. Comparison of performance among those models is difficult, because different data sets and different modeling algorithms have been used to build and evaluate the models. Specifically, one model focused on intrinsic hepatotoxicity data (Cheng and Dixon, 2003), while others utilized idiosyncratic or mostly idiosyncratic data (Egan et al., 2004; Cruz-Monteagudo et al., 2008; Matthews et al., 2009a; Ekins et al., 2010; Greene et al., 2010; Rodgers et al., 2010; Chen et al., 2011; Low et al., 2011). In addition, some models used human data (Cheng and Dixon, 2003; Egan et al., 2004; Cruz-Monteagudo et al., 2008; Matthews et al., 2009b; Ekins et al., 2010; Greene et al., 2010; Rodgers et al., 2010), whereas others used animal data (Fourches et al., 2010; Greene et al., 2010; Low et al., 2011). Whereas most models were developed to predict a general hepatotoxicity endpoint (Cheng and Dixon, 2003; Egan et al., 2004; Cruz-Monteagudo et al., 2008; Ekins et al., 2010; Fourches et al., 2010; Greene et al., 2010; Low et al., 2011; Przybylak and Cronin, 2012), three models predicted more specific endpoints, mostly elevations of liver serum enzymes (Clark et al., 2004; Matthews et al., 2009b; Rodgers et al., 2010). In addition, two models used *in vitro* data (Clark et al., 2004; Chan et al., 2007), one model used a mixture of both *in vitro* and *in vivo* data (Sakuratani et al., 2008), while the remaining models were developed based on *in vivo* data (Cheng and Dixon, 2003; Egan et al., 2004; Cruz-Monteagudo et al., 2008; Matthews et al., 2009b; Ekins et al., 2010; Greene et al., 2010; Rodgers et al. 2010; Low et al., 2011; Chen et al., 2013b).

Since intrinsic hepatotoxicity is dose dependent and often considered to be predictable, in theory it should be feasible to use *in silico* models based on the chemical structure to predict intrinsic toxicity. However, reports of the use of *in silico* models to predict intrinsic hepatotoxicity are rare, and only one model was developed strictly based on intrinsic hepatotoxicity data (Cheng and Dixon, 2003). In that study, the intrinsic hepatotoxicity data were collected from various scientific publications including books, compilations, and journal articles, which compiled a data set with 382 compounds. In that data set, 149 compounds were hepatotoxicity positive with intrinsic toxic outcomes including dose-dependent hepatocellular, cholestatic, mixed and other types of liver injury, and dose-related elevated levels of liver enzymes in more than 10% of patients. The remaining compounds were hepatotoxicity negative, that is, without apparent evidence of causing DILI in humans. A total of 25 structural descriptors were used to build a QSAR model using Monte Carlo linear regression algorithm. To improve the model's robustness, 151 different decision trees based on different training subsets were generated by using recursive partitioning with an ensemble approach. An unknown compound was predicted by each of these trees, and the final prediction activity was achieved by assessing the results from all trees. The overall concordance was 85% and 67% for leave-1%-out and leave-10%-out cross-validation with a sensitivity of 78% and 76% and a specificity of 90% and 75%, respectively (Cheng and Dixon, 2003). Although this model still may have difficulties explaining the underlying mechanisms, the ensemble method correctly identified 81% of the external validation set, therefore showing substantial promise for assessing the risk of intrinsic hepatotoxicity caused by various compounds.

It can be inferred that *in silico* models should have a high modeling performance if used to predict intrinsic drug hepatotoxicity in humans. However, since no other reports are available focusing on intrinsic hepatotoxicity, the progress made so far is disappointing. One major challenge is that drugs that reach the market have good safety profiles in clinical trials and therefore do not cause intrinsic liver injury, although they may cause idiosyncratic DILI. Hence most of the data on DILI that can be retrieved from the literature, from clinical trials, or from spontaneous reporting systems such as FAERS is not applicable to intrinsic DILI. In the future, further development and evaluation of *in silico* models that predict intrinsic hepatotoxicity should be prioritized. This will require access to additional data sets, which include high-quality hepatotoxicity data on a diverse range of compounds that cause intrinsic human liver injury.

Because of its rare occurrence and dose-independent manifestation, prediction of idiosyncratic hepatotoxicity is a huge challenge in drug discovery and development process (Chen et al., 2013b). It is expected that idiosyncratic hepatotoxicity involves multiple factors, multiple biological events, and complex mechanisms. Therefore, the final manifestation of idiosyncratic hepatotoxicity is believed to be a product of interaction of individual factors, that is, a crosstalk between chemical properties, host factors, and environmental factors, with a "multiple determinant hypothesis" (Li, 2002). This hypothesis may explain the poor performance of the current *in silico* models for prediction of idiosyncratic hepatotoxicity solely based on chemical structure. Cruz-Monteagudo

et al. developed QSAR models to predict idiosyncratic hepatotoxicity and obtained the best concordance of 86.4% in prediction based on linear discriminant analysis (LDA) (Cruz-Monteagudo et al., 2008). However, only 74 drugs (33 toxic compounds and 41 nontoxic compounds) were used to build the models, which might be insufficient for a reliable *in silico* model to screen drugs. The prediction performance of any *in silico* models should be rigorously challenged by increasing the number of compounds for validation. For example, the US FDA carried out an effort to create a human health effects database and then built QSAR models to predict hepatobiliary and urinary tract adverse effects of drugs (Clark et al., 2004; Matthews et al., 2009a, b). The data were collected from FDA postmarketing surveillance databases and the published literature, which generated a data set with about 1660 compounds. Four QSAR programs were used to construct predictive models, resulting in an average sensitivity of 39.3% and specificity of 78.4% in the internal validation procedure.

Well-annotated hepatotoxicity data are expected to improve the model performance (Chen et al., 2011). For example, based on identification of 13 hepatotoxic side effects to distinguish DILI-positive drugs from the negatives, *in silico* models for each of these side effects were developed based on structural data. A DILI prediction system was developed that combined these 13 models and provided positive predictive value of 91% (Liu et al., 2011). However, when the criteria developed by Xu et al. (2008) were used to annotate DILI drugs, a QSAR model that used a Bayesian classification approach and was based on a training set of 295 compounds (114 DILI drugs) and a test set of 237 compounds (114 DILI drugs) only correctly identified 56.0% of positive drugs and 66.7% of negative drugs, with a concordance of 59.9% in the external validation procedure (Ekins et al., 2010). Meanwhile, when based on hepatotoxicity data annotated using DILI information from FDA drug labeling, a QSAR model gave an overall prediction accuracy of 68.9% for an external validation set (Chen et al., 2013b). QSAR modeling of specific endpoints of hepatotoxicity has provided a controversial outcome, since for many compounds the predicted outcomes could not be verified due to the limited hepatotoxicity information that could be obtained from publicly available data sources (Rodgers et al., 2010). Furthermore, it remains unclear whether the chemical descriptors used to build and test QSAR models are sufficient to account for the many diverse mechanisms by which hepatotoxicity may arise, particularly idiosyncratic DILI.

DfW is a knowledge-based expert system for predicting the toxicity of a chemical from its molecular structure (Judson et al., 2003). This system is composed of structural alerts, example compounds, and rules that may each contribute to the toxicity predictions. However, DfW is designed to aid in chemical carcinogenicity risk assessment, and hepatotoxicity is not the major toxicity prediction from this system. The toxicological endpoints in the current version include carcinogenicity, mutagenicity, genotoxicity, skin sensitization, teratogenicity, irritation, respiratory sensitization, and reproductive toxicity. ToxAlerts is another expert system developed for virtual screening of chemical libraries to identify potentially toxic compounds (Sushko et al., 2012). This system is a Web-based publicly accessible platform and tightly integrated with the OCHEM (http://ochem.eu) (Sushko et al., 2011), which is a widely used platform to perform the QSAR studies online. Nearly 600 structural alerts have been developed in this system for various types of endpoints including compounds that form reactive metabolites, an important mechanism for hepatotoxicity. Importantly, this system can be used to deal with large chemical libraries against the structural alerts, generating a detailed profile of chemicals with structural alerts and endpoints, which is helpful for decision-making regarding whether a compound should be examined experimentally, validated with established QSAR models, or removed from the pipeline.

Marchant et al. reported an expert system model to detect intrinsic hepatotoxicity observed in preclinical test species in early stage of drug discovery (Marchant et al., 2009). By examining 244 compounds, the authors identified structural alerts based on the corresponding occurrence, and 74 structural alerts were collected by exploring publicly available literature and proprietary data sets. However, the prediction performance of these structural alerts was not provided. For the development of structural alerts, Greene et al. proposed a four-stage process (Greene et al., 2010). This is initiated by exploring empirical relationships between compounds and toxicological outcomes to identify the candidate structural classes, followed by a close investigation of these classes using all available sources to understand the underlying mechanisms. Next, an alert is developed only if sufficient evidence exists to rationalize a SAR model. Finally, the predictive performance of SAR models is evaluated by external validation. Using this process, the authors collected 1266 chemicals and developed SAR models for 38 classes of compounds, which were analyzed in DfW to permit clearly supported and transparent predictions. These SAR models were validated by a Pfizer-developed data set of 626 compounds, which revealed an overall concordance of 56% with specificity and sensitivity of 73% and 46%, respectively, with intrinsic hepatotoxicity observed in experimental animals.

In recent years, success has been achieved in *in silico* modeling of a variety of MIEs that describe the initial interactions between compounds and biological entities (Ellison et al., 2011). Since MIEs are a consequence of molecular interactions, in general they are reliant on molecular structure and should be more amenable to *in silico* modeling than the downstream biological pathways by which they may result in DILI (Matthews et al., 2009a). Successful predictions arising from MIE analysis could be especially valuable if used to aid compound design and prioritization if used during drug discovery.

One important MIE by which many compounds cause hepatotoxicity is via the formation of electrophilic reactive metabolites that bind covalently to the nucleophilic region of biological macromolecules (see Section 8.3.2). Although covalent binding is dependent on thermodynamic and kinetic factors, electrophilic potential may be predicted by using structural alerts (Ashby and Tennant, 1988; Ashby et al., 1989). For example, the Organization for Economic Cooperation and Development (OECD) (Q)SAR Toolbox includes an assessment of the potential electrophilic mechanisms of a compound (Enoch and Cronin, 2010; Enoch et al., 2011a). The application of this toolbox has been evaluated for predicting and profiling the carcinogenic and mutagenic potential of chemicals (Devillers and Mombelli, 2010; Mombelli and Devillers, 2010). In addition, a series of sulfur-containing chemicals were profiled for protein binding potential by using OECD QSAR Toolbox, which were experimentally verified (Richarz et al., 2014).

Mechanistic analysis is considered to be useful when defining "chemical categories" (Roberts et al., 2007; Enoch et al., 2008; Enoch et al., 2009), which are groups of chemicals participating in the same MIE via the same mechanism. Enoch and Cronin discussed five specific mechanisms of electrophilic reactivity based on chemical descriptors (Aptula and Roberts, 2006; Enoch et al., 2011b). This enabled the toxic potential of a new compound to be assessed from the known toxicity profiles of its neighbors within the same chemical category (Enoch et al., 2011b). Recently, a scheme for categorizing chemicals and developing structural alerts has also been detailed (Hewitt et al., 2013). In that study, a diverse data set with 951 compounds was utilized, and a number of structurally restricted categories were produced by using structural similarity methods. When comparing the query compound with structurally similar compounds in the same chemical category, it is important also to ensure that with similar administration routes and daily doses are used when attempting to predict hepatotoxicity.

Free radicals, as molecules with a single unpaired electron, are extremely reactive. One of the major concerns of free radicals is their ability to induce oxidative stress (Kovacic and Jacintho, 2001). Predicting the potential of a compound to generate free radicals is similar to the methodology used for prediction of covalent binding based on an understanding of a molecule's chemistry. Therefore, structural alerts can also be used to predict the potential to generate free radicals. The role of oxidative stress in hepatotoxicity has been investigated (Gibson et al., 1996; Amin and Hamza, 2005; Oh et al., 2012), and compounds with ability to cause both covalent binding and free radical formation may pose a particular hazard. However, currently it is not possible to infer whether oxidative stress is more important than covalent binding as a cause of hepatotoxicity hazard.

It is important to recognize that high-quality hepatotoxicity data, which are extremely important for the development of *in silico* models, are limited currently. Whereas intrinsic hepatotoxicity is predictable in theory, and it is known that some structural alerts are involved in several types of MIEs involved in hepatotoxicity, the ability of *in silico* models to predict idiosyncratic hepatotoxicity, based purely on molecular structure, seems to be insufficient. To improve the prediction of hepatotoxicity in the future, more efforts should be focused on understanding its mechanisms. Meanwhile, the nongenetic host factors, genetic host factors, and environmental factors should also be investigated and integrated into modeling process for predicting DILI.

8.6 SYSTEMS PHARMACOLOGY AND DILI

As has been highlighted earlier in this chapter, DILI is difficult to predict with traditional hepatocyte or animal experimental models (Chen et al., 2014; Regev, 2014). Differences in biochemistry and physiology between rat models and humans can produce divergent responses to drugs. There are a number of examples of drugs that did not cause DILI in rat or dog models but generated LFTs in clinical trials in humans (i.e., false negatives; Regev, 2013). Further complicating the challenge of translating animal data to humans, when assessing risk, is the fact that multiple mechanisms have been identified as contributors to DILI. A complete evaluation of DILI liabilities for a given compound includes assay screening for bile acid transport inhibition, mitochondrial toxicity, reactive metabolites, oxidative stress, and ER stress, in addition to others (Ramappa and Aithal, 2013; Yuan and Kaplowitz, 2013; Aleo et al., 2014; Kenna, 2014). Integrating these data to account for signals in multiple mechanistic pathways in addition to accounting for species differences presents a challenge that can be addressed by employing systems pharmacology approaches.

While multiple systems biology approaches such as toxicogenomics (Ganter et al., 2006; Blomme et al., 2009; Judson et al., 2014), QSAR modeling (Low et al., 2011), and physiologically based pharmacokinetic (PBPK) modeling (Andersen, 1995; Willmann et al., 2005; Jamei et al., 2009) have been applied to varying degrees to predict DILI, mechanistic, integrative mathematical modeling provides yet another tool. Several efforts to create mathematical models of liver injury have recently come to the forefront (Przybylak and Cronin, 2012), including the Virtual Liver Network (Kuepfer et al., 2014), the Strand model (Subramanian et al., 2008), and the EPA's Virtual Liver. An additional mechanistic mathematical model of DILI is DILIsym®. The members of the DILI-sim Initiative, which includes representatives

from 15 to 20 pharmaceutical companies, have supported and guided the development of DILIsym along with the Hamner Institutes since 2011.

DILIsym has been designed to predict DILI in multiple species, including mice, rats, dogs, and humans, including accounting for interindividual variability. Equations describing hepatocyte turnover due to apoptosis, necrosis, and proliferation are at the core of the model. Moreover, DILIsym includes mathematical representations (submodels) of many of the key mechanisms that are thought to elicit hepatocellular apoptosis and/or necrosis, including mitochondrial toxicity, bile acid transporter inhibition, ROS, and reactive metabolites. Compound (parent and/or metabolite) exposures in the liver are predicted using PBPK submodels within DILIsym. Also included is a representation of the innate immune system and its pro- and anti-inflammatory contributions to liver injury. Each submodel (individually and collectively) has been optimized based on literature from the data in both untreated and treated states. As the submodels interact, simulated liver compound concentrations can cause liver injury via one or more of the mechanisms. The drugs used to optimize DILIsym include acetaminophen (Woodhead et al., 2012), methapyrilene (Howell et al., 2012), bosentan (Woodhead et al., 2014), troglitazone (Yang et al., 2014), and others.

DILIsym provides a means for integrating laboratory screening data, as they are used as inputs for DILIsym. Inhibition constants (Ki) for BSEP, MRP3/MRP4, and NTCP describe a compound's propensity for reducing bile acid transport. Glutathione depletion and/or adduct generation in microsomes or hepatocytes helps quantify reactive metabolite generation, while additional assays can also measure ROS production in hepatocytes. All of these data can be used to determine parameter values quantifying the effect of the parent and/or metabolite on each of the mechanistic pathways in DILIsym. Mitochondrial toxicity parameters are determined by combining the measurement of changes in cellular respiration with simulations in a separate, standalone model, MITOsym® (Yang et al., 2015). Standard DMPK data can be used to define the input parameters for the PBPK submodel. Combining all these parameters provides the ability for DILIsym to evaluate the contribution of each mechanism while accounting for the amount of compound (parent and/or metabolite) in a given dosing paradigm. This provides users with the ability to integrate and evaluate the *in vitro* preclinical screening data that have been collected for a given compound.

The DILIsym systems pharmacology approach has been applied to clinical development in multiple ways. DILIsym simulations were able to predict the species differences between rats and humans with respect to liver injury following the administration of troglitazone (Yang et al., 2014) and methapyrilene (Howell et al., 2012). Employing *in vitro* and preclinical *in vivo* data inputs, DILIsym has also been used to discern the hepatotoxicity properties of one compound regioisomer from another (Howell et al., 2014a). DILIsym has additionally been used to better understand the magnitude of the injury (hepatocyte loss) in human volunteers with ALT increases following treatment (Howell et al., 2014b). Similarly, DILIsym could also be used to better understand the underlying mechanisms associated with the observed biomarker increases associated with injury.

DILIsym has recently been expanded (v4A) to include the contributions of lipotoxicity to liver injury. Lipotoxicity is the effect of lipids to elicit cytotoxicity, either via apoptosis or necrosis (Zámbó et al., 2013). Several clinical studies have implicated lipotoxicity as a primary contributor to nonalcoholic fatty liver disease and/or nonalcoholic steatohepatitis (Kotronen et al., 2008; Maximos et al., 2015), via correlations between hepatic lipids and circulating ALT. While correlating with lipotoxicity, triglycerides do not actively participate in the cytotoxicity. In fact, there is evidence to suggest that esterification of fatty acids into triglyceride may actually be protective against lipotoxicity (Li et al., 2009; Noguchi et al., 2009; Leamy et al., 2013). Saturated fatty acids (SFA) appear to play a primary role in mediating lipotoxicity (Li et al., 2009; Noguchi et al., 2009; Leamy et al., 2013; Zámbó et al., 2013). Elevated SFA levels lead to increases in cellular ROS levels (Li et al., 2009; Noguchi et al., 2009; Bao et al., 2010; Dou et al., 2011; Gentile et al., 2011; Egnatchik et al., 2014; Wang et al., 2014). When sufficiently elevated, the oxidative species lead to activation of the mitochondrial pore transition, activation of caspase enzymes in the apoptosis cascade, and reductions in mitochondrial ATP synthesis (Cortez-Pinto et al., 1999; Belosludtsev et al., 2006; Song et al., 2007; Noguchi et al., 2009; Zhang et al., 2010; Egnatchik et al., 2014). Thus, drugs that cause an increase in hepatic SFA, such as FAO inhibitors or VLDL release inhibitors, can now be evaluated for DILI risk with DILIsym.

Etomoxir is an inhibitor of FAO (via mitochondrial carnitine palmitoyltransferase inhibition) that was originally developed for the treatment of type 2 diabetes and congestive heart failure (Holubarsch et al., 2007). Etomoxir did not make it to market due to efficacy and safety issues, however. Hepatotoxicity with etomoxir in humans was observed in the "Etomoxir for the Recovery of Glucose Oxidation" (ERGO) clinical trial among 216 congestive heart failure patients (Holubarsch et al., 2007). This trial was terminated prematurely because high liver enzyme (ALT) levels were detected in several patients. After 6 weeks of treatment, four patients were found with ALT and AST elevated as high as 15× ULN. Liver enzymes returned to normal upon withdrawal of etomoxir administration. Etomoxir treatment of heart failure patients was terminated following the ERGO clinical trial (Lionetti et al., 2011). An unresolved question with this trial is whether the observed hepatotoxicity was due to FAO inhibition directly causing mitochondrial toxicity or rather a consequence of subsequent accumulation of lipids and lipotoxicity.

DILIsym was recently applied to evaluate each of these hypotheses with the simulation of 80 mg q.d. etomoxir dosing in a simulated population (SimPops™). The FAO inhibition effects were based on *in vitro* data (Agius et al., 1991). In simulations where the model parameters excluded contributions of lipotoxicity to liver injury, no simulated patients were predicted to have ALT increases. However, simulations including lipotoxicity showed an incidence of DILI that was similar to that seen in the ERGO trial, with 2/300 simulated patients predicted to have ALT >5× ULN (see Table 8.2).

Moreover, the predicted ALT increases did not occur until after 4–6 weeks of dosing, as was observed in the ERGO trial. It is reasonable to conclude from the comparison of the DILIsym simulation results with the clinical data that etomoxir caused an accumulation of lipids due to the FAO inhibition properties of the compound and lipotoxicity for the LFTs in the ERGO trial.

Systems pharmacology approaches, such as those utilized in DILIsym, can not only help integrate multiple streams of data and make predictions of DILI risk, but they can also be used to better understand the nature of liver injury when it occurs. It is perhaps the complementarity with available wet lab data that provides the greatest strength of systems pharmacology. The number of false negatives in traditional clinical development screening processes can be reduced by combining data collected in multiple species with mathematical predictions of DILI in humans, thereby enhancing patient safety.

8.7 SUMMARY

Assimilating a comprehensive approach for dealing with DILI during the drug discovery and development process is a daunting task. As one can see from the numerous

TABLE 8.2 Comparison of Observed with Simulated Incidence of DILI Following Etomoxir Treatment When Including Lipotoxicity Effects on the Liver

Hypothesis	N	Number of Patients with LFTs	Range of ALT Values	Timing of LFTs
Reported clinical data	226	4	200–600 U/L	>6 weeks
Direct mitochondrial toxicity simulation results	300	0	Normal	None
Lipotoxicity due to accumulation of lipid simulation results	300	2	200–400 U/L	>5 weeks

approaches presented here and published previously (Hornberg et al., 2013a, b), there are several ways to approach this issue, and what is considered state-of-the-art knowledge and methodologies is ever changing, making constant assimilation and renewal of new screening models difficult in organizations. Imagine the costs of rescreening hundreds or thousands of compounds to revalidate a new assay or technology platform and how rapidly one would have to do this to keep current. We have learned that there are usually trade-offs in sensitivity and specificity that can only be truly appreciated during extensive validation exercises. One usually has to accept some level of loss (e.g., sensitivity/specificity) for a perceived advantage (e.g., addressing numerous mechanisms of DILI). This just has to be accepted to gain advantages in sunk costs and time to maintain or gain the most recent scientific approaches to be applied for screening efforts. Many large pharmaceutical organizations may still only have bits and pieces of this complex screening approach or have it embedded in different parts of a large organization that make it difficult to assemble the pieces into an integrated assessment. Other smaller pharmaceutical companies may only have access to these pieces through outsourcing to contract or academic labs, where the approaches to screening in terms of assays become even more diverse and esoteric. True validation of these models is also a challenge, as typically the test drugs used for this purpose represents only a small fraction of the numerous drugs that cause DILI and only some of these exhibit clear signals in cell-based assays. Therefore the currently available data does not adequately address the broad range of chemical space and pharmacology that is encountered in practice during drug discovery. Furthermore, the currently available schemes for compound annotation may not enable the most mechanistically valid and robust discrimination between hepatotoxic and nonhepatotoxic drugs. At present, the DILIsym software provides the most comprehensive computational approach for integrating data into an overall DILI hazard and risk assessment. However, this approach is still in its infancy, and the mechanisms incorporated within the model currently are limited, as are the computational approaches used to analyze data and simulate outcomes. It also requires data inputs provided by numerous wet assays, all of which have value but also exhibit potential limitations. Furthermore, DILIsym simulations of liver injury require either direct access to the software or negotiation of a fee-for-service contract with the initiative.

Fundamentally, without this type of software approach, one still needs to assemble the basic inputs for analysis (structural alerts with some assessment of chemical metabolic reactivity and covalent binding levels, some measurement of cytotoxicity or cell injury to hepatocytes, interruption in hepatic basolateral and apical transporters, and mitochondrial injury assessment as a core) with additional approaches as necessary to capture additional mechanisms that drive DILI. At some point it needs to be acknowledged that we might never be able to assemble and adequately integrate all of the relevant mechanistic information to make a perfect prediction of hepatotoxicity *in vivo*. Currently it seems prudent to screen for the major initiating mechanisms that drive DILI and thereby to try to avoid the avoidable, as much as possible. This is the basis of the integrated approach advocated currently by many groups (Thompson et al., 2012; Morgan et al., 2013; Kenna et al., 2015; Shah et al., 2015), where it appears we can estimate with higher probability the potential of a compound to cause clinically concerning DILI in the human population (hazard detection vs. possible risk perspective). The problem as always with "predictions" here is that they are really probabilistic in nature, as opposed to true predictions, since ultimately information in the clinic is needed to make the final adjustments in the "prediction." Clinical doses and corresponding exposures of therapeutic medicines are known only after the completion of a successful phase II study, which makes any notion of a prediction really a probability assessment. And since many possible mechanisms of DILI are not covered in some of these more standard cell-based testing schemes, it is important not to make claims of safety based on a lack of response in these assays. For example, ximelagatran caused severe liver injury in humans without a strong signal in cell-based assays of cytotoxicity (Ainscow et al., 2008), emphasizing the need for developing assays that can assess additional important mechanisms (most notably immune-mediated DILI).

The challenge in this process is even more daunting when one considers that for the most part, severe manifestations of DILI are at least parts of a dual-edged sword. Drug properties/structures (that confer liabilities that drive liver injury) conspire in the host body (where there are predisposing genetic, disease, and exposure to polypharmacy and/or environmental factors) to produce severe liver injury that does not occur in other patients with similar backgrounds. The individual nature of most DILI manifestation makes it a "personalized" medicine project where more complex models are needed (please see Chapter 20), which hopefully will better represent *in vivo* physiology and complex interactions. The potential promise of induced pluripotent stem cells from sensitive patients may also provide experimental tools that can provide better insights into underlying genetic predisposition, but not necessarily epigenetic mechanisms or susceptibility factors. With time, these more sophisticated approaches may bring us closer to providing more accurate predictions of clinical DILI in terms of frequency, magnitude of the response, relative level of risk, and ultimately sensitivity to potential injury, which is so dearly sought when attempting to predict and understand safety in individual human patients.

REFERENCES

Abraham VC, Towne DL, Waring JF, Warrior U, Burns DJ (2008) Application of a high-content multiparameter cytotoxicity assay to prioritize compounds based on toxicity potential in humans. *J Biomol Screen* 13, 527–37.

Adams DH, Ju C, Ramaiah SK, Uetrecht J, Jaeschke H (2010) Mechanisms of immune-mediated liver injury. *Toxicol Sci*, 115, 307–21.

Agius L, Peak M, Sherratt SA (1991) Differences between human, rat and guinea pig hepatocyte cultures. A comparative study of their rates of beta-oxidation and esterification of palmitate and their sensitivity to R-etomoxir. *Biochem Pharmacol* 42, 1711–15.

Ainscow EK, Pilling JE, Brown NM, Orme AT, Sullivan M, Hargreaves AC, Cooke EL, Sullivan E, Carlsson S, Andersson TB (2008) Investigations into the liver effects of ximelagatran using high content screening of primary human hepatocyte cultures. *Expert Opin Drug Saf* 7, 351–65.

Akash MS, Rehman K, Chen S (2013) Role of inflammatory mechanisms in pathogenesis of type 2 diabetes mellitus. *J Cell Biochem* 114, 525–31.

Alderton W, Berghmans S, Butler P, Chassaing H, Fleming A, Golder Z, Richards F, Gardner I (2010) Accumulation and metabolism of drugs and CYP probe substrates in zebrafish larvae. *Xenobiotica* 40, 547–57.

Aleo MD, Luo Y, Swiss R, Bonin PD, Potter DM, Will Y (2014) Human drug-induced liver injury severity is highly associated with dual inhibition of liver mitochondrial function and bile salt export pump. *Hepatology* 60, 1015–22.

Ali S, Van Mil HG, Richardson MK (2011) Large-scale assessment of the zebrafish embryo as a possible predictive model in toxicity testing. *PLoS One* 6, e21076.

Amin A, Hamza AA (2005) Oxidative stress mediates drug-induced hepatotoxicity in rats: a possible role of DNA fragmentation. *Toxicology* 208, 367–75.

Andersen ME (1995) Development of physiologically based pharmacokinetic and physiologically based pharmacodynamic models for applications in toxicology and risk assessment. *Toxicol Lett* 79, 35–44.

Aptula AO, Roberts, DW (2006) Mechanistic applicability domains for nonanimal-based prediction of toxicological end points: general principles and application to reactive toxicity. *Chem Res Toxicol* 19, 1097–105.

Armstrong MJ, Carey MC (1982) The hydrophobic-hydrophilic balance of bile salts. Inverse correlation between reverse-phase high performance liquid chromatographic mobilities and micellar cholesterol-solubilizing capacities. *J Lipid Res* 23, 70–80.

Ashby J, Tennant RW (1988) Chemical structure, Salmonella mutagenicity and extent of carcinogenicity as indicators of genotoxic carcinogenesis among 222 chemicals tested in rodents by the U.S. NCI/NTP. *Mutat Res* 204, 17–115.

Ashby J, Tennant RW, Zeiger E, Stasiewicz S (1989) Classification according to chemical structure, mutagenicity to Salmonella and level of carcinogenicity of a further 42 chemicals tested for carcinogenicity by the U.S. National Toxicology Program. *Mutat Res* 223, 73–103.

Attili AF, Angelico M, Cantafora A, Alvaro D, Capocaccia L (1986) Bile acid-induced liver toxicity: relation to the hydrophobic-hydrophilic balance of bile acids. *Med Hypotheses* 19, 57–69.

Bale SS, Vernetti L, Senutovitch N, Jindal R, Hegde M, Gough A, McCarty WJ, Bakan A, Bhushan A, Shun TY, Golberg I, DeBiasio R, Usta OB, Taylor DL, Yarmush ML (2014) *In vitro* platforms for evaluating liver toxicity. *Exp Biol Med (Maywood)* 239, 1180–91.

Bao J, Scott I, Lu Z, Pang L, Dimond CC, Gius D, Sack MN (2010) SIRT3 is regulated by nutrient excess and modulates hepatic susceptibility to lipotoxicity. *Free Radic Biol Med* 49, 1230–7.

Belosludtsev K, Saris NE, Andersson LC, Belosludtseva N, Agafonov A, Sharma A, Moshkov DA, Mironova GD (2006) On the mechanism of palmitic acid-induced apoptosis: the role of a pore induced by palmitic acid and Ca2+ in mitochondria. *J Bioenerg Biomembr* 38, 113–20.

Benbow JW, Aubrecht J, Banker MJ, Nettleton D, Aleo MD (2010) Predicting safety toleration of pharmaceutical chemical leads: cytotoxicity correlations to exploratory toxicity studies. *Toxicol Lett* 197, 175–82.

Bhushan A, Senutovitch N, Bale SS, McCarty WJ, Hegde M, Jindal R, Golberg I, Berk UO, Yarmush ML, Vernetti L, Gough A, Bakan A, Shun TY, DeBiasio R, Lansing TD (2013) Towards a three-dimensional microfluidic liver platform for predicting drug efficacy and toxicity in humans. *Stem Cell Res Ther* 4 Suppl 1, S16.

Blomme EA, Yang Y, Waring JF (2009) Use of toxicogenomics to understand mechanisms of drug-induced hepatotoxicity during drug discovery and development. *Toxicol Lett* 186, 22–31.

Boelsterli UA, Lim PL (2007) Mitochondrial abnormalities—a link to idiosyncratic drug hepatotoxicity? *Toxicol Appl Pharmacol* 220, 92–107.

Bonnefont-Rousselot D, Bastard JP, Jaudon MC, Delattre J (2000) Consequences of the diabetic status on the oxidant/antioxidant balance. *Diabetes Metab* 26, 163–76.

Cargnin S, Jommi C, Canonico PL, Genazzani AA, Terrazzino S (2014) Diagnostic accuracy of HLA-B*57:01 screening for the prediction of abacavir hypersensitivity and clinical utility of the test: a meta-analytic review. *Pharmacogenomics* 15, 963–76.

Chan K, Jensen NS, Silber PM, O'Brien, PJ (2007) Structure-activity relationships for halobenzene induced cytotoxicity in rat and human hepatocytes. *Chem Biol Interact* 165, 165–74.

Chen M, Vijay V, Shi Q, Liu Z, Fang H, Tong W (2011) FDA-approved drug labeling for the study of drug-induced liver injury. *Drug Discov Today* 16, 697–703.

Chen M, Borlak J, Tong W (2013a) High lipophilicity and high daily dose of oral medications are associated with significant risk for drug-induced liver injury. *Hepatology* 58, 388–96.

Chen M, Hong H, Fang H, Kelly R, Zhou G, Borlak J, Tong W (2013b) Quantitative structure-activity relationship models for predicting drug-induced liver injury based on FDA-approved drug labeling annotation and using a large collection of drugs. *Toxicol Sci* 136, 242–9.

Chen M, Bisgin H, Tong L, Hong H, Fang H, Borlak J, Tong W (2014) Toward predictive models for drug-induced liver injury in humans: are we there yet? *Biomark Med* 8, 201–13.

Cheng A, Dixon SL (2003) *In silico* models for the prediction of dose-dependent human hepatotoxicity. *J Comput Aided Mol Des* 17, 811–23.

Chijiiwa K, Mizuta A, Ueda J, Takamatsu Y, Nakamura K, Watanabe M, Kuroki S, Tanaka M (2002) Relation of biliary bile acid output to hepatic adenosine triphosphate level and biliary indocyanine green excretion in humans. *World J Surg* 26, 457–61.

Chng HT, Ho HK, Yap CW, Lam SH, Chan EC (2012) An investigation of the bioactivation potential and metabolism profile of Zebrafish versus human. *J Biomol Screen*, 17, 974–86.

Chu X, Korzekwa K, Elsby R, Fenner K, Galetin A, Lai Y, Matsson P, Moss A, Nagar S, Rosania GR, Bai JP, Polli JW, Sugiyama Y, Brouwer KL (2013) Intracellular drug concentrations and transporters: measurement, modeling, and implications for the liver. *Clin Pharmacol Ther* 94, 126–41.

Clark RD, Wolohan PR, Hodgkin EE, Kelly JH, Sussman NL (2004) Modelling *in vitro* hepatotoxicity using molecular interaction fields and SIMCA. *J Mol Graph Model* 22, 487–97.

Cortez-Pinto H, Chatham J, Chacko VP, Arnold C, Rashid A, Diehl AM (1999) Alterations in liver ATP homeostasis in human non-alcoholic steatohepatitis: a pilot study. *JAMA* 282, 1659–64.

Cruz-Monteagudo M, Cordeiro MN, Borges F (2008) Computational chemistry approach for the early detection of drug-induced idiosyncratic liver toxicity. *J Comput Chem* 29, 533–49.

Cui Y, König J, Keppler D (2001) Vectorial transport by double-transfected cells expressing the human uptake transporter SLC21A8 and the apical export pump ABCC2. *Mol Pharmacol* 60, 934–43.

Dawson S, Stahl S, Paul N, Barber J, Kenna JG (2012) *In vitro* inhibition of the bile salt export pump correlates with risk of cholestatic drug-induced liver injury in humans. *Drug Metab Dispos* 40, 130–8.

De Bruyn T, Chatterjee S, Fattah S, Keemink J, Nicolaï J, Augustijns P, Annaert P (2013) Sandwich-cultured hepatocytes: utility for *in vitro* exploration of hepatobiliary drug disposition and drug-induced hepatotoxicity. *Expert Opin Drug Metab Toxicol* 9, 589–616.

Devillers J, Mombelli E (2010) Evaluation of the OECD QSAR Application Toolbox and Toxtree for estimating the mutagenicity of chemicals. Part 1. Aromatic amines. *SAR QSAR Environ Res* 21, 753–69.

Divakaruni AS, Rogers GW, Murphy AN (2014) Measuring mitochondrial function in permeabilized cells using the Seahorse XF analyzer or a Clark-type oxygen electrode. *Curr Protoc Toxicol* 60, 25.2.1–16.

Donato MT, Tolosa L, Jiménez N, Castell JV, Gómez-Lechón MJ (2012) High-content imaging technology for the evaluation of drug-induced steatosis using a multiparametric cell-based assay. *J Biomol Screen* 17, 394–400.

Dou X, Wang Z, Yao T, Song Z (2011) Cysteine aggravates palmitate-induced cell death in hepatocytes. *Life Sci* 89, 878–85.

Driessen M, Kienhuis AS, Vitins AP, Pennings JL, Pronk TE, Van Den Brandhof EJ, Roodbergen M, Van De Water B, Van Der Ven LT (2014) Gene expression markers in the zebrafish embryo reflect a hepatotoxic response in animal models and humans. *Toxicol Lett* 230, 48–56.

Durham SK, Pearl, GM (2001) Computational methods to predict drug safety liabilities. *Curr Opin Drug Discov Dev* 4, 110–15.

Dutu A, Borza V, Mosora N, Motocu M, Benga G (1985) ATP-ase activity of mitochondria isolated from needle-biopsy liver samples of diabetic subjects. *Med Interne* 23, 201–6.

Dykens JA, Jamieson J, Marroquin L, Nadanaciva S, Billis PA, Will Y (2008a) Biguanide-induced mitochondrial dysfunction yields increased lactate production and cytotoxicity of aerobically-poised HepG2 cells and human hepatocytes *in vitro*. *Toxicol Appl Pharmacol* 233, 203–10.

Dykens JA, Jamieson JD, Marroquin LD, Nadanaciva S, Xu JJ, Dunn MC, Smith AR, Will Y (2008b) *In vitro* assessment of mitochondrial dysfunction and cytotoxicity of nefazodone, trazodone, and buspirone. *Toxicol Sci* 103, 335–45.

Egan WJ, Zlokarnik G, Grootenhuis, PD (2004) *In silico* prediction of drug safety: despite progress there is abundant room for improvement. *Drug Discov Today Technol* 1, 381–7.

Egnatchik RA, Leamy AK, Noguchi Y, Shiota M, Young JD (2014) Palmitate-induced activation of mitochondrial metabolism promotes oxidative stress and apoptosis in H4IIEC3 rat hepatocytes. *Metabolism* 63, 283–95.

Ekins S, Williams AJ, Xu JJ (2010) A predictive ligand-based Bayesian model for human drug-induced liver injury. *Drug Metab Dispos* 38, 2302–8.

Ellison CM, Enoch SJ, Cronin MT (2011) A review of the use of *in silico* methods to predict the chemistry of molecular initiating events related to drug toxicity. *Expert Opin Drug Metab Toxicol* 7, 1481–95.

Enoch SJ, Cronin MT (2010) A review of the electrophilic reaction chemistry involved in covalent DNA binding. *Crit Rev Toxicol* 40, 728–48.

Enoch SJ, Cronin, MT, Schultz, TW, Madden, JC (2008) Quantitative and mechanistic read across for predicting the skin sensitization potential of alkenes acting via Michael addition. *Chem Res Toxicol* 21, 513–20.

Enoch SJ, Roberts DW, Cronin MT (2009) Electrophilic reaction chemistry of low molecular weight respiratory sensitizers. *Chem Res Toxicol* 22, 1447–53.

Enoch SJ, Cronin MT, Ellison CM (2011a) The use of a chemistry-based profiler for covalent DNA binding in the development of chemical categories for read-across for genotoxicity. *Altern Lab Anim* 39, 131–45.

Enoch SJ, Ellison, CM, Schultz, TW, Cronin, MT (2011b) A review of the electrophilic reaction chemistry involved in covalent protein binding relevant to toxicity. *Crit Rev Toxicol* 41, 783–802.

European Medicine Agency Committee for Human Medicinal Products (CHMP) (2012) Guideline on the Investigation of Drug Interactions CPMP/EWP/560/95/Rev. 1 Corr.* http://www.ema.europa.eu/docs/en_GB/document_library/Scientific_guideline/2012/07/WC500129606.pdf (accessed October 10, 2015)

Fahrmayr C, König J, Auge D, Mieth M, Fromm MF (2012) Identification of drugs and drug metabolites as substrates of multidrug resistance protein 2 (MRP2) using triple-transfected MDCK-OATP1B1-UGT1A1-MRP2 cells. *Br J Pharmacol* 165, 1836–47.

Fattinger K, Funk C, Pantze M, Weber C, Reichen J, Stieger B, Meier PJ (2001) The endothelin antagonist bosentan inhibits the canalicular bile salt export pump: a potential mechanism for hepatic adverse reactions. *Clin Pharmacol Ther* 69, 223–31.

Fourches D, Barnes JC, Day NC, Bradley P, Reed JZ, Tropsha A (2010) Cheminformatics analysis of assertions mined from literature that describe drug-induced liver injury in different species. *Chem Res Toxicol* 23, 171–83.

Fulda S, Gorman AM, Hori O, Samali A (2010) Cellular stress responses: cell survival and cell death. *Int J Cell Biol* 214074, doi:10.1155/2010/214074.

Funk C, Pantze M, Jehle L, Ponelle C, Scheuermann G, Lazendic M, Gasser R (2001) Troglitazone-induced intrahepatic cholestasis by an interference with the hepatobiliary export of bile acids in male and female rats. Correlation with the gender difference in troglitazone sulfate formation and the inhibition of the canalicular bile salt export pump (Bsep) by troglitazone and troglitazone sulfate. *Toxicology* 167, 83–98.

Ganter B, Snyder RD, Halbert DN, Lee MD (2006) Toxicogenomics in drug discovery and development: mechanistic analysis of compound/class-dependent effects using the DrugMatrix database. *Pharmacogenomics* 7, 1025–44.

Garside H, Marcoe KF, Chesnut-Speelman J, Foster AJ, Muthas D, Kenna JG, Warrior U, Bowes J, Baumgartner J (2014) Evaluation of the use of imaging parameters for the detection of compound-induced hepatotoxicity in 384-well cultures of HepG2 cells and cryopreserved primary human hepatocytes. *Toxicol In Vitro* 28, 171–81.

Gentile CL, Frye MA, Pagliassotti MJ (2011) Fatty acids and the endoplasmic reticulum in nonalcoholic fatty liver disease. *Biofactors* 37, 8–16.

Gerk PM, Vore M (2002) Regulation of expression of the multidrug resistance-associated protein 2 (MRP2) and its role in drug disposition. *J Pharmacol Exp Ther* 302, 407–15.

Gibson JD, Pumford NR, Samokyszyn VM, Hinson JA (1996) Mechanism of acetaminophen-induced hepatotoxicity: covalent binding versus oxidative stress. *Chem Res Toxicol* 9, 580–5.

Goldstone JV, Mcarthur AG, Kubota A, Zanette J, Parente T, Jonsson ME, Nelson DR, Stegeman JJ (2010) Identification and developmental expression of the full complement of Cytochrome P450 genes in Zebrafish. *BMC Genomics* 11, 643.

Gonzalez FJ (2012) Nuclear receptor control of enterohepatic circulation. *Comp Physiol* 2, 2811–28.

Greene N, Fisk L, Naven RT, Note RR, Patel ML, Pelletier DJ (2010) Developing structure-activity relationships for the prediction of hepatotoxicity. *Chem Res Toxicol* 23, 1215–22.

Gustafsson F, Foster AJ, Sarda S, Bridgland-Taylor MH, Kenna JG (2014) A correlation between the *in vitro* drug toxicity of drugs to cell lines that express human P450s and their propensity to cause liver injury in humans. *Toxicol Sci* 137, 189–211.

Haasio K, Nissinen E, Sopanen L, Heinonen EH (2002) Different toxicological profile of two COMT inhibitors *in vivo*: the role of uncoupling effects. *J Neural Transm* 109, 1391–401.

Hayes JD, McLellan LI (1999) Glutathione and glutathione-dependent enzymes represent a co-ordinately regulated defence against oxidative stress. *Free Radic Res* 31, 273–300.

He JH, Guo SY, Zhu F, Zhu JJ, Chen YX, Huang CJ, Gao JM, Dong QX, Xuan YX, Li CQ (2013) A zebrafish phenotypic assay for assessing drug-induced hepatotoxicity. *J Pharmacol Toxicol Methods* 67, 25–32.

Hewitt M, Enoch SJ, Madden JC, Przybylak KR, Cronin MT (2013) Hepatotoxicity: a scheme for generating chemical categories for read-across, structural alerts and insights into mechanism(s) of action. *Crit Rev Toxicol* 43, 537–58.

Hill A (2011) *Hepatotoxicity Testing in Larval Zebrafish*. John Wiley & Sons, Inc., 89–102.

Hill A, Mesens N, Steemans M, Xu JJ, Aleo MD (2012) Comparisons between *in vitro* whole cell imaging and *in vivo* zebrafish-based approaches for identifying potential human hepatotoxicants earlier in pharmaceutical development. *Drug Metab Rev* 44, 127–40.

Hinson JA, Roberts DW, James LP (2010) Mechanisms of acetaminophen-induced liver necrosis. *Handb Exp Pharmacol* 196, 369–405.

Hofmann AF, Hagey LR (2008) Bile acids: chemistry, pathochemistry, biology, pathobiology, and therapeutics. *Cell Mol Life Sci* 65, 2461–83.

Holubarsch CJ, Rohrbach M, Karrasch M, Boehm E, Polonski L, Ponikowski P, Rhein S (2007) A double-blind randomized multi-centre clinical trial to evaluate the efficacy and safety of two doses of etomoxir in comparison with placebo in patients with moderate congestive heart failure: the ERGO (etomoxir for the recovery of glucose oxidation) study. *Clin Sci (Lond)* 113, 205–12.

Hornberg JJ, Laursen M, Brenden N, Persson M, Thougaard AV, Toft DB, Mow T (2013a) Exploratory toxicology as an integrated part of drug discovery - Part I: why and how. *Drug Discov Today* 19, 1131–6.

Hornberg JJ, Laursen M, Brenden N, Persson M, Thougaard AV, Toft DB, Mow T (2013b) Exploratory toxicology as an integrated part of drug discovery. Part II: screening strategies. *Drug Discov Today* 19, 1137–44.

Howell BA, Yang Y, Kumar R, Woodhead JL, Harrill AH, Clewell HJ III, Andersen ME, Siler SQ, Watkins PB (2012) *In vitro* to *in vivo* extrapolation and species response comparisons for drug-induced liver injury (DILI) using DILIsym™: a mechanistic, mathematical model of DILI. *J Pharmacokinet Pharmacodyn* 39, 527–41.

Howell BA, Siler SQ, Watkins PB (2014a) Use of a systems model of drug-induced liver injury (DILIsym(®)) to elucidate the mechanistic differences between acetaminophen and its less-toxic isomer, AMAP, in mice. *Toxicol Lett* 226, 163–72.

Howell BA, Siler SQ, Shoda LK, Yang Y, Woodhead JL, Watkins PB (2014b) A mechanistic model of drug-induced liver injury aids the interpretation of elevated liver transaminase levels in a phase I clinical trial. *CPT Pharmacometrics Syst Pharmacol* 3, e98, doi: 10.1038/psp.2013.74.

Hsiao CJ, Younis H, Boelsterli UA (2010) Trovafloxacin, a fluoroquinolone antibiotic with hepatotoxic potential, causes mitochondrial peroxynitrite stress in a mouse model of underlying mitochondrial dysfunction. *Chem Biol Interact* 188, 204–13.

Hynes J, Swiss RL, Will Y (2012) High-throughput analysis of mitochondrial oxygen consumption. *Methods Mol Biol* 810, 59–72.

Hynes J, Nadanaciva S, Swiss R, Carey C, Kirwan S, Will Y (2013) A high-throughput dual parameter assay for assessing drug-induced mitochondrial dysfunction provides additional predictivity over two established mitochondrial toxicity assays. *Toxicol In Vitro* 27, 560–9.

Isogai S, Horiguchi M, Weinstein BM (2001) The vascular anatomy of the developing zebrafish: an atlas of embryonic and early larval development. *Dev Biol* 230, 278–301.

Jamei M, Marciniak S, Feng K, Barnett A, Tucker G, Rostami-Hodjegan A (2009) The Simcyp population-based ADME simulator. *Expert Opin Drug Metab Toxicol* 5, 211–23.

Jones HS, Panter GH, Hutchinson TH, Chipman JK (2010) Oxidative and conjugative xenobiotic metabolism in zebrafish larvae *in vivo*. *Zebrafish* 7, 23–30.

Ju C, Reilly T (2012) Role of immune reactions in drug-induced liver injury (DILI). *Drug Metab Rev*, 44, 107–15.

Judson PN, Marchant CA, Vessey JD (2003) Using argumentation for absolute reasoning about the potential toxicity of chemicals. *J Chem Inf Comput Sci* 43, 1364–70.

Judson R, Houck K, Martin M, Knudsen T, Thomas RS, Sipes N, Shah I, Wambaugh J, Crofton K (2014) *In vitro* and modelling approaches to risk assessment from the U.S. Environmental Protection Agency ToxCast programme. *Basic Clin Pharmacol Toxicol* 115, 69–76.

Kamiike W, Nakahara M, Nakao K, Koseki M, Nishida T, Kawashima Y, Watanabe F, Tagawa K (1985) Correlation between cellular ATP level and bile excretion in the rat liver. *Transplantation* 39, 50–5.

Kaplowitz N, DeLeve L (2013) *Drug-Induced Liver Disease*. Academic Press, Waltham, MA; Third Revised edition.

Kashimshetty R, Desai VG, Kale VM, Lee T, Moland CL, Branham WS, New LS, Chan EC, Younis H, Boelsterli UA (2009) Underlying mitochondrial dysfunction triggers flutamide-induced oxidative liver injury in a mouse model of idiosyncratic drug toxicity. *Toxicol Appl Pharmacol* 238, 150–9.

Kenna JG (2013) Mechanisms, pathology, and clinical presentation of hepatotoxicity of anesthetic agents. In: *Drug-Induced Liver Disease*, Eds Kaplowitz N, DeLeve LD. Academic Press, Waltham, MA, 403–20.

Kenna JG (2014) Current concepts in drug-induced bile salt export pump (BSEP) interference. *Curr Protoc Toxicol* 61, 23.7.1–15.

Kenna JG, Stahl SH, Eakins JA, Foster AJ, Andersson LC, Bergare J, Billger M, Elebring M, Elmore CS, Thompson RA (2015) Multiple compound-related adverse properties contribute to liver injury caused by endothelin receptor antagonists. *J Pharmacol Exp Ther* 352, 281–90.

Khetani SR, Kanchagar C, Ukairo O, Krzyzewski S, Moore A, Shi J, Aoyama S, Aleo M, Will Y (2013) Use of micropatterned cocultures to detect compounds that cause drug-induced liver injury in humans. *Toxicol Sci* 132, 107–17.

Köck K, Ferslew BC, Netterberg I, Yang K, Urban TJ, Swaan PW, Stewart PW, Brouwer KL (2014). Risk factors for development of cholestatic drug-induced liver injury: inhibition of hepatic basolateral bile acid transporters multidrug resistance-associated proteins 3 and 4. *Drug Metab Dispos*, 42, 665–74.

Kostrubsky SE, Strom SC, Kalgutkar AS, Kulkarni S, Atherton J, Mireles R, Feng B, Kubik R, Hanson J, Urda E, Mutlib AE (2006) Inhibition of hepatobiliary transport as a predictive method for clinical hepatotoxicity of nefazodone. *Toxicol Sci* 90, 451–9.

Kotronen A, Juurinen L, Hakkarainen A, Westerbacka J, Cornér A, Bergholm R, Yki-Järvinen H (2008) Liver fat is increased in type 2 diabetic patients and underestimated by serum alanine aminotransferase compared with equally obese nondiabetic subjects. *Diabetes Care* 31, 165–9.

Kovacic P, Jacintho, JD (2001) Mechanisms of carcinogenesis: focus on oxidative stress and electron transfer. *Curr Med Chem* 8, 773–96.

Kuepfer L, Kerb R, Henney AM (2014) Clinical translation in the virtual liver network. *CPT Pharmacometrics Syst Pharmacol* 3, e127, doi: 10.1038/psp.2014.25.

Kullak-Ublick GA, Stieger B, Meier PJ (2004) Enterohepatic bile salt transporters in normal physiology and liver disease. *Gastroenterology* 126, 322–42.

Kurbatova OV, Izmailova TD, Surkov AN, Namazova-Baranova LS, Poliakova SI, Miroshkina LV, Semenova GF, Samokhina IV, Kapustina E, Dukhova ZN, Potapov AS, Petrichuk SV (2014) Mitochondrial dysfunction in children with hepatic forms of glycogen storage disease. *Vestn Ross Akad Med Nauk* 7–8, 78–84.

Kurumiya Y, Nagino M, Nozawa K, Kamiya J, Uesaka K, Sano T, Yoshida S, Nimura Y (2003) Biliary bile acid concentration is a simple and reliable indicator for liver function after hepatobiliary resection for biliary cancer. *Surgery* 133, 512–20.

Labbe G, Pessayre D, Fromenty B (2008) Drug-induced liver injury through mitochondrial dysfunction: mechanisms and detection during preclinical safety studies. *Fundam Clin Pharmacol* 22, 335–53.

Lam P, Soroka CJ, Boyer JL (2010) The bile salt export pump: clinical and experimental aspects of genetic and acquired cholestatic liver disease. *Semin Liver Dis* 30, 125–33.

Lammert C, Einarsson S, Saha C, Niklasson A, Bjornsson E, Chalasani N (2008) Relationship between daily dose of oral medications and idiosyncratic drug-induced liver injury: search for signals. *Hepatology* 47, 2003–9.

Lammert C, Bjornsson E, Niklasson A, Chalasani N (2010) Oral medications with significant hepatic metabolism at higher risk for hepatic adverse events. *Hepatology* 51, 615–20.

Lang C, Meier Y, Stieger B, Beuers U, Lang T, Kerb R, Kullak-Ublick GA, Meier PJ, Pauli-Magnus C (2007) Mutations and polymorphisms in the bile salt export pump and the multidrug resistance protein 3 associated with drug-induced liver injury. *Pharmacogenet Genomics* 17, 47–60.

Laskin DL (2009) Macrophages and inflammatory mediators in chemical toxicity: a battle of forces. *Chem Res Toxicol* 22, 1376–85.

Laverty H, Benson C, Cartwright E, Cross M, Garland C, Hammond T, Holloway C, McMahon N, Milligan J, Park B, Pirmohamed M, Pollard C, Radford J, Roome N, Sager P, Singh S, Suter T, Suter W, Trafford A, Volders P, Wallis R, Weaver R, York M, Valentin J (2011). How can we improve our understanding of cardiovascular safety liabilities to develop safer medicines? *Br J Pharmacol*, 163, 675–93.

Leamy AK, Egnatchik RA, Young JD (2013) Molecular mechanisms and the role of saturated fatty acids in the progression of non-alcoholic fatty liver disease. *Prog Lipid Res* 52, 165–74.

Lee YH, Chung MC, Lin Q, Boelsterli UA (2008) Troglitazone-induced hepatic mitochondrial proteome expression dynamics in heterozygous Sod2(+/-) mice: two-stage oxidative injury. *Toxicol Appl Pharmacol* 231, 43–51.

Li AP (2002) A review of the common properties of drugs with idiosyncratic hepatotoxicity and the "multiple determinant hypothesis" for the manifestation of idiosyncratic drug toxicity. *Chem Biol Interact* 142, 7–23.

Li AP (2004) An integrated, multidisciplinary approach for drug safety assessment. *Drug Discov Today* 9, 687–93.

Li ZZ, Berk M, McIntyre TM, Feldstein AE (2009) Hepatic lipid partitioning and liver damage in nonalcoholic fatty liver disease: role of stearoyl-CoA desaturase. *J Biol Chem* 284, 5637–44.

Lionetti V, Stanley WC, Recchia FA (2011) Modulating fatty acid oxidation in heart failure. *Cardiovasc Res* 90, 202–9.

Liu X, Chism JP, LeCluyse EL, Brouwer KR, Brouwer KL (1999) Correlation of biliary excretion in sandwich-cultured rat hepatocytes and *in vivo* in rats. *Drug Metab Dispos* 27, 637–44.

Liu Z, Shi Q, Ding D, Kelly R, Fang H, Tong W (2011) Translating clinical findings into knowledge in drug safety evaluation—drug induced liver injury prediction system (DILIps). *PLoS Comput Biol* 7, e1002310.

Low Y, Uehara T, Minowa Y, Yamada H, Ohno Y, Urushidani T, Sedykh A, Muratov E, Kuz'min V, Fourches D, Zhu H, Rusyn I, Tropsha A (2011) Predicting drug-induced hepatotoxicity using QSAR and toxicogenomics approaches. *Chem Res Toxicol* 24, 1251–62.

Lu S, Jessen B, Strock C, Will Y (2012) The contribution of physicochemical properties to multiple *in vitro* cytotoxicity endpoints. *Toxicol In Vitro* 26, 613–20.

Marchant CA, Fisk L, Note RR, Patel ML, Suarez D (2009) An expert system approach to the assessment of hepatotoxic potential. *Chem Biodivers* 6, 2107–14.

Marroquin LD, Hynes J, Dykens JA, Jamieson JD, Will Y (2007) Circumventing the Crabtree effect: replacing media glucose with galactose increases susceptibility of HepG2 cells to mitochondrial toxicants. *Toxicol Sci* 97, 539–47.

Massart J, Begriche K, Buron N, Porceddu M, Borgne-Sanchez A, Fromenty B (2013) Drug-induced inhibition of mitochondrial fatty acid oxidation and steatosis. *Curr Pathobiol Rep* 1, 147–57.

Matthews EJ, Kruhlak NL, Benz RD, Aragones Sabate D, Marchant CA, Contrera JF (2009a) Identification of structure-activity relationships for adverse effects of pharmaceuticals in humans: part C: use of QSAR and an expert system for the estimation of the mechanism of action of drug-induced hepatobiliary and urinary tract toxicities. *Regul Toxicol Pharmacol* 54, 43–65.

Matthews EJ, Ursem CJ, Kruhlak NL, Benz RD, Sabate DA, Yang C, Klopman G, Contrera JF (2009b) Identification of structure-activity relationships for adverse effects of pharmaceuticals in humans: part B. Use of (Q)SAR systems for early detection of drug-induced hepatobiliary and urinary tract toxicities. *Regul Toxicol Pharmacol* 54, 23–42.

Maximos M, Bril F, Portillo Sanchez P, Lomonaco R, Orsak B, Biernacki D, Suman A, Weber M, Cusi K (2015) The role of liver fat and insulin resistance as determinants of plasma aminotransferase elevation in nonalcoholic fatty liver disease. *Hepatology* 61, 153–60.

Mehta R, Chan K, Lee O, Tafazoli S, O'Brien P (2008) Drug-associated mitochondrial toxicity. In: *Drug-Induced Mitochondrial Dysfunction*, Eds Dykens JA, Will Y. John Wiley & Sons, Inc., Hoboken, NJ, 71–126.

Menke AL, Spitsbergen JM, Wolterbeek AP, Woutersen RA (2011) Normal anatomy and histology of the adult zebrafish. *Toxicol Pathol* 39, 759–75.

Mesens N, Crawford AD, Menke A, Hung PD, Van Goethem F, Nuyts R, Hansen E, Wolterbeek A, Van Gompel J, De Witte P, Esguerra CV (2015) Are zebrafish larvae suitable for assessing the hepatotoxicity potential of drug candidates? *J Appl Toxicol*, doi:10.1002/jat.3091.

Messner S, Agarkova I, Moritz W, Kelm JM (2013) Multi-cell type human liver microtissues for hepatotoxicity testing. *Arch Toxicol* 87, 209–13.

Mita S, Suzuki H, Akita H, Hayashi H, Onuki R, Hofmann AF, Sugiyama Y (2006a) Vectorial transport of unconjugated and conjugated bile salts by monolayers of LLC-PK1 cells doubly transfected with human NTCP and BSEP or with rat Ntcp and Bsep. *Am J Physiol Gastrointest Liver Physiol* 290, G550–6.

Mita S, Suzuki H, Akita H, Hayashi H, Onuki R, Hofmann AF, Sugiyama Y (2006b) Inhibition of bile acid transport across Na+/taurocholate cotransporting polypeptide (SLC10A1) and bile salt export pump (ABCB 11)-coexpressing LLC-PK1 cells by cholestasis-inducing drugs. *Drug Metab Dispos* 34, 1575–81.

Mombelli E, Devillers J (2010) Evaluation of the OECD (Q)SAR Application Toolbox and Toxtree for predicting and profiling the carcinogenic potential of chemicals. *SAR QSAR Environ Res* 21, 731–52.

Monshi MM, Faulkner L, Gibson A, Jenkins RE, Farrell J, Earnshaw CJ, Alfirevic A, Cederbrant K, Daly AK, French N, Pirmohamed M, Park BK, Naisbitt DJ (2011) Human leukocyte antigen (HLA)-B*57:01-restricted activation of drug-specific T cells provides the immunological basis for flucloxacillin-induced liver injury. *Hepatology* 57, 727–39.

Morgan RE, Trauner M, van Staden CJ, Lee PH, Ramachandran B, Eschenberg M, Afshari CA, Qualls CW Jr, Lightfoot-Dunn R, Hamadeh HK (2010) Interference with bile salt export pump function is a susceptibility factor for human liver injury in drug development. *Toxicol Sci* 118, 485–500.

Morgan RE, van Staden CJ, Chen Y, Kalyanaraman N, Kalanzi J, Dunn RT II, Afshari CA, Hamadeh HK (2013) A multifactorial approach to hepatobiliary transporter assessment enables improved therapeutic compound development. *Toxicol Sci* 136, 216–41.

Nadanaciva S, Dykens JA, Bernal A, Capaldi RA, Will Y (2007) Mitochondrial impairment by PPAR agonists and statins identified via immunocaptured OXPHOS complex activities and respiration. *Toxicol Appl Pharmacol* 223, 277–87.

Nadanaciva S, Willis JH, Barker ML, Gharaibeh D, Capaldi RA, Marusich MF, Will Y (2009) Lateral-flow immunoassay for detecting drug-induced inhibition of mitochondrial DNA replication and mtDNA-encoded protein synthesis. *J Immunol Methods*, 343, 1–12.

Nadanaciva S, Murray J, Wilson C, Gebhard DF, Will Y (2011a) High-throughput assays for assessing mitochondrial dysfunction caused by compounds that impair mtDNA-encoded protein levels in eukaryotic cells. *Curr Protoc Toxicol* Chapter 3, Unit 3.11, doi:10.1002/0471140856.tx0311s48.

Nadanaciva S, Lu S, Gebhard DF, Jessen BA, Pennie WD, Will Y (2011b) A high content screening assay for identifying lysosomotropic compounds. *Toxicol In Vitro* 25, 715–23.

Nadanaciva S, Rana P, Beeson GC, Chen D, Ferrick DA, Beeson CC, Will Y (2012) Assessment of drug-induced mitochondrial dysfunction via altered cellular respiration and acidification measured in a 96-well platform. *J Bioenerg Biomembr* 44, 421–37.

Nadanaciva S, Aleo MD, Strock CJ, Stedman DB, Wang H, Will Y (2013) Toxicity assessments of nonsteroidal anti-inflammatory drugs in isolated mitochondria, rat hepatocytes, and zebrafish show good concordance across chemical classes. *Toxicol Appl Pharmacol* 272, 272–80.

Nakayama S, Atsumi R, Takakusa H, Kobayashi Y, Kurihara A, Nagai Y, Nakai D, Okazaki O (2009) A zone classification system for risk assessment of idiosyncratic drug toxicity using daily dose and covalent binding. *Drug Metab Dispos* 37, 1970–7.

Naven RT, Swiss R, Klug-McLeod J, Will Y, Greene N (2013) The development of structure-activity relationships for mitochondrial dysfunction: uncoupling of oxidative phosphorylation. *Toxicol Sci* 131, 271–8.

Nieminen AL, Saylor AK, Herman B, Lemasters JJ (1994) ATP depletion rather than mitochondrial depolarization mediates hepatocyte killing after metabolic inhibition. *Am J Physiol* 267, C67–74.

Noguchi Y, Young JD, Aleman JO, Hansen ME, Kelleher JK, Stephanopoulos G (2009) Effect of anaplerotic fluxes and amino acid availability on hepatic lipoapoptosis. *J Biol Chem* 284, 33425–36.

O'Brien PJ (2014) High-content analysis in toxicology: screening substances for human toxicity potential, elucidating subcellular mechanisms and *in vivo* use as translational safety biomarkers. *Basic Clin Pharmacol Toxicol* 115, 4–17.

O'Brien PJ, Irwin W, Diaz D, Howard-Cofield E, Krejsa CM, Slaughter MR, Gao B, Kaludercic N, Angeline A, Bernardi P, Brain P, Hougham C (2006) High concordance of drug-induced human hepatotoxicity with *in vitro* cytotoxicity measured in a novel cell-based model using high content screening. *Arch Toxicol* 80:580–604.

Oh JM, Jung YS, Jeon BS, Yoon BI, Lee KS, Kim BH, Oh SJ, Kim SK (2012) Evaluation of hepatotoxicity and oxidative stress in rats treated with tert-butyl hydroperoxide. *Food Chem Toxicol* 50, 1215–21.

Olson H, Betton G, Robinson D, Thomas K, Monro A, Kolaja G, Lilly P, Sanders J, Sipes G, Bracken W, Dorato M, Van Deun K, Smith P, Berger B, Heller A (2000) Concordance of the toxicity of pharmaceuticals in humans and in animals. *Regul Toxicol Pharmacol* 32, 56–67.

Pachkoria K., Lucena MI, Molokhia M, Cueto R, Carballo AS, Carvajal A, Andrade RJ (2007) Genetic and molecular factors in drug-induced liver injury: a review. *Curr Drug Saf* 2, 97–112.

Pack M, Solnica-Krezel L, Malicki J, Neuhauss SC, Schier AF, Stemple DL, Driever W, Fishman MC (1996) Mutations affecting development of zebrafish digestive organs. *Development* 123, 321–8.

Palmeira CM, Rolo AP (2004) Mitochondrially-mediated toxicity of bile acids. *Toxicology* 203, 1–15.

Park BK, Boobis A, Clarke S, Goldring CE, Jones D, Kenna JG, Lambert C, Laverty HG, Naisbitt DJ, Nelson S, Nicoll-Griffith DA, Obach RS, Routledge P, Smith DA, Tweedie DJ, Vermeulen N, Williams DP, Wilson ID, Baillie TA (2011). Managing the challenge of chemically reactive metabolites in drug development. *Nat Rev Drug Discov*, 10, 292–306.

Parng C, Seng WL, Semino C, Mcgrath P (2002) Zebrafish: a preclinical model for drug screening. *Assay Drug Dev Technol* 1, 41–8.

Pearl GM, Livingston-Carr S, Durham SK (2001) Integration of computational analysis as a sentinel tool in toxicological assessments. *Curr Top Med Chem* 1, 247–55.

Peraza MA, Burdick AD, Marin HE, Gonzalez FJ, Peters JM (2006) The toxicology of ligands for peroxisome proliferator-activated receptors (PPAR). *Toxicol Sci* 90, 269–95.

Pérez-Carreras M, Del Hoyo P, Martín MA, Rubio JC, Martín A, Castellano G, Colina F, Arenas J, Solis-Herruzo JA (2003) Defective hepatic mitochondrial respiratory chain in patients with nonalcoholic steatohepatitis. *Hepatology* 38, 999–1007.

Persson M, Løye AF, Mow T, Hornberg JJ (2013) A high content screening assay to predict human drug-induced liver injury during drug discovery. *J Pharmacol Toxicol Methods* 68, 302–13.

Persson M, Løye AF, Jacquet M, Mow NS, Thougaard AV, Mow T, Hornberg JJ (2014) High-content analysis/screening for predictive toxicology: application to hepatotoxicity and genotoxicity. *Basic Clin Pharmacol Toxicol* 115, 18–23.

Pessayre D, Fromenty B, Berson A, Robin MA, Letteron P, Moreau R, Mansouri A (2012) Central role of mitochondria in drug-induced liver injury. *Drug Metab Rev* 44, 34–87.

Peterson RT, Macrae CA (2012) Systematic approaches to toxicology in the zebrafish. *Annu Rev Pharmacol Toxicol*, 52, 433–53.

Pohl LR, Satoh H, Christ DD, Kenna JG (1988) The immunological and metabolic basis of drug hypersensitivities. *Annu Rev Pharmacol* 28, 367–87.

Porceddu M, Buron N, Roussel C, Labbe G, Fromenty B, Borgne-Sanchez A (2012) Prediction of liver injury induced by chemicals in human with a multiparametric assay on isolated mouse liver mitochondria. *Toxicol Sci* 129, 332–45.

Przybylak KR, Cronin MT (2012) *In silico* models for drug-induced liver injury--current status. *Expert Opin Drug Metab Toxicol* 8, 201–17.

Qiu L, Taimi M, Finley J, Aleo MD, Strock C, Gilbert J, Qin S, Will Y (2015) High-Content Imaging in Human and Rat Hepatocytes using the fluorescent dyes CLF and CMFDA is not specific enough to assess BSEP/Bsep and or MRP2/Mrp2 Inhibition by Cholestatic Drugs. *Appli In Vitro Toxicol*, 1(3), 198–212.

Rabol R, Boushel R, Almdal T, Hansen CN, Ploug T, Haugaard SB, Prats C, Madsbad S, Dela F (2010). Opposite effects of pioglitazone and rosiglitazone on mitochondrial respiration in skeletal muscle of patients with type 2 diabetes. *Diabetes Obes Metab*, 12, 806–14.

Ramaiahgari SC, den Braver MW, Herpers B, Terpstra V, Commandeur JN, van de Water B, Price LS (2014) A 3D *in vitro* model of differentiated HepG2 cell spheroids with improved liver-like properties for repeated dose high-throughput toxicity studies. *Arch Toxicol* 88, 1083–95.

Ramappa V, Aithal GP (2013) Hepatotoxicity related to anti-tuberculosis drugs: mechanisms and management. *J Clin Exp Hepatol* 3, 37–49.

Regev A (2013) How to avoid being surprised by hepatotoxicity at the final stages of drug development and approval. *Clin Liver Dis* 17, 749–67.

Regev A (2014) Drug-induced liver injury and drug development: industry perspective. *Semin Liver Dis* 34, 227–39.

Richarz AN, Schultz TW, Cronin MT, Enoch SJ (2014) Experimental verification of structural alerts for the protein binding of sulfur-containing compounds. *SAR QSAR Environ Res* 25, 325–41.

Roberts DW, Patlewicz G, Kern PS, Gerberick F, Kimber I, Dearman RJ, Ryan CA, Basketter DA, Aptula AO (2007) Mechanistic applicability domain classification of a local lymph node assay dataset for skin sensitization. *Chem Res Toxicol* 20, 1019–30.

Rodgers AD, Zhu H, Fourches D, Rusyn I, Tropsha A (2010) Modeling liver-related adverse effects of drugs using nearest neighbor quantitative structure-activity relationship method. *Chem Res Toxicol* 23, 724–32.

Rodrigues CM, Fan G, Wong PY, Kren BT, Steer CJ (1998) Ursodeoxycholic acid may inhibit deoxycholic acid-induced apoptosis by modulating mitochondrial transmembrane potential and reactive oxygen species production. *Mol Med* 4, 165–78.

Rodrigues CM, Ma X, Linehan-Stieers C, Fan G, Kren BT, Steer CJ (1999) Ursodeoxycholic acid prevents cytochrome c release in apoptosis by inhibiting mitochondrial membrane depolarization and channel formation. *Cell Death Differ* 6, 842–54.

Rodrigues AD, Lai Y, Cvijic ME, Elkin LL, Zvyaga T, Soars MG (2014). Drug-induced perturbations of the bile acid pool, cholestasis, and hepatotoxicity: mechanistic considerations beyond the direct inhibition of the bile salt export pump. *Drug Metab Dispos* 42, 566–74.

Rogers GW, Nadanaciva S, Swiss R, Divakaruni AS, Will Y (2014) Assessment of Fatty Acid Beta oxidation in cells and isolated mitochondria. *Curr Protoc Toxicol* 60, 25.3.1–19.

Rolo AP, Oliveira PJ, Moreno AJM, Palmeira CM (2000) Bile acids affect liver mitochondrial bioenergetics: possible relevance for cholestasis therapy. *Toxicol Sci* 57, 177–85.

Roth A, Singer T (2014) The application of 3D cell models to support drug safety assessment: opportunities & challenges. *Adv Drug Deliv Rev* 69–70, 179–89.

Sakuratani Y, Sato S, Nishikawa S, Yamada J, Maekawa A, Hayashi M (2008) Category analysis of the substituted anilines studied in a 28-day repeat-dose toxicity test conducted on rats: correlation between toxicity and chemical structure. *SAR QSAR Environ Res* 19, 681–96.

Sayuk GS, Elwing JE, Lisker-Melman M (2007) Hepatic glycogenosis: an underrecognized source of abnormal liver function tests? *Dig Dis Sci* 52, 936–8.

Seifter S, England S (1988) Energy Metabolism. In: *The Liver: biology and pathobiology*, Ed Arias IM. Raven Press, New York, 279–310.

Selvaraj S, Ghebremichael M, Li M, Foli Y, Langs-Barlow A, Ogbuagu A, Barakat L, Tubridy E, Edifor R, Lam W, Cheng YC, Paintsil E (2014) Antiretroviral therapy-induced mitochondrial toxicity: potential mechanisms beyond polymerase-γ inhibition. *Clin Pharmacol Ther* 96, 110–20.

Shah F, Leung L, Barton HA, Will Y, Rodrigues AD, Greene N, Aleo MD (2015) Setting Clinical exposure levels of concern for drug-induced liver injury (DILI) using mechanistic *in vitro* assays. *Toxicol Sci* 147(2):500–14.

Simon-Hettich B, Rothfuss A, Steger-Hartmann T (2006) Use of computer-assisted prediction of toxic effects of chemical substances. *Toxicology* 224, 156–62.

Sirenko O, Hesley J, Rusyn I, Cromwell EF (2014) High-content assays for hepatotoxicity using induced pluripotent stem cell-derived cells. *Assay Drug Dev Technol* 12, 43–54.

Slater TF, Delaney VB (1970) Liver adenosine triphosphate content and bile flow rate in the rat. *Biochem J* 116, 303–8.

Snyder RD (2009) An update on the genotoxicity and carcinogenicity of marketed pharmaceuticals with reference to *in silico* predictivity. *Environ Mol Mutagen* 50, 435–50.

Sodeman T, Bronk SF, Roberts PJ, Miyoshi H, Gores GJ (2000) Bile salts mediate hepatocyte apoptosis by increasing cell surface trafficking of Fas. *Am J Physiol Gastrointest Liver Physiol* 278, G992–9.

Song Z, Song M, Lee DY, Liu Y, Deaciuc IV, McClain CJ (2007) Silymarin prevents palmitate-induced lipotoxicity in HepG2 cells: involvement of maintenance of Akt kinase activation. *Basic Clin Pharmacol Toxicol* 101, 262–8.

Spivey JR, Bronk SF, Gores GJ (1993) Glycochenodeoxycholate-induced lethal hepatocellular injury in rat hepatocytes. Role of ATP depletion and cytosolic free calcium. *J Clin Invest* 92, 17–24.

Stepan AF, Walker DP, Bauman J, Price DA, Baillie TA, Kalgutkar AS, Aleo MD (2011) Structural alert/reactive metabolite concept as applied in medicinal chemistry to mitigate the risk of idiosyncratic drug toxicity: a perspective based on the critical examination of trends in the top 200 drugs marketed in the United States. *Chem Res Toxicol* 24, 1345–410.

Stieger B, Fattinger K, Madon J, Kullak-Ublick GA, Meier PJ (2000) Drug- and estrogen-induced cholestasis through inhibition of the hepatocellular bile salt export pump (Bsep) of rat liver. *Gastroenterology* 118, 422–30.

Stieger B, Meier Y, Meier PJ (2007) The bile salt export pump. *Pflugers Arch* 453, 611–20.

Strautnieks SS, Bull LN, Knisely AS, Kocoshis SA, Dahl N, Arnell H, Sokal E, Dahan K, Childs S, Ling V, Tanner MS, Kagalwalla AF, Németh A, Pawlowska J, Baker A, Mieli-Vergani G, Freimer NB, Gardiner RM, Thompson RJ (1998) A gene encoding a liver-specific ABC transporter is mutated in progressive familial intrahepatic cholestasis. *Nat Genet* 20, 233–8.

Subramanian K, Raghavan S, Rajan Bhat A, Das S, Bajpai Dikshit J, Kumar R, Narasimha MK, Nalini R, Radhakrishnan R, Raghunathan S (2008) A systems biology based integrative framework to enhance the predictivity of in vitro methods for drug-induced liver injury. *Expert Opin Drug Saf* 7, 647–62.

Sushko I, Novotarskyi S, Korner R, Pandey AK, Rupp M, Teetz W, Brandmaier S, Abdelaziz A, Prokopenko VV, Tanchuk VY, Todeschini R, Varnek A, Marcou G, Ertl P, Potemkin V, Grishina M, Gasteiger J, Schwab C, Baskin II, Palyulin VA, Radchenko EV, Welsh WJ, Kholodovych V, Chekmarev D, Cherkasov A, Aires-de-Sousa J, Zhang QY, Bender A, Nigsch F, Patiny L, Williams A, Tkachenko V, Tetko IV (2011) Online chemical modeling environment (OCHEM): web platform for data storage, model development and publishing of chemical information. *J Comput Aided Mol Des* 25, 533–54.

Sushko I, Salmina E, Potemkin VA, Poda G, Tetko IV (2012) ToxAlerts: a Web server of structural alerts for toxic chemicals and compounds with potential adverse reactions. *J Chem Inf Model* 52, 2310–16.

Susukida T, Sekine S, Ogimura E, Aoki S, Oizumi K, Horie T, Ito K (2015) Basal efflux of bile acids contributes to drug-induced bile acid-dependent hepatocyte toxicity in rat sandwich-cultured hepatocytes. *Toxicol In Vitro* 29:1454–63.

Suzuki A, Andrade RJ, Bjornsson E, Lucena MI, Lee, WM, Yuen, NA, Hunt, CM, Freston, JW (2010) Drugs associated with hepatotoxicity and their reporting frequency of liver adverse events in VigiBase: unified list based on international collaborative work. *Drug Saf* 33, 503–22.

Swift B, Pfeifer ND, Brouwer KL (2010) Sandwich-cultured hepatocytes: an in vitro model to evaluate hepatobiliary transporter-based drug interactions and hepatotoxicity. *Drug Metab Rev* 42, 446–71.

Swiss R, Niles A, Cali JJ, Nadanaciva S, Will Y (2013) Validation of a HTS-amenable assay to detect drug-induced mitochondrial toxicity in the absence and presence of cell death. *Toxicol In Vitro* 27, 1789–97.

Thompson RA, Isin EM, Li Y, Weaver R, Weidolf L, Wilson I, Claesson A, Page K, Dolgos H, Kenna JG (2011) Risk assessment and mitigation strategies for reactive metabolites in drug discovery and development. *Chem Biol Interact* 192, 65–71.

Thompson RA, Isin EM, Li Y, Weidolf L, Page K, Wilson I, Swallow S, Middleton B, Stahl S, Foster AJ, Dolgos H, Weaver R, Kenna JG (2012) In vitro approach to assess the potential for risk of idiosyncratic adverse reactions caused by candidate drugs. *Chem Res Toxicol* 25, 1616–32.

Tirmenstein MA, Hu CX, Gales TL, Maleeff BE, Narayanan PK, Kurali E, Hart TK, Thomas HC, Schwartz LW (2002) Effects of troglitazone on HepG2 viability and mitochondrial function. *Toxicol Sci* 69, 131–8.

Tolosa L, Pinto S, Donato MT, Lahoz A, Castell JV, O'Connor JE, Gomez-Lechon MJ (2012) Development of a multiparametric cell-based protocol to screen and classify the hepatotoxicity potential of drugs. *Toxicol Sci* 127, 187–98.

Trask OJ Jr, Moore A, LeCluyse EL (2014) A micropatterned hepatocyte coculture model for assessment of liver toxicity using high-content imaging analysis. *Assay Drug Dev Technol* 12, 16–27.

Trauner M, Boyer JL (2003) Bile salt transporters: molecular characterization, function, and regulation. *Physiol Rev* 83, 633–71.

Ullrich RG (2007) Idiosyncratic toxicity: a convergence of risk factors. *Annu Rev Med* 58, 17–34.

Ursem CJ, Kruhlak NL, Contrera JF, MacLaughlin PM, Benz RD, Matthews EJ (2009) Identification of structure-activity relationships for adverse effects of pharmaceuticals in humans. Part A: use of FDA post-market reports to create a database of hepatobiliary and urinary tract toxicities. *Regul Toxicol Pharmacol* 54, 1–22.

Valerio LG Jr. (2009) In silico toxicology for the pharmaceutical sciences. *Toxicol Appl Pharmacol* 241, 356–70.

Vliegenthart AD, Tucker CS, Del Pozo J, Dear JW (2014) Zebrafish as model organisms for studying drug-induced liver injury. *Br J Clin Pharmacol* 78, 1217–27.

Vogt A, Codore H, Day BW, Hukriede NA, Tsang M (2010) Development of automated imaging and analysis for zebrafish chemical screens. *J Vis Exp*, 40, 1900, doi:10.3791/1900.

Walgren JL, Mitchell MD, Thompson DC (2005) Role of metabolism in drug-induced idiosyncratic hepatotoxicity. *Crit Rev Toxicol*, 35, 325–61.

Walker SL, Ariga J, Mathias JR, Coothankandaswamy V, Xie X, Distel M, Koster RW, Parsons MJ, Bhalla KN, Saxena MT, Mumm JS (2012) Automated reporter quantification in vivo: high-throughput screening method for reporter-based assays in zebrafish. *PLoS One*, 7, e29916.

Wang W, Liu X, Gelinas D, Ciruna B, Sun Y (2007) A fully automated robotic system for microinjection of zebrafish embryos. *PLoS One*, 2, e862.

Wang C, Li H, Meng Q, Du Y, Xiao F, Zhang Q, Yu J, Li K, Chen S, Huang Z, Liu B, Guo F (2014) ATF4 deficiency protects hepatocytes from oxidative stress via inhibiting CYP2E1 expression. *J Cell Mol Med* 18, 80–90.

Warner DJ, Chen H, Cantin LD, Kenna JG, Stahl S, Walker CL, Noeske T (2012) Mitigating the inhibition of human bile salt export pump by drugs: opportunities provided by physicochemical property modulation, in silico modeling, and structural modification. *Drug Metab Dispos* 40, 2332–41.

Welch MA, Kock K, Urban TJ, Brouwer KL, Swaan PW (2015) Toward predicting drug-induced liver injury: parallel computational approaches to identify multidrug resistance protein 4 and bile salt export pump inhibitors. *Drug Metab Dispos* 43, 725–34.

Westwood FR, Bigley A, Randall K, Marsden AM, Scott RC (2005) Statin-induced muscle necrosis in the rat: distribution, development, and fibre selectivity. *Toxicol Pathol*, 33, 246–57.

Will Y, Dykens J (2014) Mitochondrial toxicity assessment in industry--a decade of technology development and insight. *Expert Opin Drug Metab Toxicol* 10,1061–7.

Williams CD, Jaeschke H (2012) Role of innate and adaptive immunity during drug-induced liver injury. *Toxicol Res* 1, 161–70.

Willmann S, Lippert J, Schmitt W (2005) From physicochemistry to absorption and distribution: predictive mechanistic modelling and computational tools. *Expert Opin Drug Metab Toxicol* 1, 159–68.

Woodhead JL, Howell BA, Yang Y, Harrill AH, Clewell HJ III, Andersen ME, Siler SQ, Watkins PB (2012) An analysis of N-acetylcysteine treatment for acetaminophen overdose using a systems model of drug-induced liver injury. *J Pharmacol Exp Ther* 342, 529–40.

Woodhead JL, Yang K, Siler SQ, Watkins PB, Brouwer KL, Barton HA, Howell BA (2014) Exploring BSEP inhibition-mediated toxicity with a mechanistic model of drug-induced liver injury. *Front Pharmacol* 5, 240.

Xie Y, McGill MR, Dorko K, Kumer SC, Schmitt TM, Forster J, Jaeschke H (2014) Mechanisms of acetaminophen-induced cell death in primary human hepatocytes. *Toxicol Appl Pharmacol* 279, 266–74.

Xu JJ, Henstock PV, Dunn MC, Smith AR, Chabot JR, de Graaf D (2008) Cellular imaging predictions of clinical drug-induced liver injury. *Toxicol Sci* 105, 97–105.

Xu JJ, Dunn MC, Smith AR, Tien ES (2012) Assessment of hepatotoxicity potential of drug candidate molecules including kinase inhibitors by hepatocyte imaging assay technology and bile flux imaging assay technology. *Methods Mol Biol* 795, 83–107.

Yamazaki M, Miyake M, Sato H, Masutomi N, Tsutsui N, Adam KP, Alexander DC, Lawton KA, Milburn MV, Ryals JA, Wulff JE, Guo L (2013) Perturbation of bile acid homeostasis is an early pathogenesis event of drug induced liver injury in rats. *Toxicol Appl Pharmacol* 268, 79–89.

Yang K, Woodhead JL, Watkins PB, Howell BA, Brouwer KL (2014) Systems pharmacology modeling predicts delayed presentation and species differences in bile acid-mediated troglitazone hepatotoxicity. *Clin Pharmacol Ther* 96, 589–98.

Yang Y, Nadanaciva S, Will Y, Woodhead JL, Howell BA, Watkins PB, Siler SQ (2015) MITOsym®: a mechanistic, mathematical model of hepatocellular respiration and bioenergetics. *Pharm Res* 32, 1975–92.

Yerushalmi B, Dahl R, Devereaux MW, Gumpricht E, Sokol RJ (2001) Bile acid-induced rat hepatocyte apoptosis is inhibited by antioxidants and blockers of the mitochondrial permeability transition. *Hepatology* 33, 616–26.

Yu K, Geng X, Chen M, Zhang J, Wang B, Ilic K, Tong W (2014) High daily dose and being a substrate of cytochrome P450 enzymes are two important predictors of drug-induced liver injury. *Drug Metab Dispos* 42, 744–50.

Yuan L, Kaplowitz N (2013) Mechanisms of drug-induced liver injury. *Clin Liver Dis* 17, 507–18.

Zámbó V, Simon-Szabó L, Szelényi P, Kereszturi E, Bánhegyi G, Csala M (2013) Lipotoxicity in the liver. *World J Hepatol* 5, 550–7.

Zhang L, Seitz LC, Abramczyk AM, Chan C (2010) Synergistic effect of cAMP and palmitate in promoting altered mitochondrial function and cell death in HepG2 cells. *Exp Cell Res* 316, 716–27.

Zhang Y, Li F, Patterson AD, Wang Y, Krausz KW, Neale G, Thomas S, Nachagari D, Vogel P, Vore M, Gonzalez FJ, Schuetz JD (2012) Abcb11 deficiency induces cholestasis coupled to impaired β-fatty acid oxidation in mice. *J Biol Chem* 287, 24784–94.

Zheng SJ, Wang P, Tsabary G, Chen YH (2004) Critical roles of TRAIL in hepatic cell death and hepatic inflammation. *J Clin Invest* 113, 58–64.

Zhu X, Kruhlak NL (2014) Construction and analysis of a human hepatotoxicity database suitable for QSAR modeling using post-market safety data. *Toxicology* 321, 62–72.

Zimmerman HJ (1999) *Hepatotoxicity: The Adverse Effects of Drugs and Other Chemicals on the Liver*. Lippincott Williams & Wilkins, Philadelphia, PA; Second edition.

9

CARDIAC

DAVID J. GALLACHER[1], GARY GINTANT[2], NAJAH ABI-GERGES[3], MARK R. DAVIES[4], HUA RONG LU[1], KIMBERLY M. HOAGLAND[5], GEORG RAST[6], BRIAN D. GUTH[6,7], HUGO M. VARGAS[5] AND ROBERT L. HAMLIN[8]

[1] *Global Safety Pharmacology, Janssen Research & Development, a division of Janssen Pharmaceutica N.V., Beerse, Belgium*

[2] *Integrative Pharmacology, Integrated Science & Technology, AbbVie, North Chicago, IL, USA*

[3] *AnaBios Corporation, San Diego, CA, USA*

[4] *QT-Informatics Limited, Macclesfield, England*

[5] *Integrated Discovery and Safety Pharmacology, Department of Comparative Biology and Safety Sciences, Amgen Inc., Thousand Oaks, CA, USA*

[6] *Drug Discovery Support, Boehringer Ingelheim Pharma GmbH & Co. KG, Biberach (Riss), Germany*

[7] *DST/NWU, Preclinical Drug Development Platform, Faculty of Health Sciences, North-West University, Potchefstroom, South Africa*

[8] *Department of Veterinary Medicine and School of Biomedical Engineering, The Ohio State University, Columbus, OH, USA*

9.1 GENERAL INTRODUCTION

Although misty in its origin, the subsequent evolution of safety pharmacology (SP) firmly established this discipline and has proven value in hazard identification and risk assessment in both drug discovery and drug development. The establishment of the SP discipline was nicely captured in the article by Bass et al. (2004). In their manuscript, Bass et al. acknowledge G. Zbinden's original concern with the "neglect of function and obsession with structure in toxicity testing" (1984), given that "these functional effects are much more frequent in clinical practice than the toxic reactions due to morphological and biochemical lesions" (1979). Consequently, he suggested that SP was born out of the fact that "organ functions can be toxicological targets in humans exposed to potential new medications and that drug effects on such organ functions (unlike organ structures) are not readily detected by routine toxicological studies." This latter statement is not surprising since toxicology studies are primarily designed for a different purpose and not for the measurement of functional parameters (e.g., respiration rate and blood pressure (BP)) that are impacted by changes in behavior and the study environment. Indeed exaggerated pharmacological effects can have long-term structural

adaptive consequences, but since these responses can have both pathological and pharmacological triggers, the SP discipline predominantly helped to develop appropriate sensitive models and study designs to discriminate between these two causes for organ damage.

Initially, pharmaceutical companies often conducted their own nonstandardized general pharmacology/SP assessment, as an ancillary part of discovery efficacy studies, to understand potential risk and benefit (Kinter et al., 1994). The conclusions of these studies were not always systematically connected with the outcome of toxicology studies, and these two assessments were often disjointed. The Japanese Ministry of Health and Welfare (JMOHW) was the first to address this limitation by formally drafting and implementing the *Japanese Guidelines for Nonclinical Studies of Drugs Manual* (Yakuji Nippo, 1995). Subsequently these guidelines became the core pharmaceutical industry reference document for the conduct of all general pharmacology/SP. Among pharmacological tests required on respiratory, gastrointestinal, renal, central and peripheral nervous systems, etc., effects on the cardiovascular (CV) system were requested and normally "from the anesthetized animal," which was in contrast to subsequent guidelines (e.g., International Conference on Harmonization (ICH) S7A).

Drug Discovery Toxicology: From Target Assessment to Translational Biomarkers, First Edition. Edited by Yvonne Will, J. Eric McDuffie, Andrew J. Olaharski, and Brandon D. Jeffy.
© 2016 John Wiley & Sons, Inc. Published 2016 by John Wiley & Sons, Inc.

During the period 1995–1998, many draft guidance documents appeared from Japan, Europe, and the United States and were discussed and debated at a meeting of the General Pharmacology/Safety Pharmacology Discussion Group (*later incorporated as the Safety Pharmacology Society in 2000*; http://www.safetypharmacology.org) in September 2000 (Bass et al., 2004). Later in 1998, the JMOHW and the Japan Pharmaceutical Manufacturers Association proposed to the ICH Steering Group the adoption of an initiative on SP. This proposal was accepted and given the designation of Topic S7 (Bass et al., 2004). Through the years 2000–2001 the ICH guideline on SP testing (ICH S7A) was rolled out across the different regulatory regions worldwide. These guidelines focused on measuring potential adverse effects on the CV, respiratory, and central nervous systems (CNS), where a transient, acute interruption in organ function could be potentially life threatening. For CV observations, the effects on BP, heart rate, and the electrocardiogram (ECG) were recommended. In contrast to the Japanese guidelines of 1995, it was preferable to use unanesthetized, unrestrained, conscious telemetered animals (e.g., dogs). Furthermore, it was advised to dose via the intended clinical route of administration up to limits showing moderate adverse effects that should not interfere with the interpretation of results (e.g., tremors during ECG recordings). In addition, unspecified *in vitro/ex vivo* evaluations for cardiac repolarization and conductance abnormalities should be "considered" for safety evaluation.

Concomitantly, during the 1990s an increasing number of non-CV therapeutics from different chemical classes were found to induce the potentially life-threatening arrhythmia called torsades de pointes (TdP), a polymorphic ventricular arrhythmia that sometimes leads to ventricular fibrillation and death. By virtue of the fact that not all incidences of drug-induced TdP will lead to death, it was discovered that this type of arrhythmia is associated with a drug-induced prolongation of the QT interval on the ECG (i.e., often expressed as QTc to reflect a correction for the physiologically induced change by heart rate itself). In addition, it was discovered that blockade of the I_{Kr} ion channel (often referred to as human ether-à-go-go-related gene (hERG) channel) underlies this prolongation of the QT interval. Indeed, the incidence of drug-induced TdP was becoming increasingly concerning to the regulatory authorities, and consequently some of drugs with this liability were removed from the market (e.g., terodiline (1991), terfenadine (1998), astemizole (1999), and grepafloxacin (1999)). Over the last few decades, serious adverse effects on CV function became one of the largest causes of withdrawal of drugs from the marketplace (Redfern et al., 2010; Laverty et al., 2011), and through this time CV SP has also evolved in parallel, in order to appropriately deal with this challenge. In response to this concern and in the absence of global regulatory guidelines, the Committee for Proprietary Medicinal Products (CPMP) from the European Agency for the Evaluation of Medicinal Products (EMEA) issued a *points to consider document* (CPMP/986/96) in December 1997 providing some advice for both preclinical and clinical safety testing for proarrhythmic potential. This document spoke of robust assessment of BP, heart rate, and the ECG including assessments of morphology changes in a nonrodent species (typically the dog), but it specified for the first time screening for potential drug effects on the myocardial action potential duration (APD) in particular at 50, 60, and 90% of repolarization, resting membrane potential, and V_{max} (the maximum rate of rise in depolarization) in an appropriate experimental model. Obviously, this needed to be conducted in a suitable species (e.g., rabbit, guinea pig, dog, pig, etc.) and isolated tissue such as the cardiac Purkinje fiber or papillary muscle preferably using different pacing frequencies to account for phenomena like use- and reverse use-dependence. This document became an important "springboard" for the ICH to accept an initiative on SP in 1998 (S7), which eventually led to the ICHS7A guideline in 2000–2001 and later to the more specific electrophysiology ICHS7B guidelines (Guidance for Industry: S7B Nonclinical Evaluation of the Potential for Delayed Ventricular Repolarization (QT Interval Prolongation) by Human Pharmaceuticals. US Department of Health and Human Services, Food and Drug Administration, Center for Drug Evaluation and Research, Center for Biologics Evaluation and Research, ICH, October 2005).

In parallel, between 1998 and 2005 there were intense scientific debates around the general practicality and overall predictability of APD assays to clearly define risks associated with new molecular entities (NMEs) (clinical drug candidates) that were moving toward phase 1 clinical testing in man. Questions arose about APD assays regarding species differences in expression of myocardial ion channels that are found in man but not in certain animals. For example, the rapidly activating repolarizing potassium ionic current I_{Kr} that is encoded by the so-called hERG is an important repolarizing ion channel in man but is not found in rats. Due to the belief that all known torsadogenic compounds had hERG ion channel blocking effects and prolonged the QT interval in man, a cellular assay exploring pharmacological concentration/response was adopted as part of the core SP battery. In contrast to the *points to consider* document (CPMP/986/96; December 1997) recommendations, AP assays were relegated to a follow-up evaluation position, should the readout between the hERG assay, *in vivo* preclinical telemetry, and clinical studies be discordant, since they could potentially help in understanding such discrepancies. Following the appearance of ICHS7B, the routine use of *in vitro* APD assays for regulatory submissions decreased, although their value continued to be appreciated by a number of pharmaceutical companies as part of their integrated risk assessment in support of applications for phase 1 clinical trials. Section 9.2 discusses the electrophysiological methods, value, and potential drawbacks in using these tests.

With 8 years' post-ICH S7B research experience, the regulatory authorities became concerned that there is too much focus preclinically on drugs effects on I_{Kr}/hERG and on QTc prolongation clinically, which we now know is only a poor surrogate marker of proarrhythmia (since many but not all hERG blockers/QTc prolongers induce TdP), and industry is not adequately addressing the endpoint of primary clinical concern, that is, the arrhythmia itself. Although the ICH S7B recommendations have helped to largely eliminate new torsadogenic drugs entering the market, there is an important limitation associated with this approach in that increases in the QTc interval are very common but not particularly specific for predicting ventricular torsadogenic risk. Indeed there are clinically marketed drugs that block I_{Kr} at therapeutic plasma exposures that are not proarrhythmic (e.g., ranolazine and verapamil; Chi, 2013a; Fulmer, 2013). Moreover, the extent of QTc prolongation, pure I_{Kr} blockers excluded, is largely drug specific, and the QTc can be prolonged by many factors not directly associated with proarrhythmia (e.g., food, temperature, drugs, autonomic disturbances, glucose/insulin levels, circadian rhythms).

Although ICH S7B undoubtedly reduced the number of QT-prolonging drugs and therefore met a primary goal of achieving potentially safer medicines, the assumption of a causative hERG–QT–arrhythmia relationship did not advance our understanding of the truly relevant factors that lead to proarrhythmia. Indeed this may have led to a negative impact in bringing forward valuable therapies for potentially life-threatening diseases and areas of unmet medical need. As a consequence, the Cardiac Safety Research Consortium, the Health and Environmental Sciences Institute (HESI), and the Food and Drug Administration (FDA) sponsored a "Think Tank" that was held at the FDA's White Oak facilities, Silver Spring, MD, on July 23, 2013 (Chi, 2013b). The goal was to define a new testing approach in which proarrhythmic risk would be primarily assessed using nonclinical *in vitro* human models, based on a mechanistic understanding of the reasons for TdP proarrhythmia. The outcome of this meeting resulted in a "White Paper" publication by Sager et al. (2014). This paper outlined a three-pronged preclinical initiative called the "Comprehensive *In Vitro* Proarrhythmia Assay" (CiPA), which consists of understanding the effects of NMEs on multiple ion channels (including I_{Kr}) using voltage clamp techniques, *in silico* prediction simulations (proarrhythmic liability) based on the ion channel effects, and an integrated human cellular study to provide confirmatory electrophysiological data (most likely involving human stem cell-derived cardiomyocytes (CMs)). We will elaborate on these topics in Sections 9.3 and 9.4, respectively.

The specific evolution of CV safety testing, the largest component of regulatory SP, in some respects was responsible for the making of SP as an independent discipline. Due to the devastating impact of CV adverse effects on the success of pharmaceutical companies, a tremendous investment (both head count and capital) has been given into this area that has helped to expand our knowledge of CV research. In recent years, many new techniques, assays, and software systems have been developed to help researchers. One of the most notable is the use of telemetry technology to allow the remote assessment of various organ functions (including CV) from the conscious laboratory animal in its natural home environment. We will provide more insight into this extremely valuable technology for safety pharmacologists in Section 9.5.

Over the last 10 years it has become clear that some drugs, in particular therapies for oncology indications, are associated with left ventricular heart dysfunction. Indeed these effects can be found with various classes of agent such as anthracyclines, alkylating agents, antimetabolites, monoclonal antibody (mAb)-based tyrosine kinase inhibitors, antimicrotubule agents, proteasome inhibitors, and small-molecule (SM) tyrosine kinase inhibitors (Yeh and Bickford, 2009). The incidence of this particular adverse effect ranges from 0.5% with imatinib mesylate (Gleevec), a SM tyrosine kinase inhibitor, up to 28% of patients with for cyclophosphamide (Cytoxan; an alkylating agent) and trastuzumab (Herceptin; a large-molecule tyrosine kinase inhibitor). In Section 9.6, we discuss various different *in vitro* and *in vivo* models that can help to identify drugs with potential negative and positive effects on cardiac contractility. One of the most commonly used surrogate markers for contractility (LVdP/dt$_{max}$: the rate of change of the left ventricular pressure with respect to time) is routinely measured *in vivo* in anesthetized and conscious animal models, but this parameter is invasive and not so easily measured in man. On the other hand, echocardiography is more routinely used in man to understand negative effects of drugs or to help diagnose patients with heart dysfunction, but this is not routinely employed within preclinical studies in industry. In order to provide a correlation between changes in ejection fraction, using echocardiography, and LVdP/dt$_{max}$, using BP data derived from catheterization of the left ventricle, Cools et al. (2014) compared these methods within studies conducted in healthy telemetered dogs. Cools et al. found that a change of 1000 mmHg/s relates to a change of ~7% in ejection fraction. This translational work now allows us to understand what change in LVdP/dt$_{max}$ within a preclinical study relates to something measurable in the clinic.

With the opportunity to have more specific targeting of therapeutics, thus potentially minimizing off-target adverse effects, there has been an increasing interest in developing mAb treatments and biologics in general. Discussion of how the CV SP assessment of conventional small molecules differs from the approach taken for biologics will be detailed later in Section 9.7.

There has been a long-standing discussion on whether safety pharmacologists should study the effects of NME's in CV disease models mimicking the patient that may eventually be treated with that compound, rather than using the genetically homogeneous animals that are routinely used in standard safety testing (Dixit and Boelsterli, 2007; Morgan et al., 2013). Although there are very few published cases where disease models have been employed in safety research (Damiano et al., 2015), the systematic use of such models in safety assessment has never really taken hold. The lack of uptake in this approach may be due to the fact that in drug discovery, an NME enters clinical studies after being tested in a number of models covering different facets of the patient situation: how can we appropriately choose which animal disease model covers the relevant patient population? Indeed it may be easier to define relevant changes in measured parameters from healthy *genetically sterile* animals and try and interpret these changes in the context of a disease-carrying patient. However, in the final subsection of this chapter, we provide a thought-provoking discussion as to why safety pharmacologists should consider embracing new approaches.

New drug-induced CV issues are currently confronting the pharmaceutical industry (e.g., drug-induced stroke and myocardial infarction). Over the last 10–15 years a number of drugs have been stopped in clinical development or have been withdrawn from some markets for being associated with an increased risk for strokes and heart attacks (e.g., torcetrapib, a CETP inhibitor for atherosclerosis; Barter et al., 2007); sibutramine, a norepinephrine/serotonin reuptake inhibitor antiobesity drug (Scheen, 2010); tegaserod, a partial $5HT_4$ agonist for irritable disease (http://www.fda.gov/drugs/drugsafety/postmarketdrugsafety informationforpatientsandproviders/ucm103223.htm); Vioxx, a COX-2 inhibitor for arthritis (Baron et al., 2008; Bottone and Barry, 2009). In most of these cases, these issues remained undetected until after completion of clinical trials, and these drugs were on the market. Due to the relatively low frequency of these events, standard pre-clinical models designed to identify acute and subacute changes had poor predictivity of what would occur in the general patient population. Indeed the potential causes underlying these effects could be multifactorial, including (i) hypercoagulability, (ii) hemodynamic changes (stasis, turbulence), and (iii) endothelial injury/dysfunction (Virchow, 1856). These three factors form part of Virchow's triad and are thought to contribute to thrombosis, and any single or combination effect could potentially be involved in causing stroke and myocardial infarction following many months of treatment with the aforementioned drugs. Although there are some public–private consortia looking into this area of research (e.g., HESI biomarker working group), advances in this area are slow and therefore are not discussed here. Our focus has been primarily given to the progress made in understanding proarrhythmic risk over the last 20 years, an area that continues to evolve. In any case, SP will need to adapt in parallel with the continually evolving innovative environment of the pharmaceutical industry where novel unexplored targets and technologies are being exploited in the treatment of diseases of unmet medical need.

9.2 CLASSICAL *IN VITRO/EX VIVO* ASSESSMENT OF CARDIAC ELECTROPHYSIOLOGIC EFFECTS

9.2.1 Introduction

While clinical ECG recordings provide globally integrated records of cardiac electrical activity from the body surface, it is well recognized that electrical control of cardiac function is more complex than information summarized from even the most sophisticated surface ECG recordings. Multiple insights regarding basic physiology and pharmacology can be gleaned from *in vitro* cardiac studies focused on subcellular, cellular, and organ levels, providing further understanding of the mechanisms responsible for beneficial as well as untoward drug effects. Perhaps the first *in vitro* recordings of cardiac electrical activity were reported in 1883 by Burdon-Sanderson and Page who reported the surprisingly long duration of ventricular AP (compared to nerve tissues) using capillary electrometers to measure field potentials from tortoise heart. Arguably, the first intracellular recordings of ventricular electrical activity were demonstrations of repetitive electrical activity of rat ventricular embryonic myocytes by Hogg et al. (1934). With the adoption of intracellular microelectrode techniques, more detailed pioneering studies of Draper and Weidmann (1951) and Hoffman and Cranefield (1960) described cardiac electrophysiology at a cellular level. Subsequent application of voltage and patch clamp techniques to isolated ventricular myocytes allowed for the measurement of native cardiac currents (see review by Varro and Papp, 1992). All four approaches (extracellular recordings (measuring field potentials), intracellular recordings (measuring action potentials), as well as voltage and patch clamp recordings (measuring whole cell electrical activity, membrane currents, as well as single channel recordings), are used today in a wide variety of cardiac preparations to define drug effects on subcellular, cellular, tissue, and organ levels and aid in the selection of novel pharmacologic therapeutics and risk–benefit assessments.

The goal of this chapter section is to highlight various nonautomated "classical" *in vitro* techniques used today to evaluate drug effects on cardiac electrophysiology. It should be recognized that *in vitro* assays typically focused on simpler, less integrated responses (e.g., ionic currents) may be better suited for hazard identification and mechanistic studies (often employed early in drug discovery), while those focused on more complex and more fully integrated responses (AP and Langendorff-perfused hearts) may be more appropriate for risk assessment. It is also recognized that the utility and expectations of studies using simpler versus more complex integrated systems are different and that both should be considered as complementary when evaluating the untoward or adverse effects of novel therapeutics on cardiac function. Finally, it must be understood that electrophysiological studies must ultimately be linked to cellular calcium handling and myocyte contractility, which influence the integrated cellular and organ response. For practical purposes, we will limit the discussion to studies evaluating acute electrophysiologic effects on ventricular myocardium of SM (nonbiologic) drugs.

9.2.2 Subcellular Techniques

Despite their simplicity, binding studies to various channels, transporters, and receptors provide a valuable early first assessment of potential cardiac effects of novel compounds. Studies evaluating the displacement of bound radiolabeled or fluorescent ligands from the Kv.11.1 (I_{Kr}) channel encoded by the hERG (Diaz et al., 2004) or thallium flux assays (Huang et al., 2010, for a comparison of two approaches). Block of this channel is well known for its ability to delay repolarization, an effect linked to TdP proarrhythmia (see Rampe and Brown, 2013 for a review). Other cardiac channels that are routinely probed include the fast sodium current (Nav1.5) and L-type calcium current (Cav1.2). Such indirect studies are also valuable in assessing potential liabilities of intracellular channels (e.g., the ryanodine channel located on the sarcoplasmic reticulum responsible for the release of calcium for cardiac contraction (Viero et al., 2012) and IP3 receptors involved in calcium signaling (Rossi et al., 2012)). More functionally based ion flux assays and voltage-sensitive dyes have also been used to assess drug effects on individual channels as well as whole-cell responses (Rezazadeh and Hesketh, 2004; Bowlby et al., 2008; Titus et al., 2009). Ideally, results obtained from more indirect binding studies should subsequently be confirmed using functional studies such as those measuring ionic currents, AP, cardiac conduction, or intervals on ECG recordings (representing the spectrum of less to more complex integrated systems). This is essential to establish whether effects on the currents are realized at expected therapeutic exposures and whether such effects modulate cardiac function.

9.2.3 Ionic Currents

The use of voltage clamp techniques to evaluate drug effects on cardiac ionic current remains the gold standard for assessing ionic current-based electrophysiologic effects. Using voltage clamp techniques, it is possible to precisely control the membrane potential (and hence activity) of voltage-dependent channels and measure drug effects on current amplitude and kinetics under defined and well-controlled conditions (Bean et al., 1983). Indeed, such studies form the basis for our understanding of the dynamics of drug–channel interactions involving channel receptors whose affinity may be modulated by the functional state of the channel (as described, respectively, by Hille (1977) and Hondeghem and Katzung (1984)) or less dynamic receptors with activity modulated by drug accessibility (Starmer et al., 1984). Such conceptual frameworks are useful in the understanding the effects of voltage- and use-dependent block of cardiac currents by drugs linked to their proarrhythmic as well as antiarrhythmic effects (Roden and Hoffman, 1985; Task Force of the Working Group on Arrhythmias of the European Society of Cardiology, 1991; Members of the Sicilian Gambit., 2001). While I_{kr} is recognized as the most common current blocked by drugs that affect delayed repolarization linked to proarrhythmia, such effects can also be initiated or modulated by other currents that may either promote or inhibit prolongation of the QT interval in humans and proarrhythmia. Thus, the integrated response of multiple ionic currents should be considered when evaluating a drugs' electrophysiologic effects (as is done by myocytes on a cellular level; see Fig. 9.1).

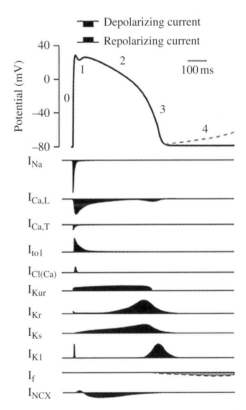

FIGURE 9.1 Schematic illustration of some key ionic currents involved in the integrated ventricular action potential. The action potential waveform can be divided into different phases, namely, phase 0 (rapid depolarization due to fast inward sodium current), phase 1 (early) repolarization, phase 2 (plateau, sustained by calcium current), and phase 3 (final repolarization resulting from activation of multiple K$^+$ currents including I_{Kr}, I_{K1}, and I_{Ks}). The contributions of various inward depolarizing currents (downward deflections) and outward repolarizing currents (upward deflections) defining the voltage time course of repolarization are shown below the action potential. From Hoekstra et al. (2012).

Earlier ionic current studies using native channels in cardiac myocytes from different species have been supplanted with studies employing transfected human channels and heterologous expression systems such as human embryonic kidney (HEK) or Chinese hamster ovary (CHO) cells. Such systems, which do not normally express these channels, provide a more convenient method to study ionic currents in the absence of other overlapping (and potentially confounding) currents, though at the possible loss of important channel subunits and modulating cellular systems. While drug effects are typically reported as IC_{50} values for current block (or sometimes current enhancement), voltage- and time-dependent characteristics of block and unblock can also be directly evaluated. Further advances in planar patch techniques have supported the development of automated patch clamp platforms to characterized drug effects on currents (Witchel, 2010; Farre and Fertig, 2014; Walsh, 2015). However, it should be recognized that experimental conditions (intracellular and extracellular solutions, room vs. physiologic temperature, quality of "seal" formation with the cells of interest (giga-ohm seals for recordings being the gold standard), nominal vs. actual concentration of test compound in well) all may influence the variability, quality, and ease of translation of results from automated patch platforms (Dankers and Moller, 2014; Elkins et al., 2013). The potency of block of I_{Kr} current in heterologous expression systems is often compared to anticipated therapeutic free plasma concentrations, thus defining a "safety margin" for QT prolongation. This approach has been shown to produce "false-positive" results (e.g., with such well-recognized drugs as verapamil, pentobarbital, ranolazine, and some additional compounds) and should be used in conjunction with other assays (Gintant, 2011) that consider multiple cardiac currents in evaluating a drugs effect on delayed repolarization. The integration of drug effects on multiple ionic currents is essential in predicting the overall effect on cardiac repolarization (Mirams et al., 2014), which is a recognized surrogate marker of proarrhythmia and forms one of the pillars of the CiPA paradigm proposed for evaluating proarrhythmic risk more directly based on mechanistic assessments (Gintant et al., 2016; Sager et al., 2014).

9.2.4 AP/Repolarization Assays

It is well recognized that the cardiac AP reflects the summation of multiple ionic currents, exchangers, and pumps that do not function in isolation and that a drug's electrophysiologic effect may be difficult to predict based on effects of a few individual ionic currents. Thus, the need for AP studies to elucidate electrophysiologic effects remains, with changes in the various phases of the AP recordings (or extracellular potentials) often providing the first suggestion of electrophysiologic activity. Such studies may employ isolated myocytes, tissues, or isolated hearts. Figure 9.2 illustrates the multiple variations in approaches possible.

While APD/repolarization assays are suggested as one approach to explore the risk for delayed repolarization (ICH S7B guidance), they are generally not widely used at present, likely due to the ease of use (and interpretation) of hERG current assays compared to AP/repolarization studies in supporting regulatory submission, lack of accepted industry best-standard/practices for APD/repolarization assays, as well as potentially important differences in ventricular repolarization across species (vs. humans). For example, it is well established that I_{Kr}/hERG plays a prominent role in

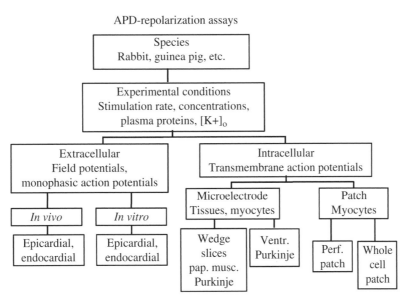

FIGURE 9.2 Multiple approaches possible for *in vitro* evaluation of electrophysiologic effects on cardiac cells and tissues.

repolarization of human but not rat ventricle, while transient outward current (Ito) plays a greater role in rat ventricular repolarization (Oudit et al., 2001). Despite these limitations, the use of experimental studies of APD/repolarization assays (reflecting integrated responses in the simplest integrated electrophysiologic system, a myocyte) is valuable in identifying potential proarrhythmic effects/mechanisms that may be further evaluated in later discovery efforts.

9.2.5   Proarrhythmia Assays

More complex preclinical assays to interrogate proarrhythmic liabilities have been developed that involve tissues and whole hearts. Such models have been informative in defining mechanisms of proarrhythmia and inform on such endpoints such as cardiac conduction, spatial and temporal dispersion of repolarization, enhanced repolarization heterogeneity leading to transmural dispersion, and triggered activity (such as early afterdepolarizations (EADs), a recognized trigger for TdP arrhythmia for drugs that delay repolarization.

One such proarrhythmia model is the rabbit wedge preparation (Wu et al., 2005; Wang et al., 2008; Liu et al., 2012). This model provides a summary risk of TdP score based on measures of QT interval prolongation, Tpeak–Tend/QT ratio, and development of EADs (with and without close coupled extrasystoles). In scoring model results, the incidence of EADs receives the greatest weight in scoring. In general, this model is able to distinguish torsadogenic versus nontorsadogenic compounds, as well as provide evidence of the safety of some hERG-blocking drugs with relatively low hERG safety margins (Liu et al., 2006, 2012). However, this model is lower throughput, technically difficult, and labor intensive and is thus more suited to later-stage drug discovery efforts.

A second recognized *in vitro* proarrhythmia model measures ventricular monophasic AP recordings from endocardial and epicardial surfaces obtained from methoxamine-pretreated Langendorff-perfused female rabbit hearts (Hondeghem et al., 2001; Hondeghem and Hoffmann, 2003). Developed as an automated platform by Hondeghem et al. (2001; Hondeghem and Hoffmann, 2003b), the SCREENIT assay provides measures of repolarization instability (enhanced by irregular stimulation patterns) in the setting of escalating drug concentrations. Using this approach, repolarization instability (defined from variations in monophasic AP changes recorded in response to stimulation rate changes) form part an overall risk assessment using "TRIaD," an acronym referring to *T*riangulation of the AP waveform, *R*everse use-dependence (greater drug-induced prolongation at slower stimulation rates), *I*nstability of repolarization, and *D*ispersion of repolarization, with all elements playing a role in delayed repolarization-induced proarrhythmia. In a validation study evaluating 55 blinded compounds (grouped according to five torsadogenic risk categories based on Redfern et al., 2003), Lawrence et al. (2006) reported that SCREENIT could predict clinical outcomes for drugs recognized for either

very high or low torsadogenic risk: the model was less precise in predicting risk for drugs with less certain proarrhythmic risk. A later study with a smaller set of nine blinded agents with varying risk of proarrhythmia correctly ranked this compound set (Steidl-Nichols et al., 2008). However, concentrations at which delayed repolarization, triangulation, instability, reverse use-dependence, and ectopic beats occurred varied based on concentration ranges used, consistent with inadequate equilibration times for some test agents.

In an interesting study, Lu et al. (2007) compared the electrophysiologic effects of four antibiotics (sparfloxacin, moxifloxacin, erythromycin, and telithromycin) across four *in vitro* assays at concentrations linked to therapeutic exposures. This study reported different orders of rankings for each compound across the four assays (functional hERG current assay, the rabbit Purkinje fiber repolarization assay, the rabbit left ventricular wedge preparation, and the rabbit Langendorff heart assay). In summary, the authors reported that the hERG current assay, Purkinje fiber, and isolated Langendorff-perfused rabbit heart could be used to assess the ability of this test set to delay repolarization, while the wedge preparation appeared to be more predictive for detecting torsadogenic risk *in vitro*. This later finding may reflect a closer resemblance of rabbit wedge preparations to a proarrhythmic substrate that supports TdP proarrhythmia.

9.2.6   Future Directions: Stem Cell-Derived CMs

Armed with the ability to reprogram cells, it is now possible to obtain human induced pluripotent stem cell-derived cardiomyocytes (hiPSC-CMs) to use as *in vitro* model systems of the human heart. The ideal hiPSC-CM would recapitulate the adult human myocyte closely both electrophysiologically (in the expression of ion channels and production of the AP) and in regard to cardiac contractility (force generation, positive staircase, etc.) while avoiding known differences compared with nonhuman species. Despite some shortcomings (a common finding for all *in vitro* models), studies using hiPSC-CMs are able to detect clinically recognized proarrhythmic drugs (Harris et al., 2013) using techniques that include extracellular field potential recordings. Indeed, the use of myocytes as a cellular model for proarrhythmia is an integral part of the CiPA initiative to directly assess proarrhythmic risk *in vitro* (Sager et al., 2014; Gintant et al., 2015). Future studies with hiPSC-CM preparations (presently focused on 2D models and evolving 3-dimensional tissue-like constructs) should ultimately replace the need for cellular and tissue studies using nonhuman sources.

9.2.7   Conclusions

We have come full circle with *in vitro* approaches to evaluate drug effects, now employing *in vitro* techniques that span subcellular-, cellular-, and organ-based approaches. Added to our armamentarium are human induced pluripotent stem cell-derived myocytes studied with newer technologies based on field potential recordings that were recognized and used as far back as 1883.

As with all *in vitro* studies, one must always be aware of difficulties in translating results obtained with simpler model systems (or more complex integrated preparations) to clinical findings. At present, one must also balance the benefits of a human-based and simple (but still evolving) myocyte preparation with the added complexity provided by *in vitro* tissue- and organ-based animal models (with potential unique pharmacological proarrhythmic substrates) compared to humans. The value of the later approach depends on the similarity of the proarrhythmic substrates in animal models versus humans. Future efforts at tissue engineering should provide a means of creating human myocardial preparations that provide the more complex proarrhythmic substrates *in vitro* that would eventually replace animal models used presently.

9.3 CARDIAC ION CHANNELS AND *IN SILICO* PREDICTION

9.3.1 Introduction

In silico modeling at a cell physiology level is becoming increasingly important in cardiac research, drug efficacy, and safety testing and has attracted attention from the pharmaceutical industry, US FDA, and other regulatory agencies. Various approaches are used to quantitatively describe the interactions of a drug with cardiac ion channels, including empirical statistical models and biophysically detailed mathematical models. Empirical statistical models relate an observation to an outcome using quantitative structure–activity relationship models (Mistry et al., 2015), while biophysically detailed mathematical models of cardiac electrical activity link molecular dynamics to biophysical models and are used to explore how a compound, interfering with specific cardiac ion channel function, may explain effects at the cell, tissue, and organ scales (reviewed by Mirams et al., 2012).

9.3.2 High-Throughput Cardiac Ion Channel Data

There are a variety of plate-based electrophysiology systems available to test for effects of drugs on cardiac ion channels overexpressed in cell lines. Conductance block data (i.e., the molar concentration of a drug producing 50% inhibition of an ionic current (IC_{50})) that are derived from these systems have enabled assessment of the key molecules responsible for the cardiac AP as a means of early cardiac safety assessment (Pollard et al., 2010). The questions therefore become: (i) What can be done to integrate these multiple experimental IC_{50} measurements? (ii) How can IC_{50}

screening variability be accounted for, and what does it mean in the physiological environment? For a very strong potency in a single ion channel assay, for example, the hERG channel (that encodes for a protein channel known as Kv.11.1, the pore-forming subunit of the rapid component of the delayed rectifier K$^+$ current that mediates the early repolarization of the AP) screening assay, the interpretation can be more straightforward, but when multiple ion channel assays give mixed effects, the conclusions are less clear, even between safety pharmacologists. There is therefore a clear requirement for a systematic (and reproducible) approach to interpreting these mixed data sources and thereby reducing bias—the assumptions of such an approach are ideally informed by but not contingent on experts in SP.

9.3.3 *In Silico* Approaches

In silico approaches for prediction of ion channel inhibition build on considerable work stemming back to the work carried out by Hodgkin and Huxley (1952a–d) and then later by, among others, Denis Noble at the University of Oxford to build models of the cardiac ion channels (reviewed by Noble et al., 2012). Generation of models and their extension thereof to study or incorporate novel experimental data has been extensive (Niederer et al., 2009). The result of this effort is that there is now a large collection of cellular models of the cardiac system that are capable of simulating how multiple ion channels act in concert to create the single cardiac cell AP. These mathematical representations have been modified to allow scaling of the ionic currents that flow through these channels based on the measured inhibitory (or agonistic) effects of drugs. Such effects are generally simulated in these models by using the form of conductance block provided by the equation for scaling currents (Eq. 9.1). Alternative approaches for representing different actions (e.g., time and voltage dependence) of drugs on the cardiac ion channels have been discussed extensively in a review by Brennan et al. (2009):

$$\text{Current}(\%) = \frac{100}{1 + \left(\text{Drug} / IC_{50}\right)^{h}} \tag{9.1}$$

Several studies have now explored how well integration of ionic current effects can translate to a measurable effect against either a biomarker of the single-cell AP (e.g., the APD at 90% repolarization (APD90); Bottino et al., 2006; Mirams et al., 2011; Davies et al., 2012) or on the ECG (e.g., the QT interval) or in proarrhythmic potential (Kramer et al., 2013; Mirams et al., 2014). In each case, a set of compounds (training set) with measured ion channel pharmacology, typically against three or more cardiac ion channels and with a corresponding effect or noneffect against a cardiac output, such as the change in AP trace, are used to calibrate the model parameters, and then an independent set of compounds (test set) are used to validate and score the model's performance in predicting an effect. A framework for assessing the predictivity of an assay or model has been described by Valentin et al. (2009), where the sensitivity (the proportion of true positive outcomes), specificity (the proportion of true negatives), and predictivity (prediction of both true positive and negative outcomes) of a model can be assessed. Performance criteria bins for model performance have been established to define a score of 65–74% as sufficient (it is not recommended to use a model that scores in the sufficient range for short-listing of compounds), 75–84% as good, and >85% as excellent (Genschow et al., 2002). The reasons for scoring <100% are likely to be numerous and may include an incomplete representation and measurement of the relevant cardiac biology, differences due to the temperature used during data generation, binding kinetics of drugs to channels, pharmacokinetic properties, and uncertainty in the experimental measurements.

Uncertainty enters the equation with respect to either a drug's effect (IC_{50}) or variability between experimental platforms, laboratories, cell types, and intra-animal/intraspecies differences. In a paper by Elkins et al. (2013), analysis of the typical variance in experimentally measured IC_{50} values of multiple ion channels at different laboratories allowed to determine when repeated ion channel screens should be performed to reduce uncertainty in a drug's action to acceptable levels to allow a meaningful interpretation of the data. While determination of the experimental uncertainty can be quantified, the uncertainty based on different

cells, between animals, and between human patients is more complex. A number of studies have therefore attempted to address this issue by not just assessing a single model structure and parameter set but by also considering many equally plausible parameter sets informed by experimental measurements of, for example, the cardiac AP. The *in silico* action potential (isAP) model (Davies et al., 2012) is an example of why a "single" model is not sufficient and why some estimate of uncertainty is required. This can be done by considering models of the same structure and yet different parameters to give a sense of robustness or confidence in the prediction of AP models. A review by Sarkar et al. (2012) tackles how parameter space across multiple cellular components, as a surrogate of interindividual variability, can be varied and how this can be equated to the experimental measurement. Figure 9.3 demonstrates how scaling of just three ionic currents in the isAP model can lead to a nonlinear output (APD90 in this case) and how different combinations of these parameters can give rise to similar APD90 results. The ensemble approaches have since been applied in increasing frequencies moving away from the idea of a single variant per single experimental observation to a multitude of equally valid variants per observed measurement (Britton et al., 2013).

The logical progression is in considering how best to improve a model's performance by considering what additional data to include. A study by Di Veroli et al. (2013) considers the kinetics of drug's block that can influence use and voltage dependence and how this may lead to the development of more predictive *in silico* models. It is

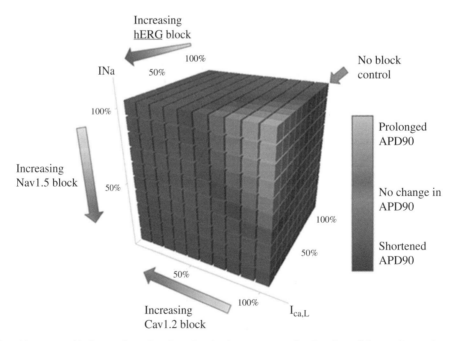

FIGURE 9.3 Predicted impact and balance of varying three key ionic currents on the duration of the cardiac action potential at 90% repolarization (APD90). (*See insert for color representation of the figure.*)

important to establish a standardized cardiac ion channel assay approach, so that consistent and reproducible electrophysiological data may be obtained and contribute to better prediction of the proarrhythmic potential of new drugs. Another factor impacting model accuracy is understanding how to account for the role of kinases, G-protein-coupled receptors, β-adrenergic receptor stimulation, drug accumulation in cardiac tissues, and trafficking of ion channel proteins to the cell surface. Furthermore, incorporating physiologically based pharmacokinetic modeling (PBPK) data and various estimates of uncertainty can be added to enhance model's performance (Hamon et al., 2015). While these approaches increase complexity in the models, other

approaches have attempted to simplify the parameters by identifying the model features that are most impactful in predicting a set of outcomes (Mistry et al., 2015). These more empirical and pragmatic approaches offer a chance to reflect on the key biological mechanisms that drive the observed cardiac effects, and offer the best chance for an early drug discovery virtual screen (Kramer et al., 2013; Mistry et al., 2015). All of these will be brought into focus with the adoption of a new, integrated nonclinical *in vitro/in silico* paradigm, the CiPA, for the assessment of a candidate drug's proarrhythmic liability, and the CiPA initiative should provide guidance on best practice to incorporate experimental findings (Sager et al., 2014; Fermini et al., 2015).

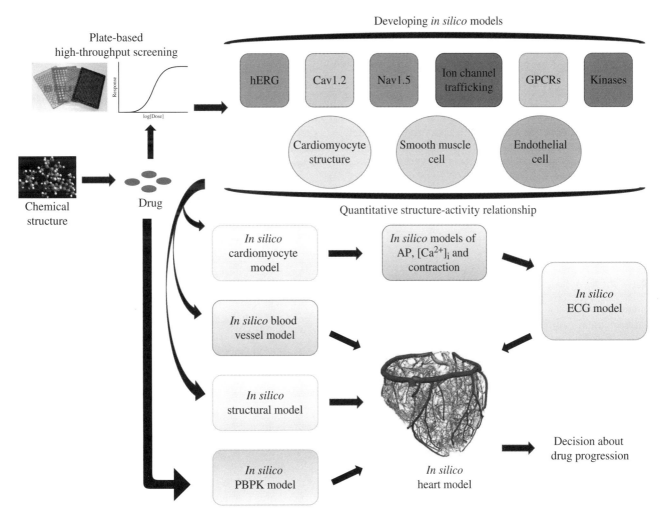

FIGURE 9.4 Proposed virtual human heart-on-a-chip risk assessment concept. Micromodels representing different cardiovascular functions can be integrated into a single microheart in a physiologically relevant manner to evaluate, early in the drug discovery process, the potential of a new chemical entity to induce cardiotoxicity. Inclusion of the drug pharmacokinetic properties would additionally allow users the ability to consider the relevant drug exposure in the heart. AP, action potential; Cav1.2, calcium ion channel protein that in humans is encoded by the CACNA1C gene; ECG, electrocardiogram; GPCRs, G-protein-coupled receptors; hERG, potassium ion channel protein that in humans is encoded by the ether-à-go-go-related gene; Nav1.5, sodium ion channel protein that in humans is encoded by the SCN5A gene; PBPK, physiologically based pharmacokinetic modeling. (*See insert for color representation of the figure.*)

Human induced pluripotent stem cell-derived cardiomyocytes are emerging as a tool for assessing the cardiotoxic potential of new drugs in higher-throughput format. While *in silico* tools are well described for predicting effects on CMs from different regions of the heart, *in silico* tools for prediction of perturbing function of human stem cell-derived CMs have yet to be developed. A recent approach to tackle this shortcoming has been to reverse engineer or retune the ionic current values from an existing *in silico* model such that the simulated AP approximates the measured recording from the human stem cell-derived CMs (Paci et al., 2012, 2013).

In addition to affecting cardiac AP, drugs have been also shown to affect the function of the heart. It is therefore essential to have a strategy that, in the context of drug discovery, allows generation of data sets that account for as many endpoints as possible and assist in selecting molecules with the lowest probability of cardiac liabilities for progression to human studies (Pollard et al., 2010). Moreover, building new *in silico* cardiac models, which allow to virtually and acutely/chronically assess the effects of drugs on other CV functions (like the QRS and PR intervals on the surface of the ECG, heart rate, BP, and contraction) and cardiac structure, and PBPK models would be indispensable to build a virtually integrated human heart-on-a-chip model that may provide a detailed analysis about the cardiac liability of new drugs (Fig. 9.4).

9.4   FROM ANIMAL *EX VIVO/IN VITRO* MODELS TO HUMAN STEM CELL-DERIVED CMs FOR CARDIAC SAFETY TESTING

9.4.1   Introduction

Although preclinical safety assessment approaches of some pharmaceutical companies currently provide sufficient assessment of drug-induced cardiac risk, evaluation is often complex and requires integration of findings from multiple *in vitro* and *in vivo* assays using nonhuman animal species, which may not all translate well to humans. Using animal *in vitro* models (isolated CMs, cardiac tissues, whole heart, etc.), several studies suggest that many factors can influence drug-induced prolongation of the QT interval/APD and potential for cardiac arrhythmias in *in vitro* cardiac tissue preparations. These factors include gender (Lu et al., 2000, 2001), pacing rate (Adamantidis et al., 1995; Lu et al., 2002, 2005, 2010), temperature, K^+ or Mg^{2+} concentrations in the perfusion solution (Roden and Hoffman, 1985; Davidenko et al., 1989; Kaseda et al., 1989), acidosis (Rozanski and Witt, 1991), species (Lu et al., 2001), and cardiac tissue subtype within a species, for example, rabbit (Lu et al., 2005, 2008). The correlation between drug responses in nonhuman animal-derived primary CMs, isolated cardiac tissue assays, or *in vivo* animal models and responses in humans is not always clear, as evidenced by the persistence of CV safety/toxicity in general as a major cause of drug attrition in early drug development and of drug withdrawal from the market (Laverty et al., 2011). Therefore, the development of human-derived *in vitro* models may help improve human safety predictivity. Furthermore, an increasing focus on the Replacement, Refinement, and Reduction (3Rs) in relation to animal studies provides an additional incentive to investigate the application of a simplified *in vitro* humanized platform to prioritize compounds before undertaking animal studies.

The limited availability of adult human cardiac tissue/cells from donors for investigation of disease mechanisms and cardiac effects of drugs remains a major challenge in the field. Human *stem cell*-derived *CMs* provide a practical alternative source of human CMs and a platform to screen for electrophysiological effects, since both *human embryonic and induced pluripotent stem cell*-derived *cardiomyocytes* (*hES*-CMs and hiPSC-CMs) possess many of the electrophysiological characteristics of primary human CMs. As such, they are increasingly used as a new source of cardiac cells for basic research, phenotypic screening, and drug safety assessment (Liang et al., 2013). Moreover, the FDA's CiPA proposal as a means to characterize proarrhythmia risk preclinically has greatly stimulated investigations into the use and validity of hES-/hiPSC-CMs for predicting cardiac safety (Sager et al., 2014).

9.4.2   Currently Available Technologies

There are several different technologies that currently use hES-/hiPSC-CMs to investigate drug-induced cardiac safety/toxicity, including (but not limited to) the technologies listed below for early drug screening:

1. Whole-cell patch clamp recordings are used to investigate voltage dependence of major cardiac ion currents (I_{Na}, I_{Ca}, I_{Kr}, and I_{Ks}) and pharmacological responses to ion channel blockers in single hiPSC-CMs, although the resting membrane potential is lower than in adult human ventricular cells due low I_{K1} current (Honda et al., 2011; Amuzescu et al., 2014). These major cardiac ion currents, and pharmacological responses to their respective blockers, in hiPSC-CMs are on the whole similar to those found in adult human CMs.

2. *Multielectrode array (MEA)*: Field potential duration (FPD) taken from hiPSC-CMs *in vitro* correlates with the QT interval in an ECG (Fig. 9.5). FPD was prolonged by I_{Kr}/hERG blockers, whereas it was shortened by an I_{Ca} blocker, indicating that these ion current components contribute to FPD in hiPSC-CMs (Braam et al., 2013; Harris et al., 2013; Navarrete et al., 2013).

3. *Calcium transient measurements*: The detection of Ca^{2+} transients is widely used in high-throughput screening assays in cells preloaded with a fluorescent

calcium-sensitive dye using, for example, FDSS-Hamamatsu (Puppala et al., 2013) or Kinetic Image Cytometer (KIC) systems (Cerignoli et al., 2012; Lu et al., 2015). Changes in Ca^{2+} transient duration closely reflect prolonged or shortened AP/FPD/QT interval (Fig. 9.5) (Ian Spencer et al., 2014).

4. Optical AP measurements using voltage-sensitive dyes (i.e., ANEPPS dyes) are also commonly used (Herron et al., 2012; Leyton-Mange et al., 2014). Changes in optical measurement of AP duration closely reflect prolonged or shortened AP/FPD/QT interval (Fig. 9.5) (Ian Spencer et al., 2014).

In many of the above assays, other parameters including beat rate, amplitude and slope of the upstroke, and incidence of arrhythmias could be also measured. However, the translational value of these parameters to *in vivo* or man still needs to be further investigated.

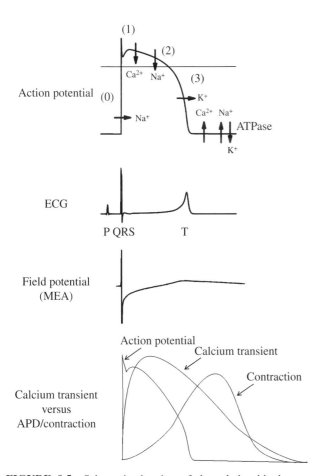

FIGURE 9.5 Schematic drawing of the relationship between cardiac action potential duration (APD) and the duration of the QT interval on the surface electrocardiogram (ECG) or the field potential duration from multielectrode array (MEA) recordings (top three parts of the figure); lower panel shows a schematic of the relationship between the calcium transient, APD, and contraction on hiPSC-CMs.

5. Impedance technologies (i.e., xCELLigence Cardio System) using hiPSC-CMs provide a relatively high-throughput assay to evaluate CM beating rate and impedance parameters such as amplitude and cell index (CI) and to detect drug-induced acute and chronic/delayed cardiac toxicity (Scott et al., 2014). The impedance technology can also be combined with MEA measurement, for example, the updated ACEA xCELLigence machine has been introduced in 2015 and is called "xCELLigence® RTCA CardioECR System."

6. Video microscopy image-based analysis (phase-contrast video microscopy) in combination with MEA or Ca^{2+} transient measurements can also be used for drug safety screening (Hayakawa et al., 2014), although this assay is still under early investigation.

9.4.3 Conclusions

Recent advances in human stem cell biology have paved the way for incorporation of human cell models into drug safety evaluation using the aforementioned assay systems. Although progressing, research in this field is mainly still at an early stage. The efficiency and reproducibility of the differentiation procedures used to generate these cells have to be improved to enable cost-effective production suitable for the pharmaceutical industry. In addition, although hiPSC-CMs appear to be useful "cells" for drug screening, caution must be exercised when interpreting findings because the quality/batch-to-batch variability of cells, different coating cell methods, and different cell providers may result in variable outcomes. Information regarding the properties of different ion channels and their response to hormones, phosphorylation, and other messengers that regulate the biology of hiPSC-CMs is only starting to emerge, and more studies will be required to establish the validity of the assays for future cardiac safety testing.

9.5 *IN VIVO* TELEMETRY CAPABILITIES AND PRECLINICAL DRUG DEVELOPMENT

9.5.1 Introduction

Telemetry technology was developed to monitor BP, body temperature, ECG, and/or numerous other physiological signals from a variety of species (rats, mice, dogs, pigs, and nonhuman primates (NHPs)) used in preclinical efficacy, toxicology, and SP evaluations during drug discovery and development. Telemetry utilizes remote monitoring methods, where biological signals from devices either chronically implanted (invasive) or worn by (noninvasive) a subject are received, processed, and transmitted wirelessly to computerized data acquisition and storage systems (Nikita, 2014). Advantages of telemetry include remote acquisition of high-fidelity signals for extended periods from conscious, unrestrained animals, where autonomic homeostatic reflexes and circadian rhythms remain intact, allowing the conduct of preclinical studies under conditions that closely resemble a clinical situation. Once data are acquired, software packages allow automated extraction of specific parameters from the physiological signals using sophisticated algorithms (i.e., marking ECG waveforms to obtain QT interval measures or derivation of left ventricular pressure parameters) or more complex analyses using modules capable of employing pattern recognition to identify signal irregularities (i.e., ECG arrhythmia) or characterize other patterns to study relevance to normal physiology (i.e., heart rate variability).

The advantages of using remote monitoring are evident when telemetry methods for measuring BP are compared with the noninvasive (tail cuff) and invasive (exteriorized saline-filled catheters) methods used prior to the availability of telemetry (Kurtz et al., 2005). Use of tail cuff methods in rats requires restraint and warming procedures to ensure proper tail blood flow to obtain a BP measurement during cuff inflation. Only "snap shots" of data are acquired during tail cuff BP measurements, making the method unsuitable for chronic studies or for evaluating the effects of circadian rhythms on drug responses. While use of an exteriorized fluid-filled arterial cannula enables continuous recording of BP, it requires tethering of the animals and poses both high infection risk and catheter patency loss, which may limit chronic BP monitoring. Notably, the dynamic frequency response for the specialized gel-filled or solid-state implantable telemetry devices is superior, so the BP signal fidelity is high over long periods of time, and these telemetry-based electronics also facilitate monitoring of left ventricular pressure to study cardiac function (Sarazan, 2014). Moreover, telemetry implants are available for concurrent assessment of other parameters, including (but not limited to) body temperature, ECG, activity, and respiratory parameters. Addition of video recording during an experiment means that concurrent evaluation of animal behavior can be included for an integrated assessment in preclinical SP studies. These advances not only increase scientific quality, but they also benefit animal welfare by refinement of experimental methods and reduction of the use of animals, consistent with the 3Rs.

9.5.2   CV SP Evaluations Using Telemetry

CV safety issues are a major cause of drug attrition in both the nonclinical and clinical phases of drug discovery and development, and an estimated 45% of drug withdrawals during the postmarketing phase are attributed to adverse CV events, such as myocardial infarction, arrhythmia, and cardiac arrest (Laverty et al., 2011; Pierson et al., 2013).

Preclinical SP studies are designed to characterize both the pharmacodynamics (PD) and pharmacokinetics (PK) associated with an adverse drug-related effect. From this PK/PD profile, an integrated safety liability assessment can be conducted before first-in-human studies, where the risk of an adverse effect occurring is evaluated, safety margins are calculated based on drug exposure, and recommendations for clinical safety monitoring are made (Pugsley et al., 2008; Vargas et al., 2008; Bass et al., 2011). Drug-induced prolongation of cardiac repolarization is one of the most serious adverse CV effects and is associated with the risk of a potentially lethal ventricular arrhythmia called "TdP" (Raehl et al., 1985; Darpo, 2001). The QT interval on the ECG represents cardiac depolarization and repolarization and is therefore assessed in preclinical and clinical studies as an indicator of proarrhythmic risk, as an increase in QT interval duration (and the heart rate-corrected QT interval (QTc interval)) occurs with prolongation of ventricular repolarization.

To improve human risk assessment, ICH S7A guidelines specifically recommend that potential new therapeutics be evaluated for CV safety in a telemetry study conducted in dogs or NHPs (US FDA S7A, 2001), as part of a comprehensive nonclinical risk assessment prior to the first-in-human clinical studies, along with an *in vitro* assessment of a drug's potential to inhibit the delayed rectifier potassium current (I_{Kr}/hERG) and other ion channels contributing to ventricular repolarization (Pugsley et al., 2008; Vargas et al., 2008). I_{Kr} is the main repolarizing current in ventricular CMs, in both humans and nonrodent preclinical species (dog, pig, and NHP), and I_{Kr} inhibition is therefore considered to be a surrogate marker of QT/QTc interval prolongation and risk for TdP (Lagrutta et al., 2008). Many reports demonstrate that dog, pig, and NHP preclinical models that employ fully implantable (Chaves et al., 2006, 2007) or wearable (Chui et al., 2009; Derakhchan et al., 2014) telemetry devices provide sensitive methods to detect QT/QTc interval prolongation and are predictive of effects in man (Ewart et al., 2014). Rats are not considered an acceptable species for *in vivo* assessment of QT/QTc interval prolongation risk in humans, since the rat has a different AP morphology, suggesting that the repolarizing currents (I_{Kr} and I_{Ks}) important in human ventricular repolarization play either a minimal role or that different repolarizing current patterns exist in the rat (i.e., Ito; Gussak et al., 2000). In addition to differences in repolarization current patterns, rats have higher heart rates and generally don't have a readily apparent ST interval, making marking of ECG waveforms for quantitative assessment challenging. While QT/QTc interval prolongation remains the primary indicator of proarrhythmic risk assessment, its association with TdP arrhythmias remains imperfect, and several alternative indices have been proposed, including evaluation of changes in T-wave morphology (Hanton et al., 2008), ventricular repolarization heterogeneity ($T_{peak} - T_{end}$ and beat-to-beat instability; Fossa and Zhou, 2010; Said et al., 2012), and the electromechanical window (van der Linde et al., 2010; Stams et al., 2014). All of these potential arrhythmia risk markers can be readily addressed using data acquired and analyzed using telemetry methods and statistical methodologies.

While arrhythmia risk assessment remains a key focus of a preclinical safety evaluation in nonrodent telemetry studies, studying the effects of drug candidates on hemodynamics (BP and heart rate) is also an important component of a CV safety evaluation (Pugsley et al., 2008; Bass et al., 2011). Controversy remains around what magnitude of change in BP and heart rate indicates a "signal of concern," and the predictivity of hemodynamic effects observed in preclinical studies to the clinic is poor compared to QT/QTc interval. However, a recent manuscript points out that there are study design differences and methodological differences in how BP is measured in clinical and preclinical studies that may be responsible for this poor predictivity (Ewart et al., 2014). Nevertheless, use of telemetry to evaluate hemodynamics remains an important part of both efficacy and safety evaluations during drug discovery and development. The telemetric approach to safety science is much broader than for cardiovascular purposes alone, and some additional opportunities in this field are mentioned in the following. These include the ability to evaluate the effects of drugs on respiratory and CNS function that is organ systems that are well known to have influences on the CV system itself.

9.5.3 Evaluation of Respiratory Function Using Telemetry

Studying effects on respiratory function is important in drug development, since a number of drugs have been reported to cause bronchoconstriction or elicit changes in ventilatory parameters, particularly respiratory rate and tidal volume. Indeed, Safety Pharmacology ICH S7A guidelines mandate evaluation of the effects of therapeutic candidates on respiratory function (US FDA S7A, 2001), and both noninvasive and fully implantable telemetry methods are available that allow for an integrated cardiorespiratory assessment in conscious animals. Fully implantable telemetry units that include an additional pressure catheter can be inserted into the thoracic cavity to obtain an intrathoracic pressure signal, from which respiratory rate and depth of respiration can be assessed (KMH personal experience). Recently, implantable telemetry that includes an impedance sensor to monitor respiratory function (tidal volume and respiratory rate) has become available, where improved calibration techniques can be conducted without invasive measures or anesthesia, since there is no movement of elastic bands common with external telemetric systems using respiratory inductance plethysmography methodologies (Authier et al., 2010; Kearney et al., 2010; Murphy et al., 2010).

9.5.4 Evaluation of CNS Using Telemetry

Seizures are reported as a frequent reason for injury or death during clinical trials (Bass et al., 2004) and can be evoked by both CNS and nonneural therapeutic agents, either directly or indirectly. Quite frequently, the first evidence of seizure activity in preclinical studies is observation of overt physical or behavioral signs that may include abnormal posture, tremors, and ataxia, which are often reported in standard neurobehavioral SP studies in rats or mice. However, confirmation of a seizure requires an evaluation of the electroencephalogram (EEG) to assess electrophysiological activity in the brain, as abnormal behaviors can manifest after drug treatment that are not seizure-related (Redfern et al., 2005; Lahunta et al., 2006; Easter et al., 2009). Distinct EEG patterns prior to seizure onset and during a seizure have been identified and used to establish clinical proconvulsive criteria, and these observations have been applied to preclinical models. Telemetric preclinical EEG monitoring methods, coupled with digitally synchronized video capture to observe animal behavior (Authier et al., 2010) and monitoring of the electromyograph (EMG) to assess skeletal muscle electrical activity, has been successfully used in follow-up preclinical studies to confirm seizure activity (Durmuller et al., 2007; van der Linde et al., 2011; Bassett et al., 2014).

9.5.5 Evaluation of Other Systems Using Telemetry

Telemetric methods have been extended to the study of other physiological systems, where the advantages of long-term monitoring in the conscious state have been leveraged in both preclinical and clinical evaluations. Some examples include:

Gastrointestinal motility: Biotelemetry capsules ("smart pill"; wireless motility capsule) have been developed for 24-h noninvasive measurement of gut intraluminal pressure and motility, as well as tracking temperature and pH gradients in the gastrointestinal tract, which can be used to evaluate gastrointestinal dysfunction (i.e., irritable bowel syndrome, dysphagia) and disease management during preclinical development and in the clinic (Lalezari, 2012; Shi et al., 2014).

Electrooculogram (EOG): The EOG measures electrical potential difference between the front and back of the eye and is used to measure eye movements (Arden and Constable, 2006). A group of investigators recently reported using telemetry-based polysomnography-derived variables, including EEG, EMG, and EOG, in a monkey model to assess responses to pharmacological agents known to disrupt sleep architecture in humans and to demonstrate clinical translation of this animal model for the study of drug-induced sleep disturbances (Authier et al., 2014). Another group also collectively monitored EEG, EMG, and EOG signals via telemetric methods to assess arousability from different sleep stages during treatment with an anti-insomnia agent in a dog model (Tannenbaum et al., 2014).

Intraocular pressure (IOP): Continuous 24-h IOP profiles are of increasing interest in the study of glaucoma, and recent reports from studies in conscious undisturbed rats (Lozano et al., 2015), rabbits (Paschalis et al., 2014), and monkeys (Downs et al., 2011) using fully implantable telemetric methods show the importance of assessing effects of antiglaucoma drugs on IOP under both diurnal and nocturnal conditions. Contact lenses with telemetry chips for continuous IOP monitoring are also being used in glaucoma patients to evaluate responses to antiglaucoma agents and monitor IOP variation associated with glaucoma progression (Lorenz et al., 2013).

9.6 ASSESSMENT OF MYOCARDIAL CONTRACTILITY IN PRECLINICAL MODELS

9.6.1 Introduction

Cardiac output is tightly regulated to provide the needed oxygen and nutrients to the various tissues of the body. Output is modulated by both neuronal and humoral inputs that can adjust the cardiac output through either changes in the heart rate or contractile force, in most cases both. There is also a dependency of the myocardial contractile force on the rate, providing a further mechanism to finely tune the relationship between oxygen consumption and oxygen delivery. It is therefore not surprising that drug exposure may have an impact on myocardial contractility. Since the CV system is modulated by both sympathetic and parasympathetic input, drugs that interact with these systems may affect myocardial function. Examples of these include sympathomimetics and adrenoreceptor blockers that either mimic or block, respectively, the endogenous sympathetic input to the heart. Other pharmacological mechanisms can also have substantial effects on myocardial contractility. For example, compounds that interact with calcium homeostasis in CMs have well-known effects on contractile function, based on the important role of calcium activation of the myofibrils. Other classes of compounds have also been shown to markedly affect myocardial contractile function, even when the mechanisms involved are not understood. Itraconazole, an antifungal medication, is a good example of a compound that was found to have a substantial negative inotropic effect on the heart through a mechanism that has not yet been elucidated (Qu et al., 2013).

Both increases and decreases in myocardial contractility (i.e., positive and negative inotropic effects) can be detrimental clinically depending upon the patient population, further emphasizing the importance of detecting such effects early in the drug discovery process. Negative inotropic effects may be contraindicated in patients with established cardiac insufficiency and could, in the worst case, drive them into acute heart failure. An increase in myocardial contractility is typically seen with exercise and is, as such, not necessarily associated with increased CV risk. However, in patients with coronary artery disease and limited coronary artery inflow, increased inotropic state and increased myocardial oxygen consumption cannot be adequately compensated by increased myocardial blood flow and oxygen delivery. In this case, an imbalance between oxygen supply and demand results in the development of myocardial ischemia, particularly in the deeper layers of the myocardium distal to the coronary stenosis. Thus, both positive and negative modulation of myocardial contractility can be clinically relevant, and drugs with these actions should be detected to assess their potential limitations in specific patient populations. As such, having preclinical assays to identify these activities early within the drug research and development process can be critically important.

9.6.2 Gold Standard Approaches

Due to the recognition that the loading conditions of the heart, meaning the pressure within the ventricle prior to contraction or "preload" as well as the pressure against which the heart pump is attempting to expel blood or "afterload," the quantitative assessment of myocardial contractility is challenging. A further consideration is that the inotropic state of the heart is intrinsically linked to the rate of contraction, such that changes in the heart rate need to be considered when making such determinations. The currently favored approach to determine the myocardial contractile state was proposed some years ago by Suga et al. (1976) based on the relationship between intraventricular BP and the volume of the left ventricle throughout the contractile cycle of the heart. Plotting pressure versus volume over time results in a "pressure–volume loop" with which a pressure- and heart rate-independent assessment of the inotropic state of the heart can be made. Despite its theoretical attractiveness, the use of this approach for screening has not found much application in drug discovery and development due to its technical challenges and rather low throughput. For the early drug discovery process, *in vitro* approaches with a medium to high throughput would be preferable. Once development candidates are identified, use of *in vivo* models to evaluate overall CV safety are typically used and should include an assessment of possible drug-induced effects on myocardial contractility.

9.6.3 *In Vitro* and *Ex Vivo* Assays

Due to its potential clinical importance, drug-induced effects on myocardial contractility need to be detected as early as possible in drug discovery in order to drive design and selection of drug candidates with low risk of CV liabilities. In order for this to be done in the early phases of a drug discovery program, there is necessity for appropriate *in vitro* assays that are capable of detecting these effects while only requiring small amounts of test article and with a reasonable throughput and cost. Some of these goals can be achieved with the use of isolated myocardial tissue from various sources and isolated hearts from various animal species and, more recently, through the use of cells or tissues engineered from human stem cells. Each of these approaches offers advantages and disadvantages that are summarized in the following.

9.6.3.1 *Isolated Tissue* Myocardial tissue harvested from an animal or even in some cases from man can be used to assess possible effects of a test article on the contractile state of the myocardium. Isolated atria, papillary muscles, trabeculae, or strips taken from papillary muscles or the ventricular wall have been used for this type of study (Toda, 1969; Brown and Erdmann, 1985; Brown et al., 1987; Wilson and Broadley, 1989; Saetrum Opgaard et al., 2000).

Isolated myocardial tissue needs to be maintained in a temperature-controlled physiological buffer system and suspended in an apparatus that allows for either an isometric (contraction at constant length) or isotonic (muscle shortening against a defined load) contraction of the muscle. Small amounts of the test article can be added to the buffer to test for changes in the contractile performance of the muscle at different drug concentrations. The studies require technical expertise, are rather low throughput, and require animals as the source of the tissue. One advantage of this approach is that the drug-induced effects can be assessed in the absence of secondary factors that may come into play in an *in vivo* system, such that direct effects on myocardial contractility can be detected and directly related to the drug concentration. Novel approaches to the performance of such studies and their evaluation have been suggested. For example, the use of a work-loop analysis appears to enhance the predictivity of drug-induced effects seen in man (Gharanei et al., 2014).

9.6.3.2 *Isolated Hearts* Similarly, an entire heart can be removed from an animal donor (e.g., rabbit) and studied in isolation via Langendorff perfusion using different modes of contraction such as isovolumetric (Qu et al., 2013) or working heart under various conditions of preload and afterload (Werchan and McDonough, 1987). This approach shares many of the characteristics of the isolated tissue approach, including the need for technical expertise to run

the study, low amounts of compound required, and the ability to examine effects over a range of drug concentrations. Once again, as an isolated organ, the heart in this type of experimental preparation is not subject to neural or humoral influences that can modify the effects caused by the test article. For all *in vitro* techniques, it is important to consider that they are usually run in a plasma protein-free environment, which may overestimate effects for highly plasma protein-bound compounds. Another important consideration that may affect predictive accuracy of *in vitro* systems is that they usually do not generate potentially active metabolites that may occur in *in vivo* test systems.

Although the relationship between drug concentration and effect in isolated tissue or heart preparations may not exactly mimic that what is seen *in vivo*, these models can still be very useful for hazard identification and prioritization of compounds when they are compared to each other.

9.6.3.3 *Novel Technologies* In recent years, there have been efforts to reduce the use of animals in SP studies, as well as an increased emphasis on using human-based tissue to minimize possible species-selective effects. The development of human induced pluripotent stem cell (iPSC)-derived cardiomyocytes offers a new approach with the potential not only of replacing animal tissue sources for such studies but also offering a test system of human origin (He et al., 2003; Hazeltine et al., 2012). Contraction of cardiomyocyte monolayers or three-dimensional cultures using these cells has emerged as a novel approach to detect drug-induced effects on myocardial contractility (Ramade et al., 2014; Rodriguez et al., 2014).

Despite their theoretical appeal, iPSC-derived CM models are still in a developmental phase and have not yet been validated enough to fully understand their advantages and possible pitfalls. In general, it appears that stem cell-derived CMs are distinct from native adult CMs and that there may be relevant differences between stem cell-derived human CMs from different origins and differentiated using distinct protocols and culture conditions. Particularly, contractility and excitation–contraction coupling are very complex physiological processes that depend on both subcellular microstructures and their adequate physiological function. A perfect match of structure and function is hard to achieve in an *ex vivo* environment (Rao et al., 2013).

9.6.4 *In Vivo* Assays

The mainstay of evaluation of drug-induced effects on myocardial contractility has been the use of *in vivo* animal models. Although the dog has been the species most commonly used for this purpose, other species (pig, NHP, rodents) have been used successfully for this purpose. Whereas this once again introduces the possibility of having species-dependent effects that could have an impact on the assessment of a given drug, overall the translation of effects seen in these *in vivo* models to that observed later in patients appears to be quite good (Steidl-Nichols et al., 2014).

9.6.4.1 Anesthetized Animal Models

9.6.4.1 Anesthetized Animal Models A mainstay of pharmacological research has been the use of anesthetized animals, in particular the dog, for assessing drug-induced effects on not only the contractile state of the heart but also effects on the heart rate and BP. Although the anesthetized state blocks most neural input to the heart, which in turn might affect myocardial contractility, animals can still be studied in a state in which their baroreceptor reflex is largely intact and changes in BP can be compensated by adjustments of heart rate. The contractility of the heart can be assessed in various ways, including the use of a catheter or pressure transducer in the left ventricular chamber. The first derivative of the resultant pressure signal represents the rate of pressure development within the heart during its contraction (systole). The maximum pressure increase, $LVdP/dt_{max}$, has been shown to be highly sensitive to changes in the contractile state of the heart although it has some dependency on both the pressure in the ventricle at the onset of the contraction (preload) and the pressure against the heart when it is ejecting blood (afterload). Alternatively, one can use echocardiography to assess left ventricular stroke volume and ejection fraction, both of which are dependent upon the inotropic state of the heart. This approach is attractive since similar measurements can be performed easily in patients in the clinic. Finally, in an anesthetized animal model, it is possible to use the relationship between left ventricular pressure and volume in an approach similar to that suggested by Suga et al. (1976). Ventricular volume can be measured either echocardiographically or using conduction catheters that record changes in overall ventricular blood volume over time. Pressure–volume relationships compared before and during drug treatment can then be used to obtain a pressure-independent assessment of contractility effects.

9.6.4.2 Conscious Animal Models In the context of SP studies, there has been a great interest in performing CV assessments under optimal physiological conditions in the conscious animal. This is specifically mentioned in ICH guidances S7A and S7B as useful for detecting hemodynamic or proarrhythmic risk of new drugs, with the thought that the conscious animal, and therefore without the presence of anesthetic agents, may provide the most relevant model for accurate prediction of effects that might be seen in patients. As with anesthetized animal models, the dog is the most commonly used species, followed by the NHP and the pig. The conscious animal, as opposed to the anesthetized animal, responds to its environment and factors other than the administration of a test article, which can have substantial impact on CV parameters, including the myocardial contractile state. Spurious noises, smells, or vibrations can

lead to excitement of the animals, which can complicate the interpretation of drug-induced changes. Thus, to be used optimally, the animals need to be highly trained to their experimental environment and need to be shielded from external stimuli. Nevertheless, some of these negative environmental factors can be reduced when employing best practices as outlined by Leishman et al. (2012). Use of fully implantable telemetry-based data acquisition systems can include a catheter or transducer in the left ventricle to derive the contractility index $LVdP/dt_{max}$, as mentioned earlier in the description of the anesthetized animal model. Due to its dependence upon heart rate and both preload (measured as end-diastolic left ventricular pressure) and afterload (assessed using aortic or arterial BP), one must simultaneously make these measurements to interpret the data. With training, the conscious animal (typically the dog) is amenable to the use of echocardiography for generating data on ventricular volume. Although echocardiography does not require that the animals be instrumented prior to the assessment, obtaining measurements of left ventricular pressure requires some sort of invasive pressure measurement.

Finally, it should be noted that there have been attempts to circumvent the more invasive approach to assess myocardial contractility through the use of the measurement of QA interval (Cambridge and Whiting, 1986). The QA interval is the time between the Q-wave of the ECG and the time of the opening of the aortic value (A), seen as the beginning of systolic arterial pressure increase. The benefit of this measurement is that only an arterial pressure measurement and an ECG are required. Whereas this has been suggested to be a useful endpoint when other approaches may not be accessible, potential disadvantages have been recognized (Norton et al., 2009).

9.6.5 Translation to Clinic

It is of interest to note that the assessment of a drug's potential to affect the myocardial contractile state is not mentioned in the ICH S7A guidance for the core battery SP studies and is mentioned only as a potential follow-up study. This should not be construed to mean that such effects are not important should they occur, but it more likely reflects the fact that the technology to generate such an assessment at the time the guidance was written was not commonly used in the pharmaceutical industry and there was some degree of uncertainty as to the robustness of an assessment based on the measurement of $LVdP/dt_{max}$. More recently, the use of a left ventricular catheter or transducer in conscious animals, using telemetry-based approaches, has become more common, and the value of detecting such effects has become well recognized, especially for internal decision-making. The robustness of the use of $LVdP/dt_{max}$ to detect changes in contractile state was also recently put to the test in a HESI-based consortium in which a standardized study design in conscious, instrumented dogs, was used in different laboratories, using similar (but not identical) experimental models (Guth et al., 2015). In this study, four agents known from their clinical use to produce significant effects on the inotropic state of the heart were tested. Two positive (amrinone and pimobendan) and two negative inotropic agents (atenolol and itraconazole) were used in this validation. The results of that study demonstrated clearly that across the laboratories involved, the drug- and dose-dependent effects of the agents on contractility parameters tested could be detected. Furthermore, the plasma drug levels in the dogs associated with the effect were very similar to those causing the effects in patients, suggesting that this model may have clinical translatability.

9.7 ASSESSMENT OF LARGE VERSUS SMALL MOLECULES IN CV SP

9.7.1 Introduction

Biotechnology-derived "large-molecule" pharmaceuticals or biopharmaceuticals (BPs) are proteinaceous molecules such as mAb, soluble/decoy receptors, hormones, enzymes, cytokines, and growth factors that are produced in various biological expression systems and are used to diagnose, treat, or prevent various diseases. Large modified antibodies, for example, antibody–drug conjugates, and small peptides are other examples of the diversity of BPs. The features of BPs that distinguish them from traditional SM therapeutics are their larger physical size and molecular weight, molecular complexity, and unique selectivity for the intended therapeutic target. Based on these properties, BPs are expected to have less off-target activities relative to SM therapeutics and consequently have a reduced risk of off-target adverse effects in humans (Table 9.1; Giezen et al., 2008; Vargas et al., 2013; Amouzadeh et al., 2015).

SP assessment of BPs is based on ICH M3(R2), S6(R1), S7A, and S7B guidelines. With regard to BPs, the M3(R2) guideline addresses only the "timing of nonclinical studies relative to clinical development" and defers to S6 guideline

for nonclinical safety assessment of biotechnology-derived drugs. The S6(R1) guideline indicates that "It is important to investigate the potential for undesirable pharmacological activity in appropriate animal models and, where necessary, to incorporate particular monitoring for these activities in the toxicity studies and/or clinical studies." Guidelines S7A and S7B specifically indicate that SP evaluation, including QTc prolongation, of BP can be performed using dedicated studies or integrated into toxicity studies. Recent review articles have detailed information on SP assessment of BP, oligonucleotides, and vaccines (Berman et al., 2014; Kim et al., 2014; Amouzadeh et al., 2015).

9.7.2 CV SP Evaluation

The nonclinical approach for CV risk assessment of new drug candidates should address potential effects on BP, heart rate, and ECG and functional impact on critical cardiac ion channels; thus a battery of in vitro and in vivo methods are used (Table 9.2). The approach for BPs and SMs are quite different. A primary parameter evaluated in the CV risk assessment of SMs is their potential inhibitory effect on hERG channel function using a voltage clamp assay (Vargas et al., 2008). Performing this assay for BPs is not considered appropriate because large proteinaceous molecules are not expected to pass though the plasma membrane to gain access to the channel pore nor are they likely to interact with the "toxin-binding site" on the extracellular surface of the hERG channel. Recent in vitro studies using anti-hERG-specific antibodies show that these antibodies do not inhibit the function of the channel because they do not bind to key epitopes near the external pore region like BeKm-1 (Qu et al., 2011). Therefore, performing a hERG assay with BPs (including antibody–drug conjugate and peptides) is considered irrelevant and is not recommended (Vargas et al., 2008).

Drug candidate SMs are profiled for secondary pharmacology activity to identify potential off-target effects and side effect liabilities (Bowes et al., 2012; Papoian et al., 2015); however, BPs are not are not routinely profiled in the

TABLE 9.1   Some Characteristics of Small Molecules and Biopharmaceuticals

Attribute	Small Molecules	Biopharmaceuticals
Modality	Synthetic chemicals	mAb, peptides, peptibodies, fusion proteins, ADC, BiTEs®, and vaccines
Molecular mass (Da)	<500–1000	1,000–150,000
Target selectivity	Low to high	High
Typical species of choice	Rodent and nonrodent (dog, NHP, minipig)	NHP
Typical dosing route	Oral	Parenteral
PK (half-life)	Short (<24 h)	Long (days to weeks)
Regulatory guidelines	ICH M3, S7A, and S7B	ICH S6R1, S7A, and S7B

ADC, antibody–drug conjugates; BiTEs, bispecific T-cell engaging antibodies; Da, dalton; ICH, International Conference on Harmonization; mAb, monoclonal antibody; NHP, nonhuman primates; PK, pharmacokinetics.

TABLE 9.2 Cardiovascular Safety Pharmacology: Comparison of Small Molecules and Biopharmaceuticals

Approaches	Small Molecules	Biopharmaceuticals
Cardiac ion channel profiling *in vitro*	Routine (hERG, NaV1.5, etc.)	Not routine
Receptor panel for off-target profiling *in vitro*	Routine	Not routine
Cardiovascular profiling *in vivo*	• Implant telemetry • Single (acute) dose	• Noninvasive telemetry integrated into toxicity studies • Repeat (chronic) dose
QT prolongation liability	• hERG potency • QT_c assessment (dog)	• QT_c assessment (NHP)
Cardiovascular risk	"On/off target"	"On target" (exaggerated pharmacology)

NHP, nonhuman primates.

same manner because of their high specificity and selectivity for their human target receptor (Table 9.2).

CV evaluation *in vivo* is another critical component in the profiling and safety evaluation of new SM and BP candidates (Leishman et al., 2012). It is common to use dog and NHP models to determine the overall effect of an SM drug on hemodynamics and cardiac electrical activity; however NHPs are primarily used in CV evaluations for BP drugs (Table 9.1). Given that BP are proteins engineered to interact with specific domains (epitopes) on human receptor targets, it is critical to use the most relevant animal species to assess CV and toxicological risks associated with the expressed therapeutic target (receptor, channel, etc.). CV assessment of BPs in rodent and canine models may not be possible because these species may not always express an orthologue of the human target or they have little to no pharmacological activity against the orthologue. Several case examples of CV telemetry evaluations in NHP demonstrate the value of this species in profiling novel SM and BP agents (Santostefano et al., 2012; Caruso et al., 2014), and a variety of invasive and noninvasive CV monitoring methods are available for routine use in this species (Heyen and Vargas, 2015). The rationale for use of NHP in CV SP studies is supported by a recent pharma-wide survey of SP practices (Authier et al., 2013).

A key limitation of typical CV telemetry studies is that they are designed primarily as acute (single dose) experiments, so functional effects that intensify or diminish (due to tolerance) with longer exposure are not evaluated systematically (Redfern et al., 2013). This limitation represents a gap in the ability to perform clinical risk assessments based on functional hazards that occur with chronic dosing. This cardiovascular safety gap can be mitigated by introducing sensitive evaluations into repeat-dose toxicity studies or performing dedicated repeat-dose CV telemetry studies. The changing landscape of CV evaluation for BPs and SMs is underscored by a recent pharmaceutical industry survey, which reported that many drug sponsors are actively using improved functional methods, like jacket-based telemetry systems for noninvasive CV monitoring, to detect functional effects after acute and chronic treatment in exploratory or investigational new drug-enabling toxicity studies (Authier et al., 2013). A recent publication reviewed a broad range of noninvasive methods for the functional assessment of acute and chronic changes in the CV system (and other organs) of nonrodents, especially the NHP (Valentin et al., 2015).

9.8 PATIENTS DO NOT NECESSARILY RESPOND TO DRUGS AND DEVICES AS DO GENETICALLY IDENTICAL, YOUNG MATURE, HEALTHY MICE!

One of the tasks of SP is to predict, in patients, adverse events produced by novel drugs and/or devices. "Prediction" is easily understood and calculated as sensitivity and specificity. "Adverse events" refer to change in any or all CV properties that (may) translate to morbidity and/or mortality. Tacit in change is that we know what magnitude of change—and for how long that change exists—makes a difference. "In patients" is precisely as it says. Tacit in that expression is that we must appreciate responses of patients with a disease for which the test article is indicated or that may have a spectrum of other confounding physiologies that may not be predicted by responses in a homogeneous population of young, healthy animals, often of a species different from that for which the test article is intended. How monumental is the extrapolation from genetically homogeneous, young mature, healthy *Mus musculus*, *Cavia porcellus*, *Canis familiaris*, or even *Macaca mullata* to a population of genetically heterogeneous *Homo sapiens* possessing a spectrum of heterogeneous physiologies and pathophysiologies?

The FDA has stated, repeatedly throughout their documents and without reference to whether or not studies might be performed on normal or disease surrogates, that preclinical studies should be conducted under conditions "as close to those in the clinic as possible." This is unambiguous and is not consistent with conducting studies on normal animals that do not have and/or could never get the disease for which a test article is indicated or that lack the receptor for which the test article is to become a drug.

The thesis of this chapter section is that the validity of extrapolation between preclinical studies and what might occur in the clinic can be improved by conducting studies in SP on animals possessing the specific indications for which the test article is intended and that—at least—represent, better, the anatomical and physiological heterogeneities of patients.

It is obvious that some surrogates for humans (Fig. 9.6) are inappropriate to use in attempts to predict events in humans. A few examples are given below:

1. Rodents do not possess the robust I_{Kr} channel; therefore toxicity manifested by effects on that channel may not be predicted in studies on rodents (Pond et al., 2000).

2. Guinea pigs have spectacular coronary artery collateral channels, and studies conducted on test articles that affect collaterals are unlikely to yield results applicable to humans (Maxwell et al., 1987).

3. Mice have absence of penetration of Purkinje fibers into the ventricular free walls (Pallante et al., 2010), and sheep have complete penetration of Purkinje fibers from endocardium to subepicardium; therefore studies exploring the role of Purkinje fibers in pathogenesis of arrhythmia are unlikely to yield information useful to humans (Hamlin and Scher, 1961).

4. Most rabbits—particularly albino rabbits—possess atropinase and will not respond to atropine; therefore studies to investigate muscarinic mechanisms using atropine as a challenge or provocateur are unlikely to yield information applicable to humans (Walsh and Schwartz-Bloom, 2004).

5. Pigs have large O waves preceding P waves generated by "peculiar" activation of the right atrium; therefore studies conducted on them to explore early SA nodal–right atrial activation might not yield information applicable to humans (Hamlin, 2010).

But—although less obvious—it is equally true that studies conducted on healthy surrogates for man may not produce information applicable to man, not only because there are species effects but also because the disease for which the humans are being treated modifies the pharmacological and/or physiological properties of the test article. The following are but a few examples out of hundreds:

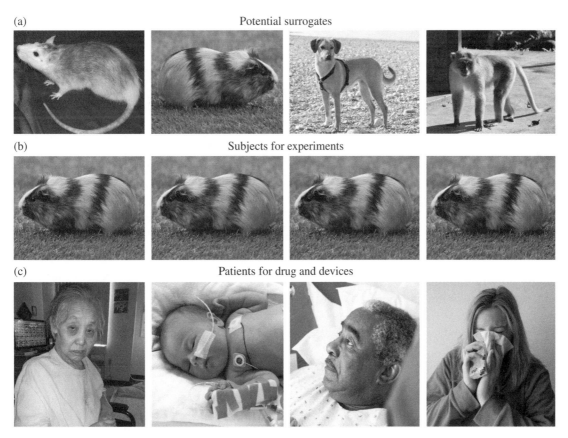

(a)   Potential surrogates

(b)   Subjects for experiments

(c)   Patients for drug and devices

FIGURE 9.6   (a) Potential surrogates for human patients. Is it reasonable to believe that studies conducted on at least 1 of these potential surrogates for man might yield data applicable to man? (b) Example of animals selected for use in safety pharmacology. Notice absence of heterogeneities of phenotypes. If this species possesses the anatomy and physiology necessary, in man, to result in an adverse response to a test article, studies conducted on it should yield applicable results. However if it does not, results applicable to man would not be expected. (c) Examples of humans, showing obvious enormous heterogeneity of phenotypes, for which we attempt to predict adverse clinical events, many of which occur because of phenotypic heterogeneities. (*See insert for color representation of the figure.*)

1. Digitalis glycosides increase myocardial contractility mildly and uniformly in normal myocardium but increase contractility in failing hearts proportional with the degree of failure and do not increase contractility in ischemic myocardium (Braunwald, 1985). These glycosides are negative lusitropes during all phases of diastole in normals but have dissimilar changes in diastolic function in heart failure and may improve ventricular diastolic filling in patients with hypertrophic cardiomyopathy by increasing left atrial function. They increase ventricular Purkinje fiber irritability only at high doses in normals but increase irritability and likelihood of arrhythmia at lower concentrations in heart failure. Additionally, they primarily vasoconstrict normal arteries while secondarily vasodilating arteries in humans with heart failure. Studies of digitalis in normals would underestimate inotropic, lusitropic, and bathmotropic effects and would err in knowledge of vascular resistance .

2. Nitroxyl (HNO) donors and sodium nitroprusside (NaNP) decrease stroke volume and cardiac output in normals but increase both in heart failure patients (del Rio et al., 2014). They decrease $dLVP/dt_{max}$ but contractility is unchanged in normals; but $dLVP/dt_{max}$ is unchanged when heart failure is present. Studies of HNO donors conducted in normals would completely miss the therapeutic benefit that is demonstrated clearly in heart failure.

3. Diuretics and angiotensin-converting enzyme (ACE) inhibitors decrease cardiac output, $dLVP/d_{tmax}$, and end-diastolic pressure–volume relationships in normals but change these parameters, little, in heart failure patients. Studies in normals depend principally on decreasing preload, and effects on patients with diseases that increase preload would be/may be missed.

4. Dofetilide slows heart rate and lengthens QT in normal, awake rabbits but rarely produces TdP (Fig. 9.7, top panel, top figure), whereas in rabbits with heart failure, a majority develop TdP (Fig. 9.7, top panel, bottom figure) (Panyasing et al., 2010). Studies on normal rabbits would have missed torsadogenicity.

5. An ACE inhibitor (enalaprilat), a mixed β- and α-blocker (carvedilol), and both together produced no blunting of the increase in heart rate in response to exercise in normal dogs (Fig. 9.7, middle panel, left) but blunted the increase in heart rate in dogs with heart failure (Fig. 9.7, middle panel, right). Studies of an ACE inhibitor, a mixed adrenergic blocker, and both together would have missed the blunting of chronotropy observed in dogs with heart failure.

6. Dofetilide given to dogs with acute heart block but not in heart failure does not produce TdP (Fig. 9.7, bottom panel, left) but when given to dogs with heart failure due to chronic AV block produces TdP (Fig. 9.7, bottom panel, right) (Dunnink et al., 2012). The torsadogenic risk of dofetilide would be greatly understated if studied in dogs without heart failure.

Before a commitment is made to use animals with disease as subjects for studies in SP, there are many issues that must be addressed:

1. Have results yielded from studies on healthy, homogeneous populations of surrogates provided the necessary/desired information? If so, there may be no need for diseased models.

2. Is use of surrogates with disease cost-effective, that is, expense versus benefit?

3. Are animal models too expensive and too challenging to produce/to house?

4. Is there a source of animal models, and how reproducible is the model?

5. Is production and use of animals with disease less humane?

6. Will regulatory agencies sanction (possibly even expect) the use of animal models with disease? Will they require use of models with specific diseases? Will they weigh more heavily, and arrive at decisions more robustly, from results of studies on models of disease than from studies on normal? Will they demand validation that studies of any test article conducted on any disease model are more predictive than studies on normal?

7. Might the pharmaceutical/device industry utilize disease models for "internal" decisions whether or not regulatory agencies endorse them?

8. Is the mechanism of disease the same in the animal model as it is in man? How similar is the animal model of disease to the actual human disease?

Animal models of disease that might be considered for use in SP should be dictated by whether or not they improve predictive value of adverse reactions in patients, but of course they may also be valuable/preferred in studies on mechanism of action and efficacy. Scientists may be guided by selection of specific models that are well known to manifest increased sensitivity for adverse reactions in the clinic (e.g., obesity, senility, heart failure, hypertrophy, diabetes). Clinicians already know, and consider in design of clinical trials, many features of subjects (e.g., age, geographical location, sex, somatotype, strain) that influence the outcome of studies. It is likely that diversity of subjects in preclinical studies is as important as diversity of subjects in clinical studies. My suspicion is that studies conducted on one mouse, one rat, one rabbit, one guinea pig, one dog, and one monkey per each of four groups (vehicle, low dose, mid

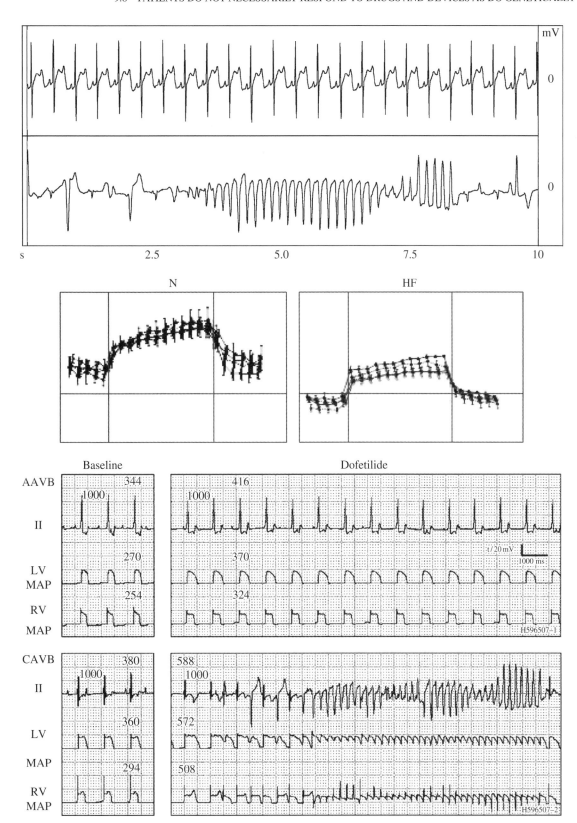

FIGURE 9.7   Examples of how responses to test articles are different whether given to normals or to animals with disease. (top panel) Rabbit (Panyasing et al., 2010) developed torsades de pointes (bottom tracing) only when in heart failure. (middle panel) Differences in heart rate responses during exercise occurred only in responses to enalaprilat and carvedilol) when dogs were in heart failure (HF). (bottom panel) Torsades de pointes developed (bottom tracing) in response to dofetilide (reference 11) only when dogs had complete AV block and were in heart failure.

dose, high dose) may be superior to conducting studies on six of any one species in each group! It is highly unlikely that at least one of those six species would not possess the polymorphism(s), present in humans, that is(are) responsible for an adverse event, should it occur!

9.8.1   Conclusions

Animal models with heart failure respond differently than normal animals. The differences may well obfuscate potentially adverse or beneficial effects. We do not know, yet, which heterogeneities are important/relevant, but we believe many recipients of test articles (e.g., antineoplastics, antibiotics, psychotropics) are aged and/or obese and that because of the prevalences of certain states of disease (e.g., heart failure, systemic arterial hypertension, diabetes, renal insufficiency) they should be considered as reasonable surrogates for man. In addition to selecting the "correct" species (i.e., at the very least the species that possesses the receptors for the test article), it is equally important to select surrogates with a disease or with diseases that is/are known to affect the pharmacology and possibly the toxicology of the test article. (Note: Statisticians do not like, but understand clearly, the concept of heterogeneity of population because of the worry of increased variability and decrease in statistical power for a given "n." But a study of torsadogenicity of dofetilide, conducted on even 1 million "identical" rats having 0 coefficients of variation for all physiological variables, would never turn up toxicity mediated by altered hERG physiology because—of course—the rat does not possess the hERG channel!)

Substituting use of disease models for healthy animals at all stages of testing may be unnecessary; however with resolution of issues 1 through 8 (see preceding text), it may be indicated for lead compounds that have "survived" routine SP/toxicology investigations and with proven efficacy.

REFERENCES

Adamantidis MM, Lacroix DL, Caron, JF, Dupuis, BA (1995). Electrophysiological and arrhythmogenic effects of the histamine type 1-receptor antagonist astemizole on rabbit Purkinje fibers: clinical relevance. *J Cardiovasc Pharmacol* 26: 319–327.

Amouzadeh HR, Engwall MJ, Vargas HM (2015). Safety pharmacology evaluation of biopharmaceuticals. *Handb Exp Pharmacol* 229: 385–404.

Amuzescu B, Scheel O, Knott T (2014). Novel automated patch-clamp assays on stem cell-derived cardiomyocytes: will they standardize in vitro pharmacology and arrhythmia research? *J Phys Chem Biophys* 4: 4.

Arden GB, Constable PA (2006). The electro-oculogram. *Prog Retin Eye Res* 25(2): 207–248.

Authier S, Haefner P, Fournier S, Troncy E, Moon LB (2010). Combined cardiopulmonary assessments with implantable telemetry device in conscious freely moving cynomolgus monkeys. *J Pharmacol Toxicol Methods* 62(1): 6–11.

Authier S, Vargas HM, Curtis MJ, Holbrook M, Pugsley MK (2013). Safety pharmacology investigations in toxicology studies: an industry survey. *J Pharmacol Toxicol Methods* 68: 44–51.

Authier S, Bassett L, Pouliot M, Rachalski A, Troncy E, Paquette D, Mongrain V (2014). Effects of amphetamine, diazepam and caffeine on polysomnography (EEG, EMG, EOG)-derived variables measured using telemetry in Cynomolgus monkeys. *J Pharmacol Toxicol Methods* 70(3): 287–294.

Baron JA, Sandler RS, Bresalier RS, Lanas A, Morton DG, Riddell R, Iverson ER, Demets DL (Nov 15, 2008). Cardiovascular events associated with rofecoxib: final analysis of the APPROVe trial. *Lancet* 372(9651): 1756–1764. Erratum in: Lancet. (Nov 15, 2008);372(9651), 1732.

Barter PJ, Caulfield M, Eriksson M, Grundy SM, Kastelein JJ, Komajda M, Lopez-Sendon J, Mosca L, Tardif JC, Waters DD, Shear CL, Revkin JH, Buhr KA, Fisher MR, Tall AR, Brewer B; ILLUMINATE Investigators (Nov 22, 2007). Effects of torcetrapib in patients at high risk for coronary events. *N Engl J Med* 357(21): 2109–2122.

Bass A, Kinter L, Williams P (2004). Origins, practices and future of safety pharmacology. *J Pharmacol Toxicol Methods* 49(3): 145–151.

Bass AS, Vargas HM, Valentin JP, Kinter LB, Hammond T, Wallis R, Siegl PK, Yamamoto K (2011). Safety pharmacology in 2010 and beyond: survey of significant events of the past 10 years and a roadmap to the immediate-, intermediate- and long-term future in recognition of the tenth anniversary of the Safety Pharmacology Society. *J Pharmacol Toxicol Methods* 64(1): 7–15.

Bassett L, Troncy E, Pouliot M, Paquette D, Ascah A, Authier S (2014). Telemetry video-electroencephalography (EEG) in rats, dogs and non-human primates: methods in follow-up safety pharmacology seizure liability assessments. *J Pharmacol Toxicol Methods* 70(3): 230–240.

Bean BP, Cohen CJ, Tsien RW (May 1983). Lidocaine block of cardiac sodium channels. *J Gen Physiol* 81(5): 613–642.

Berman CL, Cannon K, Cui Y, Kornbrust DJ, Lagrutta A, Sun SZ, Tepper J, Waldron G, Younis HS. (2014). Recommendations for safety pharmacology evaluations of oligonucleotide-based therapeutics. *Nucleic Acid Therapeut* 24: 291–301.

Bottino D, Penland RC, Stamps A, Traebert M, Dumotier B, Georgiva A, Helmlinger G, Lett GS (2006). Preclinical cardiac safety assessment of pharmaceutical compounds using an integrated systems-based computer model of the heart. *Prog Biophys Mol Biol* 90: 414–443.

Bottone FG Jr, Barry WT (June 2009). Postmarketing surveillance of serious adverse events associated with the use of rofecoxib from 1999–2002. *Curr Med Res Opin* 25(6): 1535–1550.

Bowes J, Brown AJ, Hamon J, Jarolimek W, Sridhar A, Waldron G, Whitebread S (2012). Reducing safety-related drug attrition: the use of in vitro pharmacological profiling. *Nat Rev Drug Discov* 11: 909–922.

Bowlby MR, Peri R, Zhang H, Dunlop J (Nov 2008). hERG (KCNH2 or Kv11.1) K+ channels: screening for cardiac arrhythmia risk. *Curr Drug Metab* 9(9): 965–970.

Braam SR, Tertoolen L, Casini S, Matsa E, Lu HR, Teisman A, Passier R, Denning C, Gallacher DJ, Towart R, Mummery CL (2013). Repolarization reserve determines drug responses in human pluripotent stem cell derived cardiomyocytes. *Stem Cell Res* 10: 48–56.

Braunwald E (1985). Effects of digitalis on the normal and the failing heart. *J Am Coll Cardiol* 5(Suppl A): 51A–59A. Review.

Brennan T, Fink M, Rodriguez B (2009). Multiscale modelling of drug-induced effects on cardiac electrophysiological activity. *Eur J Pharm Sci* 36: 62–77.

Britton OJ, Bueno-Orovio A, Van Ammel K, Lu HR, Towart R, Gallacher DJ, Rodriguez B (2013). Experimentally calibrated population of models predicts and explains intersubject variability in cardiac cellular electrophysiology. *Proc Natl Acad Sci U S A* 110: E2098–E2105.

Brown L, Erdmann E (1985). Concentration-response curves of positive inotropic agents before and after ouabain pretreatment. *Cardiovasc Res* 19(5): 288–298.

Brown L, Näbauer M, Erdmann E (1987). Dobutamine: positive inotropy by nonselective adrenoceptor agonism in isolated guinea pig and human myocardium. *Naunyn Schmiedebergs Arch Pharmacol* 335(4): 385–390.

Cambridge D, Whiting MV (1986). Evaluation of the QA interval as an index of cardiac contractility in anaesthetized dogs. Responses to changes in cardiac loading and heart rate. *Cardiovasc Res* 20(6): 444–450.

Caruso A, Frances N, Meille C, Greiter-Wilke A, Hillebrecht A, Lavé T (2014). Translational PK/PD modeling for cardiovascular safety assessment of drug candidates: methods and examples in drug development. *J Pharmacol Toxicol Methods* 70: 73–85.

Cerignoli F, Charlot D, Whittaker R, Ingermanson R, Gehalot P, Savchenko A, Gallacher DJ, Towart R, Price JH, McDonough PM, Mercola M (2012). High throughput measurement of Ca^{2+} dynamics for drug risk assessment in human stem cell-derived cardiomyocytes by kinetic image cytometry. *J Pharmacol Toxicol Methods* 66(3): 246–256.

Chaves AA, Keller WJ, O'Sullivan S, Williams MA, Fitzgerald LE, McPherson HE, Goykhman D, Ward PD, Hoe CM, Mixson L, Briscoe RJ (2006). Cardiovascular monkey telemetry: sensitivity to detect QT interval prolongation. *J Pharmacol Toxicol Methods* 54(2): 150–158.

Chaves AA, Zingaro GJ, Yordy MA, Bustard KA, O'Sullivan S, Galijatovic-Idrizbegovic A, Schuck H, Christian DB, Hoe CM, Briscoe RJ (2007). A highly sensitive canine telemetry model for detection of QT interval prolongation: studies with moxifloxacin, haloperidol and MK-499. *J Pharmacol Toxicol Methods* 56(2): 103–114.

Chi KR (2013a). Revolution dawning in cardiotoxicity testing. *Nat Rev Drug Discov* 12: 565–567.

Chi KR (2013b). Speedy validation sought for new cardiotoxicity testing strategy. *Nat Rev Drug Discov* 12: 655.

Chui RW, Fosdick A, Conner R, Jiang J, Bruenner BA, Vargas HM (2009). Assessment of two external telemetry systems (PhysioJacket and JET) in beagle dogs with telemetry implants. *J Pharmacol Toxicol Methods* 60(1): 58–68.

Cools F, Dhuyvetter D, Vanlommel A, Janssens S, Borghys H, Geys H, Gallacher DJ (Jan–Feb 2014). A translational assessment of preclinical versus clinical tools for the measurement of cardiac contractility: comparison of LV dP/dt(max) with echocardiography in telemetry implanted beagle dogs. *J Pharmacol Toxicol Methods* 69(1): 17–23.

Damiano BP, van der Linde HJ, Van Deuren B, Somers Y, Lubomirski M, Teisman A, Gallacher DJ (Mar–Apr 2015). Characterization of an anesthetized dog model of transient cardiac ischemia and rapid pacing: a pilot study for preclinical assessment of the potential for proarrhythmic risk of novel drug candidates. *J Pharmacol Toxicol Methods* 72: 72–84.

Danker T, Möller C (Sept 2, 2014). Early identification of hERG liability in drug discovery programs by automated patch clamp. *Front Pharmacol* 5: 203.

Darpo B (2001). Spectrum of drugs prolonging QT interval and the incidence of torsades de pointes. *Eur Heart J* 3(Suppl K): K70–K80.

Davidenko JM, Cohen L, Goodrow R, Antzelevitch C (1989). Quinidine-induced action potential prolongation, early afterdepolarizations, and triggered activity in canine Purkinje fibers. Effects of stimulation rate, potassium, and magnesium. *Circulation* 79: 674–686.

Davies MR, Mistry HB, Hussein L, Pollard CE, Valentin J-P, Swinton J, Abi-Gerges N (2012). An in silico canine cardiac midmyocardial action potential duration model as a tool for early drug safety assessment. *Am J Physiol Heart Circ Physiol* 302: H1466–H1480.

Derakhchan K, Chui RW, Stevens D, Gu W, Vargas HM (2014). Detection of QTc interval prolongation using jacket telemetry in conscious non-human primates: comparison with implanted telemetry. *Br J Pharmacol* 171(2): 509–522.

Di Veroli GY, Davies MR, Zhang H, Abi-Gerges N, Boyett MR (2013). High-throughput screening of drug-binding dynamics to HERG improves early drug safety assessment. *Am J Physiol Heart Circ Physiol* 304: H104–H117.

Diaz GJ, Daniell K, Leitza ST, Martin RL, Su Z, McDermott JS, Cox BF, Gintant GA (Nov–Dec 2004). The [3H]dofetilide binding assay is a predictive screening tool for hERG blockade and proarrhythmia: comparison of intact cell and membrane preparations and effects of altering [K+]o. *J Pharmacol Toxicol Methods* 50(3): 187–199.

Dixit R, Boelsterli UA (2007). Healthy animals and animal models of human disease(s) in safety assessment of human pharmaceuticals, including therapeutic antibodies. *Drug Discov Today* 12(7/8): 336–342.

Downs JC, Burgoyne CF, Seigfried WP, Reynaud JF, Strouthidis NG, Sallee V (2011). 24-hour IOP telemetry in the nonhuman primate: implant system performance and initial characterization of IOP at multiple timescales. *Invest Ophthalmol Vis Sci* 52(10): 7365–7375.

Draper MH Weidmann S (1951). Cardiac resting and action potentials recorded with an intracellular electrode. *J Physiol* 115: 74–94.

Dunnink A, van Opstal JM, Oosterhoff P, Winckels SK, Beekman JD, van der Nagel R, Cora Verduyn S, Vos MA (2012). Ventricular remodelling is a prerequisite for the induction of dofetilide-induced torsade de pointes arrhythmias in the anaesthetized, complete atrio-ventricular-block dog. *Europace* 14: 431–436.

Durmuller N, Guillaume P, Lacroix P, Porsolt RD, Moser P (2007). The use of the dog electroencephalogram (EEG) in safety pharmacology to evaluate proconvulsant risk. *J Pharmacol Toxicol Methods* 56(2): 234–238.

Easter A, Bell ME, Damewood JR Jr, Redfern WS, Valentin JP, Winter MJ, Fonck C, Bialecki RA (2009). Approaches to seizure risk assessment in preclinical drug discovery. *Drug Discov Today* 14(17–18): 876–884.

Elkins RC, Davies MR, Brough SJ, Gavaghan DJ, Cui Y, Abi-Gerges N, Mirams GR (2013). Variability in high-throughput ion-channel screening data and consequences for cardiac safety assessment. *J Pharmacol Toxicol Methods* 68: 112–122.

Ewart L, Aylott M, Deurinck M, Engwall M, Gallacher DJ, Geys H, Jarvis P, Ju H, Leishman D, Leong L, McMahonN, Mead A, Milliken P, Suter W, Teisman A, Van Ammel K, Vargas HM, Wallis R, Valentin JP (2014). The concordance between nonclinical and phase I clinical cardiovascular assessment from a cross-company data sharing initiative. *Toxicol Sci* 142(2): 427–435.

Farre C, Fertig N (Oct 2014). New strategies in ion channel screening for drug discovery: are there ways to improve its productivity? *Expert Opin Drug Discov* 9(10): 1103–1107.

Fermini B, Hancox JC, Abi-Gerges N, Bridgland-Taylor MH, Chaudhary KW, Colatsky T, Correll K, Crumb W, Damiano B, Erdemli G, Gintant G, Imredy J, Koerner J, Kramer J, Levesque P, Li Z, Lindqvist A, Obejero-Paz C, Rampe D, Sawada K, Strauss DG, Vandenberg JI (July 13, 2015). A new perspective in the field of cardiac safety testing through the Comprehensive in-vitro Proarrhythmia Assay (CiPA) paradigm. J Biomol Screen pii: 1087057115594589. http://www.ncbi.nlm.nih.gov/pubmed/26170255 [Epub ahead of print].

Fossa AA, Zhou M (2010). Assessing QT prolongation and electrocardiography restitution using a beat-to-beat method. *Cardiol J* 17(3): 230–243.

Fulmer T (August 2013). Rushing to abandon tQT. *Biocentury* 21(30): A1–A4.

Genschow E, Spielmann H, Scholz G, Seiler A, Brown N, Piersma A, Brady M., Clemann N, Huuskonen H, Paillard F, Bremer S, Becker K (2002). The ECVAM international validation study on in vitro embryotoxicity tests: results of the definitive phase and evaluation of prediction models. *ATLA* 30: 151–176.

Gharanei M, Hussain A, James RS, Janneh O, Maddock H (2014). Investigation into the cardiotoxic effects of doxorubicin on contractile function and the protection afforded by cyclosporine A using the work-loop assay. *Toxicol In Vitro* 28(5): 722–731.

Giezen TJ, Mantel-Teeuwisse AK, Straus SM, Schellekens H, Leufkens HG, Egberts AC (2008). Safety-related regulatory actions for biological approved in the United States and the European Union. *JAMA* 300: 1887–1896.

Gintant G (Feb 2011) An evaluation of hERG current assay performance: translating preclinical safety studies to clinical QT prolongation. *Pharmacol Ther* 129(2): 109–119.

Gintant GA, Sager P, Stockbridge N (2016). An evolving paradigm for preclinical cardiac safety testing enabling improved drug discovery. *Nat Rev Drug Discov* (in press).

Gussak I, Chaitman BR, Kopecky SL, Nerbonne JM (2000). Rapid ventricular repolarization in rodents: electrocardiographic manifestations, molecular mechanisms, and clinical insights. *J Electrocardiol* 33(2): 159–170.

Guth BD, Chiang AY, Doyle J, Engwall MJ, Guillon JM, Hoffmann P, Koener J, Mittelstadt S, Ottinger S, Pierson JB, Pugsley MK, Rossman EI, Walisser J, Sarazan RD (2015). The evaluation of drug-induced changes in cardiac inotropy in dogs: results from a HESI-sponsored consortium. *J Pharmacol Toxicol Methods* 75: 70–90.

Hamlin R (2010). QRS in pigs versus in dogs. *J Pharmacol Toxicol Methods* 62: 4–5.

Hamlin R, Scher A (1961). Ventricular activation process and genesis of QRS complex in the goat. *Am J Physiol* 200: 223–228.

Hamon J, Renner M, Jamei M, Lukas A, Kopp-Schneider A, Bois FY (2015). Quantitative invitro to invivo extrapolation of tissues toxicity. *Toxicol In Vitro* pii: S0887-2333(15): 00012-0. doi:http://dx.doi.org/10.1016/j.tiv.2015.01.011.

Hanton G, Yvon A, Racaud A (2008). Temporal variability of QT interval and changes in T wave morphology in dogs as markers of the clinical risk of drug-induced proarrhythmia. *J Pharmacol Toxicol Methods* 57(3): 194–201.

Harris K, Aylott M, Cui Y, Louttit JB, McMahon NC, Sridhar A (Aug 2013). Comparison of electrophysiological data from human-induced pluripotent stem cell-derived cardiomyocytes to functional preclinical safety assays. *Toxicol Sci* 134(2): 412–426.

Hayakawa T, Kunihiro T, Ando T, Kobayashi S, Matsui E, Yada H, Kanda Y, Kurokawa J, Furukawa T (2014). Image-based evaluation of contraction-relaxation kinetics of human-induced pluripotent stem cell-derived cardiomyocytes: correlation and complementarity with extracellular electrophysiology. *J Mol Cell Cardiol* 77: 178–191.

Hazeltine LB, Simmons CS, Salick MR, Lian X, Badur MG, Han W, Delgado SM, Wakatsuki T, Crone WC, Pruitt BL, Palecek SP (2012). Effects of substrate mechanics on contractility of cardiomyocytes generated from human pluripotent stem cells. *Int J Cell Biol* 2012. E Pub 508294.

He JQ, Ma Y, Lee Y, Thomson JA, Kamp TJ (2003). Human Embryonic stem cells develop into multiple types of cardiac myocytes. *Circ Res* 93: 32–39.

Herron TJ, Lee P, Jalife J (2012). Optical imaging of voltage and calcium in cardiac cells and tissues. *Cir Res* 110: 609–623.

Heyen JR, Vargas HM (2015). The use of nonhuman primates in cardiovascular safety assessment. In *The Nonhuman Primate in Nonclinical Drug Development and Safety Assessment*, pp. 551–572, Bluemel, J, Korte, S, Schenck, E, Weinbauer, G eds., London, San Diego, Waltham, Oxford: Elsevier.

Hille B (Apr 1977). Local anesthetics: hydrophilic and hydrophobic pathways for the drug-receptor reaction. *J Gen Physiol* 69(4): 497–515.

Hodgkin AL, Huxley AF (1952a). A quantitative description of membrane current and its application to conduction and excitation in nerve. *J Physiol* 117(4): 500–544.

Hodgkin AL, Huxley AF (1952b). The dual effect of membrane potential on sodium conductance in the giant axon of Loligo. *J Physiol* 116(4): 497–506.

Hodgkin AL, Huxley AF (1952c). The components of membrane conductance in the giant axon of Loligo. *J Physiol* 116(4): 473–496.

Hodgkin AL, Huxley AF (1952d). Currents carried by sodium and potassium ions through the membrane of the giant axon of Loligo. *J Physiol* 116(4): 449–472.

Hoekstra M, Mummery CL, Wilde AA, Bezzina CR, Verkerk AO (Aug 31, 2012). Induced pluripotent stem cell derived cardiomyocytes as models for cardiac arrhythmias. *Front Physiol* 3: 346. eCollection 2012.

Hoffman BF, Cranefield PF (1960). *Electrophysiology of the Heart.* Mount Kisco, NY: Future Publishing Co.

Hogg BM, Goss CM, Cole KS (1934). Potentials in embryo rat heart muscle cultures. *Proc Soc Exp Biol Med* 32: 304–307.

Honda M, Kiyokawa J, Tabo M, Inoue T (2011). Electrophysiological characterization of cardiomyocytes derived from human induced pluripotent stem cells. *J Pharmacol Sci* 117: 149–159.

Hondeghem LM, Hoffmann P (Jan 2003a). Blinded test in isolated female rabbit heart reliably identifies action potential duration prolongation and proarrhythmic drugs: importance of triangulation, reverse use dependence, and instability. *J Cardiovasc Pharmacol* 41(1): 14–24.

Hondeghem LM, Hoffmann P (2003b). Blinded test in isolated female rabbit heart reliably identifies action potential duration prolongation and proarrhythmic drugs: importance of triangulation, reverse use dependence, and instability. *J Cardiovasc Pharmacol* 41(1): 14–24.

Hondeghem LM, Katzung BG (1984). Antiarrhythmic agents: the modulated receptor mechanism of action of sodium and calcium channel-blocking drugs. *Annu Rev Pharmacol Toxicol* 24: 387–423.

Hondeghem LM, Carlsson L, Duker G (Apr 17, 2001). Instability and triangulation of the action potential predict serious proarrhythmia, but action potential duration prolongation is antiarrhythmic. *Circulation* 103(15): 2004–2013.

Huang XP, Mangano T, Hufeisen S, Setola V, Roth BL (Dec 2010). Identification of human Ether-à-go-go related gene modulators by three screening platforms in an academic drug-discovery setting. *Assay Drug Dev Technol* 8(6): 727–742.

Kaseda S, Gilmour RF Jr, Zipes DP (1989). Depressant effect of magnesium on early after depolarizations and triggered activity induced by cesium, quinidine, and 4-aminopyridine in canine cardiac Purkinje fibers. *Am Heart J* 118: 458–466.

Kearney K, Metea M, Gleason T, Edwards T, Atterson P (2010). Evaluation of respiratory function in freely moving Beagle dogs using implanted impedance technology. *J Pharmacol Toxicol Methods* 62(2): 119–126.

Kim TW, Kim KS, Seo JW, Park SY, Henry SP (2014). Antisense oligonucleotides on neurobehavior, respiratory and cardiovascular function and hERG channel current studies. *J Pharmacol Toxicol Methods* 69: 49–60.

Kinter LB, Gosset KA, Kerns WD (1994). Status of safety pharmacology in the pharmaceutical industry—1993. *Drug Dev Res* 32: 298–216.

Kramer J, Obejero-Paz CA, Myatt G, Kuryshev YA, Bruening-Wright A, Verducci JS, Brown AM (2013). MICE models: superior to the HERG model in predicting Torsade de Pointes. *Sci Rep* 3: 2100.

Kurtz TW, Griffin KA, Bidani AK, Davisson RL, Hall JE (2005). Recommendations for blood pressure measurement in humans and experimental animals (part 2: blood pressure measurement in experimental animals). *Hypertension* 45: 299–310.

Lagrutta AA, Trepakova ES, Salata JJ (2008). The hERG channel and risk of drug-acquired cardiac arrhythmia: an overview. *Curr Top Med Chem* 8(13): 1102–1112.

Lahunta A, Glass EN, Kent M (2006). Classifying involuntary muscle contractions. *Compend Contin Educ Pract Vet* 28: 516–530.

Lalezari D (2012). Gastrointestinal pH profile in subjects with irritable bowel syndrome. *Ann Gastroenterol* 25: 1–5.

Laverty H, Benson C, Cartwright E, Cross MJ, Garland C, Hammond T, Holloway C, McMahon N, Milligan J, Park BK, Pirmohamed M, Pollard C, Radford J, Roome N, Sager P, Singh S, Suter T, Suter W, Trafford A, Volders PGA, Wallis R, Weaver R, York M, Valentin JP (2011). How can we improve our understanding of cardiovascular safety liabilities to develop safer medicines?. *Br J Pharmacol* 163: 675–693.

Lawrence CL, Bridgland-Taylor MH, Pollard CE, Hammond TG, Valentin JP (2006). A rabbit Langendorff heart proarrhythmia model: predictive value for clinical identification of Torsades de Pointes. *Br J Pharmacol* 149(7): 845–860.

Leishman DJ, Beck TW, Dybdal N, Gallacher DJ, Guth BD, Holbrook M, Roche B, Wallis RM (May–Jun 2012). Best practice in the conduct of key nonclinical cardiovascular assessments in drug development: current recommendations from the Safety Pharmacology Society. *J Pharmacol Toxicol Methods* 65: 93–101.

Leyton-Mange JS, Mills RW, Macri VS, Jang MY, Butte FN, Ellinor PT, Milan DJ (2014). Rapid cellular phenotyping of human pluripotent stem cell-derived cardiomyocytes using a genetically encoded fluorescent voltage sensor. *Stem Cell Rep* 2: 163–170.

Liang P, Lan F, Lee AS, Gong T, Sanchez-Freire V, Wang Y, Diecke S, Sallam K, Knowles JW, Wang PJ, Nguyen PK, Bers DM, Robbins RC, Wu JC (2013). Drug screening using a library of human induced pluripotent stem cell-derived cardiomyocytes reveals disease-specific patterns of cardiotoxicity. *Circulation* 127: 1677–1691.

Liu T, Brown BS, Wu Y, Antzelevitch C, Kowey PR, Yan GX (Aug 2006). Blinded validation of the isolated arterially perfused rabbit ventricular wedge in preclinical assessment of drug-induced proarrhythmias. *Heart Rhythm* 3(8): 948–956.

Liu T, Traebert M, Ju H, Suter W, Guo D, Hoffmann P, Kowey PR, Yan GX (Oct 2012). Differentiating electrophysiological effects and cardiac safety of drugs based on the electrocardiogram: a blinded validation. *Heart Rhythm* 9(10): 1706–1715.

Lorenz K Korb C Herzog N, Vetter JM, Elflein H, Keilani MM, Pfeiffer N (2013). Tolerability of 24-hour intraocular pressure monitoring of a pressure-sensitive contact lens. *J Glaucoma* 22(4): 311–316.

Lozano DC, Hartwick AT, Twa MD (2015). Circadian rhythm of intraocular pressure in the adult rat. *Chronobiol Int* 32: 513–523.

Lu HR, Marien R, Saels A, De Clerck F (2000). Are there sex-specific differences in ventricular repolarization or in drug-induced early afterdepolarizations in isolated rabbit Purkinje fibers? *J Cardiovasc Pharmacol* 36: 132–139.

Lu HR, Remeysen P, Somers K, Saels A, De Clerck F (2001). Female gender is a risk factor for drug-induced long QT and cardiac arrhythmias in an in vivo rabbit model. *J Cardiovasc Electrophysiol* 12: 538–545.

Lu HR, Vlaminckx E, Van Ammel K, De Clerck F (2002). Drug-induced long QT in isolated rabbit Purkinje fibers: importance of action potential duration, triangulation and early afterdepolarizations. *Eur J Pharmacol* 452: 183–192.

Lu HR, Vlaminckx E, Teisman A, Gallacher DJ (2005). Choice of cardiac tissue plays an important role in the evaluation of drug-induced prolongation of the QT interval in vitro in rabbit. *J Pharmacol Toxicol Methods* 52: 90–105.

Lu HR, Vlaminckx E, Van de Water A, Rohrbacher J, Hermans A, Gallacher DJ (2007). In vitro experimental models for the risk assessment of antibiotic-induced QT prolongation. *Eur J Pharm* 577: 222–232.

Lu HR, Vlaminckx E, Gallacher DJ (2008). Choice of cardiac tissue in vitro plays an important role in assessing the risk of drug-induced cardiac arrhythmias in human: beyond QT prolongation. *J Pharmacol Toxicol Methods* 57: 1–8.

Lu HR, J Rohrbacher, E Vlaminckx, Van Ammel K, Yan GX, Gallacher DJ (2010). Predicting drug-induced slowing of conduction and pro-arrhythmia: identifying the 'bad' sodium current blockers. *Br J Pharmacol* 160: 60–76.

Lu HR, Whittaker R, Price JH, Vega R, Pfeiffer ER, Cerignoli F, Towart R, Gallacher DJ (2015). High throughput measurement of Ca++ dynamics in human stem cell-derived cardiomyocytes by kinetic image cytometry: a cardiac risk assessment characterization using a large panel of cardioactive and inactive compounds. *Toxicol Sci* pii: kfv201.

Maxwell M, Hears D, Yellon D(1987). Species variation in coronary collateral circulation during regional myocardial ischemia. *Cardiovasc Res* 21: 737–746.

Members of the Sicilian Gambit (Dec 4, 2001). New approaches to antiarrhythmic therapy, part I: emerging therapeutic applications of the cell biology of cardiac arrhythmias. *Circulation* 104(23): 2865–2873.

Mirams GR, Cui Y, Sher A, Fink M, Cooper J, Heath BM, McMahon NC, Gavaghan DJ, Noble D (2011). Simulation of multiple ion channel block provides improved early prediction of compounds' clinical torsadogenic risk. *Cardiovasc Res* 91: 53–61.

Mirams GR, Davies MR, Cui Y, Kohl P, Noble D (2012). Application of cardiac electrophysiology simulations to pro-arrhythmic safety testing. *Br J Pharmacol* 167: 932–945.

Mirams GR, Davies MR, Brough SJ, Bridgland-Taylor MH, Cui Y, Gavaghan DJ, Abi-Gerges N (Nov–Dec 2014). Prediction of Thorough QT study results using action potential simulations based on ion channel screens. *J Pharmacol Toxicol Methods* 70(3): 246–254.

Mistry HB, Davies MR, Di Veroli GY (2015). A new classifier-based strategy for in-silico ion-channel cardiac drug safety assessment. *Front Pharmacol* 6: 59.

Morgan SJ, Elangbam CS, Berens S, Janovitz E, Vitsky A, Zabaka T, Conour L (2013). Use of animal models of human disease for nonclinical safety assessment of novel pharmaceuticals. *Toxicol Pathol* 41: 508–518.

Murphy DJ, Renninger JP, Schramek D (2010). Respiratory inductive plethysmography as a method for measuring ventilatory parameters in conscious, non-restrained dogs. *J Pharmacol Toxicol Methods* 62(1): 47–53.

Navarrete EG, Liang P, Lan F, Sanchez-Freire V, Simmons C, Gong T, Sharma A, Burridge PW, Patlolla B, Lee AS, Wu H, Beygui RE, Wu SM, Robbins RC, Bers DM, Wu JC (2013). Screening drug-induced arrhythmia vents using human induced pluripotent stem cell-derived cardiomyocytes and low-impedance microelectrode arrays. *Circulation* 128: S3–S13.

Niederer SA, Fink M, Noble D, Smith NP (2009). A meta-analysis of cardiac electrophysiology computational models. *Exp Physiol* 94: 486–495.

Nikita KS (2014). Introduction to biomedical telemetry. In *Handbook of Biomedical Telemetry*, Nikita, KS, ed., John Wiley & Sons, Hoboken, NJ, pp. 1–26.

Noble D, Garny A, Noble PJ (2012). How the Hodgkin-Huxley equations inspired the cardiac physiome project. *J Physiol* 590: 2613–2628.

Norton K, Iacono G, Vezina M (2009). Assessment of the pharmacological effects of inotropic drugs on left ventricular pressure and contractility: an evaluation of the QA interval as an indirect indicator of cardiac inotropism. *J Pharmacol Toxicol Methods* 60(2): 193–197.

Oudit GY, Kassiri Z, Sah R, Ramirez RJ, Zobel C, Backx PH (2001). The molecular physiology of the cardiac transient outward potassium current (I(to)) in normal and diseased myocardium. *J Mol Cell Cardiol* 33(5): 851–872.

Paci M, Sartiani L, Del Lungo M, Jaconi M, Mugelli A, Cerbai E, Severi S (2012). Mathematical modelling of the action potential of human embryonic stem cell derived cardiomyocytes. *Biomed Eng Online* 11: 61.

Paci M, Hyttinen J, Aalto-Setala K, Severi S (2013). Computational models of ventricular- and atrial-like human induced pluripotent stem cell derived cardiomyocytes. *Ann Biomed Eng* 41: 2334–2348.

Pallante BA, Giovannone S, Fang-Yu L, Zhang J, Liu N, Kang G, Dun W, Boyden PA, Fishman GI (2010). Contactin-2 expression in the cardiac Purkinje fiber network. *Circ Arrhythm Electrophysiol* 3(2): 186–194.

Panyasing Y, Kijtawornrat A, Del Rio C, Carnes C, Hamlin RL (2010). Uni- or bi-ventricular hypertrophy and susceptibility to drug-induced torsades de pointes. *J Pharmacol Toxicol Methods* 62: 148–156.

Papoian T, Chui TJ, Elayan I, Jagadesh G, Khan I, Laniyonu AA, Li CX, Saulnier M, Simpson N, Yang B (2015). Secondary pharmacology data to assess potential off-target activity of new drugs: a regulatory perspective. *Nat Rev Drug Discov* 14: 294.

Paschalis EI, Cade F, Melki S, Pasquale LR, Dohlman CH, Ciolino JB (2014). Reliable intraocular pressure measurement using automated radio-wave telemetry. *Clin Ophthalmol* 8: 177–185.

Pierson JB, Berridge BR, Brooks MB, Dreher K, Koerner J, Schultze AE, Sarazan RD, Valentin JP, Vargas HM, Pettit SD (2013). A public–private consortium advances cardiac safety evaluation: achievements of the HESI cardiac safety technical committee. *J Pharmacol Toxicol Methods* 68: 7–12.

Pollard CE, Abi Gerges N, Bridgland-Taylor MH, Easter A, Hammond TG, Valentin JP (2010). An introduction to QT interval prolongation and non-clinical approaches to assessing and reducing risk. *Br J Pharmacol* 159: 12–21.

Pond AL, Scheve BK, Benedict AT, Petrecca K, Van Wagoner DR, Shrier A, Nerbonne JM (2000). Expression of distinct ERG proteins in rat, mouse, and human heart. Relation to functional I Kr channels. *J Biol Chem* 275: 5997–6006.

Pugsley MK, Authier S, Curtis MJ (2008). Principles of safety pharmacology. *Br J Pharmacol* 154(7): 1382–1399.

Puppala D., Collis L., Sun S, Bonato V, Chen X, Anson B, Pletcher M, Fermini B, Engle SJ (2013). Comparative gene expression profiling in human induced pluripotent stem cell derived cardiocytes and human and cynomolgus heart tissue. *Toxicol Sci* 131: 292–301.

Qu Y, Fang M, Gao B, Chui RW, Vargas HM (2011). BeKm-1, a peptide inhibitor of human ether-a-go-go-related gene potassium currents, prolongs QTc intervals in isolated rabbit heart. *J Pharmacol Exp Ther* 337: 2–8.

Qu Y, Fang M, Gao B, Amouzadeh HR, Li N, Narayanan P, Acton P, Lawrence J, Vargas HM (2013). Itraconazole decreases left ventricular contractility in isolated rabbit heart: mechanism of action. *Toxicol Appl Pharmacol* 268(2): 113–122.

Raehl CL, Patel AK, LeRoy M (1985). Drug-induced torsade de pointes. *Clin Pharm* 4: 675–690.

Ramade A, Legant WR, Picart C, Chen CS, Boudou T. Microfabrication of a platform to measure and manipulate the mechanics of engineered microtissues. *Methods Cell Biol* 2014 121: 191–211.

Rampe D, Brown AM (Jul–Aug 2013). A history of the role of the hERG channel in cardiac risk assessment. *J Pharmacol Toxicol Methods* 68(1): 13–22.

Rao C, Prodromakis T, Kolker L, Chaudhry UA, Trantidou T, Sridhar A, Weekes C, Camelliti P, Harding SE, Darzi A, Yacoub MH, Athanasiou T, Terracciano CM (2013). The effect of microgrooved culture substrates on calcium cycling of cardiac myocytes derived from human induced pluripotent stem cells. *Biomaterials* 34(10): 2399–2411.

Redfern WS, Carlsson L, Davis AS, Lynch WG, MacKenzie I, Palethorpe S, Siegl PK, Strang I, Sullivan AT, Wallis R, Camm AJ, Hammond TG (Apr 1, 2003). Relationships between preclinical cardiac electrophysiology, clinical QT interval prolongation and torsade de pointes for a broad range of drugs: evidence for a provisional safety margin in drug development. *Cardiovasc Res* 58(1): 32–45.

Redfern WS, Strang I, Storey S, Heys C, Barnard C, Lawton K, Hammond TG, Valentin JP (2005). Spectrum of effects detected in the rat functional observational battery following oral administration of non-CNS targeted compounds. *J Pharmacol Toxicol Methods* 52(1): 77–82.

Redfern WS, Ewart L, Hammond TG, Bialecki R, Kinter L, Lindgren S, Pollard CE, Roberts R, Rolf MG, Valentin JP (2010). Impact and frequency of different toxicities throughout the pharmaceutical life cycle. *The Toxicologist* 114: 1081.

Redfern WS, Ewart LC, Lainee P, Pinches M, Robinsona S, Valentina J-P (2013). Functional assessment in repeat-dose toxicity studies: the art of the possible. *Toxicol Res* 2: 209–234.

Rezazadeh S, Hesketh JC, Fedida D (Oct 2004). Rb+ flux through hERG channels affects the potency of channel blocking drugs: correlation with data obtained using a high-throughput Rb+ efflux assay. *J Biomol Screen* 9(7): 588–597.

del Rio CL, Youngblood B, Kloepfer P, Ueyama Y, George R, Wallery J, Reardon J, Hamlin RL (2014). Differential effects of a novel inodilator in conscious dogs with normal or dilated-cardiomyopathic ventricles: a look through left-ventricular pressure-volume analyses. *J Pharmacol Toxicol Methods* 70: 324.

Roden DM, Hoffman BF (Jun 1985). Action potential prolongation and induction of abnormal automaticity by low quinidine concentrations in canine Purkinje fibers. Relationship to potassium and cycle length. *Circ Res* 56(6): 857–867.

Rodriguez ML, Graham BT, Pabon LM, Han SJ, Murry CE, Sniadecki NJ (2014). Measuring the contractile forces of human induced pluripotent stem cell-derived cardiomyocytes with arrays of microposts. *J Biomech Eng* 136(136): 051005.

Rossi AM, Tovey SC, Rahman T, Prole DL, Taylor CW (Aug 2012). Analysis of IP3 receptors in and out of cells. *Biochim Biophys Acta* 1820(8): 1214–1227.

Rozanski GJ, Witt RC (1991). Early after depolarizations and triggered activity in rabbit cardiac Purkinje fibers recovering from ischemic-like conditions. Role of acidosis. *Circulation* 83: 1352–1360.

Saetrum Opgaard O, Hasbak P, de Vries R, Saxena PR, Edvinsson L (2000). Positive inotropy mediated via CGRP receptors in isolated human myocardial trabeculae. *Eur J Pharmacol* 397(2–3): 373–382.

Sager PT, Gintant G, Turner JR, Pettit S, Stockbridge N (Mar 2014). Rechanneling the cardiac proarrhythmia safety paradigm: a meeting report from the cardiac safety research consortium. *Am Heart J* 167(3): 292–300.

Said TH, Wilson LD, Jeyaraj D, Fossa AA, Rosenbaum DS (2012). Transmural dispersion of repolarization as a preclinical marker of drug-induced proarrhythmia. *J Cardiovasc Pharmacol* 60(2): 165–171.

Santostefano MJ, Kirchner J, Vissinga C, Fort M, Lear S, Pan WJ, Prince PJ, Hensley KM, Tran D, Rock D, Vargas HM, Narayanan P, Jawando R, Rees W, Reindel JF, Reynhardt K, Everds N (2012). Off-target platelet activation in macaques unique to a therapeutic monoclonal antibody. *Toxicol Pathol* 40: 899–917.

Sarazan RD (2014). Cardiovascular pressure measurement in safety assessment studies: technology requirements and potential errors. *J Pharmacol Toxicol Methods* 70(3): 210–223.

Sarkar AX, Christini DJ, Sobie EA (2012). Exploiting mathematical models to illuminate electrophysiological variability between individuals. *J Physiol* 590: 2555–2567.

Scheen AJ (2010). Cardiovascular risk-benefit profile of sibutramine. *Am J Cardiovasc Drugs* 10(5): 321–334.

Scott CW, Zhang X, Abi-Gerges N, Lamore SD, Abassi YA, Peters MF (2014). An impedance-based cellular assay using human iPSC-derived cardiomyocytes to quantify modulators of cardiac contractility. *Toxicol Sci* 142: 331–338.

Shi Q, Wang J, Chen D, Chen J, Li J, Bao K (2014). In vitro and in vivo characterization of wireless and passive micro system enabling gastrointestinal pressure monitoring. *Biomed Microdevices* 16(6): 859–868.

Spencer CI, Baba S, Nakamura K, Hua EA, Sears MAF, Chi-cheng Fu, Zhang J, Balijepalli S, Tomoda K, Hayashi Y, Lizarraga P, Wojciak J, Scheinman MM, Aalto-Setälä K, Makielski JC, January CT, Healy KE, Kamp TJ, Yamanaka S, Conklin BR (2014). Calcium transients closely reflect Prolonged action potentials in iPSC models of inherited cardiac arrhythmia. *Stem Cell Reports* 3(2): 269–281.

Stams TR, Bourgonje VJ, Beekman HD, Schoenmakers M, van der Nagel R, Oosterhoff P, van Opstal JM, Vos MA (2014). The electromechanical window is no better than QT prolongation to assess risk of Torsade de Pointes in the complete atrioventricular block model in dogs. *Br J Pharmacol* 171(3): 714–722.

Starmer CF, Grant AO, Strauss HC (Jul 1984). Mechanisms of use-dependent block of sodium channels in excitable membranes by local anesthetics. *Biophys J* 46(1): 15–27.

Steidl-Nichols JV, Hanton G, Leaney J, Liu RC, Leishman D, McHarg A, Wallis R (Jan–Feb 2008). Impact of study design on proarrhythmia prediction in the SCREENIT rabbit isolated heart model. *J Pharmacol Toxicol Methods* 57(1): 9–22.

Steidl-Nichols J, Bhatt S, Hemkens M, Heyen J, Marshall C, Li D, Flynn D, Wisialowski T, Northcott C (2014). Blood pressure and heart rate measures: how well do pre-clinical models translate? *FASEB J* 28(1): Supplement 681.7 (Abstract).

Suga H, Sagawa K, Kostiuk DP (1976). Controls of ventricular contractility assessed by pressure-volume ratio, Emax. *Cardiovasc Res* 10: 582–592.

Tannenbaum PL, Stevens J, Binns J, Savitz AT, Garson SL, Fox SV, Coleman P, Kuduk SD, Gotter AL, Marino M, Tye SJ, Uslaner JM, Winrow CJ, Renger JJ (2014). Orexin receptor antagonist-induced sleep does not impair the ability to wake in response to emotionally salient acoustic stimuli in dogs. *Front Behav Neurosci* 8: 182.

Task Force of the Working Group on Arrhythmias of the European Society of Cardiology (1991). The Sicilian gambit. A new approach to the classification of antiarrhythmic drugs based on their actions on arrhythmogenic mechanisms. *Circulation* 84: 1831–1851.

The European Agency for the Evaluation of Medicinal Products Human Medicines Evaluation Unit. *Points to Consider: The Assessment of the Potential for QT Interval Prolongation by Non-cardiovascular Medicinal Products*. The European Agency for the Evaluation of Medicinal Products Human Medicines Evaluation Unit, London, December 17, 1997 (http://www.fda.gov/ohrms/dockets/ac/03/briefing/pubs%5Ccpmp.pdf, accessed October 19, 2015).

Titus SA, Beacham D, Shahane SA, Southall N, Xia M, Huang R, Hooten E, Zhao Y, Shou L, Austin CP, Zheng W (Nov 1, 2009). A new homogeneous high-throughput screening assay for profiling compound activity on the human ether-a-go-go-related gene channel. *Anal Biochem* 394(1): 30–38.

Toda N (1969). Effects of calcium, sodium and potassium ions on contractility of isolated atria and their responses to noradrenaline. *Br J Pharmacol* 36(2): 350–367.

U.S. Food and Drug Administration, Guidance for industry (2001). S7A Safety Pharmacology Studies for Human Pharmaceuticals (http://www.fda.gov/downloads/drugs/guidancecompliance regulatoryinformation/guidances/ucm074959.pdf, accessed October 17, 2015).

Valentin JP, Bialecki R, Ewart L, Hammond T, Leishmann D, Lindgren S, Martinez V, Pollard C, Redfern W, Wallis R (2009). A framework to assess the translation of safety pharmacology data to humans. *J Pharmacol Toxicol Methods* 60: 152–158.

Valentin J-P, Delaunois A, Lainée P, Skinner M, Vargas H (2015). Functional assessments in nonhuman primate toxicology studies to support drug development. In *The Nonhuman Primate in Nonclinical Drug Development and Safety Assessment*, pp. 417–435, Bluemel, J, Korte, S, Schenck, E, Weinbauer, G eds., London, San Diego, Waltham, Oxford: Elsevier.

van der Linde HJ, Van Deuren B, Somers Y, Loenders B, Towart R, Gallacher DJ (2010). The electro-mechanical window: a risk marker for Torsade de Pointes in a canine model of drug induced arrhythmias. *Br J Pharmacol* 161(7): 1444–1454.

van der Linde HJ, Van Deuren B, Somers Y, Teisman A, Drinkenburg WH, Gallacher DJ (2011). EEG in the FEAB model: measurement of electroencephalographical burst suppression and seizure liability in safety pharmacology. *J Pharmacol Toxicol Methods* 63(1): 96–101.

Vargas HM, Bass AS, Breidenbach A, Feldman HS, Gintant GA, Harmer AR, Heath B, Hoffmann P, Lagrutta A, Leishman D, McMahon N, Mittelstadt S, Polonchuk L, Pugsley MK, Salata JJ, Valentin JP (2008). Scientific review and recommendations on preclinical cardiovascular safety evaluation of biologics. *J Pharmacol Toxicol Methods* 58(2): 72–76.

Vargas HM, Amouzadeh HR, Engwall MJ (2013). Nonclinical strategy considerations for safety pharmacology: evaluation of biopharmaceuticals. *Expert Opin Drug Saf* 12(1): 91–102.

Varró A, Papp JG (Sep 1992). The impact of single cell voltage clamp on the understanding of the cardiac ventricular action potential. *Cardioscience* 3(3): 131–144.

Viero C, Thomas NL, Euden J, Mason SA, George CH, Williams AJ (2012). Techniques and methodologies to study the ryanodine receptor at the molecular, subcellular and cellular level. *Adv Exp Med Biol* 740: 183–215.

Virchow R (1856). Thrombose und Embolie. Gefässentzündung und septische Infektion. In *Gesammelte Abhandlungen zur wissenschaftlichen Medicin (in German)*, pp. 219–732, Frankfurt am Main: Von Meidinger & Sohn.

Walsh KB (Feb 2015). Targeting cardiac potassium channels for state-of-the-art drug discovery. *Expert Opin Drug Discov* 10(2): 157–169.

Walsh CT, Schwartz-Bloom RD (2004). Factors Modifying the Effects of Drugs in Individuals. In *Pharmacology: Drug Actions and Reactions" or "Levine's Pharmacology: Drug Actions and Reactions*, 7th edition. p. 274, Boca Raton, FL: Taylor & Francis.

Wang D, Patel C, Cui C, Yan GX (Aug 2008). Preclinical assessment of drug-induced proarrhythmias: role of the arterially perfused rabbit left ventricular wedge preparation. *Pharmacol Ther* 119(2): 141–151.

Werchan PM, McDonough KH (1987). The right ventricular working heart preparation. *Proc Soc Exp Biol Med* 185(3): 339–346.

Wilson AN, Broadley KJ (1989). Analysis of the direct and indirect effects of adenosine on atrial and ventricular cardiac muscle. *Can J Physiol Pharmacol* 67(4): 294–303.

Witchel HJ (Feb 2010). Emerging trends in ion channel-based assays for predicting the cardiac safety of drugs. *IDrugs* 13(2): 90–96.

Wu Y, Carlsson L, Liu T, Kowey PR, Yan GX (Aug 2005). Assessment of the proarrhythmic potential of the novel antiarrhythmic agent AZD7009 and dofetilide in experimental models of torsades de pointes. *J Cardiovasc Electrophysiol* 16(8): 898–904.

Yakuji Nippo (1995). *Japanese Guidelines for Non-Clinical Studies of Drugs Manual*. Tokyo: Yakuji Nippo.

Yeh ETH, Bickford CL (2009). Cardiovascular complications of cancer therapy: incidence, pathogenesis, diagnosis and management. *J Am Coll Cardiol* 53(24): 2231–2247.

Zbinden G (1979). *Pharmacological Methods in Toxicology*. Elmsford, NY: Pergamon.

Zbinden G (1984). Neglect of function and obsession with structure in toxicity testing. *Proceedings of the IUPHAR 9th International Congress of Pharmacology*, 1, Scientific & Medical Division, Macmillan Press, London, UK, 43–49.

10

PREDICTIVE *IN VITRO* MODELS FOR ASSESSMENT OF NEPHROTOXICITY AND DRUG–DRUG INTERACTIONS *IN VITRO*

LAWRENCE H. LASH

Department of Pharmacology, School of Medicine, Wayne State University, Detroit, MI, USA

10.1 INTRODUCTION

10.1.1 Considerations for Studying the Kidneys as a Target Organ for Drugs and Toxic Chemicals

The kidneys are frequent targets for and exhibit adverse effects from many drugs and toxic chemicals due to their physiological functions and structure. In terms of function, the renal epithelium is exposed to any chemical in the blood by virtue of both glomerular filtration and secretion from the nearby renal circulation. A critical fact that emphasizes the unique physiology of the kidneys is that they typically comprise <1% of total body weight yet receive 20–25% of cardiac output. Hence, the kidneys are disproportionately exposed to chemicals in the blood.

Glomerular filtration results in accumulation of blood-borne chemicals in the tubular lumen, whereas secretion from the renal circulation initially results in cellular accumulation of these chemicals. In both cases, a key determinant is the presence, particularly in the proximal tubular (PT) epithelium, of a large and diverse array of plasma membrane transport proteins that mediate either the cellular uptake or efflux of various drugs or drug metabolites. These membrane transport systems on the basolateral plasma membrane (BLM) and brush-border plasma membrane (BBM; also called luminal membrane) contribute to chemical-/drug-induced nephrotoxicity by catalyzing uptake, often of charged or zwitterionic substrates, and accumulating them in the renal cells against large concentration or electrochemical gradients.

Another aspect of renal function that may significantly contribute to the kidneys as a target organ in drug or chemical exposures is the urinary concentrating mechanism that serves to reabsorb nutrients such as hexoses and amino acids; excrete waste products such as ammonia, creatinine, and other nitrogenous compounds; and regulate water and electrolyte balance. This can result in very high concentrations of chemicals in renal cells, lumens, or interstitial space. The urinary concentrating mechanism also provides an illustration of structure serving function. In this case, the renal vasculature and epithelium exist in parallel such that renal capillaries are in close proximity to reabsorptive and secretory surfaces of the tubular epithelium.

Finally, despite their primary function being considered as the site for reabsorption and excretion of drugs and chemicals, the kidneys are also important sites of metabolism for many chemicals (Lohr et al., 1998). With respect to susceptibility of the kidneys to chemically induced injury, renal metabolism does not only include reactions such as cytochrome P450 (CYP)-dependent oxidation and Phase II conjugation that enhance excretion of drugs but for some chemicals includes bioactivation reactions that result in the creation of reactive intermediates that may be cytotoxic, genotoxic, or carcinogenic (Lash, 1994). An example of this type of bioactivation pathway is that for the glutathione (GSH)-dependent metabolism of halogenated hydrocarbons such as trichloroethylene (TCE) (Lash et al., 2014a) and perchloroethylene (Perc) (Lash and Parker, 2001), which results in the formation of reactive, sulfur-containing metabolites that are both acutely cytotoxic and genotoxic.

Drug Discovery Toxicology: From Target Assessment to Translational Biomarkers, First Edition. Edited by Yvonne Will, J. Eric McDuffie, Andrew J. Olaharski, and Brandon D. Jeffy.
© 2016 John Wiley & Sons, Inc. Published 2016 by John Wiley & Sons, Inc.

Hence, there are four major factors that contribute to the frequent susceptibility of the kidneys to chemically induced injury: (i) blood flow; (ii) urinary concentrating mechanisms; (iii) the processes of filtration, absorption, and secretion; and (iv) the presence of bioactivation reactions.

Besides these four "susceptibility factors," one must consider the heterogeneous structure of the kidneys, because many drugs and chemicals act at specific sites in the tissue or particular sites are more susceptible than others to injury due to their intrinsic properties. At the level of the whole organ, the kidneys are subdivided into three regions: cortex, outer medulla, and inner medulla. The outer medulla is further divided into the outer stripe and inner stripe. The functional unit of the kidneys is the nephron, which is a series of epithelial cells of varying types that form a tube of one-layer thick cells enclosing a luminal compartment. Mammalian kidneys contain millions of nephrons, each of which leads into a common collecting duct. Nephrons are categorized as either superficial, midcortical, or juxtamedullary, depending on the location of their glomeruli and loop of Henle. The renal cortex contains glomeruli, proximal convoluted, and some proximal straight tubules, macula densa, cortical thick ascending limbs, distal convoluted tubules, connecting tubules, initial collecting ducts, interlobular arteries, and afferent and efferent capillary networks. The outer stripe of the outer medulla is located just below or inside the cortex and contains proximal straight tubules, medullary thick ascending limbs, and outer medullary collecting ducts. The inner stripe of the outer medulla is located below or inside the outer stripe and contains thin descending limbs, thick ascending limbs, and outer medullary collecting ducts. The inner medulla contains thin descending limbs, thin ascending limbs, and inner medullary collecting ducts. The renal vasculature surrounds the tubule and parallels the nephron.

The various cell populations of the nephron exhibit diverse morphological, biochemical, and functional properties (Table 10.1). Differences exist among cell types in morphology, membrane structure, energetics, and the composition of transporters and drug metabolism enzymes. As a consequence of this functional heterogeneity, exposure of the kidneys, either *in vivo* or *in vitro*, to various nephrotoxicants or to pathological conditions such as ischemia or hypoxia produces distinct patterns of cellular injury. Thus, certain cell populations are either specific targets of or are particularly susceptible to one form of injury or another. In some cases, the specificity is due to the selective presence in a given cell population of a membrane transport system or a bioactivation or detoxication enzyme. In other cases, however, susceptibility is explained by the basic biochemical function of the cell. The proximal tubules are the most common targets for chemically induced injury because they are the first nephron cell type with which filtered chemicals come in contact, they possess a large array of membrane transporters that mediate uptake and intracellular accumulation of chemicals, and they contain relatively high activities of drug metabolism enzymes that can metabolize chemicals to reactive and toxic species. Hence, when characterizing chemically induced toxicity or screening drug candidates, knowledge of the site of action or sites within the nephron that are potentially more susceptible than others is important in achieving the goals of the study.

10.1.2 Advantages and Limitations of *In Vitro* Models in General for Mechanistic Toxicology and Screening of Potential Adverse Effects

A broad range of experimental models has been developed for the study of mechanisms of chemically induced nephrotoxicity and the screening of potential new drugs for adverse

TABLE 10.1 Summary of Key Biochemical, Physiological, and Morphological Properties of Selected Nephron Cell Types

Property	Proximal Tubule	Distal Convoluted Tubule (DCT)	Medullary Thick Ascending Limb (mTAL)
Shape	Large, cuboidal	Small, cuboidal	Flattened
Localization	Cortex (S1, S2), OSOM (S2, S3)	Cortex	OM
Oxygenation	High	High	Low
Plasma membrane	High-density BB microvilli; extensive BL invaginations	Few BB microvilli; few BL invaginations	Few BB microvilli; few BL invaginations
Mitochondrial density	High	Moderate	Moderate
Energetics	High OXPHOS, FA oxidation, gluconeogenesis	Glycolysis, low OXPHOS	Glycolysis, low OXPHOS
Transport function	OA⁻, OC⁺, glucose, amino acids; water permeable	Na⁺, K⁺, H⁺, HCO₃⁻; water permeable	Divalent cations, Na⁺-K⁺-Cl⁻; low water permeability
Drug metabolism	CYP, FMO, GSH-dependent enzymes	Low activities	PGS

Modified from Lash (2012).
BB, brush border; BL, basolateral; CYP, cytochrome P450; FA, fatty acid; FMO, flavin-containing monooxygenase; OA⁻, organic anion; OC⁺, organic cation; OM, outer medulla; OSOM, outer stripe of outer medulla; OXPHOS, oxidative phosphorylation; PGS, prostaglandin synthase.

effects on kidney function. Although *in vivo* models, such as rodents, are critical in establishing target organ specificity for investigating effects involving systemic processes, such as immune-mediated toxicity, and for performing invasive studies *in vivo* that would obviously not be possible in humans, they suffer from many limitations, not the least of which is expense and the high utilization of animals. In contrast, *in vitro* models derived from the kidneys of both laboratory animals and humans have been developed and optimized over the past few decades to more accurately reflect processes that occur *in vivo*. Despite these advances, however, there are significant limitations in the use of *in vitro* rodent models to assess relevant effects for humans and to make accurate predictions for responses in humans (Steinberg and DeSesso, 1993; Pelekis and Krishnan, 1997).

Due the existence of both quantitative and qualitative species-dependent differences, use of animal models for human health risk assessment or drug screening involving the kidneys as a target organ can be limited by significant uncertainties. One of the factors that often cannot be adequately taken into account for humans using rodent-derived experimental models is the existence of genetic polymorphisms in drug metabolism enzymes and membrane transporters. Additionally, dietary, social, and environmental differences that do not exist in studies in laboratory animals also need to be taken into account to accurately assess effects in and consequent risks for humans. Accordingly, *in vitro* models have been developed that are either derived from human kidneys or that more closely mimic human kidneys.

One potential approach that has been suggested to better approximate events that occur in humans is to use nonhuman primates or *in vitro* models derived from nonhuman primate kidneys. Caution is needed with such studies, however, because while responses of nonhuman primates or renal cells derived from nonhuman primates may differ from those of rodents or renal cells derived from rodents, respectively, they may still also differ from those of humans or human-derived renal cells. For example, in a study of screening of metalloproteinase inhibitors as potential anti-inflammatory therapeutic agents, some of which were found to be nephrotoxic in rats, a different pattern of cytotoxicity was observed in primary cultures of PT cells from humans and cynomolgus monkeys (Cai et al., 2009). This suggests that species differences in drug susceptibility and mechanisms of action can also exist among different primate species.

Advantages of *in vitro* models include the ability to carefully modulate exposure conditions and dissect mechanisms of action. Unlike studies in laboratory animals, *in vitro* models enable the use of paired controls and treated samples from the same set of incubations. Another advantage specifically for studies of renal function and nephrotoxicity is that *in vitro* models can be constructed that derive from specific nephron segments, thereby allowing study of effects at the specific sites in the kidneys where they occur.

Limitations exist as well and must be considered when evaluating data from *in vitro* models. Mechanistically, one must consider that processes that underlie physiological or pathological perturbations may require systemic responses, such as might occur with immune-mediated processes as noted previously, and these may not be readily assessed in such models. Other limitations include short-term viability and the potential for artifacts that result from the isolation process. These can be mitigated by careful validation as described in the following section. Approaches to accomplish this include performing comprehensive analyses of key drug metabolism and membrane transport pathways, cellular energetics, and cellular redox status, among other parameters.

10.1.3 Types of *In Vitro* Models Available for Studying Human Kidneys

Potential *in vitro* models for studying mechanisms of toxicity or screening of drugs in human kidneys are generally similar to those used in laboratory animals such as rodents or rabbits. These include the isolated perfused kidney, renal slices, isolated perfused tubules or tubules/tubular fragments, suspensions of freshly isolated cells, primary cell cultures, and immortalized renal cell culture lines. Although all of the models are theoretically possible, the most commonly used *in vitro* models are primary cell cultures and immortalized cell lines, although the latter is more common than the former.

A distinct advantage of primary cell culture is that it is directly derived from the cell type of interest. This is important because such tissue, particularly epithelial cells, tends to exhibit a high degree of dedifferentiation when immortalized or placed in long-term culture. However, the utility of primary cell cultures is obviously dependent on the availability of fresh tissue. Despite the increasing need for kidneys for transplant due to increasing incidence of chronic kidney disease and end-stage renal failure in the US population (Atlas of Chronic Kidney Disease in the United States, 2014), availability of fresh human kidneys for research is quite reasonable. This is largely owing to the fact that criteria for kidneys being effectively used for transplant are very stringent so that many potential donor kidneys are deemed to be poor candidates for transplantation.

Few immortalized cell lines derived from human renal epithelial cells are available, particularly from normal kidney. Probably the best-known and most commonly used cell line derived from normal human kidney is the HK-2 cell, which was developed in 1994 (Ryan et al., 1994). In discussing this model elsewhere (Lash et al., 2014b), we noted that despite validation of several PT functions, immortalization of the cells by viral transduction undoubtedly will cause changes in cellular function, particularly related to processes such as cell proliferation and stress response. Additionally, a recent study (Huang et al., 2015) compared the sensitivity and consistency of biomarker responses to a battery of well-established nephrotoxicants of HK-2 cells and primary cultures of human proximal tubular (hPT) cells. The authors reported that HK-2 cells are neither sensitive nor accurate in indicating the occurrence of nephrotoxicity due to these agents, whereas the hPT primary cell cultures appear to more closely mimic what occurs *in vivo*.

As an example of the availability and use of fresh human kidneys for research, the following describes our experience in obtaining such organs over a period of a little more than 7 years (previously described in Lash et al. (2014b)). From January 2001 through March 2008, we received through material transfer agreements a total of 119 human kidneys, most recently from the International Institute for the Advancement of Medicine (IIAM, Edison, NJ; www.iiam. org). Donor demographics were as follows: 58 males (mean age $\pm$ SEM = 57.5 $\pm$ 1.9; range = 30–81) and 61 females (mean age $\pm$ SEM = 60.4 $\pm$ 1.3; range = 34–80). Of the 119 kidney donors, 10 were African Americans, 2 were Asians, 99 were Caucasians, 6 were Hispanics, 1 was Native American, and 1 was a Pacific Islander. Kidneys were perfused with Wisconsin or similar medium, kept on wet ice, and delivered to the laboratory within 15–24 h of being taken from the donor. Kidneys were certified as normal (i.e., nonneoplastic) by a pathologist and were rejected for transplant for reasons (e.g., excessive glomerular sclerosis, arterial plaque) that do not affect our ability to isolate viable PT cells.

Limitations in the number of accepted kidneys over the aforementioned time period were due to both cost and need. Because an average human kidney yields 500 million to 1 billion PT cells, this is typically more than sufficient for most sets of experiments, limiting the need for additional tissue. With regard to expense, although each kidney costs $1000–1200 (for purchase and delivery), this is not out of line considering that the material is from humans and provides 10- to 20-fold more cellular material than a typical pair of kidneys from an adult rat.

One alternative experimental model for human kidney cells that has been receiving increasing attention involves human stem cell-based approaches to generate PT-like cells (Narayanan et al., 2013; Xia et al., 2013; Kang and Han, 2014; Lam et al., 2014; Takasato et al., 2014). Although the primary focus of studies with these approaches has been on renal morphogenesis and development, human induced pluripotent stem cells have been developed and used to demonstrate protection from ischemia–reperfusion and some forms of drug-induced injury (Narayanan et al., 2013; Tiong et al., 2014). Human induced pluripotent stem cells are preferred by some investigators because of the safety and ethical concerns with the use of human embryonic stem cells. Further development of these models and validation of endpoints used to predict renal damage are still needed but may lead to their more extensive application in the study of chemically induced nephrotoxicity and drug design.

This chapter will focus on the use of hPT primary cell cultures as an *in vitro* experimental model for drug design, development, and toxicity screening. After briefly outlining the types of biological and molecular processes that are typically assessed in most toxicology and drug development studies, the remaining sections will focus on methods for hPT cell isolation and culture, validation, some specific applications of hPT primary cell cultures that are not readily feasible with other models, some specific examples of how these cells have been used to study drug-induced cytotoxicity, and a summary of key processes that are part of drug design and development that need to be quantified and how this can be accomplished with hPT primary cell cultures.

10.2 BIOLOGICAL PROCESSES AND TOXIC RESPONSES OF THE KIDNEYS THAT ARE NORMALLY MEASURED IN TOXICOLOGY RESEARCH AND DRUG DEVELOPMENT STUDIES

In the course of choosing and developing an *in vitro* experimental model for drug design and discovery studies, it is critical to identify the responses and processes that need to be determined. In this manner, the proper model can be chosen based on validation of such responses and processes. Although any *in vitro* model has limitations and can only mimic or reflect processes that occur *in vivo* to a certain extent, the responses and processes that are needed for drug design and discovery need to be relevant to those that occur *in vivo*. A summary of these parameters and comparison between how these can or cannot be measured both *in vivo* and *in vitro* are presented in Table 10.2.

One thing that is clear from the table is that *in vivo* assessments of renal function in humans are limited due to the need for them to be noninvasive. In contrast, *in vitro* cell culture models enable processes to be examined at the cellular and molecular levels where they occur. Additionally, many of the responses and processes that are measured *in vivo* can be similarly measured in the *in vitro* model. Notable among these is the measurement of sensitive biomarkers of renal injury in urine in the *in vivo* model and in the extracellular medium in the *in vitro* cell culture model. Kidney injury molecule-1 (Kim-1) is an excellent example that has been characterized by Bonventre and colleagues (Ichimura et al., 1998, 2004; Han et al., 2002; Vaidya et al., 2006; Hoffmann et al., 2010). Kim-1 is a type 1 transmembrane protein that is undetectable in normal kidney tissue but is expressed at very high levels in dedifferentiated PT epithelial cells of human and rodent kidneys after either ischemic or chemically induced injury. It appears to satisfy several of the criteria for being an ideal biomarker of effect for renal injury: Kim-1 is stable in urine for prolonged periods of time, it is specific to the kidneys, its expression increases markedly from a baseline of essentially zero, and its increased expression occurs early

TABLE 10.2   Biological Processes and Toxic Responses of the Kidneys that Are Normally Measured in Toxicology Research and Drug Development Studies: Comparison between *In Vivo* or Whole Kidney and *In Vitro* Proximal Tubule Cell Culture Models

Response/Process	*In Vivo* or Whole Kidney	*In Vitro* PT Cell Culture
Glomerular filtration	eGFR measurements	Not relevant
Urinary biomarkers	Urine: SCr, BUN, Kim-1, CysC, NGAL, miRNAs, etc.	Extracellular medium: Kim-1, CysC, NGAL, miRNAs, etc.
Transepithelial transport: Secretion or reabsorption	Lumen-to-cell-to-plasma or plasma-to-cell-to-lumen fluxes	Growth on Transwell filters or similar matrix; bidirectional transepithelial flux
Specific renal transport processes	Not readily measurable	OATs, OCTs, D-glucose, MRPs, P-gp
Renal drug metabolism	Difficult to specifically measure	Easy to measure
Mechanistic studies	Limited	Extensive

BUN, blood urea nitrogen; CysC, cystatin C; eGFR, estimated glomerular filtration rate; Kim-1, kidney injury molecule-1; MRP, multidrug resistance protein; miRNA, micro-RNA; NGAL, neutrophil gelatinase-associated lipocalin; OAT, organic anion transporter; OCT, organic cation transporter; P-gp, P-glycoprotein.

in the pathologic or toxic event, thus indicating a high degree of sensitivity. Both protein and mRNA expressions are increased in injured renal tissue.

10.3   PRIMARY CULTURES OF hPT CELLS

10.3.1   Methods for hPT Cell Isolation

hPT cells are isolated from whole human kidneys by mincing renal cortical tissue, digestion with collagenase type I, filtration through nylon sieves, and use of sterile conditions (Todd et al., 1995; Cummings and Lash, 2000; Cummings et al., 2000a). All buffers are continuously bubbled with 95% O_2/5% CO_2 and maintained at 37°C. Kidney cortex is sliced, washed with sterile PBS, and minced, and the pieces are placed in a trypsinization flask filled with 40 ml of sterile, filtered Hanks' buffer, containing 25-mM $NaHCO_3$, 25-mM HEPES, pH 7.4, 0.5-mM EGTA, 0.2% (w/v) bovine serum albumin, 50-µg/ml gentamicin, 1.3-mg/ml collagenase type I, and 0.59-mg/ml $CaCl_2$, which was filtered prior to use. Minced cortical pieces are subjected to collagenase digestion for 60 min; after which the supernatant is filtered through a 70-µm mesh filter to remove tissue fragments, centrifuged at 150 × g for 7 min, and the pellet resuspended in 1:1 Dulbecco's Modified Eagle's Medium/Ham's F12 (DMEM:F12). Yield is typically ~5 to 7 × 10⁶ cells per 1 g of human kidney cortex. Basal medium (serum-free, hormonally defined) is DMEM:F12 (1:1), and supplements include HEPES buffer (15 mM, pH 7.4), $NaHCO_3$ (20 mM), antibiotics (only through day 3 of culture; 192-IU penicillin/ml + 200-µg streptomycin sulfate/ml + 2.5-µg amphotericin B/ml), insulin (5 µg/ml = 0.87 µM), transferrin (5 µg/ml = 66 nM), sodium selenite (30 nM), hydrocortisone (100 ng/ml = 0.28 µM), epidermal growth factor (100 ng/ml = 17 nM), and 3,3′,5-triiodo-DL-thyronine (T_3) (7.5 pg/ml = 111 nM) (Cummings et al., 2000a; Lash et al., 2001, 2003, 2005).

Although these hPT cells are predominantly (≥85%) of PT origin, a small degree of contamination with other epithelial cell populations is unavoidable and, therefore, will exist. While our laboratory has typically not performed additional purification procedures with the hPT cells, options are available to further enrich the PT cell preparation. For example, density gradient centrifugation of the renal cortical cells on Percoll can be performed to obtain PT cells that are >95% of PT origin and distal tubular (DT) cells that have minimal contamination with cells of the PT region, as we have done with cortical cells from rat kidneys (Lash and Tokarz, 1989; Lash et al., 1995).

Despite some obvious advantages of using primary cultures of hPT cells, which are derived directly from human kidneys, several cautions should be noted. First, maintenance of sterility is critical to prevent contamination from bacteria. Second, use of a serum-free, hormonally defined medium is critical to maintain epithelial cell purity; whereas the typical culture medium used for established cell lines includes serum to optimize growth; its inclusion here would lead to overgrowth by the small number of fibroblasts. Finally, although the cell cultures are directly derived from PT cells and thus initially reflect all the biochemical and physiological properties of that cell population *in vivo*, it is well documented that growth of cells in culture, particularly epithelial cells, can lead to a gradual or sometimes abrupt loss of certain differentiated properties. Loss of differentiated properties may include decreased expression and/or functionality of plasma membrane transporters, decreased expression of drug metabolism enzymes, loss of BBM, and decreased mitochondrial function leading to cellular energy metabolism becoming primarily glycolytic rather than oxidative phosphorylation. Hence, it is critical to validate that key properties and functions of the hPT cell as they exist *in vivo* are maintained in primary culture.

10.3.2 Validation of hPT Primary Cell Cultures

Validation is a critical step in establishing a model as a useful tool for testing and predictions. A summary of some properties and functions that are determined as part of validation of the *in vitro* model is shown in Table 10.3. Properties that need to be assessed and validated include morphology, drug transport and metabolism, cellular energetics, and redox status. Other properties or functions, such as expression of hormone receptors, synthesis of specific gene products (e.g., erythropoietin), or response to specific hormones, may also be assessed if required for a specific experimental design. An additional, important consideration that can only be assessed with human tissue is that genetic polymorphisms exist for several key plasma membrane transporters (Lash et al., 2006) and drug metabolism enzymes (Lash et al., 2003, 2008). For example, detected levels of protein expression in hPT cells of organic anion transporters (OATs) such as OAT1 and OAT3, P-glycoprotein (P-gp or MDR1), CYPs, and flavin-containing monooxygenases (FMOs) will likely vary considerably among tissue samples from different donors. Accordingly, the appropriate validation strategy is to establish maintenance of expression levels over the course of primary culture rather than quantifying specified levels of expression.

All epithelial cells in culture grow with their luminal membrane facing upward and their basolateral membrane facing downward. To effectively measure transepithelial drug transport or transport across the BLM, the hPT cells need to be grown on a matrix that enables maintenance of epithelial polarity. An example of such a matrix is Transwell® filter plates. These plates contain compartments with raised, semipermeable membranes on which the cells attach. As epithelial cells grow with their BBM facing upward, the upper compartment of the Transwell filter plate is analogous to the tubular lumen so that addition of substrate to this compartment allows for access to the BBM. In contrast, the basolateral surface faces downward so that the lower compartment of the Transwell filter plate is analogous to the renal interstitial space and addition of substrate to this compartment allows for access to the BLM. Morphology of control hPT cells grown for 5 days is shown in Fig. 10.1. The cells exhibit the expected cuboidal shape with a single, large nucleus and numerous mitochondria.

10.3.3 Advantages and Limitations of hPT Primary Cell Cultures

The most significant advantage of using primary cultures of hPT cells as a model to screen and test drugs is that they are derived directly from human kidneys. As discussed earlier, many drugs and chemicals exhibit species-specific effects, making nonhuman systems not always accurate as an experimental model. Although many have suggested that *in vitro* models from other primates may be suitable surrogates for those from humans, our previous study (Cai et al., 2009) using primary cultures of PT cells from cynomolgus monkeys showed differences in sensitivity and patterns of response to

TABLE 10.3 Validation of Primary Cultures of hPT Cells

Property	Validated Characteristics
Morphology	1. Cuboidal shape 2. Single nucleus 3. Numerous mitochondria

Drug transport
OATs, OATPs, OCTs, P-gp,
MRPs, amino acid
transporters
	1. Transport activity
	2. Transporter expression
	  a. Total expression
	  b. Subcellular localization

Drug metabolism Phase I: CYPs, FMOs Phase II: GSTs, UGTs, SULTs	1. Activity 2. Enzyme expression
Cellular energetics	1. Mitochondrial function 2. Glycolysis rate 3. Gluconeogenesis rate 4. Adenine nucleotide charge
Redox status	1. Glutathione status 2. Lipid peroxidation

Properties of hPT cells validated in previously published studies (Cummings and Lash, 2000; Cummings et al., 2000a; Lash et al., 2001, 2003, 2005, 2007, 2008; Xu et al., 2008).

CYP, cytochrome P450; FMO, flavin-containing monooxygenase; GST, glutathione *S*-transferase; MRP, multidrug resistance-associated protein; OAT, organic anion transporter; OATP, organic anion transporting polypeptide; OCT, organic cation transporter; P-gp, P-glycoprotein; SULT, sulfotransferase; UGT, UDP-glucuronosyltransferase.

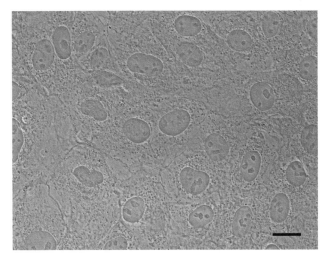

FIGURE 10.1 Morphology of hPT cells. Cells were cultured for 5 days in supplemented, serum-free, hormonally defined DMEM:F12 medium on 35 mm polystyrene dishes coated with Vitrogen (a proprietary collagen-containing mixture) until confluence. Photomicrographs were taken at 100× magnification on a Carl Zeiss confocal laser scanning microscope. Bar = 5 μm.

a series of metalloproteinase inhibitors that are nephrotoxic in rats. Other advantages include the ability to readily manipulate incubation conditions with the primary cell cultures to readily screen for therapeutic efficacy or toxicity and probe mechanisms of action.

Potential limitations include the tendency for primary cells in culture to dedifferentiate and lose functional expression of enzymes and membrane transporters during the course of culture. For example, expression of enzymes important for drug metabolism, such as CYPs, tends to readily decrease if culture conditions are not optimized. Plasma membrane transporters, such as many of the OATs, can readily internalize and become nonfunctional during the course of culture (Lash et al., 2006). Our strategy to counteract these potential problems has been to use the serum-free, hormonally defined medium that has become standard with primary culture of epithelial cells and then add other supplements to optimize maintenance of differentiated function (Lash et al., 1995; Lash, 2012).

Another potential limitation for mechanistic studies is that primary cell cultures are often difficult to transfect in comparison with immortalized cell lines. Overexpression of genes of interest or inhibition of gene expression with small-interfering RNAs (siRNAs) is an important tool for studying the function of specific proteins in a process of interest. We recently used Lipofectamine™ to transfect primary cultures of PT cells from uninephrectomized rats with cDNAs for the two mitochondrial transporters that can catalyze uptake of reduced GSH (Benipal and Lash, 2013). Expression of cDNAs for the 2-oxoglutarate carrier (*Slc25a11*) and dicarboxylate carrier (*Slc25a10*) was confirmed by real-time polymerase chain reaction. In both cases, overexpression of the two cDNAs was associated with marked increases in accumulation of GSH in the mitochondria. This experimental approach would be likely to work equally as well in primary cultures of hPT cells, although that must be directly confirmed.

10.3.4 Genetic Polymorphisms and Interindividual Susceptibility

A major difference in drug metabolism and transport studies in humans and human tissues as compared to those in experimental animals or tissues from those animals is the existence of genetic polymorphisms in humans. Such genetic polymorphisms can result in markedly different abilities to metabolize or transport drugs, subsequently resulting in altered therapeutic efficacy and/or susceptibility to toxic side effects. Although one can model such differences in experimental animals by manipulating expression of specific genes, this presumes knowledge of a specific polymorphism. In reality, numerous polymorphisms are continually being discovered so that studies in other species or tissue from other species cannot reveal such novel genetic differences.

Even with relatively limited numbers of samples (generally <10), we have detected marked variations in levels of protein expression in hPT cells for glutathione *S*-transferase alpha (GSTA) and pi (GSTP) and CYP4A11 (Cummings et al., 2000a), the FMO3 and FMO5 (Krause et al., 2003), CYP3A4 (Lash et al., 2008), and OAT1, OAT3, and P-gp (Lash et al., 2006). Hence, primary hPT cell cultures can be used to investigate the impact of several different types of polymorphisms on responses to chemical or drug exposures.

10.4 TOXICOLOGY STUDIES IN hPT PRIMARY CELL CULTURES

A wide array of mechanisms for causing cell injury and cell death by chemical or drug exposures can be studied in primary cultures of hPT cells. Figure 10.2 illustrates general responses of renal PT cells to nephrotoxic drugs. In most cases, membrane transport, intracellular metabolism to generate a reactive intermediate, and interactions with cellular macromolecules are prerequisites for cytotoxicity. The specific response to the chemical exposure can range from various forms of cytotoxicity to altered cellular regulation and is usually determined by exposure dose and time.

Responses of primary cultures of hPT cells to the nephrotoxic metabolite of TCE *S*-(1,2-dichlorovinyl)-L-cysteine (DCVC) provide an illustration of this principle. We showed that primary cultures of hPT cells incubated with DCVC undergo necrotic cell death at relatively high concentrations (≥100 μM) and long incubation times (≥24 h), whereas incubation with relatively low concentrations (<100 μM) for relatively short periods of time (≤8 h) can result in apoptosis (Lash et al., 2001). An increase in the proportion of hPT cells in S-phase (as assessed by flow cytometry) and rates of DNA synthesis (as assessed by incorporation of [³H]-dTTP) also occurred with DCVC treatment and followed similar patterns as apoptosis. This suggests that such low-dose, early responses can only occur when the hPT cell population still has adequate energy and viability.

More subtle responses, such as changes in protein and gene expression, are also observed in hPT cell cultures incubated with low micromolar concentrations of DCVC for up to 72 h (Lash et al., 2005; Lock et al., 2006). Proteins and genes identified as being significantly upregulated or downregulated include those involved in stress response and regulation of cell growth, proliferation, and apoptosis. Sublethal injury of renal proximal tubules is typically followed by repair and regeneration. Newly synthesized renal cells are not fully differentiated but must undergo a program of redifferentiation. This can be followed by immunofluorescence staining for cell surface markers such as cytokeratins (epithelial marker) and vimentin (endothelial marker). As an example, primary cultures of rat PT cells incubated for 72 h with 10-μM DCVC exhibited expression of vimentin, indicating the presence of

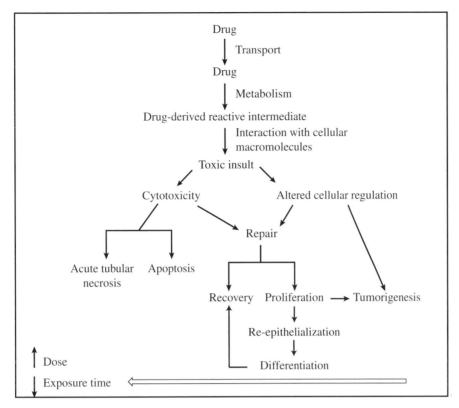

FIGURE 10.2   Scheme of responses of hPT cells to exposure to a nephrotoxic drug.

dedifferentiated cells (Cummings et al., 2000b). In contrast, control rat PT cells were negative for vimentin. Although similar measurements have not been performed with hPT cells, the same experimental approaches can be used to demonstrate both the presence of dedifferentiated cells and repair in PT cells derived from human kidneys.

Mitochondrial function, particularly in the proximal tubules, has long been known to be critical for renal function due to the high activity and number of ATP-requiring processes (Soltoff, 1986). Events in mitochondria are increasingly recognized as central to the etiology of numerous diseases and pathological states (Wallace, 2005; Pieczenik and Neustadt, 2007; Meyer et al., 2013), including those involving the kidneys (Hariri, 2009). Additionally, many chemicals, including environmental contaminants such as TCE, produce tissue injury that involves mitochondrial dysfunction as a key and early step in the mechanism of action. Mitochondria in primary cultures of hPT cells were demonstrated to be early targets for reactive metabolites generated from DCVC metabolism, and mitochondrial dysfunction was shown to be an obligatory step that precedes the diverse responses to DCVC or its sulfoxide metabolite (Lash et al., 2003; Xu et al., 2008).

Numerous methods are available to identify and quantify the extent of mitochondrial dysfunction in hPT cells, such as measurements of ATP concentrations by colorimetric or luminescence-based assays, rates of substrate-dependent oxygen consumption as measured by polarography with an oxygen-sensitive electrode, membrane potential with a fluorescent dye such as 5,5′,6,6′-tetrachloro-1,1′,3, 3′-tetraethyl-benzoimadazolylcarbocyanine iodide (JC-1), or activities of key mitochondrial enzymes such as aconitase.

An important research area of both nephrology and toxicology in general is that of biomarker discovery. Biomarkers are biological molecules that when changed in amount can be used as an indicator of some biological, pathological, or toxicological process. For these to be most useful, particularly in clinical settings in patients, biomarkers must be readily measurable in urine or blood. These biomarkers may be divided into three categories: (i) biomarkers of exposure, (ii) biomarkers of effect, and (iii) biomarkers of susceptibility. The last classification is perhaps the most valuable for human health risk assessment but is the most difficult to establish. The first two may be related or may be distinct, depending on the marker. A biomarker of exposure could be a metabolite of the drug or chemical of interest or may be a parameter that changes upon exposure to the chemical or drug. To be most useful, the biomarker of exposure should be sensitive, occur at measurable levels, and be detectable as early as possible after exposure has occurred or begun. The related biomarker of effect, in contrast, is typically not a metabolite of the chemical or drug but is some parameter that changes in response to the chemical or drug producing an adverse effect in the kidneys.

As described previously, a panel of biomarkers of effect, including Kim-1, CysC, and NGAL, has recently been approved by the US Food and Drug Administration for detection of adverse effects on kidney function (Hoffmann et al., 2010). Kim-1 is an excellent case in point and has been used in primary cultures of hPT cells to detect cytotoxicity (Huang et al., 2015). Thus, these highly sensitive and renal-specific biomarkers can be monitored in a preclinical model that is highly relevant to humans.

10.5 CRITICAL STUDIES FOR DRUG DISCOVERY IN hPT PRIMARY CELL CULTURES

This section briefly summarizes some key parameters or processes that are needed to properly use hPT primary cell cultures to pursue drug discovery and screening studies.

10.5.1 Phase I and Phase II Drug Metabolism

For primary cultures of hPT cells to be an appropriate and useful experimental model for drug discovery, it is necessary that the cells exhibit functional expression of Phase I and Phase II drug metabolism enzymes. Although the liver is quantitatively the major site of drug metabolism in the body, adverse effects of some drugs are linked specifically to renal metabolism (Lohr et al., 1998). Accordingly, determination of expression (both mRNA and protein) and activity for relevant enzymes is critical. As noted previously, key enzymes include CYPs, FMOs, GSTs, UGTs, and SULTs.

10.5.2 Membrane Transport

10.5.2.1 Organic Anion and Organic Cation Transport Pathways
Because most drugs of interest are either organic anions or organic cations, the various carriers that exist on the BLM and BBM of PT cells must be shown to be functional. As noted earlier, for many of these carriers that mediate drug uptake into the PT cell (e.g., OATs, OCTs, OATPs), it is not sufficient to demonstrate expression of protein in a total cell extract because the carriers may become internalized into vesicles and thus not be functional in membrane transport. Similarly, function of appropriate MRPs and P-gp in drug efflux is important to demonstrate so that the complete handling of the drug, including transepithelial transport, secretion, and reabsorption, is functional in the experimental model.

10.5.2.2 Experimental Models to Assess Transport and Transport Interactions between Different Drugs
hPT cells, like other epithelial cells, are polarized, which means that they grow with a distinct sidedness, that the two plasma membrane surfaces are physically separated by tight junctions and that each membrane surface has a distinctive composition of proteins including transporters. Although growth of hPT cells on solid surfaces (i.e., dishes, flasks, or multiwell plates) seems to be appropriate for many purposes and is a very common method, cells grown in this manner, while still polarized, have only their BBMs accessible to the extracellular space. To better mimic the *in vivo* polarization of their epithelial cells and assess the function of transporters on both plasma membrane surfaces, hPT cells need to be seeded and grown on a semipermeable membrane such as that provided by Transwell plates or Millicell filter inserts (Lash et al., 2006). Although cell growth on such a surface tends to be slower than that on a collagen-coated, solid surface, it is needed to have media access to the BLM.

10.5.2.3 Implications of Transport Interactions for Drug Toxicity and Therapeutic Efficacy
Uses of hPT primary cell cultures for characterizing transport of drugs and assessing potential drug–drug interactions (DDIs) at the level of competition for membrane transport are important applications of this model for drug design. Many drugs are taken off the market because of renal transport issues, which can include (i) competition with endogenous substrates, leading to disturbances in their distribution; (ii) competition with other, commonly used drugs, leading to DDIs; and (iii) high-activity or high-affinity transport leading to accumulation in the PT cells, intracellular bioactivation, and toxicity. hPT cells in primary culture are a model system that can test preclinically for these issues and avoid costly problems later.

The membrane transport data one obtains has several implications. Besides facilitating excretion or reabsorption of a drug, renal plasma membrane transport may also function to deliver drugs to bioactivation enzymes within the PT cells. Because of genetic polymorphisms in several membrane transporters or other regulatory differences, transporter function may vary, leading to differential ability to reabsorb and/or excrete both endogenous substrates (e.g., steroid hormones, bile acids, nucleotides) and xenobiotic substrates. For therapeutic agents, these differences can lead to altered drug half-life, altered efficacy, and potential toxicity.

10.6 SUMMARY AND CONCLUSIONS

10.6.1 Advantages and Limitations of Performing Studies in hPT Primary Cell Cultures

A broad array of assays can be conducted with hPT cells in primary culture that can be applied for drug design and screening as well as investigation of mechanism of action (Table 10.4). For many of these assays, commercial kits are available and assays can be conducted with cells grown on 24- or 96-well plates, thus making them somewhat high throughput and, therefore, suitable for screening a large number of derivatives and analogues.

TABLE 10.4 Panel of Assays that Can Be Conducted in Primary Cultures of hPT Cells for Drug Design and Screening and Investigation of Mechanism of Action

Parameter Measured or Assessed	Assay Methods
Morphology	1. Confocal microscopy 2. Phase-contrast microscopy
Cell viability and proliferation	1. Lactate dehydrogenase (LDH) release (necrosis) 2. 3-(4,5-dimethylthiazol-2-yl)-2,5-diphen-yltetrazolium bromide (MTT) fluorescence (viability and proliferation) 3. Propidium iodide/annexin staining, flow cytometry, and flow-activated cell sorting (FACS) analysis (necrosis, apoptosis, cell cycle status) 4. DNA and protein synthesis with radiolabeled precursors (cell death, proliferation)
Key metabolites	1. ATP (energetics) 2. Glutathione status (redox status) 3. Pyridine nucleotides (redox status)
Mitochondrial function	1. Membrane potential with fluorescent dyes 2. Oxygen consumption
Oxidative and nitrosative stress	1. Reactive oxygen species-sensitive fluorescent dyes 2. Aconitase inhibition 3. Thiol–disulfide status 4. Lipid peroxidation assays (e.g., thiobarbituric acid-reactive substance)
Membrane transport function	1. Organic anion/cation transport by OATs, OATPs, MRPs, MDR1 (P-gp) 2. D-Glucose and L-amino acid transport
Genetic effects	1. Changes in gene or protein expression 2. DNA damage
Biomarker release	1. Assays of enzymes or proteins in extracellular medium

MDR1 (P-gp), multidrug resistance protein 1 (P-glycoprotein); OAT, organic anion transporter; OATP, organic anion transporting polypeptide.

An important application for the use of hPT cells that is relevant for drug screening and development is the identification of DDIs. DDIs can be based on competition for or inhibition or induction of metabolism or membrane transport. Because of species-dependent differences in these processes, examination of DDIs in rodent models may not provide information that is readily applicable to humans. Another important application that cannot be readily modeled in nonhuman renal models is that of studying the influence of genetic polymorphisms on drug disposition and action. Such polymorphisms are known to exist for both drug metabolism enzymes such as CYPs and FMOs (Lash et al., 2008) and for membrane transporters such as the OATs and P-gp (Lash et al., 2006).

Potential limitations to the use of primary cultures of hPT cells include the need to consider the potential for genetic polymorphisms that may impact the pharmacokinetics and pharmacodynamics for a given drug under study. The other major limitation or caution is the tendency for the cellular phenotype to change during the course of culture, making the *in vitro* model no longer an accurate reflection of the *in vivo* PT cell. Both of these issues can be taken into account, but these are by no means simple problems to control in an experimental model.

In summary, primary cultures of hPT cells represent a unique and highly relevant experimental model for the study of drug metabolism, transport, mechanism of action, and nephrotoxicity. Although there exist the usual cautions with primary cell culture models, and cell culture models in general, once technique is mastered and appropriate conditions are used to minimize potential issues with the culture process, the primary hPT cells have numerous advantages over other models for obtaining information about drug action that is directly relevant to humans.

10.6.2 Future Directions

hPT cells will grow to confluence in 5–7 days, depending on the growth surface and on whether or not tissue culture dishes are first coated with collagen. While it is possible to passage hPT cells to study more complex responses that require longer periods of time, caution must be exercised in using such approaches because of the previously discussed problems with dedifferentiation that often occur with extended cell culture. We have recently addressed this possible experimental approach by growing hPT cells and passaging them for up to four generations and suggest that at least through two passages (equivalent to three generations), hPT cells represent a valid experimental model to study processes or responses that require a longer time frame than is possible with just cells in primary culture (i.e., 30–40 days vs. 5–7 days) (Lash et al., 2014b).

Although clear differences in growth rate and expression of some key regulatory proteins were observed with increasing passage number, several lines of evidence suggested that hPT cells, at least through the second passage, are a viable *in vitro* model to study cellular and molecular responses of the human kidney to chemical and other stresses that require a longer exposure or response time than is possible during the 5- to 7-day course of primary culture. Key lines of evidence for the validity of the passaged cells included maintenance of epithelial morphology, positive cytokeratin staining, and maintenance of redox status through multiple generations. Cellular proliferation rate was within 50% of that of primary cells through passage three. Beyond that stage, however, it is clear that cellular properties and responses to chemical and environmental stresses will likely exhibit qualitative differences from what are observed in primary cells.

Another important feature of any *in vitro* toxicological model and one that is certainly relevant to drug design is that responses to well-characterized chemical toxicants are qualitatively and quantitatively similar to what are observed in the *in vivo* target cell. Although some modest differences were observed, responses of primary cell cultures through second-passage cells were similar for both a specific nephrotoxicant (DCVC) and two model toxicants that act by either oxidation or alkylation of low molecular weight thiols such as GSH (*tert*-butyl hydroperoxide and methyl vinyl ketone). Thus, although additional parameters need to be studied and characterized, hPT cells from primary culture through at least the second passage appear to be a valid model for drug design and toxicology studies.

REFERENCES

Atlas of Chronic Kidney Disease in the United States. (2014) 2013 USRDS annual data report: Atlas of chronic kidney disease and end-stage renal disease in the United States. *Am. J. Kidney Dis.* 63 (Suppl. 1), A1–A7; e1–e478.

Benipal, B. and Lash, L.H. (2013) Modulation of mitochondrial glutathione status and cellular energetics in primary cultures of proximal tubular cells from remnant kidney of uninephrectomized rats. *Biochem. Pharmacol.* 85, 1379–1388.

Cai, H., Agrawal, A.K., Putt, D.A., Hashim, M., Reddy, A., Brodfuehrer, J., Surendran, N., and Lash, L.H. (2009) Assessment of the renal toxicity of novel anti-inflammatory compounds using cynomolgus monkey and human kidney cells. *Toxicology* 258, 56–63.

Cummings, B.S. and Lash, L.H. (2000) Metabolism and toxicity of trichloroethylene and *S*-(1,2-dichlorovinyl)-L-cysteine in freshly isolated human proximal tubular cells. *Toxicol. Sci.* 53, 458–466.

Cummings, B.S., Lasker, J.M., and Lash, L.H. (2000a) Expression of glutathione-dependent enzymes and cytochrome P450s in freshly isolated and primary cultures of proximal tubular cells from human kidney. *J. Pharmacol. Exp. Ther.* 293, 677–685.

Cummings, B.S., Zangar, R.C., Novak, R.F., and Lash, L.H. (2000b) Cytotoxicity of trichloroethylene and S-(1,2-dichlorovinyl)-L-cysteine in primary cultures of rat renal proximal tubular and distal tubular cells. *Toxicology* 150, 83–98.

Han, W.K., Bailly, V., Abichandani, R., Thadhani, R., and Bonventre, J.V. (2002) Kidney injury molecule-1 (KIM-1): A novel biomarker for human renal proximal tubule injury. *Kidney Int.* 62, 237–244.

Hariri, A. (2009) Mitochondrial diseases of the kidney. In: *Genetic Diseases of the Kidney* (Lifton, R.P., Somlo, S., Giebisch, G.H., and Seldin, D.W, Eds.), pp. 559–569, Elsevier, Amsterdam.

Hoffmann, D., Adler, M., Vaidya, V.S., Rached, E., Mulrane, L., Gallagher, W.M., Callanan, J.J., Gautier, J.C., Matheis, K., Staedtler, F., Dieterle, F., Brandenburg, A., Sposny, A., Hewitt, P., Ellinger-Ziegelbauer, H., Bonventre, J.V., Dekant, W.,

and Mally, A. (2010) Performance of novel kidney biomarkers in preclinical toxicity studies. *Toxicol. Sci.* 116, 8–22.

Huang, J.X., Kaelsin, G., Ranall, M.V., Blaskovich, M.A., Becker, B., Butler, M.S., Little, M.H., Lash, L.H., and Cooper, M.A. (2015) Evaluation of biomarkers for in vitro prediction of drug-induced nephrotoxicity: Comparison of HK-2, immortalized human proximal tubule epithelial, and primary cultures of human proximal tubular cells. *Pharmacol. Res. Perspect.*, 3, e00148.

Ichimura, T., Bonventre, J.V., Bailly, V., Wei, H., Hession, C.A., Cate, R.L., and Sanicola, M. (1998) Kidney injury molecule-1 (KIM-1), a putative epithelial cell adhesion molecule containing a novel immunoglobulin domain, is up-regulated in renal cells after injury. *J. Biol. Chem.* 273, 4135–4142.

Ichimura, T., Hung, C.C., Yang, S.A., Stevens, J.L., and Bonventre, J.V. (2004) Kidney injury molecule-1: A tissue and urinary biomarker for nephrotoxicant induced renal injury. *Am. J. Physiol.* 286, F552–F563.

Kang, M. and Han, Y.-M. (2014) Differentiation of human pluripotent stem cells into nephron progenitor cells in a serum and feeder free system. *PLoS One* 9, e94888.

Krause, R.J., Lash, L.H., and Elfarra, A.A. (2003) Human kidney flavin-containing monooxygenases (FMOs) and their potential roles in cysteine S-conjugate metabolism and nephrotoxicity. *J. Pharmacol. Exp. Ther.* 304, 185–191.

Lam, A.Q., Freedman, B.S., Morizane, R., Lerou, P.H., Valerius, M.T., and Bonventre, J.V. (2014) Rapid and efficient differentiation of human pluripotent stem cells into intermediate mesoderm that forms tubules expressing kidney proximal tubular markers. *J. Am. Soc. Nephrol.* 25, 1211–1225.

Lash, L.H. (1994) Role of renal metabolism in risk to toxic chemicals. *Environ. Health Perspect.* 102 (Suppl. 11), 75–79.

Lash, L.H. (2012) Human proximal tubular cells as an in vitro model for drug screening and mechanistic toxicology. AltTox Essay, http://alttox.org/ttrc/toxicity-tests/repeated-dose/way-forward/lash/ (accessed on October 12, 2015).

Lash, L.H. and Parker, J.C. (2001) Hepatic and renal toxicities associated with perchloroethylene. *Pharmacol. Rev.* 53, 177–208.

Lash, L.H. and Tokarz, J.J. (1989) Isolation of two distinct populations of cells from rat kidney cortex and their use in the study of chemical-induced toxicity. *Anal. Biochem.* 182, 271–279.

Lash, L.H., Tokarz, J.J., and Pegouske, D.M. (1995) Susceptibility of primary cultures of proximal tubular and distal tubular cells from rat kidney to chemically induced toxicity. *Toxicology* 103, 85–103.

Lash, L.H., Hueni, S.E., and Putt, D.A. (2001) Apoptosis, necrosis and cell proliferation induced by *S*-(1,2-dichlorovinyl)-L-cysteine in primary cultures of human proximal tubular cells. *Toxicol. Appl. Pharmacol.* 177, 1–16.

Lash, L.H., Putt, D.A., Hueni, S.E., Krause, R.J., and Elfarra, A.A. (2003) Roles of necrosis, apoptosis, and mitochondrial dysfunction in *S*-(1,2-dichlorovinyl)-L-cysteine sulfoxide-induced cytotoxicity in primary cultures of human renal proximal tubular cells. *J. Pharmacol. Exp. Ther.* 305, 1163–1172.

Lash, L.H., Putt, D.A., Hueni, S.E., and Horwitz, B.P. (2005) Molecular markers of trichloroethylene-induced toxicity in human kidney cells. *Toxicol. Appl. Pharmacol.* 206, 157–168.

Lash, L.H., Putt, D.A., and Cai, H. (2006) Membrane transport function in primary cultures of human proximal tubular cells. *Toxicology* 228, 200–218.

Lash, L.H., Putt, D.A., Hueni, S.E., Payton, S.G., and Zwickl, J. (2007) Interactive toxicity of inorganic mercury and trichloroethylene in rat and human proximal tubule: Effects on apoptosis, necrosis and role of glutathione status. *Toxicol. Appl. Pharmacol.* 221, 349–362.

Lash, L.H., Putt, D.A., and Cai, H. (2008) Drug metabolism enzyme expression and activity in primary cultures of human proximal tubular cells. *Toxicology* 244, 56–65.

Lash, L.H., Chiu, W.A., Guyton, K.Z., and Rusyn, I.I. (2014a) Trichloroethylene biotransformation and its role in mutagenicity, carcinogenicity and target organ toxicity Mutat. *Res. Rev.* 762, 22–36.

Lash, L.H., Putt, D.A., and Benipal, B. (2014b) Multigenerational study of chemically induced cytotoxicity and proliferation in cultures of human proximal tubular cells. *Int. J. Mol. Sci.* 15, 21348–21365.

Lock, E.A., Barth, J.L., Argraves, S.W., and Schnellmann, R.G. (2006) Changes in gene expression in human renal proximal tubule cells exposed to low concentrations of S-(1,2-dichlorovinyl)-L-cysteine, a metabolite of trichloroethylene. *Toxicol. Sci.* 216, 319–330.

Lohr, J.W., Willsky, G.R., and Acara, A. (1998) Renal drug metabolism. *Pharmacol. Rev.* 50, 107–141.

Meyer, J.N., Leung, M.C.K., Rooney, J.P., Sendoel, A., Hengartner, M.O., Kisby, G.E., and Bess, A.S. (2013) Mitochondria as a target of environmental toxicants. *Toxicol. Sci.* 134, 1–17.

Narayanan, K., Schumacher, K.M., Tasnim, F., Kandasamy, K., Schumacher, A., Ni, M., Gao, S.; Zink, D., and Ying, J.Y. (2013) Human embryonic stem cells differentiate into functional renal proximal tubular-like cells. *Kidney Int.* 83, 593–603.

Pelekis, M. and Krishnan, K. (1997) Assessing the relevance of rodent data on chemical interactions for health risk assessment purposes: A case study with dichloromethane-toluene mixture. *Regul. Toxicol. Pharmacol.* 25, 79–86.

Pieczenik, S.R. and Neustadt, J. (2007) Mitochondrial dysfunction and molecular pathways of disease. *Exp. Mol. Pathol.* 83, 84–92.

Ryan, M.J., Johnson, G., Kirk, J., Fuerstenberg, S.M., Zager, R.A., and Torok-Storb, B. (1994) HK-2: An immortalized proximal tubule epithelial cell line from normal adult human kidney. *Kidney Int.* 45, 48–57.

Soltoff, S.P. (1986) ATP and the regulation of renal cell function. *Annu. Rev. Physiol.* 48, 9–31.

Steinberg, A.D. and DeSesso, J.M. (1993) Have animal data been used inappropriately to estimate risks to humans from environmental trichloroethylene. *Regul. Toxicol. Pharmacol.* 18, 137–153.

Takasato, M., Er, P.X., Becroft, M., Vanslambrouck, J.M., Stanley, E.G., Elefanty, A.G., and Little, M.H. (2014) Directing human embryonic stem cell differentiation towards a renal lineage generates a self-organizing kidney. *Nat. Cell Biol.* 16, 118–126.

Tiong, H.Y., Huang, P., Xiong, P., Xiong, S., Li, Y., Vathsala, A., and Zink, D. (2014) Drug-induced nephrotoxicity: Clinical impact and preclinical in vitro models. *Mol. Pharm.* 11, 1933–1948.

Todd, J.H., McMartin, K.E., and Sens, D.A. (1995) Enzymatic isolation and serum-free culture of human renal cells. In: *Methods in Molecular Medicine: Human Cell Culture Protocols* (Jones, G.E., Ed.), pp. 431–435, Humana Press, Totowa, NJ.

Vaidya, V.S., Ramirez, V., Ichimura, T., Bobadilla, N.A., and Bonventre, J.V. (2006) Urinary kidney injury molecule-1: A sensitive quantitative biomarker for early detection of kidney tubular injury. *Am. J. Physiol.* 290, F517–F529.

Wallace, D.C. (2005) A mitochondrial paradigm of metabolic and degenerative diseases, aging, and cancer: A dawn for evolutionary medicine. *Annu. Rev. Genet.* 39, 359–407.

Xia, Y., Nivet, E., Sancho-Martinez, I., Gallegos, T., Suzuki, K., Okamura, D.; Wu, M.-Z., Dubova, I., Esteban, C.R., Montserrat, N., Campistol, J.M., and Izpisua Belmonte, J.C. (2013) Directed differentiation of human pluripotent cells to ureteric bud kidney progenitor-like cells. *Nat. Cell Biol.* 15, 1507–1515.

Xu, F., Papanayotou, I., Putt, D.A., Wang, J., and Lash, L.H. (2008) Role of mitochondrial dysfunction in cellular responses to S-(1,2-dichlorovinyl)-L-cysteine in primary cultures of human proximal tubular cells. *Biochem. Pharmacol.* 76, 552–567.

11

PREDICTING ORGAN TOXICITY *IN VITRO*: BONE MARROW

Ivan Rich[1] and Andrew J. Olaharski[2]

[1] *Hemogenix, Inc., Colorado Springs, CO, USA*

[2] *Toxicology, Agios Pharmaceuticals, Cambridge, MA, USA*

11.1 INTRODUCTION

Drug discovery toxicology is a step-wise process that identifies the safety profile of a therapeutic compound from target assessment to candidate nomination. Bone marrow toxicity has traditionally taken a backseat to other common and/or severe organ toxicities during this process. There are many reasons for this, the obvious being that bone marrow toxicity is commonly identified following exposures to anticancer, anti-inflammatory, and antiviral drugs and is often used as a surrogate pharmacodynamic biomarker when evaluating compound-related toxicity. Yet, it would be false to assume that only these drug classes have an effect on bone marrow cells. Bone marrow toxicity signals may lead to significant delays and, in some cases, the shuttering of a program. Even for oncology, where bone marrow toxicity is widely accepted, it is often a major limitation to performing full treatment protocols (Parchment et al., 1998; Pessina et al., 2003). Identifying hematopoietic toxicants and their potency earlier during the drug development process prior to multiple-day *in vivo* toxicology studies has the potential to save considerable time, effort, cost, and animal usage and improve the overall success of the project.

11.2 BIOLOGY OF THE HEMATOPOIETIC SYSTEM

Although the bone marrow is home to adult hematopoiesis, it is also home to another less understood system, namely, the mesenchymal stem or stromal cell (MSC) system (Delorme et al., 2006). These cells are also called multipotent mesenchymal progenitor cells. The MSC system is responsible, in part, for providing the hematopoietic microenvironment or stroma; an absolute requirement for normal hematopoiesis. The cells associated with the hematopoietic microenvironment are multifaceted and include macrophages, fibroblasts, endothelial cells, reticulum cells, and fat cells. Together, they provide the foundation for the stem cell niche. The MSC system is also responsible for adipogenesis, chondrogenesis, and osteogenesis. The association between the MSC system and hematopoiesis means that if one system is affected by toxicants, there is a high probability that the other will also be impacted.

Hematopoiesis is one of five (six if the MSC system is also included) continuously proliferating systems of the body, the others being the cells of the gastrointestinal tract, the reproductive systems, cells of the skin, and those in the axel of the eye. Cells from these organs and tissues have the capacity to continuously produce cells throughout the animal's lifetime. It is now known that definitive "stem cells," those that are committed to producing cells for a specific organ or tissue, can exist in virtually every organ and tissue and are responsible for the partial proliferation that can result in regeneration. One example would be liver regeneration (Than and Newsome, 2014). It is the presence and capability of the primary definitive stem cells that are responsible for this effect.

Primary definitive stem cells, as opposed to nondefinitive stem cells such as embryonic stem (ES) cells or induced

Drug Discovery Toxicology: From Target Assessment to Translational Biomarkers, First Edition. Edited by Yvonne Will, J. Eric McDuffie, Andrew J. Olaharski, and Brandon D. Jeffy.

pluripotent stem (iPS) cells, represent the rarest and most primitive cells in the body (Rich, 2015a). They are morphologically unidentifiable but exhibit the capacity for the greatest proliferation potential of all cell types. It is this potential for proliferation that delineates a stem cell from all other lineage-specific cells. The potential for proliferation gives the stem cell its specific characteristics. It is responsible for the ability of stem cells to self-renew. The higher the proliferation potential, the more primitive the stem cell population and the greater the self-renewal capacity. Stem cell primitiveness also directly correlates with inherent potency, which for cellular therapeutic or regenerative medicine products, also correlates with engraftment potential (Harper and Rich, 2013; Patterson et al., 2015;Rich, 2015a).

The different degrees of potential for proliferation by stem cells also cause it to be one of the most sensitive to the toxicity of drugs (Harper and Rich, 2013). Most stem cells are in a quiescent, nonproliferative state (Rich, 2015b); however, this does not mean that they are immune to potential toxicity. Small molecules may modulate quiescent cells, and for stem cells, this effect may not be evident until the cells reenter the cell cycle. Attenuations may also appear during cell development or differentiation as increased apoptosis or necrosis or other events that inhibit the system from establishing and/or maintaining homeostasis.

11.3 HEMOTOXICITY

Hemo- or hematotoxicity occurring in the bone marrow is often referred to as myelotoxicity, since the word myelo comes from the Greek, muelos, meaning "marrow." However, the term myelotoxicity is often confused with impairment to the granulocyte–macrophage (GM) lineage that causes neutropenia (Fig. 11.1). Indeed, many studies, including those performed by the European Center for the Validation of Alternative Methods (ECVAM) (Pessina et al., 2001, 2003, 2009) focused on drugs that produce myelosuppression. In addition, as part of the toxicology evaluation performed by the Toxicology and Pharmacology Branch (TPB) of the National Cancer Institute (NCI), drugs known to produce *in vivo* myelosuppression in humans, canines, and mice were retrospectively assessed *in vitro* using the *in vitro* colony-forming unit (CFU) or cell (CFC) assay specific for the GM lineage.

However, focusing just on the GM lineage is a rather myopic view of the bone marrow as a target to toxicity. Myelotoxicity has been extended to include toxicity to the megakaryopoietic lineage as a target for thrombocytopenia (Parent-Massin, 2001; Parent-Massin et al., 2010). In reality, the bone marrow is actually a target for multiple toxicities affecting multiple cells populations. These include primitive, proliferating cell populations as well as those produced

later in development. Therefore, it is insufficient to determine toxicity to one or two cell populations especially if these are (i) not the primary targets, and (ii) *in vivo* animal studies provide no information regarding these targets. Thus, *in vitro* platforms enable rapid, reliable, and cost-effective screening of drugs, compounds, and environmentally dangerous chemicals that could predict *in vivo* bone marrow toxicity prior to actually performing costly animal studies and, more importantly, before any human subjects are exposed to potentially harmful products. This implies that any *in vitro* technology used to predict toxicity must also demonstrate a high *in vivo* concordance (Olaharski et al., 2009).

In practice, such a high *in vitro* to *in vivo* concordance for myelotoxicity has been known for decades. The hematopoietic hierarchical organization (Fig. 11.1) and its regulation are, to a large part, based on animal and human perturbations (drugs, irradiation, etc.) that have been detected using *in vitro* methods such as the CFU/CFC assay. The term *hemotoxicity*, rather than myelotoxicity, will be used henceforth since the former implies that the bone marrow is the adult site for toxicity and encompasses all cells of the hematopoietic system.

11.4 MEASURING HEMOTOXICITY

The colony-forming assay was first introduced in 1966 for mouse bone marrow cells (Bradley and Metcalf, 1966; Pluznik and Sachs, 1966). It was later developed for human bone marrow cells in 1971 by Pike and Robinson. During this time, agar was used as the semisolid medium in which the cells were suspended to allow them to form colonies. Another commonality during this time was that all assays detected early progenitor cells of the GM lineage. However, also in 1971, Stephenson and colleagues developed a plasma clot assay to grow erythropoietic precursor cells designated as colony-forming unit–erythroid (CFU-E) colonies. Later, Iscove et al. (1974) developed the methylcellulose version of the CFU-E assay as well as the burst-forming unit–erythroid or BFU-E assay that detects erythropoietic progenitor cells at a similar level of development to the GM progenitors. The term colony-forming unit came from the early 1960s when it was not known whether assays detected a single cell or an aggregate (i.e., unit) of cells that formed a colony. Today, we know that that all hematopoietic colony-forming assays are clonal, meaning that each colony is derived from a single cell. The nomenclature has remained for some cell populations such as CFU-E and BFU-E, but it is now more correct to designate the colony as being derived from a CFC rather than a unit.

11.4.1 Uses of the CFC Assay

The methylcellulose version of the *in vitro* CFC assay is primarily used today for basic research, cellular therapy, and

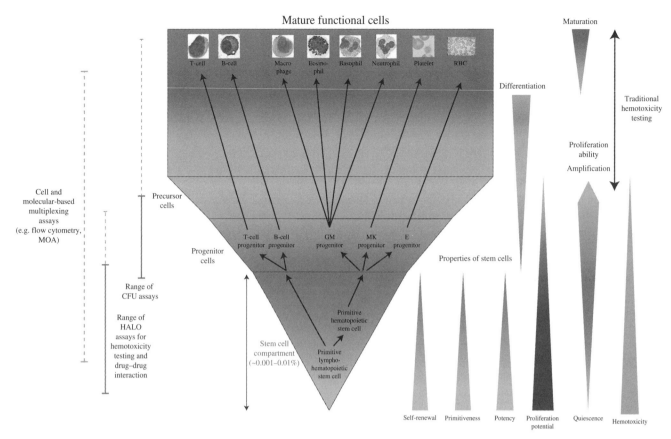

FIGURE 11.1 The organization of the hematopoietic system defines the type and severity of hemotoxicity. This diagram illustrates the organizational and hierarchical structure of the lymphohematopoietic system. This consists of a stem cell compartment that feeds into amplification compartments provided by the progenitor and precursor cells followed by the primary differentiation and maturation compartments to produce the mature functional cells. Stem cell hemotoxicity and its severity are defined by two measures of proliferation, namely, proliferation ability and proliferation potential or capacity. These, in turn, define the sensitivity of the stem cells to respond to different insults and perturbations. In general, the more primitive a stem cell, the greater its sensitivity to hemotoxic-inducing drug and compounds and the greater the severity on the system as a whole. (*See insert for color representation of the figure.*)

some toxicology applications. There are several aspects of the assay that require highlighting: Firstly, all hematopoietic stem, progenitor, and precursor cell cultures required the presence of growth factors and/or cytokines. With the exception of lineage-specific factors such as erythropoietin (EPO), the colony-stimulating factors (CSFs), and thrombopoietin (TPO), most hematopoietic growth factors and cytokines are pleiotropic. Without growth factors, primitive hematopoietic cells undergo apoptosis. Therefore, even for toxicity assays, growth factor cocktails are required to stimulate the cells to induce proliferation. Secondly, the number of colonies produced in a CFC cultures is determined by several factors including the concentration of the growth factors/cytokines used to stimulate a specific cell population, the total live cell count, and the compound concentration(s) employed. Usually the growth factor/cytokine concentrations used are optimal, meaning that the dose used produces a response on the upper plateau of the dose–response curve. However, in some cases, suboptimal concentrations are required. The cell numbers must also be optimized for a

specific experimental design. Finally, the compound dose–response must be selected to produce a range that will provide information on either the enhancing concentrations (EC) values or inhibitory concentrations (IC) values. All of these factors contribute to the specificity and sensitivity of the assay. Thirdly, although colony formation and, therefore, colony count require cell proliferation, it is false to assume that the CFC assay is a proliferation assay. Having said that, the size of the colony produced within a specific time period is an indication of the primitiveness of the cell producing the colony; the larger the colony, the more primitive the CFC. However, unless the number of cells can be counted and correlated with the size of the colony, the CFC assay cannot be used as a measure of cell proliferation. Instead, the CFC assay is used to determine differentiation and maturation ability. This is because the cells produced in the colony must differentiate and mature in order to identify the colony as being derived from a specific stem, progenitor, or precursor CFC. Therefore, the CFC assay should be used primarily to determine if a compound affects the differentiation and/or maturation processes. The types of colonies produced are dependent upon the growth factors/cytokines and their concentration. This allows the hematopoietic system to be manipulated so that the cells in different lineages can be analyzed. Very few agents affect both single lineage and only the differentiation and/or maturation process—the CFC assay can be used to investigate these processes.

11.4.2   *In Vitro/In Vivo* Concordance

Despite the limitations described in the next section, the CFC assay has played a defining role in elucidating the hierarchical structure and regulatory mechanisms governing the steady state and perturbed hematopoietic system. Investigators studying other biological systems look to the hematopoietic system and the assays used to dissect the system as a means to understand and unravel how other systems work. Although the *in vitro* CFC assay was developed in the 1960s, the use of drugs and other agents to investigate the response of the blood-forming system in different species started in earnest approximately 15 years ago. The effects seen in animals and humans could be mimicked *in vitro*, and many of the growth factors and cytokines used today were originally discovered using *in vitro* assays, which in turn, led to the discovery of specific cell populations and the organizational structure of the hematopoietic and lymphopoietic systems. Similar organizational structures are found in virtually every definitive stem cell system. Thus, *in vitro* assays such as the CFC assay can, in the majority of cases, be used to mimic the *in vivo* situation and therefore provide a high degree of concordance.

11.4.3   Limitations of the CFC Assay

The CFC assay was developed as a research tool. With the exception of incorporating recombinant growth factors and cytokines, the CFC assay has not changed since its inception. Although CCD cameras and software algorithms have been developed to count colonies, the assay relies on microscopic analysis and differentiation of colony types that is subjective and low throughput. In addition, there are no standards and controls that allow the assay to be validated according to regulatory requirements (FDA, 2001). As such, it is difficult to use the assay for GLP-compliant studies. As mentioned previously, the assay does not quantitatively measure cell proliferation and/or viability. Since many drugs and other agents target different aspects of the proliferation process, using the CFC assay provides only a retrospective indication of a response that would have occurred much earlier during cell development. This could lead to a false interpretation of results. The use of methylcellulose, a viscous semisolid medium in which the cells are suspended, together with the subjective nature of the readout, results in very high coefficients of variation that make it extremely difficult to perform interlaboratory comparisons. ECVAM has been the only

organization that ever attempted this systematically with some success. Umbilical cord blood banks that routinely use the assay as a growth/functionality assay have shown an abysmal record of variation with this assay (Spellman et al., 2011). Since methylcellulose cannot be accurately dispensed and liquid handlers are not available to dispense a viscous medium, high-throughput screening is impossible. Finally, the assay requires a steep learning curve, and proficiency in colony counting requires an excessive amount of time.

11.5   THE NEXT GENERATION OF ASSAYS

These shortcomings led several groups to consider a more robust, objective, instrument-based readout. The MTT (3-(4,5-dimethylthiazol-2-yl)-2,5,-diphenyltetrazolium bromide) colorimetric readout, based on the reduction of a tetrozolium substrate primarily by the mitochondria to a yellow formazan product, has been used for both clonal (CFU-equivalent) and liquid culture assays (Kriegler et al., 1987; Horowitz and King, 2000). Modification of the MTT assay versions laid the groundwork for the next generation of assays: A fully standardized and validated ATP bioluminescence assay platform with high-throughput capability (discussed in the following).

11.6   PROLIFERATION OR DIFFERENTIATION?

As discussed previously, one of the primary properties of stem cells is their ability to proliferate. Primitive progenitor cells have a higher rate of proliferation, which decreases as the differentiation process increases (Fig. 11.1). In contrast to stem cells, which are undifferentiated, progenitor cells have entered a specific differentiation lineage and progress to mature functional cells. Only when a stem cell crosses the "determination boundary" does it cease to be a stem cell. The processes of proliferation and differentiation, as alluded to previously, are often blurred because the processes overlap with each other. From an assay viewpoint, it is imperative to distinguish between these two biological processes because each requires a different readout. Using the incorrect readout can lead to a misinterpretation of results and false conclusions. If a drug targets the proliferation process, an assay that measures proliferation, not differentiation, must be used. In the bone marrow, both processes occur concomitantly in multiple cell types. It is therefore important to distinguish between the ability of, and potential for, proliferation and differentiation. The term "ability" or "status" is used to define a specific function of cells at a particular dose and point in time. For toxicity studies, this would be defined as the IC value. The term "potential" is used to define the capacity of cells to perform a specific function. As shown in Fig. 11.1, the proliferation ability and potential of different hematopoietic cell populations are inversely proportional to each other. It would therefore be expected that the proliferation potential of progenitor cells is less than that for stem cells, and within the stem cell compartment, the proliferation potential of primitive stem cells would be greater than that for mature stem cells. This is precisely what is found in practice (Fig. 11.2) when different hematopoietic cells populations are measured (Harper and Rich, 2013). Proliferation potential is measured by the slope of the cell dose–response linear regression curve fit; the steeper the slope, the more primitive the cell and the greater the proliferation potential. In this way, it is possible to define the cell population being measured.

11.7 MEASURING AND PREDICTING HEMOTOXICITY *IN VITRO*

Since the colony assay cannot be used to measure stem and progenitor cell proliferation parameters, a different readout was required. Originally, a methylcellulose version of the colony-forming assay was rebuilt to incorporate a adenosine triphosphate (ATP) bioluminescence readout for 96-well plates (Rich and Hall, 2005; Rich, 2007). This first-generation hematopoietic assay produced colonies that could be manually enumerated to determine cell differentiation followed by a second readout on the same cells in which the cells in the colonies were lysed to release the intracellular ATP

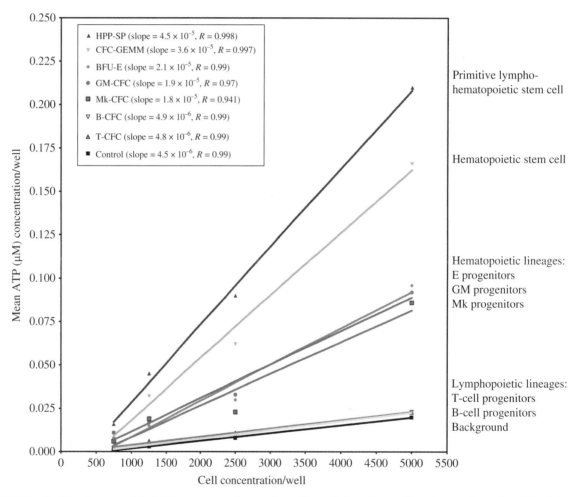

FIGURE 11.2 Proliferation potential of different stem and progenitor cell populations. Stem cells and progenitor cells within their respective compartments exhibit different degrees of proliferation potential, which is defined as the capacity of the cells to proliferate. Within the differentiation compartment of any biological system, the proliferation potential decreases as the cells differentiate and mature. Within the primitive stem and progenitor cell compartments, the degree of proliferation potential can be quantitatively measured by the slope of the cell population dose -response. As seen in this diagram, stem cells demonstrate a steeper slope than progenitor cells. This is to be expected since stem cells, by definition, have the greatest proliferation potential of all cells. However, within the stem cell compartment, primitive stem cell populations can be distinguished from more mature stem cell populations. Therefore, measurement of proliferation potential defines the degree of "stemness" or primitiveness of the cells. The properties of stem cells such as self-renewal capacity, primitiveness, and potency are all dependent on the proliferation potential. As a result, toxicity to different stem cell populations can also be dependent upon the stem cell' proliferation potential.

(iATP). The ATP acted as a limiting substrate for a luciferin/luciferase reaction to produce bioluminescence detected as light in a plate luminometer. As described previously, there is direct correlation between colony counts and the ATP concentration. This methylcellulose ATP assay (now called CAMEO-96) was validated against the Registry of Cytotoxicity (NIH, 2001). However, as mentioned previously, using methylcellulose does not allow high-throughput screening using 384-well plate formats. To do this, a new technology was required. This is known as Suspension Expansion Culture (SEC) technology. Not only did this technology allow for the first high-throughput screening of hematopoietic cells in 384-well plates, but it also enabled a more rapid completion of the assay while concomitantly increasing sensitivity fourfold. This platform, designated HALO-Tox HT, is now the basis for an advanced hemotoxicity screening platform as well as for other high-throughput, tissue-specific cytotoxicity assays (e.g., immune system, neural cells, and hepatocytes as well as other *ex vivo* primary cell types, ES, iPS, and cell lines).

During the validation of CAMEO-96 against the Registry of Cytotoxicity (NIH) and HALO against many other different compounds and drugs, it became clear that the results obtained when analyzing the stem cell response provided several paradigms for predicting effects to individual parts or the "global" lymphohematopoietic system (Rich and Hall, 2005). In general, anything that results in a decrease in hematopoietic stem cell proliferation will affect all three hematopoietic lineages (erythropoietic, myelomonocytic, and megakaryopoietic). The lineages may be affected to different degrees, but a decrease in all mature cells can be expected. If the more primitive lymphohematopoietic stem cell population is affected, not only will hematopoiesis decrease, but lymphopoiesis and the immune cell populations will also decrease. Examples of IC values for several drugs and compounds that were tested to determine the response of the primitive hematopoietic stem cell, designated as CFC-GEMM (CFC–granulocyte, erythroid, macrophage, megakaryocyte), can be found in previously published manuscripts (Rich and Hall, 2005; Olaharski et al., 2009).

11.8 DETECTING STEM AND PROGENITOR CELL DOWNSTREAM EVENTS

It is unusual that a drug is so specific that it only affects a single lineage of the blood-forming system. This is because all of the lineages are connected at the stem cell level, and if one lineage is affected to a greater degree than another, a compensatory effect occurs in the other lineages. This can be detected at both the stem and progenitor cell levels using four, five, or seven population "global" predictive hemotoxicity assays. In such an instance, one or two primitive stem cells are accompanied by either the three hematopoietic lineages and/or the two lymphopoietic lineages (B and T

lineages), providing an excellent overview of the whole system through the measurement of the response across the different lineages. The ability to multiplex and correlate different assay readouts is a particularly valuable asset for *in vitro* toxicology. For example, although HALO measures proliferation ability or potential in response to an agent, the effect on specific populations during the culture period on the differentiation and/or maturation process can be determined by multiplexing with expression of membrane markers by flow cytometry. The onset of cytotoxicity might be preceded by genotoxicity. The induction of apoptosis or specific markers of cell stress can also be studied and related to changes in DNA and/or RNA profiles. These and many other examples of multiplexing can all be performed on the same sample thereby reducing time and cost.

It is possible to home in on a specific cell population using flow cytometry and determine the MOA if a drug causes a block in differentiation and induces stress or an apoptotic event at a specific stage of development downstream from the stem and primitive progenitor cells. Alternatively, genetic and protein analysis would provide information on the gene and mRNA activity and the transcription factors involved in a specific response. In other cases, it may be necessary to determine the expression of intracellular or released growth factors and cytokines produced by cultures. The production of megakaryocytes and their development into platelets is often important, and here the combination of different flow cytometric techniques comes into play by using specific membrane expression markers combined with ploidy (DNA) analysis. These are just a few of the multiplexing capabilities that can be performed to understand both the point of action of a drug and its MOA.

11.9 BONE MARROW TOXICITY TESTING DURING DRUG DEVELOPMENT

According to the traditional drug development paradigm, the ability of a molecule to cause bone marrow toxicity would not be discovered until repeat dose testing has been performed in an *in vivo* model system. Typically the earliest *in vivo* studies performed during the drug development process are not normally toxicity studies but rather pharmacological efforts meant to understand possible efficacy in an *in vivo* model (often the mouse). While there are opportunities to harmonize efforts and gain early glimpses of possible toxicities, this is uncommon in practice. Thus, it is often not until there are potent molecules with acceptable pharmacokinetic profiles that have already exhibited some efficacy that toxicity profiling begins. While timelines vary among and even within companies, it is not unusual that 5 years will have passed from the time that a molecular target is identified that an *in vivo* toxicity study will be initiated. This is due to many reasons, but significant effort is required to identify (develop) mature chemical equity against that target ensuring that sufficient bioavailability will occur following oral dosing. This time can be increased an additional 12–18 months if the toxicity is not observed until nonrodent testing begins (such as the dog, minipig, or cynomolgus monkey). If and when bone marrow toxicity is identified, retrospective implementation of a counter screen to identify whether backup molecules also exhibit the same toxicity causes additional delays (sometimes 6–9 months), costs, and the increased use of a larger number of animals. The inclusion of an *in vitro* assay that can proactively predict *in vivo* bone marrow toxicity with a high degree of sensitivity and specificity would drastically reduce timelines, costs, and the number of animals needed to enable the start of clinical trials in man.

As introduced previously, the HALO assay has demonstrated excellent *in vitro* to *in vivo* concordance in the past, correctly identifying molecules that caused *in vivo* toxicity with approximately 80% accuracy across a structurally diverse set of 56 pharmaceutical molecules when an *in vitro* toxicity threshold of 20 µM was used (Olaharski et al., 2009). Arguably of more value was the discovery that 100% of all the molecules exhibiting an IC_{50} concentration lower than 1 µM caused bone marrow toxicity *in vivo* without a single false-positive result (Olaharski et al., 2009). The inherent value of this knowledge is that it enables decision-making. One of the biggest concerns of basing decisions off exploratory proactive toxicity testing in the pharmaceutical industry is the possibility of false-positive results. That is, a drug will test positive in an *in vitro* assay but upon *in vivo* testing is identified not to be recapitulated. Whereas it might not be as detrimental in the very early space, where chemical matter is abundant and structure activity relationships are explored, the

tolerance for potential false-positive *in vitro* results decreases immensely during lead development. In the absence of the *in vivo* result, the *in vitro* assay data will likely be used to deprioritize further development of that molecule. Stated another way, implementation of an assay with a high false-positive rate can lead to the discontinuation of a number of good drugs. Thus, the ability to state that molecules with an IC_{50} value of <1 µM are all bone marrow toxicants (and none that had IC_{50} values of <1 µM were not) is a powerful piece of information that can lead to early proactive decision-making with a very low likelihood of throwing the "baby out with the proverbial bath water." The ability to confidently implement a proactive testing scheme can greatly reduce timelines, animal usage, and attrition from *in vivo* safety findings.

11.10   PARADIGM FOR *IN VITRO* HEMOTOXICITY TESTING

Figure 11.3 demonstrates how *in vitro* hemotoxicity testing can be applied to the drug development pipeline. Including *in vitro*, human stem cell hemotoxicity screening during the ADME/Tox phase of drug development would be the most time- and cost-efficient method for assessing potential deleterious effects. Since such effects on the stem cells will be amplified to the mature functional compartment, the predictive value level for Phase 1/2 human clinical trials would be the highest. Since regulatory agencies still require preclinical animal studies on at least two different species, more predictive information can be accrued by performing species comparisons and mechanism of action (MOA) studies. In addition, the ability to use the same standardized

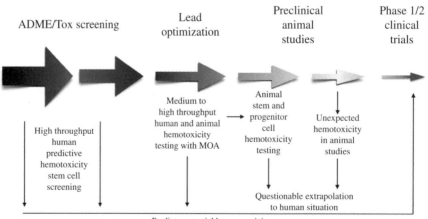

FIGURE 11.3   When to use hemotoxicity testing during drug development. Traditional hemotoxicity testing is performed during preclinical animal studies when circulating blood elements and morphologic interpretation are performed. Many unexpected observations and results are found during preclinical studies that could have been predicted prior to the onset of this time-consuming and costly stage in drug development. *In vitro* hemotoxicity screening during the lead optimization stage can alert to potential toxic and safety issues that might occur at later stages by proactively deprioritizing the further development of molecules that have a high likelihood of being hematoxic in animals and man, thereby reducing animal usage as well as time and resources.

and validated readout on multiple cell types allows a direct comparison of potential toxicity and a high degree of ranking capability. Interspecies comparisons can then be applied prior to any costly preclinical animal studies. It should be emphasized that unexpected hemotoxicity is often observed when considerable time, effort, and money have been invested into preclinical animal studies without prior *in vitro* hemotoxicity testing (often after only 7- or 14-day toxicology studies in rodents and/or the nonrodent species has been completed). At this point in drug development, the question is less about whether toxicity can be predicted but rather why an unexpected toxicity was observed and whether that toxicity is on or off target.

11.11 PREDICTING STARTING DOSES FOR ANIMAL AND HUMAN CLINICAL TRIALS

An interesting aspect of validating an assay against the Registry of Cytotoxicity is the ability to use *in vitro* IC values to estimate initial starting doses or maximum tolerated doses for either animal or human clinical studies (FDA, 2005). For many anticancer drugs, it was shown that the estimated predicted dose calculated from the IC_{20}, IC_{50}, and/or IC_{90} values corresponded, at least in the same order of magnitude, to the clinical doses used to treat various cancers in the clinic (Beveridge et al., 2004; Rich and Hall, 2005). Another approach is to identify the lethal dose in 10% (LD_{10}) of mice and generate the IC90 ratio in mouse and human CFU-GM cells *in vitro*, information which can be utilized to predict the maximum tolerated dose of a therapeutic agent in humans. The concept is that a predictive model can be created by adjusting a mouse-derived MTD for the differential interspecies sensitivity *in vitro* (Pessina et al., 2003). While admittedly a small sample size, the data supported the author's conclusions that the assay was able to accurately predict human MTD within a 4× range (20/23 molecules; 83%). Due to demonstrated high *in vitro* to *in vivo* concordance for these assays to predict bone marrow toxicity, the earlier in the drug development pipeline this type of information can be obtained, the fewer the surprises and unexpected results when the drug candidate reaches the preclinical and human clinical trial stages. This information can also be critically important for rank-ordering molecules early in development, identifying therapeutic windows and assisting teams in properly choosing which molecule to advance.

11.12 FUTURE TRENDS

The biopharmaceutical industry is open to new trends and technologies, which is driven by time and cost restraints and the need to produce new therapeutic drugs within the confines of regulatory approval. No *in vitro* or *in vivo* animal assay will predict 100% the information obtained when a drug is administered to a patient, though the ability to predict potential toxic side effects using *in vitro* surrogate assays prior to the initiation of clinical studies can be enormously beneficial.

A considerable amount of attention has been given to 3-dimensional (3D) cultures as providing an improved simulation of the tissue than 2D cultures. It may not be surprising that one of the first 2D/3D culture systems was developed for primitive hematopoietic cells as early as 1973 (Dexter et al., 1973, 1977). These types of cultures allowed an adherent layer of stromal cells to be produced upon which colonies of primitive hematopoietic stem cells could grow. The colonies were designated "cobblestone area-forming cells" (Ploemacher et al., 1991) and later long-term culture-initiating cells or LCT-IC (Liu et al., 2013). They are considered the most primitive of all *in vitro* lymphohematopoietic stem cells because they can be maintained in culture for many weeks, indicating extensive self-renewal capacity. 3D culture technology has advanced considerably, and new technology such as tissue or body-on-a-chip devices take into account many physical conditions that can influence the response of cells to different perturbations (Esch et al., 2011). Many different matrices are now used to establish 3D cultures of different tissues and organs. Nevertheless, removal of cells from the matrix to analyze their phenotype, genetic makeup, etc. is still a very difficult task that needs to be actively addressed if 3D culture systems are to provide information on potential risk and improve safety.

It is assumed that if ES and iPS cells are given the right growth factors and conditions, they will produce the desired cell type. The nondefinitive stem cells must however first produce definitive stem cells prior to producing the required functional cell type. Stem cells do not represent a single population but rather a continuum of cells each at different stages of primitiveness and exhibit slightly different properties (Fig. 11.1). Since the methods to specifically identify different stem cell populations are either unavailable or not sufficiently specific and accurate, it cannot be guaranteed that the same population is produced every time. Moreover, because the cells are present in a continuum and the properties are always in flux, cells that might be characterized as expressing the CD34 antigen at one point in time may either have a lower frequency of expression or may not be expressing the antigen at a later time point. Finally, for many cells expressing a phenotypic marker (e.g., CD34+), the marker may not be specific. Although considered a "stem cell marker," only a very small proportion of stem cells (in culture) express CD34 within a select timeframe, and the highest proportion expressed is not by stem cells but by progenitor cells. It is therefore important to understand the properties and characteristics of the target cells being used and their limitations since these will have a direct effect on the interpretation of results and conclusions drawn.

Personalized medicine is now becoming a very important issue, and assays described previously could be used to predict the response of a single patient to a drug or to investigate drug–drug interactions. One of the most "exotic" applications that can be specifically directed to personalized medicine is the field of chronotherapy, which is based upon the premise that virtually all cellular functions have an internal clock and circadian rhythm. In its simplest form, chronotherapy is the administration of drugs at a certain time of day that results in low toxicity, but high efficacy, which can improve the therapeutic ratio if the circadian rhythm of the target cells is taken into consideration. Hematopoietic stem and progenitor cells demonstrate different circadian rhythms and therefore their chronotoxicity can be widely different (Smaaland et al., 2002; Harper and Rich, 2013; Hrushesky and Rich, 2015). This is an area of toxicology that has remained underutilized, yet the clinical advantages to personalized medicine have been shown to be significant.

11.13 CONCLUSIONS

Stem cell biology, therapy, and toxicology may represent the new "hype," but it also has to be considered in the context of what is actually known. Stem cells and their use hold a great deal of promise in toxicology, but this promise should not be at the expense of lack of knowledge and false assumptions that might prove to be the undoing of a promising new drug. Understanding the biology, physiology, and interaction with other biological systems is paramount to understanding how different toxicants may deleteriously affect bone marrow.

REFERENCES

Beveridge R, Reitman JF, Fausel C, Leather H, McFarland H, Rifkin RM, Valley A. *Guide to Selected Cancer Chemotherapy Regimens and Associated Adverse Events*. Fifth Edition. Amgen, Thousand Oaks, CA, 2004.

Bradley TR, Metcalf D. The growth of mouse bone marrow cells in vitro. *Aust J Exp Biol Med Sci*. 1966; 44:287–299.

Delorme B, Chateauvieux S, Charbord P. The concept of mesenchymal stem cells. *Regen Med*. 2006; 1(4):497–509.

Dexter TM, Allen TD, Lajtha LG, Schofield R, Lord BI. Stimulation of differentiation and proliferation of hematopoietic cells in vitro. *J Cell Physiol*. 1973; 82(3):461–473.

Dexter TM, Allen TD, Lajtha LG. Conditions controlling the proliferation of hematopoietic stem cells in vitro. *J Cell Physiol*. 1977; 91(3):335–344.

ECVAM. http://ecvam-dbalm.jrc.ec.europa.eu/s_invitoxprot_res.cfm?&selTopic=&selModel=&selMaterial=&selProtocol=&selProtocolString=&selProtocolFree=&selProtocolNum=&selContactP=0&selExpSystem=&selpreview6=&selEndpoint=0&selpreview7=&selCompounds=&selpreview10=&selCasn=&selE
cn=&metStatus=0&selpreview12=&tclick=0&firstRec=61 (accessed on October 13, 2015).

Esch MB, King TL, Shuler ML. The role of body-on-a-chip devices in drug and toxicity studies. *Annu Rev Biomed Eng*. 2011; 15:55–72.

Food and Drug Administration (FDA). Guidance for industry. Bioanalytical method validation. 2001. http://www.fda.gov/downloads/Drugs/Guidances/ucm070107.pdf (accessed on October 13, 2015).

Food and Drug Administration (FDA). Guidance for industry. Estimating the maximum safe starting doses in initial clinical trials for therapeutics in adult healthy volunteers. 2005. http://www.fda.gov/downloads/Drugs/Guidances/UCM078932.pdf (accessed on October 13, 2015).

Harper H, Rich IN. Stem cell predictive hemotoxicology. In: *Stem Cell Biology in Normal Life and Diseases*, Alimoghaddam K (Ed). Intech. 2013; pp 79–108. http://dx.doi.org/10.5772/46173 (accessed on October 13, 2015).

Horowitz D, King AG. Colorimetric determination of inhibition of hematopoietic progenitor cells in soft agar. *J Immunol Methods*. 2000; 244:49–58.

Hrushesky W, Rich IN. Measuring stem cell circadian rhythm. *Methods Mol Biol*. 2015; 1235: 81–96.

Iscove NN, Sieber F, Winterhalter KH. Erythroid colony formation in cultures of mouse and human bone marrow: analysis of the requirement for erythropoietin by gel filtration and affinity chromatography on agarose-concanavalin A. *J Cell Physiol*. 1974; 83:309–320.

Kriegler AB, Bradley TR, Hodgson GS, McNiece IK. A colorimetric liquid culture assays of a growth factor for primitive murine macrophage progenitor cells. *J Immunol Methods*. 1987; 103:93–102.

Liu M, Miller CL, Eaves CJ. Human long-term culture initiating cell assay. *Methods Mol Biol*. 2013; 946:241–256.

NIH. Guidance document on using in vitro data to estimate in vivo starting doses for acute toxicity. NIH Publication No: 01-4500. 2001. http://ntp.niehs.nih.gov/iccvam/docs/acutetox_docs/guidance0801/iv_guide.pdf (accessed on October 13, 2015).

Olaharski AJ, Uppal H, Cooper M, Platz S, Zabka TS, Kolaja KL. In vitro to in vivo concordance of a high throughput assay of bone marrow toxicity across a diverse set of drug candidates. *Toxicol Lett*. 2009; 188:98–103.

Parchment RE, Gordon M, Grieshaber CK, Sessa C, Volpe D, Ghielmini M. Predicting hematological toxicity (myelosuppression) of cytotoxic drug therapy from in vitro tests. *Annals of Oncology* 1998; 9(4):357–64.

Parent-Massin D. Relevance of clonogenic assays in hematotoxicology. *Cell Biol Toxicol*. 2001; 17(2):87–94.

Parent-Massin D, Hymery N, Sibiril Y. Stem cells in myelotoxicity. *Toxicology*. 2010; 267(1–3):112–117.

Patterson J, Moore CH, Palser E, Hearn JC, Dumitru D, Harper HA, Rich IN. Detecting primitive hematopoietic stem cells in total nucleated and mononuclear cell fractions from umbilical cord blood segments and units. *J Transl Med*. 2015; 13:94.

Pessina A, Albella B, Bueren J, Brantom P, Casati S, Gribaldo L, Croera C, Gagliardi G, Foti P, Parchment R, Parent-Massin D, Sibiril Y, van Den Heuvel R. Prevalidation of a model for predicting acute neutropenia by colony forming unit granulocyte/macrophage (CFU-GM) assay. *Toxicol In Vitro*. 2001; 15:729–740.

Pessina A, Albella B, Bayo M, Bueren J, Brantom P, Casati S, Croera C, Gagliardi G, Foti P, Parchment R, Parent-Massin D, Schoeters G, Sibiril Y, Van Den Heuvel R, Gribaldo L. Application of the CFU-GM assay to predict acute drug-induced neutropenia: an international blind trial to validate a prediction model for the maximum tolerated dose (MTD) of myelosuppressive xenobiotics. *Toxicol Sci*. 2003; 75:355–367.

Pessina A, Parent-Massin D, Albella B, Van Den Heuvel R, Casati S, Croera C, Malerba I, Sibiril Y, Gomez, S, de Smedt A, Gribaldo L. Application of human CFU-Mk assay to predict potential thrombocytotoxicity of drugs. *Toxicol In Vitro*. 2009; 23:194–200.

Ploemacher RE, van der Sluijs JP, van Beurden CA, Baert MR, Chan PL. Use of limiting-dilution type long-term marrow cultures in frequency analysis of marrow-repopulating and spleen colony-forming hematopoietic stem cells in the mouse. *Blood*. 1991; 78(10):2527–2533.

Pluznik DH, Sachs L. The induction of clones of normal mast cells by a substance from conditioned medium. *Exp Cell Res*. 1966; 43:553–563.

Rich IN. High-throughput in vitro hemotoxicity testing and in vitro cross-platform comparative toxicity. *Expert Opin Drug Metab Toxicol*. 2007; 3:295–307.

Rich IN. A short primer in stem cell biology. *Methods Mol Biol*. 2015a; 1235:pp. 1–6.

Rich IN. Measurement of hematopoietic stem cell proliferation, self-rewal, and expansion potential. *Methods Mol Biol*. 2015b; 1235:pp. 7–18.

Rich IN, Hall KM. Validation and development of a predictive paradigm for hemotoxicity using a multifunctional bioluminescence colony-forming proliferation assay. *Toxicol Sci*. 2005; 87:427–441.

Smaaland R, Sothern RB, Laerum OD, Abrahamsen JF. Rhythms in human bone marrow and blood cells. *Chronobiol Int*. 2002;19(1):101–127.

Spellman S, Hurley CK, Brady C, Phillips-Johnson L, Chow R, Laughlin M, McMannis J, Reems JA, Regan D, Rubinstein P, Kurtzberg J. Guidelines for the development and validation of new potency assays for the evaluation of umbilical cord blood. *Cytotherapy*. 2011; 13(7): 848–855.

Than NN, Newsome PN. Stem cells for liver regeneration. *QJM*. 2014; 107(6):417–421.

Toxicology and Pharmacology Branch (TPB) of the National Cancer Institute (NCI). https://dtp.cancer.gov/organization/TpB/toxicology-pharmacology_primer.htm (accessed on January 4, 2016).

12

PREDICTING ORGAN TOXICITY *IN VITRO*: DERMAL TOXICITY

PATRICK J. HAYDEN[1], MICHAEL BACHELOR[1], MITCHELL KLAUSNER[1] AND HELENA KANDÁROVÁ[2]

[1] *MatTek Corporation, Ashland, MA, USA*

[2] *MatTek In Vitro Life Science Laboratories, Bratislava, Slovak Republic*

12.1 INTRODUCTION

The skin is commonly exposed to pharmaceutical active ingredients in the form of topically applied therapeutics or systemically exposed following oral, intravenous, or inhaled therapeutics administration. Cutaneous reactions are among the most prevalent adverse effects of therapeutic drugs, occurring in up to 3% of all hospitalized patients and accounting for as much as 20–30% of all reported adverse drug reactions (Svensson et al., 2001). Adverse cutaneous drug reactions range from relatively minor skin rashes to more serious toxic consequences such as Stevens–Johnson syndrome (SJS) and toxic epidermal necrolysis (TEN), which may become severe or life threatening (Svensson et al., 2001; Lee and Thompson, 2006; Farshchian et al., 2015). These cutaneous effects can impose major limitations on clinical therapeutic use. Therefore, predictive preclinical models of human cutaneous toxicity are of significant utility during the pharmaceutical development process. Preclinical models of human dermal penetration or percutaneous absorption are also important for development of topical pharmaceuticals.

Traditional cutaneous safety testing of pharmaceuticals and cosmetics has relied on the use of animal models. However, animal skin is significantly different from human skin in terms of hair and other skin appendages and does not always predict human skin effects accurately (Jírová et al., 2010; Basketter et al., 2012). In addition, questions regarding the ethics of using animals for safety testing of cosmetics have led to worldwide efforts to develop human-based *in vitro* cutaneous tests.

This chapter will cover the state-of-the-art of *in vitro* human skin models and the utility of these models for preclinical drug development applications including screening for skin sensitization, phototoxicity, irritation, genotoxicity, pigmentation effects, and percutaneous absorption. The validation status of the various models and applications and areas where further development is needed are also highlighted.

12.2 OVERVIEW OF DRUG-INDUCED ADVERSE CUTANEOUS REACTIONS

Adverse cutaneous drug reactions may result from immune-mediated processes, photochemical reactions, mechanisms that are directly related to the therapeutic activity of the drug, or idiosyncratic toxicities for which the mechanisms are poorly understood (Svensson et al., 2001; Lee and Thompson, 2006; Farshchian et al., 2015). Although a wide range of drugs can cause adverse skin reactions, antibiotics and nonsteroidal anti-inflammatory drugs are among the more frequently reported causes (Svensson et al., 2001; Lee and Thompson, 2006; Farshchian et al., 2015). The clinical morphology of adverse cutaneous drug reactions may be classified as exanthematous, morbilliform, fixed drug eruption, urticaria, SJS, and TEN (Svensson et al., 2001; Lee and Thompson, 2006; Farshchian et al., 2015). Severe cutaneous reactions such as SJS and TEN are life-threatening conditions with mortality rates as high as 30% (Svensson et al., 2001).

Drug Discovery Toxicology: From Target Assessment to Translational Biomarkers, First Edition. Edited by Yvonne Will, J. Eric McDuffie, Andrew J. Olaharski, and Brandon D. Jeffy.

Immune-mediated mechanisms are believed to be responsible for a high proportion of adverse cutaneous drug reactions. Small-molecule therapeutics are not inherently immunogenic but must first become conjugated to larger biological molecules in order to be recognized by the immune system. Thus, immunogenic drugs must possess reactive electrophilic properties or undergo bioactivation to form reactive electrophilic metabolites under biological conditions. In order for adverse immunologic reactions to occur within the skin, the protein conjugation event also likely occurs within the skin, with antigen presentation to immune cells by epidermal keratinocytes or resident dendritic cells (i.e., Langerhans cells) (Svensson et al., 2001; Khan et al., 2006).

Photochemical reactions leading to phototoxicity or photoallergy are also responsible for a significant proportion of adverse cutaneous drug reactions. Drug-induced photosensitivity is usually limited to drugs with an absorption spectrum in the UVA region (320–400 nm) of solar light. However, UVB absorption may contribute to phototoxicity of topically applied therapeutics. Following initial absorption of photons, phototoxicity can be produced by the formation of free radicals, which can form covalent adducts or lead to the oxidation of cellular molecules (type I reaction). An alternate pathway for phototoxicity involves the generation of singlet oxygen (type II reaction) from photoactivated drug molecules, with subsequent oxidation of biomolecules. Although rarely observed for systemically administered drugs (EMA, 2015), photoallergy may occur when covalently modified biomolecules produced from reactions with photoactivated drugs act as haptens to provoke an immune response.

Drugs whose mechanism of therapeutic action involves interference with specific cellular molecules or signaling pathways may also produce mechanism-based adverse side effects in tissues including skin (Rudmann, 2013). Chemotherapeutic drugs that target the EGFR or MAPK pathways are examples of drugs that commonly produce cutaneous adverse reactions including inflammatory (rash), squamoproliferative, melanocytic (vitiligo), and hair/nail effects (Borovicka et al., 2011; Fischer et al., 2013; Manousaridis et al., 2013; Rosen et al., 2013; Choi, 2014; Macdonald et al., 2015a, b).

Nonimmune-mediated idiosyncratic cutaneous reactions can arise from a variety of causes including cumulative toxicity, overdose, drug interactions, and metabolic alterations (Lee and Thompson, 2006). Viral infections can influence and increase the incidence of these cutaneous adverse drug reactions (Svensson et al., 2001; Lee and Thompson, 2006). Pharmacogenomics play an important role in individual susceptibility to adverse skin reactions as well (Lonjou et al., 2006; Tassaneeyakul et al., 2009; Aihara, 2011; Kulkantrakorn et al., 2012; Pirmohamed, 2012; Borroni, 2015).

12.3 OVERVIEW OF *IN VITRO* SKIN MODELS WITH RELEVANCE TO PRECLINICAL DRUG DEVELOPMENT

Since the initial development of methods for *in vitro* culture of human keratinocytes (Rheinwald and Green, 1975), two-dimensional monolayer cultures of submerged keratinocytes have been widely utilized for skin-related research. Primary and immortalized fibroblast and melanocyte cultures are also now widely available for *in vitro* research and toxicology testing applications (Hsu and Herlyn, 1996; Costin and Hearing, 2007). Skin-resident immune cells (e.g., Langerhans cells) are difficult to isolate and maintain in primary culture. However, monocyte-derived dendritic cells and dendritic cells derived from cord blood precursors are well established, as are a variety of immortalized dendritic-like cell lines (Ayehunie et al., 2009; Reuter et al., 2011).

Methods for *in vitro* reconstruction of three-dimensional (3D) differentiated human skin were developed during the 1980s (Bell et al., 1981). Cultured at the air–liquid interface (ALI), these *in vitro* reconstructed human epidermal (RhE) tissues display an *in vivo*-like stratified structure and functional epidermal layers including stratum corneum barrier. RhE may also be cocultured with dermal constructs containing viable fibroblasts (i.e., "full-thickness" models) as well as melanocytes (Fig. 12.1). The ALI culture format allows *in vivo*-like exposure and wounding scenarios including

(a)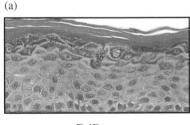

EpiDerm

(b)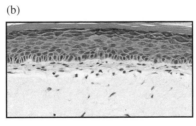

EpiDermFT

(c)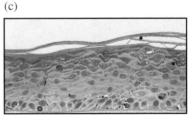

MelanoDerm

FIGURE 12.1 Examples of commercially available *in vitro* reconstructed human epidermal (RhE) tissues: (a) EpiDerm partial thickness model produced from primary keratinocytes; (b) EpiDermFT full-thickness coculture model including epidermal and dermal components; and (c) MelanoDerm coculture model of primary keratinocytes and melanocytes. Reproduced with permission of MatTek Corporation, Ashland, MA. (*See insert for color representation of the figure.*)

topical application of test materials and exposure to UV radiation. Systemic exposure is mimicked by addition of test agents to the culture medium below the constructs. When produced utilizing cells obtained from diseased donors, these cultures offer opportunities to model human diseases such as psoriasis (Chamcheu et al., 2015) and atopic dermatitis.

Explant culture of human or animal skin is also possible using ALI culture methods. An advantage of excised skin is the presence of numerous cell types and appendages such as hair and sweat glands. Cadaver skin is also often utilized for skin penetration studies. Human explant skin cultures and *in vitro* RhE models have both been shown to possess significant drug metabolizing capabilities that may affect drug activity and toxicity or promote hapten formation within cutaneous tissue (Hu et al., 2010). The disadvantages of excised skin or cadaver models are limited accessibility of the skin tissue, reproducibility issues, and lack of tissue viability and metabolism (in the case of cadaver skin).

A number of commercial 3D RhE models are now available from companies such as MatTek Corporation (EpiDerm, EpiDermFT, MelanoDerm), Henkel (Phenion FT Skin Model), EPISKIN (SkinEthic, EpiSkin), and others (Fig. 12.1). Several of these models including EpiDerm, SkinEthic, EpiSkin, and others have been validated for regulatory use in skin corrosion and irritation testing (OECD and EU Guidelines) and are partially accepted by some specific guidelines for phototoxicity testing (International Conference on Harmonisation of Technical Requirements for Registration of Pharmaceuticals for Human Use (ICH) S10) and medical device testing (ISO 10933).

Deficiencies of current commercially available *in vitro* RhE models include the lack of fatty layer, lack of skin appendages such as hair follicles and sweat glands, and lack of immune cells, innervation, or blood perfusion. However, hair follicle-containing skin models have been reported in the literature (Michel et al., 1999). In addition, it is possible to combine *in vitro* RhE skin models with hair follicles or other *in vitro* organs such as liver in perfused multiorgan chips (MOCs) (Fig. 12.2) (Ataç et al., 2013; Maschmeyer et al., 2015a, b). Due to the ongoing high level of research and development activities in the area of 3D organotypic models including skin models, further advances in complexity and functionality of *in vitro* skin models are anticipated.

12.4   SPECIFIC APPLICATIONS OF *IN VITRO* SKIN MODELS AND PREDICTIVE *IN VITRO* ASSAYS RELEVANT TO PHARMACEUTICAL DEVELOPMENT

12.4.1   Skin Sensitization

Immune-mediated mechanisms are responsible for a high proportion of adverse cutaneous drug reactions (Svensson et al., 2001; Lee and Thompson, 2006; Farshchian et al., 2015). The *in vivo* murine local lymph node assay (LLNA) has been widely utilized for skin sensitization testing but legislative and societal pressures have driven efforts to develop *in vitro* alternatives for skin sensitization testing. The European Center for the Validation of Alternative Methods (ECVAM) has validated several *in vitro* methods including the human Cell Line Activation Test (h-CLAT) (ECVAM, 2015a), the direct peptide reactivity assay (DPRA) (ECVAM, 2015b), and KeratinoSens™ assay for skin sensitization testing (ECVAM, 2015c). The ECVAM validated sensitization methods work well for materials that can be solubilized for these aqueous-based assays, and integrated testing strategies for evaluation of chemical sensitizing potential and potency

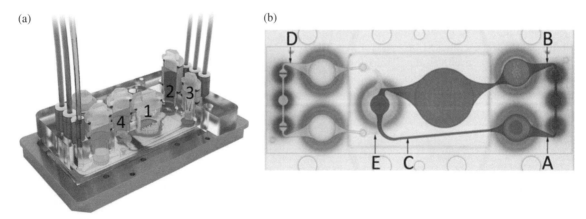

(a)   (b)

FIGURE 12.2   A microfluidic four-organ-chip device at a glance. (a) 3D view of the device comprising two polycarbonate cover plates, the PDMS–glass chip (footprint: 76 mm × 25 mm; height: 3 mm), accommodating a surrogate blood flow circuit (pink) and an excretory flow circuit (yellow). Numbers represent the four tissue culture compartments for intestine (1), liver (2), skin (3), and kidney (4) tissue. A central cross section of each tissue culture compartment aligned along the interconnecting microchannel is depicted. (b) Top view of the four-organ-chip layout illustrating the positions of three measuring spots (A, B, and C) in the surrogate blood circuit and two spots (D and E) in the excretory circuit. Figure courtesy of TissUse, Berlin, Germany, with permission. (*See insert for color representation of the figure.*)

have been described (Rovida et al., 2015). A major limitation of these assays is an inability to evaluate insoluble materials or finished formulations. For water-insoluble materials or formulations, methods that utilize 3D RhE models are in development.

A promising *in vitro* approach for skin sensitization testing that utilizes 3D RhE models are assays that measure activation of various gene expression pathways such as the antioxidant response pathway (Natsch and Emter, 2008). The SenCeeTox® assay (McKim et al., 2012) developed by CeeTox, Inc. (now Cyprotex) utilizes 3D RhE models to predict the sensitization potential and potency of test materials using a multiparametric analysis of (i) changes in the expression of 11 genes related to the Nrf2/Keap1/ARE or AhR/ARNT/XRE signaling pathways, (ii) direct reactivity by assessing glutathione (GSH) depletion, (iii) cell viability using LDH, and (iv) molecular descriptors. A proprietary algorithm is used to integrate the data and predict potency of sensitizer based on LLNA categories (non/minimal, moderate, strong, extreme).

A second approach to *in vitro* RhE skin sensitization assays that is currently under development involves measuring secretion of IL-18. IL-18 has been shown to play a key proximal role in the induction of allergic contact sensitization (Antonopoulos et al., 2008), and human keratinocytes constitutively express IL-18 mRNA and protein, which is induced following exposure to contact sensitizers (Van Och et al., 2005; Galbiati et al., 2011). Recent studies with RhE models have shown IL-18 to be a promising endpoint for RhE skin sensitization screening, including determination of sensitizer potency (Deng et al., 2011; Gibbs et al., 2013).

A summary of *in vitro* approaches to assessment of immune-mediated skin effects is shown in Table 12.1.

12.4.2 Phototoxicity

A number of *in vitro* models have been developed for assessing the phototoxic potential of chemicals. The most widely used *in vitro* phototoxicity assay is the "*in vitro* 3T3 Neutral Red Uptake Phototoxicity Test" (3T3 NRU PT). This assay was validated by ECVAM and subsequently adopted by the Organization for Economic Co-operation Development (OECD) as Test Guideline (TG) 432 (OECD,

2004a). The 3T3 NRU PT measures the viability of mouse Balb/c 3T3 cells after exposure to a chemical in the presence and absence of simulated solar light (generally 290–700 nm). Cytotoxicity (cell death) causes a reduction in the uptake of Neutral Red (a dye) by the cells. The assay endpoint is the concentration of test substance that reduces cell viability to 50% of the untreated control value (IC50 value). Drawbacks of this assay are that solubility of test materials can be a limiting factor, and it has been found to produce a high number of false-positive responses with pharmaceuticals (EMA, 2015; FDA, 2015).

In vitro 3D RhE model phototoxicity tests have been developed (Edwards et al., 1994). These assays also measure tissue viability after exposure to a chemical in the presence and absence of light. ECVAM sponsored a successful prevalidation of the EpiDerm Phototoxicity Test (EpiDerm PT) in three laboratories under blind sample conditions (Liebsch et al., 1999). A subsequent ECVAM-supported project demonstrated that the EpiDerm PT may also serve as a tool for the determination of phototoxic potency of topically applied phototoxins (Kandárová et al., 2005; Kejlová et al., 2007). The EpiDerm PT is established in many laboratories of the cosmetic industry worldwide, and the protocol has been successfully adopted for other commercial epidermal models.

Advantages of the EpiDerm PT over the 3T3 NRU PT are the 3D structure, the stratum corneum barrier, and the lipid profile that are similar to those of human epidermis. These features enable application of undiluted solid and liquid materials as well as testing of water-insoluble materials and UVB absorbing chemicals, conditions which are not possible in the 3T3 NRU PT assay due to the lack of barrier, inability of 3T3 cells to tolerate UVB and solubility issues.

The US Food and Drug Administration (FDA), the European Federation of Pharmaceutical Industries and Associations (EFPIA), and the European Medicines Agency (EMA) have recognized the usefulness of the 3T3 NRU PT and 3D RhE models for topical phototoxicity testing and have incorporated the tests into guidelines for photosafety testing of systemic and dermally applied drugs (EMA, 2015; FDA, 2015). The guidelines generally apply to new active pharmaceutical ingredients (APIs), new excipients, clinical formulations for dermal application, and photodynamic therapy products, excluding peptides, proteins, antibody drug conjugates, and oligonucleotides.

The ICH guidelines suggest tiered assessment strategies for systemic and topically applied pharmaceuticals that should be conducted prior to exposure of large numbers of human subjects (EMA, 2015; FDA, 2015). A summary of the assessment strategy recommended in the ICH guidance document is reproduced in Fig. 12.3. The initial assessment considers the photochemical properties (290–700 nm absorption spectrum) and inherent photoreactivity (e.g., UVA-induced ROS generation) of the API or excipients.

TABLE 12.1 *In Vitro* **Approaches to Assessment of Potential for Immune-Mediated Skin Effects**

	Model/Assay	References
Systemic API/medical device: H$_2$O soluble	h-CLAT	ECVAM (2015a)
	DPRA	ECVAM (2015b)
	KeratinoSens	ECVAM (2015c)
Topical API or formulation/ medical device: H$_2$O soluble or insoluble	SenCeeTox RhE	McKim (2012)
	IL-18 RhE	Deng (2011) and Gibbs (2013)

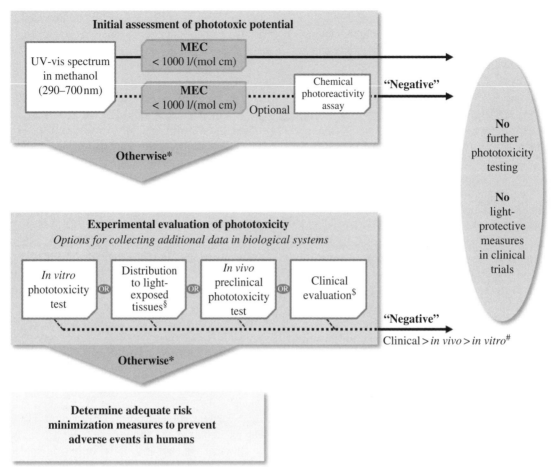

* "Otherwise": data do not support a low potential for phototoxicity or have not been generated (assay/test/evaluation not conducted)

A "negative" result in an appropriately conducted *in vivo* phototoxicity study supersedes a positive *in vitro* result. A robust clinical phototoxicity assessment indicating no concern supersedes any positive non clinical results. A positive result in an *in vitro* phototoxicity test could also, on a case-by-case basis, be negated by tissue distribution data (see text). In the United States, for products applied dermally, a dedicated clinical trial for phototoxicity on the to-be-marketed formulation can be warranted in support of product approval.

$ Clinical evaluation could range from standard reporting of adverse events in clinical studies to a dedicated clinical photosafety trial.

§ Tissue distribution is not a consideration for the phototoxicity of dermal products.

FIGURE 12.3   Outline of possible phototoxicity assessment strategies for pharmaceuticals given via systemic and dermal routes, from ICH Guideline S10, Guidance on photosafety evaluation of chemicals. EMA, 2015; FDA, 2015.

If these tests reveal a lack of phototoxic potential, no further testing is necessary. If further testing is indicated, *in vitro* models including the 3T3 NRU PT, 3D RhE testing, and *in vivo* models may be utilized. In support of efforts to reduce the use of animal testing, validated *in vitro* models should be considered before use of animal models (EMA, 2015; FDA, 2015). Currently, there are no formally validated animal models for phototoxicity, and standardized study designs for animal phototoxicity tests have not been established (EMA, 2015; FDA, 2015).

Because the 3T3 NRU PT has a high sensitivity but low specificity for predicting clinical phototoxicity, a negative result in the 3T3 NRU PT indicates that no further testing is required. However, a positive result should be confirmed by further *in vivo* or clinical testing. *In vitro* 3D RhE tests can be performed when solubility issues prevent use of the 3T3 NRU PT or for topical formulations. However, the sensitivity of the *in vitro* 3D tests has in some cases been found to be lower than *in vivo* tests, so assay conditions should consider the use of higher strength formulations or increased exposure times. Under adequate test conditions, a negative 3D RhE PT result indicates that the phototoxicity potential of the formulation can be regarded as low. In the European Union and Japan, no further phototoxicity testing is generally recommended following a negative 3D RhE PT result. However, in the United States, negative test results may require further clinical photosafety assessment (EMA, 2015).

12.4.3 Skin Irritation

Skin irritation can be produced by nonimmune-mediated idiosyncratic cutaneous reactions resulting from cumulative toxicity, overdose, drug interactions, and metabolic alterations (Lee and Thompson, 2006). RhE skin irritation tests (SIT) have been validated and adopted by OECD as TG 439 for determining the irritation hazard from topical exposure to chemicals and mixtures (Kandárová et al., 2009; OECD, 2009). TG 439 is based on the ability of irritant chemicals to penetrate the stratum corneum and produce cytotoxicity in the underlying cell layers. The TG classifies test substances as either irritating or nonirritating based on results of the MTT viability assay.

The EpiDerm SIT is endorsed in the ISO 10993-10 (ISO, 2015) for assessment of medical devices. However, since extracts from medical devices have very low irritation potential and may be underpredicted by the OECD TG 439 protocol, a validation study using a modified EpiDerm SIT (utilizing an extended exposure time and increased dose) is currently underway to address the concerns of the medical devices industry and ISO (Casas et al., 2013; Kandarova et al., 2015).

Alternate methods have been developed to evaluate the severity of skin irritation for finished cosmetic formulations, and these may be adaptable for testing topical or systemic pharmaceuticals. A study supported by the EC's Directorate-General for Research examined the concordance between human *in vivo* and *in vitro* skin irritation classifications of cosmetic products (Faller et al., 2002). In this study, 22 formulations covering a broad range of *in vivo* scores and different cosmetic product classes were tested in humans and *in vitro* with three RhE models. The mean total *in vivo* scores (human data) for the 22 coded products were best correlated to *in vitro* data from the three RhE models using an ET50 assay (ET50 = the exposure time for a test material required to reduce the tissue viability to 50%). Using the ET50 assay, correlation coefficients of $R = 0.94$, $R = 0.90$, and 0.84 were observed using EpiDerm, Cosmital Wella, and EPISKIN, respectively. This study clearly demonstrated the usefulness and relevance of RhE models for the safety assessment of topical formulations. Moreover, it demonstrated a high correlation of RhE-derived *in vitro* data with results obtained in humans. Additional adaptations of ET50- or IC50-based assays may be appropriate for irritation testing of systemically administered pharmaceuticals.

12.5 MECHANISM-BASED CUTANEOUS ADVERSE EFFECTS

Chemotherapeutic drugs that target the EGFR or MAPK pathways are examples of drugs that commonly produce cutaneous adverse reactions including inflammatory (rash), squamoproliferative, melanocytic (vitiligo), and hair/nail effects. Monolayer keratinocyte cultures have been effectively utilized to elucidate these mechanistic pathways and may also be useful for screening candidate therapeutics for potential cutaneous effects. In the case of cancer chemotherapeutics that target the EGFR, the ability to inhibit keratinocyte proliferation *in vitro* correlates with *in vivo* efficacy as well as the adverse side effect of skin rash development (Laux et al., 2006).

12.5.1 Percutaneous Absorption

Percutaneous absorption of a substance or formulation depends on a number of factors, including the area of contact, duration of exposure, lipophilicity, molecular weight, concentration of the test substance, integrity of the stratum corneum, lipid profile of the stratum corneum, and thickness of the epidermis (OECD, 2004b, c). According to OECD Guidance Document 28, *in vitro* skin models can be used for skin penetration studies in a regulatory context if data from reference chemicals are consistent with those in the published literature (OECD, 2004b).

Reconstructed human skin models have been evaluated and extensively utilized for in dermal penetration assessment (Dreher et al., 2002; Lotte et al., 2002; Shigeta et al., 2004; Hammell et al., 2005; Karande et al., 2005; Lombardi Borgia et al., 2008; Schäfer-Korting et al., 2008b; Yuan et al., 2008; Hu et al., 2009a; Kurasawa et al., 2009; Desai et al., 2013). Prevalidation (Schäfer-Korting et al., 2006) and formal validation studies (Schäfer-Korting et al., 2008a) have been performed to investigate whether commercially available RhE models are suitable for *in vitro* skin absorption testing. Substances were tested in ten laboratories under carefully controlled conditions under both infinite-dose and finite-dose protocols.

In general, permeation of test chemicals through the RhE models exceeded that of human epidermis and pig skin, but the ranking of substance permeation through the three RhE models and pig skin accurately reflected the permeation through human epidermis. The EpiDerm RhE model provided the best correlation to human epidermal sheets (HES), as demonstrated by Papp values and a correlation coefficient of $r^2 = 0.932$. These studies demonstrate that RhE models are useful alternatives to human skin for the *in vitro* assessment of chemical permeation and penetration. Advantages of the RhE models compared to heat-separated human/animal epidermal sheets are their reproducibility, commercial availability, and xenobiotic metabolism capabilities, which have been demonstrated to be very similar to *in vivo* human skin (Hu et al., 2010).

12.5.2 Genotoxicity

The timing of genotoxicity testing requirements in support of exploratory clinical trials is dependent upon the intended clinical dose to be administered. Single or repeat doses that are in the expected therapeutic use range may require prior assessment of genotoxicity potential (EMA, 2008). The ICH has issued guidance on genotoxicity testing of pharmaceuticals (FDA, 2012). The standard genotoxicity test battery generally involves assessment of mutagenicity in the bacterial reverse mutation or "Ames" test (OECD TG 471), an *in vitro* assay in mammalian cells such as the *in vitro* micronucleus test (MNvit) (OECD TG 487), and/ or *in vivo* tests (FDA, 2012). Despite a high sensitivity for identifying genotoxic chemicals, the low specificity of these *in vitro* assays (and their combination) produces large numbers of false-positive predictions (Carmichael, 2007; FDA, 2012).

In vitro 3D RhE models that closely mimic human epidermal barrier and metabolism provide a relevant approach for genotoxicity testing of topically applied chemicals and formulations or for evaluation of skin-specific genotoxicity. An advantage of RhE model assays compared to submerged 2D or suspension cell assays is that the test chemicals and formulations can be applied at final use concentrations (e.g., finished products such as creams or lotions) using realistic exposure scenarios. The barrier of the RhE models is important since topically applied materials will only be genotoxic *in vivo* if they penetrate the stratum corneum to reach the mitotically active keratinocytes present in the basal layer of the epidermis. Furthermore, because many chemicals may require metabolic activation to produce genotoxicity, the ability of RhE models to reproduce *in vivo* human skin metabolism is a key consideration for skin genotoxicity testing. RhE models have demonstrated metabolic capabilities that closely resemble human skin metabolism (Hu et al., 2009a; 2010).

Two approaches to assess genotoxicity of topically applied materials using RhE models have been developed. The most advanced approach is an adaptation of the MNT (a robust assay that is accepted by OECD as validated for numerous cell types). The *in vitro* 3D human reconstructed skin micronucleus (RSMN) assay has proven to work well with the EpiDerm model (Curren et al., 2006; Hu et al., 2009b). In a recent prevalidation study, 20 chemicals were evaluated including genotoxic rodent skin carcinogens, genotoxic rodent nonskin carcinogens, and systemic rodent carcinogens/noncarcinogens. All participating laboratories correctly identified the test chemicals as either genotoxic or nongenotoxic (Aardema et al., 2010).

A second genotoxicity test, the comet assay, detects a broad spectrum of DNA damage. A comet assay protocol developed and evaluated using the EpiDerm model demonstrated good interlaboratory reproducibility and concordance with *in vivo* data (Reus et al., 2013). Additional *in vitro* RhE comet assay development and evaluation of intra- and interlaboratory reproducibility with EpiDerm, Realskin, and the Phenion model are currently ongoing.

12.5.3 Skin Lightening/Melanogenesis

Topical formulations that modulate skin pigmentation (e.g., skin-lightening or skin-tanning agents) are used for a diverse range of dermatologic or cosmetic indications. RhEs containing functional melanocytes provide a relevant preclinical model for toxicity and efficacy evaluation of skin-lightening/ tanning formulations (Fig. 12.1). The melanocytes within the cocultures undergo spontaneous melanogenesis leading to tissues of varying pigmentation levels consistent with the melanocyte donor source (e.g., Asian, Black, Caucasian).

The ALI culture format allows for topical application of skin lighteners or self-tanning agents, providing a useful *in vitro* means to evaluate cosmetic and pharmaceutical agents designed to modulate skin pigmentation. Sensitization, phototoxicity, irritation, and genotoxicity assays as described previously can also be conducted with the melanocyte containing RhEs. Efficacy assays typically involve measuring macroscopic darkening and melanin production. RhE models containing melanocytes have found extensive utility for evaluation of pigmentation or skin lightening following treatment with topically or systemically applied cosmetic ingredients and formulations (Sugimoto et al., 2004; Nesterov et al., 2008; Choi et al., 2010; Iriyama et al., 2011; Kasraee et al., 2011; Jain et al., 2012; Kim et al., 2012; Murase et al., 2013.

12.6 SUMMARY

Cutaneous reactions are among the most prevalent adverse effects of therapeutic drugs. Ranging from relatively minor skin rashes to more serious toxic consequences such as SJS and TEN, these cutaneous effects can impose major limitations on clinical therapeutic use. Mechanisms of adverse cutaneous drug reactions may be immune mediated, photochemical, directly related to therapeutic activity, or idiosyncratic toxicities for which the mechanisms are poorly understood.

In vitro models that are relevant for assessing the potential for adverse cutaneous reactions include monolayer or suspension cultures of skin and immune cells, as well as 3D organotypic RhE models, some of which are commercially available and validated for specific regulatory applications. Deficiencies of currently available *in vitro* RhE models include a lack of fatty layer, skin appendages innervations, and blood perfusion. However, due to an ongoing high level of research and development activities, further advances in complexity and functionality of *in vitro* skin models, including coculture with other organ models on MOCs, are anticipated.

In vitro skin model systems are useful for screening of APIs or formulations for potential adverse cutaneous effects including immune-mediated effects, phototoxicity, genotoxicity, alteration in pigmentation, and irritation. 3D RhEs may also be useful for *in vitro* evaluation of percutaneous absorption. These systems and assays can be applied early in the development process prior to more costly animal or clinical studies. Some of these *in vitro* skin model systems have already been adopted by regulatory authorities and have been incorporated into regulatory guidelines. Sharing of data and experiences gained from using various skin models among stakeholders, and additional validations and qualifications of the models for use with therapeutics, will facilitate further acceptance of *in vitro* skin model data by regulatory authorities.

REFERENCES

Aardema, M.J., Barnett, B.C., Khambatta, Z., Reisinger, K., Ouedraogo-Arras, G., Faquet, B., Ginestet, A.C., Mun, G.C., Dahl, E.L., Hewitt, N.J., Corvi, R., Curren, R.D. (2010) International prevalidation studies of the EpiDerm 3D human reconstructed skin micronucleus (RSMN) assay: transferability and reproducibility. *Mutat Res*, 701 (2), 123–131.

Aihara, M. (2011) Pharmacogenetics of cutaneous adverse drug reactions. *J Dermatol*, 38 (3), 246–254.

Antonopoulos, C., Cumberbatch, M., Mee, J.B., Dearman, R.J., Wei, X.Q., Liew, F.Y., Kimber, I., Groves, R.W. (2008) IL-18 is a key proximal mediator of contact hypersensitivity and allergen-induced Langerhans cell migration in murine epidermis. *J Leukoc Biol*, 83 (2), 361–367.

Ataç, B., Wagner, I., Horland, R., Lauster, R., Marx, U., Tonevitsky, A.G., Azar, R.P., Lindner, G. (2013) Skin and hair on-a-chip: in vitro skin models versus ex vivo tissue maintenance with dynamic perfusion. *Lab Chip*, 13 (18), 3555–3561.

Ayehunie, S., Snell, M., Child, M., Klausner, M. (2009) A plasmacytoid dendritic cell (CD123+/CD11c-) based assay system to predict contact allergenicity of chemicals. *Toxicology*, 264 (1–2), 1–9.

Basketter, D., Jírova, D., Kandárová, H. (2012) Review of skin irritation/corrosion Hazards on the basis of human data: a regulatory perspective. *Interdiscip Toxicol*, 5 (2), 98–104.

Bell, E., Ehrlich, H.P., Buttle, D.J., Nakatsuji, T. (1981) Living tissue formed in vitro and accepted as skin-equivalent tissue of full thickness. *Science*, 211, 1052–1054.

Borovicka, J.H., Calahan, C., Gandhi, M., Abraham, T.S., Kwasny, M.J., Haley, A.C., West, D.P., Lacouture, M.E. (2011) Economic burden of dermatologic adverse events induced by molecularly targeted cancer agents. *Arch Dermatol*, 147 (12), 1403–1409.

Borroni, R.G. (2015) Role of dermatology in pharmacogenomics: drug-induced skin injury. *Pharmacogenomics*, 16 (4), 401–412.

Carmichael, P. (2007) False Positive Results in In Vitro Genotoxicity Testing: Future Approaches. http://www.ecopa.eu/wp-content/uploads/ecopa_annual_2007--13_Carmichael.pdf (accessed on October 14, 2015).

Casas, J.W., Lewerenz, G.M., Rankin, E.A., Willoughby, J.A. Sr., Blakeman, L.C., McKim, J.M. Jr., Coleman, K.P. (2013) In vitro human skin irritation test for evaluation of medical device extracts. *Toxicol In Vitro*, 27 (8), 2175–2183.

Chamcheu, J.C., Pal, H.C., Siddiqui, I.A., Adhami, V.M., Ayehunie, S., Boylan, B.T., Noubissi, F.K., Khan, N., Syed, D.N., Elmets, C.A., Wood, G.S., Afaq, F., Mukhtar, H. (2015) Prodifferentiation, anti-inflammatory and antiproliferative effects of delphinidin, a dietary anthocyanidin, in a full-thickness three-dimensional reconstituted human skin model of psoriasis. *Skin Pharmacol Physiol*, 28 (4), 177–188.

Choi, J.N. (2014) Dermatologic adverse events to chemotherapeutic agents, Part 2: BRAF inhibitors, MEK inhibitors, and ipilimumab. *Semin Cutan Med Surg*, 33 (1), 40–48.

Choi, W., Wolber, R., Gerwat, W., Mann, T., Batzer, J., Smuda, C., Liu, H., Kolbe, L., Hearing, V.J. (2010) The fibroblast-derived paracrine factor neuregulin-1 has a novel role in regulating the constitutive color and melanocyte function in human skin. *J Cell Sci*, 123, 3102–3111.

Costin, G.E., Hearing, V.J. (2007) Human skin pigmentation: melanocytes modulate skin color in response to stress. *FASEB J*, 21 (4), 976–994.

Curren, R.D., Mun, G.C., Gibson, D.P., Aardema, M.J. (2006) Development of a method for assessing micronucleus induction in a 3D human skin model (EpiDerm). *Mutat Res*, 607 (2), 192–204.

Deng, W., Oldach, J., Armento, A., Ayehunie, S., Kandarova, H., Letasiova, S., Klausner, M., Hayden, P.J. (2011) Il-18 secretion as a marker for identification of contact sensitizers in the Epiderm in vitro human skin model. *Toxicologist*, 120 (Suppl. 12), 550.

Desai, P.R., Shah, P.P., Hayden, P., Singh, M. (2013) Investigation of follicular and non-follicular pathways for polyarginine and oleic acid-modified nanoparticles. *Pharm Res*, 30 (4), 1037–1049.

Dreher, F., Patouillet, C., Fouchard, F., Zanini, M., Messager, A., Roguet, R., Cottin, M., Leclaire, J., Benech-Kieffer, F. (2002) Improvement of the experimental setup to assess cutaneous bioavailability on human skin models: dynamic protocol. *Skin Pharmacol Appl Skin Physiol*, 15 (Suppl 1), 31–39.

ECVAM (2015a) https://eurl-ecvam.jrc.ec.europa.eu/eurl-ecvam-recommendations/eurl-ecvam-recommendation-on-the-human-cell-line-activation-test-h-clat-for-skin-sensitisation-testing (accessed on October 14, 2015).

ECVAM (2015b) https://eurl-ecvam.jrc.ec.europa.eu/eurl-ecvam-recommendations/eurl-ecvam-recommendation-on-the-direct-peptide-reactivity-assay-dpra (accessed on October 14, 2015).

ECVAM (2015c) https://eurl-ecvam.jrc.ec.europa.eu/eurl-ecvam-recommendations/recommendation-keratinosens-skin-sensitisation (accessed on October 14, 2015).

Edwards, S.M., Donnelly, T.A., Sayre, R.M., Rheins, L.A., Spielmann, H., Liebsch, M. (1994) Quantitative in vitro assessment of phototoxicity using a human skin model, skin2. *Photodermatol Photoimmunol Photomed*, 10 (3), 111–117.

EMA (2008) Note for Guidance on Non-Clinical Safety Studies for the Conduct of Human Clinical Trials and Marketing Authorization for Pharmaceuticals. http://www.ema.europa.eu/docs/en_GB/document_library/Scientific_guideline/2009/09/WC500002941.pdf (accessed on October 14, 2015).

EMA (2015) ICH Guideline S10 Guidance on Photosafety Evaluation of Pharmaceuticals. http://www.ema.europa.eu/docs/en_GB/document_library/Regulatory_and_procedural_guideline/2012/12/WC500136404.pdf (accessed on October 14, 2015).

Faller, C., Bracher, M., Dami, N., Roguet, R. (2002) Predictive ability of reconstructed human epidermis equivalents for the assessment of skin irritation of cosmetics. *Toxicol In Vitro*, 16, 557–572.

Farshchian, M., Ansar, A., Zamanian, A., Rahmatpour-Rokni, G., Kimyai-Asadi, A., Farshchian, M. (2015) Drug-induced skin reactions: a 2-year study. *Clin Cosmet Investig Dermatol*, 8, 53–56.

FDA (2012) Guidance for Industry: S2(R1) Genotoxicity Testing and Data Interpretation for Pharmaceuticals Intended for Human Use. http://www.fda.gov/downloads/Drugs/Guidances/ucm074931.pdf (accessed on October 14, 2015).

FDA (2015) Guidance for Industry: S10 Photosafety Evaluation of Pharmaceuticals. http://www.fda.gov/downloads/drugs/guidancecomplianceregulatoryinformation/guidances/ucm337572.pdf (accessed on October 14, 2015).

Fischer, A., Rosen, A.C., Ensslin, C.J., Wu, S., Lacouture, M.E. (2013) Pruritus to anticancer agents targeting the EGFR, BRAF, and CTLA-4. *Dermatol Ther*, 26 (2), 135–148.

Galbiati, V., Mitjans, M., Lucchi, L., Viviani, B., Galli, C.L., Marinovich, M., Corsini, E. (2011) Further development of the NCTC 2544 IL-18 assay to identify in vitro contact allergens. *Toxicol In Vitro*, 25 (3), 724–732.

Gibbs, S., Corsini, E., Spiekstra, S.W., Galbiati, V., Fuchs, H.W., Degeorge, G., Troese, M., Hayden, P., Deng, W., Roggen E. (2013) An epidermal equivalent assay for identification and ranking potency of contact sensitizers. *Toxicol Appl Pharmacol*, 272 (2), 529–541.

Hammell, D.C., Stolarczyk, E.I., Klausner, M., Hamad, M.O., Crooks, P.A., Stinchcomb, A.L. (2005) Bioconversion of naltrexone and its 3-o-alkyl-ester prodrugs in a human skin equivalent. *J Pharm Sci*, 94, 828–836.

Hsu, M.Y., Herlyn, M. (1996) Cultivation of normal human epidermal melanocytes. *Methods Mol Med*, 2, 9–20.

Hu, T., Bailey, R.E., Morrall, S.W., Aardema, M.J., Stanley, L.A., Skare, J.A. (2009a) Dermal penetration and metabolism of p-aminophenol and p-phenylenediamine: application of the epiderm human reconstructed epidermis model. *Toxicol Lett*, 188, 119–129.

Hu, T., Kaluzhny, Y., Mun, G.C., Barnett, B., Karetsky, V., Wilt, N., Klausner, M., Curren, R.D., Aardema, M.J. (2009b) Intralaboratory and interlaboratory evaluation of the EpiDerm 3D human reconstructed skin micronucleus (RSMN) assay. *Mutat Res*, 673 (2), 100–108.

Hu, T., Khambatta, Z.S., Hayden, P.J., Bolmarcich, J., Binder, R.L., Robinson, M.K., Carr, G.J., Tiesman, J.P., Jarrold, B.B., Osborne, R., Reichling, T.D., Nemeth, S.T., Aardema, M.J. (2010) Xenobiotic metabolism gene expression in the EpiDerm in vitro 3D human epidermis model compared to human skin. *Toxicol In Vitro*, 24 (5), 1450–1463.

Iriyama, S., Ono, T., Aoki, H., Amano, S. (2011) Hyperpigmentation in human solar lentigo is promoted by heparanase-induced loss of heparan sulfate chains at the dermal–epidermal junction. *J Dermatol Sci*, 64, 223–228.

ISO 10993-10:2010 (2015) Biological Evaluation of Medical Devices—Part 10: Tests for Irritation and Skin Sensitization. http://www.iso.org/iso/catalogue_detail?csnumber=40884 (accessed on October 14, 2015).

Jain, P., Sonti, S., Garruto, J., Mehta, R., Banga, A.K. (2012) Formulation optimization, skin irritation, and efficacy characterization of a novel skin-lightening agent. *J Cosmet Dermatol*, 11, 101–110.

Jírová, D., Basketter, D., Liebsch, M., Bendová, H., Kejlová, K., Marriott, M., Kandárová, H. (2010) Comparison of human skin irritation patch test data with in vitro skin irritation assays and animal data. *Contact Dermatitis*, 62 (2), 109–116.

Kandárová, H., Kejlová K., Jírová D., Tharmann J., Liebsch M. (2005) ECVAM feasibility study: can the pre-validated *in vitro* skin model phototoxicity assay be upgraded to quantify phototoxic potency of topical phototoxins? *ALTEX, Special Issue*, 22, 156.

Kandárová, H., Hayden, P., Klausner, M., Kubilus, J., Sheasgreen, J. (2009) An in vitro skin irritation test (SIT) using the EpiDerm reconstructed human epidermal (RHE) model. *J Vis Exp*, (29), pii: 1366.

Kandarova, H., Letasiova, S., Milasova, T., Willoughby, J., de Jong, W., de la Fonteyne, L., Bachelor, M., Breyfogle, B., Coleman, K. (2015) Development, optimization, and standardization of an *in vitro* skin irritation test for medical devices using the reconstructed human tissue model epiderm. *Toxicologist*, 144 (1), 443.

Karande, P., Jain, A., Ergun, K., Kispersky, V., Mitragotri, S. (2005) Design principles of chemical penetration enhancers for transdermal drug delivery. *Proc Natl Acad Sci U S A*, 102 (13), 4688–4693.

Kasraee, B., Nikolic, D.S., Salomon, D., Carraux, P., Fontao, L., Piguet, V., Omrani, G.R., Sorg, O., Saurat, J.-H. (2011) Ebselen is a new skin depigmenting agent that inhibits melanin biosynthesis and melanosomal transfer. *Exp Dermatol*, 21, 19–24.

Kejlová, K., Jírová, D., Bendová, H., Kandárová, H., Weidenhoffer Z., Kolářová, H., Liebsch, M. (2007) Phototoxicity of bergamot oil assessed by in vitro techniques in combination with human patch tests. *Toxicol In Vitro*, 21, 1298–1303.

Khan, F.D., Roychowdhury, S., Gaspari, A.A., Svensson, C.K. (2006) Immune response to xenobiotics in the skin: from contact sensitivity to drug allergy. *Expert Opin Drug Metab Toxicol*, 2 (2), 261–272.

Kim, M., Park, J., Song, K., Kim, H.G., Koh, J.-S., Boo, Y.C. (2012) Screening of plant extracts for human tyrosinase inhibiting effects. *Int J Cosmet Sci*, 34, 202–208.

Kulkantrakorn, K., Tassaneeyakul, W., Tiamkao, S., Jantararoungtong, T., Prabmechai, N., Vannaprasaht, S., Chumworathayi, P., Chen, P., Sritipsukho, P. (2012) HLA-B*1502 strongly predicts carbamazepine-induced Stevens-Johnson syndrome and toxic epidermal necrolysis in Thai patients with neuropathic pain. *Pain Pract*, 12 (3), 202–208.

Kurasawa, M., Kuroda, S., Kida, N., Murata, M., Oba, A., Yamamoto, T., Sasaki, H. (2009) Regulation of tight junction permeability by sodium caprate in human keratinocytes and reconstructed epidermis. *Biochem Biophys Res Commun*, 381, 171–175.

Laux, I., Jain, A., Singh, S., Agus, D.B. (2006) Epidermal growth factor receptor dimerization status determines skin toxicity to HER-kinase targeted therapies. *Br J Cancer*, 94 (1), 85–92.

Lee, A., Thompson, J. (2006) Drug-induced skin reactions. In: Lee, A. (ed.) *Adverse Drug Reactions*, 2nd edn., Pharmaceutical Press, London, pp 125–156.

Liebsch, M., Traue, D., Barrabas, C., Spielmann, H., Gerberick, G.F., Cruse, L., Diembeck, W., Pfannenbecker, U., Spieker, J., Holzhütter, H.G., Brantom, P., Aspin, P., Southee, J. (1999) Prevalidation of the EpiDerm phototoxicity test. In: Clark, D., Lisansky, S., Macmillan, R.(eds.) *Alternatives to Animal Testing II: Proceedings of the Second International Scientific Conference Organised by the European Cosmetic Industry, Brussels, Belgium*,CPL Press, Newbury, UK, pp. 160–166.

Lombardi Borgia, S., Schlupp, P., Mehnert, W., Schafer-Korting, M. (2008) In vitro skin absorption and drug release—a comparison of six commercial prednicarbate preparations for topical use. *Eur J Pharm Biopharm*, 68, 380–389.

Lonjou, C., Thomas, L., Borot, N., Ledger, N., de Toma, C., LeLouet, H., Graf, E., Schumacher, M., Hovnanian, A., Mockenhaupt, M., Roujeau, J.C.; RegiSCAR Group. (2006) A marker for Stevens-Johnson syndrome …: ethnicity matters. *Pharmacogenomics J*, 6 (4), 265–268.

Lotte, C., Patouillet, C., Zanini, M., Messager, A., Roguet, R. (2002) Permeation and skin absorption: reproducibility of various industrial reconstructed human skin models. *Skin Pharmacol Appl Skin Physiol*, 15 (suppl 1), 18–30.

Macdonald, J.B., Macdonald, B., Golitz, L.E., LoRusso, P., Sekulic, A. (2015a) Cutaneous adverse effects of targeted therapies: part I: inhibitors of the cellular membrane. *J Am Acad Dermatol,* 72 (2), 203–218.

Macdonald, J.B., Macdonald, B., Golitz, L.E., LoRusso, P., Sekulic, A. (2015b) Cutaneous adverse effects of targeted therapies: part II: inhibitors of intracellular molecular signaling pathways. *J Am Acad Dermatol,* 72 (2), 221–236.

Manousaridis, I., Mavridou, S., Goerdt, S., Leverkus, M., Utikal, J. (2013) Cutaneous side effects of inhibitors of the RAS/RAF/MEK/ERK signalling pathway and their management. *J Eur Acad Dermatol Venereol*, 27 (1), 11–18.

Maschmeyer, I., Hasenberg, T., Jaenicke, A., Lindner, M., Lorenz, A.K., Zech, J., Garbe, L.A., Sonntag, F., Hayden, P., Ayehunie, S., Lauster, R., Marx, U., Materne, E.M. (2015a) Chip-based human liver-intestine and liver-skin co-cultures—a first step toward systemic repeated dose substance testing in vitro. *Eur J Pharm Biopharm*, 95, 77–87.

Maschmeyer, I., Lorenz, A.K., Schimek, K., Hasenberg, T., Ramme, A.P., Hübner, J., Lindner, M., Drewell, C., Bauer, S., Thomas, A., Sambo, N.S., Sonntag, F., Lauster, R., Marx, U. (2015b) A four-organ-chip for interconnected long-term co-culture of human intestine, liver, skin and kidney equivalents. *Lab Chip*, 15 (12), 2688–2699.

McKim, J.M. Jr., Keller, D.J., Gorski, J.R. (2012) An in vitro method for detecting chemical sensitization using human reconstructed skin models and its applicability to cosmetic, pharmaceutical and medical device safety testing. *Cutan Ocul Toxicol,* 31 (4), 292–305.

Michel, M., L'Heureux, N., Pouliot, R., Xu, W., Auger, F.A., Germain, L. (1999) Characterization of a new tissue-engineered human skin equivalent with hair. *In Vitro Cell Dev Biol Anim*, 35 (6), 318–326.

Murase, D., Hachiya, A., Takano, K., Hicks, R., Visscher, M.O., Kitahara, T., Hase, T., Takema, Y., Yoshimori, T. (2013) Autophagy has a significant role in determining skin color by regulating melanosome degradation in keratinocytes. *J Investig Dermatol*, 133 (10), 2416–2424.

Natsch, A., Emter, R. (2008) Skin sensitizers induce antioxidant response element dependent genes: application to the in vitro testing of sensitisation potential of chemicals. *Toxicol Sci* 102, 110–119.

Nesterov, A., Zhao, J., Minter, D., Hertel, C., Ma, W., Abeysinghe, P., Hong, M., Jia, Q. (2008) 1-(2,4-dihydroxyphenyl)-3-(2,4-dimethoxy-3-methyl-phenyl)propane, a novel tyrosinase inhibitor with strong depigmenting effects. *Chem Pharm Bull*, 56 (9) 1292–1296.

OECD (2004a) OECD Guideline for Testing of Chemicals In Vitro 3T3 NRU Phototoxicity Test. http://ntp.niehs.nih.gov/iccvam/suppdocs/feddocs/oecd/oecdtg432-508.pdf (accessed on October 14, 2015).

OECD (2004b). Guidance Document for the Conduct of Skin Absorption Studies. OECD Series on Testing and Assessment. No. 28. Paris: OECD.

OECD (2004c). OECD Guidelines for the Testing of Chemicals. Test No. 427. Skin absorption: *In vivo* method. Paris: OECD.

OECD (2009). OECD Guideline for Testing of Chemicals, No. 439. In Vitro Skin Irritation: Reconstructed Human Epidermis Test Method. Paris: OECD.

Pirmohamed, M. (2012) Genetics and the potential for predictive tests in adverse drug reactions. *Chem Immunol Allergy*, 97, 18–31.

Reus, A.A., Reisinger, K., Downs, T.R., Carr, G.J., Zeller, A., Corvi, R., Krul, C.A., Pfuhler, S. (2013) Comet assay in reconstructed 3D human epidermal skin models—investigation of intra- and inter-laboratory reproducibility with coded chemicals. *Mutagenesis*, 28 (6), 709–720.

Reuter, H., Spieker, J., Gerlach, S., Engels, U., Pape, W., Kolbe, L., Schmucker, R., Wenck, H., Diembeck, W., Wittern, K.P., Reisinger, K., Schepky, A.G. (2011) In vitro detection of contact allergens: development of an optimized protocol using human peripheral blood monocyte-derived dendritic cells. *Toxicol In Vitro*, 25 (1), 315–323.

Rheinwald, J.G., Green, H. (1975) Serial cultivation of strains of human epidermal keratinocytes: the formation of keratinizing colonies from single cells. *Cell*, 6 (3), 331–343.

Rosen, A.C., Case, E.C., Dusza, S.W., Balagula, Y., Gordon, J., West, D.P., Lacouture, M.E. (2013) Impact of dermatologic adverse events on quality of life in 283 cancer patients: a questionnaire study in a dermatology referral clinic. *Am J Clin Dermatol*, 14 (4), 327–333.

Rovida, C., Alépée, N., Api, A.M., Basketter, D.A., Bois, F.Y., Caloni, F., Corsini, E., Daneshian, M., Eskes, C., Ezendam, J., Fuchs, H., Hayden, P., Hegele-Hartung, C., Hoffmann, S., Hubesch, B., Jacobs, M.N., Jaworska, J., Kleensang, A., Kleinstreuer, N., Lalko, J., Landsiedel, R., Lebreux, F., Luechtefeld, T., Locatelli, M., Mehling, A., Natsch, A., Pitchford, J.W., Prater, D., Prieto, P., Schepky, A., Schüürmann, G., Smirnova, L., Toole, C., van Vliet, E., Weisensee, D., Hartung, T. (2015) Integrated testing strategies (ITS) for safety assessment. *ALTEX,* 32 (1), 25–40.

Rudmann, D.G. (2013) On-target and off-target-based toxicologic effects. *Toxicol Pathol*, 41 (2), 310–314.

Schäfer-Korting, M., Bock, U., Gamer, A., Haberland, A., Haltner-Ukaomadu, E., Kaca, M., Kamp, H., Kietzmann, M., Korting, H.C., Krachter, H.-U., Lehr, C.M., Liebsch, M., Mehling, A., Netzlaff, F., Niedorf, F., Rübbelke, M.K., Schäfer, U., Schmidt, E., Schreiber, S., Schroder, K.-R., Spielmann, H., Vuia, A. (2006) Reconstructed human epidermis for skin absorption testing: results of the German prevalidation study. *ATLA,* 34, 283–294.

Schäfer-Korting, M., Bock, U., Diembeck, W., Düsing, H.-J., Gamer, A., Haltner-Ukomadu, E., Hoffmann, C., Kaca, M., Kamp, H., Kersen, S., Kietzmann, M., Korting, H.C., Krächter, H.-U., Lehr, C.-M., Liebsch, M., Mehling, A., Müller-Goymann, C., Netzlaff, F., Niedorf, F., Rübbelke, M.K., Schäfer, U., Schmidt, E., Schreiber, S., Spielmann, H., Vuia, A., Weimer, M. (2008a) The use of reconstructed human epidermis for skin absorption testing: results of the validation study. *ATLA*, 36, 161–187.

Schäfer-Korting, M., Mahmoud, A., Borgia, S.L., Brüggener, B., Kleuser, B., Schreiber, S., Mehnert, W. (2008b) Reconstructed epidermis and full-thickness skin for absorption testing: influence of the vehicles used on steroid permeation. *ATLA*, 36, 441–452.

Shigeta, Y., Imanaka, H., Yonezawa, S., Oku, N., Baba, N., Miakino, T. (2004) Suppressed permeation of linoleic acid in a liposomal formulation through reconstructed skin tissue. *Biol Pharm Bull*, 27, 879–882.

Sugimoto, K., Nishimura, T., Nomura, K., Sugimoto, K., Kurikia, T. (2004) Inhibitory effects of á-arbutin on melanin synthesis in cultured human melanoma cells and a three-dimensional human skin model. *Biol Pharm Bull*, 27 (4), 510–514.

Svensson, C.K., Cowen, E.W., Gaspari, A.A. (2001) Cutaneous drug reactions. *Pharmacol Rev*, 53 (3), 357–379.

Tassaneeyakul, W., Jantararoungtong, T., Chen, P., Lin, P.Y., Tiamkao, S., Khunarkornsiri, U., Chucherd, P., Konyoung, P., Vannaprasaht, S., Choonhakarn, C., Pisuttimarn, P., Sangviroon, A., Tassaneeyakul, W. (2009) Strong association between HLA-B*5801 and allopurinol-induced Stevens-Johnson syndrome and toxic epidermal necrolysis in a Thai population. *Pharmacogenet Genomics*, 19 (9), 704–709.

Van Och, F.M., Van Loveren, H., Van Wolfswinkel, J.C., Machielsen, A.J., Vandebriel, R.J. (2005) Assessment of potency of allergenic activity of low molecular weight compounds based on IL-1alpha and IL-18 production by a murine and human keratinocyte cell line. *Toxicology*, 210 (2–3), 95–109.

Yuan, J.S., Ansari, M., Samaan, M., Acosta, E.J. (2008) Linker-based lecithin microemulsions for transdermal delivery of lidocaine. *Int J Pharm*, 349, 130–143.

13

IN VITRO METHODS IN IMMUNOTOXICITY ASSESSMENT

Xu Zhu and Ellen Evans

Immunotoxicology Center of Emphasis, Drug Safety Research and Development, Pfizer Inc., Groton, CT, USA

13.1 INTRODUCTION AND PERSPECTIVES ON IN VITRO IMMUNOTOXICITY SCREENING

One aspect of safety assessment that is growing in importance is the determination of the impact of pharmaceuticals on the immune system. The immune system is very complex due to the evolution of multiple pathways, allowing each species to respond to a variety of organisms; those organisms which are typically encountered and coevolved with the host species, as well as organisms which are novel to the host species or individual. There is also a built-in redundancy in the immune system to account for the fact that survival is dependent on the ability to ward off invaders. Because of this complexity and redundancy, there is no single *in vitro* screening paradigm that will reliably identify the spectrum of potential adverse effects associated with pharmaceuticals which impact the immune system. In fact, regulatory guidance (ICH S-8) relies on *in vivo* standard toxicity studies (STS) as the primary screen for unintended effects on the immune system (ICH, 2005). STS, which are required for registration, include many endpoints which assess immune status, including clinical signs, hematology, serum chemistry (particularly albumin and globulins) (Evans, 2008), and histopathology of bone marrow, thymus, and peripheral lymphoid tissues as well as tissues which may be targets of immunostimulation, inflammation, or infection. It is not practical, and is rarely necessary, to perform functional tests prior to STS to screen for unintended or unexpected effects on the immune system, particularly when there are no holistic, high-throughput screening tools available. In fact, the most commonly used broad tool for immunotoxicity testing is an *in vivo* assay, the T-cell-dependent antibody response (TDAR), which is in widespread use in the chemical industry to identify unintended immunosuppression. In pharmaceutical testing, TDAR is primarily used as a follow-on study to determine if there is a functional defect in immunity when findings in STS suggest immunotoxicity but no specific mechanism is identified or suspected. This assay covers multiple immune functions in that it requires intact, coordinated antigen-presenting, T-cell, and B-cell function to produce antibodies against a particular antigen and can measure both primary and recall responses to the antigen. However, it does not assess innate immunity or cytotoxic T lymphocyte function. The TDAR also requires time (it is typically conducted with 28 days of dosing) and enough animals to overcome variability (Lebrec et al., 2014) and allow determination of dose response, limiting its usefulness as a screening assay.

When is it appropriate to conduct *in vitro* assays for immunotoxicity? Every molecule, target, and indication should be evaluated on a case-by-case basis, both for the need to conduct assays and the selection of appropriate assays. The ICH S8 guidance suggests that the decision to conduct specific assays for immunotoxicity should be made based on a "Weight of Evidence (WOE)" review (ICH, 2005; Spanhaak, 2006). Factors that are taken into consideration are findings from STS, pharmacologic properties, intended patient population, structural similarities to known immunotoxicants, drug disposition, and clinical findings suggestive of effects on the immune system. The assays can be used to predict potential liabilities early in a program or understand

Drug Discovery Toxicology: From Target Assessment to Translational Biomarkers, First Edition. Edited by Yvonne Will, J. Eric McDuffie, Andrew J. Olaharski, and Brandon D. Jeffy.

mechanism of action. In some cases, they may be appropriate for screening molecules in a series or targets with known immunotoxicity liabilities (ICH, 2005; Kawabata and Evans, 2012).

For unexpected immunotoxicity seen in STS, *in vitro* assays may be conducted as a follow-on to TDAR results or instead of a TDAR assay, depending on the nature of the findings. For example, in the case of bone marrow suppression, a TDAR would not be a logical functional assay; a more appropriate approach would be bone marrow culture evaluating the affected line(s), particularly using human cells in comparison with those of the toxicology species to determine translatability and relative magnitude of the finding. In the case of a bacterial infection in an *in vivo* study, neutrophil function could be evaluated. Decisions should be scientifically based—trying to apply a broad screening approach is likely to result in unnecessary testing and may miss the actual defect responsible for the finding.

For molecules that are intended or expected to impact the immune system, some early characterization of immune effects may be warranted to identify liabilities and understand biological relevance of intended toxicology species and translatability of findings to humans. Evaluations should be conducted to determine target engagement and presence of target on analogous cells. However, particularly in the case of the immune system, these assessments are not enough to ensure biological relevance of a toxicology species, as there are significant species differences in immune responses, which are often not well characterized, and downstream biology or distribution of target in tissues or on cells may vary (Pallardy and Hunig, 2010; Evans, 2014) and may not be fully characterized. There is increasing emphasis on using human cells to assess the effects of immunomodulatory drugs *in vitro* (Kawabata and Evans, 2012). Functional and mechanistic studies focus on mechanism of action and breadth of impact (is only the intended pathway affected, or are additional pathways affected which might counter or exacerbate intended immunomodulation?), and the assays are often developed de novo to answer specific questions unique to the molecule or target.

13.2   OVERVIEW OF THE IMMUNE SYSTEM

This overview provides a superficial description of the immune system, focusing primarily on those aspects which may be impacted by pharmaceuticals unintentionally. For a more thorough review of the immune system, there are some excellent books on the subject (Murphy, 2014; Abbas et al., 2015). The immune system is traditionally categorized into two compartments based on how antigens are recognized: "innate" immunity, which is nonspecific and does not require prior exposure, and "acquired" immunity, in which exposure to an antigen brings about future recognition of that antigen and responses involving antibody production (humoral immunity) and/or specific cell-mediated immunity. The initial response is referred to as a primary immune response. Upon subsequent exposure to the same antigen, secondary responses occur, which tend to be more rapid and are of greater magnitude than the primary response.

Immune cells are not confined to a single site within the body. The bone marrow and thymus are generally considered primary lymphatic organs and sites of production for B and T cells, respectively. In addition, bone marrow is the primary site of hematopoiesis in health, producing granulocytes, erythrocytes, platelets, and monocytes, all of which can be involved in immune responses. Secondary lymphoid tissues include lymph nodes, tonsils, spleen, Peyer's patches, and mucosa-associated lymphoid tissue (MALT), which are responsible for filtering the contents of the extracellular fluids, that is, lymph and blood. The secondary lymphoid tissues are major sites of antigen presentation and lymphocyte activation. Various types of immune cells also reside in key organs including brain, heart, lung, liver and kidneys in order to protect the body against infectious organisms and other invaders.

As the immune system evolved as a means to protect the body from infection and eliminate cells (e.g., virus containing, senescent, or tumor cells), various mechanisms (e.g., anti-inflammatory cytokines, regulatory T cells) also evolved to limit or mitigate the damage to normal tissues caused by an immune response or prevent attack on normal cells. Uncontrolled immunostimulation may result in circulatory collapse or damage to tissues, which may be life threatening.

Some of the mechanisms by which immunostimulation can be adverse are through mast cell degranulation, discussed as part of Type I hypersensitivity reactions in the next paragraph, inflammatory cytokine release, and complement activation. Cytokines serve many purposes, including calling in immune and inflammatory cells, activating and stimulating proliferation of cells, and perpetuating immune and inflammatory responses. Some cytokines, such as interleukin 10 (IL-10), have dampening effects on inflammation. Cytokines and complement are involved in many immune/inflammatory processes, both "acquired" and "innate," regardless of initiating cause. The complement system consists of both soluble proteins and proteins bound to cell membranes. There are three potential pathways (classical, lectin, and alternative) in which the complement cascade becomes activated to ultimately form complement split products including C4a, C3a, and C5a, resulting in the release of mediators which can cause anaphylaxis, call in inflammatory cells, and trigger neutrophil activation and degranulation, resulting in tissue destruction and amplification of the response. For the patient, these events cause changes in blood pressure, flushing, rash, and dyspnea, as well as other clinical signs associated with inflammation and cytokine release (Ricklin and Lambris, 2007; Moghimi et al., 2011).

Hypersensitivity is a form of immunostimulation in which normal immune responses are triggered inappropriately or are directed at inappropriate targets. Hypersensitivity reactions involving acquired immune responses traditionally have been classified into four categories (Gell and Coombs, 1968; Pichler, 2003). Type I reactions are immediate hypersensitivity reactions, which involve immunoglobulin E (IgE) and mediators such as histamine released from mast cells and basophils. Consequences include vasodilation, increased vessel wall permeability, itching, and in extreme cases, systemic anaphylaxis. Type II reactions are classic autoimmune reactions, mediated by immunoglobulin G (IgG) or immunoglobulin M (IgM) antibodies, where the antibody (Ab) binds to host cells, forming an antigen/antibody (Ag/Ab) complex (also referred to as "immune complex") and activating complement, resulting in cellular destruction and surrounding tissue damage. In Type III reactions, immune complexes form in circulation, deposit in tissues, and activate complement. Such activation further attracts neutrophils and monocytes, which causes tissue destruction and inflammation. Type IV reactions are also referred to as "delayed-type hypersensitivity reactions" because they typically take several days to manifest clinically. These reactions are mediated by sensitized T lymphocytes rather than antibodies.

The effects of drugs on the immune system can impact the drug development process when those effects pose a potential risk to human safety. Adverse effects of drugs include immunosuppression (excessive dampening of immune responses) and excessive immunostimulation, including hypersensitivity, and development of *in vitro* methods has focused on these areas. There are still unmet needs in prediction of hypersensitivity and autoimmunity, likely due to the multifactorial nature of these syndromes and lack of mechanistic understanding in the scientific community. Although there is still a lack of validated *in vitro* tests to assess immunotoxicity, there is ongoing research in this area, and those assays that do exist have utility in understanding mechanisms of toxicity and identifying potential hazards in a human-relevant system.

Some examples of *in vitro* assays for immunotoxicity assessment follow (Table 13.1), but this list is not intended to be comprehensive. It should be borne in mind that this is a rapidly growing area of research, and many of the assays employed in immunotoxicity or immunosafety evaluation of pharmaceuticals are fit-for-purpose assays based on mechanism or target and are beyond the scope of this chapter. Evaluation of effects on individual cell types and mechanisms (e.g., neutrophil, macrophage, cytotoxic T lymphocyte, and NK function) will not be discussed, as these are rarely employed unless driven by STS findings or intended mechanism of action.

TABLE 13.1 Examples of *In Vitro* Testing Assay for Immunotoxicity Assessment

Assay	Endpoint	Screening Purpose	References
Human lymphocyte activation assay (HuLA)	Proliferation, flu antigen-specific IgG	Immunosuppression	Collinge et al. (2010)
Cytokine release assay	Cytokines	Immunostimulation	Vidal et al. (2010)
Fcγ receptor binding assay	Binding affinities of drug to FcγR	Binding suggests possibility of ADCC; follow-on ADCC studies conducted if positive	Brennan et al. (2010), Evans (2014), and CHMP (2009)
Complement binding assay	Binding affinities of drug to C1q component of complement	Binding suggests possibility of CDC; follow-on CDC studies conducted if positive	Brennan et al. (2010), Evans (2014), and CHMP (2009)
Complement activation CH50	Complement-mediated hemolysis	Immunostimulation	Pham et al. (2014)
Complement activation ELISA, Western blot	Complement activation fragments	Immunostimulation	Dobrovolskaia et al. (2008)
Complement WIESLAB assay	Complement pathway	Immunostimulation	EuroDiagnostica (www.eurodiagnostica.com)
T-cell priming culture	Proliferation markers, cytokine secretion, and T-cell phenotype	Hypersensitivity	Faulkner et al. (2012)
Dendritic cell and keratinocyte activation	Activation markers, cytokine secretion	Chemical contact sensitization	Basketter and Maxwell (2007) and Teunis et al. (2013)
Mast cell assay	Release of mediators	Pseudoallergy	Toyoguchi et al. (2000) and Hohman and Dreskin (2001)

13.3   EXAMPLES OF *IN VITRO* APPROACHES

13.3.1   Acquired Immune Responses

An alternative to the traditional *in vivo* TDAR assay is the recently developed *in vitro* human lymphocyte activation (HuLA) assay using human peripheral blood mononuclear cells (PBMCs), which measures secondary influenza-specific responses. This assay is sensitive to, and can differentiate the responses of, a number of known immunosuppressive compounds with different mechanisms of action at concentrations within their respective therapeutic ranges. Various endpoints can be evaluated, including proliferation and flu antigen-specific antibody (IgM and IgG)-secreting cells (Collinge et al., 2010).

Other *in vitro* assays have been developed, which use animal cells, for example, an assay developed by Fischer et al.(2011), which uses rodent blood cells and splenocytes exposed to sheep red blood cells, and the Mishell–Dutton assay, which is an *in vitro* immunization culture system measuring an primary antibody responses of mouse splenocytes (Michell and Dutton, 1967). Although the HuLA assay of Collinge et al. (2010) only measures the secondary response, it has advantages over these *in vitro* methods in that it uses human cells and a human disease-relevant antigen.

13.3.2   Fcγ Receptor and Complement Binding

If a monoclonal antibody bound to its intended target on a cell surface interacts with Fcγ receptors (FcγRs) on NK or other effector cells, the target cells may be killed via antibody-dependent cell-mediated cytotoxicity (ADCC). Similarly, binding to the C1q component of complement may result in complement activation and complement-dependent cytotoxicity (CDC) of the target cells. In both of these scenarios, cytokines are released and inflammatory responses are triggered, exacerbating the toxicity. The potential for molecules to bind to FcγR and C1q should be determined, particularly when target cell killing is not the desired effect, and ADCC or CDC conducted as a follow-up for positive results (CHMP, 2009; Brennan et al., 2010; Evans, 2014).

13.3.3   Assessment of Hypersensitivity

Adverse drug reactions (ADRs) are considered the sixth leading cause of death in the United States and cost between 30 and 130 billion annually (White et al., 1999). Hypersensitivity reactions are thought to be responsible for about 6–10% of ADRs (Thong and Tan, 2011). Manifestations of drug hypersensitivity include contact hypersensitivity, respiratory hypersensitivity, systemic hypersensitivity, and autoimmunity. Pseudoallergy is also included in the category of "hypersensitivity," although the inciting causes are different (Descotes and Choquet-Kastylevsky, 2001). Unlike the other hypersensitivities, which are considered immune-mediated because immune cells initiate the responses, pseudoallergic events, also called nonimmune hypersensitivities, do not require prior exposure and result from direct action of the drug on an immune cell or complement. The mechanisms for drug-induced pseudoallergy include cytokine release, histamine release from mast cells, alterations in arachidonic acid metabolite pathways, and/or direct complement activation. Despite the differences in initiating events, clinical signs associated with pseudoallergic reactions are similar to those observed with immune-mediated hypersensitivities because the effector mechanisms are the same.

13.3.3.1   Drug Allergy (Gell and Coombs Types I, III, and IV Hypersensitivity)   The mechanisms of drug-induced immune-mediated hypersensitivity are not fully understood, but there are several hypotheses. The hapten hypothesis proposes that small molecule drugs or their metabolites are able to bind covalently to proteins and form haptens, which could induce an immune response (Landsteiner and Jacobs, 1935; Mitchell et al., 1973). The pharmacological interaction (p-i) hypothesis proposed by Pichler suggests that the parent drug binds reversibly to the complex formed by the complex of MHC II on antigen-presenting cells (APCs) and the T-cell receptor on T cells to initiate an immune response (Pichler, 2002). Finally, according to the danger hypothesis proposed by Polly Matzinger, injured tissue produces danger signals that activate APCs, leading to upregulation of costimulatory molecules and induction of an immunological response (Matzinger, 1994).

Several *in vivo* methods exist and are used as tools to determine the potential for contact, and potentially respiratory, hypersensitivity (e.g., the local lymph node (LLNA) (Kimber et al., 2002) and popliteal lymph node (PLNA) (Pieters, 2001) assays); and the mouse drug allergy model, which is a modification of the LLNA, demonstrates promise as a screening tool in that a number of drugs which cause idiosyncratic hypersensitivity reactions in humans were positive in this assay (Whritenour et al., 2014; Zhu et al., 2014). However, there are no fully predictive animal models of drug allergy in humans.

Great efforts have been made by multiple groups to develop *in vitro* methods and assays to identify drug candidates or chemicals with the potential to induce allergic reactions. In 2012, Faulkner et al. reported their *in vitro* T-cell culture system. In this method, peripheral blood obtained from healthy volunteers was used to prepare a coculture of T cells and dendritic cells with the drug tested, and responses to drug-specific stimulation were measured by examining the changes in proliferation markers, cytokine secretion, and T-cell phenotype. This assay appears to be a promising tool for predicting drug allergy potential and is currently under further investigation and development (Faulkner et al., 2012). Other *in vitro* approaches involving dendritic cells and keratinocytes have been used for identifying contact allergens and irritants. While these assays may have promise in predicting contact sensitization for chemicals, the use of such assays to predict reactions with systemically administered drugs requires further investigation (Basketter and Maxwell, 2007; Galbiati et al., 2011; Teunis et al., 2013).

13.3.3.2 Autoimmunity (Gell and Coombs Type II Hypersensitivity)

Autoimmunity represents a process when a specific adaptive immune response is mounted against self-antigens. Antibodies or cytotoxic T cells directed against self-antigens or haptens formed by drug bound to self-proteins can lead to direct tissue damage, immune complex deposition with complement activation, or stimulation of target function (Descotes, 1990; Knowles et al., 2000). Autoimmunity can be organ specific or systemic. The mechanisms that lead to autoimmune diseases remain largely unknown, but reviews of adverse reactions suggest that drugs can induce autoimmune phenomena, which are usually systemic rather than organ specific (Uetrecht, 2005; Uetrecht and Naisbitt, 2013). Some of the most common types of drug-induced autoimmune disorders include lupus-like syndrome, hemolytic anemia, thrombocytopenia, liver injury, and vasculitis (Rose and Bhatia, 1995; Uetrecht and Naisbitt, 2013).

Although drug-induced autoimmunity is a very important health issue and can significantly impact human safety, the multifactorial nature of these reactions and poor mechanistic understanding have hampered efforts to develop predictive tools. There is a great need to develop *in vitro* models to detect autoimmunity potential. As far as *in vivo* tools, the PLNA and its modification, reporter antigen-PLNA (RA-PLNA) have been proposed as having utility in autoimmunity prediction (Descotes and Verdier, 1995; Pieters and Albers, 1999; Pieters, 2007). However, the value of the assays for autoimmunity prediction has not been validated and they are not in wide use for this purpose in drug development. Methods such as screening for autoantibody production in preclinical toxicology studies have also been proposed and showed some promise for predicting autoimmune effects associated with certain protein drugs (Wierda et al., 2001).

13.3.3.3 Pseudoallergy

Mast Cell Degranulation Mast cells play a role in Type I hypersensitivity reactions, but they can also be stimulated by pharmaceuticals in an IgE-independent manner due to direct interactions with mast cell membranes or surface receptors (de Weck, 1984). The reaction *in vivo* typically occurs at maximum drug concentration, or very soon after an IV infusion, and tends to diminish with subsequent exposures to the drug. Pseudoallergy is a rare event, and routine screening is neither warranted nor simple. However, if a potential liability is identified, a series of compounds can be evaluated using an *in vitro* assay in which mast cells are exposed to the drug, and histamine (or other mediator) release is measured (Toyoguchi et al., 2000; Hohman and Dreskin, 2001; Evans, 2014). This assay can be used to characterize findings in preclinical studies using mast cells from the preclinical species and compared with results using human mast cells to determine translatability and relative risk.

Cytokine Release One *in vitro* assay that is in wide use for immunotoxicity screening, particularly for monoclonal antibodies (mAbs) with targets on immune cells, is the cytokine release assay (CRA). Cytokines released by activated cells *in vivo* may trigger a systemic inflammatory response, which can be manifested by mild systems of fever and malaise, hypotension, or life-threatening multiorgan failure. When the target of a pharmaceutical drug is present on an immune cell, the ability of the drug to stimulate cytokine release is often assessed using human cells (whole blood or PBMCs) incubated with the drug, either in solution or with cells and/or drug immobilized to promote cross-linking of the mAbs. Subsequently, a panel of cytokines (e.g., IL-1, TNF, IFNγ, IL-6, IL-10) is measured. There is, however, very little understanding of the translatability of the data from the *in vitro* system to human patients. Therefore, the assay is useful primarily as hazard identification for cytokine release potential, as opposed to determining exposure margins or predicting clinical consequences (Vidal et al., 2010).

Complement Activation Complement activation has been described as a potential liability with liposomes, Cremophor, nanoparticles and oligonucleotides (Huttel et al., 1980; Szebeni et al., 2003; Advani et al., 2005; Salvador-Morales et al., 2006; Moghimi et al., 2011; Henry et al., 2014; Pham et al., 2014), and for molecules or solvents in these categories, screening is warranted. In the CH50 assay, complement activation is measured *in vitro* by incubating the test molecule with human (or other species) serum and exposing the mixture to antibody-sensitized sheep red blood cells. Complement-mediated hemolysis is then determined. If the test molecule activates complement, the red blood cells will not be lysed because the available complement will have been activated and consumed (Pham et al., 2014). A simpler, more sensitive method of assessing likelihood of complement activation is incubation of serum or plasma with drug and measuring activation fragments via enzyme-linked immunoassay (ELISA). Most commercial kits have been developed to measure human complement fragments, and may not all cross-react with preclinical species, but there is one kit (Quidel Corporation) that works for additional species by allowing the activated complement to be converted to human SC5b-9 (soluble membrane attack complex), which is then measured by a human-specific ELISA. Another assay, the WIESLAB® system (EuroDiagnostica), identifies which complement pathway(s) are activated by a drug candidate, and complement cleavage products can also be identified by Western blot using fragment-specific antibodies (Dobrovolskaia et al., 2008).

13.3.4 Immunogenicity of Biologics

Immunogenicity is referred to as the ability of a substance to elicit an immune response (Murphy, 2011). In drug development, both low molecular weight compounds and large molecules can be immunogenic via different mechanisms. Small-molecule drugs, for example, penicillin, may covalently bind to proteins forming hapten–protein complexes and then become immunogenic. Peptide- or protein-based biotherapeutics often have more complicated structures, which can be considered as "foreign" or "nonself" by the immune system, and they may also lead to the formation of antidrug antibodies (ADA), which could potentially induce severe adverse reactions, such as complement activation, immune complex disease, and IgE-mediated anaphylactic reactions (Chung et al., 2008; Evans, 2014). In addition to molecule structure and inherent immunogenicity potential, immunogenicity may be related to several other factors, such as patient factors (underlying disease, genetic background), route of administration, impurities, and manufacturing process. Characterization of immunogenicity is crucial in the drug development and marketing of therapeutic proteins. In addition to potential toxicities associated with ADA, the pharmacokinetic and pharmacodynamics profile may be altered significantly, resulting in diminished or prolonged exposure or diminished efficacy due to neutralization of the therapeutic.

Evaluation of protein drugs for immunogenic potential is difficult in a nonclinical setting. As human proteins are inevitably immunogenic in animals, *in vivo* models using mice, rats, rabbits, and cynomolgus monkeys are usually of limited value for the evaluation of human immunogenicity. Therefore, research is ongoing to develop *in vitro* predictive assays for immunogenicity in the context of ADA production. These methods may also prove useful to evaluate the allergenic potential of protein drugs, although they have not been extensively validated in this regard (Kawabata et al., 1996; Wierda et al., 2001).

Some efforts in the development of *in vitro* assays for immunogenicity prediction have focused on approaches such as identifying problematic sequences, measuring T-cell responses, and evaluating responses of antigen-presenting DCs to drug. *In silico* T-cell epitope prediction methods and HLA-binding affinity assays have been used to identify sequences within the therapeutic protein that bind to specific MHC alleles (Tong et al., 2007) because there is some evidence of a genetic component to immunogenicity and allergic reactions. Biotherapeutics with immunogenic potential may induce cellular immune responses involving the primary activation of T lymphocytes (Gorbachev and Fairchild, 2001), which can be measured in various ways, including the identification of activation markers using immunophenotyping. A few assays have been developed to expand the antigen-specific T cells from the naïve pool in human peripheral blood samples and examine their activation (Dai and Streilein, 1998; Rustemeyer et al., 1999). In addition, measuring proliferation of T cells and their cytokine production has also been proposed. However, currently, the assays discussed in this section have not been validated for immunogenicity or immunotoxicity assessment, and this remains an unmet need.

13.3.5 Immunotoxicity Due to Myelotoxicity

Molecules that damage or affect normal bone marrow functions may induce immunotoxicity as well, since bone marrow is the site of hematopoiesis. Clinical manifestations of drug-induced myelotoxicity include peripheral pancytopenia if stem cells or all hematopoietic lineages are involved or anemia, leukopenia, and/or thrombocytopenia, if only specific lineages are affected (Carey, 2003; Evans, 2008). Typical bone marrow culture assays focus on granulocyte, monocyte, megakaryocyte, and/or erythroid lineages. Lymphopoiesis can be evaluated by other functional assays such as lymphocyte proliferation assays, which may include assessment of proliferation of individual lymphoid subsets. For more details on assessment of bone marrow toxicity, please see Chapter IV.11.

13.4 CONCLUSIONS

While there are many *in vitro* assays available to assess immune function, and therefore immunotoxicity potential, it is not practical to recommend routine screening for immunotoxicity. There are no assays available that cover the spectrum of potential immunotoxicity effects. The nature of the immune system, with its many different cell types, tissues, and available responses, precludes the establishment of a reliable, inclusive screening paradigm for unexpected immunotoxicity, which can be broadly applied or provide adequate substitutes for available *in vivo* methods such as STS immune system endpoints and TDAR. However, *in vitro* methods to assess the impact of the drug on immunity do have a solid place in drug development, particularly in the development of molecules with known potential for impact on the immune system and those for which a WOE review suggests a potential liability. These methods are very useful in early screening to assess pharmacology, species relevance, and unacceptable immunotoxicity and may be used to rank potential drug candidates or targets in terms of safety liability. The decision to use *in vitro* assays and the selection of appropriate *in vitro* assays should always be made on a case-by-case basis and be driven by scientific review of the characteristics of the molecule, the biology of the target, potential redundant or compensatory pathways, the indication, and the patient population.

REFERENCES

Abbas, A., Lichtman, A. & Pillai, S. eds. (2015). *Cellular and Molecular Immunology*, Elsevier, Canada.

Advani, R., Lum, B.L., Fisher, G.A., Halsey, J., Geary, R.S., Holmlund, J.T., Kwoh, T.J., Dorr, F.A., & Sikic, B.I. (2005). A phase I trial of aprinocarsen (ISIS 3521/LY900003), an antisense inhibitor of protein kinase C-alpha administered as a 24-hour weekly infusion schedule in patients with advanced cancer. *Investigational New Drugs*, 23, 467–477.

Basketter, D. & Maxwell, G. (2007). In vitro approaches to the identification and characterization of skin sensitizers. *Cutaneous and Ocular Toxicology*, 26, 359–373.

Brennan, F.R., Morton, L.D., Spindeldreher, S., Kiessling, A., Allenspach, R., Hey, A., Muller, P.Y., Frings, W., & Sims, J. (2010). Safety and immunotoxicity assessment of immunomodulatory monoclonal antibodies. *MAbs*, 2, 233–255.

Carey, P.J. (2003). Drug-induced myelosuppression: diagnosis and management. *Drug Safety*, 26, 691–706.

CHMP (2009). Guideline on development, production, characterisation and specifications for monoclonal antibodies and related products, 40CFR Part 261., US Government Printing Office, Washington, DC.

Chung, C.H., Mirakhur, B., Chan, E., Le, Q.T., Berlin, J., Morse, M., Murphy, B.A., Satinover, S.M., Hosen, J., Mauro, D., Slebos, R.J., Zhou, Q., Gold, D., Hatley, T., Hicklin, D.J., & Platts-Mills, T.A. (2008). Cetuximab-induced anaphylaxis and IgE specific for galactose-alpha-1,3-galactose. *The New England Journal of Medicine*, 358, 1109–1117.

Collinge, M., Cole, S.H., Schneider, P.A., Donovan, C.B., Kamperschroer, C., & Kawabata, T.T. (2010). Human lymphocyte activation assay: an in vitro method for predictive immunotoxicity testing. *Journal of Immunotoxicology*, 7, 357–366.

Dai, R. & Streilein, J.W. (1998). Naive, hapten-specific human T lymphocytes are primed in vitro with derivatized blood mononuclear cells. *The Journal of Investigative Dermatology*, 110, 29–33.

Descotes, J. (1990). *Drug-Induced Immune Diseases* Elsevier, Amsterdam.

Descotes, J. & Choquet-Kastylevsky, G. (2001). Gell and Coombs's classification: is it still valid? *Toxicology*, 158, 43–49.

Descotes, J. & Verdier, F. (1995). Popliteal lymph node assay In *Methods in Immunotoxicology*, 1, (Burleson, G.R., Dean, J.H., & Munson, A.E. eds.), pp. 189–196, Wiley-Liss, New York.

Dobrovolskaia, M.A., Aggarwal, P., Hall, J.B., & McNeil, S.E. (2008). Preclinical studies to understand nanoparticle interaction with the immune system and its potential effects on nanoparticle biodistribution. *Molecular Pharmaceutics*, 5, 487–495.

Evans, E.W. (2008). Clinical pathology as crucial insight into immunotoxicity testing In *Immunotoxicology Strategies for Pharmaceutical Safety Assessment* (Herzyk, D.J. & Bussiere, J.L. eds.), pp. 13–26. John Wiley & Sons, Inc., Hoboken, NJ.

Evans, E.W. (2014). Regulatory forum commentary: is unexpected immunostimulation manageable in pharmaceutical development? *Toxicologic Pathology*, 42, 1053–1057.

Faulkner, L., Martinsson, K., Santoyo-Castelazo, A., Cederbrant, K., Schuppe-Koistinen, I., Powell, H., Tugwood, J., Naisbitt, D.J., & Park, B.K. (2012). The development of in vitro culture methods to characterize primary T-cell responses to drugs. *Toxicological Sciences*, 127, 150–158.

Fischer, A., Koeper, L.M., & Vohr, H.W. (2011). Specific antibody responses of primary cells from different cell sources are able to predict immunotoxicity in vitro. *Toxicology In Vitro*, 25, 1966–1973.

Galbiati, V., Mitjans, M., Lucchi, L., Viviani, B., Galli, C.L., Marinovich, M., & Corsini, E. (2011). Further development of the NCTC 2544 IL-18 assay to identify in vitro contact allergens. *Toxicology In Vitro*, 25, 724–732.

Gell, P.G.H. & Coombs, R.R.A. (1968). *Clinical Aspects of Immunology*, Blackwell Scientific, Oxford/Edinburgh.

Gorbachev, A.V. & Fairchild, R.L. (2001). Induction and regulation of T-cell priming for contact hypersensitivity. *Critical Reviews in Immunology*, 21, 451–472.

Henry, S.P., Jagels, M.A., Hugli, T.E., Manalili, S., Geary, R.S., Giclas, P.C., & Levin, A.A. (2014). Mechanism of alternative complement pathway dysregulation by a phosphorothioate oligonucleotide in monkey and human serum. *Nucleic Acid Therapeutics*, 24, 326–335.

Hohman, R.J. & Dreskin, S.C. (2001). Measuring degranulation of mast cells. *Current Protocols in Immunology*. doi:10.1002/0471142735.im0726s08.

Huttel, M.S., Schou Olesen, A., & Stoffersen, E. (1980). Complement-mediated reactions to diazepam with Cremophor as solvent (Stesolid MR). *British Journal of Anaesthesia*, 52, 77–79.

ICH. (2005). Harmonised Tripartite Guideline: Immunotoxicity Studies for Human Pharmaceuticals S8 In: International Conference on Harmonisation of Technical Requirements for Registration of Pharmaceuticals for Human Use, 15 September 2005.

Kawabata, T.T. & Evans, E.W. (2012). Development of immunotoxicity testing strategies for immunomodulatory drugs. *Toxicologic Pathology*, 40, 288–293.

Kawabata, T.T., Babcock, L.S., & Horn, P.A. (1996). Specific IgE and IgG1 responses to subtilisin Carlsberg (Alcalase) in mice: development of an intratracheal exposure model. *Fundamental and Applied Toxicology*, 29, 238–243.

Kimber, I., Dearman, R.J., Basketter, D.A., Ryan, C.A., & Gerberick, G.F. (2002). The local lymph node assay: past, present and future. *Contact Dermatitis*, 47, 315–328.

Knowles, S.R., Uetrecht, J., & Shear, N.H. (2000). Idiosyncratic drug reactions: the reactive metabolite syndromes. *Lancet*, 356, 1587–1591.

Landsteiner, K. & Jacobs, J. (1935). Studies on the sensitization of animals with simple chemical compounds. *The Journal of Experimental Medicine*, 61, 643–656.

Lebrec, H., Molinier, B., Boverhof, D., Collinge, M., Freebern, W., Henson, K., Mytych, D.T., Ochs, H.D., Wange, R., Yang, Y., Zhou, L., Arrington, J., Christin-Piche, M.S., & Shenton, J. (2014). The T-cell-dependent antibody response assay in non-clinical studies of pharmaceuticals and chemicals: study design, data analysis, interpretation. *Regulatory Toxicology and Pharmacology*, 69, 7–21.

Matzinger, P. (1994). Tolerance, danger, and the extended family. *Annual Review of Immunology*, 12, 991–1045.

Michell, R.I. & Dutton, R.W. (1967). Immunization of dissociated mouse spleen cell cultures from normal mice. *Journal of Experimental Medicine*, 126, 423–442.

Mitchell, J.R., Jollow, D.J., Gillette, J.R., & Brodie, B.B. (1973). Drug metabolism as a cause of drug toxicity. *Drug Metabolism and Disposition*, 1, 418–423.

Moghimi, S.M., Andersen, A.J., Ahmadvand, D., Wibroe, P.P., Andresen, T.L., & Hunter, A.C. (2011). Material properties in complement activation. *Advanced Drug Delivery Reviews*, 63, 1000–1007.

Murphy, K.M. (2011). *Janeway's Immunobiology*, Garland Science, New York.

Murphy, K. ed. (2014). *Janeway's Immunobiology*, Garland Science, New York.

Pallardy, M. & Hunig, T. (2010). Primate testing of TGN1412: right target, wrong cell. *British Journal of Pharmacology*, 161, 509–511.

Pham, C.T., Thomas, D.G., Beiser, J., Mitchell, L.M., Huang, J.L., Senpan, A., Hu, G., Gordon, M., Baker, N.A., Pan, D., Lanza, G.M., & Hourcade, D.E. (2014). Application of a hemolysis assay for analysis of complement activation by perfluorocarbon nanoparticles. *Nanomedicine*, 10, 651–660.

Pichler, W.J. (2002). Pharmacological interaction of drugs with antigen-specific immune receptors: the p-i concept. *Current Opinion in Allergy and Clinical Immunology*, 2, 301–305.

Pichler, W.J. (2003). Delayed drug hypersensitivity reactions. *Annals of Internal Medicine*, 139, 683–693.

Pieters, R. (2001). The popliteal lymph node assay: a tool for predicting drug allergies. *Toxicology*, 158, 65–69.

Pieters, R. (2007). Detection of autoimmunity by pharmaceuticals. *Methods*, 41, 112–117.

Pieters, R. & Albers, R. (1999). Screening tests for autoimmune-related immunotoxicity. *Environmental Health Perspectives*, 107 Suppl 5, 673–677.

Ricklin, D. & Lambris, J.D. (2007). Complement-targeted therapeutics. *Nature Biotechnology*, 25, 1265–1275.

Rose, N.R. & Bhatia, S. (1995). Autoimmunity: animal models of human autoimmune disease In *Methods in Immunotoxicology*, 2, (Burleson, G.R., Dean, J.H., & Munson, A.E. eds.), pp. 427–445, Wiley-Liss, New York.

Rustemeyer, T., De Ligter, S., Von Blomberg, B.M., Frosch, P.J., & Scheper, R.J. (1999). Human T lymphocyte priming in vitro by haptenated autologous dendritic cells. *Clinical and Experimental Immunology*, 117, 209–216.

Salvador-Morales, C., Flahaut, E., Sim, E., Sloan, J., Green, M.L., & Sim, R.B. (2006). Complement activation and protein adsorption by carbon nanotubes. *Molecular Immunology*, 43, 193–201.

Spanhaak, S. (2006). The ICH S8 immunotoxicity guidance. Immune function assessment and toxicological pathology: autonomous or synergistic methods to predict immunotoxicity? *Experimental and Toxicologic Pathology*, 57, 373–376.

Szebeni, J., Baranyi, L., Savay, S., Milosevits, J., Bodo, M., Bunger, R., & Alving, C.R. (2003). The interaction of liposomes with the complement system: in vitro and in vivo assays. *Methods in Enzymology*, 373, 136–154.

Teunis, M., Corsini, E., Smits, M., Madsen, C.B., Eltze, T., Ezendam, J., Galbiati, V., Gremmer, E., Krul, C., Landin, A., Landsiedel, R., Pieters, R., Rasmussen, T.F., Reinders, J., Roggen, E., Spiekstra, S., & Gibbs, S. (2013). Transfer of a two-tiered keratinocyte assay: IL-18 production by NCTC2544 to determine the skin sensitizing capacity and epidermal equivalent assay to determine sensitizer potency. *Toxicology In Vitro*, 27, 1135–1150.

Thong, B.Y. & Tan, T.C. (2011). Epidemiology and risk factors for drug allergy. *British Journal of Clinical Pharmacology*, 71, 684–700.

Tong, J.C., Tan, T.W., & Ranganathan, S. (2007). Methods and protocols for prediction of immunogenic epitopes. *Briefings in Bioinformatics*, 8, 96–108.

Toyoguchi, T., Ebihara, M., Ojima, F., Hosoya, J., Shoji, T., & Nakagawa, Y. (2000). Histamine release induced by antimicrobial agents and effects of antimicrobial agents on vancomycin-induced histamine release from rat peritoneal mast cells. *The Journal of Pharmacy and Pharmacology*, 52, 327–331.

Uetrecht, J. (2005). Current trends in drug-induced autoimmunity. *Autoimmunity Reviews*, 4, 309–314.

Uetrecht, J. & Naisbitt, D.J. (2013). Idiosyncratic adverse drug reactions: current concepts. *Pharmacological Reviews*, 65, 779–808.

Vidal, J.M., Kawabata, T.T., Thorpe, R., Silva-Lima, B., Cederbrant, K., Poole, S., Mueller-Berghaus, J., Pallardy, M., & Van der Laan, J.W. (2010). In vitro cytokine release assays for predicting cytokine release syndrome: the current state-of-the-science. Report of a European Medicines Agency Workshop. *Cytokine*, 51, 213–215.

de Weck, A.L. (1984). Pathophysiologic mechanisms of allergic and pseudo-allergic reactions to foods, food additives and drugs. *Annals of Allergy*, 53, 583–586.

White, T.J., Arakelian, A., & Rho, J.P. (1999). Counting the costs of drug-related adverse events. *PharmacoEconomics*, 15, 445–458.

Whritenour, J., Cole, S., Zhu, X., Li, D., & Kawabata, T.T. (2014). Development and partial validation of a mouse model for predicting drug hypersensitivity reactions. *Journal of Immunotoxicology*, 11, 141–147.

Wierda, D., Smith, H.W., & Zwickl, C.M. (2001). Immunogenicity of biopharmaceuticals in laboratory animals. *Toxicology*, 158, 71–74.

Zhu, X., Cole, S.H., Kawabata, T.T., & Whritenour, J. (2014). Characterization of the draining lymph node response in the mouse drug allergy model: a model for drug hypersensitivity reactions. *Journal of Immunotoxicology*, 12, 376–384.

14

STRATEGIES AND ASSAYS FOR MINIMIZING RISK OF OCULAR TOXICITY DURING EARLY DEVELOPMENT OF SYSTEMICALLY ADMINISTERED DRUGS

CHRIS J. SOMPS[1], PAUL BUTLER[2], JAY H. FORTNER[3], KERI E. CANNON[4] AND WENHU HUANG[5]

[1] Investigative Toxicology, Drug Safety Research and Development, Pfizer Inc., Groton, CT, USA

[2] Global Safety Pharmacology, Drug Safety Research and Development, Pfizer Inc., La Jolla, CA, USA

[3] Veterinary Science & Technology, Comparative Medicine, Pfizer Inc., Groton, CT, USA

[4] Toxicology, Halozyme Therapeutics, Inc., San Diego, CA, USA

[5] General Toxicology, Drug Safety Research and Development, Pfizer Inc., La Jolla, CA, USA

14.1 INTRODUCTION

Ocular toxicity accounts for a significant level of safety-related attrition during the development of systemically administered, non-oculotherapeutic drugs. For example, retinal toxicity was reported to cause ~7% of the new drug failures for Bristol–Myers Squibb between 1993 and 2006 (Car, 2006). Ocular toxicity accounted for a similar attrition rate within Pfizer from 2006 to 2010 and was the fourth most common cause of attrition during that time period. Although it is not a life-threatening toxicity, there is little tolerance for adverse visual system effects among drug developers, payers, physicians, or patients (Brick, 1995; Fraunfelder, 2001). Typically, oculotoxic drugs are identified during standard animal testing using postmortem histopathology, potentially after a significant investment of preclinical development time (3–6 years) and resource (hundreds of millions of dollars) (Kramer et al., 2007; Brock et al., 2013; DiMasi et al., 2015). High-risk compounds need to be identified much earlier in the drug development process, giving the medicinal chemist and toxicologist an opportunity to avoid or minimize safety risks when investments are still small and alternative chemical matter is still abundant (Kramer et al., 2007). In order to identify drug candidates with a risk for ocular toxicity prior to significant investment, a combination of in silico, in vitro, and higher-throughput

in vivo assays and screening strategies should be implemented. This chapter outlines internal decision-making strategies and tools that pharmaceutical companies may use to better predict and manage ocular safety concerns early in the development of systemically administered, non-oculotherapeutic drugs. In silico and in vitro tools and strategies have been discussed previously (Somps et al., 2009) and are only briefly summarized in the next section. Our primary focus in this chapter is on the use of in vivo models during early discovery toxicology assessments and in particular on in-life evaluations, which can be completed faster and more efficiently than postmortem analyses. Additionally, we identify and discuss gaps, emerging strategies, and future needs for improved detection of ocular toxicity risk early in the drug discovery process.

14.2 IN SILICO AND IN VITRO TOOLS AND STRATEGIES

Clearly, if a new drug is not intended to treat an ocular disease, the best strategy for avoiding ocular toxicity is to avoid exposure to the eye all together. However, our ability to predict ocular penetrance of systemically administered drugs remains relatively limited. Assuming a systemically administered drug and/or metabolite gets into an ocular

Drug Discovery Toxicology: From Target Assessment to Translational Biomarkers, First Edition. Edited by Yvonne Will, J. Eric McDuffie, Andrew J. Olaharski, and Brandon D. Jeffy.

compartment, the potential for ocular toxicity will then be determined by the presence or absence of the intended drug target in that compartment, activity at any unintended molecular targets (off-target activity), or general physicochemical properties of the drug and its metabolites. This information can often be obtained using *in silico* strategies, for example, mining existing literature and publically available databases, and confirmed with specific *in vitro* studies.

Often the first step in predicting ocular toxicity is determining the expression of the target in the eye, both in healthy individuals as well as in the intended patient population. It is also important to understand any species differences in ocular expression so that appropriate toxicology animal models can be chosen that optimize translation to humans. Any information on the ocular phenotypes associated with genetic variants of the target, including gene knockout or overexpressing animal models, as well as information on target biology and/or pharmacology (e.g., with marketed drugs) will further aid in predicting an ocular safety risk. For example, the potential drug target aldehyde dehydrogenase 1 family, member A1 (ALDH1A1) carries a risk of ocular toxicity based on expression in the lens (Cooper et al., 1993), reports of cataracts in ALDH1A1 knock-out (KO) mice (Lassen et al., 2007), and a putative role of this ALDH isoform in protecting the lens from oxidative stress (Choudhary et al., 2005; Chen et al., 2013). Follow-up *in vitro* and *ex vivo* studies in human lens epithelial cells and rat lens explants with pharmacological and molecular tools (Somps et al., 2009) confirmed this risk and helped inform a Pfizer internal decision not to develop a drug targeting this dehydrogenase.

Off-target hits are probably the most difficult to predict. However, close-in analogues, or isoforms, of an intended molecular target are often known, and their expression in the eye may lead to unintended pharmacology or toxicity. For example, phosphodiesterase type 5 (PDE5) inhibitors, marketed for the treatment of erectile dysfunction, are associated with visual disturbance (Kerr and Danesh-Meyer, 2009), due to the unintended activity of these inhibitors at a similar phosphodiesterase, PDE6, a critical enzyme in the visual transduction cascade (Chabre and Deterre, 1989; Yarfitz and Hurley, 1994). While not sufficiently adverse to significantly impact drug development or marketing, an earlier appreciation of this off-target effect of PDE5 inhibitors may have enabled drug developers to avoid visually relevant activity at PDE6 altogether.

Chemical structure and physicochemical properties of compounds and their metabolites are also important to predicting toxicity (Waring et al., 2015), including ocular toxicity. Some drugs or their metabolites may not act on a specific molecular target in the eye, but rather cause nonspecific effects like disrupting critical membrane structural organization in sensitive ocular compartments like the lens leading to cataracts (Jacob et al., 2013). Additionally, drugs that bind to melanin may be retained at higher concentrations in the retinal

pigment epithelium and may also increase the risk of retinal toxicity via either direct or photosensitizing mechanisms (Dayhaw-Barker, 2002).

In vitro assays, cell culture systems or tissue explants, can be used to further understand potential risks identified using *in silico* prediction approaches (e.g., Toimela et al., 1995). Molecular manipulations of targets, for example, target down- or upregulation combined with gene expression profiling in cell culture systems can provide confirmation and mechanistic insight into target modulation risks (Liebler and Guengerich, 2005; Simic et al., 2005). Due to the variety and complexity of ocular tissues and cell types, higher-throughput screening assays for ocular toxicity that use simple cell cultures are not typically applied in a prospective manner in a drug development setting. More detailed examples of *in silico* and *in vitro* assays used to identify and characterize ocular toxicity in drug discovery have been described elsewhere (Somps et al., 2009).

Often the best way to predict ocular toxicity risk prospectively, or to confirm predictions based on *in silico* and *in vitro* models, is to evaluate new drug candidates in a whole animal model. The question then becomes how best to do this quickly and efficiently in a discovery toxicology environment. In the remainder of this chapter, we highlight *in vivo* models and in-life assessment strategies that facilitate early detection of ocular toxicity risk, including higher-throughput behavioral models and noninvasive ophthalmic examinations. We also review emerging tools and strategies and identify gaps and future needs for improved detection of ocular toxicity early in the development of non-oculotherapeutic drugs.

14.3 HIGHER-THROUGHPUT *IN VIVO* TOOLS AND STRATEGIES

14.3.1 Ocular Reflexes and Associated Behaviors

The optokinetic response (OKR) and optomotor response (OMR) are commonly used to detect alterations in visual function and performance in a variety of animal models, as well as in human subjects. The OKR, also called optokinetic nystagmus (OKN), is triggered when the entire visual field drifts across the retina and produces eye rotation in the same direction and velocity as the image (Mitchiner et al., 1976; Iwashita et al., 2001; Douglas et al., 2005; Cahill and Nathans, 2008). This rotation reduces the movement of the image on the retina and is intermittently interrupted by saccades or rapid rotations of the eye in the opposite direction (Mitchiner et al., 1976; Iwashita et al., 2001; Douglas et al., 2005; Cahill and Nathans, 2008). Saccades (defined as a rapid jerky movement of the eye especially as it jumps from fixation on one point to another) function to reset the eye position for a new steady rotation period (Mitchiner et al., 1976; Iwashita et al., 2001; Douglas et al., 2005; Cahill and

Nathans, 2008). An OMR is also initiated when the visual field moves across the retina and consists of involuntary movements of the head and/or body to assist in image stabilization (Kretschmer et al., 2013) and encompasses head movements initiated via the vestibulo-ocular reflex (VOR). These head movements stimulate the vestibular organs, which then send signals to the brain stem to generate compensatory eye movements in the opposite direction of the image motion (Iwashita et al., 2001; Cahill and Nathans, 2008). Both OKR and VOR collectively function to stabilize images across the retina over a wide spectrum of head and body movements (Iwashita et al., 2001; Cahill and Nathans, 2008). In the following sections we describe how these responses can be used to detect ocular toxicity in zebrafish and rodents.

14.3.1.1 Zebrafish as a Model for Assessing Ocular Toxicity

The zebrafish is a small tropical fish that has become extremely popular with geneticists and molecular biologists given that the organization of the genome and the genetic pathways controlling signal transduction and development are highly conserved between zebrafish and humans. Over the past decade, zebrafish have become an established model to study a variety of human diseases (Zon and Peterson, 2005; Phillips and Westerfield, 2014). Zebrafish have recently been utilized in drug discovery programs (Kari et al., 2007) and for the assessment of safety liabilities (Barros et al., 2008).

Zebrafish are an important vertebrate model for studying vision-related disorders especially as the visual system is fundamentally similar to humans (Goldsmith and Harris, 2003). For instance, the zebrafish retina has a cone-dense retina giving the fish a color-rich vision similar to humans, providing a potential advantage over testing compounds for effects on visual function in nocturnal rodents, which have rod-dominant retinas. The visual system development is very rapid as zebrafish larvae depend on their sense of vision to evade predators and to catch prey. By 5 days postfertilization (dpf) the visual system is well developed when assessed using electrophysiological, morphological, or behavioral criteria (Bilotta and Saszik, 2001). The uses and value of zebrafish as a model of ocular disease have been recently reviewed (Gestri et al., 2012; Chhetri et al., 2014), highlighting the fish's contribution to the understanding of human ocular disease as well as screening approaches to identify novel therapeutics.

Several papers have recently been published on the validation and use of zebrafish (*Danio rerio*) larvae as early safety pharmacology screens in visual safety assessment (Berghmans et al., 2008; Richards et al., 2008). Importantly, zebrafish larvae are amenable to medium-to-high-throughput *in vivo* screening (Redfern et al., 2011). The larvae are small enough (<1 mm in diameter) to be easily distributed and maintained in 96-well plates, which enables drug screening approaches using only milligrams of test article. This is facilitated by the fact that the larvae are dimethyl sulfoxide (DMSO) tolerant (1%) and can readily absorb compounds from the water through their skin, and by 4–5 dpf the onset of regular and coordinated gut peristaltic waves occur (Holmberg et al., 2004) thus allowing oral ingestion. This time period also coincides with maturation of absorption through the protogills (Rombough, 2002). However, despite these different routes for compound uptake, nonuniform and unpredictable compound exposures have been reported (Fleming and Alderton, 2013). Ways around these issues have been suggested and include establishing the pharmacokinetics of a wide range of drugs with different physicochemical properties in zebrafish larvae to establish rules for zebrafish uptake (Berghmans et al., 2008).

To assess visual function in zebrafish, behavioral assays using either OKR or OMR have been developed that take advantage of inherent visual reflexes in the fish. Both assays have been successfully employed to detect mutant zebrafish with defects in the visual system (reviewed in Neuhauss (2003)) as well as assessing the effect of compounds on visual function (Berghmans et al., 2008; Richards et al., 2008). More recently, the systems for quantifying the visualmotor response (VMR), the locomotor movements of zebrafish larvae in response to a light stimulus, have been developed (Deeti et al., 2014) and validated with a number of compounds. The VMR assay allows for the simultaneous monitoring of individual larvae in wells of a 96-well plate in response to lights being turned "on" and "off." A brief outline of the zebrafish OKR, OMR, and VMR assays, methodology, and the effects of tested compounds on visual function follows.

Assessing Oculotoxicity in Zebrafish Using the OKR   For assessing the effects of compounds on the OKR, the maximum tolerated concentration (MTC) for each compound first needs to be determined in the larvae to identify toxic concentrations and ensure that the response is not compromised by general toxicity (Hutchinson et al., 2009). Once the MTC has been established, the larvae are exposed to test compounds for 5 days (from 3 to 8 dpf). In order to measure the OKR in compound or vehicle-treated zebrafish larvae, the larvae (typically $n = 10$/group) need to be immobilized to restrain the body while maintaining the ability for eye movement and are then exposed to a visual stimulus that usually consists of alternate black and white stripes that rotate around the larvae (Fig. 14.1) (Brockerhoff, 2006; Alderton, 2012). The larvae respond by moving their eyes in the direction of the moving stripes (both clockwise and counterclockwise) with eye movements being recorded and the numbers of saccades determined. If the vehicle-treated larvae control scores <20 saccades, the experiment is considered invalid (Richards et al., 2008).

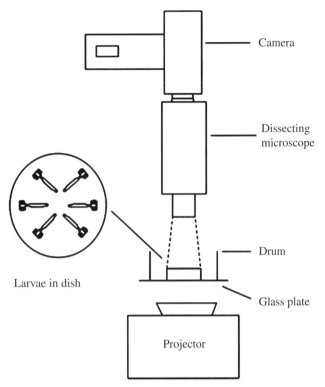

FIGURE 14.1 Diagram of the OKR apparatus. A projector plays a movie of black and white grating on to the inside of the drum, which surrounds a dish containing the 8-dpf zebrafish larvae immobilized in methylcellulose (inset). A camera records movies of the larvae for subsequent analysis of eye position and quantification of the number of saccades for the eye of each larva. Alderton (2012), figure 15.4. Reproduced with permission of John Wiley & Sons.

The OKR has been extensively used for laboratory screening of visual function following genetic manipulations and has been used to identify and characterize many zebrafish mutants that are associated with vision defects, for example, defect in lens development (Chhetri et al., 2014). To date, only two studies have been published on the effect of drugs on visual function using the OKR assay. Richards et al. (2008) used the OKR assay to follow up on visual function effects detected in primary screening in the OMR assay. More recently, Deeti et al. (2014) showed that 9 out of 10 negative controls (comprising antibiotics and neuromodulators) had no effect on the OKR, while 5 out of 6 oculotoxic compounds showed adverse effects on the response. Pfizer has internal data, in collaboration with an external partner, which showed that the OKR assay had the same sensitivity as the OMR assay with seven out of nine compounds evaluated showing the expected results on visual function (Butler personal communication). Chhetri et al. (2014) proposed that the OKR assay results in more robust, reliable, and quantifiable behavioral data compared to the other behavioral assays described in the following sections. However, the assay does have the disadvantage of being more technically

challenging and has a lower screening throughput than the OMR or VMR assays.

Assessing Oculotoxicity in Zebrafish Using the OMR In zebrafish, the OMR can be elicited by moving horizontal black and white stripes below long transparent chambers in which the larvae swim (Fig. 14.2) (Richards et al., 2008). Once the MTC has been established, the larvae are exposed to test compounds for 5 days (from 3 to 8 dpf) and then transferred to a transparent, multilane acrylic block containing embryo medium with 10 larvae per lane and with two groups of vehicle controls in each assay. A black and white stripe movie is played for a certain length of time (e.g., 45 s), and the position of the larvae to the end of the channel is recorded (see Fig. 14.2). Control (wild-type) zebrafish swim toward the perceived motion and gather at one end of the channel, while zebrafish with ocular defects swim in random patterns. At the end of the run, the number of larvae in a "pass area" (which typically represents the furthest 25% of the channel) is calculated and the mean percentage in the pass area of a number of runs (e.g., 6) obtained. If either of the vehicle control groups score <60% overall, then the assay is considered invalid (Richards et al., 2008).

In a validation study undertaken by Richards et al. (2008), the authors reported on the results of the blinded testing of 27 compounds in an OMR assay. The compounds tested caused adverse effects on the visual system via a variety of mechanisms. Thirteen out of nineteen positive controls tested produced the expected oculotoxic effects, while six out of eight negative compounds were correctly predicted. This study gave an overall predictivity of 70% for the OMR assay with a sensitivity of 68% and specificity of 75%. These findings are supported by another similar validation study (Berghmans et al., 2008) that found that seven out of nine compounds tested produced the expected effects on zebrafish visual function (sensitivity of 78%). Overall, these results suggest that the OMR assay in zebrafish could be useful in predicting the adverse effects of drugs on visual function in humans and would support its potential as a screen for frontloading safety pharmacology assessments of this endpoint *in vivo*. However, this assay is known to be sensitive to the generation of false positive data for compounds that impair locomotor activity without affecting vision (Richards et al., 2008).

Assessing Oculotoxicity in Zebrafish Using the VMR The VMR assesses the locomotor movements of zebrafish larvae in response to lights being turned on and off (Emran et al., 2008; Yin et al., 2012). In brief, individual larvae (5 dpf that have been previously treated for 48 h with either vehicle or selected compounds) are placed in embryo medium in a well of a 96-well clear microtiter plate (12 fish/treatment group). The plate is then placed inside a recording chamber

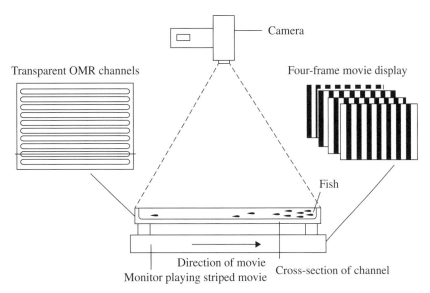

FIGURE 14.2 Diagram of the OMR apparatus, viewed from above and in cross section. Richards et al. (2008). Reproduced with permission of Elsevier.

and the light driving parameters set so that there is an initial stabilization period (30 min) followed by four periods where the light changes from on to off and vice versa in 20-min intervals. Threshold detection values are set to detect burst and freeze activity and locomotor behavior (using an infrared tracking system) of each larvae measured over the time course of the study. The effect of drug treatment on the VMR is quantified by comparing the increased or decreased activity in response to lights being off or on (Deeti et al., 2014).

Using the VMR assay, Deeti et al. (2014) reported that 9 out of 10 negative control drugs (comprising antibiotics and neuromodulators) had no effect on the VMR, while five of the six oculotoxicants tested showed adverse effects on zebrafish visual behavior giving an overall assay sensitivity of 83% (false negative rate of 17%), a specificity of 100%, and positive predictive value of 100%, a significant improvement over the OMR. These validation data therefore appear very promising especially since the assay can be assessed in a higher-throughput 96-well format to detect effects on visual function. Additional studies, including the testing of a much larger compound validation set, need to be undertaken by other laboratories to confirm these data. If confirmed, then the VMR endpoint in zebrafish larvae would be a viable ocular toxicity screen, and the assay could become a second tier screen for programs with potential ocular safety concerns (based on prior knowledge of target or compound-related risks) and reduce the number of rodent toxicology studies required to identify a new chemical entity with reduced risk of ocular toxicity.

Overall, published validation studies suggest that the zebrafish models described to assess visual function do appear to have the required sensitivity and specificity to

detect the oculotoxic effects of drugs known to cause ocular toxicity in humans. The use of zebrafish larvae would therefore appear to have utility as an early safety screen to predict the oculotoxicity profile of novel drugs, with the potential to bridge *in vitro* to *in vivo* studies.

14.3.1.2 Rodents as a Model for Assessing Ocular Toxicity

Several different approaches are available for assessing the OKR and OMR in rodents, but all rely on a similar premise that consists of rotating a pattern of vertical stripes, sinusoidal grid, or random dots across the visual field of the animal (Kretschmer et al., 2013). Traditionally, OKR assessments used large mechanical drums or cylinders and turntables to rotate the images around the animal (Mitchiner et al., 1976; Precht and Cazin, 1979; Fuller, 1985; Iwashita et al., 2001; Cahill and Nathans, 2008). However, with the advent of virtual imaging, newer systems utilize liquid crystal displays (LCD) or other digital techniques to project a computer-generated image of rotating vertical stripes (Prusky et al., 2002, 2004; Douglas et al., 2005; Kretschmer et al., 2013). The use of digital images to stimulate an OKR has several advantages over its counterpart. First, the velocity, spatial frequency, and contrast can be easily and rapidly adjusted (Prusky et al., 2004; Douglas et al., 2005; Kretschmer et al., 2013). Second, the rapid appearance of the stimulus often captures the attention of the animal (Douglas et al., 2005). Third, the appearance of the stimulus can be repeated multiple times within a brief period of time (i.e., a few seconds) in order to confirm the presence of an appropriate OKR (Douglas et al., 2005). However, one disadvantage to the use of LCDs for stimulation is that visual acuity may be underestimated (Kretschmer et al., 2013). The backlight of LCDs lacks UV light that stimulates

UV-sensitive blue cones of the eye. Therefore, the rodent retina is not optimally stimulated by LCDs (Kretschmer et al., 2013).

More recently, OMR assessments have become more popular than the traditional OKR measurements. This is in part due to the stressful procedure of head restraint necessary for OKR assessments. Assays can now incorporate OMR measurements in addition to determining OKR thresholds, which allows for a more complete assessment of visual processing (Abdeljalil et al., 2005; Douglas et al., 2005; Kretschmer et al., 2013). The setup for a typical OMR assessment consists of a square chamber comprised of four LCD screens facing inward and projecting the stimulus pattern (Fig. 14.3) (Redfern et al., 2011). In the middle of the chamber, there is a platform for the animal to be placed on. A camera mounted above the platform captures images of the animal and its movements in response to the rotating stimulus pattern (Douglas et al., 2005). These virtual systems are advantageous because they allow the animal to freely move about on the platform instead of being restrained. The system can then be kept centered on the head by monitoring the position of the freely moving animal in order to obtain OMR measurements (Douglas et al., 2005). When compared to visual acuity measurements obtained by assessing the OKR in animals with fixated heads, the visual acuity measurements obtained solely from OMR movements were only slightly lower (Kretschmer et al., 2013). Therefore, it may be possible to just use head tracking movements in the OMR arenas as a surrogate for OKR changes (Kretschmer et al., 2013). Finally, both OKR and OMR assessments can be scored manually, via automated scoring system or both.

Thus far, OKR and OMR assessments have been used primarily as part of the screening process for genetically engineered rodents. However, more recent efforts have been focused on the use of these assessments as preclinical safety evaluations. Redfern et al. (2011) demonstrated the suitability of measuring OMR over time and possible inclusion on toxicology studies as early assessments for ocular toxicity when exposed to sodium iodate. However, more research is needed to determine the suitability of these assessments for early drug discovery and their translatability to the clinic.

14.3.2 Noninvasive Ophthalmic Examinations

The routine ophthalmic examination is an efficient and effective technique to identify ocular toxicity (Munger and Collins, 2013; Wilkie, 2014). Long a required evaluation in Good Laboratory Practice (GLP) safety studies, these examinations can be readily applied to any mammalian *in vivo* study in which ocular effects are anticipated or observed. Minimal pharmacologic intervention is required to examine the eye, and most typical laboratory animal species require only manual restraint. The examinations are not invasive, and the same animal or cohort of animals can be examined

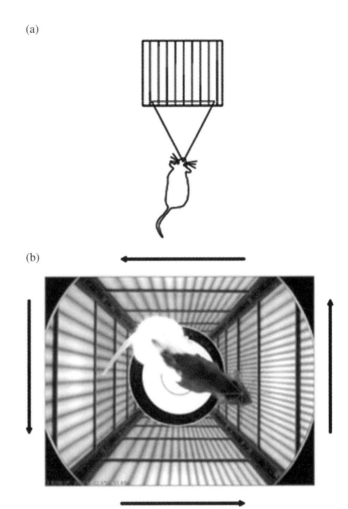

FIGURE 14.3 (a) Visual acuity is measured in cycles/degree, which takes into account the distance from the eyes to the image. In the cartoon this is simply the number of sine wave cycles in the grating display divided by the angle of view. (b) Image from the camera above a rat in the OptoMotry© chamber. The cursor positioned between the eyes of the rat can be seen, as can the vertical gratings displayed on the four monitors. The position of the cursor determines the center of the "virtual cylinder." Note that the gratings nearest the eyes are closer together than more distant gratings so as to maintain the eyes at the center of the virtual cylinder. The horizontal "lines" are the edges of the mirror above the animal, with multiple reflections by the mirror beneath the platform. The arrows indicate the direction of movement of the gratings (in this example, anticlockwise); the rat will head track in the same direction. Redfern et al. (2011, pp. 102–114). Reproduced with permission of Elsevier.

multiple times in one study. Using a proper lexicon of ophthalmic terminology, the results can be promptly recorded and shared with excellent understanding of what was observed; ophthalmic imaging is not required.

Histopathology of the eye is a complementary procedure to the routine ophthalmic examination. Microscopic evaluation of ocular tissue permits identification of specific cell types, a level of detail generally beyond the clinical ophthalmic examination. However, the intensive resource demand during processing, limited tissue available on one glass slide, and high magnification used during examination affect the applicability of this procedure for high-throughput screening. Moreover, the clinical ophthalmic examination allows evaluation of more territory within the eye that typically will inform the selection of specific sites for subsequent histopathologic evaluation. Therefore, ocular histopathology complements the ophthalmic examination but cannot replace it.

Pharmacologic dilation of the pupil (mydriasis) is mandatory for adequate evaluation of the lens and fundus, including retina, optic disc, and macula (if present) (Taradach and Greaves, 1984; Munger and Collins, 2013). Tropicamide 1% administered topically to the eye 10–20 min prior to the examination provides excellent dilation for the procedure and resolves within hours. Rodents with densely pigmented eyes may require two instillations of the mydriatic separated by 15–20 min. Atropine also provides dilation, however its duration of action is excessive and, in albino species, could increase the risk of light-induced retinal degeneration (Williams, 2002). The examiner must be aware that once the pupil is dilated, the iris will be mostly hidden and cannot be adequately evaluated, and a decision must be made whether to include a premydriatic examination to allow proper observation of the iris.

Intraocular pressure (IOP) measurement can be applied to laboratory animals (Williams, 2007; Coster et al., 2008; Munger and Collins, 2013). The tonometry procedure requires topical corneal anesthesia. Proparacaine hydrochloride 0.5%, administered as a single drop to each eye immediately before measurement, provides adequate desensitization for this brief procedure. Applanation tonometry requires physical contact between the instrument and the corneal surface; a gentle and consistent touch with the tonometer is essential. Unlike tonometry using fixed, tabletop equipment with adult humans, this procedure in laboratory animals demands multiple, consecutive measures that are averaged and displayed by the equipment. Tonometers that rely on air pulses can cause startle reflex activity that interferes with consistent measurement.

14.3.2.1 Ophthalmic Equipment

Correct ophthalmic equipment is essential for the examination (Kuhlman et al., 1992; Saint-Macary and Berthoux, 1994; Hubert et al., 1999; Williams, 2007; Coster et al., 2008; Munger and Collins, 2013; Wilkie, 2014). Although the direct ophthalmoscope is a standard tool, indirect ophthalmoscopy is more flexible and provides a better field of view due to the many handheld lenses that are available for this technique. The examiner can elect a more panoramic view of the fundus for quick scanning and reserve the higher magnification lenses for particular areas of interest. Use of the indirect ophthalmoscope also keeps the examiner's face away from the animal subject, a consideration with larger species. The other essential for

ophthalmic examinations is the handheld slit lamp biomicroscope (SLBM). The SLBM is an easily portable, low-power microscope with an attached, focused, and variable light source. It can be applied to many different aspects of the examination including pupillary light reflex, undilated iris review, corneal fluorescein staining, and detailed examination of the cornea, anterior chamber, and lens. Handheld tonometers that are appropriate for larger species (eyeball size similar to human) and, separately, for rodents are available. The cost of this equipment is considerable, however with proper care it will last for a long time and is applicable across all species of interest.

14.3.2.2 Ophthalmic Examination Procedure

The basic ophthalmic examination procedure is similar for all species and is best performed in a dimly lit room. It can be tailored to fit the needs of the study and the particular species. Rodents can be examined conscious, and skilled technical assistance for manual restraint of the animals is required (Taradach and Greaves, 1984; Hubert et al., 1999; Williams, 2007; Coster et al., 2008). Large numbers of animals can be examined quickly with the indirect ophthalmoscope. The examiner may opt to forego the SLBM and utilize backlighting from the fundic reflex, seen through the indirect headset without a handheld lens, for evaluation of the clear ocular tissues. This technique permits detection of very small changes in transparency, however it is most effective with albino species. Most laboratory dogs are quite tractable and are also examined while conscious with manual restraint. Basic functional testing in the dog is easy and rapid. Pupillary light reflex and menace reflex testing permit evaluation of the entire visual system within seconds. Rabbits (Holve et al., 2011) and pigs (Saint-Macary and Berthoux, 1994) can also be examined conscious, however the typical requirement for mechanical restraint and the need for the examiner to encourage cooperation from the animal result in a slower pace of examination.

Examination of the nonhuman primate requires sedation (Kuhlman et al., 1992). The highly mobile primate eye and their anxious nature when handled mandate a chemically restrained animal for an adequate examination. Human safety considerations also are a factor. Ketamine hydrochloride works well due to its safety and controllable duration. The time to onset of mydriasis following tropicamide administration must be balanced against the time to the animal's recovery from ketamine's sedative effect. A refinement is the addition of a water-based lubricant eye drop at the time of tropicamide administration. This helps protect the cornea from desiccation while the monkey is sedated and may also aid in achieving prompt and satisfactory dilation by increasing tropicamide residence time in the precorneal tear film. Working with a seasoned technical staff appropriately trained for ketamine injection, the examiner can evaluate many primates in a reasonable time.

14.3.2.3 Trained Examiner Selecting the individual to perform ophthalmic examinations is critical (Wilkie, 2014). Mere familiarity with proper use of the equipment and a single species is not sufficient. Ideally the examiner has extensive experience with the species and strain of laboratory animal to be evaluated. Although the mammalian eye has a highly conserved, basic architecture, each species has definite characteristics that set its eye clearly apart from all others. Normal lens size and shape, optic disc (optic nerve head) shape and placement, retinal blood vessel pattern, pigmentation, and presence of a tapetum or macula are examples of the many ocular tissues that vary among the species. Within a species, normal variation among individuals can be significant, and the examiner must be prepared to reliably distinguish the abnormal against this background following a brief observation. Unlike most other evaluations, the only ophthalmic data collected is the examiner's immediate interpretation; peer review or quality control of the observations at a later date is not possible.

In summary, the routine ophthalmic examination is readily applicable to the drug discovery environment. It permits rapid ocular evaluation with no loss of animal subjects. The required equipment is an expensive investment but will not wear out or become obsolete. However, care must be applied when identifying a qualified examiner. Observations can be performed at any interval required, allowing the same animal to be compared to itself. Reversibility is readily evaluated by ophthalmic examination performed once dosing with the test article has been completed. While many ocular adverse effects are fully or partially reversible, loss of retinal neurons and rupture of lens cell membranes are not. Any ocular toxicity, and especially if it is nonreversible, can be challenging to risk manage. Earlier detection permits informed decision-making and potential saving of tremendous resources for the project team.

14.4 STRATEGIES, GAPS, AND EMERGING TECHNOLOGIES

14.4.1 Strategic Deployment of *In Silico, In Vitro,* and *In Vivo* Tools

As mentioned in the introduction of this chapter, the best way to reduce the risk of ocular toxicity when developing a nonocular drug is to minimize exposure to the eye as much as possible; in other words develop a drug that has limited ability to cross the blood–ocular or various diffusional barriers. However, while there have been significant advances in understanding and modeling the various blood–ocular and diffusional barriers to the eye (Hornof et al., 2005; Toda et al., 2011), such tools are still a long way from being able to reliably predict *in vivo* ocular penetrance of candidate drugs. *In silico* and *in vitro* models that accurately predict whether or not a systemically administered molecule will get into the various ocular compartments are needed. Ideally such models would enable cross-species comparisons, and thus confidence that a lack of ocular penetrance in nonclinical models would translate to humans.

Until the molecular characteristics that determine ocular exposure are better understood and taken into consideration during screening and selection of non-oculotherapeutic drug candidates, drug developers can strategically deploy a series of *in silico* and *in vitro* assays like those described previously (Somps et al., 2009), as well as higher-throughput *in vivo* assays, in the early discovery phase of drug development as outlined in Figure 14.4. The goal of deploying such assays is to triage compounds for risk of ocular toxicity ahead of more time-consuming and expensive *in vivo* studies, for example, GLP toxicology studies or specialty ocular assessments like electroretinography (ERG). Although *in silico* and *in vitro* assays are most often used to identify a risk or potential hazard, higher-throughput in-life bioassays as described in this chapter can be used to get an early estimate of the therapeutic index. Drug candidates that cause ocular effects only at drug levels that far exceed those needed to treat a particular disease, or levels unlikely to ever occur in patients, may be chosen to move forward in development.

If the drug candidate gets into, or is assumed to enter, one or more ocular compartments, the drug developer will need to be able to predict if that candidate has the potential to alter the normal function of the cells and tissues of that ocular compartment. This in turn will depend on whether (i) the intended target of the candidate drug is expressed in ocular tissues, (ii) high homology analogues of the intended target are expressed in these tissues leading to off-target effects, and (iii) the drug structure itself, for example, physical/chemical properties or structural subcomponents of the parent structure or metabolites carry risk for adverse ocular tissue interactions. Clearly, any additional information published on the effects of modulating the amount or activity of a target or close target analogue expressed in the eye using chemical or molecular tools, or through naturally occurring genetic variants, will also inform the ocular toxicity risk. Public and commercial databases are available for determining expression of targets and their close homologues (Rung and Brazma, 2013) across various tissues, and gene and protein expression profile portals, for example, BioGPS and Protein Atlas, facilitate access to annotated data. However, current expression data sets are largely based on human tissues and thus do not enable cross-species comparisons nor, typically, do they provide the resolution needed to understand expression at the level of different ocular compartments. Additionally, structural alert databases for ocular toxicity are still in their infancy (Greene, 2007; Somps et al., 2009; Solimeo et al., 2012). Knowledge of the expression of druggable targets in ocular compartments including cornea, iris, ciliary body, lens, retina, retinal pigment epithelium, and

Discovery toxicology ocular safety decision tree

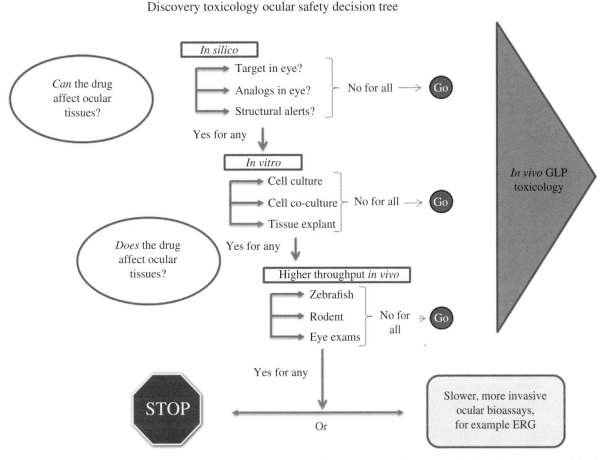

FIGURE 14.4   Decisions start with the application of less expensive and faster *in silico* and *in vitro* tools. Based on outcomes of the *in silico* and/or *in vitro* studies, drug developers may choose to continue further testing in higher-throughput *in vivo* assays or move into standard GLP toxicology studies or progressively more invasive and slower ocular evaluations like electroretinography.

choroid, across species including mouse, rat, dog, nonhuman primate, rabbit, and human, are needed, as are better and more complete structure–activity relationship (SAR) databases for ocular toxicity.

If there are reasons to be concerned that a drug candidate may cause adverse ocular effects, those concerns can be evaluated in tissue or cell type-specific *in vitro* assays. Isolated lenses from rodents, for example, have been used to understand the potential of a candidate drug to cause lenticular opacities and cataracts (Walsh Clang and Aleo, 1997; Aleo et al., 2000, 2005; Sampath et al., 2012). Similarly, cell culture models of ocular epithelia (e.g., retinal and lenticular), either primary or transformed cells, have been used to determine the potential of systemic drugs to cause ocular toxicity (Mannerstrom et al., 2002; Verdugo-Gazdik et al., 2006). Models developed to support ocular irritancy studies (Wilson et al., 2015) can be adapted to investigate risk of corneal/conjunctival toxicities associated with systemically delivered drugs. However, the predictivity of these models is often unsatisfactory due to the limited number of interacting cell types and the lack of systemically

derived influences. Additionally, lack of readily available human ocular tissues makes it difficult to predict the translation of any finding to human risk. New and better models of the eye, for example, 2D co-culture systems (German et al., 2008; Amirpour et al., 2013), 3D ocular organoids (Eiraku et al., 2011; Nakano et al., 2012), and the rapid advances in human stem cell-derived ocular cells (Nakano et al., 2012; Ramsden et al., 2013), hold the promise for greatly improved *in vitro* models for detecting ocular toxicity risk.

If *in silico* or *in vitro* tools predict an increased risk for ocular toxicity, these risks can be further evaluated in higher-throughput *in vivo* screens like those described in this chapter. However, in-life behavioral assessments or clinical eye examinations may not always be sufficiently sensitive for detecting ocular toxicity. For example, MNU-induced retinal toxicity in the rat was not detected with funduscopic eye examinations but was readily detected as photoreceptor loss using standard H&E staining (Liu et al., 2004). Similarly, the false negative rate in the zebrafish or rodent behavioral assays can be as high as one in three. Finally, in-life behavioral testing and ophthalmic examinations, while faster and less labor-intensive than terminal histopathology or more rigorous bioassays like ERG, can still be time consuming and expensive to use early in the drug discovery process. There remains additional need for easy-to-use, yet sensitive and predictive, in-life biomarkers for early detection of ocular toxicity. To address this need we are currently pursuing peripherally available biomarkers of ocular toxicity in the blood.

14.4.2 Emerging Biomarkers of Retinal Toxicity

Nucleic acid-based microRNA (or miRNA) is an emerging class of biomarkers that has been expanding rapidly in the past decade, with increasing application in disease diagnosis and in safety assessments (Siddeek et al., 2014; Marrone et al., 2015). MicroRNAs comprise a class of small, noncoding endogenous single-stranded RNAs that contain about 19–24 nucleotides and regulate gene expression and associated protein levels. Thus they can also be used to elucidate mechanisms of ocular toxicity. The total number of unique, mature human miRNAs has been predicated to be around 3000 (Friedlander et al., 2014). miRNA has the characteristics of an ideal biomarker, including specificity with restricted tissue/organ expression, stability in a readily accessible peripheral compartment like the blood, small sample volume requirements, ability to withstand long-term storage and freeze–thaw cycles, and robust analytic methods for rapid, simple, and quantitative detection (Mitchell et al., 2008).

A number of ocular specific miRNAs have been detected by miRNA array analysis and validated with Northern blot and *in situ* hybridization (Ryan et al., 2006). Several miRNAs highly enriched in the mammalian eye were also reported with distinct distributions in the retina (e.g., miR-124a, miR-181, miR-9, miR-183, miR-182, and miR-204) (Karali et al., 2007). Recently, we have further evaluated these retina-expressed miRNAs as possible toxicity biomarkers using retinal explant, organotypic cultures treated with known retinal toxicants. The data generated using the organotypic explant model revealed the miR-183/miR-182/miR-96 cluster as possible biomarkers of retinal tissue damage caused by exposure to retinal toxicants (Peng et al., 2014).

Promising miRNA biomarkers were further validated in animal models for their *in vivo* translatability and potential utility in exploratory toxicology screening studies. One of the small-molecule compounds that we used was an early generation of a pan-CDK inhibitor, which causes retinal degeneration or atrophy in rodents following intravenous delivery (Illanes et al., 2006). Treatment with this retinal toxicant produced changes in the circulating levels of the miR-183/miR-182/miR-96 miRNA cluster, which correlated well to the severity of retina injury as evaluated by an *in vivo* ophthalmic exam, visual functional analysis of ERG, and robust histopathology analysis (Peng et al., 2014).

Further validation of these potential retinal miRNA biomarkers will be needed in additional toxicological species, for example, dog and monkey, following treatments with a variety of retinal toxicants, as well as possible studies in laser-induced choroid neovascularization (CNV) and streptozotocin-induced diabetic retinopathy ocular disease models (Rittenhouse et al., 2014). Understanding the temporal and spatial relationships between retinal injuries and circulating miRNAs will be critical to fully understand the specificity, sensitivity, predictivity, and reliability of these biomarkers. Further evaluation of the translatability of circulating miRNA toxicity biomarkers in humans will be exciting as the next tier of retinal miRNA biomarker research. Finally, in addition to retinal miRNA biomarkers, there is the need to establish and validate other ocular tissue-specific miRNA biomarkers, for cornea, iris/ciliary body, and lens.

14.5   SUMMARY

Ocular toxicity accounts for a significant amount of safety-related attrition during the development of systemically administered drugs. This attrition often occurs after a significant investment of time and resource in a drug's development. High-risk compounds need to be identified much earlier in the drug development process, giving pharmaceutical developers an opportunity to avoid or minimize safety risks when investments are still small. By strategically combining *in silico*, *in vitro*, and higher-throughput *in vivo* models, along with working to improve these models with, for example, miRNA biomarkers, drug developers will be able to identify and manage ocular safety risk much earlier in a drug's development timeline.

REFERENCES

Abdeljalil J, Hamid M, Abdel-Mouttalib O, Stephane R, Raymond R, Johan A, Jose S, Pierre C, Serge P (2005) The optomotor response: a robust first-line visual screening method for mice. *Vision Research* 45:1439–1446.

Alderton W (2012) Assessment of the effects of function in Larval Zebrafish. In: *Zebrafish: Methods for Assessing Drug Safety and Toxicity* (McGrath, P., ed), pp. 191–203 Hoboken, NJ: John Wiley & Sons, Inc.

Aleo MD, Avery MJ, Beierschmitt WP, Drupa CA, Fortner JH, Kaplan AH, Navetta KA, Shepard RM, Walsh CM (2000) The use of explant lens culture to assess cataractogenic potential. *The Annals of the New York Academy of Sciences* 919:171–187.

Aleo MD, Doshna CM, Navetta KA (2005) Ciglitazone-induced lenticular opacities in rats: in vivo and whole lens explant culture evaluation. *The Journal of Pharmacology and Experimental Therapeutics* 312:1027–1033.

Amirpour N, Nasr-Esfahani MH, Esfandiari E, Razavi S, Karamali F (2013) Comparing three methods of co-culture of retinal pigment epithelium with progenitor cells derived human embryonic stem cells. *International Journal of Preventive Medicine* 4:1243–1250.

Barros TP, Alderton WK, Reynolds HM, Roach AG, Berghmans S (2008) Zebrafish: an emerging technology for in vivo pharmacological assessment to identify potential safety liabilities in early drug discovery. *British Journal of Pharmacology* 154:1400–1413.

Berghmans S, Butler P, Goldsmith P, Waldron G, Gardner I, Golder Z, Richards FM, Kimber G, Roach A, Alderton W, Fleming A (2008) Zebrafish based assays for the assessment of cardiac, visual and gut function—potential safety screens for early drug discovery. *Journal of Pharmacological and Toxicological Methods* 58:59–68.

Bilotta J, Saszik S (2001) The zebrafish as a model visual system. *International Journal of Developmental Neuroscience* 19:621–629.

Brick DC (1995) Medication errors result in costly claims for ophthalmologists. *Survey of Ophthalmology* 40:232–236.

Brock WJ, Somps CJ, Torti V, Render JA, Jamison J, Rivera MI (2013) Ocular toxicity assessment from systemically administered xenobiotics: considerations in drug development. *International Journal of Toxicology* 32:171–188.

Brockerhoff SE (2006) Measuring the optokinetic response of zebrafish larvae. *Nature Protocols* 1:2448–2451.

Cahill H, Nathans J (2008) The optokinetic reflex as a tool for quantitative analyses of nervous system function in mice: application to genetic and drug-induced variation. *PloS One* 3:e2055.

Car BD (2006) Enabling technologies in reducing drug attrition due to safety failures. *American Drug Discovery* 1:53–56.

Chabre M, Deterre P (1989) Molecular mechanism of visual transduction. *European Journal of Biochemistry* 179:255–266.

Chen Y, Thompson DC, Koppaka V, Jester JV, Vasiliou V (2013) Ocular aldehyde dehydrogenases: protection against ultraviolet damage and maintenance of transparency for vision. *Progress in Retinal and Eye Research* 33:28–39.

Chhetri J, Jacobson G, Gueven N (2014) Zebrafish—on the move towards ophthalmological research. *Eye (London, England)* 28:367–380.

Choudhary S, Xiao T, Vergara LA, Srivastava S, Nees D, Piatigorsky J, Ansari NH (2005) Role of aldehyde dehydrogenase isozymes in the defense of rat lens and human lens epithelial cells against oxidative stress. *Investigative Ophthalmology & Visual Science* 46:259–267.

Cooper DL, Isola NR, Stevenson K, Baptist EW (1993) Members of the ALDH gene family are lens and corneal crystallins. *Advances in Experimental Medicine and Biology* 328:169–179.

Coster ME, Stiles J, Krohne SG, Raskin RE (2008) Results of diagnostic ophthalmic testing in healthy guinea pigs. *Journal of the American Veterinary Medical Association* 232:1825–1833.

Dayhaw-Barker P (2002) Retinal pigment epithelium melanin and ocular toxicity. *International Journal of Toxicology* 21:451–454.

Deeti S, O'Farrell S, Kennedy BN (2014) Early safety assessment of human oculotoxic drugs using the zebrafish visualmotor response. *Journal of Pharmacological and Toxicological Methods* 69:1–8.

DiMasi JA, Grabowski G, Hansen RW (2015) Cost of developing a new drug. *New England Journal of Medicine* 372:1972.

Douglas RM, Alam NM, Silver BD, McGill TJ, Tschetter WW, Prusky GT (2005) Independent visual threshold measurements in the two eyes of freely moving rats and mice using a virtual-reality optokinetic system. *Visual Neuroscience* 22:677–684.

Eiraku M, Takata N, Ishibashi H, Kawada M, Sakakura E, Okuda S, Sekiguchi K, Adachi T, Sasai Y (2011) Self-organizing optic-cup morphogenesis in three-dimensional culture. *Nature* 472:51–56.

Emran F, Rihel J, Dowling JE (2008) A behavioral assay to measure responsiveness of zebrafish to changes in light intensities. *Journal of Visualized Experiments* (20):pii:923.

Fleming A, Alderton WK (2013) Zebrafish in pharmaceutical industry research: finding the best fit. *Drug Discovery Today: Disease Models* 10:e43–e50.

Fraunfelder FT (2001) *Drug-Induced Ocular Side Effects*. Boston, MA: Butterworth-Heinemann.

Friedlander MR, Lizano E, Houben AJ, Bezdan D, Banez-Coronel M, Kudla G, Mateu-Huertas E, Kagerbauer B, Gonzalez J, Chen KC, LeProust EM, Marti E, Estivill X (2014) Evidence for the biogenesis of more than 1,000 novel human microRNAs. *Genome Biology* 15:R57.

Fuller JH (1985) Eye and head movements in the pigmented rat. *Vision Research* 25:1121–1128.

German OL, Buzzi E, Rotstein NP, Rodriguez-Boulan E, Politi LE (2008) Retinal pigment epithelial cells promote spatial reorganization and differentiation of retina photoreceptors. *Journal of Neuroscience Research* 86:3503–3514.

Gestri G, Link BA, Neuhauss SC (2012) The visual system of zebrafish and its use to model human ocular diseases. *Developmental Neurobiology* 72:302–327.

Goldsmith P, Harris WA (2003) The zebrafish as a tool for understanding the biology of visual disorders. *Seminars in Cell & Developmental Biology* 14:11–18.

Greene N (2007) *Computational Models to Predict Toxicity*. Oxford: Elsevier Ltd.

Holmberg A, Schwerte T, Pelster B, Holmgren S (2004) Ontogeny of the gut motility control system in zebrafish Danio rerio embryos and larvae. *The Journal of Experimental Biology* 207:4085–4094.

Holve DL, Mundwiler KE, Pritt SL (2011) Incidence of spontaneous ocular lesions in laboratory rabbits. *Comparative Medicine* 61:436–440.

Hornof M, Toropainen E, Urtti A (2005) Cell culture models of the ocular barriers. *European Journal of Pharmaceutics and Biopharmaceutics* 60:207–225.

Hubert MF, Gerin G, Durand-Cavagna G (1999) Spontaneous ophthalmic lesions in young Swiss mice. *Laboratory Animal Science* 49:232–240.

Hutchinson TH, Bogi C, Winter MJ, Owens JW (2009) Benefits of the maximum tolerated dose (MTD) and maximum tolerated concentration (MTC) concept in aquatic toxicology. *Aquatic Toxicology* 91:197–202.

Illanes O, Anderson S, Niesman M, Zwick L, Jessen BA (2006) Retinal and peripheral nerve toxicity induced by the administration of a pan-cyclin dependent kinase (cdk) inhibitor in mice. *Toxicologic Pathology* 34:243–248.

Iwashita M, Kanai R, Funabiki K, Matsuda K, Hirano T (2001) Dynamic properties, interactions and adaptive modifications of vestibulo-ocular reflex and optokinetic response in mice. *Neuroscience Research* 39:299–311.

Jacob RF, Aleo MD, Self-Medlin Y, Doshna CM, Mason RP (2013) 1,2-naphthoquinone stimulates lipid peroxidation and cholesterol domain formation in model membranes. *Investigative Ophthalmology & Visual Science* 54:7189–7197.

Karali M, Peluso I, Marigo V, Banfi S (2007) Identification and characterization of microRNAs expressed in the mouse eye. *Investigative Ophthalmology & Visual Science* 48:509–515.

Kari G, Rodeck U, Dicker AP (2007) Zebrafish: an emerging model system for human disease and drug discovery. *Clinical Pharmacology and Therapeutics* 82:70–80.

Kerr NM, Danesh-Meyer HV (2009) Phosphodiesterase inhibitors and the eye. *Clinical & Experimental Ophthalmology* 37:514–523.

Kramer JA, Sagartz JE, Morris DL (2007) The application of discovery toxicology and pathology towards the design of safer pharmaceutical lead candidates. *Nature Reviews Drug Discovery* 6:636–649.

Kretschmer F, Kretschmer V, Kunze VP, Kretzberg J (2013) OMR-arena: automated measurement and stimulation system to determine mouse visual thresholds based on optomotor responses. *PloS One* 8:e78058.

Kuhlman SM, Rubin LF, Ridgway RL (1992) Prevalence of ophthalmic lesions in wild-caught cynomolgus monkeys. *Progress in Veterinary & Comparative Ophthalmology* 2:20–28.

Lassen N, Bateman JB, Estey T, Kuszak JR, Nees DW, Piatigorsky J, Duester G, Day BJ, Huang J, Hines LM, Vasiliou V (2007) Multiple and additive functions of ALDH3A1 and ALDH1A1: cataract phenotype and ocular oxidative damage in Aldh3a1(-/-)/Aldh1a1(-/-) knock-out mice. *The Journal of Biological Chemistry* 282:25668–25676.

Liebler DC, Guengerich FP (2005) Elucidating mechanisms of drug-induced toxicity. *Nature Reviews Drug Discovery* 4:410–420.

Liu CN, Simic D, Baltrukonis D, Doshna C, Render J, Fortner J, Verdugo ME, Weisenburger WP, Somps C (2004) Early MNU-induced retinal toxicity detected by electroretinography and visual acuity in funduscopically normal rats. *Investigative Ophthalmology and Visual Science* 45:804.

Mannerstrom M, Zorn-Kruppa M, Diehl H, Engelke M, Toimela T, Maenpaa H, Huhtala A, Uusitalo H, Salminen L, Pappas P, Marselos M, Mantyla M, Mantyla E, Tahti H (2002) Evaluation of the cytotoxicity of selected systemic and intravitreally dosed drugs in the cultures of human retinal pigment epithelial cell line and of pig primary retinal pigment epithelial cells. *Toxicology In Vitro* 16:193–200.

Marrone AK, Beland FA, Pogribny IP (2015) The role for microRNAs in drug toxicity and in safety assessment. *Expert Opinion on Drug Metabolism & Toxicology* 11:601–611.

Mitchell PS, Parkin RK, Kroh EM, Fritz BR, Wyman SK, Pogosova-Agadjanyan EL, Peterson A, Noteboom J, O'Briant KC, Allen A, Lin DW, Urban N, Drescher CW, Knudsen BS, Stirewalt DL, Gentleman R, Vessella RL, Nelson PS, Martin DB, Tewari M (2008) Circulating microRNAs as stable blood-based markers for cancer detection. *Proceedings of the National Academy of Sciences of the United States of America* 105:10513–10518.

Mitchiner JC, Pinto LH, Vanable JW, Jr. (1976) Visually evoked eye movements in the mouse (Mus musculus). *Vision Research* 16:1169–1171.

Munger RJ, Collins MA (2013) Assessment of ocular toxicity potential: basic theory and techniques. In: *Assessing Ocular Toxicity in Laboratory Animals* (AB, W. and Collins M, eds), pp. 23–52 New York: Springer.

Nakano T, Ando S, Takata N, Kawada M, Muguruma K, Sekiguchi K, Saito K, Yonemura S, Eiraku M, Sasai Y (2012) Self-formation of optic cups and storable stratified neural retina from human ESCs. *Cell Stem Cell* 10:771–785.

Neuhauss SC (2003) Behavioral genetic approaches to visual system development and function in zebrafish. *Journal of Neurobiology* 54:148–160.

Peng Q, Huang W, John-Baptiste A (2014) Circulating microRNAs as biomarkers of retinal toxicity. *Journal of Applied Toxicology* 34:695–702.

Phillips JB, Westerfield M (2014) Zebrafish models in translational research: tipping the scales toward advancements in human health. *Disease Models & Mechanisms* 7:739–743.

Precht W, Cazin L (1979) Functional deficits in the optokinetic system of albino rats. *Experimental Brain Research* 37:183–186.

Prusky GT, Harker KT, Douglas RM, Whishaw IQ (2002) Variation in visual acuity within pigmented, and between pigmented and albino rat strains. *Behavioural Brain Research* 136:339–348.

Prusky GT, Alam NM, Beekman S, Douglas RM (2004) Rapid quantification of adult and developing mouse spatial vision using a virtual optomotor system. *Investigative Ophthalmology & Visual Science* 45:4611–4616.

Ramsden CM, Powner MB, Carr AJ, Smart MJ, da Cruz L, Coffey PJ (2013) Stem cells in retinal regeneration: past, present and future. *Development* 140:2576–2585.

Redfern WS, Storey S, Tse K, Hussain Q, Maung KP, Valentin JP, Ahmed G, Bigley A, Heathcote D, McKay JS (2011) Evaluation of a convenient method of assessing rodent visual function in safety pharmacology studies: effects of sodium iodate on visual acuity and retinal morphology in albino and pigmented rats and mice. *Journal of Pharmacological and Toxicological Methods* 63:102–114.

Richards FM, Alderton WK, Kimber GM, Liu Z, Strang I, Redfern WS, Valentin JP, Winter MJ, Hutchinson TH (2008) Validation of the use of zebrafish larvae in visual safety assessment. *Journal of Pharmacological and Toxicological Methods* 58:50–58.

Rittenhouse KD, Johnson TR, Vicini P, Hirakawa B, Kalabat D, Yang AH, Huang W, Basile AS (2014) RTP801 gene expression is differentially upregulated in retinopathy and is silenced by PF-04523655, a 19-Mer siRNA directed against RTP801. *Investigative Ophthalmology & Visual Science* 55:1232–1240.

Rombough P (2002) Gills are needed for ionoregulation before they are needed for O(2) uptake in developing zebrafish, Danio rerio. *The Journal of Experimental Biology* 205:1787–1794.

Rung J, Brazma A (2013) Reuse of public genome-wide gene expression data. *Nature Reviews Genetics* 14:89–99.

Ryan DG, Oliveira-Fernandes M, Lavker RM (2006) MicroRNAs of the mammalian eye display distinct and overlapping tissue specificity. *Molecular Vision* 12:1175–1184.

Saint-Macary G, Berthoux C (1994) Ophthalmic observations in the young Yucatan micropig. *Laboratory Animal Science* 1 44:334–337.

Sampath S, McLean LA, Buono C, Moulin P, Wolf A, Chibout SD, Pognan F, Busch S, Shangari N, Cruz E, Gurnani M, Patel P, Reising A (2012) The use of rat lens explant cultures to study the mechanism of drug-induced cataractogenesis. *Toxicological Sciences* 126:128–139.

Siddeek B, Inoubli L, Lakhdari N, Rachel PB, Fussell KC, Schneider S, Mauduit C, Benahmed M (2014) MicroRNAs as potential biomarkers in diseases and toxicology. *Mutation Research, Genetic Toxicology and Environmental Mutagenesis* 764–765:46–57.

Simic D, Liu CN, Baltrukonis D, Somps CJ (2005) Outward current in SRA 01/04 human lens epithelial cells is reduced following BK channel knockdown with RNAi. *Investigative Ophthalmology and Visual Science* 46:1859.

Solimeo R, Zhang J, Kim M, Sedykh A, Zhu H (2012) Predicting chemical ocular toxicity using a combinatorial QSAR approach. *Chemical Research in Toxicology* 25:2763–2769.

Somps CJ, Greene N, Render JA, Aleo MD, Fortner JH, Dykens JA, Phillips G (2009) A current practice for predicting ocular toxicity of systemically delivered drugs. *Cutaneous and Ocular Toxicology* 28:1–18.

Taradach C, Greaves P (1984) Spontaneous eye lesions in laboratory animals: incidence in relation to age. *Critical Reviews in Toxicology* 12:121–147.

Toda R, Kawazu K, Oyabu M, Miyazaki T, Kiuchi Y (2011) Comparison of drug permeabilities across the blood-retinal barrier, blood-aqueous humor barrier, and blood-brain barrier. *Journal of Pharmaceutical Sciences* 100:3904–3911.

Toimela T, Tahti H, Salminen L (1995) Retinal pigment epithelium cell culture as a model for evaluation of the toxicity of tamoxifen and chloroquine. *Ophthalmic Research* 27 Suppl 1:150–153.

Verdugo-Gazdik ME, Simic D, Opsahl AC, Tengowski MW (2006) Investigating cytoskeletal alterations as a potential marker of retinal and lens drug-related toxicity. *Assay and Drug Development Technologies* 4:695–707.

Walsh Clang CM, Aleo MD (1997) Mechanistic analysis of S-(1,2-dichlorovinyl)-L-cysteine-induced cataractogenesis in vitro. *Toxicology and Applied Pharmacology* 146:144–155.

Waring MJ, Arrowsmith J, Leach AR, Leeson PD, Mandrell S, Owen RM, Pairaudeau G, Pennie WD, Pickett SD, Wang J, Wallace O, Weir A (2015) An analysis of the attrition of drug candidates from four major pharmaceutical companies. *Nature Reviews Drug Discovery* 14:475–486.

Wilkie DA (2014) The ophthalmic examination as it pertains to general ocular toxicology: basic and advanced techniques and species-associated findings. In: *Ocular Pharmacology and Toxicology Methods in Pharmacology and Toxicology* (BC, G., ed), pp. 143–203 New York: Springer-Verlag.

Williams DL (2002) Ocular disease in rats: a review. *Veterinary Ophthalmology* 5:183–191.

Williams DL (2007) Rabbit and rodent ophthalmology. *The European Journal of Companion Animal Practice* 17:242–252.

Wilson SL, Ahearne M, Hopkinson A (2015) An overview of current techniques for ocular toxicity testing. *Toxicology* 327C:32–46.

Yarfitz S, Hurley JB (1994) Transduction mechanisms of vertebrate and invertebrate photoreceptors. *The Journal of Biological Chemistry* 269:14329–14332.

Yin J, Brocher J, Linder B, Hirmer A, Sundaramurthi H, Fischer U, Winkler C (2012) The 1D4 antibody labels outer segments of long double cone but not rod photoreceptors in zebrafish. *Investigative Ophthalmology & Visual Science* 53:4943–4951.

Zon LI, Peterson RT (2005) In vivo drug discovery in the zebrafish. *Nature Reviews Drug Discovery* 4:35–44.

15

PREDICTING ORGAN TOXICITY *IN VIVO*—CENTRAL NERVOUS SYSTEM

GREET TEUNS[1] AND ALISON EASTER[2]

[1] *Janssen Research & Development, Janssen Pharmaceutica N.V., Beerse, Belgium*

[2] *Biogen Inc., Cambridge, MA, USA*

15.1 INTRODUCTION

The central nervous system (CNS) adverse events account for ~14 and 21% of attrition for safety reasons in preclinical drug discovery and early clinical development, respectively (Redfern et al., 2010), and there is significant resource expended in predicting potential CNS adverse effects in preclinical development. In agreement with this, together with the cardiovascular and respiratory systems, the CNS is recognized as a key organ system in the International Conference on Harmonisation of Technical Requirements for Registration of Pharmaceuticals for Human Use and Safety Pharmacology Studies for Human Pharmaceuticals S7A (ICH S7A). Consequently, potential CNS adverse events must assessed in preclinical models prior to administration in humans. Traditionally, this is assessed by monitoring general behavioral changes *in vivo* using models such as the Irwin assay (Irwin, 1968) or functional observational battery (FOB) (Moser et al., 1995) as a primary screen. In addition, a number of animal models designed to assess specific CNS effects (e.g., seizure, abuse liability, drug dependence) may be used. More recently, there has been significant effort invested in investigating methods suitable for use in earlier phases of drug discovery when it is not always feasible to run resource-intensive *in vivo* assays. Here we overview the *in vitro* and *in vivo* models currently used in general CNS preclinical risk assessment and also emerging methods, which may supplement the more standard approaches. In addition we review the specialized models available for assessment of two specific CNS ADRs, seizure,

and drug abuse liability. Similar models exist for a number of other CNS ADRs, for example, cognition deficits, psychiatric effects, and effects on sleep architecture but will not be discussed here.

15.2 MODELS FOR ASSESSMENT OF CNS ADRs

15.2.1 *In Vivo* Behavioral Batteries

To date neurotoxicity testing is classically based on neurobehavioral assessments and on morphological and histopathological endpoints. To assess the behavioral, neurologic, and autonomic status in laboratory animals, the modified Irwin's test or the FOB are employed (Irwin, 1968; Moser, 2000). The rat is usually chosen as preferred species, and the drug candidate is tested at three dosages in a single-dose administration of which the low dose lays within the therapeutic range and the high dose is situated in a toxic range (ICH S7A, 2000) or, in case of impeding physicochemical properties, limited to the maximum feasible dose.

Both tests include a series of noninvasive observational (in cage) and interactive measurements to evaluate the neurofunctional integrity of the rat. Locomotor activity, behavioral changes, motor–affective (e.g., touch escape) and sensory (e.g., startle response) reflex responses, coordination and gait, CNS excitation (e.g., tremors), and autonomic features as pupil size and body temperature are evaluated (ICH, 2000). As a standard, these measurements are executed at several time points after the single-dose administration

Drug Discovery Toxicology: From Target Assessment to Translational Biomarkers, First Edition. Edited by Yvonne Will, J. Eric McDuffie, Andrew J. Olaharski, and Brandon D. Jeffy.

and are recorded based on onset, duration, and severity. Permanent or delayed neurotoxicity can be determined by including an additional observational time point at 7 days postdose.

In particular the modified Irwin's test enables to distinguish between most classes of psychoactive, neurologic, and autonomic drugs and even between drug candidates of the same class of drugs. With this test it is also feasible to differentiate between the different strains of rodents.

15.2.2 *In Vitro* Models

15.2.2.1 *General Considerations*
The ideal *in vitro* model will have low resource requirements, high throughput, and be predictive of *in vivo* and clinical endpoints. However, due to complexity of the CNS, this has proved difficult to model *in vitro*. Consequently, in comparison to other organ systems, there are few *in vitro* assays commonly used in preclinical risk assessment. The major issue is that neuronal cell cultures do not reflect the complexity of the intact CNS. This consists of large populations of neurons, which form vast neural networks; these are supported by nonexcitable support cells such as glial cells. In an added level of complexity, there are multiple types of both neurons and glia, and the exact proportions vary widely in different regions of the CNS. For these reasons, it is very difficult to recapitulate even one anatomical structure of the CNS in an *in vitro* assay for drug screening. An additional consideration when using *in vitro* CNS assays is that these assays do not take into account blood–brain barrier (BBB) function. Even if compounds have significant *in vitro* CNS activity, this may not be physiologically relevant if CNS exposure is limited by poor BBB penetration *in vivo*. For this reason, in order to avoid high numbers of false positives, *in vitro* CNS data should be interpreted in the context of CNS exposure, which itself may be measured or predicted based on physicochemical properties.

15.2.2.2 *Pharmacological Profiling*
The CNS expresses a wide range of drug targets including ion channels, G-protein coupled receptors, enzymes, and transporters. Off-target pharmacological modulation of any of these could potentially cause adverse effects on CNS function. In the early phases of drug discovery, pharmacological profiling can be used to give an overall assessment of compound activity at these targets (Bowes et al., 2012). This is generally achieved using high-throughput radioligand binding assays although functional assays may also be used to establish agonist or antagonist activity. In order to assess potential CNS effects, lists of targets known to be expressed in the CNS or to be involved in specific CNS ADRs can be used. For instance, lists of targets associated with seizure liability (Easter et al., 2009) and abuse liability (FDA, Guidance for Industry: Assessment of Abuse Potential of Drugs) have been proposed.

However, given the lack of understanding of the precise molecular mechanisms underlying CNS function, it is often difficult to predict the functional consequences of off-target binding. If necessary, activity at CNS-expressed targets can trigger testing in follow-up assays to investigate the potential physiological consequences of this pharmacological modulation.

15.2.2.3 *Neurotoxicity Assays*
The pharmacological profiling assays described previously typically use recombinant cell lines overexpressing the molecular target of interest and simply reflect compound binding to these targets. Another strategy for CNS risk assessment is to culture neuronal cells and assess general effects on cell function. Various *in vitro* neurotoxicity assays have been described and are available at contract research organizations. These typically utilize cultured neuronal-derived cell lines or primary cells and use general cytotoxicity endpoints such as proliferation and apoptosis (Moors et al., 2009; Culbreth et al., 2012); more neuron-specific endpoints such as neurite outgrowth have also been used (Radio et al., 2010). However, it is often difficult to distinguish general cytotoxicity from specific neurotoxicity, and these assays are not widely used as a primary screen.

15.2.2.4 *Electrophysiological Assays*
Brain slices maintain some of the neuronal circuitry and electrophysiological properties of the intact brain (Lynch and Schubert, 1980) and are a useful tool to investigate how molecules can modulate brain function, either to drive efficacy (Cho et al., 2007) or to induce CNS adverse events (Fountain et al., 1992). Electrophysiological measurement of electrical activity from cell cultures or brain slices has been a significant area of interest in development of new assays to better assess CNS adverse effects without performing complex and expensive *in vivo* studies. Initial work focused on *ex vivo* brain slice preparations, and these can be prepared from multiple species (usually rodent) and brain areas. In terms of preclinical safety screening, the hippocampal brain slice assay has been used most frequently, typically in the assessment of seizure liability (Easter et al., 2009, see also Section 15.3.2). There is also potential utility in other areas such as cognitive deficits or sedative effects. Memory formation in rodents, for example, during exploratory behaviors, has been shown to be associated with theta rhythm neuronal firing (Buzsaki et al., 1989); similar stimulation patterns have been shown to induce long-term potentiation (LTP) *in vitro* brain slices. Therefore, it appears that some of the circuitry involved in memory formation *in vivo* is retained *in vitro* suggesting that brain slice electrophysiology may be a useful tool to investigate drug-induced cognitive impairment. In agreement with this, benzodiazepines, which are well known to cause cognition deficits clinically, have been shown to decrease the magnitude of

LTP in hippocampal brain slices (del Cerro et al., 1992). Brain slice electrophysiology has also been used to investigate mechanisms of drug-induced sleep disruption. It has long been known that rhythmic network-driven oscillatory activity can be recorded from *in vitro* brain slices, and the frequency of this activity is similar to the EEG frequencies associated with sleep. For example, theta rhythm activity has been recorded from cortical brain slices and anesthetic agents including benzodiazepines, and barbiturates depress the action potential firing underlying this oscillatory network activity (Lukatch et al., 2005). Despite these promising data these techniques have not been widely used in screening for CNS ADRs, and the low throughput and high resource requirements of brain slice electrophysiology mean these assays are not suitable for use in screening of large numbers of compounds and use has been largely limited to investigational/mechanistic studies.

15.2.2.5 Newer Technologies A more recent development is the use of multielectrode array (MEA) recording, which can be used to record electrical activity from neuronal cultures, which have been shown to form functional networks *in vitro*. MEAs typically have 64 recording sites, and multiple arrays can be combined to increase throughput (Johnstone et al., 2010). In addition to increasing experimental throughput, a further advantage of using cell cultures is that, unlike brain slices, they can be maintained in long-term culture facilitating investigation of the effects of chronic drug exposure. More recently, the plate-based MEA assay systems amenable to higher-throughput screening have been developed, which is of particular interest in terms of drug screening. The potential value of these systems in neurotoxicity screening has been assessed using large validation sets consisting of positive and negative control compounds with known *in vivo* effects (DeFranchi et al., 2011; McConnell et al., 2012; Valdivia et al., 2014). These studies used primary cultures of neuronal cells; more recently it has been demonstrated that human iPS-derived neuronal cultures can also be used (Odawara et al., 2014). Although more commonly used in assessment of cardiac risk, there is significant interest in using this technology in the area of seizure risk assessment and MEA-based assays using neuronal cultures are available at contract research organizations. An area of increasing interest is microphysiological systems or "organs-on-chips"; these microfluidic devices allow culture of different cell types in an environment where the *in vivo* environment is more closely replicated (Bhatia and Ingber, 2014). This recreation of tissue- and organ-level physiology in an *in vitro* system is not possible using conventional cell culture techniques and is an exciting development for CNS research. As technology and scientific understanding advances, it is hoped that these assays may start to bridge the gap between simple cell culture assays and complex *in vivo* assays.

15.3 SEIZURE LIABILITY TESTING

15.3.1 Introduction

Drug-induced seizures are a serious, potentially life-threatening CNS ADR that can result in the failure of drugs to be licensed for use or withdrawn from the market. Drug-induced seizure is associated with a wide variety of drug classes, including non-CNS targeted therapies (Zaccara et al., 1990). Perhaps surprisingly, it has been estimated that around 65% of marketed compounds associated with convulsion or seizure clinically are non-CNS targeted (Easter et al., 2009). The term seizure refers to a period of rhythmic, synchronized abnormal neuronal activity, which may result in a number of symptoms including visual disturbances, tingling, mood changes, or the more obvious events typical of a classic tonic–clonic convulsion. This latter event is characterized by muscle rigidity followed by large amplitude rhythmic jerking movements. It is important to note that seizure activity is not always followed by the behavioral changes that define a convulsion, meaning it can be difficult to detect in standard behavioral studies. To make a definitive diagnosis of drug-induced seizure, recording of abnormal electrographic activity is required. However, in most studies this is absent and the terms seizure and convulsion are often used interchangeably to refer to the behavioral consequences of the seizure. Despite the potentially serious consequences of drug-induced seizures/convulsions, there are no regulatory guidelines describing how this should be addressed in preclinical studies. CNS safety pharmacology studies performed as part of regulatory submissions are only mandated to assess general behavioral effects as outlined in the ICH S7A guidelines. However, a number of optional studies may be used to address this issue. Such studies are usually undertaken if there are risk flags that would indicate seizure potential during early discovery, such as (i) existing data suggesting the compounds from that chemical or drug class are seizurogenic, (ii) evidence or knowledge that the compound penetrates the BBB, (iii) pharmacological profile indicating that the compound interacts with molecular targets associated with seizure risk, and/or (iv) any *in vivo* behavioral observation suggesting convulsions or other abnormal CNS effects (Easter et al., 2009). In contrast to most CNS adverse events, *in vitro* assays have been relatively well utilized in this area although *in vivo* models are generally considered the best understood and most commonly used models in assessment of seizure liability.

15.3.2 Medium/High Throughput Approaches to Assess Seizure Liability of Drug Candidates

15.3.2.1 In Vivo Zebra Fish Larvae The zebra fish larvae (*Danio rerio*; Zf) have been assessed as a medium/high-throughput *in vivo* model for detection of seizure liability (Baraban et al., 2005; Winter et al., 2008). Exposure of zebra fish larvae to a variety of convulsant compounds such as pentylenetetrazole (PTZ) induces abnormal behavioral activity in a concentration-dependent manner. Specifically, a large increase in swimming speed/activity in combination with rapid circling, followed by a loss of posture at higher concentrations, suggests a clonus-like convulsion. Furthermore, electrophysiological recordings in adult Zf revealed electrographic discharges consistent with ictal events. This convulsant activity can be blocked by exposure to antiepileptic drugs (Berghmans et al., 2007). Of relevance for drug screening, Zf larvae can be kept in plate-based wells and locomotor activity recorded using automated video tracking, meaning this assay is of potential value in early drug discovery phases.

15.3.2.2 Electrophysiological Recording from Brain Slices and Cultured Cells

As discussed previously, brain slice electrophysiology is of potential value in assessment of a number of CNS ADRs, including seizure liability (see Section 15.2.2.4). To date, most work in this area has focused on the hippocampal brain slice largely due to a defined cytoarchitecture that makes it amenable to electrophysiological recording. The hippocampus is also strongly linked to partial seizures, including temporal lobe epilepsy (Schwartzkroin, 1994). Using single-microelectrode field potential recording of evoked synaptic activity from hippocampal brain slices, it has been shown that a number of standard seizurogenic compounds, including PTZ (Leweke et al., 1990; Rostampour et al., 2002), cause strong excitatory effects resulting in increased firing in neuronal cell populations. In addition, similar effects have been seen with a number of drugs associated with seizure clinically, including antibiotics (Schmuck et al., 1998), antipsychotics (Oliver et al., 1982), and antidepressants (Luchins et al., 1984). Similar results have been reported using multislice recording systems, which allow increased experimental throughput although this is still low relative to cell-based in vitro assays (Easter et al., 2007). Although most of the work in this area has focused on hippocampal brain slices, compounds may elicit seizure in a number of different brain regions. In agreement with this, seizure-like activity has also been recorded from neocortical (Voss and Sleigh, 2010) and thalamocortical (Gibbs et al., 2002) brain slices. For this reason, there may be utility in investigating the use of nonhippocampal brain slice preparations; the most appropriate preparation is likely to be dependent on the exact mechanism of drug-induced seizure. An additional development is the use of MEAs, which consist of a grid of several recording sites, which allow the simultaneous recording of electrical activity across a brain slice (Gonzalez-Sulser et al., 2011). The previous techniques may also be adapted to recording from cell populations, either primary cell cultures, stem cell-derived neuronal cultures (Illes et al., 2007), or mixed organotypic cultures (Berdichevsky et al., 2009).

15.3.3 In Vivo Approaches to Assess Seizure Liability of Drug Candidates

There are two main types of in vivo seizure models, precipitant models and electroencephalogram (EEG) recording. These are generally used where there is known target-related risk or to follow-up on behavioral observations seen in other in vivo studies.

15.3.3.1 Precipitant Models

Quantitative precipitant challenge assays classically applied in antiepileptic efficacy testing have been widely used to assess proconvulsant risk in preclinical studies (Porsolt et al., 2002). The most commonly used precipitants are chemoconvulsants and electrical stimulation, although other precipitants have been identified (e.g., sound, shaking). Of these precipitant models, the chemoconvulsant $GABA_A$ antagonist PTZ is the most commonly used. In each method, prior to administering the test compound, the sensitivity of the animal to precipitant-evoked seizures is established. The test compound is then dosed followed by challenge with the precipitant. A proconvulsant compound is associated with a lowering of the threshold to the precipitant challenge. Although precipitant models do not measure the direct convulsant effect of test compounds when administered alone, many proconvulsant compounds identified using precipitant models, such as venlafaxine, fentanyl, or isoniazid, also induce convulsions when administered alone at higher doses. Thus, the PTZ assay can be used as an indirect preclinical measure of convulsant risk. Importantly, not all convulsant compounds cause proconvulsant effects; the antidepressant bupropion that has a well-established seizurogenic potential in the clinic does not alter seizure threshold in the PTZ assay (Tutka et al., 2004).

15.3.3.2 EEG Recording

The EEG is the definitive seizure detection assay both preclinically and clinically. During EEG recording, electrical activity is measured directly using electrodes attached to the surface of the head. This method can also be utilized in preclinical species (rodent or nonrodent). In the case of rodent studies, electrodes are usually surgically implanted several days prior to the study start. An EEG recording represents the activity of a large population of neurons that span the cerebral cortex. Seizure activity is characterized by repeated spike–wave activity, which is due to abnormal synchronized firing (Tatum et al., 2008); this EEG biomarker translates across all clinical species. These spikes are thought to represent the summated synchronous firing of pyramidal neurons that span the cerebral cortex of all mammals. Regardless of where or how a seizure is initiated, drug-induced, physical trauma or epilepsy, once it generalizes to include the cerebral cortex, the spike–wave morphology is easily recognizable. There are other EEG waveforms such as paroxysmal spikes, fast ripples, or postictal depression that can serve to confirm the ictal nature of observed EEG abnormalities. In many cases, these changes in electrical activity precede the behavioral manifestation of seizure, the tonic–clonic convulsion. For this reason, it is useful to make video recordings of animal behavior so any electrographic events can be matched to behavioral observations. EEGs can help determine, beyond behavioral ambiguities, if a given drug causes seizure and at what doses or exposures. For these reasons, EEG is the gold standard model for assessment of drug-induced seizure. EEG recordings can be performed in a standalone study or can be added to general toxicology studies; this is particularly valuable where effects are only seen following repeat dosing of the test compound. Although EEG recording gives the definitive diagnosis of drug-induced seizure, this is a resource-intensive assay and is generally only used at the later phases of drug discovery or in drug development and is not amenable to high-throughput screening.

15.4 DRUG ABUSE LIABILITY TESTING

15.4.1 Introduction

The nonmedical use of prescription drugs has evolved toward a severe health issue worldwide. To date, prescription drugs are the second of abused substances after marijuana in teens in the United States (White House ONDCP, 2014) and include pain killers (opioids), stimulants (ADHD medication), and antidepressants.

The concomitant risk for dependence behavior or addiction when abusing prescription drugs has led to the release of stringent guidances by the drug licensing authorities (EMA, 2006; ICH, 2009; FDA, 2010, 2015), forcing the Pharma Industry to formally investigate novel CNS-active drug candidates for their potential to induce physical and/or psychological dependence.

To date the preclinical platform for the translational research of the abuse potential of novel CNS-active drug candidates comprises dedicated behavioral animal tests, each investigating different aspects of abuse potential: the nonprecipitated withdrawal test studies the physical dependence potential of a CNS-active drug candidate and is a mandatory test. The evaluation of discriminative stimulus properties of drug candidates, including drug profiling and functional resemblance to known psychoactive reference drugs, is executed through the drug discrimination learning (DDL) test. Rewarding and reinforcing properties are determined directly via the intravenous self-administration (IVSA) model or indirectly via the conditioned place preference (CPP) test (if the drug candidate formulation does not allow intravenous administration). Reinforcement is a term used to describe the relationship between the behavior (i.e., drug taking) and the consequences of that behavior (i.e., drug effect) (Bozarth, 1987). A positive reinforcement changes the environment by adding a stimulus that increases the likelihood of the behavior to recur in the future.

For each test a known psychoactive drug of abuse with similar pharmacological properties as the drug candidate needs to be included, as results are only relevant in relation to those of the chosen reference drug. These reference drugs are known to interfere with neurotransmitter systems involved in the neuronal pathway associated with substance abuse (Camí and Farré, 2003). The systems comprise dopamine and dopamine transporters (cocaine), mu opioids (morphine), serotonin (LSD), cannabinoids (THC), GABA (benzodiazepines), acetylcholine (nicotine), N-methyl-D-aspartate (NMDA) (ketamine), and norepinephrine. However the complexity of testing new CNS-active drug candidates for possible abuse potential often lies within the off-target pharmacotoxicological profile, as the mechanism of action of the majority of these drug candidates is distinct from that of the known classes of abused drugs (EMA, 2006; FDA, 2010). This implies that the construct of the preclinical abuse liability assessment, in particular with regard to the relevance of the reference drug versus the drug candidate, is not straightforward and might even impose more studies with different reference drugs, which may be distinct from the pharmacology but will rather apply to the same therapeutic class. The behavioral side effects (preclinical and/or clinical) caused by a novel CNS-active drug candidate may also justify the selection of a reference drug with a similar behavioral profile. An illustration hereof is the choice of chlordiazepoxide (CIV scheduled drug) if the drug candidate exhibits sedation although no affinity/similarity to the pharmacological (benzodiazepine) or therapeutic indication (anxiety treatment) of chlordiazepoxide is present.

A proper dose range of the drug candidate must be selected in terms of preclinically testing a minimally effective dose and a supratherapeutic dose, the latter being a severalfold of the human efficacious dose (expressed as the maximum plasma exposure, C_{eff} Hu) to meet the requirements of the regulatory guidances (EMA, 2006; ICH, 2009; FDA, 2010). This involves thorough knowledge of the toxicological and toxicokinetic profile of the drug candidate.

In addition, kinetics need to be determined in the preclinical abuse liability studies. A CNS-active drug candidate with a kinetic profile demonstrating a short half-life might be subject to potential abuse.

With respect to animal welfare, care must be taken not to interact with the animal's comfort and health status (USDA Animal Welfare Act, 1996).

The translational approach of a preclinical abuse liability assessment involves a thorough understanding and knowledge of the (neuro) pharmacology and the toxicological profile of the drug candidate and of the concomitant psychoactive reference drug(s), and the evaluation of the biological relevance of the outcome of the preclinical tests must be carefully determined. As such the preclinical abuse liability assessment provides robust and predictive preclinical data on the abuse potential of new medicines with a novel mechanism of action (Horton et al., 2013).

15.4.2 Preclinical Models to Test Abuse Potential of CNS-Active Drug Candidates

According to the regulatory requirements of drug licensing authorities, the rat is considered the preferred species to perform drug abuse liability studies (EMA, 2006; ICH, 2009; FDA, 2010).

The studies are executed following the GLP requirements (OECD, 1998; FDA, 2014).

15.4.2.1 Nonprecipitated Withdrawal Test
Physical dependence is defined as a state of adaptation of the body to the administered drug substance, manifested by a drug class-specific withdrawal syndrome that can be produced by abrupt cessation, rapid dose reduction, decreasing blood level of the drug, and/or administration of an antagonist (AAPM, 2001). The presence of physical dependence only is not equivalent to addiction as it may occur with the chronic use of any substance, legal or illegal, even when taken as prescribed (NIDA, 2012). However, physical dependence can lead to craving for the drug to relieve the withdrawal symptoms.

The nonprecipitated withdrawal test studies the physical dependence potential of a CNS-active drug candidate in view of a positive reference drug relevant to the drug candidate and includes a dosing phase followed by abrupt termination of drug treatment. The dosages should allow sufficient exposure to obtain adequate receptor occupancy and/or brain penetration. Preferably the clinical administration route is utilized. The duration of the repeated dose phase must allow a neuroadaptation of the body to the drug candidate. The length of the withdrawal period is dependent on the half-life of the drug candidate but does usually not exceed 1 week.

At various daily time points during specified days of both the repeated dose phase and the withdrawal period, the rats are observed for changes in behavior, and body temperature, body weight, and food intake are recorded (Teuns, 2015; Table 15.1).

Albeit the nonprecipitated withdrawal test is not as complex as the tests involved in the investigation of reinforcing and rewarding properties of a drug candidate, many factors have to be taken into account to ensure a proper study design and hence a correct interpretation of the results.

15.4.2.2 DDL Test
The DDL test is considered a sensitive test to profile the discriminative stimulus effects of new CNS-active drug candidates (stimulus generalization) relative to those of known psychoactive drugs in view of their receptor-binding profile (functional resemblance) or interaction with new neurotransmitter pathways. This stimulus generalization procedure can be executed using a variety of known substances of abuse (cues), which display their existing intrinsic stimulus effects via different receptors (Colpaert et al., 1975; Meert et al., 1990).

If the drug candidate has no pharmacological similarity with known drugs of abuse, which are acting via the previously mentioned neurotransmitters, a scheduled comparator drug approved for the same disease condition can be considered. For first-in-class drug candidates, it might also be appropriate to select a reference drug, which is primarily based upon the clinical signs observed in preclinical and/or clinical trials.

A two-lever food-reinforced DDL procedure includes a food reward training, a drug discrimination training with a psychoactive drug, and a final test phase (Fig. 15.1).

During the food reward training, animals are learned to press levers up to 10 lever presses for 1 food reward (fixed ratio of reinforcement 10 or FR10). The subsequent drug discrimination training implies the process to learn to discriminate between saline and a training drug, based upon the subjective effects of the training drug as experienced by the animal, and made visible through the correct levers presses for food rewarding (food reinforcement).

During the final test phase, the training drug or saline is replaced by single escalating doses of the drug candidate (including a control/vehicle substitution without active compound). If the animals experience the drug candidate as having comparable interoceptive cues as the training drug, animals will select the drug lever; subsequent bar pressing of this drug lever is rewarded with food pellets (FR10 schedule). If the animals do not associate the cueing effects with the training drug or if no effects are present, the saline lever is selected and subsequent pressing on that lever is reinforced (FR10 schedule).

Points of attention in the DDL test include the possibility to retest a certain test dose, sham treatment to avoid conditioning effects due to the chosen clinical route of the drug candidate versus the common injectable (IP or SC) route of the psychoactive training drugs, and careful consideration of possible responses due to active excipients used as vehicle. Using the clinical route and the feasibility to test different types of formulations in this model adds to the opportunity to use a broad test dose range.

15.4.2.3 IVSA Test
The IVSA paradigm, using a fixed-ratio schedule of reinforcement, is considered highly predictive for abuse liability testing of new CNS-active drug candidates within drug development. As required by the drug licensing authorities, FR10 (10 lever presses before an intravenous dose is released via self-administration) has to be obtained in the IVSA paradigm (FDA Dialogue Session, 2010). However, some of the known psychoactive drugs cannot be used in the general FR10 schedule (e.g., morphine). Lower (individual) FRs may then be justified.

TABLE 15.1   List of Specific Behavioral Observations with Concomitant Scores and/or Grades

Observations	Scores	
Lethality	P,D,I or −	Preterminally killed, dead, incidental dead, or absent
Sniffing	+/−	Present/absent
Abnormal licking	+/−	Present/absent
Abnormal preening	+/−	Present/absent
Grooming	+/−	Present/absent
Wet dog shakes	+/−	Present/absent
Twitches	0/4	0: no; 1: slight; 2: mild; 3: obvious; 4: severe
Tremors	0/4	0: no; 1: slight; 2: mild; 3: obvious; 4: severe
		Specific tremor : X if present
Convulsions	+/−	Present/absent
Specific convulsions		X if present:
Clonic-type convulsions		Asymmetrical
		Running excitement
		Champing
		Popcorn
		Asphyxial
Tonic-type convulsions		Opisthotonus
		Emprosthotonus
Miscellaneous-type convulsions		Rock and roll
		Sitting up
		Praying
Locomotor activity	0/4	0: no; 1: slight; 2: mild; 3: obvious; 4: severe
Abnormal biting	+/−	Present/absent
Restlessness	+/−	Present/absent
Writhing	+/−	Present/absent
Body carriage		
Straub tail	0/4	0: flat; 1: horizontal stretched; 2: 30° upward;
		3: 60° upward; 4: 90° upward
Dyspnoea	+/−	Present/absent
Arousal		
Excitation	0/4	0: no; 1: slight; 2: mild; 3: obvious; 4: severe
Jumping	+/−	Present/absent
Sedation	0/4	0: no; 1: slight; 2: mild; 3: obvious; 4: severe
Alertness	+/−	Present/absent
Narrowing palpebral fissure	0/4	0: completely closed
		1: narrowed; no reaction to stimuli
		2: narrowed; half-closed eyelids
		3: slightly narrowed; open after stimulus
		4: open
Defecation	−2/2	−2: no feces
		−1: decreased feces
		0: normal
		1: increased feces
		2: diarrhea
Lacrimation	+/−	Present/absent
Catalepsy	+/−	Present/absent
Abnormal gait		
Ataxia	0/4	0: no; 1: slight; 2: mild; 3: obvious; 4: severe
Tiptoe gait	+/−	Present/absent
Shuffling movements	+/−	Present/absent
Salivation	+/−	Present/absent

Taken with permission of G. Teuns (2015).

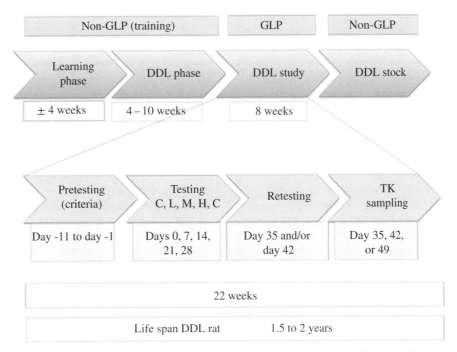

FIGURE 15.1 Outline of a full drug discrimination learning study. The training phases preceding the study and the maintenance of the stock animals are not under GLP regulation.

The basis of the IVSA rat model includes (i) a surgical procedure providing rats with an IV indwelling catheter; (ii) a training phase during which the animals are trained to IV self-administer a psychoactive drug at a fixed ratio of reinforcement; and (iii) the full test, during which the drug candidate is presented for IV self-administration at different dosages. One potential test design consists of five phases during which the IV self-administration of the reference drug (phases 1, 3 and 5) is alternated with the IV self-administration of saline (phase 2) and the IVSA of three to four dosages of the drug candidate and its concomitant control/vehicle solution (phase 4).

The administration of drugs in an IVSA paradigm is not under control of the investigator but relies on the rat's behavior to self-administer psychoactive drugs or drug candidates (Fig. 15.2).

The choice of a relevant reinforcer as reference drug in an IVSA study is based upon the characteristics of the drug candidate in view of its pharmacological or therapeutic class. Hallucinogenic drugs like LSD are not reliably self-administered in rats, and testing drug candidates with these type of reference drugs to investigate rewarding and/or reinforcing properties will need to be executed with other preclinical models like, for example, the CPP test (Bardo, 2000) or the DDL.

Other major issues to encounter in the IVSA paradigm are the solubility of the drug candidate in order to enable IVSA, and its concentration in the solution to fulfill the requirement of testing a multiple of the human efficacious dose (C_{eff} Hu).

Testing a dose range of a substance with reinforcing properties will result in a typically bell-shaped dose–response

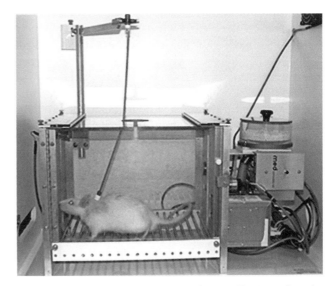

FIGURE 15.2 Setting of the IVSA equipment. (*See insert for color representation of the figure.*)

curve and not in a linear dose–response as classically seen in toxicity studies (Van Ree et al., 1999; Piazza et al., 2000). This indicates the importance of a well-considered dose selection for the outcome of this type of studies.

Another IVSA procedure is the progressive ratio reinforcement schedule, which compels increasing lever presses from the animals before an intravenous dose is released via self-administration. This procedure enables to define the so-called breakpoint at which an animal will no longer press the required amount of lever presses to obtain an intravenous dose. This breaking point is considered a measure for relative reinforcing efficacy (Richardson and Roberts, 1996). The value of progressive ratio assessments is realized in the ability of this schedule of reinforcement to provide a quantitative assessment of the reinforcing effects of novel drug candidates. While the FR self-administration assessment provides a qualitative assessment of reinforcement (i.e., a drug is self-administered or not), the PR assessment provides a quantitative assessment of reinforcement important for the comparison of relative reinforcing efficacy between reference drugs and drug candidates.

An alternative method to investigate reinforcing properties of a drug candidate is the intracranial self-administration test (Goeders and Smith, 1987; McBride et al., 1999; Collins et al., 2012; Bauer et al., 2013). The intracranial method is also an invasive method, and the exact location of the probe is crucial but also limiting the outcome. Most often intracranial electrodes are implanted in the medial forebrain bundle or in the ventral tegmental area (VTA). This method is not considered a standard procedure to determine the abuse potential of drug candidates in drug development but can possess value in certain situations and may be promising as future research on optogenetic technologies rather than electrical stimulation evolves (Negus and Miller, 2014).

15.4.2.4 *CPP Test*

The drug-induced CPP test is, in contrast to the IVSA model, a noninvasive and short-lasting method to measure the drug reward and, indirectly, the reinforcing properties of a drug candidate in naïve, untrained animals.

This model also allows all routes of administration, which makes the clinical route (often oral administration) applicable for testing. Solutions as well as suspensions can be administered and a dose range up to a (sub)toxic level can be reached, including a dose that is a multiple of the human efficacious dose, as required by the drug licensing authorities (EMA, 2006; ICH, 2009; FDA, 2010).

The general outline of our CPP model comprises three distinct phases: preconditioning (habituation and pretest), conditioning, and postconditioning (i.e., the test phase) (Tzschentke, 1998, 2007; Bardo, 2000; Prus et al., 2009). A two- or three-compartment test box can be employed for testing. The habituation phase, part of the preconditioning, enables animals to become accustomed to the different environments of the test box. At pretest, animals can again freely move, but the time spent in each compartment is recorded during this phase. In the unbiased test model, the individual place preference at pretest is not taken into account for pairing drug or water/saline to a specific compartment, and animals are assigned to the drug-paired or nondrug-paired environment for conditioning using a body weight-based stratified randomization procedure prior to starting the habituation phase.

During the conditioning phase animals are treated daily, receiving the test compound or water/saline on alternate days. On those days that the test compound is administered, animals are placed in the drug-paired compartment, whereas on the days that water/saline is administered, animals are placed in the nondrug-paired compartment.

At posttest, neither test compound nor water is administered, and animals are given free choice to the different compartments. The time spent in each compartment is recorded. Classically, a drug-induced CPP or positive reinforcing effect is obtained if the animals spent more time in the drug-paired compartment than in the nondrug-paired compartment at posttest.

The length of the conditioning phase, and subsequently the number of drug pairings is, among others, considered a critical variable to test drug candidates with a new mechanism of action, which may possess only weak reinforcing properties (Fig. 15.3).

The drug licensing authorities consider the CPP as a valuable but less robust model for testing reinforcing properties of a drug candidate, but it can replace the IVSA paradigm in case of insolubility of the drug candidate (FDA, 2010).

Besides the CPP test, which is based on the rewarding properties of a drug, a conditioned place aversion (CPA) test is also often described in the literature (Cunningham et al., 2006). In the latter, rats will spend less time in the drug-paired compartment at posttest. A CPA is often related but not limited to drugs that produce aversive effects. Examples hereof described in the literature include naloxone (Cunningham et al., 2006) and lithium chloride (Prus et al., 2009).

15.4.2.5 *Conclusion*

To date no *in vitro* models are available to study the physical dependence potential and the rewarding and reinforcing properties of novel CNS-active drug candidates in development. The preclinical abuse liability assessment represents the integrated data of *in vivo* animal studies to predict the abuse potential of CNS-active drug candidates.

15.5 GENERAL CONCLUSIONS

15.5.1 *In Vitro*

As with all drug-induced adverse events, there is a drive to develop predictive *in vitro* models to enable assessment of CNS ADRs in the early phases of drug discovery. However, due to the complex structure of the CNS, it has proved difficult to model in cell-based systems, and efforts in this area have been of limited success. For this reason, *in vitro* CNS assays are rarely used as frontline screens in drug safety screening, and use has been mainly limited to mechanistic studies, particularly using electrophysiological techniques.

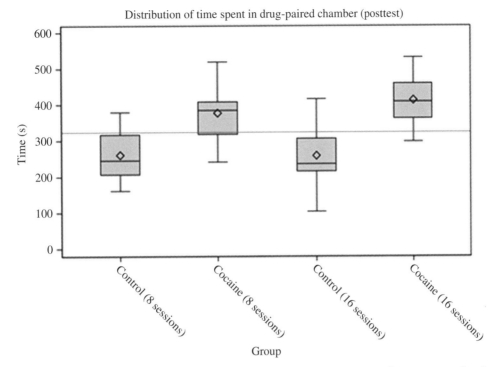

FIGURE 15.3   Box plot visualization of the distribution of the time spent in the drug-paired chamber at posttest after 8 or 16 drug pairings with cocaine (10 mg/kg subcutaneous administration), respectively. Control = saline 0.9% (subcutaneous administration).

As scientific knowledge increases and technology advances, particularly in the area of 3D cell cultures and microphysiological systems, it is hoped that more predictive *in vitro* assays will be developed. However, these techniques are in the early phases of development, and significant validation work is required before these are used routinely in preclinical assessment of drug safety.

15.5.2   *In Vivo*

The preclinical translational platform for predicting drug-induced neurotoxicity is still limited to behavioral assessments and morphological and histopathological endpoints. Testing the neurofunctional integrity of laboratory animals is a sensitive method but does not distinguish between physiological, pharmacological, or nonnervous system effects induced by the drug candidate. As such considerable scientific knowledge of the pharmacology and the toxicology of the drug candidate and a broad behavioral expertise are needed to correctly interpret the results and to assess the relevance hereof.

Conventional *in vivo* seizurogenic models like the PTZ model, with its pros and cons, might be replaced by the *in vivo* zebra fish larvae model in future as a medium/high-throughput *in vivo* model for detection of seizure liability of drug candidates in early discovery.

The preclinical abuse liability testing assessment comprising the investigation of the drug-induced physical dependence potential and the rewarding and reinforcing properties of CNS-active drug candidates relative to known psychoactive drugs of abuse is considered a predictive *in vivo* platform.

However as the mechanisms of action of novel drug candidates are usually very different from those of the reference drugs of abuse, the translational approach requires a multidisciplinary knowledge and expertise at the level of, *among others*, pharmacology and toxicology.

In conclusion, it can be stated that the nonclinical risk assessment of drug-induced neurotoxicity is still driven by *in vivo* investigations due to the complexity of the CNS, although promising *in vitro* alternatives are being developed as early screening models.

REFERENCES

AAPM American Academy of Pain Medicine, APS American Pain Society, and ASAM American Society of Addiction Medicine (2001) Consensus Document. Definitions Related to the Use of Opioids for the Treatment of Pain.

Baraban SC Taylor MR, Castro PA, Baier H (2005) Pentylenetetrazole induced changes in zebrafish behavior, neural activity and c-fos expression. *Neuroscience* 131:759–768.

Bardo M (2000) Conditioned place preference: what does it add to our preclinical understanding of drug reward? *Psychopharmacology (Berl)* 153(1):31–43.

Bauer C, Banks M, Blough B, Negus S (2013) Use of intracranial self-stimulation to evaluate abuse-related and abuse-limiting effects of monoamine releasers in rats. *Br J Pharmacol* 168(4):850–862.

Berdichevsky Y, Sabolek H, Levine JB, Staley KJ, Yarmush ML (2009) Microfluidics and multielectrode array-compatible organotypic slice culture method. *J Neurosci Methods* 178:59–64.

Berghmans S, Hunt J, Roach A, Goldsmith P (2007) Zebrafish offer the potential for a primary screen to identify a wide variety of potential anticonvulsants. *Epilepsy Res* 75(1):18–28.

Bhatia S, Ingber D (2014) Microfluidic organs-on-chips. *Nat Biotechnol* 8:760–772.

Bowes J, Brown A, Hamon J, Jarolimek W, Sirdhar A, Waldron G, Whitebread S (2012) Reducing safety-related drug attrition: the use of in vitro pharmacological profiling. *Nat Rev Drug Discov* 11:909–922.

Bozarth M (1987) An Overview of Assessing Drug Reinforcement. *Methods of Assessing the Reinforcing Properties of Abused Drugs* (pp. 635–658). M.A. Bozarth (Ed.), New York: Springer-Verlag.

Buzsaki G, Bickford RG, Ryan LJ, Young S, Prohaska O, Mandel RJ, Gage FH (1989) Multisite recording of brain field potentials and unit activity in freely moving rats. *J Neurosci Methods* 28:209–217.

Camí J, Farré M (2003) Drug addiction. *N Engl J Med* 349:975–986.

del Cerro S, Jung M, Lynch G (1992) Benzodiazepines block long-term potentiation in slices of hippocampus and piriform cortex. *Neuroscience* 49:1–6.

Cho S, Wood A, Bowlby MR (2007) Brain slices as models for neurodegenerative disease and screening platforms to identify novel therapeutics. *Curr Neuropharmacol* 5:19–33.

Collins A, Pogun S, Nesil T, Kanit L (2012) Oral nicotine self-administration in rodents. *J Addict Res Ther* Suppl 2: 004, 1–19.

Colpaert F, Niemegeers C, Janssen P (1975) The narcotic cue: evidence for the specificity of the stimulus properties of narcotic drugs. *Arch Int Pharmacodyn Ther* 218:268–276.

Culbreth ME, Harrill JA, Freudenrick TM, Mundy WR, Shafer TJ (2012) Comparison of chemical-induced changes in proliferation and apoptosis in human and mouse neuroprogenitor cells. *Neurotoxicology* 33:1499–1510.

Cunningham CL, Gremel CM, Groblewski PA (2006) Drug-induced conditioned place preference and aversion in mice. *Nat Protoc* 1(4):1662–1670.

DeFranchi E, Novellino A, Whelan M, Vogel S, Ramirez T, van Ravenzwaay B, Landsiedel R (2011) Feasibility assessment of microelectrode chip assay as a method of detecting neurotoxicity in vitro. *Front Neuroeng* 4(6):1–12.

Easter A, Sharp TH, Valentin J-P, Pollard CE (2007) Pharmacological validation of a semi-automated in vitro hippocampal slice assay for the assessment of seizure liability. *J Pharmacol Toxicol Methods* 56:223–233.

Easter A, Bell ME, Damewood JR Jr, Redfern WS, Valentin JP, Winter MJ, Fonck C, Bialecki RA (2009) Approaches to seizure risk assessment in preclinical drug discovery. *Drug Discov Today* 14:876–884.

EMA: EMEA/CHMP/SWP/94227/2004. Adopted by CHMP (2006) Guideline on the Non-Clinical Investigation of the Dependence Potential of Medicinal Products.

FDA (Revised 2014) CFR Title 21 Part 58 Good Laboratory Practice for Nonclinical Laboratory Studies.

FDA at the Dialogue Session on Abuse Potential (2010) Comments by CSS at F2F Meeting in Bethesda, MD. Public information.

FDA U.S. Department of Health and Human Services Food and Drug Administration—Center for Drug Evaluation and Research (CDER) (2010) Guidance for Industry. Assessment of Abuse Potential of Drugs. Draft Guidance.

FDA U.S. Department of Health and Human Services Food and Drug Administration—Center for Drug Evaluation and Research (CDER) (2015). Guidance for Industry Abuse-Deterrent Opioids—Evaluation and Labeling.

Fountain SB, Ting YL, Teyler TJ (1992) The in vitro hippocampal slice preparation as a screen for neurotoxicity. *Toxicol In Vitro* 6:77–87.

Gibbs JW III, Zhang YF, Ahmed HS, Coulter DA (2002) Anticonvulsant actions of lamotrigine on spontaneous thalamo-cortical rhythms. *Epilepsia* 43:342–349.

Goeders N, Smith J (1987) Intracranial self-administration methodologies. *Neurosci Biobehav Rev* 11(3):319–329.

Gonzalez-Sulser A, Wang J, Motamedi GK, Avoli M, Vicini S, Dzakpasu R (2011) The 4-aminopyridine in vitro epilepsy model analyzed with a perforated multi-electrode array. *Neuropharmacology* 60:1142–1153.

Horton D, Potter D, Mead A (2013) A translational pharmacology approach to understanding the predictive value of abuse potential assessments. *Behav Pharmacol* 24:410–436.

Illes S, Fleischer W, Siebler M, Hartung H-P, Dihne M (2007) Development and pharmacological modulation of embryonic stem cell-derived neuronal network activity. *Exp Neurol* 207:171–176.

International Conference on Harmonisation of Technical Requirements for Registration of Pharmaceuticals for Human Use (2000) ICH Harmonised Tripartite Guideline for Safety Pharmacology Studies for Human Use. Recommended for Adoption at Step 4 of the ICH Process on November 8, 2000.

International Conference on Harmonisation of Technical Requirements for Registration of Pharmaceuticals for Human Use (2009). ICH Harmonised Tripartite Guideline. Guidance of Nonclinical Safety Studies for the Conduct of Human Clinical Trials and Marketing Authorization for Pharmaceuticals. M3(R2). Current step 4-version.

Irwin S (1968) Comprehensive observational assessment: Ia. A systemic, quantitative procedure for assessing the behavioural and physiologic state of the mouse. *Psychopharmacologia* 13:222–257.

Johnstone AFM, Gross GW, Weiss DG, Schroeder OH-, Gramowski A, Shafer TJ (2010) Microelectrode arrays: a physiologically based neurotoxicity testing platform for the 21st century. *Neurotoxicology* 31:331–350.

Leweke FM, Louvel J, Rausche G, Heinemann U (1990) Effects of pentetrazol on neuronal activity and on extracellular calcium concentration in rat hippocampal slices. *Epilepsy Res* 6:187–198.

Luchins DJ, Oliver AP, Wyatt RJ (1984) Seizures with antidepressants: an in vitro technique to assess relative risk. *Epilepsia* 25:25–32.

Lukatch HS, Kiddoo CE, Maciver MB (2005) Anesthetic-induced burst suppression EEG activity requires glutamate-mediated excitatory synaptic transmission. *Cereb Cortex* 15:1322–1331.

Lynch G, Schubert P (1980) The use of in vitro brain slices for multidisciplinary studies of synaptic function. *Annu Rev Neurosci* 3:1–22.

McBride W, Murphy J, Ikemoto S (1999) Localization of brain reinforcement mechanisms: intracranial self-administration and intracranial place-conditioning studies. *Behav Brain Res* 101:129–152.

McConnell E, McClain M, Ross J, LeFew W, Shafer T (2012) Evaluation of multi-well microelectrode arrays for neurotoxicity screening using a chemical training set. *Neurotoxicology* 33(5):1048–1057.

Meert T, De Haes P, Vermote P, Janssen P (1990) A pharmacological validation of ritanserin and risperidone in the drug discrimination test procedure in the rat. *Drug Dev Res* 19:353–373.

Moors M, Rockel TD, Abel J, Cline JE, Gassmann K, Schreiber T, Schuwald J, Weinmann N, Fritsche E (2009) Human neurospheres as three-dimensional cellular systems for developmental neurotoxicity testing. *Environ Health Perspect* 117:1131–1138.

Moser VC (2000) The functional observational battery in adult and developing rats. *Neurotoxicology* 21(6):989–996.

Moser VC, Cheek BM, McPhail RC (1995) A multidisciplinary approach to toxicological screening. neurobehavioral toxicology. *J Toxicol Environ Health* 45:173–210.

Negus S, Miller L (2014) Intracranial self-stimulation to evaluate abuse potential of drugs. *Pharmacol Rev* 66(3):869–917.

NIDA National Institute on Drug Abuse (2012) The Science of Drug Abuse and Addiction. In Media Guide. Retrieved from: http://www.drugabuse.gov/publications/media-guide/science-drug-abuse-addiction (accessed October 16, 2015).

Odawara A, Saitoh Y, Alhebshi A, Gotoh M, Suzuki I (2014) Long-term electrophysiological activity and pharmacological response of a human induced pluripotent stem cell-derived neuron and astrocyte co-culture. *Biochem Biophys Res Commun* 443(4):1176–1181.

Oliver AP, Luchins DJ, Wyatt RJ (1982) Neuroleptic-induced seizures: an in vitro technique for assessing relative risk. *Arch Gen Psychiatry* 39:206–209.

Organisation for Economic Co-operation and Development (OECD) (1998) Principles of Good Laboratory Practice. Number 1.

Piazza P, Deroche-Gamonent V, Rouge-Pont F, Le Moal M (2000) Vertical shifts in self-administration dose–response functions predict a drug-vulnerable phenotype predisposed to addiction. *J Neurosci* 20(11):4226–4232.

Porsolt RD, Lemaire M, Durmuller N, Roux S (2002) New perspectives in CNS safety pharmacology. *Fundam Clin Pharmacol* 16(3):197–207.

Prus A, James J, Rosecrans J (2009) Conditioned Place Preference. *Methods of Behavior Analysis in Neuroscience*, 2nd edition. Frontiers in Neuroscience (pp. 59–77). J.J. Buccafusco (Ed.), Boca Raton, FL: CRC Press.

Radio NM, Freudenrich TM, Robinete BL, Crofton KM, Mundy WR (2010) Comparison of PC12 and cerebellar granule cell cultures for evaluating neurite outgrowth using high content analysis. *Neurotoxicol Teratol* 31:25–35.

Redfern WS, Ewart L, Hammond TG, Bialecki R, Kinter L, Lindgren S, Pollard CE, Roberts, R, Rolf MG, Valentin J-P (2010) Impact and frequency of different toxicities throughout the pharmaceutical life cycle. *Toxicologist* 114(S1):1081.

Richardson N, Roberts D (1996) Progressive ratio schedules in drug self-administration studies in rats: a method to evaluate reinforcing efficacy. *J Neurosci Methods* 66:1–11.

Rostampour M, Fathollahi Y, Semnanian S, Hajizadeh S, Mirnajafizadeh J, Shafizadeh M (2002) Cysteamine pretreatment reduces pentylenetetrazol-induced plasticity and epileptiform discharge in the CA1 region of rat hippocampal slices. *Brain Res* 955:98–103.

Schmuck G, Schurmann A, Schluter G (1998) Determination of the excitatory potencies of fluoroquinolones in the central nervous system by an in vitro model. *Antimicrob Agents Chemother* 42:1831–1836.

Schwartzkroin PA (1994) Role of the hippocampus in epilepsy. *Hippocampus* 4:239–242.

Tatum WO, Husain AM, Benbadis SR (2008) *Handbook of EEG Interpretation*. New York: Demos Medical Publishing.

Teuns G (2015) Assessing Physical Dependence. *Preclinical Assessment of Abuse Potential for New Pharmaceuticals* (pp. 101–127). C.G. Markgraf; T.J. Hudzik; D.C. Compton (Ed.), Cambridge, MA: Elsevier.

Tutka P, Barczynski B, Wielosz M (2004) Convulsant and anticonvulsant effects of bupropion in mice. *Eur J Pharmacol* 499:117–120.

Tzschentke T (1998) Measuring reward with the conditioned place preference paradigm: a comprehensive review of drug effects, recent progress and new issues. *Prog Neurobiol* 56(6):613–672.

Tzschentke T (2007) Measuring reward with the conditioned place preference (CPP) paradigm: update of the last decade. *Addict Biol* 12(3–4):227–462.

U.S. Department of Agriculture's (USDA) Animal Welfare Act (9 CFR Parts 1, 2 and 3) (1996) *The Guide for the Care and Use of Laboratory Animals*. Washington, DC: National Academy Press.

Valdivia P, Martin M, LeFew W, Ross J, Houch K, Shafer T (2014) Multi-well microelectrode array recordings detect neuroactivity of ToxCast compounds. *Neurotoxicology* 44:204–217.

Van Ree J, Gerrits M, Vanderschuren L (1999) Opioids, reward and addiction: an encounter of biology, psychology, and medicine. *Pharmacol Rev* 51:341–396.

Voss LJ, Sleigh JW (2010) Stability of brain neocortical slice seizure-like activity during low-magnesium exposure: measurement and effect of artificial cerebrospinal fluid temperature. *J Neurosci Methods* 192:214–218.

White House Office of National Drug Control Policy (ONDCP), U.S (2014) Research on Prescription Drug Abuse Centers for Disease Control and Prevention (CDC). http://www.white house.gov/ondcp/prescription-drug-abuse. (accessed October 16 2015).

Winter MJ Redfern WS, Hayfield AJ, Owen SF, Valentin JP, Hutchinson TH (2008) Validation of a larval zebrafish locomotor assay for assessing the seizure liability of early-stage development drugs. *J Pharmacol Toxicol Methods* 57:176–187.

Zaccara G, Muscas GC, Messori A (1990) Clinical features, pathogenesis and management of drug-induced seizures. *Drug Saf* 5:109–151.

16

BIOMARKERS, CELL MODELS, AND *IN VITRO* ASSAYS FOR GASTROINTESTINAL TOXICOLOGY

ALLISON VITSKY[1] AND GINA M. YANOCHKO[2]

[1] *Biomarkers, Drug Safety Research and Development, Pfizer, Inc., La Jolla, CA, USA*

[2] *Investigative Toxicology, Drug Safety Research and Development, Pfizer, Inc., La Jolla, CA, USA*

16.1 INTRODUCTION

The gastrointestinal (GI) tract is commonly affected by xenobiotics. Not only is it the first site of interaction with orally administered compounds, but it is an intricate system that includes many different (often interreliant) tissue types, an array of microbiota that are constantly interacting with a local immune system, as well as a continuous barrage of ingesta. As a result of this complexity of structure and function, this organ can be impaired by numerous mechanisms. GI toxicity is particularly common for certain classes of pharmaceuticals such as antibiotics, NSAIDS, and cancer therapies and reportedly occurs mainly as a result of four main reasons: primary pharmacology, impairment of GI defenses, direct injury, and changes in microbiome (Norman and Hawkey, 2010).

GI toxicity is often accompanied by clinical signs. 5–25% patients receiving antibiotics for bacterial infection, for instance, develop antibiotic-associated diarrhea (AAD) (Walk and Young, 2008), an issue that arises due to several mechanisms. Disruption of the gut microbiota can result in the loss of bacterial populations with metabolic consequences such as loss of carbohydrate metabolism (examples include clindamycin, ampicillin, and metronidazole) (Hogenauer et al., 1998; Bartlett, 2002). 10–20% of cases of AAD are caused by overgrowth of pathogenic bacteria, such as *C. difficile*, as a result of gut microbiota disruption (e.g., with cephalosporins, aminopenicillins, and clindamycin) (Hogenauer et al., 1998; Bartlett, 2002). Antibiotics can also have direct effects on the intestinal mucosa related to allergic or toxic reactions (neomycin) or cause pharmacologic effects on motility (erythromycin, amoxicillin; Hogenauer et al., 1998; Bartlett, 2002).

Cytotoxic therapies are also notably associated with GI toxicity, as they do not discriminate between highly proliferative cancer cells and highly proliferative normal cells, such as those within the small intestinal crypts. Furthermore, the high turnover/metabolic rate of many cells within the GI tract makes them exquisitely sensitive to hypoxia, so these tissues can also be affected secondarily when there is a vascular effect following xenobiotic administration. Adverse GI clinical signs during cancer chemotherapy include nausea, vomiting, and diarrhea (secretory, osmotic, malabsorption, exudative, and dysmotility-related) (Richardson and Dobish, 2007; Davila and Bresalier, 2008; Loriot et al., 2008; Boussios et al., 2012). Though there are conventional treatments available for certain GI toxicities, dose reductions and delays or cessation of treatment due to dose-limiting GI toxicity can have impact on a number of factors: treatment success (due to dose reductions), changes in the pharmacokinetics and pharmacodynamics of orally dosed drugs due to destruction of the GI tract, increased treatment costs due to the need for additional treatments, and decreased quality of life for patients experiencing severe symptoms (Richardson and Dobish, 2007; Loriot et al., 2008). Since GI toxicity from cytotoxic chemotherapy is due to the lack of specificity, advances in cancer treatments with molecularly targeted agents were hoped to decrease GI adverse events. However, many molecularly targeted cancer chemotherapies including monoclonal antibodies and small molecules have

Drug Discovery Toxicology: From Target Assessment to Translational Biomarkers, First Edition. Edited by Yvonne Will, J. Eric McDuffie, Andrew J. Olaharski, and Brandon D. Jeffy.

high incidence of GI toxicity. Targeted epidermal growth factor inhibitors (erlotinib, gefitinib, lapatinib, and HKI-272) have reported diarrhea (multiple grades) in phase I–III studies of 40–84% (Loriot et al., 2008). This high incidence is not restricted to epidermal growth factor receptor (EGFR) inhibition; the incidence of diarrhea is high for other molecularly targeted agents: the vascular endothelial growth factor receptor 3 (VEGFR3) inhibitors sorafenib and sunitinib have 33 and 20% incidence of grade 2–3 diarrhea; flavopiridol, a pan-cyclin-dependent kinase (pan-CDK) inhibitor, has 50% incidence of diarrhea; and the proteasome inhibitor bortezomib had ~30% incidence (Loriot et al., 2008).

Olson et al. (2000) compiled data from 12 major pharmaceutical companies to analyze the concordance between animal and human toxicity findings for compounds that had gone through clinical development. GI toxicity in humans accounted for 10% of project terminations across all therapeutic areas. The authors note that despite being the second most common human toxicity, it had the lowest rate for reason of termination. The therapeutic areas found to have the most frequent occurrence of GI toxicity were anti-infectives, anti-inflammatories, and anticancer with nearly 40% of GI toxicities in these areas attributed to the primary pharmacology of the compound. This group reported an 85% correlation of observed GI toxicity in preclinical animal studies with observed human GI toxicity, suggesting that animal models can accurately predict the occurrence of GI toxicity in humans. These findings support our qualitative experience that GI toxicity has been generally accepted as a toxicity that can be carried along through the drug discovery process and as Olson et al. put it, "a nuisance" toxicity, at least for certain therapeutic areas (Olson et al., 2000). However, patient tolerance for GI adverse events, even in the oncology therapeutic area, is declining.

Because of the complex nature of the GI system, a single chapter cannot reasonably provide a comprehensive list of all possible dosing outcomes that are possible or biomarkers and testing systems that could be utilized in a drug development setting. Rather, it is our intention to deliver examples of some of the major characteristics of this organ system that will help to highlight pertinent examples of GI toxicity seen during preclinical testing, followed by some of the tools that can be used to detect and characterize these situations.

16.2 ANATOMIC AND PHYSIOLOGIC CONSIDERATIONS

16.2.1 Oral Cavity

Although the complicated structure of the oral cavity includes many types of hard and soft tissues, the primary target tends to be the lining epithelium, which both protects the tissues from mechanical trauma and acts as a barrier against local microflora (Squier and Kremer, 2001). Disruption of this cell layer presenting as oral mucositis and/or ulceration is one of the most commonly reported outcomes of xenobiotic administration (Chaveli-Lopez, 2014), and in fact, erosions and ulcerations are a common manifestation of GI toxicity throughout the entire GI tract (Parfitt and Driman, 2007), most notably seen with chemotherapeutic compounds. Previously thought to be due to widespread necrosis of rapidly dividing cell populations, that is, basal epithelial cells, the development of oral mucositis is now proposed to occur through a multistep process. Simplistically, this process involves complex interplay between cell damage and resultant generation of reactive oxygen species, the activation of (among others) the NF-κB pathway, cytokine release, and the presence and proliferation of local microorganisms (Sonis, 2009). Complicating matters is the presence of secretory IgA within the saliva, a decrease of which (as may occur with xerostomia or salivary gland alterations) can result in imbalance of oral microorganisms and predisposition to the development of such lesions (Marcotte and Lavoie, 1998). Likely further reflecting the interrelationship of many biological factors are less frequent instances of epithelial hyperplasia affecting one or more oral epithelia, reported both preclinically (Vitsky et al., 2013) and clinically (Villalón et al., 2009).

16.2.2 Esophagus

The esophagus, an epithelial-lined muscular tube, allows passage of food to the stomach (Gelberg, 2014). Similar to the oral cavity, the epithelium is a common site of test article-associated lesions, although most drug-related tissue damage is thought to be associated with direct or prolonged contact of orally administered compounds with the esophageal mucosa (Jasperson, 2000). It is clear, however, that systemic factors are capable of playing a part in esophageal lesion development, as esophagitis has also been reported preclinically following intraperitoneal compound administration (Vitsky et al., 2013).

16.2.3 Stomach

Multiple cell types make up the gastric mucosa, including cells that function in digestion and nutrient absorption and as a barrier against low luminal pH and the resident bacterial flora, in lowest numbers here compared to rest of GI tract (Guarner and Malagelada, 2003). While direct effects of xenobiotics administered orally have been reported, both as erosive/ulcerative gastritis and exacerbated esophageal reflux (Peter et al., 1998; Lowe et al., 2000), the mucosa and other gastric tissues are also prone to damage following vascular insult (Hanton et al., 2008). Other compounds, notably nonsteroidal anti-inflammatory agents, not only diminish blood flow, resulting in ischemic risk, but also directly induce additional biochemical damage by uncoupling oxidative phosphorylation within the mitochondria of the mucosal lining cells (Somasundaram et al., 1995).

16.2.4 Small and Large Intestine

As previously mentioned, unselective toxicity of highly proliferating cells has historically been one of the most common reasons for preclinical toxicity, and the small intestinal crypts are often one of the foci of such an event (Phillips and Sternberg, 1975; Moore, 1979; Moore, 1986). Destruction of the proliferating components of the GI tract eventually leads to alteration of the crypt–villus axis leading to imbalance between the secretory and absorptive gut components (Viele, 2003). The variety and overlap of clinical signs and histologic lesions can be significant and at times have been recognized mainly by their mimicry of noniatrogenic disease states, such as tissue dysplasia and celiac-like disease (McCarthy et al., 2015). Other toxic effects occur due to dysmotility (Sellers and Morton, 2014), metaplasia (Milano et al., 2004), modification of intestinal enzymes (Cain et al., 1968), or deposition of nonabsorbable compounds (Arnold et al., 2013), and still others arise from alteration of vascular supply (Daniele et al., 2001; Klestov et al., 2001; Hass et al., 2007; Salk et al., 2013).

The introduction of biologics and molecular-targeted compounds may have resulted in fewer indiscriminate negative effects but has unfortunately not ameliorated the presence of GI toxicity. Targeted epidermal growth factor inhibitors (i.e., erlotinib, gefitinib, lapatinib, and HKI-272) have reported diarrhea (multiple grades) in phase I–III studies of 40–84% (Loriot et al., 2008). This outcome is not restricted to EGFR inhibition; the incidence of diarrhea is also high for other molecularly targeted agents: the VEGFR3 inhibitors sorafenib and sunitinib have 33 and 20% incidence, respectively, of grade 2–3 diarrhea; the pan-CDK inhibitor flavopiridol has 50% incidence of diarrhea; and the proteasome inhibitor bortezomib had ~30% incidence of diarrhea (Loriot et al., 2008).

Mechanisms of GI toxicity from molecularly targeted agents include changes in ionic secretion (EGFR inhibitors), direct mucosal damage (VEGFR inhibitors, flavopiridol, imatinib), changes in microbiome (mammalian target of rapamycin (mTOR) inhibitors), and autonomic nerve dysfunction (bortezomib) (Loriot et al., 2008). These mechanisms are often related to the antitumor action of the drug either due to direct pharmacological action on targets that are expressed both in tumor cells and in the normal epithelium of the GI tract or because the drug has activity at multiple targets (i.e., imatinib, sorafenib, or sunitinib) (Keefe and Gibson, 2007; Loriot et al., 2008).

Although the importance of intact mucosal epithelium's protective role against the luminal microbiota is well known, recent works detail a far more complicated interaction than simply barrier versus invader (Hooper and MacPherson, 2010; Duerr and Hornef, 2012; Chassaing and Gewirtz, 2014), detailing emergent information about the important role of the microbiome and intestinal immunity in prevention and formation of disease, as well as the utility of nutrition to dampen xenobiotic effects through modulating the interaction between the host, microbiome, and xenobiotic (Xue et al., 2011). The intestinal immune system plays a tenuous role in maintaining overall enteric health, protecting the body against a large, diverse population of resident microorganisms (not to mention a constant influx of new organisms), ensuring a measured, appropriate response that limits microbial invasion while avoiding overreaction. Simplistically, this is accomplished by minimizing interactions between enteric microorganisms and mucosal epithelium (mucus layer, secreted IgA, and antimicrobial peptides such as lectins), reacting to bacteria that break this barrier (resident macrophages within the intestine and lymphocytes serving adaptive or innate immune functions), and limiting the spread of any successful invaders (anatomic boundaries, resident macrophages in other organs, and mesenteric lymph nodes (Hooper and MacPherson, 2010). Disruption of the balance of this complex system has on occasion resulted in inflammation, infection, and neoplasia in both preclinical species (Morris et al., 2010) and human patients (Dulai et al., 2014; Venditti et al., 2015).

16.3 GI BIOMARKERS

In preclinical studies, histopathology has traditionally been the primary method of demonstrating GI injury. While this remains the most reliable way to detect and characterize tissue damage at the conclusion of a study, the need for real-time detection remains. The literature abounds on biomarkers of GI disease, though most are nonspecific and/or insensitive. The field is improving, however, and we describe several biomarkers here that have some possible or demonstrated utility in preclinical species. Many orally administered probe tests, such as the sucrose breath test, are not considered to have utility in preclinical studies due to the requirement for radioactive tracers and are therefore not covered.

16.3.1 Biomarkers of Epithelial Mass, Intestinal Function, or Cellular Damage

16.3.1.1 Citrulline Citrulline, an intermediate in the urea cycle, is the nitrogen end product of glutamine metabolism. Since the small intestinal epithelium is the predominant site of glutamine uptake and metabolism, it is therefore the principal source of circulating citrulline (van de Poll et al., 2007), so damage to this tissue would be expected to diminish normal blood levels. This is supported by reports of reduced citrulline concentrations that correlate with diminished intestinal mass, induced by surgical (so-called "short bowel" syndrome), HIV-associated, myeloablative, and radiation-induced mechanisms (Crenn et al., 2000; Lutgens et al., 2003; Lutgens et al., 2005; Papadia et al., 2007; Picot et al., 2010). Although historically, enzymatic assays have lacked sensitivity, more recent assays utilizing high-performance lipid chromatography–mass spectrometry have reportedly improved sensitivity (Crenn et al., 2000). In rats dosed with known GI toxicants, decreases in circulating citrulline have correlated well with small intestinal mucosal lesions (John-Baptiste et al., 2012; Vitsky et al., 2013), indicating it may hold promise as a sensitive and specific biomarker of injury, though further investigations are needed to determine predictive functionality.

16.3.1.2 Hydroxyproline Both increased collagenase and decreased collagen have been reported in ulcerated stomach tissues (Hasebe et al., 1987). Through capillary electrophoresis–mass spectrometry-based metabolic profiling of rats administered NSAIDs (aspirin, ibuprofen) or ethanol or subjected to environmental stress, a Japanese group recently identified hydroxyproline, a modified amino acid found in collagen, as a potential biomarker of gastric ulceration (Takeuchi et al., 2013, 2014a). In these investigations, decreases in serum hydroxyproline were present in all test groups and correlated with gastric ulceration. Importantly, in an additional study, rats codosed with aspirin and either omeprazole or famotidine exhibited neither decreased serum hydroxyproline nor ulcer formation, in contrast to aspirin-dosed rats, which as in other studies showed decreases in serum hydroxyproline that correlated with gastric ulceration (Takeuchi et al., 2014b). While additional work is needed to determine specificity of this parameter as well as the utility of this biomarker in other species, it may hold promise in rodent preclinical studies.

16.3.1.3 Trefoil Factors Trefoil factors are cysteine-rich, mucin-associated peptides (MAPs) that are richly expressed in the GI tract and upregulated in inflammatory and erosive/ulcerative conditions. Trefoil factor (TF) 1 is reportedly upregulated in ulcerated rat gastric mucosa (Ulaganathan et al., 2001), and TF 1 and TF2 (measured in serum via immunoassay) have been shown to be increased in human patients with inflammatory bowel disease (Vestergaard et al., 2004; Aamann et al., 2014).

16.3.1.4 Diamine Oxidase Diamine oxidase (DAO, histaminase) is a degradative enzyme of the polyamine metabolic pathway that is expressed in mature epithelium lining the rodent intestinal villi (Luk et al., 1980). Diminished blood concentrations, measurable by tritiated water assay, have been shown to correlate well with tissue expression (likewise decreased following injury resulting in loss of mature epithelial cells) and presence and extent of intestinal mucosal lesions (Luk et al., 1980; Moriyama et al., 2006). Normally, circulating DAO levels are very low, which creates difficulty when attempting to detect decreases within blood compartments (serum, plasma) as seen during investigations with various known intestinal toxicants in rats (John-Baptiste et al., 2012). Circulating levels can be enhanced by the *in vivo* administration of histamine (Luk et al., 1983; Moriyama et al., 2006); although such an addition is not possible during many preclinical safety studies, it might be possible for investigative studies or other types of examinations.

16.3.1.5 Intestinal Fatty Acid-Binding Proteins Fatty acid-binding proteins (FABP) are small (~15 kD) cytoplasmic proteins expressed in tissues that participate in high levels of fatty acid metabolism. Intestinal FABP (I-FABP), one of several FABP subtypes, is primarily expressed in the intestinal epithelium, with highest expression in the small intestine (Pelsers et al., 2003). Studies examining mesenteric infarction in humans have demonstrated a correlation between increases in circulating I-FABP and severity and extent of intestinal lesions (Kanda et al., 1996; Wiercinska-Drapalo et al., 2008). In the neonatal rat, increased intestinal tissue expression was noted in a model of infectious necrotizing enterocolitis (Gonçalves et al., 2015), and increases in circulating I-FABP, measurable by ELISA, have been reported in a rat ischemia–perfusion model (Evennett et al., 2014).

16.3.1.6 MicroRNAs MicroRNAs (miRNAs) are endogenous small (~19–23 nucleotides) noncoding RNAs that serve to negatively regulate gene expression through interaction with messenger RNA (mRNA). Unlike mRNAs, however, miRNAs have a long *in vivo* half-life and notable stability *in vitro*; furthermore, they generally exhibit a high degree of cross-species homology (Mikaelian et al., 2013). Following cell damage, miRNAs are released into the surrounding environment and as a result have garnered attention as emerging biomarkers for a variety of tissue disturbances. Recent literature describes expression patterns specific to various segments of the GI tract (Fassan et al., 2011), and it seems likely that novel miRNA biomarkers could emerge, possibly even with the ability to pinpoint both location and extent of GI tissue damage. In a series of investigative rat studies, miR194 and miR215 have been shown to have utility as a sensitive biomarker of intestinal mucosal damage (Yang et al., 2013; Kalabat et al., 2015).

16.3.2 Biomarkers of Inflammation

Since these biomarkers may be altered due to inflammation, they are not always considered specific for preclinical GI toxicities manifesting with alternate primary lesions (e.g., crypt epithelial apoptosis/necrosis associated with antiproliferative compound administration). However, since inflammation can occur as a secondary change in these cases due to mucosal barrier disruption (Groschwitz and Hogan, 2009), they may play a role in detecting xenobiotic-related GI lesions.

16.3.2.1 Calprotectin

Calprotectin, a calcium-binding protein found within neutrophils and monocytes, makes up 60% of the cytoplasmic protein within neutrophils and acts to fight several types of microorganisms. It is resistant to bacterial degradation and is reported to be quite stable at room temperature (Røseth et al., 1992). It is evenly distributed within the feces, and fecal concentrations have been demonstrated to correlate with the presence and endoscopic severity of inflammatory bowel syndrome (Lehmann et al., 2015; Menees et al., 2015). Recently it was reported that fecal calprotectin concentrations measured by ELISA correlated with intestinal injury in a rat model of necrotizing enterocolitis (Saglam et al., 2015), although, in a rat investigative study in which the primary lesion was mucosal atrophy, calprotectin did not correlate with severity of histologic findings (John-Baptiste et al., 2012).

16.3.2.2 CD64

CD64, the high-affinity Fc receptor (FcgRI) for IgG1 and IgG3, is primarily expressed on mononuclear phagocytes. The minimal expression present on resting neutrophils is upregulated when myeloid cells in the bone marrow are exposed to G-CSF and IFN-gamma, and the receptor then can exert a number of immune functions, such as internalization of immune complexes and crosslinking C-reactive protein and serum amyloid P. Flow cytometric analysis of CD64 has been reported as a useful tool in human medicine, with values differentiating between gastroenteritides of various causes (Tillinger et al., 2009).

16.4   CELL MODELS OF THE GI TRACT

The challenges for applying cell models of the GI tract for efficacy or toxicology studies are severalfold and similar to developing cell models of other organ systems: (i) Is a simple 2D cell culture model sufficient, or is an advanced culture model needed? (ii) Can a relatively low-throughput organoid model be used as a screening tool, and how can it be adapted to fulfill that need? (iii) Is cell-type or species specificity in the model required? (iv) What is the relevance of *in vitro* data to *in vivo* preclinical findings and further to human clinical findings? Cell-based models of the GI tract, as with other organ systems, have been widely used to understand and model efficacy and toxicity of compounds during the drug development process. These models vary greatly in their complexity with models of increasing complexity (i.e., three-dimensional (3D) organoids) becoming more accessible and thus more popular.

16.4.1 Cell Lines and Primary Cells

The simplest models are cell lines or primary cells isolated from the GI tract grown as monolayers in standard 2D multiwell cell culture plates. The more commonly used and widely available GI cell lines are Caco2, T84, HT-29, and IEC-6 (Fig. 16.1; ATCC, Manassas, VA; HTB-37, CCL-248, HTB-38, and CRL-1592 respectively). Additional cell lines have been generated but are less commonly used or are difficult to obtain by for-profit pharmaceutical companies (Sambruy et al., 2001; Cencič and Langerholc, 2010; Langerholc et al., 2011). IEC lines are nontransformed rat epithelial cell lines isolated from the rat small intestine and other locations (Quaroni et al., 1979). The widely used human cell lines such as Caco-2, T84, and HT-29 were derived from colon adenocarcinoma and thus have some limitations when trying to determine the mechanism of toxicity to normal GI tract for compounds in the drug development pipeline. The use of cell lines for *in vitro* toxicity assessment have several limitations, namely, they often lack appropriate cell-type-specific differentiation markers, they are tumorigenic, or they lack polarization and structural complexity of the GI tract (such as a crypt–villus axis).

In order to avoid limitations associated with cell lines (dedifferentiation, lack of cell-type specificity) researchers have looked to primary cell cultures of the GI tract. A number of culture and isolation techniques have been described (Lawson et al., 1982; Quaroni et al., 1991; Evans et al., 1992, 1994; Kaeffer, 2002). There have historically been significant drawbacks to the use of primary cultures of the GI tract, namely, poor survival *in vitro*, low cell yields, contamination from fibroblasts, and difficulty maintaining the functional differentiation for extended periods of time (Pageot et al., 2000; Kaeffer, 2002). Commercial sources of primary cells from rats, dogs, and humans are available (Lonza Group Ltd., Walkersville, MD; Creative Bioarray, Shirley, NY; Cell Biologics, Chicago, IL; CHI Scientific, Maynard, MA). Common molecular markers of differentiation used to characterize epithelial cell models of the GI tract include lactase, sucrase-isomaltase, and dipeptidyl peptidase-4 (Sanderson et al., 1996; Simon-Assmann et al., 2007). Functional endpoints are also useful markers of differentiation including transepithelial electrical resistance (TEER), inhibition of fluorescently conjugated dextran, transport, and ionic or acid secretion (Sanderson et al., 1996; Sambruy et al., 2001; Gunzel et al., 2010; Srinivasan et al., 2015). Culture conditions can alter the expression of differentiation markers (Simon-Assmann et al., 2007), and there have been advances in our understanding of the growth factor requirements for maintaining and differentiating primary cells in culture, particularly in relation to the need for Wnt signaling (Sancho et al., 2003; Ootani et al., 2009; Sato et al., 2009; Gon et al., 2013; Krausova and Korinek, 2014).

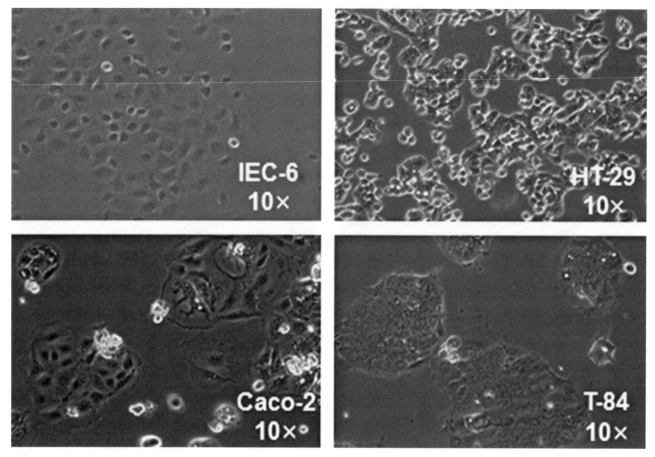

FIGURE 16.1 Morphology of commonly used gastrointestinal cell lines in monolayer culture. Representative gastrointestinal cell lines were cultured on 75 cm² cell culture flasks with appropriate growth media.

16.4.2 Induced Pluripotent Stem Cells

Induced pluripotent stem cells (iPSCs) offer a leap forward in our ability to model the GI tract *in vitro*: the cells are of human origin and thus provide researcher with a human cell-based tool, and they have the capacity for long-term regeneration, providing consistency and reducing the need for multiple primary cell isolations (Robinton and Daley, 2012; Finkbeiner and Spence, 2013; Hynds and Giangreco, 2013). In 2009, Spence et al. reported the first system for inducing human pluripotent stem cells (PSCs) to an intestinal lineage. The process entails four stages of growth factor induction resulting in cultures containing the four main cell types lining the GI tract: enterocytes, Paneth cells, goblet cells, and enteroendocrine cells (Spence et al., 2011). Kauffman et al. (2013) compared commercially available primary human intestinal cells (Lonza Group Ltd) with human PSCs induced along an intestinal lineage as a monolayer on transwell inserts and Caco-2 cells for markers of enterocyte differentiation, TEER, FITC–dextran permeability, and P-glycoprotein transport. Compared to Caco-2 cells, the iPSC-derived intestinal cells had equivalent or better expression of differentiation markers for several cell lineages including enterocytes (E-cadherin, villin), stem cells (Lgr5, ASCL2), Paneth cells (lysozyme), and enteroendocrine cells (chromogranin A). TEER values for iPSC-derived intestinal cells were ~1000 Ω cm², similar to Caco-2 cells (Fig. 16.2; Tanoue et al., 2008; Ferruzza et al., 2013; Srinivasan et al., 2015) that coincided with low permeability to FITC–dextran (Kauffman et al., 2013). The primary human intestinal cells were also similar in gene expression patterns to Caco-2 and iPSC-derived intestinal cells. The TEER from primary cells was higher than for iPSC-derived intestinal cells or Caco-2 (~1500–2500 vs. ~1000 Ω cm²), and they also prevented FITC–dextran permeability (Kauffman et al., 2013).

16.4.3 Coculture Systems

Combining cell types *in vitro* by coculturing is one way to increase the long-term viability as well as complexity of *in vitro* models, and to this end, cocultures of intestinal epithelial cells with myofibroblasts, immune cells, or bacteria have been utilized to model the interactions between epithelial cells and others within the GI tract. Susewind et al. (2015) investigated the effect of coculturing Caco-2 cells

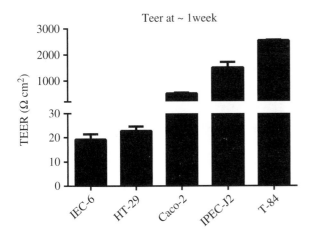

FIGURE 16.2 Transepithelial electrical resistance (TEER) of some commonly used gastrointestinal cell lines. Representative gastrointestinal cell lines were cultured on transwell filters with appropriate growth media. TEER values were obtained with REMS by World Precision Instruments (Sarasota, FL) after 7 days in culture.

in a transwell system with two immune cell populations (macrophages, THP-1 cells and dendritic cells, MUTZ-3). The response to exposure to silver, titanium dioxide, and gold nanoparticles on cytotoxicity, TEER, and IL-8 induction was measured with and without IL-1β-induced inflammation. TEER values were similar for Caco-2 monolayers and cocultures. Differences in IL-8 secretion and gene expression were evident with silver nanoparticles, but significant increases were observed only with coculture conditions (Susewind et al., 2015). Lahar et al. (2011) cocultured intestinal subepithelial myofibroblasts from mouse and human as a feeder layer to prolong human intestinal epithelial cell growth *in vitro* and *in vivo*. When grown in isolation, in the absence of a feeder layer, the primary human intestinal cells grew for 2–3 days, but their culture time could be prolonged for 50–60 days with the presence of mouse or fetal human intestinal myofibroblasts (Lahar et al., 2011). Interestingly, the cocultures with human fetal myofibroblasts were less dependent on exogenous growth factors for prolonged growth *in vitro* (Lahar et al., 2011). Primary cell "kits" of human intestinal myofibroblasts and epithelial cells are commercially available from Lonza Group Ltd. Cocultures of GI epithelial cells with cells of the immune system can be used to model the crosstalk between epithelial cells, the immune system, and gut microbiota. When cocultured with PBMC in a transwell system and stimulated with enteropathogenic and commensal bacteria, Caco-2 cells had differential levels of cytokine induction in response to pathogenic and commensal bacteria only when cultured in the presence of PBMC in the basal chamber (Haller et al., 2000). When cocultured with Caco-2 cells in a transwell system, dendritic cells had modified phenotypes (decreased MHC class II and CD86 expression), were less sensitive to

stimulation with Toll-like receptor ligands, had decreased production of the inflammatory cytokines (IL-8 and IL-10), and increased TGF-β (Butler et al., 2006).

16.4.4 3D Organoid Models

Advances in our understanding of the molecular mechanisms of differentiation and of 3D culture techniques have led to significant improvement in our ability to culture, maintain, and model the GI tract *in vitro* (Fig. 16.3). In 2009, two labs (Ootani et al., 2009; Sato et al., 2009) reported long-term culture and passage of intestinal Lgr5+ stem cells from the mouse as well as organoid development from these cells. Though each lab used slightly different techniques, both models required intact Wnt and Notch signaling for longevity and supported the growth of multiple cell types of the intestine (enterocytes, goblet cells, Paneth cells, and enteroendocrine cells). The 3D organoids reported by Sato et al. (2009) were generated from single Lgr5+ stem cells grown and propagated in matrigel, which contained a central lumen, lined by villus-like epithelial-lined structures with crypt-like domains at the tips. Ootani et al. (2009) started with mouse neonatal intestinal tissue and used a collagen matrix with an air–liquid interface. Though it takes a four-stage process to derive cells of intestinal lineage from human iPSCs, intestinal organoids derived from iPSCs have similar properties to those obtained from primary tissue or single Lgr5+ stem cells including growth of the four main cell types, polarized epithelium, villus-like structures, and proliferative crypt-like domains with cells that express stem cell markers such as Lgr5+, SOX9, and ASCL2 (Spence et al., 2011).

Intestinal organoids appear to be a useful *in vitro* model of the GI tract for toxicology applications. Exposure of intestinal organoids to the gamma-secretase inhibitor dibenzazepine caused goblet cell metaplasia consistent with preclinical and clinical experience with gamma-secretase inhibition (Ootani et al., 2009). Bromodomain and extraterminal (BET) domain proteins are chromatin reader proteins that are important targets for inflammation and oncology and are involved in homeostasis of the stem cell component of the GI tract (Bolden et al., 2014). Wistar Han rats dosed for 5 days with a Brd4/BET domain inhibitor exhibited inappetance, body weight loss, and duodenal villous atrophy, though mice and humans seem to tolerate Brd4 inhibition (Bolden et al., 2014; Wagoner et al., 2015). Intestinal organoids exposed to the Brd4/BET domain inhibitor exhibited loss of the fast-cycling intestinal stem cell component, confirming the *in vivo* findings (Wagoner et al., 2015). Furthermore, there was species specific sensitivity to Brd4/BET domain inhibition with dogs and rats being more sensitive than mice and humans (Wagoner et al., 2015). These results confirm the *in vivo* preclinical and clinical experience with Brd4/BET inhibitors and support the use of

Going from *this...*   to *this*

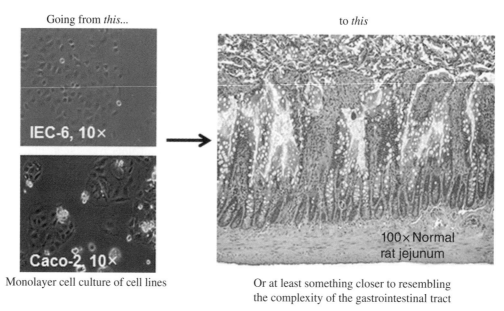

Monolayer cell culture of cell lines   Or at least something closer to resembling
                                              the complexity of the gastrointestinal tract

FIGURE 16.3   Challenges with *in vitro* models of the GI tract. Successive improvements in cell isolations and culture and understanding the molecular mechanisms of differentiation and microfluidic technology are improving our ability to model the GI tract *in vitro*.

intestinal organoids as a useful *in vitro* model for GI toxicity. Additionally, this work demonstrates how intestinal organoid culture significantly improves our ability to understand differences in species-specific toxicity and clinical translatability. Though obtaining primary tissue from multiple species can be difficult, the possibility of long-term propagation of intestinal organoids limits the need for multiple primary cell isolations, thus reducing the number of animals needed for research.

Gastric organoid models from primary cells and iPSCs have also been described. Gastric spheroids derived from different regions of the human stomach can maintain their region-specific differences (antrum vs. corpus) and be propagated in culture for more than 9 months (~20 passages; Schlaermann et al., 2014). These spheroids can then be differentiated into organoid cultures that exhibit folded, gland-like structures and expression of E-cadherin, β-catenin, and mucin 5AC (MUC5AC) and a small population of proliferating cells marked by Ki-67 (Schlaermann et al., 2014). In order to use established functional endpoints for gastric epithelia and monitor the signaling pathways activated by *Helicobacter pylori* infection, the authors isolated cells from 3D organoids and plated them in 2D culture. The cells retained their differentiated characteristics and responded to *H. pylori* infection with cytotoxin-associated gene A (CagA) translocation and phosphorylation and morphological changes consistent with infection (Schlaermann et al., 2014). McCracken et al. (2014) established human iPSC-derived organoids with different functional domains of the stomach (fundus and antrum) by differential growth factor signaling: posterior foregut specification required Wnt, fibroblast growth factor 4 (FGF4), NOG, and retinoic acid, while

further antral specification required retinoic acid, NOG, and EGF with additional EGF required for differentiated organoid growth. The differentiated organoids contained a variety of cell lineages including LGFR5 and SOX9-expressing cells in proliferative zones, endocrine cells (gastrin, ghrelin, somatostatin, and serotonin positive), and mucous cells, and they had RNA-seq profiles similar to human fetal stomach. Importantly, the gastric organoids responded to injection of *H. pylori* with appropriate pathophysiological response of CagA translocation to epithelial cells and CagA/c-Met receptor complex with accompanying c-Met phosphorylation. For both these organoid models (hIPSC- or primary cell-derived), the response of the organoids with the appropriate receptor stimulation and morphological changes in response to infections suggests that the gastric organoid model can improve the translatability of *in vitro* model systems and be a useful toxicological tool.

The presence of multiple differentiated cell types, morphological compartments reminiscent of structural domains *in vivo* (crypt–villus axes), relevant biological signaling and response to pathogens, and long-term propagation suggest that organoid models represent a significant improvement in our ability to model the GI tract *in vitro*. However, organoid models do have limitations: (i) Gastric organoids lack acidic luminal environment, and parietal cell lineage is apparently difficult to recapitulate. (ii) Lack of complete reconstruction of all GI cell types, such as microbiota and the enteric nervous system. (iii) Lack of physiologically relevant mechanical cues. (iv) Accessibility of the luminal compartment for drug exposure and sampling. (v) Can we adequately model species-related differences, for example, the nonglandular portion of the rodent stomach? (vi) Their complexity and low-throughput limit their utility for screening and application to established functional endpoints such as TEER, FITC permeability, epithelial transport, and high-content assays.

16.4.5   Organs-on-a-Chip

Microfluidic cell culture devices have the potential to further improve our ability to model the GI tract beyond improvements with 3D organoid models and address some of the limitations with organoid models. The term "organ-on-a-chip" refers to cell culture systems using organ-specific cells (i.e., Caco-2 cells as a model for enterocytes) grown on engineered microfluidic devices that were originally fabricated using methods similar to those for computer microchip manufacturing (Bhatia and Ingber, 2014). Organs-on-chips have several advantages over organoid models because they are completely engineered: (i) microsensor integration (e.g., inclusion of TEER as an endpoint), (ii) ability to provide controlled fluid flow and shear stress, (iii) cell patterning that can be controlled, (iv) porous substrates that can be integrated allowing for cell–cell and tissue–tissue interfaces and assessment of barrier function, and (v) incorporation of complex mechanical microenvironment with cyclic mechanical strain (thus mimicking peristaltic movements of the intestine; Bhatia and Ingber, 2014).

To date, GI-on-a-chip models have utilized Caco-2 cells (Kimura et al., 2008; Mahler et al., 2009; Esch et al., 2012; Kim et al., 2012; Kim and Ingber, 2013). Two papers from the Ingber lab described the culture of Caco-2 cells in two-chambered, linear microfluidic devices that integrated vacuum chambers to apply both shear stress (flow = 30 μl/h, 0.02 dyn/cm^2) and cyclic mechanical strain to mimic peristalsis (10%, 0.15 Hz; Kim et al., 2012; Kim and Ingber, 2013). Under these conditions Caco-2 cells developed complex morphological structures with basally located crypt-like domains that vertically extended into villus-like structures lined with epithelial cells. Importantly and even though the starting material were Caco-2 cells, all four main cell types of the intestine—enterocytes, enteroendocrine cells, Paneth cells, and goblet cells—were present in the cultures in ratios and locations consistent with *in vivo* localization. For example, Paneth cells were located in crypt-like domains, which also contained the greatest percentage of proliferating cells, while enterocyte and enteroendocrine and goblet cells were located in villus-like regions. A number of cell biological factors were improved under these culture conditions as compared to Caco-2 cells grown in static conditions in transwell chambers: increased cytochrome P450 3A4 activity, cells sixfold taller in size, improved glucose uptake, and increased paracellular permeability. Even more important, Kim et al. (2012) showed once differentiated, Caco-2-on-a-chip cultures could be cocultured in the presence of an endogenous human gut bacteria species

(*Lactobacillus rhamnosus*) without loss of viability of either human or bacterial cells and in fact, coculture in the presence of shear stress and cyclic mechanical strain improved the TEER of differentiated Caco-2 in agreement with *in vitro* and *in vivo* probiotic exposure. Another interesting aspect of these results is that the only difference in culture conditions besides the presence of collagen I and matrigel extracellular matrix was the shear stress and mechanical strain. Recapitulation of multiple GI cell types and complex architecture in organoids requires multiple stages of growth factor induction—an important component of which is intact Wnt signaling (Ootani et al., 2009; Sato et al., 2009). The apparent lack of the need for exogenously added Wnt ligands suggests that the combination of shear stress and cyclic mechanical strain may induce Wnt signaling pathways. Further investigation into the process of differentiation in this system will provide much needed information to understand the mechanisms of differentiation induced by shear stress and mechanical strain and be useful to compare with the differentiation processes identified to date for 3D organoid structures.

16.5   CELL-BASED *IN VITRO* ASSAYS FOR SCREENING AND MECHANISTIC INVESTIGATIONS TO GI TOXICITY

Many endpoints of GI tract function have been paired with the use of cell lines and primary cells to model and screen toxicity of drugs in development. Despite the limitations of both cell lines and primary cells, they can be useful as simple models and have the added advantage of being high throughput. Early target assessment or hypotheses from *in vivo* findings can aid in understanding of the need for simple or complex models as well as aid in the selection of an appropriate cell line for *in vitro* mechanistic and screening toxicology studies. Common questions we ask when defining the appropriateness of an *in vitro* model are: Is the target expressed in the cell line? Are we derisking a lesion that appears to be regionally specific and thus has the potential need for cell-type specificity and primary cell isolation? Or is the hypothesized mechanism sufficiently general that we could use a nonspecific and simple cell model? For example, IEC-6 cells, derived from the crypt cell population of the small intestine, could be a useful model for screening or mechanistic work for targets expressed in or lesions identified in the proliferating compartments of the GI tract but might be a poor model for identifying toxicity mechanisms related to changes in barrier function due to effects on tight or adherens junctions as they do not form a resistant monolayer in culture (Fig. 16.2; Bastian et al., 1999; Puthia et al., 2006). In the latter situation, it might be appropriate to use Caco-2 or IPEC-J2 cells to evaluate the effect of one or a series of compounds on epithelial barrier integrity as they do form high-resistance monolayers with intact tight and adherens junction signaling (Fig. 16.2). The three most common *in vitro* endpoints used for screening and mechanistic investigation of GI toxicity are cell viability, cell migration, and epithelial barrier integrity. These assays are particularly suited for enterocyte-based toxicity mechanisms and can be applied to any primary cell or line.

16.5.1   Cell Viability

Cell viability assays, though very simple, can be a powerful screening tool for programs with compounds that cause lesions to the proliferative regions of the intestine. They have the disadvantage of not providing much mechanistic information but can be a useful preliminary screen for large sets of compounds particularly if they can be combined with data sets containing different chemical series and different target activities to determine whether the toxicity is on or off target. A variety of kits for cell viability are commercially available, and there is extensive literature comparing their utility (Kepp et al., 2011).

16.5.2   Cell Migration

Migration is an important cell biology mechanism in the GI tract since the first step in wound restitution is the migration of cells on the margins of a wound over the exposed surface of the lamina propria to cover the wound and also as new cells generated in proliferative zones migrate up villi or out of gastric pits to repopulate the surface. Cell migration can be measured *in vitro* with a scratch wound assay, transmembrane/Boyden chamber assay, cell exclusion assay, or microfluidic chamber assays (Hulkower and Herber, 2011). In a scratch wound assay, a cell monolayer is "wounded" or scratched, and the movement of cells into the cleared area is monitored over time. Cell migration from the apical to basal chamber of a transwell system is measured in a Boyden chamber migration assay usually with the application of a growth stimulus in the basal chamber as the stimulus to migrate. Cell exclusion assays are similar to scratch wound assay in that the movement of cells into a cleared area is measured; the difference is that instead of wounding the monolayer, a physical barrier is placed into the well prior to plating the cells, and after application of the appropriate stimulus, the barrier is removed, revealing a clear area into which cells can migrate (Hulkower and Herber, 2011). Histology and immunofluorescence microscopy can be used for assessing cell migration after wounding in novel cell models of the GI tract (Ayehounie et al., 2014).

16.5.3   Barrier Integrity

The GI surface is lined by a single layer of epithelial cells. This cellular barrier coupled with the mucus layer provides the protective barrier between the luminal contents of the GI tract and the underlying tissues and access to the systemic circulation. Molecularly, the epithelial barrier is maintained by tight and adherens junctions and desmosomes between enterocytes (reviewed in Peterson and Artis, 2014). Epithelial barrier function is typically measured with cells grown on transwell inserts and quantified by TEER (or TER) that measures the transport of ions across a monolayer or by assessing the status of paracellular pathway by passage of fluorescently conjugated molecules, such as dextran, from the apical to the basal chamber of transwell cultures (Blikslager et al., 2007). TEER can also be measured with a Ussing chamber with pieces of tissue from the GI tract (Srinivasan et al., 2015).

There are, of course, many other *in vitro*, *ex vivo*, and *in vivo* endpoints that can be employed depending on target cell type, understanding of potential for GI effects related to the primary pharmacology of the target or lesions identified from preclinical models, such as fecal microflora composition, charcoal transit time, ileum contraction, or quantification of short-circuit current with Ussing chambers. For excellent review of endpoints for assessment of GI toxicity, the reader is referred to Kapp (2008).

16.6   SUMMARY/CONCLUSIONS/CHALLENGES

The high degree of concordance of GI tract effects in response to individual xenobiotics is well documented across species (Olson et al., 2000; Greaves et al., 2004; Horner et al., 2013); however, preclinical challenges remain. Extensive differences in anatomy, physiology, and biochemistry across species (Kararli, 1995) can affect not only drug absorption for orally administered compounds but also the response of the GI tract components to them, a factor that could directly impact the translatability of biomarkers. The challenge deepens when we attempt to translate *in vitro* assay findings across multiple species. Recent advances in our ability to culture more complex *in vitro* models of the GI tract coupled with the application of currently accepted functional endpoints should improve *in vitro* to *in vivo* translation. Additionally, continual improvement in our understanding of the interaction between epithelial cells and gut microbiome has deepened our understanding of GI toxicology and provides hope of decreasing GI side effects for newly developed drugs.

REFERENCES

Aamann, L., Vestergaard, E.M., and Grønbæk, H. 2014. Trefoil factors in inflammatory bowel disease. *World J Gastroenterol*, 20(12): 3223–3230.

Arnold, C.A., Limketkai, B.N., and Liu, T.C., Montgomery, E., Nazari, K., Torbenson, M.S., Yearsley, M.M., and Lam-Himlin, D. 2013. Renvela crystals in the gastrointestinal tract: a new entity. *Mod Pathol*, 4: 455–458.

Ayehounie, S., Stevens, Z., Landry, T., Cataldo, A., Armento, A., Klausner, M., and Hayden, P. 2014. Organotypic human small intestine tissue to assess epithelial restitution. 17th Annual AAPS NERDG Meeting, May 2014, Farmington, CT.

Bartlett, J.G. 2002. Antibiotic-associated diarrhea. *N Engl J Med*, 346(5): 334–339.

Bastian, S.E.P., Walton, P.E., Ballard, F.J., and Belford, D.A. 1999. Transport of IGF-1 across epithelial cell monolayers. *J Endocrinol*, 162: 361–369.

Bhatia, S.N. and Ingber, D.E. 2014. Microfluidic organs-on-chips. *Nat Biotechnol*, 8: 760–772.

Blikslager, A.T., Moeser, A.J., Gookin, J.L., Jones, S.L., and Odle, J. 2007. Restoration of barrier function in injured intestinal mucosa. *Physiol Rev*, 87: 545–564.

Bolden, J.E., Tasdemir, N., Dow, L.E., van Es, J.H., Wilkinson, J.E., Zhao, Z., Clevers, H., and Lowe, S.W. 2014. Inducible in vivo silencing of Brd4 identifies potential toxicities of sustained BET protein inhibition. *Cell Rep*, 8(6): 1919–1929.

Boussios, S., Penteroudakis, G., Katsanos, K., and Pavlidis, N. 2012. Systemic treatment-induced gastrointestinal toxicity: incidence, clinical presentation and management. *Ann Gastroenterol*, 25(2): 106–118.

Butler, M., Ng, C.-Y., van Heel, D.A., Lombardi, G., Lechler, R., Playford, R.J., and Ghosh, S. 2006. Modulation of dendritic cell phenotype and function in an in vitro model of the intestinal epithelium. *Eur J Immunol*, 36: 864–874.

Cain, G.D., Reiner, E.B., and Patterson, M. 1968. Effects of neomycin on disaccharidase activity of the small bowel. *Arch Intern Med*, 122(4): 311–314.

Cencič, A. and Langerholc, T. 2010. Functional cell models of the gut and their applications in food microbiology—a review. *Int J Food Microbiol*, 141: S4–S14.

Chassaing, B. and Gewirtz, A.T. 2014. Gut microbiota, low grade inflammation and metabolic syndrome. *Toxicol Pathol*, 42: 49–53.

Chaveli-Lopez, C. 2014. Oral toxicity produced by chemotherapy: a review. *Oral Med and Pathol*, 6(1): 81–90.

Crenn, P., Coudray–Lucas, C., Thuillier, F., Cynober, L., and Messing, B. 2000. Postabsorptive plasma citrulline concentration is a marker of absorptive enterocyte mass and intestinal failure in humans. *Gastroenterology*, 119(6): 1496–1505.

Daniele, B., Rossi, G.B., Losito, S., Gridelli, C., and de Bellis, M. 2001. Ischemic colitis associated with paclitaxel. *J Clin Gastroenterol*, 33: 159–160.

Davila, M. and Bresalier, R.S. 2008. Gastrointestinal complications of oncologic therapy. *Nat Clin Pract Gastroenterol Hepatol*, 5(12): 682–696.

Duerr, C.U. and Hornef, M.W. 2012. The mammalian intestinal epithelium as integral player in the establishment and maintenance of host-microbial homeostasis. *Semin Immunol*, 24(1):25–35.

Dulai PS, Thompson KD, Blunt HB, Dubinsky MC, and Siegel CA. 2014. Risks of serious infection or lymphoma with anti-tumor necrosis factor therapy for oediatric inflammatory bowel disease: a systematic review. *Clin Gatroenterol Hepatol*, 12: 1443–1451.

Esch, M.B., Sung, J.H., Yang, J., Yu, C., Yu, J., March, J.C., and Shuler, M.L. 2012. On chip porous polymer membranes for integration of gastrointestinal tract epithelium with microfluidic "body-on-a-chip" devices. *Biomed Microdevices*, 14: 895–906.

Evans, G.S., Flint, N., Somers, A.S., Eyden, B., and Potten, C.S. 1992. The development of a method for the preparation of rat intestinal epithelial cell primary cultures. *J Cell Sci*, 101: 219–231.

Evans, G.S., Flint, N., and Potten, C.S. 1994. Primary cultures for studies of cell regulation and physiology in intestinal epithelium. *Annu Rev Physiol*, 56: 399–417.

Evennett, N., Cerigioni, E., Hall, N.J., Pierro, A., and Eaton, S. 2014. Smooth muscle actin as a novel serologic marker of severe intestinal damage in rat intestinal ischemia–reperfusion and human necrotising enterocolitis. *J Surg Res*, 191(2): 323–330.

Fassan, M., Croce, C.M., and Rugge, M. 2011. miRNAs in precancerous lesions of the gastrointestinal tract. *World J Gastroenterol*, 28: 5231–5239.

Ferruzza, S., Rossi, C., Sambuy, Y., and Scarino, M.L. 2013. Serum-reduced and serum-free media for differentiation of Caco-2 cells. *ALTEX*, 30(2): 159–168.

Finkbeiner, S.R. and Spence, J. R. 2013. A gutsy task: generating intestinal tissue from human pluripotent stem cells. *Dig Dis Sci*, 58: 1176–1184.

Gelberg, H.B. 2014. Comparative anatomy, physiology, and mechanisms of disease production in the esophagus, stomach, and small intestine. *Toxicol Pathol*, 42: 54–66.

Gon, H., Fumoto, K., Ku, Y., Matsumoto, S., and Kikuci, A. 2013. Wnt5a signaling promotes apical and basolateral polarization of single epithelial cells. *Mol Biol Cell*, 24: 3764–3774.

Gonçalves, F.L., Soares, L.M., Figueira, R.L., Simões, A.L., Gallindo, R.M., and Sbragia, L. 2015. Evaluation of the expression of I-FABP and L-FABP in a necrotizing enterocolitis model after the use of *Lactobacillus acidophilus*. *J Pediatr Surg*, 50(4): 543–549.

Greaves, P., Williams, A., and Eve, M. 2004. First dose of potential new medicines to humans: how animals help. *Nat Rev Drug Discov*, 3(3): 226–236.

Groschwitz, K.R. and Hogan, S.P. 2009. Intestinal barrier function: molecular regulation and disease pathogenesis. *J Allergy Clin Immunol*, 124(1), 3–20.

Guarner, F. and Malagelada, J.R. 2003. Gut flora in health and disease. *Lancet*, 361(9356), 512–519.

Gunzel, D., Krug, S.M., Rosenthal, R., and Fromm, M. 2010. Biophysical methods to study tight junction permeability. *Curr Top Membr*, 65: 40–78.

Haller, D., Blum, S., Bode, C., Hammes, W.P., and Schiffrin, E.J. 2000. Activation of human peripheral blood mononuclear cells by nonpathogenic bacteria in vitro: evidence of NK cells as primary targets. *Infect Immun*, 68(2): 752–759.

Hanton, G., Sobry, C., Daguès, N., Provost, J.P., Le Net, J.L., Comby, P., and Chevalier, S. 2008. Characterisation of the vascular and inflammatory lesions induced by the PDE4 inhibitor CI-1044 in the dog. *Toxicol Lett*, 179(1): 15–22.

Hasebe, T., Harasawa, S., Miwa, T., Shibata, T., and Inayama, S. 1987. Collagen and collagenase in ulcer tissue-1. The healing process of acetic acid ulcers in rats. *Tokai J Exp Clin Med*, 12(3): 147–158.

Hass, D.J., Kozuch, P., and Brandt, L.J. 2007. Pharmacologically mediated colon ischemia. *Am J Gastroenterol*, 102(8): 1765–1780.

Hogenauer, C., Hammer, H.F., Krejs, G.J., and Reisinger, E.C. 1998. Mechanisms and management of antibiotic-associated diarrhea. *Clin Infect Dis*, 27: 702–710.

Hooper, L.V. and MacPherson, A.J. 2010. Immune Adaptations that Maintain Homeostasis with the Intestinal Microbiota. *Nat Rev Immunol*, 10: 159–169.

Horner, S., Ryan, D., Robinson, S., Callander, R., Stamp, K., and Roberts, R.A. 2013. Target organ toxicities in studies conducted to support first time in man dosing: an analysis across species and therapy areas. *Regul Toxicol Pharmacol*, 65(3): 334–343.

Hulkower, K.I. and Herber, R.L. 2011. Cell migration and invasion assays as tools for drug discovery. *Pharmaceutics*, 3: 107–124.

Hynds, R.E. and Giangreco, A. 2013. Concise review: the relevance of human stem-cell derived organoid models for epithelial translational medicine. *Stem Cells*, 31: 417–422.

Jasperson, D. 2000. Drug-induced oesophageal disorders: pathogenesis, incidence, prevention, and management. *Drug Saf*, 22(3): 237–249.

John-Baptiste, A., Huang, W., Kindt, E., Wu, A., Vitsky, A., Scott, W., Gross, C., Yang, A.H., Schaiff, W.T., Ramaiah, S.K. 2012. Evaluation of potential gastrointestinal biomarkers in a PAK4 inhibitor-treated preclinical toxicity model to address unmonitorable gastrointestinal toxicity. *Toxicol Pathol*, 40(3): 482–490.

Kaeffer, B. 2002. Mammalian intestinal epithelial cells in primary culture: a mini-review. *In Vitro Cell Dev Biol Anim*, 38:123–134.

Kalabat, D., Kindt, E., Vitsky, A., Scott, W., Huang, W., and Yang, A.H. 2015. Identification and evaluation of novel microRNA biomarkers in biofluids for drug-induced intestinal toxicity, manuscript in preparation.

Kanda, T., Fujii, H., Tani, T., Murakami, H., Suda, T., Sakai, Y., Ono, T., and Hatakeyama, K. 1996. Intestinal fatty acid–binding protein is a useful diagnostic marker for mesenteric infarction in humans. *Gastroenterology*, 110: 339–343.

Kapp, R.W. Jr. 2008. Gastrointestinal toxicology. In Hayes, A.W. (Ed), *Principles and Methods of Toxicology*, 1541–1584. CRC Press, Boca Raton, FL.

Kararli, T.T. 1995. Comparison of the gastrointestinal anatomy, physiology, and biochemistry of humans and commonly used laboratory animals. *Biopharm Drug Dispos*, 16(5): 351–380.

Kauffman, A.L., Gyurdieva, A.V., Mabus, J.R., Ferguson, C., Yan, Z., and Hornby, P.J. 2013. Alternative functional in vitro models of human intestinal epithelia. *Front Pharmacol*, 4: 79, doi: 10.3389/fphar.2013.00079.

Keefe, D.M.K. and Gibson, R.J. 2007. Mucosal injury from targeted anti-cancer therapy. *Support Care Cancer*, 15: 483–490.

Kepp, O., Galluzzi, L., Lipinski, M., Yuan, J., and Kroemer, G. 2011. Cell death assays for drug discovery. *Nat Rev Drug Discov*, 10, 221–237.

Kim, H.J. and Ingber, D.E. 2013. Gut-on-a-chip microenvironment induces human intestinal cells to undergo villus differentiation. *Integr Biol*, 5: 1130–1140.

Kim, H.J., Huh, D., Hamilton, G., and Ingber, D.E. 2012. Human gut-on-a-chip inhabited by microbial flora that experiences intestinal peristalsis-like motions and flow. *Lab Chip*, 12: 2165–2174.

Kimura, H., Yamamoto, T., Sakai, H., Sakai, Y., and Fujii, T. 2008. An integrated microfluidic system for long-term perfusion culture and on-line monitoring of intestinal tissue models. *Lab Chip*, 8: 741–746.

Klestov, A., Kubler, P., and Meulet, J. 2001. Recurrent ischaemic colitis associated with pseudoephedrine use. *Intern Med J*, 31(3): 195–196.

Krausova, M. and Korinek, V. 2014. Wnt signaling in adult intestinal stem cells and cancer. *Cell Signal*, 26(3): 570–579.

Lahar, N., Lei, N.Y., Wang, J., Jabaji, Z., Tung, S.C., Joshi, V., Lewis, M., Stelzner, M., Martin, M.G., and Dunn, J.C.Y. 2011. Intestinal subepithelial myofibroblasts support in vitro and in vivo growth of human small intestinal epithelium. *PLoS One*, 6(11): e26898–e84651.

Langerholc, T., Maragkoudakis, P.A., Wollgast, J., Gradisnik, L., and Cencič, A. 2011. Novel and established intestinal cell line models—an indispensable tool in food science and nutrition. *Trends Food Sci Technol*, 22(S1): S11–S20.

Lawson, A.J., Smit, R.A., Jeffers, N.A., and Osborne, J.W. 1982. Isolation of rat intestinal crypt cells. *Cell Tissue Kinet* 15: 69–80.

Lehmann, F.S., Burri, E., and Beglinger, C. 2015. The role and utility of faecal markers in inflammatory bowel disease. *Ther Adv Gastroenterol*, 8(1): 23–36.

Loriot, Y., Perlemuter, G., Malka, D., Penault-Lorca, F., Boige, V., Deutsch, E., Massard, C., Armand, J. P., and Soria, J.-C.. 2008. Drug insight: gastrointestinal and hepatic adverse effects of molecular-targeted agents in cancer therapy. *Nat Clin Pract*, 5(5): 268–278.

Lowe, C.E., Depew, W.T., Vanner, S.J., Paterson, W.G., and Meddings, J.B. 2000. Upper gastrointestinal toxicity of alendronate. *Am J Gastroenterol*, 95: 624–640.

Luk, G.D., Bayless, T.M., and Baylin, S.B. 1980. Diamine oxidase (histaminase). A circulating marker for rat intestinal mucosal maturation and integrity. *J Clin Invest*, 66(1): 66–70.

Luk, G.D., Bayless, T.M., and Baylin, S.B. 1983. Plasma postheparin diamine oxidase. Sensitive provocative test for quantitating length of acute intestinal mucosal injury in the rat. *J Clin Invest*, 71(5): 1308–1315.

Lutgens, L.C., Deutz, N.E., Gueulette, J., Cleutjens, J.P., Berger, M.P., Wouters, B.G., von Meyenfeldt, M.F., and Lambin, P. 2003. Citrulline: a physiologic marker enabling quantitation and monitoring of epithelial radiation-induced small bowel damage. *Int J Radiat Oncol Biol Phys*, 57(4): 1067–1074.

Lutgens, L.C., Blijlevens, N., Deutz, N.E., Donnelly, J.P., Lambin, P., and de Pauw, B.E. 2005. Monitoring myeloablative therapy-induced small bowel toxicity by serum citrulline concentration. *Cancer*, 103(1): 191–199.

Mahler, G.J., Esch, M.B., Glahn, R.P., and Shuler, M.L. 2009. Characterization of a gastrointestinal tract microscale cell culture analog used to predict drug toxicity. *Biotechnol Bioeng*, 104(1): 193–205.

Marcotte, H. and Lavoie, M.C. 1998. Oral microbial ecology and the role of salivary immunoglobulin A. *Microbiol Mol Biol Rev*, 62(1): 71–109.

McCarthy, A.J., Lauwers, G.Y., and Sheahan, K. 2015. Iatrogenic pathology of the intestines. *Histopathology*, 66(1): 15–28.

McCracken, K.W., Cata, E.M., Crawford, C.M., Sinagoga, K.L., Schumacher, M., Rockich, B.E., Tsai, Y.-H., Mayhew, C.N., Spence, J.R., Zavros, Y., and Wells, J.M. 2014. Modelling human development and disease in pluripotent stem-cell derived gastric organoids. *Nature*, 516(18): 400–404.

Menees, S., Powell, C., Kurlander, J., Goel, A., and Chey, W.D. 2015. A meta-analysis of the utility of C-reactive protein, erythrocyte sedimentation rate, fecal calprotectin, and fecal lactoferrin to exclude inflammatory bowel disease in adults with IBS. *Am J Gastroenterol*, 110: 444–454.

Mikaelian, I., Scicchitano, M., Mendes, O., Thomas, R.A., and LeRoy, B.E. 2013. Frontiers in preclinical safety biomarkers: MicroRNAs and messenger RNA. *Toxicol Pathol*, 41: 18.

Milano, J., McKay, J., Dagenais, C., Foster-Brown, L., Pognan, F., Gadient, R., Jacobs, R.T., Zacco, A., Greenberg, B., and Ciaccio, P.J. 2004. Modulation of notch processing by γ-secretase inhibitors causes intestinal goblet cell metaplasia and induction of genes known to specify gut secretory lineage differentiation. *Toxicol Sci*, 82(1): 341–358.

Moore, J.V. 1979. Ablation of murine jejunal crypts by alkylating agents. *Br J Cancer*, 39(2): 175.

Moore, J.V. 1986. The "gastrointestinal syndrome" after chemotherapy: inferences from mouse survival time, and from histologically-and clonogenically-defined cell death in intestinal crypts. *Br J Cancer*, Supplement 7: 16.

Moriyama, K., Kouchi, Y., Morinaga, H., Irimura, K., Hayashi, T., Ohuchida, A., Goto, T., Yoshizawa, Y. 2006. Diamine oxidase, a plasma biomarker in rats to GI tract toxicity of oral fluorouracil anti-cancer drugs. *Toxicology*, 217(2–3):233–239.

Morris, D.L., O'Neil, S.P., Devraj, R.V., Portanova, J.P., Gilles, R.W., Gross, C.J., Curtiss, S.W., Komocsar, W.J., Garner, D.S., Happa, F.A., Kraus, L.J., Nikula, K.J., Monahan, J.B., Selness, S.R., Galluppi, G.R., Shevlin, K.M., Kramer, J.A., Walker, J.K., Messing, D.M., Anderson, D.R., Mourey, R.J., Whiteley, L.O., Daniels, J.S., Yang, J.Z., Rowlands, P.C., Alden, C.L., Davis, J.W. II, and Sagartz, J.E. 2010. Acute lymphoid and gastrointestinal toxicity induced by selective p38alpha map kinase and map kinase-activated protein kinase-2 (MK2) inhibitors in the dog. *Toxicol Pathol* 38(4): 606–618.

Norman, A. and Hawkey, C.J. 2010. Drug-induced gastrointestinal disorders, *Medicine*, 39(3): 162–168).

Olson, H., Betton, G., Robinson, D., Thomas, K., Monro, A., Kolaja, G., Lilly, P., Sanders, J., Sipes, G., Bracken, W., Dorato, M., Van Deun, K., Smith, P., Berger, B., and Heller, A. 2000. Concordance of the toxicity of pharmaceuticals in humans and in animals. *Regul Toxicol Pharmacol*, 32(1): 56–67.

Ootani, A., Li, X., Sangiorgi, E., Ho, Q.T., Ueno, H., Toda, S., Sugihara, H., Fujimoto, K., Weissman, I.L., Capecchi, M.R., and Kuo, C.J. 2009. Sustained in vitro intestinal epithelial culture within a Wnt-dependent stem cell niche. *Nat Med*, 15: 701–706.

Pageot, L.-P., Perreault, N., Basora, N., Francoeur, C., Magny, P., and Beaulieu, J.-F. 2000. Human cell models to study small intestinal functions: recapitulation of the crypt-villus axis. *Microsc Res Tech*, 49: 394–406.

Papadia, C., Sherwood, R.A., Kalantzis, C., Wallis, K., Volta, U., Fiorini, E., and Forbes, A. 2007. Plasma citrulline concentration: a reliable marker of small bowel absorptive capacity independent of intestinal inflammation. *Am J Gastroenterol*, 102(7): 1474–1482.

Parfitt, J.R. and Driman, D.K. 2007. Pathological effects of drugs on the gastrointestinal tract: a review. *Hum Pathol*, 38: 527–536.

Pelsers, M.M., Namiot, Z., Kisielewski, W., Namiotd, A., Januszkiewicze, M., Hermensf, W.T., and Glatza, J.F.C. 2003. Intestinal-type and liver-type fatty acid-binding protein in the intestine. Tissue distribution and clinical utility. *Clin Biochem*, 36: 529–535.

Peter, C.P., Kindt, M.V., and Majka, J.A. 1998. Comparative study of potential for bisphosphonates to damage gastric mucosa in rats. *Dig Dis Sci*, 43(5): 1009–1015.

Peterson, L.W. and Artis, D. 2014. Intestinal epithelial cells: regulators of barrier function and immune homeostasis. *Nat Rev Immunol*, 14: 141–153.

Phillips, F.S. and Sternberg, S.S. 1975. The lethal actions of antitumor agents in proliferating cell systems in vivo. *Am J Pathol*, 81(1): 205.

Picot, D., Garin, L., Trivin, F., Kossovsky, M.P., Darmaun, D., and Thibault, R. 2010. Plasma citrulline is a marker of absorptive small bowel length in patients with transient enterostomy and acute intestinal failure. *Clin Nutr* 29: 235–242.

Puthia, M.K., Sio, S.W., Lu, J., and Tan, K.S. 2006. Blastocystis ratti induces contact-independent apoptosis, F-actin rearrangement, and barrier function disruption in IEC-6 cells. *Infect Immun*, 74(7): 4114–4123.

Quaroni, A., Wands, J., Trelstad, R.L., and Isselbacher, K.J. 1979. Epithelioid cell cultures from rat small intestine. Characterization by morphologic and immunologic criteria. *J Cell Biol*, 80(2): 248–265.

Quaroni, A., Calnek, D., Quaroni, E., and Chandler, J.S. 1991. Keratin expression in rat intestinal crypt and villus cells: analysis with a panel of monoclonal antibodies. *J Biol Chem* 266(18): 11923–11931.

Richardson, G. and Dobish, R. 2007. Chemotherapy induced diarrhea. *J Oncol Pharm Pract*, 12: 181–198.

Robinton, D.A. and Daley, G.Q. 2012. The promise of induced pluripotent stem cells in research and therapy. *Nature*, 481: 295–305.

Røseth, A.G., Fagerhol, M.K., Aadland, E., and Schjønsby, H. 1992. Assessment of the neutrophil dominating protein calprotectin in feces. A methodologic study. *Scand J Gastroenterol*, 27(9): 793–798.

Saglam, C., Kesik, V., Caliskan, B., Avci, T., Agilli, M., Yigit, N., Babacan, O., Korkmazer, N., Atas, E., and Gulgun, M. 2015. Small intestinal lactoferrin and calprotectin levels in different stages of necrotizing enterocolitis in a rat model. *Adv Med Sci*, 60(2): 199–203.

Salk, A., Stobaugh, D.J., Deepak, P., and Ehrenpreis, E.D. 2013. Ischaemic colitis in rheumatoid arthritis patients receiving tumour necrosis factor-α inhibitors: an analysis of reports to the US FDA adverse event reporting system. *Drug Saf*, 36: 329–334.

Sambruy, Y., Ferruzza, S., Ranaldi, G., and De Angelis, I. 2001. Intestinal cell culture models: applications in toxicology and pharmacology. *Cell Biol Toxicol*, 17: 301–317.

Sancho, E., Batlle, E., and Clevers, H. 2003. Live and let die in the intestinal epithelium. *Curr Opin Cell Biol*, 15: 763–770.

Sanderson, I.R., Ezzell, R.M., Kedinger, M., Erlanger, M., Xu, Z.-X., Pringault, E., Leon-Robine, S., Louvard, D., and Walker, W.A. 1996. Human fetal enterocytes in vitro: modulation of the phenotype by extracellular matrix. *Proc Natl Acad Sci U S A*, 93: 7717–7722.

Sato, T., Vries, R.G., Snippert, H.J., van de Wetering, M., Barker, N., Stange, D.E., van Es, J.H., Abo, A., Kujala, P., Peters, P.J., and Clevers, H. 2009. Single Lgr5 stem cells build crypt-villus structures in vitro without a mesenchymal niche. *Nature*, 459: 262–265.

Schlaermann, P., Toelle, B., Berger, H., Schmidt, S.C., Glanemann, M., Ordemann, J., Bartfeld, S., Mollenkopf, H.J., and Meyer, T.F. 2014. A novel human gastric primary cell culture system for modelling *Helicobacter pylori* infection in vitro. *Gut*, gutjnl-2014-307949, doi: 10.1136/gutjnl-2014-307949.

Sellers, R.S. and Morton, D. 2014. The colon: from banal to brilliant. *Toxicol Pathol*, 42: 67–81.

Simon-Assmann, P., Turck, N., Sidhoum-Jenny, M., Gradwohl, G., and Kedinger, M. 2007. In vitro models of intestinal epithelial cell differentiation. *Cell Biol Toxicol*, 23: 241–256.

Somasundaram, S., Hayllar, H., Rafi, S., Wrigglesworth, J.M., MacPherson, A.J.S., and Bjarnason, I. 1995. The biochemical basis of NSAID drug-induced damage to the gastrointestinal tract: a review and hypothesis. *Scand J Gastroenterol*, 30: 289–299.

Sonis, S.T. 2009. Mucositis: the impact, biology and therapeutic opportunities of oral mucositis. *Oral Oncol*, 45: 1015–1020.

Spence, J.R., Mayhew, C.N., Rankin, S.A., Kuhar, M.F., Vallance, J.E., Tolle, K., Hoskins, E.E., Kalinichenko, V.V., Wells, S.I., Zorn, A.M., Shroyer, N.F., and Wells, J.M. 2011. Directed differentiation of human pluripotent stem cells into intestinal tissue in vitro. *Nature*, 470: 105–109.

Squier, C.A. and Kremer, M.J. 2001. Biology of oral cavity and esophagus. *J Natl Cancer Inst Monogr*, 29: 7–15.

Srinivasan, B., Kolli, A.R., Esch, M.B., Abaci, H.E., Shuler, M.L., and Hickman, J.J. 2015. TEER measurement techniques for in vitro barrier model systems. *J Lab Autom*, 20(2): 107–126.

Susewind, J., de Souza Carvalho-Wodarz, C., Repnik, U., Collnot, E.-M., Schneider-Daum, N., Griffiths, G.W., and Lehr, C.-M. 2015. A 3D co-culture of three human cell lines to model the inflamed intestinal mucosa for safety testing of nanomaterials. *Nanotoxicology*, doi:10.3109/17435390.2015.1008065.

Takeuchi, K., Ohishi, M., Ota, S., Suzumura, K., Naraoka, H., Ohata, T., Seki, J., Miyamae, Y., Honma, M., and Soga, T. 2013. Metabolic profiling to identify potential serum biomarkers for gastric ulceration induced by nonsteroid anti-inflammatory drugs. *J Proteome Res*, 12(3): 1399–1407.

Takeuchi, K., Endo, K., Suzumura, K., Naraoka, H., Ohata, T., Seki, J., Miyamae, Y., Honma, M., and Soga, T. 2014a. Metabolomic analysis of the effects of omeprazole and famotidine on aspirin-induced gastric injury. *Metabolomics*, 10(5): 995–1004.

Takeuchi, K., Ohishi, M., Endo, K., Suzumura, K., Naraoka, H., Ohata, T., Seki, J., Miyamae, Y., Honma, M., and Soga, T. 2014b. Hydroxyproline, a serum biomarker candidate for gastric ulcer in rats: a comparison study of metabolic analysis of gastric ulcer models induced by ethanol, stress, and aspirin. *Biomark Insights*, 9: 61.

Tanoue, T., Nishitani, Y., Kanazawa, K., Hashimoto, T., and Mizuno, M. 2008. In vitro model to estimate gut inflammation using co-cultured Caco-2 and RAW264.7 cells. *Biochem Biophys Res Commun*, 374: 565–569.

Tillinger, W., Jilch, R., Jilma, B., Brunner, H., Koeller, U., Lichtenberger, C., Waldhör, T., and Reinisch, W. 2009. Expression of the high-affinity IgG receptor FcRI (CD64) in patients with inflammatory bowel disease: a new biomarker for gastroenterologic diagnostics. *Am J Gastroenterol*, 104: 102–109.

Ulaganathan, M., Familari, M., Yeomans, N.D., Giraud, A.S., and Cook, G.A. 2001. Spatio-temporal expression of trefoil peptide following severe gastric ulceration in the rat implicates it in late-stage repair processes. *J Gastroenterol Hepatol* 16(5): 506–512.

Van de Poll, M.C.G., Ligthart-Melis, G.C., Boelens, P.G., Deutz, N.E., van Leeuwen, P.A., and Dejong, C.H. 2007. Intestinal and hepatic metabolism of glutamine and citrulline in humans. *J Physiol*, 518(2): 819–827.

Venditti, O., DeLisi, D., Caricato, M., Caputo, D., Capolupo, G.T., Taffon, C., Pagliara, E., Battisi, S., Frezza, A.M., Muda, A.O., Tonini, G., and Santini, D. 2015. Ipilimumab and immune-mediated adverse events: a case report of anti-CTLA4-induced ileitis. *BMC Cancer*, 15: 87.

Vestergaard, E.M., Brynskov, J., Ejskjaer, K., Clausen, J.T., Thim, L., Nexø, E., and Poulsen, S.S. 2004. Immunoassays of human trefoil factors 1 and 2: measured on serum from patients with inflammatory bowel disease. *Scand J Clin Lab Invest*, 64: 146–156.

Viele, C.S. 2003. Overview of chemotherapy-induced diarrhea. *Semin Oncol Nurs*, 19(Supp 3): 2–5.

Villalón, G, Martín, J.M., Pinazo, M.I., Calduch, L., Alonso, V., and Jordá, E. 2009. Focal acral hyperpigmentation in a patient undergoing chemotherapy with capecitabine. *Am J Clin Dermatol*, 10(4): 261–263.

Vitsky, A., Huang, W., Kindt, E., Yang, A.H., Scott, W., Kalabat, D., Sace, D., and John-Baptiste, A. 2013. Evaluation of potential biomarkers of gastrointestinal toxicity in preclinical safety studies. Abstr STP Meeting, Toxicologic Pathology, June 2013, Portland, OR.

Wagoner, M., Kelsal, J., Hattersley, M., Hickling, K., Pederson, J., Harris, J., Keirstead, N., Heathcote, D., Newham, P. 2015. Bromodomain and extraterminal (BET) domain inhibitors induce a loss of intestinal stem cells and villous atrophy. *Toxicol Lett*, 229: S75–S76.

Walk, S.T. and Young, V.B. 2008. Emerging insights into antibiotic-associated diarrhea and clostridium difficile infection through the lens of microbial ecology. *Interdiscip Perspect Infect Dis*. Article ID 125081. doi:10.1155/2008/125081.

Wiercinska-Drapalo, A., Jaroszewicz, J., Siwak, E., Pogorzelska, J., and Prokopowicz, D. 2008. Intestinal fatty acid binding protein (I-FABP) as a possible biomarker of ileitis in patients with ulcerative colitis. *Regul Pept*, 147: 25–28.

Xue, H., Sawyer, M.B., Wischmeyer, P.E., and Baracos, V.E. 2011. Nutrition modulation of gastrointestinal toxicity related to cancer chemotherapy: from preclinical findings to clinical strategy. *J Parenter Enteral Nutr* 15(1): 74–90.

Yang, A.H., Kalabat, D., Vitsky, A., Scott, W., Kindt, E., John-Baptiste, A., and Huang, W. 2013. MicroRNA biomarkers of gastrointestinal toxicity in tissues and biofluids. Abstr STP Meeting, Toxicologic Pathology, June 2013, Portland, OR.

17

PRECLINICAL SAFETY ASSESSMENT OF DRUG CANDIDATE-INDUCED PANCREATIC TOXICITY: FROM AN APPLIED PERSPECTIVE

KARRIE A. BRENNEMAN[1], SHASHI K. RAMAIAH[2] AND LAUREN M. GAUTHIER[3]

[1] Toxicologic Pathology, Drug Safety Research and Development, Pfizer Inc., Andover, MA, USA

[2] Biomarkers, Drug Safety Research and Development, Pfizer Inc., Cambridge, MA, USA

[3] Investigative Toxicology, Drug Safety Research and Development, Pfizer Inc., Andover, MA, USA

17.1 DRUG-INDUCED PANCREATIC TOXICITY

17.1.1 Introduction

Pancreatic toxicity is rarely encountered in preclinical safety assessment of drug candidates and, as a consequence, is an infrequent cause of attrition in drug development programs. In a recent survey from AstraZeneca of 1-month preclinical toxicity studies evaluating small-molecule drug candidates, pancreas was identified as a target organ in 6 of 77 rodent and 5 of 75 nonrodent studies (Horner et al., 2013). Despite these low incidences, serious concern is warranted when pancreas is identified as a target organ, because of the lack of specific and sensitive clinically monitorable biomarkers and the potentially serious clinical complications of undetected drug-induced injury, most notably acute and chronic pancreatitis (AP and CP, respectively), pancreatic neoplasia, and type 1 diabetes. Additional hurdles are the lack of clinically relevant animal models and *in vitro* screening tools that adequately model the human disease process to enable drug discovery programs. When observed, toxicity may directly affect the parenchymal cells of the exocrine or the endocrine pancreas or both as well as nonparenchymal tissue components, such as the vasculature or innervation. Yet, pancreatic toxicity in animal models and/or animal-derived *in vitro* platforms may not translate to humans.

17.1.1.1 Exocrine The exocrine portion of the pancreas is predominantly comprised of acinar cells (Fig. 17.1) that produce precursors of digestive enzymes (proenzymes) that are stored in zymogen granules and, with appropriate stimulation, are secreted into pancreatic ducts that carry them to the lumen of the digestive tract, where they become activated. Within the duodenum, the intestinal brush border enzyme enteropeptidase (enterokinase) converts the proenzyme trypsinogen to its active form trypsin. In turn, trypsin can autocatalytically activate trypsinogen to trypsin as well as convert other pancreatic proenzymes, such as proelastase, procarboxypeptidase, chymotrypsinogens, and others, to their active forms. When this autocatalytic cascade of digestive enzyme activation occurs within acinar cells, it is thought to be central to the pathogenesis of AP. Drug-induced toxicity directly affecting acinar cells may result in a variety of degenerative changes (e.g., zymogen granule depletion, vacuolization, autophagy), apoptosis, or even necrosis, often leading to the premature release and activation of digestive enzymes within the pancreas and secondarily inciting widespread necrosis, ischemic injury, and inflammation characteristic of AP. With longer duration of injury, atrophy, and fibrosis as well as proliferative changes including hyperplasia, preneoplastic lesions and neoplasia may also occur (Cattley et al., 2013; Wallig and Sullivan, 2013).

Drug Discovery Toxicology: From Target Assessment to Translational Biomarkers, First Edition. Edited by Yvonne Will, J. Eric McDuffie, Andrew J. Olaharski, and Brandon D. Jeffy.

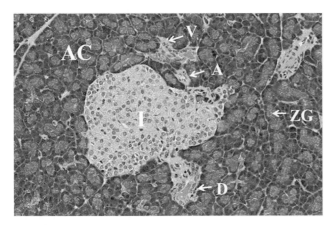

FIGURE 17.1 Microscopic appearance of the normal rat pancreas. An islet of endocrine cells is surrounded by acinar cells of the exocrine pancreas in a normal rat. A, arteriole; AC, acinar cells; D, ductule; I, islet; V, venule; ZG, zymogen granules. (H&E stain; original magnification 200×) (*See insert for color representation of the figure.*)

17.1.1.2 Endocrine The endocrine pancreas consists of four main types of peptide hormone-secreting cells (alpha, beta, gamma, and delta) clustered within the islets (Fig. 17.1) that are scattered throughout the exocrine pancreas. As a brief overview, alpha cells secrete glucagon, which elevates blood glucose levels; beta cells secrete insulin, which decreases blood glucose levels; delta cells secrete somatostatin, which inhibits insulin and glucagon secretion; and gamma cells secrete pancreatic polypeptide, which is involved in the regulation of both endocrine and exocrine secretions. In response to exposure to drug candidates, islet cells may experience a similar range of alterations (i.e., from degeneration, necrosis, and/or inflammation to atrophy, fibrosis, and/or proliferation) as those observed in acinar cells. Microscopically observable alterations may or may not be accompanied by functional perturbations in hormone production and/or secretion (Chandra et al., 2013; Rosol et al., 2013).

17.1.1.3 Nonparenchymal Toxicity may also affect ductal epithelial cells as well nonparenchymal tissue elements, such as the vasculature or innervation, possibly leading to secondary injury of parenchymal cell populations. The microvasculature of the pancreas has the unique feature that the blood supplies of the endocrine and exocrine pancreas are interconnected via a portal system of small capillaries that carry the peptide hormones secreted by the islets directly to the peri-islet acinar tissue allowing for dynamic and homeostatic metabolic regulation. Drug candidates have been reported to injure the microvasculature at the endocrine–exocrine interface in the rat resulting in both acute and chronic injury to islets and surrounding acinar tissue, which may progress to necrosis and atrophy of whole pancreatic lobules (Brenneman et al., 2014).

17.1.2 Drug-Induced Pancreatic Exocrine Toxicity in Humans: Pancreatitis

Drug-induced injury to the exocrine pancreas in humans most commonly presents as mild to moderate symptoms of AP (abdominal pain, nausea, and vomiting, among others) that subside with drug withdrawal and supportive therapy; however, severe and even fatal cases can occur (Jones et al., 2015). The pathogenesis and classifications of AP in humans have been reviewed previously. AP is commonly characterized by the onset of reversible parenchymal and peripancreatic fat necrosis with associated inflammation, and cases of AP may be subclassified as interstitial edematous AP or necrotizing AP (based upon the extent of inflammation and necrosis) or by severity. Significant necrosis or organ failure is rarely evident in mild AP, whereas severe AP is characterized by extensive necrosis that may lead to intrapancreatic vasculopathy (i.e., thrombosis and intraparenchymal hemorrhage) and even multiple organ failure (Jones et al., 2015). CP is defined as prolonged inflammation of the pancreas associated with irreversible destruction of exocrine parenchyma, fibrosis, and, in the later stages, concurrent destruction of endocrine parenchyma (Hruban, 2015). Rare cases of drug-induced CP have been reported, sometimes with the functional irreversible consequence of pancreatic exocrine insufficiency (PEI) as has recently been reported for the tyrosine kinase inhibitor sorafenib (Lee and Dalia, 2012; Hescot et al., 2013).

Drug-induced AP (DIAP) is rarely reported in humans (~0.1–5% of all cases of AP), likely because of the dual challenges of diagnosing pancreatitis as well as making a definitive association between drug exposure and AP in humans (Hung and Lanfranco, 2014; Tenner, 2014). Ideally, a drug is classified as toxic to the pancreas only when it induces AP documented by clinical findings and specific laboratory tests, if the condition improves when the drug is discontinued, and worsens or reoccurs upon rechallenge. A drug that meets these criteria would be classified as Ia in the five-category (Ia, Ib, II, III, and IV), evidence-based, classification system established in 2007 to categorize the strength of the association between the occurrence of AP and drug administration (Badalov et al., 2007). However, more commonly, the association of AP with drug administration is less stringent, based upon only one or few case reports, and suffers from inadequate evidence to prove a causal association (Tenner, 2014). A recent example of unsupported causal associations in the literature was reports of incretin mimetics causing DIAP that were subsequently cleared of this risk when the FDA and EMA's retrospective analysis of clinical data failed to demonstrate an increased risk for AP beyond that typical of the diabetic patient population nor any evidence of pancreatic toxicity in preclinical studies (Tatarkiewicz et al., 2013; Egan et al., 2014; Hung and Lanfranco, 2014).

Yet, the number of drugs reported to be associated with pancreatitis in human patients is rising rapidly, even in the context of the accuracy of these associations being actively questioned (Hung and Lanfranco, 2014; Tenner, 2014). A survey from 1989 reported that only eight drugs had fulfilled the criteria necessary to be categorized as toxic to the pancreas. These included azathioprine, estrogens, furosemide, methyldopa, pentamidine, procainamide, sulfonamides, and thiazide diuretics (Scarpelli, 1989). In a more recent survey from 2008, published reports of DIAP existed for at least 40 of the top 200 most prescribed medications (Kaurich, 2008). A 2014 publication states that over 100 drugs have been reported to cause AP in the scientific literature (Tenner, 2014). And yet another publication states that 525 different drugs have been reported in the WHO database to induce AP as an adverse reaction (Nitsche et al., 2010).

17.1.3 Mechanisms of Drug-Induced Pancreatic Toxicity

17.1.3.1 Exocrine Even though a large number of drugs are reported to be associated with DIAP in humans, the mechanism of toxicity is often not well understood, and, as a consequence, it is often difficult to determine if the effects are intrinsic for all drugs within a class, since the mechanism of toxicity may or may not be related to the drug's expected pharmacology. In fact, the majority of cases of DIAP reported have an idiosyncratic nature in that the adverse effect on the pancreas occurs at a low incidence, is unpredictable, is often not directly related to the pharmacodynamic mechanism of the drug, and is theorized to be mediated by immunologic or cytotoxic effects triggered by the drug or its metabolites (Hung and Lanfranco, 2014). The mechanisms of action for DIAP that have been reported are based upon theories extracted from case reports, case–control studies, animal studies, and other experimental data. Potential mechanisms of toxicity for DIAP are highly varied and include structural causes (i.e., cholestatic liver injury, spasm of the sphincter of Oddi, pancreatic duct constriction or obstruction), cytotoxicity, metabolic effects, accumulation of toxic intermediate metabolites, and hypersensitivity reactions (Table 17.1) (Underwood and Frye, 1993; Badalov et al., 2007; Nitsche et al., 2010; Hung and Lanfranco, 2014; Jones et al., 2015). Other mechanisms of DIAP are associated with secondary effects of drugs, such as hypertriglyceridemia, chronic hypercalcemia, or gallstone formation, which are risk factors for AP. As examples of secondary effects causing pancreatic toxicity, localized angioedema of the pancreatic duct has been implicated in the case of angiotensin-converting enzyme (ACE) inhibitors, and hypertriglyceridemia and arteriolar thrombosis have been theorized as mechanisms of toxicity for estrogens and hormone replacement therapy (Badalov et al., 2007; Kaurich, 2008; Jones et al., 2015).

17.1.3.2 Endocrine Many drugs may predispose to or induce diabetes in humans, as well as cause deterioration of glucose control if administered to those with preexisting diabetes by affecting the secretion of insulin, increasing insulin resistance, or both. These drugs range from weakly diabetogenic medications, for example, antihypertensives and statins, to medications that are more strongly diabetogenic,

TABLE 17.1   Potential Mechanisms of Drug-Induced Acute Pancreatitis (DIAP) in Humans

Potential Mechanisms		Select Drugs Associated with Acute Pancreatitis
Structural	Cholestatic liver injury	Azathioprine, cytarabine
	Spasm of the sphincter of Oddi	Codeine, erythromycin, opioids
	Duct obstruction	Angiotensin-converting enzyme (ACE) inhibitors, for example, enalapril (angioedema) and ceftriaxone (gallstones)
	Duct constriction	Sulindac
Cytotoxicity		Acetaminophen, didanosine, isoniazid, L-asparaginase, mesalamine, metronidazole, nucleoside reverse transcriptase inhibitors (NRTIs), pentamidine, protease inhibitors, sitagliptin, statins, valproic acid
Metabolic	Hypertriglyceridemia	Beta-blockers, clomiphene, corticosteroids, dibenzodiazepine-derived atypical antipsychotics (clozapine, olanzapine, and quetiapine), estrogens, furosemide, isotretinoin, propofol, protease inhibitors, retinoid derivatives, tamoxifen, thiazides
	Hypercalcemia	IV calcium, thiazides, vitamin D
Accumulation of toxic intermediate metabolites		Statins, tetracycline
Hypersensitivity reactions		Aminosalicylates (sulfasalazine and mesalazine), azathioprine/6-MP, sulfonamides
Vascular	Thrombosis	Estrogens/hormone replacement therapy
	Microvascular injury	Angiotensin II (AT) receptor blockers, contrast agent (iopamidol), sorafenib

From Badalov et al. (2007), Hung and Lanfranco, (2014), Jones et al. (2015), Nitsche et al. (2010), and Pezelli et al. (2010).

for example, steroids, antipsychotics, and many immunosuppressive agents, such as cyclosporine. Alloxan and streptozotocin are both well-known direct beta-cell toxins that induce beta-cell necrosis via production of reactive oxygen species (Szkudelski, 2001). Interestingly, after it was discovered that streptozotocin causes beta-cell toxicity, the drug, which was initially developed for use as an antibiotic, was repurposed and is now utilized to induce various animal models of type 1 and type 2 diabetes as well as an effective treatment for metastatic β-cell neoplasms in human patients, whose tumors cannot be surgically resected.

17.2 PRECLINICAL EVALUATION OF PANCREATIC TOXICITY

17.2.1 Introduction

The potential risk for pancreatic toxicity for a drug candidate may first be recognized within the context of a preclinical toxicity study. However, currently, the risk for drug-induced pancreatic toxicity is often recognized during the conduct of clinical trials or even after drug approval and widespread use in a large population of human patients with various risk factors and comorbidities that may predispose them to drug-induced pancreatic injury. Unfortunately, retrospective analysis of the preclinical data for marketed drugs may not reveal evidence of similar types of pancreatic injury liabilities. The preclinical species most commonly reported to be affected is the rat, likely because it is the species most frequently utilized in preclinical toxicity and investigative studies. However, the value of assessing preclinical pancreatic toxicity in the rat has been questioned, because the anatomy, physiology, and molecular cell biology of the rat pancreas differ substantially from those in humans probably more so than any other preclinical species (Case, 2006). Pancreatic toxicity has also been reported in the mouse and occasionally in the dog, nonhuman primate (NHP), and other species.

The initial hazard identification of pancreatic toxicity in a preclinical study is always concerning because of the potentially serious human health risks, which may include fulminant, life-threatening cases of DIAP or a lifelong illness, such as CP, type 1 diabetes, or pancreatic neoplasia. Because both the exocrine and the endocrine pancreas have enormous reserve capacities and there are limited sensitive and specific biomarkers to detect pancreatic injury in both animals and humans, there is always the concern that subclinical DIAP may initially go undetected in humans. Further, AP may progress until a threshold of severity is evident in minimally or asymptomatic healthy human volunteers exposed to drug candidates with unknown risks and/or in previously healthy patients that receive standard-of-care agents with marked

pancreatic liabilities (Wallig and Sullivan, 2013). This is particularly concerning given the potential for long periods of latency (weeks to months) before drug-induced pancreatic toxicity can be definitively diagnosed. For example, the first cases of drug-induced CP were reported in the 1950s and were associated with chlorthalidone and cortisone (Hung and Lanfranco, 2014). These same factors make the detection of pancreatic toxicants difficult in preclinical safety assessments.

17.2.2 Risk Management and Understanding the Potential for Clinical Translation

Drug candidate-related pancreatic toxicity identified in preclinical models may or may not be predictive of a relevant human health risk. Thus, to ensure safety within the context of a drug development program, further investigations may be warranted to better understand the underlying mechanism and potential translatability to humans. These investigations often consist of parallel efforts to better characterize the lesion including its temporal pathogenesis, explore potential underlying cellular and molecular mechanisms of toxicity, and evaluate the possible role of the intended molecular target, that is, on-target versus off-target toxicity. Other contributing factors that may also be investigated include but are not limited to pancreas-selective drug accumulation, local metabolic activation, and genetic predispositions. With additional characterization, *in vitro* and/or *in vivo* screening tools may evolve to enable selection of drug candidates free of the pancreatic liabilities and/or aid the identification of predictive biomarkers that may be utilized for drug candidate selection as well as monitoring for potential safety signals throughout the drug development process.

The potential for clinical translatability of preclinical findings of pancreatic toxicity needs to be determined on a case-by-case basis. Since pancreatic toxicity is a known risk factor for many approved drugs, it is clear that, if first identified early on in a drug development program, the risk does not necessarily preclude the further development of the drug candidate. However, such risk would need to be fully understood mechanistically, so that its potential for extrapolation to humans could be determined, successfully monitored and managed in the clinic, and balanced relative to the desired benefit of the proposed drug candidate for the patient population.

An example of successful risk management for preclinical pancreatic toxicity is the drug development program for an antisclerostin mAb for sclerosteosis, in which repeat-dose toxicology studies showed peri-islet hemorrhage in the pancreas of the rat but not the NHP. Subsequent investigations determined that the mechanism of toxicity was a rat-specific drug-induced thrombocytopenia that led to a hypocoagulable state and secondary pancreatic hemorrhage.

The species-specific nature of these reversible toxicological findings combined with the ease of clinical monitoring with routine platelet counts allowed for the safe initiation of clinical trials (Rudmann et al., 2012). Similarly, a novel cholecystokinin (CCK)-1 receptor agonist being explored for the treatment of obesity successfully reached the clinic, because its preclinical pancreatic toxicity of necrotizing pancreatitis was shown to be rodent specific. The toxicity did not occur in the cynomolgus macaque and was not observed in a 24-week clinical trial in which patients were closely monitored for pancreatic structural abnormalities using abdominal ultrasound and magnetic resonance imaging (Myer et al., 2014). In contrast, the potent diabetogenic effect of cyclosporine, a calcineurin inhibitor immunosuppressive agent, is successfully modeled in Wistar rats and translatable to humans. After 3 weeks of cyclosporine administration, the Wistar rat showed impaired glucose tolerance and decreased insulin production and beta-cell mass (Hahn et al., 1986). The risk of this reversible drug-induced toxicity is considered acceptable for the organ transplant patient population and is successfully monitored using clinically relevant changes in blood glucose concentrations (Räkel and Karelis, 2011).

17.2.3 Interspecies and Interstrain Differences in Susceptibility to Pancreatic Toxicity

The lack of correlation between preclinical pancreatic toxicity findings and safety findings in humans, when it occurs, may be related to interspecies and/or interstrain differences in susceptibility to the mechanism of pancreatic toxicity along with other factors. There are numerous anatomical and functional differences that exist between humans and the preclinical species and also between strains or sexes of various preclinical models, most notably the rat. Although in a general sense, both macroscopically and microscopically, the pancreata of humans and the most common preclinical species have many similarities, they also have many major and minor differences that may determine the variability in response to drug candidates. Such similarities and differences are reviewed extensively in the literature, and readers are referred to these reviews (Case, 2006; Greaves, 2012; Cattley et al., 2013; Chandra et al., 2013; Rosol et al., 2013; Wallig and Roussex, 2013; Wallig and Sullivan, 2013). The translatability of preclinical pancreatic toxicity may hinge upon whether or not a key component of the underlying mechanism(s) of toxicity is present in the preclinical species evaluated as well as humans.

Select major and minor differences exist between species at the gross anatomical, cellular, and molecular levels. For example, the pancreatic ducts in rodents open into the common bile duct, whereas, in the dog and human, the pancreatic duct or ducts open directly into the duodenum (Greaves, 2012). In the rat, part of the acinar blood supply bypasses the islets and provides blood directly to the acinar capillary network; this is unlike the vasculature of the dog and primate in which most arterial branches to acinar cells first pass through the islets (Greaves, 2012). In rodents, the pancreatic islet has a central core of insulin-secreting beta cells surrounded by a thin mantle zone of alpha, delta, and gamma cells, whereas, in NHPs, alpha cells predominate centrally and are surrounded by peripheral zone of beta cells. In contrast, human islets are made up of a highly varied mixture of hormone-secreting cells without evidence of a core or mantle zone (Fig. 17.2) (Steiner et al., 2010; Chandra et al., 2013).

The distribution of drug-metabolizing enzymes (i.e., cytochrome P450 and glutathione-*S*-transferase isozymes) also varies between species such that most enzymes are expressed in the mouse, NHP, and human, although several are absent in rats and dogs (Ulrich et al., 2002). Interspecies differences in glucose transporter-2 (GLUT2) expression do well to explain the greater susceptibility of rat beta islet cells to injury with streptozotocin and alloxan compared with human islet cells. Both of these beta-cell toxins rely upon GLUT2 for intracellular uptake, and lower GLUT2 expression in human islets results in lower toxin uptake at a given exposure level, compared with rat islets, and therefore less toxicity (Eizirik et al., 1994; DeVos et al., 1995). Differential expression of GLUT2 also explains the greater susceptibility of old world NHPs to streptozotocin-induced type 1 diabetes versus new world NHPs (Kramer et al., 2009).

Differences between rat strains in their susceptibility to pancreas-specific venular permeability induced by intravenous monastral blue B administration may play a role in differences in their susceptibilities to development of islet inflammation and type 1 diabetes (Desemone et al., 1989). Rat strain, species, and sex differences were also demonstrated for susceptibility to azaserine-induced atypical acinar cell nodules (AACN) that are preneoplastic lesions that may develop into adenomas and carcinomas. The W/LEW and Wistar rat strains were highly responsive to nodule induction with 100% and 90% induction rates, respectively, whereas F344 rats were less susceptible and developed about 10% as many AACN as the Wistar rats (Roebuck and Longnecker, 1977).

Given these many differences between preclinical species and humans as well as the potential for mechanistic complexity in drug-induced pancreatic injury, derisking preclinical pancreatic toxicity is challenging. Currently, some of the biggest hurdles for risk management are the unclear relevance of preclinical pancreatic histopathology findings to human health risk, the lack of clinically monitorable pancreas-specific biomarkers, the lack of relevant animal models of human pancreatic sequelae, and the need for advanced *in vitro* models that better replicate conditions in humans (Wallig and Rousseaux, 2013).

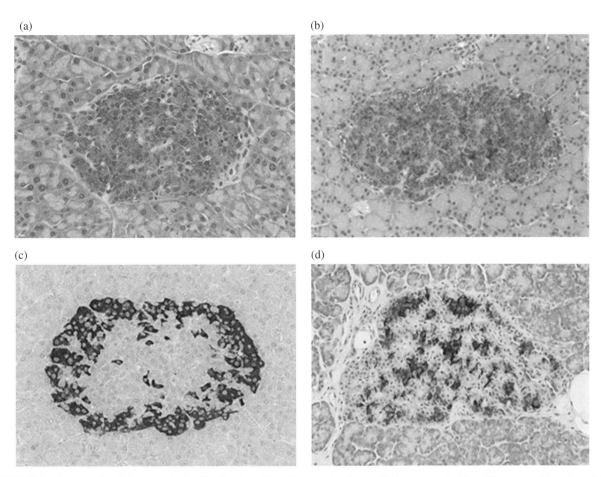

FIGURE 17.2 Interspecies differences in the distribution of insulin-secreting beta cells in pancreatic islets. The mouse (a) and rat (b) have a central core of beta cells, the cynomolgus macaque (c) has a peripheral zone of beta cells, and the human (d) has beta cells scattered throughout the islet. (Insulin immunohistochemistry; original magnification 200×) (*See insert for color representation of the figure.*)

17.3 PRECLINICAL PANCREATIC TOXICITY ASSESSMENT: *IN VIVO*

17.3.1 Routine Assessment

In preclinical toxicity studies, drug candidate-induced pancreatic injury may be associated with clinical observations such as decreased activity, decreased body weight gain, body weight loss, decreased food consumption, or possibly alterations in the feces due to maldigestion in the case of chronic injury with atrophy where PEI may have developed. However, similar to humans, pancreatic toxicity in preclinical models may often be clinically silent, especially if the injury is of lesser severity and/or slowly progressive. Drug candidate-induced pancreatic injury is most commonly detected by routine histopathologic evaluation of hematoxylin and eosin (H&E)-stained tissue sections, as part of the evaluation of pathology endpoints for a dose range-finding exploratory or regulatory toxicity study (Fig. 17.3). The lesions, if of sufficient size, number, and color contrast to normal pancreatic tissue, may also be detectable macroscopically. Routine tissue sampling consists of a single representative cross section that, in the case of rodents, usually is harvested from the tail region of the pancreas adjacent to the spleen. In nonrodents, a single section of the tail or body is routinely examined. If the lesion is diffuse or multifocal and widely distributed throughout the pancreas, a single section may be adequate to assess toxicity. If the lesions are few and/or locally distributed, additional sections may be needed to ensure sampling of the affected areas. Pancreatic weights can also be evaluated at necropsy to quantify loss or gain of parenchymal cell populations or other tissue changes.

Microscopic findings in the pancreas of preclinical species may be attributed to drug candidate administration if they occur exclusively in test article-dosed animals, have a dose-related pattern of occurrence, and/or have an increased incidence and/or severity compared with vehicle control-dosed animals, especially if the incidence is also greater than that of relevant historical controls. Drug candidate-induced lesions are often distinguishable from common background lesions in the pancreas of the preclinical species; however,

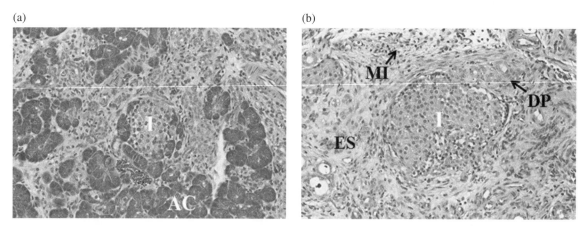

FIGURE 17.3   Microscopic changes in the rat pancreas induced by cyanohydroxybutene (CHB). CHB is a known acinar cell toxicant used to model exocrine injury in rodents. Progressive acinar cell degeneration, necrosis, and atrophy with mixed cell inflammatory infiltrate and ductular proliferation ranging from moderate (a) to severe (b) are evident in routine H&E-stained sections 7 days following a single subcutaneous dose. AC, acinar cells; DP, ductular proliferation; ES, edematous stroma; I, islet; MI, mixed cell inflammatory infiltrate. (H&E; original magnification 200×) (*See insert for color representation of the figure.*)

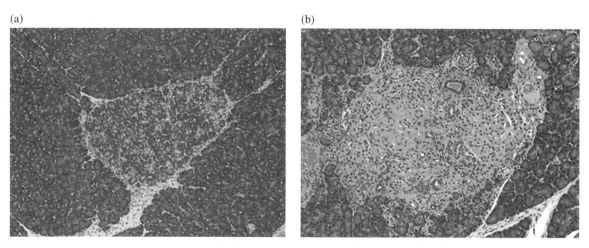

FIGURE 17.4   Common background microscopic findings in the rat pancreas. Focal acinar (lobular) atrophy (a) and islet fibrosis with inflammation and/or hemorrhage (b) may be observed at a low incidence in normal, vehicle control-dosed rats. Similar lesions may also be induced by drug candidate administration but are observed at a greater incidence and/or severity compared with controls. (H&E; original magnification 50× and 100×, respectively.) (*See insert for color representation of the figure.*)

drug candidates may also exacerbate common background findings by increasing the incidence and/or severity or decreasing the age of onset. In the rat, common background findings in the exocrine pancreas include focal ductular obstruction by inspissated material, focal acinar (lobular) atrophy, mononuclear infiltrates, and scattered apoptotic acinar cells. Another common background lesion that may affect the endocrine pancreas of the rat is spontaneous peri-islet hemorrhage and fibrosis, which is most commonly reported in the Sprague-Dawley strain (Dillberger, 1994; Imaoka et al., 2007) (Fig. 17.4). Zymogen granule depletion in acinar cells may be a direct consequence of drug candidate administration or a secondary effect of decreased food consumption and/or body weight loss (Fig. 17.5). Similarly, vacuolization of acinar cells may be drug candidate induced

or artifactual, that is, related to how the tissue was handled and/or processed. Gentle tissue handling and rapid tissue preservation are recommended for optimal microscopic evaluation as well as tissue-based molecular analyses due to the enzyme content of the tissue and the tendency toward rapid autolysis. For example, RNA isolation protocols for the rodent pancreas have modifications to further inhibit the abundant ribonucleases present in pancreatic tissue, so that high-quality RNA can be extracted for expression analysis (Azevedo-Pouly et al., 2014).

17.3.2   Specialized Techniques

If preclinical pancreatic toxicity is identified, the first step to better understanding the pathogenesis of the lesion is often a

(a)　　　　　　　　　　　　　　　　(b)

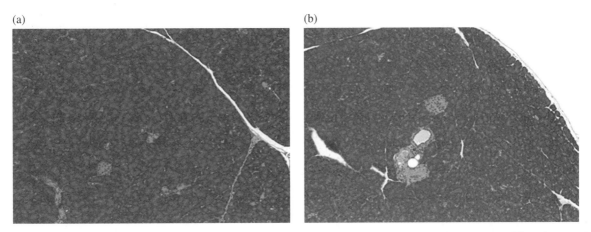

FIGURE 17.5 Zymogen granulation depletion in acinar cells. Zymogen granulation depletion is observed as a diffuse decrease in cytoplasmic eosinophilic (pink) staining within acinar cells (b) compared with the normal rat pancreas (a). This change may be directly related to drug candidate administration or secondary to decreased food consumption and/or body weight loss. (H&E; original magnification 50×) (*See insert for color representation of the figure.*)

time-course study that allows the pathologist to evaluate the location and character of the acute injury, the time of onset and to watch the lesion progression over time, including an assessment of reversibility. For these investigative studies, the pancreas is often more extensively sampled than in routine toxicology studies. In the case of rodents, the entire pancreas can be sampled and embedded flat to be evaluated microscopically *in toto* to maximize the ability of the pathologist to detect infrequent lesions. The whole tissue preparation also provides an opportunity to compare the distribution of lesions across different regions of the pancreas (head, body, and tail), that is, so-called lesion mapping, which may implicate a particular tissue component as the site of origin for the pathologic process. For example, the tail of the rodent pancreas has a greater density of islets and may be a more common site for lesions originating in or near islets. In the case of large animals, representative sections can be evaluated from different anatomical regions as well as from sites of gross abnormalities.

Specialized histochemical, immunohistochemical (IHC), and other tissue section-based molecular techniques (e.g., *in situ* hybridization (ISH)) can be used to better characterize the nature of the injury, cell types affected and its relationship to test article distribution and/or its intended target. The distribution of digestive enzymes or their proenzymes in the zymogen granules of acinar cells can be demonstrated using IHC markers for trypsin, chymotrypsin, carboxypeptidases A and B, lipase, amylase, elastase, DNase, and RNase (Cattley et al., 2013). Cytokeratin IHC markers can be used to differentiate epithelial components of the pancreas, most notably the cells of the ductal system. Amylase and MIST1 have used been as IHC markers to confirm acinar cell differentiation experimentally. Trypsinogen activation peptide (TAP), a peptide released from trypsinogen when trypsin is activated, can be used as an IHC marker for intracellular activation

of trypsinogen (Cattley et al., 2013). IHC identification of peptide hormones to visualize subtypes of endocrine cells within the islets may be critical to characterizing injury and loss or, alternatively, proliferation of particular subpopulations of hormone-secreting cells. Immunohistochemistry along with histomorphometry via image analysis as well as stereological approaches can be used to identify and quantify changes in specific cell populations or other tissue features relative to the tissue area or volume, respectively. Transmission electron microscopy (TEM) is a valuable tool for characterizing the earliest stages of cellular injury at the ultrastructural level. Frozen tissue samples may be used for molecular analyses to assess exposure-related changes in gene expression or protein levels as well as drug tissue concentration to determine if the drug candidate is selectively taken up by and concentrated in the pancreas.

17.4 PANCREATIC BIOMARKERS

17.4.1 Introduction

The lack of sensitive and specific biomarkers for exocrine and endocrine pancreatic injury is a major hurdle for drug candidate screenings, preclinical assessments in animals/*in vitro* platforms, and clinical monitoring of potential drug-induced pancreatic injury. In regard to exocrine toxicity, the most widely used traditional biomarkers are digestive enzymes such as amylase and lipase, which originate from the exocrine pancreas. These enzymes may leak from injured acinar cells or obstructed ducts into the vasculature, where they are subsequently measured, representing minimally invasive, conventional biomarkers. In regard to endocrine toxicity, the first indicators of drug candidate- or drug-induced injury are usually related to impaired blood sugar regulation reflected by changes in blood glucose (increased) and/or insulin concentrations (often evident as early and progressive increases followed by decreases at later stages). Although these biomarkers can be useful and indicative of drug-induced effects, they often lack adequate specificity and/or sensitivity.

Even in the absence of reliable biomarkers to accurately assess pancreatic toxicity, there is not consistent and reliable exchange of biomarker knowledge across pharmaceutical companies to drive biomarker discovery, development, and implementation. To address these biomarker gaps, a cross-industry working group is being established (c.2015) within the Predictive Safety Testing Consortium (PSTC; http://c-path.org/programs/pstc) to explore the development of potential new and improved biomarkers for drug-induced pancreatic toxicity.

The biomarker section will focus on exocrine injury biomarkers and endocrine functional biomarkers that are relevant to drug discovery and development. For exocrine biomarkers, the section is divided into traditional, exploratory, and emerging pancreatic biomarkers. Traditional biomarkers, as defined in this manuscript, are those parameters for which assay methodologies are currently available and can be added to a preclinical safety study protocol if pancreas is identified or suspected as a likely target organ of toxicity. Exploratory biomarkers are those that are sparingly used in human and veterinary diagnostic medicine and are adapted in preclinical drug discovery settings with additional fit-for-purpose assay development. Emerging biomarkers are those not currently used in veterinary or human medicine, but that are being explored in the pharmaceutical industry due to specific pancreatic injury biomarker needs. Generally, there is minimal prospective knowledge on the value of emerging pancreatic biomarkers with most biomarker knowledge being generated using biomarker discovery approaches. While serum cytokines, chemokines, acute-phase proteins, and complement products may also provide added value for detecting pancreatitis, the specificity of these biomarkers to the pancreas is questionable and thus beyond the scope of this discussion. For each of these biomarker sections, background, biomarker value, and current gaps will be discussed.

17.4.2 Exocrine Injury Biomarkers in Humans and Preclinical Species

17.4.2.1 Traditional Biomarkers: Serum Amylase and Lipase

Background Serum biomarkers for exocrine pancreatic injury in humans and preclinical species traditionally include amylase and lipase. In terms of assay methodology, serum amylase and lipase use an activity-based endpoint, which utilizes the substrate (e.g., diacylglycerol is the lipolytic substrate for serum lipase) to detect the serum activity levels. The assays have been used for over a decade in diagnostic and experimental settings. Short-duration elevations in serum amylase and lipase can be observed following acute acinar injury in humans and preclinical species, and these assays are commonly used in a diagnostic setting to evaluate for pancreatic injury. Although amylase and lipase can be measured, they are not routinely assessed in preclinical toxicology studies, and thus experience with these analytes is generally limited.

Current Gaps Although commonly used in a diagnostic setting to test for pancreatic injury in humans, serum amylase and lipase have limited utility for detecting drug candidate-induced pancreatic injury due to their poor sensitivity, lack of specificity, and short duration of clinically relevant elevations even within the context of acute widespread acinar cell necrosis (Banks and Freeman, 2006). As a consequence, serum levels of amylase and lipase can be within normal ranges in cases of AP, or, if elevated during the course of AP, the magnitude of the increase does not necessarily correlate with the severity of the AP. The sensitivity of these biomarkers may be further reduced within the context of chronic disease or decreased food consumption and/or body weight loss, which may reduce pancreatic stores of enzymes and the likelihood that enzyme leakage into the serum could be detected if acinar injury should occur. In general, lipase is considered to be more sensitive for detecting pancreatic injury than amylase, especially in humans. Also in humans, significant increases in serum amylase and lipase (>3 times the upper limit of normal), if present, may contribute to a diagnosis of AP; however, presenting signs such as abdominal pain and imaging modalities using a combination of ultrasound and computed tomography (CT) are often more relevant for diagnosing AP (Banks and Freeman, 2006).

Increases in serum amylase are not pancreas specific and may also be caused by injury to the liver, intestine, and/or salivary gland (depending upon species) or related to decreased renal clearance due to underlying kidney disease (Simpson et al., 1991; Steiner, 2003). It is worth noting that there are interspecies differences in regard to the tissue distribution of the isozymes of amylase. There are four different amylase isozymes that can be found in the pancreas (high concentrations in all species), liver (rat, mouse, and dogs), intestine (duodenum and ileum, especially in the rat), and salivary gland (rat and mouse have levels comparable to pancreas, whereas dogs and NHPs have insignificant amounts) (Jacobs, 1989; Hokari et al., 2003). The presence of amylase in multiple tissues in preclinical species is also reflected by the large normal reference range (e.g., in rats of ~900–2000 U/l) compared with serum lipase (which has a normal reference range in rats of 6–26 U/l). Similar to amylase isozymes, increases in serum total lipase isoforms are also not pancreas specific and may be related to injury to the gastric mucosa or decreased renal clearance (Lassen, 2006a).

Because these enzymes are not specific to the pancreas and are rapidly cleared by the kidneys, serum amylase and lipase are now considered to be of very little value in the diagnosis of DIAP. However, in suspected cases of pancreatic injury where elevations in amylase or lipase do occur, the lack of tissue specificity can be addressed by the use of assays specific for the pancreatic form of each enzyme (i.e., pancreatic (or p-)amylase and p-lipase, respectively).

17.4.2.2 Exploratory Biomarkers: Trypsin-Like Immunoreactivity, Pancreatic Lipase Immunoreactivity, and Trypsinogen Activation Peptides

Background Trypsinogen is synthesized exclusively by the acinar cells of the pancreas, and measurement of this zymogen by a trypsin-like immunoreactivity (TLI) assay provides an indirect index of pancreatic damage. TLI detects both trypsinogen and trypsin forms and hence the use of the term TLI to describe the total concentration of these two immunoreactive species. It should be noted that the active enzyme (trypsin) is only present in the serum when there is pancreatic inflammation. As a consequence, TLI concentration has been shown to be useful for the diagnosis of AP, although it has a low sensitivity (Steiner, 2003).

The pancreatic lipase immunoreactivity (PLI) assay is a measure of lipase and its related precursors. While the function of pancreatic lipase is the same as any other lipase in the body, the molecular structure is unique. For this reason, a radioimmunoassay (RIA) and subsequently an enzyme-linked immunosorbent assay (ELISA) were developed and validated by the Westbrook, ME, US-based manufacturer, IDEXX Laboratories, Inc., for dogs and cats to measure canine and feline PLI (i.e., cPLI and fPLI, respectively). Relative to TLI, species-specific PLI assays have better sensitivity and specificity for detection of pancreatitis (Xenoulis, 2015). Based upon the available data, PLI appears to be the most sensitive laboratory test available for the diagnosis of pancreatitis. Although rat-reactive commercial PLI antibody is available, there is as yet a significant need to develop and validate an ELISA assay that can reliably detect the antigen in preclinical species.

TAP are small peptides that are released when trypsinogen is activated to trypsin. Under physiologic conditions, trypsinogen activation occurs in the intestinal lumen and is mediated by enteropeptidase. Within the intestinal brush border membranes, TAP is quickly degraded by peptidases. In AP, trypsinogen is prematurely activated within pancreatic acinar cells, and TAP is released into the peripheral circulation and potentially can be detected in serum and urine (Sáez et al., 2005). Significant increases in plasma and/or urine TAP concentrations have been reported in feline, canine, and human patients with AP (Johnson et al., 2004). In humans, there is enough data in the literature to show

serum and urine TAP to be excellent markers for the detection of AP (Sáez et al., 2005). Similar to TLI and PLI, there are limited reagents available for accurate detection of TAP by immunoassays in the serum of preclinical species, especially in rodents.

Current Gaps Serum PLI appears to be promising as a biomarker of acute exocrine injury with AP being associated with increased PLI and not being significantly affected by renal function. Unlike traditional biomarkers, the reagent availability is limited for testing TLI, PLI, and TAP, which have not been tested in rodents and/or nonrodents using well-characterized biofluid samples, especially from cases of DIAP. Most of the available reagents are for companion animals, such as dogs and cats. Even with the use of reliable reagents for detecting elevations in TLI, there are some gaps with serum TLI for diagnosis of AP or CP in veterinary diagnostic settings. For example, serum TLI can be elevated in sick or malnourished dogs without evidence of pancreatitis, and TLI is increased in only 30–40% of cats and dogs with pancreatitis showing lack of specificity (Steiner, 2003). Compared to TLI and PLI, TAP has a circulating half-life of <8 min and elevations can be easily missed in a terminal or single blood collection sample. For this reason, in humans with AP, the measurement of urine TAP (normalized to creatinine) is thought to have an advantage over measurement in serum with a prolonged half-life of over 5 h in urine (Sáez et al., 2005).

17.4.2.3 Emerging Biomarkers: Peptides and MicroRNAs

There are multiple candidate emerging biomarkers reported in the literature related to zymogen activation (e.g., chymotrypsinogen, trypsinogens 1 and 2), proteases (procarboxypeptidase B, carboxypeptidases 1 and 2, elastase 3B), and fibrotic events related to matrix degradation (TIMP-1, MMPs, collagen formation markers). Most notably, peptide biomarkers and microRNAs (miRNAs) have been discussed as emerging pancreas-specific biomarkers.

Peptide Biomarkers: Background Two relevant published papers are discussed, which have identified peptide signatures generated in mice and rats using the pancreatic toxicant cyanohydroxybutene (CHB). The concept is based on the increased release of peptidases (carboxypeptidase A, chymotrypsin, and elastase) during pancreatitis. Using a liquid chromatography–mass spectrometry (LC–MS) approach with trypsin digestion and the addition of internal standards and peptide quantitation, two distinct peptide signatures (RA1609 and RT2864 peaks) were generated from trypsin digestion of albumin and carboxypeptidase A digestion of trypsin III, respectively. Following induction of pancreatic injury, decreases in the RA1609 peak and increases in the RT2864 peak have been observed in at least three different species (mouse, rat, and human). The peptide signatures changed immediately following injury and tracked with acinar cell injury even after the initial degranulation of zymogen granules (Walgren et al., 2007a, b).

In a cerulein-treated rat model, Walgren et al. (2007a, b) showed that amylase and lipase increased significantly as early as ~8 h after cerulein administration, while the RA1609 peak decreased and the RT2864 peak increased both at 8 and 24 h after cerulein injection. In a dose–response study using CHB as a model toxicant in rats with apoptosis and zymogen granule depletion occurring at multiple dose levels (50–200 mg/kg), lipase levels increased significantly at 150 and 200 mg/kg, while the RA1609 peak decreased and the RT2864 peak increased at both 150 and 200 mg/kg of CHB confirming the value of the peaks in specifically detecting pancreatic injury. This rat-derived data was further confirmed in 98 human-derived samples. In this specific pancreatitis human subject cohort, the RT2864 peak identified patients with 85% sensitivity and 94% specificity, and the RA1623/1641 peaks with 100% sensitivity and 88% specificity.

miRNA Biomarkers: Background   miRNA biomarkers are small, 18–25 nucleotide, noncoding RNA sequences that regulate gene expression at the posttranscriptional level by binding to and inhibiting translation of target messenger RNAs (Panarelli and Yantiss, 2011). Evaluation of circulating miRNAs is the most recent effort to identify novel biomarkers in preclinical safety as miRNAs often display tissue-specific expression and may be released from the tissues into circulation during organ-specific injury induced by drug candidates. One distinct gap that serum miRNA analysis potentially may address is tissue specificity due to the identification of pancreas-specific miRNAs. Another advantage with this technology is the potential for interspecies translatability in addition to a longer postinjury duration of detectable alterations due to the stability of miRNAs.

Plasma miR-216a elevation was detected by Kong et al. (2010) in a rat pancreatic injury model 24 h postadministration of L-arginine. In a similar rat pancreatic injury model, Endo et al. (2013) reported that both miR-216a and miR-216b were significantly increased 24 h after L-arginine injection. Using laser microdissection, Endo et al. (2013) concluded that miR-216a and miR-216b were predominantly expressed in acinar cells of the pancreas as compared to pancreatic islets. Usborne et al. (2014) also examined miR-216a and miR-375 as potential biomarkers in two rat models of AP and reported that miR-216a and miR-375 were significantly elevated in serum 24 h postdosing of CHB and concluded that pancreas-enriched miRNAs hold promise as novel serum-based biomarkers for exocrine pancreatic injury. Goodwin et al. (2014) completed an evaluation of miR-216a and miR-217 as potential biomarkers of acute

pancreatic injury in rats and mice. Both miRNAs showed time- and dose-dependent responses to pancreatic injury induced by L-arginine and pancreatic duct ligation and displayed wider dynamic ranges of response than serum amylase or lipase.

Current Gaps   Although the peptide signatures have been identified using several pancreatic toxicants, such peptides have not been tested prospectively when pancreatic injury has occurred. Although the LC–MS assay development is feasible for identifying the signature, the methodology and workflow need to be established to confirm the approach. Also, it will be important to test the presence of these peptides across other species such as in dogs or NHPs to determine if the biomarkers are translatable. With regard to miRNA, the potential utility of miRNAs as plasma-based biomarkers of acute pancreatic injury has been evaluated in several studies, but much of this work has focused on animal models induced by chemical compounds that may not be relevant to preclinical safety studies. Additional baseline data and evidence of specificity are needed under a wider range of drug-induced injury and disease states as well as in other species including humans.

17.4.3   Endocrine/Islet Functional Biomarkers for Humans and Preclinical Species

In contrast with pancreatic exocrine injury biomarkers discussed in the previous section, there are currently no injury markers reported for the endocrine pancreas or islets. Available islet-specific biomarkers routinely used to understand islet health in general include parameters that test the functionality of islets relative to the synthesis of insulin and concurrent regulation of blood glucose concentrations.

17.4.3.1   Glucose Tolerance Test and Hormones   Insulin and glucagon, derived from beta and alpha pancreatic islet cells, respectively, are the two most important hormones involved in glucose homeostasis (Lassen, 2006b). In humans and preclinical species, a glucose tolerance test (GTT) is a commonly used functional assay to assess carbohydrate metabolism and indirectly pancreatic hormone function, that is, a GTT determines the rate at which glucose is cleared from the blood. An oral GTT (OGTT) is performed by orally administering glucose and subsequently measuring glucose, insulin, and/or glucagon levels over time (e.g., at 30, 60, 90, and 120 min postadministration of drug). A prolonged elevation (>120 min) in both plasma glucose and insulin constitutes impaired glucose tolerance and insulin resistance. OGTT is highly sensitive, and this procedure is used across species to understand insulin sensitivity (Lassen, 2006b). In addition to measurement of insulin, by-products of insulin synthesis, such as C-peptide, can be assessed. C-peptide is formed from the cleavage of proinsulin following its transport from the endoplasmic reticulum to the Golgi and subsequent packaging into granules. As such, both C-peptide and insulin can be used to evaluate endocrine function (Wu et al., 2012).

When evaluating potential islet pancreatic toxicants, considerations for timing of glucose challenge should be based on peak concentration of the toxicant achieved after single or multiple doses. Concurrent measurement of pancreatic hormones such as insulin, C-peptide, and glucagon in addition to serum glucose may be helpful in the interpretation of the ability of the endocrine pancreas to maintain glucose homeostasis. It should be noted that there are no strict interpretation guidelines on what magnitude of change constitutes a meaningful change when there is pancreatic islet damage.

17.4.4 A Note on Biomarkers of Vascular Injury Relevant to the Pancreas

In addition to exocrine and endocrine components, vascular injury can also be observed within the pancreas (Brenneman et al., 2014). Although there are some pancreas-specific microvasculature patterns of expression, there are no biomarkers that can be assessed in serum that are specific to pancreatic microvasculature, that is, acinar and islet capillaries. In the absence of concurrent vascular damage in nonpancreatic tissues, changes in vascular injury biomarkers in the blood can be interpreted as likely originating from the pancreas. In this context, endothelial microparticles (EMPs) have gained attention and could be valuable as a biomarker (Helbing et al., 2014). Microparticles are very small vesicles (0.1–1 μm in size) originating for the plasma membrane via blebbing, wherein a portion of the cell membrane is released into circulation. Since microparticles are part of the plasma membrane, the receptors and cytosolic molecules, including RNA, are maintained within the microparticle and can be assessed using flow cytometry in a time-course fashion with concurrent controls. This technology is still in the exploratory stages and needs additional assay validation and qualification (linkage to the vascular damage) before this assay is available for prospective biomarker applications.

17.4.5 Author's Opinion on the Strategy for Investments to Address Pancreatic Biomarker Gaps

It is currently well known that there is a lack of sensitive and specific pancreatic biomarkers available for use in drug development research. This is especially true in the preclinical setting, because these biomarkers are not routinely evaluated and preclinical pancreatic toxicity occurs infrequently. While exploratory biomarkers such as TLI, PLI, and TAP have been sporadically tested, there is great value in investing in the development of reliable reagents across preclinical species to accurately assess their value in well-characterized biological samples. Testing species specificity, when reagents are available, in addition to developing new species-specific reagents for these assays is an approach that could quickly lead to prioritization of assays using well-annotated pancreatic injury samples. Practically speaking, there are bottlenecks in the development of reliable immunoassay reagents across species. To address this, investment in miRNA and the LC–MS approach for peptide biomarkers makes practical biomarker development sense. In addition to targeted miRNA assessment based on the literature, broad miRNA profiling of pancreas (exocrine and islet) and serum simultaneously in a relevant pancreatic injury model would provide an unbiased approach to biomarker discovery. A broad protein biomarker discovery approach may be challenging due to reproducibility issues as a result of significant proteolysis within the pancreas compared to other organs, such as the liver or kidney. However, testing well-annotated samples with the serum peptide biomarkers already reported in the literature is feasible and can be executed within reasonable time and costs. Investigators will have the option to utilize newer platforms available for broad biomarker profiling and discovery (e.g., next-generation sequencing, aptamer technology, and multiple reaction monitoring LC–MS), but such approaches will generally need significant funding and time. In addition, a small set of biomarkers is what is needed to address acute pancreatic injury rather than a larger panel of exploratory biomarker candidates typically used to understand novel toxicity of a drug candidate.

17.5 PRECLINICAL PANCREATIC TOXICITY ASSESSMENT: *IN VITRO*

17.5.1 Introduction to Pancreatic Cell Culture

When preclinical pancreatic toxicity is recognized from in-life, anatomic, and/or clinical pathology findings during the conduct of an *in vivo* toxicity study, *in vitro* toxicity assessment is a common component of the derisking efforts. The primary advantage of *in vitro* work is that it allows one to model drug candidate exposure and cellular response and explore possible mechanisms of toxicity within a simplified, defined, and controlled environment. When investigating pancreatic toxicity, this approach may work well for drug candidate-induced injury of an abundant, isolatable, and culturable tissue component, such as acinar cells or islets, when it is directly injured by the administered drug candidate. However, if the mechanism of toxicity involves more than a single tissue component or has a more complex pathogenesis, as it may in the pancreas, given its complex anatomy and physiology, this approach may be inadequate to model the *in vivo* toxicity.

With the pancreas functioning as an interdependent dual system, as both an exocrine and endocrine gland, efforts have been made to design *in vitro* models that isolate cells from both systems. However, there are limited knowledge and literature pertaining to the isolation of live pancreatic cell subsets, mechanistic modeling, and reliable high-throughput screening for *in vitro* assessments of toxicity mainly due to the complex homeostatic regulatory systems involved. Due to the proteolytic enzyme content of the pancreas, cell isolations and downstream *in vitro* work can often prove to be difficult, but are not impossible with the correct protocol. Acini and islets can both be isolated by perfusion of the common bile duct or direct infiltration of the pancreatic lobules with collagenase and a trypsin inhibitor (to prevent autodigestion), followed by manual dissociation and size separation into the smaller acinar cells and larger whole islets that contain multiple subtypes of hormone-secreting cells. Unlike the mechanical isolation of many other parenchymal cell populations, a 3-day incubation period is recommended to allow for recovery of the cells from the isolation process. After isolation, the whole islets can be further subdivided into individual cells by further dissociation and flow cytometric cell purification.

17.5.2 Modeling *In Vivo* Toxicity *In Vitro*, Testing Translatability, and *In Vitro* Screening Tools

Once the tissue component of interest is successfully isolated, the goal of pancreatic *in vitro* work is to define models that correctly mimic the *in vivo* environment where the toxicity occurred, test for possible translatability of mechanisms between preclinical species and humans, and confirm or predict mechanisms of toxicity as a screening tool. The *in vitro* methods described in this chapter emphasize these goals, highlighting the endpoints available for assessing drug candidate-induced injury and dysfunction unique to the pancreas.

17.5.2.1 *Exocrine*

Drug candidate-induced toxicity of the exocrine pancreas is mainly associated with degenerative changes, such as autophagy, degranulation, and degeneration and/or necrosis of the acinar cells. Autophagy is a basic catabolic mechanism that involves the uptake into vacuoles and degradation of unnecessary or dysfunctional cellular components, including zymogen granules, through the actions of lysosomes. Autophagy may have a physiological role in that it promotes cell survival during starvation by maintaining cellular energy levels; however, autophagy is also a hallmark of sublethal injury to acinar cells and may precede acinar cell necrosis in drug candidate-induced injury (Wallig and Sullivan, 2013). Autophagic vacuoles can be observed in histologic sections at the light microscopic and ultrastructural levels (Fig. 17.6) as well as in cultured acinar cells. Colocalization with Beclin-1, an autophagosomal protein, and autophagosomal-integrated proteins, VMP1 (Vaccaro et al., 2003) and/or LC3 (Zhang et al., 2014), may be used to confirm the autophagic origin of these vacuoles. The cells can also be lysed to probe via Western blot for these autophagic markers (Meyer et al., 2013).

Acinar cell degranulation results from signal transduction in the cells during a stress response that releases previously inactive proenzymes as secretory enzymes (such as trypsin, chymotrypsin, amylase, and lipase) from zymogen granules (Ji et al., 2009). Release of amylase and lipase from acinar cells during a stress response can be monitored *in vitro* by measuring the amount of enzyme secretions released into the cell culture medium using a chemistry analyzer (e.g., Siemens ADVIA 1800) or ELISA. Although *in vivo* increases in serum activity levels of amylase and lipase are not always indicative of pancreatic toxicity due to their possibly originating from multiple tissues, *in vitro* measurements of

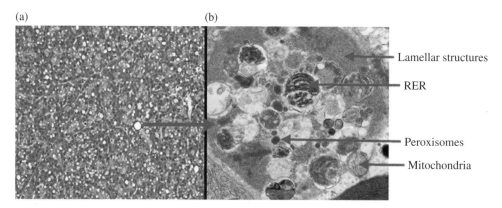

(a)　　　　　(b)

Lamellar structures

RER

Peroxisomes

Mitochondria

FIGURE 17.6 Drug candidate-induced pancreatic acinar cell vacuolization due to autophagy in rats. Drug candidate administration was associated with vacuolization of acinar cells in H&E-stained sections of the rat pancreas (a), which correlated ultrastructurally with autophagosomes filled with degenerating organelles (b). Autophagy is a hallmark of sublethal injury to acinar cells and may precede acinar cell necrosis. (*See insert for color representation of the figure.*)

amylase and lipase from culture media directly assess the effect of drug candidate exposure on the cultured acinar cells. Cerulein, an oligopeptide that can stimulate digestive enzyme secretions, is often used as a positive control comparator for activation of amylase and lipase secretion from acinar cells (Kim, 2008). The activation and conversion of trypsinogen to trypsin or other activated proteases can be monitored *in vitro* by using fluorogenic substrates. Upon enzymatic cleavage, the nonfluorescent substrate is converted resulting in an increase in fluorescence. The substrate can be used to continuously measure enzyme activity in acinar cell extracts and purified enzyme preparations using a fluorometer or fluorescence microplate reader or can be used for intracellular protease assays with analysis by flow cytometry or fluorescence microscopy (Halangka et al., 1997).

Necrosis can occur from overactivation of autophagy and subsequent degranulation. If the cell death or toxicity finding is believed to be drug candidate related, acinar cells can be cultured and treated with the drug candidate of interest in a high-throughput format and the viability of the cells can be assessed by measuring levels of ATP relative to a control over the course of the treatment.

17.5.2.2 *Endocrine*

Islets, as whole islets or dissociated cells, can be evaluated for drug candidate-induced toxicity through assessment of insulin regulation, glucose tolerance, and/or cell death. Effects of drug candidate exposure on islets can be monitored by ELISA for insulin secretion, glucose uptake, and, similar to acinar cells, cytotoxicity by measuring ATP levels. To monitor insulin regulation, cultured islets can be treated with the drug candidate of choice and exposed to physiological levels of glucose. Upon exposure to glucose, the amount of insulin secreted from the islets can be quantitated in the culture medium via ELISA as measurement of insulin regulation and overall cell functionality (Kooptiwut et al., 2002). Glucose is the primary source of energy for most cells, and its uptake into islets, being the first rate-limiting step in glucose metabolism, is highly regulated and, if altered, may indicate drug-induced toxicity. Cells exhibiting insulin resistance show diminished glucose uptake in response to insulin stimulation (Schwartz et al., 2013).

Glucose uptake can be measured using the glucose analogue 2-deoxyglucose (2-DG), which is taken up by cells and phosphorylated by hexokinase to 2-DG6P (Sigma Aldrich). 2-DG6P cannot be further metabolized and accumulates in cells in a manner directly proportional to the glucose uptake and is measured by a coupled enzymatic assay by generating a fluorometric product.

17.5.2.3 *Microvascular*

In addition to analysis of endocrine and exocrine cell systems in the pancreas, additional cell types may play a role in the development of a drug candidate-induced pancreatic toxicity. The dense microvasculature of the pancreas plays a key role within the islet and as an interface between the islet and the surrounding acinar tissue. With an interdependent physical and functional relationship with beta cells, islet endothelial cells are involved not only in the delivery of oxygen and nutrients to endocrine cells but also induce insulin gene expression during islet development, affect adult beta cell function, promote beta cell proliferation, and produce a number of vasoactive, angiogenic substances and growth factors (Zanone et al., 2008).

In vitro, the capillary endothelial cells of the pancreas are still attached to the outer rim of whole islets postisolation and, with conditioned medium, may be coaxed to proliferate outward onto extracellular matrix-coated plastic plates for further mechanistic studies and functional analyses. The isolated capillary endothelial cells can be characterized by examining α-1 proteinase inhibitor and nephrin expression via immunocytochemistry or flow cytometry (Zanone et al., 2008). Not only do these endothelial cells play a role in drug candidate-induced effects on insulin secretion and glucose regulation of islets, but they also act as the regulators of leukocyte recruitment into the islets. The pancreatic microvasculature is therefore likely to play a role in the altered physiology of the endocrine and exocrine pancreas if affected by drug candidate exposure.

In vitro investigations with acinar, islet, or endothelial cells individually may be beneficial for screening of drug candidates for direct, single-cell-type, cytotoxic effects, but it is important to note that toxicities with a mechanism involving interactions between more than one cell type concurrently (e.g., endothelial/endocrine interactions) cannot be definitively assessed in a single-cell culture system. More complicated organotypic models may be needed if the drug candidate-induced toxicity is mediated through multicell type injury and response.

17.5.3 Case Study 1: Drug Candidate-Induced Direct Acinar Cell Toxicity *In Vivo* with Confirmation of Toxicity and Drug Candidate Screening *In Vitro*

An advantage of *in vitro* methods is that they are usually amenable to miniaturization and automation, yielding high-throughput screening methods for screening drug candidates. A success story for this approach is a preclinical toxicity evaluation in the rat with a direct acinar toxicant that resulted in a rapid onset of macroscopic and microscopic lesions (necrosis progressing to atrophy) diffusely affecting the pancreas with concurrent increases in traditional serum biomarkers, including amylase and lipase, indicating acinar cell injury. Acinar cells were isolated, and the cytotoxicity profile of the acinar toxicant was consistent *in vitro* with the necrosis observed *in vivo*. Amylase, lipase, and activation of trypsin were assessed *in vitro* and demonstrated an increase in the amount of digestive enzymes secreted via ELISA or

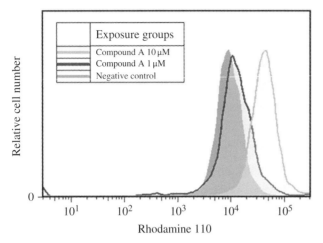

FIGURE 17.7 *In vitro* assessment of drug candidate-induced trypsin activation related to autophagy. Autophagy-related activation of trypsinogen to trypsin was measured in drug candidate-exposed (compound A) acinar cells at 1 and 10 μM concentrations by using fluorogenic substrates. Upon enzymatic cleavage, the nonfluorescent substrate was converted to a fluorescent substrate in the drug candidate-exposed cells resulting in an increase in fluorescence compared with the negative control.

flow cytometry in the acinar toxicant-exposed cells (Fig. 17.7). Due to this correlation with *in vivo* findings and successful modeling of toxicity, the *in vitro* models were used to screen many candidates and identified a go-forward drug candidate free of toxicity both in the *in vitro* model and ultimately in preclinical *in vivo* toxicity studies. This example shows that there are *in vitro* assays using a single cell type, that is, one relevant cell population that is the direct target of toxicity, available that can effectively predict *in vivo* pancreatic toxicities and can be used as highly efficient tools for drug candidate screening.

17.5.4 Case Study 2: Drug Candidate-Induced Microvascular Injury at the Exocrine–Endocrine Interface in the Rat with Unsuccessful Confirmation of Toxicity *In Vitro* and No Pancreas-Specific Monitorable Biomarkers Identified

A case of drug candidate-induced injury to the microvasculature of the endocrine–exocrine portal system in the rat was reported with gross and microscopic lesions that presented as minimal to mild, multifocal, degenerative, and inflammatory foci (Brenneman et al., 2014). The lesions (peri-islet fibrin exudation, hemorrhage, and inflammation that extended into surrounding acinar tissue as necrosis and progressed to islet fibrosis and lobular atrophy) had some early features that were similar to the common background lesion (i.e., spontaneous peri-islet hemorrhage and fibrosis) in the Sprague-Dawley rat (Dillberger, 1994; Imaoka et al., 2007), although they were clearly drug candidate related due

to a dose-related increase in incidence and severity compared with controls.

To further characterize the toxicity and investigate its possible mechanism, a 14-day time-course study in the rat was conducted to allow for the observation of the lesion development and progression (Brenneman et al., 2014). Light microscopic (i.e., histochemical and IHC with quantification) and ultrastructural (i.e., TEM) examination showed that the injury appeared to originate as degeneration and loss of endothelial cells in the capillaries at the exocrine–endocrine interface. This injury was not associated with any increases in traditional or exploratory serum biomarkers of exocrine injury nor any evidence of endocrine dysfunction as assessed via an OGTT; however, it was associated with an increase in EMPs that was not pancreas specific.

In an effort to derisk the preclinical pancreatic toxicity, several *in vitro* approaches were used to attempt to identify the mechanism of toxicity. Initially, rat primary acinar cells were isolated and treated with a series of chemically similar test articles to screen for cytotoxicity. Only a few of the test articles tested were associated with acinar cell cytotoxicity, which was not predictive of *in vivo* toxicity nor was it associated with elevations in amylase and lipase in the cell culture medium, indicating a false-positive outcome. Islets were isolated and insulin regulation and glucose tolerance were assessed again with no changes related to the *in vivo* toxicity. With this work, it was determined that endothelial cell cultures from the islet periphery or alternatively a three-dimensional (3D) model including multiple cell types would be needed to further explore possible mechanisms of toxicity; in the interim, 14-day *in vivo* toxicity studies in the rat were utilized for drug candidate screening.

17.5.5 Emerging Technologies/Gaps: Organotypic Models

The lesson learned from these case studies was that *in vitro* models can be used as screening tools when investigating direct injury to a relevant single cell type, but traditional two-dimensional (2D) cell culture models often bear little physical, molecular, or physiological similarity to the pancreas *in vivo*. This highlights a need to continue efforts to define more complex multicell systems, such as organotypic and *ex vivo* models, for assessment of drug candidate-induced pancreatic toxicity.

The inability of 2D cultures to fully recapitulate the organizational structure and subcellular processes of a complex microtissue environment has led to recent advancements in pancreatic organotypic cultures. These 3D cultures have been designed in multiple formats (e.g., organoids and organ-on-a-chip) to mimic the complex spatial morphology of an organ and to allow biologically relevant cell–cell and cell–matrix interactions. Due in part to the inability of researchers to create an effective 2D model of diabetes that translates *in vivo*, 3D models of the pancreas currently focus primarily on recapitulating the metabolic processes of islets (Boj et al., 2015). With the recent advances in organotypic technology, pancreas-specific 3D models would greatly improve the tools available for screening, identifying, and derisking potential drug candidate-induced pancreatic toxicities *in vitro*.

Pancreatic organoids are multicell microtissues formed by adhesive forces with self-produced cell matrices, providing an isolated yet physiologically relevant environment for various cell types to be screened (Hynds and Giangreco, 2013). Recently, the model of organoids has been applied to the development of artificial or pseudoislets made from multiple pancreatic cell lines, typically comprised of α, β, and pancreatic epithelial-like cell types (Jo et al., 2014). These artificial islets have been shown to express and secrete pancreatic hormones such as insulin, glucagon, and somatostatin, demonstrating that they may be an ideal model for prediction of drug candidate-induced pancreatic toxicity and underlying mechanisms of diabetes mellitus. The artificial islets and other organoid models have been developed to enhance functionality of the target cell types through cell–cell contact, but it is also important to use models that mimic the mechanics of how the pancreas operates, making microfluidic models ideal for pancreatic toxicity assessment.

Culture of cells using various microfluidic devices, such as organ-on-a-chip, is becoming more common within research to mimic the interactions of drugs between endothelium and other various cell types in a microvascular environment. Pancreatic islets are highly vascularized *in vivo* and, within this tissue microenvironment, endothelial and islet cells often have 1:1 association (Sankar et al., 2011). *In vitro*, this ratio is maintained immediately after cell isolation due to endothelial cells remaining attached to the islets, but time in culture causes the endothelial cells to deteriorate, losing density and branched morphology. It is believed that the loss in viability and functionality of the endothelial cells is due to the absence of blood flow and the limited diffusion of media into the interior of the tissue. To improve flow of media into the tissue, microfluidic models have been developed to culture islets with a variable range of flow rates. Culturing the islets in these models with media flowing enhances the viability and functionality of both the endothelial and islet cells compared to classically cultured islets, making this model a technological innovation for *in vitro* toxicity assessments in which islet and endothelial interactions are key.

In contrast with these new 3D coculture technologies, *ex vivo* or *in situ* perfusion methods, using the rat pancreas, are a unique opportunity to evaluate the effects of drug candidates or explore possible mechanisms of toxicity in the whole pancreas (Ross, 1972). This procedure is technically challenging and involves cannulating the arterial blood supply to the pancreas and perfusing with an oxygenated physiologic solution mimicking blood flow to maintain viability during drug exposure. The venous outflow of the pancreas as well as the major pancreatic ducts can be cannulated to collect samples for analysis of drug effects on glucose as well as hormone and digestive enzyme secretions. In this model, effects on blood flow and pressure can also be monitored (Hillaire-Buys et al., 1985). This isolated whole-organ perfusion system has advantages compared to the *in vitro* cell isolation methods, because it maintains the complex anatomical and functional relationships of the whole organ including the endocrine and exocrine portions and their interconnected vasculatures (Ross, 1972).

As previously stated, the primary advantage of *in vitro* work is that it allows one to model drug candidate exposure and cellular response and explore possible mechanisms of toxicity within a simplified, defined, and controlled environment. This can be accomplished by using primary cells in a 2D culture, but due to the intricate anatomy and physiology of the pancreas, the isolated 2D single-cell system has significant limitations. Great advancements have been made with the use of organotypic and microfluidic *in vitro* as well as whole-organ perfusion model systems for attempting to simulate *in vivo* toxicity, and with similar assay endpoints as 2D cultures, organotypic cultures may become the new screening paradigm for assessment of drug candidate-induced pancreatic toxicity.

17.6 SUMMARY AND CONCLUSIONS

The reporting of drug-induced pancreatic toxicity in humans is rapidly on the rise in recent decades, most commonly presents as AP, and its association with drug exposure is often weak, idiosyncratic and based upon only one or a few case reports. The mechanism of toxicity is often not known, and the likelihood that a preclinical species will model the same outcome and mechanism of injury is also often not known, although it does occasionally occur. Given these circumstances, any preclinical pancreatic toxicity, although of questionable relevance for human health risk assessment, is concerning. Because, if it does translate to humans, the injury may be clinically silent for a long time and could progress to serious consequences including AP and CP, which predispose to pancreatic cancer and/or type 1 diabetes. Currently, some of the biggest hurdles for risk management of preclinical pancreatic toxicity are the lack of understanding of the relevance of preclinical findings for human health risk, the lack of clinically monitorable biomarkers, the lack of relevant animal models of human pancreatic disease, and the need to develop *in vitro* models that replicate the *in vivo* clinical situation. Future cross-industry efforts will focus on filling these gaps by sharing learnings from firsthand experiences derisking preclinical pancreatic toxicity for drug development programs.

ACKNOWLEDGMENTS

The authors would like to thank Jianying Wang for her contributions to the discussion of miRNA biomarkers and Robert Mauthe for his review of the manuscript.

REFERENCES

Azevedo-Pouly, A. C. P., Elgamal, O. A., and Schmittgen, T. D. (2014): RNA isolation from mouse pancreas: A ribonuclease-rich tissue. *J. Vis. Exp.*, (90), e51777, doi:10.3791/51779.

Badalov, N., Baradarian, R., Iswara, K., Li, J., Steinberg, W., and Tenner, S. (2007): Drug-induced AP: An evidence-based review. *Clin. Gastroenterol. Hepatol.*, 5, 648–661.

Banks P. A. and Freeman M. L. (2006): Practice guidelines in acute pancreatitis. *Am. J. Gastroenterol.* 101, 2379–2400.

Boj, S. F., Hwang, C-I., Baker, L. A., Chio, I. I. C., Engle, D. D., Corbo, V., Jager, M., Ponz-Sarvise, M., Tiriac, H., Spector, M. S., Gracanin, A., Oni, T., Yu, K. H., van Boxtel, R., Huch, M., Rivera, K. D., Wilson, J. P., Feigin, M. E., Öhlund, D., Handly-Santana, A., Ardito-Abraham, C. M., Ludwig, M., Elyada, E., Alagesan, B., Biffi, G., Yordanov, G. N., Delcuze, B., Creighton, B., Wright, K., Park, Y., Morsink, F. H. M., Molenaar, I. Q., Borel Rinkes, I. H., Cuppen, E., Hao, Y., Jin, Y., Nijman, I. J., Iacobuzio-Donahue, C., Leach, S. D., Pappin, D. J., Hammell, M., Klimstra, D. S., Basturk, O., Hruban, R. H., Offerhaus, G. J., Vries, R. G. J., Clevers, H., and Tuveson, D. A. (2015): Organoid models of human and mouse ductal pancreatic cancer. *Cell*, 160(1–2), 324–338.

Brenneman, K. A., Ramaiah, S. K., Rohde, C. M., Messing, D. M., O'Neil, S. P., Gauthier, L. M., Stewart, Z. S., Mantena, S. R., Shevlin, K. M., Leonard, C.G., Sokolowski, S. A., Lin, H., Carraher, D. C., Jesson, M. I., Tomlinson, L., Zhan, Y., Bobrowski, W. F., Bailey, S. A., Vogel, W. M., Morris, D. L., Whiteley, L. O., and Davis, J. W. II (2014): Mechanistic investigations of test article-induced pancreatic toxicity at the endocrine-exocrine interface in the rat. *Toxicol. Pathol.*, 42, 229–242.

Case, R.M. (2006): Is the rat pancreas an appropriate model of the human pancreas? *Pancreatology*, 6, 180–190.

Cattley, R. C., Popp, J. A., and Vonderfecht S. L. (2013): Liver, gallbladder and exocrine pancreas. In: *Toxicologic Pathology: Preclinical Safety Assessment* (Sahota, P. S., Popp, J. A., Hardisty, J. F., and Gopinath, C., eds.), pp. 345–366, CRC Press, Boca Raton, FL.

Chandra, S., Hoenerhoff, M. J., and Peterson, R. (2013): Endocrine glands. In: *Toxicologic Pathology: Preclinical Safety Assessment* (Sahota, P.S., Popp, J. A., Hardisty, J. F., and Gopinath, C., eds.), pp. 692–716, CRC Press, Boca Raton, FL.

De Vos, A., Heimberg, J., Quartier, E., Huypens, P., Bouwens, L., Pipeleers, D., and Schuit, F. (1995): Human and rat beta cells differ in glucose transporter but not glucokinase expression. *J. Clin. Invest.*, 96, 2489–2495.

Desemone, J., Majno, G., Joris, I., Handler, E. S., Rossini, A. A., and Mordes J. P. (1989): Morphological and physiological characteristics of pancreas-specific venular permeability induced by Monastral blue B. *Exp. Mol. Pathol.*, 52, 141–153.

Dillberger, J. E. (1994): Age-related pancreatic islet changes in Sprague-Dawley rats. *Toxicol. Pathol.*, 22, 48–55.

Egan, A. G., Blind, E., Dunder, K., de Graeff, P. A., Hummer, B. T., Bourcier, T., and Rosebraugh, C. (2014): Pancreatic safety of incretin-based drugs—FDA and EMA assessment. *N. Engl. J. Med.*, 370, 794–795.

Eizirik, D. L., Pipeleers, D. G., Ling, Z., Welsh, N., Hellerström, C., and Andersson, A. (1994): Major species differences between humans and rodents in the susceptibility to pancreatic ß-cell injury. *Proc. Natl. Acad. Sci. U. S. A.*, 91, 9253–9256.

Endo K., Weng J., Kito N., Fukushima Y., and Iwai N. (2013): miR-216a and miR-216b as markers for acute phased pancreatic injury. *Biomed. Res.*, 34(4), 179–188.

Goodwin D., Rosenzweig B., Zhang J., Stewart S., Thompson K., and Rouse R. (2014): Evaluation of miR-216a and miR-217 as potential biomarkers of acute pancreatic injury in rats and mice. *Biomarkers*, 19, 517–529.

Greaves, P. (2012): Exocrine pancreas. In: *Histopathology of Preclinical Toxicity Studies: Interpretation and Relevance in Drug Safety Studies*, 4th ed., p. 489, Academic Press, London, UK.

Hahn, H. J., Laube, F., Lucke, S., Klöting, I., Kohnert, K. D., and Warzock, R. (1986): Toxic effects of cyclosporine on the endocrine pancreas of Wistar rats. *Transplantation*, 41(1), 44–47.

Halangka,W., Stürzebecherb, J., Matthias, B., Schulze, H., and Lipperta, H. (1997): Trypsinogen activation in rat pancreatic acinar cells hyperstimulated by caerulein. *Biochim. Biophys. Acta*, 1362(2–3), 243–251.

Helbing, T., Olivier, C., Bode, C., Moser, M., and Diehl P. (2014): Role of microparticles in endothelial dysfunction and arterial hypertension. *World J. Cardiol.*, 26(6), 1135–1139.

Hescot, S., Vignaus, O., and Goldwasser F. (2013): Pancreatic atrophy—A new late toxic effect of sorafenib. *N. Engl. J. Med.* 369(15), 1475–1476.

Hillaire-Buys, D., Gross, R., Blayac, J.-P., Ribes, G., and Loubatières-Mariani, M.-M. (1985): Effects of α-adrenoceptor agonists and antagonists on insulin secreting cells and pancreatic blood vessels: Comparative study. *Eur. J. Pharmacol.*, 117, 253–257.

Hokari, S., Miura, K., Koyama, I., Kobayashi, M., Matsunaga, T., Iino, N., and Komoda, T. (2003): Expression of alpha-amylase isozymes in rat tissues. *Comp. Biochem. Physiol. B Biochem. Mol. Biol.*, 135, 63–69.

Horner, S., Ryan, D., Robinson, S., Callander, R., Stamp, K., and Roberts, R. A. (2013): Target organ toxicities in studies conducted to support first time in man dosing: An analysis across species and therapy areas. *Regul. Toxicol. Pharmacol.*, 65, 334–343.

Hruban, R. H. (2015): The pancreas. In: *Robbins and Cotran the Pathological Basis of Disease* (Kumar, V., Abbas, A. K., and Aster, J. C., eds.), pp. 884–889, Elsevier Saunders, Philadelphia, PA.

Hung, W. Y. and Lanfranco, O. A. (2014): Contemporary review of drug-induced pancreatitis: A different perspective. *World J. Gastrointest. Pathophysiol.*, 5(4), 405–415.

Hynds, R. and Giangreco, A. (2013): The relevance of human stem cell-derived organoid models for epithelial translational medicine. *Stem Cells*, 31(3), 417–422.

Imaoka, M., Satoh, H., and Furuhama, K. (2007): Age- and sex-related differences in spontaneous hemorrhage and fibrosis of the pancreatic islets in Sprague-Dawley rats. *Toxicol. Pathol.*, 35, 388–394.

Jacobs, R. M. (1989): The origins of canine serum amylase and lipase. *Vet. Pathol.*, 26, 525–527.

Ji, B., Gaiser, S., Chen, X., Ernst, S., and Logsdon, C. (2009): Intracellular trypsin induces pancreatic acinar cell death but not NF-κB activation. *J. Biol. Chem.*, 284(26), 17488–17498.

Jo, Y. H., Jang, I. J., Nemeno, J. G., Lee, S., Kim, B. Y., Nam, B. M., Yang, W., Lee, K. M., Kim, H., Takebe, T., Kim, Y. S., and Lee, J. I. (2014): Artificial islets from hybrid spheroids of three pancreatic cell lines. *Transplant. Proc.*, 46(4), 1156–1160.

Johnson, C. D., Lempinen, M., Imrie, C. W., Puolakkainen, P., Kemppainen, E., Carter, R., and McKay C. (2004): Urinary trypsinogen activation peptide as a marker of severe acute pancreatitis. *Br. J. Surg.* 91, 1027–1033.

Jones, M. R., Hall, O. H., Kaye, A. M., and Kaye, A. D. (2015): Drug-induced acute pancreatitis: A review. *Ochsner J.*, 15, 45–51.

Kaurich, T. (2008): Drug-induced acute pancreatitis. *Proc. (Baylor Univ. Med. Cent.)*, 21(1), 77–81.

Kim, H. (2008): Cerulein pancreatitis: Oxidative stress, inflammation and apoptosis. *Gut Liver*, 2(2), 74–80.

Kong, X. Y., Du Y. Q., Li L., Liu, J. Q., Wang, G. K., Zhu, J. Q., Man, X. H., Gong, Y. F., Xiao, L. N., Zheng, Y. Z., Deng, S. X., Gu, J. J., and Li, Z. S. (2010): Plasma miR-216a as a potential marker of pancreatic injury in a rat model of acute pancreatitis. *World J. Gastroenterol.*, 16, 4599–4604.

Kooptiwut, S., Zraika, S., Thorburn, A., Dunlop, M., Darwiche, R., Kay, T., Proietto, J., and Andrikopoulos, S. (2002): Comparison of insulin secretory function in two mouse models with different susceptibility to β-cell failure. *Endocrinology*, 143(6), 2085–2092.

Kramer, J., Moeller, E. L., Hachey, A., Mansfield, K. G., and Wachtman, L. (2009): Differential expression of GLUT2 in pancreatic islets and kidneys of new and old world nonhuman primates. *Am. J. Physiol. Regul. Integr. Comp. Physiol.*, 296, R786–R793.

Lassen, D. E. (2006a): Laboratory evaluation of exocrine pancreas. In: *Veterinary Hematology and Clinical Chemistry* (Thrall, M., Baker, D. C., Campbell, T. W., DeNicola, D., Fettman, M. J., Lasse, E. D., Rebar, A., and Weiser, G., eds.). pp. 377–385. Blackwell Publishing Ltd., Oxford, UK.

Lassen, D. E. (2006b): Laboratory evaluation of endocrine pancreas and of glucose metabolism. In: *Veterinary Hematology and Clinical Chemistry* (Thrall, M., Baker, D. C., Campbell, T. W., DeNicola, D., Fettman, M. J., Lasse, E. D., Rebar, A., and Weiser, G., eds.). pp 431–443. Blackwell Publishing Ltd., Oxford, UK.

Lee, S. C. and Dalia, S. M. (2012): Drug-induced chronic pancreatitis. *Med. Health Rhode Island*, 95(1), 19–20.

Meyer, G., Czompa, A., Reboul, C., Csepanyi, E., Czegledi, A., Bak, I., Balla, G., Balla, J., Tosaki, A., and Lekli, I. (2013): The cellular autophagy markers Beclin-1 and LC3B-II are increased during reperfusion in fibrillated mouse hearts. *Curr. Pharm. Des.*, 19(39), 6912–6918.

Myer, J. R., Romach, E. H., and Elangbam, C. S. (2014): Species- and dose-specific pancreatic responses and progression in single- and repeat-dose studies with GI181771X: A novel cholecystokinin 1 receptor agonist in mice, rats and monkeys. *Toxicol. Pathol.*, 42, 260–274.

Nitsche, C. J., Jamieson, N., Lerch, M. M., and Mayerle, J. V. (2010): Drug-induced pancreatitis. *Best Pract. Res. Clin. Gastroenterol.*, 24, 143–155.

Panarelli, N. C. and Yantiss, R. K. (2011): MicroRNA expression in selected carcinomas of the gastrointestinal tract. *Pathol. Res. Int.*, 2011, 1–10.

Pezelli, R., Corinaldesi, R., and Morselli-Labate, A. M. (2010): Tyrosine kinase inhibitors and acute pancreatitis. *Int. J. Pancreatol.*, 11(3), 291–293.

Räkel, A. and Karelis, A. D. (2011): New-onset diabetes after transplantation: Risk factors and clinical impact. *Diabetes Metab.*, 37, 1–14.

Roebuck, B. D. and Longnecker, D. S. (1977): Species and rat strain variation in pancreatic nodule induction by azaserine. *J. Natl. Cancer Inst.*, 59(4), 1273–1277.

Rosol, T. J., DeLellis, R. A., Harvey, P. W., and Sutcliffe, C. (2013): Endocrine pancreas. In: *Haschek and Rousseaux's Handbook of Toxicologic Pathology* (Haschek, W., Rousseaux, C., and Wallig, M., eds.), pp. 2475–2492, Elsevier/Academic Press, Amsterdam.

Ross, B. D. (1972): Endocrine organs. In: *Perfusion Techniques in Biochemistry: A Laboratory Manual in the Use of Isolated Perfused Organs in Biochemical Experimentation*. pp. 321–355, Clarendon Press, Oxford, UK.

Rudmann, D. G., Page, T. J., Vahle, J. L., Chouinard, L., Haile, S., Poitout, F., Baskin, G., Lambert, A.-J., Walker, P., Glazier, G., Awori, M., and Bernier, L. (2012): Rat-specific decreases in platelet count caused by a humanized monoclonal antibody against sclerostin. *J. Toxicol. Sci.*, 125(2), 586–594.

Sáez, J., Martínez, J., Trigo, C., Sánchez-Payá, J., Compañy, L., Laveda, R., Griñó, P., García, C., and Pérez-Mateo, M. (2005): *World J. Gastroenterol.*, 13, 7261–7265.

Sankar, K. S., Green, B. J., Crocker, A. R., Verity, J. E., Altamentova, S. M., and Rocheleau, J. V. (2011): Culturing pancreatic islets in microfluidic flow enhances morphology of the associated endothelial cells. *PLoS One*, 6(9), e24904.

Scarpelli, D. G. (1989): Toxicology of the pancreas. *Toxicol. Appl. Pharmacol.*, 101(3), 543–554.

Schwartz, M. W., Seeley, R. J., Tschöp, M. H., Woods, S. C., Morton, G. J., Myers, M. G., and D'Alessio, D. (2013): Cooperation between brain and islet in glucose homeostasis and diabetes. *Nature*, 503, 59–66.

Simpson, K. W., Simpson, J. W., Lake, S., Morton, D. B., and Batt, R. M. (1991): Effect of pancreatectomy on plasma activities of amylase, isoamylase, lipase and trypsin-like-immunoreactivity in dogs. *Res. Vet. Sci.*, 51, 78–82.

Steiner, J. M. (2003): Diagnosis of pancreatitis. *Vet. Clin. North Am. Small Anim. Pract.*, 33, 1181–1195.

Steiner, D. J., Kim, A., Miller, K., and Hara, M. (2010): Pancreatic islet plasticity: Interspecies comparison of islet architecture and composition. *Islets*, 2(3), 135–145.

Szkudelski, T. (2001): The mechanism of alloxan and streptozotocin action in B cells of the rat pancreas. *Physiol. Res.*, 50, 536–546.

Tatarkiewicz, K., Belanger, P., Gu, G., Parkes, D., and Roy, D. (2013): No evidence of drug-induced pancreatitis in rats treated with exenatide for 13 weeks. *Diabetes Obes. Metab.*, 15(5), 417–426.

Tenner, S. (2014): Drug-induced acute pancreatitis: Does it exist? *World J. Gastrointest. Pathophysiol.*, 20(44), 16529–16534.

Ulrich, A. B., Standop, J., Schmied, B. M., Schneider, M. B., Lawson, T. A., and Pour, P. M. (2002): Species differences in the distribution of drug-metabolizing enzymes in the pancreas. *Toxicol. Pathol.*, 30(2), 247–253.

Underwood, T.W. and Frye, C. B. (1993): Drug-induced pancreatitis. *Clin. Pharm.*, 12(6), 440–448.

Usborne, A. L., Smith, A.T., Engle, S. K., Watson, D. E., Sullivan, J. M., and Walgren, J. L. (2014): Biomarkers of exocrine pancreatic injury in 2 rat acute pancreatitis models. *Toxicol. Pathol.* 42, 195–203.

Vaccaro, M. I., Grasso, D., Ropolo, A., Iovanna, J. L., Cerquetti, M. C. (2003): VMP1 expression correlates with acinar cell cytoplasmic vacuolization in arginine-induced acute pancreatitis. *Pancreatology*, 3(1), 69–74.

Walgren, J. L., Mitchell, M. D., Whiteley, L. O., and Thompson, D. C. (2007a): Identification of novel peptide safety markers for exocrine pancreatic toxicity induced by cyanohydroxybutene. *Toxicol. Sci.*, 96, 174–183.

Walgren, J. L., Mitchell, M. D., Whiteley, L. O., and Thompson, D. C. (2007b): Evaluation of two novel peptide safety markers for exocrine pancreatic toxicity. *Toxicol. Sci.*, 96, 174–183.

Wallig, M. A. and Rousseaux, C. (2013): Endocrine pancreas. In: *Haschek and Rousseaux's Handbook of Toxicologic Pathology* (Haschek, W., Rousseaux, C., and Wallig, M., eds.), pp. 2475–2492, Academic Press, London, UK.

Wallig, M. A. and Sullivan, J. M. (2013): Exocrine pancreas. In: *Haschek and Rousseaux's Handbook of Toxicologic Pathology* (Haschek, W., Rousseaux, C., and Wallig, M., eds.), pp. 2361–2390, Academic Press, London, UK.

Wu, D., Yue, F., Zou, C., Chan, P., and Zhang, Y. A. (2012): Analysis of glucose metabolism in cynomolgus monkeys during aging. *Biogerontology*, 13, 147–155.

Xenoulis, P. G. (2015): Diagnosis of pancreatitis in dogs and cats. *J. Small Anim. Pract.* 56, 13–26.

Zanone, M. M., Favaro, E., and Camussi, G. (2008): From endothelial to beta cells: Insights into pancreatic islet microendothelium. *Curr. Diabetes Rev.*, 4(1), 1–9.

Zhang, L., Zhang, J., Shea, K., Xu, L., Tobin, G., Knapton, A., Sharron, S., and Rouse, R. (2014): Autophagy in pancreatic acinar cells in caerulein-treated mice: Immunolocalization of related proteins and their potential as markers of pancreatitis. *Toxicol. Pathol.*, 42(2), 435–457.

PART V

ADDRESSING THE FALSE NEGATIVE SPACE—INCREASING PREDICTIVITY

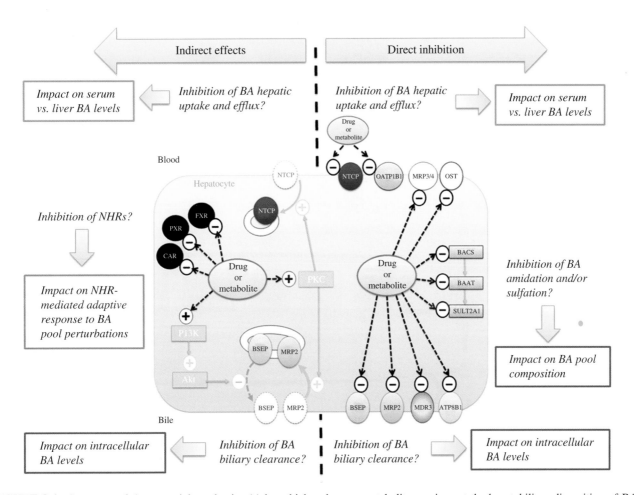

FIGURE 8.6 Summary of the potential mechanism(s) by which a drug or metabolite can impact the hepatobiliary disposition of BAs (Rodrigues et al. (2014), figure 3, pp. 566–574. Reproduced with permission of ASPET).

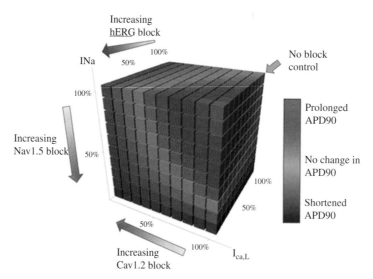

FIGURE 9.3 Predicted impact and balance of varying three key ionic currents on the duration of the cardiac action potential at 90% repolarization (APD90).

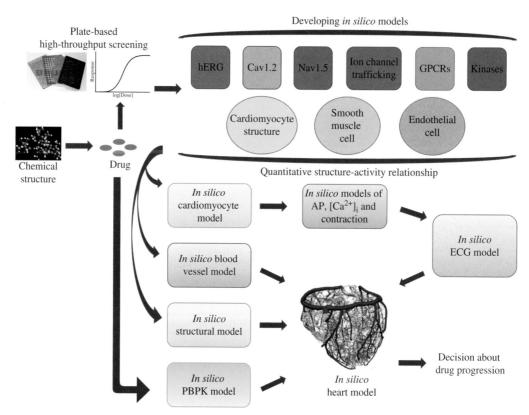

FIGURE 9.4 Proposed virtual human heart-on-a-chip risk assessment concept. Micromodels representing different cardiovascular functions can be integrated into a single microheart in a physiologically relevant manner to evaluate, early in the drug discovery process, the potential of a new chemical entity to induce cardiotoxicity. Inclusion of the drug pharmacokinetic properties would additionally allow users the ability to consider the relevant drug exposure in the heart. AP, action potential; Cav1.2, calcium ion channel protein that in humans is encoded by the CACNA1C gene; ECG, electrocardiogram; GPCRs, G-protein-coupled receptors; hERG, potassium ion channel protein that in humans is encoded by the ether-à-go-go-related gene; Nav1.5, sodium ion channel protein that in humans is encoded by the SCN5A gene; PBPK, physiologically based pharmacokinetic modeling.

FIGURE 9.6 (a) Potential surrogates for human patients. Is it reasonable to believe that studies conducted on at least 1 of these potential surrogates for man might yield data applicable to man? (b) Example of animals selected for use in safety pharmacology. Notice absence of heterogeneities of phenotypes. If this species possesses the anatomy and physiology necessary, in man, to result in an adverse response to a test article, studies conducted on it should yield applicable results. However if it does not, results applicable to man would not be expected. (c) Examples of humans, showing obvious enormous heterogeneity of phenotypes, for which we attempt to predict adverse clinical events, many of which occur because of phenotypic heterogeneities.

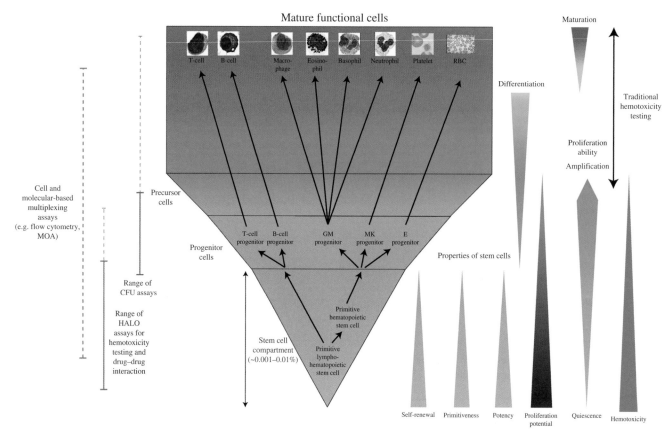

FIGURE 11.1 The organization of the hematopoietic system defines the type and severity of hemotoxicity. This diagram illustrates the organizational and hierarchical structure of the lymphohematopoietic system. This consists of a stem cell compartment that feeds into amplification compartments provided by the progenitor and precursor cells followed by the primary differentiation and maturation compartments to produce the mature functional cells. Stem cell hemotoxicity and its severity are defined by two measures of proliferation, namely, proliferation ability and proliferation potential or capacity. These, in turn, define the sensitivity of the stem cells to respond to different insults and perturbations. In general, the more primitive a stem cell, the greater its sensitivity to hemotoxic-inducing drug and compounds and the greater the severity on the system as a whole.

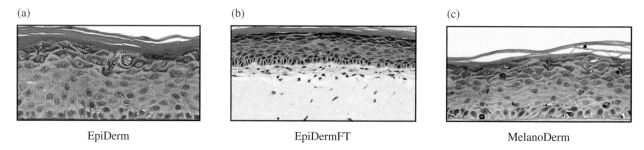

FIGURE 12.1 Examples of commercially available *in vitro* reconstructed human epidermal (RhE) tissues: (a) EpiDerm partial thickness model produced from primary keratinocytes; (b) EpiDermFT full-thickness coculture model including epidermal and dermal components; and (c) MelanoDerm coculture model of primary keratinocytes and melanocytes. Reproduced with permission of MatTek Corporation, Ashland, MA.

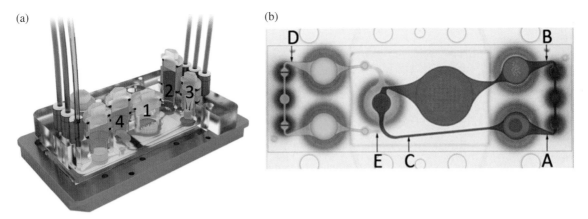

FIGURE 12.2 A microfluidic four-organ-chip device at a glance. (a) 3D view of the device comprising two polycarbonate cover plates, the PDMS–glass chip (footprint: 76 mm×25 mm; height: 3 mm), accommodating a surrogate blood flow circuit (pink) and an excretory flow circuit (yellow). Numbers represent the four tissue culture compartments for intestine (1), liver (2), skin (3), and kidney (4) tissue. A central cross section of each tissue culture compartment aligned along the interconnecting microchannel is depicted. (b) Top view of the four-organ-chip layout illustrating the positions of three measuring spots (A, B, and C) in the surrogate blood circuit and two spots (D and E) in the excretory circuit. Figure courtesy of TissUse, Berlin, Germany, with permission.

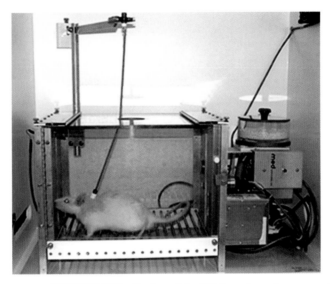

FIGURE 15.2 Setting of the IVSA equipment.

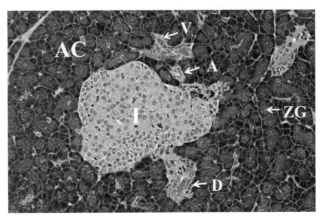

FIGURE 17.1 Microscopic appearance of the normal rat pancreas. An islet of endocrine cells is surrounded by acinar cells of the exocrine pancreas in a normal rat. A, arteriole; AC, acinar cells; D, ductule; I, islet; V, venule; ZG, zymogen granules. (H&E stain; original magnification 200×)

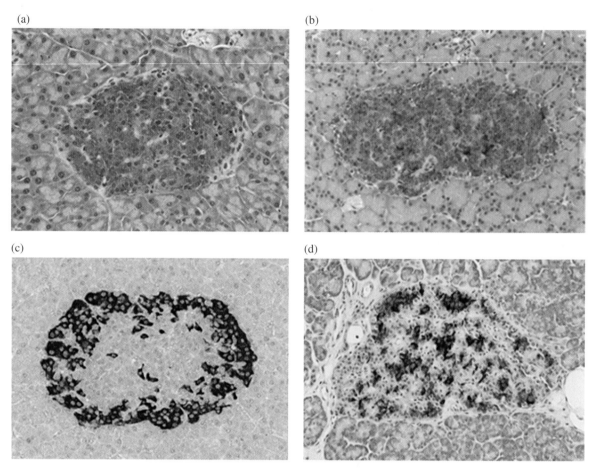

FIGURE 17.2 Interspecies differences in the distribution of insulin-secreting beta cells in pancreatic islets. The mouse (a) and rat (b) have a central core of beta cells, the cynomolgus macaque (c) has a peripheral zone of beta cells, and the human (d) has beta cells scattered throughout the islet. (Insulin immunohistochemistry; original magnification 200×)

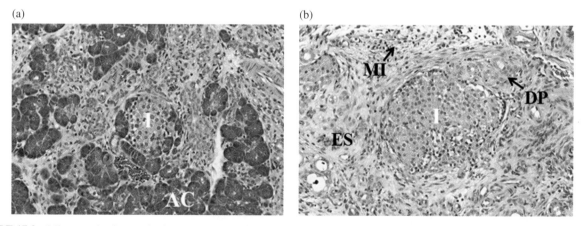

FIGURE 17.3 Microscopic changes in the rat pancreas induced by cyanohydroxybutene (CHB). CHB is a known acinar cell toxicant used to model exocrine injury in rodents. Progressive acinar cell degeneration, necrosis, and atrophy with mixed cell inflammatory infiltrate and ductular proliferation ranging from moderate (a) to severe (b) are evident in routine H&E-stained sections 7 days following a single subcutaneous dose. AC, acinar cells; DP, ductular proliferation; ES, edematous stroma; I, islet; MI, mixed cell inflammatory infiltrate. (H&E; original magnification 200×)

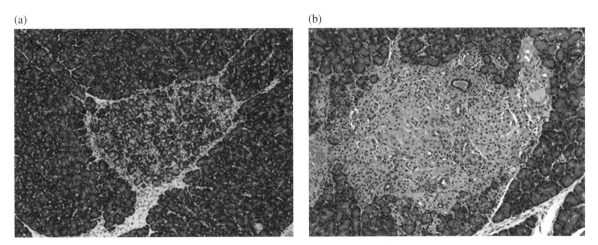

FIGURE 17.4 Common background microscopic findings in the rat pancreas. Focal acinar (lobular) atrophy (a) and islet fibrosis with inflammation and/or hemorrhage (b) may be observed at a low incidence in normal, vehicle control-dosed rats. Similar lesions may also be induced by drug candidate administration but are observed at a greater incidence and/or severity compared with controls. (H&E; original magnification 50× and 100×, respectively.)

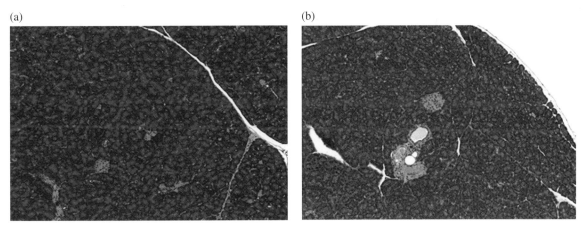

FIGURE 17.5 Zymogen granulation depletion in acinar cells. Zymogen granulation depletion is observed as a diffuse decrease in cytoplasmic eosinophilic (pink) staining within acinar cells (b) compared with the normal rat pancreas (a). This change may be directly related to drug candidate administration or secondary to decreased food consumption and/or body weight loss. (H&E; original magnification 50×)

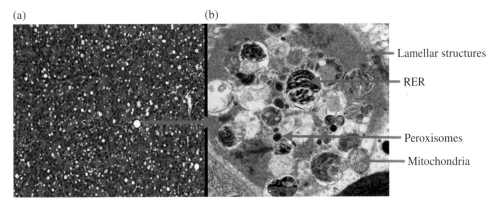

FIGURE 17.6 Drug candidate-induced pancreatic acinar cell vacuolization due to autophagy in rats. Drug candidate administration was associated with vacuolization of acinar cells in H&E-stained sections of the rat pancreas (a), which correlated ultrastructurally with autophagosomes filled with degenerating organelles (b). Autophagy is a hallmark of sublethal injury to acinar cells and may precede acinar cell necrosis.

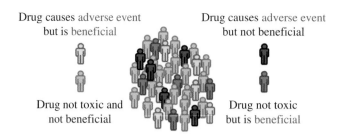

FIGURE 20.1 Response to a given drug is often variable, in which a patient population with a similar diagnosis may actually represent a heterogeneous group of responders and nonresponders. Pharmacogenetic tests have the potential to identify each subset of patients in order to more efficiently tailor drug therapy.

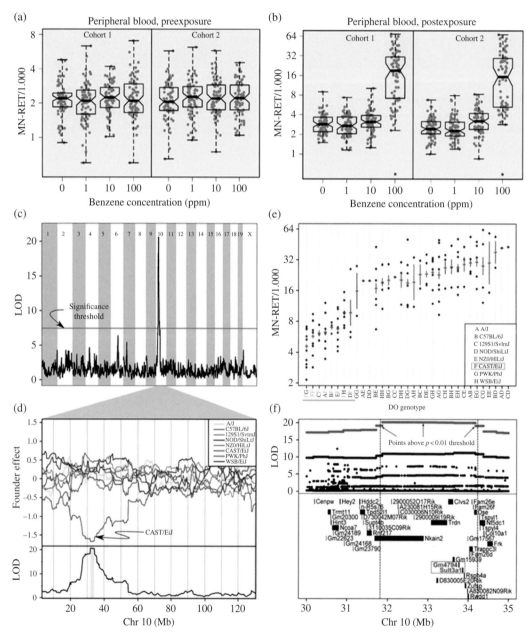

FIGURE 20.4 (a) Preexposure blood MN-RETs. Boxes represent the median and interquartile range, and whiskers cover the entire data range. (b) Benzene postexposure blood MN-RETs. There was a high degree of concordance between cohorts 1 and 2. (c) Plot of the LOD at each marker; red line indicates the permutation-derived significance threshold. (d) Founder effect plots (top) and the LOD score (bottom) on Chr 10; DO mice that have the CAST/EiJ founder haplotype (green) are associated with lower MN-RETs. (e) MN-RET values by DO genotype at the marker with the maximum LOD score on Chr 10 (31.868 Mb). Data points indicate bone marrow MN-RET values for individual DO mice, and red lines denote mean ±SE. Genotypes are listed on the x-axis, with each DO founder represented by a letter; genotypes containing the CAST/EiJ allele are shown in green. (f) Association mapping within the Chr 10 QTL interval. (f, top) Each data point shows the LOD score at one SNP; red data points indicate scores above the significance threshold. (f, bottom) Genes in the QTL interval. Dashed vertical lines denote the QTL support interval. *Sult3a1* is highlighted in red. Adapted from French et al. (2015).

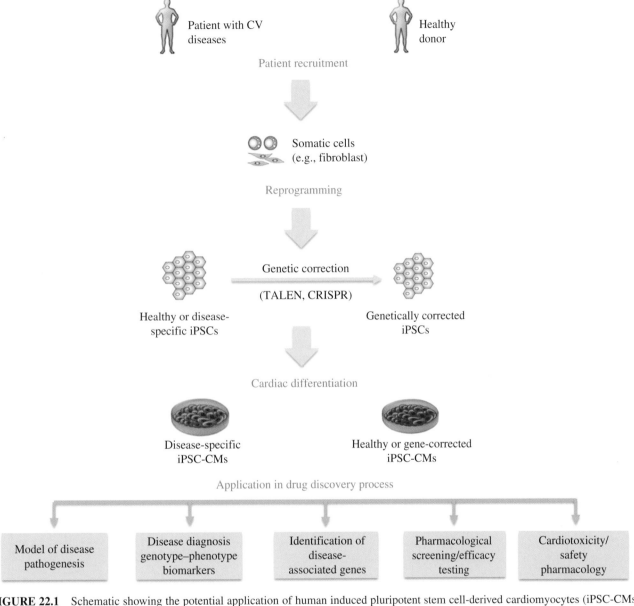

FIGURE 22.1 Schematic showing the potential application of human induced pluripotent stem cell-derived cardiomyocytes (iPSC-CMs) in disease modeling, cardiotoxicity/safety pharmacology assessment, and drug discovery. Somatic cells (e.g., fibroblasts, blood cells, etc.) can be obtained from healthy donors and patients with cardiovascular diseases (e.g., arrhythmia, cardiomyopathies, etc.) followed by efficient reprogramming into well-characterized libraries of iPSCs. Early integration of such disease-specific cardiomyocytes derived from the standardized iPSC library can augment the identification of disease-relevant NCEs or "disease-specific hits" and provide human biology-relevant cardiac cell type to assess the potential cardiotoxicity (e.g., hERG blocking, QT prolongation, or effects on other cardiac ion channels orchestrating the AP waveform) of NCEs early on in drug discovery. Strategic inclusion of iPSC-CMs is expected to reduce the false positives and false negatives associated with the traditional drug discovery process and may help reduce the high attrition rate at costly clinical phases of drug development. In addition, genes identified by disease modeling studies could be corrected using recently developed gene-editing technologies such as transcription activator-like effector nucleases (TALENs) and clustered regularly interspaced short palindromic repeats (CRISPRs) to generate genome-corrected iPSCs for regenerative therapies.

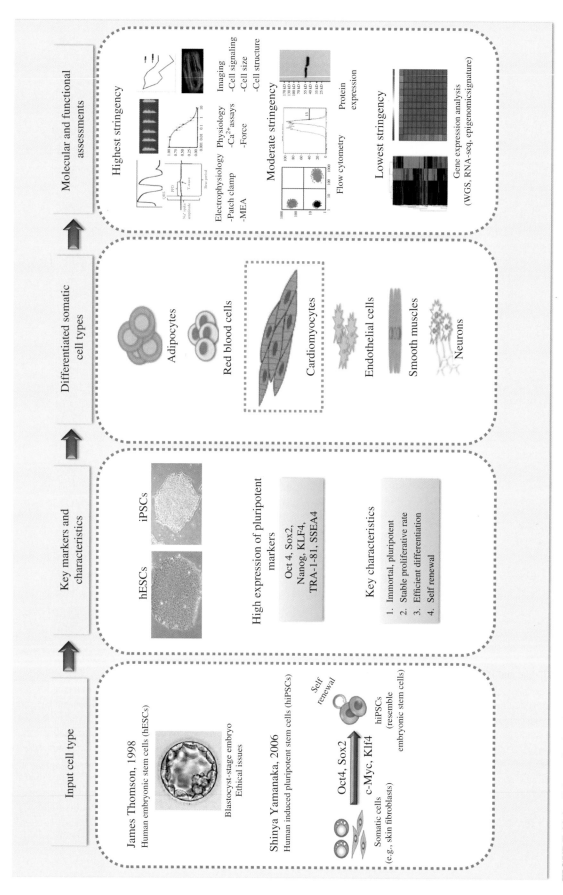

FIGURE 22.2 Schematic showing a battery of *in vitro* assays or descriptors aimed at characterizing the electrophysiology, physiology, cell surface marker expression, and gene expression analysis of the cardiomyocytes obtained from human pluripotent stem cell (hPSCs), which include embryonic stem cells (ESCs) and induced pluripotent stem cells (iPSCs). The assays or descriptors are stratified based on the relative stringency of the data obtained by a particular assay or technique. This strategy can be used to delineate the quantitative and qualitative differences between diseased and control cellular phenotypes, and intra- and interline variability associated with the currently available reprogramming and cardiac differentiations methods. These platforms can be adopted as a standard for quality control in mass production of hPSC-CMs in drug discovery and regenerative therapy.

	1 Step	Multistep

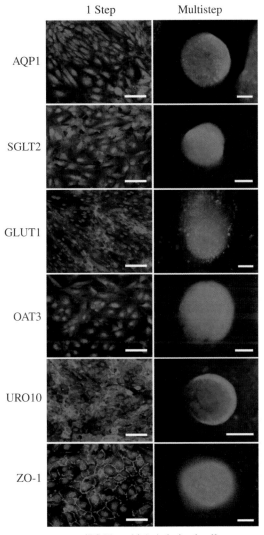

AQP1

SGLT2

GLUT1

OAT3

URO10

ZO-1

iPS(Foreskin)-4-derived cells

FIGURE 23.2 HPTC-like and PTC-like cells generated with the currently most rapid protocols. iPS(Foreskin)-4 hiPSC were differentiated with a 1-step protocol (Kandasamy et al., 2015) or a multistep protocol using CHIR, FGF2, and RA (Lam et al., 2014) (see also Fig. 23.1). Both protocols resulted in the generation of cells expressing PTC markers (listed on the left) that were detected by immunostaining (immunostaining and cell nuclei:gray). HPTC-like cells generated with the 1-step protocol (left) formed a differentiated simple epithelium when cultivated in multiwell plates (similar to HPTC; the similarity of these cells and HPTC was confirmed by various other methods; for details see text). In contrast, PTC-like cells formed round 3D cell aggregates (right). Features of individual HPTC-like cells cultivated in multiwell plates (left) can be easily addressed by imaging and such cells would be suitable for High-content screering HCS. HPTC-like cells (left) were fixed on day 8 and PTC-like cells (right) were fixed on day 11 or 12 after seeding. Scale bars: 100 μm.

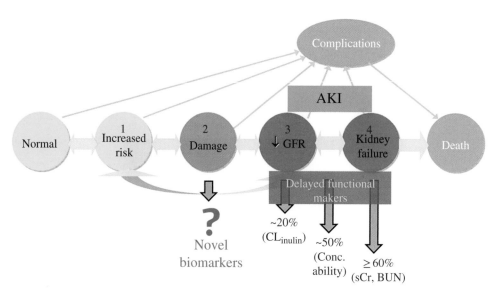

FIGURE 28.1 Insensitive, delayed, and nonspecific traditional kidney injury/functional markers on a conceptual framework of acute kidney injury (AKI), which is divided into several stages from the increased risk to damage, followed by a decrease in glomerular filtration rate (GFR), with progressing to kidney failure and death. Early, sensitive, and specific AKI biomarkers are not available or well qualified for AKI detection and safety assessment. ? indicates there are no biomarkers available for early AKI detection. Adapted from Lameire et al. (2008, pp. 392–402).

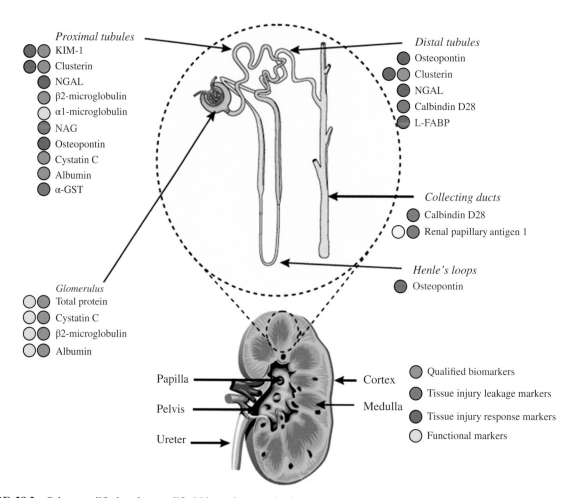

FIGURE 28.2 Select qualified and unqualified biomarkers used when monitoring for drug-induced kidney injury to specific nephron segments in rats. Adapted from Bonventre et al. (2010).

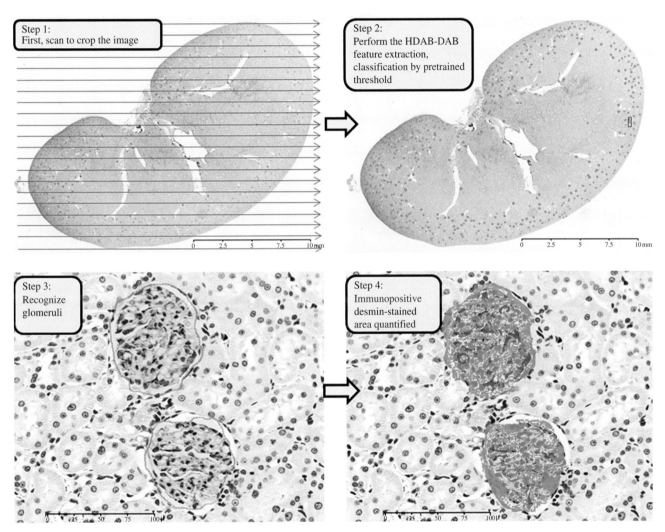

FIGURE 36.1 Fully automated computational image analysis workflow. First, the image of a whole tissue section (e.g., kidney) is scanned and subsequent subimages are processed (cropped). Second, the HDAB-DAB extraction feature combined with a mean and poly-local linear filter supports image classification using the pretrained linear threshold. This tool allows for candidate cropped subimage selection; the images reveal positively stained morphological structures (e.g., glomeruli and nonglomerular tissues—*see inset of recognized glomeruli*). Lastly, the algorithm automatically recognizes and quantifies the immunopositively stained area and intensity (such as the desmin immunostaining shown).

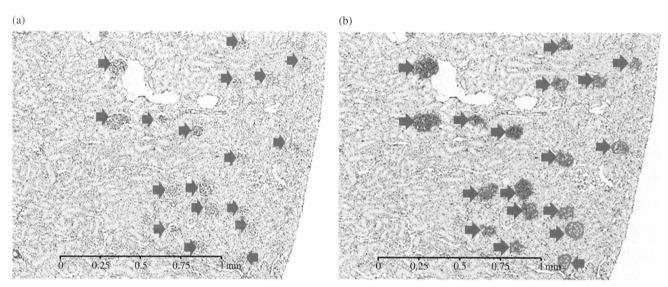

FIGURE 36.2 Desmin immunoreactivity in outer cortical (green arrows) and juxtamedullary (red arrows) glomeruli from a DOX-treated rat (a) and quantification step for assessing areas of enhanced desmin immunereactivity in damaged podocytes of juxtamedullary glomeruli (b). Glomeruli are recognized (green color); desmin-positive-labeled podocytes (red color).

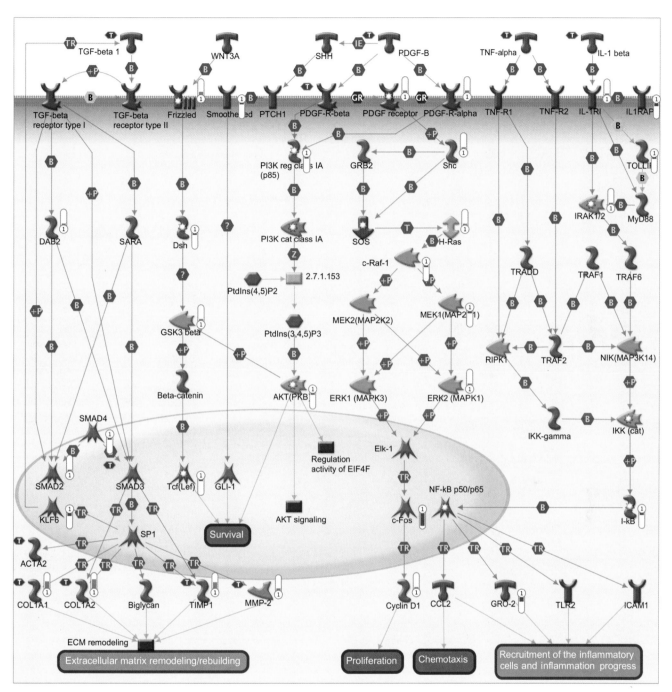

FIGURE 45.1 Canonical pathway map image representing processes involved in stellate cell activation and the development liver fibrosis. Map is from the Systems Toxicology Module of MetaCore (Thomson Reuters Inc.), full legend available at https://ftp.genego.com/files/ MC_legend.pdf. Compounds are represented by hexagons, proteins by solid shapes representing different classes of protein, and enzymatic reactions by gray rectangles. Protein–protein, compound–protein, and compound–reaction interactions are shown as unidirectional arrows and a mechanism of interaction represented by letters in hexagonal boxes over the arrows. Red color of interactions and mechanisms represents a negative effect (inhibition, downregulation) and green a positive effect (activation, upregulation). "Thermometers" represent differential gene expression data for severe alcoholic hepatitis versus normal liver (data taken from Affò et al., 2013). Red color fill represents upregulation of the gene for the adjacent protein, blue—downregulation. Magnitude of the fill is semiquantitative (relative to the highest differential expression level on the map). Purple hexagons containing a white "T" represent map objects that correspond to known biomarkers of drug-induced liver fibrosis in the Systems Toxicology Module of MetaCore database.

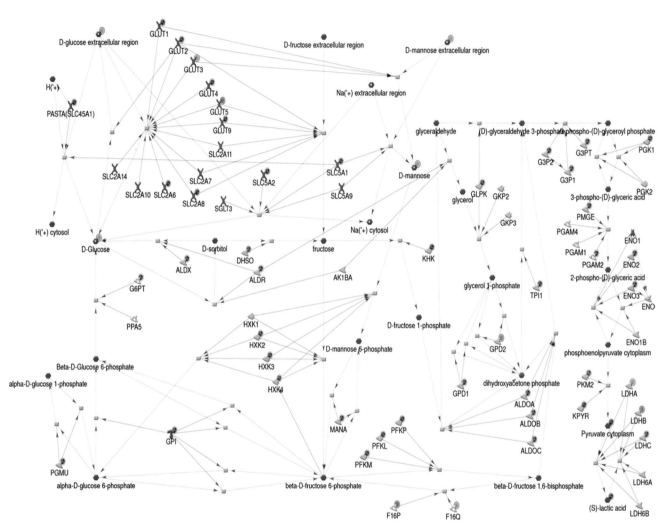

FIGURE 45.2 Dynamic metabolite and transcript changes in glycolysis and glucose transport after treatment of rats with cisplatin or gentamicin. Urine metabolites and kidney transcripts are indicated by a filled circle on the upper right-hand corner of each network object; a red dot in the circle indicates upregulation, whereas a blue dot indicates downregulation from the vehicle control group. Mixed colors indicate differential regulation patterns across dose and time groups. Small-molecule metabolites are represented by purple hexagons, and reactions are represented by gray rectangles. Gene/protein objects are represented by other various shapes and colors depending on their functional annotations. Green arrows represent activation, red arrows represent inhibition, and gray arrows represent other types of unspecified interactions (e.g., molecular transport or complex component binding). Xu et al. (2008). Reproduced with permission of American Chemical Society.

18

ANIMAL MODELS OF DISEASE FOR FUTURE TOXICITY PREDICTIONS

Sherry J. Morgan[1] and Chandikumar S. Elangbam[2]

[1] Preclinical Safety, AbbVie, Inc., North Chicago, IL, USA

[2] Pathophysiology, Safety Assessment, GlaxoSmithKline, Research Triangle Park, NC, USA

18.1 INTRODUCTION

Animal models play a crucial role and are fundamentally required to advance our scientific understanding and development of novel therapies for various human diseases and pathologies. Identification and characterization of novel drug targets also require nonclinical testing in appropriate animal models that are relevant to human pathologies and clinical outcomes. When utilized appropriately, with full awareness of each animal model's limitations, and in combination with other resources (e.g., human cells and tissues, biomarkers), they help to identify altered molecular targets and pathways for better understanding of disease pathogenesis and subsequent development of new therapeutic agents and their potential safety liabilities. Understanding mechanisms of potential safety liabilities/toxicities will help to assess the risk and the relevance of the nonclinical data to humans. However, because of inherent ethnic and genetic variability and possible environmental and epigenetic effects on specific genes and pathologic responses in humans, it is unlikely to ever be replicated in a single inbred animal strain.

Animal models of human disease are commonly utilized in early discovery studies, either in elucidation of the pathogenesis of the disease or the potential for a particular compound to provide a therapeutic benefit. However, animal models of human disease are relatively rarely utilized in nonclinical safety testing; instead, conventional (healthy) rodent and nonrodent models are typically utilized. This chapter will outline some of the available models of human disease by an organ/system approach. While clinical adverse drug reactions may be apparent in either a healthy or diseased population, others may be more frequently observed in a diseased population. From a clinical perspective, adverse reactions affecting the hepatic, cardiovascular (CV), neurological, and gastrointestinal system are the most frequent adverse events resulting in termination of compound development (Olson et al., 2000; Stevens and Baker, 2009). As the models are discussed, it will become apparent that no one animal model currently depicts all facets or the heterogeneity of the human disease. Selection of an inappropriate or poorly understood model of human disease may result in substantial risks to the development of a compound. Alternatively, selection of an appropriate model and careful development of a targeted study plan may result in enhanced understanding of the actual risks to the human population under consideration.

As will be discussed in the following sections, there are various diseased animal models, including spontaneous, chemically/surgically induced, or genetically engineered models for academic and pharmaceutical research applications. However, details of each existing or newer models are outside the scope of this chapter. Key perspectives discussed include strengths and weaknesses of commonly used animal models, potential causes/factors for poor translation of safety liabilities from animals to humans, and their application in the nonclinical mechanistic and safety studies.

Drug Discovery Toxicology: From Target Assessment to Translational Biomarkers, First Edition. Edited by Yvonne Will, J. Eric McDuffie, Andrew J. Olaharski, and Brandon D. Jeffy.
© 2016 John Wiley & Sons, Inc. Published 2016 by John Wiley & Sons, Inc.

18.2 HEPATIC DISEASE MODELS

18.2.1 Hepatic Toxicity: Relevance to Drug Attrition

Along with CV toxicity, hepatic toxicity is a primary reason for drug withdrawals during clinical testing (Stevens and Baker, 2009). In contrast to the CV system, the liver does have a considerable capacity for regeneration and/or repair. However, overwhelming insult and/or a host of factors may influence the ability of the liver to respond to injury and thus have a deleterious effect on survival. As is the case with CV toxicity, drug-induced liver injury (DILI) is can also be of concern, with DILI considered to be the most frequent single cause of safety-related drug marketing withdrawals over the past 50 years (Temple and Himmel, 2002).

The liver plays a central role in biotransformation/detoxification as well as metabolism (amino acids, proteins, carbohydrates, hormones, lipids and lipoproteins, porphyrin, vitamins, bile acids, bilirubin). Thus, an effect on function of the liver may result in important ramifications on other organ systems. Intrinsic hepatic toxicity is considered to be predictable, dose and time dependent, occurring in most if not all subjects, and generally reproducible in animals. As expected, this type of toxicity is relatively easy to predict/ monitor in nonclinical species and/or in humans. In contrast to intrinsic toxicity, idiosyncratic hepatic toxicity is considered unpredictable and dose and time independent, occurring sporadically and often not apparent until after monitoring in a large number of exposed individuals. While relatively rare, idiosyncratic toxicity can have important ramifications for a pharmaceutical compound, ranging from limiting use to withdrawal from the market (Kaplowitz 2001; Li, 2002; Kaplowitz, 2005).

Due to the high rate of attrition attributed to hepatotoxicity and the potential for serious consequences with significant hepatotoxicity, there are three regulatory guidance documents that specifically address nonclinical and/or clinical evaluation of new clinical entities (NCEs) in respect to hepatotoxicity. The EMEA guidance (Non-clinical Guidance on Drug-Induced Hepatotoxicity) (EMEA, 2010) provides recommendations for nonclinical safety evaluation of medicinal products that are systemically absorbed, excluding DNA-reactive (cytotoxic) anticancer products. The FDA guidance (Drug-Induced Liver Injury: Premarketing Clinical Evaluation) (FDA, 2009) provides recommendations for the clinical safety assessment of the potential of a drug to cause severe liver injury (i.e., irreversible liver failure that is fatal or requires liver transplantation). The guidance notes that most drugs that cause severe DILI do so infrequently, and thus careful evaluation of the signals that indicate that a drug has the potential for severe DILI is paramount. A guidance from Canada (Recommendations from the Scientific Advisory Panel Subgroups on Hepatotoxicity: Hepatotoxicity of Health Products) (Health Canada, 2012) provides a detailed classification for hepatotoxicity, considering multiple critical facets (type of injury, predictability/reproducibility in a nonclinical setting, and clinical significance).

18.2.2 Hepatic Toxicity: Reasons for Poor Translation from Animal to Human

There are likely a multitude of reasons for the low predictability of hepatic toxicities. Differences in metabolism (production) or route of elimination (e.g., renal vs. hepatic) of toxic metabolites between nonclinical species and humans may play a major role in the production of hepatic toxicity. Even if the production and elimination of toxic metabolites are similar between species, differences in hepatic transporters between may play a role in DILI from either an uptake or efflux perspective. From the standpoint of efflux, various xenobiotics may result in transporters such as BSEP or MDR3 function, leading to cholestasis with hepatocellular accumulation of bile salts. With concomitant medication administration, the effect on transporters may be of even more importance as a xenobiotic may result in increased transporter-mediated uptake of a hepatotoxic constituent, leading to enhanced toxicity. Alternatively, inhibition of hepatocellular uptake may constitute an approach for treatment of poisoning with some hepatotoxins (e.g., phalloidin, microcystin-LR). Obviously, functional differences between a normal and abnormal liver may play a major role, with preexisting liver disease in a clinical setting predisposing a liver to injury from the standpoint of decreased functional mass (either hepatocellular or biliary) or blood supply. Finally, there is a relatively high incidence of idiosyncratic (immune-mediated) hepatotoxicity in a clinical setting (Thiele, 2007), which, by nature, is not readily predictable.

While mild elevations of serum transaminases may be a signal for severe DILI, they may also be generated by compounds with low potential for causing injury or an impact on hepatic function. Although central goals of nonclinical and clinical investigations are to avoid toxicities, including hepatotoxicity, efforts should be made to understand mechanisms and thus place the risk of an apparent/potential toxicity into perspective prior to making a decision on whether or not to continue nonclinical or clinical investigations of an NCE in order to prevent premature discontinuation of a potentially beneficial NCE.

18.2.3 Available Hepatic Models to Predict Hepatic Toxicity or Understand Molecular Mechanisms of Toxicity: Advantages and Limitations

There are multitudes of naturally occurring diseases of the liver, including infectious, inflammatory, neoplastic, and metabolic disorders. Naturally occurring diseases for which there are potential models include models for liver cancer, fatty liver disease, fibrosis, bile duct loss, viral-induced inflammation (hepatitis B/hepatitis C), and idiosyncratic hepatitis. Examples of selected animal models of human disease will be briefly summarized.

18.2.3.1 Liver Cancer Models Hepatocellular carcinoma (HCC) is the most common form of primary liver cancer and the third leading cause of cancer-related death. By the time HCC is diagnosed, therapeutic options are limited and survival after diagnosis is poor. Therefore, improved abilities to prevent, diagnose, and treat HCC are needed.

There are a wide variety of methodologies that have been employed in developing rodent models of HCC. The genotoxic drug diethylnitrosamine (DEN) has been widely utilized to induce liver cancer in mice. DEN undergoes metabolic activation by P450 enzymes and, if administered to mice <2 weeks of age, acts as a complete carcinogen. If administered later, tumor promotion is required by either coadministration of xenobiotics (phenobarbital, carbon tetrachloride) or dietary manipulation (high-fat diet). Activation of the specific pathways in this model exhibits similarities with the human counterpart, supporting the usefulness of this model (Lee et al., 2004; Stahl et al., 2005; Calvisi et al., 2006; Bakiri and Wagner, 2013). A common predisposing factor for human HCC is infection with hepatitis B virus (HBV) or hepatitis C virus (HCV). While mice are resistant to infection with both of these viruses, transgenic mice have been developed that have viral DNA sequences integrated into the host genome. The use of these mice has enhanced understanding of the initiation and promotion of HCC by these viruses (Bakiri and Wagner, 2013). While beyond the scope of this overview, a host of other genetic manipulations have been undertaken in mouse models to enable a better understanding of the role of oncogenes as well as functional aspects (Mdr2 gene, Aox gene) in the development of HCC (Bakiri and Wagner, 2013). Xenograft models in which human tumors are developed in immunocompromised mice have been core to the identification of therapeutic regimes for treatment of various human cancers for many years. However, there is poor correlation between new chemical entities that appear to be effective in a discovery setting compared to those that will be successful clinically. The reasons for this lack of correlation are complex and poorly understood, but lack of replication of tumor–stroma interactions may play a role because human cells may fail to respond to signals from the mouse stroma (Bakiri and Wagner, 2013). Some models have been developed to assist in optimizing xenograft models, including generation of "humanized" recipient mouse strains expressing human ligands, such as HGF (Zhang et al., 2005; Bakiri and Wagner, 2013) and mouse cell lines that express human version of P450 metabolizing enzymes (Cheung and Gonzalez, 2008; Bakiri and Wagner, 2013).

It should also be kept in mind that in many instances (as with viral hepatitis and nonalcoholic steatohepatitis (NASH)), inflammation and/or fibrosis is a predisposing factor and it is feasible that incorporating such facets into xenograft models may result in a closer parallel to human HCC (Bakiri and Wagner, 2013).

18.2.3.2 Fatty Liver Disease Nonalcoholic fatty liver disease (NAFLD) is rapidly becoming the most common liver disease worldwide, with a prevalence of ~20–30% in the general population of Western countries. NAFLD is characterized by macrovesicular steatosis or fatty liver. While NAFLD is primarily limited to obese individuals (present in 80–90% of obese individuals), it has also been associated with some surgical procedures, metabolic conditions (present in 30–50% of diabetics and up to 90% of patients with hyperlipidemia), and medications. Some patients with NAFLD develop a more serious syndrome termed NASH. It is estimated that NASH affects ~2–3% of the general population of Western countries. Features of NASH include hepatocellular necrosis or ballooning change of the cytoplasm (with or without Mallory's hyaline droplets), inflammation, megamitochondria, iron deposition, as well as fibrosis in some patients. Ultimately, NASH progress to cirrhosis and HCC (Cave et al., 2007; Bellentani et al., 2010; Kopec and Burns, 2011). Currently, there are no reliable noninvasive biomarkers for NAFLD or NASH, and diagnosis requires a liver biopsy. Also, there are few effective treatment regimens other than controlled weight loss. Thus, the spectrum of NAFLD/NASH provides an opportunity for a model of human disease in both elucidation of effective diagnostic aids and therapeutic regimens.

Rodent models of fatty liver disease employ dietary and/ or genetic manipulation. Efforts are underway to elucidate the ideal combination of diet, exercise, and genetic background that reliably results in a syndrome that adequately represents the human counterpart of NAFLD/ NASH. While either high-fat or high-carbohydrate diets tend to elicit fatty change that may be analogous to NAFLD, a diet that is high in both fat and carbohydrates is typically required to result in a microscopic change of steatohepatitis that may be analogous to NASH. Hepatic microscopic changes tend to vary with specific dietary manipulations, with more severe changes being induced with diets that are high in unsaturated fats, trans fats, and fructose. Diets deficient in specific macromolecules (choline deficient or choline–methionine deficient) may result in steatohepatitis that is not associated with a metabolic syndrome (high-calorie diets). While the lack of association with a metabolic syndrome precludes them from being considered as an ideal model of human NAFLD or NASH from a response to therapy perspective, the rapid onset of steatohepatitis that may result from manipulation of choline/choline–methionine levels has resulted in a model that may be of considerable benefit in understanding certain aspects of the pathogenesis of the human disease. While wild-type rodents may develop fatty liver disease if overfed (intragastric infusion of a liquid diet that is analogous to a conventional chow diet) or with dietary manipulation as outlined above, some genetically modified rodents may develop fatty liver when fed a high amount of a conventional chow diet. Specific models include the leptin-deficient (*ob/ob*) mouse, cholecystokinin-1 (CCK1) receptor-deficient (Otsuka Long-Evans Tokushima Fatty (OLETF)) rats, and Alms1-deficient (*foz/foz*) mice. Leptin deficiency in *ob/ob* mice results in broad phenotypic abnormalities that may limit their usefulness as a human counterpart for NAFLD/NASH. However, the metabolic changes in the OLETF rat are entirely attributable to excessive excess food intake as they have a defect in the satiety signal mediated by CCK1 along with a lack of CCK1-mediated inhibition of the brain orexigenic neuropeptide Y. The *foz/foz*-deficient mouse develops hyperphagia at approximately 2 months of age. Feeding a conventional chow diet results in a limited steatosis, whereas feeding a high-fat diet results in a more severe liver disease than would be noted in wild-type mice (Maher, 2011).

18.2.3.3 Liver Fibrosis

Liver fibrosis is a common end result of inflammation and/or necrosis. While the liver does have considerable regenerative capabilities, cytokine release associated with the inflammatory/necrotic process can lead to fibrosis that can have a deleterious effect on hepatic function, not only from the aspect of decreased hepatic functional mass but also from the standpoint of compromising blood supply. Animal models of hepatic fibrosis would be valuable from the standpoint of facilitating the development of noninvasive biomarkers as well as development of interventional agents.

Hepatic fibrosis has been induced in nonclinical models by chemical methods including carbon tetrachloride (CCl_4), thioacetamide (TAA), and dimethylnitrosamine (DMN) or DEN. The fibrosis induced by repeated dose of CCl_4 is typically reversible after discontinuation of the compound, whereas the fibrosis induced by TAA and DMN/DEN typically is not as readily reversible and, with DMN/DEN, may continue to progress after compound administration has been discontinued. Not surprisingly, these compounds with continued injury and fibrosis may ultimately develop hepatic neoplasms (HCC and/or cholangiocarcinoma). Thus, these models may play a dual role (investigations of fibrosis and hepatic neoplasia).

Surgical models also exit for evaluation of hepatic fibrosis. The most common method employed is ligation of the common bile duct in rats or mice. The subsequent response involves not only fibrosis but also proliferation of biliary epithelial cells, oval cells, portal inflammation, and, of course, cholestasis. With time, there is progression to biliary cirrhosis and hepatic failure. Relatively high mortality rates due to rupture of the gallbladder (mice) or biliary cyst (rat) and subsequent leakage of bile limit the usefulness of this model (Liu et al., 2013).

18.2.3.4 Primary Biliary Cirrhosis/Bile Duct Loss

Primary biliary cirrhosis (PBC) is most common in middle-aged women and is of unknown etiology (Perez-Osterreicher et al., 2011) and is characterized by the destruction of intrahepatic biliary epithelial cells (primarily interlobular bile ducts) and a progressive portal lymphoplasmacytic inflammatory response with peribiliary and periportal fibrosis in the later stages (Liu et al., 2013). While the specific etiology is unknown, it is considered to be an autoimmune disorder of both a humoral and cellular immunological basis (Liu et al., 2013). Animal models for PBC include both spontaneous mouse models and xenobiotic models. Spontaneous mouse models include NOD.c3c4 mice (generated by introgression of B6- and B10-derived insulin-dependent diabetes regions into nonobese diabetic (NOD) mice), which are subsequently protected from diabetes. These mice develop a spontaneous autoimmune biliary change with features similar to the human disorder with the exception of biliary cysts in the mouse but not the human. Another spontaneous model is the dnTGF-βRII knockout mouse, which is of high value, not only because of the microscopic similarities to the human disease but also because of the low interanimal variability, which is critical in understanding the potential response to a therapeutic agent (Liu et al., 2013). A third spontaneous model is the Mdr2 knockout mouse, which develops progressive loss of small bile ducts, presumably due to inability to secrete phospholipid into the bile (Smit et al., 1993; Leveille-Webster, 1994). Xenobiotic-induced models of PBC include dietary administration of 3,5-diethoxycarbonyl-1, 4-dihydrocollidine (DDC) in mice. Dietary administration of DDC results in cholangitis with periductal fibrosis (Fickert et al., 2007).

18.2.3.5 Viral-Induced Inflammation

It is estimated that more than 500 million individuals worldwide are infected with HBV or HCV. Many infections are undetected since the initial signs may be very mild/nonspecific. While patients may spontaneously clear the virus, others develop a chronic infection that may progress to cirrhosis, liver failure, or HCC (Kuehn, 2009).

Investigation of the pathogenesis of HBV and HCV has been complicated because the only recognized model of HCV infection is the chimpanzee. This species only rarely develops viral-associated chronic liver disease. Because of limitations of this model (not ideal model, low availability), other models have been developed to provide insight into the pathogenesis of these viral infections. For HBV, naturally occurring animal diseases in other nonclinical species exist, which bear similarities to human HBV infection (Eastern American woodchuck and woodchuck HBV, Peking duck and duck HBV). For HCV, numerous transgenic mouse models that overexpress one or more HCV proteins have been generated to study the effects of specific viral proteins on hepatic pathology. Investigative studies with these models has revealed important insights into the pathogenesis of the hepatic toxicity associated with the virus, including a link between the HCV core protein expression and hepatic steatosis and HCC (Liu et al., 2013).

The development of effective models of animal disease of HCV to evaluate potential therapeutic agents is very complex due to the extremely narrow species trophism. While the reasons behind the species trophism are not completely understood, nearly all the HBV- or HCV-permissive rodent models require xenografting of human hepatocytes and a permanent lack of rejection of these engrafted cells. A detailed discussion of the various types of rodent models utilized in understanding the pathogenesis of HBV/HCV and potential for therapy/vaccine development is beyond the scope of this chapter and can be found in recent reviews (Liu et al., 2013).

18.2.3.6 Idiosyncratic Toxicity While idiosyncratic drug-induced liver injury (IDILI) toxicity is a relatively rare occurrence, it may result in significant consequences for both the individuals involved and the future marketability of the compound that associated with the toxicity. Since idiosyncratic toxicities in general are typically rare events in a clinical setting, they are not readily predicable in a nonclinical setting utilizing conventional animal models. Oftentimes, idiosyncratic reactions are not evident until large numbers of patients are exposed to the drug (e.g., postmarketing). Idiosyncratic adverse drug reactions that target the liver are a major cause for drugs being withdrawn from the market or having restrictions placed on use (Kaplowitz, 2001; Kaplowitz, 2005).

Identification of suitable models of idiosyncratic drug reactions would be a valuable resource in guiding development of compounds and/or identification of patients who may be at risk for development of an idiosyncratic response. Characteristics of an ideal animal model of IDILI include the production of liver injury in a large fraction of the nonclinical species and sufficient similarities to the human condition (clinical and anatomic pathology, mode of toxicity, temporal features, and risk factors). Additionally, the test system should be capable of distinguishing between drugs that cause human IDILI and those that do not (e.g., positive and negative predictive capabilities) and optimally should be a relatively inexpensive species/strain (Roth and Ganey, 2011).

While the mechanistic basis of idiosyncratic reactions is not well understood, attempts to develop animal models of IDILI have focused on several potential hypotheses for mechanisms that are outlined in Table 18.1. Possible models

TABLE 18.1   Possible Mechanisms of Idiosyncratic Drug-Induced Liver Injury and Models

Mechanism/ Hypothesis	Comment	Possible Model(s)
Drug disposition polymorphism	Related to elimination, drug-metabolizing enzymes, transporters	Transgenic mice (knockout/knock-in for drug-metabolizing enzymes/transporters Chimeras with transplanted human hepatocytes
Adaptive immunity (hapten)	Example: severe hepatotoxicity in humans with halothane	Currently no suitable animal models available
Mitochondrial dysfunction	May consider *in vitro* studies	Few *in vivo* models: (i) fialuridine in woodchucks with woodchucks, (ii) panadiplon in rabbits, and (iii) valproic acid in jvs+/− mice, troglitazone, nimesulide, and flutamide in superoxide dismutase 2(SOD2)-deficient mice
Inflammatory stress	Patients with underlying inflammatory conditions (arthritis, atherosclerosis, asthma, etc.) may be more susceptible to inflammatory stress	Several drugs that cause human IDILI also cause liver injury upon cotreatment with small nontoxic dose of LPS → hepatocellular necrosis, neutrophil infiltration Of interest to note that drugs most often associated with IDILI are those used in inflammatory conditions— antibiotics and anti-inflammatory agents
	Endotoxin (LPS) and other bacterial/viral products may precipitate inflammatory stress	Very complicated model—ability to produce characteristic lesion is very "compound specific" as it pertains to timing of LPS and other factors
Failure to adapt	Mild injury in majority of patients may progress to more fulminant effect in some	Currently no suitable animal models available
Multiple determinants	Example: inflammatory stress+drug disposition polymorphism	Female Balb/cJ mice fasted overnight+halothane → severe hepatic lesion; male Balb/cJ mice fasted overnight+LPS+halothane → similar change

that may be helpful in the elucidation of the pathogenesis of the hypothesis are included (Roth and Ganey, 2011). While progress has been made in the elucidation of the pathogenesis and potential for prediction of IDILI, there is still a considerable unmet need for animal models and IDILI.

18.3 CARDIOVASCULAR DISEASE MODELS

18.3.1 Cardiac Toxicity: Relevance to Drug Attrition

Cardiac toxicity is a major cause of drug attrition during the clinical development and postapproval drug withdrawal of new drugs (Laverty et al., 2011). The US Department of Health and Human Services estimates that nearly 1 million patients experience adverse drug reactions each year (Classen et al., 1997). The major causes of drug attrition are lack of efficacy (accounting for ~30% of failures) and safety (toxicology and clinical safety accounting for ~30% of failures) (Ferri et al., 2013). According to the analysis conducted by Stevens and Baker (2009), safety liabilities related to the CV system account for 45% of the total postapproval drug withdrawal from the market. A higher incidence of potentially severe CV adverse drug reactions is detected in the later stages of clinical trials (phase II and onward) where drug candidates are tested over longer periods of time in larger patient populations, which is unacceptable due to the risks for patients (Laverty et al., 2011). In contrast, cardiac toxicity in phase I clinical trials accounts for only 9% of total drug attrition where the drug candidates are tested in fewer patients and for shorter duration (Sibille et al., 1998), indicating there is a need for more accurately predicting the risk of drug-induced cardiac liabilities in nonclinical and early clinical stages. The main reasons for the high rates of CV adverse drug reactions in patients are due to the limited capacity of nonclinical screening assays to detect cardiac toxicity and probably due in part to the strict application of the minimal requirements defined in the ICH guidelines (International Safety Pharmacology Guidelines) (ICH-S7A, 2000; ICH-S7B, 2005). Over the last decade, CV risk assessments and strategies have tended to focus on identifying compounds with the ability to alter cardiac electrophysiology (e.g., QT interval prolongation, arrhythmias). Mounting evidence suggests that nonclinical assays are effective in identifying drug candidates with a risk for QT interval prolongation in humans (Wallis, 2010). However, CV-related attrition is not limited to altered electrophysiological parameters. The safety community as a whole should recognize that CV-related drug attrition can include deleterious effects other than hERG blockade (Peters et al., 2014) and can arise as a result of multiple mechanisms and risk factors contributing to adverse outcomes, such as congestive heart failure (HF), stroke, and myocardial necrosis/ischemia. In the case of necrosis/ischemia, results from nonclinical

models have not translated into the clinic. Nonclinical studies failed to predict the elevated CV risks (e.g., thrombotic events, HF, and hypertension) of rofecoxib, torcetrapib, trastuzumab, and antineoplastic tyrosine kinase inhibitors in patients (Jnni et al., 2004; Burnier, 2005; Kerkela et al., 2006; Chu et al., 2007; Chien and Rugo, 2010). Consequently, both the FDA and pharmaceutical industry have mandated early nonclinical drug screening guidelines to detect potential cardiac toxicity of new drug candidates to minimize cardiac risks in humans (ICH-S7A, 2000; Fermini and Fossa, 2003). Despite such guidelines, drug-induced cardiac toxicity has resulted in numerous preventable patient deaths and the costly withdrawal of pharmacological products from the market (Roden, 2001; De Bruin et al., 2005; van Noord et al., 2011). Early detection and characterization of cardiac toxicity risks (in the nonclinical stage of drug development) are clearly the ideal strategy of new drug candidates prior to the clinical trials. Addressing the safety issues early will help sponsors select new drug candidates with the highest probability of success and avoid failure prior to investment in clinical studies and putting patients at risk.

18.3.2 Cardiac Toxicity: Reasons for Poor Translation from Animal to Human

The various nonclinical animal models to assess CV drug safety and efficacy do not always predict clinical outcomes in humans. Safety-related attrition of candidate drugs during the discovery, nonclinical, and clinical development phases is one of the major contributors to the current poor research and development productivity of the pharmaceutical industry and rising cost of developing new drugs today. This is due, at least in part, to inadequacies of appropriate animal models for nonclinical studies. The CV system of each species used in nonclinical research has evolved differently in order to meet the unique demands of a given species, and to date, there is no perfect animal model of the human CV function and structure. Although the basic principles of cardiac excitation and contraction are relatively conserved, there are significant differences between the human and animal CV system that must be taken into consideration, particularly in the case of small rodents. For example, rodent hearts have adapted to function at relatively high heart rates (up to 800 beats per min) compared to humans (72 beats per min) (Ostergaard et al., 2010). Hearts of smaller animal species need to contract and relax more rapidly than larger species in order to maintain cardiac output at these higher heart rates. Furthermore, a mouse can increase its cardiac output by typically no more than one-third, whereas humans can increase cardiac output as high as 10-fold in highly trained individuals. Humans have a greater ability than rodents to increase their cardiac output during exercise by changing both heart rate and stroke volume (i.e., the cardiac reserve of humans is much greater than rodents). Similarly, the resting heart rate of a rabbit is considerably higher than humans. Although the heart rate reserve in rabbits is greater than the heart rate reserve of mice (~30–40%) and rats (~40–50%), it is still less than the reserve (~140–170%) in humans (Bolter and Atkinson, 1988; Flamm et al., 1990; Stratton et al., 1994; Desai et al., 1997; Fewell et al., 1997; Lujan et al., 2012). These differences can affect outcomes of studies and experiments that involve physical exercise/training, and therefore, caution should be observed when relating and translating studies and mechanisms of disease or drug-induced toxicity from small rodents or rabbits to humans. Large animal models, such as canines, are better suited for addressing these issues since the resting heart rate is closer to that of the human.

The use of animal hearts as surrogate models for the human heart is also limited because of different electrophysiological properties in murine, canine, and rabbit cardiomyocytes and fails to respond to certain drugs in a similar manner to humans (Zicha et al., 2003; Akar et al., 2004; Davis et al., 2011). The ventricular action potential duration is much shorter in small rodents than that of humans (Nerbonne, 2004; Glukhov et al., 2010a, 2010b; Fedorov et al., 2011). In contrast to human cardiomyocytes, mouse ventricular action potentials have rapid repolarization and lack of prominent plateau phase (Nerbonne, 2004; Salama and London, 2007). There are also differences between small rodents and humans at the cardiomyocyte myofilaments: (i) myosin heavy chain (MHC) (small rodent ventricular cardiomyocytes predominately express fast α-MHC (N94–100%) compared to slow β-MHC (N90–95%) in humans) (Hamilton and Ianuzzo, 1991; Miyata et al., 2000; Reiser et al., 2001; Alpert et al., 2002; Wang et al., 2002; Krenz et al., 2003; Lemon et al., 2011) and (ii) ventricular titin (sarcomeric protein) isoforms (human ventricles express both compliant N2BA and stiff N2B titin isoforms with a relative N2BA : N2B ratio of ~0.4–1.2, whereas mouse and rat predominately express N2B titin and considerably lower N2BA titin isoform with N2BA : N2B ratio ranges of ~0.05–0.25 in mouse and ~0.06–0.1 in rat) (Cazorla et al., 2000; Neagoe et al., 2003; Makarenko et al., 2004; Nagueh et al., 2004; Opitz and Linke, 2005). There is clearly an inverse relationship between body weight and heart rate across animal species because of the size-dependent differences in metabolic demand (Speakman, 2005; Ostergaard et al., 2010). In small animals, higher cardiac contractile activity and greater oxygen consumption in small animals lead to increased free radical generation, reduced contraction–relaxation times, and differences in calcium handling and myofilament protein isoforms (Bassani et al., 1994; Speakman, 2005). For example, in rodents, 92% of cytosolic calcium is sequestrated by the sarco-/endoplasmic reticulum calcium ATPase (SERCA), and the remaining 8% is extruded by the sarcolemmal sodium/calcium exchange (NCX) (Bassani et al.,

1994). In contrast, SERCA and NCX in humans account for 76 and 24% of calcium normalization, respectively (Piacentino et al., 2003). Similar differences between rabbit and human myocardium also exist, including atrial transient outward (Ito) current, ventricular expression levels of some potassium ion channel subunits, relative expression levels of cardiac titin N2B and N2BA isoforms, and left ventricular wall motion (Fermini et al., 1992; Wang et al., 1999; Cazorla et al., 2000; Neagoe et al., 2003; Zicha et al., 2003; Makarenko et al., 2004; Nagueh et al., 2004; Jung et al., 2012). The myocardial contractile kinetics of rodents and rabbits are faster than that of humans, and therefore, depending on the particular CV process being studied, caution needs to be taken when trying to extrapolate studies from rodents or rabbits to humans. Despite the high degree of similarity among human, canine, and porcine myocardium, some differences have been reported among these species as well (e.g., kinetics of Ito current inactivation and recovery and/or Purkinje fiber distribution) (Lacroix et al., 2002; Li et al., 2003; Akar et al., 2004; Schultz et al., 2007). It is believed that a major factor contributing to the limited success of predicting clinical outcome using nonclinical models is due to limited understanding of the translatability across model systems and species. Furthermore, the use of large animals (primarily pigs and dogs and more exceptionally sheep and nonhuman primates) is limited by multiple factors but not limited to their relatively high cost compared to rodents, the increased amount of test article required for dosing, increased and complex maintenance requirements, necessity for skilled personnel, and increased social ethical concerns.

18.3.3 Available CV Models to Predict Cardiac Toxicity or Understand Molecular Mechanisms of Toxicity: Advantages and Limitations

Several diseased animal models mimic human CV diseases such as hypertension, HF, atherosclerosis, and stroke (Guihaire et al., 2013). These models have been used historically, and many of them continue to be used, while recent technologies now facilitate development of many newer and refined animal models, including surgical (e.g., aortic or pulmonary artery banding, pulmonary shunts for pulmonary hypertension, cardiac hypertrophy, HF), drug-induced (e.g., monocrotaline, angioproliferative agents for pulmonary hypertension), chronic hypoxia (for pulmonary hypertension), and genetically engineered models by altering the genetic composition of an animal by mutating, deleting, or overexpressing a targeted gene. Examples of commonly utilized animal models of CV disease in the early drug discovery include apolipoprotein E-deficient and low-density lipoprotein receptor-deficient mice for atherosclerosis (Zhang et al., 1992; Sanan et al., 1998), Dahl salt-sensitive (Dahl/SS) and spontaneously hypertensive rats (SHRs) for hypertension (Okamoto and Aoki, 1963; Okamoto, 1969;

Yamori et al., 1972; Rapp and Dene, 1985; Doi et al., 2000; Qu et al., 2000; Horgan et al., 2014), *Scn5a* mouse models for cardiac ion channelopathies and arrhythmias (Charpentier et al., 2004, 2008; Nerbonne, 2004; McCauley and Wehrens, 2009; Huang et al., 2012), Watanabe heritable hyperlipidemic and St. Thomas' Hospital rabbits for hyperlipidemia (Watanabe, 1980; Nordestgaard and Lewis, 1991; Beaty et al., 1992), and bone morphogenetic peptide receptor type 2 knockout model and Fawn-hooded rats for pulmonary hypertension and right HF (West et al., 2004; Bonnet et al., 2006; Guihaire et al., 2013; Kapourchali et al., 2014). Most of these models have at least some, often severe, limitations and deviations from the human disease. For example, the Dahl/SS rat progresses rapidly to cardiac hypertrophy and is therefore unsuitable as a chronic disease model (Horgan et al., 2014). Likewise, 57% of male SHRs at the age of 18–24 months have been shown to progress from hypertension-induced compensated hypertrophy to decompensated HF, with 13% survival without HF, and 30% died or killed due to noncardiac reasons (e.g., stroke, tumor, or debilitation). Another limitation of the SHR model is diastolic dysfunction, with the onset varying from 16 to 32 weeks (Bing et al., 1995; Sun et al., 2008). In the investigation of histopathologic effects in a SHR model, use of a heterogeneous population of animals may result in exclusion of potentially beneficial compounds as it may appear that the compound exhibits no benefit or, alternatively, exacerbates hypertension-associated histopathologic changes. In the latter case, the apparent histopathologic changes may simply be secondary to preexisting heterogeneity in manifestation of the disease. The inability to develop a model with sufficient homogeneity with respect to the disease may hinder the use of the model for both drug efficacy and toxicity studies. In conclusion, an improved understanding of the mechanisms of toxicity can be ascertained only through the use of appropriate animal models with full understanding of model's limitations and confounding factors, including survival, and existing pathologies that obscures the ability to reliably detect drug-induced alterations.

18.4 NERVOUS SYSTEM DISEASE MODELS

18.4.1 Nervous System Toxicity: Relevance to Drug Attrition

It is estimated that neurological effects account for 22% of adverse drug reactions in a clinical setting and are among the top three (along with hepatic and CV) reasons for termination of drug development during clinical testing (Olson et al., 2000; Stevens and Baker, 2009). Animal models of human neurological diseases are commonly utilized in discovery evaluation in order to evaluate response to potential therapeutic agents and/or to elucidate pathogenesis. Some of the major nervous system models of human disease will be briefly outlined in order to provide a background on some of the available models.

18.4.2 Nervous System Toxicity: Reasons for Poor Translation from Animal to Human

Some neurological effects are not likely to be detected in nonclinical species irrespective of whether one considers conventional animal models (healthy rodent/nonrodent) or models of human disease. There are inherent and obvious reasons for difficulties in translation of adverse clinical signs from animals to humans. Many of the adverse events encountered (amnesia, mild anxiety, emotional lability, dizziness or paresthesia without ambulatory disturbances, headache, risk-taking behavior, etc.) in humans would not be readily detected in nonclinical species. Also, some patient populations may be at an enhanced risk for neurological effects due to preexisting disease (compromise of blood–brain barrier due to injury, neoplasia). For some clinical disease indications, it may be helpful to consider evaluation of a compound in an animal model of neurological disease, such as with compounds intended to treat lysosomal storage diseases involving the central nervous system.

18.4.3 Available Nervous System Models to Predict Nervous System Toxicity or Understand Molecular Mechanisms of Toxicity: Advantages and Limitations

There are multitudes of naturally occurring diseases of the central nervous system, including degenerative, traumatic, neoplastic, and metabolic disorders. Naturally occurring diseases for which there are potential models include models for include Parkinson's disease (PD), Huntington's disease (HD), Alzheimer's disease (AD), cerebral ischemia, miscellaneous neoplastic conditions, and lysosomal storage diseases. Examples of selected animal models of human disease will be briefly summarized.

18.4.3.1 Parkinson's Disease Models Parkinson's disease (PD) is a progressive neurodegenerative disease that results in deficits in motor function related to loss of dopaminergic neurons in the substantia nigra with resultant decrease in dopamine input to the basal ganglia. Microscopic characteristics of PD include the development of cytoplasmic Lewy bodies, abnormal aggregates of α-synuclein, and other proteins (Shah et al., 2010). Patients exhibit tremors and rigidity. Critical aspects of an animal model of PD include normal levels of functioning dopaminergic neurons in the substantia nigra at birth with substantial selective loss with age, easily detectable and measureable motor deficits, development of Lewy bodies, and a short disease course (Beal, 2001). Some of the more common models of PD are outlined in Table 18.2. Models of PD are currently utilized to

TABLE 18.2 Animal Models of Parkinson's Disease

Model	Selective Loss of Dopaminergic Neurons	Detectable, Measureable Motor Deficits	Lewy Bodies	Short Course	Comments
Reserpine model	No morphologic changes	Yes	No	Yes	Helpful in evaluation of pathophysiology Not always predictive of positive response in humans
Methamphetamine model	No morphologic changes	No	No	Yes	Similar to above
6-Hydroxydopamine	Yes—very acute	Yes	No	Yes	Typically unilateral injection—contralateral side of brain as negative control
1-Methyl 4-phenyl-1,2,3, 6-tetrahydropyridine (MPTP)	Yes	Yes	No	Yes	One of best models to date Metabolite (MPP⁺) selectively taken up by dopaminergic neurons
Rotenone	Yes	Yes	Yes	Yes (also progressive)	Advantage—progressive rather than acute Disadvantage—inconsistent

elucidate aspect of the pathogenesis and evaluate potential therapeutic agents for therapy. While such models are not likely to be utilized in drug safety evaluation from a full development standpoint, it may be helpful to evaluate selected tissues from these models as an add-on in early discovery studies.

18.4.3.2 Huntington's Disease

Huntington's disease (HD) is an autosomal dominant neurologic disease characterized by neuronal death in selected portions of the brain (especially the striatum, with major effects on the caudate nucleus and putamen). Patients exhibit chorea (abnormal, involuntary muscle movements), loss of motor incoordination, bradykinesia, and dystonia. Transgenic mouse models have been utilized for investigation of the pathogenesis/pathophysiology of HD (Shah et al., 2010).

18.4.3.3 Alzheimer's Disease

Alzheimer's disease (AD) is a neurodegenerative disease that is a common cause of dementia in the adult population. Microscopic changes are primarily found in regions of the brain involving memory and cognition, including the hippocampus, amygdala, neocortex, and basal forebrain cholinergic system (Morrison and Hof, 1997). Characteristic changes include tau-containing neurofibrillary tangles (NFT) and β-amyloid plaques (Wong et al., 2002). The plaques are composed of a polypeptide derivative of amyloid precursor protein (APP) (Glenner and Wong, 1984).

Available models include either nontransgenic (spontaneous) or transgenic models. The nontransgenic models include primate models (deposit amyloid beta (Aβ) in the brain parenchyma and cerebral vascular walls), canine models (deposit Aβ without NFT), and other models including rabbits and rodents (Jay et al., 2011). Because of the wide variation in the extent of disease in these nontransgenic/spontaneous models and the difficulty in obtaining sufficient animals for investigation, it is unlikely that they will play a major role in drug safety evaluation. In contrast, transgenic models, especially mouse models that focus on the overexpression of mutant gene products associated familial AD (FAD) (early onset) have been useful in the identification of potential susceptibility genes, disease modifiers, and therapeutic agents (Rockenstein et al., 1995; Gotz et al., 2004). While there are numerous transgenic models of AD, no one model is an ideal model, with all of the salient features of AD. The two most commonly employed transgenic mouse models are the APP and the tau mouse model. The APP transgenic mouse models expresses mutations similar to those found in the APP gene of human FAD patients with plaque formation and age-associated memory loss (Shah et al., 2010). Expression of these transgenes results in Aβ deposits in the neural parenchyma, cerebral vascular walls, or both. However, these models do not develop NFT and typically do not exhibit neuronal loss

(Jay et al., 2011). The tau transgenic mouse models overexpress the human tau isoform and also exhibit AD-like symptoms (Shah et al., 2010). In these models, human tau protein is expressed and hyperphosphorylated, but the mice do not develop Aβ deposits.

One well-characterized transgenic mouse model of AD is the PDAPP (platelet-derived growth factor promoter expressing APP) transgenic mouse. This model overexpresses human APP and, beginning at 7–8 months of age, depots Aβ in the hippocampus and, to a lesser extent, in the cerebral parenchyma and vasculature. While this model features some of the critical aspects of AD, PDAPP mice do not develop prominent neuronal degeneration or NFT (Jay et al., 2011).

As is the case with HD models, it is not likely that use of models of AD will become salient components of safety evaluation, but toxicologists and pathologists supporting discovery/nonclinical efficacy investigations may be called upon to provide input on study design/interpretation of results. In particular, if adverse effects are encountered in nonclinical efficacy studies, understanding the pathobiology of the effect (e.g., potential consequences of rapid Aβ removal) would be critical in assessment of the go/no-go determination for the compound.

18.4.3.4 Lysosomal Storage Diseases

There are multiple lysosomal storage diseases in humans for which the central nervous system is the most biologically and clinically significant. Naturally occurring large animal models exist for several of these diseases (GM_1, GM_2, Gaucher's disease, globoid cell leukodystrophy, Niemann–Pick disease (NPD), type II glycogenosis, or Pompe's disease). However, use of large animal models for efficacy or in drug safety testing is not optimal due to the difficulty in obtaining sufficient numbers of animals for evaluation and/or the difficulty in providing sufficient compound for testing, particularly in an early discovery phase. Instead, transgenic mouse models are the preferred species for evaluation of efficacy or safety due to their size, availability of sufficient number of animals (litter size), and relative homogeneity in disease state between animals.

While several of the human lysosomal storage diseases do not have a suitable transgenic mouse model counterpart, two disease models are available that provide considerable similarity to their human counterpart (model of NPD and Pompe's disease) (Horinouchi et al., 1995; Bijvoet et al., 1999). Specifics on the use of one of these models in drug safety testing will be provided in the following.

Niemann-Pick disease is an autosomal recessive disorder in which patients accumulate sphingomyelin and, to a lesser extent, cholesterol and other phospholipids within lysosomes due to a deficiency of acid sphingomyelinase deficiency. While the monocyte/macrophage cell lineage is extensively affected, it is the accumulation within other cell types, including hepatocytes, endothelial cells, pulmonary epithelial cells, and neurons, that are of more concern. The acid sphingomyelinase knockout (ASMKO) mouse accumulates sphingomyelin in a tissue distribution similar to that of humans with NPD. While the animals are normal at birth, by 2 months there is considerable tissue accumulation of sphingomyelin, and by 4 months, neurological clinical signs (ataxia, tremors) are evident. The animals typically die by 6–8 months of age (Murray et al., 2014).

An extremely relevant manuscript outlining nonclinical safety evaluation of recombinant human acid sphingomyelinase (rhASM) has recently been published (Murray et al., 2014). The salient feature of the nonclinical safety evaluation was the difference in tolerability of the compound between conventional animal models (wild-type mouse, rat, dog, cynomolgus monkey) and the ASMKO mouse model. While very well tolerated in the conventional animal models, high dosages of rhASM resulted in dose-dependent toxicity in the ASMKO mouse, characterized by CV shock, systemic inflammation, and death. The investigators concluded that the adverse effects in the transgenic mouse model were a reflection of rapid breakdown of sphingomyelin into ceramide or other toxic downstream metabolites. The use of this model was critical to providing recommendations for dosing regimens for human clinical trials.

18.4.3.5 *Diabetic Neuropathy*

Numerous animal models are available to facilitate investigation of the pathobiology and potential therapeutic interventions for diabetic neuropathy. In general, models may be induced (streptozotocin (STZ) or diet), spontaneous, or transgenic mice. Details on the scope and suitability of the extensive number of available models is outside of the scope of this chapter, but the details including advantages/disadvantages of specific models can be found in a recent review article (Islam, 2013).

18.5 GASTROINTESTINAL INJURY MODELS

18.5.1 Gastrointestinal (GI) Toxicity: Relevance to Drug Attrition

The adverse drug reactions due to GI causes represent ~67% of adverse drug reactions proclaimed on drug labels and account for 23% of adverse events encountered in phase I studies (Sibille et al., 1998; Redfern et al., 2002). According to the 1998–2008 drug withdrawal data, the primary reasons for drug withdrawals in the United States, Europe, and Asia were CV (33%), liver (29%), psychiatric/addiction (8%), GI (2%), muscle (2%), and others (22%) including renal dysfunction, accelerated carcinogenicity or death, mutagenesis, severe drug–drug interactions with ethanol, hypersensitivity, and pulmonary hypertension (MacDonald and Robertson, 2009). In the oncology therapeutic area, GI toxicity continues to be a significant dose-limiting safety concern because of anticancer agent's targeting of highly proliferative cells, such as epithelial cells lining of the GI tract. GI tract is the initial entry site of the body of oral drugs or compounds and is susceptible to toxicity of a wide variety of drug classes including antipsychotics, anticancers, antidepressants, antibacterials, and anti-inflammatories. GI mucosa is also a site for metabolic activation or detoxification of compounds that can increase or decrease a compound's toxicity. The GI toxicity can lead to widespread systemic effects, including nutrient malabsorption and resultant malnutrition and starvation. GI toxicity is also complicated by gut microbes (termed "microbiota"), which are involved in the biotransformation of xenobiotics and complex interactions with host mucosal epithelial and immune cells and physiological processes such as digestion, energy homeostasis, and development of gut-associated lymphoid tissues (Bakhtiar et al.,2013). Furthermore, altered GI function not only affects pharmacokinetic profiles but also results in dose-limiting toxicity of therapeutic agents. As a result of this complex pathophysiology, there are no GI biomarkers that can accurately predict the morphological evidence of injury.

18.5.2 Gastrointestinal Toxicity: Reasons for Poor Translation from Animal to Human

Major contributing factors for poor translation of GI toxicity from animal models to humans are complex GI pathophysiology involving muscle, nerve, epithelial and other cell types, microbial biotransformation, and lack of predictive biomarkers. In addition, findings from healthy animals with controlled dietary regimen and age range are unlikely to translate to patients with varied ages, gender, lifestyles, genetic backgrounds, and preexisting diseases. Although the basic functions of GI are similar between humans and laboratory animals, there are numerous anatomical and functional differences (Sanger et al., 2011). Differences in physiological factors, such as pH, bile, pancreatic juice, and mucus and fluid volume and content can result in alteration of dissolution rates, solubility, transit times, and membrane transport of drug molecules. It is important to understand that these species-specific differences significantly impact the translational value of animal to human GI pathophysiology. Details on the anatomical differences between humans and commonly used laboratory animals can be found in various reviews (Kararli, 1995; Sanger et al., 2011). In addition to rodent's inability to vomit, there are also marked differences between humans and rodents in gastric electrophysiology, hormonal effects, and genetics (Table 18.3). In mouse and guinea pig, the gastric fundus is relatively quiescent, whereas in humans slow-wave activities are ~5 cycles per min (cpm) in the fundus and corpus and >7 cpm in the antrum (Rhee et al., 2011). Large rises in blood vasopressin were noted in humans during nausea, whereas rodents given the same

TABLE 18.3 GI Tract: Major Anatomical and Functional Differences between Humans and Rats

GI Tract	Rats	Humans
Stomach		
Proximal stomach	Lined by stratified squamous epithelium (known as forestomach in rodents)	No forestomach
Ability to secrete acid in proximal stomach	No (forestomach)	Yes
Absolute surface area (m^2)	0.000624	0.053
Small intestine		
Length (m)	1.25 (83% of total)	6.80 (81% of total)
Absolute surface area (m^2)	1	200
Large intestine		
Length (m)	0.25 (17% of total)	1.55 (19% of total)
Absolute surface area (m^2)	0.034	0.35
Total intestinal length (m)	1.50	8.35
Microbial fermentation in large cecum	Yes	No
Response to an emetic stimulus		
Vomit	No	Yes
Plasma vasopressin	No increase	Large increase
Plasma oxytocin	Large increase	No increase
Distinct molecular differences		
Functional motilin system	No	Yes
5-Hydroxytryptamine (5-HT)3 receptor subunits	5-HT3A and B only	5-HT3A, B, C, D, and E
Melanin-concentrating hormone receptor type 2	No	Yes
Neuropeptide Y receptor type 6	Yes	No

m, meter.

emetogenic stimuli failed to raise blood vasopressin but showed a rise in oxytocin (Sanger et al., 2011). Recently, it has shown that the GI hormone motilin (a hormone released from the upper gut of humans during fasting to stimulate hunger) is functionally absent in rodents (He et al., 2010; Sanger et al., 2011). Similarly, certain 5-HT3 receptor subunits are absent in rodents but expressed in humans and other species capable of emesis (Holbrook et al., 2009). In addition, there are species-related differences in GI receptors and/or sensitivity to certain drugs or chemicals, including cannabinoid-1 receptor antagonist, opioids, histamine, and protease-activated receptors (Breunig et al., 2007; Sanger, 2007; Benko et al., 2010; Mueller et al., 2011). It is therefore critical to fully understand these species-related differences in terms of how these might impact on translating mechanisms of drug/compound-induced findings from rodents to humans.

18.5.3 Available Gastrointestinal Animal Models to Predict Gastrointestinal Toxicity or Understand Molecular Mechanisms of Toxicity: Advantages and Limitations

The use of animal models has added our understanding of various human diseases, including inflammatory bowel disease (IBD), carcinogenesis, tumor biology, and the impact of specific molecular events on various human cancers. Mechanistic insights learned from such models permit us to design or refine novel therapeutic strategies and to characterize the efficacy and safety of such novel treatments. Although animal models have shown significant promise in the drug development efforts for diseases, such as cancer and IBD, these models have been generally disappointing for certain complex disorders, such as irritable bowel syndrome. Various animal models have been established to study a variety of human GI diseases and cancers, but the details of each model are outside the scope of this chapter. Selected examples of the most commonly used GI animal models are described in the following.

18.5.3.1 GI Injury and Ulcer Models Acute or chronic GI injury models are induced by various compounds or agents in rodents. Intragastric administrations of concentrated ethanol (50–100%), sodium hydroxide (NaOH), acetic acid, and hydrochloric acid (HCl) have been shown to induce mucosal injury in fasted rats. Of these models, ethanol and HCL models are relevant to humans, mimicking alcohol consumption and overproduction of HCL in the stomach. GI ulcer models include administration of nonsteroidal anti-inflammatory drugs (NSAIDs) such as

aspirin and indomethacin. Stomach ulcers are induced by single oral dose of aspirin or indomethacin in fasted rats (Whittle, 1977; Guth et al., 1979), whereas small intestinal ulcers are induced by daily subcutaneous injections of indomethacin in nonfasted rats (Del Soldato, 1984). In addition, histamine, cholinergic drugs, and pentagastrin have been shown to induce duodenal ulcers in rats and guinea pigs (Robert et al., 1970; Cho and Pfeiffer, 1981). These models have been used to study the pathogenesis of acute gastric injury or ulceration and for developing therapeutic strategies (Szabo and Cho, 1988a, b; Cho et al., 1990). However, these animal models are rarely used in safety studies.

18.5.3.2 GI Disease Models: *Idiopathic IBD* Idiopathic IBD is a set of two diseases, those being Crohn's disease (CD) and ulcerative colitis (UC). Both CD and UC are chronic inflammatory disease with different predilection sites; CD may affect any portion of the GI from the esophagus to anus with higher prevalence in distal small intestine and colon, whereas UC is limited to the colon and rectum. The exact cause of IBD is unknown; however, animal models have led to the understanding of pathogenesis and environmental triggers that contribute to IBD in humans. More than sixty distinct animal models of IBD were developed, and approximately forty of these are currently used in experimental research studies (Mizoguchi et al., 2003; Mizoguchi, 2012). These animal models are classified as (i) chemically induced, (ii) genetically engineered (i.e., knockout and transgenic models), (iii) spontaneous (inbred), and (iv) immune mediated (transfer of activated immunocytes in immunodeficient mice—severe combined immunodeficiency (SCID) mice or Rag knockout mice). Of these, chemically induced and genetically engineered models of IBD (Table 18.4) are most frequently used in nonclinical efficacy and pharmacokinetic studies but rarely for toxicity studies. Dextran sulfate sodium (DSS)- and trinitrobenzene sulfonic acid (TNBS)-induced colitis and interleukin-10 (IL-10) knockout models are used to test the pharmacology and efficacy of probiotics such as *Bifidobacterium lactis* and drugs, including biopharmaceuticals (Claes et al., 2010; Kim et al., 2010).

18.5.3.3 GI Cancer Models Colorectal cancer (CRC) is a major human cancer with the highest death rates in the United States, Australia, New Zealand, and Eastern European countries. Many factors play a role in the pathogenesis of CRC, including diet, genetics, and diseases such as UC (Derry et al. 2013). Based on the potential mechanisms involved in humans and search for specific etiologies, several animal models have been developed. Animal models for CRC fit into three broad categories: (i) spontaneous, (ii) chemically or environmentally induced, and (iii) cancers induced by genetic manipulation.

Spontaneous CRC in Animals (Dogs and Monkeys) Many similarities exist between dog and human CRC, such as location (occur more commonly in large intestine than small intestine), morphology (pedunculated adenomas and cytoplasmic and nuclear accumulation of β-catenin), and tendency to progress to malignancy (Church et al., 1987; Valerius et al., 1997; McEntee and Brenneman, 1999). The dog has been chosen as an attractive model for oncology research by the National Cancer Institute Comparative Oncology Program. Despite all of the similarities, the utility of dogs for CRC research is severely limited by the low prevalence of the disease in the pet dog population (<1%) (Schaffer and Schiefer, 1968; Church et al., 1987). Among nonhuman primates, the cotton-top tamarin (*Saguinus oedipus*) with idiopathic UC has been shown to develop colonic adenocarcinoma (as high as 39%) and resultant death in 5–7 years (Lushbaugh et al., 1978; Cheverud et al., 1993). The cause is unknown, but there is evidence that environmental stress and luminal microbes may play a role in the pathogenesis of colonic adenocarcinoma in cotton-top tamarin primates (Wood et al., 2000; Mansfield et al., 2001). However, the widespread use of this model is limited due to cost, long latency of carcinogenesis, and ethical concerns. Spontaneous GI cancers are rare in rodents (Chandra et al., 2010), but Western diet-induced rodent models of spontaneous CRC have been described. Rats and mice given diet with common features of Western diet (i.e., increased fat and decreased calcium and vitamin D) or modified Western diet with lower fiber, folate, methionine, and choline content for 12 weeks to 2 years have developed hyperplastic/dysplastic colonic epithelium leading to invasive colonic adenocarcinoma (Newmark et al., 1990, 2001; Richter et al., 1995; Risio et al., 1996). These animal models appear to capture much of the complexity that underlines CRC in humans. However, there are deficiencies, such as lack of information on molecular mutations and use of diet with deficient calcium level (Johnson and Fleet, 2013).

Chemically or Environmentally Induced CRC Rodents (rats and mice) have been employed for chemically induced CRC models. A large number of chemicals are known to cause cancers of the small and large intestines in rodents (Ward and Treuting, 2014). The most commonly used chemicals to induce and promote CRC in rats and mice are 1,2-dimethylhydrazine (DMH) and its metabolite, azoxymethane (AOM) (Bissahoyo et al., 2005; Rosenberg et al., 2009). AOM and DMH are alkylating agents that are administered by either subcutaneous or intraperitoneal injections for 30 weeks or more to induce CRC in rats. Although the CRCs induced by AOM or DMH in rats have many similarities to human CRC, there are noticeable differences between the two. AOM or DMH-induced CRCs in rats develop in the background of flat mucosa without polyp formation, whereas many human CRCs arise from

TABLE 18.4 Idiopathic Inflammatory Bowel Disease Models

Model	Pathogenesis	Advantages	Limitations
Chemically induced			
Dextran sulfate sodium (DSS)	Chelation of divalent cations (e.g., Ca^{2+}, Mg^{2+}) required for IEC tight Junctions	Relatively cheap, easy to use, highly reproducible, reliable for both acute and recurrent inflammatory episodes, extensive published data, and used commonly for testing new therapeutic agents	DSS-induced colitis depends on dosage used, duration, manufacturer of DSS, mice strain (C3H/HeJ and Balb/c mouse strains are more susceptible), sex (male mice are more susceptible) and microbial environment (germ-free versus specific pathogen-free), and self-limiting inflammation with highly variable severity from mouse to mouse and from region to region
Dinitrobenzene sulfonic acid (DNBS)/ethanol; trinitrobenzene sulfonic acid (TNBS)/ethanol; oxazolone	Haptenization of colonic proteins	Similar to above	TNBS/DNBS-induced colitis depends on mouse strain (i.e., SJL/J, C3HeJ, and Balb/c are susceptible, whereas C57BL/6 and DBA/2 are highly resistant), site (restricted to the injection site), dose, and source and may cause unacceptably high mortality Oxazolone-induced colitis—intrarectal administration required; most conventional mouse strains were resistant to oxazolone-induced colitis; skin presensitization with oxazolone needed for reproducible colitis
Spontaneous			
SAMP1/Yit, SAMP1/YitFc	Sublines of SAM mice from the AKR/J background	SAMP1/YitFc—spontaneous inflammation and fibrosis, no need for stimulus or additional manipulations, and GI findings that resemble Crohn's disease SAMP1/Yit—transmural inflammatory lesions in the terminal ileum with 100% penetrance by 30 weeks of age	SAMP1—early signs of senescence (amyloidosis, alopecia, and osteoporosis) SAMP1/YitFc—low breeding rate, difficult to use on a large scale, and large colonies required for routine experimentation, not commercially available
C3H/HeJBir mouse substrain	A substrain of TLR-4 lacking, C3H/HeJ mice unresponsive to LPS	Inflammation mainly involves the cecum and right colon at 3–6 weeks of age and resolves spontaneously by 12 weeks of age	C3H/HeJBir—greatly influenced by housing conditions and enteric microflora
Immune mediated			
Transfer of activated immunocytes in immunodeficient mice—SCID mice or Rag knockout mice	Repletion of lymphopenic RAG1/2$^{-/-}$ or SCID mice of variable genetic backgrounds with naive CD4+ CD45RB^{high} T cells	Reliable chronic models, high reproducibility of T cell-dependent chronic intestinal inflammation (well-defined mechanism), commercially available, suitable for testing target-specific drug candidates	Expensive and laborious
Genetically engineered			
IL-10 knockout mice (IL-10$^{-/-}$) most commonly used	Disruption of the IL-10 gene	Reliable chronic models for testing target-specific new therapeutic agents, commercially available with new updated models, clinical signs derived from endogenous mechanisms, and valuable means for studying pathophysiology	Expensive to establish and maintain these models, low penetrance of the phenotype in some mouse strains, heterogeneity of disease onset in individual mice, lack of insight into the diverse mechanisms (e.g., epigenetic, transcriptional, systemic), and the lack of knowledge for compensatory mechanism
STAT3 knockout mice (STAT3$^{-/-}$)	Disruption of the STAT3 gene in neutrophils and macrophages) by Cre/loxP recombination	Similar to above	Similar to above
IL-7 transgenic mice (IL-7 tg)	Enhanced IL-7 production by mucosal T cells	Similar to above	Similar to above
STAT4 transgenic mice (STAT4 tg)	STAT4 overexpression in CD4+ T cells	Similar to above	Similar to above

adenomatous polyps (Ward, 1974). The incidence of metastasis is relatively low in AOM- or DMH-induced CRC in rats compared to CRC patients where the metastatic rate to regional lymph node is ~50% (Rosenberg, et al., 2009). Multiple administrations and longer treatment duration (40 weeks or longer) are required for mice to develop CRC. However, mice given AOM followed by multiple cycles of DSS developed CRC within a relatively short term (Tanaka et al., 2003; Rosenberg et al., 2009). The majority of these tumors show mutations of the β-catenin gene (Ctnnb1), resulting in stabilization of β-catenin (Ctnnb1) and increased WNT signaling to drive carcinogenesis (Yamada et al., 2000). The AOM/DSS mouse model of colitis-associated CRC is the most frequently used in preclinical testing of chemopreventive or therapeutic compounds, such as cyclooxygenase-2 inhibitors, peroxisome proliferator-activated receptors, plant-derived compounds, and probiotics (Kim et al., 2010; De Robertis et al., 2011; Tanaka 2012).

Other chemical inducers of CRC in rodents include 2-amino-1-methyl-6-phenylimidazo[4,5-*b*]pyridine (PhIP) and nitroso compounds such as *N*-methyl-*N*-nitro-*N*-nitrosoguanidine (MNNG) and *N*-methyl-*N*-nitrosourea (MNU). Among these chemicals, PhIP, a heterocyclic amine generated during the cooking of meat and fish at a high temperature, is highly relevant to human cancer because of an epidemiologic link between PhIP from cooked meat to increased CRC risk in humans (Sinha and Rothman, 1999; Sinha et al., 1999). PhIP is a colon cancer mutagen in rats, but in mice, it induces aberrant crypt foci but not colon cancer (Esumi et al., 1989; Tudek et al., 1989; Nakagama et al., 2005). Rats treated with PhIP had a 50% incidence of colon cancers, and cotreating PhIP with DSS or treating APC-min mice with PhIP has been shown to enhance carcinogenesis (Hasegawa et al., 1993; Tanaka et al., 2005; Andreassen et al., 2006). In PhIP-induced cancers, β-catenin (Ctnnb1) and Apc gene mutations are common compared to rare mutations of Kras and Tp53 genes (Dashwood et al., 1998; Tsukamoto et al., 2000). The susceptibility to PhIP-induced colon cancer is also dependent on the strain of rat: BUF/Nac rats are highly susceptible; F344, Wistar, and Brown Norway (BN) rats are moderately sensitive; and ACI rats are relatively resistant to PhIP-induced formation of aberrant crypt foci (Ishiguro et al., 1999). In the case of nitroso compounds, both MNNG and MNU are direct-acting carcinogens. Oral administration of MNNG or MNU in rodents causes neoplasia of various organs, including the stomach, small intestine, large intestine, kidney, skin, lung, and thymus (Koestner et al., 1977; Maekawa et al., 1988). However, when administered via the rectum, MNU or MNNG reproducibly induces a high incidence of colon cancer in rodents (Narisawa and Weisburger, 1975; Narisawa et al., 1976; Qin et al., 1994). The molecular profile is not completely understood, but rat colon cancers induced by MNU or MNNG have been shown to contain Kras and Apc mutations. Major confounding variables for nitroso compound-induced model are high incidence of extracolonic neoplasia and early deaths.

Inflammation-Induced CRC Chronic inflammation via complex mechanisms involving production of chemokines, growth factors, and reactive oxygen species, activation of immune cells, and DNA damage in the setting of proliferation is associated with an increased risk of developing neoplasia at the site of inflammation (Coussens and Werb, 2002; Ohshima et al., 2003; Macarthur et al., 2004; Ohshima et al., 2005). Numerous human cancers are associated with chronic inflammation from infections, chemical exposure, and inflammatory diseases such as UC and CD. In UC, duration and severity of inflammation correlate with the cumulative probability of CRC development, ranging from 2% at 10 years of colitis to 18% after 30 years (Eaden et al., 2001). CD patients with a similar extent and severity of colitis as UC patients have similar risks for development of CRC (Zisman and Rubin, 2008). However, CD patients with the most severe inflammation in the small intestine have a 12- to 60-fold increased risk of developing small bowel adenocarcinoma (Bernstein et al., 2001). A classic example of chemically induced inflammation causing cancer is DSS-induced chronic colitis in rodents. DSS is not a direct carcinogen, but administration causes chronic inflammation through activation of macrophages and release of proinflammatory cytokines. DSS-induced inflammation subsequently leads to CRC. As described earlier, cotreating DSS with either AOM or PhIP enhances the development of CRC in rodents. Similarly, chronic colitis induced by DNBS, TNBS, or oxazolone can result in the development of CRC in rodents. These chemicals haptenize proteins autologous or derived from the microflora, causing them to be immunogenic and resultant priming of antigen-specific immune cells (Neurath et al., 1996; Qiu et al., 1999; Arita et al., 2005). These activated immune cells produce proinflammatory cytokines, which then enhance inflammation and development of carcinogenesis (Schiechl et al., 2011). Newer inflammation-induced colon cancer models have been developed using immunodeficient mice infected with *Helicobacter* species (Maggio-Price et al., 2005, 2006, 2009; Erdman and Poutahidis 2010).

Cancers Induced by Genetic Manipulation

ADENOMATOUS POLYPOSIS COLI RODENT MODEL Adenomatous polyposis coli (APC) is a dual-function tumor suppressor gene and encodes a protein that has been implicated in a variety of cellular functions including cellular proliferation, differentiation, cytoskeleton regulation, migration, and apoptosis (Neufeld, 2009; Minde et al., 2011; Perez-Sayans et al., 2012). Mechanistically, APC regulates levels of β-catenin, an important mediator of

Wnt/β-catenin signaling pathway. The Wnt/β-catenin signaling pathway plays an important role in the development of normal intestinal epithelium as well as colorectal carcinomas (Oshima et al., 1997). In humans, ~55% of CRCs occur in the descending colon and rectum, and of these, more than 80% have inactivated APC and 50% have β-catenin mutations without APC mutations. Most CRCs arise from benign adenomatous polyps. The progression of polyps to invasive or metastatic cancer (i.e., adenoma-to-carcinoma progression), also known as the Vogelstein model (Fearon and Vogelstein, 1990), involves the stepwise accumulation of multiple mutations in three additional pathways, such as MAPK pathway (i.e., KRAS-activating mutations), TGFβ-induced differentiation pathway (e.g., loss of SMAD2 and SMAD4 genes), and DNA repair pathways (i.e., inactivation of TP53 gene). In contrast to adenoma-to-carcinoma progression, human cancers in the ascending colon account for 20%, but they are characterized by genetic lesions in DNA mismatch repair genes. Mutations in the DNA mismatch repair genes cause alterations of microsatellites, resulting in microsatellite instability (Geiersbach and Samowitz, 2011). Germ line mutations of two major mismatch repair genes, such as MutS homologue 2 (MSH2) and MutL homologue 1 (MLH1), are important for the initiation of carcinogenesis in ~90% of hereditary Lynch syndrome (also known as hereditary nonpolyposis CRC) cases (Geiersbach and Samowitz, 2011).

Rodent models with various Apc mutations have been the workhorse of preclinical CRC research. Since the development of the first mouse model with a germ line Apc mutation in the early 1990s, over 43 different Apc rodent mutants have been generated (Johnson and Fleet, 2013; Zeineldin and Neufeld, 2013). All characterized motifs in human APC are conserved in murine Apc with 87.9% identical proteins and 91.9% similar amino acids. The first phenotype of these mutant mice had multiple intestinal neoplasms or "Min." Mice homozygous for Apc^{Min} died early in embryonic stage, whereas heterozygous Apc$^{Min/+}$ developed multiple tumors in small and large intestines but greater than 10-fold more tumors in the small intestine. Similarly, F344-*Pirc* (polyposis in the rat colon) rat model with APC mutation developed tumors throughout the intestine (Amos-Landgraf et al., 2007). APC rodent models are generated using various technologies, such as chemical mutagenesis screen (e.g., F344-*Pirc*, Apc$^{Min/+}$), insertion of an antibiotic-resistant gene (Apc^{1309}, Apc^{1638N}, and Apc^{1638T}), and Cre/lox-induced excision (e.g., Apc$^{1322T/+}$ and ApcΔ^{e1-15}). Details of these models are summarized in several excellent reviews (Johnson and Fleet, 2013; Washington et al., 2013; Zeineldin and Neufeld, 2013; Fleet, 2014). Apc rodent models have been used for elucidating the effect of various environmental and genetic factors on intestinal tumorigenesis and for testing potential chemopreventive and therapeutic agents. There are limitations of the APC rodent model.

Apc$^{Min/+}$ mice typically develop many more small intestinal tumors than colon tumors (Moser et al., 1990; Moser et al., 1992). Other complications of Apc$^{Min/+}$ mice include frequent hemorrhages at tumor sites, anemia, and early deaths (Moser et al., 1990). Mice used to model colon cancer originating from defects in DNA mismatch repair and microsatellite instability include Mlh1$^{-/-}$ (null MutL homologue 1) and Msh2$^{-/-}$ (null MutS homologue 2) mice. Mlh1$^{-/-}$ and Msh2$^{-/-}$ mice develop both lymphoid and intestinal tumors and succumb primarily to lymphomas after 6 months of age (Reitmair et al., 1996; Edelmann et al., 1999). Genetically modified mouse models with alterations in multiple pathways of CRC have been developed (Johnson and Fleet, 2013; Washington et al., 2013) but the details on these models are outside of scope of this chapter.

XENOGRAFTS (COLON CANCER CELL LINES AND PATIENT-DERIVED SPECIMENS)   Human cancer xenograft models have been established in immunodeficient athymic nude mice and SCID mice using cancer tissues obtained from patients or from colon cancer cell lines. Although xenograft models using colon cancer cell lines are easy to establish, these models are insufficient and often misleading for screening anticancer effects of new therapeutic compounds (Sausville and Burger, 2006; Monsma et al., 2012; Siolas and Hannon, 2013). To address these limitations, patient-derived xenograft (PDX) models have been developed. Recently, PDX models have gained momentum by direct use of surgically resected human tumor specimens, thereby retaining patient's tumor heterogeneity (Talmadge et al., 2007; Rubio-Viqueira and Hidalgo, 2009; Williams et al., 2013). The establishment of PDX models involves transplantation of resected tumor specimens to immunodeficient mice by subcutaneous implantation or into the kidney capsule (O'Brien et al., 2007; Ricci-Vitiani et al., 2007; Burgenske et al., 2014). Alternatively, PDX models can be established by surgical implantation of tumor specimens in the colon serosa of immunodeficient mice (Rashidi et al., 2000; Puig et al., 2013). PDX models, therefore, offer a relevant method to study tumor biology, including initiation and progression of metastasis and for assessing the effects of therapeutic intervention. These models have been used for developing gene signature patterns that predict tumor response to cytotoxic agents (Fiebig et al., 2007). The gene signature patterns generated from the PDX models represent the main genetic characteristics of CRC in humans (Julien et al., 2012). PDX mice models also facilitate in the discovery of predictive biomarkers of drug sensitivity and resistance. However, there are several factors to hinder the use of PDX mice models, such as high cost, availability, requirement of special technical skills, long latency periods after engraftment, and variable engraftment rates (Dangles-Marie et al., 2007; Siolas and Hannon, 2013).

18.6 RENAL INJURY MODELS

18.6.1 Renal Toxicity: Relevance to Drug Attrition

Drug-induced renal toxicity represents one of the main reasons for drug attrition and continues to be an ongoing challenge for the pharmaceutical industry and for patient's safety and welfare (Lin and Will, 2012). The kidney plays an important role in the elimination of drugs and metabolites via excretion but also as a site for metabolic activation and/or detoxification of drugs/chemicals. Similar to liver, drug-induced injury via metabolic activation occurs in kidneys. Moreover, the kidney is uniquely susceptible to drug-induced toxicity because of its disproportionately high blood flow and anatomical and functional complexity. When compared to the liver with an enormous capacity for cellular regeneration and repair, kidney cells have a very limited regenerative/repair capacity (Betton et al., 2005; Pereira et al., 2012). A variety of drugs cause ~20% of acute renal failure in humans. But in older adults, the incidence of drug-induced nephrotoxicity may be as high as 66% (Naughton, 2008) and include drug classes, such as antibiotics, NSAIDs, and anticancer and immunosuppressive drugs.

18.6.2 Renal Toxicity: Reasons for Poor Translation from Animal to Human

Animal models have greatly enhanced the understanding of renal injury in humans because of many similarities between rodents and humans in anatomical and physiological characteristics, including absorption, distribution, metabolism, and excretion (ADME) of the test material (parent or metabolites). Despite many similarities in ADME, there are variations in pathophysiology between humans and animals that can have a significant impact on the animal to human translation. For example, promising treatment outcomes of acute renal failure in animal models have not translated successfully to humans. It is known that rodent's ability to concentrate urine is twice that of humans. The osmotic ratios (urine/plasma) are 8.9 in rats and 4.2 in humans; and the urine specific gravity ranges from 1.050 to 1.062 in rats and 1.003 to 1.030 in humans (Bivin et al., 1979; Christiansen et al., 1997; Kasiske and Keane, 2000). This is due to the anatomical differences between humans and rats. The relative medullary thickness of rat kidney is substantially greater than that of human kidney (~0.6–1.4 $\times 10^6$ nephrons in each human kidney vs. 30,000 nephrons in each adult rat kidney) (Nielsen et al., 2012). Other structural differences include unilobular/unipapillary kidney in rodents, rabbits, and dogs compared to multipapillary kidney in humans, pigs, and sheep and the gender-related renal anatomical dimorphism in mice (Giraud et al., 2011; Treuting and Dintzis, 2012). These species-specific differences may profoundly impact the interpretation of renal injury. For example, rat models are

able to closely recapitulate the major pathological manifestations of human renal diseases, whereas mice are notoriously resistant to development of renal lesions after injury (Breyer et al., 2005). Sex-related susceptibility of certain drugs/chemicals to kidney injury/tumor is well known. Male and female (to a lesser extent) rats produce a rat-specific protein, called α-2 microglobulin, which has been shown to interact with the metabolites of certain chemicals. α-2 Microglobulin is synthesized in large quantities in the male rat liver and to a lesser extent in certain glands of both male and female rats (Lock and Hard, 2010). Following the glomerular filtration, α-2 microglobulin is reabsorbed and degraded by proteolytic enzymes in the proximal tubules. Excessive α-2 microglobulin accumulation is responsible for the kidney tumors in male rats (e.g., Fischer 344, Sprague-Dawley (S-D), Buffalo, or BN) after chronic exposure to a number of hydrocarbons, including decalin and d-limonene (Hard et al., 1993). The NCI-black-Reiter (NBR), the only known rat strain, is not susceptible to the development of kidney tumors because of the lack of α-2 microglobulin (male NBR do not produce significant amounts of α-2 microglobulin) (Chatterjee et al., 1989). Similarly, kidney tumors are not seen in female rats or mice (both sexes) because of the lack of this unique urinary protein. It does not occur in any species including guinea pigs, hamsters, dogs, and nonhuman primates (Borghoff et al., 1990). Based on the sex and species-specific findings, scientific groups and regulatory agencies have concluded that kidney tumors in male rats are not relevant to humans, as humans do not produce α-2 microglobulin.

There are marked differences in the renal metabolism among various animal species and strains that may account for the reported diversity of renal injury and susceptibility. The rat kidney contains glutamine synthetase (an enzyme that converts ammonium glutamate to glutamine) but absent in dog and human kidneys (Lemieux et al., 1976). Various rat strains have different sulfate conjugation steps in the kidney for phenolic drugs and phenolic monoamines (phenol sulfotransferase). For example, the induction of renal phenol sulfotransferase by dexamethasone was slight (23% increase in activity) in F344 rats but marked (317% rise in activity) in S-D rats (Maus et al., 1982). Similar to the liver, the kidney contains microsomal oxidase enzymes (Mitchell et al., 1977), and activation of a specific cytochrome P450 isozyme enhances the drug-induced nephrotoxicity in certain rat strains. Compared with S-D, F344 rats are highly susceptible to acetaminophen-induced nephrotoxicity, and this effect may be due to the activation of a specific P450 isozyme in F344 rats but not in S-D rats (Newton et al., 1983). Another contributing factor for acetaminophen-induced differences between F344 and S-D rat strains is the ability of parent compound/metabolites to reach the kidney (i.e., pharmacokinetics). Young F344 (2–3-month-old) rats given acetaminophen showed reduced excretion of parent compound or metabolite (p-aminophenol) and associated renal dysfunction and morphologic changes. S-D rats, on the contrary, appeared resistant to acetaminophen-induced nephrotoxicity because of the enhanced acetaminophen (parent and metabolites) excretion (Tarloff et al., 1989). Similarly, S-D rats excrete greater concentrations of urinary tobramycin, an aminoglycoside antibiotic, over a 24-h period when compared with F344 rats. The decreased ability to excrete tobramycin contributed to enhanced nephrotoxicity in F344 rats (Reinhard et al., 1994). Goodrich and Hottendorf (1995) reported that S-D rats given 90 mg/kg dose of tobramycin failed to markedly alter renal function in S-D rats, whereas F344 rats given 30 mg/kg dose showed marked renal dysfunction and injury. Another antibiotic, gentamicin has been shown to cause similar strain-related nephrotoxic effects. F344 rats given 40 mg/kg gentamicin exhibited nephrotoxicity in F344 rats, whereas 70–100 mg/kg gentamicin was required to induce an equivalent adverse effect in S-D rats. Male F344 rats were more susceptible to gentamicin-induced nephrotoxicity than that of females. It should be noted that male F344 rats are more prone to severe chronic progressive nephropathy (CPN) than other strains, and CPN is known to interfere the pharmacokinetics as well as the toxicity of various drugs (Goodman et al., 1994).

One of the major complications that hampered the use of rodents, particularly for longer-duration toxicity studies, is CPN, a rodent-specific, age-related renal disease. In S-D rats, CPN is first detected at 2–3 months of age and then progressed to severe renal dysfunction and death from renal failure at <18 months (Palm, 1998; Hard and Khan 2004; Travlos et al., 2007). Reduced or loss of cortical mass observed in advanced CPN has been shown to enhance renal injury by increasing delivery of toxicants to the remaining functioning nephrons. CPN also affects baseline clinical pathology parameters and may obscure biomarkers of renal injury. Furthermore, the key morphologic features involving tubular basement membranes in rats with CPN do not recapitulate key hallmarks of kidney senescence and chronic renal failure in humans (Trevisan et al., 2010). Furthermore, different strains and genetic makeups of the same animal species often display tremendous heterogeneity in developing renal lesions after injury. Therefore, several critical factors, such as animal species, strain, gender, and age, may play a crucial role in determining the study outcomes and their relevance to humans.

18.6.3 Available Renal Models to Predict Renal Toxicity or Understand Molecular Mechanisms of Toxicity: Advantages and Limitations

Among various models, rodents are most frequently used as preferred animal models for understanding the complex pathophysiological responses associated with the development and progression of kidney diseases in humans. These models not only provide opportunities to improve treatment strategies but also explore the mechanisms of drug-induced nephrotoxicity, including genetic and environmental factors. A number of animal models are available to study human diseases such as acute kidney injury/disease and chronic kidney disease (CKD). A detailed review of these models is beyond the scope of this chapter. Examples of selected animal models of human disease are described below.

18.6.3.1 Acute Kidney Injury or Disease Models   Acute kidney injury (AKI), characterized by tubular injury, inflammation, and vascular damage, is a common clinical problem with high morbidity and mortality rates in hospitalized patients. Although AKI is a well-recognized complication of nephrotoxic drugs, there are many systemic diseases that may lead to AKI in humans including diabetes, hypertension, and autoimmune diseases (Ramesh and Reeves, 2003; Jiang et al., 2007). Most of AKI pathophysiology knowledge has been derived from animal models. Based on major causes of clinical AKI, three types of animal models have been developed such as ischemic/reperfusion (I/R), toxin/drug, and sepsis-induced models to mimic vascular compromise, drug toxicity, and infection, respectively.

Ischemic/reperfusion model, induced by clamping of renal artery, is the most commonly used animal model for investigational studies. I/R models were initially developed in dogs, pigs, and rabbits because of their large size and easy to handle, including surgery (Kaboth, 1965; Baker et al., 2006; Balasubramanian et al., 2012; Lee et al., 2012). Rodents (rats and mice) became the most popular I/R model in recent years. In rat model, large areas of renal cortex are essentially no longer perfused due to clamping of renal artery, which in turn results in cortical necrosis. However, it should be noted that human AKI differs considerably from I/R animal model. The blood flow in human AKI never fully ceases, and the damage consisting of primarily focal necrotic areas with apoptosis and desquamation of viable cells occurs mainly from focal mismatches of oxygen delivery by impaired microcirculation and increased demand due to cellular stress (Gobe et al., 1999). However, the detailed histopathology of human AKI is largely unknown because of the rarity of biopsies performed in the acute phase of injury. As a result, the extent of tubular damage (proximal vs. distal) and cell types affected still remained controversial (Heyman et al., 2010). Although I/R models elicit AKI, there are marked differences in the susceptibility to ischemic AKI among different mouse strains and even different colonies of the same strain. For example, Swiss (National Institutes of Health) and 129/Sv mice have been shown to be resistant or less sensitive to ischemic AKI than C57BL/6 and BALB/c mice (Burne et al., 2000; Lu et al., 2012). Similarly, AKI is also affected by animal age and sex. Aged rats (60–65 weeks old) are more susceptible to ischemic AKI than young rats of 6–7 weeks (Kusaka et al., 2012).

Other less commonly used animal models include toxic injury (e.g., cisplatin and folic acid (FA)) and sepsis model induced by cecal ligation and puncture (Roncal et al., 2007; Izuwa et al., 2009; Singh et al., 2012). Cisplatin-induced renal toxicity is primarily associated with impaired glomerular filtration and proximal tubular injury that progresses to AKI (Jones et al., 1985). For cisplatin-induced AKI, males are more sensitive than females (Park et al., 2004; Wei et al., 2005). FA-induced AKI is associated with the rapid tubular appearance of FA crystals and subsequent tubular necrosis, epithelial regeneration, and fibrosis (Mullin et al., 1976). Sepsis-induced AKI models are used to study the pathophysiology of AKI in critically ill patients (Chertow et al., 2005). However, these models are difficult to interpret because of highly complex mechanisms leading to AKI and are rarely used in safety studies of an investigational drug.

18.6.3.2 Chronic Kidney Disease Models

Chronic kidney disease (CKD) in humans results from recurrent or progressive injuries in glomeruli, tubules, interstitium, and/or vasculature and is contributed by a variety of etiologies such as genetic, autoimmune, infectious, environmental, diets, and drugs. AKI is considered a trigger for subsequent development of CKD and eventually end-stage renal disease (ESRD) in humans (Bucaloiu et al., 2012; Rifkin et al., 2012). Many animal models have been developed to study pathogenesis, mechanisms, and effects of therapeutic interventions. However, these animal models do not exactly mimic human diseases, and most of them are strain, gender, or age dependent. Although not perfect, the careful use of disease animal models offers the opportunity to focus on individual mechanisms and potential for developing new therapeutics. The following are selected examples of CKD animal models.

Diabetic Nephropathy Diabetes-associated kidney disease (also known as diabetic nephropathy (DN)) is one of the leading causes of CKD and ESRD in humans. Approximately 40% of both insulin-dependent type 1 and type 2 diabetics have advanced CKD or ESRD (Kong et al. 2013; Kaur et al., 2014). Rodent models of DN include artificially induced, spontaneous, and genetically engineered (knockout and transgenic). These models are frequently used for understanding the complex pathogenesis of DN (Chatzigeorgiou et al., 2009; Kaur et al., 2014). STZ and alloxan are widely used for artificially inducing type 1 diabetes mellitus (T1DM) in rodents, and the kidney changes, such as mesangial expansion and glomerular scarring, are characteristic of human DN (Leiter, 1982; Leiter, 1985; Sun et al., 2006; Tesch and Allen, 2007). S-D, Wistar-Kyoto (WKY), or spontaneously hypertensive rats have shown to develop DN in 4–8 weeks after single intraperitoneal injection of STZ (40–65 mg/kg dose) (Casey et al., 2005; Shah and Singh 2006; Gojo et al., 2007;

Budhiraja and Singh 2008; Haidara et al., 2009). Similarly, rats develop DN after single dose of alloxan (120 mg/kg dose) intravenous or intraperitoneal injection. However, alloxan-treated animals show severe polydipsia, hyperglycemia, glycosuria, hyperlipidemia, and polyphagia and develop various complications such as DN, neuropathy, cardiomyopathy, and retinopathy. Furthermore, the incidence of alloxan-induced ketosis and mortality is higher than that of STZ. Therefore, STZ is preferred over alloxan for inducing DN (Bailey and Day 1989; Young et al., 1990; Katovitch et al., 1991; Pele-Tounian et al., 1998). Although STZ-induced DN has been widely used, there is potential for nonspecific renal toxicity that may complicate the interpretation.

For type 2 diabetes mellitus (T2DM)-induced DN, high-fat diet is widely used to induce insulin resistance and obesity and DN in C57BL6 mice (diet-induced obese mice) (Surwit et al., 1988, 1995; Petro et al., 2004). Obese Zucker rats are an inbred strain with mild glucose intolerance and peripheral insulin resistance similar to that found in humans with type 2 diabetes. Similarly, multiple low-dose injections of STZ have been shown to induce DN in nephrectomized (5/6 or uninephrectomized) rats with most features of human DN from T2DM including hyperglycemia, hypoinsulinemia, hyperlipidemia, hypertension, and microalbuminuria, followed by overt albuminuria, mesangial expansion, and terminal glomerular sclerosis (Sugano et al., 2006; Kong et al., 2013). Genetic animal models serve as an essential experimental tool for investigating the molecular mechanisms and genetic susceptibility in the development of DN. There are numerous genetic models that mimic human DN such as insulin-2 Akita mice (mutant mouse model of T1DM from Jackson Laboratories), Db/db mice, Goto–Kakizaki rats, Zucker diabetic fatty rats, KK mice, FVB/NJ (Swiss) mice, ROP (ragged/+, oligosyndactyly/+, pintail/+) mice, ICER 1cTg (inducible cAMP early repressor transgenic) mice, and human RAGE (the receptor for advanced glycation endproducts) gene overexpressed mice (Mathews, 2002; Kaur et al., 2014).

Hypertensive Nephropathy Long-standing hypertension may develop hypertensive nephropathy (HN), a frequent cause of uremia and death in humans. The exact mechanism of HN is not fully understood. But HN might be the result of long-standing hypertension with subsequent injury to vessel wall and associated pathologies (glomerulonephritis, fibrinoid arteriopathy, and arteriosclerosis). Animal models to study human HN include Dahl salt-sensitive rat, stroke-prone spontaneous hypertensive rat, and ren2 transgenic rat (Yang et al., 2010).

Immune-Mediated Glomerular Disease Animal models have been crucial to the understanding of immune-mediated glomerular disease in humans. Several animal models have developed using injection of preformed antibodies to glomerular antigens, active immunization against glomerular antigens, and immunization against nonrenal circulating antigens. The most commonly used animal models include anti-Thy1 nephritis, nephrotoxic nephritis, and spontaneous lupus glomerulonephritis (Yang et al., 2010; Becker and Hewitson, 2013). Anti-Thy1 nephritis is induced by injecting an antibody to Thy-1, an antigen found on glomerular mesangial cells. In rats, anti-Th1 antibody reacts with mesangial cells and causes mesangiolysis, fibrin deposition with monocyte/macrophage infiltration, and subsequent mesangial cell proliferation and mesangial matrix expansion (Fujimoto et al., 1964; Mosley et al., 2000; Becker and Hewitson, 2013). In nephrotoxic nephritis, mice or rats are injected first with rabbit (or sheep) IgG and then with rabbit (or sheep) antiserum to glomerular basement membrane. The resulting glomerular disease mimics human immune complex glomerulonephritis (Christensen et al., 2006; Fu et al., 2007). Likewise, the pathology of lupus glomerulonephritis has been extensively studied in inbred mouse strains, such as MRL lpr/lpr, BXSB, NZB x NZW F1, NZM2328, and NZM2410 (Bagavant et al., 2004). These mouse strains are genetically susceptible to systemic lupus erythematosus (SLE) and develop autoimmune responses and renal pathology. Additionally, procainamide, hydralazine, gold salts, and D-penicillamine have been shown induce to SLE in BN rats (Foster, 1999). There are also transgenic models based on other target antigens (e.g., phospholipase A2 receptor, podocyte antigens) under development (Farquhar et al., 1995; Glassock, 2009; Becker and Hewitson, 2013).

18.7   RESPIRATORY DISEASE MODELS

18.7.1   Respiratory Toxicity: Relevance to Drug Attrition

Adverse reactions affecting the respiratory system are not considered to be frequent cause of drug attrition (Olson et al., 2000). The reason for the low rate of attrition is not clear, but could be in part due to the high prediction of risks for effects on the respiratory system in conventional nonclinical safety evaluation (Schein et al., 1970; Fletcher, 1978). While not a common cause of drug attrition, there is a high unmet need for effective therapies to treat a variety of acute and chronic respiratory disorders. Animal models of human respiratory disease are commonly utilized in discovery evaluation in order to evaluate response to potential therapeutic agents and/or to elucidate pathogenesis. While prediction of pulmonary toxicity is typically satisfactory with conventional animal models (healthy rodent/nonrodent), in pulmonary changes seen in these models, it may be appropriate to consider evaluation of the compound in an animal model of respiratory disease prior to proceeding to humans with significant respiratory compromise, particularly if

effects in the respiratory tract have been identified with conventional animal models. Some of the major respiratory models of human disease will be briefly outlined in order to provide a background on this important therapeutic area.

18.7.2   Respiratory Toxicity: Reasons for Adequate Translation from Animal to Human

Prior to administration of an NCE in humans, the compound's effect on respiratory function is evaluated in a nonclinical setting (typically in a rodent) as per ICH S7A. Simple visual inspection is not considered adequate, so quantitative measurements (tidal volume and minute volume) in addition to respiratory rate are recorded. While quantitative assessment is typically conducted with single-dose administration, microscopic evaluation of the lung, trachea, and nasal passages are conducted as part of the repeat-dose administration toxicology studies.

18.7.3   Available Respiratory Models to Predict Respiratory Toxicity or Understand Molecular Mechanisms of Toxicity: Advantages and Limitations

There are numerous naturally occurring diseases of the respiratory system that may affect either the upper or lower respiratory tract. The models for some of the major respiratory disorders (chronic obstructive pulmonary disease (COPD), acute respiratory distress syndrome (ARDS), asthma, fibrotic lung disease) will be briefly summarized.

18.7.3.1   Chronic   Obstructive   Pulmonary   Disease Chronic obstructive pulmonary disease (COPD) is a major global health problem and is predicted to become the third most common cause of death by 2020 (Murray and Lopez, 1997; Chung et al., 2002; Lopez, et al., 2006). An excellent review of available models of COPD has been provided by Groneberg and Chung (2004) and will be briefly summarized in the following. Animal models may be "induced" (e.g., tobacco smoke, nitrogen dioxide, or sulfur dioxide), or they may consist of genetically targeted mice. As compared to asthma, relatively few animal models have been developed. Genetically targeted mice have also been utilized as models of human COPD. COPD is characterized by progressive airflow obstruction of peripheral airways associated with inflammation, emphysema, and mucus production. One of the limitations in developing a suitable animal model for COPD is the difficulty in producing small airway damage/obstruction because small animals (e.g., rodents) have relatively few levels of airway branching as compared to the human. None of the models are ideal as none consistently produce all of the salient features of human COPD. Thus, the selection of the mode should be based on the specific question to be answered. Some of the more common models of COPD are presented in Table 18.5.

TABLE 18.5 Models of Chronic Obstructive Pulmonary (COPD) Disease[a]

Target	Comments	Advantages/Disadvantages
Inhalation model—tobacco smoke	Guinea pig, rabbit, dog, rat, and mouse; guinea pig is especially susceptible, susceptibility in mice varies with strain	Rat appears to be very resistant to emphysema type of lesions but do develop corticosteroid resistance While there are differences from human (lesion and mediators), it still provides useful mechanism to investigate pathogenesis
Inhalation model—sulfur dioxide	Chronic injury and repair in the rat and guinea pig	Mucous hypersecretion present, but airway inflammation is confined to large airways
Inhalation model—nitrogen dioxide	Considerable interspecies variability—guinea pigs more sensitive than mice/rats; variability in susceptibility within mouse strains (genetic component)	Can lead to death in nonclinical species (pulmonary edema/hemorrhage, pleural effusion)
Inhalation model—oxidant stimuli (ozone) and particles (silica/coal dust, diesel exhaust particles)	May also have pulmonary tumor development with diesel exhaust particles	
Tissue-degrading approaches	Proteinases—human neutrophil elastase, porcine pancreatic elastase, papain → emphysema	Simple and rapid to produce but does not have inflammatory component of COPD
Gene targeting approaches	Gene depletion and overexpression in mice—ID function/role of different genes involved—integrins, MMPs, elastase, PDGF-A, and FGFR	May combine these models with induced models (e.g., tobacco smoke)

[a] From Groneberg and Chung (2004).

18.7.3.2 Acute Respiratory Distress Syndrome

An excellent review of animal models of acute lung injury has been provided by Matute-Bello et al. (2008) and will be summarized in the following. Microscopic characteristics of acute lung injury in humans include neutrophilic alveolitis, injury of the alveolar epithelium and endothelium, edema, hyaline membrane formation, microvascular thrombi, and, in the later stage, fibrin deposition with ultimate repair with fibrosis. Different animal models have been utilized to investigate the pathogenesis and potential therapeutic regimens for the disorder. Most models are based on reproducing the known risk factors for ARDS in humans, including sepsis, lipid embolism secondary to bone fracture, acid aspiration, and ischemia/reperfusion of pulmonary or distal vascular beds. Due to the complexity of ARDS and progression from an acute to chronic state, finding a single animal model that satisfies all characteristics of the disorder is not yet possible. Other complicating factors that play a role in the identification of ideal animal model for ARDS include differences in Toll-like receptors, the mononuclear phagocyte system, nitric oxide, and chemokines/chemokine receptors between animals and humans. Some of the primary models and their technical issues/limitations are presented in Table 18.6.

18.7.3.3 Asthma

An excellent review of animal models of asthma has been provided by Zosky and Sly and will be summarized in the following (Zosky and Sly, 2007). Animal models of human asthma have been in used for over 100 years (Mosmann and Coffman, 1989; Karol, 1994) in order to understand the pathogenesis and potential for therapeutic agent intervention. Use of these models has been of central importance in the elucidation of the mechanisms of development and progression of asthma in humans, including the identification of the relative importance of Th1 and Th2 in mouse models of allergic airway inflammation (Mosmann and Coffman, 1989).

The mouse has commonly been utilized as a model for human asthma. The mouse that can be sensitized to a variety of antigens has proven to be a valuable resource in understanding many components of the pathogenesis of asthma in humans. However, the mouse model is not a perfect model as there are differences in degranulation of eosinophils and release of serotonin from mast cells between mice and humans in the mouse model of asthma compared to human asthma. Transgenic mice may be a valuable asset to utilize in understanding pathways and potential realms for therapeutic intervention.

The rat is also a common model for human asthma and has a size advantage over the mouse, resulting in enhanced ability for collection of blood and bronchoalveolar fluid. The BN rat is the most commonly utilized strain due to its pronounced IgE and inflammatory response to challenge following sensitization (Hylkema et al., 2002). As is the case with the mouse, rats are not considered a perfect model due to differences in the release of serotonin from mast cells between rats and humans.

While guinea pigs are one of the oldest animal models of human asthma, mechanistic studies in this species are limited

TABLE 18.6 Models of Acute Lung Injury[a]

Target	Example and Mechanism	Technical Issues/Limitations
Alveolar epithelium	Oleic acid—attempt to reproduce lung injury from lipid embolism	Insoluble in water; requires IV administration
	Endothelial cell necrosis	Lesions are multifocal
	Endotoxin (LPS)—endothelial cell apoptosis	Highly variable species response to LPS (differs between animals with and without pulmonary intravascular macrophages); also varies between strains (BALB/c mice sensitive)
		LPS preps vary in purity
		LPS by itself results in changes less severe than effects of live bacteria in the lung
	Acid aspiration—low pH (e.g., 1.5) fluid instillation	Narrow range between injurious and noninjurious acid concentrations
		Not like "real aspiration" since with aspiration there are food particles, bacteria, etc.
	Hyperoxia—damage from reactive oxygen species (free radicals) → necrosis and apoptosis	Relevance to humans unknown as humans exposed to 100% oxygen for up to 3 days do not develop lung injury
	Surfactant depletion by saline lavage → alveolar collapse, enhanced mechanical injury to alveolar walls, impaired alveolar host defenses	In adult humans, depletion of surfactant is typically a result of rather than a cause of acute lung injury
		Requires general anesthesia and intubation
	Mechanical ventilation—direct damage due to mechanical stretch/activation of intracellular pathways; role of "stretch" different in normal versus inflamed lungs	Technically challenging
		Requires anesthesia and intubation
	Intratracheal bleomycin	Species- and strain-related differences in susceptibility (high expression of bleomycin hydrolase)
Capillary endothelium and alveolar epithelium	Ischemia/reperfusion—as noted following lung transplantation (not related to rejection) and other causes of ischemia/reperfusion	Some technical challenges—severity worse when the lungs remain deflated during ischemic period and when pulmonary and bronchial circulations are occluded together
		Models of non-pulmonary beds is the only stimulus, typically requires secondary stimulus (e.g., LPS) to produce significant lesion
	Sepsis—administration of live bacteria (IV or pulmonary instillation) or creation of endogenous infection (e.g., cecal puncture)	With exception of pulmonary instillation, typically less severe than human disease

[a] From Matute-Bello et al. (2008).

by the low number of inbred strains and the lack of guinea pig-specific reagents. The guinea pig is the most widely used model for contact hypersensitivity to chemical irritants and proteins and has been used as a screening model for drugs that act through particular pathways considered to be relevant to human asthma. However, as with other models, care must be taken to avoid broad generalization and comparisons between the guinea pig and human. For example, some pharmacologic agents (methacholine, histamine, allergen) result in similar airway responses in the guinea pig and human, whereas others (leukotrienes) are considerably different.

The dog has also been used as a model for human asthma. Genetics plays an important role in dogs, similar to the situation in humans, and recent models have employed

selective breeding of dogs with high IgE titers. While dogs can be sensitized with a variety of antigens, one of the more useful models may be the aerosolized ragweed extract model, which results in long-term increased IgE, eosinophilia, and airway responsiveness.

Sheep have also been utilized as a model for human asthma with the primary inciting allergen being inoculation with *Ascaris suum*. As is the case with dogs and humans, there is considerable variability between sheep with the degree of response to allergen exposure. While a potential useful model, the size and cost of the animals may be a challenge.

As indicated earlier, there is no perfect animal model of human asthma. None of the animal models completely mirror the spectrum of changes noted in humans, particularly in respect to an ability to produce both an acute and a chronic/continued response (several animal models develop tolerance). Also, while some therapeutic agents that have shown benefit in animal models have translated into benefit in human asthma, there are many instances in which models have not been predicted for therapeutic agent benefit.

18.7.3.4 Fibrotic Lung Disease

As is the case with the other models of human respiratory disease models, excellent reviews of animal models of fibrotic lung disease are available in the literature. A recent publication from Moore et al. will be briefly outlined in the following (Moore et al., 2013).

Interstitial pulmonary fibrosis (IPF) has numerous potential inciting causes, including occupational or medical exposure, as a result of genetic defects, posttrauma, or acute lung injury, or the origin may be unknown. As is the case with animal models in general, no one animal model of pulmonary fibrosis fully recapitulates the human syndrome, either in respect to progression of disease or capturing all salient histopathologic features. Additionally, since there are so many different potential inciting causes of IPF in humans, resulting in considerable variability in histopathology in the human syndrome, making capturing all of the features in one animal model is even more difficult. Despite the inherent shortcomings, animal models of pulmonary fibrosis have considerable value in the elucidation of the pathogenesis/evaluation of potential therapeutic agents.

There are several animal models of IPF that can be induced by administration of a variety of agents. The most common are asbestos, silica, bleomycin, and fluorescein isothiocyanate (FITC). The route of administration is dependent upon the agent and may be intratracheal (asbestos, silica, bleomycin, FITC), aerosol/inhalation (asbestos, silica, bleomycin), pharyngeal (silica), or other parenteral routes including intraperitoneal/subcutaneous/intravenous (bleomycin). Care must be taken to avoid contamination including baking of asbestos/silica to avoid LPS-related lesions. Rodents are the most commonly employed model for all of these agents. Considerable progress has been achieved in better understanding the models (i.e., different strains of rodents develop more significant histopathologic changes for the different inciting agents, and older rodents typically develop more severe changes as compared to younger animals of the same strain with the same inciting agent).

An additional approach that has been of value is a more targeted approach—the cytokine overexpression models. In these models, the overexpression of cytokines, including TGF-β, TGF-α, IL-13, TNF-α, and IL-1β, is employed to develop a model that results in pulmonary fibrosis.

Other animal models have been employed to more closely mimic IPF that can follow an acute lung injury. In this syndrome, the pulmonary fibrosis occurs as a result of a severe lung injury (acute lung injury/ARDS), resulting in severe complications, with secondary pulmonary hypertension and

high mortality. The animal models include acid instillation, radiation, and lung contusion.

Because none of the animal models described previously fully recapitulate the pathogenesis/pathology of IPF, attempts are ongoing to "humanize" a model for IPF. Administration of human IPF fibroblasts, but not fibroblasts from normal human lungs into immunodeficient mice, results in focal alveolar remodeling (Pierce et al., 2007; Trujillo et al., 2010). The cells also tend to activate murine epithelial cells and fibroblasts with resultant remodeling. Use of this model provides a unique opportunity to evaluate the potential for therapeutic intervention.

18.8 CONCLUSION

Animal models that mimic the clinical disease in humans play a crucial role and are fundamentally required to advance our scientific understanding and development of novel therapies for various human diseases and pathologies. Identification and characterization of novel drug targets also require nonclinical testing in appropriate animal models that are relevant to human pathologies and clinical outcomes. In spite of undeniable differences between humans and animals, animal models still help to identify altered molecular targets and pathways for better understanding of disease pathogenesis and subsequent development of new therapeutic agents and their potential safety liabilities. Understanding mechanisms of potential safety liabilities/toxicities will help to assess the risk and the relevance of the nonclinical data to humans. The strategy to accurately detect the target organ toxicity using animal models depends in part on the similarity in mechanisms of toxicity across species. For example, the majority of untoward effects with cytotoxic anticancer drugs are readily predicted in preclinical safety studies because of their similar basic mechanism of action. Traditional animal models, such as rats, mice, dogs, and nonhuman primates, are generally considered "predictive" of human toxicity in the 30-day or less nonclinical safety testing studies that support phase I clinical trials (Olson et al., 2000).

Unfortunately, prediction of toxicities that will be encountered in humans is not perfect. Species differences in metabolic pathways exist, and these differences can cause qualitative and quantitative differences in exposure to metabolites and result in species-specific toxicity of certain drugs (e.g., Efavirenz—nephrotoxic in rats but not in monkeys or humans). Nonclinical safety testing studies are typically undertaken in "healthy juvenile" animals, although new drugs are often intended for diseased patient populations, which may encounter particular risks, especially when associated with long-duration exposures. Young, healthy animals may not consistently translate subtle or chronically progressive risks to patients with preexisting disease. Consequently, traditional animal models in nonclinical safety testing may be less effective at modeling and/or predicting drug-associated safety concerns in diseased patient populations. In some instances, diseased animal models may provide a greater understanding of safety risks of new drug candidates in nonclinical or clinical development, with their greatest value being in targeted or hypothesis-driven studies to help understand the mechanism of a particular toxicity. Utilizing appropriate diseased models may allow for more robust evaluation of toxicity, particularly in respect to off-target toxicity, and may provide greater predictivity to clinical outcomes. This is particularly relevant for evaluating compounds with new safety challenges, including immune responses, exaggerated pharmacology, tissue cross-reactivity, and species-specific responses. For example, as outlined in the section on disease models of the nervous system, significant target organ toxicity (e.g., CV shock, systemic inflammation, and death) was present in transgenic (ASMKO) mice administered rhASM but not in healthy conventional species, such as mice, rats, and dogs. Similarly, as outlined in the section on cardiovascular disease models, SHRs are much more

sensitive than normotensive rats to the cardiotoxic effects of doxorubicin, highlighting the roles of hypertension in cardiac toxicity and its relevance to human risk.

Despite the potential attributes of animal models of human disease, diseased animal models have been primarily utilized to study the pathogenesis of a disease in academic laboratories or to test the efficacy of new therapeutic agents in pharmaceutical industries. Such models have rarely been adopted in the nonclinical toxicity studies of new therapeutic agents mainly due to the perception (real or otherwise) about the lack of translatability of these models to humans that has already been mentioned. A number of critical issues may arise in drug toxicity studies due to insufficient knowledge of model-specific background pathology and sensitivity and functional/anatomical differences among species. It is also important to determine whether a given group of animals modeling a specific disease can be maintained over the intended duration of efficacy and safety studies. Many genetically engineered or surgically induced models may not have sufficient long-term survivability, which may substantially limit their use, particularly for safety studies. Similar

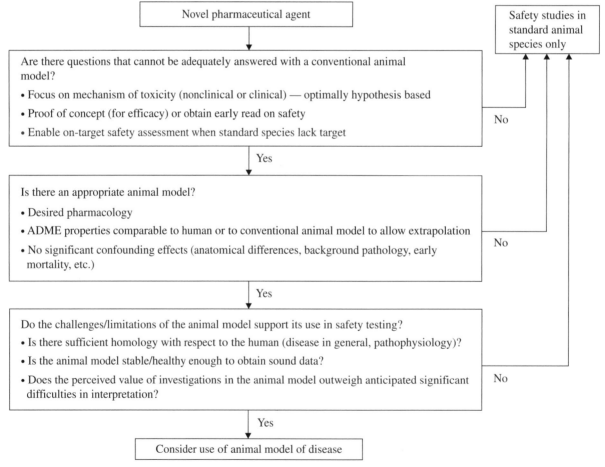

FIGURE 18.1 Decision tree—selection of animal disease models for conducting nonclinical safety studies. Morgan et al. (2013, pp. 408–418). Reproduced with permission from SAGE Publications.

considerations would be apparent for animal models with naturally occurring or implanted tumors as control and low-dose animals may have survival issues, precluding long-term nonclinical safety testing. It is also important to consider the life span differences between humans and model animals in that relatively short-lived animals may not be the right model to accurately mimic human diseases/disorders that are manifested late in life.

In summary, the use of diseased animal models should be reserved for answering specific hypothesis-driven questions in order to understand the pathogenesis and/or relevance of toxicities noted in conventional animals or to understand the pathogenesis of a new toxicity noted in a clinical setting. The selection of the "right" animal models is crucial, and animal models selected must be relevant to humans in regard to ADME and underlying toxicity mechanisms. Critical steps prior to considering animal models of disease include (i) degree of homogeneity with respect to the human disease, (ii) rigorous characterization—similar clinical and histological features—so that any experimental findings in animals would be relevant and applied to humans and (iii) reproducible so that various laboratories could make valid and scientific comparison of findings or study outcomes. The flow chart below provides suggestions of factors to take into consideration before proceeding with nonclinical safety testing in an animal model of human disease (Fig. 18.1).

REFERENCES

Akar, F.G., Wu, R.C., Deschenes, I., Armoundas, A.A., Piacentino, V. III, Houser, S.R., and Tomaselli, G.F. (2004): Phenotypic differences in transient outward K+ current of human and canine ventricular myocytes: insights into molecular composition of ventricular Ito. *Am. J. Physiol. Heart Circ. Physiol.* 286, H602–H609.

Alpert, N.R., Brosseau, C., Federico, A., Krenz, M., Robbins, J., and Warshaw, D.M. (2002): Molecular mechanics of mouse cardiac myosin isoforms. *Am. J. Physiol. Heart Circ. Physiol.* 283, H1446–H1454.

Amos-Landgraf, J.M., Kwong, L.N., Kendziorski, C.M., Reichelderfer, M., Torrealba, J., Weichert, J., Haag, J.D., Chen, K.S., Waller, J.L., Gould, M.N., and Dove, W.F. (2007): A target-selected Apc-mutant rat kindred enhances the modeling of familial human colon cancer. *Proc. Natl. Acad. Sci. U. S. A.*, 104, 4036–4041.

Andreassen, A., Vikse, R., Mikalsen, A., Adamovic, T., Steffensen, I.L., Hjertholm, H., Levan, G., and Alexander, J. (2006): 2-Amino-1-methyl-6-phenylimidazo[4,5-b]pyridine (PhIP) induces genetic changes in murine intestinal tumours and cells with ApcMin mutation. *Mutat. Res.* 604, 60–70.

Arita, M., Yoshida, M., Hong, S., Tjonahen, E., Glickman, J.N., Petasis, N.A., Blumberg, R.S., and Serhan, C.N. (2005): Resolvin E1, an endogenous lipid mediator derived from omega-3 eicosapentaenoic acid, protects against 2,4,6-trinitrobenzene sulfonic acid-induced colitis. *Proc. Natl. Acad. Sci. U. S. A.* 102, 7671–7676.

Bagavant, H., Deshmukh, U.S., Gaskin, F., and Fu, S.M. (2004): Lupus glomerulonephritis revisited 2004: autoimmunity and end-organ damage. *Scand. J. Immunol.* 60, 52–63.

Bailey, C.J. and Day, C. (1989): Traditional plant medicines as treatments for diabetes. *Diabetes Care* 12, 553–564.

Baker, R.C., Armstrong, M.A., Young, I.S., McClean, E., O'Rourke, D., Campbell, F.C., D'Sa, A.A., and McBride, W.T. (2006): Methylprednisolone increases urinary nitrate concentrations and reduces subclinical renal injury during infrarenal aortic ischemia reperfusion. *Ann. Surg.* 244, 821–826.

Bakhtiar, S.M., LeBlanc, J.G., Salvucci, E., Ali, A., Martin, R., Langella, P., Chatel, J.M., Miyoshi, A., Bermudez-Humaran, L.G., and Azevedo, V. (2013): Implications of the human microbiome in inflammatory bowel diseases. *FEMS Microbiol. Lett.* 342, 10–17.

Bakiri, L. and Wagner, E.F. (2013): Mouse models for liver cancer. *Mol. Oncol.* 7, 206–223.

Balasubramanian, S., Jansen, M., Valerius, M.T., Humphreys, B.D., and Strom, T.B. (2012): Orphan nuclear receptor Nur77 promotes acute kidney injury and renal epithelial apoptosis. *J. Am. Soc. Nephrol.* 23, 674–686.

Bassani, J.W., Bassani, R.A., and Bers, D.M. (1994): Relaxation in rabbit and rat cardiac cells: species-dependent differences in cellular mechanisms. *J. Physiol.* 476, 279–293.

Beal, M.F. (2001): Experimental models of Parkinson's disease. *Nat. Rev. Neurosci.* 2, 325–334.

Beaty, T.H., Prenger, V.L., Virgil, D.G., Lewis, B., Kwiterovich, P.O., and Bachorik, P.S. (1992): A genetic model for control of hypertriglyceridemia and apolipoprotein B levels in the Johns Hopkins colony of St. *Thomas Hospital rabbits. Genetics* 132, 1095–1104.

Becker, G.J. and Hewitson, T.D. (2013): Animal models of chronic kidney disease: useful but not perfect. *Nephrol. Dial. Transplant.* 28, 2432–2438.

Bellentani, S., Scaglioni, F., Marino, M., and Bedogni, G. (2010): Epidemiology of non-alcoholic fatty liver disease. *Dig. Dis.* 28, 155–161.

Benko, R., Molnar, Z., Nemes, D., Dekany, A., Kelemen, D., Illenyi, L., Cseke, L., Papp, A., Varga, G., and Bartho, L. (2010): Unexpected insensitivity of the cholinergic motor responses to morphine in the human small intestine. *Pharmacology* 86, 145–148.

Bernstein, C.N., Blanchard, J.F., Kliewer, E., and Wajda, A. (2001): Cancer risk in patients with inflammatory bowel disease: a population-based study. *Cancer* 91, 854–862.

Betton, G.R., Kenne, K., Somers, R., and Marr, A. (2005): Protein biomarkers of nephrotoxicity; a review and findings with cyclosporin A, a signal transduction kinase inhibitor and N-phenylanthranilic acid. *Cancer Biomark.* 1, 59–67.

Bijvoet, A.G.A., Van Hirtum, H., Vermey, M., Val Leenen, D., Ven Der Ploeg, A.T., Mooi, W.J., and Reuser, A.J. (1999): Pathological features of glycogen storage disease type II highlighted in the knockout mouse model. *J. Pathol.* 189, 416–424.

Bing, O.H., Brooks, W.W., Robinson, K.G., Slawsky, M.T., Hayes, J.A., Litwin, S.E., Sen, S., and Conrad, C.H. (1995):

The spontaneously hypertensive rat as a model of the transition from compensated left ventricular hypertrophy to failure. *J. Mol. Cell. Cardiol.* 27, 383–396.

Bissahoyo, A., Pearsall, R.S., Hanlon, K., Amann, V., Hicks, D., Godfrey, V.L., and Threadgill, D.W. (2005): Azoxymethane is a genetic background-dependent colorectal tumor initiator and promoter in mice: effects of dose, route, and diet. *Toxicol. Sci.* 88, 340–345.

Bivin, W.S., Crawford, M.P., and Brewer, N.R. (1979): Morphophysiology. In: *The Laboratory Rat.* pp. 73–103, Academic Press, New York.

Bolter, C.P. and Atkinson, K.J. (1988): Maximum heart rate responses to exercise and isoproterenol in the trained rat. *Am. J. Physiol.* 254, R834–R839.

Bonnet, S., Michelakis, E.D., Porter, C.J., Andrade-Navarro, M.A., Thebaud, B., Bonnet, S., Haromy, A., Harry, G., Moudgil, R., McMurtry, M.S., Weir, E.K., and Archer, S.L. (2006): An abnormal mitochondrial-hypoxia inducible factor-1alpha-Kv channel pathway disrupts oxygen sensing and triggers pulmonary arterial hypertension in fawn hooded rats: similarities to human pulmonary arterial hypertension. *Circulation* 113, 2630–2641.

Borghoff, S.J., Short, B.G., and Swenberg, J.A. (1990): Biochemical mechanisms and pathobiology of alpha 2u-globulin nephropathy. *Annu. Rev. Pharmacol. Toxicol.* 30, 349–367.

Breunig, E., Michel, K., Zeller, F., Seidl, S., Weyhern, C.W., and Schemann, M. (2007): Histamine excites neurones in the human submucous plexus through activation of H1, H2, H3 and H4 receptors. *J. Physiol.* 583, 731–742.

Breyer, M.D., Bottinger, E., Brosius, F.C. III, Coffman, T.M., Harris, R.C., Heilig, C.W., and Sharma, K. (2005): Mouse models of diabetic nephropathy. *J. Am. Soc. Nephrol.* 16, 27–45.

Bucaloiu, I.D., Kirchner, H.L., Norfolk, E.R., Hartle, J.E. 2nd., and Perkins, R.M. (2012): Increased risk of death and de novo chronic kidney disease following reversible acute kidney injury. *Kidney Int.* 81, 477–485.

Budhiraja, S. and Singh, J. (2008): Protein kinase C beta inhibitors: a new therapeutic target for diabetic nephropathy and vascular complications. *Fundam. Clin. Pharmacol.* 22, 231–240.

Burgenske, D.M., Monsma, D.J., Dylewski, D., Scott, S.B., Sayfie, A.D., Kim, D.G., Luchtefeld, M., Martin, K.R., Stephenson, P., Hostetter, G., Dujovny, N., and MacKeigan, J.P. (2014): Establishment of genetically diverse patient-derived xenografts of colorectal cancer. *Am J Cancer Res.* 19, 824–837.

Burne, M.J., Haq, M., Matsuse, H., Mohapatra, S., and Rabb, H. (2000): Genetic susceptibility to renal ischemia reperfusion injury revealed in a murine model. *Transplantation* 69, 1023–1025.

Burnier, M., 2005: The safety of rofecoxib. *Expert Opin. Drug Saf.* 4, 491–499.

Calvisi, D.F., Ladu, S., Gordon, A., Farina, M., Conner, E.A., Lee, J.S., Factor, V.M., and Thorgeirsson, S.S. (2006): Ubiquitous activation of Ras and Jak/Stat pathways in human HCC. *Gastroenterology* 13, 1117–1128.

Casey, R.G., Joyce, M., Roche-Nagle, G., Chen, G., and Bouchier-Hayes, D. (2005): Pravastatin modulates early diabetic nephropathy in an experimental model of diabetic renal disease. *J. Surg. Res.* 123, 176–181.

Cave, M., Deaciuc, I., Mendez, C., Song, Z., Joshi-Barve, S., Barve, S., and McClain, C. (2007): Nonalcoholic fatty liver disease: predisposing factors and the role of nutrition. *J. Nutr. Biochem.* 18, 184–195.

Cazorla, O., Freiburg, A., Helmes, M., Centner, T., McNabb, M., Wu, Y., Trombitas, K., Labeit, S., and Granzier, H. (2000): Differential expression of cardiac titin isoforms and modulation of cellular stiffness. *Circ. Res.* 86, 59–67.

Chandra, S.A., Nolan, M.W., and Malarkey, D.E. (2010): Chemical carcinogenesis of the gastrointestinal tract in rodents: an overview with emphasis on NTP carcinogenesis bioassays. *Toxicol. Pathol.* 38, 188–197.

Charpentier, F., Demolombe, S., and Escande, D. (2004): Cardiac channelopathies: from men to mice. *Ann. Med.* 36(Suppl. 1), 28–34.

Charpentier, F., Bourge, A., and Merot, J. (2008): Mouse models of SCN5A-related cardiac arrhythmias. *Prog. Biophys. Mol. Biol.* 98, 230–237.

Chatterjee, B., Demyan, W.F., Song, C.S., Garg, B.D., and Roy, A.K. (1989): Loss of androgenic induction of a 2u-globulin gene family in the liver of NIH black rats. *Endocrinology* 125, 1385–1388.

Chatzigeorgiou, A., Halapas, A., Kalafatakis, K., and Kamper, E. (2009): The use of animal models in the study of diabetes mellitus. *In Vivo* 23, 245–258.

Chertow, G.M., Burdick, E., Honour, M., Bonventre, J.V., and Bates, D.W. (2005): Acute kidney injury, mortality, length of stay, and costs in hospitalized patients. *J. Am. Soc. Nephrol.* 16, 3365–3370.

Cheung, C. and Gonzalez, F.J. (2008): Humanized mouse lines and their application for prediction of human drug metabolism and toxicological risk assessment. *J. Pharmacol. Exp. Ther.* 327, 288–299.

Cheverud, J.M., Tardif, S., Henke, M.A., and Clapp, N.K. (1993): Genetic epidemiology of colon cancer in the cotton-top tamarin (*Saguinus oedipus*). *Hum. Biol.* 65, 1005–1012.

Chien, A.J. and Rugo, H.S. (2010): The cardiac safety of trastuzumab in the treatment of breast cancer. *Expert Opin. Drug Saf.* 9, 335–346.

Cho, C.H. and Pfeiffer, C.J. (1981): Gastrointestinal ulceration in the guinea pig in response to dimaprit, histamine, and H1- and H2-blocking agents. *Dig. Dis. Sci.* 26, 306–311.

Cho, C.H., Chen, B.W., Hui, W.M., Luk, C.T., and Lam, S.K. (1990): Endogenous prostaglandins: its role in gastric mucosal blood flow and ethanol ulceration in rats. *Prostaglandins* 40, 397–403.

Christensen, M., Su, A.W., Snyder, R.W., Greco, A., Lipschutz, J.H., and Madaio, M.P. (2006): Simvastatin protection against acute immune-mediated glomerulonephritis in mice. *Kidney Int.* 69, 457–463.

Christiansen, T., Rasch, R., Stodkilde-Jorgensen, H., and Flyvbjerg, A. (1997): Relationship between MRI and morphometric kidney measurements in diabetic and non-diabetic rats. *Kidney Int.* 51, 50–56.

Chu, T.F., Rupnick, M.A., Kerkela, R., Dallabrida, S.M., Zurakowski, D., Nguyen, L., Woulfe, K., Pravda, E., Cassiola, F., Desai, J., George, S., Morgan, J.A., Harris, D.M., Ismail, N.S., Chen, J.H., Schoen, F.J., Van den Abbeele, A.D., Demetri, G.D., Force, T., and Chen, M.H. (2007): Cardiotoxicity associated with tyrosine kinase inhibitor sunitinib. *Lancet* 370, 2011–2019.

Chung, F., Barnes, N., Allen, M., Angus, R., Corris, P., Knox, A., Miles, J., Morice, A., O'Reilly, J., and Richardson, M. (2002): Assessing the burden of respiratory disease in the UK. *Respir. Med.* 96, 963–975.

Church, E.M., Mehlhaff, C.J., and Patnaik, A.K. (1987): Colorectal adenocarcinoma in dogs: 78 cases (1973–1984). *J. Am. Vet. Med. Assoc.* 191, 727–730.

Claes, I.J., Lebeer, S., Shen, C., Verhoeven, T.L., Dilissen, E., De Hertogh, G., Bullens, D.M., Ceuppens, J.L., Van Assche, G., Vermeire, S., Rutgeerts, P., Vanderleyden, J., and De Keersmaecker, S.C. (2010): Impact of lipoteichoic acid modification on the performance of the probiotic Lactobacillus rhamnosus GG in experimental colitis. *Clin. Exp. Immunol.* 162, 306–314.

Classen, D.C., Pestotnik, S.L., Evans, R.S., Lloyd, J.F., and Burke, J.P. (1997): Adverse drug events in hospitalized patients. Excess length of stay, extra costs, and attributable mortality. *JAMA* 277, 301–306.

Coussens, L.M. and Werb, Z. (2002): Inflammation and cancer. *Nature* 420, 860–867.

Dangles-Marie, V., Pocard, M., Richon, S., Weiswald, L.B., Assayag, F., Saulnier, P., Judde, J.G., Janneau, J.L., Auger, N., Validire, P., Dutrillaux, B., Praz, F., Bellet, D., and Poupon, M.F. (2007): Establishment of human colon cancer cell lines from fresh tumors versus xenografts: comparison of success rate and cell line features. *Cancer Res.* 67, 398–407.

Dashwood, R.H., Suzui, M., Nakagama, H., Sugimura, T., and Nagao, M. (1998): High frequency of beta-catenin (ctnnb1) mutations in the colon tumors induced by two heterocyclic amines in the F344 rat. *Cancer Res.* 58, 1127–1129.

Davis, R.P., van den Berg, C.W., Casini, S., Braam, S.R., and Mummery, C.L. (2011): Pluripotent stem cell models of cardiac disease and their implication for drug discovery and development. *Trends Mol. Med.* 17, 475–484.

De Bruin, M.L., Pettersson, M., Meyboom, R.H., Hoes, A.W., and Leufkens, H.G. (2005): Anti-HERG activity and the risk of drug-induced arrhythmias and sudden death. *Eur. Heart J.* 26, 590–597.

De Robertis, M., Massi, E., Poeta, M.L., Carotti, S., Morini, S., Cecchetelli, L., Signori, E., and Fazio, V.M. (2011): The AOM/DSS murine model for the study of colon carcinogenesis: from pathways to diagnosis and therapy studies. *J Carcinog.* 10, 9.

Del Soldato, P. (1984): Nonsteroidal anti-inflammatory drugs and their actions in the intestine—pharmacological, emotional, dietary and microbial factors. In: *Advances in Inflammation Research* (Weissman, G., ed.), Vol. 6, pp. 89–101, Raven Press, New York.

Derry, M.M., Raina, K., Agarwal, C., and Agarwal, R. (2013): Identifying molecular targets of lifestyle modifications in colon cancer prevention. *Front Oncol.* 3, 119.

Desai, K.H., Sato, R., Schauble, E., Barsh, G.S., Kobilka, B.K., and Bernstein, D. (1997): Cardiovascular indexes in the mouse at rest and with exercise: new tools to study models of cardiac disease. *Am. J. Physiol.* 272, H1053–H1061.

Doi, R., Masuyama, T., Yamamoto, K., Doi, Y., Mano, T., Sakata, Y., Ono, K., Kuzuya, T., Hirota, S., Koyama, T., Miwa, T., and Hori, M. (2000): Development of different phenotypes of hypertensive heart failure: systolic versus diastolic failure in Dahl salt-sensitive rats. *J. Hypertens.* 18, 111–120.

Eaden, J.A., Abrams, K.R., and Mayberry, J.F. (2001): The risk of colorectal cancer in ulcerative colitis: a metaanalysis. *Gut* 48, 526–535.

Edelmann, W., Yang, F., Kuraguchi, M., Heyer, J., Lia, M., Kneitz, B., Fan, K.H., Brown, A.M.C., Lipkin, M., and Kucherlapati, R. (1999): Tumorigenesis in Mlh1 and Mlh1/Apc1638N mutant mice. *Cancer Res.* 59, 1301–1307.

EMEA (2010): European Medicines Evaluation Agency. Committee for Medicinal Products for Human Use (CHMP). Reflection paper on non-clinical evaluation of drug-induced liver injury (DILI). EMEA/CHMP/SWP/150115/2006.

Erdman, S.E. and Poutahidis, T. (2010): Roles for inflammation and regulatory T cells in colon cancer. *Toxicol. Pathol.* 38, 76–87.

Esumi, H., Ohgaki, H., Kohzen, E., Takayama, S., and Sugimura, T. (1989): Induction of lymphoma in CDF1 mice by the food mutagen, 2-amino-1-methyl-6-phenylimidazo[4,5-b]pyridine. *Jpn. J. Cancer Res.* 80, 1176–1178.

Farquhar, M.G., Saito, A., Kerjaschki, D., and Orlando, R.A. (1995): The Heymann nephritis antigenic complex: megalin (gp330) and RAP. *J. Am. Soc. Nephrol.* 6, 35–47.

FDA (2009): Guidance for Industry: Drug-Induced Liver Injury: Premarketing Clinical Evaluation. *Release date* July 2009.

Fearon, E.R. and Vogelstein, B. (1990): A genetic model for colorectal tumorigenesis. *Cell* 61, 759–767.

Fedorov, V.V., Glukhov, A.V., Ambrosi, C.M., Kostecki, G., Chang, R., Janks, D., Schuessler, R.B., Moazami, N., Nichols, C.G., and Efimov, I.R. (2011): Effects of KATP channel openers diazoxide and pinacidil in coronary-perfused atria and ventricles from failing and non-failing human hearts. *J. Mol. Cell. Cardiol.* 51, 215–225.

Fermini, B. and Fossa, A.A. (2003): The impact of drug-induced QT interval prolongation on drug discovery and development. *Nat. Rev. Drug Discov.* 2, 439–447.

Fermini, B., Wang, Z., Duan, D., and Nattel, S. (1992): Differences in rate dependence of transient outward current in rabbit and human atrium. *Am. J. Physiol.* 263, H1747–H1754.

Ferri, N., Siegl, P., Corsini, A., Herrmann, J., Lerman, A., and Benghozi, R. (2013): Drug attrition during pre-clinical and clinical development: understanding and managing drug-induced cardiotoxicity. *Pharmacol. Ther.* 138, 470–484.

Fewell, J.G., Osinska, H., Klevitsky, R., Ng, W., Sfyris, G., Bahrehmand, F., and Robbins, J. (1997): A treadmill exercise regimen for identifying cardiovascular phenotypes in transgenic mice. *Am. J. Physiol.* 273, H1595–H1605.

Fickert, P., Stoger, U., Fuchsbichler, A., Moustafa, T., Marschall, H.U., Denk, H., and Trauner, M. (2007): A new xenobiotic-induced mouse model of sclerosing cholangitis and biliary fibrosis. *Am. J. Pathol.* 171, 525–536.

Fiebig, H.H., Schuler, J., Bausch, N., Hofmann, M., Metz, T., and Korrat, A. (2007): Gene signatures developed from patient tumor explants grown in nude mice to predict tumor response to 11 cytotoxic drugs. *Cancer Genomics Proteomics* 4, 197–209.

Flamm, S.D., Taki, J., Moore, R., Lewis, S.F., Keech, F., Maltais, F., Ahmad, M., Callahan, R., Dragotakes, S., Alpert, N., and Strauss, H.W. (1990): Redistribution of regional and organ blood volume and effect on cardiac function in relation to upright exercise intensity in healthy human subjects. *Circulation* 81, 1550–1559.

Fleet, J.C. (2014): Animal models of gastrointestinal and liver diseases. New mouse models for studying dietary prevention of colorectal cancer. *Am. J. Physiol. Gastrointest. Liver Physiol.* 307, G249–G259.

Fletcher, A.P. (1978): Drug safety tests and subsequent clinical experience. *J. R. Soc. Med.* 71, 693–696.

Foster, M.H. (1999): Relevance of systemic lupus erythematosus nephritis animal models to human disease. *Semin. Nephrol.* 19, 12–24.

Fu, Y., Du, Y., and Mohan, C. (2007): Experimental anti-GBM disease as a tool for studying spontaneous lupus nephritis. *Clin. Immunol.* 124, 1090–1118.

Fujimoto, T., Okada, M., Kondo, Y., and Tada, T. (1964): The nature of masugi nephritis; histo- and immunopathological studies. *Acta Pathol. Jpn.* 14, 275–310.

Geiersbach, K.B. and Samowitz, W.S. (2011): Microsatellite instability and colorectal cancer. *Arch. Pathol. Lab. Med.* 135, 1269–1277.

Giraud, S., Favreau, F., Chatauret, N., Thuillier, R., Maiga, S., and Hauet, T. (2011): Contribution of large pig for renal ischemia-reperfusion and transplantation studies: the preclinical model. *J. Biomed. Biotechnol.* 2011, 532127.

Glassock, R.J. (2009): Human idiopathic membranous nephropathy—a mystery solved? *N. Engl. J. Med.* 361, 81–83.

Glenner, G. and Wong, C. (1984): Alzheimer's disease: initial report of the purification and characterization of a novel cerebrovascular amyloid protein. *Biochem. Biophys. Res. Commun.* 120, 885–890.

Glukhov, A.V., Fedorov, V.V., Lou, Q., Ravikumar, V.K., Kalish, P.W., Schuessler, R.B., Moazami, N., and Efimov, I.R. (2010a): Transmural dispersion of repolarization in failing and nonfailing human ventricle. *Circ. Res.* 106, 981–991.

Glukhov, A.V., Flagg, T.P., Fedorov, V.V., Efimov, I.R., and Nichols, C.G. (2010b): Differential K (ATP) channel pharmacology in intact mouse heart. *J. Mol. Cell. Cardiol.* 48, 152–160.

Gobe, G., Willgoss, D., Hogg, N., Schoch, E., and Endre, Z. (1999): Cell survival or death in renal tubular epithelium after ischemia-reperfusion injury. *Kidney Int.* 56, 1299–1304.

Gojo, A., Utsunomiya, K., Taniguchi, K., Yokota, T., Ishizawa, S., Kanazawa, Y., Kurata, H., and Tajima, N. (2007) The Rho-kinase inhibitor, fasudil, attenuates diabetic nephropathy in streptozotocin-induced diabetic rats. *Eur. J. Pharmacol.* 568, 242–247.

Goodman, D.G., Boorman, G.A., and Strandberg, J.D. (1994): Selection and use of the B6C3F1 mouse and F344 rat in long-term bioassays for carcinogenicity. In: *Handbook of Carcinogen Testing*, 2nd ed. (Milman, H.A., and Weisburger, E.K., eds.), pp. 347–390, Noyes, Park Ridge, NJ.

Goodrich, J.A. and Hottendorf, G.H. (1995): Tobramycin gender-related nephrotoxicity in Fischer but not Sprague-Dawley rats. *Toxicol. Lett.* 75, 127–131.

Gotz, J., Streffer, J., David, D., Schild, A., Hoerndli, F., Pennanen, L., Kurosinski, P., and Chen, F. (2004): Transgenic animal models of Alzheimer's disease and related disorders: histopathology, behavior and therapy. *Mol. Psychiatry* 9, 664–683.

Groneberg, D.A. and Chung, K.F. (2004): Models of chronic obstructive pulmonary disease. *Respir. Res.* 4, 18–34.

Guihaire, J., Bogaard, H.J., Flecher, E., Noly, P.E., Mercier, O., Haddad, F., and Fadel, E. (2013): Experimental models of right heart failure: a window for translational research in pulmonary hypertension. *Semin. Respir. Crit. Care Med.* 34, 689–699.

Guth, P.H., Aures, D., and Paulsen, G. (1979): Topical aspirin plus HCl gastric lesions in the rat. Cytoprotective effect of prostaglandin, cimetidine, and probanthine. *Gastroenterology* 76, 88–93.

Haidara, M.A., Mikhailidis, D.P., Rateb, M., Ahmed, Z.A., Yassin, H.Z., Ibrahim, I.M., and Rashed, L.A. (2009): Evaluation of the effect of oxidative stress and vitamin E supplementation on renal function in rats with streptozotocin-induced Type 1 diabetes. *J. Diabet. Complicat.* 23, 130–136.

Hamilton, N. and Ianuzzo, C.D. (1991): Contractile and calcium regulating capacities of myocardia of different sized mammals scale with resting heart rate. *Mol. Cell. Biochem.* 106, 133–141.

Hard, G.C. and Khan, K.N. (2004): A contemporary overview of chronic progressive nephropathy in the laboratory rat, and its significance for human risk assessment. *Toxicol. Pathol.* 32, 171–180.

Hard, G.C., Rodgers, I.S., Baetcke, K.P., Richards, W.L., McGaughy, R.E., and Valcovic, L.R. (1993): Hazard evaluation of chemicals that cause accumulation of alpha 2u-globulin, hyaline droplet nephropathy, and tubule neoplasia in the kidneys of male rats. *Environ. Health Perspect.* 99, 313–349.

Hasegawa, R., Sano, M., Tamano, S., Imaida, K., Shirai, T., Nagao, M., Sugimura, T., and Ito, N. (1993): Dose-dependence of 2-amino-1-methyl-6-phenylimidazo[4,5-b]-pyridine (PhIP) carcinogenicity in rats. *Carcinogenesis* 14, 2553–2557.

He, J., Irwin, D.M., Chen, R., and Zhang, Y.P. (2010): Stepwise loss of motilin and its specific receptor genes in rodents. *J. Mol. Endocrinol.* 44, 37–44.

Health Canada (2012): Recommendations from the Scientific Advisory Panel Sub-groups on Hepatotoxicity: Hepatotoxicity of Health Products. *Release date April* 18, 2012.

Heyman, S.N., Rosenberger, C., and Rosen, S. (2010): Experimental ischemia reperfusion: biases and myths-the proximal vs. distal hypoxic tubular injury debate revisited. *Kidney Int..* 77, 9–16.

Holbrook, J.D., Gill, C.H., Zebda, N., Spencer, J.P., Leyland, R., Rance, K.H., Trinh, H., Balmer, G., Kelly, F.M., Yusaf, S.P., Courtenay, N., Luck, J., Rhodes, A., Modha, S., Moore, S.E.,

Sanger, G.J., and Gunthorpe, M.J. (2009): Characterisation of 5-HT3C, 5-HT3D and 5-HT3E receptor subunits: evolution, distribution and function. *J. Neurochem.* 108, 384–396.

Horgan, S., Watson, C., Glezeva, N., and Baugh, J. (2014): Murine models of diastolic dysfunction and heart failure with preserved ejection fraction. *J. Card. Fail.* 20, 984–995.

Horinouchi, K., Erlich, S., Perl, D.P., Ferlinz, K., Bisgaier, C.L., Sandhoff, K., Desnick, R.J., Stewart, C.L., and Schuchman, E.H. (1995): Acid sphingomyelinase deficient mice: a model of types A and B Niemann-Pick disease. *Nat. Genet.* 10, 288–293.

Huang, C.L., Lei, L., Matthews, G.D., Zhang, Y., and Lei, M. (2012): Pathophysiological mechanisms of sino-atrial dysfunction and ventricular conduction disease associated with Scn5a deficiency: insights from mouse models. *Front. Physiol.* 3, 234.

Hylkema, M.N., Hoekstra, M.O., Luinge, M., and Timens, W. (2002): The strength of the ova-induced airway inflammation in rats is strain dependent. *Clin. Exp. Immunol.* 129, 390–396.

International Conference on Harmonisation (ICH) of Technical Requirements for Registration of Pharmaceuticals for Human use (2000): S7A—Safety Pharmacological Studies for Human Pharmaceuticals. Available at http://www.ich.org/products/guidelines/safety/article/safety-guidelines.html (accessed on October 16, 2015).

International Conference on Harmonisation (ICH) of Technical Requirements for Registration of Pharmaceuticals for Human use (2005): S7B—The Nonclinical Evaluation of the Potential for Delayed Ventricular Repolarization. Available at http://www.ich.org/products/guidelines/safety/article/safety-guidelines.html (accessed on October 16, 2015).

Ishiguro, Y., Ochiai, M., Sugimura, T., Nagao, M., and Nakagama, H. (1999): Strain differences of rats in the susceptibility to aberrant crypt foci formation by 2-amino-1-methyl-6-phenylimidazo-[4,5-b]pyridine: no implication of Apc and Pla2g2a genetic polymorphisms in differential susceptibility. *Carcinogenesis* 20, 1063–1068.

Islam, M.S. (2013): Animal models of diabetic neuropathy: progress since 1960s. *J Diabetes Res.* 2013, 1–9.

Izuwa, Y., Kusaba, J., Horiuchi, M., Aiba, T., Kawasaki, H., and Kurosaki, Y. (2009): Comparative study of increased plasma quinidine concentration in rats with glycerol and cisplatin induced acute renal failure. *Drug Metab. Pharmacokinet.* 24, 451–457.

Jay, G.W., Demattos, R.B., Weinstein, E.J., Philbert, M.A., Pardo, I.D., and Brown, T.P. (2011): Animal models for neural disease. *Toxicol. Pathol.* 39, 167–169.

Jiang, M., Wei, Q., Pabla, N., Dong, G., Wang, C.Y., Yang, T., Smith, S.B., and Dong, Z. (2007): Effects of hydroxyl radical scavenging on cisplatin-induced p53 activation, tubular cell apoptosis and nephrotoxicity. *Biochem. Pharmacol.* 73, 1499–1510.

Jnni, P., Nartey, L., Reichenbach, S., Sterchi, R., Dieppe, P.A., and Egger, M. (2004): Risk of cardiovascular events and rofecoxib: cumulative meta-analysis. *Lancet* 364, 2021–2029.

Johnson, R.L. and Fleet, J.C. (2013): Animal models of colorectal cancer. *Cancer Metastasis Rev.* 32, 39–61.

Jones, T.W., Chorpa, S., Kaufman, J.S., Flamenbaum, W., and Trump, B.F. (1985): Cis-diamminedichloroplatinum (II)-induced acute renal failure in the rat: enzyme histochemical studies. *Toxicol. Pathol.* 13, 296–305.

Julien, S., Merino-Trigo, A., Lacroix, L., Pocard, M., Goere, D., Mariani, P., Landron, S., Bigot, L., Nemati, F., Dartigues, P., Weiswald, L.B., Lantuas, D., Morgand, L., Pham, E., Gonin, P., Dangles-Marie, V., Job, B., Dessen, P., Bruno, A., Pierre, A., De The, H., Soliman, H., Nunes, M., Lardier, G., Calvet, L., Demers, B., Prevost, G., Vrignaud, P., Roman-Roman, S., Duchamp, O., and Berthet, C. (2012): Characterization of a large panel of patient-derived tumor xenografts representing the clinical heterogeneity of human colorectal cancer. *Clin. Cancer Res.* 18, 5314–5328.

Jung, B., Odening, K.E., Dall'Armellina, E., Foll, D., Menza, M., Markl, M., and Schneider, J.E. (2012): A quantitative comparison of regional myocardial motion in mice, rabbits and humans using in-vivo phase contrast CMR. *J. Cardiovasc. Magn. Reson.* 14, 87.

Kaboth, U. (1965): Comparative functional and morphological studies on the Ischemia-damaged rat kidney. *Z. Gesamte Exp. Med.* 138, 561–580.

Kaplowitz, N. (2001): Drug-induced liver disorders: implications for drug development and regulation. *Drug Saf.* 24, 483–490.

Kaplowitz, N. (2005): Idiosyncratic drug hepatotoxicity. *Nat. Rev. Drug Discov.* 4(6), 489–499.

Kapourchali, F.R., Surendiran, G., Chen, L., Uitz, E., Bahadori, B., and Moghadasian, M.H. (2014): Animal models of atherosclerosis. *World J Clin Cases.* 2, 126–132.

Kararli, T.T. (1995): Comparison of the gastrointestinal anatomy, physiology, and biochemistry of humans and commonly used laboratory animals. *Biopharm. Drug Dispos.* 16, 351–380.

Karol, M.H. (1994): Animal models of occupational asthma. *Eur. Respir. J.* 7, 555–568.

Kasiske, B.L. and Keane, W.F. (2000): Laboratory assessment of renal disease: clearance, urinalysis, and renal biopsy. *Biopharm. Drug Dispos.* 16, 351–380.

Katovitch, M.J., Meldrum, M.J., and Vasselli, J.R. (1991): Beneficial effects of dietary acarbose in the streptozotocin induced diabetic rat. *Metabolism* 40, 1275–1282.

Kaur, M., Bedi, O., Sachdeva, S., Reddy, B.V., and Kumar, P. (2014): Rodent animal models: from mild to advanced stages of diabetic nephropathy. *Inflammopharmacology* 22, 279–293.

Kerkela, R., Grazette, L., Yacobi, R., Iliescu, C., Patten, R., Beahm, C., Walters, B., Shevtsov, S., Pesant, S., Clubb, F.J., Rosenzweig, A., Salomon, R.N., Van Etten, R.A., Alroy, J., Durand, J.B., and Force, T. (2006): Cardiotoxicity of the cancer therapeutic agent imatinib mesylate. *Nat. Med.* 12, 908–916.

Kim, S.W., Kim, H.M., Yang, K.M., Kim, S.A., Kim, S.K., An, M.J., Park, J.J., Lee, S.K., Kim, T.I., Kim, W.H., and Cheon, J.H. (2010): *Bifidobacterium lactis* inhibits NF-kappaB in intestinal epithelial cells and prevents acute colitis and colitis-associated colon cancer in mice. *Inflamm. Bowel Dis.* 16, 1514–1525.

Koestner, A.W., Ruecker, F.A., and Koestner, A. (1977): Morphology and pathogenesis of tumors of the thymus and stomach in Sprague-Dawley rats following intragastric administration of methyl nitrosourea (MNU). *Int. J. Cancer* 20, 418–426.

Kong, L.L., Wu, H., Cui, W.P., Zhou, W.H., Luo, P., Sun, J., Yuan, H., and Miao, L.N. (2013): Advances in murine models of diabetic nephropathy. *J. Diabetes Res.* 2013(6), 1–10.

Kopec, K.L. and Burns, D. (2011): Nonalcoholic fatty liver disease: a review of the spectrum of disease, diagnosis and therapy. *Nutr. Clin. Pract.* 26, 565–576.

Krenz, M., Sanbe, A., Bouyer-Dalloz, F., Gulick, J., Klevitsky, R., Hewett, T.E., Osinska, H.E., Lorenz, J.N., Brosseau, C., Federico, A., Alpert, N.R., Warshaw, D.M., Perryman, M.B., Helmke, S.M., and Robbins, J. (2003): Analysis of myosin heavy chain functionality in the heart. *J. Biol. Chem.* 278, 17466–17474.

Kuehn BM (2009). Silent epidemic of viral hepatitis may lead to boom in serious disease. *JAMA* 302, 1949–1950.

Kusaka, J., Koga, H., Hagiwara, S., Hasegawa, A., Kudo, K., and Noguchi, T.P. (2012): Age-dependent responses to renal ischemia-reperfusion injury. *J. Surg. Res.* 172, 153–158.

Lacroix, D., Gluais, P., Marquie, C., D'Hoinne, C., Adamantidis, M., and Bastide, M. (2002): Repolarization abnormalities and their arrhythmogenic consequences in porcine tachycardia-induced cardiomyopathy. *Cardiovasc. Res.* 54, 42–50.

Laverty, H., Benson, C., Cartwright, E., Cross, M., Garland, C., Hammond, T., Holloway, C., McMahon, N., Milligan, J., Park, B., Pirmohamed, M., Pollard, C., Radford, J., Roome, N., Sager, P., Singh, S., Suter, T., Suter, W., Trafford, A., Volders, P., Wallis, R., Weaver, R., York, M., and Valentin, J. (2011): How can we improve our understanding of cardiovascular safety liabilities to develop safer medicines? *Br. J. Pharmacol.* 163, 675–693.

Lee, J.S., Shu, I.S., Mikaelyan, A., Calvisi, D.F., Heo, J., Reddy, J.K., and Thorgeirsson, S.S. (2004): Application of comparative functional genomics to identify best-fit mouse models to study human cancer. *Nat. Genet.* 36, 1306–1311.

Lee, H.T., Park, S.W., Kim, M., Ham, A., Anderson, L.J., Brown, K.M., D'Agati, V.D., and Cox, G.N. (2012): Interleukin-11 protects against renal ischemia and reperfusion injury. *Am. J. Physiol. Renal Physiol.* 302, 1166–1175.

Leiter, E.H. (1982): Multiple low-dose streptozotocin-induced hyperglycemia and insulitis in C57BL mice: influence of inbred background, sex, and thymus. *Proc. Natl. Acad. Sci. U. S. A.* 79, 630–634.

Leiter, E.H. (1985): Differential susceptibility of BALB/c sublines to diabetes induction by multi-dose streptozotocin treatment. *Curr. Top. Microbiol. Immunol.* 122, 78–85.

Lemieux, G., Baverel, G., Vinay, P., and Wadoux, P. (1976): Glutamine synthetase and glutamyltransferase in the kidney of man, dog, and rat. *Am. J. Physiol.* 231, 1068–1073.

Lemon, D.D., Papst, P.J., Joly, K., Plato, C.F., and McKinsey, T.A. (2011): A high-performance liquid chromatography assay for quantification of cardiac myosin heavy chain isoform protein expression. *Anal. Biochem.* 408, 132–135.

Leveille-Webster, C.R. and Arias, M. (1994): Mdr 2 knockout mice link biliary phospholipid deficiency with small bile duct destruction. *Hepatology* 19, 1528–1534.

Li, A.P. (2002): A review of the common properties of drugs with idiosyncratic hepatotoxicity and the "multiple determinant hypothesis" for the manifestation of idiosyncratic drug toxicity. *Chem. Biol. Interact.* 142, 7–23.

Li, G.R., Du, X.L., Siow, Y.L., K, O., Tse, H.F., and Lau, C.P. (2003): Calcium-activated transient outward chloride current and phase 1 repolarization of swine ventricular action potential. *Cardiovasc. Res.* 58, 89–98.

Lin, Z. and Will, Y. (2012): Evaluation of drugs with specific organ toxicities in organ-specific cell lines. *Toxicol. Sci.* 126, 114–127.

Liu, Y., Meyer, C., Xu, C., Weng, H., Hellerbrand, C., ten Dijke, P., and Dooley, S. (2013): Animal models of chronic liver diseases. *Am. J. Physiol. Gastrointest. Liver Physiol.* 304, G449–G468.

Lock, E.A. and Hard, G.C. (2010): Alpha2u-globulin nephropathy and chronic progressive nephropathy as modes of action for renal tubule tumor induction in rats, and their possible interaction. In: *Cancer Risk Assessment—Chemical Carcinogenesis, Hazard Evaluation, and Risk Quantification*, (Hsu, C.H. and Stedeford, T. Eds.), pp. 248–255, John Wiley & Sons, Ltd., Oxford, England.

Lopez, A.D., Shibuya, K., Rao, C., Mathers, C.D., Hansell, A.L., Held, L.S., Schmid, V., and Buist, S. (2006): Chronic obstructive pulmonary disease: current burden and future predictions. *Eur. Respir. J.* 27, 297–412.

Lu, X., Li, N., Shushakova, N., Schmitt, R., Menne, J., Susnik, N., Meier, M., Leitges, M., Haller, H., Gueler, F., and Rong, S. (2012): C57BL/6 and 129/Sv mice: genetic difference to renal ischemia-reperfusion. *J. Nephrol.* 5, 738–743.

Lujan, H.L., Janbaih, H., Feng, H.Z., Jin, J.P., and DiCarlo, S.E. (2012): Ventricular function during exercise in mice and rats. *Am. J. Physiol. Regul. Integr. Comp. Physiol.* 302, R68–R74.

Lushbaugh, C.C., Humason, G.L., Swartzendruber, D.C., Richter, C.B., and Gengozian, N. (1978): Spontaneous colonic adenocarcinoma in marmosets. *Primates Med.* 10, 119–134.

Macarthur, M., Hold, G.L., and El-Omar, E.M. (2004): Inflammation and cancer II. Role of chronic inflammation and cytokine gene polymorphisms in the pathogenesis of gastrointestinal malignancy. *Am. J. Physiol. Gastrointest. Liver Physiol.* 286, G515–G520.

MacDonald, J.S. and Robertson, R.T. (2009): Toxicity testing in the 21st century: a view from the pharmaceutical industry. *Toxicol. Sci.* 110, 40–46.

Maekawa, A., Onodera, H., Kanno, J., Furuta, K., Nagaoka, T., Todate, A., Matsushima, Y., Oh-hara, T., and Kawazoe, Y. (1988): Carcinogenicity and organ specificity of N-trimethylsilylmethyl-N-nitrosourea (TMSMNU), N-neopentyl-N-nitrosourea (neoPNU), and N-methyl-N-nitrosourea (MNU) in rats. *J. Cancer Res. Clin. Oncol.* 114, 473–476.

Maggio-Price, L., Bielefeldt-Ohmann, H., Treuting, P., Iritani, B.M., Zeng, W., Nicks, A., Tsang, M., Shows, D., Morrissey, P., and Viney, J.L. (2005): Dual infection with *Helicobacter bilis* and *Helicobacter hepaticus* in p-glycoprotein-deficient mdr1a/ mice results in colitis that progresses to dysplasia. *Am. J. Pathol.* 166, 1793–1806.

Maggio-Price, L., Treuting, P., Zeng, W., Tsang, M., Bielefeldt-Ohmann, H., and Iritani, B.M. (2006): Helicobacter infection is required for inflammation and colon cancer in SMAD3-deficient mice. *Cancer Res.* 66, 828–838.

Maggio-Price, L., Treuting, P., Bielefeldt-Ohmann, H., Seamons, A., Drivdahl, R., Zeng, W., Lai, L., Huycke, M., Phelps, S., Brabb, T., and Iritani, B.M. (2009): Bacterial infection of Smad3/Rag2 double-null mice with transforming growth factor-beta dysregulation as a model for studying inflammation-associated colon cancer. *Am. J. Pathol.* 174, 317–329.

Maher, J.J. (2011): New insights from rodent models of fatty liver disease. *Antioxid. Redox Signal.* 15, 535–550.

Makarenko, I., Opitz, C.A., Leake, M.C., Neagoe, C., Kulke, M., Gwathmey, J.K., del Monte, F., Hajjar, R.J., and Linke, W.A. (2004): Passive stiffness changes caused by upregulation of compliant titin isoforms in human dilated cardiomyopathy hearts. *Circ. Res.* 95, 708–716.

Mansfield, K.G., Lin, K.C., Xia, D., Newman, J.V., Schauer, D.B., MacKey, J., Lackner, A.A., and Carville, A. (2001): Enteropathogenic *Escherichia coli* and ulcerative colitis in cotton-top tamarins (*Saguinus oedipus*). *J. Infect. Dis.* 184, 803–807.

Mathews, C. (2002): Rodent models for the study of type 2 diabetes in children (juvenile diabesity). *Pediatr. Diabetes* 3, 163–173.

Matute-Bello, G., Frevert, C.W., and Martin, T.R. (2008): Animal models of acute lung injury. *Am J Physiol Lung Cell Mol Physiol.* 295, L379–L399.

Maus, T.P., Pearson, R.K., Anderson, R.J., Woodson, L.C., Reiter, C., and Weinshilboum, R.M. (1982): Rat phenol sulfotransferase. Assay procedure, developmental changes, and glucocorticoid regulation. *Biochem. Pharmacol.* 31, 849–856.

McCauley, M.D. and Wehrens, X.H.T. (2009): Animal models of arrhythmogenic cardiomyopathy. *Dis. Model. Mech.* 2, 563–570.

McEntee, M.F. and Brenneman, K.A. (1999): Dysregulation of beta-catenin is common in canine sporadic colorectal tumors. *Vet. Pathol.* 36, 228–236.

Minde, D.P., Anvarian, Z., Rudiger, S.G., and Maurice, M.M. (2011): Messing up disorder: how do missense mutations in the tumor suppressor protein APC lead to cancer? *Mol. Cancer.* 10, 101.

Mitchell, J.R., Mcmurtry, R.J., Statham, C.N., and Nelson, S.D. (1977): Molecular basis for several strain and diet on the promoting effects of sodium L-ascorbate in two-stage urinary bladder carcinogenesis in rats. *Cancer Res.* 47, 3492–3495.

Miyata, S., Minobe, W., Bristow, M.R., and Leinwand, L.A. (2000): Myosin heavy chain isoform expression in the failing and nonfailing human heart. *Circ. Res.* 86, 386–390.

Mizoguchi, A. (2012): Animal models of inflammatory bowel disease. *Prog. Mol. Biol. Transl. Sci.* 105, 263–320.

Mizoguchi, A., Mizoguchi, E., and Bhan, A.K. (2003): Immune networks in animal models of inflammatory bowel disease. *Inflamm. Bowel Dis.* 9, 246–259.

Monsma, D.J., Monks, N.R., Cherba, D.M., Dylewski, D., Eugster, E., Jahn, H., Srikanth, S., Scott, S.B., Richardson, P.J., Everts, R.E., Ishkin, A., Nikolsky, Y., Resau, J.H., Sigler, R., Nickoloff, B.J., and Webb, C.P. (2012): Genomic characterization of explant tumorgraft models derived from fresh patient tumor tissue. *J. Transl. Med.* 10, 125.

Moore, B.B., Lawson, W.E., Our, T.D., Sisson, T.H., Raghavendran, K., and Hogaboam, C.M. (2013): Animal models of fibrotic lung disease. *Am. J. Respir. Cell Mol. Biol.* 49, 167–179.

Morgan, S.J., Elangbam, C.S., Berens, S., Janovitz, E., Vitsky, A., Zabka, T., and Conour, L. (2013): Use of animal models of human disease for nonclinical safety assessment of novel pharmaceuticals. *Toxicol Pathol.* 41, 508–418.

Morrison, J. and Hof, P. (1997): Life and death of neurons in the aging brain. *Science* 278, 412–429.

Moser, A.R., Pitot, H.C., and Dove, W.F. (1990): A dominant mutation that predisposes to multiple intestinal neoplasia in the mouse. *Science* 247, 322–324.

Moser, A.R., Dove, W.F., Roth, K.A., and Gordon, J.I. (1992): The Min (multiple intestinal neoplasia) mutation: its effect on gut epithelial cell differentiation and interaction with a modifier system. *J. Cell Biol.* 116, 1517–1526.

Mosley, K., Collar, J., and Cattell, V. (2000): Mesangial cell necrosis in Thy 1 glomerulonephritis—an ultrastructural study. *Virchows Arch.* 436, 567–573.

Mosmann, T.R. and Coffman, R.L. (1989): Th1 and Th2-cells: different patterns of lymphokine secretion lead to different functional properties. *Annu. Rev. Immunol.* 7, 145–173.

Mueller, K., Michel, K., Krueger, D., Demir, I.E., Ceyhan, G.O., Zeller, F., Kreis, M.E., and Schemann, M. (2011): Activity of protease-activated receptors in the human submucous plexus. *Gastroenterology* 141, 2088–2097.

Mullin, E.M., Bonar, R.A., and Paulson, D.F. (1976): Acute tubular necrosis. An experimental model detailing the biochemical events accompanying renal injury and recovery. *Investig. Urol.* 13, 289–294.

Murray, C.J.L. and Lopez, A.D. (1997): Global mortality, disability and the contribution of risk factors: global burden of disease study. *Lancet* 349, 1436–1442.

Murray, J.M., Thompson, A.M., Vitsky, A., Hawes, M., Chuang, W., Pacheco, J., Wilson, S., McPherson, J.M., Thurberg, B.L., Karey, K.P., and Andrews, L. (2014): Nonclinical safety assessment of recombinant human sphingomyelinase (rhASM) for the treatment of acid sphingomyelinase deficiency: the utility of animal models of disease in the toxicological evaluation of potential therapeutics. *Mol. Genet. Metab.* 114, 217–225.

Nagueh, S.F., Shah, G., Wu, Y., Torre-Amione, G., King, N.M., Lahmers, S., Witt, C.C., Becker, K., Labeit, S., and Granzier, H.L. (2004): Altered titin expression, myocardial stiffness, and left ventricular function in patients with dilated cardiomyopathy. *Circulation* 110, 155–162.

Nakagama, H., Nakanishi, M., and Ochiai, M. (2005): Modeling human colon cancer in rodents using a foodborne carcinogen, PhIP. *Cancer Sci.* 96, 627–636.

Narisawa, T. and Weisburger, J.H. (1975): Colon cancer induction in mice by intrarectal instillation of N-methylnitosorurea (38498). *Proc. Soc. Exp. Biol. Med.* 148, 166–169.

Narisawa, T., Wong, C.Q., Maronpot, R.R., and Weisburger, J.H. (1976): Large bowel carcinogenesis in mice and rats by several intrarectal doses of methylnitrosourea and negative effect of nitrite plus methylurea. *Cancer Res.* 36, 505–510.

Naughton, C.A. (2008): Drug-induced nephrotoxicity. *Am. Fam. Physician* 78, 743–750.

Neagoe, C., Opitz, C.A., Makarenko, I., and Linke, W.A. (2003): Gigantic variety: expression patterns of titin isoforms in striated muscles and consequences for myofibrillar passive stiffness. *J. Muscle Res. Cell Motil.* 24, 175–189.

Nerbonne, J.M. (2004): Studying cardiac arrhythmias in the mouse—a reasonable model for probing mechanisms? Trends Cardiovasc. *Med.* 14, 83–93.

Neufeld, K.L. (2009): Nuclear APC. *Adv. Exp. Med. Biol.* 656, 13–29.

Neurath, M.F., Fuss, I., Kelsall, B., Meyer zum Buschenfelde, K.H., and Strober, W. (1996): Effect of IL-12 and antibodies to IL-12 on established granulomatous colitis in mice. *Ann. N. Y. Acad. Sci.* 795, 368–370.

Newmark, H.L., Lipkin, M., and Maheshwari, N. (1990): Colonic hyperplasia and hyperproliferation induced by a nutritional stress diet with four components of Western-style diet. *J. Natl. Cancer Inst.* 82, 491–496.

Newmark, H.L., Yang, K., Lipkin, M., Kopelovich, L., Liu, Y., Fan, K., and Shinozaki, H. (2001): A Western-style diet induces benign and malignant neoplasms in the colon of normal C57Bl/6 mice. *Carcinogenesis* 22, 1871–1875.

Newton, J.F., Yoshimoto, M., Bernstein, J., Rush, G.F., and Hook, J.B. (1983): Acetaminophen nephrotoxicity in the rat. II. Strain differences in nephrotoxicity and metabolism of p-aminophenol, a metabolite of acetaminophen. *Toxicol. Appl. Pharmacol.* 69, 307–318.

Nielsen, S., Kwon, T.-H., Fenton, R.A., and Praetorious, J. (2012): Anatomy of the kidney. In: *Brenner and Rector's Kidney*, (Taal, M.W., Chertow, G.M., Marsden, P.A., Yu, A.S.L., Brenner, B.M. eds.), pp. 31–93, Elsevier, Philadelphia.

van Noord, C., Sturkenboom, M.C., Straus, S.M., Witteman, J.C., and Stricker, B.H. (2011): Non-cardiovascular drugs that inhibit hERG-encoded potassium channels and risk of sudden cardiac death. *Heart* 97, 215–220.

Nordestgaard, B.G. and Lewis, B. (1991): Intermediate density lipoprotein levels are strong predictors of the extent of aortic atherosclerosis in the St. *Thomas's Hospital rabbit strain.* *Atherosclerosis* 87, 39–46.

O'Brien, C.A., Pollett, A., Gallinger, S., and Dick, J.E. (2007): A human colon cancer cell capable of initiating tumour growth in immunodeficient mice. *Nature* 445, 106–110.

Ohshima, H., Tatemichi, M., and Sawa, T. (2003): Chemical basis of inflammation-induced carcinogenesis. *Arch. Biochem. Biophys.* 417, 3–11.

Ohshima, H., Tazawa, H., Sylla, B.S., and Sawa, T. (2005): Prevention of human cancer by modulation of chronic inflammatory processes. *Mutat. Res.* 591, 110–122.

Okamoto, K. (1969): Spontaneous hypertension in rats. *Int. Rev. Exp. Pathol.* 7, 227–270.

Okamoto, K. and Aoki, K. (1963): Development of a strain of spontaneously hypertensive rats. *Jpn. Circ. J.* 27, 282–293.

Olson, H., Betton, G., Robinson, D., Thomas, K., Monro, A., Kolaja, G., Lilly, P., Sanders, J., Sipes, G., Bracken, W., Dorato, M., Van Deun, K., Smith, P., Berger, B., and Heller, A. (2000): Concordance of the toxicity of pharmaceuticals in humans and in animals. *Regul. Toxicol. Pharmacol.* 32, 56–67.

Opitz, C.A. and Linke, W.A. (2005): Plasticity of cardiac titin/connectin in heart development. *J. Muscle Res. Cell Motil.* 26, 333–342.

Oshima, H., Oshima, M., Kobayashi, M., Tsutsumi, M., and Taketo, M.M. (1997): Morphological and molecular processes of polyp formation in Apc(delta716) knockout mice. *Cancer Res.* 57, 1644–1649.

Ostergaard, G., Hansen, H.N., and Ottesen, J.L. (2010): Physiological, hematological, and clinical chemistry parameters, including conversion factors. In: *Handbook of Laboratory Animal Science, Volume I: Essential Principles and Practices* Vol. 1, 3rd ed. (Hau, J., and Schapiro, S.J., eds.), pp. 667–707, CRC Press, Boca Raton, FL.

Palm, M. (1998): The incidence of chronic progressive nephrosis in young Sprague-Dawley rats from two different breeders. *Lab. Anim* 32, 477–482.

Park, K.M., Kim, J.I., Ahn, Y., Bonventre, A.J., and Bonventre, J.V. (2004): Testosterone is responsible for enhanced susceptibility of males to ischemic renal injury. *J. Biol. Chem.* 279, 52282–52292.

Pele-Tounian, A., Wang, X., Rondu, F., Lamouri, A., Touboul, E., Marc, S., and Ktorza, A. (1998): Potent antihyperglycaemic property of a new imidazoline derivative S-22068 (PMS 847) in a rat model of NIDDM. *Br. J. Pharmacol.* 124, 1591–1596.

Pereira, C.V., Nadanaciva, S., Oliveira, P.J., and Will, Y. (2012): The contribution of oxidative stress to drug-induced organ toxicity and its detection in vitro and in vivo. *Expert Opin. Drug Metab. Toxicol.* 8, 219–237.

Perez-Osterreicher, M., Osterreicher, C.H., and Trauner, M. (2011): Fibrosis in autoimmune and cholestatic liver disease. *Best Pract. Res. Clin. Gastroenterol.* 25, 245–258.

Perez-Sayans, M., Suarez-Penaranda, J.M., Herranz-Carnero, M., Gayoso-Diz, P., Barros-Angueira, F., Gandara-Rey, J.M., and Garcia-Garcia, A. (2012): The role of the adenomatous polyposis coli (APC) in oral squamous cell carcinoma. *Oral Oncol.* 48, 56–60.

Peters, M.F., Lamore, S.D., Guo, L., Scott, C.W., and Kolaja, K.L. (2014): Human stem cell-derived cardiomyocytes in cellular impedance assays: bringing cardiotoxicity screening to the front line. *Cardiovasc. Toxicol.* doi:10.1007/s12012-014-9268-9.

Petro, A.E., Cotter, J., Cooper, D.A., Peters, J.C., Surwit, S.J., and Surwit, R.S. (2004): Fat, carbohydrate, and calories in the development of diabetes and obesity in the C57BL/6J mouse. *Metabolism* 53, 454–457.

Piacentino, V. III, Weber, C.R., Chen, X., Weisser-Thomas, J., Margulies, K.B., Bers, D.M., and Houser, S.R. (2003): Cellular basis of abnormal calcium transients of failing human ventricular myocytes. *Circ. Res.* 92, 651–658.

Pierce, E.M., Carpenter, K., Jakubzick, C., Kunkel, S.L., Flaherty, K.R., Martinez, F.J., and Hogaboam, C.M. (2007): Therapeutic targeting of CC ligand 21 or CC chemokine receptor 7 abrogates pulmonary fibrosis induced by the adoptive transfer of human pulmonary fibroblasts to immunodeficient mice. *Am. J. Pathol.* 170, 1152–1164.

Puig, I., Chicote, I., Tenbaum, S.P., Arques, O., Herance, J.R., Gispert, J.D., Jimenez, J., Landolfi, S., Caci, K., Allende, H., Mendizabal, L., Moreno, D., Charco, R., Espin, E., Prat, A., Elez, M.E., Argiles, G., Vivancos, A., Tabernero, J., Rojas, S., and Palmer, H.G. (2013): A personalized preclinical model to evaluate the metastatic potential of patient-derived colon cancer initiating cells. *Clin. Cancer Res.* 19, 6787–6801.

Qin, X., Zarkovic, M., Nakatsuru, Y., Arai, M., Oda, H., and Ishikawa, T. (1994): DNA adduct formation and assessment of aberrant crypt foci in vivo in the rat colon mucosa after treatment with N-methyl-nitrosourea. *Carcinogenesis* 15, 851–855.

Qiu, B.S., Vallance, B.A., Blennerhassett, P.A., and Collins, S.M. (1999): The role of CD4+ lymphocytes in the susceptibility of mice to stress-induced reactivation of experimental colitis. *Nat. Med.* 5, 1178–1182.

Qu, P., Hamada, M., Ikeda, S., Hiasa, G., Shigematsu, Y., and Hiwada, K. (2000): Time-course changes in left ventricular geometry and function during the development of hypertension in Dahl salt-sensitive rats. *Hypertens. Res.* 23, 613–623.

Ramesh, G. and Reeves, W.B. (2003): TNFR2-mediated apoptosis and necrosis in cisplatin-induced acute renal failure. *Am. J. Physiol. Renal Physiol.* 285, F610–F618.

Rapp, J.P. and Dene, H. (1985): Development and characteristics of inbred strains of Dahl salt-sensitive and salt-resistant rats. *Hypertension* 7, 340–349.

Rashidi, B., Gamagami, R., Sasson, A., Sun, F.X., Geller, J., Moossa, A.R., and Hoffman, R.M. (2000): An orthotopic mouse model of remetastasis of human colon cancer liver metastasis. *Clin. Cancer Res.* 6, 2556–2561.

Redfern, W.S., Wakefield, I.D., Prior, H., Pollard, C.E., Hammond, T.G., and Valentin, J.P. (2002): Safety pharmacology–a progressive approach. *Fundam. Clin. Pharmacol.* 16, 161–173.

Reinhard, M.K., Bekersky, H., Sanders, T.W., Jr., Hamis, B.J., and Hottendorf, G.H. (1994): Effects of polyaspartic acid on pharmacokinetics of tobramycin in two strains of rat. *Antimicrob. Agents Chemother.* 38, 79–82.

Reiser, P.J., Portman, M.A., Ning, X.H., and Schomisch Moravec, C. (2001): Human cardiac myosin heavy chain isoforms in fetal and failing adult atria and ventricles. *Am. J. Physiol. Heart Circ. Physiol.* 280, H1814–H1820.

Reitmair, A.H., Redston, M., Cai, J.C., Chuang, T.C., Bjerknes, M., Cheng, H., Hay, K., Gallinger, S., Bapat, B., and Mak, T.W. (1996): Spontaneous intestinal carcinomas and skin neoplasms in Msh2-deficient mice. *Cancer Res.* 56, 3842–3849.

Rhee, P.L., Lee, J.Y., Son, H.J., Kim, J.J., Rhee, J.C., Kim, S., Koh, S.D., Hwang, S.J., Sanders, K.M., and Ward, S.M. (2011): Analysis of pacemaker activity in the human stomach. *J. Physiol.* 589, 6105–6118.

Ricci-Vitiani, L., Lombardi, D.G., Pilozzi, E., Biffoni, M., Todaro, M., Peschle, C., and De Maria, R. (2007): Identification and expansion of human colon-cancer-initiating cells. *Nature* 445, 111–115.

Richter, F., Newmark, H.L., Richter, A., Leung, D., and Lipkin, M. (1995): Inhibition of Western-diet induced hyperproliferation and hyperplasia in mouse colon by two sources of calcium. *Carcinogenesis* 16, 2685–2689.

Rifkin, D.E., Coca, S.G., and Kalantar-Zadeh, K. (2012): Does AKI truly lead to CKD? *J. Am. Soc. Nephrol.* 23, 979–984.

Risio, M., Lipkin, M., Newmark, H., Yang, K., Rossini, F.P., Steele, V.E., Boone, C.W., and Kelloff, G.J. (1996): Apoptosis, cell replication, and Western-style diet-induced tumorigenesis in mouse colon. *Cancer Res.* 56, 4910–4916.

Robert, A., Stout, T.J., and Dale, J.E. (1970): Production by secretagogues of duodenal ulcers in the rat. *Gastroenterology* 59, 95–102.

Rockenstein, E., McConlogue, L., Tan, H., Power, M., Masliah, E., and Mucke, L. (1995): Levels and alternative splicing of amyloid beta protein precursor (APP) transcripts in brains of APP transgenic mice and humans with Alzheimer's disease. *Biol. Chem.* 270, 28257–28267.

Roden, D.M. (2001): Pharmacogenetics and drug-induced arrhythmias. *Cardiovasc. Res.* 50, 224–231.

Roncal, C.S., Mu, W., Croker, B., Reungjui, S., Ouyang, X., Tabah-Fisch, I., Johnson, R.J., and Ejaz, A.A. (2007): Effect of elevated serum uric acid on cisplatin induced acute renal failure. *Am. J. Physiol. Renal Physiol.* 292, 116–122.

Rosenberg, D.W., Giardina, C., and Tanaka, T. (2009): Mouse models for the study of colon carcinogenesis. *Carcinogenesis* 30, 183–196.

Roth, R.A. and Ganey, P. (2011): Animal models of idiosyncratic drug-induced liver injury—current status. *Crit. Rev. Toxicol.* 41, 723–739.

Rubio-Viqueira, B. and Hidalgo, M. (2009): Direct in vivo xenograft tumor model for predicting chemotherapeutic drug response in cancer patients. *Clin. Pharmacol. Ther.* 85, 217–221.

Salama, G. and London, B. (2007): Mouse models of long QT syndrome. *J. Physiol.* 578, 43–53.

Sanan, D.A., Newland, D.L., Tao, R., Marcovina, S., Wang, J., Mooser, V., Hammer, R.E., and Hobbs, H.H. (1998): Low density lipoprotein receptor-negative mice expressing human apolipoprotein B-100 develop complex atherosclerotic lesions on a chow diet: no accentuation by apolipoprotein(a). *Proc. Natl. Acad. Sci. U. S. A.* 95, 4544–4549.

Sanger, G.J. (2007): Endocannabinoids and the gastrointestinal tract: what are the key questions? *Br. J. Pharmacol.* 152, 663–670.

Sanger, G.J., Holbrook, J.D., and Andrews, P.L.R. (2011): The translational value of rodent gastrointestinal functions: a cautionary tale. *Trends Pharmacol. Sci.* 32, 402–409.

Sausville, E.A. and Burger, A.M. (2006): Contributions of human tumor xenografts to anticancer drug development. *Cancer Res.* 66, 3351–3354.

Schaffer, E. and Schiefer, B. (1968): Incidence and types of canine rectal carcinomas. *J. Small Anim. Pract.* 9, 491–496.

Schein, P.S., Davis, R.D., Carter, S., Newman, J., Schein, D.R., and Rall, D.P. (1970): The evaluation of anticancer drugs in dogs and monkeys for prediction of qualitative toxicities in man. *Clin. Pharmacol. Ther.* 11, 3–40.

Schiechl, G., Bauer, B., Fuss, I., Lang, S.A., Moser, C., Ruemmele, P., Rose-John, S., Neurath, M.F., Geissler, E.K., Schlitt, H.J., Strober, W., and Fichtner-Feigl, S. (2011): Tumor development in murine ulcerative colitis depends on MyD88 signaling of colonic F4/80+CD11b(high)Gr1(low) macrophages. *J. Clin. Invest.* 121, 1692–1708.

Schultz, J.H., Volk, T., Bassalay, P., Hennings, J.C., Hubner, C.A., and Ehmke, H. (2007): Molecular and functional characterization of Kv4.2 and KChIP2 expressed in the porcine left ventricle. *Pflugers Arch.* 454, 195–207.

Shah, D.I. and Singh, M. (2006): Possible role of exogenous cAMP to improve vascular endothelial dysfunction in hypertensive rats. *Fundam. Clin. Pharmacol.* 20, 595–604.

Shah, A., Garzon-Muvdi, T., Mahajan, R., Duenas, V.J., and Quinones-Hinojosa, A. (2010): Animal models of neurological disease. *Adv. Exp. Med. Biol.* 671, 23–40.

Sibille, M., Deigat, N., Janin, A., Kirkesseli, S., and Durand, D.V. (1998): Adverse events in phase-I studies: a report in 1015 healthy volunteers. *Eur. J. Clin. Pharmacol.* 54, 13–20.

Singh, A.P., Junemann, A., Muthuraman, A., Jaggi, A.S., Singh, N., Grover, K., and Dhawan, R. (2012): Animal models of acute renal failure. *Pharmacol. Rep.* 64, 31–44.

Sinha, R. and Rothman, N. (1999): Role of well-done, grilled red meat, heterocyclic amines (HCAs) in the etiology of human cancer. *Cancer Lett.* 143, 189–194.

Sinha, R., Chow, W.H., Kulldorff, M., DeNobile, J., Butler, J., Garcia-Closas, M., Weil, R., Hoover, R.N., and Rothman, N. (1999): Well-done, grilled red meat increases the risk of colorectal adenomas. *Cancer Res.* 59, 4320–4324.

Siolas, D. and Hannon, G.J. (2013): Patient-derived tumor xenografts: transforming clinical samples into mouse models. *Cancer Res.* 73, 5315–5319.

Smit, J.J, Schinkel, A.H., Oude Elferink, R.P.J., Groen, A.K., Wagenaar, B., von Deemter, L., Mol, C.A., Ottenhoff, R., van der Lugt, N.M., and van Roon, M.A. (1993): Homozygous disruption of the murine mdr2 p-glycoprotein gene leads to a complete absence of phospholipid from bile and to liver disease. *Cell* 75, 451–462.

Speakman, J.R. (2005): Body size, energy metabolism and lifespan. *J. Exp. Biol.* 208, 1717–1730.

Stahl, S., Ittrich, C., Marx-Stoleting, P., Kohle, C., Altug-Teber, O., Reiss, O., Bonin, M., Jobst, J., Kaiser, S., Buchmann, A., and Schwarz, M. (2005): Genotype-phenotype relationships in hepatocellular tumors from mice and man. *Hepatology* 42, 353–361.

Stevens, J.L. and Baker, T.K. (2009): The future of drug safety testing: expanding the view and narrowing the focus. *Drug Discov. Today* 14, 162–167.

Stratton, J.R., Levy, W.C., Cerqueira, M.D., Schwartz, R.S., and Abrass, I.B. (1994): Cardiovascular responses to exercise. Effects of aging and exercise training in healthy men. *Circulation* 89, 1648–1655.

Sugano, M., Yamato, H., Hayashi, T., Ochiai, H., Kakuchi, J., Goto, S., Nishijima, F., Iino, N., Kazama, J.J., Takeuchi, T., Mokuda, O., Ishikawa, T., and Okazaki, R. (2006): High-fat diet in low-dose-streptozotocin-treated heminephrectomized rats induces all features of human type 2 diabetic nephropathy: a new rat model of diabetic nephropathy. *Nutr. Metab. Cardiovasc. Dis.* 16, 477–484.

Sun, S., Wang, Y., Li, Q., Tian, Y., Liu, M., and Yu, Y. (2006): Effects of benazepril on renal function and kidney expression of matrix metalloproteinase-2 and tissue inhibitor of metalloproteinase-2 in diabetic rats. *Chin Med J (Engl)* 119, 814–821.

Sun, Y., Liu, G., Song, T., Liu, F., Kang, W., Zhang, Y., and Ge, Z. (2008): Upregulation of GRP78 and caspase-12 in diastolic failing heart. *Acta Biochim. Pol.* 55, 511–516.

Surwit, R.S., Kuhn, C.M., Cochrane, C., McCubbin, J.A., and Feinglos, M.N. (1988): Diet-induced type II diabetes in C57BL/6J mice. *Diabetes* 37, 1163–1167.

Surwit, R.S., Feinglos, M.N., Rodin, J., Sutherland, A., Petro, A.E., Opara, E.C., Kuhn, C.M., and Rebuffe-Scrive, M. (1995): Differential effects of fat and sucrose on the development of obesity and diabetes in C57BL/6J and A/J mice. *Metabolism* 44, 645–651.

Szabo, S. and Cho, C.H. (1988a): From cysteamine to MPTP: structure-activity studies with duodenal ulcerogens. *Toxicol. Pathol.* 16, 205–212.

Szabo, S. and Cho, C.H. (1988b): Animal models for studying the role of eicosanoids in peptic ulcer disease. In: *Eicosanoids and the Gastrointestinal Tract* (Hillier, K., ed.), Vol. 2, 75–102, Springer, the Netherlands.

Talmadge, J.E., Singh, R.K., Fidler, I.J., and Raz, A. (2007): Murine models to evaluate novel and conventional therapeutic strategies for cancer. *Am. J. Pathol.* 170, 793–804.

Tanaka, T. (2012): Preclinical cancer chemoprevention studies using animal model of inflammation-associated colorectal carcinogenesis. *Cancers* 4, 673–700.

Tanaka, T., Kohno, H., Suzuki, R., Yamada, Y., Sugie, S., and Mori, H. (2003): A novel inflammation-related mouse colon carcinogenesis model induced by azoxymethane and dextran sodium sulfate. *Cancer Sci.* 94, 965–973.

Tanaka, T., Suzuki, R., Kohno, H., Sugie, S., Takahashi, M., and Wakabayashi, K. (2005): Colonic adenocarcinomas rapidly induced by the combined treatment with 2-amino-1-methyl-6-phenylimidazo[4,5-b]pyridine and dextran sodium sulfate in male ICR mice possess beta-catenin gene mutations and increases immunoreactivity for beta-catenin, cyclooxygenase-2 and inducible nitric oxide synthase. *Carcinogenesis* 26, 229–238.

Tarloff, J.B., Goldstein, R.S., Morgan, D.G., and Hook, J.B. (1989): Acetaminophen and p-aminophenol nephrotoxicity in aging male Sprague-Dawley and Fischer 344 rats. *Fundam. Appl. Toxicol.* 12, 78–91.

Temple, R.J. and Himmel, M.H. (2002): Safety of newly approved drugs: implications for prescribing. *JAMA* 287, 2273–2275.

Tesch, G.H. and Allen, T.J. (2007): Rodent models of streptozotocin-induced diabetic nephropathy. *Nephrology (Carlton)* 12, 261–266.

Thiele, D.L. (2007): Immunological mechanisms in drug-induced liver injury. In *Drug-Induced Liver Disease*, 2nd ed. (Kaplowitz, N., and DeLeve, L.D., eds.), pp. 115–123, Informa Health Care, New York.

Travlos, G.S., Betz, L., and Hard, G.C. (2007): Influence of diet or route of exposure on chronic progressive nephropathy, kidney weight, blood urea nitrogen and creatinine concentrations in 13-week toxicity studies for control F344/N male rats [abstract]. *Toxicol. Pathol.* 35, 190–191.

Treuting, P.M. and Dintzis, S. (2012): *Comparative Anatomy and Histology: A Mouse and Human Atlas.* Academic Press, Elsevier, London.

Trevisan, A., Nicolli, A., and Chiara, F. (2010): Are rats the appropriate experimental model to understand age-related renal drug metabolism and toxicity? *Expert Opin. Drug Metab. Toxicol.* 6, 1451–1459.

Trujillo, G., Meneghin, A., Flaherty, K.R., Sholl, L.M., Myers, J.L., Kazerooni, E.A., Gross, B.H., Oak, S.R., Coelho, A.L., Evanoff, H., Day, E., Toews, G.B., Joshi, A.D., Schaller, M.A., Waters, B., Jarai, G., Westwick, J., Kunkel, S.L., Martinez, F.J., and Hogaboam, C.M. (2010): TLR9 differentiates rapidly from slowly progressing forms of idiopathic pulmonary fibrosis. *Sci. Transl. Med.* 2, 57–82.

Tsukamoto, T., Tanaka, H., Fukami, H., Inoue, M., Takahashi, M., Wakabayashi, K., and Tatematsu, M. (2000): More frequent beta-catenin gene mutations in adenomas than in aberrant crypt foci or adenocarcinomas in the large intestines of 2-amino-1-methyl-6-phenylimidazo[4,5-b]pyridine (PhIP)-treated rats. *Jpn. J. Cancer Res.* 91, 729–796.

Tudek, B., Bird, R.P., and Bruce, W.R. (1989): Foci of aberrant crypts in the colons of mice and rats exposed to carcinogens associated with foods. *Cancer Res.* 49, 1236–1240.

Valerius, K.D., Powers, B.E., McPherron, M.A., Hutchison, J.M., Mann, F.A., and Withrow, S.J. (1997): Adenomatous polyps and carcinoma in situ of the canine colon and rectum: 34 cases (1982–1994). *J. Am. Anim. Hosp. Assoc.* 33, 156–160.

Wallis, R.M. (2010): Integrated risk assessment and predictive value to humans of non-clinical repolarization assays. *Br. J. Pharmacol.* 159, 115–121.

Wang, Z., Feng, J., Shi, H., Pond, A., Nerbonne, J.M., and Nattel, S. (1999): Potential molecular basis of different physiological properties of the transient outward K+ current in rabbit and human atrial myocytes. *Circ. Res.* 84, 551–561.

Wang, J., Liu, X., Ren, B., Rupp, H., Takeda, N., and Dhalla, N.S. (2002): Modification of myosin gene expression by imidapril in failing heart due to myocardial infarction. *J. Mol. Cell. Cardiol.* 34, 847–857.

Ward, J.M. (1974): Morphogenesis of chemically induced neoplasms of the colon and small intestine in rats. *Lab. Invest.* 30, 505–513.

Ward, J.M. and Treuting, P.M. (2014): Rodent intestinal epithelial carcinogenesis: pathology and preclinical models. *Toxicol. Pathol.* 42, 148–161.

Washington, M.K., Powell, A.E., Sullivan, R., Sundberg, J.P., Wright, N., Coffey, R.J., and Dove, W.F. (2013): Pathology of rodent models of intestinal cancer: progress report and recommendations. *Gastroenterology* 144, 705–717.

Watanabe, Y. (1980): Serial inbreeding of rabbits with hereditary hyperlipidemia (WHHL-rabbit). *Atherosclerosis* 36, 261–268.

Wei, Q., Wang, M.H., and Dong, Z. (2005): Differential gender differences in ischemic and nephrotoxic acute renal failure. *Am. J. Nephrol.* 25, 491–499.

West, J., Fagan, K., Steudel, W., Fouty, B., Lane, K., Harral, J., Hoedt-Miller, M., Tada, Y., Ozimek, J., Tuder, R., and Rodman, D.M. (2004): Pulmonary hypertension in transgenic mice expressing a dominant-negative BMPRII gene in smooth muscle. *Circ. Res.* 94, 1109–1114.

Whittle, B.J. (1977): Mechanisms underlying gastric mucosal damage induced by indomethacin and bile-salts, and the actions of prostaglandins. *Br. J. Pharmacol.* 60, 455–460.

Williams, S.A., Anderson, W.C., Santaguida, M.T., and Dylla, S.J. (2013): Patient-derived xenografts, the cancer stem cell paradigm, and cancer pathobiology in the 21st century. *Lab. Invest.* 93, 970–982.

Wong, P.C., Cai, H., Borchelt, D.R., and Price, D.L. (2002): Genetically engineered mouse models of neurodegenerative diseases. *Nat. Neurosci.* 5, 633–639.

Wood, J.D., Peck, O.C., Tefend, K.S., Stonerook, M.J., Caniano, D.A., Mutabagani, K.H., Lhotak, S., and Sharma, H.M. (2000): Evidence that colitis is initiated by environmental stress and sustained by fecal factors in the cotton-top tamarin (*Saguinus oedipus*). *Dig. Dis. Sci.* 45, 385–393.

Yamada, Y., Yoshimi, N., Hirose, Y., Kawabata, K., Matsunaga, K., Shimizu, M., Hara, A., and Mori, H. (2000): Frequent beta-catenin gene mutations and accumulations of the protein in the putative preneoplastic lesions lacking macroscopic aberrant crypt foci appearance, in rat colon carcinogenesis. *Cancer Res.* 60, 3323–3327.

Yamori, Y., Ooshima, A., and Okamoto, K. (1972): Genetic factors involved in spontaneous hypertension in rats an analysis of F 2 segregate generation. *Jpn. Circ. J.* 36, 561–568.

Yang, H.C., Zuo, Y., and Fogo, A.B. (2010): Models of chronic kidney disease. *Drug Discov. Today Dis. Models* 7, 13–19.

Young, D.A., Ho, R.S., Bell, P.A., Cohen, D.K., McIntosh, R.H., and Nadelson, J. (1990): Inhibition of hepatic glucose production by SDZ 51641. *Diabetes* 39, 1408–1413.

Zeineldin, M. and Neufeld, K.L. (2013): More than two decades of Apc modeling in rodents. *Biochim. Biophys. Acta* 1836, 80–89.

Zhang, S.H., Reddick, R.L., Piedrahita, J.A., and Maeda, N. (1992): Spontaneous hypercholesterolemia and arterial lesions in mice lacking apolipoprotein E. *Science* 258, 468–471.

Zhang, Y.W., Su, Y., Lanning, N., Gustafson, M., Shinomiya, N., Zhao, P., Cao, B., Tsarfaty, G., Wang, L.M., Hay, R., and Vande Woude, G.F. (2005): Enhanced growth of human met-expressing xenografts in a new strain of immunocompromised mice transgenic for human hepatocytes growth factor/scatter factor. *Oncogene* 24, 101–106.

Zicha, S., Moss, I., Allen, B., Varro, A., Papp, J., Dumaine, R., Antzelevich, C., and Nattel, S. (2003): Molecular basis of species-specific expression of repolarizing K+ currents in the heart. *Am. J. Physiol. Heart Circ. Physiol.* 285, H1641–H1649.

Zisman, T.L. and Rubin, D.T. (2008): Colorectal cancer and dysplasia in inflammatory bowel disease. *World J. Gastroenterol.* 14, 2662–2669.

Zosky, G.R. and Sly, P.D. (2007): Animal models of asthma. *Clin. Exp. Allergy* 7, 973–988.

19

THE USE OF GENETICALLY MODIFIED ANIMALS IN DISCOVERY TOXICOLOGY

DOLORES DIAZ AND JONATHAN M. MAHER

Discovery Toxicology, Safety Assessment, Genentech, Inc., South San Francisco, CA, USA

19.1 INTRODUCTION

Genetically engineered mouse models are widely used in discovery toxicology for a variety of purposes, including assessing the safety of the intended pharmacological target, investigating mechanisms of toxicity (i.e., on-target vs. off-target effects), and aiding in the assessment of the disposition and metabolism of pharmaceuticals in establishing the toxicokinetics of a drug. An important consideration when generating knockout (KO) models to infer the safety of a particular pharmaceutical target is the translation and relevance of the KO mouse phenotype to the human condition and therefore the human relevance of the mouse findings. Although it would be simplistic to assume that every finding in a KO rodent phenotype is relevant to human risk, the relatively strong correlation observed between KO phenotypes and human pharmacology strongly suggests that KO phenotypes should in general be considered relevant to human safety when attempting to understand pharmacologic modulation of a target. A retrospective evaluation of the KO phenotypes for the targets of the 100 best-selling drugs, covering ~43 mammalian targets, indicates that these phenotypes correlated well with known drug pharmacological effects. In particular, of these 43 mammalian targets, 34 have been knocked out and 29 of the resulting KO phenotypes approximate the therapeutic effect of pharmaceutical intervention, providing in most cases a direct correlation between KO phenotype and the therapeutic effect of the drug (Zambrowicz and Sands, 2003). This correlation was observed despite the limitations of potential gene compensation, embryonic lethality, developmental issues, and differences between mouse and human physiology.

Several genetically modified mouse and rat models are available and can be leveraged in safety investigations and derisking, mostly for exploration of target safety. The evolution and advances in KO technologies have allowed for faster, more affordable, and more sophisticated models to be created and studied. Each one of these models has advantages and limitations (Table 19.1), and it is important to understand these factors in order to select the most appropriate model for the proper interpretation of the data generated. Additionally, in certain cases, several models can be used in a complementary fashion to obtain a comprehensive picture of target safety.

The most basic KO model is a constitutive KO, in which the target gene is disrupted in all cells of the body and at all stages of development; in some cases these models offer limited value for safety assessment, since gene disruption during development can be embryonic lethal. More relevant models for safety are conditional KOs, which are mainly divided into two types. These include tissue-specific KOs, in which the gene is inactivated in particular cell type by a tissue-specific promoter, and inducible KOs, in which the gene of interest is inactivated at a given time point by use of a tamoxifen- or tetracycline-driven Cre–lox system. These models are not mutually exclusive, and tissue-specific, inducible KOs can be engineered in all cells of the body or in a particular tissue type (Robbins, 1993; Bronson and Smithies, 1994).

Drug Discovery Toxicology: From Target Assessment to Translational Biomarkers, First Edition. Edited by Yvonne Will, J. Eric McDuffie, Andrew J. Olaharski, and Brandon D. Jeffy.
© 2016 John Wiley & Sons, Inc. Published 2016 by John Wiley & Sons, Inc.

TABLE 19.1 Genetic Models in Discovery Toxicology

	Advantages	Limitations
Germ line KOs	• Rapid model generation • Complete KO in all tissues • Can be leveraged for developmental assessment	• Can be embryonic lethal • Potential confounding developmental effects, which can over or underestimate toxicity • Potential scaffolding effects can result in overestimation of toxicities
Germ line—tissue-specific KO	• Can bypass embryonic lethality if the KO gene is not critical for development in the particular tissue • Useful when interrogating the role of a gene in a particular tissue, including its role in development	• Can be embryonic lethal if KO tissue has a critical role in development • Potential confounding developmental effects • Potential scaffolding effects can result in overestimation of toxicities • Assessment limited to tissue with KO gene • Potential scaffolding effects
Conditional KO in adulthood	• More relevant to pharmacological inhibition in adults • Flexibility in timing of the KO	• Potential incomplete KO • Limitations in obtaining KO in brain • Potential confounding effects of tamoxifen if used • Potential scaffolding effects
Conditional KO in adulthood—tissue specific	• More relevant to pharmacological inhibition in adults • Useful when interrogating the role of a gene in a particular tissue in isolation	• Assessment limited to tissue with KO gene • Potential incomplete KO • Limitations in obtaining KO in brain • Potential confounding effects of tamoxifen if used • Potential scaffolding effects
Germ line KI (inactive drug-binding site)	• Complete KO in all tissues • Preserves scaffolding function and more relevant to pharmacological inhibition.	• Can be embryonic lethal • Potential confounding developmental effects
Conditional KI on adulthood (inactive drug-binding site)	• Complete KO in all tissues • Most relevant model to mimic pharmacological effects in adults	• Potential incomplete KO • Limitation in obtaining KO in brain • Potential confounding effects of tamoxifen

Knock-in models are models in which a gene sequence has been inserted into a specific gene locus. A subset of knock-in models are represented by point mutation models, in which a small DNA modification of typically <3 base pairs is introduced into a gene of interest. This selective alteration of a single amino acid results in the modification or inactivation of the function of the protein in a targeted fashion, without significantly compromising the protein's tridimensional structure or affecting scaffolding function. Targeted models such as kinase-dead or ligand-binding domain-dead knock-in models are typically thought to be the most representative genetic models to determine on-target effects that might be observed with a small-molecule inhibitor of the protein (Sacca et al., 2010; Doyle et al., 2012). Similarly, inducible versions of targeted knock-in models are viable options, although they add complexity and decrease the success rate of generating the model.

Nuclease-based systems (like the CRISPR/Cas9 system) represent a relatively recent development and a significant improvement over traditional embryonic stem (ES) cell gene targeting. Nucleases create specific double-stranded breaks at the target locus, which trigger DNA repair mechanisms that facilitate constitutive KOs or knock-ins (Jinek et al., 2012; Wang et al., 2013). The main advantages of nuclease systems are speed and the potential for efficient multifunctional gene targeting, which can be particularly time consuming to generate with traditional methods that require multiple crossings.

19.2 LARGE-SCALE GENE TARGETING AND PHENOTYPING EFFORTS

Since gene targeting is such a valuable tool to elucidate and study gene function, efforts have been made to use this approach on a large scale. The most concerted and systematic gene targeting and phenotyping effort in mice has been undertaken by the International Knockout Mouse Consortium (IKMC). The IKMC set out to create a library of mouse ES cell lines representing each one of the ~20,000 genes in the C57BL/6 mouse genome (most commonly used mouse strain and strain for which the whole genomic sequence is available). The IKMC is comprised of several US and ex-US programs, including the Knockout Mouse Project (KOMP, which is a trans-NIH initiative), European Conditional Mouse Mutagenesis Program (EUCOMM), and the Canadian North American Conditional Mouse Mutagenesis Program (NorCOMM). With ES cells for over 14,000 gene KOs completed, the international scientific community has embarked on the second phase of this project: generating and phenotyping mice for each targeted gene, with the goal of reaching 5000 phenotyped KO genes by 2016. The International Mouse Phenotyping Consortium (IMPC) has as a goal to characterize the phenotype of every mouse generated by the IKMC and to preserve these mice in repositories and make them available to the scientific community (https://www.mousephenotype.org/; this website includes a useful searchable database, with mouse phenotype information when available).

A more limited large-scale phenotypic screen was conducted by Tang et al. (2010) by generating mouse KOs for 472 genes encoding secreted and transmembrane proteins and following-up with thorough phenotyping of the resulting mice, revealing a large array of neurological, immunological, metabolic, and cardiac functions for these genes (Tang et al., 2010). An interesting observation from this study was the lack of correlation between phenotype and tissue expression, even after excluding all of the secreted protein coding genes from the analysis. This finding emphasizes the need for caution when attempting to link gene function to gene expression patterns and the potential for overlooking valuable information with tissue-specific KO generation or targeted analysis of mutant mice.

19.3   USE OF GENETICALLY MODIFIED ANIMAL MODELS IN DISCOVERY TOXICOLOGY

When tackling a novel target in drug discovery, a benign phenotype of a relevant genetically engineered model (a KO for antagonists/inhibitors or an overexpressing transgenic for agonists/inducers) offers initial reassurance that the target can be potentially tractable from a safety perspective. A significant percentage of gene mutations and KOs result in embryonic lethality (Papaioannou and Behringer, 2012). Embryo lethal phenotypes do not necessarily mean that a target is not tractable in the adult organism, since developmental processes that might be sensitive to target KO might not be relevant in adulthood. An illustrative example is provided by 3-hydroxy-3-methylgluraryl-coenzyme A (HMG-CoA), the target of the widely used statin drugs, for which genetic KOs are embryonic lethal (Ohashi et al., 2003). On the other hand, careful analysis of these KOs, although challenging (since death can be the extreme manifestation of a primary defect that can be quite subtle), it can potentially provide useful

information about target safety (Turgeon and Meloche, 2009; Papaioannou and Behringer, 2012). As an example of a useful analysis of an embryo lethal phenotype, KO of erythropoietin or its receptor in mice resulted in identical embryonic lethal phenotypes, in which death occurred at embryonic day 13. A further study of the embryos indicated that there was a failure to produce red blood cells and a failure of fetal liver erythropoiesis (Wu et al., 1995). This is a good example in which an embryonic lethal phenotype still produced relevant information related to the gene function and pharmaceutical utility of the target. Lastly, embryo lethal phenotypes can also inform the assessment of the potential developmental toxicity of a drug.

In some cases, even when full KOs are embryonic lethal, heterozygous mice (mice carrying one mutated and one wild-type allele) might be viable, and characterization of these mice can provide useful and relevant safety information. Bcl-xL is a prosurvival member of the Bcl-2 family that regulates developmentally programmed and stress-induced cell death, and which has been investigated as a potential oncology target. Bcl-xL also has an important role in platelet survival, and although Bcl-xL$^{-/-}$ mice die at midgestation, heterozygous $+/-$ mice develop normally. Hematological characterization of these mice reveals platelet counts significantly lower (~20% lower) than those of wild-type counterparts (Mason et al., 2007). This phenotype is consistent with a decrease in platelet counts observed as a result of pharmacological inhibition of Bcl-xL in mice (Mason et al., 2007) and rats (Leverson et al., 2015), a finding that also translated to cancer patients (Tse et al., 2008; Schoenwaelder et al., 2011). The knowledge of the platelet effects in the Bcl-xL heterozygous mice provided evidence that the preclinical and clinical platelet effects were target related; this type of evidence can be very helpful in the discovery space when elucidating mechanisms of toxicity.

One can also speculate that heterozygous KO mice might be more relevant and representative of partial pharmacological target modulation (i.e., 50% target inhibition), but even when these mice can be useful for interrogating dose–response relationships, they can also underestimate safety issues, especially in situations where robust target modulation (i.e., >50%) is needed for efficacy.

In addition to potentially overpredicting the toxicity of molecules, the reverse and sometimes overlooked possibility is that germ line KOs can also underpredict toxicity if there are compensatory mechanisms at play during development. An example of this are the serotonin receptor 5-HT1B KOs, which exhibit an increased locomotor response to cocaine (instead of the expected decrease in activity that is triggered by 5-HT1B antagonists), or the finding that mice lacking the dopamine transporter gene (the main binding site for cocaine) still show dependence behavior toward cocaine. These paradoxical effects are believed to be the result of alterations in other brain systems that have occurred in response to the absence of the 5-HT1B receptor during development (Pich and Epping-Jordan, 1998; Rocha et al., 1998; Scearce-Levie et al., 1999). In the discovery space one should therefore consider the potential for underpredicting potential safety liabilities when characterizing embryonic KOs.

Tissue-specific KOs can provide a strategy to bypass embryonic lethality and allow the investigation of gene function within particular tissues of interest. An example is provided by the insulin receptor KOs, which die within 4–5 days after birth due to severe ketosis (Accili et al., 1996; Joshi et al., 1996), but for which several tissue-specific KOs, including muscle, liver, pancreas, brain, and brown fat, enabled the understanding of the role of the insulin receptor in these tissues (Bruning et al., 1998; Kulkarni et al., 1999; Bruning et al., 2000; Michael et al., 2000; Guerra et al., 2001). Although there are instances in which tissue-specific KOs might be useful in discovery toxicology, their usefulness is limited by the restricted nature of their targeted assessment, and more general KOs that mimic the effects of a drug in whole organisms are usually preferable.

In general, the preferred genetic model to assess target safety is one that most closely resembles pharmacological modulation of the intended target. This can differ depending on the drug or drug target, therefore careful consideration must be given when selecting KO models. Typically the most relevant model would be an inducible knock-in model where gene deletion (for inhibitors) is triggered during adulthood, and the deletion is restricted to the binding domain of the target, still allowing expression of the inactive (for inhibitors) target protein. A model like this would bypass any developmental phenotypic influence (in some cases including embryonic lethality), as well as preserve the scaffolding functions of the protein and as such, more closely resemble pharmacological engagement in an adult organism. An example of an embryonic lethal phenotype that is bypassed with an inducible KO model in adulthood is illustrated by B-Raf, a widely pursued oncology target. B-Raf germ line KO mice die in midgestation due to vascular defects (Wojnowski et al., 1997; Galabova-Kovacs et al., 2006). However, mice where the B-Raf gene has been knocked out in adulthood are viable and healthy and appear histologically normal (Blasco et al., 2011).

If a molecule targets two different isoforms of a protein, either by design or by the inability to obtain selectivity among closely related isoforms, the most relevant KOs for safety evaluation should be dual KOs of the targeted isoforms. An example that illustrates the complexities of drug target KOs and that incorporates embryo lethal phenotypes, tissue-specific KO, and dual isoform KOs is provided by MEK KOs.

One MEK inhibitor, Trametinib, has been approved, and several others are in clinical trials for oncology indications (Akinleye et al., 2013); these inhibitors target both MEK1 and MEK2 isoforms. MEK2 KO mice are phenotypically normal (Belanger et al., 2003), whereas MEK1 KO mice die

at midgestation due to abnormal development and reduced placental vascularization (Giroux et al., 1999), therefore a dual KO model that would recapitulate the pharmacological inhibition is not possible by simply crossing the two strains. Skin-specific dual KO mice were obtained by crossing MEK2 KO mice with skin-specific MEK1 KO mice. These mice died within 24 h of birth due to skin defects resulting in compromised barrier function. This skin phenotype is consistent with one of the most characteristic (and in some cases dose limiting) toxicities of MEK inhibitors in preclinical models and humans, which is severe skin rash (Choi, 2014; Jain et al., 2014; Chung and Reilly, 2015). These complex models that are able to address pharmacological relevance by knocking out multiple targets, and bypass embryonic lethality with tissue-specific KOs, can be potentially helpful in elucidating mechanisms of toxicity in the discovery space (i.e., evaluating whether a tissue-specific toxicity is on or off target).

Double KO mice can also be relevant models to investigate the safety of potential drug combinations. For example, Mcl-1 and Bcl-xL are antiapoptotic proteins that are putative drug targets for oncology, and in particular the potential combination of Mcl-1 and Bcl-xL inhibitors would therefore be of theoretical therapeutic interest. Both proteins also seem to play important roles in liver homeostasis and these roles appear to be cooperative and nonredundant, as illustrated by the increased severity of mouse phenotypes with increased number of ablated alleles for these two genes. Bcl-xL$^{+/+}$+Mcl-1$^{-/-}$ mice are viable but have increased hepatocyte apoptosis (Hikita et al., 2009; Vick et al., 2009); a similar phenotype is observed in Bcl-xL$^{-/-}$+Mcl-1$^{+/+}$ (Takehara and Takahashi, 2003) and Bcl-xL$^{+/-}$+Mcl-1$^{+/-}$ (Hikita et al., 2009). Homozygous null mice for both alleles die soon after birth of liver failure. In addition, Mcl-1$^{-/-}$ mice dosed with Bc-xL inhibitor ABT-737 die within 2 h of dosing of fulminant liver failure (Hikita et al., 2009). These experimental data strongly suggests that the combination of Bcl-xL and Mcl-1 inhibitors is unlikely to be tolerated and caution should be taken in pursuing such a combined therapeutic approach. This example illustrates how knowledge of target KO phenotypes can effectively be used in the discovery space to inform potential success of certain drug combinations.

Given the utility of genetic models in assessing the safety of the target and the long timeline for model generation and characterization (typically 9–12 months), it is prudent to trigger the generation of these models early in the life cycle of a project, ideally in the research space or the early hit-to-lead phase. Even when the findings might not be decisional for program fate, the information can be very useful in flagging safety concerns that might need further characterization in safety studies or in contributing to the mechanistic understanding of toxicity findings when they arise in animal studies or even in humans.

One of the most useful applications of KO models in discovery toxicology is in the determination of mechanism of toxicity and, in particular, in differentiating on-target from off-target drivers of toxicity. This differentiation is typically critical for decision-making in the discovery space, since off-target findings can usually be solvable through structural modifications, but on-target findings are typically considered unavoidable. In this regard, one of the most useful criteria in the weight of evidence to indicate that a particular toxicity is on target is consistency with the phenotype of KO mice.

This strategy was nicely illustrated in the case of LRRK2 inhibitor toxicity, where two structurally distinct and highly selective LRRK2 small-molecule inhibitors caused concerning lung toxicity in cynomolgus monkeys. The finding, accumulation of lamellar bodies in type II pneumocytes, was morphologically identical to that seen in LRRK2 KO mice (Herzig et al., 2011; Fuji et al., 2015), which supported the conclusion that this finding was target related.

Interestingly, this finding was not observed in mice or rats treated with the same small-molecule LRRK2 inhibitors at plasma exposures that matched or exceeded those achieved cynomolgus monkeys (Fuji et al., 2015). The implications of this observation are important in discovery toxicology, since absence of the finding in rodents would initially suggest that the phenotype of the KO mice was not recapitulated with pharmacological inhibition, and one would be tempted to conclude that this phenotype was perhaps a worst-case scenario related to developmental inhibition or scaffolding function that was not applicable to the intended drug. However, the finding emerged in the monkey study, raising the possibility that rodents might need a higher threshold for pharmacological triggering of this particular finding compared to monkey. Interestingly, these two inhibitors caused no apparent renal toxicity in rodents or monkeys, despite the presence of renal findings in the mouse and rat LRRK2 KOs (Ness et al., 2013; Fuji et al., 2015), which illustrates further the complexities in understanding the translation and relevance of KO mouse phenotypes.

This example emphasizes the importance of not discounting phenotypic findings, even if these findings are not recapitulated in rodent toxicology studies as the findings may still remain relevant in other species or with increased target engagement.

Another useful application of KOs in elucidating mechanisms of toxicity consists of dosing the KO with the molecule of interest and, if the toxicity persists in the absence of the target, this is a strong indication of an off-target driver. A good example of this approach is provided by studies conducted with polo-like kinase 2 (PLK2) inhibitors. PLK2 is a kinase that phosphorylates synuclein, and it is considered a putative target for the treatment of Parkinson's disease. This kinase also has a demonstrated role in the attachment of kinetochores to microtubules during mitosis, which raises a concern for potential genotoxicity. Although selective PLK2

inhibitors were not genotoxic *in vitro*, one of these inhibitors caused a significant increase in micronuclei in rats. *In vivo* genotoxicity based on the potential biological function would suggest PLK2 is an intractable target for a CNS indication, which led to further investigation using KO models. These KO mice demonstrated that the genotoxicity (increase in micronucleated reticulocytes) observed *in vivo* with a selective PLK2 inhibitor was likely unrelated to PLK2 pharmacological activity and therefore likely off target, since a similar increase in micronucleated reticulocytes was observed in both wild type and PLK2 KO mice after treatment with a PLK2 small-molecule inhibitor. Similar plasma exposures to the inhibitor were also confirmed in both wild type and KO mice, ruling out confounding effects related to pharmacokinetics. This KO model was therefore useful in demonstrating that PLK2 (as a drug target) is not linked to genotoxicity, which provided reassurance about the tractability of this target for Parkinson's disease from a safety standpoint (Fitzgerald et al., 2013).

Another example of dosing a KO to confirm that a toxicity finding is off target was provided by May et al. (2011), where dosing BACE1 KO mice (strain BACE1 tm1Pcw) daily for 9 weeks with the SM BACE1 inhibitor LY2811376 revealed accumulation of autofluorescent material and degenerative changes in the retinal epithelium, which resembled findings observed in rats given similar doses of LY2811376 for 3 months. Although the findings in the rat study resulted in clinical discontinuation of LY2811376, the KO mouse data provided evidence that this toxicity was likely off target and therefore potentially solvable through chemical modifications (May et al., 2011).

Alternatively, if an observed toxicity is absent in a KO dosed with the molecule of interest, this implicates the target as responsible or contributing to the toxicity. This concept is more applicable to agonist molecules or inducers, but it could also theoretically be applicable to antagonists/inhibitors when using germ line KOs in which developmental compensation has masked the pharmacological effects of the target. An example of this is provided by glucagon-like peptide receptor (GLP-1R) agonists. GLP-1R agonists are associated with the development of thyroid C-cell tumors after lifetime exposure in rodents; these tumors are preceded by increase plasma calcitonin, thought to be mediated via GLP-1R, since these findings are not present when treating GLP-1R KO mice (Madsen et al., 2012).

Finally, to illustrate the complexity of interpreting information obtained from GEMM models, the following is an example that at first glance might seem paradoxical of how the phenotype of a kinase-dead gene can be considerably worse than the phenotype of the completely absent gene. Receptor-interacting protein kinase 1 (RIPK1) and 3 are involved in the triggering of "necroptosis," a programmed form of necrosis or inflammatory cell death. RIPK3 KO mice are viable, but mice in which the catalytic domain of

RIPK3 is mutated and inactivated during development die around embryonic day 11.5 with abnormal yolk sac vasculature. In addition, mice in which the catalytic domain of RIPK3 is mutated and inactivated in adulthood also die after 6–7 days due to intestinal toxicity (Newton et al., 2014). Kinase-dead RIPK3 is believed to interact with RIPK1 to promote apoptotic cell death in the intestine; it is unclear why this catalytically inactive mutant engages RIPK1, FADD, and caspase-8 but wild-type RIKP3 does not. This paper underlies the potential limited therapeutic benefit of RIPK3 inhibitors because of the possibility that they could cause apoptotic cell death, but leaves the door open for therapeutic inhibitors of RIPK1, given that inhibition of its kinase activity has not shown to have deleterious consequences in the short term. This example illustrates the benefit of exploring several mouse models in order to fully appreciate the complexities of the biology as it relates to safety implications in order to better inform potential target safety derisking in the discovery space.

Given that rats are the most commonly used rodent tox species, generating and characterizing the phenotypes of rat KOs can be extremely helpful in derisking programs and in interpreting toxicity findings of compounds. Although there are currently few published examples of phenotyped rat KOs, the pharmaceutical industry is moving in this direction, and we expect to see an increased number of examples of rat models in the future. A published KO rat example was presented by Ness et al. (2013), where they describe a detailed phenotypic study of LRRK2 KO rats, which have renal, metabolic, and immunological perturbations suggestive of potentially complex biological safety implications for this target (Ness et al., 2013).

In summary, in this chapter we have outlined how genetically modified animal models, mostly mouse KO models, can be leveraged in the discovery toxicology space, particularly for providing insights into target safety and investigating mechanisms of toxicity. These models, along with a solid understanding of their limitations and with thoughtful interpretation, can be powerful tools in early derisking, and research teams can benefit greatly from incorporating GEMMs into their safety paradigms.

19.4 THE USE OF GENETICALLY MODIFIED ANIMALS IN PHARMACOKINETIC AND METABOLISM STUDIES

In addition to the role of genetically modified animals in target safety assessment and investigating mechanisms of toxicity, a key use of these animal models is the assessment of drug metabolism and toxicokinetics of experimental therapeutics, with drug exposure playing an important role in overall safety. The properties of absorption, distribution, metabolism, and elimination (ADME) for small-molecule

therapeutics are dictated primarily by a limited number of key nuclear receptors, enzymes, and transporters. The development and early adoption of KO mice for these key genes in the pharmaceutical industry have led to a clear understanding of the strengths and weaknesses of each of these models. One of the first reported experiments conducted in the field was targeted disruption of the drug-metabolizing enzyme cytochrome P450 1A2 in mice (CYP1A2) (Liang et al., 1996). When the muscle relaxant zoxazolamine was administered, the lack of CYP1A2 led to markedly decreased drug metabolism, causing decreased clearance and prolonged sleep paralysis in the CYP1A2 KO mice (Pineau et al., 1995). In the 20 years since those initial studies took place, the increase in the number of new genetically modified animal models and experimental designs has steadily climbed. The newest technologies such as double and triple transporter KO models, humanized P450 mice, KO rats, and humanized liver models have become powerful tools in the assessment of human pharmacokinetics. The utility of these models in assessing the PK properties in early drug development will be highlighted with several examples (Table 19.2).

19.4.1 Drug Metabolism

There are a multitude of enzymes that make up xenobiotic detoxification, which is divided into phase I and phase II metabolism. The first step in the process is an oxidation reaction, which is typically conducted by the cytochrome P450 system. Whereas over 57 P450s exist in humans, only a selected number play a prominent role in drug metabolism (Guengerich, 2006). Eight of the most common CYPs that are often evaluated in human reaction phenotyping include CYP1A1/2, CYP2B6, CYP2C9, CYP2C19, CYP2D6, CYP2E1, and CYP3A4. Of these, CYP3A4 is responsible for nearly 50% of the metabolism of all marketed drugs and has been the most studied enzyme of the CYP family. The ease and availability of *in vitro* profiling allow for an early understanding of the metabolism and key drivers of clearance for an investigational drug. Once these parameters are determined for a particular drug or compound, GEMM models have potential utility in the assessment of exaggerated pharmacology or increased safety risk in the clinical setting in terms of both human genetic variability (e.g., polymorphisms) or drug–drug interactions (DDI). It is also critical to understand the variability in genetic background and the limitations of each model, which can be significant.

Unfortunately not only do rodents and humans differ in the CYP orthologues expressed, but significant differences can be observed in terms of functional inhibition, substrate selectivity, and enzyme induction. Notably the mouse has a significant expansion in the number of P450s expressed, with more than twice the number of Cyps than in humans (Muruganandan and Sinal, 2008). This has limited the use of

TABLE 19.2 Genetic Models for Pharmacokinetics and Metabolism

	Key Issues	Utility in Drug Discovery and Development
KO mice	• Marked species differences in substrate specificity, activity, and induction limit translation • Compensation by other enzymes, transporters, or receptors upon deletion of single gene • Functional redundancy limits interpretation	• P450s: KO models generally not useful due to species differences • Transporters: compound transporter KO models can be used to assess contributions to CNS penetration and oral bioavailability • Nuclear receptors: compound PXR/CAR and AhR mice useful for some induction liabilities where species-specific activation is not observed
Humanized transgenic models	• Random insertion leading to disruption of bystander gene • Transgene insertion is in poor genomic location • Copy number not ideal • May not have deletion of murine orthologues • Variable expression in tissues • Tg construct might not contain all response elements needed for proper expression • Interplay between humanized protein and murine environment	• In general the models have limited utility with the existence of knock-in humanized models
Knock-in humanized models	• Technically challenging • Construct might not contain all response elements needed for proper expression • Interplay between humanized protein and murine environment causing hybrid metabolic profile	• Models with humanization of key human P450s (CYP3A, CYP2D) can aid in human clearance predictions with *in vivo* context (i.e., protein binding) • Modeling human induction liabilities (i.e., rifampicin)
Humanized liver models	• Degree and consistency of human hepatocyte colonization • Animal health and histopathology • Severely immunocompromised • Unable to recapitulate extrahepatic metabolism, potentially leading to a hybrid metabolic profile and disposition • Compensatory effects on extrahepatic tissues	• Detection of human-predominant toxicities (i.e., fialuridine, bosentan) • Modeling human induction liabilities (i.e., rifampicin) • Prediction of human PK where *in vitro* models are limited
KO rats	• Marked species differences in substrate specificity, activity, and induction limit translation • Compensation by other enzymes, transporters, or receptors upon deletion • Functional redundancy limits interpretation • Limited number of available models and compound mutants • No current availability of humanized rat models	• Significant practical advantages of rats over mice in blood sampling and manipulation • Often species of choice for toxicity studies, avoiding extrapolation from mouse • Fewer animals needed than mouse due to repeat sampling • Future development of humanized rat models would likely eclipse mouse models

P450 KO mice due to the concerns about lack of human translation. In particular the most striking species differences in both expression and P450 metabolism are in the CYP2D6/Cyp2D and CYP3A4/Cyp3a families, which together metabolize >70% of marketed drugs (Gonzalez and Yu, 2006). One example is debrisoquine, an antihypertensive agent that is often utilized for CYPD26 phenotyping. In mice, no formation of the 4-hydroxy metabolite is observed, suggesting that mice cannot adequately recapitulate the human condition (Bogaards et al., 2000). Similarly for the probe substrate coumarin, the differences in CYP2D6

hydroxylase activity are responsible for 200-fold higher exposures in human than in rat (Pearce et al., 1992; Yamazaki et al., 1994; Bogaards et al., 2000). The CYP2C family also plays a role in metabolism of some pharmaceutical agents, and a Taconic Cyp2C KO model, with deletion of the 14 members of the murine 2C family, was utilized to examine a DDI interaction of the Cyp3a substrate midazolam (MDZ, victim) and the Cyp3a inhibitor, troleandomycin (TAO; perpetrator) (Grimsley et al., 2013). Although deletion of the Cyp2C loci markedly increased the exposure of MDZ versus WT, addition of the TAO inhibitor should theoretically have effectively caused synergistic increases in parent MDZ drug levels when both CYP3A/2C metabolism was compromised. However, there was little difference in the ratio of increase between WT and KO, casting some doubt on the utility of this model to evaluate human DDI prediction.

Although other P450s are generally less involved in drug metabolism compared to CYP3A4, CYP2D6, and CYP2C9/19, some of these KO mice models have been proven useful in understanding DDI and toxicity potential. For example, the human CYP2E1 is an enzyme that is relatively more conserved among various species compared to other relevant P450s. Cyp2e1 KO mice were able to highlight the contributions of the enzyme to acetaminophen toxicity, which has been shown to be of predominant importance in causing formation of the reactive NAPQI metabolite in humans (Lee et al., 1996; Manyike et al., 2000; Cheung et al., 2005). For other substrates, the relevance of murine Cyp2e1 KO is relatively high and fits with the Guengerich schema where the cross-species extrapolations for CYP2E1 are the highest among the P450s (Guengerich, 1997). Another example is the CYP1A1/2 KO mouse model. These enzymes are known to metabolize an overall small percentage of marketed drugs, but numerous examples of CYP1A2 substrates, such as a subset of antidepressants and antipsychotics, exist. In general, Cyp1a1/2 KO mice have contributed significantly to the understanding of bioactivation of polyaromatic hydrocarbons, whose mechanism of carcinogenesis is relevant in humans (Kleiner et al., 2004). Conversely, despite the existence of some isolated examples like the use of Cyp1a2 KO in assessing caffeine clearance (Buters et al., 1996), their utility in assessing pharmaceutical exposures in humans has been limited. In both cases for Cyp2e1 and Cyp1a2, once the murine Cyp was deleted, the human orthologue was introduced and quickly became the more relevant model.

The development of humanized P450 models was an important landmark in creating relevant preclinical models for modeling human drug metabolism. Several studies have utilized Tg overexpression of human CYPs on an intact murine Cyp background, including several notable examples. These include this first humanized P450 mouse model, which expressed human CYP2D6 on top of the existing murine Cyp2d family (Corchero et al., 2001; Miksys et al., 2005). Similar examples were short to follow and included CYP1A1, CYP1A2, CYP2E1, CYP2D6, CYP3A4, and CYP3A7 (Gonzalez, 2007). To reduce confounding factors, the best humanized models typically utilize genomic deletion of all mouse orthologues to reduce background contributions of murine genes, a technically challenging approach that has become feasible only in recent years.

Perhaps the most notable example is the mouse Cyp3a family. In mouse, there are eight *Cyp3a* genes and three pseudogenes, all of which are spread across 7 MB of genomic DNA and the majority of which are clustered in a 0.8 MB region (van Herwaarden et al., 2007). *Flox-p* sites were introduced upstream and downstream of this region so this entire region could be deleted when a tissue-specific Cre recombinase was introduced, such as an albumin (liver) or villin-driven (intestine) construct. These mice could be crossed with Cyp3a13 KO mice to create a mouse with complete deletion of the Cyp3a murine family (van Herwaarden et al., 2007). Assessment of docetaxel, a chemotherapeutic agent with a narrow therapeutic index, was conducted in Cyp3a family KO mice, showing an almost sevenfold increase in exposure versus WT, which was attenuated when the human CYP3A4 was introduced. One of the inherent weaknesses (or strengths) of the model is that CYP3A4 has activities in both liver and intestine that have important contributions to the overall pharmacokinetics of any drug, thus one important consideration when utilizing mice on this background is the tissue-specific transgene that drives the deletion of these loci. This was elegantly demonstrated in the same Tg-Cyp3A4 human model previously, which could delineate the contributions of liver versus intestine CYP3A4 metabolism for triazolam (van Waterschoot et al., 2009). This model revealed that, surprisingly, intestinal CYP3A4 was the key driver of exposure rather than hepatic CYP3A4.

In general, mouse "knock-in" models are preferred to transgenic rescue and humanization. Knock-in techniques rely on homologous recombination of a specific locus to ensure positional location into the genome. In the case of knock-in humanized mice, the genome will be overwritten with a modified construct that has significant homology with the mouse but contains the human P450 of interest. During ES cell selection, the construct will, with low frequency, be inserted into one locus. Once the ES cell of interest has been selected, it will be incorporated into a blastocyst, and the resultant mice, assuming germ line transmission, will be bred until homozygous mice are generated. Conversely, in the case of transgenic mice, empirical selection of a transgenic line that phenotypically recapitulates an expected phenotype is the ultimate goal. This can be tricky, and some of the pitfalls have been reviewed elsewhere (Matthaei, 2007). Similarly, knock-in humanized CYP3A4/3A7 mice were first reported in 2011 (Hasegawa et al., 2011a). Although murine Cyp3a13 was not deleted in this model, this enzyme was believed to not play a major role in metabolism and was poorly expressed in liver and small intestine. The tissue distribution and phenotypic aspects of the mice are highly consistent with successful development of a knock-in model and do not require crossing the mice with transgenic Cre-expressing systems to generate deletion in a tissue of interest, as in other advanced models (van Waterschoot et al., 2009). Furthermore, these knock-in mice have been crossed with other existing model systems, including CYP2D6 and CYP3C9 mice, as well as regulatory nuclear receptors such as PXR and CAR such that induction can be assessed. In essence, the key drivers in metabolic clearance (~70% of marketed drugs) are almost completely incorporated into one existing mouse model. With each additional humanization and crossing, these mouse models have improved incrementally toward recapitulating human metabolism over the last few decades.

In addition to phase I metabolism, some phase II enzymes, which typically conjugate and make substrates more hydrophilic, can contribute significantly to metabolism and clearance of drugs. In particular, the UDP-glucuronosyl-transferases (UGTs), sulfotransferases (SULTs), and glutathione S-transferases (GSTs) are notable examples of phase II enzymes. In general, phase II enzymes are less involved in clinical DDIs and toxicity, which has been speculated to be due to higher turnover rates than phase I enzymes. As a representative example for phase II enzymes, one such clinically relevant DDI is irinotecan (topoisomerase inhibitor) and erlotinib (epidermal growth factor receptor inhibitor), which lead to significant toxicity in combination (Liu et al., 2010). Similarly UGT1A1 plays a role in bilirubin glucuronide conjugation, and significant increases in bilirubin levels are associated with hepatotoxicity (Temple, 2006). Although poor glucuronidation is a risk factor for drug-induced liver injury (Raza et al., 2013), the main concern is cholestatic liver injury, which is diagnostically measured by total bilirubin levels. Thus understanding whether a small molecule inhibits UGT activity or bile flow is an important distinction. To this end, animal models exist and have been utilized for many years in mechanistic studies. In murine KO models, disruption of exon 4 of Ugt1a led to premature lethality due to hyperbilirubinemia and kernicterus (Nguyen et al., 2008), which makes this model too severe to be helpful. Instead of Ugt KO mouse models, the field has relied heavily on a spontaneous mutation in Wistar rats (GUNN, 1938). Notable examples in investigating hyperbilirubinemia findings in Gunn rats include mechanistic studies demonstrating UGT inhibition by antiviral drug cocktails (Zucker et al., 2001; Kempf et al., 2006). In mice, recent humanized knock-in models deleted the murine Ugt1a family and expressed *UGT1A1*28*, which markedly decreased phenotypic UGT1A1 activity to a level consistent with human patients that have Gilbert's syndrome. Patients with the *UGT1A1*28* genotype and Gilbert's disease are known to be susceptible to toxicity upon exposure to some drugs (Cai et al., 2010). Considering the viability and clinical relevance of this model, this could be another valuable tool in evaluating hyperbilirubinemia and clinical DDIs.

19.4.2 Drug Transporters

In addition to drug-metabolizing enzymes, transporters play an important role in drug disposition and clearance. There are over 48 active ATP-binding cassette transporters in humans, and many more passive or facilitative transporters that play key roles in transporting endogenous and xenobiotic substrates. Key roles of transporters in drug pharmacokinetics

are (i) mediating uptake from the gut of orally administered compounds, (ii) preventing or allowing entry of drugs into the central nervous system via blood–brain barrier (BBB), (iii) mediating renal secretion, and (iv) facilitating biliary clearance. To address the relevance of transporters in pharmaceutical development, an International Transporter Consortium was formed. The group has outlined the relevance of the various models used to address safety and clinical DDIs, including highlighting the most relevant animal models in the field in a publication that is recommended as an important reference (International Transporter et al., 2010).

Much like the P450 families, rodent species have undergone significant expansion of many of the transporter families, and there are notable differences in orthologues expressed and substrate specificities. In fact, the most studied human transporter, P-glycoprotein (PgP; MDR1), actually has two isoforms in mice (Mdr1a/Mdr1b). Significant overlap in substrate specificity has also been observed in different species for many key transporters. This level of functional redundancy has led to subtle phenotypes in KO models, confounding the interpretation that the "one drug, one transporter" assumption is too simplistic to be practically applied. To this end, the field has moved to the creation of compound transporter KO models.

Although there are numerous transporter GEMM models, the most utilized to date are the Mdr1a/Mdr1b (PgP) KO mice (Schinkel et al., 1997). PgP plays a crucial role in blocking the effective uptake of substrates in the BBB and gut, and potent substrate specificity of a drug for PgP can greatly limit systemic or brain exposures. In some cases, a lack of specificity is desired to aid in better CNS exposures or to gain better oral bioavailability of a drug. In other cases, as with some chemotherapeutic agents, limiting brain disposition is desired, and one discovery safety strategy utilized is to improve the efflux transporter substrate specificity to avoid any potential CNS-related liabilities. A very thorough study that utilized 34 CNS active and 8 non-CNS active compounds was conducted in the Mdr1a/Mdr1b KO mice to examine brain-to-plasma (B : P) ratios over a time course (Doran et al., 2005). Deletion of PgP increased brain exposures for most compounds, with the exception of a few non-CNS compounds like quinidine and amiodarone (with large B : P increases). However, for the vast majority of compounds, the B : P ratio increased only by a factor of 2. Furthermore, ivermectin, a compound that shows a 100-fold increase in B : P ratio in the Mdr1a deficient mice which results in lethality, has never been reported to have caused CNS-related toxicity in humans (Kalvass et al., 2013). These observations seem to cast some doubt on the translation of these preclinical models in terms of safety, suggesting that DDIs at the BBB are unlikely to cause significant clinical risk.

Another important efflux transporter that has significant substrate specificity and functional redundancy with PgP is

BCRP. A triple KO Mdr1a/b/Bcrp mouse was engineered in order to determine if exacerbated or even synergistic accumulation in the CNS might be driven by dual inhibition of both PgP and BCRP. Recent reports have suggested that combined deletion of the three transporters significantly increased CNS exposures for rivaroxaban, as well as decreased overall clearance of this anticoagulant (Gong et al., 2013). It is likely that this model represents one of the better models for understanding BBB penetration, since it lacks the compensatory actions of Bcrp. However, in general, the transporter field lacks the level of GEMM model sophistication of the P450 field, and models such as knock-in humanized mice could greatly advance the landscape of the transporter field but are not available to date for evaluating the CNS liabilities of potentially small-molecule therapeutics.

In recent years, technological advances in zinc finger, TALEN, and CRISPR genome editing has allowed for development of KO rat models. The rat is the preferred rodent safety model given their practical advantages, including easier PK sampling, larger blood volumes, ease of cannulation, and a more extensive understanding of background incidence of histopathology findings. As transporters have a notable role in drug disposition, a large number of transporter KO rats are available, including Mdr1a (PgP), Bcrp, and Mrp2 (Chu et al., 2012; Zamek-Gliszczynski et al., 2012). Where comparisons of drug disposition have been made between Mdr1 KO mice and rats, the correlations have been excellent (Bundgaard et al., 2012). Although the rat models offer the ability to understand the mechanistic contributions of individual transporters in isolation, the lack of compound transporter models has limited their utility as tools in drug development. Although some compound KO rat models are available such as Mdr1a/b and Mdr1a/Bcrp KO rats, the zinc finger technology requires crossbreeding and takes an ever-increasing amount of work to get full deletion of multiple transporters. However, with the onset of newer technologies like CRISPR and TALENs, this process may be significantly sped up so that more sophisticated models can be rapidly developed; these technological advances should hopefully contribute to the development of more relevant transporter rat models in early safety evaluations.

19.4.3 Nuclear Receptors and Coordinate Induction

There are a few key nuclear receptors that can significantly affect the regulation of phase I and II metabolism as well as uptake and efflux transporter expression. This is sometimes referred to as coordinated regulation due to the ability of a single receptor to modulate uptake, metabolism, and efflux via a single mechanism. When such events occur, they can lead to DDIs where the effects of one drug (perpetrator) can decrease the exposure and efficacy of another drug (victim).

This type of phenomenon was illustrated in some patients taking St John's wort as a herbal antidepressant concomitantly with contraceptives, leading to unexpected pregnancies due to the decrease in exposure and half-life of the contraceptive (Hall et al., 2003). Furthermore, a drug can also decrease its own exposure in a process known as autoinduction. These types of interactions are often referred to as induction liabilities, and clinical pharmacology guidelines outline screening for induction of CYP1A2, CYP2B6, and CYP3A4, which are the P450s that have the most potential for marked induction. The nuclear receptors that modulate the regulation of these genes are the aryl hydrocarbon receptor (AhR), the constitutive androstane receptor (CAR), and the pregnane X Receptor (PXR), and high-throughput screens are often applied in early drug discovery and development to interrogate whether potent activation of these receptors occurs. In general, the availability of these receptor assays and the growing quality of primary plated hepatocytes from human and preclinical species allow for the interrogation of a potential induction liability. Although the translation of these *in vitro* assays is very good, questions can remain about the quantitative translation of exposure loss clinically. The rodent P450s and nuclear receptor pathways involved in induction are generally functionally conserved, but the substrate specificity and potency for a drug is often markedly different, as is the P450 orthologue being upregulated. In such cases, recent advances in GEMM models allow for an improved context in using an *in vivo* system to interrogate an induction liability without costly additions to clinical trials.

Nuclear receptor KO mouse models have been utilized for many years, with the Ahr KO mouse first developed in 1995 and demonstrating the *in vivo* phenotype of knocking out the dioxin receptor (Fernandez-Salguero et al., 1995). Similarly Pxr KO and hPXR Tg-humanized mice quickly followed, showing an induction profile that was responsive to human-specific ligands such as rifampicin (Xie et al., 2000). Rat KO models have also been developed, including AhR/CAR/PXR triple KO models. However, because of the large differences in species specificity and the cross activation of receptors such as PXR and CAR, more advanced models have been developed in recent years. Due to the large number of CYP3A4 substrates and the inducibility of this enzyme, a humanized mouse model was engineered to contain the two major nuclear receptors that regulate CYP3A4, PXR, and CAR. The DDI potential was examined in this hPXR/hCAR/hCYP3A4/3A7 mouse, with triazolam (TRZ) as the victim, and with the CAR/PXR activators rifampicin, sulfinpyrazone, and pioglitazone coadministered (Hasegawa et al., 2011a). Decreases in AUCs of TRZ were greater than 90% with coadministration of rifampicin at 10 mg/kg/day, demonstrating the potent effects of an induction liability on compound disposition. To date, similar models with hAHR/hCYP1A1/hCYP1A2 and hCAR/hPXR/

hCYP2B6 have not been reported, likely due to their lesser role in overall metabolism and induction by drugs. Further refinement of these models will likely occur over the next few years, resulting in models that are more predictive of human PK modeling and exposure projections.

19.4.4 Humanized Liver Models

Humanized models can be helpful in understanding the contributions of individual, sets, or families of genes. However, these models cannot fully recapitulate the functions of a human cell type or an organ. The ADME properties of xenobiotic handling and the regulation of these systems is an intricate process, and humanized metabolism in phase I, for example, may negatively or positively affect murine phase II metabolism and efflux transport. Similarly, marked species differences in activation of nuclear receptors can lead to discrepant effects on induction of many P450s and transporters. Thus, the "Holy Grail" has been to create preclinical models that have complete recapitulation of key drug-metabolizing tissues, with the most important cell type being the hepatocyte. To this end, several humanized hepatocyte models have been attempted, with various degrees of success. Although these models have some interesting properties, there are also significant drawbacks that are important to discuss.

Several different types of mouse models with humanized livers have been developed, and all of them include two main components. The first is that the murine hepatocytes have to be damaged in a targeted fashion, leading to eventual hepatocyte cell death. Three main systems have evolved to accomplish this task, most notably (i) a transgenic albumin-driven uroplasminogen activator (uPA), (ii) a transgenic albumin-driven thymidine kinase (TK) system, and (iii) a tyrosine catabolic enzyme fumarylacetoacetate hydrolase (Fah) system. The uPA system, which was the first one to be developed, has shown notable drawbacks. The uPA relies on plasminogen activation, which is suppressive of hepatocyte growth factor activity and overall hepatocyte function and ultimately leads to cell death (Mars et al., 1993). Plasminogen activation also causes notable extrahepatic effects, including difficulty in breeding, renal injury, and complexity in the protocol and time frame when conducting *in vivo* studies (Hasegawa et al., 2011b). Furthermore, to increase the viability of these mice, an anticoagulant (i.e., Futhan) is typically administered, which could confound metabolism and PK studies. In some cases, examples of selective pressure and transgenic inactivation have been observed, which led to retention of viable murine hepatocytes (Rhim et al., 1995). Due to some of the constraints with the uPA model, other systems have been developed, like the TK and Fah models. The TK model is a well-established, efficient system for xenografts (Heyman et al., 1989). When ganciclovir is administered, tissue-specific expression of the TK enzyme leads to cell death in any cell type expressing the TK

construct, thus the albumin-Cre transgene drives murine hepatocyte death. In the case of Fah mice, supplementation with 2-(2-nitro-4-trifluoromethylbenzoyl)-1,3-cyclohexanedione (NTBC) is required to prevent the onset of tyrosinemia, which is the overall mechanism of murine cell death when NTBC is withheld. In these two later systems, the timing for the reconstitution of the liver by splenic transplantation of human hepatocytes can be controlled through chemical induction or supplementation.

The second main step in the generation of mice with humanized livers is that the mice have to be immunocompromised to ensure tolerance to the engraftment with human hepatocytes. Several types of mice have emerged as viable options for engraftment. The first are severe combined immunodeficiency (SKID) mice, which lack functional T and B cells. These mice are typically kept on a NOD/LtJ background, which increases the reconstitution index (RI) of an engrafted tissue or cell type due to further diminishment of immune function (Ito et al., 2002). A variant of this model that is gaining traction as a next-generation model is one that incorporates an IL-2rg mutation. These mice, deemed NOG mice, have less NK cell activity and a longer lifespan due in part to the decrease in thymic lymphomas (Shultz et al., 2005). An alternative variant, the Rag2$^{-/-}$/Il2rg$^{-/-}$ (RG) mice, similarly lack T- and B-cell function and were empirically selected to have suitable characteristics as a host model (Hasegawa et al., 2011b). Thus, the SKID, NOG, and RG models have all been shown to have RIs above 90% with relevant human albumin levels and serve as viable models for *in vivo* studies with nearly complete humanization of the liver. It is also important to understand that this field is still evolving and that improved combinations of these models could be created in the future.

In terms of safety, these humanized liver models have been shown to replicate some human toxicities that had not been observed in preclinical species. Two recently published examples illustrate the potential value of humanized liver models in safety assessment. Fialuridine is a nucleoside analogue that was developed for hepatitis B and that caused severe drug-induced liver injury in clinical trials, but no hepatotoxicity in preclinical species, despite being dosed at 1000-fold higher levels compared to clinical doses (Richardson et al., 1994). Several lines of evidence implicated mitochondrial toxicity as the cause of injury, in whole or part caused by significant differences in the expression and function of the equilibrative nucleoside transporter 1 (hENT1) (Lee et al., 2006). As this is an atypical transporter that is not usually examined in drug discovery and development, there would be little reason to suspect such a large, species-dependent toxicity. The humanized TK-NOG mice were able to recapitulate the fialuridine hepatotoxicity toxicity observed in humans in a dose-responsive manner. Hepatotoxicity was not observed in control mice or in humanized mice dosed with sofosbuvir, a fialuridine analogue that had shown no evidence of clinical DILI (Xu et al., 2014). Similar work by Xu and colleagues demonstrated that bosentan, an endothelin antagonist that showed clinical hepatotoxicity but limited evidence of preclinical hepatotoxicity, caused liver injury in the TK-NOG model that was similar to the cholestatic effects observed clinically (Xu et al., 2015).

Although the assessments of the overall capacity of these models to replicate human hepatotoxicity are still in progress, these emerging examples suggest that these models could be very valuable, particularly in cases of potential cholestatic liver injury due to transporter inhibition, an application where preclinical studies have shown limited predictivity of DILI (Morgan et al., 2013).

For hepatic metabolism, each of the humanized liver models generally recapitulates the human metabolism well. For example, the phase I metabolism (and in some cases, phase II metabolism and transporter activity) has been evaluated for expression, phenotypic activity, and induction (Azuma et al., 2007; Strom et al., 2010; Hasegawa et al., 2011a; Ohtsuki et al., 2014). In virtually all the published reports, the humanized liver had a phenotype that was consistent with the donor hepatocyte lot. The return of the hepatocytes to their *in vivo* host environment with murine nonparenchymal stroma restores the metabolic profile in a manner that has been unattainable in 2D hepatocyte cell cultures, and it provides the added context of relevant exposures and drug protein binding. This has led to some novel uses of these mice, including *in vivo* recapitulation of disproportionally formed human metabolites that are poorly detected in *in vitro* systems, as well as understanding the potential for covalent adduct formation *in vivo* (Kamimura et al., 2010). The predictivity of *in vitro* metabolism is generally adequate given that microsomes and cell culture hepatocytes are 73–81% predictive of human metabolism (Dalvie et al., 2009). However, in specific instances, these mice may be ideal for addressing some of the "metabolites in safety testing" (MIST) guidelines where *in vitro* models are inadequate.

Despite the recapitulation of several human-predominant toxicities, several outstanding questions and potential limitations remain about the overall utility of these models. First, it is important to note that these animals are severely immunocompromised. This could significantly affect the interpretation of a safety study if the liver injury is potentially immune mediated or alter expression of enzymes or transporters. Similarly, the complex paracrine and endocrine signals that can significantly affect toxicity and drug metabolism are still not well understood in these models. For example, the interactions of murine stellate cells, endothelial cells, and cholangiocytes with the xeno-hepatocytes may not be fully recapitulated, and thus some drug toxicities may be diminished because of the lack of interplay between these cell types. Furthermore, there are significant species differences in hormonal regulation, and the murine endocrine system likely has regulatory impact on the human metabolic enzyme profile. This fits into the larger picture that humanizing the liver likely leads to compensatory effects on murine gene expression in other tissues and effects in the overall clearance of the drug in a manner that is difficult to predict and that might be different than in humans. Fortunately with the data that exists to date and given the large variability in the human population, these unknowns do not seem to be major drivers in the ultimate interpretation of the data, but the caveats should be considered on a case-by-case basis.

19.5 CONCLUSIONS

The utility of GEMM models in pharmacokinetics and metabolism and in assessing the safety profile of novel small-molecule therapeutics has dramatically evolved in the past few years, and the field is currently in flux. The creation of compound humanized knock-in and murine humanized liver models has reached a point where the utility of these models is approaching general acceptance and potential adoption in cases where standard *in vitro* models have failed. However, many of the most advanced models have not yet been thoroughly validated and their integration into the drug discovery and development has been limited. With careful examination of the pros and cons and clear experimental rationale, the utility of advanced GEMM models can be cost-effective and provide unparalleled clinical translation compared with existing options. Furthermore, technological advances in genome editing will likely result in GEMM models that have even better translatability and adoption in drug discovery and development paradigms.

REFERENCES

Accili, D., Drago, J., Lee, E. J., Johnson, M. D., Cool, M. H., Salvatore, P., Asico, L. D., Jose, P. A., Taylor, S. I., and Westphal, H. (1996). Early neonatal death in mice homozygous for a null allele of the insulin receptor gene. *Nat Genet* 12, 106–109.

Akinleye, A., Furqan, M., Mukhi, N., Ravella, P., and Liu, D. (2013). MEK and the inhibitors: from bench to bedside. *J Hematol Oncol* 6, 27.

Azuma, H., Paulk, N., Ranade, A., Dorrell, C., Al-Dhalimy, M., Ellis, E., Strom, S., Kay, M. A., Finegold, M., and Grompe, M. (2007). Robust expansion of human hepatocytes in Fah(−/−)/Rag2(−/−)/Il2rg (−/−) mice. *Nat Biotechnol* 25, 903–910.

Belanger, L. F., Roy, S., Tremblay, M., Brott, B., Steff, A. M., Mourad, W., Hugo, P., Erikson, R., and Charron, J. (2003). Mek2 is dispensable for mouse growth and development. *Mol Cell Biol* 23, 4778–4787.

Blasco, R. B., Francoz, S., Santamaria, D., Canamero, M., Dubus, P., Charron, J., Baccarini, M., and Barbacid, M. (2011). c-Raf, but not B-Raf, is essential for development of K-Ras oncogene-driven non-small cell lung carcinoma. *Cancer Cell* 19, 652–663.

Bogaards, J. J., Bertrand, M., Jackson, P., Oudshoorn, M. J., Weaver, R. J., van Bladeren, P. J., and Walther, B. (2000). Determining the best animal model for human cytochrome P450 activities: a comparison of mouse, rat, rabbit, dog, micropig, monkey and man. *Xenobiotica* 30, 1131–1152.

Bronson, S. K. and Smithies, O. (1994). Altering mice by homologous recombination using embryonic stem cells. *J Biol Chem* 269, 27155–27158.

Bruning, J. C., Michael, M. D., Winnay, J. N., Hayashi, T., Horsch, D., Accili, D., Goodyear, L. J., and Kahn, C. R. (1998). A muscle-specific insulin receptor knockout exhibits features of the metabolic syndrome of NIDDM without altering glucose tolerance. *Mol Cell* 2, 559–569.

Bruning, J. C., Gautam, D., Burks, D. J., Gillette, J., Schubert, M., Orban, P. C., Klein, R., Krone, W., Muller-Wieland, D., and Kahn, C. R. (2000). Role of brain insulin receptor in control of body weight and reproduction. *Science* 289, 2122–2125.

Bundgaard, C., Jensen, C. J. N., and Garmer, M. (2012). Species comparison of in vivo P-glycoprotein-mediated brain efflux using mdr1a-deficient rats and mice. *Drug Metab Dispos* 40, 461–466.

Buters, J. T., Tang, B. K., Pineau, T., Gelboin, H. V., Kimura, S., and Gonzalez, F. J. (1996). Role of CYP1A2 in caffeine pharmacokinetics and metabolism: studies using mice deficient in CYP1A2. *Pharmacogenetics* 6, 291–296.

Cai, H., Nguyen, N., Peterkin, V., Yang, Y. S., Hotz, K., La Placa, D. B., Chen, S., Tukey, R. H., and Stevens, J. C. (2010). A humanized UGT1 mouse model expressing the UGT1A1*28 allele for assessing drug clearance by UGT1A1-dependent glucuronidation. *Drug metab Dispos* 38, 879–886.

Cheung, C., Yu, A. M., Ward, J. M., Krausz, K. W., Akiyama, T. E., Feigenbaum, L., and Gonzalez, F. J. (2005). The cyp2e1-humanized transgenic mouse: role of cyp2e1 in acetaminophen hepatotoxicity. *Drug Metab Dispos* 33, 449–457.

Choi, J. N. (2014). Dermatologic adverse events to chemotherapeutic agents, part 2: BRAF inhibitors, MEK inhibitors, and ipilimumab. *Semin Cutan Med Surg* 33, 40–48.

Chu, X., Zhang, Z., Yabut, J., Horwitz, S., Levorse, J., Li, X. Q., Zhu, L., Lederman, H., Ortiga, R., Strauss, J., Li, X., Owens, K. A., Dragovic, J., Vogt, T., Evers, R., and Shin, M. K. (2012). Characterization of multidrug resistance 1a/P-glycoprotein knockout rats generated by zinc finger nucleases. *Mol Pharmacol* 81, 220–227.

Chung, C. and Reilly, S. (2015). Trametinib: a novel signal transduction inhibitor for the treatment of metastatic cutaneous melanoma. *Am J Health-Syst Pharm* 72, 101–110.

Corchero, J., Granvil, C. P., Akiyama, T. E., Hayhurst, G. P., Pimprale, S., Feigenbaum, L., Idle, J. R., and Gonzalez, F. J. (2001). The CYP2D6 humanized mouse: effect of the human CYP2D6 transgene and HNF4alpha on the disposition of debrisoquine in the mouse. *Mol Pharmacol* 60, 1260–1267.

Dalvie, D., Obach, R. S., Kang, P., Prakash, C., Loi, C. M., Hurst, S., Nedderman, A., Goulet, L., Smith, E., Bu, H. Z., and Smith, D. A. (2009). Assessment of three human in vitro systems in the generation of major human excretory and circulating metabolites. *Chem Res Toxicol* 22, 357–368.

Doran, A., Obach, R. S., Smith, B. J., Hosea, N. A., Becker, S., Callegari, E., Chen, C., Chen, X., Choo, E., Cianfrogna, J., Cox, L. M., Gibbs, J. P., Gibbs, M. A., Hatch, H., Hop, C. E., Kasman, I. N., Laperle, J., Liu, J., Liu, X., Logman, M., Maclin, D., Nedza, F. M., Nelson, F., Olson, E., Rahematpura, S., Raunig, D., Rogers, S., Schmidt, K., Spracklin, D. K., Szewc, M., Troutman, M., Tseng, E., Tu, M., Van Deusen, J. W., Venkatakrishnan, K., Walens, G., Wang, E. Q., Wong, D., Yasgar, A. S., and Zhang, C. (2005). The impact of P-glycoprotein on the disposition of drugs targeted for indications of the central nervous system: evaluation using the MDR1A/1B knockout mouse model. *Drug Metab Dispos* 33, 165–174.

Doyle, A., McGarry, M. P., Lee, N. A., and Lee, J. J. (2012). The construction of transgenic and gene knockout/knockin mouse models of human disease. *Transgenic Res* 21, 327–349.

Fernandez-Salguero, P., Pineau, T., Hilbert, D. M., McPhail, T., Lee, S. S., Kimura, S., Nebert, D. W., Rudikoff, S., Ward, J. M., and Gonzalez, F. J. (1995). Immune system impairment and hepatic fibrosis in mice lacking the dioxin-binding Ah receptor. *Science* 268, 722–726.

Fitzgerald, K., Bergeron, M., Willits, C., Bowers, S., Aubele, D. L., Goldbach, E., Tonn, G., Ness, D., and Olaharski, A. (2013). Pharmacological inhibition of polo like kinase 2 (PLK2) does not cause chromosomal damage or result in the formation of micronuclei. *Toxicol Appl Pharmacol* 269(1), 1–7.

Fuji, R. N., Flagella, M., Baca, M., MA, S. B., Brodbeck, J., Chan, B. K., Fiske, B. K., Honigberg, L., Jubb, A. M., Katavolos, P., Lee, D. W., Lewin-Koh, S. C., Lin, T., Liu, X., Liu, S., Lyssikatos, J. P., O'Mahony, J., Reichelt, M., Roose-Girma, M., Sheng, Z., Sherer, T., Smith, A., Solon, M., Sweeney, Z. K., Tarrant, J., Urkowitz, A., Warming, S., Yaylaoglu, M., Zhang, S., Zhu, H., Estrada, A. A., and Watts, R. J. (2015). Effect of selective LRRK2 kinase inhibition on nonhuman primate lung. *Sci Transl Med* 7, 273ra215.

Galabova-Kovacs, G., Matzen, D., Piazzolla, D., Meissl, K., Plyushch, T., Chen, A. P., Silva, A., and Baccarini, M. (2006). Essential role of B-Raf in ERK activation during extraembryonic development. *Proc Natl Acad Sci U S A* 103, 1325–1330.

Giroux, S., Tremblay, M., Bernard, D., Cardin-Girard, J. F., Aubry, S., Larouche, L., Rousseau, S., Huot, J., Landry, J., Jeannotte, L., and Charron, J. (1999). Embryonic death of Mek1-deficient mice reveals a role for this kinase in angiogenesis in the labyrinthine region of the placenta. *Curr Biol* 9, 369–372.

Gong, I. Y., Mansell, S. E., and Kim, R. B. (2013). Absence of both MDR1 (ABCB1) and breast cancer resistance protein (ABCG2) transporters significantly alters rivaroxaban disposition and central nervous system entry. *Basic Clin Pharmacol Toxicol* 112, 164–170.

Gonzalez, F. J. (2007). CYP3A4 and pregnane X receptor humanized mice. *J Biochem Mol Toxicol* 21, 158–162.

Gonzalez, F. J. and Yu, A. M. (2006). Cytochrome P450 and xenobiotic receptor humanized mice. *Annu Rev Pharmacol Toxicol* 46, 41–64.

Grimsley, A., Gallagher, R., Hutchison, M., Pickup, K., Wilson, I. D., and Samuelsson, K. (2013). Drug-drug interactions and metabolism in cytochrome P450 2C knockout mice: application to troleandomycin and midazolam. *Biochem Pharmacol* 86, 529–538.

Guengerich, F. P. (1997). Comparisons of catalytic selectivity of cytochrome P450 subfamily enzymes from different species. *Chem Biol Interact* 106, 161–182.

Guengerich, F. P. (2006). Cytochrome P450s and other enzymes in drug metabolism and toxicity. *AAPS J* 8, E101–E111.

Guerra, C., Navarro, P., Valverde, A. M., Arribas, M., Bruning, J., Kozak, L. P., Kahn, C. R., and Benito, M. (2001). Brown adipose tissue-specific insulin receptor knockout shows diabetic phenotype without insulin resistance. *J Clin Invest* 108, 1205–1213.

Gunn, C. H. (1938). Hereditary Acholuric Jaundice: in a new mutant strain of rats. *J Hered* 29, 137–139.

Hall, S. D., Wang, Z., Huang, S. M., Hamman, M. A., Vasavada, N., Adigun, A. Q., Hilligoss, J. K., Miller, M., and Gorski, J. C. (2003). The interaction between St John's wort and an oral contraceptive. *Clin Pharmacol Ther* 74, 525–535.

Hasegawa, M., Kapelyukh, Y., Tahara, H., Seibler, J., Rode, A., Krueger, S., Lee, D. N., Wolf, C. R., and Scheer, N. (2011a). Quantitative prediction of human pregnane X receptor and cytochrome P450 3A4 mediated drug-drug interaction in a novel multiple humanized mouse line. *Mol Pharmacol* 80, 518–528.

Hasegawa, M., Kawai, K., Mitsui, T., Taniguchi, K., Monnai, M., Wakui, M., Ito, M., Suematsu, M., Peltz, G., Nakamura, M., and Suemizu, H. (2011b). The reconstituted "humanized liver" in TK-NOG mice is mature and functional. *Biochem Biophys Res Commun* 405, 405–410.

van Herwaarden, A. E., Wagenaar, E., van der Kruijssen, C. M., van Waterschoot, R. A., Smit, J. W., Song, J. Y., van der Valk, M. A., van Tellingen, O., van der Hoorn, J. W., Rosing, H., Beijnen, J. H., and Schinkel, A. H. (2007). Knockout of cytochrome P450 3A yields new mouse models for understanding xenobiotic metabolism. *J Clin Invest* 117, 3583–3592.

Herzig, M. C., Kolly, C., Persohn, E., Theil, D., Schweizer, T., Hafner, T., Stemmelen, C., Troxler, T. J., Schmid, P., Danner, S., Schnell, C. R., Mueller, M., Kinzel, B., Grevot, A., Bolognani, F., Stirn, M., Kuhn, R. R., Kaupmann, K., van der Putten, P. H., Rovelli, G., and Shimshek, D. R. (2011). LRRK2 protein levels are determined by kinase function and are crucial for kidney and lung homeostasis in mice. *Hum Mol Genet* 20, 4209–4223.

Heyman, R. A., Borrelli, E., Lesley, J., Anderson, D., Richman, D. D., Baird, S. M., Hyman, R., and Evans, R. M. (1989). Thymidine kinase obliteration: creation of transgenic mice with controlled immune deficiency. *Proc Natl Acad Sci U S A* 86, 2698–2702.

Hikita, H., Takehara, T., Shimizu, S., Kodama, T., Li, W., Miyagi, T., Hosui, A., Ishida, H., Ohkawa, K., Kanto, T., Hiramatsu, N., Yin, X. M., Hennighausen, L., Tatsumi, T., and Hayashi, N. (2009). Mcl-1 and Bcl-xL cooperatively maintain integrity of hepatocytes in developing and adult murine liver. *Hepatology* 50, 1217–1226.

International Transporter, C., Giacomini, K. M., Huang, S. M., Tweedie, D. J., Benet, L. Z., Brouwer, K. L., Chu, X., Dahlin, A., Evers, R., Fischer, V., Hillgren, K. M., Hoffmaster, K. A., Ishikawa, T., Keppler, D., Kim, R. B., Lee, C. A., Niemi, M., Polli, J. W., Sugiyama, Y., Swaan, P. W., Ware, J. A., Wright, S. H., Yee, S. W., Zamek-Gliszczynski, M. J., and Zhang, L.

(2010). Membrane transporters in drug development. *Nat Rev Drug Discov* 9, 215–236.

Ito, M., Hiramatsu, H., Kobayashi, K., Suzue, K., Kawahata, M., Hioki, K., Ueyama, Y., Koyanagi, Y., Sugamura, K., Tsuji, K., Heike, T., and Nakahata, T. (2002). NOD/SCID/γ mouse: an excellent recipient mouse model for engraftment of human cells. *Blood* 100, 3175–3182.

Jain, N., Curran, E., Iyengar, N. M., Diaz-Flores, E., Kunnavakkam, R., Popplewell, L., Kirschbaum, M. H., Karrison, T., Erba, H. P., Green, M., Poire, X., Koval, G., Shannon, K., Reddy, P. L., Joseph, L., Atallah, E. L., Dy, P., Thomas, S. P., Smith, S. E., Doyle, L. A., Stadler, W. M., Larson, R. A., Stock, W., and Odenike, O. (2014). Phase II study of the oral MEK inhibitor selumetinib in advanced acute myelogenous leukemia: a University of Chicago phase II consortium trial. *Clin Cancer Res* 20, 490–498.

Jinek, M., Chylinski, K., Fonfara, I., Hauer, M., Doudna, J. A., and Charpentier, E. (2012). A programmable dual-RNA-guided DNA endonuclease in adaptive bacterial immunity. *Science* 337, 816–821.

Joshi, R. L., Lamothe, B., Cordonnier, N., Mesbah, K., Monthioux, E., Jami, J., and Bucchini, D. (1996). Targeted disruption of the insulin receptor gene in the mouse results in neonatal lethality. *EMBO J* 15, 1542–1547.

Kalvass, J. C., Polli, J. W., Bourdet, D. L., Feng, B., Huang, S. M., Liu, X., Smith, Q. R., Zhang, L. K., Zamek-Gliszczynski, M. J., and International Transporter, C. (2013). Why clinical modulation of efflux transport at the human blood–brain barrier is unlikely: the ITC evidence-based position. *Clin Pharmacol Ther* 94, 80–94.

Kamimura, H., Nakada, N., Suzuki, K., Mera, A., Souda, K., Murakami, Y., Tanaka, K., Iwatsubo, T., Kawamura, A., and Usui, T. (2010). Assessment of chimeric mice with humanized liver as a tool for predicting circulating human metabolites. *Drug Metab Pharmacokinet* 25, 223–235.

Kempf, D. J., Waring, J. F., Morfitt, D. C., Werner, P., Ebert, B., Mitten, M., Nguyen, B., Randolph, J. T., DeGoey, D. A., Klein, L. L., and Marsh, K. (2006). Practical preclinical model for assessing the potential for unconjugated hyperbilirubinemia produced by human immunodeficiency virus protease inhibitors. *Antimicrob Agents Chemother* 50, 762–764.

Kleiner, H. E., Vulimiri, S. V., Hatten, W. B., Reed, M. J., Nebert, D. W., Jefcoate, C. R., and DiGiovanni, J. (2004). Role of cytochrome p4501 family members in the metabolic activation of polycyclic aromatic hydrocarbons in mouse epidermis. *Chem Res Toxicol* 17, 1667–1674.

Kulkarni, R. N., Bruning, J. C., Winnay, J. N., Postic, C., Magnuson, M. A., and Kahn, C. R. (1999). Tissue-specific knockout of the insulin receptor in pancreatic beta cells creates an insulin secretory defect similar to that in type 2 diabetes. *Cell* 96, 329–339.

Lee, S. S., Buters, J. T., Pineau, T., Fernandez-Salguero, P., and Gonzalez, F. J. (1996). Role of CYP2E1 in the hepatotoxicity of acetaminophen. *J Biol Chem* 271, 12063–12067.

Lee, E. W., Lai, Y., Zhang, H., and Unadkat, J. D. (2006). Identification of the mitochondrial targeting signal of the human equilibrative nucleoside transporter 1 (hENT1): implications for interspecies differences in mitochondrial toxicity of fialuridine. *J Biol Chem* 281, 16700–16706.

Leverson, J. D., Phillips, D. C., Mitten, M. J., Boghaert, E. R., Diaz, D., Tahir, S. K., Belmont, L. D., Nimmer, P., Xiao, Y., Ma, X. M., Lowes, K. N., Kovar, P., Chen, J., Jin, S., Smith, M., Xue, J., Zhang, H., Oleksijew, A., Magoc, T. J., Vaidya, K. S., Albert, D. H., Tarrant, J. M., La, N., Wang, L., Tao, Z. F., Wendt, M. D., Sampath, D., Rosenberg, S. H., Tse, C., DC, S. H., Fairbrother, W. J., Elmore, S. W., and Souers, A. J. (2015). Exploiting selective BCL-2 family inhibitors to dissect cell survival dependencies and define improved strategies for cancer therapy. *Sci Transl Med* 7, 279ra240.

Liang, H. C., Li, H., McKinnon, R. A., Duffy, J. J., Potter, S. S., Puga, A., and Nebert, D. W. (1996). Cyp1a2(−/−) null mutant mice develop normally but show deficient drug metabolism. *Proc Natl Acad Sci U S A* 93, 1671–1676.

Liu, Y., Ramírez, J., House, L., and Ratain, M. J. (2010). The UGT1A1*28 polymorphism correlates with erlotinib's effect on SN-38 glucuronidation. *Eur J Cancer* 46, 2097–2103.

Madsen, L. W., Knauf, J. A., Gotfredsen, C., Pilling, A., Sjogren, I., Andersen, S., Andersen, L., de Boer, A. S., Manova, K., Barlas, A., Vundavalli, S., Nyborg, N. C., Knudsen, L. B., Moelck, A. M., and Fagin, J. A. (2012). GLP-1 receptor agonists and the thyroid: C-cell effects in mice are mediated via the GLP-1 receptor and not associated with RET activation. *Endocrinology* 153, 1538–1547.

Manyike, P. T., Kharasch, E. D., Kalhorn, T. F., and Slattery, J. T. (2000). Contribution of CYP2E1 and CYP3A to acetaminophen reactive metabolite formation. *Clin Pharmacol Ther* 67, 275–282.

Mars, W. M., Zarnegar, R., and Michalopoulos, G. K. (1993). Activation of hepatocyte growth factor by the plasminogen activators uPA and tPA. *Am J Pathol* 143, 949–958.

Mason, K. D., Carpinelli, M. R., Fletcher, J. I., Collinge, J. E., Hilton, A. A., Ellis, S., Kelly, P. N., Ekert, P. G., Metcalf, D., Roberts, A. W., Huang, D. C., and Kile, B. T. (2007). Programmed anuclear cell death delimits platelet life span. *Cell* 128, 1173–1186.

Matthaei, K. I. (2007). Genetically manipulated mice: a powerful tool with unsuspected caveats. *J Physiol* 582, 481–488.

May, P. C., Dean, R. A., Lowe, S. L., Martenyi, F., Sheehan, S. M., Boggs, L. N., Monk, S. A., Mathes, B. M., Mergott, D. J., Watson, B. M., Stout, S. L., Timm, D. E., Smith Labell, E., Gonzales, C. R., Nakano, M., Jhee, S. S., Yen, M., Ereshefsky, L., Lindstrom, T. D., Calligaro, D. O., Cocke, P. J., Greg Hall, D., Friedrich, S., Citron, M., and Audia, J. E. (2011). Robust central reduction of amyloid-beta in humans with an orally available, non-peptidic beta-secretase inhibitor. *J Neurosci* 31, 16507–16516.

Michael, M. D., Kulkarni, R. N., Postic, C., Previs, S. F., Shulman, G. I., Magnuson, M. A., and Kahn, C. R. (2000). Loss of insulin signaling in hepatocytes leads to severe insulin resistance and progressive hepatic dysfunction. *Mol Cell* 6, 87–97.

Miksys, S. L., Cheung, C., Gonzalez, F. J., and Tyndale, R. F. (2005). Human CYP2D6 and mouse CYP2Ds: organ distribution in a humanized mouse model. *Drug Metab Dispos* 33, 1495–1502.

Morgan, R. E., van Staden, C. J., Chen, Y., Kalyanaraman, N., Kalanzi, J., Dunn, R. T., II, Afshari, C. A., and Hamadeh, H. K. (2013). A multifactorial approach to hepatobiliary transporter assessment enables improved therapeutic compound development. *Toxicol Sci* 136, 216–241.

Muruganandan, S. and Sinal, C. J. (2008). Mice as clinically relevant models for the study of cytochrome P450-dependent metabolism. *Clin Pharmacol Ther* 83, 818–828.

Ness, D., Ren, Z., Gardai, S., Sharpnack, D., Johnson, V. J., Brennan, R. J., Brigham, E. F., and Olaharski, A. J. (2013). Leucine-rich repeat kinase 2 (LRRK2)-deficient rats exhibit renal tubule injury and perturbations in metabolic and immunological homeostasis. *PLoS One* 8, e66164.

Newton, K., Dugger, D. L., Wickliffe, K. E., Kapoor, N., de Almagro, M. C., Vucic, D., Komuves, L., Ferrando, R. E., French, D. M., Webster, J., Roose-Girma, M., Warming, S., and Dixit, V. M. (2014). Activity of protein kinase RIPK3 determines whether cells die by necroptosis or apoptosis. *Science* 343, 1357–1360.

Nguyen, N., Bonzo, J. A., Chen, S., Chouinard, S., Kelner, M. J., Hardiman, G., Belanger, A., and Tukey, R. H. (2008). Disruption of the ugt1 locus in mice resembles human Crigler-Najjar type I disease. *J Biol Chem* 283, 7901–7911.

Ohashi, K., Osuga, J., Tozawa, R., Kitamine, T., Yagyu, H., Sekiya, M., Tomita, S., Okazaki, H., Tamura, Y., Yahagi, N., Iizuka, Y., Harada, K., Gotoda, T., Shimano, H., Yamada, N., and Ishibashi, S. (2003). Early embryonic lethality caused by targeted disruption of the 3-hydroxy-3-methylglutaryl-CoA reductase gene. *J Biol Chem* 278, 42936–42941.

Ohtsuki, S., Kawakami, H., Inoue, T., Nakamura, K., Tateno, C., Katsukura, Y., Obuchi, W., Uchida, Y., Kamiie, J., Horie, T., and Terasaki, T. (2014). Validation of uPA/SCID mouse with humanized liver as a human liver model: protein quantification of transporters, cytochromes P450, and UDP-glucuronosyltransferases by LC-MS/MS. *Drug Metab Dispos* 42, 1039–1043.

Papaioannou, V. E. and Behringer, R. R. (2012). Early embryonic lethality in genetically engineered mice: diagnosis and phenotypic analysis. *Vet Pathol* 49, 64–70.

Pearce, R., Greenway, D., and Parkinson, A. (1992). Species differences and interindividual variation in liver microsomal cytochrome P450 2A enzymes: effects on coumarin, dicumarol, and testosterone oxidation. *Arch Biochem Biophys* 298, 211–225.

Pich, E. M. and Epping-Jordan, M. P. (1998). Transgenic mice in drug dependence research. *Ann Med* 30, 390–396.

Pineau, T., Fernandez-Salguero, P., Lee, S. S., McPhail, T., Ward, J. M., and Gonzalez, F. J. (1995). Neonatal lethality associated with respiratory distress in mice lacking cytochrome P450 1A2. *Proc Natl Acad Sci U S A* 92, 5134–5138.

Raza, A., Vierling, J., and Hussain, K. B. (2013). Genetics of drug-induced hepatotoxicity toxicity in Gilbert's syndrome. *Am J Gastroenterol* 108, 1936–1937.

Rhim, J. A., Sandgren, E. P., Palmiter, R. D., and Brinster, R. L. (1995). Complete reconstitution of mouse liver with xenogeneic hepatocytes. *Proc Natl Acad Sci U S A* 92, 4942–4946.

Richardson, F. C., Engelhardt, J. A., and Bowsher, R. R. (1994). Fialuridine accumulates in DNA of dogs, monkeys, and rats following long-term oral administration. *Proc Natl Acad Sci U S A* 91, 12003–12007.

Robbins, J. (1993). Gene targeting. The precise manipulation of the mammalian genome. *Circ Res* 73, 3–9.

Rocha, B. A., Scearce-Levie, K., Lucas, J. J., Hiroi, N., Castanon, N., Crabbe, J. C., Nestler, E. J., and Hen, R. (1998). Increased vulnerability to cocaine in mice lacking the serotonin-1B receptor. *Nature* 393, 175–178.

Sacca, R., Engle, S. J., Qin, W., Stock, J. L., and McNeish, J. D. (2010). Genetically engineered mouse models in drug discovery research. *Methods Mol Biol* 602, 37–54.

Scearce-Levie, K., Chen, J. P., Gardner, E., and Hen, R. (1999). 5-HT receptor knockout mice: pharmacological tools or models of psychiatric disorders. *Ann N Y Acad Sci* 868, 701–715.

Schinkel, A. H., Mayer, U., Wagenaar, E., Mol, C. A., van Deemter, L., Smit, J. J., van der Valk, M. A., Voordouw, A. C., Spits, H., van Tellingen, O., Zijlmans, J. M., Fibbe, W. E., and Borst, P. (1997). Normal viability and altered pharmacokinetics in mice lacking mdr1-type (drug-transporting) P-glycoproteins. *Proc Natl Acad Sci U S A* 94, 4028–4033.

Schoenwaelder, S. M., Jarman, K. E., Gardiner, E. E., Hua, M., Qiao, J., White, M. J., Josefsson, E. C., Alwis, I., Ono, A., Willcox, A., Andrews, R. K., Mason, K. D., Salem, H. H., Huang, D. C., Kile, B. T., Roberts, A. W., and Jackson, S. P. (2011). Bcl-xL-inhibitory BH3 mimetics can induce a transient thrombocytopathy that undermines the hemostatic function of platelets. *Blood* 118, 1663–1674.

Shultz, L. D., Lyons, B. L., Burzenski, L. M., Gott, B., Chen, X., Chaleff, S., Kotb, M., Gillies, S. D., King, M., Mangada, J., Greiner, D. L., and Handgretinger, R. (2005). Human lymphoid and myeloid cell development in NOD/LtSz-scid IL2R gamma null mice engrafted with mobilized human hemopoietic stem cells. *J Immunol* 174, 6477–6489.

Strom, S. C., Davila, J., and Grompe, M. (2010). Chimeric mice with humanized liver: tools for the study of drug metabolism, excretion, and toxicity. *Methods Mol Biol* 640, 491–509.

Takehara, T. and Takahashi, H. (2003). Suppression of Bcl-xL deamidation in human hepatocellular carcinomas. *Cancer Res* 63, 3054–3057.

Tang, T., Li, L., Tang, J., Li, Y., Lin, W. Y., Martin, F., Grant, D., Solloway, M., Parker, L., Ye, W., Forrest, W., Ghilardi, N., Oravecz, T., Platt, K. A., Rice, D. S., Hansen, G. M., Abuin, A., Eberhart, D. E., Godowski, P., Holt, K. H., Peterson, A., Zambrowicz, B. P., and de Sauvage, F. J. (2010). A mouse knockout library for secreted and transmembrane proteins. *Nat Biotechnol* 28, 749–755.

Temple, R. (2006). Hy's law: predicting serious hepatotoxicity. *Pharmacoepidemiol Drug Saf* 15, 241–243.

Tse, C., Shoemaker, A. R., Adickes, J., Anderson, M. G., Chen, J., Jin, S., Johnson, E. F., Marsh, K. C., Mitten, M. J., Nimmer, P., Roberts, L., Tahir, S. K., Xiao, Y., Yang, X., Zhang, H., Fesik, S., Rosenberg, S. H., and Elmore, S. W. (2008). ABT-263: a potent and orally bioavailable Bcl-2 family inhibitor. *Cancer Res* 68, 3421–3428.

Turgeon, B. and Meloche, S. (2009). Interpreting neonatal lethal phenotypes in mouse mutants: insights into gene function and human diseases. *Physiol Rev* 89, 1–26.

Vick, B., Weber, A., Urbanik, T., Maass, T., Teufel, A., Krammer, P. H., Opferman, J. T., Schuchmann, M., Galle, P. R., and Schulze-Bergkamen, H. (2009). Knockout of myeloid cell leukemia-1 induces liver damage and increases apoptosis susceptibility of murine hepatocytes. *Hepatology* 49, 627–636.

Wang, H., Yang, H., Shivalila, C. S., Dawlaty, M. M., Cheng, A. W., Zhang, F., and Jaenisch, R. (2013). One-step generation of mice carrying mutations in multiple genes by CRISPR/Cas-mediated genome engineering. *Cell* 153, 910–918.

van Waterschoot, R. A., Rooswinkel, R. W., Sparidans, R. W., van Herwaarden, A. E., Beijnen, J. H., and Schinkel, A. H. (2009). Inhibition and stimulation of intestinal and hepatic CYP3A activity: studies in humanized CYP3A4 transgenic mice using triazolam. *Drug Metab Dispos* 37, 2305–2313.

Wojnowski, L., Zimmer, A. M., Beck, T. W., Hahn, H., Bernal, R., Rapp, U. R., and Zimmer, A. (1997). Endothelial apoptosis in Braf-deficient mice. *Nat Genet* 16, 293–297.

Wu, H., Liu, X., Jaenisch, R., and Lodish, H. F. (1995). Generation of committed erythroid BFU-E and CFU-E progenitors does not require erythropoietin or the erythropoietin receptor. *Cell* 83, 59–67.

Xie, W., Barwick, J. L., Downes, M., Blumberg, B., Simon, C. M., Nelson, M. C., Neuschwander-Tetri, B. A., Brunt, E. M., Guzelian, P. S., and Evans, R. M. (2000). Humanized xenobiotic response in mice expressing nuclear receptor SXR. *Nature* 406, 435–439.

Xu, D., Nishimura, T., Nishimura, S., Zhang, H., Zheng, M., Guo, Y. Y., Masek, M., Michie, S. A., Glenn, J., and Peltz, G. (2014). Fialuridine induces acute liver failure in chimeric TK-NOG mice: a model for detecting hepatic drug toxicity prior to human testing. *PLoS Med* 11, e1001628.

Xu, D., Wu, M., Nishimura, S., Nishimura, T., Michie, S. A., Zheng, M., Yang, Z., Yates, A. J., Day, J. S., Hillgren, K. M., Takeda, S. T., Guan, Y., Guo, Y., and Peltz, G. (2015). Chimeric TK-NOG mice: a predictive model for cholestatic human liver toxicity. *J Pharmacol Exp Ther* 352, 274–280.

Yamazaki, H., Mimura, M., Sugahara, C., and Shimada, T. (1994). Catalytic roles of rat and human cytochrome P450 2A enzymes in testosterone 7 alpha- and coumarin 7-hydroxylations. *Biochem Pharmacol* 48, 1524–1527.

Zambrowicz, B. P. and Sands, A. T. (2003). Knockouts model the 100 best-selling drugs—will they model the next 100? *Nat Rev Drug Discov* 2, 38–51.

Zamek-Gliszczynski, M. J., Bedwell, D. W., Bao, J. Q., and Higgins, J. W. (2012). Characterization of SAGE Mdr1a (P-gp), Bcrp, and Mrp2 knockout rats using loperamide, paclitaxel, sulfasalazine, and carboxydichlorofluorescein pharmacokinetics. *Drug Metab Dispos* 40, 1825–1833.

Zucker, S. D., Qin, X., Rouster, S. D., Yu, F., Green, R. M., Keshavan, P., Feinberg, J., and Sherman, K. E. (2001). Mechanism of indinavir-induced hyperbilirubinemia. *Proc Natl Acad Sci U S A* 98, 12671–12676.

20

MOUSE POPULATION-BASED TOXICOLOGY FOR PERSONALIZED MEDICINE AND IMPROVED SAFETY PREDICTION

ALISON H. HARRILL

Department of Environmental and Occupational Health, Regulatory Sciences Program, The University of Arkansas for Medical Sciences, Little Rock, AR, USA

20.1 INTRODUCTION

Despite rising drug development costs that now average over a billion dollars (US) for a single drug, the number of approvals for new drugs has changed little over the past decade (Allison, 2012). In part owing to a lack of return on investment, there were many pharmaceutical mergers and tens of thousands of layoffs in the industry from 2007 to 2012. While many reasons have been attributed to the lack of apparent productivity, a major culprit is a low probability of success, which has been estimated as low as 10% for investigational new drugs entering phase II clinical trials (Arrowsmith, 2012).

There is an urgent need for new tools and approaches that can accurately identify human-relevant safety concerns for chemical and pharmaceutical agents. Rapid, high-capacity assay systems have received much attention in recent years owing to the growing rate at which new chemical agents are introduced—approximately 700 new chemicals are introduced into commerce each year. Such efforts include ToxCast, Tox21, ILSI-HESI Risk21 Project, and DARPA initiatives to develop "tissues on a chip." However, the major challenge in toxicology today is the ability to extrapolate risk from experimental systems to human populations. Systems toxicology, or the integration of traditional toxicology approaches with the development and implementation of transcriptomics, proteomics, and metabolomics,

offers the potential to more accurately predict adverse health effects in humans. Yet, while significant resources have been devoted to develop *in vitro* or *in silico* approaches that may afford greater throughput, little effort has been undertaken to improve the whole animal studies that remain a necessary component for bringing pharmaceutical agents to market.

20.2 PHARMACOGENETICS AND POPULATION VARIABILITY

Interindividual variability in response to drugs has gained attention in recent years as a target for improving understanding of drug responses. There are many factors that can affect an individual's response, such as demographic and environmental factors that include gender, age, dietary intake, comorbid conditions, and concomitant drug or complementary alternative medicine use.

It is now well accepted that genetic variants affect responses to drugs and chemicals (Kalow, 1965; Burns, 1968). One of the earliest studies in this field compared plasma drug half-lives in identical and fraternal twin pairs and demonstrated that greater differences existed between the fraternal twins (Vesell and Page, 1968). In recent years, several more monogenic pharmacogenetic traits have been reported (Nebert et al., 2008). One example

Drug Discovery Toxicology: From Target Assessment to Translational Biomarkers, First Edition. Edited by Yvonne Will, J. Eric McDuffie, Andrew J. Olaharski, and Brandon D. Jeffy.
© 2016 John Wiley & Sons, Inc. Published 2016 by John Wiley & Sons, Inc.

of such a genotype–phenotype association is the finding that *apolipoprotein E* (*APOE*)-4/4 allele carriers respond poorly to conventional Alzheimer's disease treatments (Cacabelos, 2007).

Successes in this field have led to important regulatory action by the FDA that has allowed drugs to remain on the market due to the availability of genetic testing. Examples include assigning the dose of 6-mercaptopurine based on the genotype of thiopurine *S*-methyltransferase (*TPMT*) (Yates et al., 1997) and the dose of warfarin based on the genotypes of vitamin K epoxide reductase complex (*VKORC1*) (Rost et al., 2004) and cytochrome P450 2C9 (*CYP2C9*) (Higashi et al., 2002). In fact, warfarin represents one of the first drugs for which toxicogenetic information (by VKORC1 haplotyping) better explained the dose variance than pharmacokinetic-based pharmacogenomic data (by *CYP2C9* genotyping) (Weinshilboum and Wang, 2006). The data on warfarin, therefore, represents a shift from monogenic toxicogenetic testing to a polygenic model, which will be encountered with increasing frequency as research in this field expands to include a greater variety of pharmaceuticals.

The wealth of information on genetic polymorphisms now available through the Human Genome and HapMap Projects has led to a dramatic increase in studies that seek to connect genetic variants with toxicity and pharmacologic phenotypes. Because the base pair sequence variation among individuals averages to be about 1 in 500–1000 base pairs (Venter et al., 2001), it is reasonable to expect that a significant number of genes will contain polymorphisms that contribute to disease and that many will play a role in adverse drug responses. Most of the current research focuses on the association between phenotype/disease and single nucleotide polymorphisms (SNPs). SNPs are an attractive choice for biomarkers of adverse responses because, unlike other factors that contribute to a toxicity phenotype such as age, comorbidity, and environment, an individual's genetic code remains largely stable throughout their lifetime. Genetic testing offers the potential for replacing empirical dose adjustment based upon therapeutic assessment of pharmacologic or toxic effect after initial dosing. In addition, predictive genetic tests may have value in the drug development process by rescuing drugs that failed phase III clinical trials or following market approval due to toxicity within a subset of patients (Fig. 20.1) (Weiss et al., 2008). A key example of this concept is the genetic testing available to patients with HIV who are prescribed the drug abacavir, in which screening for major histocompatibility complex, class I, B (HLA-B)*5701 reduces the risk of hypersensitivity reactions (Mallal et al., 2008).

A promising new alternative is a translational pharmacogenetic approach in which genetically diverse mouse populations serve as a surrogate for human populations. The major advantage of this approach is the potential for

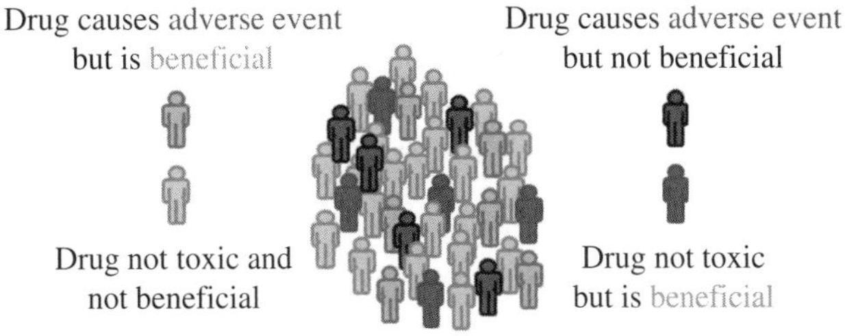

FIGURE 20.1 Response to a given drug is often variable, in which a patient population with a similar diagnosis may actually represent a heterogeneous group of responders and nonresponders. Pharmacogenetic tests have the potential to identify each subset of patients in order to more efficiently tailor drug therapy. (*See insert for color representation of the figure.*)

inclusion of sensitive mice that may react more "like" humans at a given dose. In addition, mouse populations that have recently become available have greater randomization of polymorphisms and a higher frequency of rare alleles, which translates to a much smaller number of animals needed for genetic analysis as compared to a human study. In addition, pharmacogenetic studies to identify safety biomarkers are often unsuited for clinical studies due to ethical concerns. As will be discussed throughout this chapter, it is common for genetically different mice to exhibit vastly different responses to a given dose of a drug. Thus, while much recent light has been shed on poor performance of animal research data to translate to human discoveries, some of the lack of translatability may be due to the limited genetic contexts that are typically used in animal studies. By using a mouse population rather than a single strain, researchers can refine the central experimental question from "can mice react like humans?" to "which mouse reacts most like humans?."

Throughout the chapter, several terms are utilized to describe analysis of genetic information and the results of such investigations. Table 20.1 defines each term for ease of reference throughout the text. Much of this chapter focuses on experimentation that facilitates a hypothesis-generating approach known as genome-wide association (GWA). In these studies, researchers seek to identify regions of the genome containing sequence variants that, in essence, distinguish responder individuals from nonresponder individuals (with a caveat that many mapped traits are quantitative and continuous rather than binary as in the following example). A simplified schematic of a GWA study is presented in Figure 20.2. In this cartoon, the bottom half of the figure indicates the significance score at any given locus and within each mouse chromosome. In plots that indicate genome-wide significance, higher scores represent a more significant association because the log of odds (LOD or $-\log(p)$) is plotted on the y-axis. In the example shown in Figure 20.2, regions with low significance scores, such as those indicated

TABLE 20.1 Pharmacogenetic Terms and Definitions

Term (Aliases)	Definition
Allele (variant, polymorphism)	A variant form of a gene, which can encompass genetic sequence variation or insertion/deletion of DNA bases. Sometimes called a mutation; however, that term is avoided here because of its special association in genetic toxicology studies. An allele is major when it occurs with greater frequency in the population as compared to the minor, less frequent, allele
Genomic mapping (QTL mapping, genome-wide association (GWA), genotype–phenotype association analysis)	Statistical techniques that that utilize genotypic information and quantitative trait measurements to identify QTL
Haplotype	A set of gene variants that tend to be inherited together
Pharmacogenetics (pharmacogenomics, toxicogenomics)	The evaluation of drug effects and their relationship to genetic factors (here, only DNA sequence is considered)
Quantitative trait loci (QTL)	A section of DNA (the locus) that is associated with variation in a phenotype (the quantitative trait)
Quantitative trait gene (QTG)	A gene encoded by DNA within a QTL interval that is suspected to contain a candidate pharmacogenetic marker
Transcriptomics	Analysis of global mRNA transcript expression using microarrays

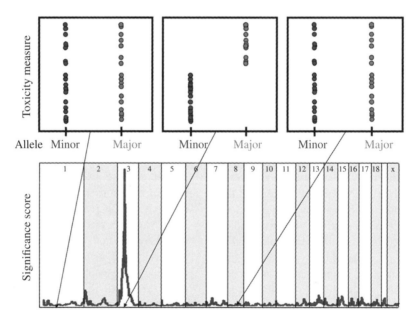

FIGURE 20.2 Illustration of QTL mapping. (Top) Circles indicate phenotype values for tested mice, with the values separated by mice harboring either the major or minor allele at the indicated locus. Panels correspond to gene regions as indicated by the black line. (Bottom) Manhattan plot showing location of a QTL on Chr 3. Reproduced with permission of Daniel Gatti, The Jackson Laboratory.

on chromosomes 1 and 8, have no association with the trait; thus, when the allelic patterns are examined, there is no discernment between responder and nonresponder individuals. In contrast, the region on chromosome 3 is highly significant, and there is an allelic pattern that can distinguish the responders. Such a region would likely contain a quantitative trait gene (QTG) that may be useful as a pharmacogenetic marker.

20.3 RODENT POPULATIONS ENABLE A POPULATION-BASED APPROACH TO TOXICOLOGY

In the past, the standard model for dissection of complex traits in mammals was a genetic cross of two inbred mouse strains. Within a cross, mating of two inbred strains results in an F1 generation for which offspring were mated to each other (called an intercross) or back to a progenitor strain (called a backcross) (Flint and Eskin, 2012). The offspring of the F2 generation were then phenotyped and genotyped in order to perform a linkage analysis that enabled determination of gene regions containing variation that influenced the trait. These gene regions are termed "quantitative trait loci (QTL)." A drawback of the described approach is a lack of resolution to identify the causative gene, because the QTL regions identified in two-way crosses of this nature are often large (spanning tens of megabases) and thus contain hundreds of candidate genes (Flint and Eskin, 2012). Owing to a lack of resolution, determination of specific genetic drivers using previous methods has been costly and time consuming and has resulted in identification of few specific disease drivers. The inefficiency of this method spurred research and development into mouse population models that could provide advantages toward the identification of specific QTGs, whose variants could serve as biomarkers of a genetic condition or risk factor (Fig. 20.3).

The genetic variation within panels of inbred mouse strains, while greater than that in recombinant inbred (RI) lines, can be somewhat limited due to the fact that the majority of classical inbred strains are derived from *Mus musculus* (*M.m.*) *domesticus*. To address this concern, the Collaborative Cross (CC) mouse strains and Diversity Outbred (DO) stock were recently developed and designed to incorporate large genetic variation (Churchill et al., 2004). To develop this resource, a controlled breeding program was designed to randomize genetic elements among the progeny derived from eight parental strains. A recent study demonstrated that the genetic variation present in the CC and DO

represents the optimal polymorphism architecture for the study of systems biology when compared to historical RI lines or to panels of classical inbred strains and was demonstrated to be more reflective of the genetic variation expected in natural populations (Roberts et al., 2007). Further characterization of this resource may improve the translational ability of pharmacogenetic studies in rodents.

20.3.1 Mouse Diversity Panel

Largely driven by advances in high-density genotyping, the mouse diversity panel (MDP) was one of the first animal models put forward as a method for narrowing QTL intervals that might have utility for personalized medicine. A well-designed MDP consists of classical laboratory and wild-derived inbred mouse strains. While the exact strain composition may vary, the majority of mouse panels consist of 20–40 inbred strains, many of which are classical laboratory strains derived from a common set of genetic founders. Common laboratory strains were largely derived from a *M.m. musculus* or *M.m. domesticus* subspecific origin. Owing to the particulars of mouse strain breeding in the twentieth century (reviewed elsewhere (Beck et al., 2000)), there is a lack of sufficient genetic diversity among classical laboratory strains, which can result in large genetic regions containing no genetic diversity in which to query for a genetic risk factor (these regions are called "identical by descent"). It is generally accepted that inclusion of wild-derived strains (such as CAST/EiJ or MSM/Ms for example, which are derived from a different subspecific origin (*M.m. castaneous* and *M.m. molossinus*)), as well as mice that were

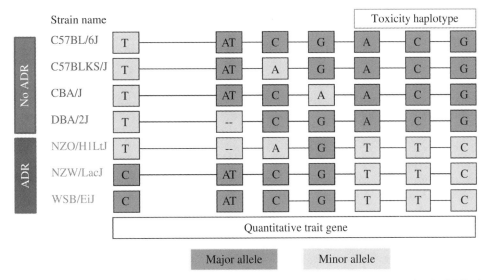

FIGURE 20.3 This simple example demonstrates an association with a genetic sequence haplotype within this QTG—in this case the TTC haplotype—with an increased risk of an adverse drug reaction (ADR). This example shows fewer strains than would be necessary to power a QTL mapping study. From this example, you may observe that having only a small number of allelic patterns at any locus (columns) will result in a large number of false positive associations just by chance. This is one reason why a large number of genetic contexts are better suited for QTL mapping.

more recently derived from the wild—such as WSB/EiJ—can reduce the number and size of intervals that are identical by descent. However, even with inclusion of the wild-derived inbred strains, there remains a disadvantage of a weighted population structure that requires statistical methodology to overcome. An additional disadvantage is that there are many gene variants that are more frequently inherited together or in high linkage disequilibrium (LD). The consequence for genetic mapping is that multiple gene regions across the genome may appear to be significant, but it is difficult to determine the "true positive" peak that would be ideal to focus downstream QTG validation studies.

Despite several limitations for genomic mapping analyses, MDP studies have yielded important insights into toxicity outcomes associated with pharmaceutical drugs and chemicals. An exemplar case is a study conducted with acetaminophen, which is associated with liver toxicity in the case of overdose. Surprisingly, a small percentage of individuals who were taking therapeutic doses of acetaminophen were found to experience liver injury in a controlled clinical study (maximum dose of 4 g/day, although the guidelines were changed in 2011 to reduce this to 3 g/day for most products) (Physicians Education Gram, 2011). This study, published in 2006, demonstrated a remarkable incidence of transient alanine aminotransferase (ALT) increases in healthy volunteers administered the maximum therapeutic dose of acetaminophen, with a higher percentage of responders being of Hispanic ethnicity (Watkins et al., 2006). Following oral administration of a high dose of acetaminophen to MDP strains, it was observed that genetic background greatly influenced the toxicity response and that the response was likely unrelated to differences in drug metabolism (Harrill et al., 2009b; Howell et al., 2012). Genomic mapping identified more than one QTL; however, targeted resequencing of candidate genes in DNA from clinical studies confirmed a positive association with a variant in the gene *Cd44* in both mice and humans. Confidence in the association was strengthened via observation of increased toxicity in *Cd44*-knockout mice (on a C57BL/6J background) and independent verification of an enrichment of the risk allele in clinical cases of acute acetaminophen-induced liver failure (Harrill et al., 2009b; Court et al., 2014).

The MDP approach has also been used successfully to investigate mechanisms of drug toxicity. An example of this approach is a recent investigation into the ketolide antibiotic PF-04287881, which was abandoned from development following observation of liver biomarker elevations in clinical trials (Mosedale et al., 2014). Susceptible inbred strains with liver injury in response to the drug were found to exhibit phospholipidosis in Kupffer cells. This finding was not entirely surprising given that PF-04287881 is a cationic amphiphilic drug (CAD) and that phospholipidosis has been associated with previously developed CADs, including antiarrhythmics, antidepressants, and other antibiotics (Halliwell, 1997). Once susceptible and resistant strains

were identified using biochemical and histopathological criteria, a subset of four strains was identified for global transcriptomic analysis using liver tissue. Pathway analysis of transcripts that were altered by the drug and that were unique in susceptible strains indicated that PF-04287881-induced phospholipidosis was mediated by alterations in phospholipid metabolism and lysosomal function and, furthermore, that the liver injury due to PF-04287881 was linked to changes in expression of genes involved in protein degradation, leading to accumulation of oxidized proteins.

MDP models have also been utilized for identification of genetic factors that influence positive response to drug therapy. An exemplar case is a study which investigated behavioral, neurochemical, and transcriptomic responses to chronic exposure to fluoxetine. In that study, it was observed that genetic background greatly influenced the therapeutic response, with 5/30 strains identified as negative responders, 13/30 strains identified as positive responders to the medication, and the remainder of the strains identified as nonresponders (Benton et al., 2012). Genetic association was found with the gene encoding cellular proliferation/adhesion molecule 1 (*Cadm1*), which may indicate a potential role for neuro/gliogenesis in depression.

Mouse embryonic fibroblasts (MEFs) that have been generated from MDP strains may have utility to investigate drug-induced safety risks. However, to date there has been a single study reported, which screened 65 compounds across 32 strains (Suzuki et al., 2014), and the relationship between *in vitro* assay findings and *in vivo* responses was not investigated across compounds. A success of the study was that a QTG was identified that was associated with rotenone toxicity. This gene, *Cybb*, was subsequently validated experimentally both *in vitro* and *in vivo*, providing a promising test case for utilizing cell-based screens for pharmacogenetic safety studies.

Despite some successes with MDP mice, it can be difficult to identify QTGs using classical inbred strains. Genetic variation in classical inbred strains, such as C57BL/6J and DBA/2J, is both limited and unevenly distributed across the genome (Yang et al., 2007, 2011). In addition, because these strains were derived from a limited founder pool, significant bottlenecking has occurred among the strains, resulting in extensive long-range LD (Petkov et al., 2005; Cheng et al., 2010; Collaborative Cross Consortium, 2012). LD occurs when alleles of two or more genes are inherited together more often than would be expected by chance. The high degree of LD present in MDP strains can therefore confound attempts to use this model for association mapping, making it difficult to discern which gene regions contains the QTG that influences the trait.

20.3.2 CC Mice

While there are many approaches to identify QTL, all involve a population of individuals with a measurable phenotype, a database of genotypic variation present within that population, and statistical measures that serve to link the magnitude of the phenotype with a specific genotype or polymorphism. Classical approaches have often sought to utilize the genotypic and phenotypic diversity present in F2 or backcross mouse populations; however, the genetic variability in these populations is limited. Due to the relatively low number of recombination events in F2 and backcross populations, identification of precise QTL locations is often more difficult, although the statistical power to detect a QTL interval is high. RI lines offer the advantage of fixed genomes, but these lines can be expensive to acquire and maintain.

Considered a "next-generation" genetic resource population, the CC resource consists of a genetically diverse panel of RI (CC-RI) mice (Iraqi et al., 2014). The CC population was conceived as a community resource for systems genetics. The CC-RI lines were created via a systematic breeding program derived from an initial cross of eight inbred founder mouse strains. Of the founder strains, five are classically derived strains of the *M.m. domesticus* subspecies (129S1/SvImJ, A/J, C57BL/6J, NOD/ShiLtJ, and NZO/H1LtJ). The remainder of the founder strains represents *M. m. castaneus* (CAST/EiJ) and wild-derived *M.m musculus* and *M. m. domesticus* mice (PWK/PhJ and WSB/EiJ, respectively) (Collaborative Cross Consortium, 2012). All CC lines were independently bred in a scheme that combined the genetic variation present in the eight founder strains into CC lines over three generations followed by inbreeding of each breeding funnel to homozygosity (Collaborative Cross Consortium, 2012). Currently (as of 2015), independent sets of CC lines are generated and maintained at three centers: Chapel Hill in the United States (CC-UNC), Tel Aviv in Israel (CC-TAU), and Perth in Australia (CC-GND). CC lines are considered complete once they have reached 98% homozygosity and are considered distributable for researchers at 90% homozygosity, which is verified using genetic information obtained via specialized genotyping arrays for this purpose (Yang et al., 2009). Owing to the availability of genotyping and bioinformatics resources for CC and DO mice, users will have access to an unprecedented level of detail on the genetic sequences of the CC lines (Welsh et al., 2012).

The CC mice have been utilized in several efforts to identify models of disease (particularly for which animal models are lacking or have historically been reported as ineffective using common rodent strains). For example, research of therapies for Ebola virus infection using the mouse-adapted strain of Ebola virus (MA-EBOV) was historically restricted to macaques, guinea pigs, and Syrian hamsters, owing to a failure of common mouse strains to reproduce the hemorrhagic hallmarks of the human disease pathogenesis. In a landmark study utilizing 47 recombinant inbred intercross (RIX) lines of CC mice (CC-RIX), researchers found that genetic background played an important role in disease pathogenesis, with a phenotypic outcomes across CC lines ranging from complete resistance to lethal disease or severe pathology that was consistent with Ebola hemorrhagic fever (EHF) (Rasmussen et al., 2014). In addition to these observations, it was observed that select CC-RIX lines experienced lethality without symptoms of EHF. Thus, a screening strategy to identify CC lines that exhibit pathology consistent with human disease may advance therapeutic development.

In addition, spontaneous disease observed in CC lines may provide unique opportunities for identifying mouse models of human medical conditions. One such possibility is the study of inflammatory bowel disease (IBD), an immune-mediated condition that is modulated by aberrant responses to intestinal microflora under certain host genetic and environmental contexts. While rodent models of IBD had existed, prior models were reliant upon interventions to induce the disease that consist of either chemical induction or introduction of an infectious agent to induce the disease. A CC line that has been identified as a model for IBD was discovered fortuitously when it was observed that there was a high frequency of rectal prolapse in the CC011/Unc line (Rogala et al., 2014). Following observance of the prolapse, animals generally maintained a good body condition; however, the prolapsed tissue, because necrotic or ulcerated, resulted in deteriorating condition that necessitated euthanasia for humane reasons in the affected animals. Affected animals in the line were found to have hallmark features of colitis by histological assessment of gastrointestinal tissues, and no coincident pathogenic infection was found.

20.3.3 DO Mice

DO mice are a complementary or "sister" mouse population model that was derived through the CC development pipeline. DO mice are a heterogeneous stock derived from the same eight founder strains as the CC. Independent lineages (144 in total) were initially selected from the CC breeding colony while the lines were still segregating, and these mice were utilized to seed the DO population (Chesler et al., 2008). The DO mice are currently maintained as a randomized breeding colony of 175 breeding pairs. Each DO mouse harbors a high level of heterozygosity, with the population at large providing a vast array of allelic combinations. This level of heterozygosity is maintained by randomly selecting a male and female from each first litter and assigning to a new breeding pair to generate the next generation. Such a mating scheme doubles the effective population size, minimizes genetic drift, and minimizes selection on the allele frequencies within the population (Rockman and Kruglyak, 2008). DO mice are robust and breed efficiently, averaging 7 pups (±2.4 SD) in first litters (Churchill et al., 2012).

The utility of using DO mice to model human disease and population-level responses has been demonstrated in recent manuscripts. As with CC mice, DO mice have been employed as a tool to study a variety of human conditions, including pain sensitivity (Recla et al., 2014), development of atherosclerosis (Smallwood et al., 2014), susceptibility to *Mycobacterium tuberculosis* infection (Harrison et al., 2014), and neurobehavioral traits (Logan et al., 2013), including addiction to drugs of abuse (Dickson et al., 2015).

There is growing enthusiasm for utilizing DO mice to set exposure thresholds for chemicals that may be occupational hazards. A landmark study that utilized the DO for this purpose investigated whether current guidelines set for benzene exposures in an occupational setting were adequately protecting both sensitive and resistant workers (French et al., 2015). Typical experiments to identify the "benchmark dose (BMD)" or point of departure for adverse effects are conducted in genetically homogeneous B6C3F1/J mice. In this study, a dose-dependent increase in benzene-induced chromosomal damage was observed, with a wide degree of susceptibility reported across a cohort of DO mice (Fig. 20.4a, b). Furthermore, the BMD limit was an order of magnitude lower in the DO mice as compared to B6C3F1/J mice, suggesting that the increased genetic diversity in the DO could be exploited to identify a more conservative concentration threshold that could protect more workers from the adverse effects of benzene exposure.

In addition to determining safety thresholds for benzene, genomic mapping was performed to identify QTGs that were associated with benzene toxicity, as measured by the micronuclei frequency of reticulocytes from peripheral blood and bone marrow. The investigators identified a highly significant peak (LOD > 20) on chromosome 10 that contained several genes (Fig. 20.4c, d). Analysis of the effects of founder haplotypes within the QTL interval enabled determination that DO mice who had inherited the CAST/EiJ founder haplotype were less sensitive to benzene versus DO mice without the haplotype (Fig. 20.4d, e). *Sult3a1* was identified as a candidate gene in the region because it contained alleles that are unique to the CAST/EiJ haplotype (Fig. 20.4f). Interestingly, examination of mRNA expression patterns of this gene in the eight founder strains led to the discovery that CAST/EiJ mice have a greater expression of the gene and the genome harbors a copy number expansion of the gene. Thus, DO mice that harbor the CAST/EiJ haplotype likely have a greater capacity to detoxify benzene, providing a foundation for resistance to benzene-induced chromosomal damage. While this study involved inhalation of a chemical, rather than administration of a pharmaceutical therapy, the workflow utilized in the benzene study can provide a template for gaining mechanistic insight into outcomes associated with drugs.

20.4 APPLICATIONS FOR PHARMACEUTICAL SAFETY SCIENCE

20.4.1 Personalized Medicine: Development of Companion Diagnostics

The majority of the pharmacogenetic biomarkers in use today have been identified through retrospective analyses. It has not yet become routine to collect genetic information or other molecular data from clinical trial subjects in prospective trials. Therefore, a major challenge to the industry is to identify predictive biomarkers prospectively or codevelop these biomarkers as a *companion diagnostic* that can facilitate development of the therapeutic agent. Companion diagnostics are defined as *in vitro* diagnostic (IVD) tests that ensure the safety and effectiveness of therapeutic products (FDA, 2011; Philip et al., 2011). In effect, the IVD may be essential for the product to meet labeled safety and efficacy claims, and this situation has become more common (FDA, 2011). Such tests distinguish sensitive subpopulations who will benefit from treatment or who may be at a greater risk for certain severe adverse side effects. As such, these new technologies allow for personalization of medical therapy by tailoring the drug treatment to the individual patient's genetic predisposition.

Because pharmacogenetic analysis necessitates inclusion of a large number of individuals in most cases, there are benefits to utilizing a translational pharmacogenetic approach that first utilizes mice. In situations for which safety concerns necessitate development of an IVD to "rescue" a new drug, animal studies are attractive where ethical concerns may prevent further clinical testing. For example, one might consider initiating a safety pharmacogenetic study upon observation of a serious ADR in either late-stage clinical trials or upon entering the marketplace.

20.4.2 Biomarkers of Sensitivity

Mouse population models may also be utilized to identify biomarkers that are associated with toxicity apart from genetic sequence variation. An example of this approach is a recent transcriptomic study that was conducted using a panel of 35 inbred strains (Harrill et al., 2009a). In that study, strains were identified as either sensitive or resistant to a high oral dose of acetaminophen. Global mRNA transcript expression was measured in livers extracted from all of the strains and the data were analyzed such that strain (genetic background), treatment (acetaminophen or vehicle), and liver injury score (% necrosis) were used as factors in an ANCOVA model (analysis of covariance). This analysis allowed for discernment of transcripts with expression that changed: (i) with treatment, but not with strain; (ii) with strain, but not with treatment; and (iii) with treatment and strain and that varied with the amount of necrosis. The 26

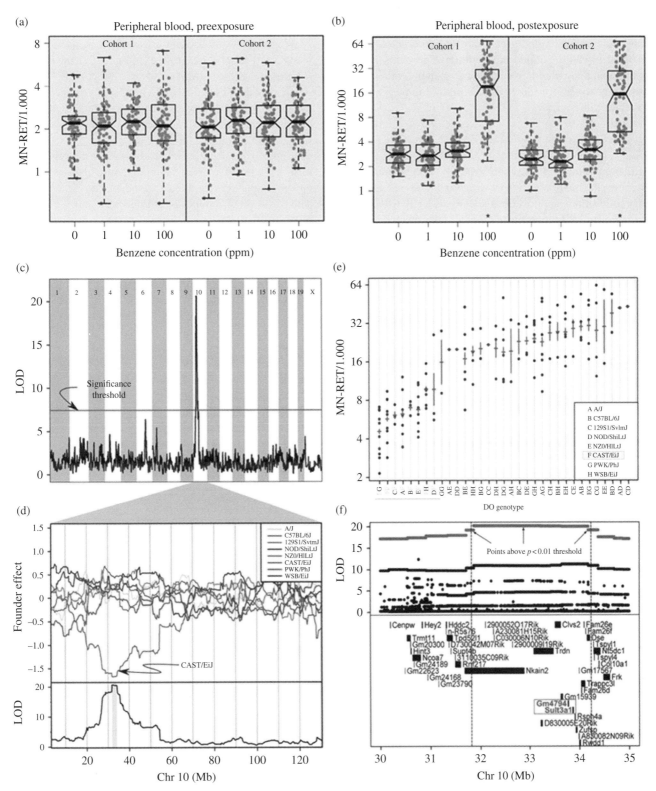

FIGURE 20.4 (a) Preexposure blood MN-RETs. Boxes represent the median and interquartile range, and whiskers cover the entire data range. (b) Benzene postexposure blood MN-RETs. There was a high degree of concordance between cohorts 1 and 2. (c) Plot of the LOD at each marker; red line indicates the permutation-derived significance threshold. (d) Founder effect plots (top) and the LOD score (bottom) on Chr 10; DO mice that have the CAST/EiJ founder haplotype (green) are associated with lower MN-RETs. (e) MN-RET values by DO geno-type at the marker with the maximum LOD score on Chr 10 (31.868 Mb). Data points indicate bone marrow MN-RET values for individual DO mice, and red lines denote mean ± SE. Genotypes are listed on the x-axis, with each DO founder represented by a letter; genotypes con-taining the CAST/EiJ allele are shown in green. (f) Association mapping within the Chr 10 QTL interval. (f, top) Each data point shows the LOD score at one SNP; red data points indicate scores above the significance threshold. (f, bottom) Genes in the QTL interval. Dashed vertical lines denote the QTL support interval. *Sult3a1* is highlighted in red. Adapted from French et al. (2015). (*See insert for color repre-sentation of the figure.*)

transcripts in the latter group mostly represented genes involved in cell death and proliferation. This same experimental paradigm could be extended to metabolomic or proteomic analyses to identify accessible biomarkers that correlate with toxicity.

20.4.3 Mode of Action

A mouse population-based approach may also be employed to inform the mode of action for drug toxicity. A recent demonstration of this application is a study that utilized a panel of inbred strains to investigate the molecular mechanisms of genetic sensitivity to isoniazid-induced steatosis (Church et al., 2014). The study was conducted as a consortia effort by members of the Health and Environmental Sciences Institute (HESI)'s Application of Genomics to Mechanism-based Risk Assessment Technical Committee. In that study, a combined transcriptomic, metabolomic, and pharmacogenetic analysis was utilized to provide evidence for a novel hypothesis that isoniazid increases the capacity for formation of lipid droplets while concurrently reducing the capacity for exporting stored fat from the hepatocytes in sensitive strains.

An important aspect of this approach is that a subset of sensitive and resistant strains may be selected for downstream analysis after an initial screen of a larger strain panel. By selecting strains with certain characteristics, it is possible to reduce costs and complexity of targeted experiments to elucidate toxicity mechanisms.

20.5 STUDY DESIGN CONSIDERATIONS FOR GENOMIC MAPPING

Use of mouse populations in drug safety assessment is largely experimental at this stage, and, therefore, there is a lack of regulatory guidance on the conduct of such a study. In lieu of firm guidelines, general considerations for study designs founded on the past decade of research investigations will be presented.

20.5.1 Dose Selection

As the rat tends to be a more frequent model for safety studies, dose selection for pharmacogenetic studies in mice can present a challenge. Depending on the study objective, there are multiple criteria that may come into play upon selecting a dose for such a study, and no clear guidelines are yet available from the regulatory agencies. A second challenge is that the rodent (and other mammalian species) data may not indicate any potential for the ADR subsequently observed in the clinic. The potential for a positive outcome rests, in part, on a hypothesis that a mouse population-based approach will include some genetically sensitive individuals that will manifest the toxicity of concern.

Where *in vivo* mouse data are available in classical strains or stocks, it may be ideal to use a dose that is consistent with (or close to) the no observed adverse effect level (NOAEL). Selecting a dose near the NOAEL for safety studies will help to minimize any off-target adverse events that might occur in a mouse population study. However, where there are indications from the mouse studies of the human-relevant ADR, a dose approximating the lowest adverse effect level (LOAEL) may be utilized instead. Where mouse data are lacking entirely, it may be beneficial to conduct a dose range-finding study in a small number of animals to identify a dose that will limit undesired adverse effects. Because mouse population studies require a large number of animals, conducting this step on the front end may produce cost savings down the road. Finally, it may be useful to consider pharmacokinetic data in mice and in humans to enable selection of a dose that is within a reasonable order of magnitude to approximate the clinical exposure data.

However, there is a caveat that the relative sensitivities of genetically different mice may lead some animals to manifest side effects that were wholly unanticipated by either the classical strain or clinical studies. As more studies are conducted, estimates of the frequency of these events will become more readily available.

20.5.2 Model Selection

20.5.2.1 DO The likelihood of severe morbidity and mortality may be greater in CC-RI lines or in inbred strains of an MDP, owing to a lack of hybrid vigor. One example of this phenomenon occurred in the MDP investigation of isoniazid-induced hepatic steatosis. In that study, drug-treated members of two strains (P/J and WSB/EiJ) had to be excluded from subsequent analyses owing to mortality that was not associated with liver injury (Church et al., 2014). For these reasons, the DO mice may be preferable to use for safety pharmacogenetic studies, particularly in cases where the mode of action is not understood. In addition, owing to high fecundity of DO mice, these mice may be used preferentially over the CC strains for reproductive and developmental studies for which it is desirable to determine the genetic underpinning of certain traits. In addition, because the DO mice breed well and are bred in three large breeding waves per year, it is more straightforward to obtain a large cohort of similarly aged DO mice as compared to CC lines.

20.5.2.2 CC The CC lines may also be used in safety studies; however, the current limitations in availability of animals may limit use. The CC-RI lines have great potential to be utilized for efficacy studies involving certain disease states; however, as more phenotypic information is collated from this population, information regarding spontaneous disease will become more readily available. A positive aspect of the CC lines is the reproducibility of members within a line, providing the ability to repeat exposures or test multiple doses in the same genetic context. This also means that if a certain CC line is identified as a model, there is a potential to test next-in-class drugs with fewer animals used than would be required for a typical pharmacogenetics study.

20.5.2.3 *MDP*

As discussed previously, there are drawbacks to using multiple classical inbred lines for mapping studies, but there are some positive aspects to this population. The first is the availability of historical data for many strains (e.g., C57BL/6J or Balb/cJ) that may be useful to replicate. In addition, there is the availability of MEF lines for many strains that may be an attractive choice for *in vitro* analyses while the CC and DO MEF and stem cell lines are still under development (Suzuki et al., 2014). It is important to note that strains even with a similar name (e.g., C57BL/6N and C57BL/6J) can have genetic sequence differences owing to genetic drift that has occurred where the population is housed (Festing, 2010). There is also the potential for genetic contamination, spontaneous mutations, and genetic drift to occur even in an established line if the strain has been maintained over many generations (Wiles and Taft, 2010). Therefore, it is important to have consistency when obtaining inbred strains. For these reasons, some vendors now "reset" strains after a defined number of breeding generations using cryopreserved embryos derived from the original stock.

20.5.3 Sample Size

The numbers of animals used for a study is highly dependent upon the goals of the study, magnitude and direction of the measured effect, expected variation in the effect, desired significance level, and power (i.e., the probability of finding an effect—usually ~80%) (Charan and Kantharia, 2013). For the purposes of this chapter, sample sizes for pharmacogenetic studies will be the major focus.

20.5.3.1 *CC*

For the majority of studies conducted to date, panels of CC lines have been the standard model. Simulation analysis has proposed that an "idealized minimum" number of animals for a genetic study involving CC lines is 128 lines. For a QTL with additive effects, using 128 strains enables detection of a major QTL with an effect size of 0.25 and 90% power (Tsaih et al., 2005). For comparison, the same statistical model indicates that power is reduced to 60% when the CC panel is reduced to 64 strains. Another recent retrospective analysis included metrics of heritability, and the genetic coefficient of variation confirmed that it may be possible to identify a strong QTL that maps to a resolution as narrow as 1 Mb with as few as 100 CC lines (Iraqi et al., 2014).

It should be mentioned that there are at least two types of genetic studies that have been proposed for CC lines. The first type involves genomic mapping of traits measured directly in the standard CC lines as described previously. A variant of this analysis that has been proposed is called a RIX (formally known as a *diallel cross*), which is comprised of F1 progeny of a panel of CC lines. By generating RIX lines, investigators can evaluate parent-of-origin effects, as well as determine whether a gene variant is dominant or recessive (it should be noted that these effects may also be evaluated in the DO). An advantage of utilizing RIX lines is that the mice are no longer fully inbred, which can improve the vitality of the animals by introducing hybrid vigor. While the combinatorial effect of a RIX cross increases the number of unique recombinant genomes available (Threadgill et al., 2002), the increase does not overcome the limitation of having a small panel of RI lines (Tsaih et al., 2005). An advantage of a RIX cross is that the frequency of false positive associations that result from nonsyntenic lineage is reduced in genomic mapping analysis. However, studies that investigated the utility of using RIX lines versus traditional RI panels have concluded that there appears to be little advantage of RIX crosses for genomic mapping analysis (Tsaih et al., 2005).

20.5.3.2 *DO*

DO mice typically require greater numbers of animals for a genome-wide association study (GWAS) as compared to CC mice. The reason is that DO mice have a high level of heterozygosity, whereas CC mice are largely homozygous at every locus. A major consideration for the sample size is that the magnitude of the effect of a gene variant is often not known *a priori*. Therefore, an experimental strategy could be to start with fewer numbers of animals, analyze the GWAS data, and then add more to the study if the necessary power was not achieved. This strategy is more amenable to DO mice at the present time, although it is expected that more CC lines will become available in the future to bolster experimentation.

For DO mice, simulated power analyses indicate that as few as 200 mice can be utilized to detect QTL regions with large effects. In contrast, loci that account for <5% of trait variance may require up to 1000 mice (Gatti et al., 2014b). However, it is unlikely that loci with very minor effects will have utility in personalized medicine, meaning that achieving power to detect such an effect may be unproductive. On the other hand, for projects where it is desirable to explain as much of the trait variance as possible (e.g., for a neurobehavioral outcome), larger numbers of animals may be preferable. In routine studies to identify pharmacogenetic markers for susceptibility to drug toxicity, a rule of thumb is to start with 400 DO mice. Power simulations indicate that with 400 mice, QTL can be detected (at $\alpha = 0.05$) that account for 10% of the phenotypic variance with 80% power. Based on existing data from DO mice at generations 7 and 8 of outbreeding, 400 mice provide a median recombination block width of 0.33 Mb (Gatti et al., 2014b). This resolution is fine enough to map loci down to a handful of genes. A caveat is that, in practice, the local LD structure will also influence the QTL width.

20.5.3.3   MDP   The rule of thumb for MDP sample sizes is "the more strains the better." In the past, successful studies have been completed using between 34 and 36 inbred strains (Harrill et al., 2009c).

20.5.4   Phenotyping

The phenotype (or experimental outcome) measured is an important consideration in the design of studies involving mouse populations. Because mice of various genetic contexts will likely exhibit different outcomes, it is important to choose a phenotype for which sampling and measurement error can be minimized. The more precise the quantitatively measured outcome is, the less noise there will be in the data set and the better the estimate of variance will be. This is because the quantitative phenotype information will be utilized as a starting input for the genome-wide association models. Examples of ideally suited phenotypes can include, but are not limited to, quantitative pathology scores, organ weights, or quantitative tissue markers. Special consideration should be given to utilizing blood-based leakage biomarkers as there may be differences in clearance rates among genetically diverse animals that could affect endpoint measures.

The phenotype–genotype association will inevitably influence the QTL that is identified by mapping. Assuming litter size (phenotype A) is measured in a set of 200 DO mice, it would be expected that the QTL identified by GWA for phenotype A would be different from the QTL identified for anxiety (phenotype B), unless the two traits are governed by the same set of genes. This distinction becomes important because it may be necessary to conduct a study in which many biomarkers and outcomes are measured. Once genotypic data are collected, it is straightforward to map all of the phenotypes collected from the study. Phenotypes that are strongly correlated have a greater probability of being modulated by the same QTGs.

20.5.5   Genome-Wide Association Analysis

A variety of public data analysis packages are available for genomic mapping applications in both CC and DO populations. Several methodologies also exist for QTL mapping, but discussion of the analysis particulars is beyond the scope of this chapter. Here, a few of the analysis packages that have been utilized in recent investigations are presented.

20.5.5.1   CC   The UNC Computational Genetics group has developed a suite of tools for viewing and analyzing genetic data from the CC lines. One such tool is TreeQA, which is a tree-based association mapping method that can incorporate evolutionary history of the genome into the analysis. Essentially, the algorithm utilizes local phylogenies constructed in genomic regions that exhibit no evidence of historical recombination (Pan et al., 2009). Other packages that have been utilized for the analysis include HAPPY. HBREM (Vered et al., 2014), which employs a logistic regression model to fit covariates. Residuals from the model are then used as the response variable for QTL mapping using linear regression, with the Bayesian random effects model HBREM used to estimate individual haplotype effects (Durrant and Mott, 2010).

20.5.5.2   DO   The currently used software is a Bioconductor package called DOQTL, which can be utilized in R (Gatti et al., 2014a, b). DOQTL provides a suite of tools for processing DO genotype data, reconstructing individual genomes by inferring sequences using haplotypes from the founder strains, and performing QTL analysis. In addition, the package allows for assessment of founder haplotype effects on the QTL interval.

20.5.5.3   MDP   MDPs can be composed of many different strains; however, the choice of strains should fall within those for which genetic sequence data are available and with good coverage of polymorphisms throughout the genome. Traditional inbred *M. m. musculus* strains have minimal levels of intrastrain polymorphisms. Thus, it is prudent to select panels that include *M. m. musculus* strains and *M. m. domesticus* strains. While the inclusion of *M. m. castaneous* alleles in the CC and DO populations is an advantage, it may be somewhat detrimental to a mapping study to include an *M. m. castaneous* or, similarly, a wild-derived line in a MDP. This is because inclusion of these strains affects the population structure such that many false positives may arise simply due to the genetic divergence of these lines.

In part due to population structure across inbred strains, a popular statistical method for QTL mapping for this population is Efficient Mixed-Model Association (EMMA), available as an R package. EMMA corrects for confounding population structure and genetic relatedness of the strains, thus removing the potential for findings that may be false positive associations (Kang et al., 2008). Another model that has been used previously is SNPster, which assesses significance for 3-SNP windows across the genome as an inferred haplotype block (McClurg et al., 2006, 2007). SNPster works best when there is even and close spacing of known polymorphisms throughout the genome; otherwise QTL analysis may result in identifying very large genomic regions in the association analysis.

20.5.6   Candidate Gene Analysis

Following QTL analysis, confirmation of candidate QTGs is the next logical experimental step. Because the main objective in pharmacogenetics is translation of results to humans, an ideal confirmatory study would involve sequencing of the candidate gene in human subjects (cases and controls, matched for ethnicity). A higher frequency of the variant in the toxicity cases and an increased odds ratio for mutation carriers may signify utility for a pharmacogenetic test.

The selection of follow-up studies depends, in part, on the phenotype and what is known about the function of the QTG from gene and pathway databases. In addition, it may depend on whether the QTG is suspected to underlie a monogenic trait or whether multiple QTL peaks were identified that rather suggest a polygenic trait. Where a QTG is suspected of having a large effect on the phenotype, follow-up studies in either gene knockout mice or CRISPR-edited mice may be useful to confirm the relative importance of the gene to the response. In the case of a QTG identified via a DO study, it may be handy to perform follow-up analyses in selected CC lines that contain the variant(s) of interest.

In cases where the trait is polygenic, it may be useful to consider a pathway-based approach. Wang et al. recently proposed a method for which the power to detect causal mechanisms of disease may be more robust because multiple contributing factors are considered together, as opposed to focusing on a few SNPs with the highest association score (Wang et al., 2007). This approach, which combines GWA with the gene set enrichment algorithm, identified a number of pathways that may be associated with Parkinson disease and age-related macular degeneration.

20.5.7 Cost Considerations

Because the costs of animals can vary, discussion of cost considerations is limited to generalizations of the current circumstances for the different models (as of 2016).

20.5.7.1 CC The major cost advantages to using CC lines are the ready availability of genotypic and phenotypic information for each line and the ability to use fewer mice to detect a potential QTL. Because each CC line is inbred, genetic sequence information is fixed, thus the animals do not need to be genotyped by the end user. Genetic sequence information for each CC line may be downloaded from http://csbio.unc.edu/CCstatus/index.py?run=Pseudo.

As previously mentioned, the various CC lines are maintained at three different sites, which are geographically distant from each other. The Collaborative Cross Consortium has made available both CC lines that are complete and lines that are "distributable" (98 and 90% homozygosity, respectively) (Welsh et al., 2012). Distributed CC lines fall under a Conditions of Use (COU) agreement that limits research to internal use and prevents redistribution to other institutions without prior permission. The current list of distributable lines, including breeding performance and

physical characteristics, is available online at http://csbio.unc.edu/CCstatus/index.py. There is an animal cost plus freight charges per animal ordered. The CC mice are generally distributed as breeding trios, with the end user responsible for generating the required number of offspring for analysis. Therefore, the user must factor in caging costs for breeding pairs, offspring, and the variable time that it will take to generate enough offspring from all of the CC lines purchased. Regular CC users may not find the need to maintain a breeding colony a disadvantage as this may afford future cost savings. However, users with sporadic or one-off projects may find this to be a disadvantage of the current CC distribution system.

20.5.7.2 DO There is also an animal cost plus freight charges per animal for DO mice, which are available from the Jackson Laboratory, Bar Harbor, ME. Mice are available from three breeding waves distributed throughout the year. An advantage is that users can obtain hundreds of mice at a time, allowing for completion of experiments within a single order. Major costs associated with using the DO for pharmacogenetic experiments is the cost of purchasing the animals and the need to genotype every individual animal. This is accomplished by submitting DNA samples to a contract facility to analyze using the GigaMUGA (Mouse Universal Genotyping Array) for a per sample fee. The GigaMUGA is essentially the same array that is used to genotype CC lines, although many of the lines were genotyped using the previous version (MegaMUGA—778,000 markers). The GigaMUGA has ~150,000 markers, in addition to probes that can detect known structural variants including insertions, deletions, and duplications.

20.5.7.3 MDP Costs for MDP mice are highly dependent upon the number of strains and the number of animals within a strain that are desired. Per strain costs can vary widely depending on the demand for a given strain and how difficult a strain may be to breed and maintain. However, as with the CC, because each inbred strain is a renewable resource, strains with polymorphism data available do not have to be genotyped for each study. Whole genome sequencing data are available for many strains from the Mouse Genomes Project of the Wellcome Trust Sanger Institute (http://www.sanger.ac.uk/resources/mouse/genomes/).

20.5.8 Health Status

20.5.8.1 CC The current distribution center for CC lines in the United States is a university setting at UNC. CC mice are housed in facilities representing two tiers of pathogen control—either specific pathogen free (SPF) or barrier. As of 2012, the SPF facility exhibited negative serology for the following common pathogens: EDIM, TMEV GDVII, MHV, Mycoplasma pulmonis, MPV, MVM, Parvo NS-1,

PVM, and Sendai. Additionally, some are tested for CAR bacillus, Ectromelia, LCMV, MAD1, MAD2, mCMV, Polyoma, and REO3. The barrier facility is designed to exclude all of the pathogens in the SPF as well as MNV, PVM, *Pasteurella pneumotropica*, and *Helicobacter*. There is an ongoing effort to rederive all of the CC lines through *in vitro* fertilization and c-section such that all of the lines will be able to be housed in the barrier facility.

Mice are also available from Tel Aviv University (TAU). Currently at TAU, all CC mice are maintained in a conventional facility, while the rederived lines are maintained in individual ventilation cages (IVC) and housed in a SPF facility. CC lines at TAU undergo routine microbiology monitoring according to the Federation of European Laboratory Animal Science Association (FELASA) recommendations (Nicklas et al., 2002).

20.5.8.2 DO Because DO mice are maintained by a mouse distributor, there are strict exclusion criteria for animal housing. Exclusion criteria and reporting requirements can be found by accessing http://jaxmice.jax.org/health/agents_list.html. Production facility reports for DO mice include monitoring data for common mouse pathogens. Current health reports may be accessed at http://jaxmice.jax.org/strain/009376.html under the "Technical Support" drop down menu.

20.5.8.3 MDP The health status of inbred strains will vary based on the source, which can be an academic center or vendor. It will be necessary to check the health status with each supplier prior to ordering.

20.6 SUMMARY

The next wave of pharmacogenomic and genetic research is the individualization of medicine and the ability to tailor drug therapy for specific patients. This area will be greatly facilitated by the development of easy and inexpensive whole-genome genotyping on an individual basis. Because genome association using inbred, complex RI lines, or outbred stocks is an emerging technique, we expect that the experimental pipelines will be refined to facilitate personalized assessment of efficacy and toxicity. However, the availability of controlled mouse genetic breeding programs, namely, the CC and DO mice, will enable a superior precision in genetic mapping as compared to previous resources and will revolutionize the ability to detect the genetic control of drug responses in individuals. Translational research bridging rodent and human drug responses will be the key to the success of this field, and great emphasis should be given to collaboration across laboratories and disciplines. This effort will require team research within centers and consortia in order to be fully realized.

REFERENCES

Allison, M., 2012. Reinventing clinical trials. *Nat Biotechnol* 30, 41–49.

Arrowsmith, J., 2012. A decade of change. *Nat Rev Drug Discov* 11, 17–18.

Beck, J.A., Lloyd, S., Hafezparast, M., Lennon-Pierce, M., Eppig, J.T., Festing, M.F., Fisher, E.M., 2000. Genealogies of mouse inbred strains. *Nat Genet* 24, 23–25.

Benton, C.S., Miller, B.H., Skwerer, S., Suzuki, O., Schultz, L.E., Cameron, M.D., Marron, J.S., Pletcher, M.T., Wiltshire, T., 2012. Evaluating genetic markers and neurobiochemical analytes for fluoxetine response using a panel of mouse inbred strains. *Psychopharmacology (Berl)* 221, 297–315.

Burns, J.J., 1968. IV. Pharmacogenetics and drug toxicity. Variation of drug metabolism in animals and the prediction of drug action in man. *Ann N Y Acad Sci* 151, 959–967.

Cacabelos, R., 2007. Pharmacogenetic basis for therapeutic optimization in Alzheimer's disease. *Mol DiagnTher* 11, 385–405.

Charan, J., Kantharia, N.D., 2013. How to calculate sample size in animal studies? *J Pharmacol Pharmacother* 4, 303–306.

Cheng, R., Lim, J.E., Samocha, K.E., Sokoloff, G., Abney, M., Skol, A.D., Palmer, A.A., 2010. Genome-wide association studies and the problem of relatedness among advanced intercross lines and other highly recombinant populations. *Genetics* 185, 1033–1044.

Chesler, E.J., Miller, D.R., Branstetter, L.R., Galloway, L.D., Jackson, B.L., Philip, V.M., Voy, B.H., Culiat, C.T., Threadgill, D.W., Williams, R.W., Churchill, G.A., Johnson, D.K., Manly, K.F., 2008. The collaborative cross at Oak Ridge national laboratory: developing a powerful resource for systems genetics. *Mamm Genome* 19, 382–389.

Church, R.J., Wu, H., Mosedale, M., Sumner, S.J., Pathmasiri, W., Kurtz, C.L., Pletcher, M.T., Eaddy, J.S., Pandher, K., Singer, M., Batheja, A., Watkins, P.B., Adkins, K., Harrill, A.H., 2014. A systems biology approach utilizing a mouse diversity panel identifies genetic differences influencing isoniazid-induced microvesicular steatosis. *Toxicol Sci* 140, 481–492.

Churchill, G.A., Airey, D.C., Allayee, H., Angel, J.M., Attie, A.D., Beatty, J., Beavis, W.D., Belknap, J.K., Bennett, B., Berrettini, W., Bleich, A., Bogue, M., Broman, K.W., Buck, K.J., Buckler, E., Burmeister, M., Chesler, E.J., Cheverud, J.M., Clapcote, S., Cook, M.N., Cox, R.D., Crabbe, J.C., Crusio, W.E., Darvasi, A., Deschepper, C.F., Doerge, R.W., Farber, C.R., Forejt, J., Gaile, D., Garlow, S.J., Geiger, H., Gershenfeld, H., Gordon, T., Gu, J., Gu, W., de Haan, G., Hayes, N.L., Heller, C., Himmelbauer, H., Hitzemann, R., Hunter, K., Hsu, H.C., Iraqi, F.A., Ivandic, B., Jacob, H.J., Jansen, R.C., Jepsen, K.J., Johnson, D.K., Johnson, T.E., Kempermann, G., Kendziorski, C., Kotb, M., Kooy, R.F., Llamas, B., Lammert, F., Lassalle, J.M., Lowenstein, P.R., Lu, L., Lusis, A., Manly, K.F., Marcucio, R., Matthews, D., Medrano, J.F., Miller, D.R., Mittleman, G., Mock, B.A., Mogil, J.S., Montagutelli, X., Morahan, G., Morris, D.G., Mott, R., Nadeau, J.H., Nagase, H., Nowakowski, R.S., O'Hara, B.F., Osadchuk, A.V., Page, G.P., Paigen, B., Paigen, K., Palmer, A.A., Pan, H.J., Peltonen-Palotie, L., Peirce, J., Pomp, D., Pravenec, M., Prows, D.R., Qi, Z., Reeves, R.H., Roder, J.,

Rosen, G.D., Schadt, E.E., Schalkwyk, L.C., Seltzer, Z., Shimomura, K., Shou, S., Sillanpaa, M.J., Siracusa, L.D., Snoeck, H.W., Spearow, J.L., Svenson, K., Tarantino, L.M., Threadgill, D., Toth, L.A., Valdar, W., de Villena, F.P., Warden, C., Whatley, S., Williams, R.W., Wiltshire, T., Yi, N., Zhang, D., Zhang, M., Zou, F., 2004. The collaborative cross, a community resource for the genetic analysis of complex traits. *Nat Genet* 36, 1133–1137.

Churchill, G.A., Gatti, D.M., Munger, S.C., Svenson, K.L., 2012. The Diversity Outbred mouse population. *Mamm Genome* 23, 713–718.

Collaborative Cross Consortium, 2012. The genome architecture of the collaborative cross mouse genetic reference population. *Genetics* 190, 389–401.

Court, M.H., Peter, I., Hazarika, S., Vasiadi, M., Greenblatt, D.J., Lee, W.M., Acute Liver Failure Study Group, 2014. Candidate gene polymorphisms in patients with acetaminophen-induced acute liver failure. *Drug Metab Dispos* 42, 28–32.

Dickson, P.E., Ndukum, J., Wilcox, T., Clark, J., Roy, B., Zhang, L., Li, Y., Lin, D.T., Chesler, E.J., 2015. Association of novelty-related behaviors and intravenous cocaine self-administration in Diversity Outbred mice. *Psychopharmacology (Berl)* 232, 1011–1024.

Durrant, C., Mott, R., 2010. Bayesian quantitative trait locus mapping using inferred haplotypes. *Genetics* 184, 839–852.

FDA, 2011. In Vitro Companion Diagnostic Devices: Guidance for Industry and Food and Drug Administration Staff.

Festing, M.F., 2010. Improving toxicity screening and drug development by using genetically defined strains. *Methods Mol Biol* 602, 1–21.

Flint, J., Eskin, E., 2012. Genome-wide association studies in mice. *Nat Rev Genet* 13, 807–817.

French, J.E., Gatti, D.M., Morgan, D.L., Kissling, G.E., Shockley, K.R., Knudsen, G.A., Shepard, K.G., Price, H.C., King, D., Witt, K.L., Pedersen, L.C., Munger, S.C., Svenson, K.L., Churchill, G.A., 2015. Diversity Outbred mice identify population-based exposure thresholds and genetic factors that influence benzene-induced genotoxicity. *Environ Health Perspect* 123, 237–245.

Gatti, D.M., Broman, K., Shabalin, A., 2014a. DOQTL: Genotyping and QTL Mapping in DO Mice. Bioconductor.

Gatti, D.M., Svenson, K.L., Shabalin, A., Wu, L.Y., Valdar, W., Simecek, P., Goodwin, N., Cheng, R., Pomp, D., Palmer, A., Chesler, E.J., Broman, K.W., Churchill, G.A., 2014b. Quantitative trait locus mapping methods for Diversity Outbred mice. *G3 (Bethesda)* 4, 1623–1633.

Halliwell, W.H., 1997. Cationic amphiphilic drug-induced phospholipidosis. *Toxicol Pathol* 25, 53–60.

Harrill, A.H., Ross, P.K., Gatti, D.M., Threadgill, D.W., Rusyn, I., 2009a. Population-based discovery of toxicogenomics biomarkers for hepatotoxicity using a laboratory strain diversity panel. *Toxicol Sci* 110, 235–243.

Harrill, A.H., Watkins, P.B., Su, S., Ross, P.K., Harbourt, D.E., Stylianou, I.M., Boorman, G.A., Russo, M.W., Sackler, R.S., Harris, S.C., Smith, P.C., Tennant, R., Bogue, M., Paigen, K., Harris, C., Contractor, T., Wiltshire, T., Rusyn, I., Threadgill, D.W., 2009b. Mouse population-guided resequencing reveals that variants in CD44 contribute to acetaminophen-induced liver injury in humans. *Genome Res* 19, 1507–1515.

Harrison, D.E., Astle, C.M., Niazi, M.K., Major, S., Beamer, G.L., 2014. Genetically diverse mice are novel and valuable models of age-associated susceptibility to Mycobacterium tuberculosis. *Immun Ageing* 11, 24.

Higashi, M.K., Veenstra, D.L., Kondo, L.M., Wittkowsky, A.K., Srinouanprachanh, S.L., Farin, F.M., Rettie, A.E., 2002. Association between CYP2C9 genetic variants and anticoagulation-related outcomes during warfarin therapy. *JAMA* 287, 1690–1698.

Howell, B.A., Yang, Y., Kumar, R., Woodhead, J.L., Harrill, A.H., Clewell, H.J., III, Andersen, M.E., Siler, S.Q., Watkins, P.B., 2012. In vitro to in vivo extrapolation and species response comparisons for drug-induced liver injury (DILI) using DILIsym: a mechanistic, mathematical model of DILI. *J Pharmacokinet Pharmacodyn* 39, 527–541.

Iraqi, F.A., Athamni, H., Dorman, A., Salymah, Y., Tomlinson, I., Nashif, A., Shusterman, A., Weiss, E., Houri-Haddad, Y., Mott, R., Soller, M., 2014. Heritability and coefficient of genetic variation analyses of phenotypic traits provide strong basis for high-resolution QTL mapping in the Collaborative Cross mouse genetic reference population. *Mamm Genome* 25, 109–119.

Kalow, W., 1965. Contribution of hereditary factors to the response to drugs. *Fed Proc* 24, 1259–1265.

Kang, H.M., Zaitlen, N.A., Wade, C.M., Kirby, A., Heckerman, D., Daly, M.J., Eskin, E., 2008. Efficient control of population structure in model organism association mapping. *Genetics* 178, 1709–1723.

Logan, R.W., Robledo, R.F., Recla, J.M., Philip, V.M., Bubier, J.A., Jay, J.J., Harwood, C., Wilcox, T., Gatti, D.M., Bult, C.J., Churchill, G.A., Chesler, E.J., 2013. High-precision genetic mapping of behavioral traits in the Diversity Outbred mouse population. *Genes Brain Behav* 12, 424–437.

Mallal, S., Phillips, E., Carosi, G., Molina, J.M., Workman, C., Tomazic, J., Jagel-Guedes, E., Rugina, S., Kozyrev, O., Cid, J.F., Hay, P., Nolan, D., Hughes, S., Hughes, A., Ryan, S., Fitch, N., Thorborn, D., Benbow, A., 2008. HLA-B*5701 screening for hypersensitivity to abacavir. *N Engl J Med* 358, 568–579.

McClurg, P., Pletcher, M.T., Wiltshire, T., Su, A.I., 2006. Comparative analysis of haplotype association mapping algorithms. *BMC Bioinformatics* 7, 61.

McClurg, P., Janes, J., Wu, C., Delano, D.L., Walker, J.R., Batalov, S., Takahashi, J.S., Shimomura, K., Kohsaka, A., Bass, J., Wiltshire, T., Su, A.I., 2007. Genomewide association analysis in diverse inbred mice: power and population structure. *Genetics* 176, 675–683.

Mosedale, M., Wu, H., Kurtz, C.L., Schmidt, S.P., Adkins, K., Harrill, A.H., 2014. Dysregulation of protein degradation pathways may mediate the liver injury and phospholipidosis associated with a cationic amphiphilic antibiotic drug. *Toxicol Appl Pharmacol* 280, 21–29.

Nebert, D.W., Zhang, G., Vesell, E.S., 2008. From human genetics and genomics to pharmacogenetics and pharmacogenomics: past lessons, future directions. *Drug Metab Rev* 40, 187–224.

Nicklas, W., Baneux, P., Boot, R., Decelle, T., Deeny, A.A., Fumanelli, M., Illgen-Wilcke, B., FELASA, 2002. Recommendations for the health monitoring of rodent and rabbit colonies in breeding and experimental units. *Lab Anim* 36, 20–42.

Pan, F., McMillan, L., Pardo-Manuel De Villena, F., Threadgill, D., Wang, W., 2009. TreeQA: quantitative genome wide association mapping using local perfect phylogeny trees. *Pac Symp Biocomput*, 415–426.

Petkov, P.M., Graber, J.H., Churchill, G.A., DiPetrillo, K., King, B.L., Paigen, K., 2005. Evidence of a large-scale functional organization of mammalian chromosomes. *PLoS Genet* 1, e33.

Philip, R., Carrington, L., Chan, M., 2011. US FDA perspective on challenges in co-developing in vitro companion diagnostics and targeted cancer therapeutics. *Bioanalysis* 3, 383–389.

Physicians Education Gram, 2011. Acetaminophen and Hepatotoxicity—New Max Dose of 3gm/Day. Pro Pharma Pharmaceutical Consultants, Inc., Northridge, CA.

Rasmussen, A.L., Okumura, A., Ferris, M.T., Green, R., Feldmann, F., Kelly, S.M., Scott, D.P., Safronetz, D., Haddock, E., LaCasse, R., Thomas, M.J., Sova, P., Carter, V.S., Weiss, J.M., Miller, D.R., Shaw, G.D., Korth, M.J., Heise, M.T., Baric, R.S., de Villena, F.P., Feldmann, H., Katze, M.G., 2014. Host genetic diversity enables Ebola hemorrhagic fever pathogenesis and resistance. *Science* 346, 987–991.

Recla, J.M., Robledo, R.F., Gatti, D.M., Bult, C.J., Churchill, G.A., Chesler, E.J., 2014. Precise genetic mapping and integrative bioinformatics in Diversity Outbred mice reveals Hydin as a novel pain gene. *Mamm Genome* 25, 211–222.

Roberts, A., Pardo-Manuel, D.V., Wang, W., McMillan, L., Threadgill, D.W., 2007. The polymorphism architecture of mouse genetic resources elucidated using genome-wide resequencing data: implications for QTL discovery and systems genetics. *Mamm Genome* 18, 473–481.

Rockman, M.V., Kruglyak, L., 2008. Breeding designs for recombinant inbred advanced intercross lines. *Genetics* 179, 1069–1078.

Rogala, A.R., Morgan, A.P., Christensen, A.M., Gooch, T.J., Bell, T.A., Miller, D.R., Godfrey, V.L., de Villena, F.P., 2014. The collaborative cross as a resource for modeling human disease: CC011/Unc, a new mouse model for spontaneous colitis. *Mamm Genome* 25, 95–108.

Rost, S., Fregin, A., Ivaskevicius, V., Conzelmann, E., Hortnagel, K., Pelz, H.J., Lappegard, K., Seifried, E., Scharrer, I., Tuddenham, E.G., Muller, C.R., Strom, T.M., Oldenburg, J., 2004. Mutations in VKORC1 cause warfarin resistance and multiple coagulation factor deficiency type 2. *Nature* 427, 537–541.

Smallwood, T.L., Gatti, D.M., Quizon, P., Weinstock, G.M., Jung, K.C., Zhao, L., Hua, K., Pomp, D., Bennett, B.J., 2014. High-resolution genetic mapping in the Diversity Outbred mouse population identifies Apobec1 as a candidate gene for atherosclerosis. *G3 (Bethesda)* 4, 2353–2363.

Suzuki, O.T., Frick, A., Parks, B.B., Trask, O.J., Jr., Butz, N., Steffy, B., Chan, E., Scoville, D.K., Healy, E., Benton, C., McQuaid, P.E., Thomas, R.S., Wiltshire, T., 2014. A cellular genetics approach identifies gene-drug interactions and pinpoints drug toxicity pathway nodes. *Front Genet* 5, 272.

Threadgill, D.W., Hunter, K.W., Williams, R.W., 2002. Genetic dissection of complex and quantitative traits: from fantasy to reality via a community effort. *Mamm Genome* 13, 175–178.

Tsaih, S.W., Lu, L., Airey, D.C., Williams, R.W., Churchill, G.A., 2005. Quantitative trait mapping in a diallel cross of recombinant inbred lines. *Mamm Genome* 16, 344–355.

Venter, J.C., Adams, M.D., Myers, E.W., Li, P.W., Mural, R.J., Sutton, G.G., Smith, H.O., Yandell, M., Evans, C.A., Holt, R.A., Gocayne, J.D., Amanatides, P., Ballew, R.M., Huson, D.H., Wortman, J.R., Zhang, Q., Kodira, C.D., Zheng, X.H., Chen, L., Skupski, M., Subramanian, G., Thomas, P.D., Zhang, J., Gabor Miklos, G.L., Nelson, C., Broder, S., Clark, A.G., Nadeau, J., McKusick, V.A., Zinder, N., Levine, A.J., Roberts, R.J., Simon, M., Slayman, C., Hunkapiller, M., Bolanos, R., Delcher, A., Dew, I., Fasulo, D., Flanigan, M., Florea, L., Halpern, A., Hannenhalli, S., Kravitz, S., Levy, S., Mobarry, C., Reinert, K., Remington, K., Abu-Threideh, J., Beasley, E., Biddick, K., Bonazzi, V., Brandon, R., Cargill, M., Chandramouliswaran, I., Charlab, R., Chaturvedi, K., Deng, Z., Di, F.V., Dunn, P., Eilbeck, K., Evangelista, C., Gabrielian, A.E., Gan, W., Ge, W., Gong, F., Gu, Z., Guan, P., Heiman, T.J., Higgins, M.E., Ji, R.R., Ke, Z., Ketchum, K.A., Lai, Z., Lei, Y., Li, Z., Li, J., Liang, Y., Lin, X., Lu, F., Merkulov, G.V., Milshina, N., Moore, H.M., Naik, A.K., Narayan, V.A., Neelam, B., Nusskern, D., Rusch, D.B., Salzberg, S., Shao, W., Shue, B., Sun, J., Wang, Z., Wang, A., Wang, X., Wang, J., Wei, M., Wides, R., Xiao, C., Yan, C., Yao, A., Ye, J., Zhan, M., Zhang, W., Zhang, H., Zhao, Q., Zheng, L., Zhong, F., Zhong, W., Zhu, S., Zhao, S., Gilbert, D., Baumhueter, S., Spier, G., Carter, C., Cravchik, A., Woodage, T., Ali, F., An, H., Awe, A., Baldwin, D., Baden, H., Barnstead, M., Barrow, I., Beeson, K., Busam, D., Carver, A., Center, A., Cheng, M.L., Curry, L., Danaher, S., Davenport, L., Desilets, R., Dietz, S., Dodson, K., Doup, L., Ferriera, S., Garg, N., Gluecksmann, A., Hart, B., Haynes, J., Haynes, C., Heiner, C., Hladun, S., Hostin, D., Houck, J., Howland, T., Ibegwam, C., Johnson, J., Kalush, F., Kline, L., Koduru, S., Love, A., Mann, F., May, D., McCawley, S., McIntosh, T., McMullen, I., Moy, M., Moy, L., Murphy, B., Nelson, K., Pfannkoch, C., Pratts, E., Puri, V., Qureshi, H., Reardon, M., Rodriguez, R., Rogers, Y.H., Romblad, D., Ruhfel, B., Scott, R., Sitter, C., Smallwood, M., Stewart, E., Strong, R., Suh, E., Thomas, R., Tint, N.N., Tse, S., Vech, C., Wang, G., Wetter, J., Williams, S., Williams, M., Windsor, S., Winn-Deen, E., Wolfe, K., Zaveri, J., Zaveri, K., Abril, J.F., Guigo, R., Campbell, M.J., Sjolander, K.V., Karlak, B., Kejariwal, A., Mi, H., Lazareva, B., Hatton, T., Narechania, A., Diemer, K., Muruganujan, A., Guo, N., Sato, S., Bafna, V., Istrail, S., Lippert, R., Schwartz, R., Walenz, B., Yooseph, S., Allen, D., Basu, A., Baxendale, J., Blick, L., Caminha, M., Carnes-Stine, J., Caulk, P., Chiang, Y.H., Coyne, M., Dahlke, C., Mays, A., Dombroski, M., Donnelly, M., Ely, D., Esparham, S., Fosler, C., Gire, H., Glanowski, S., Glasser, K., Glodek, A., Gorokhov, M., Graham, K., Gropman, B., Harris, M., Heil, J., Henderson, S., Hoover, J., Jennings, D., Jordan, C., Jordan, J., Kasha, J., Kagan, L., Kraft, C., Levitsky, A., Lewis, M., Liu, X., Lopez, J., Ma, D., Majoros, W., McDaniel, J., Murphy, S., Newman, M., Nguyen, T., Nguyen, N., Nodell, M., 2001. The sequence of the human genome. *Science* 291, 1304–1351.

Vered, K., Durrant, C., Mott, R., Iraqi, F.A., 2014. Susceptibility to Klebsiella pneumonaie infection in collaborative cross mice is a complex trait controlled by at least three loci acting at different time points. *BMC Genomics* 15, 865.

Vesell, E.S., Page, J.G., 1968. Genetic control of dicumarol levels in man. *J Clin Invest* 47, 2657–2663.

Wang, K., Li, M., Bucan, M., 2007. Pathway-based approaches for analysis of genomewide association studies. *Am J Hum Genet* 81, 1278–1283.

Watkins, P.B., Kaplowitz, N., Slattery, J.T., Colonese, C.R., Colucci, S.V., Stewart, P.W., Harris, S.C., 2006. Aminotransferase elevations in healthy adults receiving 4 grams of acetaminophen daily: a randomized controlled trial. *JAMA* 296, 87-93.

Weinshilboum, R.M., Wang, L., 2006. Pharmacogenetics and pharmacogenomics: development, science, and translation. *Annu Rev Genomics Hum Genet* 7, 223–245.

Weiss, S.T., McLeod, H.L., Flockhart, D.A., Dolan, M.E., Benowitz, N.L., Johnson, J.A., Ratain, M.J., Giacomini, K.M., 2008. Creating and evaluating genetic tests predictive of drug response. *Nat Rev Drug Discov* 7, 568–574.

Welsh, C.E., Miller, D.R., Manly, K.F., Wang, J., McMillan, L., Morahan, G., Mott, R., Iraqi, F.A., Threadgill, D.W., de Villena, F.P., 2012. Status and access to the collaborative cross population. *Mamm Genome* 23, 706–712.

Wiles, M.V., Taft, R.A., 2010. The sophisticated mouse: protecting a precious reagent. *Methods Mol Biol* 602, 23–36.

Yang, H., Bell, T.A., Churchill, G.A., Pardo-Manuel de Villena, F., 2007. On the subspecific origin of the laboratory mouse. *Nat Genet* 39, 1100–1107.

Yang, H., Ding, Y., Hutchins, L.N., Szatkiewicz, J., Bell, T.A., Paigen, B.J., Graber, J.H., de Villena, F.P., Churchill, G.A., 2009. A customized and versatile high-density genotyping array for the mouse. *Nat Methods* 6, 663–666.

Yang, H., Wang, J.R., Didion, J.P., Buus, R.J., Bell, T.A., Welsh, C.E., Bonhomme, F., Yu, A.H., Nachman, M.W., Pialek, J., Tucker, P., Boursot, P., McMillan, L., Churchill, G.A., de Villena, F.P., 2011. Subspecific origin and haplotype diversity in the laboratory mouse. *Nat Genet* 43, 648–655.

Yates, C.R., Krynetski, E.Y., Loennechen, T., Fessing, M.Y., Tai, H.L., Pui, C.H., Relling, M.V., Evans, W.E., 1997. Molecular diagnosis of thiopurine S-methyltransferase deficiency: genetic basis for azathioprine and mercaptopurine intolerance. *Ann Intern Med* 126, 608–614.

PART VI

STEM CELLS IN TOXICOLOGY

21

APPLICATION OF PLURIPOTENT STEM CELLS IN DRUG-INDUCED LIVER INJURY SAFETY ASSESSMENT

CHRISTOPHER S. PRIDGEON, FANG ZHANG, JAMES A. HESLOP, CHARLOTTE M.L. NUGUES, NEIL R. KITTERINGHAM, B. KEVIN PARK AND CHRISTOPHER E.P. GOLDRING

MRC Centre for Drug Safety Science, Department of Molecular and Clinical Pharmacology, The Institute of Translational Medicine, University of Liverpool, Liverpool, UK

21.1 THE LIVER, HEPATOCYTES, AND DRUG-INDUCED LIVER INJURY

The liver is the major hub of xenobiotic metabolism; it responds to a wide variety of xenobiotic compounds and converts them into nontoxic by-products. Hepatocytes make up the majority of cells in the liver (75–80%) and contain many liver-enriched enzymes involved in xenobiotic metabolism, most notably transporters and the cytochrome P450s (CYPs). The liver is strategically positioned as the first-pass organ after absorption from the gastrointestinal tract and therefore receives the highest dose of absorbed xenobiotics. In addition to xenobiotic metabolism the liver is also responsible for secreting serum proteins such as albumin and α1-antitrypsin and producing urea. The liver is also implicated in the regulation of blood glucose; glycogen is stored and released from the liver under the control of insulin and glucagon secreted from the pancreas.

The liver is divided into lobes: each of these lobes is made up of smaller lobules that are roughly hexagonal in cross section and fit together in a lattice not unlike a honeycomb. Blood enters the lobule through the portal triad (the perivenous region); this blood then flows along the lobule via the sinusoid and drains through the central vein (the pericentral region). Along the sinusoid, hepatocytes are tightly packed together where they perform xenobiotic metabolism and other functions during the passage of blood. These hepatocytes exhibit a spectrum of phenotypes along the sinusoid specializing in different areas of xenobiotic metabolism. For example, expression of CYP2E1, responsible for the metabolism of most notably paracetamol and ethanol, is restricted to pericentral regions of the liver (Dicker and Cederbaum, 1991). This explains the localized pericentral damage seen during paracetamol overdose and presents a challenge to models of drug-induced liver injury (DILI), which must be able to detect toxicity to all zones of the liver *in vitro*, where the zonation of the lobule is not present (Fig. 21.1 and Table 21.1).

During the course of xenobiotic metabolism, compounds may be converted into chemically reactive metabolites that can damage cellular macromolecules. When this damage results in perturbations to the cell or in cell death, it may be a key step in the onset of DILI. DILI is a major cause of morbidity and mortality estimated to be responsible for 6.5% of all hospital admissions, 0.15% of which culminate in fatality (Pirmohamed et al., 2004); this also represents a large financial burden on the National Health Service costing an estimated £466 million ($734.3 million USD) per year (Pirmohamed et al., 2004). DILI is an enormous issue for pharmaceutical companies not only due to attrition of drugs during development but also when products must be withdrawn from the market postapproval due to unpredicted safety concerns causing large financial loss.

DILI is divided by type of occurrence into predictable and idiosyncratic. Predictable DILI is the most common form and follows dose–response curves and is typically present in all patients during overdose. The archetypical example of predictable DILI is that of paracetamol hepatotoxicity and is the leading cause of acute liver failure in

Drug Discovery Toxicology: From Target Assessment to Translational Biomarkers, First Edition. Edited by Yvonne Will, J. Eric McDuffie, Andrew J. Olaharski, and Brandon D. Jeffy.

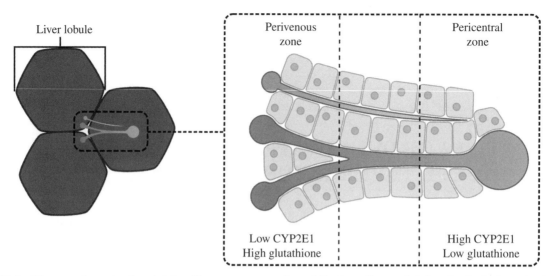

FIGURE 21.1 Diagram of zonation liver lobules. Hepatocytes vary substantially depending on their location in the liver. A prime example is the expression of CYP2E1 in pericentral regions of the liver lobule that can functionally affect cells during paracetamol overdose. An *in vitro* model should either be able to predict toxicity from any zone of the liver or be sufficiently characterized that the zone of the liver best represented is known.

both the United Kingdom and the United States (Fontana, 2008; Lancaster et al., 2015). Normally paracetamol is metabolized without ill effect, though during overdose the toxic metabolite NAPQI accumulates and depletes hepatic glutathione, after which NAPQI is able to bind to cellular macromolecules, causing hepatotoxicity. Although treatments such as *n*-acetylcysteine are available for paracetamol overdose, these are only effective if treatment commences rapidly after overdose, and if overdose progresses to acute liver failure, the only lifesaving treatment is liver transplant.

Idiosyncratic DILI encompasses the unpredictable reactions to a drug that do not necessarily follow a dose–response pattern and do not occur in every member of a population. The causes of idiosyncratic DILI are not fully understood though several mechanisms are thought to be implicated. Interindividual genetic variation and variable immune activation play a role (Hussaini and Farrington, 2014; Metushi et al., 2015) though other mechanisms have also been highlighted such as perturbation to mitochondrial function, disruption of glucose homeostasis, and other more general perturbations sensitizing cells to damage through secondary mechanisms (Ulrich et al., 2001). Idiosyncratic DILI is very rare, affecting ~19 per 100,000 individuals (Hussaini and Farrington, 2014); this makes it extremely difficult to predict during clinical trials due to the low frequency of susceptible variant genotypes in the population. Despite the low incidence, idiosyncratic DILI is a major safety concern, not only because of the danger to the few individuals susceptible to idiosyncratic DILI but also because of the attrition of potentially useful drugs because of unexpected toxicity.

21.2 CURRENT MODELS OF DILI

Plainly DILI is a major cause for concern toward which efforts to detect potential hepatotoxicants before they reach later stages of development should be made. During development, molecules must be tested using *in vitro* models before progressing to animal and human trials. There are several *in vitro* models available including primary human hepatocytes, immortalized cell lines and animal models, and, the main subject of this chapter, pluripotent stem cell-derived hepatocyte-like cells (HLCs).

21.2.1 Primary Human Hepatocytes

Freshly isolated primary human hepatocytes are considered the gold standard for toxicity screening due to their high metabolic relevance. Primary human hepatocytes are isolated from human liver tissue obtained from surgical resections, which is digested enzymatically to release the cells from the extracellular matrix (ECM). Hepatocytes are purified from the nonparenchymal cells (NPCs) and plated out ready for testing. Despite their high metabolic relevance, primary human hepatocytes are not an ideal model; they are mired by several major flaws: dedifferentiation, scarcity, and interdonor variation.

Dedifferentiation begins as soon as the tissue becomes ischemic during resection and is worsened during isolation by induction of ischemia–reperfusion injury when the tissue is reperfused and use of collagenase that is typically contaminated with lipopolysaccharide producing an inflammatory stress response (Wang et al., 1998; Elaut et al., 2006). Dedifferentiation is characterized by the loss of

TABLE 21.1 Comparison of iPSC HLC Differentiation Methods and Their Efficiency

Reference	Method of stem cell differentiation			Differentiation efficiency	Phase I and II enzyme activity		
	Stem cell type	Culture format	Differentiation factors	% ALB +ve HLCs (assay)	Enzyme (assay method)	% hPH comparator	Other comparators
Chen et al. (2012)	hESC (H9), hiPSC (hFb derived)	Monolayer	AA, ITS, HGF, Wnt3A, OSM, DMSO, DEX	—	CYP3A4 (bioluminescence)	100	hiPSC
Cayo et al. (2012)	hiPSC (hFb derived)	Monolayer	AA, BMP4, FGF2, OSM, B27	—	—	—	—
Schwartz et al. (2012)	hiPSC (hFb derived)	Monolayer	AA, BMP4, FGF2, HGF, OSM	80 (ICC)	—	—	—
Takayama et al. (2012)	hESC (H9), hiPSC (hFb derived)	Monolayer	AA, SOX17, HEX, BMP4, FGF4, HNF4α, HGF, OSM, DEX	—	CYP3A4, CYP2C9, CYP1A2 (fluorescence)	100, >10, <1	—
Choi et al. (2013)	hiPSC (from AAT-deficient patients)	Monolayer	B27, AA, FGF4, HGF, OSM, DEX	—	CYP3A4, CYP2D6, CYP2C19, CYP1A2 (bioluminescence)	80, 70, 90, 90	—
Ramasamy et al. (2013)	hESC (H1)	Monolayer and 3D culture in AlgiMatrix plate	AA, DMSO, HGF, OSM	—	CYP3A4 (bioluminescence)	—	HepG2
Gieseck et al. (2014)	hiPSC (hFb derived)	Monolayer, 3D single cell or clump culture in RAFT system	AA, FGF2, BMP4, LY-294002	—	CYP3A4 for 2D day 35 (HPLC-MS), CYP3A4 for 3D day 45 (bioluminescence)	4, 25	—
Jia et al. (2014)	hiPSC (from urine cells of HA patient)	Monolayer, EB formation	AA, FGF4, BMP2, HGF, KGF, OSM, DEX	64 (FACS)	—	—	—
Avior et al. (2015)	hESC (13)	Monolayer	AA, B27, Wnt3A, HGF, DMSO, DEX, OSM, FGF2, LCA, MK4	83 (FACS)	CYP3A4 and CYP1A2, CYP2E1 and CYP2C9 (fluorescence)	30, 8	HepG2, hESC without LCA/MK4
Chien et al. (2015)	hiPSC (from dental pulp stromal cells)	Coculture with MEF, EB formation	BSA, AA, FGF4, BMP2, HGF, KGF, OSM, DEX, B27, miR122	—	—	—	—

AA, activin A; AAT, α1-antitrypsin; ALB, albumin; BMP4, bone morphogenetic protein 4; BSA, bovine serum albumin; CYP, cytochrome P450; DEX, dexamethasone; DMSO, dimethyl sulfoxide; EB, embryoid body; FACS, fluorescence-activated cell sorting; fFb, fetal fibroblast; FGF2/4, fibroblast growth factor 2/4; HA, hemophilia A; hESC, human embryonic stem cell; HEX, haematopoietically expressed homeobox; HGF, hepatocyte growth factor; hiPSC, human induced pluripotent stem cell; HLCs, hepatocyte-like cells; HNF4α, hepatocyte nuclear factor 4α; hPH, human primary hepatocyte; HPLC-MS, high-performance liquid chromatography–mass spectrometry; ICC, immunocytochemistry; ITS, insulin–transferrin–selenium; KGF, keratinocyte growth factor; LCA, lithocholic acid; LY-294002, 2-morpholin-4-yl-8-phenylchromen-4-one; MEF, mouse embryonic fibroblasts; MK4, vitamin K2; miR-122, microRNA-122; OSM, oncostatin M; RAFT, real architecture for 3D systems; SOX17, SRY (sex determining region Y)-box 17; Wnt3A, wingless-type MMTV integration site family, member 3A.

mature phenotype and regression toward a more fetal-like phenotype; xenobiotic metabolism enzymes and transporters are particularly affected showing manyfold reduced expression within hours after isolation (Elaut et al., 2006). The characteristic cuboidal morphology with defined cell junctions is also rapidly lost within a few days in culture. Loss of xenobiotic metabolism capabilities reduces the physiological relevance of the cells and precludes chronic studies, a severe limitation when DILI often occurs after chronic administration of a compound (Hussaini and Farrington, 2014).

Genetic differences between individuals are reflected in the differing expression of xenobiotic metabolism enzymes between donors. Isolated hepatocytes reflect the phenotype of that individual; variations in CYPs and transferases are common and can also be more frequent in certain ethnicities (Fakunle and Loring, 2012). Interdonor variation can also affect other aspects of cell culture, such as amenability to plating. Most importantly in terms of toxicity testing, interdonor variability makes experimental repeats using primary hepatocytes extremely challenging.

Scarcity arises as a result of the dependency on collection of tissue from surgery that is amenable to cell isolation. These surgeries can be sporadic, and institutions that have the necessary links to healthcare to obtain surgical samples are not commonplace, heightening scarcity. Although liver tissue used for hepatocyte isolation is taken from areas of healthy liver, it should be noted that most donor livers have been exposed to cancer or other insults or in some cases rejected for transplantation, which is a consideration that should be made when using primary hepatocytes.

Cryopreservation of hepatocytes allows their storage and transport rather than immediate use after isolation. However there is an associated loss of phenotype with this technique, making cryopreserved hepatocytes less physiologically relevant than primary human hepatocytes (Terry et al., 2006).

21.2.2   Murine Models

Animal models, rodents in particular, are used to predict hepatotoxicity. However, due to differences in metabolism between species, conventional animal models cannot be as accurately predictive of human hepatotoxicity as primary human hepatocytes. For example, rodents are much more effective at detoxifying arsenic than humans (Wang et al., 2006), and animal testing of fialuridine was not predictive of severe hepatotoxicity in humans leading to five deaths during clinical trials (McKenzie et al., 1995). Despite these limitations, the use of conventional animals during safety assessment is commonplace.

Recent technological developments show promise to make animal testing more physiologically relevant to humans. Humanized TK-NOG mice have recently been produced using both primary human hepatocytes and induced pluripotent stem cells (iPSCs) to produce functional human organs that are more physiologically relevant to humans for drug testing (Hasegawa et al., 2011; Takebe et al., 2013). These humanized organs were demonstrated to be functionally human by metabolism of debrisoquine mediated by CYP2D6, which is enriched in humans but lacking in murine liver (Hasegawa et al., 2011). Similar liver repopulation results have been generated in uPA transgenic and fah$^{-/-}$ mice though with less success (Dandri et al., 2001; Bissig et al., 2010).

These models offer metabolically relevant models for drug testing and also offer some advantages over primary human hepatocytes, namely, that compounds can be tested over a longer time scale than is possible with primary human hepatocytes; in addition, if the mice are repopulated using iPSC HLCs derived from multiple donors, the variability between those donors could be recapitulated in the animals, giving better prediction of idiosyncratic DILI. However it should be noted that these animal models are costly and technically complex to produce and do not help to reduce the number of animals used in animal testing. Furthermore, while these animals may have humanized livers, they are not always completely transformed, and the rest of the system remains murine and is therefore not necessarily directly relevant to or predictive of human toxicity.

21.2.3   Cell Lines

Currently, the most commonly used cell models for hepatotoxicity testing are immortalized cell lines derived from hepatic cancers; by far the most common of these is the ubiquitous HepG2 cell line derived from the hepatocarcinoma of a 15-year-old Caucasian male. HepG2 cells are a poor approximation of *in vivo* hepatocytes from which they are derived, expressing vanishingly low abundances of CYPs and poor expression or other hepatic-specific markers (Wilkening et al., 2003; Guo et al., 2011). Without expression of these proteins, HepG2 cells are unable to properly recapitulate the xenobiotic metabolism that occurs *in vivo* and are more suited to the detection of general cellular toxicity. HepG2s however are commonly used in big pharma as the cell model of choice for hepatotoxicity testing due to their ease of culture, availability, and long-established history of use and amenability to high-throughput testing.

Huh7 cells are another similar hepatoma-derived cell line initially obtained from a 57-year-old Japanese male. Huh7 cells show a marginally more hepatocyte-like phenotype than HepG2 cells showing increased expression of CYP1A2 and CYP2D6 when compared (Guo et al., 2011). Additionally, a recent paper has shown that over time in culture, confluent Huh7 cells improve in hepatic phenotype most notably in the expression of CYP3A4 (Sivertsson et al., 2013).

Although there are other immortalized cell lines that may be used for hepatotoxicity testing such as HepaRG cells, in each case there are disadvantages that are unlikely to ever be resolved due to the static nature of such lines, most prominently their lack of physiological relevance compared with the gold standard model, primary human hepatocytes (Guo et al., 2011). A recent study has shown, among several hepatoma cell lines commonly used for hepatotoxicity testing (including HepG2 and Huh-7), gene expression of 251 drug-metabolizing enzymes and transporters correlated poorly to pooled primary human hepatocytes from four donors; Huh-7 cells achieved a correlation of 0.71 and HepG2 cells of only 0.6. Also of note during this study is that CYP2E1 and CYP3A4 were undetected in both Huh-7 and HepG2 cells, while CYP1A2 and CYP2D6 were detected at extremely low levels (Guo et al., 2011).

Other cell lines available for hepatotoxicity modeling include the HepaRG, which is more metabolically relevant than HepG2 cells, but are also more difficult to culture and are supplied terminally differentiated, increasing costs.

21.2.4 Stem Cell Models

Stem cells are a promising method of producing a theoretically inexhaustible supply of hepatocytes from either embryonic or adult cells that can be used for toxicity testing or transplant. Embryonic stem cells (ESCs) are phenotypically limited compared with iPSCs that can be derived from any individual. Given the wide array of phenotypes required to accurately predict idiosyncratic DILI, the wide array of phenotypes offered by iPSCs is a clear advantage. Therefore ESCs will not be the focus of this chapter; instead it will focus on the development and use of iPSCs. iPSCs are derived from cells from adult donors and reverted to a pluripotent state, that is, able to enter any of the three germ layers. The iPSCs are subsequently differentiated into the cell type of interest such as HLCs (iPSC HLCs).

iPSCs were first produced in two landmark papers by the Yamanaka group in Japan (Takahashi and Yamanaka, 2006; Takahashi et al., 2007). These iPSCs were first generated from murine cells and subsequently from human cells using the now ubiquitous OSKM factors (Oct3/4, Sox2, Klf4, and c-Myc) also referred to as the Yamanaka factors. These factors are expressed in ESCs, and their expression in adult cells reverts the cells to a pluripotent state.

In the original papers from the Yamanaka group, transgenes were retrovirally delivered and integrated randomly into the genome. Retroviral transduction is an efficient method of producing reprogrammed cells (compared with other methods of transduction) with 0.1 and 0.01% efficiency in mouse embryonic fibroblasts and human fibroblasts, respectively (González et al., 2011). While the high reprogramming efficiency of retroviral transduction is convenient, random integration of potent oncogenes like c-Myc makes

tumorigenicity a concern in transplanted cells. In addition, random insertion into the genome can cause undesirable effects on expression of other genes. As such, multiple alternative methods for generating iPSCs have emerged. Alternative methods of iPSC generation may be integrative or nonintegrative. Integrative methods insert directly into the genome, while nonintegrative methods are either transiently expressed without genome integration or utilize methods other than transgene expression to obtain pluripotency, for example, direct introduction of proteins or use of small molecules.

21.2.4.1 *Integrative Methods* The success of retroviral delivery as a method to produce iPSCs has led to the investigation of other viruses including lentiviruses, which have produced similar effects (Blelloch et al., 2007; Yu et al., 2007). In addition, transfection of linear DNA has been demonstrated as an alternative method to viruses. Combination of integrative methods with advances in transgenic technology such as the use of tet-inducible and polycistronic vectors and Cre/loxP gene excision has formed the backbone of advances in integrative methods (Hasegawa et al., 2007; Kaji et al., 2009; González et al., 2011). In general terms, use of smaller vectors and fewer transductions (i.e., polycistronic vectors) in combination with tighter control over expression of these transgenes (e.g., tet repression and similar mechanisms) with removal after reprogramming (Cre/LoxP) allows induction of pluripotency with greater finesse.

Close control of the potent oncogenes associated with reprogramming is necessary if the iPSCs are to be used for transplant into humans. However, due to the random integration of transgenes and the potential oncogenic effects, integrative methods are generally considered unsuitable for use in human transplants; even after excision of the transgenes by methods such as Cre/LoxP, a genomic scar remains, which, if integrated in an important gene, could have deleterious or oncogenic effects.

Transposon-based gene delivery is an alternative method to produce iPSCs. Transposons are discrete DNA fragments whose mobility relies on the activity of a transposase. The enzyme, which is encoded by the transposon itself, binds at or near inverted terminal repeats and displaces the transposon from its genomic site. Following excision, the transposase catalyzes the integration of the transposon at a new genomic location (Guo et al., 2011). The inherent properties of transposons to move freely between or within genomic sites make them an attractive tool for gene delivery (Sivertsson et al., 2013). PiggyBac transposons were isolated from *Trichoplusia ni* and were found to be active in many species, including humans (Blelloch et al., 2007). They were used to incorporate reprogramming factors and subsequently generate iPSCs (Yu et al., 2007). The piggyBac transposon system consists of the piggyBac transposase encoded by plasmid and a donor plasmid containing the DNA sequence of interest flanked by terminal repeats recognized by the transposase (Hasegawa et al., 2007). This system offers several advantages over other integrative reprogramming methods. It is less immunogenic than viruses, and, unlike the Cre-excisable linear transgenes, its removal is seamless (González et al., 2011). However, as for all other integrative systems, it requires an excision step following cell reprogramming. Furthermore, the transposase not only excises the transposon from its genomic location but also mediates its relocation at another site. The latter event can therefore complicate the excision process (Yu et al., 2007).

21.2.4.2 *Nonintegrative Methods*

The alternative to integrative methods are nonintegrative methods where no modifications to the genomic DNA are made. These methods hold promise as the expression of transgenes is transient and can be cleared from the cell after reprogramming with greatly reduced oncogenic risk due to aberrant expression of transgenes. Multiple methods have been attempted to produce integration-free iPSCs including integration-deficient viruses (Stadtfeld et al., 2008; Zhou and Freed, 2009), episomal delivery (Okita et al., 2008; Gonzalez et al., 2009; Jia et al., 2010), RNA delivery (Warren et al., 2010), protein delivery (Zhou et al., 2009), and small-molecule mimics (Kim et al., 2009). Generally, nonintegrative methods have much lower efficiency than their integrative counterparts and can take much longer to achieve pluripotency (González et al., 2011).

The first integration-free iPSCs were produced using a replication-defective adenoviral vector, where OSKM factors were delivered by separate vectors (Stadtfeld et al., 2008). F-deficient Sendai virus has also been used to deliver transgenes to a wide array of host cells where it replicates as negative-sense ssRNA, the virus is "diluted out" after multiple passages, and the loss of the virus can be confirmed by PCR (Fusaki et al., 2009; Nishimura et al., 2011).

Episomal delivery uses transfection of several plasmids or a single polycistronic plasmid expressing OSKM factors. These plasmids are simpler to produce than viral particles and are able to produce integration-free iPSCs. However, integration is a concern and not all iPSCs generated by this method will be integration-free. It has been shown that as many as 92% of iPSCs generated by episomal transfection had integrated plasmid DNA (Gonzalez et al., 2009) requiring laborious screening to detect integration-free clones. Over time in culture, nonreplicating plasmids are diluted out and lost. In order to eliminate the requirement for serial transfections, self-replicating plasmids have been produced, which can be removed through their dependence on drug selection (Yu et al., 2007). The use of self-replicating plasmids is highly inefficient, presumably due to the large plasmid size, generating 3–6 colonies per million cells transfected (González et al., 2011).

Direct delivery of synthetic mRNA encoding OSKM and Lin28 (a translational enhancer that binds let-7, thereby regulating self-renewal of stem cells (Shyh-Chang and Daley, 2013; The UniProt Consortium, 2015)) has been shown to produce highly efficient transfections in neonatal fibroblasts, with ~2% of cells reprogrammed (Warren et al., 2010). Extensive modifications of the synthetic mRNA in order to avoid targeting from the innate immune response of the target cells are required, including phosphatase treatment, substitution of cytosine to 5-methylcytosine and uridine to pseudouridine, and inclusion of B18R in the culture medium (Warren et al., 2010). The high gene doses involved in this technique are cause for concern with regard to genomic instability arising from high expression of c-Myc (González et al., 2011). In a similar vein, delivery of OSKM proteins fused to proteins mediating their transduction has been shown to produce iPSCs albeit with extremely slow kinetics and at low efficiencies (Zhou et al., 2009); in addition the reproducible production and purification of sufficient quantities of protein are challenging, somewhat hamstringing this technique.

Reprogramming has recently been demonstrated in murine cells using a cocktail of small molecules (Hou et al., 2013). This approach shows great promise in that genetic manipulation of the cells is entirely avoided with no risky expression of potent oncogenes. However, the efficiency of reprogramming is low and thus far has not been replicated in human cells.

21.3 USES OF iPSC HLCs

iPSC HLCs offer an alternative to immortalized cell lines due to the ability for unlimited expansion. While it is true that iPSC HLCs currently exhibit a phenotype closer to fetal than adult liver (Baxter et al., 2015), they also show potential for increasing physiological relevance as differentiation protocols become more advanced, a trait that cannot be matched using established hepatoma-derived cell lines. It has been show that incorporation of iPSC HLC-derived liver buds into mice leads to a maturation of those cells (Takebe et al., 2013); this indicated that iPSC HLCs are capable of expressing a more mature phenotype that *in vitro* methods are currently incapable of producing. Furthermore, iPSC HLCs may be produced from a wide array of donors while retaining the phenotype of that donor (Takayama et al., 2014); in this manner iPSC HLCs can model the effects of drug interactions in many phenotypes, reducing the likelihood of idiosyncratic ADRs, particularly if donors are intelligently screened and selected to give a broad panel of the phenotypes present in the human population.

The unlimited expansion potential of iPSC HLCs makes them suitable replacements for immortalized cell lines in high-throughput screening. Furthermore, the immature phenotype of iPSC HLCs with sufficient characterization to determine exactly what stage of development is most closely represented may be exploited in embryotoxicity testing. Indeed, murine models (using 3T3 fibroblasts and ESCs) are currently used to test embryotoxicity of all new compounds (Höfer et al., 2004; Kuske et al., 2012), and use of a human model would likely increase physiological relevance and improve predictivity of human toxicity.

In addition to showing clear advantages over immortalized cell lines, iPSC HLCs also show advantages over human hepatocytes. Despite being the gold standard, primary hepatocytes are mired by several issues discussed earlier in this chapter. Conversely, iPSC HLCs can be stably maintained in culture for extended periods of time allowing for repeat experimentation on near identical cells, something that is not possible with primary human hepatocytes, cryopreserved or otherwise. Furthermore, primary hepatocytes are subject to rapid dedifferentiation, which is not the case for iPSC HLCs, allowing for study over longer periods of time and possibly for detection of chronic toxicity.

In order to improve iPSC HLCs, focus should be given to the improvement of the phenotype of these cells with comparison to primary human hepatocytes (Fig. 21.2). This may be broken down into pharmacological relevance, that is,

the expression of xenobiotic metabolism enzymes and transporters; physiological relevance, that is, the production of proteins typical to hepatocytes such as albumin and alpha-fetoprotein; and toxicological relevance, that is, behavior of the cell when stressed including cell death when appropriate.

In addition, incorporation of NPCs into these models including immune cells that are known to play a role in idiosyncratic ADRs will help to produce more physiologically relevant models of hepatotoxicity, as it is important to appreciate that not all hepatotoxicity arises from toxicity to hepatocytes.

21.4 CHALLENGES OF USING IPSCs AND NEW DIRECTIONS FOR IMPROVEMENT

iPSCs resolve the controversial issue of using embryonic tissue and allow for genetic and phenotypic diversity in pharmacological and toxicological evaluations, therefore showing potential for high-throughput screening of drug candidates, as well as representing the genetic variability in different populations and in diseased or healthy cohorts. However, the application of iPSCs in drug screening is restricted by several problems: reprogramming efficiency is

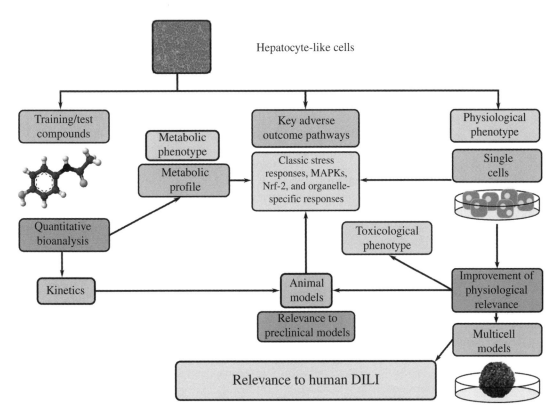

FIGURE 21.2 Roadmap toward producing stem cell-derived models to improve mechanistic understanding and prediction of human DILI. Physiological, pharmacological, and toxicological characterization of iPSC HLCs is necessary in order to fully exploit the strengths of these cells. This will include the use of toxicity/stress reporters, and a panel of test compounds, thereby defining what toxicological purpose each line is fit for.

often low, and there are concerns regarding the genomic authenticity of the iPSCs when exogenous sequences were introduced by viral transfection as the typical approach for cellular reprogramming (Kia et al., 2013). Oncogenes may be switched on during reprogramming. Alternative methods for iPSC generation have been discussed earlier in this chapter, and a reprogramming approach that permits high-efficiency transformation without compromising genomic integrity has yet to be found.

There is difficulty in controlling differentiation efficiencies, hence inconsistency in gene profile and functionality of HLCs between different studies, or different batches of experiments using the same iPSC line. The heterogeneous cell population generated during differentiation is particularly problematic for hepatotoxicity testing as typical markers such as ATP and LDH are not hepatocyte specific; therefore it does not always reflect the genuine response of HLCs to testing compounds. Discovery of hepatocyte-specific markers may help resolve this problem. For example, detection of the liver-enriched biomarker microRNA-122 (miR-122) in iPSC HLCs after exposed to toxic compounds was able to show cytotoxic sensitivity similar to ATP and LDH but only in differentiated HLCs rather than iPSCs (Fontana, 2008).

Thus far, differentiation of iPSCs to a mature HLC phenotype that is comparable to primary hepatocytes has proven to be challenging. HLCs generated in different studies appeared to be more fetal-like. Small molecules identified by high-throughput screening demonstrated their potential for HLC maturation (increased albumin secretion, CYP activity, and gene expression of ABC transporters), although the mechanisms are unclear (Bohme et al., 1994).

21.4.1 Complex Culture Systems

Advances in complex culture systems may improve the application of iPSC in drug safety assessment. Many studies compared conventional 2D monolayer culture with complex 3D and coculture systems for phenotype and functionality. Compared to 2D, 3D or coculture systems generally produced more physiologically relevant characteristics: better inflammatory pathways (more *in vivo*-like cytokine profiles in response to toxins) (Martin et al., 2004), better metabolic and pharmacological capacity (increased albumin secretion and phase I and II enzyme activity) (Marx-Stoelting et al., 2009; Musunuru, 2013), and formation of more *in vivo*-like cell morphology, polarization, and organization (Seiler and Spielmann, 2011; Tomizawa et al., 2013). These characteristics were achieved by re-creating physiological microenvironment (heterogeneous cell types, spatial architecture, mechanical and chemical composition of ECM) that improves cell–cell and cell–ECM interactions and restores molecular and stromal signaling (e.g., cytokines, integrin, adhesion molecules) (Astashkina and Grainger, 2014).

The interaction of hepatocytes with other NPCs such as Kupffer cells, endothelial cells, and stellate cells is crucial to maintaining the pharmacological responses of the liver. These NPCs activate inflammatory pathways and the release of growth factors and reactive oxygen species (ROS) in response to toxins; therefore they are very important for inducing and regulating DILI. Efforts have been made to develop complex culture systems, with the aim of maintaining hepatic cells for longer *in vitro* without significant depletion of the normal hepatocyte functions (Roth and Singer, 2014). Examples of these culture approaches include micropatterned coculture (MPCC) (Ulvestad et al., 2013), 3D spheroids (Wang et al., 1998; Ulrich et al., 2001; Daly et al., 2009; Yusa et al., 2011), bioreactors (Feng et al., 2009; Nishimura et al., 2011), and microfluidic dynamic perfusion system (Lei et al., 2012; Li et al., 2015).

21.4.2 Coculture

Coculture models combine different cell types in the same system, initially in 2D transwells, but recently such models have also become available in 3D defined structures (Ding et al., 2013; Shyh-Chang and Daley, 2013). Primary human hepatocytes cultured with NPCs such as Kupffer or stellate cells showed improved function and prolonged culture time. MPCCs were developed with microtechnology as a stable and functional hepatic model for drug screening. Primary human hepatocytes were seeded to micropatterned ECM in multiwell elastomeric device and were surrounded by fibroblast cells. This culture system maintained hepatic phenotype and functions (phase I and II enzyme activity, gene profile, canalicular transport, and albumin secretion) for up to 6 weeks (Khetani and Bhatia, 2008). A recent study showed better prediction of hepatotoxicity by iPSCs in MPCC than in conventional culture, with sensitivity and specificity comparable to primary human hepatocytes (Ware et al., 2015). iPSC HLCs in MPCC were treated with 47 drugs (37 known hepatotoxins and 10 Non-liver toxins). TC_{50} values (concentration that decreased the measured endpoints by 50%) for assays of albumin, urea, and ATP were used for binary decisions of toxicity. iMPCCs showed a 65% sensitivity for hepatotoxins (24 out of 37) compared with PHH-MPCCs at 70% and conventional confluent culture at 35%. All three models correctly identified all 10 nonliver toxins. Nagamoto showed improved maturation of iPSC-derived HLCs when cocultured with Swiss 3T3 cells using the cell sheet engineering technology (upregulation of CYP450 gene expression and albumin secretion). This study highlighted the importance of type I collagen in hepatic maturation (Ding et al., 2013).

21.4.3 3D Culture

3D culture allows formation of microtissue when hepatocytes alone or in combination with NPCs are seeded in specialized matrices. Examples of this type of culture system include hanging drop plates (InSphero, Schlieren, Switzerland), alginate scaffolds, or porous nylon scaffolds, which direct the assembly of spheroids. Ramasamy et al. (2013) used sequential 2D and 3D culture for differentiation of hESC-derived HLCs (Kim et al., 2009). ESCs were differentiated in monolayer to definitive endoderm cells first and then inoculated to 3D AlgiMatrix scaffold and treated with Rho-associated kinase inhibitor, which formed spheroids of various sizes depending on the seeding density. These spheroids showed better markers of differentiation and maturation such as apolipoprotein F, tryptophan dioxygenase, and CYP3A4 in gene profile compared with 2D differentiation. Medium-sized spheroids formed from a lower seeding density showed higher CYP3A4 activity. Gieseck et al. reported an improvement in cell maturation by transferring fully differentiated iPSC HLCs in 2D culture to 3D *Real Architecture for 3D Tissue* (RAFT, TAP Biosystems, Royston, United Kingdom) matrix as single cells or small epithelial clumps (Gieseck et al., 2014). RAFT creates cell-seeded hydrogels by mixing cells with neutralized collagen in a 96-well format. Compared to 2D culture, cells in 3D clumps showed a more hepatocyte-like gene profile of hepatic markers, and CYP3A4 activity was maintained for 75 days. Polarization and bile canaliculi formation were also induced in 3D clump culture, suggesting the importance of maintaining cell–cell junctions. Zhang et al. reported a method for generating more hepatic-like iPSC HLC spheroids in 3D micropatterned plate compared with 2D culture; this may be a powerful tool for high-throughput screening (Warren et al., 2010).

21.4.4 Perfusion Bioreactors

Perfusion bioreactors are dynamic culture systems that aim to restore cellular mechanosensitivity in physiological environment. Cells are often seeded to bioscaffolds and continuously perfused with culture medium driven by various types of pumps (e.g., pneumatic, peristaltic, gear, or propeller) at a flow rate that mimics hepatic circulation. This allows continuous cell exposure to nutrients and oxygen, as well as *in vivo*-like shear stress, which are important conditions for cell expansion and maturation (Feng et al., 2009). Microfluidic chips enable the modeling of *in vivo* flow and shear stress *in vitro*. Cells are seeded into biocoated scaffolds in 12 wells and perfused via pneumatic micropump-controlled flow, mimicking the architecture of hepatic sinusoids. The LiverChip system has been claimed to enhance maturation of iPSC HLCs and maintain functional hepatocytes with improved viability and longevity and predict and distinguish hepatotoxicity of known hepatotoxic compounds from less toxic analogues (Krahenbuhl et al., 1994; Lei et al., 2012; Kamalian et al., 2015).

Although dynamic perfusion culture has not been widely applied in studies of iPSC differentiation, it has shown potential in other hepatic models. Clark et al. (2014) developed a model to identify new therapeutic strategies for metastasis by incorporating the hepatic niche (fresh human hepatocytes and NPCs) and breast cancer cells in a microfluidic cell culture system incorporating flow (Li et al., 2015). Darnell et al. (2011) cultured HepaRG cells in a 3D multicompartment capillary membrane bioreactor under dynamic conditions (Li et al., 2015). These cells showed stable P450 metabolic capacity over several weeks, as well as polarity of transporter expression and bile canalicular formation.

In summary, the use of complex culture systems in iPSC HLC differentiation has demonstrated notable advantages in cell maturity and longevity (e.g. ,albumin and urea secretion maintained for months, gene profile shifting from fetal to mature hepatocytes, sustained and inducible CYP activity over long term). However, more efforts are required to develop methodology to produce more stable and homogeneous iPSC HLCs, as well as to decipher the mechanisms and key signaling pathways involved in hepatocyte differentiation and maturation. Furthermore, the development of reliable, practical, and standardized culture systems that are amenable to high-throughput screening should also be considered. Nonetheless, the application of iPSC HLCs in complex culture systems shows great potential for DILI safety assessment, particularly in high-throughput screening and chronic toxicity assessment.

Other novel approaches may have an impact on application of iPSC HLCs. For example, Shan et al. (2013) identified small molecules of functional and proliferative hits via high-throughput screening using human hepatocytes cocultured with fibroblasts (Bohme et al., 1994). Two of these hits, FH1 and FPH1, showed the ability of promoting differentiation of iPSCs toward a hepatic lineage and maturation of iPSC HLCs toward adult hepatocytes (gene profile level of phase I–III enzyme expression comparable to adult hepatocytes, increased albumin secretion, and CYP activity and decreased alpha-fetoprotein expression in treated iPSC HLCs). A recent study delivered miR-122, a liver-enriched microRNA, complexed with polyurethane-graft-short-branch polyethylenimine copolymer (PU-PEI) (in nanostructured amphiphatic carboxymethyl-hexanoyl chitosan), to iPSCs and enhanced the maturation of iPSC HLCs. The miR122-iPSC HLCs expressed higher levels of gene and protein markers for hepatic function compared to miR-scrambled iPSC HLCs (Takebe et al., 2013).

21.5 ALTERNATE USES OF HLCs IN TOXICITY ASSESSMENT

It is clear that iPSC HLCs currently lack many of the key qualities required to be a one-for-all model of hepatotoxicity; however, that does not preclude the use of iPSC HLCs in modeling other less metabolically dependent endpoints.

In short, it is important to find what purpose iPSC HLCs are currently fit for.

One such use is as a model of mitochondrial perturbation. Recent work has shown that HepG2s may provide a model of mitochondrial perturbation that is more amenable to mechanistic studies than human hepatocytes (Kamalian et al., 2015). However, HepG2s are genotypically abnormal and lack physiological translatability; therefore, iPSC HLCs may provide a more physiologically relevant model of this perturbation. Research has shown that iPSC HLCs, like hepatocytes, but unlike pluripotent stem cells, can survive in galactose/ornithine-based culture that forces cells to use oxidative phosphorylation in the absence of glucose (Tomizawa et al., 2013), suggestive of a bioenergetic switch during differentiation to a situation that is more similar to hepatocytes. The use of human iPSC HLCs derived from patients with Alper's syndrome, a disease that increases susceptibility to valproic acid-derived hepatotoxicity, was successfully used to model this increased toxicity compared to control iPSC HLCs (Li et al., 2015). Furthermore, the mechanism of toxicity, more frequent bursts of superoxide generation, was delineated using these cells; therefore, in cases where toxicity is not dependent on metabolism, iPSC HLCs are valuable models of toxicity.

Further, DILI-induced cholestasis may also be modeled. Many drugs associated with cholestasis, such as cyclosporin A, rifampicin, and glibenclamide, have been found to be a competitive inhibitor of the bile salt export pump (BSEP) (Bohme et al., 1994; Byrne et al., 2002; Leuthold et al., 2009), preventing the export of bile salts and resulting in their cytotoxic accumulation and consequently DILI (Krahenbuhl et al., 1994). iPSC HLCs have been shown to have some degree of BSEP activity (Ulvestad et al., 2013) and may therefore be able to triage compounds with cholestatic liability through inhibition of this mechanism.

The advent of gene modulation technology, in the form of zinc fingers, TALENs, and CRISPR/Cas9, also provides the capacity to achieve mechanistic studies of hepatotoxicity. Yusa et al. reported that such work was feasible in the context of gene therapy, demonstrating that the defective gene in the alpha-1-antitrypsin deficiency could be corrected using zinc-finger nucleases in iPSCs and be functional in the resulting HLCs (Yusa et al., 2011). CRISPR/Cas9 knockout technology has also been successfully applied to pluripotent stem cells (Ding et al., 2013), and the differentiation of these cells to HLCs may provide human in vitro disease models or equivalents of ex vivo cells from knockout mice that are commonly used to understand the mechanisms that underlie hepatotoxicity of a given compound (Musunuru, 2013).

The "capturing" of the patient genotype is a fundamental advantage of iPSC HLCs over more traditional and hESC-derived models of hepatotoxicity. DILI is often idiosyncratic in nature and can occur in only very small numbers of patients, in the case of abacavir and flucloxacillin, which are immune mediated and associated with specific HLA types (Martin et al., 2004; Daly et al., 2009); iPSCs allow for the selection of patients with risk-associated HLA types. The testing of these HLA-typed HLCs with immune cells with the same HLA type may provide a truly unique model of DILI. Such investigation would currently be dependent on the toxicological profile of the drug, that is, does it require metabolism before HLA presentation? Nevertheless, HLA-associated toxicity is currently only feasible in iPSC-based models, and the development of a human in vitro system able to screen for immune-mediated idiosyncratic DILI would represent a *paradigm shift* in hepatotoxicity modeling.

Furthermore, once metabolically competent HLCs are established, simpler iPSC HLC panel-based screening representing the major genotypic variations present in the population (e.g., CYP2D6 null genotype (Takayama et al., 2014)) would enhance current compound screening models utilizing the restricted genotypic range of animal strains and cell lines.

iPSCs and iPSC HLCs are an exciting model for the prediction of toxicity. Although current iPSC HLCs are not sufficiently mature to fully replace other models of hepatotoxicity, protocols for the derivation of mature cells are constantly improving and in coming years will likely represent a sufficiently mature cell type. In the meantime, there are uses for iPSC HLCs that can exploit their immature phenotype such as embryotoxicity testing and testing of compounds where metabolic activation is not essential. iPSCs represent a unique and dynamic model for the prediction of toxicity that hold great potential for future improvement, particularly relative to other more established models such as cell lines and simple culture of primary cells.

REFERENCES

Astashkina, A. & Grainger, D. W. 2014. Critical analysis of 3-D organoid in vitro cell culture models for high-throughput drug candidate toxicity assessments. *Advanced Drug Delivery Reviews*, 69–0, 1–18.

Avior, Y., Levy, G., Zimerman, M., Kitsberg, D., Schwartz, R., Sadeh, R., Moussaieff, A., Cohen, M., Itskovitz-Eldor, J., & Nahmias, Y. 2015. Microbial-derived lithocholic acid and vitamin K drive the metabolic maturation of pluripotent stem cells-derived and fetal hepatocytes. *Hepatology*, 62, 265–78.

Baxter, M., Withey, S., Harrison, S., Segeritz, C.-P., Zhang, F., Atkinson-Dell, R., Rowe, C., Gerrard, D. T., Sison-Young, R., Jenkins, R., Henry, J., Berry, A. A., Mohamet, L., Best, M., Fenwick, S. W., Malik, H., Kitteringham, N. R., Goldring, C. E., Piper Hanley, K., Vallier, L., & Hanley, N. A. 2015. Phenotypic and functional analyses show stem cell-derived hepatocyte-like cells better mimic fetal rather than adult hepatocytes. *Journal of Hepatology*, 62, 581–9.

Bissig, K.-D., Wieland, S. F., Tran, P., Isogawa, M., Le, T. T., Chisari, F. V., & Verma, I. M. 2010. Human liver chimeric mice provide a model for hepatitis B and C virus infection and treatment. *The Journal of Clinical Investigation*, 120, 924–30.

Blelloch, R., Venere, M., Yen, J., & Ramalho-Santos, M. 2007. Generation of induced pluripotent stem cells in the absence of drug selection. *Cell Stem Cell*, 1, 245–7.

Bohme, M., Muller, M., Leier, I., Jedlitschky, G., & Keppler, D. 1994. Cholestasis caused by inhibition of the adenosine triphosphate-dependent bile salt transport in rat liver. *Gastroenterology*, 107, 255–65.

Byrne, J. A., Strautnieks, S. S., Mieli-Vergani, G., Higgins, C. F., Linton, K. J., & Thompson, R. J. 2002. The human bile salt export pump: characterization of substrate specificity and identification of inhibitors. *Gastroenterology*, 123, 1649–58.

Cayo, M. A., Cai, J., Delaforest, A., Noto, F. K., Nagaoka, M., Clark, B. S., Collery, R. F., Si-Tayeb, K., & Duncan, S. A. 2012. JD induced pluripotent stem cell-derived hepatocytes faithfully recapitulate the pathophysiology of familial hypercholesterolemia. *Hepatology*, 56, 2163–71.

Chen, Y. F., Tseng, C. Y., Wang, H. W., Kuo, H. C., Yang, V. W., & Lee, O. K. 2012. Rapid generation of mature hepatocyte-like cells from human induced pluripotent stem cells by an efficient three-step protocol. *Hepatology*, 55, 1193–203.

Chien, Y., Chang, Y. L., Li, H. Y., Larsson, M., Wu, W. W., Chien, C. S., Wang, C. Y., Chu, P. Y., Chen, K. H., Lo, W. L., Chiou, S. H., Lan, Y. T., Huo, T. I., Lee, S. D., & Huang, P. I., 2015. Synergistic effects of carboxymethyl-hexanoyl chitosan, cationic polyurethane-short branch PEI in miR122 gene delivery: accelerated differentiation of iPSCs into mature hepatocyte-like cells and improved stem cell therapy in a hepatic failure model. *Acta Biomater.* 13, 228–44. doi:10.1016/j.actbio.2014.11.018.

Choi, S. M., Kim, Y., Shim, J. S., Park, J. T., Wang, R. H., Leach, S. D., Liu, J. O., Deng, C., Ye, Z., & Jang, Y. Y. 2013. Efficient drug screening and gene correction for treating liver disease using patient-specific stem cells. *Hepatology*, 57, 2458–68.

Clark, A.M., Wheeler, S. E., Taylor, D. P., Pillai, V. C., Young, C. L., Prantil-Baun, R., Nguyen, T., Stolz, D. B., Borenstein, J. T., Lauffenburger, D. A., Venkataramanan, R., Griffith, L. G., & Wells, A. 2014. A microphysiological system model of therapy for liver micrometastases. *Experimental Biology and Medicine*, 239(9), 1170–9.

Daly, A. K., Donaldson, P. T., Bhatnagar, P., Shen, Y., Pe'er, I., Floratos, A., Daly, M. J., Goldstein, D. B., John, S., Nelson, M. R., Graham, J., Park, B. K., Dillon, J. F., Bernal, W., Cordell, H. J., Pirmohamed, M., Aithal, G. P., & Day, C. P. 2009. HLA-B*5701 genotype is a major determinant of drug-induced liver injury due to flucloxacillin. *Nature Genetics*, 41, 816–9.

Dandri, M., Burda, M. R., Torok, E., Pollok, J. M., Iwanska, A., Sommer, G., Rogiers, X., Rogler, C. E., Gupta, S., Will, H., Greten, H., & Petersen, J. 2001. Repopulation of mouse liver with human hepatocytes and in vivo infection with hepatitis B virus. *Hepatology*, 33, 981–8.

Darnell, M., Schreiter, T., Zeilinger, K., Urbaniak, T., Söderdahl, T., Rossberg, I., Dillnér, B., Berg, A.L., Gerlach, J.C., & Andersson, T. B. 2011. Cytochrome P450-dependent metabolism in HepaRG cells cultured in a dynamic three-dimensional bioreactor. *Drug Metabolism and Disposition*, 39(7), 1131–8.

Dicker, E. & Cederbaum, A. I. 1991. Increased oxidation of dimethylnitrosamine in pericentral microsomes after pyrazole induction of cytochrome P-4502E1. *Alcoholism, Clinical and Experimental Research*, 15, 1072–6.

Ding, Q., Regan, S. N., Xia, Y., Oostrom, L. A., Cowan, C. A., & Musunuru, K. 2013. Enhanced efficiency of human pluripotent stem cell genome editing through replacing TALENs with CRISPRs. *Cell Stem Cell*, 12, 393–4.

Elaut, G., Henkens, T., Papeleu, P., Snykers, S., Vinken, M., Vanhaecke, T., & Rogiers, V. 2006. Molecular mechanisms underlying the dedifferentiation process of isolated hepatocytes and their cultures. *Current Drug Metabolism*, 7, 629–60.

Fakunle, E. S. & Loring, J. F. 2012. Ethnically diverse pluripotent stem cells for drug development. *Trends in Molecular Medicine*, 18, 709–16.

Feng, B., Ng, J. H., Heng, J. C. & Ng, H. H. 2009. Molecules that promote or enhance reprogramming of somatic cells to induced pluripotent stem cells. *Cell Stem Cell*, 4, 301–12.

Fontana, R. J. 2008. Acute liver failure including acetaminophen overdose. *The Medical Clinics of North America*, 92, 761–94.

Fusaki, N., Ban, H., Nishiyama, A., Saeki, K., & Hasegawa, M. 2009. Efficient induction of transgene-free human pluripotent stem cells using a vector based on Sendai virus, an RNA virus that does not integrate into the host genome. *Proceedings of the Japan Academy, Series B Physical and Biological Sciences*, 85, 348–62.

Gieseck, R. L., 3rd, Hannan, N. R., Bort, R., Hanley, N. A., Drake, R. A., Cameron, G. W., Wynn, T. A., & Vallier, L. 2014. Maturation of induced pluripotent stem cell derived hepatocytes by 3D-culture. *PLoS One*, 9, e86372.

Gonzalez, F., Barragan Monasterio, M., Tiscornia, G., Montserrat Pulido, N., Vassena, R., Batlle Morera, L., Rodriguez Piza, I., & Belmonte, J. C. I. 2009. Generation of mouse-induced pluripotent stem cells by transient expression of a single nonviral polycistronic vector. *Proceedings of the National Academy of Sciences of the United States of America*, 106, 8918–22.

González, F., Boué, S., & Belmonte, J. C. I. 2011. Methods for making induced pluripotent stem cells: reprogramming à la carte. *Nature Reviews Genetics*, 12, 231–42.

Guo, L., Dial, S., Shi, L., Branham, W., Liu, J., Fang, J.-L., Green, B., Deng, H., Kaput, J., & Ning, B. 2011. Similarities and differences in the expression of drug metabolizing enzymes between human hepatic cell lines and primary Human Hepatocytes. *Drug Metabolism and Disposition*, 39(3), 528–38.

Hasegawa, K., Cowan, A. B., Nakatsuji, N., & Suemori, H. 2007. Efficient multicistronic expression of a transgene in human embryonic stem cells. *Stem Cells*, 25, 1707–12.

Hasegawa, M., Kawai, K., Mitsui, T., Taniguchi, K., Monnai, M., Wakui, M., Ito, M., Suematsu, M., Peltz, G., Nakamura, M., & Suemizu, H. 2011. The reconstituted "humanized liver" in TK-NOG mice is mature and functional. *Biochemical and Biophysical Research Communications*, 405, 405–10.

Höfer, T., Gerner, I., Gundert-Remy, U., Liebsch, M., Schulte, A., Spielmann, H., Vogel, R., & Wettig, K. 2004. Animal testing and alternative approaches for the human health risk assessment under the proposed new European chemicals regulation. *Archives of Toxicology*, 78, 549–64.

Hou, P., Li, Y., Zhang, X., Liu, C., Guan, J., Li, H., Zhao, T., Ye, J., Yang, W., Liu, K., Ge, J., Xu, J., Zhang, Q., Zhao, Y., & Deng, H. 2013. Pluripotent stem cells induced from mouse somatic cells by small-molecule compounds. *Science*, 341, 651–4.

Hussaini, S. H. & Farrington, E. A. 2014. Idiosyncratic drug-induced liver injury: an update on the 2007 overview. *Expert Opinion on Drug Safety*, 13, 67–81.

Jia, F., Wilson, K. D., Sun, N., Gupta, D. M., Huang, M., Li, Z., Panetta, N. J., Chen, Z. Y., Robbins, R. C., Kay, M. A., Longaker, M. T., & Wu, J. C. 2010. A nonviral minicircle vector for deriving human iPS cells. *Nature Methods*, 7, 197–9.

Jia, B., Chen, S., Zhao, Z., Liu, P., Cai, J., Qin, D., Du, J., Wu, C., Chen, Q., Cai, X., Zhang, H., Yu, Y., Pei, D., Zhong, M., & Pan, G. 2014. Modeling of hemophilia A using patient-specific induced pluripotent stem cells derived from urine cells. *Life Sciences*, 108, 22–9.

Kaji, K., Norrby, K., Paca, A., Mileikovsky, M., Mohseni, P., & Woltjen, K. 2009. Virus-free induction of pluripotency and subsequent excision of reprogramming factors. *Nature*, 458, 771–5.

Kamalian, L., Chadwick, A. E., Bayliss, M., French, N. S., Monshouwer, M., Snoeys, J., & Park, B. K. 2015. The utility of HepG2 cells to identify direct mitochondrial dysfunction in the absence of cell death. *Toxicology in Vitro*, 29, 732–40.

Khetani, S. R. & Bhatia, S. N. 2008. Microscale culture of human liver cells for drug development. *Nature Biotechnology*, 26, 120–6.

Kia, R., Sison, R. L., Heslop, J., Kitteringham, N. R., Hanley, N., Mills, J. S., Park, B. K., & Goldring, C. E. 2013. Stem cell-derived hepatocytes as a predictive model for drug-induced liver injury: are we there yet? *British Journal of Clinical Pharmacology*, 75, 885–96.

Kim, D., Kim, C. H., Moon, J. I., Chung, Y. G., Chang, M. Y., Han, B. S., Ko, S., Yang, E., Cha, K. Y., Lanza, R., & Kim, K. S. 2009. Generation of human induced pluripotent stem cells by direct delivery of reprogramming proteins. *Cell Stem Cell*, 4, 472–6.

Krahenbuhl, S., Talos, C., Fischer, S., & Reichen, J. 1994. Toxicity of bile acids on the electron transport chain of isolated rat liver mitochondria. *Hepatology*, 19, 471–9.

Kuske, B., Pulyanina, P. Y., & Zur Nieden, N. I. 2012. Embryonic stem cell test: stem cell use in predicting developmental cardiotoxicity and osteotoxicity. *Methods in Molecular Biology*, 889, 147–79.

Lancaster, E., Hiatt, J., & Zarrinpar, A. 2015. Acetaminophen hepatotoxicity: an updated review. *Archives of Toxicology*, 89, 193–9.

Lei, F., Haque, R., Xiong, X., & Song, J. 2012. Directed differentiation of induced pluripotent stem cells towards T lymphocytes. *Journal of Visualized Experiments*, (63) e3986.

Leuthold, S., Hagenbuch, B., Mohebbi, N., Wagner, C. A., Meier, P. J., & Stieger, B. 2009. Mechanisms of pH-gradient driven transport mediated by organic anion polypeptide transporters. *American Journal of Physiology - Cellular Physiology*, 296, C570–C582.

Li, S., Guo, J., Ying, Z., Chen, S., Yang, L., Chen, K., Long, Q., Qin, D., Pei, D., & Liu, X. 2015. Valproic acid-induced hepatotoxicity in alpers syndrome is associated with mitochondrial

permeability transition pore opening-dependent apoptotic sensitivity in an induced pluripotent stem cell model. *Hepatology*, 61, 1730–9.

Martin, A. M., Nolan, D., Gaudieri, S., Almeida, C. A., Nolan, R., James, I., Carvalho, F., Phillips, E., Christiansen, F. T., Purcell, A. W., Mccluskey, J., & Mallal, S. 2004. Predisposition to abacavir hypersensitivity conferred by HLA-B*5701 and a haplotypic Hsp70-Hom variant. *Proceedings of the National Academy of Sciences*, 101, 4180–5.

Marx-Stoelting, P., Adriaens, E., Ahr, H. J., Bremer, S., Garthoff, B., Gelbke, H. P., Piersma, A., Pellizzer, C., Reuter, U., Rogiers, V., Schenk, B., Schwengberg, S., Seiler, A., Spielmann, H., Steemans, M., Stedman, D. B., Vanparys, P., Vericat, J. A., Verwei, M., Van Der Water, F., Weimer, M. & Schwarz, M. 2009. A review of the implementation of the embryonic stem cell test (EST). The report and recommendations of an ECVAM/RePorTect Workshop. *Alternatives to Laboratory Animals*, 37, 313–28.

Mckenzie, R., Fried, M. W., Sallie, R., Conjeevaram, H., Di Bisceglie, A. M., Park, Y., Savarese, B., Kleiner, D., Tsokos, M., Luciano, C., Pruett, T., Stotka, J. L., Straus, S. E., & Hoofnagle, J. H. 1995. Hepatic failure and lactic acidosis due to fialuridine (FIAU), an investigational nucleoside analogue for chronic hepatitis B. *New England Journal of Medicine*, 333, 1099–105.

Metushi, I. G., Hayes, M. A., & Uetrecht, J. 2015. Treatment of PD-1(−/−) mice with amodiaquine and anti-CTLA4 leads to liver injury similar to idiosyncratic liver injury in patients. *Hepatology*, 61, 1332–42.

Musunuru, K. 2013. Genome editing of human pluripotent stem cells to generate human cellular disease models. *Disease Models & Mechanisms*, 6, 896–904.

Nishimura, K., Sano, M., Ohtaka, M., Furuta, B., Umemura, Y., Nakajima, Y., Ikehara, Y., Kobayashi, T., Segawa, H., Takayasu, S., Sato, H., Motomura, K., Uchida, E., Kanayasu-Toyoda, T., Asashima, M., Nakauchi, H., Yamaguchi, T., & Nakanishi, M. 2011. Development of defective and persistent sendai virus vector: a unique gene delivery/expression system ideal for cell reprogramming. *Journal of Biological Chemistry*, 286, 4760–71.

Okita, K., Nakagawa, M., Hyenjong, H., Ichisaka, T., & Yamanaka, S. 2008. Generation of mouse induced pluripotent stem cells without viral vectors. *Science*, 322, 949–53.

Pirmohamed, M., James, S., Meakin, S., Green, C., Scott, A. K., Walley, T. J., Farrar, K., Park, B. K., & Breckenridge, A. M. 2004. Adverse drug reactions as cause of admission to hospital: prospective analysis of 18 820 patients. *BMJ*, 329, 15–9.

Ramasamy, T. S., Yu, J. S., Selden, C., Hodgson, H., & Cui, W. 2013. Application of three-dimensional culture conditions to human embryonic stem cell-derived definitive endoderm cells enhances hepatocyte differentiation and functionality. *Tissue Engineering Part A*, 19(3–4), 360–7.

Roth, A. & Singer, T. 2014. The application of 3D cell models to support drug safety assessment: opportunities & challenges. *Advanced Drug Delivery Reviews*, 69–70, 179–89.

Schwartz, R. E., Trehan, K., Andrus, L., Sheahan, T. P., Ploss, A., Duncan, S. A., Rice, C. M., & Bhatia, S. N. 2012. Modeling hepatitis C virus infection using human induced pluripotent stem cells. *Proceedings of the National Academy of Sciences of the United States of America*, 109, 2544–8.

Seiler, A. E. M. & Spielmann, H. 2011. The validated embryonic stem cell test to predict embryotoxicity in vitro. *Nature Protocols*, 6, 961–78.

Shan, J., Schwartz, R. E., Ross, N. T., Logan, D. J., Thomas, D., Duncan, S. A., North, T. E., Goessling, W., Carpenter, A. E., & Bhatia, S. N. 2013. Identification of small molecules for human hepatocyte expansion and iPS differentiation. *Nature chemical biology*, 9(8), 514–20.

Shyh-Chang, N. & Daley, G. Q. 2013. Lin28: primal regulator of growth and metabolism in stem cells. *Cell Stem Cell*, 12, 395–406.

Sivertsson, L., Edebert, I., Palmertz, M. P., Ingelman-Sundberg, M., & Neve, E. P. A. 2013. Induced CYP3A4 expression in confluent Huh7 hepatoma cells as a result of decreased cell proliferation and subsequent pregnane X receptor activation. *Molecular Pharmacology*, 83, 659–70.

Stadtfeld, M., Nagaya, M., Utikal, J., Weir, G., & Hochedlinger, K. 2008. Induced pluripotent stem cells generated without viral integration. *Science (New York, N.Y.)*, 322, 945–9.

Takahashi, K. & Yamanaka, S. 2006. Induction of pluripotent stem cells from mouse embryonic and adult fibroblast cultures by defined factors. *Cell*, 126, 663–76.

Takahashi, K., Tanabe, K., Ohnuki, M., Narita, M., Ichisaka, T., Tomoda, K., & Yamanaka, S. 2007. Induction of pluripotent stem cells from adult human fibroblasts by defined factors. *Cell*, 131, 861–72.

Takayama, K., Inamura, M., Kawabata, K., Katayama, K., Higuchi, M., Tashiro, K., Nonaka, A., Sakurai, F., Hayakawa, T., Furue, M. K., & Mizuguchi, H. 2012. Efficient generation of functional hepatocytes from human embryonic stem cells and induced pluripotent stem cells by HNF4alpha transduction. *Molecular Therapy*, 20, 127–37.

Takayama, K., Morisaki, Y., Kuno, S., Nagamoto, Y., Harada, K., Furukawa, N., Ohtaka, M., Nishimura, K., Imagawa, K., Sakurai, F., Tachibana, M., Sumazaki, R., Noguchi, E., Nakanishi, M., Hirata, K., Kawabata, K., & Mizuguchi, H. 2014. Prediction of interindividual differences in hepatic functions and drug sensitivity by using human iPS-derived hepatocytes. *Proceedings of the National Academy of Sciences*, 111, 16772–7.

Takebe, T., Sekine, K., Enomura, M., Koike, H., Kimura, M., Ogaeri, T., Zhang, R.-R., Ueno, Y., Zheng, Y.-W., Koike, N., Aoyama, S., Adachi, Y., & Taniguchi, H. 2013. Vascularized and functional human liver from an iPSC-derived organ bud transplant. *Nature*, 499, 481–4.

Terry, C., Dhawan, A., Mitry, R. R., & Hughes, R. D. 2006. Cryopreservation of isolated human hepatocytes for transplantation: state of the art. *Cryobiology*, 53, 149–59.

The Uniprot Consortium 2015. UniProt: a hub for protein information. *Nucleic Acids Research*, 43, D204–D212.

Tomizawa, M., Shinozaki, F., Sugiyama, T., Yamamoto, S., Sueishi, M., & Yoshida, T. 2013. Survival of primary human hepatocytes and death of induced pluripotent stem cells in media lacking glucose and arginine. *PLoS One*, 8, e71897.

Ulrich, R. G., Bacon, J. A., Brass, E. P., Cramer, C. T., Petrella, D. K., & Sun, E. L. 2001. Metabolic, idiosyncratic toxicity of drugs: overview of the hepatic toxicity induced by the anxiolytic, panadiplon. *Chemico-Biological Interactions*, 134, 251–70.

Ulvestad, M., Nordell, P., Asplund, A., Rehnström, M., Jacobsson, S., Holmgren, G., Davidson, L., Brolén, G., Edsbagge, J., Björquist, P., Küppers-Munther, B., & Andersson, T. B. 2013. Drug metabolizing enzyme and transporter protein profiles of hepatocytes derived from human embryonic and induced pluripotent stem cells. *Biochemical Pharmacology*, 86, 691–702.

Wang, H., Gao, X., Fukumoto, S., Tademoto, S., Sato, K., & Hirai, K. 1998. Post-isolation inducible nitric oxide synthase gene expression due to collagenase buffer perfusion and characterization of the gene regulation in primary cultured murine hepatocytes. *The Journal of Biochemistry*, 124, 892–9.

Wang, A., Holladay, S. D., Wolf, D. C., Ahmed, S. A., & Robertson, J. L. 2006. Reproductive and developmental toxicity of arsenic in rodents: a review. *International Journal of Toxicology*, 25, 319–31.

Ware, B. R., Berger, D. R., & Khetani, S. R. 2015. Prediction of drug-induced liver injury in micropatterned co-cultures containing iPSC-derived human hepatocytes. *Toxicological Sciences*, 145, 252–62.

Warren, L., Manos, P. D., Ahfeldt, T., Loh, Y. H., Li, H., Lau, F., Ebina, W., Mandal, P. K., Smith, Z. D., Meissner, A., Daley, G. Q., Brack, A. S., Collins, J. J., Cowan, C., Schlaeger, T. M., & Rossi, D. J. 2010. Highly efficient reprogramming to pluripotency and directed differentiation of human cells with synthetic modified mRNA. *Cell Stem Cell*, 7, 618–30.

Wilkening, S., Stahl, F., & Bader, A. 2003. Comparison of primary human hepatocytes and hepatoma cell line Hepg2 with regard to their biotransformation properties. *Drug Metabolism and Disposition*, 31, 1035–42.

Yu, J., Vodyanik, M. A., Smuga-Otto, K., Antosiewicz-Bourget, J., Frane, J. L., Tian, S., Nie, J., Jonsdottir, G. A., Ruotti, V., Stewart, R., Slukvin, II, & Thomson, J. A. 2007. Induced pluripotent stem cell lines derived from human somatic cells. *Science*, 318, 1917–20.

Yusa, K., Rashid, S. T., Strick-Marchand, H., Varela, I., Liu, P. Q., Paschon, D. E., Miranda, E., Ordonez, A., Hannan, N. R., Rouhani, F. J., Darche, S., Alexander, G., Marciniak, S. J., Fusaki, N., Hasegawa, M., Holmes, M. C., Di Santo, J. P., Lomas, D. A., Bradley, A., & Vallier, L. 2011. Targeted gene correction of alpha1-antitrypsin deficiency in induced pluripotent stem cells. *Nature*, 478, 391–4.

Zhang, R. R., Takebe, T., Miyazaki, L., Takayama, M., Koike, H., Kimura, M., Enomura, M., Zheng, Y. W., Sekine, K., & Taniguchi, H., 2014. Efficient hepatic differentiation of human induced pluripotent stem cells in a three-dimensional microscale culture. *Methods Mol. Biol.* 1210, 131–41. doi:10.1007/978-1-4939-1435-7_10.

Zhou, W. & Freed, C. R. 2009. Adenoviral gene delivery can reprogram human fibroblasts to induced pluripotent stem cells. *Stem Cells*, 27, 2667–74.

Zhou, H., Wu, S., Joo, J. Y., Zhu, S., Han, D. W., Lin, T., Trauger, S., Bien, G., Yao, S., Zhu, Y., Siuzdak, G., Scholer, H. R., Duan, L., & Ding, S. 2009. Generation of induced pluripotent stem cells using recombinant proteins. *Cell Stem Cell*, 4, 381–4.

22

HUMAN PLURIPOTENT STEM CELL-DERIVED CARDIOMYOCYTES: A NEW PARADIGM IN PREDICTIVE PHARMACOLOGY AND TOXICOLOGY

Praveen Shukla, Priyanka Garg and Joseph C. Wu

Stanford Cardiovascular Institute, Institute for Stem Cell Biology and Regenerative Medicine, Department of Medicine, Division of Cardiology, Stanford University School of Medicine, Stanford, CA, USA

22.1 INTRODUCTION

New drug discovery and development is a multidisciplinary, multiyear endeavor requiring allocation of dedicated resources aimed at reducing the ever-increasing burden of human diseases on society. The average time of developing a single drug from inception to launch is in the range of 10–15 years, costing ~$800 million to $5 billion (Munos and Chin, 2009; Desmond-Hellmann, 2013; Engle and Puppala, 2013). At present, many pharmaceutical research and development (R&D) organizations devote more than $5 billion/year to R&D, with over $30 billion/year of cumulative spending, equivalent to the total NIH budget of ~$30 billion (Munos and Chin, 2009; Desmond-Hellmann, 2013; Engle and Puppala, 2013; Hanna, 2015). Despite such focused and concerted efforts mounted by pharmaceutical companies, the number of drugs that fail in clinical trials (i.e., attrition rate) or are withdrawn from the market due to previously unknown negative side effects attributed to off- and on-target toxicity is on rise (Hutchinson and Kirk, 2011; Ferri et al., 2013; Mordwinkin et al., 2013). With respect to cardiovascular diseases, only 20% of agents earn FDA approval for human use following costly and lengthy clinical trial; for anticancer drugs, the approval rate falls to a meager 5% (Hutchinson and Kirk, 2011). Although the high attrition rate inherent to the drug discovery process can be attributed to several factors (e.g., scientific, technical, and managerial issues), off-target cardiotoxicity appears to be the most common cause of delay both in gaining drug approval and postmarketing withdrawal of drugs (Force and Kolaja, 2011;

Ferri et al., 2013; Mordwinkin et al., 2013). For instance, drug-induced cardiotoxicity manifested as idiosyncratic precipitation of life-threatening cardiac arrhythmia (i.e., torsades de pointes (TdP)) has caused withdrawal of several drugs across different therapeutic segments (Force and Kolaja, 2011; Ferri et al., 2013; Mordwinkin et al., 2013). Moreover, based on the potential risk of cardiovascular toxicity, the FDA has issued a "black box warning" for more than 100 drugs (Mordwinkin et al., 2013), thereby reducing drug developers' revenue and return on the investment. Not surprisingly, a meta-analysis shows a good correlation between FDA issued black box warnings and postmarketing withdrawal of drug, as 30% of all drugs withdrawn for market carried a black box warning (Wang et al., 2010).

Given the unpredictability and rarity of such unexpected drug-induced cardiotoxicity (e.g., the estimated incidence of adverse cardiac events with cisapride is about 1 in 111,000 prescriptions) (Malik and Camm, 2001), regulatory agencies in the United States, Europe, and Japan have taken a strong initiative and devised the ICH S7A and S7B guidelines to test all new chemical entities (NCEs) for their ability to inhibit the human *Ether-à-go-go-Related* Gene (hERG) channel current and thereby cause cardiotoxicity (ICH, 2005). Presently, the preclinical testing paradigm relies heavily on the use of *in vitro* cell lines such as Chinese hamster ovary (CHO) and human embryonic kidney (HEK) cells overexpressing hERG channels, *ex vivo* tissue preparations such as isolated arterially perfused left ventricular rabbit wedge preparations, and *in vivo* studies such as chronic dog atrioventricular (AV) block models (Mandenius et al., 2011).

Drug Discovery Toxicology: From Target Assessment to Translational Biomarkers, First Edition. Edited by Yvonne Will, J. Eric McDuffie, Andrew J. Olaharski, and Brandon D. Jeffy.

However, the high costs associated with these assays, their poor predictability owing to interspecies differences in cardiac electrophysiology and human biology (Kaese and Verheule, 2012), and a high probability of discarding NCEs (due to false positives), which otherwise could have been drugs with superior efficacy and safety profiles, remain big challenges that need immediate attention from all stakeholders involved in the drug discovery process.

The advent of human pluripotent stem cells (hPSCs), which include both human embryonic stem cell (ESCs) (Thomson et al., 1998) and human induced pluripotent stem cells (iPSCs) (Takahashi et al., 2007; Yu et al., 2007), provides an unprecedented opportunity to generate potentially limitless quantities of hPSC-derived somatic cell types (including human cardiomyocytes) for studying disease mechanisms, identifying novel drug targets, and accelerating drug screening (Burridge et al., 2012; Matsa et al., 2014). Of hPSCs, iPSCs carry distinctive advantages over ESCs as (i) they obviate the ethical issues associated with the use of human embryos and (ii) multiple lines can be established from both healthy controls and patients who have a disease of interest with known onset and severity (Wilson and Wu, 2015). The unique ability of hPSCs to self-renew and remain pluripotent (i.e., the ability to differentiate into virtually any cell of the human body, including neurons, cardiomyocytes, and hepatocytes (Murry and Keller, 2008; Inoue et al., 2014)), coupled with the recent refinement and improvement in reprogramming and cardiac differentiation strategies, has dramatically expanded the pharmaceutical industry's interest in using hPSC technology in preclinical drug screening and toxicity testing.

In this chapter, we review the recent advances and refinements in efficient and large-scale generation and characterization of iPSC-derived cardiomyocytes (iPSC-CMs), with an emphasis on the progress made in recapitulating clinically relevant cardiovascular diseases using iPSC-CMs. In addition, we discuss the potential benefit of integrating iPSC technology into early stages of preclinical drug screening, safety pharmacology, and toxicological testing by reducing the new drug attrition rate. We conclude with a discussion of current challenges before the iPSC technology can be fully exploited for a myriad of applications, including disease modeling, drug target identification, drug screening, and toxicological testing (Fig. 22.1).

22.2 ADVENT OF hPSCs: REPROGRAMMING AND CARDIAC DIFFERENTIATION

22.2.1 Reprogramming

Since the original discovery of iPSCs by Takahashi and Yamanaka, who described the induction of a pluripotent state (first in mice (Takahashi and Yamanaka, 2006) and then in

humans (Takahashi et al., 2007)) by forced ectopic overexpression of a mere four pluripotent transcription factors/core genes (*Oct4*, *Sox2*, *Klf4*, and *c-Myc*), considerable progress has been made in improving reprogramming methods. Initial use of genome-integrating viral vectors such as retroviruses (Takahashi and Yamanaka, 2006; Takahashi et al., 2007; Huangfu et al., 2008; Lowry et al., 2008) and lentiviruses (Yu et al., 2007; Stadtfeld et al., 2008; Sommer et al., 2009; Anokye-Danso et al., 2011) for reprogramming had several limitations such as incomplete proviral silencing, slow kinetics, and genomic integration of exogenous sequences with a potential risk of insertional mutagenesis and/or reactivation of viral transgenes, raising safety concerns related to immunogenicity and the risk of tumorigenesis (Zhao et al., 2011). In addition, early iPSCs are suboptimal for therapeutic use because the integrated viruses may affect the phenotype of their cells derived for disease modeling and drug testing. Such limitations led to the development of several other reprogramming methods, including excisable PiggyBac transposons (Woltjen et al., 2009; Somers et al., 2010), nonintegrating DNA-based (e.g., adenoviral vectors, mini-circle) (Yu et al., 2007; Okita et al., 2008; Sommer et al., 2009; Zhou and Freed, 2009; Jia et al., 2010; Si-Tayeb et al., 2010), and DNA-free (e.g., proteins, modified mRNA, microRNA) methods (Fusaki et al., 2009; Kim et al., 2009; Zhou et al., 2009; Warren et al., 2010; Miyoshi et al., 2011; Nishimura et al., 2011). Induction of pluripotency by use of cell-permeable, nonimmunogenic, small chemical compounds has also been successfully demonstrated (Ichida et al., 2009; Maherali and Hochedlinger, 2009; Itzhaki et al., 2011; Hou et al., 2013). Despite their varied success in generating iPSCs, often without a genetic footprint, these methods remain extremely labor intensive and inefficient (ranging from 0.001 to 0.2%) regardless of the method used.

22.2.2 Cardiac Differentiation

The versatile ability of iPSCs to differentiate into all somatic cell types of the body, including iPSC-CMs, promises an unrestricted access to healthy and disease-specific iPSC-CMs for a variety of applications (Murry and Keller, 2008; Burridge et al., 2012; Inoue et al., 2014; Matsa et al., 2014). However, to meet these goals, an acute need to develop more cost-effective, easily scalable, and robust differentiation approaches to produce iPSC-CMs in sufficient quality and quantity for downstream analysis must be met. To this end, significant progress has been made in refining and improving cardiac differentiation strategies that can be grouped into three distinct categories (Matsa et al., 2014): (1) embryoid body (EB) methods, (2) monolayer differentiation, and (3) suspension culture systems. Independent of the methodology adopted, the *in vitro* differentiation of hPSCs to CMs relies on imitating the sequential stages of embryonic cardiac development and involves (i) a transition from hPSCs to a

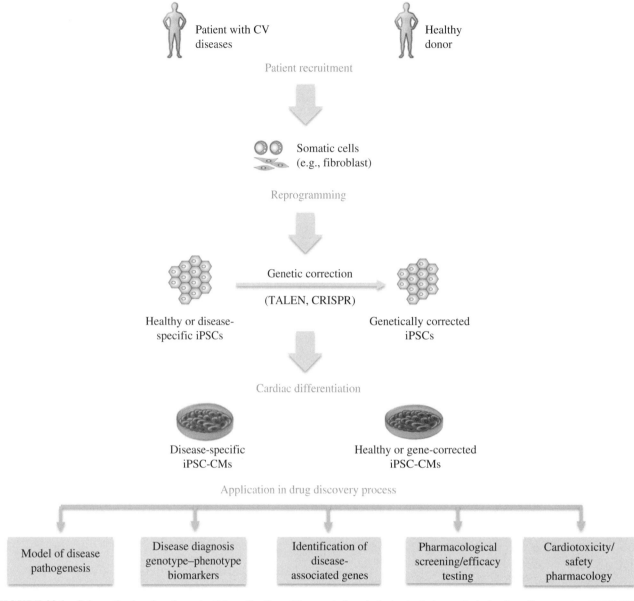

FIGURE 22.1   Schematic showing the potential application of human induced pluripotent stem cell-derived cardiomyocytes (iPSC-CMs) in disease modeling, cardiotoxicity/safety pharmacology assessment, and drug discovery. Somatic cells (e.g., fibroblasts, blood cells, etc.) can be obtained from healthy donors and patients with cardiovascular diseases (e.g., arrhythmia, cardiomyopathies, etc.) followed by efficient reprogramming into well-characterized libraries of iPSCs. Early integration of such disease-specific cardiomyocytes derived from the standardized iPSC library can augment the identification of disease-relevant NCEs or "disease-specific hits" and provide human biology-relevant cardiac cell type to assess the potential cardiotoxicity (e.g., hERG blocking, QT prolongation, or effects on other cardiac ion channels orchestrating the AP waveform) of NCEs early on in drug discovery. Strategic inclusion of iPSC-CMs is expected to reduce the false positives and false negatives associated with the traditional drug discovery process and may help reduce the high attrition rate at costly clinical phases of drug development. In addition, genes identified by disease modeling studies could be corrected using recently developed gene-editing technologies such as transcription activator-like effector nucleases (TALENs) and clustered regularly interspaced short palindromic repeats (CRISPRs) to generate genome-corrected iPSCs for regenerative therapies. (*See insert for color representation of the figure.*)

mesoderm progenitor cells; (ii) a transition from mesodermal progenitor cells to precardiac mesoderm cells; (iii) specification of the precardiac mesoderm to the cardiac mesoderm; (iv) specification in which the cardiac mesoderm further develops into cardiac progenitor cells; (v) transition from cardiac progenitor cells to immature cardiomyocytes; and (vi) maturation of the CMs measured by time in culture postcardiac specification, with days 30–40 being the most common time point for hPSC-CM characterization and drug testing (Burridge et al., 2012).

The EB method was the first to be used. This method entails the spontaneous appearance of contracting areas containing cardiomyocytes in suspension-grown EBs (Kehat et al., 2001). Although invaluable in demonstrating early proof of principle, EB methods are time consuming, highly inefficient, and unpredictable and require the use of complex media. Eventual refinement of EB methods and close simulation and modulation of key cardiac lineage developmental stages led to the monolayer-based differentiation. This method is more predictable and results in significantly higher yield but still requires complex undefined medium components, including some of animal origin (Burridge et al., 2012, 2014). Moreover, intra- and interline variability of cardiac differentiation reported between different research laboratories remains a concern. The recent development of a chemically defined cardiac differentiation medium (Burridge et al., 2014), coupled with successful demonstration of good manufacturing practices (GMP)-compliant stirred spinner flask suspension culture (Chen et al., 2012; Lei and Schaffer, 2013; Kempf et al., 2014) to produce large quantities of hPSC-CMs, is expected to accelerate the use of hPSC-CMs in therapeutic and drug discovery applications.

22.3 iPSC-BASED DISEASE MODELING AND DRUG TESTING

Cellular modeling of human heart disease has been hampered by severe limitations, including inaccessibility of human cardiac tissue, difficult donor consent, limited tissue sample quantities, and low proliferative capacity of freshly dissociated human cardiomyocytes (Brandenburger et al., 2012; Matsa et al., 2014). These problems have led to the extensive use of surrogates *in vitro* (i.e., HEK or CHO overexpressing hERG channel) and *in vivo* model systems (i.e., rodents and large animals). These models have been useful in gaining insight in the pathogenesis of cardiovascular diseases, with the aforementioned limitations. Mounting evidence in the literature points toward the utility of iPSCs in recapitulating the cellular phenotypes of several cardiovascular diseases, including cardiac channelopathies and cardiomyopathies (Table 22.1) (Matsa et al., 2014). The current iPSC-based disease modeling paradigm differentiates CMs from several iPSC clones derived from somatic cells of an affected member with known mutation(s) causing a particular cardiac disorder. Subsequently, a versatile battery of functional, biochemical, and gene expression assays with a range of stringencies is used to thoroughly characterize the cellular disease phenotype and assess drug responses (Fig. 22.2).

The *KCNQ1* and *KCNH2* genes code for two major cardiac repolarizing currents, and mutations in these channels have been shown to cause long QT syndrome (LQTS) type 1 and type 2, respectively (Modell and Lehmann, 2006). iPSC-CMs carrying mutations in these cardiac potassium channels show abnormal prolongation of AP duration and increased arrhythmogenicity when presented with β-adrenergic stimulation or blockade of potassium channels (Moretti et al., 2010; Itzhaki et al., 2011; Matsa et al., 2011). By contrast, the pharmacological blockade of either the β-adrenergic receptors or calcium channels rescues this phenotype (Moretti et al., 2010; Itzhaki et al., 2011; Matsa et al., 2011). Similarly, iPSC-CMs are also used to model the LQTS type 3, which has a relatively lower prevalence and is precipitated by bradycardia (Churko et al., 2013; Terrenoire et al., 2013). Heterologous expression studies identified the mutations in the cardiac sodium ion channel-encoding gene *SCN5A*, which cause an increase in inward depolarizing late sodium current during phases I–III of repolarization and are manifested as LQTS. iPSC-CMs from LQTS type three patients exhibit marked prolongation of the AP duration at lower beat frequencies and elevation in late sodium current, which can be rescued by pacing at higher beat frequencies and treatment with the sodium channel blocker, mexiletine (Churko et al., 2013; Terrenoire et al., 2013). iPSC-CMs carrying a specific mutation in L-type calcium ($Ca_V1.2$) gene also recapitulated the LQTS type 8 (Timothy syndrome) disease phenotype, which is characterized by excessive Ca^{2+} influx leading to prolonged action potentials. Roscovitine, a compound that increases the voltage-dependent inactivation of $Ca_V1.2$, restored the electrical and Ca^{2+} signaling defect of iPSC-CMs from Timothy syndrome patients (Yazawa et al., 2011). Furthermore, iPSC-CMs have been successfully used to demonstrate the underlying calcium-signaling defects and arrhythmias associated with catecholaminergic polymorphic ventricular tachycardia (CPVT) types 1 and 2, along with the pharmacological approaches to rescue the disease phenotype (Carvajal-Vergara et al., 2010; Itzhaki et al., 2012) (Table 22.1).

In addition, iPSC-CMs have been used to model structural heart diseases, including familial dilated cardiomyopathy (DCM) (Sun et al., 2012) and hypertrophic cardiomyopathy (HCM) (Lan et al., 2013) caused by mutations in cardiac troponin T (cTnT) and myosin heavy chain 7 (MYH7), respectively. In both studies, iPSC-CMs recapitulated the disease phenotype at the cellular level with pharmacological rescue by chronic treatment with β-adrenergic blocker (propranolol or metoprolol) or the L-type Ca^{2+} channel blocker (verapamil) (Sun et al., 2012; Lan et al., 2013) (Table 22.1). Collectively, these studies reinforce the ability of iPSC-CMs to model channelopathies and cardiomyopathies as well as to test the clinical efficacy of drugs.

Despite the rapid and impressive progress in iPSC disease modeling, there are several limitations that need to be addressed, including lack of appropriate controls and modeling of diseases without a known genetic cause. Current iPSC-based disease modeling studies (Table 22.1) use age-matched, unaffected cells within the same family pedigree as controls, but these are suboptimal because of differences in the individuals

TABLE 22.1 Cardiovascular Disease Modeling and Drug Testing Using hPSC-Derived Cardiomyocytes

Cardiovascular Diseases	Mutation	hPSC Cell Type	Major Functional Assays	Disease Phenotype/Proposed Mechanism(s)	Pharmacological Rescue	References
Long QT syndrome						
				hPSC-based disease modeling		
Long QT-1	*KCNQ1* R190Q	CMs	Standard patch clamp	Significant and abnormal reduction in outward I_K current with consequential reduction in repolarization velocity; anomalous subcellular localization of R190Q KCNQ1	β-Adrenergic blockade abrogates isoproterenol-induced EAD	Moretti et al. (2010)
Long QT-2	*KCNH2* A614V	CMs	Standard patch clamp MEA	Reduction of potassium current I_{Kr}; APD prolongation precipitating into spontaneous EADs	Abrogation of EADs by mechanistic pharmacology involving blocking calcium, potassium, and late sodium currents with nifedipine, pinacidil, and ranolazine, respectively	Itzhaki et al. (2011)
Long QT-2	*KCNH2* G1681A	CMs	Standard patch clamp, MEA	Reduction of potassium current I_{Kr}; APD prolongation precipitating into spontaneous EADs	β-Adrenergic blockade by nadolol and propranolol abrogates isoproterenol-induced EAD Potassium channel activators like nicorandil and PD118057 rectifies APD prolongation and reduces EADs	Matsa et al. (2011)
Long QT-8 (Timothy syndrome)	*Cav1.2* G406R	CMs	Standard patch clamp, calcium imaging	Excessive Ca^{2+} influx coupled with abnormal Ca^{2+} transients results in APD prolongation and irregular contractions/electrical activity	Roscovitine rescues the disease phenotype by improving Ca^{2+} channel inactivation and thus influx of excessive Ca^{2+}	Yazawa et al. (2011)
Long QT-3	*SCN5A* V1763M *SCN5A* F1473C K897T	CMs	Standard patch clamp	An abnormal increase in late Na^+ current caused by either delayed channel inactivation or a faster recovery from inactivation results in increase in intracellular Na^+ loading manifested as APD prolongation	Pharmacological blockade of $Na_v1.5$ with mexiletine abrogates both the pathogenic increase in late Na^+ current and APD prolongation specific to LQT3 iPSC-CMs	Churko et al. (2013); Terrenoire et al. (2013)
Catecholaminergic polymorphic ventricular tachycardia (CPVT)						
CPVT-1	*RyR2* S406L	CMs	Standard patch clamp, calcium imaging	Catecholamine-induced increases in diastolic Ca^{2+} concentrations, reduced sarcoplasmic reticulum Ca^{2+} loading coupled with increased frequency and duration of elementary Ca^{2+} release precipitating into spontaneous delayed after depolarizations (DADs)	Restoration of normal Ca^{2+} spark with consequential reduction in arrhythmogenicity by dantrolene	Jung et al. (2012)

Disease	Gene/Mutation	Cell type	Method	Phenotype	Pharmacology/Treatment	Reference
CPVT-1	*RyR2* M4109R P2328S	CMs	Standard patch clamp, calcium imaging	Independent replication of observations noted above along with observations of early after depolarizations (EADs)	Mechanistic pharmacology approach involving flecainide, thapsigargin, and β-blockers rescued disease phenotype by improving Ca^{2+} transient anomalies	Itzhaki et al. (2012); Kujala et al. (2012)
CPVT-2	*CASQ2* D307H	CMs	Standard patch clamp, calcium imaging	Structural changes: immature cardiomyocytes with less organized myofibrils, enlarged sarcoplasmic reticulum cisternae, and reduced number of caveolae. Functional changes: increases in spontaneous appearance of oscillatory arrhythmic prepotentials and after contractions secondary to abnormal elevation of diastolic Ca^{2+}	ND	Novak et al. (2012)
Cardiomyopathies						
DCM	*TNNT2* R173W	CMs	Standard patch clamp, calcium imaging, MEA	Abnormal sarcomeric α-actinin distribution coupled with altered regulation of Ca^{2+} manifesting as reduction of beat rate and weaker contraction	β-Adrenergic blockade with metoprolol improves abnormal sarcomeric organization	Sun et al. (2012)
HCM	*MYH7* c.1988GNA; p.R663H	CMs	Standard patch clamp, calcium imaging, MEA	Cellular hypertrophy characterized by calcineurin–NFAT activation and upregulation of hypertrophic transcription factors. Dysregulation of Ca^{2+} cycling manifesting as contractile arrhythmia	Mechanistic pharmacology approach involving propranolol, verapamil, nifedipine, diltiazem, lidocaine, mexiletine, ranolazine rescued disease phenotype	Lan et al. (2013)
ARVD/C	*PKP2* Plakophilin-2	CMs	Calcium imaging	Exaggerated lipogenesis and apoptosis in mutant PKP2 iPSC-CMs and Ca^{2+}-handling deficits in case of iPSC-CMs with a homozygous PKP2 mutation	ND	Kim et al. (2013)
LEOPARD syndrome	*PTPN11* T468M	CMs	NA	CMs are larger, have a higher degree of sarcomeric organization and preferential localization of NFATC4 in the nucleus compared to normal CMs	ND	Carvajal-Vergara et al. (2010)
Hypoplastic left heart syndrome (HLHS)	NA	CMs	Standard patch clamp, calcium imaging	A lower level of myofibrillar organization, persistence of a fetal gene expression pattern, changes in commitment to ventricular versus atrial lineages and display different calcium transient patterns and electrophysiological properties versus control	ND	Jiang et al. (2014)

(Continued)

TABLE 22.1 (Continued)

Cardiovascular Diseases	Mutation	hPSC Cell Type	Major Functional Assays	Disease Phenotype/Proposed Mechanism(s)	Pharmacological Rescue	References
Barth syndrome	*Tafazzin* BTH-H *TAZ*	CMs	Heart-on-chip technology Stress measurements	Structural sarcomeric abnormalities manifested as weakened contraction and abnormal decrease in systolic stresses	Mechanistic pharmacology approach involving mitoTEMPO, an antioxidant, bromoenol lactone, inhibitor of phospholipase A$_2$, linoleic acid (LA), a fatty acid precursor of mature cardiolipin	Wang et al. (2014b)
Genome-edited iPSC-based disease modeling						
LQT-1	KCNQ1 (R190Q G269S G345E)	CMs	Standard patch clamp, calcium imaging	CMs displayed long QT syndrome phenotype and significant prolongation of the action potential duration compared with the unedited control cells	Pharmacological rescue of disease phenotype by a calcium channel blocker, nifedipine, and a K$_{ATP}$ channel opener, pinacidil	Wang et al. (2014a)
LQT-2	KCNH2 (A614V)					

APD, action potential duration; ARVD/C, arrhythmogenic right ventricular dysplasia/cardiomyopathy; CM, cardiomyocyte; CPVT, catecholaminergic polymorphic ventricular tachycardia; DAD, delayed after depolarizations; DCM, dilated cardiomyopathy; EAD, early after depolarization; HCM, hypertrophic cardiomyopathy; I$_{Kr}$, rectifier potassium current; I$_{Na}$, sodium current; MEA, multielectrode array; TNNT2, troponin T type 2.

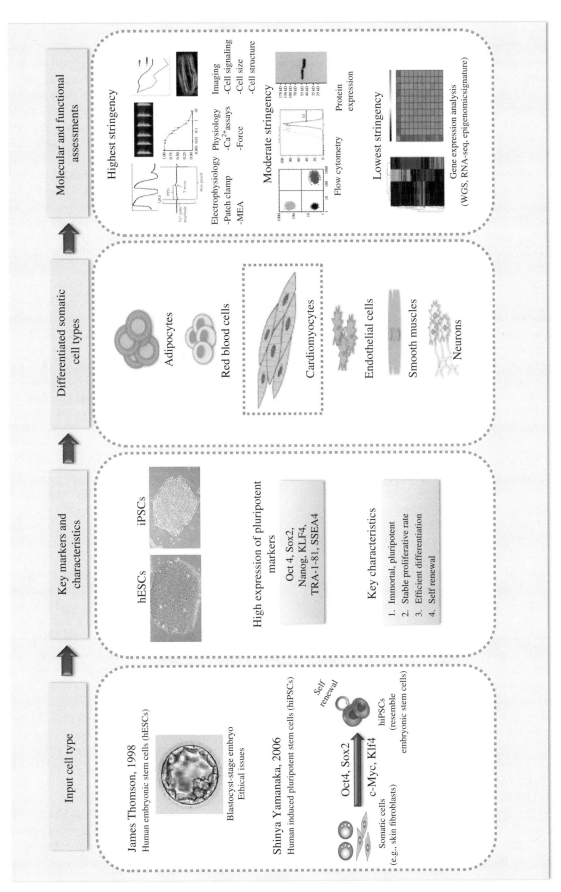

FIGURE 22.2 Schematic showing a battery of *in vitro* assays or descriptors aimed at characterizing the electrophysiology, physiology, cell surface marker expression, and gene expression analysis of the cardiomyocytes obtained from human pluripotent stem cell (hPSCs), which include embryonic stem cells (ESCs) and induced pluripotent stem cells (iPSCs). The assays or descriptors are stratified based on the relative stringency of the data obtained by a particular assay or technique. This strategy can be used to delineate the quantitative and qualitative differences between diseased and control cellular phenotypes, and intra- and interline variability associated with the currently available reprogramming and cardiac differentiations methods. These platforms can be adopted as a standard for quality control in mass production of hPSC-CMs in drug discovery and regenerative therapy. (*See insert for color representation of the figure.*)

genetic backgrounds and other confounding factors, including (i) unmatched age, gender, and ethnicity between the patients and control individuals; (ii) differences in methodology used to induce pluripotency (e.g., lentivirus vs. RNA transfection); and (iii) differences in passage number and adaptation to culture of the iPSC lines (Wang et al., 2014a). In addition, even in studies in which healthy siblings have been used as controls for disease patients, this would only account for an average of ~50% of the genome that is shared between any siblings, and the relevant phenotypic differences could be the result of DNA variants in the other ~50% of the genome, rather than the purported disease-associated mutations (Musunuru, 2013). The very processes of generating, expanding, and passaging iPSC lines have also been shown to cause accumulation of a variety of genetic alterations such as copy number variations (CNVs) to chromosomal amplifications, deletions, and rearrangements (Hussein et al., 2011; Pera, 2011). In addition, there are concerns that potential somatic cell epigenetic memory from the original tissue source of the reprogrammed cells can affect the differentiation efficiency (Kim et al., 2010; Musunuru, 2013; Sanchez-Freire et al., 2014). Thus, there is a need to develop strategies to minimize confounding variables so that the disease-causing mutations are the only differences between diseased and control iPSCs.

The recent development in genome-editing technology allows the efficient introduction of specific genetic alterations into the genomes of iPSCs with the unedited cells serving as an isogenic control (Hsu et al., 2014). This strategy promises to eliminate, or at least mitigate, all of the other confounding variables, allowing investigators to directly connect genotype to phenotype to establish causality. A recent study by Wang et al. (Wang et al., 2014a) successfully applied genome-editing technology to introduce the variants of the ion channel genes *KCNQ1* and *KCNH2* with dominant negative mutations, causing LQTS type 1 and type 2, respectively, followed by pharmacological rescue of disease phenotype by a calcium channel blocker (nifedipine) and a K_{ATP} channel opener (pinacidil) (Wang et al., 2014a). So far, iPSC-CMs have only been applied to model disorders for which mutations in particular genes have been identified. Future studies are likely to demonstrate the potential of these iPSC systems in delineating the pathophysiological mechanisms for diseases for which no gene mutations are yet known (Sallam et al., 2015).

22.4 TRADITIONAL TARGET-CENTRIC DRUG DISCOVERY PARADIGM

The traditional drug discovery paradigm is heavily reliant on testing a library of NCEs in a simplified, scalable cell-based assay involving overexpression of a well-characterized target in an immortal heterologous expression system (Fig. 22.3). Although useful, such assay systems lack the human disease-specific subcellular signaling and disease interactome, making them less optimal in identification and nomination of lead compounds with suboptimal efficacy. The subsequent progression and testing of such lead compounds in more complex and costly whole animal models, where significant interspecies differences exist both in terms of human physiology and disease pathophysiology (Kaese and Verheule, 2012), further contribute to advancement of molecules with a higher propensity for failure in clinical trials. In addition, a number of cardiovascular and noncardiovascular drugs with effective therapeutic efficacy have been withdrawn from the market due to unpredictable life-threatening cardiotoxicity (Mordwinkin et al., 2013). This further highlights the suboptimal predictability of preclinical safety pharmacological models to accurately and reliably identify the potential cardiotoxicity of NCEs. For example, ICH guidelines require testing all NCEs for their ability to inhibit whole-cell hERG current in an *in vitro* assay involving overexpression of the hERG channel in HEK or CHO cells (Cavero and Crumb, 2005). However, this cell-based assay system does not simulate human cardiomyocyte physiology as it lacks all other cardiac ion channels orchestrating the cardiac AP waveform. Therefore, taken together, there is an immediate need to develop and integrate better predicative human disease-relevant model systems across different stages of drug discovery.

22.5 iPSC-BASED DRUG DISCOVERY PARADIGM

The recent discovery followed by tremendous progress made in generating iPSCs from both healthy and diseased subjects of all ages and ethnic backgrounds offers an unprecedented opportunity to integrate iPSC-based human disease-relevant cellular models in drug discovery process (Murry and Keller, 2008; Burridge et al., 2012; Inoue et al., 2014; Matsa et al., 2014). The early integration of such models will enable not only the identification of human disease-specific hits (as opposed to target-specific hits inherent to traditional drug discovery paradigm) but will also allow to serve as a better predictor of potential cardiotoxicity of NCEs (Fig. 22.3).

At present, NCEs are not exposed to healthy human subjects or relevant patient population until phases I and II of clinical trials, thereby significantly contributing to the high attrition rate associated with the target-centric drug discovery paradigm. Early integration of iPSC-based models in the current drug discovery paradigm may enable identification and validation of high-fidelity drug targets and optimization of lead compounds with superior pharmacological efficacy while providing clinically relevant assessment of cardiotoxicity. This will also boost drug development pipelines for pharmaceutical developers and reduce reliance on animal models.

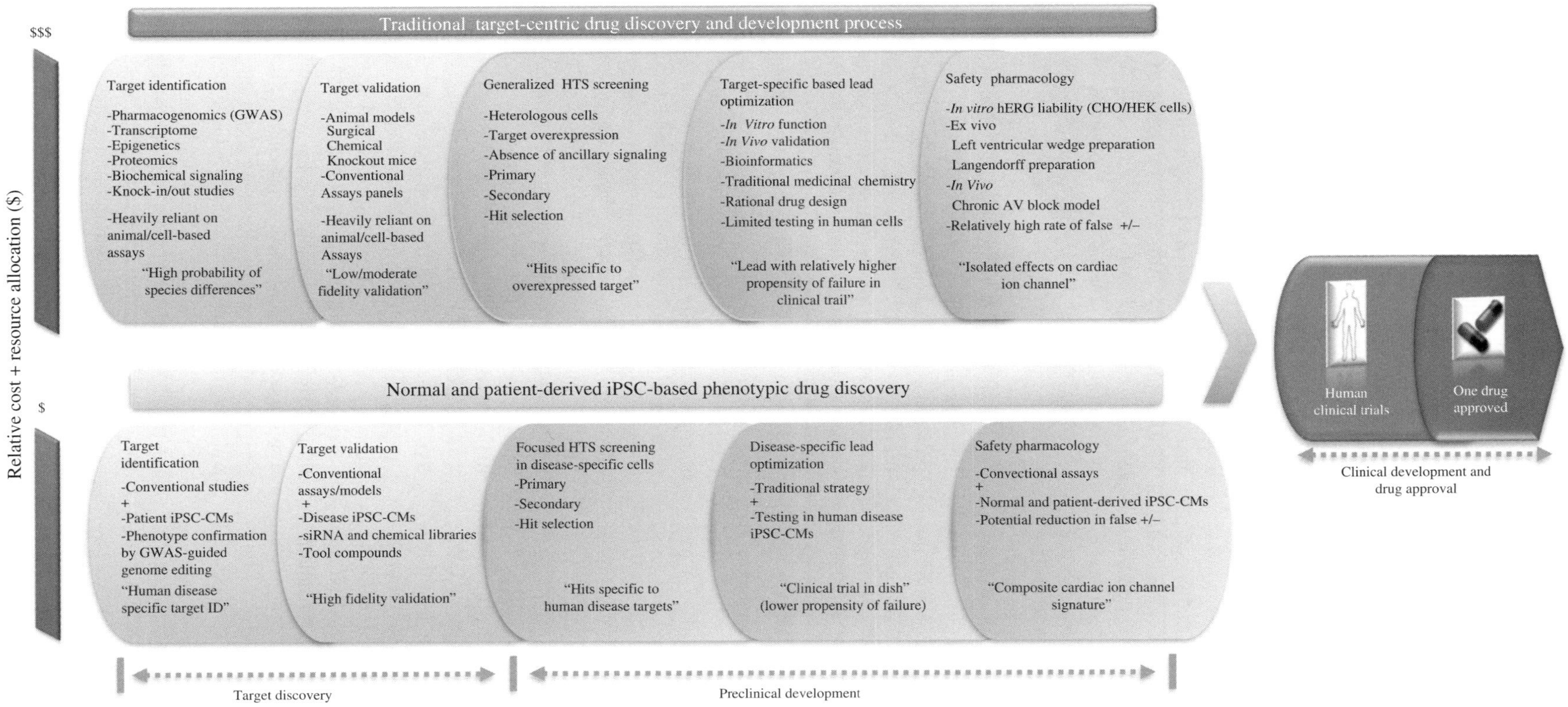

FIGURE 22.3 Schematic showing the strategies and assays used during preclinical phases of traditional target-centric versus iPSC-based phenotypic drug discovery paradigm. The target-centric approach relies heavily on identification of a single disease-specific target, followed by screening of large chemical libraries against that identified target overexpressed in a heterologous expression system. Either genetically engineered or chemically induced disease animal models are then used to validate the efficacy of a lead compound at a systems biology level. The therapeutic or pharmacological efficacy of the lead compound is not tested in target patient populations until a phase II of clinical trial. Similarly, assessment of cardiotoxicity of lead compound relies on use of either overexpressed hERG channels in immortal cell lines like HEK and CHO, followed by testing in more thorough *ex vivo* and *in vivo* models, which suffer from considerable interspecies variations. Because iPSCs carry a unique potential of differentiating into all somatic cell types of the body, including iPSC-CMs, these cells can complement the existing strategies and assays commonly used in a target-centric drug discovery paradigm. The iPSC-based models will allow interrogation of new chemical moieties in a human disease-relevant cell system that otherwise would not be exposed to patients until phase I of clinical trials. Since iPSC-CMs express all the cardiac ion channels underlying the cardiac action potential signature, the assessment of new chemical entities (NCEs) in iPSC-CMs will provide a composite cardiac ion channel signature of the NCEs, thereby potentially reducing false positives and false negatives associated with the current battery of test as outlined by ICH S7B guidelines. Early integration of iPSC-based *in vitro* model systems is expected to reduce the total relative funding and resources currently allocated for discovering new drugs, from target identification to approval.

22.5.1 Target Identification and Validation: "Clinical Trial in a Dish"

Numerous studies have demonstrated the ability of patient-specific iPSC-CMs to faithfully recapitulate the cellular phenotypes of several cardiac diseases and to predict pharmacological efficacy and cardiotoxicity of known and investigational drugs (Table 22.1). Retrospective interrogation can now be used to identify disease-specific novel drug targets, including but not limited to signaling molecules, enzymes, receptors, and ion channels. A comprehensive approach that includes knockdown or knock-in of identified targets can be achieved by standard molecular and cellular biology techniques, which can demonstrate the pathogenic link of drug targets that cause the disease under investigation (Table 22.1). Similarly, for monogenic disorders, genome-editing technology can be used to introduce a disease-causing mutation in iPSC-CMs to establish the genotype–phenotype relationship (Wang et al., 2014a). Subsequently, the routine assays of target-centric drug discovery can be used to further validate the targets identified in iPSC-CM-based models, including RNAi knockdown, pharmacological inhibition, and activation of targets in small animal models, and genetically engineered mice models can be used to validate the target at the whole animal or systems biology level.

Perhaps the most important application of iPSC models is in a phenotypic drug screening process in which the exact molecular and structural identity of a drug target is not yet fully established. In such complex multifactorial diseases with unknown causative molecular mechanism(s), recapitulation of the disease phenotype or pathology by iPSC-CMs would facilitate the identification of lead compounds in a target-agnostic drug screening fashion, that is, compounds will be screened and identified based on their ability to revert or rescue the overall cellular disease phenotype independent of the drug target (Grskovic et al., 2011) (Fig. 22.3). In addition, the identification of several disease-specific genetic perturbations (e.g., CNVs) by a growing number of powerful genome-wide association studies (GWAS) (Arking and Chakravarti, 2009; Ndiaye et al., 2011), coupled with the availability of RNAi libraries, provides another interesting and promising venue for identifying previously unknown mechanisms of disease and novel drug targets (Iorns et al., 2007). A custom-made RNAi knockdown library can be assessed in patient-specific iPSC-CMs carrying gene candidates identified by GWAS and expression profiling studies to rescue the disease phenotype, thus potentially unraveling the potential therapeutic targets that can be exploited to develop drugs using rational drug design approaches. Disease-specific iPSC-CMs can then aid in testing the chemical moieties generated for their pharmacological efficacy and cardiotoxicity.

Drug repositioning or repurposing is another promising concept wherein an FDA-approved drug with a demonstrated record of efficacy and safety is tested for treating new diseases, thereby reducing the overall cost associated with new drug discovery while expediting the delivery of therapeutics to patients (Ashburn and Thor, 2004). Since most of the FDA-approved drugs were developed before the advent of iPSC technology, iPSC-based disease models can now be used to screen these drugs for potential therapeutic benefits in other diseases. This represents a major opportunity for industrial and academic collaboration that is now increasingly being reorganized (Oprea et al., 2012). The commercial availability of nearly all 3400 FDA-approved drugs (Chong et al., 2006) via chemical libraries such as the Johns Hopkins Clinical Compound Collections, MicroSource Spectrum Collection, Prestwick Collection, and the Sigma-Aldrich Library of Pharmacologically Active Compounds (Grskovic et al., 2011) is expected to foster greater collaboration aimed at discovering new therapeutic modalities of existing drugs aided by iPSC technology-based drug screening.

22.5.2 Safety Pharmacology and Toxicological Testing

Idiosyncratic and unpredictable adverse drug reactions including organ-specific toxicity (e.g., cardiotoxicity) remain one of the largest contributors to the high attrition rate observed with the drug discovery process and the ongoing decline in pharmaceutical R&D pipeline (Kola and Landis, 2004; Force and Kolaja, 2011; Hutchinson and Kirk, 2011). Drug-induced cardiotoxicity involves a diverse range of mechanisms, including a specific ion channel block (i.e., hERG) causing delayed repolarization and an extremely rare yet potentially life-threatening polymorphic ventricular tachycardia from TdP, while biochemical toxicities manifest as cell death, abnormal cardiac function, and even heart failure and death (Roden, 2004; Rochette et al., 2015). Finding physiologically relevant, reproducible, and reliable cell models that can be used to detect these endpoints at preclinical and discovery stages has been a challenge, as a majority of toxicity testing is done either by using immortalized cancer cell lines, primary explanted somatic cells, or live animals. Interestingly, TdP is not readily observed during clinical trials or postmarketing surveillance with most drugs. For example, the estimated incidence of adverse cardiac events with cisapride is about 1 in 111,000 prescriptions (Malik and Camm, 2001). Clinical observation of drug-induced TdP raised a serious medical concern and resulted in concerted efforts to delineate the underlying causative mechanism. The drug-induced TdP is attributed to an unintended and nonspecific inhibition of hERG-encoded rapid delayed rectifier repolarizing current (I_{Kr}), with a consequential reduction in repolarization reserve of cardiomyocytes, prolongation of AP duration, and precipitation of TdP (Roden, 2004). Spontaneous and unpredictable development of TdP is observed not only with antiarrhythmic drugs but also with different therapeutic and pharmacological classes

TABLE 22.2 Mechanistic Classes of Drugs Known to Perturb Cardiac Action Potential

Mechanism(s)	Drugs	Effect on Cardiac AP	References
hERG blockers	Terfenadine Sotalol	Prolongation of cardiac AP and QT interval leading to TdP	Monahan et al. (1990); Crumb et al. (1995); Yap and Camm (2003); Kannankeril et al. (2010)
Reduction in hERG channel trafficking to cellular membrane	Pentamidine	No acute effects on hERG current but reduces hERG channel trafficking to cell membrane only on prolonged exposure resulting in prolonged AP (false negative)	Wharton et al. (1987); Cordes et al. (2005); Kuryshev et al. (2005)
I_{Ks}, carried by the KvLQT1/minK channel blockers	Chromanol 293B	Inhibition of I_{Ks} reduces the cardiac repolarization reserve leading to prolongation of AP duration (false negative)	Peng et al. (2010); Abassi et al. (2012)
Cardiac sodium channel (NaV1.5) blockers	Quinidine	Inhibition of $Na_V1.5$ current that governs the upstroke velocity of phase 0 of the cardiac AP significantly reduces the upstroke velocity and prolongs the AP duration, independent of hERG blockade (false negative)	Roden and Hoffman (1985); Ducroq et al. (2007)
Increase in cardiac sodium channel ($Na_V1.5$) current	Alfuzosin	Delays cardiac repolarization by increasing sodium current leading to increased plateau potential and prolongation of AP duration and QT interval (false negative)	Lacerda et al. (2008); Liang et al. (2013)
Dual channel blockers	Verapamil	Verapamil-induced prolongation of AP duration due to its hERG blocking properties is negated by its inhibitory effect on cardiac calcium channels (false positive)	Zhang et al. (1999); Redfern et al. (2003)

of drugs (Mordwinkin et al., 2013). This led to an initial proposal and then widespread adoption of ICH S7A and S7B guidelines by leading regulatory agencies mandating testing of all NCEs for their ability to inhibit hERG currents *in vitro* (a pharmacodynamic endpoint, as opposed to histopathological endpoints used in traditional toxicity studies), followed by thorough clinical QT studies in healthy volunteers (Cavero and Crumb, 2005).

Since the early integration of hERG inhibition assay (and other *ex vivo* and *in vivo* assays) into preclinical stages of drug discovery (Lawrence et al., 2008), no new drug has been withdrawn due to unpredictable cardiotoxicity. However, these assays are not without limitations, which include high cost, use of large numbers of animals, low throughput, and, more importantly, a high rate of false-positive and false-negative hits reflecting poor predictability (Lawrence et al., 2008). As discussed previously, CHO or HEK cells overexpressing a single ion channel (i.e., hERG) do not recapitulate the characteristic cardiac AP, a result of the concerted action of multiple ion channels (i.e., tightly coordinated voltage-dependent opening/closing of different sodium, calcium, and potassium channels) (Hoekstra et al., 2012). An early reliance on cell-based hERG assay may miss the NCEs that could prolong cardiac AP without directly affecting hERG currents, resulting in increased propensity of false-positive and false-negative hits (Table 22.2). The probability of such NCEs to undergo subsequent testing in much more complex, time-consuming, and costly *ex vivo* and *in vivo* preparations

to reveal their true likelihood of cardiac AP perturbations is very low, because these compounds will be eliminated early in typical high-throughput screening (HTS) processes (Hoekstra et al., 2012).

The development of iPSC-CMs as models for drug-induced cardiotoxicity testing offers a promising alternative that is more physiologically relevant, more predictive, and more time and cost efficient. The iPSC-CMs may greatly augment our ability to detect (and avoid) compounds that may provoke TdP arrhythmias at the early lead optimization phase, reduce the occurrence of false-positive hits (thereby saving several promising NCEs that otherwise would have been discarded early in development), and endorse those compounds with minimal torsadogenic risks. Adoption of iPSC-CMs could also potentially reduce the number of NCEs with questionable or conflicting proarrhythmic signals obtained with simpler preclinical assays (e.g., CHO/HEK cells overexpressing only hERG channels). A retrospective detection of human cardiotoxicity by iPSC-CMs of a series of compounds that did not show any cardiotoxicity in animal- and cell-based models, but were later withdrawn during phase I clinical trials due to unexpected cardiotoxicity, would greatly increase the confidence of the pharmaceutical industry in the predictive power of iPSC-CM-based assays, thereby expediting the widespread adoption of the model to pharmaceutical use. In addition, iPSC-CM-based cellular models can be used to assess drug-induced cardiotoxicity independent of hERG channel blockade by assessing the release of cardiac troponins (cTnI and cTnT) induced by cytotoxic antibiotics such as doxorubicin and daunorubicin, cardiomyocyte hypertrophy phenotypes marked by cell elongation, loss of membrane integrity or apoptosis (McKinsey and Olson, 2005), or perturbation in calcium handling. Similarly, several AP properties indicative of cardiotoxicity involving ion channels other than hERG can be analyzed (Ma et al., 2011). For example, inhibition of I_{Na} current in phase 0 results in a reduction in upstroke velocity, whereas altered AP duration (APD_{30}) can report on I_{Ca} function in phase II, and APD_{40-90} (known as triangulation) on I_K in phase III.

There is well-documented variability in patient susceptibility to both the therapeutic and toxicological effects of drugs, and libraries of iPSCs derived from such cohorts of patients can be used to recapitulate the heterogeneity of the patient population and delineate the molecular signature underpinning the variable human drug responses. To this end, a proof-of-concept study demonstrated the differential sensitivity of iPSC-CMs derived from patients with different cardiac disorders such as LQTS, HCM, and DCM to cisapride, a known hERG blocker whose tendency to induce electrophysiological perturbations was not detected until after FDA approval (Liang et al., 2013). Use of such cardiac disease-specific panels of iPSC-CMs will not only result in better prediction of cardiotoxicity but will also help in risk stratification based on the patient response and reduction in late-stage clinical failures (Engle and Puppala, 2013).

An increasing number of studies have demonstrated the feasibility of differentiating iPSCs into different cell types of the human body such as dermal, neuronal, gastrointestinal, blood, pancreatic, and hepatic cells (Murry and Keller, 2008; Inoue et al., 2014). Interestingly, the degree of concordance between traditional animal toxicology aimed at assessing the organ-specific toxicity (e.g., gastrointestinal, cardiovascular, hematological, and hepato-/nephrotoxicity) of new drugs is in the range of <10 to ~70% of all human adverse effects (Olson et al., 2000; Greaves et al., 2004). Therefore, the inclusion of different iPSC-derived somatic cell types in preclinical toxicological studies could potentially augment the degree of concordance between preclinical toxicological data and that observed in humans. A retrospective evaluation of compounds that have shown an organ-specific toxicity either in advance stages of preclinical toxicity testing or in the clinic using iPSC-based cellular models could be done to demonstrate the superior predictability of iPSCs in toxicological testing.

Taken together, the clinical importance of safety pharmacological and toxicological studies is very obvious. However, there is also a substantial economic consequence should adverse effects be found that limit the clinical utility of a given drug, thereby restricting its market potential (e.g., FDA black box warning) or even requiring its removal from the market altogether (Mordwinkin et al., 2013). Thus, the early inclusion of iPSC-CMs (and other organ-specific somatic cell types) from healthy subjects and patients with various cardiovascular diseases may serve as a better phenotypic predictive model in safety pharmacology and toxicological assessment (Fig. 22.3) and may decrease the economic and human costs associated with drug withdrawal that is caused by unpredictable and untoward cardiotoxicity.

22.6 LIMITATIONS AND CHALLENGES

The preceding discussion demonstrates that in the past decade, a concerted and collective effort undertaken by funding agencies, academia, and industry has resulted in significant advances in our understanding of the developmental biology of iPSCs. These include methods of reprogramming iPSCs from a variety of somatic cell sources, as well as differentiation into specific cell lineages (Burridge et al., 2012). There has also been significant progress in utilizing these cells, particularly hPSC-CMs, to recapitulate disease phenotypes *in vitro*, and as human biology-relevant novel platforms to model diseases and assess the pharmacological efficacy and cardiotoxicity of both established and investigational drugs (Table 22.1). Despite these exciting steps forward, there are still major challenges associated with the widespread use of hPSC technology in drug discovery and toxicological testing, including (i) improving reprogramming efficiency, (ii) minimizing intra- and interline variability as well as genomic instability or abnormal karyotypes, (iii) reducing the high costs associated with scalable generation of hPSC derivatives to meet HTS requirements, and (iv) driving maturity of the hPSC-CMs (in terms of structure, calcium cycling, and electrophysiology) to a more adult phenotype.

With respect to reprogramming, major strides have been made in the use of nonintegrating and small-molecule-based methods. Regardless of the methods used, however, the overall reprogramming efficiency remains <1%, thus significantly increasing the time, manpower, and costs associated with this process. Moreover, there are considerable intra- and interline variations among different iPSC lines owing to residual somatic cell epigenetic memory, thus requiring subculturing before they can be differentiated efficiently into a specific cell lineage. The presence of karyotypic abnormality following long-term culture and repeated cell passaging is another challenge. Going forward, there is a need to develop more efficient, time- and cost-effective reprogramming processes to enable seamless reprogramming of somatic cell samples from large cohort of healthy and patient populations. Encouragingly, there has been a significant improvement in the cardiac differentiation efficiency of hPSCs with small-molecule-based methods yielding large quantities of high-purity hPSC-CMs under chemically defined media condition

(Burridge et al., 2014). In addition, strategies to grow hPSCs in stirred suspension systems have been successful, thereby reducing the user input while increasing the overall yield (Chen et al., 2012).

Independent of reprogramming and cardiac differentiation methods and hPSC lines, the resultant hPSC-CMs possess properties that are more akin to the human fetal CM phenotype than adult CMs (Mummery et al., 2012; Keung et al., 2014; van den Heuvel et al., 2014; Yang et al., 2014). hPSC-CMs are smaller in size and have irregular shapes (round/oval rather than rodlike shapes) (Mummery et al., 2012). In contrast to adult human cardiomyocytes, hPSC-CMs also show unorganized sarcomeres, absence of T-tubules (resulting in much slower excitation–contraction coupling), and a depolarized resting membrane potential manifested as spontaneous beating (i.e., automaticity) of ventricular- and atrial-like hPSC-CMs. By contrast, freshly dissociated human adult CMs are quiescent and beat or contract in response to stimuli (Yang et al., 2014). This increased automaticity of hPSC-CMs can be attributed to the presence of large funny current (I_f) or pacemaker current, plus underexpression of the inward rectifier potassium channel (I_{K1}), with a consequential large inward sodium current (Satin et al., 2004; Yang et al., 2014). Moreover, present cardiac differentiation methods produce a mixed population of ventricular-, atrial-, and nodal-like cells with ventricular-like CMs being the predominant cell type. This represents a potential limitation in modeling cardiac diseases that affect a specific cardiomyocyte type, especially if the AP is not the readout assay of the cellular phenotype, or in high-throughput drug screening. As mentioned previously, even fully differentiated hPSC-CMs contain a spectrum of APs with continuous ranges of properties instead of three highly distinct subpopulations of cardiac AP types. This heterogeneity corresponds to various stages of development and maturity as corroborated by single cell transcriptional profiling, which is also suggestive of significant intercell line variability (Narsinh et al., 2011; Yang et al., 2014). Currently, there are no effective and validated strategies to produce a pure population of either ventricular-, atrial-, or nodal-like CMs to model specific cardiac diseases (e.g., atrial fibrillation, ventricular arrhythmias, etc.). To this end, recent studies have shown a small-molecule-based differentiation of hPSCs toward a highly enriched population of ventricular-like CMs (Karakikes et al., 2014), an enhanced proportion of nodal-like cells (Zhu et al., 2010), and increased atrial versus ventricular specification during cardiac ESC differentiation (Zhang et al., 2011), potentially opening the door to modeling cardiac diseases specific to a subtype of CMs.

Several studies have devised strategies to increase the maturity of hPSC-CMs, such as extended time in culture, overexpressing a deficient ion channel (e.g., I_{K1}) (Lieu et al., 2013) or cardiac specific microRNA (e.g., miR-1), tissue engineering approaches (including modulating substrate stiffness, mechanical loading or stretching, 3D culture, and electrically pacing) (Nunes et al., 2013), or pharmacological and neurohormonal treatment (e.g., adrenergic stimulation and triiodothyronine) (Yang et al., 2014). These approaches promise to significantly increase the overall maturity of hPSC-CMs, thereby increasing the validity of hPSC-CM-based cell models in simulating human diseases in a dish. A detailed discussion of maturation strategy is beyond the scope of this chapter, and interested readers are encouraged to refer to other reviews (Matsa et al., 2014; Yang et al., 2014).

Lastly, at present the high cost of generating, maintaining, and differentiating desired cell type is at least two to three orders of magnitude higher than the cost for immortal cell lines currently used in HTS by pharmaceutical R&D (McGivern and Ebert, 2014). The peculiar nature of costly media components needed to support both the undifferentiated state and cardiac differentiation of hPSCs, and the average time needed to generate and establish multiple iPSC lines followed by culture of differentiated derivatives, could take upward of several months, requiring large amounts of media and dedicated skilled personnel. As discussed previously, the recent development of stirred suspension culture systems holds great promise to reduce the time and effort needed for scaling of hPSC cultures while reducing the maintenance associated with error-prone, user-dependent adherent culture systems. This, coupled with development of simplified, low-cost chemically defined cardiac differentiation medium (Burridge et al., 2012), will eventually reduce the overall cost associated with hPSC technology while increasing the scalability and yield to levels amenable to HTS specifications.

22.7 CONCLUSIONS AND FUTURE PERSPECTIVE

Considerable progress has been made in understanding the fundamentals of stem cell biology and human genetics, allowing efficient generation of iPSCs and derivatives like iPSC-CMs. The methods of somatic cell reprogramming and cardiac differentiation have improved dramatically, including development of footprint-free, small-molecule-based reprogramming, as well as GMP-compliant, scalable, suspension culture systems for hPSCs and their derivatives. Furthermore, the versatile ability of patient-specific iPSC-CMs in recapitulating the cardiac disease phenotype and predicting the therapeutic effects and cardiotoxicity of known and investigational agents has been well demonstrated and validated. However, before the full potential of iPSC-CMs can be realized in drug screening and toxicological testing, challenges pertaining to scalable generation of iPSC-CMs with consistent quality and purity, as well as the relative immaturity of iPSC-CMs, need to be addressed. The efficiency and reproducibility of the differentiation protocols must be improved to enable cost-effective production suitable for industry. Thus far, low-throughput assays (Fig. 22.2) have been used to characterize the iPSC-CMs, and efforts are underway to develop automated high-throughput platforms for large-scale drug screening. In summary, the advent of iPSC technology provides an unprecedented opportunity to generate virtually limitless quantities of patient-specific iPSC-CMs for studying cardiac disease mechanisms, drug screening, and toxicological testing (Fig. 22.1).

ACKNOWLEDGMENTS

We would like to thank Blake Wu and Joseph Gold for critical reading of the manuscript. This work is supported by research grants from American Heart Association 13EIA14420025, NIH R01 HL113006, NIH R01 HL123968, and NIH R01 HL126527 (JCW). Due to space limitations, we are unable to include all of the important citations relevant to this subject; we apologize to those investigators whose work was omitted here.

REFERENCES

Y. A. Abassi, B. Xi, N. Li, W. Ouyang, A. Seiler, M. Watzele, R. Kettenhofen, H. Bohlen, A. Ehlich, E. Kolossov, Dynamic monitoring of beating periodicity of stem cell-derived cardiomyocytes as a predictive tool for preclinical safety assessment. *Br J Pharmacol* 165, 1424–1441 (2012).

F. Anokye-Danso, C. M. Trivedi, D. Juhr, M. Gupta, Z. Cui, Y. Tian, Y. Zhang, W. Yang, P. J. Gruber, J. A. Epstein, Highly efficient miRNA-mediated reprogramming of mouse and human somatic cells to pluripotency. *Cell Stem Cell* 8, 376–388 (2011).

D. E. Arking, A. Chakravarti, Understanding cardiovascular disease through the lens of genome-wide association studies. *Trends Genet* 25, 387–394 (2009).

T. T. Ashburn, K. B. Thor, Drug repositioning: identifying and developing new uses for existing drugs. *Nat Rev Drug Discov* 3, 673–683 (2004).

M. Brandenburger, J. Wenzel, R. Bogdan, D. Richardt, F. Nguemo, M. Reppel, J. Hescheler, H. Terlau, A. Dendorfer, Organotypic slice culture from human adult ventricular myocardium. *Cardiovasc Res* 93, 50–59 (2012).

P. W. Burridge, G. Keller, J. D. Gold, J. C. Wu, Production of de novo cardiomyocytes: human pluripotent stem cell differentiation and direct reprogramming. *Cell Stem Cell* 10, 16–28 (2012).

P. W. Burridge, E. Matsa, P. Shukla, Z. C. Lin, J. M. Churko, A. D. Ebert, F. Lan, S. Diecke, B. Huber, N. M. Mordwinkin, Chemically defined generation of human cardiomyocytes. *Nat Methods* 11, 855–860 (2014).

X. Carvajal-Vergara, A. Sevilla, S. L. D'Souza, Y.-S. Ang, C. Schaniel, D.-F. Lee, L. Yang, A. D. Kaplan, E. D. Adler, R. Rozov, Patient-specific induced pluripotent stem-cell-derived models of LEOPARD syndrome. *Nature* 465, 808–812 (2010).

I. Cavero, W. Crumb, ICH S7B draft guideline on the non-clinical strategy for testing delayed cardiac repolarisation risk of drugs: a critical analysis. *Expert Opin Drug Saf* 4, 509–530 (2005).

V. C. Chen, S. M. Couture, J. Ye, Z. Lin, G. Hua, H.-I. P. Huang, J. Wu, D. Hsu, M. K. Carpenter, L. A. Couture, Scalable GMP compliant suspension culture system for human ES cells. *Stem Cell Res* 8, 388–402 (2012).

C. R. Chong, X. Chen, L. Shi, J. O. Liu, D. J. Sullivan, A clinical drug library screen identifies astemizole as an antimalarial agent. *Nat Chem Biol* 2, 415–416 (2006).

J. M. Churko, G. L. Mantalas, M. P. Snyder, J. C. Wu, Overview of high throughput sequencing technologies to elucidate molecular pathways in cardiovascular diseases. *Circ Res* 112, 1613–1623 (2013).

J. S. Cordes, Z. Sun, D. B. Lloyd, J. A. Bradley, A. C. Opsahl, M. W. Tengowski, X. Chen, J. Zhou, Pentamidine reduces hERG expression to prolong the QT interval. *Br J Pharmacol* 145, 15–23 (2005).

W. J. Crumb, B. Wible, D. J. Arnold, J. P. Payne, A. M. Brown, Blockade of multiple human cardiac potassium currents by the antihistamine terfenadine: possible mechanism for terfenadine-associated cardiotoxicity. *Mol Pharmacol* 47, 181–190 (1995).

S. Desmond-Hellmann, The Cost of Creating a New Drug Now $5 Billion, Pushing Big Pharma to Change. *Forbes 11th of August*, (2013).

J. Ducroq, R. Printemps, S. Guilbot, J. Gardette, C. Salvetat, M. Le Grand, Action potential experiments complete hERG assay and QT-interval measurements in cardiac preclinical studies. *J Pharmacol Toxicol Methods* 56, 159–170 (2007).

S. J. Engle, D. Puppala, Integrating human pluripotent stem cells into drug development. *Cell Stem Cell* 12, 669–677 (2013).

N. Ferri, P. Siegl, A. Corsini, J. Herrmann, A. Lerman, R. Benghozi, Drug attrition during pre-clinical and clinical development: understanding and managing drug-induced cardiotoxicity. *Pharmacol Ther* 138, 470–484 (2013).

T. Force, K. L. Kolaja, Cardiotoxicity of kinase inhibitors: the prediction and translation of preclinical models to clinical outcomes. *Nat Rev Drug Discov* 10, 111–126 (2011).

N. Fusaki, H. Ban, A. Nishiyama, K. Saeki, M. Hasegawa, Efficient induction of transgene-free human pluripotent stem cells using a vector based on Sendai virus, an RNA virus that does not integrate into the host genome. *Proc Jpn Acad Ser B Phys Biol Sci* 85, 348–362 (2009).

P. Greaves, A. Williams, M. Eve, First dose of potential new medicines to humans: how animals help. *Nat Rev Drug Discov* 3, 226–236 (2004).

M. Grskovic, A. Javaherian, B. Strulovici, G. Q. Daley, Induced pluripotent stem cells—opportunities for disease modelling and drug discovery. *Nat Rev Drug Discov* 10, 915–929 (2011).

M. Hanna, Matching taxpayer funding to population health needs. *Circ Res* 116, 1296–1300 (2015).

N. H. van den Heuvel, T. A. van Veen, B. Lim, M. K. Jonsson, Lessons from the heart: mirroring electrophysiological characteristics during cardiac development to in vitro differentiation of stem cell derived cardiomyocytes. *J Mol Cell Cardiol* 67, 12–25 (2014).

M. Hoekstra, C. L. Mummery, A. A. Wilde, C. R. Bezzina, A. O. Verkerk, Induced pluripotent stem cell derived cardiomyocytes as models for cardiac arrhythmias. *Front Physiol* 3, 346 (2012).

P. Hou, Y. Li, X. Zhang, C. Liu, J. Guan, H. Li, T. Zhao, J. Ye, W. Yang, K. Liu, Pluripotent stem cells induced from mouse somatic cells by small-molecule compounds. *Science* 341, 651–654 (2013).

P. D. Hsu, E. S. Lander, F. Zhang, Development and applications of CRISPR-Cas9 for genome engineering. *Cell* 157, 1262–1278 (2014).

D. Huangfu, K. Osafune, R. Maehr, W. Guo, A. Eijkelenboom, S. Chen, W. Muhlestein, D. A. Melton, Induction of pluripotent stem cells from primary human fibroblasts with only Oct4 and Sox2. *Nat Biotechnol* 26, 1269–1275 (2008).

S. M. Hussein, N. N. Batada, S. Vuoristo, R. W. Ching, R. Autio, E. Närvä, S. Ng, M. Sourour, R. Hämäläinen, C. Olsson, Copy number variation and selection during reprogramming to pluripotency. *Nature* 471, 58–62 (2011).

L. Hutchinson, R. Kirk, High drug attrition rates—where are we going wrong? *Nat Rev Clin Oncol* 8, 189–190 (2011).

ICH. Guideline on S7B, The non-clinical evaluation of the potential for delayed ventricular repolarization (Qt Interval Prolongation) by human pharmaceuticals. (http://www.ich.org/products/guidelines/safety/article/safety-guidelines.html, accessed October 21, 2015) (2005).

J. K. Ichida, J. Blanchard, K. Lam, E. Y. Son, J. E. Chung, D. Egli, K. M. Loh, A. C. Carter, F. P. Di Giorgio, K. Koszka, A small-molecule inhibitor of Tgf-β signaling replaces Sox2 in reprogramming by inducing Nanog. *Cell Stem Cell* 5, 491–503 (2009).

H. Inoue, N. Nagata, H. Kurokawa, S. Yamanaka, iPS cells: a game changer for future medicine. *EMBO J* 33, 409–417 (2014).

E. Iorns, C. J. Lord, N. Turner, A. Ashworth, Utilizing RNA interference to enhance cancer drug discovery. *Nat Rev Drug Discov* 6, 556–568 (2007).

I. Itzhaki, L. Maizels, I. Huber, L. Zwi-Dantsis, O. Caspi, A. Winterstern, O. Feldman, A. Gepstein, G. Arbel, H. Hammerman, Modelling the long QT syndrome with induced pluripotent stem cells. *Nature* 471, 225–229 (2011).

I. Itzhaki, L. Maizels, I. Huber, A. Gepstein, G. Arbel, O. Caspi, L. Miller, B. Belhassen, E. Nof, M. Glikson, Modeling of catecholaminergic polymorphic ventricular tachycardia with patient-specific human-induced pluripotent stem cells. *JACC* 60, 990–1000 (2012).

F. Jia, K. D. Wilson, N. Sun, D. M. Gupta, M. Huang, Z. Li, N. J. Panetta, Z. Y. Chen, R. C. Robbins, M. A. Kay, A nonviral minicircle vector for deriving human iPS cells. *Nat Methods* 7, 197–199 (2010).

Y. Jiang, S. Habibollah, K. Tilgner, J. Collin, T. Barta, J. Y. Al-Aama, L. Tesarov, R. Hussain, A. W. Trafford, G. Kirkwood, An induced pluripotent stem cell model of hypoplastic left heart syndrome (HLHS) reveals multiple expression and functional differences in HLHS-derived cardiac myocytes. *Stem Cells Transl Med* 3, 416–423 (2014).

C. B. Jung, A. Moretti, M. Mederos y Schnitzler, L. Iop, U. Storch, M. Bellin, T. Dorn, S. Ruppenthal, S. Pfeiffer, A. Goedel, Dantrolene rescues arrhythmogenic RYR2 defect in a patient-specific stem cell model of catecholaminergic polymorphic ventricular tachycardia. *EMBO Mol Med* 4, 180–191 (2012).

S. Kaese, S. Verheule, Cardiac electrophysiology in mice: a matter of size. *Front Physiol* 3, 345 (2012).

P. Kannankeril, D. M. Roden, D. Darbar, Drug-induced long QT syndrome. *Pharmacol Rev* 62, 760–781 (2010).

I. Karakikes, G. D. Senyei, J. Hansen, C.-W. Kong, E. U. Azeloglu, F. Stillitano, D. K. Lieu, J. Wang, L. Ren, J.-S. Hulot, Small molecule-mediated directed differentiation of human embryonic stem cells toward ventricular cardiomyocytes. *Stem Cells Transl Med* 3, 18–31 (2014).

I. Kehat, D. Kenyagin-Karsenti, M. Snir, H. Segev, M. Amit, A. Gepstein, E. Livne, O. Binah, J. Itskovitz-Eldor, L. Gepstein, Human embryonic stem cells can differentiate into myocytes with structural and functional properties of cardiomyocytes. *J Clin Invest* 108, 407 (2001).

H. Kempf, R. Olmer, C. Kropp, M. Rückert, M. Jara-Avaca, D. Robles-Diaz, A. Franke, D. A. Elliott, D. Wojciechowski, M. Fischer, Controlling expansion and cardiomyogenic differentiation of human pluripotent stem cells in scalable suspension culture. *Stem Cell Rep* 3, 1132–1146 (2014).

W. Keung, K. R. Boheler, R. A. Li, Developmental cues for the maturation of metabolic, electrophysiological and calcium handling properties of human pluripotent stem cell-derived cardiomyocytes. *Stem Cell Res Ther* 5, 17 (2014).

D. Kim, C.-H. Kim, J.-I. Moon, Y.-G. Chung, M.-Y. Chang, B.-S. Han, S. Ko, E. Yang, K. Y. Cha, R. Lanza, Generation of human induced pluripotent stem cells by direct delivery of reprogramming proteins. *Cell Stem Cell* 4, 472 (2009).

K. Kim, A. Doi, B. Wen, K. Ng, R. Zhao, P. Cahan, J. Kim, M. Aryee, H. Ji, L. Ehrlich, Epigenetic memory in induced pluripotent stem cells. *Nature* 467, 285–290 (2010).

C. Kim, J. Wong, J. Wen, S. Wang, C. Wang, S. Spiering, N. G. Kan, S. Forcales, P. L. Puri, T. C. Leone, Studying arrhythmogenic right ventricular dysplasia with patient-specific iPSCs. *Nature* 494, 105–110 (2013).

I. Kola, J. Landis, Can the pharmaceutical industry reduce attrition rates? *Nat Rev Drug Discov* 3, 711–716 (2004).

K. Kujala, J. Paavola, A. Lahti, K. Larsson, M. Pekkanen-Mattila, M. Viitasalo, A. M. Lahtinen, L. Toivonen, K. Kontula, H. Swan, Cell model of catecholaminergic polymorphic ventricular tachycardia reveals early and delayed afterdepolarizations. *PLoS One* 7, e44660 (2012).

Y. A. Kuryshev, E. Ficker, L. Wang, P. Hawryluk, A. T. Dennis, B. A. Wible, A. M. Brown, J. Kang, X.-L. Chen, K. Sawamura, Pentamidine-induced long QT syndrome and block of hERG trafficking. *J Pharmacol Exp Ther* 312, 316–323 (2005).

A. E. Lacerda, Y. A. Kuryshev, Y. Chen, M. Renganathan, H. Eng, S. J. Danthi, J. W. Kramer, T. Yang, A. M. Brown, Alfuzosin delays cardiac repolarization by a novel mechanism. *J Pharmacol Exp Ther* 324, 427–433 (2008).

F. Lan, A. S. Lee, P. Liang, V. Sanchez-Freire, P. K. Nguyen, L. Wang, L. Han, M. Yen, Y. Wang, N. Sun, Abnormal calcium handling properties underlie familial hypertrophic cardiomyopathy pathology in patient-specific induced pluripotent stem cells. *Cell Stem Cell* 12, 101–113 (2013).

C. Lawrence, C. Pollard, T. Hammond, J. P. Valentin, In vitro models of proarrhythmia. *Br J Pharmacol* 154, 1516–1522 (2008).

Y. Lei, D. V. Schaffer, A fully defined and scalable 3D culture system for human pluripotent stem cell expansion and differentiation. *Proc Natl Acad Sci U S A* 110, E5039–E5048 (2013).

P. Liang, F. Lan, A. S. Lee, T. Gong, V. Sanchez-Freire, Y. Wang, S. Diecke, K. Sallam, J. W. Knowles, P. J. Wang, Drug screening using a library of human induced pluripotent stem cell–derived cardiomyocytes reveals disease-specific patterns of cardiotoxicity. *Circulation* 127, 1677–1691 (2013).

D. K. Lieu, J.-D. Fu, N. Chiamvimonvat, K. W. C. Tung, G. P. McNerney, T. Huser, G. Keller, C.-W. Kong, R. A. Li, Mechanism-based facilitated maturation of human pluripotent stem cell-derived cardiomyocytes. *Circ Arrhythm Electrophysiol* 6, 191–201 (2013).

W. Lowry, L. Richter, R. Yachechko, A. Pyle, J. Tchieu, R. Sridharan, A. Clark, K. Plath, Generation of human induced pluripotent stem cells from dermal fibroblasts. *Proc Natl Acad Sci U S A* 105, 2883–2888 (2008).

J. Ma, L. Guo, S. J. Fiene, B. D. Anson, J. A. Thomson, T. J. Kamp, K. L. Kolaja, B. J. Swanson, C. T. January, High purity human-induced pluripotent stem cell-derived cardiomyocytes: electro-physiological properties of action potentials and ionic currents. *Am J Physiol Heart Circ Physiol* 301, H2006–H2017 (2011).

N. Maherali, K. Hochedlinger, Tgfβ signal inhibition cooperates in the induction of iPSCs and replaces Sox2 and cMyc. *Curr Biol* 19, 1718–1723 (2009).

M. Malik, A. J. Camm, Evaluation of drug-induced QT interval prolongation: implications for drug approval and labelling. *Drug Saf* 24, 323–351 (2001).

C. F. Mandenius, D. Steel, F. Noor, T. Meyer, E. Heinzle, J. Asp, S. Arain, U. Kraushaar, S. Bremer, R. Class, Cardiotoxicity testing using pluripotent stem cell-derived human cardiomyocytes and state-of-the-art bioanalytics: a review. *J Appl Toxicol* 31, 191–205 (2011).

E. Matsa, D. Rajamohan, E. Dick, L. Young, I. Mellor, A. Staniforth, C. Denning, Drug evaluation in cardiomyocytes derived from human induced pluripotent stem cells carrying a long QT syndrome type 2 mutation. *Eur Heart J* 32, 952–962 (2011).

E. Matsa, P. W. Burridge, J. C. Wu, Human stem cells for modeling heart disease and for drug discovery. *Sci Transl Med* 6, 239ps236 (2014).

J. V. McGivern, A. D. Ebert, Exploiting pluripotent stem cell technology for drug discovery, screening, safety, and toxicology assessments. *Adv Drug Deliv Rev* 69, 170–178 (2014).

T. A. McKinsey, E. N. Olson, Toward transcriptional therapies for the failing heart: chemical screens to modulate genes. *J Clin Invest* 115, 538 (2005).

N. Miyoshi, H. Ishii, H. Nagano, N. Haraguchi, D. L. Dewi, Y. Kano, S. Nishikawa, M. Tanemura, K. Mimori, F. Tanaka, Reprogramming of mouse and human cells to pluripotency using mature microRNAs. *Cell Stem Cell* 8, 633–638 (2011).

S. M. Modell, M. H. Lehmann, The long QT syndrome family of cardiac ion channelopathies: a HuGE review. *Genet Med* 8, 143–155 (2006).

B. P. Monahan, C. L. Ferguson, E. S. Killeavy, B. K. Lloyd, J. Troy, L. R. Cantilena, Torsades de pointes occurring in association with terfenadine use. *JAMA* 264, 2788–2790 (1990).

N. M. Mordwinkin, P. W. Burridge, J. C. Wu, A review of human pluripotent stem cell-derived cardiomyocytes for high-throughput drug discovery, cardiotoxicity screening, and publication standards. *J Cardiovasc Transl Res* 6, 22–30 (2013).

A. Moretti, M. Bellin, A. Welling, C. B. Jung, J. T. Lam, L. Bott-Flügel, T. Dorn, A. Goedel, C. Höhnke, F. Hofmann, Patient-specific induced pluripotent stem-cell models for long-QT syndrome. *N Engl J Med* 363, 1397–1409 (2010).

C. L. Mummery, J. Zhang, E. S. Ng, D. A. Elliott, A. G. Elefanty, T. J. Kamp, Differentiation of human embryonic stem cells and induced pluripotent stem cells to cardiomyocytes a methods overview. *Circ Res* 111, 344–358 (2012).

B. H. Munos, W. W. Chin, A call for sharing: adapting pharmaceutical research to new realities. *Sci Transl Med* 1, 9cm8 (2009).

C. E. Murry, G. Keller, Differentiation of embryonic stem cells to clinically relevant populations: lessons from embryonic development. *Cell* 132, 661–680 (2008).

K. Musunuru, Genome editing of human pluripotent stem cells to generate human cellular disease models. *Dis Model Mech* 6, 896–904 (2013).

K. H. Narsinh, N. Sun, V. Sanchez-Freire, A. S. Lee, P. Almeida, S. Hu, T. Jan, K. D. Wilson, D. Leong, J. Rosenberg, Single cell transcriptional profiling reveals heterogeneity of human induced pluripotent stem cells. *J Clin Invest* 121, 1217 (2011).

N. C. Ndiaye, M. A. Nehzad, S. El Shamieh, M. G. Stathopoulou, S. Visvikis-Siest, Cardiovascular diseases and genome-wide association studies. *Clin Chim Acta* 412, 1697–1701 (2011).

K. Nishimura, M. Sano, M. Ohtaka, T. Furuta, Y. Umemura, Y. Nakajima, Y. Ikehara, T. Kobayashi, H. Segawa, S. Takayasu, Development of defective and persistent Sendai virus vector a unique gene delivery/expression system ideal for cell reprogramming. *J Biol Chem* 286, 4760–4771 (2011).

A. Novak, L. Barad, N. Zeevi-Levin, R. Shick, R. Shtrichman, A. Lorber, J. Itskovitz-Eldor, O. Binah, Cardiomyocytes generated from CPVTD307H patients are arrhythmogenic in response to β-adrenergic stimulation. *J Cell Mol Med* 16, 468–482 (2012).

S. S. Nunes, J. W. Miklas, J. Liu, R. Aschar-Sobbi, Y. Xiao, B. Zhang, J. Jiang, S. Massé, M. Gagliardi, A. Hsieh, Biowire: a platform for maturation of human pluripotent stem cell-derived cardiomyocytes. *Nat Methods* 10, 781–787 (2013).

K. Okita, M. Nakagawa, H. Hyenjong, T. Ichisaka, S. Yamanaka, Generation of mouse induced pluripotent stem cells without viral vectors. *Science* 322, 949–953 (2008).

H. Olson, G. Betton, D. Robinson, K. Thomas, A. Monro, G. Kolaja, P. Lilly, J. Sanders, G. Sipes, W. Bracken, Concordance of the toxicity of pharmaceuticals in humans and in animals. *Regul Toxicol Pharmacol* 32, 56–67 (2000).

T. I. Oprea, J. E. Bauman, C. G. Bologa, T. Buranda, A. Chigaev, B. S. Edwards, J. W. Jarvik, H. D. Gresham, M. K. Haynes, B. Hjelle, Drug repurposing from an academic perspective. *Drug Discov Today Ther Strat* 8, 61–69 (2012).

S. Peng, A. E. Lacerda, G. E. Kirsch, A. M. Brown, A. Bruening-Wright, The action potential and comparative pharmacology of stem cell-derived human cardiomyocytes. *J Pharmacol Toxicol Methods* 61, 277–286 (2010).

M. F. Pera, Stem cells: the dark side of induced pluripotency. *Nature* 471, 46–47 (2011).

W. Redfern, L. Carlsson, A. Davis, W. Lynch, I. MacKenzie, S. Palethorpe, P. Siegl, I. Strang, A. Sullivan, R. Wallis, Relationships between preclinical cardiac electrophysiology, clinical QT interval prolongation and torsade de pointes for a broad range of drugs: evidence for a provisional safety margin in drug development. *Cardiovasc Res* 58, 32–45 (2003).

L. Rochette, C. Guenancia, A. Gudjoncik, O. Hachet, M. Zeller, Y. Cottin, C. Vergely, Anthracyclines/trastuzumab: new aspects of cardiotoxicity and molecular mechanisms. *Trends Pharmacol Sci* 36, 326–348 (2015).

D. M. Roden, Drug-induced prolongation of the QT interval. *N Engl J Med* 350, 1013–1022 (2004).

D. M. Roden, B. F. Hoffman, Action potential prolongation and induction of abnormal automaticity by low quinidine concentrations in canine Purkinje fibers. Relationship to potassium and cycle length. *Circ Res* 56, 857–867 (1985).

K Sallam, Y Li, P. T. Sager, S.R. Houser, J. C. Wu, Finding the rhythm of sudden cardiac death: new opportunities using induced pluripotent stem cell–derived cardiomyocytes. *Circ Res* 116, 1989–2004 (2015).

V. Sanchez-Freire, A. S. Lee, S. Hu, O. J. Abilez, P. Liang, F. Lan, B. C. Huber, S.-G. Ong, W. X. Hong, M. Huang, J. C. Wu, Effect of human donor cell source on differentiation and function of cardiac induced pluripotent stem cells. *JACC* 64, 436–448 (2014).

J. Satin, I. Kehat, O. Caspi, I. Huber, G. Arbel, I. Itzhaki, J. Magyar, E. A. Schroder, I. Perlman, L. Gepstein, Mechanism of spontaneous excitability in human embryonic stem cell derived cardiomyocytes. *J Physiol* 559, 479–496 (2004).

K. Si-Tayeb, F. K. Noto, A. Sepac, F. Sedlic, Z. J. Bosnjak, J. W. Lough, S. A. Duncan, Generation of human induced pluripotent stem cells by simple transient transfection of plasmid DNA encoding reprogramming factors. *BMC Dev Biol* 10, 81 (2010).

A. Somers, J. C. Jean, C. A. Sommer, A. Omari, C. C. Ford, J. A. Mills, L. Ying, A. G. Sommer, J. M. Jean, B. W. Smith, Generation of transgene-free lung disease-specific human induced pluripotent stem cells using a single excisable lentiviral stem cell cassette. *Stem Cells* 28, 1728–1740 (2010).

C. A. Sommer, M. Stadtfeld, G. J. Murphy, K. Hochedlinger, D. N. Kotton, G. Mostoslavsky, Induced pluripotent stem cell generation using a single lentiviral stem cell cassette. *Stem Cells* 27, 543–549 (2009).

M. Stadtfeld, K. Brennand, K. Hochedlinger, Reprogramming of pancreatic β cells into induced pluripotent stem cells. *Curr Biol* 18, 890–894 (2008).

N. Sun, M. Yazawa, J. Liu, L. Han, V. Sanchez-Freire, O. J. Abilez, E. G. Navarrete, S. Hu, L. Wang, A. Lee, A. Pavlovic, S. Lin, R. Chen, R. J.Hajjar, M. P. Snyder, R. E. Dolmetsch, M. J. Butte, E. A. Ashley, M. T. Longaker, R. C. Robbins, J. C. Wu, Patient-specific induced pluripotent stem cells as a model for familial dilated cardiomyopathy. *Sci Transl Med* 4, 130ra147 (2012).

K. Takahashi, S. Yamanaka, Induction of pluripotent stem cells from mouse embryonic and adult fibroblast cultures by defined factors. *Cell* 126, 663–676 (2006).

K. Takahashi, K. Tanabe, M. Ohnuki, M. Narita, T. Ichisaka, T. Tomoda, S. Yamanaka, Induction of pluripotent stem cells from adult human fibroblasts by defined factors. *Cell* 131, 861–872 (2007).

C. Terrenoire, K. Wang, K. W. C. Tung, W. K. Chung, R. H. Pass, J. T. Lu, J.-C. Jean, A. Omari, K. J. Sampson, D. N. Kotton, Induced pluripotent stem cells used to reveal drug actions in a long QT syndrome family with complex genetics. *J Gen Physiol* 141, 61–72 (2013).

J. A. Thomson, J. Itskovitz-Eldor, S. S. Shapiro, M. A. Waknitz, J. J. Swiergiel, V. S. Marshall, J. M. Jones, Embryonic stem cell lines derived from human blastocysts. *Science* 282, 1145–1147 (1998).

L. M. Wang, M. Wong, J. M. Lightwood, C. M. Cheng, Black box warning contraindicated comedications: concordance among three major drug interaction screening programs. *Ann Pharmacother* 44, 28–34 (2010).

Y. Wang, P. Liang, F. Lan, H. Wu, L. Lisowski, M. Gu, S. Hu, M. A. Kay, F. D. Urnov, R. Shinnawi, Genome editing of isogenic human induced pluripotent stem cells recapitulates long QT phenotype for drug testing. *JACC* 64, 451–459 (2014a).

G. Wang, M. L. McCain, L. Yang, A. He, F. S. Pasqualini, A. Agarwal, H. Yuan, D. Jiang, D. Zhang, L. Zangi, Modeling the mitochondrial cardiomyopathy of Barth syndrome with induced pluripotent stem cell and heart-on-chip technologies. *Nat Med* 20, 616–623 (2014b).

L. Warren, P. D. Manos, T. Ahfeldt, Y.-H. Loh, H. Li, F. Lau, W. Ebina, P. K. Mandal, Z. D. Smith, A. Meissner, Highly efficient reprogramming to pluripotency and directed differentiation of human cells with synthetic modified mRNA. *Cell Stem Cell* 7, 618–630 (2010).

J. M. Wharton, P. A. Demopulos, N. Goldschlager, Torsade de pointes during administration of pentamidine isethionate. *Am J Med* 83, 571–576 (1987).

K. D. Wilson, J. C. Wu, Induced pluripotent stem cells. *JAMA* 313, 1613–1614 (2015).

K. Woltjen, I. P. Michael, P. Mohseni, R. Desai, M. Mileikovsky, R. Hämäläinen, R. Cowling, W. Wang, P. Liu, M. Gertsenstein, piggyBac transposition reprograms fibroblasts to induced pluripotent stem cells. *Nature* 458, 766–770 (2009).

X. Yang, L. Pabon, C. E. Murry, Engineering adolescence maturation of human pluripotent stem cell–derived cardiomyocytes. *Circ Res* 114, 511–523 (2014).

Y. G. Yap, A. J. Camm, Drug induced QT prolongation and torsades de pointes. *Heart* 89, 1363–1372 (2003).

M. Yazawa, B. Hsueh, X. Jia, A. M. Pasca, J. A. Bernstein, J. Hallmayer, R. E. Dolmetsch, Using induced pluripotent stem cells to investigate cardiac phenotypes in Timothy syndrome. *Nature* 471, 230–234 (2011).

J. Yu, M. A. Vodyanik, K. Smuga-Otto, J. Antosiewicz-Bourget, J. L. Frane, S. Tian, J. Nie, G. A. Jonsdottir, V. Ruotti, R. Stewart, Induced pluripotent stem cell lines derived from human somatic cells. *Science* 318, 1917–1920 (2007).

S. Zhang, Z. Zhou, Q. Gong, J. C. Makielski, C. T. January, Mechanism of block and identification of the verapamil binding domain to HERG potassium channels. *Circ Res* 84, 989–998 (1999).

Q. Zhang, J. Jiang, P. Han, Q. Yuan, J. Zhang, X. Zhang, Y. Xu, H. Cao, Q. Meng, L. Chen, Direct differentiation of atrial and ventricular myocytes from human embryonic stem cells by alternating retinoid signals. *Cell Res* 21, 579–587 (2011).

T. Zhao, Z.-N. Zhang, Z. Rong, Y. Xu, Immunogenicity of induced pluripotent stem cells. *Nature* 474, 212–215 (2011).

W. Zhou, C. R. Freed, Adenoviral gene delivery can reprogram human fibroblasts to induced pluripotent stem cells. *Stem Cells* 27, 2667–2674 (2009).

H. Zhou, S. Wu, J. Y. Joo, S. Zhu, D. W. Han, T. Lin, S. Trauger, G. Bien, S. Yao, Y. Zhu, Generation of induced pluripotent stem cells using recombinant proteins. *Cell Stem Cell* 4, 381–384 (2009).

W.-Z. Zhu, Y. Xie, K. W. Moyes, J. D. Gold, B. Askari, M. A. Laflamme, Neuregulin/ErbB signaling regulates cardiac subtype specification in differentiating human embryonic stem cells. *Circ Res* 107, 776–786 (2010).

23

STEM CELL-DERIVED RENAL CELLS AND PREDICTIVE RENAL *IN VITRO* MODELS

JACQUELINE KAI CHIN CHUAH, YUE NING LAM, PENG HUANG AND DANIELE ZINK

Institute of Bioengineering and Nanotechnology, The Nanos, Singapore

23.1 INTRODUCTION

Liver, heart, and kidney are major target organs for drug-induced toxicity. Drug-induced nephrotoxicity (DIN) can lead to acute kidney injury (AKI) or chronic kidney disease (Guo and Nzerue, 2002; Choudhury and Ahmed, 2006). Some drugs, such as aristolochic acid, can induce both, depending on the dose (Yang et al., 2012). DIN accounts for about 20% of hospital- or community-acquired cases of AKI (Naughton, 2008). In fact, many marketed drugs were found to cause nephrotoxicity, including antibiotics, anticancer drugs, immunosuppressants, nonsteroidal anti-inflammatory drugs (NSAIDs), and radiographic contrast agents (Guo and Nzerue, 2002; Choudhury and Ahmed, 2006; Naughton, 2008; Tiong et al., 2014). The frequent use of drugs that can cause nephrotoxicity in ICUs is one major reason why among ICU patients the rate of AKI is high and ranges between 30 and 60% (Bellomo, 2011). This is associated with a rate of mortality of 40–70% when dialysis is required (Uchino et al., 2005; Tolwani, 2012).

Development of less nephrotoxic drugs is difficult. Nephrotoxicity is typically detected only late during drug development and accounts for 2% of drug attrition during preclinical studies and 19% during phase III (Redfern, 2010). A central problem is the lack of preclinical models with high predictivity. Animal models are affected by interspecies variability. One reason for this is the different patterns of drug transporters and drug-metabolizing enzymes expressed in human and animal kidneys (Lohr et al., 1998; Tiong et al., 2014). The situation is also difficult with respect

to *in vitro* models. Validated or accepted renal *in vitro* models are currently not available. Renal *in vitro* models were usually used for mechanistic studies on few selected compounds (a comprehensive review on renal *in vitro* models that were published during this millennium is provided in Tiong et al. (2014)). Thus, due to the low numbers of compounds tested, the predictivity of these models could not be determined. So far, only six *in vitro* models have been tested with >10 compounds (Duff et al., 2002; Wu et al., 2009; Lin and Will, 2012; Li et al., 2013, 2014; Kandasamy et al., 2015). The three most recently developed models predicted renal proximal tubular toxicity in humans with high accuracy (Li et al., 2013, 2014; Kandasamy et al., 2015), and two of these predictive models were based on stem cell-derived renal cells. These very recent developments will be discussed in the following. Before stem cell-based approaches and predictive models will be discussed, we will briefly describe renal drug handling and the various cell types involved.

One reason why the kidney is so frequently affected by drug-induced toxicity lies in the fact that the kidney is an excretory organ that is highly exposed to all circulating substances. The blood flow to the kidneys is high and the kidneys receive ~25% of the resting cardiac output (Guo and Nzerue, 2002). Ultrafiltration of the blood occurs at the glomerulus. The filtration barrier consists of the glomerular endothelial cells, the glomerular basement membrane, and the epithelial podocytes (Patrakka and Tryggvason, 2010). Their foot processes form a slit diaphragm that contains specialized proteins such as podocin and nephrin (NPHS1) (Mundel and Shankland, 2002). Podocytes are damaged by a

Drug Discovery Toxicology: From Target Assessment to Translational Biomarkers, First Edition. Edited by Yvonne Will, J. Eric McDuffie, Andrew J. Olaharski, and Brandon D. Jeffy.

variety of small-molecule drugs that are filtered by the glomerulus, such as the antibiotic puromycin aminonucleoside (Marshall et al., 2006) and the anticancer drug doxorubicin (Tacar et al., 2013).

Together, the glomerulus and the so-called renal tubule form the functional unit of the kidney, the nephron. Next to the glomerulus is the so-called renal proximal tubule (PT), which comes first into contact with the glomerular filtrate and reabsorbs ~95% of its water. Water reabsorption in the kidney is facilitated by aquaporins (AQP), and AQP1 is expressed in the PT (Maunsbach et al., 1997). The PT is lined by polarized proximal tubular cells (PTC), which form a simple epithelium. PTC possess an apical (luminal) brush border, through which they reabsorb electrolytes, glucose, proteins, and peptides from the filtrate. In addition, PTC control the pH of blood and urine through bicarbonate reabsorption and ammoniagenesis, and they produce the most active form of vitamin D (Brenner, 2008).

After passing through the PT, the filtrate flows through the loop of Henle and the distal tubules into the collecting ducts. The more distal parts of the nephron and the collecting ducts are mainly involved in further concentrating the urine and in regulating the electrolyte balance. Their contributions are hormone regulated and variable. Some drugs primarily damage the distal tubules and collecting ducts. This applies, for instance, to lithium (Choudhury and Ahmed, 2006).

The main renal target for compound-induced toxicity is the PTC due to its roles in active transport, metabolism, secretion, and reabsorption of endogenous and xenobiotic organic compounds and drugs. Many drugs are not efficiently cleared by ultrafiltration at the glomerulus, and PTC make major contributions to their clearance by active transport and secretion. Uptake of compounds from the blood is achieved through different organic anion and cation transporters, which are expressed at the basolateral membranes. The most important basolateral uptake transporters are the organic anion transporters (OAT) 1 and 3 and the organic cation transporter (OCT) 2 (Burckhardt and Burckhardt, 2003; Sekine et al., 2006; Burckhardt, 2012; Morrissey et al., 2012, 2013). OCT2 is important, for instance, for uptake of cisplatin, which has dose-limiting nephrotoxicity (Ciarimboli, 2014).

After uptake, organic compounds and their metabolites are further transported into the filtrate by efflux transporters expressed at the apical sides of the PTC, such as P-glycoprotein (multidrug resistance (MDR) 1), multidrug resistance-associated protein (MRP) 2 and 4, multidrug and toxin extrusion (MATE) 1, and others (Imaoka et al., 2007; Yonezawa and Inui, 2011; Morrissey et al., 2012, 2013). In fact, PTC do not only actively transport organic compounds, but they also metabolize them and express a large variety of drug-metabolizing enzymes, such as cytochrome P450 (CYP) enzymes, UDP-glucuronosyltransferases, or enzymes involved in glutathione metabolism (Lohr et al., 1998; Lash et al., 2008). For instance, conversion of the anticancer drug

ifosfamide into metabolites by CYP3A4 and 2B6 seems to be critical for PTC injury by this anticancer drug (Aleksa et al., 2005). Peptide transporters (PEPT1 and PEPT2) and the endocytotic megalin/cubilin receptor complex mediate drug uptake at the apical side of PTC. Accumulation of aminoglycoside antibiotics in PTC results from the presence of the megalin/cubilin complex (Quiros et al., 2011).

Aminoglycoside antibiotics (gentamicin, tobramycin) are directly toxic for PTC, as well as a large variety of other drugs including other antibiotics (cephalosporin antibiotics, tetracycline), anticancer drugs (cisplatin, ifosfamide), immunosuppressants (cyclosporine A, tacrolimus), radiocontrast agents, and Chinese herbal drugs (Guo and Nzerue, 2002; Choudhury and Ahmed, 2006; Naughton, 2008; Yang et al., 2012; Tiong et al., 2014). Directly toxic for PTC are also a variety of fungal toxins (citrinin, ochratoxin A) (Lockard et al., 1980; Pfohl-Leszkowicz and Manderville, 2007; Pfohl-Leszkowicz et al., 2007; Reddy and Bhoola, 2010), as well as herbicides (paraquat) (Vaziri et al., 1979; Gil et al., 2005) and environmental toxicants (heavy metals) (Madden and Fowler, 2000; Huang et al., 2009; Nordberg et al., 2012). It is worth mentioning that the compounds listed here are some examples. More complete lists of compounds that are directly toxic for PTC can be found in the referenced review articles. Due to its prominent role in water reabsorption, compound concentrations drastically increase in the PT. This enhances the toxicity of drugs that are directly toxic for PTC and can also lead to crystal formation by some compounds or their metabolites (ethylene glycol, acyclovir) (Guo and Nzerue, 2002; Hovda et al., 2010; Snellings et al., 2013). Indirect PTC damage by crystal formation would be difficult to predict by cell-based *in vitro* assays, whereas direct toxicity can be addressed.

As PTC play a prominent role in drug transport and metabolism and are a frequent target of direct toxic effects, this cell type is of major interest with respect to the development of renal *in vitro* models for drug safety screening. It would be preferred to use human PTC for such applications, due to the different expression patterns of drug transporters and drug-metabolizing enzymes in animal and human PTC (see Tiong et al., 2014, and references therein). Various human and nonhuman PTC models have been developed (Ryan et al., 1994; Bens and Vandewalle, 2008; Brown et al., 2008; Wieser et al., 2008; Lash et al., 2006, 2008; Tiong et al., 2014; see also Chapter 10).

The first *in vitro* model that predicted PT toxicity in humans with high accuracy was based on human primary renal proximal tubular cells (HPTC) (Li et al., 2013; Table 23.1). In comparison to the human and porcine PT cell lines HK-2 and LLC-PK1 use of HPTC resulted in improved predictivity. However, during this work substantial interdonor variability was observed with respect to HPTC (Li et al., 2013). Also other problems typically associated with primary cells remained unsolved, such as the cell sourcing

TABLE 23.1 Predictive Renal *In Vitro* Models

	Li et al. (2013)	Li et al. (2014)	Kandasamy et al. (2015)
Cell type	HPTC	hESC-derived HPTC-like cells	hiPSC-derived HPTC-like cells
Endpoint	IL6 and IL8 mRNA upregulation	IL6 and IL8 mRNA upregulation	IL6 and IL8 mRNA upregulation
Balanced accuracy (manual analysis)	0.81	0.76	Not determined (ND)
Balanced accuracy (automated classification)	0.99 (Su et al., 2014) (41 compounds)	ND	1.00 (training set) 0.87 (test set) (30 compounds)
Injury mechanisms	ND	ND	Correctly identified

problem and functional changes during passaging. Therefore, stem cell-based approaches were most attractive. The problem was that until recently no protocols existed for the differentiation of stem cells into more mature renal cells under *in vitro* conditions. Nevertheless, this situation has changed now. In the next sections we will provide a detailed description of the current methods for the generation of stem cell-derived renal cells. We will then describe predictive renal *in vitro* models, with an emphasis on stem cell-based renal models.

23.2 PROTOCOLS FOR THE DIFFERENTIATION OF PLURIPOTENT STEM CELLS INTO CELLS OF THE RENAL LINEAGE

23.2.1 Earlier Protocols and the Recent Race

First promising results on differentiating pluripotent stem cells (PSC) into cells of the renal lineage were obtained with a method using a combination of retinoic acid (RA; 0.1 μM), activin A (10 ng/ml), and bone morphogenetic protein (BMP)7 (50 ng/ml) (Kim and Dressler, 2005). Based on previous knowledge on vertebrate embryo development, RA and activin A were selected to stimulate the expression of early intermediate mesoderm (IM) markers. The IM is an early embryonic precursor of the kidney, gonads, and adrenal glands (Davies, 2002; Mugford et al., 2008b). BMP7 was selected because it is essential for kidney development and stimulates tubulogenesis in kidney organ cultures (Patel and Dressler, 2005; Simic and Vukicevic, 2005). With the combination of these three factors, mouse embryonic stem cell (mESC)-derived embryoid bodies (EB) could be induced to express markers that are expressed in the IM (Kim and Dressler, 2005). IM markers that were expressed by the induced embryoid bodies included paired box gene (PAX)2, Wilms' tumor (WT)1, and glial cell-derived neurotrophic factor (GDNF). In addition, induction of epithelial markers was observed, suggesting the generation of renal epithelial precursor cells *in vitro* (Kim and Dressler, 2005). This was in agreement with the finding that almost all of the cells from such treated embryoid bodies integrated into tubular epithelia when injected into developing kidney rudiments.

An alternative approach used only activin A to differentiate mESC-derived embryoid bodies into renal progenitor populations expressing PAX2, WT1, and other characteristic markers (Vigneau et al., 2007). After enrichment for cells expressing brachyury (T), which is expressed in early embryonic mesendoderm and is a key regulator of mesoderm formation (Technau, 2001; Martin and Kimelman, 2010), and injection of the cells into newborn mouse kidneys, stable integration into renal tubules was observed (Vigneau et al., 2007). Together, these earlier studies suggested that mesoderm induction is a critical step in the differentiation of PSC into renal progenitor cells. This is in agreement with the fact that the kidney is a mesoderm-derived organ (Davies, 2002; Vainio and Lin, 2002).

Upregulation of some renal markers could be also achieved *in vitro* in other studies with mESC (Kramer et al., 2006; Bruce et al., 2007; Fuente Mora et al., 2012) and human ESC (hESC) (Batchelder et al., 2009). Work performed with human induced pluripotent stem cells (hiPSC; derived from normal human kidney mesangial cells) suggested that directed differentiation into kidney podocyte-like cells could be achieved by mechanically cutting hiPSC colonies into pieces and cultivating them in suspension for 3 days in medium that was also supplemented in this case with activin A (10 ng/ml), BMP7, (15 ng/ml) and RA (0.1 μm) (Song et al., 2012). Subsequently, the cells were plated onto gelatin-coated plates where they were cultivated for another 7–8 days in the presence of RA, activin A, and BMP7 (Fig. 23.1). After 10 days the cells could be grown in normal cell culture medium without differentiation factors.

In contrast to primary human podocytes, the hiPSC-derived podocyte-like cells could be further expanded *in vitro* for extended time periods (Song et al., 2012). At day 10 of differentiation, the cells had cytoplasmic extensions and their morphology resembled the appearance of cultured human podocytes. In addition, they expressed the podocyte markers podocin and synaptopodin and also expression of WT1 and PAX2 was observed. The hiPSC-derived cells displayed also functional characteristics of podocytes and showed a contractile response to angiotensin II and endocytosis of albumin. Further, the cells integrated into WT1-positive regions in an embryonic kidney reaggregation assay (Song et al., 2012).

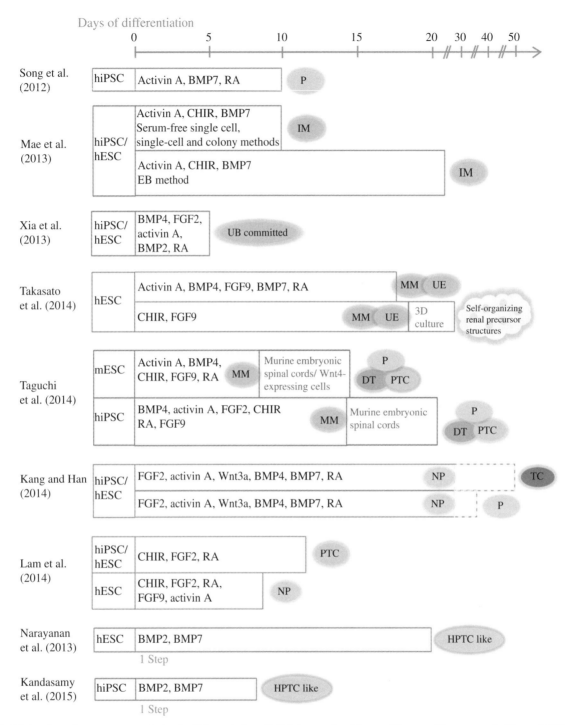

FIGURE 23.1 Overview over protocols for the differentiation of PSCs into cells of the renal lineage. The time scale (days of differentiation) is provided at the top (note that the scale changes after 20 days). The references and PSC types used by the different protocols are listed on the left. Growth factors and special materials or conditions used are indicated (the acronym CHIR is used for CHIR99021). All methods are based on multistep protocols that apply different factors and combinations thereof at different time points during the differentiation period (for details see text), with exception of the two 1-step protocols listed at the bottom. In these cases the only step is applying the BMP2/BMP7-containing differentiation medium on day 1 (see also Fig. 23.3). The renal progenitor structures and cell types generated by each protocol are indicated: embryonic precursor-like structures (IM, MM, UE, and UB-committed cells), podocyte-like cells (P), distal tubular (DT)-like cells, proximal tubular cell (PTC)-like and HPTC-like cells, renal tubular-like cells (TC), nephron progenitor structures (NP), and self-organizing precursor-like structures consisting of various cell types.

A race on the generation of stem cell-derived renal cells started in 2013. In January 2013 a protocol for the differentiation of hiPSC or hESC into IM was published (Mae et al., 2013). Odd-skipped related (OSR)1 was used as the main IM marker, and up to 90% OSR1⁺ cells were obtained. More differentiated renal cell types were only obtained at low frequency, which was not sufficient for use of these cells in any application, including *in vitro* toxicology (Mae et al., 2013). However, briefly afterward the first protocol for the differentiation of human pluripotent stem cells (hPSC) into mature renal cells was published (Narayanan et al., 2013). This protocol was based on hESC, and the differentiated hESC-derived cells exhibited many features of HPTC and were therefore called HPTC-like cells. The hESC-derived HPTC-like cells were then directly used for the development of the first stem cell-based renal *in vitro* model for the prediction of DIN (Li et al., 2014). (This *in vitro* model will be described in more detail in the following.) Later in 2013 the race on the generation of stem cell-derived renal cells continued and various alternative protocols were developed (Lam et al., 2014; Xia et al., 2013; Kang and Han, 2014; Taguchi et al., 2014; Takasato et al., 2014) (Fig. 23.1).

Importantly, all of these protocols (Lam et al., 2014; Xia et al., 2013; Kang and Han, 2014; Taguchi et al., 2014; Takasato et al., 2014) aimed mainly at applications in regenerative medicine. Goal was the development of multistep protocols that mimicked the various stages of embryonic kidney development. In contrast, the protocol developed by Narayanan et al. (2013) is a simple single-step protocol and the intention was not to mimic embryonic kidney development. This protocol (Narayanan et al., 2013) aimed at *in vitro* applications of stem cell-derived renal cells, in particular in *in vitro* toxicology, and a predictive *in vitro* model has been developed based on these cells (Li et al., 2014). Applications based on cells derived from alternative protocols mimicking embryonic kidney development (Lam et al., 2014; Xia et al., 2013; Kang and Han, 2014; Taguchi et al., 2014; Takasato et al., 2014) have not been developed yet.

23.2.2 Protocols Designed to Mimic Embryonic Kidney Development

23.2.2.1 Embryonic Kidney Development and Marker Expression Patterns Before protocols designed to mimic embryonic kidney development will be discussed in more detail, we will give a brief outline of embryonic kidney development and the markers used in various protocols. As mentioned, the kidney is a mesoderm-derived organ. The first embryonic precursor structure of mesoderm and endoderm is the primitive streak (PS), through which prospective mesodermal cells ingress during gastrulation (Tam and Loebel, 2007; Arnold and Robertson, 2009). Cells derived from a part of the posterior PS will generate the IM after

gastrulation, which is located between the somites and the lateral plate mesoderm in the early embryo. Part of the IM then gives rise to the metanephric mesenchyme (MM) and the ureteric bud (UB), from which the ureteric epithelium (UE) is derived (Davies, 2002; Vainio and Lin, 2002; Mugford et al., 2008a, b). The adult kidney is derived from the MM and the UB. The UB invades the MM and performs branching morphogenesis. The branched UB-derived structures generate the collecting duct system of the mature kidney. The tips of the branched UB/UE interact with the surrounding MM, which induces a mesenchymal-to-epithelial transition. The MM-derived epithelial precursor cells generated in these ways give rise to the epithelial parts of the nephron, which consist of the different types of renal tubular cells and the glomerular epithelial cells including podocytes (Davies, 2002; Dressler, 2002; Vainio and Lin, 2002).

Protocols for the directed differentiation of PSC into renal cells that are designed to mimic the different stages of embryonic renal development define these stages by the marker expression patterns of the cells. Widely used IM markers are OSR1 and PAX2. In fact, these markers are first expressed in the IM during embryonic development (Patel and Dressler, 2004; Mugford et al., 2008b). However, they remain to be expressed during subsequent stages of kidney development in metanephric renal precursor structures (Dressler et al., 1990; Dressler and Douglass, 1992; Eccles et al., 1992; Mugford et al., 2008b). The same applies to HOXD11, which specifies metanephric development within the IM (Mugford et al., 2008a), and is usually used as MM marker. Also other widely used MM markers like WT1 and GDNF are expressed throughout large parts of kidney development (Vainio and Lin, 2002; Kreidberg, 2010). Of note, these genes remain to be expressed even in the adult kidney, and WT1 controls gene expression in mature podocytes (Orth et al., 2000; Kann et al., 2015).

Further, renal progenitor markers become reexpressed in cell types from the adult kidney where they are usually not expressed after injury *in vivo* and during *in vitro* cultivation. Thus, WT1 and PAX2, which are not expressed in PTC of adult human kidneys, are expressed *in vitro* in primary human renal proximal tubular cells (Elberg et al., 2008; Li et al., 2013). PAX2 reexpression after injury *in vivo* and under *in vitro* conditions seems to be related to epidermal growth factor (EGF)-dependent cell proliferation (Liu et al., 1997; He et al., 2013). Renal epithelial cell media contain EGF and cell proliferation is desirable under *in vitro* conditions, as otherwise not enough cells for *in vitro* assays and screening can be obtained.

Therefore, in our opinion it is difficult to conclude from marker expression patterns of PSC-derived renal cells which developmental stage and renal embryonic structure they actually represent. Hence, in order to address the nature of the stem cell-derived renal cells and their potential applicability, a rigorous functional testing is obligatory. In the following paragraphs we will describe the various protocols for the differentiation of PSCs into renal (progenitor) cells that were designed to mimic embryonic development and the features of the derived cells.

23.2.2.2 Differentiation of hPSC into IM-Like Cells

As mentioned previously, a protocol for the differentiation of hPSC into presumable IM-like cells was published in early 2013 (Mae et al., 2013) (Fig. 23.1). This protocol was established with an OSR1-green fluorescence protein (GFP) knock-in hiPSC line. It was shown that 100 ng/ml activin A and 100 ng/ml Wnt3a were potent inducers of mesendodermal cells, in agreement with previous reports (D'Amour et al., 2006; Gadue et al., 2006). Mae et al. (2013) found that these factors induced the early mesendodermal marker T, as well as other mesendodermal markers including goosecoid (GSC) (Tada et al., 2005) and MIXL1 (Mix/Bix family transcription factor) (Pereira et al., 2011; Pulina et al., 2014). It was also found that CHIR99021 was more efficient than Wnt3a (Mae et al., 2013). CHIR99021 is a specific inhibitor of glycogen synthase kinase 3β that activates canonical Wnt signaling. In addition, the results revealed that BMP7 was the most potent inducer of OSR1 expression. Based on these and additional findings, a protocol was developed that comprised two different differentiation stages. Stage 1 comprised 2 days of treatment with 100 ng/ml activin A and 3 μM CHIR99021. Stage 2 comprised 4–20 days of treatment with 100 ng/ml BMP7 and 3 μM CHIR99021. Different variants of the protocol were developed with different media supplements and initial seeding of single cells, colonies, or EB. In most cases stage 2 comprised 8 days (Fig. 23.1).

The method was validated with different hiPSC and hESC lines. The changes in gene expression patterns observed by Mae et al. (2013) during the differentiation period are characteristic and are a common feature of all protocols, including those that were published later (see following text): (i) the mesendodermal marker T was transiently upregulated early (here after 2 days) and (ii) early upregulation of T was followed by upregulation of OSR1, PAX2, and WT1.

The OSR1+ cells generated by Mae et al. (2013) could be further differentiated into cells expressing markers typical for various renal structures, gonads and adrenal cortex, consistent with the idea that these cells had properties similar to those of early embryonic urogenital precursor cells. More differentiated renal cell types were obtained at frequencies that were too low for further applications (Mae et al., 2013).

23.2.2.3 Generation of UB-Committed Progenitor-Like Cells

A protocol developed by Xia et al. (2013) resulted in the generation of UB-committed renal progenitor-like cells (Fig. 23.1). Briefly, hiPSC or hESC were treated for 2 days with BMP4 and fibroblast growth factor (FGF)2, which resulted in downregulation of stemness markers and transient early induction of T (Xia et al., 2013). This was

followed by upregulation of OSR1, PAX2, and other IM and renal lineage markers during a subsequent 2-day treatment with activin A, BMP2, and RA. Expression profiling suggested a marker expression pattern that was more consistent with the UB lineage. Reaggregation organ culture experiments based on murine kidneys confirmed integration into UB structures. hiPSC derived from patients with polycystic kidney disease were also differentiated with the 4d *in vitro* protocol, and also these cells subsequently integrated into UB structures in chimeric murine/human kidney reaggregation cultures. This system may be of interest for disease-specific models.

As outlined previously, the UB gives rise to the collecting duct system. The collecting duct system is relatively rarely affected by DIN. Most frequently this presents as renal papillary necrosis due to excessive abuse of NSAIDs or alcohol (Kincaid-Smith, 1986; Segasothy et al., 1994). Fluid and electrolyte imbalances induced by halogenated anesthetics seem to be partly related to fluoride ion toxicity in collecting duct cells (Cittanova et al., 1996). Thus, protocols generating UB cells may be of limited interest with respect to *in vitro* toxicology, apart for the fact that the 4-day *in vitro* protocol (Xia et al., 2013) generated immature UB progenitor cells that were different from mature UB cells.

23.2.2.4 Generation of a Self-Organizing Mix of Different Renal Precursor Structures

Methods for generating a mix of self-organizing different cell types derived from UB- and MM-like structures were developed by Takasato et al. (2014) (Fig. 23.1). Using hESC, PS-like cells (expressing T and the PS marker MIXL1 (Pereira et al., 2011; Pulina et al., 2014)) were first induced by treatment with either BMP4/activin A or CHIR99021. This was followed by exposure to FGF9, which is expressed in IM (Colvin et al., 1999). Upregulation of OSR1, PAX2, and LHX1 suggested induction of IM-like cells by FGF9. Further treatment with FGF9, which also maintains mouse nephron progenitors *in vitro* (Barak et al., 2012), in combination with the previously identified (Kim and Dressler, 2005) nephrogenic factors BMP7 and RA, resulted in cells expressing SIX2 (SIX homeobox 2), WT1, GDNF, and HOXD11 (Takasato et al., 2014). Combined expression of these markers is first observed in MM and persists throughout subsequent stages of renal development. SIX2 defines the epithelial precursors of the nephron (Kobayashi et al., 2008).

In addition to markers specific for MM and MM-derived structures, also expression of UE genes (C-RET (Pachnis et al., 1993) and HOXB7 (Srinivas et al., 1999)) and of the epithelial marker E-cadherin (E-CAD) was observed. Formation of epithelial structures was promoted by RA, in agreement with previous findings (Kim and Dressler, 2005). RT-PCR analysis after an extended differentiation period of 22 days showed the expression of podocyte markers (synaptopodin (SYNPO), NPHS1, and WT1), PTC markers (AQP1 and the amino acid transporter SLC3A1), and collecting duct genes (AQP2 and SCNNB1) (Takasato et al., 2014).

To further address generation of different renal cell types and their potential interactions, an 18-day protocol was used that first induced cells with CHIR99021 (8 μM, 2 days) and FGF9 (200 ng/ml; after day 2 until day 12), followed by withdrawal of growth factors. Time-course RT-PCR analysis revealed an early transient pulse of T, followed by the upregulation of renal progenitor markers. On day 18 E-CAD$^+$ UE-like structures were surrounded by clumps of mesenchyme expressing MM markers. Cells induced with this protocol integrated into all major compartments of the developing kidney in a reaggregation assay with mouse embryonic kidneys (Takasato et al., 2014).

While most experiments with the 18-day CHIR99021–FGF9 protocol were performed with monolayers, cells were also harvested after 18 days, pelleted, and differentiated for another 4 days in 3D culture. Histological analysis revealed the formation of E-CAD$^+$ tubules expressing either UE or PTC markers (Takasato et al., 2014). The tubular structures were surrounded by cells expressing MM (WT1, PAX2) and renal vesicle markers (jagged (JAG)1, E-CAD$^+$; the renal vesicle is an epithelial precursor of the nephron). Together, the results suggested simultaneous induction of UB- and MM-derived structures and self-organization into renal precursor structures. While this approach is highly interesting with respect to regenerative medicine, the mix of various renal precursor structures may be less suitable for compound screening.

23.2.2.5 Complex Protocols for the Generation of Murine and Human Renal Precursor Structures

The strategy of Taguchi et al. (2014) was to reevaluate first the developmental origins of metanephric progenitors and to apply the information on cell characteristics and molecular clues then for the development of protocols for the directed differentiation of murine and human PSC into MM-derived structures. This was achieved by extensive isolation and sorting of cells from the MM region of developing mouse embryos and by further characterizing the marker expression patterns of these cells and their response to growth factors. Based on these findings complex multistep protocols were developed (Fig. 23.1). Most of the work on PSC differentiation was performed with mESC. In a first step comprising 2 days EB were generated, which were then induced with activin A during the following day. Subsequently, EB were treated with BMP4/10 μM CHIR99021, which resulted in an early transient peak of T at day (d) 4.5 (also posterior mesoderm markers were expressed at this time point). EB were further treated until d 8.5 with different combinations of activin A, BMP4, RA, FGF9, and CHIR99021 at various concentrations. This resulted in upregulation of markers specific for IM and MM (and structures derived thereof) like OSR1, WT1, PAX2, HOXA11, SIX2, GDNF, and others. Most of these

markers were fully upregulated at d 6.5. The whole protocol comprised 8.5 days and the results suggested generation of metanephric nephron progenitors (Taguchi et al., 2014).

Cocultivation of such progenitors with murine embryonic spinal cords or murine Wnt4-expressing cells resulted in tubulogenesis and formation of glomerulus-like structures. Multiple markers specific for renal vesicles, proximal and distal tubules, and glomerular podocytes were expressed in the respective structures. This implied nephron formation. In subsequent experiments induced EB were transplanted together with spinal cords beneath the kidney capsule of immunodeficient mice. Again, formation of nephrons with renal tubules and glomeruli was observed after 1 week. Importantly, the glomeruli became vascularized and connected to the host circulation (Taguchi et al., 2014). The blood vessels seemed to be host derived. These results open exciting perspectives for regenerative medicine.

Subsequently, a 14-day multistep protocol was developed for the induction of hiPSC-derived EB, followed by 8 days of coculture with spinal cords (Fig. 23.1). The initial 14-day protocol involved a first step of EB formation during treatment with BMP4, followed by five steps where activin A, BMP4, FGF2, CHIR99021, RA, and FGF9 were applied in different combinations and at various concentrations. Induced EB harvested after 14 days expressed OSR1, WT1, PAX2, SIX2, SALL1 (vertebrate homologue of spalt (SAL); SALL1 is essential during kidney development and for UB invasion (Nishinakamura and Takasato, 2005)), HOXA10, and HOXA11, indicative of the formation of MM-like structures or derivatives. The various markers were expressed in about 20–70% of the EB cells, depending on the actual marker assessed. Also here nephrogenesis was induced in cocultures with spinal cord. Expression of WT1/NPHS1, SAL1/CDH6, and PAX2/E-CAD was interpreted as being consistent with formation of podocytes and proximal and distal tubules (Taguchi et al., 2014). However, the majority of these markers are not very specific for particular renal structures or cell types and functional tests were not performed. In summary, the 22-day protocol based on hiPSC and spinal cord coculture resulted in the formation of 3D tissue structures expressing renal/renal progenitor markers.

23.2.2.6 Generation of Podocyte- or Renal Tubular-Like Cells with Protocols Covering Extended Time Periods

Another protocol for the stepwise differentiation of hESC or hiPSC into renal progenitor cells was developed by Kang and Han (2014) (Fig. 23.1). After adjustment for feeder-free conditions for 4 days, the cells were induced in a multistep process with activin A, Wnt3a, BMP4, FGF2, BMP7, and RA in various combinations. The whole procedure comprised about 30 days. Also here the typical pattern of changes in gene expression was observed: initial downregulation of stemness markers, followed by an early pulse of T (in combination with MIXL1 and EOMES; interpreted as PS stage), and subsequent upregulation of OSR1, PAX2, WT1, and EYA1 (interpreted as IM) and GDNF, SIX2, HOXD11, and SALL1 (interpreted as nephron progenitors). These cells could be further differentiated into presumable renal tubular-like cells (positive for AQP1, CD13 (aminopeptidase N), alkaline phosphatase (ALP), and mucin (MUC)1) or podocyte-like cells (positive for SYN, NPHS1, and podocalyxin-like). The final differentiation steps comprised 7 days in case of podocyte-like cells or involved treatment with hepatocyte growth factor (HGF) during a time period of 21 days in case of renal tubular-like cells. Thus, the complete protocol comprised up to 50 days. Besides characterizing marker expression patterns, no further functional tests had been performed.

23.2.3 Rapid and Efficient Methods for the Generation of Proximal Tubular-Like Cells

All of the protocols discussed, so far, are associated with a variety of issues limiting their applicability. One problem is long differentiation periods of up to 50 days (Fig. 23.1). Also, early precursor cells (Mae et al., 2013; Xia et al., 2013) or a mix of different renal (precursor) cell types (Taguchi et al., 2014; Takasato et al., 2014) are not suitable for compound screening, unless effects on renal development are specifically addressed. Further, 3D tissue structures, as generated by some protocols, are usually difficult to handle and are not suitable for current high-content screening (HCS) technology. Finally, the use of ill-defined compounds, like murine spinal cord (Taguchi et al., 2014), is problematic. As outlined previously, the main goal of the protocols discussed, so far, were applications in regenerative medicine. Therefore, it is not surprising that the results were exciting in terms of potential applications in regenerative medicine but suboptimal with respect to applications in compound screening. The following paragraphs will describe more rapid protocols and protocols that have been shown to be compatible with compound screening.

23.2.3.1 A Method for Generating Proximal Tubular-Like Cells by Recapitulating Different Steps of Embryonic Development
A relatively rapid and simple protocol for the directed differentiation of hPSC into proximal tubular-like cells had been developed by Lam et al. (2014). This work was performed with hESC and hiPSC. Cells that expressed PTC markers, and in addition also markers for other renal cell types, were obtained after a differentiation period of about 11–12 days (Fig. 23.1). The protocol comprised three different steps. hPSC were first treated with CHIR99021 (5 µm). Cells expressing T, MIXL1, and other PS markers, as well as N-cadherin (N-CAD), were obtained with almost 100% efficiency. These results were in agreement with the results of other protocols described in the previous sections, which demonstrated efficient meso-endoderm induction by CHIR99021 and transient early

expression of T. The expression of PS markers peaked between 36 and 48 h and pluripotency markers were down-regulated in parallel.

A variety of protocols with different factors applied for various time periods were then tested. It was found that exposure to CHIR99021 for 36 h and subsequent treatment with FGF2 (100 ng/ml) and RA (1 µm) for 3 days generated cells expressing PAX2 and LHX1 (lim homeobox 1; expressed in IM and adjacent lateral plate mesoderm (Dressler, 2009)) with an efficiency of about 70–80% (Lam et al., 2014). These PAX2$^+$ LHX1$^+$ cells, which expressed also other IM markers, were considered to represent an IM-like state. After subsequent growth factor withdrawal and further cultivation in serum-free medium for 6 days, the cells could be differentiated into proximal tubular-like cells that were stained by Lotus tetragonolobus lectin (LTL) and expressed the PTC markers N-CAD, kidney-specific cadherin (KSP-CAD), AQP1, megalin (MEG), and the primary ciliary protein polycystin-2. The cells expressed also markers characteristic for various other types of mature renal epithelial cells, such as nephrin and synaptopodin (podocytes), uromodulin (UMOD; loop of Henle), and AQP2 (collecting duct) at varying amounts in a time-dependent manner.

Of note, this protocol resulted in the formation of round laminin-bounded structures expressing these markers, which the authors described as tubular structures (Lam et al., 2014). When we applied the protocol developed by Lam et al. (2014) to a different hiPSC line (iPS(Foreskin)-4), we observed the formation of marker-expressing round 3D aggregates (Fig. 23.2). Cells that had gone through the entire differentiation procedure (referred to as day 9 cells by Lam et al. (2014) based on counting from the start of FGF2/RA treatment; note that the entire procedure is longer and comprises 11–12 days (Fig. 23.1)) integrated into the metanephric interstitium and into laminin-bounded structures when the reaggregation assay was performed with dissociated murine embryonic kidneys (Lam et al., 2014).

Also, by treating CHIR99021-, FGF2-, and RA-induced PAX$^+$ LHX1$^+$ cells with FGF9 and activin A, differentiation into SIX2$^+$ SALL1$^+$ cap mesenchyme nephron progenitor-like cells could be achieved (Fig. 23.1). These cells could be stained with LTL in response to Wnt signaling (Lam et al., 2014).

23.2.3.2 Simple 1-Step Protocols
The currently most rapid and simple protocol for the generation of proximal tubular-like cells uses hiPSC (Kandasamy et al., 2015) (Fig. 23.1). One week after seeding the cells are differentiated and ready for compound screening (Fig. 23.3). The differentiation protocol comprises one single step, which is the application of the differentiation medium (Fig. 23.3). The differentiation medium consists of commercial renal epithelial cell growth medium (REGM) supplemented with BMP2 (10 ng/ml) and BMP7 (2.5 ng/ml). hiPSC-derived

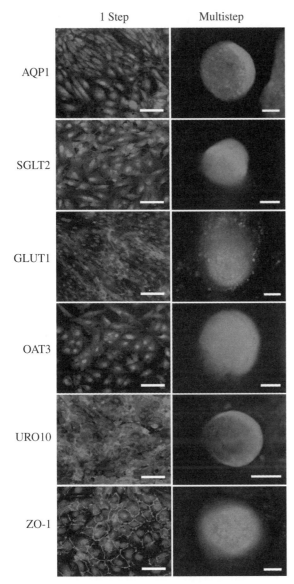

iPS(Foreskin)-4-derived cells

FIGURE 23.2 HPTC-like and PTC-like cells generated with the currently most rapid protocols. iPS(Foreskin)-4 hiPSC were differentiated with a 1-step protocol (Kandasamy et al., 2015) or a multistep protocol using CHIR, FGF2, and RA (Lam et al., 2014) (see also Fig. 23.1). Both protocols resulted in the generation of cells expressing PTC markers (listed on the left) that were detected by immunostaining (immunostaining and cell nuclei: gray). HPTC-like cells generated with the 1-step protocol (left) formed a differentiated simple epithelium when cultivated in multiwell plates (similar to HPTC; the similarity of these cells and HPTC was confirmed by various other methods; for details see text). In contrast, PTC-like cells formed round 3D cell aggregates (right). Features of individual HPTC-like cells cultivated in multiwell plates (left) can be easily addressed by imaging and such cells would be suitable for high-content screening HCS. HPTC-like cells (left) were fixed on day 8 and PTC-like cells (right) were fixed on day 11 or 12 after seeding. Scale bars: 100 μm. (*See insert for color representation of the figure.*)

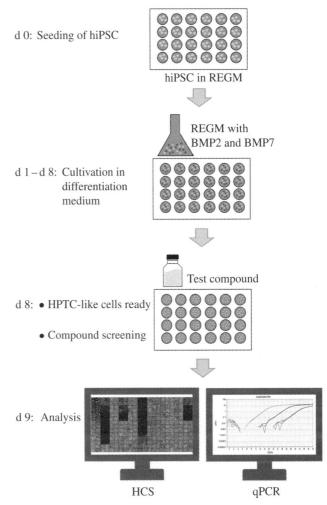

FIGURE 23.3 Flow chart of the hiPSC-based predictive renal *in vitro* model. hiPSC are seeded on day (d) 0 into multiwell plates and are cultivated from d 1 to d 8 in renal epithelial cell growth medium (REGM) supplemented with BMP2 and BMP7. HPTC-like cells are ready on d 8 (see also Figs. 23.1 and 23.2). As HPTC-like cells are highly pure, the same multiwell plate can be continuously used for cell differentiation and compound treatment on d 8 and subsequent analysis of the results on d 9. By using a qPCR-based method for determining drug-induced increase of IL6 and IL8 expression, a test balanced accuracy of 87% could be achieved when 30 compounds were screened (Kandasamy et al., 2015; see also Table 23.1).

cells differentiated in this way were used for the development of a predictive renal *in vitro* model. This model predicts nephrotoxicity in humans with a test balanced accuracy of 87% (Kandasamy et al., 2015) (Table 23.1). Before the hiPSC-derived cells and the respective *in vitro* model will be described in more detail, we will briefly summarize how the differentiation protocol was developed with hESC. The hESC-based protocol was the first method for the differentiation of hPSC into renal tubular cells (Narayanan et al., 2013) and with the hESC-derived cells the first stem cell-based renal *in vitro* model had been developed (Li et al., 2014) (Table 23.1).

A hESC-Based Protocol   When the hESC-based protocol was developed, first the impact of different extracellular matrix coatings was compared, and Matrigel gave best results (Narayanan et al., 2013). hESC seeded on Matrigel were then cultivated in REGM supplemented with various concentrations of BMP2 and/or BMP7. BMP2 is important for renal development *in vivo* and for renal tubulogenesis *in vitro* (Piscione et al., 1997; Grisaru et al., 2001; Piscione et al., 2001; Davies, 2002). BMP7 is essential for kidney development and the kidney is the major source of BMP7 in the adult human body (Gould et al., 2002; Patel and Dressler, 2005; Simic and Vukicevic, 2005). Supplementation with 10 ng/ml BMP2 and 2.5 ng/ml BMP7 gave best results in terms of differentiating hESC into AQP1-expressing cells (Narayanan et al., 2013). Additional supplementation with RA and activin A did not significantly improve the results. hESC were cultivated on Matrigel with REGM supplemented with BMP2/BMP7 for 20 days, and this was the whole differentiation protocol (Fig. 23.1). Of note, this 1-step protocol was not designed to recapitulate embryonic kidney development, which distinguishes this protocol from all other methods developed, so far.

After 20 days of differentiation, 17 PTC markers (AQP1, γ-glutamyl transferase (GGT), CD13, 25-hydroxyvitamin D$_3$ 1α-hydroxylase (VIT D3), MEG, Na$^+$/K$^+$ ATPase, glucose transporter 5 (GLUT 5), sodium-dependent glucose cotransporter 2 (SGLT2), Na$^+$ HCO3$^-$ cotransporter 1 (NBC1), OAT1, OAT3, OCT1, organic cation/carnitine transporter (OCTN2), MDR1, PEPT1, PEPT2, and KSP-CAD) and four markers specific for other renal cell types were assessed by qPCR, immunoblotting, and immunostaining. The results revealed that the expression patterns of the hESC-derived cells were similar to those of HPTC, and therefore the stem cell-derived cells were called HPTC-like cells. It should be noted that both, HPTC and HPTC-like cells, also expressed markers characteristic for other renal cell types: AQP3 (collecting ducts), NCCT (thiazide-sensitive sodium-chloride cotransporter; distal tubule (Simon et al., 1996)), NKCC2 (Na$^+$/K$^+$/2Cl$^-$ cotransporter; thick ascending limp of Henle's loop (Carota et al., 2010)), and PODXL (podocytes). All of these markers were expressed at higher levels in HPTC isolated from adult human kidneys (Narayanan et al., 2013), and expression of such markers seems to be an artifact associated with *in vitro* cultivation. Coimmunostaining confirmed that PTC markers and markers specific for other renal cell types were coexpressed in the same cells, ruling out the possibility that these mixed marker expression patterns were due to the presence of different renal cell types (Narayanan et al., 2013). (Note that also Lam et al. (2014) observed expression of markers characteristic for other renal cell types in proximal tubular-like cells.)

Also renal progenitor markers like PAX2 and WT1 and the mesenchymal marker VIM were expressed by HPTC and HPTC-like cells (Narayanan et al., 2013). Reexpression of PAX2, WT1, and VIM by primary human and animal renal proximal tubular cells cultivated *in vitro* has been consistently observed (Wallin et al., 1992; Liu et al., 1997; Weiland et al., 2007; Elberg et al., 2008; Li et al., 2013). It is important to note that PTC are not locked in a terminally differentiated epithelial state but can easily adopt more mesenchymal and progenitor-like states. This seems to be important for renal tubular repair after injury, which appears to be accomplished by surviving PTC rather than by presumable kidney stem cells (Witzgall et al., 1994; Bonventre, 2003; Humphreys et al., 2011; Kusaba et al., 2014). Also such surviving PTC typically express VIM, WT1, and PAX2 (Imgrund et al., 1999; Jiang et al., 2013; Kusaba et al., 2014) after injury and are in a proliferative state. Interestingly, also expression of PAX2 *in vitro* seems to be related to proliferation (Liu et al., 1997). Furthermore, injured PTC *in vivo* (Ichimura et al., 1998, 2004; Mishra et al., 2003; Vanmassenhove et al., 2013; Kusaba et al., 2014) as well as primary human and animal PTC *in vitro* (Weiland et al., 2007; Li et al., 2013) express typical injury markers such as kidney injury molecule (KIM)-1 and neutrophil gelatinase-associated lipocalin (NGAL), further supporting the idea that such cells are in a comparable functional state.

Therefore, differences between properties of PTC *in vivo* in the normal adult kidney and *in vitro* properties of PTC and stem cell-derived proximal tubular-like cells may not just reflect dedifferentiation or lack of full terminal differentiation. Rather, altered marker expression patterns seem to be related to the biology of these cells and their response to altered conditions. It is important to keep in mind that human and animal PTC are probably always different under *in vitro* conditions (compared to the *in vivo* situation), unless the *in vivo* situation can be exactly mimicked. This is also important with respect to the interpretation of the differentiation status of stem cell-derived proximal tubular-like cells, as the expression of progenitor markers and markers specific for other renal cell types is normal under *in vitro* conditions, also in HPTC. For practical reasons we suggest to always compare stem cell-derived cells carefully to respective primary cells cultivated *in vitro*, which gives a realistic impression of the best possible differentiation state that can be achieved under the conditions used. In addition, rigorous functional tests (e.g., of transporter functions) are often more informative than marker expression patterns in terms of the applicability of stem cell-derived cells for drug screening.

As mentioned previously, the marker expression patterns of hESC-derived HPTC-like cells were similar to those of HPTC cultivated *in vitro* (Narayanan et al., 2013). In addition, hESC-derived HPTC-like cells formed simple epithelia with tight junctions and displayed a polarized morphology with an apical brush border (Narayanan et al., 2013). Both are typical features of proximal tubular cells *in vivo* and *in vitro*. Further, HPTC-like cells and formed tubules *in vitro* and AQP1-expressing tubules were also generated *in vivo* after implantation. The cells displayed brush border enzyme activity and responsiveness to parathyroid hormone, and these typical PTC functions were maintained under bioreactor conditions. In addition, the cells performed pH-dependent ammoniagenesis, which is important for controlling the pH *in vivo*, and displayed water permeability. Most of the cells integrated into tubular epithelia when injected into mouse embryonic kidneys (Narayanan et al., 2013). Altogether, these results showed that the hESC-derived HPTC-like cells displayed many morphological and functional features that are characteristic for PTC.

A Simple 1-Step Protocol for the Differentiation of hiPSC into HPTC-Like Cells within 8 Days A disadvantage of the 1-step protocol described previously was that it was performed with hESC, which is associated with ethical and legal issues. Another point of concern was the long differentiation period of 20 days. Therefore, follow-up work was performed with hiPSC. The same 1-step protocol was applied (seeding on Matrigel in multiwell plates and cultivation in REGM supplemented with BMP2/BMP7) and the results showed that HPTC-like cells were obtained within only 8 days (Kandasamy et al., 2015) (Fig. 23.1). These cells had a purity of >90% as determined by fluorescence-activated cell sorting (FACS) with various PTC markers. Therefore, no harvesting or purification procedures were required. As the cells were already available in multiwell plates, these plates could be directly used for compound screening on day 8 (Fig. 23.3). Based on this method a very rapid and efficient hiPSC-based *in vitro* model was developed that predicted DIN with a test balanced accuracy of 87% (Kandasamy et al., 2015). This model will be described in detail in the following. First, the characteristics of the hiPSC-derived HPTC-like cells will be described here in more detail.

The analysis of gene expression patterns during hiPSC differentiation revealed a remarkable result. As mentioned, the 1-step method was not designed for recapitulating the different steps of embryonic kidney development. Still, the changes in gene expression patterns were similar compared to the other protocols discussed previously: (i) initial downregulation of stemness markers, (ii) a transient peak of T (in this case on day 3), and (iii) subsequent upregulation of OSR1 and other IM and renal (progenitor) markers (Kandasamy et al., 2015). T-expressing mesendoderm can be induced from hPSC by a wide range of conditions and factors, and most protocols discussed here used CHIR99021 or activin A for the initial induction of this state (Fig. 23.1). The 1-step protocol did not include any of these factors but used BMP2. Also BMP2 is an inducer of T and in addition promotes expression of IM genes like OSR1 and PAX2 (Vidricaire et al., 1994; James and Schultheiss, 2005). Other

methods induced IM gene expression by alternative factors such as FGFs. The functional overlap between the different factors used in different protocols as well as the remarkable flexibility with respect to the induction of different developmental stages from hPSC might explain why similar changes in gene expression could be induced with different protocols and factors. In addition to BMP2, the 1-step protocol included BMP7, which is essential for kidney development. This factor was used by various other protocols to promote renal differentiation after transient induction of T (Fig. 23.1).

When the 1-step protocol was applied to hiPSC, nephron progenitor markers (SIX2, WT1, GDNF, and HOXD11) and PTC markers (AQP1, GGT, and KSP-CAD) were upregulated together after day 7 (Kandasamy et al., 2015). HPTC showed a similar expression pattern of nephron progenitor and PTC markers as hiPSC-derived cells after day 7. As mentioned, the expression of progenitor markers in mature PTC is normal under *in vitro* conditions.

Day 8 cells were further characterized, and expression and correct subcellular localization of PTC markers (AQP1, SGLT2, GLUT1, OAT3, CD13, urothelial glycoprotein (URO10), and zonula occludens (ZO)-1) were confirmed by immunostaining (Kandasamy et al., 2015; Fig. 23.2). FACS analysis revealed PTC marker expression in >90% of d 8 cells (Kandasamy et al., 2015).

Further, gene expression in HPTC and HPTC-like cells was compared by qPCR using 31 markers. The results confirmed expression of the kidney-specific marker KSP-CAD and of the phosphate transporter SLC34A1 (Kandasamy et al., 2015), which is only expressed in fully differentiated PTC *in vivo* (Biber et al., 2009). Further, a variety of PTC-specific transporters and drug transporters were expressed (NBC1, SGLT2, GLUT5, OAT1, OAT3, OCT2, OCTN2, MDR1, MEG, PEPT1), as well as other PTC (AQP1, CD13, GGT, VIT D3, and Na$^+$/K$^+$ ATPase) and epithelial (ZO-1, E-CAD, N-CAD) markers. GGT activity was confirmed by functional assays. In addition, hiPSC-derived HPTC-like d 8 cells expressed nephron progenitor (PAX2, WT1) and injury (NGAL, KIM, VIM) markers, as well as markers for other renal cell types (PODXL, NCCT, NKCC2, UMOD, AQP3) (Kandasamy et al., 2015). HPTC showed similar gene expression patterns, and expression of nephron progenitor and injury markers and markers specific for other renal cell types is a typical *in vitro* artifact (see section "A hESC-Based Protocol").

Morphological analysis revealed that hiPSC-derived HPTC-like d 8 cells formed simple epithelia with tight junctions, and also dome formation was observed. The cells displayed a polarized phenotype with an apical brush border and generated tubules *in vitro*. Functional studies revealed characteristic cellular responses to the nephrotoxicants citrinin and rifampicin, which could be specifically inhibited with probenecid or cimetidine (Kandasamy et al., 2015). These compounds inhibit the uptake transporters for citrinin (OAT1 and OAT3) and rifampicin (OCT2). Together, these results suggest specific transporter-mediated drug uptake and proper response of the cells to nephrotoxicants. The results were in accordance with expression of OAT1 and OAT3 and OCT2 as determined by qPCR and activity of OCT2 as confirmed by functional assays (Kandasamy et al., 2015).

In summary, the results showed that on day 8 of differentiation hiPSC-derived cells had a phenotype similar to HPTC when the 1-step protocol was applied. The results on drug transporter expression and activity, transporter-mediated drug uptake, and responses to nephrotoxicants made this cell type particularly interesting for applications in *in vitro* safety screening. Compared to the alternative protocol for the rapid generation of proximal tubular-like cells (Lam et al., 2014), the 1-step protocol had the following advantages: faster (8 days vs. 11–12 days), more simple (1 step vs. 3 steps), lower initial seeding density of hiPSC (8,000 cells/cm² vs. 40,000 cells/cm²), high purity of HPTC-like cells of >90% (the purity of proximal tubular-like cells was not determined by Lam et al. (2014)), and formation of a simple epithelium that is suitable for HCS (the protocol established by Lam et al. (2014) results in the formation of 3D structures (Fig. 23.2)).

hiPSC- and hESC-derived HPTC-like cells generated by the 1-step protocol have been used for the development of *in vitro* models for the prediction of DIN (Li et al., 2014; Kandasamy et al., 2015). These are the only applications of stem cell-derived renal cells that have been developed, so far. The *in vitro* models will be described in the following section.

23.3 RENAL *IN VITRO* MODELS FOR DRUG SAFETY SCREENING

23.3.1 Microfluidic and 3D Models and Other Models that have been Tested with Lower Numbers of Compounds

A wealth of studies on DIN has been performed with renal *in vitro* models. Such studies mainly addressed mechanistic aspects and used only very limited numbers of compounds (<10). Therefore, the predictivity of these models could not be determined. (A comprehensive review on renal *in vitro* models published in this millennium is provided in Tiong et al. (2014). This review gives a detailed overview on cell types used, cultivation conditions (2D or 3D, static or microfluidic), endpoints, and numbers of compounds tested (see also Chapter 10)). Although microfluidic and 3D models are currently thought to be most promising (Davies, 2014; Kelly et al., 2013; Huang et al., 2014), a general problem with such models is that compound screening is relatively inefficient and also not compatible with current methods for high-throughput screening (HTS). Hence, existing 3D (Astashkina et al., 2012a, b; DesRochers et al., 2013) and microfluidic (Jang et al., 2013) models have only been tested, so far, with very limited numbers of compounds and their predictivity is unclear.

Most research with renal *in vitro* models has been performed with static monolayer cultures of immortalized cell lines from humans or animals or primary animal cells (Huang et al., 2014; Tiong et al., 2014). Some studies have been performed with human primary cells (Carvalho et al., 2002; Li et al., 2003, 2006; McGoldrick et al., 2003; Konigs et al., 2009) (see also Chapter 10), which included one microfluidic model (Jang et al., 2013). Here, HPTC were cultivated in a microfluidic chip on a porous membrane. This microfluidic model has been tested with one single compound (cisplatin), and the effects were determined by measuring lactate dehydrogenase (LDH) release and apoptosis. Cisplatin-induced toxicity was more pronounced under static (control) conditions. Amelioration of cisplatin-induced effects by blocking transporter-mediated uptake and cell recovery after cisplatin exposure was more pronounced under microfluidic conditions (Jang et al., 2013).

Also the renal 3D models, which have been described recently, have been tested with <10 compounds and their predictivity is unclear. One of these renal 3D models is based on hydrogel-embedded PT isolated from murine kidneys (Astashkina et al., 2012a). Thus, two of the main potential advantages of *in vitro* models do not apply here, which are (1) use of human cells and (2) avoidance of animal experiments. The four compounds tested with this model included doxorubicin, which gave in addition to cisplatin the most consistent positive results (cytokine release was measured as endpoint). However, in humans doxorubicin toxicity usually does not affect the PT. Doxorubicin has substantial hepato- and cardiotoxicity, but only minor toxic effects on the kidney, which are mainly limited to the glomerulus (Tacar et al., 2013).

Another 3D model had been developed with immortalized human renal cortical cells embedded into a mix of Matrigel and rat tail collagen (DesRochers et al., 2013). These cells displayed some features of PTC but performed also functions that are not typical for PTC. For instance, the cells were responsive to antidiuretic hormone, whereas they did not respond to parathyroid hormone. The effects of cisplatin, gentamicin, and doxorubicin were tested by measuring cell viability, LDH release, and increase of KIM-1 and NGAL. In comparison to 2D cultures, the 3D model did not clearly improve the results. Also here positive responses to doxorubicin were observed, and in this case the 3D model was more sensitive than 2D cultures to this usually non-PT-damaging drug.

23.3.2 *In Vitro* Models that have been Tested with Higher Numbers of Compounds and the First Predictive Renal *In Vitro* Model

Six renal *in vitro* models have been tested with more than 10 compounds (Duff et al., 2002; Wu et al., 2009; Lin and Will, 2012; Li et al., 2013, 2014; Kandasamy et al., 2015) and in four cases performance metrics and predictivity were determined (Lin and Will, 2012; Li et al., 2013, 2014; Kandasamy et al., 2015). The first study where a larger number of compounds have been tested was a prevalidation study funded by the European Centre for the Validation of Alternative Test Methods (ECVAM, competence inherited by the EURL ECVAM) (Duff et al., 2002). Fifteen compounds were tested with 2D models based on canine MDCK or porcine LLC-PK1 cells. Transepithelial electrical resistance and fluorescein isothiocyanate flux were measured as endpoints, and the lowest compound concentrations with significant effects were determined. No further validation work was performed on this model.

A multiplexed assay was developed with the human renal proximal tubular cell line HK-2. Here, 11 compounds were tested and positive and negative results for all compounds were compiled (Wu et al., 2009). The largest number of compounds was screened in a study that addressed organ-specific toxicity with HepG2 (hepatocellular carcinoma), H9c2 (embryonic myocardium), and NRK-52E (rat PT cell line) cells (Lin and Will, 2012). Six hundred twenty-one compounds were screened that were either nephro-, hepato-, or cardiotoxic or not toxic in humans. The ATP content of the cells was measured as endpoint. The results revealed that with this model no accurate prediction of organ-specific toxicity could be achieved, although the predictivity improved significantly when human C_{max} values were included (Lin and Will, 2012).

23.3.2.1 Prediction of Renal PT Toxicity with HPTC

HPTC were used for the development of the first predictive renal *in vitro* model (Li et al., 2013) (Table 23.1). Three batches of HPTC derived from different donors were screened with 41 compounds. Human (HK-2) and porcine (LLC-PK1) immortalized renal proximal tubular cell lines were used for comparison. All of the 41 compounds were either widely used drugs or environmental toxicants and industrial chemicals with well-characterized effects on human kidneys. Thus, the *in vitro* results could be directly compared to human clinical data. The compounds were classified into three groups: (1) nephrotoxicants that are directly toxic for PTC in human kidneys, (2) nephrotoxicants that damage the kidney in different ways and are not directly toxic for PTC in humans, and (3) compounds that are not nephrotoxic in humans.

The *in vitro* assay was based on HPTC seeded into uncoated multiwell plates that were cultivated for 3 days until the cells had formed a differentiated simple epithelium. As endpoint the expression levels of interleukin (IL)6 and IL8 were determined by qPCR. In the first analysis performed on the expression data (Li et al., 2013), a result was classified as positive if the drug-induced increase of at least one of these markers (IL6 or IL8) was above a certain threshold. When this thresholding procedure was applied, the median values (three batches of HPTC) of all major performance metrics

(sensitivity, specificity, balanced accuracy, positive predictive value (PPV), negative predictive value (NPV), and area under the curve (AUC) of the receiver operating characteristic (ROC) curves) ranged between 0.76 and 0.85. In comparison to HPTC the predictivity was lower when HK-2 or LLC-PK1 cells were used. Especially the sensitivity was low when cell lines were used with values of only 0.50 (HK-2) and 0.64 (LLC-PK1), respectively (Li et al., 2013).

23.3.3 Stem Cell-Based Predictive Models

23.3.3.1 A hESC-Based Model

After the protocol for generating hESC-derived HPTC-like cells had been established (described in section "A hESC-Based Protocol"), a 2D *in vitro* model based on this cell type was developed (Li et al., 2014) (Table 23.1). This hESC-based model was tested with the same set of 41 compounds, which had been used for testing the HPTC-based model (see Section 23.3.2.1). Also the hESC-based model used drug-induced increases in IL6 and/or IL8 determined by qPCR as endpoint. When analyzed with the thresholding procedure, the values obtained for sensitivity and specificity were 0.68 and 0.84, respectively. Also for the hESC-based model case, all major performance metrics were determined and an AUC/ROC value of 0.8 was obtained.

As similar compounds were screened with HPTC and hESC-derived HPTC-like cells, the results were directly comparable. In comparison to HPTC, the sensitivity of hESC-derived HPTC-like cells was relatively low. The mean and median values were 0.77 in case of HPTC (three batches), whereas a value of 0.68 was obtained with respect to hESC–HPTC-like cells. The lower sensitivity of hESC-derived HPTC-like cells was probably due to lower expression levels of some transporters and receptors like MEG (Li et al., 2014), which are important for the uptake of PT-specific nephrotoxicants. The specificity was similar in case of HPTC and hESC-derived HPTC-like cells (0.84). Results obtained with a second independently differentiated batch of hESC-derived HPTC-like cells revealed that positive and negative results were reproducible (Li et al., 2014).

In addition to the mRNA levels of IL6 and IL8, also established endpoints were measured with the best batch of HPTC (HPTC1, HPTC displayed interdonor variability) and hESC-derived HPTC-like cells. These endpoints included cell viability, ATP depletion, GSH depletion, and LDH leakage. In all cases the predictivity was substantially lower and the balanced accuracy ranged between 0.60 and 0.69, depending on the endpoint. In contrast, the respective values for balanced accuracy were 0.90 (HPTC1) and 0.76 (hESC-derived HPTC-like cells) when IL6/IL8 upregulation was used as endpoint. These results show that not only the selection of the cell model is important, but also the endpoints must be carefully considered. In the cases discussed here, good predictivity was only obtained when IL6/IL8 expression was measured, whereas suitable cell models in combination with widely used endpoints resulted in compromised predictivity.

The work on HPTC also revealed that kidney injury markers like VIM, KIM-1, NGAL, and IL18, which are upregulated after kidney injury *in vivo*, were not consistently upregulated *in vitro* after exposure to nephrotoxicants (Li et al., 2013). This was in agreement with other recent results ((DesRochers et al., 2013) and Predict-IV project, third, fourth and fifth Periodic Reports http://www.predict-iv.toxi. uni-wuerzburg.de/periodic_reports/). The results can be probably explained by the fact that kidney injury markers are already expressed *in vitro* in untreated cells, at least in human and animal primary PTC and hPSC-derived HPTC-like cells (Weiland et al., 2007; Elberg et al., 2008; Li et al., 2013, 2014; Kandasamy et al., 2015). This is likely related to the altered biology of these cells under *in vitro* conditions (see section "A hESC-Based Protocol"). The difficulties of successfully using potential novel biomarkers for kidney injury as endpoints in *in vitro* models show how important it is to understand how the cells are different under *in vitro* conditions. It is also important to note that predictive models can still be successfully developed if cellular differences are carefully considered and appropriate endpoints are chosen.

23.3.3.2 Prediction of Drug-Induced PT Toxicity and Injury Mechanisms with an hiPSC-Based Model and Machine Learning Methods

The weak points of the HPTC- and hESC-based models described previously (Sections 23.3.2.1 and 23.3.3.1) were the data analysis procedures. In order to improve result classification, the raw data obtained with three batches of HPTC and the IL6/IL8-based model (Li et al., 2013) were reanalyzed by machine learning (Su et al., 2014). Random forest (RF), support vector machine (SVM), k-NN, and Naïve Bayes classifiers were tested. Best results were obtained with the RF classifier and the mean values (three batches of HPTC) ranged between 0.99 and 1.00 with respect to sensitivity, specificity, balanced accuracy, and AUC/ROC (Su et al., 2014). Thus, excellent predictivity could be obtained by combining the IL6/IL8-based model with automated classification by machine learning.

The most recently developed model combined the rapid and simple 1-step protocol for the differentiation of hiPSC into HPTC-like cells (section "A Simple 1-Step Protocol for the Differentiation of hiPSC into HPTC-Like Cells within 8 Days") with the IL6/IL8-based assay and data analyses by machine learning (Kandasamy et al., 2015) (Fig. 23.3 and Table 23.1). Thirty compounds were screened. As no harvesting of the cells was required due to the high purity of HPTC-like cells derived with this protocol, the same plate in which the cells were differentiated could be used for compound screening on the evening of day 8 (Fig. 23.3). On the next morning the lysates could be harvested for qPCR analysis. Analysis of the results was performed with the RF classifier and 10-fold cross-validation. The values for sensitivity, specificity, and balanced accuracy of the test set all ranged between 0.85 and 0.89 (Kandasamy et al., 2015).

Further, drug-induced pathways and injury mechanisms were addressed by automated cellular imaging using hiPSC-derived HPTC-like cells. Drug-induced mechanisms and pathways, such as NF-κB activation, reactive oxygen species generation, and DNA damage by nephrotoxicants, were determined, and the results obtained with hiPSC-derived HPTC-like cells were in concordance with human clinical data (Kandasamy et al., 2015). These results show that hiPSC-derived HPTC-like cells can be used for the accurate prediction of drug-induced PT toxicity and the determination of underlying injury mechanisms and cellular responses.

23.4 ACHIEVEMENTS AND FUTURE DIRECTIONS

During the last 3 years, major achievements have been made in the field of *in vitro* nephrotoxicology. These include the development of various protocols for the differentiation of hPSC into different renal cell types, the establishment of predictive *in vitro* methods, and the development of the first models based on hPSC-derived renal cells that predict nephrotoxicity in humans with high accuracy. The current predictive models (Table 23.1) give a binary result. They can predict with high accuracy whether a drug will be nephrotoxic or not in humans, but they do not predict the dose. Assays that give a yes/no response can be usefully applied during early stages of development in order to flag all compounds that will be nephrotoxic in humans with a high probability. Such compounds can then either be excluded, or nephrotoxic effects can be more carefully monitored during subsequent steps of drug development if it should be decided to follow up on a specific compound.

Nevertheless, it would be desirable to predict also the dose response in humans. A key to this problem may be the expression levels of drug transporters and drug-metabolizing enzymes, which are usually different *in vitro* compared to the *in vivo* situation. Given these differences it is not surprising that the dose response is usually different *in vitro*. A wealth of expression data and *in vitro* dose–response data obtained from the same cell types and models have been generated during the development of the current predictive models. Hence, it should be possible to link these data and develop probably more sophisticated models on this basis. In addition, it would desirable to include more renal cell types into *in vitro* models, as not all nephrotoxic compounds are toxic for the proximal tubular cells.

Also, the current *in vitro* models for the prediction of PT toxicity give limited mechanistic insights. First steps toward developing models that combine high predictivity with mechanistic insights have been made with respect to the model based on hiPSC-derived HPTC-like cells (Table 23.1).

It would be important to further continue the development of such models.

With the hiPSC-based model drug mechanisms have been addressed by using automated cellular imaging, and this model, as well as other multiwell plate-based 2D models, would be suitable for HCS. Nevertheless, so far, no predictive HCS-based renal model has been established. Such models would be essential for the screening of larger compound libraries.

An interesting question is whether development of 3D and microfluidic renal models should have high priority. Usually, such models are favored based on the assumption that cell performance should be more *in vivo*-like. With respect to renal cells and PTC, this has not been convincingly demonstrated, so far. It is worth mentioning that PTC form a simple epithelium *in vivo* and are not arranged into 3D structures as hepatocytes, for instance. Also, 3D models are usually complicated, have a low reproducibility, and are not suitable for current HCS techniques. Such difficulties can be avoided with 2D models and also a good predictivity can be obtained. Of note, all current predictive renal models are 2D models (Table 23.1). Important aspects during the development of these models were the careful characterization of primary and stem cell-derived renal cells and an understanding of how the cells are different compared to the *in vivo* situation. This knowledge was used in order to identify suitable endpoints that still work under the altered *in vitro* conditions. The results have shown that not only cell performance but also careful selection of endpoints that are suitable for *in vitro* applications are crucial for obtaining high predictivity.

In the future, an exciting new area will be the establishment of hiPSC lines from patients affected by DIN and the further characterization of the underlying causes by using such cell models, as well as the development of patient-specific renal models using hiPSC-based technology. The establishment of the current hiPSC-based methods for the generation of renal cells and the further development of such methods will provide the basis for future work in this exciting new area.

ACKNOWLEDGMENTS

This work was supported by a grant from the Joint Council Office (Agency for Science, Technology and Research, Singapore) Development Program and the Institute of Bioengineering and Nanotechnology (Biomedical Research Council, Agency for Science, Technology and Research).

NOTES

While this article has been processed, based on the work by Lam et al. (2014) and Takasato et al. (2014) protocols have been established for the generation of kidney organoids from hESC and hiPSC (Freedman et al., 2015; Morizane et al., 2015; Takasato et al., 2015). Such self-organizing organoids contain different kinds of renal cells organized into structures resembling kidney tissue. Potential applications would be in regenerative medicine and compound safety screening. Organoids have been treated with one (cisplatin) or two (cisplatin and gentamicin) nephrotoxic compounds and KIM-1 induction and apoptosis has been observed. No not nephrotoxic control compounds have been used and the predictive performance of these models is unknown.

Also, a high content screening (HCS) platform for the nephrotoxicity prediction of xenobiotics with diverse chemical structures has been established (Su et al., 2015) while this article has been processed. This HCS platform is based on automated imaging in combination with phenotypic profiling and machine learning and has been pre-validated with 44 compounds including drugs, industrial chemicals, environmental toxicants and other compounds. It has a test balanced accuracy of 82% (HPTC) or 89% (immortalized human PTC) with respect to predicting nephrotoxicity in humans. This HCS platform could be also combined with stem cell-derived renal cells.

REFERENCES

Aleksa K, Matsell D, Krausz K, Gelboin H, Ito S, Koren G. 2005. Cytochrome P450 3A and 2B6 in the developing kidney: implications for ifosfamide nephrotoxicity. *Pediatr Nephrol* 20(7):872–885.

Arnold SJ, Robertson EJ. 2009. Making a commitment: cell lineage allocation and axis patterning in the early mouse embryo. *Nat Rev Mol Cell Biol* 10(2):91–103.

Astashkina AI, Mann BK, Prestwich GD, Grainger DW. 2012a. A 3-D organoid kidney culture model engineered for high-throughput nephrotoxicity assays. *Biomaterials* 33(18):4700–4711.

Astashkina AI, Mann BK, Prestwich GD, Grainger DW. 2012b. Comparing predictive drug nephrotoxicity biomarkers in kidney 3-D primary organoid culture and immortalized cell lines. *Biomaterials* 33(18):4712–4721.

Barak H, Huh SH, Chen S, Jeanpierre C, Martinovic J, Parisot M, Bole-Feysot C, Nitschke P, Salomon R, Antignac C, Ornitz DM, Kopan R. 2012. FGF9 and FGF20 maintain the stemness of nephron progenitors in mice and man. *Dev Cell* 22(6): 1191–1207.

Batchelder CA, Lee CC, Matsell DG, Yoder MC, Tarantal AF. 2009. Renal ontogeny in the rhesus monkey (Macaca mulatta) and directed differentiation of human embryonic stem cells towards kidney precursors. *Differentiation* 78(1):45–56.

Bellomo R. 2011. Acute renal failure. *Semin Respir Crit Care Med* 32(5):639–650.

Bens M, Vandewalle A. 2008. Cell models for studying renal physiology. *Pflugers Arch* 457(1):1–15.

Biber J, Hernando N, Forster I, Murer H. 2009. Regulation of phosphate transport in proximal tubules. *Pflugers Arch* 458(1): 39–52.

Bonventre JV. 2003. Dedifferentiation and proliferation of surviving epithelial cells in acute renal failure. *J Am Soc Nephrol* 14(Suppl 1):S55–S61.

Brenner BM. 2008. *Brenner and Rector's The Kidney*. Philadelphia: Saunders Elsevier.

Brown CD, Sayer R, Windass AS, Haslam IS, De Broe ME, D'Haese PC, Verhulst A. 2008. Characterisation of human tubular cell monolayers as a model of proximal tubular xenobiotic handling. *Toxicol Appl Pharmacol* 233(3):428–438.

Bruce SJ, Rea RW, Steptoe AL, Busslinger M, Bertram JF, Perkins AC. 2007. In vitro differentiation of murine embryonic stem cells toward a renal lineage. *Differentiation* 75(5):337–349.

Burckhardt G. 2012. Drug transport by organic anion transporters (OATs). *Pharmacol Ther* 136(1):106–130.

Burckhardt BC, Burckhardt G. 2003. Transport of organic anions across the basolateral membrane of proximal tubule cells. *Rev Physiol Biochem Pharmacol* 146:95–158.

Carota I, Theilig F, Oppermann M, Kongsuphol P, Rosenauer A, Schreiber R, Jensen BL, Walter S, Kunzelmann K, Castrop H. 2010. Localization and functional characterization of the human NKCC2 isoforms. *Acta Physiol (Oxf)* 199(3):327–338.

Carvalho M, Hawksworth G, Milhazes N, Borges F, Monks TJ, Fernandes E, Carvalho F, Bastos ML. 2002. Role of metabolites in MDMA (ecstasy)-induced nephrotoxicity: an in vitro study using rat and human renal proximal tubular cells. *Arch Toxicol* 76(10):581–588.

Choudhury D, Ahmed Z. 2006. Drug-associated renal dysfunction and injury. *Nat Clin Pract Nephrol* 2(2):80–91.

Ciarimboli G. 2014. Membrane transporters as mediators of cisplatin side-effects. *Anticancer Res* 34(1):547–550.

Cittanova ML, Lelongt B, Verpont MC, Geniteau-Legendre M, Wahbe F, Prie D, Coriat P, Ronco PM. 1996. Fluoride ion toxicity in human kidney collecting duct cells. *Anesthesiology* 84(2):428–435.

Colvin JS, Feldman B, Nadeau JH, Goldfarb M, Ornitz DM. 1999. Genomic organization and embryonic expression of the mouse fibroblast growth factor 9 gene. *Dev Dyn* 216(1):72–88.

D'Amour KA, Bang AG, Eliazer S, Kelly OG, Agulnick AD, Smart NG, Moorman MA, Kroon E, Carpenter MK, Baetge EE. 2006. Production of pancreatic hormone-expressing endocrine cells from human embryonic stem cells. *Nat Biotechnol* 24(11):1392–1401.

Davies JA. 2002. Morphogenesis of the metanephric kidney. *ScientificWorldJournal* 2:1937–1950.

Davies J. 2014. Engineered renal tissue as a potential platform for pharmacokinetic and nephrotoxicity testing. *Drug Discov Today* 19:725–729.

DesRochers TM, Suter L, Roth A, Kaplan DL. 2013. Bioengineered 3D human kidney tissue, a platform for the determination of nephrotoxicity. *PLoS One* 8(3):e59219.

Dressler G. 2002. Tubulogenesis in the developing mammalian kidney. *Trends Cell Biol* 12(8):390–395.

Dressler GR. 2009. Advances in early kidney specification, development and patterning. *Development* 136(23):3863–3874.

Dressler GR, Douglass EC. 1992. Pax-2 is a DNA-binding protein expressed in embryonic kidney and Wilms tumor. *Proc Natl Acad Sci U S A* 89(4):1179–1183.

Dressler GR, Deutsch U, Chowdhury K, Nornes HO, Gruss P. 1990. Pax2, a new murine paired-box-containing gene and its

expression in the developing excretory system. *Development* 109(4):787–795.

Duff T, Carter S, Feldman G, McEwan G, Pfaller W, Rhodes P, Ryan M, Hawksworth G. 2002. Transepithelial resistance and inulin permeability as endpoints in in vitro nephrotoxicity testing. *Altern Lab Anim* 30(Suppl 2):53–59.

Eccles MR, Wallis LJ, Fidler AE, Spurr NK, Goodfellow PJ, Reeve AE. 1992. Expression of the PAX2 gene in human fetal kidney and Wilms' tumor. *Cell Growth Differ* 3(5):279–289.

Elberg G, Guruswamy S, Logan CJ, Chen L, Turman MA. 2008. Plasticity of epithelial cells derived from human normal and ADPKD kidneys in primary cultures. *Cell Tissue Res* 331(2):495–508.

Freedman BS, Brooks CR, Lam AQ, Fu H, Morizane R, Agrawal V, Saad AF, Li MK, Hughes MR, Werff RV, Peters DT, Lu J, Baccei A, Siedlecki AM, Valerius MT, Musunuru K, McNagny KM, Steinman TI, Zhou J, Lerou PH, Bonventre JV. 2015. Modelling kidney disease with CRISPR-mutant kidney organoids derived from human pluripotent epiblast spheroids. *Nat Commun* 6:8715.

Fuente Mora C, Ranghini E, Bruno S, Bussolati B, Camussi G, Wilm B, Edgar D, Kenny SE, Murray P. 2012. Differentiation of podocyte and proximal tubule-like cells from a mouse kidney-derived stem cell line. *Stem Cells Dev* 21:296–307.

Gadue P, Huber TL, Paddison PJ, Keller GM. 2006. Wnt and TGF-beta signaling are required for the induction of an in vitro model of primitive streak formation using embryonic stem cells. *Proc Natl Acad Sci U S A* 103(45):16806–16811.

Gil HW, Yang JO, Lee EY, Hong SY. 2005. Paraquat-induced Fanconi syndrome. *Nephrology (Carlton)* 10(5):430–432.

Gould SE, Day M, Jones SS, Dorai H. 2002. BMP-7 regulates chemokine, cytokine, and hemodynamic gene expression in proximal tubule cells. *Kidney Int* 61(1):51–60.

Grisaru S, Cano-Gauci D, Tee J, Filmus J, Rosenblum ND. 2001. Glypican-3 modulates BMP- and FGF-mediated effects during renal branching morphogenesis. *Dev Biol* 231(1):31–46.

Guo X, Nzerue C. 2002. How to prevent, recognize, and treat drug-induced nephrotoxicity. *Cleve Clin J Med* 69(4):289–290, 293–284, 296–287 passim.

He S, Liu N, Bayliss G, Zhuang S. 2013. EGFR activity is required for renal tubular cell dedifferentiation and proliferation in a murine model of folic acid-induced acute kidney injury. *Am J Physiol Renal Physiol* 304(4):F356–F366.

Hovda KE, Guo C, Austin R, McMartin KE. 2010. Renal toxicity of ethylene glycol results from internalization of calcium oxalate crystals by proximal tubule cells. *Toxicol Lett* 192(3):365–372.

Huang M, Choi SJ, Kim DW, Kim NY, Park CH, Yu SD, Kim DS, Park KS, Song JS, Kim H, Choi BS, Yu IJ, Park JD. 2009. Risk assessment of low-level cadmium and arsenic on the kidney. *J Toxicol Environ Health A* 72(21–22):1493–1498.

Huang JX, Blaskovich MA, Cooper MA. 2014. Cell- and bio-marker-based assays for predicting nephrotoxicity. *Expert Opin Drug Metab Toxicol* 10(12):1621–1635.

Humphreys BD, Czerniak S, DiRocco DP, Hasnain W, Cheema R, Bonventre JV. 2011. Repair of injured proximal tubule does not involve specialized progenitors. *Proc Natl Acad Sci U S A* 108(22):9226–9231.

Ichimura T, Bonventre JV, Bailly V, Wei H, Hession CA, Cate RL, Sanicola M. 1998. Kidney injury molecule-1 (KIM-1), a putative epithelial cell adhesion molecule containing a novel immunoglobulin domain, is up-regulated in renal cells after injury. *J Biol Chem* 273(7):4135–4142.

Ichimura T, Hung CC, Yang SA, Stevens JL, Bonventre JV. 2004. Kidney injury molecule-1: a tissue and urinary biomarker for nephrotoxicant-induced renal injury. *Am J Physiol Renal Physiol* 286(3):F552–F563.

Imaoka T, Kusuhara H, Adachi M, Schuetz JD, Takeuchi K, Sugiyama Y. 2007. Functional involvement of multidrug resistance-associated protein 4 (MRP4/ABCC4) in the renal elimination of the antiviral drugs adefovir and tenofovir. *Mol Pharmacol* 71(2):619–627.

Imgrund M, Grone E, Grone HJ, Kretzler M, Holzman L, Schlondorff D, Rothenpieler UW. 1999. Re-expression of the developmental gene Pax-2 during experimental acute tubular necrosis in mice 1. *Kidney Int* 56(4):1423–1431.

James RG, Schultheiss TM. 2005. Bmp signaling promotes intermediate mesoderm gene expression in a dose-dependent, cell-autonomous and translation-dependent manner. *Dev Biol* 288(1):113–125.

Jang KJ, Mehr AP, Hamilton GA, McPartlin LA, Chung S, Suh KY, Ingber DE. 2013. Human kidney proximal tubule-on-a-chip for drug transport and nephrotoxicity assessment. *Integr Biol (Camb)* 5(9):1119–1129.

Jiang YS, Jiang T, Huang B, Chen PS, Ouyang J. 2013. Epithelial-mesenchymal transition of renal tubules: divergent processes of repairing in acute or chronic injury? *Med Hypotheses* 81(1):73–75.

Kandasamy K, Chuah JK, Su R, Huang P, Eng KG, Xiong S, Li Y, Chia CS, Loo LH, Zink D. 2015. Prediction of drug-induced nephrotoxicity and injury mechanisms with human induced pluripotent stem cell-derived cells and machine learning methods. *Scientific Reports* 5:12337.

Kang M, Han YM. 2014. Differentiation of human pluripotent stem cells into nephron progenitor cells in a serum and feeder free system. *PLoS One* 9(4):e94888.

Kann M, Ettou S, Jung YL, Lenz MO, Taglienti ME, Park PJ, Schermer B, Benzing T, Kreidberg JA. 2015. Genome-wide analysis of Wilms' tumor 1-controlled gene expression in podocytes reveals key regulatory mechanisms. *J Am Soc Nephrol* 26:2097–2104.

Kelly EJ, Wang Z, Voellinger JL, Yeung CK, Shen DD, Thummel KE, Zheng Y, Ligresti G, Eaton DL, Muczynski KA, Duffield JS, Neumann T, Tourovskaia A, Fauver M, Kramer G, Asp E, Himmelfarb J. 2013. Innovations in preclinical biology: ex vivo engineering of a human kidney tissue microperfusion system. *Stem Cell Res Ther* 4(Suppl 1):S17.

Kim D, Dressler GR. 2005. Nephrogenic factors promote differentiation of mouse embryonic stem cells into renal epithelia. *J Am Soc Nephrol* 16(12):3527–3534.

Kincaid-Smith P. 1986. Renal toxicity of non-narcotic analgesics. At-risk patients and prescribing applications. *Med Toxicol* 1(Suppl 1):14–22.

Kobayashi A, Valerius MT, Mugford JW, Carroll TJ, Self M, Oliver G, McMahon AP. 2008. Six2 defines and regulates a multipotent self-renewing nephron progenitor population throughout mammalian kidney development. *Cell Stem Cell* 3(2):169–181.

Konigs M, Mulac D, Schwerdt G, Gekle M, Humpf HU. 2009. Metabolism and cytotoxic effects of T-2 toxin and its metabolites on human cells in primary culture. *Toxicology* 258(2–3):106–115.

Kramer J, Steinhoff J, Klinger M, Fricke L, Rohwedel J. 2006. Cells differentiated from mouse embryonic stem cells via embryoid bodies express renal marker molecules. *Differentiation* 74(2–3):91–104.

Kreidberg JA. 2010. WT1 and kidney progenitor cells. *Organogenesis* 6(2):61–70.

Kusaba T, Lalli M, Kramann R, Kobayashi A, Humphreys BD. 2014. Differentiated kidney epithelial cells repair injured proximal tubule. *Proc Natl Acad Sci U S A* 111(4):1527–1532. Erratum in *Proc Natl Acad Sci U S A.* 2014, 111(15): 5754.

Lam AQ, Freedman BS, Morizane R, Lerou PH, Valerius MT, Bonventre JV. 2014. Rapid and efficient differentiation of human pluripotent stem cells into intermediate mesoderm that forms tubules expressing kidney proximal tubular markers. *J Am Soc Nephrol* 25(6):1211–1225.

Lash LH, Putt DA, Cai H. 2006. Membrane transport function in primary cultures of human proximal tubular cells. *Toxicology* 228(2–3):200–218.

Lash LH, Putt DA, Cai H. 2008. Drug metabolism enzyme expression and activity in primary cultures of human proximal tubular cells. *Toxicology* 244(1):56–65.

Li W, Choy DF, Lam MS, Morgan T, Sullivan ME, Post JM. 2003. Use of cultured cells of kidney origin to assess specific cytotoxic effects of nephrotoxins. *Toxicol In Vitro* 17(1):107–113.

Li W, Lam M, Choy D, Birkeland A, Sullivan ME, Post JM. 2006. Human primary renal cells as a model for toxicity assessment of chemo-therapeutic drugs. *Toxicol In Vitro* 20(5):669–676.

Li Y, Oo ZY, Chang SY, Huang P, Eng KG, Zeng JL, Kaestli AJ, Gopalan B, Kandasamy K, Tasnim F, Zink D. 2013. An in vitro method for the prediction of renal proximal tubular toxicity in humans. *Toxicol Res* 2(5):352–362.

Li Y, Kandasamy K, Chuah JKC, Lam YN, Toh WS, Oo ZY, Zink D. 2014. Identification of nephrotoxic compounds with embryonic stem cell-derived human renal proximal tubular-like cells. *Mol Pharm* 11(7):1982–1990.

Lin Z, Will Y. 2012. Evaluation of drugs with specific organ toxicities in organ-specific cell lines. *Toxicol Sci* 126(1):114–127.

Liu S, Cieslinski DA, Funke AJ, Humes HD. 1997. Transforming growth factor-beta 1 regulates the expression of Pax-2, a developmental control gene, in renal tubule cells. *Exp Nephrol* 5(4):295–300.

Lockard VG, Phillips RD, Wallace Hayes A, Berndt WO, O'Neal RM. 1980. Citrinin nephrotoxicity in rats: a light and electron microscopic study. *Exp Mol Pathol* 32(3):226–240.

Lohr JW, Willsky GR, Acara MA. 1998. Renal drug metabolism. *Pharmacol Rev* 50(1):107–141.

Madden EF, Fowler BA. 2000. Mechanisms of nephrotoxicity from metal combinations: a review. *Drug Chem Toxicol* 23(1):1–12.

Mae S, Shono A, Shiota F, Yasuno T, Kajiwara M, Gotoda-Nishimura N, Arai S, Sato-Otubo A, Toyoda T, Takahashi K, Nakayama N, Cowan CA, Aoi T, Ogawa S, McMahon AP, Yamanaka S, Osafune K. 2013. Monitoring and robust induction of nephrogenic intermediate mesoderm from human pluripotent stem cells. *Nat Commun* 4:1367.

Marshall CB, Pippin JW, Krofft RD, Shankland SJ. 2006. Puromycin aminonucleoside induces oxidant-dependent DNA damage in podocytes in vitro and in vivo. *Kidney Int* 70(11):1962–1973.

Martin BL, Kimelman D. 2010. Brachyury establishes the embryonic mesodermal progenitor niche. *Genes Dev* 24(24): 2778–2783.

Maunsbach AB, Marples D, Chin E, Ning G, Bondy C, Agre P, Nielsen S. 1997. Aquaporin-1 water channel expression in human kidney. *J Am Soc Nephrol* 8(1):1–14.

McGoldrick TA, Lock EA, Rodilla V, Hawksworth GM. 2003. Renal cysteine conjugate C-S lyase mediated toxicity of halogenated alkenes in primary cultures of human and rat proximal tubular cells. *Arch Toxicol* 77(7):365–370.

Mishra J, Ma Q, Prada A, Mitsnefes M, Zahedi K, Yang J, Barasch J, Devarajan P. 2003. Identification of neutrophil gelatinase-associated lipocalin as a novel early urinary biomarker for ischemic renal injury. *J Am Soc Nephrol* 14(10):2534–2543.

Morrissey KM, Wen CC, Johns SJ, Zhang L, Huang SM, Giacomini KM. 2012. The UCSF-FDA TransPortal: a public drug transporter database. *Clin Pharmacol Ther* 92(5):545–546.

Morrissey KM, Stocker SL, Wittwer MB, Xu L, Giacomini KM. 2013. Renal transporters in drug development. *Annu Rev Pharmacol Toxicol* 53:503–529.

Morizane R, Lam AQ, Freedman BS, Kishi S, Valerius MT, Bonventre JV. 2015. Nephron organoids derived from human pluripotent stem cells model kidney development and injury. *Nat Biotechnol* 33(11):1193–1200.

Mugford JW, Sipila P, Kobayashi A, Behringer RR, McMahon AP. 2008a. Hoxd11 specifies a program of metanephric kidney development within the intermediate mesoderm of the mouse embryo. *Dev Biol* 319(2):396–405.

Mugford JW, Sipila P, McMahon JA, McMahon AP. 2008b. Osr1 expression demarcates a multi-potent population of intermediate mesoderm that undergoes progressive restriction to an Osr1-dependent nephron progenitor compartment within the mammalian kidney. *Dev Biol* 324(1):88–98.

Mundel P, Shankland SJ. 2002. Podocyte biology and response to injury. *J Am Soc Nephrol* 13(12):3005–3015.

Narayanan K, Schumacher KM, Tasnim F, Kandasamy K, Schumacher A, Ni M, Gao S, Gopalan B, Zink D, Ying JY. 2013. Human embryonic stem cells differentiate into functional renal proximal tubular-like cells. *Kidney Int* 83(4):593–603.

Naughton CA. 2008. Drug-induced nephrotoxicity. *Am Fam Physician* 78(6):743–750.

Nishinakamura R, Takasato M. 2005. Essential roles of Sall1 in kidney development. *Kidney Int* 68(5):1948–1950.

Nordberg G, Jin T, Wu X, Lu J, Chen L, Liang Y, Lei L, Hong F, Bergdahl IA, Nordberg M. 2012. Kidney dysfunction and cadmium exposure--factors influencing dose-response relationships. *J Trace Elem Med Biol* 26(2–3):197–200.

Orth SR, Ritz E, Suter-Crazzolara C. 2000. Glial cell line-derived neurotrophic factor (GDNF) is expressed in the human kidney and is a growth factor for human mesangial cells. *Nephrol Dial Transplant* 15(5):589–595.

Pachnis V, Mankoo B, Costantini F. 1993. Expression of the c-ret proto-oncogene during mouse embryogenesis. *Development* 119(4):1005–1017.

Patel SR, Dressler GR. 2004. Expression of Pax2 in the intermediate mesoderm is regulated by YY1. *Dev Biol* 267(2):505–516.

Patel SR, Dressler GR. 2005. BMP7 signaling in renal development and disease. *Trends Mol Med* 11(11):512–518.

Patrakka J, Tryggvason K. 2010. Molecular make-up of the glomerular filtration barrier. *Biochem Biophys Res Commun* 396(1):164–169.

Pereira LA, Wong MS, Lim SM, Sides A, Stanley EG, Elefanty AG. 2011. Brachyury and related Tbx proteins interact with the Mixl1 homeodomain protein and negatively regulate Mixl1 transcriptional activity. *PLoS One* 6(12):e28394.

Pfohl-Leszkowicz A, Manderville RA. 2007. Ochratoxin A: an overview on toxicity and carcinogenicity in animals and humans. *Mol Nutr Food Res* 51(1):61–99.

Pfohl-Leszkowicz A, Tozlovanu M, Manderville R, Peraica M, Castegnaro M, Stefanovic V. 2007. New molecular and field evidences for the implication of mycotoxins but not aristolochic acid in human nephropathy and urinary tract tumor. *Mol Nutr Food Res* 51(9):1131–1146.

Piscione TD, Yager TD, Gupta IR, Grinfeld B, Pei Y, Attisano L, Wrana JL, Rosenblum ND. 1997. BMP-2 and OP-1 exert direct and opposite effects on renal branching morphogenesis. *Am J Physiol* 273(6 Pt 2):F961–F975.

Piscione TD, Phan T, Rosenblum ND. 2001. BMP7 controls collecting tubule cell proliferation and apoptosis via Smad1-dependent and -independent pathways. *Am J Physiol Renal Physiol* 280(1):F19–F33.

Pulina MV, Sahr KE, Nowotschin S, Baron MH, Hadjantonakis AK. 2014. A conditional mutant allele for analysis of Mixl1 function in the mouse. *Genesis* 52(5):417–423.

Quiros Y, Vicente-Vicente L, Morales AI, Lopez-Novoa JM, Lopez-Hernandez FJ. 2011. An integrative overview on the mechanisms underlying the renal tubular cytotoxicity of gentamicin. *Toxicol Sci* 119(2):245–256.

Reddy L, Bhoola K. 2010. Ochratoxins-food contaminants: impact on human health. *Toxins (Basel)* 2(4):771–779.

Redfern WS. 2010. Impact and frequency of different toxicities throughout the pharmaceutical life cycle. *The Toxicologist* 114(S1):1081.

Ryan MJ, Johnson G, Kirk J, Fuerstenberg SM, Zager RA, Torok-Storb B. 1994. HK-2: an immortalized proximal tubule epithelial cell line from normal adult human kidney. *Kidney Int* 45(1):48–57.

Segasothy M, Samad SA, Zulfigar A, Bennett WM. 1994. Chronic renal disease and papillary necrosis associated with the long-term use of nonsteroidal anti-inflammatory drugs as the sole or predominant analgesic. *Am J Kidney Dis* 24(1):17–24.

Sekine T, Miyazaki H, Endou H. 2006. Molecular physiology of renal organic anion transporters. *Am J Physiol Renal Physiol* 290(2):F251–F261.

Simic P, Vukicevic S. 2005. Bone morphogenetic proteins in development and homeostasis of kidney. *Cytokine Growth Factor Rev* 16(3):299–308.

Simon DB, Nelson-Williams C, Bia MJ, Ellison D, Karet FE, Molina AM, Vaara I, Iwata F, Cushner HM, Koolen M, Gainza FJ, Gitleman HJ, Lifton RP. 1996. Gitelman's variant of Bartter's syndrome, inherited hypokalaemic alkalosis, is caused by mutations in the thiazide-sensitive Na-Cl cotransporter. *Nat Genet* 12(1):24–30.

Snellings WM, Corley RA, McMartin KE, Kirman CR, Bobst SM. 2013. Oral reference dose for ethylene glycol based on oxalate crystal-induced renal tubule degeneration as the critical effect. *Regul Toxicol Pharmacol* 65(2):229–241.

Song B, Smink AM, Jones CV, Callaghan JM, Firth SD, Bernard CA, Laslett AL, Kerr PG, Ricardo SD. 2012. The directed differentiation of human iPS cells into kidney podocytes. *PLoS One* 7(9):e46453.

Srinivas S, Goldberg MR, Watanabe T, D'Agati V, al-Awqati Q, Costantini F. 1999. Expression of green fluorescent protein in the ureteric bud of transgenic mice: a new tool for the analysis of ureteric bud morphogenesis. *Dev Genet* 24(3–4):241–251.

Su R, Li Y, Zink D, Loo LH. 2014. Supervised prediction of drug-induced nephrotoxicity based on interleukin-6 and -8 expression levels. *BMC Bioinformatics* 15(Suppl. 16):S16.

Su R, Xiong S, Zink D, Loo LH. 2015. High-throughput imaging-based nephrotoxicity prediction for xenobiotics with diverse chemical structures. *Arch Toxicol*. Epub ahead of print DOI: 10.1007/s00204-015-1638-y.

Tacar O, Sriamornsak P, Dass CR. 2013. Doxorubicin: an update on anticancer molecular action, toxicity and novel drug delivery systems. *J Pharm Pharmacol* 65(2):157–170.

Tada S, Era T, Furusawa C, Sakurai H, Nishikawa S, Kinoshita M, Nakao K, Chiba T. 2005. Characterization of mesendoderm: a diverging point of the definitive endoderm and mesoderm in embryonic stem cell differentiation culture. *Development* 132(19):4363–4374.

Taguchi A, Kaku Y, Ohmori T, Sharmin S, Ogawa M, Sasaki H, Nishinakamura R. 2014. Redefining the in vivo origin of metanephric nephron progenitors enables generation of complex kidney structures from pluripotent stem cells. *Cell Stem Cell* 14(1):53–67.

Takasato M, Er PX, Becroft M, Vanslambrouck JM, Stanley EG, Elefanty AG, Little MH. 2014. Directing human embryonic stem cell differentiation towards a renal lineage generates a self-organizing kidney. *Nat Cell Biol* 16(1):118–126.

Takasato M, Er PX, Chiu HS, Maier B, Baillie GJ, Ferguson C, Parton RG, Wolvetang EJ, Roost MS, Chuva de Sousa Lopes SM, Little MH. 2015. Kidney organoids from human iPS cells contain multiple lineages and model human nephrogenesis. *Nature* 526(7574):564–568.

Tam PP, Loebel DA. 2007. Gene function in mouse embryogenesis: get set for gastrulation. *Nat Rev Genet* 8(5):368–381.

Technau U. 2001. Brachyury, the blastopore and the evolution of the mesoderm. *Bioessays* 23(9):788–794.

Tiong HY, Huang P, Xiong S, Li Y, Vathsala A, Zink D. 2014. Drug-induced nephrotoxicity: clinical impact and preclinical in vitro models. *Mol Pharm* 11(7):1933–1948.

Tolwani A. 2012. Continuous renal-replacement therapy for acute kidney injury. *N Engl J Med* 367(26):2505–2514.

Uchino S, Kellum JA, Bellomo R, Doig GS, Morimatsu H, Morgera S, Schetz M, Tan I, Bouman C, Macedo E, Gibney N, Tolwani A, Ronco C. 2005. Acute renal failure in critically ill patients: a multinational, multicenter study. *JAMA* 294(7):813–818.

Vainio S, Lin Y. 2002. Coordinating early kidney development: lessons from gene targeting. *Nat Rev Genet* 3(7):533–543.

Vanmassenhove J, Vanholder R, Nagler E, Van Biesen W. 2013. Urinary and serum biomarkers for the diagnosis of acute kidney injury: an in-depth review of the literature. *Nephrol Dial Transplant* 28(2):254–273.

Vaziri ND, Ness RL, Fairshter RD, Smith WR, Rosen SM. 1979. Nephrotoxicity of paraquat in man. *Arch Intern Med* 139(2):172–174.

Vidricaire G, Jardine K, McBurney MW. 1994. Expression of the Brachyury gene during mesoderm development in differentiating embryonal carcinoma cell cultures. *Development* 120(1):115–122.

Vigneau C, Polgar K, Striker G, Elliott J, Hyink D, Weber O, Fehling HJ, Keller G, Burrow C, Wilson P. 2007. Mouse embryonic stem cell-derived embryoid bodies generate progenitors that integrate long term into renal proximal tubules in vivo. *J Am Soc Nephrol* 18(6):1709–1720.

Wallin A, Zhang G, Jones TW, Jaken S, Stevens JL. 1992. Mechanism of the nephrogenic repair response. Studies on proliferation and vimentin expression after 35S-1,2-dichlorovinyl-L-cysteine nephrotoxicity in vivo and in cultured proximal tubule epithelial cells. *Lab Invest* 66(4):474–484.

Weiland C, Ahr HJ, Vohr HW, Ellinger-Ziegelbauer H. 2007. Characterization of primary rat proximal tubular cells by gene expression analysis. *Toxicol In Vitro* 21(3):466–491.

Wieser M, Stadler G, Jennings P, Streubel B, Pfaller W, Ambros P, Riedl C, Katinger H, Grillari J, Grillari-Voglauer R. 2008. hTERT alone immortalizes epithelial cells of renal proximal tubules without changing their functional characteristics. *Am J Physiol Renal Physiol* 295(5):F1365–F1375.

Witzgall R, Brown D, Schwarz C, Bonventre JV. 1994. Localization of proliferating cell nuclear antigen, vimentin, c-Fos, and clusterin in the postischemic kidney. Evidence for a heterogenous genetic response among nephron segments, and a large pool of mitotically active and dedifferentiated cells. *J Clin Invest* 93(5):2175–2188.

Wu Y, Connors D, Barber L, Jayachandra S, Hanumegowda UM, Adams SP. 2009. Multiplexed assay panel of cytotoxicity in HK-2 cells for detection of renal proximal tubule injury potential of compounds. *Toxicol In Vitro* 23(6):1170–1178.

Xia Y, Nivet E, Sancho-Martinez I, Gallegos T, Suzuki K, Okamura D, Wu MZ, Dubova I, Esteban CR, Montserrat N, Campistol JM, Izpisua Belmonte JC. 2013. Directed differentiation of human pluripotent cells to ureteric bud kidney progenitor-like cells. *Nat Cell Biol* 15(12):1507–1515.

Yang L, Su T, Li XM, Wang X, Cai SQ, Meng LQ, Zou WZ, Wang HY. 2012. Aristolochic acid nephropathy: variation in presentation and prognosis. *Nephrol Dial Transplant* 27(1): 292–298.

Yonezawa A, Inui K. 2011. Importance of the multidrug and toxin extrusion MATE/SLC47A family to pharmacokinetics, pharmacodynamics/toxicodynamics and pharmacogenomics. *Br J Pharmacol* 164(7):1817–1825.

PART VII

CURRENT STATUS OF PRECLINICAL *IN VIVO* TOXICITY BIOMARKERS

24

PREDICTIVE CARDIAC HYPERTROPHY BIOMARKERS IN NONCLINICAL STUDIES

STEVEN K. ENGLE

Lilly Research Laboratories, Division of Eli Lilly and Company, Lilly Corporate Center, Indianapolis, IN, USA

24.1 INTRODUCTION TO BIOMARKERS

A biomarker is a measurable substance whose concentration is reflective of an ongoing disease or toxicity in an organism (Aronson, 2005). They are applied in preclinical drug discovery toxicology to detect test article-related tissue injury and may be used in conjunction with histopathological examination of tissues to provide an additional, quantitative measure of toxicity and assist in the selection of drug candidates more likely to be tolerated in humans. Blood-based biomarkers offer the advantage of being nonterminal and thus allow serial monitoring of toxicity. Practical application of biomarkers in drug development requires the commercial availability of assays (e.g., enzyme-linked immunosorbent assay (ELISA), enzyme activity assay, mass spectrometry) or the creation of custom assays. Many well-established biomarkers, such as cardiac troponin I (cTnI) or creatine kinase–myocardial band (CK-MB), are "leakage" markers that are allowed into circulation when a cardiac cell membrane is disrupted and its contents leak out, often preceding cell death. Secreted factors, such as cytokines and hormones, may also be used as safety biomarkers to provide information about ongoing processes that may lead to injury or disruption of homeostasis and anatomical adaptation, such as pathological or physiological cardiac hypertrophy. Biomarkers may be applied at any stage of drug development, from efficacy studies in animal models (often mice), to safety studies in rats, dogs, nonhuman primates, and guinea pigs, to clinical studies and even

postmarketing. Their use has been encouraged by the US Food and Drug Administration to reduce the time and cost of drug development (Amur et al., 2008).

24.2 CARDIOVASCULAR TOXICITY

Cardiovascular (CV) toxicity is a safety liability in drug development and is responsible for significant attrition during nonclinical and clinical investigation, as well as the addition of black box warnings and withdrawal of drugs from the market after approval (Olson et al., 2000; Lasser et al., 2002; Stevens and Baker, 2009; Laverty et al., 2011). Cardiac toxicity may include arrhythmias, such as *torsades de pointes*, caused by the prolongation of QT interval, cardiomyocyte degeneration, hemodynamic alterations, and cardiac hypertrophy. Prediction of arrhythmia *in vitro* may be achieved using binding profiles against major cardiac ion channels (Fermini et al., 2016), while *in vivo* detection of arrhythmia requires electrocardiography (ECG) and can be achieved in preclinical (i.e., rat, dog, nonhuman primate, and guinea pig) safety assessment studies using implantable telemetry (electrodes surgically implanted that measure electrical activity, blood pressure, and heart rate) or jacketed external telemetry (JET) (Leishman et al., 2012; Derakhchan et al., 2014).

Prolongation of QT interval may be predicted *in vitro* by patch clamp electrophysiology in mammalian cells transfected with the human *ether-à-go-go-related* gene (hERG), a subunit of the I_{Kr} (cardiac rapidly activating delayed rectifier potassium channel), or by measuring a compound's ability

Drug Discovery Toxicology: From Target Assessment to Translational Biomarkers, First Edition. Edited by Yvonne Will, J. Eric McDuffie, Andrew J. Olaharski, and Brandon D. Jeffy.

to inhibit astemizole binding in hERG-transfected cells (Chiu et al., 2004). Myocardial cell degeneration and necrosis can be detected through histological examination of heart tissue sections and in blood using cell membrane leakage biomarkers such as CK-MB or cardiac troponins I and T (cTnI and cTnT) in the preclinical and clinical settings (Berridge et al., 2009). Apple et al. (2008) have reported on analytical characteristics of commercial cTnI immunoassays in serum from rats, dogs, and monkeys with acute myocardial injury.

Changes in blood pressure or heart rate can be detected using a sphygmomanometer (blood pressure cuff) or the aforementioned implanted telemetry devices. Cardiac hypertrophy can be detected through measurement of heart weights or by using echocardiography (ultrasound). More recently, prediction of cardiac hypertrophy has called for the discovery and validation of next-generation tools; therefore, the detection of cardiac hypertrophy and accompanying changes in blood pressure using blood-based biomarkers will be the focus of this chapter.

24.3 CARDIAC HYPERTROPHY

Cardiac hypertrophy is an adaptive response to a variety of factors, including, but not limited to, increased blood pressure or plasma volume overload (hypervolemia), cardiomyocyte injury, or structural defects and is a risk factor for drug-induced cardiac morbidity and mortality through heart failure (Levy et al., 1990). Fundamentally, cardiac hypertrophy is a compensatory increase in cardiac mass in response to increased workload, much like the response of skeletal muscle to effective resistance weight training wherein increases in skeletal muscle mass become evident (Houston, 1999). Cardiac hypertrophy is characterized at the cellular level by increased cell size, enhanced protein synthesis, fibrosis, and heightened organization of the sarcomere (Frey and Olson, 2003). In "physiologic" hypertrophy, enlargement of the heart serves to normalize workload and output, such as in elite athletes, in whom repeated, maximal increases in CV workload during training lead to increased cardiac mass, or in response to hypervolemia and increased metabolic demands during pregnancy. Physiologic hypertrophy rarely progresses to heart failure and is reversible (Baggish and Wood, 2011; Li et al., 2012). In chronic pathological conditions, such as hypertension, compensatory hypertrophic changes may eventually progress to decompensated heart failure, characterized by declining cardiac output, arrhythmia, increased apoptosis, fibrosis, chamber enlargement, and/or increased wall thickness (Frey et al., 2004). Long-standing cardiac hypertrophy of pathological origin may ultimately result in heart failure, the inability of the heart to supply sufficient blood flow to meet physiological needs (Hunter and Chien, 1999).

Cardiac hypertrophy may be eccentric, concentric, or a combination thereof (Gaasch and Zile, 2011). Enlarged cardiac chambers (increased volume), especially the left ventricle (LV), with free walls and septum of normal (during compensated phase) to somewhat decreased thickness (during decompensated heart failure), are characteristic of eccentric hypertrophy and volume overload. During eccentric hypertrophy, the sarcomere is lengthened, increasing cardiac mass without thickening the walls of the heart. Conservation of chamber volume or a slight reduction with thicker walls is characteristic of concentric hypertrophy and pressure overload (Van Vleet and Ferrans, 1995). During concentric hypertrophy, the sarcomere is thickened, thus increasing the thickness of the heart's walls and/or septum. Cardiac hypertrophy is induced by a variety of factors, including persistent arterial hypertension and increased ventricular wall stress (Krauser and Devereaux, 2006).

Clinical conditions that may result in cardiac hypertrophy include valvular regurgitation, stenosis, and components of the metabolic syndrome including obesity, diabetes mellitus, and neurohormonal imbalances (Artham et al., 2009). During diastole, the LV fills with blood after the mitral valve opens, allowing blood to flow from the left atrium into the LV while the aortic valve remains closed. When the mitral valve closes at the end of diastole, the amount of blood in the LV is referred to as end-diastolic volume (EDV) and is known as preload. If the aortic valve fails to close completely during diastole, blood is allowed to flow backward from the aorta into the LV (valvular regurgitation) as it is being filled with blood from the left atrium, thus increasing EDV and wall stress, creating a volume overload and causing eccentric hypertrophy in the LV.

Chronic hypertension, such as during obesity or in neurohormonal imbalances (Crowley et al., 2006), results in increased pressure in the aorta (pressure overload). The aortic valve opens at the start of systole and blood is forced from the LV into the aorta as the heart contracts as a result of higher pressure in the ventricle than the aorta. When the aortic valve closes at the end of systole, the remaining volume of blood in the LV is known as end-systolic volume (ESV). The resistance to blood flow from the LV is a result of pressure in the aorta and is known as afterload. When afterload is increased as a result of hypertension, the amount of blood ejected from the LV with each contraction, or stroke volume ($SV = EDV - ESV$), is reduced and ESV is increased, increasing wall stress and resulting in concentric hypertrophy of the LV.

In both previously mentioned examples, a fundamental property of the myocardium, described by the Frank–Starling mechanism, comes into play. The Frank–Starling mechanism holds that the force of contraction generated by the heart is increased as the amount of blood filling the LV is increased and the myocardium is stretched (Moss and Fitzsimons, 2002; Vincent, 2008). In the example of valvular regurgitation, excess EDV is the result of regurgitation from the aortic valve during diastole, and in hypertension, increased ESV is added to normal venous return. Either way, the volume of the ventricle is increased, causing greater stretch of the myocardium, and greater force is generated during contraction. In short, the heart is forced to work harder to maintain cardiac output (SV×HR) and grows in order to generate adequate force to move increased blood volume from the LV during systole.

24.4 DIAGNOSIS OF CARDIAC HYPERTROPHY

Cardiac hypertrophy can be accurately assessed using echocardiography (Devereux et al., 1986). Increased dimensions of the heart, such as increased chamber volume or wall thickness, can be visualized with echocardiography and changes in cardiac function, such as increased EDV or reduced ejection fraction (EF=(SV/EDV)×100), can be calculated. This allows detection of pathologies that can lead to cardiac hypertrophy before cardiac mass is increased, potentially after only one or two test article administrations in the drug development scenario. After sufficient increases in mass have occurred, abnormalities in electrical conduction may be evident, although ECG is a relatively insensitive method for detection of cardiac hypertrophy (Edhouse et al., 2002). Additionally, systemic changes that can lead to cardiac hypertrophy, such as hypertension, can be readily detected using a sphygmomanometer; however, echocardiography and blood pressure measurements are not commonly incorporated into nonclinical drug development studies, especially in rodents. Blood pressure cuffs can be used to noninvasively measure BP in rodents, dogs, or monkeys; however, such measurements are often highly variable due to the stress induced during the handling and restraint of animals, similar to "white coat syndrome" seen in patients (Buñag and Butterfield, 1982; Chester et al., 1992; Pickering, 1998).

The most accurate and representative method to collect hemodynamic data during animal studies involves surgical implantation of pressure transducers into a major artery, often the descending aorta, and the transmission of data using radiotelemetry devices implanted in the abdomen (Whitesall et al., 2004; Duan et al., 2007). After a period of recovery from the surgery, blood pressure, heart rate, and ECG can be continuously measured in conscious, freely moving animals without the presence of investigators in the animal room. The reduction in stress involved in the acquisition of data using this method allows a more accurate assessment of whether a test article has unintended and undesired effects on blood pressure, heart rate, or electrical conduction. Radiotelemetry, while desirable for the reasons mentioned previously, is costly and adds time to the already lengthy process of drug discovery. Yet, collection of hemodynamic data using radiotelemetry or

JET technology may be actualized during the compound lead optimization process when a test article becomes a candidate for human development or as an add-on when "good laboratory practice" (GLP) studies are conducted to determine whether the compound is suitable for advancing to first- in-human studies (Pritchard et al., 2003). This was first suggested by the original guideline governing safety pharmacology (U.S. Food and Drug Administration, Guidance for Industry, 2001) where "the use of new technologies and methodologies in accordance with sound scientific principles is encouraged" and was further reinforced by revisions to ICH guideline S6 (R1) (International Conference on Harmonisation, 2011) stating that safety pharmacology "may be investigated in separate studies or incorporated in the design of toxicity studies" for biotechnology-derived products.

Similarly, echocardiography, while sensitive and definitive in the assessment of cardiac dysfunction leading to cardiac hypertrophy, and measurement of cardiac hypertrophy itself, adds time and expense and requires specialized equipment and training not typically incorporated into early safety studies. Instead, detection of cardiac hypertrophy in nonclinical studies relies on heart weight measurement and routine heart histopathology. This requires sufficient duration of treatment to produce a measurable change in heart weight, and the short duration of initial toxicology studies (e.g., 1–4 days in rats) makes observation of these changes relatively uncommon until later in the development process. Histological observations consistent with cardiac hypertrophy (increase in number or size of myofibers and changes in their organization) are even less common in toxicology studies of <1-month duration. Thus, prediction of cardiac hypertrophy may be delayed until longer-term preclinical toxicology studies (i.e., 1–3 months) are conducted. For these reasons, the use of blood-based biomarkers predictive of cardiac hypertrophy or reflective of ongoing cardiac dysfunction that could lead to hypertrophy can be beneficial in the selection of safer compounds earlier in the drug development process.

24.5 BIOMARKERS OF CARDIAC HYPERTROPHY

Ideally, a biomarker of cardiac hypertrophy would be predictive, that is, its concentration in serum or plasma would be altered in advance of eventual increases in heart weight, and specific for the heart in both humans and animals commonly used in safety assessment. This would allow early detection of cardiac risk in short-duration toxicology studies (e.g., 1–4 days), serial monitoring in longer-duration toxicology studies, or pharmacology studies in order to enable the prediction of cardiac hypertrophy liabilities in humans. In practice, biomarkers are not always specific for a certain organ or type of toxicity and prediction of an adverse outcome such as cardiac hypertrophy may be best achieved through the use of multiple biomarkers reflecting the systemic and structural changes leading to increased cardiac mass and heart failure (Zile et al., 2011). Published reports of the use of biomarkers of cardiac hypertrophy in preclinical drug development are scarce and significant work remains to be done to understand the relevance and value of potential biomarkers of cardiac hypertrophy in laboratory animals. Currently, there is no "gold standard" for the prediction of cardiac hypertrophy but the following section will explore some biomarkers that have shown promise in humans and/or animals and merit further study.

Clinically, cardiac hypertrophy is assessed as part of a spectrum of changes leading to heart failure, including inflammation, extracellular matrix (ECM) remodeling, neurohormonal changes, and hemodynamic stress. Biomarkers of inflammation, such as C-reactive protein (CRP) and tumor necrosis factor-α (TNF-α), have proven useful in risk stratification in patients with heart failure (Braunwald, 2008). CRP, an acute-phase protein secreted in response to tissue injury, exerts adverse effects on the vasculature, including reduced nitric oxide synthesis and increased endothelin-1 (ET-1) production, with the net effect of increased blood pressure (Venugopal et al., 2005). In human hemodialysis patients, CRP has been found to be an independent predictor of LV hypertrophy and the pathogenic role of CRP in the development of cardiac hypertrophy has also been demonstrated in a pressure overload model in mice (Nagai et al., 2011; Monfared et al., 2013). However, CRP is secreted in mice in trace amounts and is not considered a major acute-phase protein and care must be taken when extrapolating findings to larger species (Pepys and Hirschfield, 2003; Torzewski et al., 2014).

Secretion of TNF-α from cardiomyocytes has been demonstrated in mice after administration of endotoxin and the ability of TNF-α to induce hypertrophic growth has been observed in cultured adult feline cardiac myocytes and neonatal rat cardiac myocytes (Giroir et al., 1992; Yokoyama et al., 1997; Condorelli et al., 2002). Increased circulating TNF-α concentrations have been reported in humans with severe heart failure and in mice with cardiac hypertrophy induced by pressure overload caused by aortic banding (Levine et al., 1990; Sun et al., 2007). Infusion of TNF-α has been associated with LV dysfunction in dogs (Eichenholz et al., 1992; Pagani et al., 1992). Additional cytokines implicated in cardiac hypertrophy and heart failure include ET-1, interleukin-6 (IL-6), and cardiotrophin-1 (CT-1) (Seta et al., 1996; López et al., 2009). Biomarkers of inflammation offer insight into ongoing injury processes that may exacerbate hemodynamic and cellular changes, leading to cardiac hypertrophy and heart failure. They are relevant in multiple animal species used in preclinical safety assessment as well as humans (Watterson et al., 2009); however, these markers are not specific for inflammation in the heart and could signal injury in multiple tissues. Nevertheless, inclusion of one or more in a panel of markers may aid in the prediction of cardiac toxicity in animal studies.

Matrix metalloproteinases (MMP), tissue inhibitors of metalloproteinase (TIMP), and collagen propeptides (PIIINP) are also indicative of remodeling that occurs during cardiac hypertrophy, but are also nonspecific and consistent with more advanced hypertrophy than occurs in short-duration animal studies. A panel consisting of MMP-7, MMP-9, TIMP-1, PIIINP, and NTproBNP has shown promise in predicting the presence of cardiac hypertrophy in humans with hypertension and may prove useful in animal studies; however, further research is needed (Zile et al., 2011). Osteoprotegerin (OPG), a member of the TNF receptor superfamily, has shown promise as a biomarker of cardiac hypertrophy; however, little is known regarding the utility of OPG in the context of nonclinical safety studies (Coutinho et al., 2011; Koyama et al., 2014).

Vasoactive peptides, such as adrenomedullin (AM), ET-1, and natriuretic peptides, have shown relevance in humans and animals for predicting pathological changes in the CV system and playing an active role in the development of cardiac hypertrophy. AM, a peptide hormone, is secreted from cardiac myocytes and has vasodilatory, antiapoptotic, antihypertrophic, angiogenic, antifibrotic, natriuretic, diuretic, and inotropic activities (Autelitano et al., 2001; Yanagawa and Nagaya, 2007). Increased circulating concentrations of AM have been observed in hypertensive human patients with increased LV mass and ventricular expression of AM was correlated with cardiac mass in rats with volume- or pressure overload-induced cardiac hypertrophy caused by aortocaval shunt or aortic banding (Morimoto et al., 1999; Coutinho et al., 2011; Yoshihara et al., 2005). Increased plasma concentrations of AM have also been reported in dogs with heart failure as a result of rapid ventricular pacing and were found to correlate with increased left ventricular mass, diastolic function, cardiac output, and ejection fraction (Jougasaki et al., 1997).

ET-1 and angiotensin II (Ang II) are peptide hormones that cause increased blood pressure and have been associated with cardiac hypertrophy. ET-1 is secreted from vascular endothelial cells and has been associated with inflammation, fibrosis, cardiac and vascular hypertrophy, cell proliferation, and vasoconstriction (Beghetti et al., 2005). Application of ET-1 in vitro has caused hypertrophy of cardiac myocytes from rabbits and rats (Irukayama-Tomobe et al., 2004; Bupha-Intr et al., 2012). Increased concentrations of circulating ET-1 have also been reported in human patients with cardiac hypertrophy (Hasegawa et al., 1996; Ogino et al., 2004; Coutinho et al., 2011). Plasma ET-1 concentrations were also increased in dogs with heart failure and correlated with increased left ventricular dimensions (Prosek et al., 2004).

Natriuretic peptides (NPs) have shown potential as biomarkers of hemodynamic stress and cardiac hypertrophy when used in short-duration rodent toxicology studies (Engle et al., 2011; Engle and Watson, 2016). Atrial natriuretic peptide (ANP) and brain natriuretic peptide (BNP) are hormones produced and secreted by cardiomyocytes and are important regulators of cardiac growth, blood volume, and blood pressure. The presence of a hormone in cardiomyocytes was first demonstrated in 1981 by the observation that intravenous injection of atrial extracts from normal rats produced natriuresis and diuresis by a factor that was subsequently identified as ANP (de Bold et al., 1981). BNP was identified after isolation from porcine brain and was noted for its similarity to ANP (Sudoh et al., 1988). ANP and BNP are constitutively synthesized and secreted by atrial cardiomyocytes but are also expressed and secreted by ventricular cardiomyocytes in cases of left ventricular (LV) dysfunction and cardiac hypertrophy (Yasue et al., 1994; McGrath and De Bold, 2005; Engle et al., 2010; Oyama et al., 2013). NPs are secreted in response to a variety of factors, most prominently myofiber stretch resulting from increased cardiac wall pressure or afterload (e.g., high blood pressure or plasma volume expansion) but also during myocardial ischemia, and in response humoral stimuli such as ET-1, Ang II, TNF-α, IL-1, IL-6, glucocorticoids, sex steroid hormones, and thyroid hormones (Ruskoaho, 1992; Thibault et al., 1999; Clerico et al., 2006). These various mechanisms present confounding factors when interpreting natriuretic peptide data; however, when applied in the acute setting in animal studies during safety testing, they can be useful triggers for more intense CV safety investigations, such as blood pressure or imaging studies.

ANP and BNP are stored as prohormones in vesicles within atrial cardiomyocytes (Klein et al., 1993). Upon stimulation, proANP and proBNP are proteolytically cleaved to yield the amino-terminal fragments NTproANP and NTproBNP and the active C-terminal NP hormones ANP and BNP (Potter et al., 2006). The active hormones are secreted in equimolar amounts with their respective N-terminal fragments and bind primarily to the NP receptors NPR-A and NPR-B, which are transmembrane guanylyl cyclases. A third NP receptor, NPR-C, has no known enzymatic activity and acts to clear ANP and BNP from circulation.

NPs are active in several tissues, including the heart, kidneys, and vasculature. Studies in mice have shown that ANP (Potter et al., 2006) and BNP (Jensen et al., 1998) inhibit cardiac hypertrophy and the proliferation of cardiac fibroblasts. Kidney function is regulated by NPs through increased glomerular filtration rate, inhibition of sodium and water reabsorption, and inhibition of renin secretion, resulting in natriuresis, diuresis, and downstream inhibition of the renin–angiotensin–aldosterone system. NP signaling causes relaxation of smooth muscle, resulting in dilation of blood vessels and airways, and increased vascular permeability. Taken together, these effects cause reduced plasma volume and decreased blood pressure (Jensen et al., 1998, Levin et al., 1998; Potter et al., 2006).

ANP and particularly BNP have proven to be prognostic biomarkers of heart failure in both clinical and veterinary settings and are associated most frequently with LV dysfunction and increased cardiac mass but are also relevant in the context of pulmonary embolism, pulmonary hypertension, and chronic atrial fibrillation (Lerman et al., 1993; Lloyd et al., 1994; Dickstein et al., 1995; Muders et al., 1997; Jortani et al., 2004; Clerico et al., 2006; Tang, 2007; Oyama et al., 2013). N-terminal proANP and NTproBNP are also predictive of left ventricular hypertrophy (LVH) and systolic dysfunction in humans (Schirmer and Omland, 1999; Galasko et al., 2007). Expression of ANP precursor (NPPA) and BNP precursor (NPPB) was upregulated in human cardiac ventricles with pathological hypertrophy (Schirmer and Omland, 1999; Booz, 2005). Increased plasma concentrations of ANP and BNP have been found in human patients with heart disease (Sagnella, 1998; Doust et al., 2004). Their utility in detecting cardiac stress and dysfunction in humans make NPs candidates for noninvasive translational cardiac biomarkers for use during drug safety studies in rats and other veterinary species.

ANP and BNP have very short half-lives, on the order of minutes (Kimura et al., 2007). Due to their important physiological roles, they are rapidly cleared from circulation through receptor-mediated clearance and proteolytic degradation, resulting in relatively low concentrations that are challenging to measure. N-terminal proANP and NTproBNP have no known physiological function or receptor and are cleared through renal excretion, leading to longer half-lives in blood and greater circulating concentrations while reflecting the overall secretion of the active C-terminal hormones (Thibault et al., 1988; Clerico et al., 2006). Higher concentrations and longer half-lives make the measurement N-terminal fragments less sensitive to the timing of sample collection and lead to more robust changes in cases of compound-induced cardiac stress.

ANP concentrations are approximately ten times higher than BNP in healthy human subjects (Potter et al., 2006). During congestive heart failure, BNP concentrations may be elevated by as much as 200–300-fold, compared to a more modest 10–20-fold increase in ANP. This suggests that ANP may be more relevant to normal physiology and immediate response to perturbations of homeostasis, while BNP is more involved in chronic disease. Thus, during short-duration safety studies in healthy animals, concentrations of ANP and NTproANP are more likely to show significant changes than BNP or NTproBNP. In longer-duration studies, BNP or NTproBNP may reach greater concentrations as significant LV dysfunction, cardiac hypertrophy, and induction of NPPB expression in the ventricles occur. In addition, proANP amino acid sequences between mice, rats, monkeys, dogs, and humans are relatively well conserved exhibiting >80% homology, whereas rat and human proBNP are <40% homologous (Greenberg et al., 1984; Argentin et al., 1985; Kojima et al., 1989; Kambayashi et al., 1990). The homology of pro-ANP across species allows greater assay cross-reactivity among animals typically used in toxicology studies.

24.6 CASE STUDIES

Increased circulating NTproANP concentrations have been observed in the context of pressure overload-induced cardiac hypertrophy, as early as 24 h after aortic banding and prior to measurable increases in heart weight (Colton et al., 2011). Plasma NTproANP concentrations in rats have also been observed to increase in the context of volume overload-induced cardiac hypertrophy after administration of a peroxisome proliferator-activated receptor agonist. Increases in NTproANP were observed as early as 48 h after initiation of dosing and were well correlated with final heart weights after 28 days (Engle et al., 2010). Additionally, BNP concentrations were increased in rats after 4 days of administration of rosiglitazone (Fig. 24.1). Rosiglitazone, a thiazolidinedione, has been associated with increases in cardiotoxicity and has carried a "black box" warning since 2007 (Nissen and Wolski, 2007). Thiazolidinediones in general have been linked to fluid retention and congestive heart failure (Nesto et al., 2004).

NPs and other safety biomarkers can be applied even earlier in drug development while pharmacology is still being assessed (Engle et al., 2009). Thanks to cross-species sequence homology and conservation of function, NTproANP can be measured in mice using the same assay used for rat samples. NTproANP concentrations measured in mouse

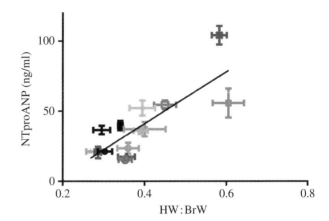

FIGURE 24.2 NTproANP and relative heart weights in mice. Group mean NTproANP concentrations in plasma versus group mean heart weight : brain weight (HW : BrW) in mice after 28 days of twice weekly administration of test articles. Each symbol is a different compound ($n = 6$ mice per group). NTproANP was measured after 3 days and relative heart weights after 28 days. NTproANP was measured Meso Scale Discovery's rat NTproANP enzyme-linked electrochemiluminescence assay. $R^2 = 0.64$.

plasma as early as 52 hours after first dose administration have been observed to correlate with increased heart weights observed after 28 days in pharmacology studies (Fig. 24.2).

24.7 CONCLUSION

Cardiac hypertrophy, as with all toxicities, is best mitigated by early detection, allowing screening of additional test articles and selection of molecules more likely to be tolerated in humans. Further work remains to be done in order to identify the most useful biomarkers for prediction of cardiac hypertrophy in laboratory animals. Markers of inflammation, ECM remodeling, neurohormonal changes, and hemodynamic stress have shown promise in human studies and some animal studies; their inclusion in subsequent animal studies will help to understand their value as nonclinical CV safety biomarkers. The application of safety biomarkers *in vivo*, such as NPs, as soon as animal studies are conducted allows a more thorough exploration of structure–activity relationships and target related toxicities and selection of safer molecules for continued development. NPs, while complex in their physiological roles, have shown promise as cardiac safety biomarkers in rodent studies. The correlation between NTproANP concentrations taken after only 2 or 3 days with heart weights after a month suggests this marker may be predictive of adverse increases in final heart weight after long-term studies in mice and rats, and possibly larger species (Engle et al., 2010; Colton et al., 2011). This allows the investigator to screen for a chronic outcome in a short study with only a few doses of test article, thus increasing throughput and improving the chances of success in finding

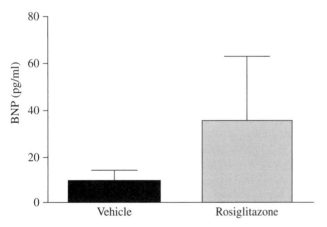

FIGURE 24.1 BNP concentrations in rat plasma. BNP concentrations in plasma from rats administered rosiglitazone (80 mg/kg) by oral gavage for 4 days ($n = 10$). BNP was measured using Meso Scale Discovery's rat BNP enzyme-linked electrochemiluminescence assay 4 h after the fourth daily dose.

a drug candidate during nonclinical studies. Early detection of changes in NP concentrations as a result of compound administration may trigger a more in-depth examination in subsequent studies, including implanted telemetry or imaging such as echocardiography or MRI, or selection of a different molecule without effects on NP concentrations. Additionally, the relevance of NPs to clinical medicine allows the possibility for translational use of NPs from early discovery through clinical development.

REFERENCES

Amur, S., Frueh, F.W., Lesko, L.J., Huang, S.M. (2008). Integration and use of biomarkers in drug development, regulation and clinical practice: a US regulatory perspective. *Biomark. Med.* 2(3):305–311.

Apple, F.S., Marakami, M.M., Ler, R., Walker, D., York, M.; HESI Technical Committee of Biomarkers Working Group on Cardiac Troponins. (2008). Analytical characteristics of commercial cardiac troponin I and T immunoassays in serum from rats, dogs, and monkeys with induced acute myocardial injury. *Clin. Chem.* 54(12):1982–1989.

Argentin, S., Nemer, M., Drouin, J., Scott, G.K., Kennedy, B.P., Davies, P.L. (1985). The gene for rat atrial natriuretic factor. *J. Biol. Chem.* 260(8):4568–4571.

Aronson, J.K. (2005). Biomarkers and surrogate endpoints. *Br. J. Clin. Pharmacol.* 59(5):491–494.

Artham, S.M., Lavie, C.J., Milani, R.V., Patel, D.A., Verma, A., Ventura H.O. (2009). Clinical impact of left ventricular hypertrophy and implication for regression. *Prog. Cardiovasc. Dis.* 52:153–167.

Autelitano, D.J., Ridings, R., Tang, F. (2001). Adrenomedullin is a regulated modulator of neonatal cardiomyocyte hypertrophy in vitro. *Cardiovasc. Res.* 51(2):255–264.

Baggish, A.L., Wood, M.J. (2011). Athlete's heart and cardiovascular care of the athlete: scientific and clinical update. *Circulation* 123(23):2723–2735.

Beghetti, M., Black, S.M., Fineman, J.R. (2005). Endothelin-1 in congenital heart disease. *Pediatr. Res.* 57(5 Pt 2):16R–20R.

Berridge, B.R., Pettit, S., Walker, D.B., Jaffe, A.S., Schultze, A.E., Herman, E., Reagan, W.J., Lipshultz, S.E., Apple, F.S., York, M.J. (2009). A translational approach to detecting drug-induced cardiac injury with cardiac troponins: consensus and recommendations from the Cardiac Troponins Biomarker Working Group of the Health and Environmental Sciences Institute. *Am. Heart J.* 158(1):21–29.

de Bold AJ, Borenstein HB, Veress AT, Sonnenberg H. (1981). A rapid and important natriuretic response to intravenous injection of atrial myocardial extracts in rats. *Life Sci.*, 28:89–94.

Booz, G.W. (2005). Putting the brakes on cardiac hypertrophy exploiting the NO-cGMP counter-regulatory system. *Hypertension.* 45:341–346.

Braunwald, E. (2008). Biomarkers in heart failure. *N. Engl. J. Med.* 358(20):2148–2159.

Buñag, R.D., Butterfield, J. (1982). Tail-cuff blood pressure measurement without external preheating in awake rats. *Hypertension.* 4(6):898–903.

Bupha-Intr, T., Haizlip, K.M., Janssen, P.M.L. (2012). Role of endothelin in the induction of cardiac hypertrophy in vitro. *PLoS One.* 7(8):e43179.

Chester, A.E., Dorr, A.E., Lund, K.R., Wood, L.D. (1992). Noninvasive measurement of blood pressure in conscious cynomolgus monkeys. *Fundam. Appl. Toxicol.* 19(1):64–68.

Chiu, P.J., Marcoe, K.F., Bounds, S.E., Lin, C.H., Feng, J.J., Lin, A., Cheng, F.C., Crumb, W.J., Mitchell, R. (2004). Validation of a [³H]astemizole binding assay in HEK293 cells expressing HERG K⁺ channels. *J. Pharmacol. Sci.* 95(3):311–319.

Clerico A, Recchia FA, Passino C, Emdin M. 2006. Cardiac endocrine function is an essential component of the homeostatic regulation network: physiological and clinical implications. *Am. J. Physiol. Heart Circ. Physiol.* 290:H17–H29.

Colton, H.M., Stokes, A.H., Yoon, L.W., Quaile, M.P., Novak, P.J., Falls, J.G., Kimbrough, C.L., Cariello, N.F., Jordan, H.L., Berridge, B.R. (2011). An initial characterization of N-terminal-proatrial natriuretic peptide in serum of Sprague Dawley rats. *Toxicol. Sci.* 120(2):262–268.

Condorelli, G., Morisco, C., Latronico, M.V., Claudio, P.P., Dent, P., Tsichlis, P., Condorelli, G., Frati, G., Drusco, A., Croce, C.M., Napoli, C. (2002). TNF-alpha signal transduction in rat neonatal cardiac myocytes: definition of pathways generating from the TNF-alpha receptor. *FASEB J.* 16(13):1732–1737.

Coutinho, T., Al-Omari, M., Mosley, T.H. Jr., Kullo, I.J. (2011). Biomarkers of left ventricular hypertrophy and remodeling in blacks. *Hypertension* 58(5):920–925.

Crowley, S.D., Gurley, S.B., Herrera, M.J., Ruiz, P., Griffiths, R., Kumar, A.P., Kim, H.S., Smithies, O., Le, T.H., Coffman, T.M. (2006). Angiotensin II causes hypertension and cardiac hypertrophy through its receptors in the kidney. *Proc. Natl. Acad. Sci.* 103(47):17985–17990.

Derakhchan, K., Chui, R.W., Stevens, D., Gu, W., Vargas, H.M. (2014) Detection of OTc interval prolongation using jacket telemetry in conscious non-human primates: comparison with implanted telemetry. *Br. J. Pharmacol.* 171(2):509–522.

Devereux, R.B., Alonso, D.R., Lutas, E.M., Gottlieb, G.J., Campo, E., Sachs, I., Reicheck, N. (1986). Echocardiographic assessment of left ventricular hypertrophy: Comparison to necropsy findings. *Am. J. Cardiol.* 57(6):450–458.

Dickstein, K., Larsen, A.I., Bonarjee, V., Thoresen, M., Aarsland, T., Hall, C. (1995). Plasma proatrial natriuretic factor is predictive of clinical status in patients with congestive heart failure. *Am. J. Cardiol.* 76(10):679–683.

Doust, J.A., Glasziou, P.P., Pietrzak, E., Dobson, A.J. (2004). A systematic review of the diagnostic accuracy of natriuretic peptides for heart failure. *Arch. Intern. Med.* 164:1978–1984.

Duan, S.Z., Ivashchenko, C.Y., Whitesall, S.E., D'Alecy, L.G., Mortensen, R.M. (2007). Direct monitoring pressure overload predicts cardiac hypertrophy in mice. *Physiol. Meas.* 28(11):1329–1339.

Edhouse, J., Thakur, R.K. and Khalil, J.M. (2002). Conditions affecting the left side of the heart. *BMJ.* 324(7348):1264–1267.

Eichenholz, P.W., Eichacker, P.Q., Hoffman, W.D., Banks, S.M., Parrillo, J.E., Danner, R.L., Natanson, C. (1992). Tumor necrosis factor challenges in canines: patterns of cardiovascular dysfunction. *Am. J. Physiol.* 263:H668–H675.

Engle, S.K., Watson, D.E. (2016). Natriuretic peptides as cardiovascular safety biomarkers in rats: comparison with blood pressure, heart rate, and heart weight. *Toxicol. Sci.* 149(2):458–472.

Engle, S.K., Huber, L., Sall, D., Uhlik, M., Watson, D.E. (2011). Use of natriuretic peptides to detect compound related changes in blood pressure in rodents. *Toxicologist CD.* 120(2):202.

Engle, S.K., Jordan, W.J., Pritt, M.L., Chiang, A.Y., Davis, M.A., Zimmermann, J.L., Rudmann, D.G., Heinz-Taheny, K.M., Irizarry, A.R., Yamamoto, Y., Mendel, D., Schultze, A.E., Cornwell, P.D., Watson, D.E. (2009). Qualification of cardiac troponin I concentration in mouse serum using isoproterenol and implementation in pharmacology studies to accelerate drug development. *Toxicol. Pathol.* 37, 617–628.

Engle, S.K., Solter, P.F., Credille, K.M., Bull, C.M., Adams, S., Berna, M.J., Schultze, A.E., Rothstein, E.C., Cockman, M.D., Pritt, M.L., Liu, H., Lu, Y., Chiang, A.Y., Watson, D.E. (2010). Detection of left ventricular hypertrophy in rats administered a peroxisome proliferator activated receptor dual agonist using natriuretic peptides and imaging. *Toxicol. Sci.* 114(2):183–192.

Fermini, B., Hancox, J.C., Abi-Gerges, N., Bridgland-Taylor, M., Chaudhary, K.W., Colatsky, T., Correll, K., Crumb, W., Damiano, B., Erdemli, G., Gintant, G., Imredy, J., Koerner, J., Kramer, J., Levesque, P., Li, Z., Lindqvist, A., Obejero-Paz, C.A., Rampe, D., Sawada, K., Strauss, D.G., Vandenberg, J.I. (2016). A new perspective in the field of cardiac safety testing through the comprehensive in vitro proarrhythmia assay paradigm. *J. Biomol. Screen.* 21(1):1–11.

Frey, N., Olson, E.N. (2003). Cardiac hypertrophy: the good, the bad, and the ugly. *Annu. Rev. Physiol.* 65:45–79.

Frey, N., Katus, H.A., Olson, E.N., Hill, J.A. (2004). Hypertrophy of the heart: a new therapeutic target? *Circulation* 109(13):1580–1589.

Gaasch, W.H., Zile, M.R. (2011). Left ventricular structural remodeling in health and disease: with special emphasis on volume, mass, and geometry. *J. Am. Coll. Cardiol.* 58(17):1733–1740.

Galasko, G., Collinson, P.O., Barnes, S.C., Gaze, D., Lahiri, A., Senior, R. (2007). Comparison of the clinical utility of atrial and B type natriuretic peptide measurement for the diagnosis of systolic dysfunction in a low-risk population. *J. Clin. Pathol.* 60:570–572.

Giroir, B.P., Johnson, J.H., Brown, T., Allen, G.L., Beutler, B. (1992). The tissue distribution of tumor necrosis factor biosynthesis during endotoxemia. *J. Clin. Invest.* 90(3):693–698.

Greenberg, B.D., Bencen, G.H., Seilhamer, J.J., Lewicki, J.A., Fiddes, J.C. (1984). Nucleotide sequence of the gene encoding human atrial natriuretic factor precursor. *Nature* 312(5995): 656–658.

Hasegawa, K., Fujiwara, H., Koshiji, M., Inada, T., Ohtani, S., Doyama, K., Tanaka, M., Matsumori, A., Fujiwara, T., Shirakami, G., Hosoda, K., Nakao, K., Sasayama, S. (1996). Endothelin-1 and its receptor in hypertrophic cardiomyopathy. *Hypertension.* 27:259–264.

Houston, M.E. (1999). Gaining weight: the scientific basis of increasing skeletal muscle mass. *Can. J. Appl. Physiol.* 24(4):305–316.

Hunter, J.J., Chien, K.R. (1999). Signaling pathways for cardiac hypertrophy and failure. *N. Engl. J. Med.* 341:1276–1283.

International Conference on Harmonisation of Technical Requirements for Registration of Pharmaceuticals for Human Use. (2011). Preclinical Safety Evaluation of Biotechnology-Derived Pharmaceuticals S6(R1). http://www.ich.org/fileadmin/Public_Web_Site/ICH_Products/Guidelines/Safety/S6_R1/Step4/S6_R1_Guideline.pdf last accessed 11 January 2016.

Irukayama-Tomobe, Y., Miyauchi, T., Sakai, S., Kasuya, Y., Ogata, T., Takanashi, M., Iemitsu, M., Sudo, T., Goto, K., Yamaguchi, I. (2004). Endothelin-1-induced cardiac hypertrophy is inhibited by activation of peroxisome proliferator-activated receptor-alpha partly via blockade of c-Jun NH2-terminal kinase pathway. *Circulation.* 109(7):904–910.

Jensen K.T., Carstens J, Pedersen E.B.. (1998). Effect of BNP on renal hemodynamics, tubular function and vasoactive hormones. *Am. J. Physiol.* 274(1):F63–F72.

Jortani, S.A., Prabhu, S.D., Valdes, R. Jr. (2004). Strategies for developing biomarkers of heart failure. *Clin. Chem.* 50(2): 265–278.

Jougasaki, M., Stevens, T.L., Borgeson, D.D., Luchner, A., Redfield, M.M., Burnett, J.C., Jr. (1997). Adrenomedullin in experimental congestive heart failure: cardiorenal activation. *Am. J. Physiol.* 273(4 Pt 2):R1392–R1399.

Kambayashi, Y., Nakao, K., Mukoyama, M., Saito, Y., Ogawa, Y., Shiono, S., Inouye, K., Yoshida, N., Imura, H. (1990). Isolation and sequence determination of human brain natriuretic peptide in human atrium. *FEBS Lett.* 259(2):341–345.

Kimura, K., Yamaguchi, Y., Horii, M., Kawata, H., Yamamoto, H., Uemura, S., Saito, Y. (2007). ANP is cleared much faster than BNP in patients with congestive heart failure. *Eur. J. Clin. Pharmacol.* 63:699–702.

Klein, R.M., Kelley, K.B., Merisko-Liversidge, E.M. (1993). A clathrin-coated vesicle-mediated pathway in atrial natriuretic peptide (ANP) secretion. *J. Mol. Cell. Cardiol.* 25(4):437–452.

Kojima, M., Minamino, N., Kangawa, K., Matsuo, H. (1989). Cloning and sequence analysis of cDNA encoding a precursor for rat brain natriuretic peptide. *Biochem. Biophys. Res. Commun.* 159(3):1420–1426.

Koyama, S., Tsuruda, T., Ideguchi, T., Kawagoe, J., Onitsuka, H., Ishikawa, T., Date, H., Hatakeyama, K., Asada, Y., Kato, J., Kitamura, K. (2014). Osteoprotegerin is secreted into the coronary circulation: a possible association with the renin-angiotensin system and cardiac hypertrophy. *Horm. Metab. Res.* 46(8):581–586.

Krauser, D.G., Devereaux, R.B. (2006). Ventricular hypertrophy and hypertension: prognostic elements and implications for management. *Herz* 31:305–316.

Lasser, K.E., Allen, P.D., Woolhandler, S.J., Himmelstein, D.U., Wolfe, S.M., Bor, D.H. (2002). Timing of new black box warnings and withdrawals for prescription medications. *JAMA* 287(17):2215–2220.

Laverty, H., Benson, C., Cartwright, E., Cross, M., Garland, C., Hammond, T., Holloway, C., McMahon, N., Milligan, J., Park, B., Pirmohamed, M., Pollard, C., Radford, J., Roome, N., Sager, P., Singh, S., Suter, T., Suter, W., Trafford, A., Volders, P., Wallis,

R., Weaver, R., York, M., Valentin, J. (2011). How can we improve our understanding of cardiovascular safety liabilities to develop safer medicines? *Br J Pharmacol.* 163(4):675–693.

Leishman D.J., Beck, T.W., Dybdal, N. Gallacher, D.J., Guth, B.D., Holbrook, M., Roche, B., Wallis, R.M. (2012). Best practice in the conduct of key nonclinical cardiovascular assessments in drug development: current recommendations from the safety pharmacology society. *J. Pharmacol. Toxicol. Methods* 65(3):93–101.

Lerman, A., Gibbons, R.J., Rodeheffer, R.J., Bailey, K.R., McKinley, L.J., Heublein, D.M., Burnett, J.C. Jr. (1993). Circulating N-terminal atrial natriuretic peptide as a marker for symptomless left-ventricular dysfunction. *Lancet* 341(8853): 1105–1109.

Levin, E.R., Gardner, D.G., Samson, W.K. (1998). Natriuretic peptides. *N. Engl. J. Med.* 339(5):321–328.

Levine, B., Kalman, J., Mayer, L., Fillit, H.M., Packer, M. (1990). Elevated circulating levels of tumor necrosis factor in severe chronic heart failure. *N. Engl. J. Med.* 323(4):236–241.

Levy, D., Garrison, R.J., Savage, D.D., Kannel, W.B., Castelli, W.P. (1990). Prognostic implications of echocardiographically determined left ventricular mass in the Framingham Heart Study. *N. Engl. J. Med.* 322:1561–1566.

Li, J., Umar, S., Amjedi, M., Iorga, A., Sharma, S., Nadadur, R.D., Regitz-Zagrosek, V., Eghbali, M. (2012). New frontiers in heart hypertrophy during pregnancy. *Am. J. Cardiovasc. Dis.* 2(3):192–207.

Lloyd, M.A., Grogan, M., Sandberg, S.M., Smith, H.C., Gersh, B.J., Edwards, W.D., Edwards, B.S. (1994). Presence of atrial natriuretic factor in ventricular tissue in tachycardia-induced cardiomyopathy. *Am. J. Cardiol.* 73(13):984–986.

López, B., González, A., Querejeta, R., Barba, J., Díez, J. (2009). Association of plasma cardiotrophin-1 with stage C heart failure in hypertensive patients: potential diagnostic implications. *J. Hypertens.* 27(2):418–424.

McGrath, M.F., de Bold, A.J. (2005) Determinants of natriuretic peptide gene expression. *Peptides* 26:933–943.

Monfared, A., Salari, A., Kazemnezhad, E., Lebadi, M., Khosravi, M., Mehrjardi, N.K., Rahimifar, S., Amini, N. (2013). Association of left ventricular hypertrophy with high-sensitive C-reactive protein in hemodialysis patients. *Int. Urol. Nephrol.* 45(6):1679–1689.

Morimoto, A., Nishikimi, T., Yoshihara, F., Horio, T., Nagaya, N., Matsuo, H., Dohi, K., Kangawa, K. (1999). Ventricular adrenomedullin levels correlate with the extent of cardiac hypertrophy in rats. *Hypertension.* 33(5):1146–1152.

Moss, R.L., Fitzsimons, D.P. (2002). Frank-starling relationship: long on importance, short on mechanism. *Circ. Res.* 90:11–13.

Muders, F., Kromer, E.P., Griese, D.P., Pfeifer, M., Hense, H.W., Riegger, G.A., Elsner, D. (1997). Evaluation of plasma natriuretic peptides as markers for left ventricular dysfunction. *Am. Heart J.* 134(3):442–449.

Nagai, T., Anzai, T., Kaneko, H., Mano, Y., Anzai, A., Maekawa, Y., Takahashi, T., Meguro, T., Yoshikawa, T., Fukuda, K. (2011). C-reactive protein overexpression exacerbates pressure overload-induced cardiac remodeling through enhanced inflammatory response. *Hypertension.* 57(2):208–215.

Nesto, R.W., Bell, D., Bonow, R.O., Fonseca, V., Grundy, S.M., Horton, E.S., Le Winter, M., Porte, D., Semenkovich, C.F., Smith, S., Young, L.H., Kahn, R. (2004). Thiazolidinedione use, fluid retention, and congestive heart failure: a consensus statement from the American Heart Association and American Diabetes Association. *Diabetes Care* 27(1):256–263.

Nissen, S.E., Wolski, K. (2007). Effect of rosiglitazone on the risk of myocardial infarction and death from cardiovascular causes. *N. Engl. J. Med.* 356(24):2457–2471.

Ogino, K., Ogura, K., Kinugawa, T., Osaki, S., Kato, M., Furuse, Y., Kinugasa, Y., Tomikura, Y., Igawa, O., Hisatome, I., Shigemasa, C. (2004). Neurohumoral profiles in patients with hypertrophic cardiomyopathy: differences to hypertensive left ventricular hypertrophy. *Circ. J.* 68:444–450.

Olson, H., Betton, G., Robinson, D., Thomas, K., Monro, A., Kolaja, G., Lilly, P., Sanders, J., Sipes, G., Bracken, W., Dorato, M., Van Deun, K., Smith, P., Berger, B., Heller, A. (2000). Concordance of the toxicity of pharmaceuticals in humans and in animals. *Regul. Toxicol. Pharmacol.* 32(1):56–67.

Oyama, M.A, Boswood, A, Connolly, D.J., Ettinger, S.J., Fox, P.R., Gordon, S.G., Rush, J.E., Sisson, D.D., Stepien, R.L., Wess, G., Zannad, F. (2013) Clinical usefulness of an assay for measurement of circulating N-terminal pro–B-type natriuretic peptide concentration in dogs and cats with heart disease. *J. Am. Vet. Med. Assoc.* 243(1):71–82.

Pagani, F.D., Baker, L.S., His, C., Knox, M., Fink, M.P., Visner, M.S. (1992). Left ventricular systolic and diastolic dysfunction after infusion of tumor necrosis factor-alpha in conscious dogs. *J Clin Invest.* 90:389–398.

Pepys, M.B., Hirschfield, G.M. (2003). C-reactive protein: a critical update. *J. Clin. Invest.* 111(12):1805–1812.

Pickering, T.G. (1998). White coat hypertension: time for action. *Circulation* 98(18):1834–1836.

Potter, L.R., Abbey-Hosch, S., Dickey, D.M. (2006). Natriuretic peptides, their receptors, and cyclic guanosine monophosphate-dependent signaling functions. *Endocr. Rev.* 27(1):47–72.

Pritchard, J.F., Jurima-Romet, M., Reimer, M.L.J., Mortimer, E., Rolfe, B., Cayen, M.N. (2003). Making better drugs: decision gates in non-clinical drug development. *Nat. Rev. Drug Discov.* 2(7):542–553.

Prosek, R., Sisson, D.D., Oyama, M.A., Biondo, A.W., Solter, P.F. (2004). Plasma endothelin-1 immunoreactivity in normal dogs and dogs with acquired heart disease. *J. Vet. Intern. Med.* 18(6):840–844.

Ruskoaho, H. (1992). Atrial natriuretic peptide: synthesis, release, and metabolism. *Pharmacol. Rev.* 44(4):479–602.

Sagnella, G.A., (1998). Measurement and significance of circulating natriuretic peptides in cardiovascular disease. *Clin. Sci.* 95, 519–529.

Schirmer, H., Omland, T. (1999). Circulating N-terminal pro-atrial natriuretic peptide is an independent predictor of left ventricular hypertrophy in the general population. *Eur. Heart J.* 20:755–763.

Seta, Y., Shan, K., Bozkurt, B., Oral, H., Mann, D.L. (1996). Basic mechanisms in heart failure: the cytokine hypothesis. *J. Card. Fail.* 2(3):243–249.

Stevens, J.L., Baker, T.K. (2009). The future of drug safety testing: expanding the view and narrowing the focus. *Drug Discov. Today* 14(3–4):162–167.

Sudoh, T., Kangawa, K., Minamino, N., Matsuo, H. (1988). A new natriuretic peptide in porcine brain. *Nature*, 332:78–81.

Sun, M., Chen, M., Dawood, F., Zurawska, U., Li, J.Y., Parker, T., Kassiri, Z., Kirshenbaum, L.A., Arnold, M., Khokha, R., Liu, P.P. (2007). Tumor necrosis factor-alpha mediates cardiac remodeling and ventricular dysfunction after pressure overload state. *Circulation.* 115(11):1398–1407.

Tang, W.H. (2007). B-type natriuretic peptide: a critical review. *Congest. Heart Fail.* 13(1):45–52.

Thibault, G., Murthy, K.K., Gutkowska, J., Seidah, N.G., Lazure, C., Chrétien, M., Cantin, M. (1988). NH2-terminal fragment of rat pro-atrial natriuretic factor in the circulation: identification, radioimmunoassay and half-life. *Peptides* 9:47–53.

Thibault, G., Amiri, F., Garcia, R. (1999). Regulation of natriuretic peptide secretion by the heart. *Annu. Rev. Physiol.* 61:193–217.

Torzewski, M., Waqar, A.B., Fan, J. (2014). Animal models of C-reactive protein. *Mediators Inflamm.* 2014:683598. Epub Apr 24.

U.S. Food and Drug Administration, Guidance for Industry (2001). S7A safety pharmacology studies for human pharmaceuticals. http://www.fda.gov/downloads/drugs/guidancecompliance regulatoryinformation/guidances/ucm074959.pdf Last accessed 11 January 2016.

Van Vleet, J.F., Ferrans, V.J. (1995). Pathology of the cardiovascular system. In: *Thomson's Special Veterinary Pathology* (Carlton, W.W., McGavin, M.D., Eds.), 2nd ed., pp. 189. Mosby, St. Louis.

Venugopal, S.K., Deveraj, S., Jialal, I. (2005). Effect of C-reactive protein on vascular cells: evidence for a proinflammatory, pro-atherogenic role. *Curr. Opin. Nephrol. Hypertens.* 14:33–37.

Vincent, J. (2008). Understanding cardiac output. *Crit. Care.* 12:174.

Watterson, C., Lanevschi, A., Horner, J., Louden, C. (2009). A comparative analysis of acute-phase proteins as inflammatory biomarkers in preclinical toxicology studies: implications for preclinical to clinical translation. *Toxicol. Pathol.* 37(1):28–33.

Whitesall, S.E., Hoff, J.B., Vollmer, A.P., D'Alecy, L.G. (2004). Comparison of simultaneous measurement of mouse systolic arterial blood pressure by radiotelemetry and tail-cuff methods. *Am. J. Physiol. Heart Circ. Physiol.* 286(6):H2408–H2415.

Yanagawa, B., Nagaya, N. (2007). Adrenomedullin: molecular mechanisms and its role in cardiac disease. *Amino Acids.* 32(1):157–164.

Yasue, H., Yoshimura, M., Sumida, H., Kikuta, K., Kugiyama, K., Jougasaki, M., Ogawa, H., Okumura, K., Mukoyama, M., Nakao, K. (1994). Localization and mechanism of secretion of B-type natriuretic peptide in comparison with those of A-type natriuretic peptide in normal subjects and patients with heart failure. *Circulation.* 90(1):195–203.

Yokoyama, T., Nakano, M., Bednarczyk, J.L., McIntyre, B.W., Entman, M., Mann, D.L. (1997). Tumor necrosis factor-alpha provokes a hypertrophic growth response in adult cardiac myocytes. *Circulation.* 95(5):1247–1252.

Yoshihara, F., Nishikimi, T., Okano, I., Hino, J., Horio, T., Tokudome, T., Suga, S., Matsuoka, H., Kangawa, K., Kawano, Y. (2005). Upregulation of intracardiac adrenomedullin and its receptor system in rats with volume overload-induced cardiac hypertrophy. *Regul. Pept.* 127(1–3):239–244.

Zile, M.R., Desantis, S.M., Baicu, C.F., Stroud, R.E., Thompson, S.B., McClure, C.D., Mehurg, S.M., Spinale, F.G. (2011). Plasma biomarkers that reflect determinants of matrix composition identify the presence of left ventricular hypertrophy and diastolic heart failure. *Circ. Heart. Fail.* 4(3):246–256.

25

VASCULAR INJURY BIOMARKERS

Tanja S. Zabka[1] and Kaïdre Bendjama[2]

[1] *Development Sciences-Safety Assessment, Genentech Inc., South San Francisco, CA, USA*

[2] *Transgene, Illkirch-Graffenstaden, France*

25.1 HISTORICAL CONTEXT OF DRUG-INDUCED VASCULAR INJURY AND DRUG DEVELOPMENT

Drug-induced vascular injury (DIVI) is a finding of concern in preclinical safety testing and can lead to significant delays or project termination because of the poor ability to noninvasively monitor the finding, variability in and sometimes infrequent and sporadic manifestation across preclinical species, and uncertainty of translation to humans (Morton et al., 2014). Historically, DIVI was associated with vasoactive mechanisms of small molecules (Mikaelian et al., 2014) that manifest in animals but not in humans with compounds such as caffeine (Johansson, 1981), theobromine (Collins et al., 1988), the phosphodiesterase IV (PDE4) inhibitor, apremilast (Kavanaugh et al., 2014), and minoxidil (Sobota, 1989). More recently, other mechanisms of DIVI in preclinical species were recognized for small molecules (Mikaelian et al., 2014), biotherapeutics (Frazier et al., 2015), or antisense oligonucleotides (Engelhardt et al., 2015). These other mechanisms of DIVI manifest with different histomorphologic features and in different and often specific vascular beds. The distribution, localization, and morphologic similarities of lesions associated with DIVI and sporadic spontaneous vasculitides in animals complicate interpretations of drug-related events. Further, DIVI usually is not associated with alteration of preclinical cardiovascular safety pharmacology and toxicity endpoints.

In humans, DIVI has been associated with most pharmaceutical classes of drug and often is not predicted from preclinical studies (Doyle and Cuellar, 2003; Taborda et al., 2013). Two categories of drug-associated vasculitis in human were proposed recently by the International Chapel Hill Consensus Conference on Nomenclature of Vasculitides (ICHCCNV) (Jennette et al., 2013), namely, antineutrophil cytoplasmic antibody (ANCA)-associated vasculitis (AAV) and immune complex small vessel vasculitis. The manifestation of DIVI limited to the skin is reported most commonly; however, it also may affect the visceral vasculature, most commonly the kidney, in conjunction with or independently of the skin with potentially life-threatening consequences (Doyle and Cuellar, 2003; Radic et al., 2012; Jennette et al., 2013; Taborda et al., 2013). DIVI is not easily differentiated clinically and/or pathologically from idiopathic vasculitides, which encompass an array of pathological entities affecting various vascular beds and are associated with histologic manifestations that largely overlap with those observed in preclinical DIVI (Radic et al., 2012; Jennette et al., 2013). DIVI has been a focus for biomarker development through joint strategies of two consortia, the Critical Path Institute's Predictive Safety Testing Consortium (PSTC; Mikaelian et al., 2014) and the Innovative Medicines Initiative Joint Undertaking (IMI JU) SAFE-T Consortium (Bendjama et al., 2014).

The mechanisms of preclinical DIVI can be grouped broadly into hemodynamic alterations through vasodilation or vasoconstriction, direct endothelial cell stimulation through on-target and off-target interactions and associated secondary lesions due to modulation of inflammatory mechanisms, and immune complex formation and complement activation, such

Drug Discovery Toxicology: From Target Assessment to Translational Biomarkers, First Edition. Edited by Yvonne Will, J. Eric McDuffie, Andrew J. Olaharski, and Brandon D. Jeffy.

as through haptenic binding, antidrug antibody formation or nonendogenous immune drug aggregate formation, disrupted regulation of the complement pathway, and stimulation of the innate immune system (Engelhardt et al., 2015; Frazier et al., 2015; Mikaelian et al., 2014). The broad categories of histomorphologic outcomes of DIVI are degeneration and necrosis, hypertrophy and hyperplasia, and inflammation (mixed or mononuclear leukocyte cell types) that occur in various combinations and can affect the endothelium (*tunica intima*), vascular smooth muscle (*tunica media*) and/or perivascular tissue (*tunica adventitia*) of the large (i.e., elastic) arteries, medium (i.e., muscular and distributing) arteries, capillaries, veins, and lymphatics (Bendjama et al., 2014; Mikaelian et al., 2014; Engelhardt et al., 2015). These categories are captured in the PSTC histomorphologic lexicon for biomarker qualification in preclinical DIVI models (Mikaelian et al., 2014) as well as in the current terminology guidelines for histologic diagnosis of VI set out by the International Harmonization of Nomenclature and Diagnostic Criteria (INHAND) and the Standard Exchange of Nonclinical Data (SEND) initiatives (Mann et al., 2012).

The categories for outcomes of DIVI are used to group phenotypic vascular findings according to histomorphologic similarities and candidate sensitive and specific biomarker response independent of etiology, mechanisms, and sites of injury (Zhang et al., 2012; Bendjama et al., 2014).

Accordingly, one could envision a panel of biomarkers that would include components common to vascular endothelial cells, vascular smooth muscle cells, and related tissue inflammation to reflect histomorphologic outcomes following compound administration to preclinical species and humans. Such a panel would comprise factors that reflect different stages of DIVI and relevant vascular components. For example, administration of PDE4 inhibitors to rats provides a model for investigating different stages of DIVI (Zhang et al., 2008) as described in Figure 25.1 with administration of SCH 351591 by oral gavage at 4.5 mg/kg/day for 4 days. Such panels are currently under qualification per a joint strategy supported by the PSTC in preclinical species using vasotoxicants and by the SAFE-T in human populations having existing vasculitides with the rationale that overlapping biology and histomorphology will reveal overlapping biomarker signatures.

25.2 CURRENT STATE OF DIVI BIOMARKERS

Most currently reported candidate biomarkers are soluble proteins that are intended to reflect the histopathologic processes occurring during DIVI in rats (Mikaelian et al., 2014) and VI in humans (Bendjama et al., 2014) (Table 25.1). Most of these markers are cytokines or acute phase proteins

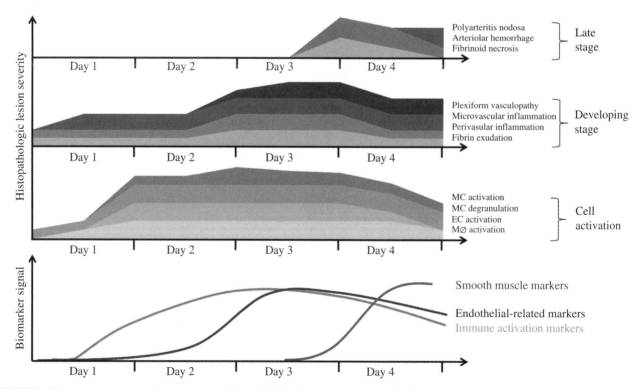

FIGURE 25.1 Stages in the development of preclinical DIVI. Zhang and colleagues (2008) used the term "polyarteritis nodosa" in this manuscript to reflect the microscopic finding of "necrotizing inflammation, fibrous thickening of arterial walls, and nodular fibroblastic proliferation in the adventitia."

TABLE 25.1 Examples of Drug-Induced Vascular Injury Identified in Preclinical Safety Studies According to the Preclinical Species, Mechanism of Toxicity, Vascular Beds Affected, and Microscopic Findings

Drug	Pharmacologic Mechanism	Affected Vessels	Morphology	References
Rats				
Phosphodiesterase 3 inhibitors	Vasodilator/inotrope/chronotrope	Coronary arteries	Medial smooth muscle cell necrosis, later mixed inflammation	Joseph et al. (1996, 1997) and Isaacs et al. (1989)
Phosphodiesterase 4 inhibitors	Vasodilator/inotrope	Small/medium-sized arteries, venules, and capillaries; multiple organs	Medial smooth muscle cell necrosis, large numbers of neutrophils	Hanton et al. (2008)
Minoxidil	Vasodilator/K+ channel opener	Coronary arteries (primarily right branches)	Medial smooth muscle cell necrosis, inflammation, hemorrhage	Mesfin et al. (1989)
CI-947	Vasodilator/adenosine agonist	Coronary arteries (arteries in other organs also affected)	Medial smooth muscle cell necrosis, hemorrhage, inflammation	Metz et al. (1991) and Enerson et al. (2006)
Endothelin receptor antagonists	Vasodilator	Coronary arteries (primarily extramural branches; small arteries in other organs sometimes affected)	Segmental medial smooth muscle cell necrosis, hemorrhage, mixed inflammation	Albassam et al. (2001), Louden et al. (1998), Doherty and Uprichard (1998), and Jones et al. (2003)
Hydralazine	Vasodilator, raised cGMP	Coronary arteries	Medial smooth muscle cell necrosis	Mesfin et al. (1987)
Digoxin	Inotrope	Coronary arteries (mainly small intramural branches)	Medial smooth muscle cell necrosis (fibrinoid), inflammation, hemorrhage	Bourdois et al. (1982) and Teske et al. (1976)
Inotropic amines (dopamine, isoproterenol)	Inotrope (adrenergic)	Coronary arteries (mainly small intramural branches)	Medial smooth muscle cell necrosis, fibroplasia, mild mononuclear inflammation	Sandusky et al. (1990)
Dogs				
β-Adrenoreceptor agonists	Positive inotropic vasodilating agent; β-adrenoreceptors coupled through G protein signals to increase cAMP	Medium-sized, muscular mesenteric arteries	Medial smooth muscle cell necrosis and hemorrhage	Rona (1985) and Greaves (2000)
Dopamine receptor agonists (e.g., Fenoldopam)	Positive inotropic vasodilating agent; DA1 receptor which acts to increase intracellular cAMP	Medium-sized, muscular mesenteric arteries	Medial smooth muscle cell necrosis and hemorrhage	Yuhas et al. (1985) and Kerns et al. (1989)
Phosphodiesterase inhibitors (PDE3 and PDE4 inhibitors)	Positive inotropic vasodilating agent; inhibits degradation of intracellular cAMP	Medium-sized mesenteric arteries	Segmental medial necrosis and hemorrhage	Isaacs et al. (1989) and Zhang et al. (2008)
Inhibitors of ATP-sensitive K+ channel (e.g., Minoxidil)	Positive inotropic vasodilating agent; increases permeability of K+ ions into the cell	Medium-sized mesenteric arteries	Segmental medial necrosis, hemorrhage, and perivascular inflammation	Joseph (2000)

(Continued)

TABLE 25.1 (*Continued*)

Drug	Pharmacologic Mechanism	Affected Vessels	Morphology	References
Adrenaline	Vasoconstrictor and –dilator; binds to α and β adrenergic receptors	Systemic small- and medium-sized arteries	Medial necrosis and hemorrhage; smooth muscle cells degeneration/hypertrophy	Greaves (2000)
Monocrotaline pyrrole	Vasoconstrictor; acts on endothelin receptors and DNA in endothelial cells	Small- and medium-sized intraacinar pulmonary arteries	Fibrinoid change and smooth muscle cells hypertrophy	Miyauchi et al. (1993)
Endothelin receptor antagonists	Vasodilator	Coronary arteries (right), coronary groove, and atrium	Medial hemorrhage and necrosis, with mixed inflammatory cell infiltrates in the adventitia and media	McDuffie et al (2006)
Swine				
ATP-sensitive K+ channel (e.g., Minoxidil)	Positive inotropic vasodilating agent; increases permeability of K+ ions into the cell	Coronary arteries (left branches, with <3 layers of medial smooth muscle)	Medial necrosis, hemorrhage, and perivascular inflammation	Herman et al. (1989)
Phosphodiesterase 4 inhibitors	Positive inotropic vasodilating agent;	Small- and medium-sized arteries of kidneys, stomach, intestine, and mesenteric nodes	Segmental vasculitis	Vogel et al. (1999)
Nonhuman primates				
Phosphodiesterase 4 inhibitors	Vasodilator/inotrope	Small- to medium-sized arteries in many organs; including coronary arteries	Medial fibrinoid necrosis, edema, hemorrhage, and surrounding inflammation	Losco et al. (2004)
Minoxidil (marmoset)	Vasodilator/K+ channel opener	Coronary arteries in left and right atrium	Adventitial fibroplasia, medial arterial hemorrhage	Hanton et al. (2008)
CI-947 (cynomolgus)	Vasodilator/adenosine agonist	Coronary arteries (arteries in other organs also affected)	Medial smooth muscle cell necrosis, mixed transmural inflammation	Albassam et al. (1998)
Endothelin receptor antagonists (cynomolgus)	Vasodilator	Coronary arteries and epididymis	Medial smooth muscle cell necrosis, mixed inflammation	Albassam et al. (1999)

related to the inflammatory processes observed in DIVI, proteins that are specific of the different components of the vascular wall such as smooth muscle proteins, or markers related to endothelial cell activation and/or injury. The candidate biomarkers in the rat were selected based on literature review and results from exploratory studies using different classes of vasotoxicants, including a tissue gene list generated from laser capture microdissection of endothelial cells and vascular smooth muscle cells (Dalmas et al., 2011) and refined from studies using vasoactive compounds and recombinant interleukin 2. The candidate biomarkers in man are driven by data from healthy volunteers, patients with acute flares of vasculitides or balloon angioplasty surgery, and patients in remission (Bendjama et al., 2014). The overlap between the rat panel and the human panel is limited by reagent availability and species differences in biology. For example, C-reactive protein is a candidate biomarker in humans and select lower species (e.g., canine) but is not a good marker of inflammation in rat. Of note, there is no candidate biomarker of vascular smooth muscle injury being tested in the rat due to poor reagent availability; however, this may be bridged using immunohistochemistry for smooth muscle markers that have shown changes in tissue with DIVI (Wiener et al., 1996; Albassam et al., 1999, 2001; Slim et al., 2003) to predict changes in circulating smooth muscle biomarkers in human should DIVI translate. Despite the limited overlap of biomarkers between the rat and human panels, the panels share biologic similarities when assessed in Ingenuity Pathway Analysis (Keirstead et al., 2015), which models and analyzes complex biological and chemistry systems, suggesting the different markers reflect similar pathophysiology. Given the multiple tissue and pathophysiologic mechanisms involved in DIVI, there is a general consensus that no single biomarker will achieve significant sensitivity and specificity in detecting DIVI. Accordingly, research has focused on interrogating panels of biomarkers, sometimes supported by predictive modeling using machine learning algorithms. The reader can refer to review articles from the PSTC, FDA, IMI SAFE-T, and others for a detailed discussion on preclinical and translational DIVI markers (Kerns et al., 2005; Louden et al., 2006; Zhang et al., 2012; Bendjama et al., 2014; Mikaelian et al., 2014).

As novel assays become available, the overlap between the clinical and nonclinical biomarkers may further increase. For instance, DIVI and some vasculitides have been associated with changes in levels of circulating endothelial microparticles (EMPs). EMPs are a type of small membrane vesicle (0.1–1 mm in diameter) that is released from cells in response to activation, injury, inflammation, and/or apoptosis and can serve as a means of intercellular communication (Rautou et al., 2011). The use of EMPs for biomarker qualification in drug development may gain broader acceptance with the advent of newer technologies such as flow cytometry-based rare cell sorting systems. Another potential rich source of DIVI biomarkers is circulating microRNAs (miRs). miRs are small noncoding RNAs that can modulate targeted pathways through posttranscriptional regulation of gene expression and thus play critical roles in diverse biological and pathological processes, including those associated with cardiovascular disease (Sen et al., 2009; Creemers et al., 2012; Zampetaki and Mayr, 2012). In studies performed in rat using recombinant interleukin 2 and vasomodulators to propose candidate protein circulating biomarkers, there were changes consistent among studies in certain circulating miRs, such as miR-21 and -132 (Thomas et al., 2012; Keirstead et al., 2015). As for EMPs, however, standardization and performance of methods will be critical to biomarker qualification of miRs intended for use in drug development and is being developed currently through PSTC. Lastly, although there is limited supportive literature, circulating mRNAs may be another source of biomarkers in the future.

Imaging modalities, such as ultrasound, computed tomography, magnetic resonance imaging, and positron emission tomography, provide additional potential means to assay for DIVI. Ultrasound is highly amenable for monitoring DIVI in preclinical and clinical drug development, because it is relatively inexpensive, noninvasive, widely available, translational, and does not require various tracers. Ultrasound provides functional evaluations of local blood flow and shear stress that may be associated with DIVI and may not be apparent by routine safety pharmacology evaluations of systemic arterial and venous blood pressure (Swanson et al., 2014). The more recent development of high-frequency ultrasound instruments provides a tool to quantify subtle hemodynamic changes, including the more recent measurement of flow-mediated dilation to indicate endothelial function and the ability to predict microscopic VI (Harris et al., 2010; Swanson et al., 2014). Although the small vascular beds (<0.2 mm) are too small to be directly monitored by ultrasonography, the hemodynamic changes in large vascular beds may inform on DIVI in the smaller caliber vessels. Ultrasound also can be used for molecular imaging of conjugated ligands or antibodies against vascular targets expressed by endothelial cells (such as VCAM-1, P-selectin, and ICAM1) (Ellegala et al., 2003; Weller et al., 2003; Kaufmann et al., 2007, 2010); however, this *in vivo* imaging modality currently is not approved for human use (Chadderdon and Kaul, 2010).

The tissue-based modalities, although not useful as monitoring tools for clinical DIVI, are important in biomarker qualification for drug development to understand the distribution of candidate circulating biomarkers in healthy and injured tissue as well as mechanism of toxicity (Dunstan et al., 2011; Mikaelian et al., 2014). Such modalities are being used with DIVI biomarkers, including immunohistochemistry (Wiener et al., 1996; Albassam et al.,

1999, 2001; Frazier et al., 2015; Hughes et al., 2015; Keirstead et al., 2015), *in situ* hybridization (Thomas et al., 2012; Keirstead et al., 2015), and mRNA and miR expression from laser capture microdissected vascular compartments (Dalmas et al., 2011; Thomas et al., 2012). These techniques have the advantage of being performed on formalin-fixed paraffin-embedded tissues, which enables biomarker localization to a specific cell type and evaluation relative to the histopathology. In contrast, tools such as western blot, qRT-PCR, and ELISA from homogenized tissues cannot be related directly to the histopathology. Importantly then, IHC/ISH and targeted mRNA and miR expression techniques may also be used to inform of a potential change in a circulating biomarker that is available as a monitoring tool in humans but not in the preclinical species in which DIVI was identified. This strategy is being used by PSTC to compensate for the absence of reagents for circulating vascular smooth muscle biomarkers. As a potential drawback, IHC and ISH can be less sensitive and target specific as well as not amenable to absolute quantitation compared to other techniques.

25.3 CURRENT STATUS AND FUTURE OF *IN VITRO* SYSTEMS TO INVESTIGATE DIVI

In vitro systems, when applied appropriately, are valuable tools in drug development for purposes of identifying risk for DIVI based on safety target assessments, rank ordering of molecular series, understanding mechanism of toxicity, identifying potential biomarkers, and demonstrating relevance of preclinical DIVI to humans. Potential endpoints for *in vitro* systems include cellular mRNA and miR expression and biochemical measurements; evaluation of the cell culture media for biochemical, protein, and genomic biomarkers; functional measurements for permeability and transcellular resistance; and evaluation of cellular morphology by gross visualization and microscopy. Ideally, the *in vitro* systems are established separately for both preclinical species and humans to enable a direct comparison to *in vivo* preclinical safety studies to support relevance of the *in vitro* system as well as predict its translation to humans. Vascular beds in different organs often have unique endothelial cell populations and respond differently to stimuli (Mikaelian et al., 2014); therefore, to ensure relevance of the *in vitro* system, it is important to consider whether the endothelial cells used in the *in vitro* system have expression profiles that correspond to the vascular bed affected *in vivo*. Recently, primary endothelial cells from 25 tissue types of male and female rats and from 14 tissue types of 23 human donors demonstrated similar gene expression profiles within species (Snyder et al., 2014), which may allay some concerns around endothelial cell heterogeneity. This work also demonstrated 40% similarity of the measured genes

between species, which underscores that DIVI identified in preclinical species may not always be relevant to humans depending on the mechanism of toxicity. It will be interesting to see further profiling of endothelial cells from different tissues in response to DIVI compounds, as well as introduce them into the newer flow- and 3D-based cell culture systems.

Traditionally, cell culture systems were comprised of endothelial cells alone or in coculture with vascular smooth muscle cells or pericytes grown on a flat two-dimensional (2D) surface and bathed in media, in which important physiologic conditions were not recapitulated (Wong et al., 2012). As an improvement, laminar flow was introduced into a vascular surrogate system comprised of endothelial and smooth muscle cells to incorporate the influence of normal hemodynamics on vascular health as well as the mechanistic component of altered hemodynamic forces associated with some drugs that cause DIVI. For example, endothelial cells exposed to physiologic shear stress *in vitro* have phenotypes more similar to endothelial cells *in vivo*. There are new three-dimensional (3D) cell culture systems that incorporate flow into an intricate microfluidic system or a high-throughput microchip platform comprised of a perfusate, multiple cells types, and physiologically meaningful microenvironments including factors in circulation and tissue gradients (Wong et al., 2012). In some models, endothelial cells and vascular smooth muscle cells are allowed to cluster and form a tube that more closely reflects normal vascular architecture including diffusion barriers and bifurcations, which enables meaningful visualization of cell interactions and movement to and from the vascular space. The intravascular space within the tube can be perfused to mimic drug exposure, mechanisms that require the interaction of blood-borne cells (such as leukocytes) and factors (such as growth factors), and mechanical forces. Finally, a specific organ microenvironment, including cellular inflammatory infiltrates, potentially can be introduced in the "perivascular" matrix and can be perfused separately. Work is under way to interrogate these new flow- and 3D vascular- based systems in the setting of DIVI, including assessment of interspecies (rat, nonhuman primate, and human) and intervascular bed (such as mesenteric and larger arteries) comparisons in relation to vasoreactivity, permeability, cell death, vascular remodeling, and molecular signaling mechanisms. For example, IL-2-induced pulmonary edema was recapitulated using a lung microchip with subsequent evaluation of therapeutic intervention, including angiopoieitin-1 and transient potential vanilloid 4 ion channel inhibitor (Huh et al., 2015). These *in vitro* platforms, despite some continued limitations, constitute an exciting opportunity to inform biomarker discovery, investigate DIVI mechanisms of toxicity, screen compounds with a DIVI liability, and predict DIVI in the clinical setting (Wong et al., 2012).

25.4 INCORPORATION OF *IN VITRO* AND *IN VIVO* TOOLS IN PRECLINICAL DRUG DEVELOPMENT

In preclinical drug development, the incorporation of non-routine safety biomarkers and *in vitro* tools for DIVI is driven by the safety assessment of the drug target in which a strong liability for DIVI is predicted or by the microscopic or safety pharmacologic evaluation from preclinical safety studies in which a signal of vascular injury is identified. The latter is the more common scenario that requires cessation of program development or the implementation of a derisking strategy that demonstrates one of several outcomes: (i) the toxicity is monitorable, reversible, and/or occurs at acceptable margins from drug levels where efficacy is expected (i.e., safety margin); (ii) the toxicity is not relevant to humans; or (iii) the toxicity is off target, so synthesis of alternative molecular structures will circumvent the toxicity. The PSTC and SAFE-T consortia efforts to identify noninvasive circulating biomarkers of DIVI are directed at establishing a monitoring paradigm for preclinical and clinical studies to facilitate safe and faster drug development. Such biomarkers also may be used on an individual program basis if strong supportive data can be generated from *in vitro* and/or *in vivo* studies in preclinical species, and the biomarkers are demonstrated to translate to humans through *in vitro* studies or retrospective data in relevant clinical populations. Demonstrating human relevance requires understanding the mechanism of toxicity and ideally uses a human *in vitro* (or *ex vivo*) system that reflects the proposed mechanism; however, without an acceptable safety margin and monitoring tool, this approach still may not enable clinical development. Finally, demonstrating off-target activity entails (i) understanding the mechanism of toxicity to separate it from the desired pharmacology; (ii) an efficient screening paradigm to rapidly identify non-DIVI chemical series; or (iii) demonstrating nonefficacious molecules will drive the same toxicity (i.e., toxicity is not related to potency and thus not to target engagement). Ideally, an *in vitro* system is used for these purposes to reduce animal use and would be in the same preclinical species to confirm relevance of the system for the DIVI finding *in vivo*.

25.5 DIVI CASE STUDY

DIVI was encountered in preclinical drug development for a therapeutic antibody (an anti-inflammatory agent) in which the *in vitro* and *in vivo* investigations demonstrated compound-related toxicity (Wang, 2015, personal communication). In a multidose safety study performed in cynomolgus monkeys, which included implanted telemetry for safety pharmacology evaluation, the antibody caused unanticipated vascular injury characterized by systemic hypotension and acute hemorrhage throughout the gastrointestinal

tract without evidence of vascular wall damage or an inflammatory response microscopically. Mechanistic investigations were based on evaluation of *in vivo* monkey serum and tissue samples (endpoints of circulating cytokine, chemokine, and complement levels and IHC staining) and *in vitro* human endothelial cell systems (endpoints of chemokine and cytokine levels in supernatant, imaging of nitric oxide production with and without target receptor blocking and of mitochondrial membrane potential, immunofluorescence cellular assay for reduced glutathione quantitation, western blot for expression of nitric oxide pathway components, and fluorescent dye assay for endothelial cell permeability). These investigations excluded the typical mechanisms of antibody-induced vascular injury (i.e., cytokine, proinflammatory, or complement mediated) and instead suggested an unanticipated off-target, promiscuous effect of the antibody on endothelial cells. This effect was to activate nitric oxide synthase with consequent downstream effects of nitric oxide production, increased mitochondrial membrane depolarization, glutathione depletion, increased paracellular permeability, and vasodilation. In addition, the antibody reduced myosin light chain phosphorylation in a human vascular smooth muscle cell *in vitro* system, which would translate to smooth muscle relaxation and vasodilation *in vivo*. This mechanism was confirmed in an *ex vivo* study utilizing segments of monkey aorta and femoral arteries. In a final single-dose monkey study, sampling from a portal vein catheter confirmed antibody-related acute local and slightly delayed systemic elevations in total circulating nitric oxide, corresponding directly to the onset and duration of systemic hypotension and microscopic evidence of acute gastrointestinal pathology. This series of *in vivo*, *ex vivo*, and *in vitro* investigations using human- and monkey-based systems with tissue- and fluid-based biomarkers provided mechanistic insight demonstrating an off-target effect but of direct relationship to the antibody and suggested human relevance. This mechanistic and translational information, in addition to the narrow therapeutic index and acute onset of the severe toxicity (preventing a safe monitoring plan), precluded further development of the therapeutic antibody.

REFERENCES

Albassam, M. A., Smith, G. S., and Macallum, G. E. (1998). Arteriopathy induced by an adenosine agonist-antihypertensive in monkeys. *Toxicol Pathol* 26, 375–80.

Albassam, M. A., Metz, A. L., Gragtmans, N. J., King, L. M., Macallum, G. E., Hallak, H., and McGuire, E. J. (1999). Coronary arteriopathy in monkeys following administration of CI-1020, an endothelin A receptor antagonist. *Toxicol Pathol* 27, 156–64.

Albassam, M. A., Metz, A. L., Potoczak, R. E., Gallagher, K. P., Haleen, S., Hallak, H., and McGuire, E. J. (2001). Studies on coronary arteriopathy in dogs following administration of CI-1020, an endothelin A receptor antagonist. *Toxicol Pathol* 29, 277–84.

Bendjama, K., Guionaud, S., Aras, G., Arber, N., Badimon, L., Bamberger, U., Bratfalean, D., Brott, D., David, M., Doessegger, L., Firat, H., Gallas, J. F., Gautier, J. C., Hoffmann, P., Kraus, S., Padro, T., Saadoun, D., Szczesny, P., Thomann, P., Vilahur, G., Lawton, M., and Cacoub, P. (2014). Translation strategy for the qualification of drug-induced vascular injury biomarkers. *Toxicol Pathol* 42, 658.

Bourdois, P., Dancla, J.-L., Faccini, J., Nachbaur, J., and Monro, A. (1982). The sub-acute toxicology of digoxin in dogs; clinical chemistry and histopathology of heart and kidneys. *Arch Toxicol* 51, 273–83.

Chadderdon, S. M. and Kaul, S. (2010). Molecular imaging with contrast enhanced ultrasound. *J Nucl Cardiol* 17, 667–77.

Collins, J. J., Elwell, M. R., Lamb, J. C. IV, Manus, A. G., Heath, J. E., and Makovec, G. T. (1988). Subchronic toxicity of orally administered (gavage and dosed-feed) theophylline in Fischer 344 rats and B6C3F1 mice. *Fundam Appl Toxicol* 11, 472–84.

Creemers, E. E., Tijsen, A. J., and Pinto, Y. M. (2012). Circulating MicroRNAs novel biomarkers and extracellular communicators in cardiovascular disease? *Circ Res* 110, 483–95.

Dalmas, D. A., Scicchitano, M. S., Mullins, D., Hughes-Earle, A., Tatsuoka, K., Magid-Slav, M., Frazier, K. S., and Thomas, H. C. (2011). Potential candidate genomic biomarkers of drug induced vascular injury in the rat. *Toxicol Appl Pharmacol* 257, 284–300.

Doherty, A. M., and Uprichard, A. C. (1998). Discovery and development of an endothelin A receptor-selective antagonist PD 156707. *Pharm Biotechnol* 11, 81–112.

Doyle, M. K. and Cuellar, M. L. (2003). Drug-induced vasculitis. *Expert Opin Drug Saf* 2, 401–9.

Dunstan, R. W., Wharton, K. A., Quigley, C., and Lowe, A. (2011). The use of immunohistochemistry for biomarker assessment: Can it compete with other technologies? *Toxicol Pathol* 39, 988–1002.

Ellegala, D. B., Leong-Poi, H., Carpenter, J. E., Klibanov, A. L., Kaul, S., Shaffrey, M. E., Sklenar, J., and Lindner, J. R. (2003). Imaging tumor angiogenesis with contrast ultrasound and microbubbles targeted to alpha(v)beta3. *Circulation* 108, 336–41.

Enerson, B. E., Lin, A., Lu, B., Zhao, H., Lawton, M. P., and Floyd, E. (2006). Acute drug-induced vascular injury in beagle dogs: pathology and correlating genomic expression. *Toxicol Pathol* 34, 27–32.

Engelhardt, J. A., Fant, P., Guionaud, S., Henry, S. P., Leach, M. W., Louden, C., Scicchitano, M. S., Weaver, J. L., Zabka, T. S., and Frazier, K. S. (2015). Scientific and Regulatory Policy Committee Points-to-consider Paper*: Drug-induced vascular injury associated with nonsmall molecule therapeutics in preclinical development: Part 2. Antisense oligonucleotides. *Toxicol Pathol* 43, 935–44.

Frazier, K. S., Engelhardt, J., Fant, P., Guionaud, S., Henry, S. P., Leach, M., Louden, C., Scicchitano, M. S., Weaver, J. L., and Zabka, T. S. 2015. Drug-induced vascular injury associated with

non-small molecule therapeutics in preclinical development: Part I. Biotherapeutics. *Toxicol Pathol* 43, 915–34.

Greaves, P. (2000). Patterns of cardiovascular pathology induced by diverse cardioactive drugs. *Toxicol Lett* 112–113, 547–52.

Hanton, G., Sobry, C., Dagues, N., Rochefort, G. Y., Bonnet, P., and Eder, V. (2008). Cardiovascular toxicity of minoxidil in the marmoset. *Toxicol Lett* 180, 157–65.

Harris, R. A., Nishiyama, S. K., Wray, D. W., Richardson, R. S. (2010). Ultrasound assessment of flow-mediated dilation. *Hypertension* 55, 1075–85.

Herman, E. H., Ferrans, V. J., Young, R. S., and Balazs, T. (1989). A comparative study of minoxidil-induced myocardial lesions in beagle dogs and miniature swine. *Toxicol Pathol* 17, 182–92.

Hughes, A., Dalmas, D., Thomas, R., Hughes-Earle, A., King, N., Zabka, T.S., Enerson, B., and Kambara, T. (2015). The Predictive Safety Testing Consortium Vascular Injury Working Group: Immunohistochemical Characterization of Candidate Biomarkers of Drug-induced Vascular Injury in Control Rat Tissues. *Society of Toxicology Annual Meeting*, March, San Diego, CA.

Huh, D., Leslie, D. C., Matthews, B. D., Fraser, J. P., Jurek, S., Hamilton, G. A., Thorneloe, K. S., McAlexander, M. A., and Ingber, D. E. (2015) A human disease model of drug toxicity-induced. *Sci Transl Med* 4, 47.

Isaacs, K. R., Joseph, E. C., and Betton, G. R. (1989). Coronary vascular lesions in dogs treated with phosphodiesterase III inhibitors. *Toxicol Pathol* 17, 153–63.

Jennette, J. C., Falk, R. J., Bacon, P. A., Basu, N., Cid, M. C., Ferrario, F., Flores-Suarez, L. F., Gross, W. L., Guillevin, L., Hagen, E. C., Hoffman, G. S., Jayne, D. R., Kallenberg, C. G., Lamprecht, P., Langford, C. A., Luqmani, R. A., Mahr, A. D., Matteson, E. L., Merkel, P. A., Ozen, S., Pusey, C. D., Rasmussen, N., Rees, A. J., Scott, D. G., Specks, U., Stone, J. H., Takahashi, K., and Watts, R. A. (2013). 2012 revised international Chapel Hill consensus conference nomenclature of vasculitides. *Arthritis Rheum* 65, 1–11.

Johansson, S. (1981). Cardiovascular lesions in Sprague-Dawley rats induced by long-term treatment with caffeine. *Acta Pathol Microbiol Scand A* 89, 185–91.

Jones, H. B., Macpherson, A., Betton, G. R., Davis, A. S., Siddall, R., and Greaves, P. (2003). Endothelin antagonist-induced coronary and systemic arteritis in the beagle dog. *Toxicol Pathol* 31, 263–72.

Joseph, E. C. (2000). Arterial lesions induced by phosphodiesterase III (PDE III) inhibitors and DA1 agonists. *Toxicol Lett* 112–113, 537–46.

Joseph E.C., Mesfin G., Kerns WD (1997). Pathogenesis of arterial lesions caused by vasoactive compounds in laboratory animals. In: Cardiovascular Toxicology, Bishop SP, Kerns WD (eds). Pergamon, New York, pp 279–307.

Joseph, E. C., Rees, J. A., and Dayan, A. D. (1996). Mesenteric arteriopathy in the rat induced by phosphodiesterase III inhibitors: an investigation of morphological, ultrastructural, and hemodynamic changes. *Toxicol Pathol* 24, 436–50.

Kaufmann, B. A., Sanders, J. M., Davis, C., Xie, A., Aldred, P., Sarembock, I. J., and Lindner, J. R. (2007). Molecular imaging of inflammation in atherosclerosis with targeted ultrasound detection of vascular cell adhesion molecule-1. *Circulation* 116, 276–84.

Kaufmann, B. A., Carr, C. L., Belcik, J. T., Xie, A., Yue, Q., Chadderdon, S., Caplan, E. S., Khangura, J., Bullens, S., Bunting, S., and Lindner, J. R. (2010). Molecular imaging of the initial inflammatory response in atherosclerosis: Implications for early detection of disease. *Arterioscler Thromb Vasc Biol* 30, 54–9.

Kavanaugh, A., Mease, P. J., Gomez-Reino, J. J., Adebajo, A. O., Wollenhaupt, J., Gladman, D. D., Lespessailles, E., Hall, S., Hochfeld, M., Hu, C., Hough, D., Stevens, R. M., and Schett, G. (2014). Treatment of psoriatic arthritis in a phase 3 randomised, placebo-controlled trial with apremilast, an oral phosphodiesterase 4 inhibitor. *Ann Rheum Dis* 73, 1020–6.

Keirstead, N., Bertinetti-Lapatki, C., Knapp, D., Albassam, M., Hughes, V., Hong, F., Roth, A., and Mikaelian, I. (2015). Temporal patterns of novel circulating biomarkers in IL-2-mediated vascular injury in the rat. *Toxicol Pathol*, 43, 984–94.

Kerns, W. D., Arena, E., Macia, R. A., Bugelski, P. J., Matthews, W. D., and Morgan, D. G. (1989). Pathogenesis of arterial lesions induced by dopaminergic compounds in the rat. *Toxicol Pathol* 17, 203–13.

Kerns, W., Schwartz, L., Blanchard, K., Burchiel, S., Essayan, D., Fung, E., Johnson, R., Lawton, M., Louden, C., MacGregor, J., Miller, F., Nagarkatti, P., Robertson, D., Snyder, P., Thomas, H., Wagner, B., Ward, A., Zhang, J.; Expert Working Group on Drug-Induced Vascular Injury. (2005). Drug-induced vascular injury—A quest for biomarkers. *Toxicol Appl Pharmacol* 203, 62–87.

Losco, P. E., Evans, E. W., Barat, S. A., Blackshear, P. E., Reyderman, L., Fine, J. S., Bober, L. A., Anthes, J. C., Mirro, E. J., and Cuss, F. M. (2004). The toxicity of SCH 351591, a novel phosphodiesterase-4 inhibitor, in Cynomolgus monkeys. *Toxicol Pathol* 32, 295–308.

Louden, C., Nambi, P., Branch, C., Gossett, K., Pullen, M., Eustis, S., and Solleveld, H. A. (1998). Coronary arterial lesions in dogs treated with an endothelin receptor antagonist. *J Cardiovasc Pharmacol* 31 Suppl 1, S384–S385.

Louden, C., Brott, D., Katein A, Kelly T., Gould, S., Jones, H., Betton, G., Valetin, J. P., and Richardson, R. J. (2006). Biomarkers and mechanisms of drug-induced vascular injury in non-rodents. *Toxicol Pathol* 34(1), 19–26.

Mann, P. C., Vahle, J., Keenan, C. M., Baker, J. F., Bradley, A. E., Goodman, D. G., Harada, T., Herbert, R., Kaufmann, W., Kellner, R., Nolte, T., Rittinghausen, S., and Tanaka, T. (2012). International harmonization of toxicologic pathology nomenclature: An overview and review of basic principles. *Toxicol Pathol* 40, 7S–13.

McDuffie, J. E., Yu, X., Sobocinski, G., Song, Y., Chupka, J., and Albassam, M. (2006). Acute coronary artery injury in dogs following administration of CI-1034, an endothelin A receptor antagonist. *Cardiovasc Toxicol* 6, 25–38.

Mesfin, G. M., Shawaryn, G. G., and Higgins, M. J. (1987). Cardiovascular alterations in dogs treated with hydralazine. *Toxicol Pathol* 15, 409–16.

Mesfin, G. M., Piper, R. C., DuCharme, D. W., Carlson, R. G., Humphrey, S. J., and Zins, G. R. (1989). Pathogenesis of cardiovascular alterations in dogs treated with minoxidil. *Toxicol Pathol* 17, 164–81.

Metz, A. L., Dominick, M. A., Suchanek, G., and Gough, A. W. (1991). Acute cardiovascular toxicity induced by an adenosine agonist-antihypertensive in beagles. *Toxicol Pathol* 19, 98–107.

Mikaelian, I., Cameron, M., Dalmas, D. A., Enerson, B. E., Gonzalez, R. J., Guionaud, S., Hoffmann, P. K., King, N. M., Lawton, M. P., Scicchitano, M. S., Smith, H. W., Thomas, R. A., Weaver, J. L., Zabka, T. S.; Vascular Injury Working Group of the Predictive Safety, C. (2014). Nonclinical safety biomarkers of drug-induced vascular injury: Current status and blueprint for the future. *Toxicol Pathol* 42, 635–57.

Miyauchi, T., Yorikane, R., Sakai, S., Sakurai, T., Okada, M., Nishikibe, M., Yano, M., Yamaguchi, I., Sugishita, Y., and Goto, K. (1993). Contribution of endogenous endothelin-1 to the progression of cardiopulmonary alterations in rats with monocrotaline-induced pulmonary hypertension. *Circ Res* 73, 887–97.

Morton, D., Houle, C. D., and Tomlinson, L. (2014). Perspectives on drug-induced vascular injury. *Toxicol Pathol* 42, 633–4.

Radic, M., Martinovic Kaliterna, D., and Radic, J. (2012). Drug-induced vasculitis: A clinical and pathological review. *Neth J Med* 70, 12–7.

Rautou, P. E., Vion, A. C., Amabile, N., Chironi, G., Simon, A., Tedgui, A., and Boulanger, C. M. (2011). Microparticles, vascular function, and atherothrombosis. *Circ Res* 109, 593–606.

Rona, G. (1985). Catecholamine cardiotoxicity. *J Mol Cell Cardiol* 17, 291–306.

Sandusky, G. E., Means, J. R., and Todd, G. C. (1990). Comparative cardiovascular toxicity in dogs given inotropic agents by continuous intravenous infusion. *Toxicol Pathol* 18, 268–78.

Sen, C. K., Gordillo, G. M., Khanna, S., and Roy, S. (2009). Micromanaging vascular biology: Tiny microRNAs play big band. *J Vasc Res* 46, 527–40.

Slim, R., Song, Y., Albassam, M., and Dethloff, L. (2003). Apoptosis and nitrative stress associated with phosphodiesterase inhibitor-induced mesenteric vasculitis in rats. *Toxicol Pathol* 31, 638–45.

Snyder, C., Kauss, A., Doudement, E. Musinipally, V., Ngo, T., Zabka, T., Uppal, H., Misner, D., Dambach, D., and Pai, R. (2014). Comparative Analysis of Endogenous Gene Expression Levels Across Human and Rat Primary Endothelial Cells. *Annual Meeting of American College of Toxicology*, Orlando, Florida, November 9–12, 2014.

Sobota, J. T. (1989). Review of cardiovascular findings in humans treated with minoxidil. *Toxicol Pathol* 17, 193–202.

Swanson, T. A., Conte, T., Deeley, B., Portugal, S., Kreeger, J. M., Obert, L. A., Joseph, E. C., Wisialowski, T. A., Sokolowski, S. A., Rief, C., Nugent, P., Lawton, M. P., and Enerson, B. E. (2014). Hemodynamic correlates of drug-induced vascular injury in the rat using high-frequency ultrasound imaging. *Toxicol Pathol* 42, 784.

Taborda, L., Amaral, B., and Isenberg, D. (2013). Drug-induced vasculitis. *Adverse Drug React Bull* 279, 1075–8.

Teske, R. H., Bishop, S. P., Righter, H. F., and Detweiler, D. K. (1976). Subacute digoxin toxicosis in the beagle dog. *Toxicol Appl Pharmacol* 35, 283–301.

Thomas, R. A., Scicchitano, M. S., Mirabile, R. C., Chau, N. T., Frazier, K. S., and Thomas, H. C. (2012). microRNA changes in rat mesentery and serum associated with drug-induced vascular injury. *Toxicol Appl Pharmacol* 262, 310–20.

Vogel, B., Kolopp, M., Fraissinette, A. d. B. d., Singer, T., Ettlin, R., Birnbaum, I., and Cordier (1999). A Minipigs in safety assessment of a phosphodiesterase type IV inhibitor for the treatment of inflammatory skin diseases. *Toxicologist* 48, 263–74.

Weller, G. E., Lu, E., Csikari, M. M., Klibanov, A. L., Fischer, D., Wagner, W. R., and Villanueva, F. S. (2003). Ultrasound imaging of acute cardiac transplant rejection with microbubbles targeted to intercellular adhesion molecule-1. *Circulation* 108, 218–24.

Wiener, J., Lombardi, D. M., Su, J. E., and Schwartz, S. M. (1996). Immunohistochemical and molecular characterization of the differential response of the rat mesenteric microvasculature to angiotensin-II infusion. *J Vasc Res* 33, 195–208.

Wong, H. K., Chan, J. M., Kamm, R. D., and Tien, J. (2012) Microfluidic models of vascular functions. *Annu Rev Biomed Eng* 14, 205–30.

Yuhas, E. M., Morgan, D. G., Arena, E., Kupp, R. P., Saunders, L. Z., and Lewis, H. B. (1985). Arterial medial necrosis and hemorrhage induced in rats by intravenous infusion of fenoldopam mesylate, a dopaminergic vasodilator. *Am J Pathol* 119, 83–91.

Zampetaki, A. and Mayr, M. (2012). MicroRNAs in vascular and metabolic disease. *Circ Res* 110, 508–22.

Zhang, J., Snyder, R. D., Herman, E. H., Knapton, A., Honchel, R., Miller, T., Espandiari, P., Goodsaid, F. M., Rosenblum, I. Y., Hanig, J. P., Sistare, F. D., and Weaver, J. L. (2008). Histopathology of vascular injury in Sprague-Dawley rats treated with phosphodiesterase IV inhibitor SCH 351591 or SCH 534385. *Toxicol Pathol* 36, 827–39.

Zhang, J., Hanig, J. P., and De Felice, A. F. (2012). Biomarkers of endothelial cell activation: Candidate markers for drug-induced vasculitis in patients or drug-induced vascular injury in animals. *Vascul Pharmacol* 56, 14–25.

26

NOVEL TRANSLATIONAL BIOMARKERS OF SKELETAL MUSCLE INJURY

PETER M. BURCH[1] AND WARREN E. GLAAB[2]

[1] *Investigative Pathology, Drug Safety Research and Development, Pfizer Inc., Groton, CT, USA*

[2] *Systems Toxicology, Investigative Laboratory Sciences, Safety Assessment, Merck Research Laboratories, West Point, PA, USA*

26.1 INTRODUCTION

Numerous therapeutics, both in the clinic and preclinical developments, have been noted to cause skeletal muscle injury by a variety of mechanisms (reviewed in Mastaglia and Needham (2012)). In some cases, this adverse event has been severe enough to result in the removal of drugs, most notably cerivastatin, from the market (Davidson, 2002). The symptoms reported can often be vague, such as fatigue and muscle pain or weakness, but can result in or progress to life-threatening conditions such as rhabdomyolysis (Valiyil and Christopher-Stine, 2010; Mor et al., 2011). The myopathy is often resolved by discontinuing the medication, making a prompt diagnosis crucial (Mor et al., 2011). For many drugs, however, the incidence and severity of myopathy can be highly variable, making diagnosis and discontinuing treatment difficult (Valiyil and Christopher-Stine, 2010). Unfortunately, these events can be equally difficult to diagnose in the animal models used for drug development as it is in the clinic. This is in part because of a lack of rigorously tested biomarkers with the specificity and/or sensitivity to routinely detect and monitor muscle injury in preclinical safety studies. Recent advances have expanded the options for noninvasive biomarkers of skeletal muscle injury. Here we briefly review the potential mechanisms of drug-induced skeletal muscle injury and describe the recent advances in the identification of biomarkers with evidence of utility in drug development. Finally, we discuss where gaps remain and possible future directions for the field.

26.2 OVERVIEW OF DRUG-INDUCED SKELETAL MUSCLE INJURY

Skeletal muscle is a highly vascularized and metabolically active tissue that can account for almost half of the mammalian body mass. Skeletal muscle is composed of bundles of multinucleated myocytes or myofibers that contain the myofibrils. The myofibrils are rodlike filaments composed of actin, myosin, troponin, and numerous other cytoskeletal and regulatory proteins, which support the structure of striated muscle and allow it to contract in response to a stimulus from the neuromuscular junctions (Schiaffino and Reggiani, 2011). These muscle fibers can be classified into different types based on metabolic activity, protein expression, and morphological and physiological characteristics (Schiaffino and Reggiani, 2011). The predominate muscle fiber types span a continuum from slow-twitch, oxidative fibers (also called type 1 fibers); to fast-twitch, oxidative (type 2A) fibers; to fast-twitch, glycolytic (type 2B) fibers (Schiaffino and Reggiani, 2011). The abundance of each fiber type found in a particular muscle group depends on the primary use of that muscle and has been well characterized in both preclinical species and humans (Armstrong and Phelps, 1984; Schiaffino and Reggiani, 2011).

Skeletal muscle is susceptible to injury by a broad range of pharmaceuticals. A number of recent reviews have provided examples of drugs that cause skeletal muscle injury and the known or suspected mechanism (Bannwarth, 2007; Kuncl, 2009; Mastaglia and Needham, 2012; Jones et al., 2014). Broadly classified, the underlying mechanisms

Drug Discovery Toxicology: From Target Assessment to Translational Biomarkers, First Edition. Edited by Yvonne Will, J. Eric McDuffie, Andrew J. Olaharski, and Brandon D. Jeffy.

of drug-induced myopathy include impairment of metabolism, mitochondrial dysfunction, necrosis, and impairment of autophagy (Jones et al., 2014). Additionally, drugs known to cause peripheral neurotoxicity can affect the neuromuscular junctions resulting in neuromyopathies (Bannwarth, 2002). Often multiple mechanisms can be ascribed to a single therapeutic shown to induce muscle damage (Jones et al., 2014).

Statins, cholesterol-lowering drugs, are perhaps the best studied and most instructive examples of drug-induced myopathy. Although the incidence of statin-induced muscle injury is low and varies depending on the type of stain prescribed, all members of this drug class have this liability (Tomaszewski et al., 2011). Symptoms can range from mild muscle pain and weakness to rhabdomyolysis (Banach et al., 2015). Preclinical and clinical studies have shown that the muscle injury caused by statins can be potentiated by interactions with other drugs such as fibrates and cyclosporine A (Smith et al., 1991; Valiyil and Christopher-Stine, 2010; Banach et al., 2015). Multiple possible mechanisms for the observed myotoxicity of statins have been identified including impaired muscle energy metabolism, mitochondrial dysfunction, and necrosis (Jones et al., 2014). It has also been shown that statin myotoxicity can result in a necrotizing autoimmune myopathy, which has been associated with the production of autoantibodies to hydroxymethylglutaryl coenzyme A reductase, the enzyme inhibited by statins (Mammen et al., 2011). Statin-induced myopathy is instructive in the demonstration that a single agent can act through multiple possible mechanisms to cause muscle injury of variable severity.

Drugs that cause unintended muscle injury can also exhibit fiber-type specificity (Bakhtiar, 2008; Kuncl, 2009). Again the statins serve as an instructive example. Preclinical studies in rats have shown that statin-induced muscle injury predominately affects the fast glycolytic fibers (Smith et al., 1991; Westwood et al., 2005). Conversely, fibrates have been shown to exhibit a type 1 fiber specificity both for efficacy and myotoxicity (De Souza et al., 2006). The differences in protein expression, metabolism, and mitochondrial function and number between the fiber types have been implicated in the differences seen in the sensitivity to particular drugs (Sirvent et al., 2008; Kuncl, 2009; Anadón et al., 2014). Identifying a fiber-type specific effect is currently done by immunohistochemistry, which is labor intensive and limits the feasibility of assessing routinely in preclinical studies. Further, a muscle biopsy would be required in clinical studies, which is impractical in most situations.

The tools available to identify and monitor drug-induced muscle injury (defined as myocyte degeneration and necrosis) both in preclinical and in clinical drug development are very limited. In preclinical studies, histopathology is the primary method used to routinely identify muscle injury. Numerous enzymatic assays have routinely been used as markers of skeletal muscle damage, including aspartate aminotransferase (AST), aldolase, and lactate dehydrogenase (LDH) (Brancaccio et al., 2010). While all of these markers can be sensitive indicators of tissue damage, they lack specificity for skeletal muscle tissue as they are expressed of multiple cell types and are used as indicators of injury to liver (e.g., AST) or, in the case of LDH, multiple tissues. Because of this, such parameters are typically not used to specifically detect skeletal muscle injury in preclinical drug development.

Total serum creatine kinase (CK) is the clinical chemistry test that is used to detect and monitor muscle injury (Brancaccio et al., 2010) but is not run routinely in most labs. The physiological role of CK is to utilize adenosine triphosphate (ATP) to catalyze the reversible conversion of creatine to phosphocreatine and serves as an important mechanism for ATP generation in skeletal muscle during periods of high energy demand (Schiaffino and Reggiani, 2011). This test is based on measuring the enzymatic activity of CK present in the serum after release from damaged tissue (Brancaccio et al., 2010). While this widely available test is useful for detecting muscle injury, it also lacks tissue specificity because total CK is also abundant in cardiac tissue. Cytosolic CK is a dimer composed of either the CK, muscle type (CKM) or CK, brain type (CKB). The total serum CK assay measures the enzymatic activity of all CK isoenzymes including the CK-MM homodimer, predominately expressed in skeletal muscle; the CK-MB heterodimer, principally released from cardiomyocytes; and the CK-BB homodimer, expressed in the brain and at lower levels in numerous other tissues (Ozawa et al., 1999; Brancaccio et al., 2007). Therefore, as has been noted by other researchers and clinicians, total serum CK cannot be used to independently distinguish between skeletal muscle and cardiac damage or can be elevated due to medical conditions not directly related to muscle disease (Walker, 2006). Total serum CK activity is also affected by numerous other factors including the age of the subject, the level of physical activity, and the amount of muscle mass, as well as factors that affect CK enzyme activity such as the instability of the CK dimer and short half-life in plasma, serum glutathione levels, and treatment with certain drugs (Szasz et al., 1978; Gunst et al., 1998; Rosalki, 1998), making it a less than ideal biomarker of skeletal muscle injury.

The lack of tissue specificity of the total serum CK assay and the reliance on histopathology to definitively diagnose drug-induced skeletal muscle injury in preclinical drug development have highlighted the need to identify better biomarkers. Beyond the characteristics desired in all safety biomarkers (reviewed in Robinson et al. (2008)), the ideal skeletal muscle injury biomarker should have a number of specific characteristics. First, it should be specific for skeletal muscle tissue injury and, most importantly, allow the discrimination between cardiac and skeletal muscle damage. It should be at least as sensitive as or more sensitive than total serum CK and correlate to the severity of damage. The biomarker should also ideally discriminate between muscle injury and muscle atrophy since age- and disease-related muscle wasting could adversely affect the interpretation of a biomarker signal, as is the case for CK (Rosalki, 1998; Brancaccio et al., 2010).

26.3 NOVEL BIOMARKERS OF DRUG-INDUCED SKELETAL MUSCLE INJURY

Several novel biomarkers of skeletal muscle degeneration/necrosis have been proposed, each with supporting literature references (Table 26.1). These skeletal muscle biomarkers include skeletal troponin I (sTnI), myosin light chain 3 (Myl3), creatine kinase M isoform (CKM, protein assay, recognizes the CKMM homodimer), fatty acid-binding protein 3 (FABP3), parvalbumin, urinary myoglobin, and select microRNAs (miRs) such as muscle-specific miRNAs miR1 and miR-133.

26.3.1 Skeletal Troponin I (sTnI)

Considering the successful application of cardiac troponin assays for monitoring cardiac tissue injury both preclinical and clinically, there has been significant interest determining the utility of skeletal muscle troponins as specific biomarkers of skeletal muscle injury (Brancaccio et al., 2010; Tonomura et al., 2012). Troponin is a heterotrimeric protein composed of a calcium-responsive (C), inhibitory (I), and tropomyosin-binding (T) subunit (Schiaffino and Reggiani, 2011) that regulates the interaction of myosin with actin during muscle contraction. The skeletal troponins are exclusively expressed in the skeletal muscle (Schiaffino and Reggiani, 2011) bolstering the possibility that they could be used as specific

injury biomarkers like the cardiac troponins. Immunoassays that can accurately measure skeletal troponin I in serum and plasma and that do not cross-react with the cardiac troponins have been developed (Takahashi et al., 1996; Tonomura et al., 2012). Furthermore, there are distinct fast-twitch and slow-twitch muscle fiber-specific isoforms of the skeletal troponins (Schiaffino and Reggiani, 2011), suggesting they could be used to develop biomarker assays for fiber-type-specific injury. sTnI is anticipated to increase sensitivity and especially specificity when compared to established biomarkers of skeletal muscle injury.

26.3.2 Creatine Kinase M (CKM)

As described previously, cytosolic CK is a dimeric protein with subunits coded by the CKM (muscle) and CKB (brain) genes (Brancaccio et al., 2010). The subunits are differentially expressed, giving rise to three cytosolic CK isoforms: MM, BB, and MB. In the skeletal muscle, 90% of CK is the CK-MM homodimer (CKM), whereas in the heart, CK consists mostly of CK-MB and to a lesser extent CK-MM (Apple et al., 1984; Brancaccio et al., 2010). While measuring total serum CK activity does not discriminate between isoforms, measuring CKM protein (i.e., CK-MM homodimer) using immunological techniques can provide much higher selectivity for CK released into circulation due to skeletal muscle damage, as well as resolve the instability/half-life liability of the enzymatic assay.

26.3.3 Myosin Light Chain 3 (Myl3)

Myl3 is an essential light chain of the myosin molecule expressed predominantly in the cardiac and skeletal muscle (Schiaffino and Reggiani, 2011). Myosin is a six-subunit mechanochemical enzyme that consists of two heavy-chain

TABLE 26.1 Circulating Biomarkers of Drug-Induced Skeletal Muscle Injury with Evidence of Potential Utility in Drug Development

Biomarker	Source	Assay	Reference(s)
Protein			
Creatine kinase	Serum	Enzymatic	Reviewed in Brancaccio et al. (2010)
Fatty acid-binding protein 3	Serum or plasma	Immunoassay	Pritt et al. (2008), Tonomura et al. (2009), and Tonomura et al. (2012)
Myosin light chain 3	Serum or plasma	Immunoassay	Tonomura et al. (2012)
Skeletal troponin I	Serum or plasma	Immunoassay	Sun et al. (2010), Vassallo et al. (2009), and Tonomura et al. (2012)
Creatine kinase, M-type	Serum or plasma	Immunoassay	Tonomura et al. (2012)
Myosin light chain 1	Serum or plasma	Immunoassay	Tonomura et al. (2009)
Myoglobin	Urine	Immunoassay	Vassallo et al. (2009)
MicroRNA			
miR-1	Plasma	qRT-PCR	Nishimura et al. (2015)
miR-133	Plasma	qRT-PCR	Laterza et al. (2009) and Nishimura et al. (2015)
miR-206	Plasma	qRT-PCR	Nishimura et al. (2015)

subunits and four light-chain subunits. The light-chain subunits consist of two regulatory light chains with phosphorylation sites (encoded by the MYL2 genes) and two essential light chains (encoded by the MYL3 genes). Following damage to muscle tissue, the constituent subunits of myosin become dissociated, and Myl3 is released into the bloodstream. Some reports have focused on determining the utility of Myl3 as a biomarker of cardiomyocyte injury (Lee and Vasan, 2005; Berna et al., 2007). More recently, however, investigators tested a series of skeletal and cardiac muscle toxicants in rats and demonstrated the utility of Myl3 as a quantitative serum biomarker of skeletal muscle injury (Tonomura et al., 2009; Tonomura et al., 2012). Given its abundant expression in type 1 skeletal muscle, it has also been noted that Myl3 may also be useful as a circulating surrogate for injury to this tissue type (Berna et al., 2007).

26.3.4 Fatty Acid-Binding Protein 3

Fatty acid-binding protein 3 (FABP3) is a small (14.5 kDa) cytoplasmic protein that plays a permissive role in transport/mobilization of fatty acids within the cellular environment (Thumser et al., 2014). FABP3 binds both saturated and polyunsaturated fatty acids with high affinity (Kd 2–60 nM). It is thought to shuttle fatty acids from the plasma membrane to intracellular sites of usage including the β-oxidation machinery. FABP3 has also been shown to be expressed in the skeletal muscle and increases in response to physiological conditions that increase fatty acid demand/availability, such as testosterone, endurance training, and nutritional state (Glatz et al., 2003). Therefore, FABP3 has been proposed as a biomarker of both cardiac and skeletal muscle damage. Several studies have linked circulating levels of FABP3 with the incidence and severity of skeletal muscle necrosis, with high performance including sensitivity, positive and negative predictivity, and false-negative rate (Pritt et al., 2008). A direct correlation between circulating concentrations of FABP3 and PPARα-agonist-induced type 1 muscle fiber toxicity was observed in rats (Pritt et al., 2008), and in a study in rats treated with the acetylcholinesterase inhibitor carbofuran (CAF), FABP3 correlated well skeletal muscle degeneration (Tonomura et al., 2009). In another rat study testing a panel of myotoxicants, FABP3 was found to be a sensitive marker for both cardiac and musculoskeletal toxicity. However, FABP3 was not able to differentiate between these two types of injury (Tonomura et al., 2012), demonstrating that this candidate biomarker is not a skeletal muscle-specific biomarker.

26.3.5 Parvalbumin

Parvalbumin is an intracellular calcium-binding protein in the EF-hand superfamily of calcium-binding proteins including troponin C, calmodulin, S100B, and calbindin (Yanez et al., 2012). Parvalbumin is a cytosolic protein that functions in the skeletal muscle to shuttle calcium from the cytosol to intracellular stores to accelerate the relaxation of fast-twitch fibers (Maughan et al., 2005; Arif, 2009). It is predominantly expressed in type 2 fast-twitch, glycolytic muscle fibers and is barely detectable in slow-twitch, oxidative fibers, suggesting the potential utility of parvalbumin as a marker of (Schmitt and Pette, 1991; Schiaffino and Reggiani, 2011). Parvalbumin is also expressed in a number of tissues and organs at considerably lower levels, including GABAergic neurons, heart, distal convoluted tubules in kidneys, ameloblasts, bone, prostate, seminal vesicle, testis, and ovary (Berchtold et al., 1984). Expression of parvalbumin is modulated by the effects of innervation, muscle workload, and dietary creatine and decreases in the fast-twitch muscle with age in rat (Cai et al., 2001; Gallo et al., 2008). Based on the biology of parvalbumin, there has been considerable interest in its potential to serve as a muscle injury biomarker but no evidence in the literature to date of its utility.

26.3.6 Myoglobin

Myoglobin is a heme protein found in both cardiac and skeletal muscles, with highest concentrations in type 1 and 2A skeletal muscle fibers (Weber et al., 2007). It is a single-chain globular protein of 153 amino acids, containing a central heme (iron-containing porphyrin) prosthetic group around which the remaining apoprotein folds (Ordway and Garry, 2004). Because of its low molecular weight, myoglobin is rapidly released from damaged muscle tissue into the systemic circulation, exhibiting increased serum concentrations within 0.5–2 h after acute muscle injury. The released myoglobin is filtered by the kidneys and, in large amounts, can be toxic to the renal tubular epithelium, potentially causing acute renal failure (Boutaud and Roberts, 2011). The rapid and variable clearance of myoglobin from the circulation often renders single point analysis of either blood or urine concentrations inadequate for assessing muscle injury; therefore, overnight urine collections have proven more useful (Vassallo et al., 2009). In addition, increased circulating myoglobin levels are not specific for cardiac or skeletal muscle injury and may due to result from renal failure or dysfunction leading to a marked decrease in renal filtration. Myoglobin from skeletal and cardiac muscle is immunologically identical and cannot be distinguished from each other by immunological methods.

26.3.7 MicroRNAs

miRNAs are endogenous, small (~22 nucleotides), noncoding RNAs that downregulate gene expression (Bartel, 2004). Some miRNAs are produced at extremely high copy numbers within cells in a tissue-specific manner (Lagos-Quintana et al., 2002; Ason et al., 2006), and such miRNAs have recently been reported to be remarkably stable in plasma

(Mitchell et al., 2008; Wang et al., 2009). MiRNAs have also been shown to be highly conserved across species, making them attractive translational tissue toxicity biomarkers. Candidate muscle-specific miRNAs include miR-1 and miR-133. These miRNAs have been shown to be sensitive in detecting drug-induced muscle injury in rats and, more importantly, have demonstrated greater specificity relative to drug-induced liver injury in these studies (Laterza et al., 2009). Although miRNAs have only recently been explored as tissue-specific accessible biomarkers, they hold promise of being superior relative to other protein biomarkers due to their tissue abundance, specificity, and stability in plasma.

26.4 REGULATORY ENDORSEMENT

Beginning in 2004, the Food and Drug Administration (FDA), the European Medicines Agency (EMA), and the Pharmaceuticals and Medical Devices Agency (PDMA) established a framework to encourage collaboration among the regulatory agencies, academia, and the pharmaceutical industry to identify and validate novel biomarkers (reviewed in Goodsaid and Papaluca (2010)). The goal of this evolving process is to review, endorse, and ultimately qualify translational biomarkers to improve nonclinical and clinical drug development decision making (Goodsaid and Papaluca, 2010). The first successful use of this process was the qualification by the FDA and EMA of seven novel kidney injury biomarkers proposed by the Predictive Safety Testing Consortium (PSTC) Nephrotoxicity Working Group, a public–private partnership of member pharmaceutical companies organized by the Critical Path Institute (Dieterle et al., 2010). This review and qualification by the regulatory agencies is intended to encourage the application and further development of these biomarkers for nonclinical and clinical drug development (Goodsaid and Papaluca, 2010).

Similar to the process used for the qualification of the kidney injury biomarkers, the PSTC's Skeletal Myopathy Working Group (SKMWG) investigated sTnI, CKM, FABP3, Myl3, parvalbumin, and urinary myoglobin as skeletal muscle degeneration and/or necrosis biomarkers that were potentially more sensitive and specific than those currently in use, specifically enzymatic CK and AST, for nonclinical drug development. The PSTC is unique in that it allows for pharmaceutical companies to collaborate on the generation and analysis of robust biomarker data packages and receive guidance from the regulatory agencies prior to submission to streamline the identification and validation of relevant biomarkers (Goodsaid et al., 2007). The member companies of the SKMWG contributed and analyzed data from eighteen studies of drug-induced skeletal muscle injury and 16 studies involving drug-induced injury of tissues other than skeletal muscle to test the sensitivity and specificity of the proposed biomarkers. During this process, both parvalbumin and urinary myoglobin failed to

outperform CK or AST in studies with skeletal muscle degeneration/necrosis. This does not mean, however, they could not be useful biomarkers of skeletal muscle injury for specific applications. The outcome of the review of the data package submitted by the working group to the EMA and FDA was a Letter of Support from the regulatory agencies, endorsing and encouraging the use and further study of sTnI, CKM, FABP3, and Myl3 as serum or plasma biomarkers of skeletal muscle injury in conjunction with total serum CK or AST in nonclinical and early clinical drug development (European Medicines Agency, 2015; Food and Drug Administration, 2015).

26.5 GAPS AND FUTURE DIRECTIONS

Future directions include the translation and qualification of next-generation biomarkers described here for use in detecting and monitoring drug-induced skeletal muscle injury in humans. The skeletal muscle proteins discussed here display a very similar tissue distribution in humans and serve the same role in skeletal muscle biology as they do in the rat (Schiaffino and Reggiani, 2011), providing some evidence to expect them to perform similarly as safety biomarkers. There is also evidence from the clinical literature, primarily studying cases of inherited muscle disease or exercise-induced muscle injury, showing the elevated circulating levels of CKM (Apple et al., 1988; Lo et al., 2010; Ayoglu et al., 2014; Hathout et al., 2014), Myl3 (Ayoglu et al., 2014), FABP3 (Sorichter et al., 1998; Pelsers et al., 2005; Hathout et al., 2014), and sTnI (Takahashi et al., 1996; Sorichter et al., 1997; Simpson et al., 2002, 2005; Matziolis et al., 2011; Foster et al., 2012; Chapman et al., 2013) by a variety of analytical methods. Similarly, clinical studies of patients with inherited muscle disorders have provided evidence of the potential utility of mIRNA-1, miRNA-133, and miRNA-206 in monitoring disease progression (Cacchiarelli et al., 2011; Zaharieva et al., 2013; Hu et al., 2014; Li et al., 2014).

There remain, however, a number of deficiencies in the currently identified biomarkers to detect or monitor skeletal muscle injury. First, the biomarkers described here have been developed specifically for the detection of myocyte degeneration and necrosis. This means that, like total serum CK or AST, the biomarker assays are detecting either protein or miRNA released from the muscle fiber into circulation after damage to the sarcolemma (Brancaccio et al., 2010). But degeneration or necrosis is not the only potential injury to skeletal muscle encountered in drug development programs. Drug-induced muscle atrophy or wasting, best exemplified by the skeletal muscle effects of long-term glucocorticoids treatment (Jones et al., 2014), is also a potential liability to a drug development program. These adverse events rarely result in myocyte degeneration or elevated total serum CK levels (Bannwarth, 2002), and, therefore, the biomarkers described here would be expected to be of limited utility.

A recent study in a mouse model of Huntington's disease did show elevated serum levels of sTnI, Myl3, and FABP3 concomitant with skeletal muscle atrophy (Magnusson-Lind et al., 2014), supporting further investigation of the utility of these markers for detecting drug-induced atrophy. Also, circulating levels of potential biomarkers principally originating from the skeletal muscle tissue could be affected by age- or disease-related changes in muscle mass. This potential issue of a change in muscle mass affecting circulating protein levels has been noted for total serum CK (Brancaccio et al., 2010). This will be of particular concern in human translation of these biomarkers and will have to be carefully controlled for in qualification studies.

Identifying biomarkers that can distinguish between fast-twitch and slow-twitch fiber injuries could also provide a significant advantage to drug development programs where there is a need to monitor a specific type of drug-induced injury or better understand the mechanism of injury. This may be possible with two of the biomarkers described here. The fast-twitch and slow-twitch fiber-type isoforms of sTnI described earlier are analytically distinguishable (Simpson et al., 2005; Sun et al., 2010). Additionally, the higher expression of Myl3 at higher levels in type 1 fibers (Berna et al., 2007; Tonomura et al., 2012) may prove advantageous in monitoring slow-twitch fiber injury. Further studies and data analysis of fiber-type specific injuries will be needed, however, to prove the utility of the markers for this purpose.

Recent advances in the identification of circulating biomarkers of muscle diseases may aid in filling the gaps in drug-induced muscle injury biomarkers cited here. Multiple studies using proteomics approaches to identify proteins elevated in the serum or plasma of muscular dystrophy patients have identified numerous potential biomarkers of muscle injury (Ayoglu et al., 2014; Hathout et al., 2014). While most of these potential biomarkers, like CK, are predominately expressed in the muscle, some were not and could be used to monitor aspects of the muscle pathology. For example, matrix metalloproteinase-9 (MMP-9) was identified as a biomarker of disease progression in Duchenne muscular dystrophy and serves a role in the inflammatory response and remodeling that accompany muscle regeneration after injury (Nadarajah et al., 2011). Circulating levels of other microRNAs, including miRNA-208 and miRNA-499, have also been reported to be elevated in muscular dystrophy patients and could potentially serve as translational biomarkers of muscle injury (Li et al., 2014). Additionally, numerous circulating biomarkers and imaging approaches have been investigated as biomarkers for monitoring the muscle loss associated with sarcopenia and cachexia (Goodwin, 2011; Nedergaard et al., 2013). Further investigation of these muscle disease biomarkers could prove them useful for monitoring drug-induced muscle injury and potentially minimize or eliminate some of the drawbacks associated with those currently in use.

26.6 CONCLUSIONS

Efforts to date have clearly identified a number of novel candidate biomarkers of skeletal muscle injury that offer advantages over the standard clinical chemistry tests for total serum CK or AST. The availability of analytical tests in a standard immunoassay format for the peptide markers or quantitative real-time PCR assays for the miRNA makes it feasible to add these biomarkers to drug development studies with minimal assay development time. These biomarkers also exhibit high homology across species, making them attractive translational skeletal muscle biomarkers. Additionally, the regulatory agencies' review and endorsement of sTnI, Myl3, FABP3, and CKM as biomarkers of skeletal muscle injury have the potential to reduce the time and cost of drug development programs that require specific assays to monitor skeletal muscle degeneration or necrosis and further promote the translation of these novel biomarkers into clinical development to monitor drug-induced skeletal muscle injury and better ensure patient safety.

REFERENCES

Anadón, A., V. Castellano, and M. R. Martínez-Larrañaga (2014). Biomarkers in drug safety evaluation. In: *Biomarkers in Toxicology*. R. C. Gupta, ed. Boston, MA, Academic Press, 923–945.

Apple, F. S., M. A. Rogers, W. M. Sherman, D. L. Costill, F. C. Hagerman, and J. L. Ivy (1984). "Profile of creatine kinase isoenzymes in skeletal muscles of marathon runners." *Clin Chem* 30(3): 413–416.

Apple, F. S., Y. Hellsten, and P. M. Clarkson (1988). "Early detection of skeletal muscle injury by assay of creatine kinase MM isoforms in serum after acute exercise." *Clin Chem* 34(6): 1102–1104.

Arif, S. H. (2009). "A Ca(2+)-binding protein with numerous roles and uses: parvalbumin in molecular biology and physiology." *Bioessays* 31(4): 410–421.

Armstrong, R. B. and R. O. Phelps (1984). "Muscle fiber type composition of the rat hindlimb." *Am J Anat* 171(3): 259–272.

Ason, B., D. K. Darnell, B. Wittbrodt, E. Berezikov, W. P. Kloosterman, J. Wittbrodt, P. B. Antin, and R. H. Plasterk (2006). "Differences in vertebrate microRNA expression." *Proc Natl Acad Sci U S A* 103(39): 14385–14389.

Ayoglu, B., A. Chaouch, H. Lochmuller, L. Politano, E. Bertini, P. Spitali, M. Hiller, E. H. Niks, F. Gualandi, F. Ponten, K. Bushby, A. Aartsma-Rus, E. Schwartz, Y. Le Priol, V. Straub, M. Uhlen, S. Cirak, P. A. t Hoen, F. Muntoni, A. Ferlini, J. M. Schwenk, P. Nilsson, and C. Al-Khalili Szigyarto (2014). "Affinity proteomics within rare diseases: a BIO-NMD study for blood biomarkers of muscular dystrophies." *EMBO Mol Med* 6(7): 918–936.

Bakhtiar, R. (2008). "Biomarkers in drug discovery and development." *J Pharmacol Toxicol Methods* 57(2): 85–91.

Banach, M., M. Rizzo, P. P. Toth, M. Farnier, M. H. Davidson, K. Al-Rasadi, W. S. Aronow, V. Athyros, D. M. Djuric, M. V. Ezhov, R. S. Greenfield, G. K. Hovingh, K. Kostner, C. Serban, D. Lighezan, Z. Fras, P. M. Moriarty, P. Muntner, A. Goudev, R. Ceska, S. J. Nicholls, M. Broncel, D. Nikolic, D. Pella, R. Puri, J. Rysz, N. D. Wong, L. Bajnok, S. R. Jones, K. K. Ray, and D. P. Mikhailidis (2015). "Statin intolerance—an attempt at a unified definition. Position paper from an International Lipid Expert Panel." *Arch Med Sci* 11(1): 1–23.

Bannwarth, B. (2002). "Drug-induced myopathies." *Expert Opin Drug Saf* 1(1): 65–70.

Bannwarth, B. (2007). "Drug-induced musculoskeletal disorders." *Drug Saf* 30(1): 27–46.

Bartel, D. P. (2004). "MicroRNAs: genomics, biogenesis, mechanism, and function." *Cell* 116(2): 281–297.

Berchtold, M. W., M. R. Celio, and C. W. Heizmann (1984). "Parvalbumin in non-muscle tissues of the rat. Quantitation and immunohistochemical localization." *J Biol Chem* 259(8): 5189–5196.

Berna, M. J., Y. Zhen, D. E. Watson, J. E. Hale, and B. L. Ackermann (2007). "Strategic use of immunoprecipitation and LC/MS/MS for trace-level protein quantification: myosin light chain 1, a biomarker of cardiac necrosis." *Anal Chem* 79(11): 4199–4205.

Boutaud, O. and L. J. Roberts, II (2011). "Mechanism-based therapeutic approaches to rhabdomyolysis-induced renal failure." *Free Radic Biol Med* 51(5): 1062–1067.

Brancaccio, P., N. Maffulli, and F. M. Limongelli (2007). "Creatine kinase monitoring in sport medicine." *Br Med Bull* 81–82: 209–230.

Brancaccio, P., G. Lippi, and N. Maffulli (2010). "Biochemical markers of muscular damage." *Clin Chem Lab Med* 48(6): 757–767.

Cacchiarelli, D., I. Legnini, J. Martone, V. Cazzella, A. D'Amico, E. Bertini, and I. Bozzoni (2011). "miRNAs as serum biomarkers for Duchenne muscular dystrophy." *EMBO Mol Med* 3(5): 258–265.

Cai, D. Q., M. Li, K. K. Lee, K. M. Lee, L. Qin, and K. M. Chan (2001). "Parvalbumin expression is downregulated in rat fast-twitch skeletal muscles during aging." *Arch Biochem Biophys* 387(2): 202–208.

Chapman, D. W., J. A. Simpson, S. Iscoe, T. Robins, and K. Nosaka (2013). "Changes in serum fast and slow skeletal troponin I concentration following maximal eccentric contractions." *J Sci Med Sport* 16: 82–85.

Davidson, M. H. (2002). "Controversy surrounding the safety of cerivastatin." *Expert Opin Drug Saf* 1(3): 207–212.

De Souza, A. T., P. D. Cornwell, X. Dai, M. J. Caguyong, and R. G. Ulrich (2006). "Agonists of the peroxisome proliferator-activated receptor alpha induce a fiber-type-selective transcriptional response in rat skeletal muscle." *Toxicol Sci* 92(2): 578–586.

Dieterle, F., F. Sistare, F. Goodsaid, M. Papaluca, J. S. Ozer, C. P. Webb, W. Baer, A. Senagore, M. J. Schipper, J. Vonderscher, S. Sultana, D. L. Gerhold, J. A. Phillips, G. Maurer, K. Carl, D. Laurie, E. Harpur, M. Sonee, D. Ennulat, D. Holder, D. Andrews-Cleavenger, Y. Z. Gu, K. L. Thompson, P. L. Goering,

J. M. Vidal, E. Abadie, R. Maciulaitis, D. Jacobson-Kram, A. F. Defelice, E. A. Hausner, M. Blank, A. Thompson, P. Harlow, D. Throckmorton, S. Xiao, N. Xu, W. Taylor, S. Vamvakas, B. Flamion, B. S. Lima, P. Kasper, M. Pasanen, K. Prasad, S. Troth, D. Bounous, D. Robinson-Gravatt, G. Betton, M. A. Davis, J. Akunda, J. E. McDuffie, L. Suter, L. Obert, M. Guffroy, M. Pinches, S. Jayadev, E. A. Blomme, S. A. Beushausen, V. G. Barlow, N. Collins, J. Waring, D. Honor, S. Snook, J. Lee, P. Rossi, E. Walker, and W. Mattes (2010). "Renal biomarker qualification submission: a dialog between the FDA-EMEA and Predictive Safety Testing Consortium." *Nat Biotechnol* 28(5): 455–462.

European Medicines Agency (2015, March 6). "Letter of Support for Skeletal Muscle Injury Biomarkers." from http://www.ema. europa.eu/docs/en_GB/document_library/Other/2015/03/ WC500184458.pdf (accessed October 20, 2015).

Food and Drug Administration (2015). "Drug Development Tools (DDT) Letters of Support." from http://www.fda.gov/Drugs/ DevelopmentApprovalProcess/ucm434382.htm (accessed October 20, 2015).

Foster, G. E., J. Nakano, A. W. Sheel, J. A. Simpson, J. D. Road, and W. D. Reid (2012). "Serum skeletal troponin I following inspiratory threshold loading in healthy young and middle-aged men." *Eur J Appl Physiol* 112: 3547–3558.

Gallo, M., I. MacLean, N. Tyreman, K. J. Martins, T. Syrotuik, T. Gordon, and C. T. Putman (2008). "Adaptive responses to creatine loading and exercise in fast-twitch rat skeletal muscle." *Am J Physiol Regul Integr Comp Physiol* 294(4): R1319–R1328.

Glatz, J. F., F. G. Schaap, B. Binas, A. Bonen, G. J. van der Vusse, and J. J. Luiken (2003). "Cytoplasmic fatty acid-binding protein facilitates fatty acid utilization by skeletal muscle." *Acta Physiol Scand* 178(4): 367–371.

Goodsaid, F. and M. Papaluca (2010). "Evolution of biomarker qualification at the health authorities." *Nat Biotechnol* 28(5): 441–443.

Goodsaid, F. M., F. W. Frueh, and W. Mattes (2007). "The Predictive Safety Testing Consortium: a synthesis of the goals, challenges and accomplishments of the critical path." *Drug Discov Today Technol* 4(2): 47–50.

Goodwin, D. W. (2011). "Imaging of skeletal muscle." *Rheum Dis Clin North Am* 37(2): 245–251, vi–vii.

Gunst, J. J., M. R. Langlois, J. R. Delanghe, M. L. De Buyzere, and G. G. Leroux-Roels (1998). "Serum creatine kinase activity is not a reliable marker for muscle damage in conditions associated with low extracellular glutathione concentration." *Clin Chem* 44(5): 939–943.

Hathout, Y., R. L. Marathi, S. Rayavarapu, A. Zhang, K. J. Brown, H. Seol, H. Gordish-Dressman, S. Cirak, L. Bello, K. Nagaraju, T. Partridge, E. P. Hoffman, S. Takeda, J. K. Mah, E. Henricson, and C. McDonald (2014). "Discovery of serum protein biomarkers in the mdx mouse model and cross-species comparison to Duchenne muscular dystrophy patients." *Hum Mol Genet* 23: 6458–6469.

Hu, J., M. Kong, Y. Ye, S. Hong, L. Cheng, and L. Jiang (2014). "Serum miR-206 and other muscle-specific microRNAs as non-invasive biomarkers for Duchenne muscular dystrophy." *J Neurochem* 129(5): 877–883.

Jones, J. D., H. L. Kirsch, R. L. Wortmann, and M. H. Pillinger (2014). "The causes of drug-induced muscle toxicity." *Curr Opin Rheumatol* 26(6): 697–703.

Kuncl, R. W. (2009). "Agents and mechanisms of toxic myopathy." *Curr Opin Neurol* 22(5): 506–515.

Lagos-Quintana, M., R. Rauhut, A. Yalcin, J. Meyer, W. Lendeckel, and T. Tuschl (2002). "Identification of tissue-specific microRNAs from mouse." *Curr Biol* 12(9): 735–739.

Laterza, O. F., L. Lim, P. W. Garrett-Engele, K. Vlasakova, N. Muniappa, W. K. Tanaka, J. M. Johnson, J. F. Sina, T. L. Fare, F. D. Sistare, and W. E. Glaab (2009). "Plasma MicroRNAs as sensitive and specific biomarkers of tissue injury." *Clin Chem* 55(11): 1977–1983.

Lee, D. S. and R. S. Vasan (2005). "Novel markers for heart failure diagnosis and prognosis." *Curr Opin Cardiol* 20(3): 201–210.

Li, X., Y. Li, L. Zhao, D. Zhang, X. Yao, H. Zhang, Y. C. Wang, X. Y. Wang, H. Xia, J. Yan, and H. Ying (2014). "Circulating muscle-specific miRNAs in Duchenne muscular dystrophy patients." *Mol Ther Nucleic Acids* 3: e177.

Lo, K. R., S. M. Hurst, K. R. Atkinson, T. Vandenbogaerde, C. M. Beaven, and J. R. Ingram (2010). "Development and validation of a sensitive immunoassay for the skeletal muscle isoform of creatine kinase." *J Sci Med Sport* 13(1): 117–119.

Magnusson-Lind, A., M. Davidsson, E. Silajdzic, C. Hansen, A. C. McCourt, S. J. Tabrizi, and M. Bjorkqvist (2014). "Skeletal muscle atrophy in R6/2 mice—altered circulating skeletal muscle markers and gene expression profile changes." *J Huntington's Dis* 3(1): 13–24.

Mammen, A. L., T. Chung, L. Christopher-Stine, P. Rosen, A. Rosen, K. R. Doering, and L. A. Casciola-Rosen (2011). "Autoantibodies against 3-hydroxy-3-methylglutaryl-coenzyme A reductase in patients with statin-associated autoimmune myopathy." *Arthritis Rheum* 63(3): 713–721.

Mastaglia, F. L. and M. Needham (2012). "Update on toxic myopathies." *Curr Neurol Neurosci Rep* 12(1): 54–61.

Matziolis, D., G. Wassilew, P. Strube, G. Matziolis, and C. Perka (2011). "Differences in muscle trauma quantifiable in the laboratory between the minimally invasive anterolateral and transgluteal approach." *Arch Orthop Trauma Surg* 131(5): 651–655.

Maughan, D. W., J. A. Henkin, and J. O. Vigoreaux (2005). "Concentrations of glycolytic enzymes and other cytosolic proteins in the diffusible fraction of a vertebrate muscle proteome." *Mol Cell Proteomics* 4(10): 1541–1549.

Mitchell, P. S., R. K. Parkin, E. M. Kroh, B. R. Fritz, S. K. Wyman, E. L. Pogosova-Agadjanyan, A. Peterson, J. Noteboom, K. C. O'Briant, A. Allen, D. W. Lin, N. Urban, C. W. Drescher, B. S. Knudsen, D. L. Stirewalt, R. Gentleman, R. L. Vessella, P. S. Nelson, D. B. Martin, and M. Tewari (2008). "Circulating microRNAs as stable blood-based markers for cancer detection." *Proc Natl Acad Sci U S A* 105(30): 10513–10518.

Mor, A., R. L. Wortmann, H. J. Mitnick, and M. H. Pillinger (2011). "Drugs causing muscle disease." *Rheum Dis Clin North Am* 37(2): 219–231, vi.

Nadarajah, V. D., M. van Putten, A. Chaouch, P. Garrood, V. Straub, H. Lochmuller, H. B. Ginjaar, A. M. Aartsma-Rus, G. J. van Ommen, J. T. den Dunnen, and P. A. t Hoen (2011). "Serum matrix metalloproteinase-9 (MMP-9) as a biomarker for monitoring disease progression in Duchenne muscular dystrophy (DMD)." *Neuromuscul Disord* 21(8): 569–578.

Nedergaard, A., M. A. Karsdal, S. Sun, and K. Henriksen (2013). "Serological muscle loss biomarkers: an overview of current concepts and future possibilities." *J Cachexia Sarcopenia Muscle* 4(1): 1–17.

Nishimura, Y., C. Kondo, Y. Morikawa, Y. Tonomura, M. Torii, J. Yamate, and T. Uehara (2015). "Plasma miR-208 as a useful biomarker for drug-induced cardiotoxicity in rats." *J Appl Toxicol* 35(2): 173–180.

Ordway, G. A. and D. J. Garry (2004). "Myoglobin: an essential hemoprotein in striated muscle." *J Exp Biol* 207(Pt 20): 3441–3446.

Ozawa, E., Y. Hagiwara, and M. Yoshida (1999). "Creatine kinase, cell membrane and Duchenne muscular dystrophy." *Mol Cell Biochem* 190(1–2): 143–151.

Pelsers, M. M., W. T. Hermens, and J. F. Glatz (2005). "Fatty acid-binding proteins as plasma markers of tissue injury." *Clin Chim Acta* 352(1–2): 15–35.

Pritt, M. L., D. G. Hall, J. Recknor, K. M. Credille, D. D. Brown, N. P. Yumibe, A. E. Schultze, and D. E. Watson (2008). "Fabp3 as a biomarker of skeletal muscle toxicity in the rat: comparison with conventional biomarkers." *Toxicol Sci* 103(2): 382–396.

Robinson, S., R. Pool, R. B. Giffin, and Institute of Medicine (U.S.). Forum on Drug Discovery Development and Translation (2008). *Emerging Safety Science: Workshop Summary*. Washington, DC, National Academies Press.

Rosalki, S. B. (1998). "Low serum creatine kinase activity." *Clin Chem* 44(5): 905.

Schiaffino, S. and C. Reggiani (2011). "Fiber types in mammalian skeletal muscles." *Physiol Rev* 91(4): 1447–1531.

Schmitt, T. L. and D. Pette (1991). "Fiber type-specific distribution of parvalbumin in rabbit skeletal muscle. A quantitative microbiochemical and immunohistochemical study." *Histochemistry* 96(6): 459–465.

Simpson, J. A., R. Labugger, G. G. Hesketh, C. D'Arsigny, D. O'Donnell, N. Matsumoto, C. P. Collier, S. Iscoe, and J. E. Van Eyk (2002). "Differential detection of skeletal troponin I isoforms in serum of a patient with rhabdomyolysis: markers of muscle injury?" *Clin Chem* 48(7): 1112–1114.

Simpson, J. A., R. Labugger, C. Collier, R. J. Brison, S. Iscoe, and J. E. Van Eyk (2005). "Fast and slow skeletal troponin I in serum from patients with various skeletal muscle disorders: a pilot study." *Clin Chem* 51(6): 966–972.

Sirvent, P., J. Mercier, and A. Lacampagne (2008). "New insights into mechanisms of statin-associated myotoxicity." *Curr Opin Pharmacol* 8(3): 333–338.

Smith, P. F., R. S. Eydelloth, S. J. Grossman, R. J. Stubbs, M. S. Schwartz, J. I. Germershausen, K. P. Vyas, P. H. Kari, and J. S. MacDonald (1991). "HMG-CoA reductase inhibitor-induced myopathy in the rat: cyclosporine A interaction and mechanism studies." *J Pharmacol Exp Ther* 257(3): 1225–1235.

Sorichter, S., J. Mair, A. Koller, W. Gebert, D. Rama, C. Calzolari, E. Artner-Dworzak, and B. Puschendorf (1997). "Skeletal troponin I as a marker of exercise-induced muscle damage." *J Appl Physiol* 83(4): 1076–1082.

Sorichter, S., J. Mair, A. Koller, M. M. Pelsers, B. Puschendorf, and J. F. Glatz (1998). "Early assessment of exercise induced skeletal muscle injury using plasma fatty acid binding protein." *Br J Sports Med* 32(2): 121–124.

Sun, D., D. Hamlin, A. Butterfield, D. E. Watson, and H. W. Smith (2010). "Electrochemiluminescent immunoassay for rat skeletal troponin I (Tnni2) in serum." *J Pharmacol Toxicol Methods* 61(1): 52–58.

Szasz, G., W. Gerhardt, and W. Gruber (1978). "Creatine kinase in serum: 5. Effect of thiols on isoenzyme activity during storage at various temperatures." *Clin Chem* 24(9): 1557–1563.

Takahashi, M., L. Lee, Q. Shi, Y. Gawad, and G. Jackowski (1996). "Use of enzyme immunoassay for measurement of skeletal troponin-I utilizing isoform-specific monoclonal antibodies." *Clin Biochem* 29(4): 301–308.

Thumser, A. E., J. B. Moore, and N. J. Plant (2014). "Fatty acid binding proteins: tissue-specific functions in health and disease." *Curr Opin Clin Nutr Metab Care* 17(2): 124–129.

Tomaszewski, M., K. M. Stepien, J. Tomaszewska, and S. J. Czuczwar (2011). "Statin-induced myopathies." *Pharmacol Rep* 63(4): 859–866.

Tonomura, Y., Y. Mori, M. Torii, and T. Uehara (2009). "Evaluation of the usefulness of biomarkers for cardiac and skeletal myotoxicity in rats." *Toxicology* 266(1–3): 48–54.

Tonomura, Y., S. Matsushima, E. Kashiwagi, K. Fujisawa, S. Takagi, Y. Nishimura, R. Fukushima, M. Torii, and M. Matsubara (2012). "Biomarker panel of cardiac and skeletal muscle troponins, fatty acid binding protein 3 and myosin light chain 3 for the accurate diagnosis of cardiotoxicity and musculoskeletal toxicity in rats." *Toxicology* 302(2–3): 179–189.

Valiyil, R. and L. Christopher-Stine (2010). "Drug-related myopathies of which the clinician should be aware." *Curr Rheumatol Rep* 12(3): 213–220.

Vassallo, J. D., E. B. Janovitz, D. M. Wescott, C. Chadwick, L. J. Lowe-Krentz, and L. D. Lehman-McKeeman (2009). "Biomarkers of drug-induced skeletal muscle injury in the rat: troponin I and myoglobin." *Toxicol Sci* 111(2): 402–412.

Walker, D. B. (2006). "Serum chemical biomarkers of cardiac injury for nonclinical safety testing." *Toxicol Pathol* 34(1): 94–104.

Wang, K., S. Zhang, B. Marzolf, P. Troisch, A. Brightman, Z. Hu, L. E. Hood, and D. J. Galas (2009). "Circulating microRNAs, potential biomarkers for drug-induced liver injury." *Proc Natl Acad Sci U S A* 106(11): 4402–4407.

Weber, M. A., R. Kinscherf, H. Krakowski-Roosen, M. Aulmann, H. Renk, A. Kunkele, L. Edler, H. U. Kauczor, and W. Hildebrandt (2007). "Myoglobin plasma level related to muscle mass and fiber composition: a clinical marker of muscle wasting?" *J Mol Med (Berl)* 85(8): 887–896.

Westwood, F. R., A. Bigley, K. Randall, A. M. Marsden, and R. C. Scott (2005). "Statin-induced muscle necrosis in the rat: distribution, development, and fibre selectivity." *Toxicol Pathol* 33(2): 246–257.

Yanez, M., J. Gil-Longo, and M. Campos-Toimil (2012). "Calcium binding proteins." *Adv Exp Med Biol* 740: 461–482.

Zaharieva, I. T., M. Calissano, M. Scoto, M. Preston, S. Cirak, L. Feng, J. Collins, R. Kole, M. Guglieri, V. Straub, K. Bushby, A. Ferlini, J. E. Morgan, and F. Muntoni (2013). "Dystromirs as serum biomarkers for monitoring the disease severity in Duchenne muscular dystrophy." *PLoS One* 8(11): e80263.

27

TRANSLATIONAL MECHANISTIC BIOMARKERS AND MODELS FOR PREDICTING DRUG-INDUCED LIVER INJURY: CLINICAL TO *IN VITRO* PERSPECTIVES

DANIEL J. ANTOINE

MRC Centre for Drug Safety Science and Department of Molecular and Clinical Pharmacology, The Institute of Translational Medicine, University of Liverpool, Liverpool, UK

27.1 INTRODUCTION

Drug-induced liver injury (DILI) represents a significant ADR for both currently used medicines and is a significant impediment to the development of new therapies. It is a major human health concern as it is a leading cause of patient morbidity and mortality. It has been widely cited that of the 10,000 documented human medicines, more than 1,000 are associated with liver injury (Lee, 2003). The overall incidence of DILI in the general population has been estimated to range from 10 to 15 cases per 100,000 patient years with the incidence of DILI resulting from an individual drug used in clinical practice ranging from 1 in 10,000 to 1 in 1,000,000 patient-years (Sgro et al., 2002; Meier et al., 2005). Although DILI accounts for <1%, hospitalized patients show phenotypic signs of DILI such as jaundice (Sgro et al., 2002; Vuppalanchi et al., 2007); it is an adverse event that most frequently results in regulatory actions leading to *black box warnings* and/or removal of drugs from the market. In the clinic, DILI accounts for more than 50% of acute liver failure (ALF) cases; and improved detection of DILI before overt liver failure occurs has been the subject of translational biomarker investigations (Lee, 2003).

Drug attrition due to DILI occurs in all phases of the development pipeline, during preclinical testing and due to toxicological liabilities identified post drug launch. In cases where the frequency is remarkable in either preclinical animal models or humans, DILI is considered "intrinsic," defined as DILI due to direct hepatocellular damage (Corsini et al., 2012).

However, another concerning manifestation, termed "idiosyncratic DILI," occurs rarely in susceptible human subjects following exposure to regulatory agency-approved therapeutic doses, but the mechanisms responsible for this clinical phenomenon remain unclear (Sgro et al., 2002).

Due to the low incidence and multiple contributing factors, confident diagnosis of DILI in humans can only be attained once other possible causes of liver injury have been excluded, which may delay the discontinuation of offending drug(s). When DILI is suspected in a patient treated with multiple medications, it may be impossible to rapidly identify the specific drug responsible, and this may force the physician to unnecessarily stop medications, potentially placing the patient at increased risk from the disease(s) being treated. Furthermore, it is almost impossible to identify which patients are susceptible to DILI during the early phase of harm. Therefore, there is a need to develop new and improved DILI biomarkers that can either confidently establish the diagnosis of DILI, identify the specific drug(s) that may cause DILI, and predict progressive and reversible DILI.

The presentation of DILI (clinical and histological) can mimic most types of naturally occurring liver diseases. Idiosyncratic hepatocellular liver injury is generally of greatest concern because it can develop quickly and become life threatening prior to evidence of jaundice. Regardless of type, detection of DILI relies upon a small number of routine laboratory tests. However, traditional biomarkers of DILI lack specificity and sensitivity, thus contributing to the poor prediction of liver toxicities in all species. The assessment of

Drug Discovery Toxicology: From Target Assessment to Translational Biomarkers, First Edition. Edited by Yvonne Will, J. Eric McDuffie, Andrew J. Olaharski, and Brandon D. Jeffy.

the potential for new chemical entities to elicit clinical hepatotoxicity is heavily dependent upon the histopathological evaluation of hepatotoxicity endpoints in preclinical species coupled with the quantitative assessment of circulating proteins that are enriched in representative hepatic tissue samples (Antoine et al., 2009a; Moggs et al., 2012). However, when clinical trials are performed, current preclinical testing regimens at best successfully correlate to clinical adverse hepatic events in about 50% of cases (Olson et al., 2000). In addition, liver biopsies are not routinely taken from clinical trial subjects or patients with overt DILI, leading to incomplete assessment of the mechanisms of liver injury for a given drug.

The lack of qualified mechanistic biomarkers has resulted major challenges to investigate the true extent and diagnosis of DILI (Aithal et al., 2011). The development and qualification of sensitive and specific hepatocellular-specific biomarkers of injury that hold translation between preclinical and clinical studies are urgently required to accelerate the pace of drug development. Improved DILI biomarkers may additionally enable patient and/or DILI-specific treatment stratification for marketed therapeutics. The potential for novel biomarkers to provide enhanced understanding of the fundamental mechanisms that result in clinical DILI is becoming increasingly recognized.

In 2001, the US National Institute of Health (NIH) defined a biomarker (i.e., oligonucleotide, protein, or metabolite) as a characteristic that is objectively measured and evaluated as an indicator of normal biological process, a pathogenic process, or a pharmacological response to a therapeutic intervention (Biomarkers Definitions Working Group, 2001). Biomarkers are classified by the US Food and Drug Administration (FDA) as either exploratory, probable valid, or known valid. A valid biomarker is further defined as a biomarker that is measured in an analytical test system with well-established performance characteristics and for which there is an established scientific framework or body of evidence that elucidates the physiological, toxicological, pharmacological, or clinical significance of the test result (Ratner, 2005). Since then, recommendations have been set to avoid confusion that the term "validation" should refer to the technical characterization and documentation of methodological performances and the term "qualification" should refer to the evidentiary process of linking a biomarker to a clinical endpoint or biological process (Matheis et al., 2011).

However, the development and clinical integration of potential hepatic biomarkers over the past 60 years have revealed only a limited number of candidates (Antoine et al., 2009a; Matheis et al., 2011; Moggs et al., 2012) compared to research focus placed on drug efficacy (Watkins, 2011). Furthermore, the delayed qualification and ultimate scientific acceptance of a potential DILI biomarker have been hindered by what has been previously thought of as the competing interest between the various stakeholders. Safety assessment

within drug development has traditionally focused on reliable clinical-to-preclinical concordance. Low baseline variability, specificity, and rapid endpoint analysis are sought after by clinicians, and the ability to provide enhanced mechanistic understanding about toxicological processes is warranted. Public and private consortia have evolved to meet these gaps; these consortia consist of leading academic groups, large pharmaceutical companies, small-to-medium enterprises, and clinical centers of excellence such as the Predictive Safety Testing Consortium (PSTC) (http://www.c-path.org/pstc.cfm) and the Safer and Faster Evidence-based Translation (SAFE-T) consortium (www.imi-safe-t.eu). Various consortia efforts, coupled with active engagements by regulatory authorities as external members and/or advisors, provide an opportunity for collaborative efforts to foster the successful identification, validation, and qualification of novel translational safety biomarkers (TSBM) for DILI (Matheis et al., 2011).

27.2 DRUG-INDUCED TOXICITY AND THE LIVER

The first-pass exposure to drugs administered orally and the high capacity for xenobiotic metabolism are considered significant reasons why the liver is a target for drug toxicity, and this aspect has been widely reviewed (Park et al., 2005). Toxic metabolites can be generated and accumulate within hepatocytes as a result of oxidation–reduction (phase I metabolism), conjugation (phase II metabolism), and the saturation of transporter activities (phase III metabolism), which normally serve to remove or detoxify xenobiotics (Park et al., 2005, 2011). The metabolic capacity of the liver coupled to the portal blood supply and the resident immune system contribute to hepatic susceptibility to drug toxicity.

The most frequent cause of DILI in the Western world is a result of acetaminophen (APAP, paracetamol) overdose, which accounted for 38,000 emergency hospital admissions in the financial year 2010–2011 in England alone (NHS, 2011). Between 1998 and 2007, almost half (46%) of reported cases of ALF had been attributed to APAP with DILI resulting from other drugs accounting for 11% (Lee et al., 2008). APAP hepatotoxicity has been widely studied and is reproducible in animal models. Therefore, due to difficulties in studying clinical idiosyncratic DILI in animal models, APAP hepatotoxicity represents an excellent paradigm to identify new biomarkers and to understand mechanisms of clinical-induced liver stress in the exploratory setting, which has clear clinical significance (Kaplowitz, 2005; Williams and Jaeschke, 2012). The biochemical basis of APAP hepatotoxicity is well defined through cytochrome P450 (CYP2E1, 1A2, and 2D6)-mediated reactive metabolite formation (N-acetyl-p-benzoquinone imine) and hepatic GSH depletion (Mitchell et al., 1973a, b). Covalent adduction to cellular proteins then propagates injury through a number of mechanisms, including oxidative stress, which ultimately leads to hepatocyte necrosis and sterile inflammation (Jollow et al., 1973) and has been recently reviewed (Potter et al., 1973).

Clinically, antidote treatment with acetylcysteine (AC) is time-consuming and results in significant bed occupancy in hospitals (around 47,000 bed days per year in England). Furthermore, adverse reactions to AC are common in around half of all identified patients. The majority of APAP-overdosed patients that present to hospital early, soon after overdose, and before acute liver injury (ALI) can be diagnosed, or confidently excluded, using current blood-based biomarkers such as alanine aminotransaminase (ALT) (Antoine et al., 2013a; Dear et al., 2013). Therefore, decisions regarding the initiation of antidote treatment are predominately based on the dose of APAP ingested and a timed blood APAP concentration (Ferner et al., 2011). This time delay coupled with the early clinical uncertainty of the presence of liver injury prevents hepatotoxicity management from being individualized, potentially leading to patients being over treated with a time-consuming and potentially harmful antidote or undertreated ALI. Novel biomarkers that detect ALI at the earliest possible time point are required both from the view point of idiosyncratic hepatotoxicity and patient risk resulting from known hepatotoxins (Dear and Antoine, 2014). Moreover, given the multistep and multicellular process of DILI, panels of biomarkers that have potential to provide insights into the underlying mechanistic basis of ALI are increasingly being recognized as fundamental to efforts in translational research and patient treatment stratification.

27.3   CURRENT STATUS OF BIOMARKERS FOR THE ASSESSMENT OF DILI

To date, only a relatively small number of blood-based tests are used to assess DILI in humans while the assessment of DILI in preclinical drug development is heavily dependent upon hepatic histological interpretation (Moggs et al., 2012). These tests primarily consist of the determination of serum total bilirubin (TBL) concentration and the activity of the enzymes alkaline phosphatase (ALP), aspartate aminotransferase (AST), and alanine aminotransferase (ALT). Elevations in the activity of these enzymes may indicate injury to biliary cells or hepatocytes, while elevations of TBL concentration represent declining hepatic function (Antoine et al., 2009a). Although the assay of ALT has become the primary screening tool to detect DILI, it is not without uncertainties. Changes in these enzymatic activities are not specific for DILI since they can occur in a number of disease processes, including viral hepatitis, fatty liver disease, and liver cancer (Ozer et al., 2008). Nor are elevations in the aminotransferases unique to liver injury since increases in ALT and AST concentrations in circulation can also result from myocardial

damage, muscle damage, or extreme exercise. Elevations in ALP concentration are not specific for biliary injury as the marker may also be attributed to hyperthyroidism or bone disease. Furthermore, the analytical methods used to quantify ALT activity have not been standardized, and a robust definition of normal reference ranges has not been agreed upon; these ranges inevitably depend upon the population group defined as normal and other preanalytical variables (e.g., assay) that often inherently vary within and/or between laboratories. Although ALT activity is regarded as generally sensitive for detecting liver injury when it occurs, it is not sensitive with respect to temporal kinetics. Furthermore, ALT activity has often been described as having little prognostic value due to the fact that an ALT elevation represents probable injury to the liver post remarkable evidence of damage. From a regulatory point of view, elevations in ALT activity are also troublesome with respect to establishing liver safety during drug treatment. However, frequent and relatively large elevations in ALT activity are associated with treatments that do not pose a clinical liver safety issues, including drugs such as heparins and tacrine (Watkins et al., 1994; Harrill et al., 2012). The challenge is to identify or assess current biomarkers (on their own or in combination) that can distinguish between benign elevations in ALT activity and the potential for a serious DILI outcome.

ALT activity is often combined with the liver-specific assessment of TBL as part of Hy's law. Hy Zimmerman first noted that a patient who presents with jaundice as a result of hepatocellular DILI has at least a 10% chance of developing ALF, regardless of which drug has caused hepatocellular injury (Zimmerman, 1968). Hy's law is currently the only accepted regulatory model to assess significant, acute DILI (Watkins et al., 2011; Senior, 2012). The FDA has developed a liver safety data management tool—evaluation of drug-induced serious liver injury (*eDISH*), which involves data visualization by plotting the peak serum ALT versus the peak serum TBL for each subject in a clinical trial (Watkins et al., 2011). Although eDISH has revolutionized the standardization and transparent means of displaying and organizing relevant liver safety data from a clinical trial, it remains limited by its reliance on Hy's law. TBL most likely rises once there has only been a substantial loss of functioning hepatocytes, placing the patient a greater risk of liver failure. Therefore, serum TBL in this setting is not a biomarker that predicts severe hepatotoxicity potentials but instead serves as confirmation that severe hepatotoxicity has occurred. An ideal biomarker would predict the liver safety profile of the drug and the patient's biomarker response(s) before liver injury has progressed to the point of clinically relevant TBL elevations. Therefore, the diagnosis of a Hy's law case is often delayed, and death of the patient may still occur despite discontinuation of drug(s) even after relatively subtle liver toxicity signals (Watkins, 2011). This further highlights the need for improvement and modification to

Hy's law criteria where novel biomarkers can fill this capabilities gap. Due to the difficulties in defining a true Hy's law case, defined by these biochemical characteristics, an international expert working group of clinicians and scientists reviewed current terminology and diagnostic criteria for DILI based on currently used clinical chemistry parameters to enable uniformed criteria to define a case as DILI and to characterize the spectrum of clinical patterns encompassing it (Zhang et al., 2011). Currently, a liver biopsy is still the definitive form of hepatic tissue injury diagnosis. Consequently, with the exception of APAP (hallmarks such as APAP adducts; *N*-acetylcysteine), there is no specific noninvasive diagnostic DILI measure, treatment, and/or prevention of DILI except for the early withdrawal of a drug in the case of suspected DILI. Therefore, there has been considerable remit to identify and develop new biomarkers that can inform the mechanistic basis of DILI and provide potential measures for patient management.

27.4 NOVEL INVESTIGATIONAL BIOMARKERS FOR DILI

The identification and eventual qualification of sensitive and specific DILI biomarkers that hold translational promise between preclinical and clinical studies is urgently required to actualize improved safety screening in drug development and sensitive clinical diagnosis of DILI for patient treatment stratification. An added benefit of novel biomarkers would be to provide enhanced understanding of the fundamental mechanisms that result in clinical DILI.

Significant resources have been directed toward the qualification of next-generation DILI biomarkers, and a number of public–private consortia have been developed, namely, the Predictive Safety Testing Consortium (PSTC) in the United States and the Safer And Faster Evidence-based Translation (SAFE-T) consortium in Europe, which are currently funding prospective preclinical and clinical DILI biomarker studies (Matheis et al., 2011). To synergize efforts and minimize overlap, these consortia collaborate closely in what has been described as a highly efficient scientific "meta-consortium" on a global scale (Zhang et al., 2011).

The strategies for biomarker discovery broadly fall into two categories, *unbiased analysis* and *targeted analysis*. Unbiased approaches driven by omic technologies to integrate biological samples are an excellent mechanism to identify novel biomarkers and develop testable theories and have been reviewed with the specific focus on hepatotoxicity (Coen, 2010; van Summeren et al., 2012). Furthermore efforts have been reported to describe their utility to define proteomic profiles that relate to idiosyncratic hepatotoxicity (Bell et al., 2012). However, these strategies are very important but are often scientifically challenging because of the paucity of well-defined clinical samples for a particular drug

and because of heterogeneous sample sets and disease manifestations. The alternative is to use model systems to investigate target analytes in suitable biofluids as biomarkers where the chemistry of the drug/molecule reflects the biology of the cell/tissue/organ and where it is possible to investigate the mechanism of appearance from the subcellular milieu into relevant biological matrices and subsequent clearance. This concept forms the basis of the Innovative Medicines Initiative Mechanism-Based Integrated Systems for the Prediction of Drug-Induced Liver Injury (MIP-DILI) consortium (www.mip-dili.eu). Here, within this review, we focus on specific mechanism-based biomarkers that have shown evidence of utility in both preclinical and clinical DILI studies and are of interest to aforementioned biomarkers and ongoing consortia activities.

Notably, progress has been made in the development of novel biomarkers for renal drug safety evaluations (Bonventre et al., 2010; Vaidya et al., 2010), which have been qualified for use by various regulatory authorities (FDA, EMA, and Pharmaceuticals and Medical Devices Agency (PMDA), Japan), and the lessons learned from these efforts can apply to the liver, in particular the concept where newer biomarkers are used to complement existing ones and not to replace them to inform stakeholders including medicinal chemists, clinicians, toxicologists, regulators, and the patient/public interests. Novel biomarkers of DILI that have shown clinical utility will be discussed later; these parameters have enhanced sensitivity and specificity for liver injury or offer mechanistic insights into the pathogenic processes that result in DILI.

27.4.1 Glutamate Dehydrogenase

Glutamate dehydrogenase (GLDH) is an enzyme present in matrix-rich mitochondria (liver) and not in crista-rich mitochondria (cardiac and skeletal muscle) (Feldman, 1989). It is important to note that while GLDH is also expressed in the brain and in the kidney, its release from these tissues enters the cerebrospinal fluid and tubular lumen respectively, rather than the blood (Feldman, 1989; Stonard, 1996). GLDH is a key enzyme in amino acid oxidation and urea production that is highly conserved across species, making it an attractive biomarker candidate (Schmidt and Schmidt, 1988). It is considered "relatively" liver specific and may serve an indicator for leakage of hepatic mitochondrial contents into the circulation (Antoine et al., 2009a). GLDH localization within the liver is regional, with higher concentrations present in the centrilobular area, the region of metabolic activation and site of tissue damage (e.g., during APAP hepatotoxicity). GLDH use as a DILI biomarker is well documented and appears more sensitive and indicative of DILI than other cytosolic enzymes (Schomaker et al., 2013). A recent study in rats subjected to multiple liver injury modalities indicated that GLDH increases were up to 10-fold greater and 3-fold more persistent than ALT elevations (O'Brien et al., 2002). Because

GLDH is localized to the mitochondrial matrix and its relative large size is approximately 330 kDa, release of GLDH into the circulation is delayed during hepatocellular necrosis when compared to cytosolic enzymes like the aminotransferases. This property may contribute to increased specificity of GLDH to indicate hepatocellular necrosis. GLDH is additionally elevated in ALI models as it is elevated in blood in both preclinical models and clinical cases of DILI and liver impairment (McGill et al., 2012; Antoine et al., 2013a, b; Schomaker et al., 2013), highlighting its potential as a translational biomarker. Circulating GLDH has also been shown to rise in healthy volunteers treated with heparins and cholestyramine, treatments that are not associated with clinically important liver injury (Harrill et al., 2012; Singhal et al., 2013), and specific recommendations have been made when drawing conclusions on data regarding sample type and specimen preparation (Jaeschke and McGill, 2013). Measurement of GLDH alone may or may not be value added in distinguishing benign elevations in ALT from those that portent severe DILI potential.

27.4.2 Acylcarnitines

Due to the fact that most large enzymes used to investigate mitochondrial dysfunction track changes in ALT activity (McGill et al., 2012), a promising approach to the identification of biomarkers of injury that are useful at earlier time points may involve metabolomics. In general, metabolic intermediates are much smaller than proteins and more likely to cross cell membranes and enter circulation before the development of liver injury. In 2009, Chen et al. (2009) measured increased levels of acylcarnitines in serum from APAP-treated mice. Acylcarnitines are derivatives of long-chain fatty acids, which are required for transport of these fatty acids into mitochondria for β-oxidation. First, a coenzyme A (CoA) group is attached in a reaction catalyzed by acyl-CoA synthetase. The CoA group is then displaced by carnitine through the action of carnitine palmitoyl transferase I (CPT I), forming an acylcarnitine that can enter the mitochondrial matrix through facilitated diffusion with the help of a carnitine–acylcarnitine translocase (CACT). Because acylcarnitines are broken down within mitochondria by carnitine palmitoyl transferase II (CPT II) and beta-oxidation, mitochondrial dysfunction may result in their accumulation. It was been shown that increases in these fatty acid–carnitine conjugates rise in the serum of mice treated with APAP (mitochondrial-dependent hepatocyte death), but not with mice treated with furosemide (which has been shown to cause liver injury without primarily affecting mitochondrial function) (McGill et al., 2014). Therefore, circulating acylcarnitines have potential as specific biomarkers of mitochondrial dysfunction. It is important to note that acylcarnitines have been shown to "not be elevated" in patients with APAP overdose (McGill et al., 2014). This is most likely due in part to the standard-of-care treatment

N-acetylcysteine (AC). However, it might be useful to measure acylcarnitines in other forms of liver injury or APAP-treated patients who present to the hospital prior to the initiation of anecdotal therapy using AC (Antoine et al., 2013a; Ferner et al., 2011; Dear et al., 2013).

27.4.3 High-Mobility Group Box-1 (HMGB1)

HMGB1 is a chromatin-binding protein that is passively released by cells undergoing necrosis where it acts as a damage-associated molecular pattern (DAMP) molecule by linking cell death to the activation of an immune response by targeting Toll-like receptors and the receptor for advanced glycation end products (RAGE) (Wang et al., 1999; Scaffidi et al., 2002; Yang et al., 2013). HMGB1 has activity at the intersection between infectious and sterile inflammation. It is also actively secreted as a cytokine by innate immune cells in a hyperacetylated form (Lu et al., 2012; Nystrom et al., 2013), and its biological function is highly dependent upon and is regulated by posttranslational redox modifications of three key cysteine residues (Venereau et al., 2012; Yang et al., 2012). Furthermore, a recently defined nomenclature has been developed to identify these functional relevant isoforms (Antoine et al., 2014). Acetylation of lysine residues is also important for the active release of HMGB1 from immune cells and for the release in cell death mechanisms such as pyroptosis or apoptosis associated with inflammatory response (Bonaldi et al., 2003; Lu et al., 2012; Nystrom et al., 2013; Lu et al., 2014). HMGB1 is an informative and early serum-based indicator of cell death processes in preclinical models of APAP poisoning (Antoine et al., 2010, 2012) and in the clinic (Antoine et al., 2013a; Dear et al., 2013). Circulating levels of total and acetylated HMGB1 displayed different temporal profiles, which, in mouse models of APAP toxicity, correlate with the onset of necrosis and inflammation, respectively (Antoine et al., 2009b). Serum levels of total HMGB1 correlate strongly with ALT activity and prothrombin time in patients with established ALI following APAP overdose (Antoine et al., 2012). The prognostic utility of acetylated HMGB1 has also been demonstrated in clinical DILI. In patients with established ALI following APAP overdose, elevations in acetylated HMGB1 associate with a poor prognosis and outcome (Antoine et al., 2012). Elevations in serum HMGB1 and a secondary rise in acetylated HMGB1 were also observed during treatment of healthy volunteers with heparins (Harrill et al., 2012). As well as being an important biomarker of APAP toxicity, conditional knockout animals for HMGB1 and novel therapeutic targeting of these signaling pathways have demonstrated its importance in the pathogenesis of the disease (Huebener et al., 2015; Yang et al., 2015). Furthermore, HMGB1 as a mechanistic player and biomarker has also been demonstrated in alcoholic liver disease (Ge et al., 2014) and preclinical and clinical cholestasis (Woolbright et al., 2013).

27.4.4 Keratin-18 (K18)

K18 is a type I intermediate filament protein expressed in epithelial cells and is responsible for cell structure and integrity (Ku et al., 2007). Caspase-mediated cleavage of K18 is an early event in cellular structural rearrangement during apoptosis (Caulin et al., 1997). Caspases 3, 7, and 9 have been implicated in the cleavage of K18 at the C-terminal DALD/S motif. Full-length K18 is released passively during necrotic cell death, whereas fragmented K18 is released with apoptosis (Schutte et al., 2004). The use of immunoassays directed toward the recognition of caspase-cleaved K18 (apoptosis) and full-length K18 (necrosis) has been reported clinically as biomarkers for the therapeutic drug monitoring of chemotherapeutic agents and for the quantification of apoptosis during liver disorders such as nonalcoholic steatohepatitis (NASH) and hepatitis C infection (Wieckowska et al., 2006; Cummings et al., 2008) and mutations in K18 predispose toward ALF and hepatotoxicity (Ku et al., 1996; Strnad et al., 2010). Circulating necrosis K18 and apoptosis K18 have been shown to represent indicators of hepatic necrotic and apoptotic events in a mouse model of APAP-induced liver injury (Antoine et al., 2009b) and during heparin-induced hepatocellular injury in humans (Harrill et al., 2012). The prognostic utility of K18 has also been demonstrated in clinical DILI and ALI (Bechmann et al., 2010; Antoine et al., 2012). In patients with established ALI following APAP overdose, elevations in absolute levels of necrosis K18 associate with a poor prognosis (KCC) and outcome and a total percentage of K18 attributed to apoptosis as associates with improved survival (Antoine et al., 2012). Interestingly, in the first blood sample taken at the point of admission following APAP overdose, when currently used markers of liver injury remained within the normal range and prior to antidote treatment, K18 and also miR-122 and HMGB1 were significantly elevated in the group of patients that subsequently went on to develop liver injury, even in patients that presented <8h post overdose (Antoine et al., 2013a). Furthermore the values of K18, HMGB1, and miR-122 at presentation correlated with the peak ALT activity and peak INR recorded during the hospital admission. Interestingly, these data are also supported by a recently published case report highlighting that life-threatening hepatotoxicity following APAP overdose could have potentially been avoided if these biomarkers had been measured (Dear et al., 2013). These data demonstrate for the first time in humans that these novel biomarkers represent more sensitive biomarkers of DILI in a temporal sense compared to currently used indicators and can be used to aid treatment stratification and identify risk.

27.4.5 MicroRNA-122 (miR-122)

MicroRNAs are small noncoding RNAs ~22–25 nucleotides in length, which predominantly serve to negatively regulate posttranscriptional gene expression. Circulating microRNAs are stable and provide disease state biomarkers spanning diverse therapeutic areas and have been associated with a wide range of tissue-specific toxicities (Starkey Lewis et al., 2012). Many microRNA species show a high degree of organ specificity and cross species conservation, which makes them attractive candidates as translational safety biomarkers (Zen and Zhang, 2012). MicroRNA-122 (miR-122) represents 75% of the total hepatic miRNA content and exhibits exclusive hepatic expression. miR-122 has been shown to be a serum biomarker of APAP-induced ALI in mice, which was more sensitive with respect to dose and time than ALT (Wang et al., 2009). The improved tissue specificity of miR-122 versus ALT is supported by the observation that clinical ALT elevations associated with muscle injury are not accompanied by concomitant elevations in miR-122 (Zhang et al., 2010). miR-122 has also been previously shown to serve as a clinical indicator of heparin-induced hepatocellular necrosis (Aithal et al., 1999; Harrill et al., 2012). Moreover, as observed in mice, miR-122 is elevated in blood following APAP overdose in man and correlates strongly with ALT activity in patients with established ALI. Furthermore miR-122 has been shown to represent a more sensitive biomarker of APAP hepatotoxicity in humans compared to routine clinical chemistry parameters (Antoine et al., 2013a; Dear et al., 2013). In these investigations, elevated miR-122 was observed in patients who present to hospital with normal liver function test values within the normal range but then later develop ALI compared to those that did not develop ALI following APAP overdose. Furthermore, lessons from healthy volunteer studies have also shown that increases in serum livers of miR-122 are associated with individuals that develop liver injury despite taking the therapeutic dose and that miR-122 rise at time points 24h before ALT activity (Thulin et al., 2014). In these APAP overdose studies, miR-122 correlated strongly with peak ALT levels (Starkey Lewis et al., 2011). Interestingly, serum miR-122 levels in APAP-ALI patients who satisfied King's College Criteria (KCC) for liver transplantation were higher than those who did not satisfy KCC. However, this did not reach statistical significance, potentially due to small patient numbers (Starkey Lewis et al., 2011). Further prospective and longitudinal biomarker studies in ALI patients will be required to determine whether miR-122 can provide added clinical prognostic value. The translational value of miR-122 as a sensitive circulating biomarker has also been demonstrated in an APAP overdose model in zebra fish (Vliegenthart et al., 2014). This represents an important observation for translational research and data interpretation given the increasing utility of this organism for earlier drug development studies. Despite the advantages of miR-12, future efforts should be coordinated to develop cross laboratory-validated methods for miRNA isolation and quantification as well as develop a consensus on normalization standards (Starkey Lewis et al., 2012).

27.5 *IN VITRO* MODELS AND THE PREDICTION OF HUMAN DILI

It is evident from the drug development failures of the past 30 years that only through a better mechanistic understanding of specific DILI will we improve our ability to predict clinical hepatotoxicity. Current *in vitro* models used for the prediction of DILI often suffer from uninformative, inappropriate, and poorly translatable end-point measurements, and this may lead to unreliable inferences being drawn. Moreover, there is currently a significant disconnect between *in vitro* DILI endpoints and those used to assess hepatotoxicity in a clinical setting through either traditional endpoints or novel biomarkers.

In vitro models used for early screening for potential hepatotoxic liability are predominantly based on single cell systems or on human tissue preparations. Liver homogenates, microsomes, and slices can all be prepared from human tissue obtained as surgical by-products, which are more readily available as a resource than in the past. While such models provide valuable insights into the metabolism and covalent binding of new compounds, it is difficult to estimate the propensity for cytotoxicity directly, and consequently most attention is now focussed on the use of intact cell models.

The three cell types currently used for *in vitro* hepatotoxicity testing are (i) primary human hepatocytes (PHHs), (ii) transformed liver cell lines, and (iii) induced pluripotent stem (iPS) cell-derived hepatocytes. PHHs are considered the gold standard model, but the scarcity of suitable tissue and limited life span of the cells in culture is a major issue for their routine use in toxicity screening assays. Furthermore, primary hepatocytes show a marked propensity to dedifferentiate under culture conditions, a feature that is particularly marked with respect to the expression of cytochrome P450 (CYP) enzymes, which decline dramatically almost immediately following isolation. Consequently, immortalized human hepatocyte cell lines, such as HepG2, HuH7, and HepaRG cells, have become popular due to the limitations of PHHs and because of interdonor variability of the primary cells. HepG2 cells exhibit a few of the characteristics of PHHs including expression of some metabolic enzymes and nuclear receptors and are used typically to screen the cytotoxicity potential of new chemical entities at the lead generation stage (Gerets et al., 2012). However, the low levels of CYPs in HepG2 cells are inappropriate for use in metabolite toxicity testing, which reduces their predictive power considerably, although culturing HepG2 cells in 3D spheroids can improve metabolic competence. To address this limitation, the HepaRG cell line was developed (Guillouzo et al., 2007) to provide for higher CYP expression and inducibility. Proteomic comparison of these cells with cryopreserved human hepatocytes, however, suggests that the HepaRG cells do not have uniformly high CYP levels, although the

CYP3A4 level is comparable to primary cells (University of Liverpool and the IMI MIP-DILI consortium, unpublished data), and these cells are generally less sensitive to DILI compounds than PHHs, based on cytotoxicity endpoints.

iPS cells and human embryonic stem cells (hESCs) offer exciting opportunities to generate all the cells of the liver, with a genetically identical background of interest. However, the phenotypes of these cells lack many mature hepatocyte features suggesting that the differentiation process and culture systems used require optimization (Kia et al., 2013). It is also unknown whether specific phenotypic and functional characteristics unique to the donor are maintained or lost in culture.

Traditional *in vitro* models based on 2D monolayer cell cultures are considered to be poorly representative of the *in vivo* environment due to their inability to form a normal morphology and do not fully replicate hepatocyte function *in vivo* (Godoy et al., 2013). Consequently, significant efforts have been undertaken to develop 3D models, which improve cell–cell and cell–matrix interactions and hepatocyte polarization, such as those described later. Such models offer the potential to better reflect the architecture of the liver and to develop more appropriate and relevant endpoints.

Spheroid (3D) culture systems hold promise with regard to improvements in hepatic phenotype. They are traditionally formed using hanging drop culture and can be both cocultured with other nonparenchymal liver cells, such as stellate cells, and miniaturized relatively easily (Riccalton-Banks et al., 2003; Abu-Absi et al., 2004; Thomas et al., 2005; Inamori et al., 2009; Godoy et al., 2013). Systems employing HepaRG cells, HepG2 cells, PHHs, and human liver cells have been shown to be stable for up to 28 days (Tostões et al., 2012; Gunness et al., 2013; Ramaiahgari et al., 2014) and suitable for high-throughput screening. However, spheroids lack a perfusion mechanism, which means that, as with standard 2D cultures, the media remains static and this does not accurately reflect the *in vivo* situation.

A recently described multicellular model that may offer promise is the formation of liver buds from pluripotent stem cells cultured with endothelial cells and mesenchymal stem cells (Takebe et al., 2013). This coculture system self-organizes into a 3D bud of cells with a complex vasculature and shows improved hepatic function compared to 2D controls. Despite this, the authors report a lack of biliary cell formation suggesting that this model, while being a step forward, requires independent reproduction and further development before fully recapturing the complex architecture of the liver.

Perfusion bioreactors present a highly complex and sophisticated approach for modeling drug-induced hepatotoxicity by offering opportunities to combine a multicellular and 3D hepatic cell system with structural and perfusion capabilities that better mimic the physiological conditions of the liver (Zeilinger et al., 2011). Human liver cell cultures in 3D formats placed in hollow fiber bioreactors have been shown to maintain drug-metabolizing (CYP) activities and gene expression during prolonged (3–4-week) cultures (Ramaiahgari et al., 2014). These models are capable of producing more reliable data, which may lend themselves to measurement of endpoints and markers with greater translational relevance for detecting of DILI. For an excellent review on bioreactor technologies in relation to liver cell culture (see reference Ebrahimkhani et al., 2014). There are however a number of disadvantages to using this technology. For example, the fibers that attempt to emulate the complex structural components of the liver are also thought to interact with certain classes of drugs, making them ineffective for modeling DILI in some instances (JJS, 2012; Ebrahimkhani et al., 2014). The major drawbacks to bioreactors are the large quantities of cells required and the high skill levels and hands-on time needed to use these systems, and these are therefore only truly usable and cost-effective when employed with cell lines such as HepaRG (Cerec et al., 2007; Darnell et al., 2011, 2012).

Currently, it is clear that there is no *in vitro* model (and therefore no endpoint or marker) that can recapitulate *in vitro*, the role of the adaptive immune system in DILI (Uetrecht and Naisbitt, 2013). While there has been some progress at modeling the interactions of the innate immune system with hepatocytes through coculture with either monocytes or Kupffer cells, this is still a work in progress and needs to be refined based on further elucidation of immune mechanisms underlying human DILI.

In summary, while the last 10 years has seen considerable developments and innovations in the use of *in vitro* toxicity models, particularly for predicting acute drug toxicity, there is still no single system that would predict acute hepatotoxicity induced by an agent such as APAP. Prediction of chronic hepatotoxicity is even less advanced, and until it is possible to incorporate components of both the innate and adaptive immune systems into the models, it is unlikely that these types of toxicity will be routinely detected preclinically.

27.6 CONCLUSIONS AND FUTURE PERSPECTIVES

Significant recent progress has been made and clinical utility has been shown regarding "mechanism-based" biomarkers such as acylcarnitines, HMGB1, K18, GLDH, and highly liver-specific markers such as miR-122. These candidate translational biomarkers may indicate progressive DILI, shed light on mechanistic aspects of clinical DILI, and/or can predict patient prognosis. To date, the vast majority of clinical DILI biomarker data have been obtained from studies of APAP-induced liver injury and have not been assessed in rare cases of idiosyncratic DILI. Yet, a clear knowledge gap still exists regarding the identification and development of

biomarkers that predict serious DILI and reflect hepatic regenerative processes. Therefore, appropriate and well-annotated tissue banks may enable ongoing efforts by the US-based Drug-Induced Liver Injury Network (DILIN) (Hoofnagle, 2004) and the Spanish Hepatotoxicity Registry (Andrade et al., 2006), which have been collecting serum and urine from all subjects in their registry. In scope are comprehensive histological evaluations of novel biomarkers (Kleiner et al., 2014). These subjects were enrolled only after the diagnosis of DILI is established; therefore, the utility of biospecimens collected will probably be limited to studies of biomarkers for diagnosis and management of DILI. The qualification of biomarkers that will predict individual subject susceptibility, or that can accurately and safely assess the liver liability of a new drug in development, will require the collection of biospecimens before the start of treatment and during treatment prior to the onset of clinically overt DILI. This effort will require the adoption of standardized liver safety databases, standardized protocols for biospecimen collection and storage, and the initiation of large prospective clinical trials, involving diverse disease populations and treatment with many different drugs. This should now become a high priority within the pharmaceutical industry.

The lessons learned from the preclinical qualification of renal safety biomarkers have demonstrated that it is also clear that no single biomarker will be the answer and that a panel approach of novel biomarkers alongside a more intelligent use of currently used biomarkers represents the way forward to inform all stakeholders. Moreover, novel translational biomarkers that reflect the mechanistic basis of DILI are fundamental to efforts in translational research. Despite significant progress in preclinical renal injury biomarker qualification, to date, clinical biomarker qualification studies are ongoing to achieve this objective.

Defining the context of use for novel biomarkers in man represents an important area of collaborative research interest. Understanding reference ranges for novel DILI biomarkers in preclinical species and their evaluation in diverse healthy human populations and liver disease cohorts is an important area of investigation and question to address. Further areas of research focus should also be targeted toward the generation of robust cross species bioanalytical assays that are standardized or point-of-care tests in parallel with a comprehensive understanding of cross species differences in biomarker expression, mechanisms of release, and clearance, distribution, and kinetics.

It is also important to understand the cost-effectiveness of a new biomarker and the added value when moving from an experimental tool to the clinical setting. Defining whether a biomarker is fit for the purpose of characterizing DILI, understanding the right biomarker assays to leverage, and defining at what particular time for an individual patient or at what stage during drug development are critical regarding the utilization of DILI biomarkers.

The qualification of novel DILI biomarkers will require application to biospecimens obtained from many different patient populations treated with many different drugs, both those that cause clinically important DILI and those that cause elevations in traditional liver chemistries but do not cause clinically important liver injury. It is important that pharmaceutical companies start now to archive samples and link these specimens to the relevant liver safety data. Ideally, liver safety data management tools should be standardized across the industry to facilitate the precompetitive collaborations on biomarker validation and qualification, such as eDISH (Watkins et al., 2011). Formal biomarker validation and qualification will warrant significant time to obtain regulatory-endorsed exploratory status via Letters of Support.

Serum ALT activity is the primary method to assess ALI associated with drugs. Despite not formally qualified against human histology for DILI, ALT has been widely regarded. Several limitations have been associated with ALT activity (Senior, 2012; Antoine et al., 2013b).

Over the past 5 years, a paradigm shift toward the thorough elevation of hepatic *in vitro* models has shown that currently available *in vitro* models and, moreover, the endpoints in use with these models lack sufficient sensitivity and specificity to allow meaningful *a priori* risk assessment of the hepatotoxic potential of candidate drugs in human. Considering the multiple molecular initiating events associated with drug liver injury, it is not realistic to suggest a single assay or test will capture all of the potential mechanisms presently described and allow an informed decision on the DILI risk potential of candidate drugs. Single endpoint cytotoxicity assays have poor concordance with *in vivo* preclinical and clinical readouts, most likely reflecting the measurement of a late event in the pathologic process of liver injury (Xu et al., 2004). In contrast to cytotoxicity endpoints, panels of molecular cell-based probes permit, by use of high-content cellular imaging, the concomitant time-resolved high-content analysis (HCA) of hepatocellular stress response and biochemical endpoints associated with molecular initiating events in liver injury (O'Brien et al., 2006; Xu et al., 2008). HCA permits a combination of endpoints, including cell loss, nuclear size, mitochondrial potential, DNA damage, apoptosis, GSH depletion, and reactive oxygen species (ROS) (Lang et al., 2006; Xu et al., 2008; Wink et al., 2014). Clearly, the success of HCA depends on the physiological relevance of the cell models that are used, and therefore HCA, while serving as a tool to flag candidate drugs for potential DILI risk, is limited by the availability of fully phenotypically and physiologically characterized cell systems. Even with fully functional *in vitro* models, there is always a level of uncertainty regarding the "predictivity" of the system in advance of human trials until efficacious doses, exposure, and the integration of the whole human response are present.

REFERENCES

Abu-Absi, S.F., L.K. Hansen, and W.S. Hu, Three-dimensional co-culture of hepatocytes and stellate cells. *Cytotechnology*, 2004. 45(3): p. 125–40.

Aithal, G.P., M.D. Rawlins, and C.P. Day, Accuracy of hepatic adverse drug reaction reporting in one English health region. *BMJ*, 1999. 319(7224): p. 1541.

Aithal, G.P., et al., Case definition and phenotype standardization in drug-induced liver injury. *Clin Pharmacol Ther*, 2011. 89(6): p. 806–15.

Andrade, R.J., et al., Outcome of acute idiosyncratic drug-induced liver injury: long-term follow-up in a hepatotoxicity registry. *Hepatology*, 2006. 44(6): p. 1581–8.

Antoine, D.J., et al., Mechanism-based bioanalysis and biomarkers for hepatic chemical stress. *Xenobiotica*, 2009a. 39(8): p. 565–77.

Antoine, D.J., et al., High-mobility group box-1 protein and keratin-18, circulating serum proteins informative of acetaminophen-induced necrosis and apoptosis in vivo. *Toxicol Sci*, 2009b. 112(2): p. 521–31.

Antoine, D.J., et al., Diet restriction inhibits apoptosis and HMGB1 oxidation and promotes inflammatory cell recruitment during acetaminophen hepatotoxicity. *Mol Med*, 2010. 16(11–12): p. 479–90.

Antoine, D.J., et al., Molecular forms of HMGB1 and keratin-18 as mechanistic biomarkers for mode of cell death and prognosis during clinical acetaminophen hepatotoxicity. *J Hepatol*, 2012. 56(5): p. 1070–9.

Antoine, D.J., et al., Mechanistic biomarkers provide early and sensitive detection of acetaminophen-induced acute liver injury at first presentation to hospital. *Hepatology*, 2013a. 58(2): p. 777–87.

Antoine, D.J., et al., Are we closer to finding biomarkers for identifying acute drug-induced liver injury? *Biomark Med*, 2013b. 7(3): p. 383–6.

Antoine, D.J., et al., A systematic nomenclature for the redox states of high mobility group box (HMGB) proteins. *Mol Med*, 2014. 20(1): p. 135–7.

Bechmann, L.P., et al., Cytokeratin 18-based modification of the MELD score improves prediction of spontaneous survival after acute liver injury. *J Hepatol*, 2010. 53(4): p. 639–47.

Bell, L.N., et al., Serum proteomic profiling in patients with drug-induced liver injury. *Aliment Pharmacol Ther*, 2012. 35(5): p. 600–12.

Biomarkers Definitions Working Group, Biomarkers and surrogate endpoints: preferred definitions and conceptual framework. *Clin Pharmacol Ther*, 2001. 69(3): p. 89–95.

Bonaldi, T., et al., Monocytic cells hyperacetylate chromatin protein HMGB1 to redirect it towards secretion. *EMBO J*, 2003. 22(20): p. 5551–60.

Bonventre, J.V., et al., Next-generation biomarkers for detecting kidney toxicity. *Nat Biotechnol*, 2010. 28(5): p. 436–40.

Caulin, C., G.S. Salvesen, and R.G. Oshima, Caspase cleavage of keratin 18 and reorganization of intermediate filaments during epithelial cell apoptosis. *J Cell Biol*, 1997. 138(6): p. 1379–94.

Cerec, V., et al., Transdifferentiation of hepatocyte-like cells from the human hepatoma HepaRG cell line through bipotent progenitor. *Hepatology*, 2007. 45(4): p. 957–67.

Chen, C., et al., Serum metabolomics reveals irreversible inhibition of fatty acid beta-oxidation through the suppression of PPARalpha activation as a contributing mechanism of acetaminophen-induced hepatotoxicity. *Chem Res Toxicol*, 2009. 22(4): p. 699–707.

Coen, M., A metabonomic approach for mechanistic exploration of pre-clinical toxicology. *Toxicology*, 2010. 278(3): p. 326–40.

Corsini, A., et al., Current challenges and controversies in drug-induced liver injury. *Drug Saf*, 2012. 35(12): p. 1099–117.

Cummings, J., et al., Preclinical evaluation of M30 and M65 ELISAs as biomarkers of drug induced tumor cell death and antitumor activity. *Mol Cancer Ther*, 2008. 7(3): p. 455–63.

Darnell, M., et al., Cytochrome P450-dependent metabolism in HepaRG cells cultured in a dynamic three-dimensional bioreactor. *Drug Metab Dispos*, 2011. 39(7): p. 1131–8.

Darnell, M., et al., In vitro evaluation of major in vivo drug metabolic pathways using primary human hepatocytes and HepaRG cells in suspension and a dynamic three-dimensional bioreactor system. *J Pharmacol Exp Ther*, 2012. 343(1): p. 134–44.

Dear, J.W. and D.J. Antoine, Stratification of paracetamol overdose patients using new toxicity biomarkers: current candidates and future challenges. *Expert Rev Clin Pharmacol*, 2014. 7(2): p. 181–9.

Dear, J.W., et al., Letter to the editor: early detection of paracetamol toxicity using circulating liver microRNA and markers of cell necrosis. *Br J Clin Pharmacol*, 2013. doi:10.1111/bcp.12214.

Ebrahimkhani, M.R., et al., Bioreactor technologies to support liver function in vitro. *Adv Drug Deliv Rev*, 2014. 69–70: p. 132–57.

Feldman, B.F., Cerebrospinal Fluid, in *Clinical Biochemistry of Domestic Animals*, J. Kaneko, Editor. 1989, Academic Press: San Diego. p. 835–65.

Ferner, R.E., J.W. Dear, and D.N. Bateman, Management of paracetamol poisoning. *BMJ*, 2011. 342: p. d2218.

Ge, X., et al., High mobility group box-1 (HMGB1) participates in the pathogenesis of alcoholic liver disease (ALD). *J Biol Chem*, 2014. 289(33): p. 22672–91.

Gerets, H.H., et al., Characterization of primary human hepatocytes, HepG2 cells, and HepaRG cells at the mRNA level and CYP activity in response to inducers and their predictivity for the detection of human hepatotoxins. *Cell Biol Toxicol*, 2012. 28(2): p. 69–87.

Godoy, P., et al., Recent advances in 2D and 3D in vitro systems using primary hepatocytes, alternative hepatocyte sources and non-parenchymal liver cells and their use in investigating mechanisms of hepatotoxicity, cell signaling and ADME. *Arch Toxicol*, 2013. 87(8): p. 1315–530.

Guillouzo, A., et al., The human hepatoma HepaRG cells: a highly differentiated model for studies of liver metabolism and toxicity of xenobiotics. *Chem Biol Interact*, 2007. 168(1): p. 66–73.

Gunness, P., et al., 3D organotypic cultures of human HepaRG cells: a tool for in vitro toxicity studies. *Toxicol Sci*, 2013. 133(1): p. 67–78.

Harrill, A.H., et al., The effects of heparins on the liver: application of mechanistic serum biomarkers in a randomized study in healthy volunteers. *Clin Pharmacol Ther*, 2012. 92(2): p. 214–20.

Hoofnagle, J.H., Drug-induced liver injury network (DILIN). *Hepatology*, 2004. 40(4): p. 773.

Huebener, P., et al., The HMGB1/RAGE axis triggers neutrophil-mediated injury amplification following necrosis. *J Clin Invest*, 2015. 125(2): p. 539–50.

Inamori, M., H. Mizumoto, and T. Kajiwara, An approach for formation of vascularized liver tissue by endothelial cell-covered hepatocyte spheroid integration. *Tissue Eng Part A*, 2009. 15(8): p. 2029–37.

Jaeschke, H. and M.R. McGill, Serum glutamate dehydrogenase—biomarker for liver cell death or mitochondrial dysfunction? *Toxicol Sci*, 2013. 134(1): p. 221–2.

JJS, C., The hollow fiber infection model for antimicrobial pharmacodynamics and pharmacokinetics. *Adv Pharmacoepidem Drug Saf*, 2012. S1: p. 007.

Jollow, D.J., et al., Acetaminophen-induced hepatic necrosis. II. Role of covalent binding in vivo. *J Pharmacol Exp Ther*, 1973. 187(1): p. 195–202.

Kaplowitz, N., Idiosyncratic drug hepatotoxicity. *Nat Rev Drug Discov*, 2005. 4(6): p. 489–99.

Kia, R., et al., Stem cell-derived hepatocytes as a predictive model for drug-induced liver injury: are we there yet? *Br J Clin Pharmacol*, 2013. 75(4): p. 885–96.

Kleiner, D.E., et al., Hepatic histological findings in suspected drug-induced liver injury: systematic evaluation and clinical associations. *Hepatology*, 2014. 59(2): p. 661–70.

Ku, N.O., et al., Susceptibility to hepatotoxicity in transgenic mice that express a dominant-negative human keratin 18 mutant. *J Clin Invest*, 1996. 98(4): p. 1034–46.

Ku, N.O., et al., Keratins let liver live: mutations predispose to liver disease and crosslinking generates Mallory-Denk bodies. *Hepatology*, 2007. 46(5): p. 1639–49.

Lang, P., et al., Cellular imaging in drug discovery. *Nat Rev Drug Discov*, 2006. 5(4): p. 343–56.

Lee, W.M., Drug-induced hepatotoxicity. *N Engl J Med*, 2003. 349(5): p. 474–85.

Lee, W.M., et al., Acute liver failure: summary of a workshop. *Hepatology*, 2008. 47(4): p. 1401–15.

Lu, B., et al., Novel role of PKR in inflammasome activation and HMGB1 release. *Nature*, 2012. 488(7413): p. 670–4.

Lu, B., et al., JAK/STAT1 signaling promotes HMGB1 hyperacetylation and nuclear translocation. *Proc Natl Acad Sci U S A*, 2014. 111(8): p. 3068–73.

Matheis, K., et al., A generic operational strategy to qualify translational safety biomarkers. *Drug Discov Today*, 2011. 16(13–14): p. 600–8.

McGill, M.R., et al., The mechanism underlying acetaminophen-induced hepatotoxicity in humans and mice involves mitochondrial damage and nuclear DNA fragmentation. *J Clin Invest*, 2012. 122(4): p. 1574–83.

McGill, M.R., et al., Circulating acylcarnitines as biomarkers of mitochondrial dysfunction after acetaminophen overdose in mice and humans. *Arch Toxicol*, 2014. 88(2): p. 391–401.

Meier, Y., et al., Incidence of drug-induced liver injury in medical inpatients. *Eur J Clin Pharmacol*, 2005. 61(2): p. 135–43.

Mitchell, J.R., et al., Acetaminophen-induced hepatic necrosis. I. Role of drug metabolism. *J Pharmacol Exp Ther*, 1973a. 187(1): p. 185–94.

Mitchell, J.R., et al., Acetaminophen-induced hepatic necrosis. IV. Protective role of glutathione. *J Pharmacol Exp Ther*, 1973b. 187(1): p. 211–7.

Moggs, J., et al., Investigative safety science as a competitive advantage for Pharma. *Expert Opin Drug Metab Toxicol*, 2012. 8(9): p. 1071–82.

NHS, H.E.S, http://www.hesonline.nhs.uk/Ease/servlet/Content Server?siteID=1937. 2011 (accessed October 22, 2015).

Nystrom, S., et al., TLR activation regulates damage-associated molecular pattern isoforms released during pyroptosis. *EMBO J*, 2013. 32(1): p. 86–99.

O'Brien, P.J., et al., Advantages of glutamate dehydrogenase as a blood biomarker of acute hepatic injury in rats. *Lab Anim*, 2002. 36(3): p. 313–21.

O'Brien, P.J., et al., High concordance of drug-induced human hepatotoxicity with in vitro cytotoxicity measured in a novel cell-based model using high content screening. *Arch Toxicol*, 2006. 80(9): p. 580–604.

Olson, H., et al., Concordance of the toxicity of pharmaceuticals in humans and in animals. *Regul Toxicol Pharmacol*, 2000. 32(1): p. 56–67.

Ozer, J., et al., The current state of serum biomarkers of hepatotoxicity. *Toxicology*, 2008. 245(3): p. 194–205.

Park, B.K., et al., The role of metabolic activation in drug-induced hepatotoxicity. *Annu Rev Pharmacol Toxicol*, 2005. 45: p. 177–202.

Park, B.K., et al., Drug bioactivation and protein adduct formation in the pathogenesis of drug-induced toxicity. *Chem Biol Interact*, 2011. 192(1–2): p. 30–6.

Potter, W.Z., et al., Acetaminophen-induced hepatic necrosis. 3. Cytochrome P-450-mediated covalent binding in vitro. *J Pharmacol Exp Ther*, 1973. 187(1): p. 203–10.

Ramaiahgari, S.C., et al., A 3D in vitro model of differentiated HepG2 cell spheroids with improved liver-like properties for repeated dose high-throughput toxicity studies. *Arch Toxicol*, 2014. 88(5): p. 1083–95.

Ratner, M., FDA pharmacogenomics guidance sends clear message to industry. *Nat Rev Drug Discov*, 2005. 4(5): p. 359.

Riccalton-Banks, L., et al., Long-term culture of functional liver tissue: three-dimensional coculture of primary hepatocytes and stellate cells. *Tissue Eng*, 2003. 9(3): p. 401–10.

Scaffidi, P., T. Misteli, and M.E. Bianchi, Release of chromatin protein HMGB1 by necrotic cells triggers inflammation. *Nature*, 2002. 418(6894): p. 191–5.

Schmidt, E.S. and F.W. Schmidt, Glutamate dehydrogenase: biochemical and clinical aspects of an interesting enzyme. *Clin Chim Acta*, 1988. 173(1): p. 43–55.

Schomaker, S., et al., Assessment of emerging biomarkers of liver injury in human subjects. *Toxicol Sci*, 2013. 132(2): p. 276–83.

Schutte, B., et al., Keratin 8/18 breakdown and reorganization during apoptosis. *Exp Cell Res*, 2004. 297(1): p. 11–26.

Senior, J.R., Alanine aminotransferase: a clinical and regulatory tool for detecting liver injury-past, present, and future. *Clin Pharmacol Ther*, 2012. 92(3): p. 332–9.

Sgro, C., et al., Incidence of drug-induced hepatic injuries: a French population-based study. *Hepatology*, 2002. 36(2): p. 451–5.

Singhal, R., et al., Mechanistic biomarkers of liver safety and benign elevations in serum aminotransferases—a study in healthy volunteer treated with cholestyramine. *BMC Pharmacology and Toxicology*, 2014. 15:42.

Starkey Lewis, P.J., et al., Circulating microRNAs as potential markers of human drug-induced liver injury. *Hepatology*, 2011. 54(5): p. 1767–76.

Starkey Lewis, P.J., et al., Serum microRNA biomarkers for drug-induced liver injury. *Clin Pharmacol Ther*, 2012. 92(3): p. 291–3.

Stonard, M.D., Assessment of Nephrotoxicity, in *Animal Clinical Chemistry*, G.O. Evans, Editor. 1996, Taylor & Francis: London. p. 87–9.

Strnad, P., et al., Keratin variants predispose to acute liver failure and adverse outcome: race and ethnic associations. *Gastroenterology*, 2010. 139(3): p. 828–35, 835 e1–3.

Takebe, T., et al., Vascularized and functional human liver from an iPSC-derived organ bud transplant. *Nature*, 2013. 499(7459): p. 481–4.

Thomas, R.J., et al., The effect of three-dimensional co-culture of hepatocytes and hepatic stellate cells on key hepatocyte functions in vitro. *Cells Tissues Organs*, 2005. 181(2): p. 67–79.

Thulin, P., et al., Keratin-18 and microRNA-122 complement alanine aminotransferase as novel safety biomarkers for drug-induced liver injury in two human cohorts. *Liver Int*, 2014. 34: p. 367–78.

Tostões, R.M., et al., Human liver cell spheroids in extended perfusion bioreactor culture for repeated-dose drug testing. *Hepatology*, 2012. 55(4): p. 1227–36.

Uetrecht, J. and D.J. Naisbitt, Idiosyncratic adverse drug reactions: current concepts. *Pharmacol Rev*, 2013. 65(2): p. 779–808.

Vaidya, V.S., et al., Kidney injury molecule-1 outperforms traditional biomarkers of kidney injury in preclinical biomarker qualification studies. *Nat Biotechnol*, 2010. 28(5): p. 478–85.

Van Summeren, A., et al., Proteomics in the search for mechanisms and biomarkers of drug-induced hepatotoxicity. *Toxicol In Vitro*, 2012. 26(3): p. 373–85.

Venereau, E., et al., Mutually exclusive redox forms of HMGB1 promote cell recruitment or proinflammatory cytokine release. *J Exp Med*, 2012. 209(9): p. 1519–28.

Vliegenthart, A.D., et al., Retro-orbital blood acquisition facilitates circulating microRNA measurement in zebrafish with paracetamol hepatotoxicity. *Zebrafish*, 2014. 11: p. 219–26.

Vuppalanchi, R., S. Liangpunsakul, and N. Chalasani, Etiology of new-onset jaundice: how often is it caused by idiosyncratic drug-induced liver injury in the United States? *Am J Gastroenterol*, 2007. 102(3): p. 558–62; quiz 693.

Wang, H., et al., HMG-1 as a late mediator of endotoxin lethality in mice. *Science*, 1999. 285(5425): p. 248–51.

Wang, K., et al., Circulating microRNAs, potential biomarkers for drug-induced liver injury. *Proc Natl Acad Sci U S A*, 2009. 106(11): p. 4402–7.

Watkins, P.B., Drug safety sciences and the bottleneck in drug development. *Clin Pharmacol Ther*, 2011. 89(6): p. 788–90.

Watkins, P.B., et al., Hepatotoxic effects of tacrine administration in patients with Alzheimer's disease. *JAMA*, 1994. 271(13): p. 992–8.

Watkins, P.B., et al., Evaluation of drug-induced serious hepatotoxicity (eDISH): application of this data organization approach to phase III clinical trials of rivaroxaban after total hip or knee replacement surgery. *Drug Saf*, 2011. 34(3): p. 243–52.

Wieckowska, A., et al., In vivo assessment of liver cell apoptosis as a novel biomarker of disease severity in nonalcoholic fatty liver disease. *Hepatology*, 2006. 44(1): p. 27–33.

Williams, C.D. and H. Jaeschke, Role of innate and adaptive immunity during drug-induced liver injury. *Toxicol Res*, 2012. 1: p. 161–70.

Wink, S., et al., Quantitative high content imaging of cellular adaptive stress response pathways in toxicity for chemical safety assessment. *Chem Res Toxicol*, 2014. 27: p. 338–55.

Woolbright, B.L., et al., Plasma biomarkers of liver injury and inflammation demonstrate a lack of apoptosis during obstructive cholestasis in mice. *Toxicol Appl Pharmacol*, 2013. 273(3): p. 524–31.

Xu, J.J., D. Diaz, and P.J. O'Brien, Applications of cytotoxicity assays and pre-lethal mechanistic assays for assessment of human hepatotoxicity potential. *Chem Biol Interact*, 2004. 150(1): p. 115–28.

Xu, J.J., et al., Cellular imaging predictions of clinical drug-induced liver injury. *Toxicol Sci*, 2008. 105(1): p. 97–105.

Yang, H., et al., Redox modification of cysteine residues regulates the cytokine activity of high mobility group box-1 (HMGB1). *Mol Med*, 2012. 18(1): p. 250–9.

Yang, H., et al., The many faces of HMGB1: molecular structure-functional activity in inflammation, apoptosis, and chemotaxis. *J Leukoc Biol*, 2013. 93: p. 865–73.

Yang, H., et al., MD-2 is required for disulfide HMGB1-dependent TLR4 signaling. *J Exp Med*, 2015. 212(1): p. 5–14.

Zeilinger, K., et al., Scaling down of a clinical three-dimensional perfusion multicompartment hollow fiber liver bioreactor developed for extracorporeal liver support to an analytical scale device useful for hepatic pharmacological in vitro studies. *Tissue Eng Part C Methods*, 2011. 17(5): p. 549–56.

Zen, K. and C.Y. Zhang, Circulating microRNAs: a novel class of biomarkers to diagnose and monitor human cancers. *Med Res Rev*, 2012. 32(2): p. 326–48.

Zhang, Y., et al., Plasma microRNA-122 as a biomarker for viral-, alcohol-, and chemical-related hepatic diseases. *Clin Chem*, 2010. 56(12): p. 1830–8.

Zhang, X., et al., Involvement of the immune system in idiosyncratic drug reactions. *Drug Metab Pharmacokinet*, 2011. 26(1): p. 47–59.

Zimmerman, H.J., The spectrum of hepatotoxicity. *Perspect Biol Med*, 1968. 12(1): p. 135–61.

PART VIII

KIDNEY INJURY BIOMARKERS

28

ASSESSING AND PREDICTING DRUG-INDUCED KIDNEY INJURY, FUNCTIONAL CHANGE, AND SAFETY IN PRECLINICAL STUDIES IN RATS

YAFEI CHEN

Mechanistic & Investigative Toxicology, Discovery Sciences, Janssen Research & Development, L.L.C., San Diego, CA, USA

28.1 INTRODUCTION

Drug-induced kidney injury (DIKI) or nephrotoxicity contributes to ~20% of all cases of acute kidney injury (AKI) in clinical practice (Luyckx and Naicker, 2008). In preclinical safety assessment of candidate drugs, DIKI represents one of the main toxicities causing drug failure (~9%) or adverse drug reactions (~21%) in the late development stage (Naughton, 2008; Redfern et al., 2010). Current regulatory standards for the clinical diagnosis of nephrotoxicity and kidney disease primarily rely on the detection of elevated serum creatinine (sCr) and/or blood urea nitrogen (BUN) concentrations. Preclinical investigation of drug candidate effects on renal function such as global glomerular filtration rate (GFR) and concentrating ability in rats is recommended by the International Conference on Harmonization (ICH) 7A safety pharmacology guideline when renal safety is suspect based on "cause for concern." Routine urinalysis including urine volume, specific gravity, osmolality, pH, fluid/electrolyte balance, proteins, and cytology and standard blood chemistry determinations including sCr, BUN, and various blood-based proteins are recommended as standard end points to be measured (Anonymous, 2001). However, all of these biomarkers are dependent upon a decrement in overall kidney function, may not be specific to the kidney, and do not distinguish between drug-induced direct tubule epithelial damage, kidney tissue stress-repair responses, or other mechanisms (e.g., prerenal hemodynamic changes, postrenal obstruction) that can cause diminished global GFR and/or tubular reabsorption/secretion processes (Lameire et al.,

2008; Olson et al., 2009). The current gold standard parameters are sCr and BUN, which are insensitive and inadequate to detect especially mild to moderate or transient renal injury after acute nephrotoxic drug insult (Kellum et al., 2002; Bonventre et al., 2010). Recent studies have demonstrated that increased sCr and BUN concentrations may be predictive of kidney injury only when nearly or more than 50% of the functional nephron capacity has been lost and the kidneys are unable to regulate fluid and electrolyte homeostasis (Lameire et al., 2008; Olson et al., 2009; Bonventre et al., 2010). In addition, an increase in the sCr and BUN concentrations represents a delayed indication of a functional decrement in GFR that lags behind morphological changes in the kidney during the early stage of AKI. Furthermore, an elevated sCr and BUN may be unrelated to drug-induced nephrotoxicity due to physiological states and multiple nonrenal factors such as dehydration or muscle damage (Harpur et al., 2011; Rouse et al., 2012). Taken together, given the limitations of traditional kidney injury indices (Fig. 28.1), more sensitive and nephron segment-specific biomarkers for detection of DIKI in preclinical and clinical research, specifically, novel biomarkers with the ability to effectively measure subtle effects on nephron integrity prior to notable decline of kidney function and tissue damage, are needed.

The early detection of DIKI using more sensitive and nephron site-specific urinary protein biomarkers suggests a promising approach for preclinical evaluation of compounds and safety monitoring in clinical trials (Vaidya et al., 2008; Olson et al., 2009; Bonventre et al., 2010; Urbschat et al., 2011). These small urinary proteins have been extensively

Drug Discovery Toxicology: From Target Assessment to Translational Biomarkers, First Edition. Edited by Yvonne Will, J. Eric McDuffie, Andrew J. Olaharski, and Brandon D. Jeffy.

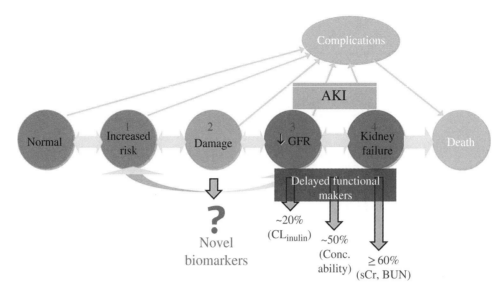

FIGURE 28.1 Insensitive, delayed, and nonspecific traditional kidney injury/functional markers on a conceptual framework of acute kidney injury (AKI), which is divided into several stages from the increased risk to damage, followed by a decrease in glomerular filtration rate (GFR), with progressing to kidney failure and death. Early, sensitive, and specific AKI biomarkers are not available or well qualified for AKI detection and safety assessment. ? indicates there are no biomarkers available for early AKI detection. Adapted from Lameire et al. (2008, pp. 392–402). (*See insert for color representation of the figure.*)

investigated over the past decade and are emerging as new-generation biomarkers for early identification of DIKI. The US Food and Drug Administration (US FDA), European Medicines Agency (EMA), and Japanese Pharmaceuticals and Medical Devices Agency (PMDA) have supported the qualification of eight novel urinary DIKI biomarkers used for preclinical safety evaluation in rats (Dieterle et al., 2010a; Ward et al., 2013; Charlton et al., 2014). These new renal safety biomarkers were qualified by the Critical Path Institute's Predictive Safety Testing Consortium (PSTC), Nephrotoxicity Working Group (Dieterle et al., 2010a; Ward et al., 2013) and the International Life Sciences Institute Health and Environmental Safety Institute (ILSI-HESI), Biomarkers of Nephrotoxicity Project Group (Harpur et al., 2011) for monitoring compound-induced renal injury in rats. The qualified DIKI urinary biomarkers, including total protein, albumin, kidney injury molecule-1 (KIM-1), clusterin, β2-microglobulin, cystatin C, trefoil factor-3 (TFF3), and renal papillary antigen-1 (RPA-1), may outperform the traditional biomarkers for particular uses in the context of nonclinical drug development to help guide safety assessments for regulatory decision-making (Tonomura et al., 2010). Other renal injury biomarkers including urinary neutrophil gelatinase-associated lipocalin (NGAL), osteopontin (OPN), α-glutathione S-transferase (α-GST), mu gamma glutathione S-transferase (GSTYb1), N-Acetyl-β-D-glucosaminidase (NAG), L-type fatty acid binding protein (L-FABP), interleukin-18 (IL-18), and calbindin D28 have been reported and recognized as candidate nephrotoxicity biomarkers (Haase et al., 2009; Bonventre et al., 2010; Lin et al., 2015; Xin et al., 2008), as listed in Table 28.1. The FDA also

TABLE 28.1   Regulatory Approved and Candidate Biomarkers of Acute Kidney Injury

Regulatory Agencies-Approved Novel Rat DIKI Biomarkers	Candidate DIKI Biomarkers (Unapproved by Regulatory Agencies)
Total protein	Urinary neutrophil gelatinase-associated lipocalin (NGAL)
Albumin	Osteopontin (OPN)
Kidney injury molecule-1 (KIM-1)	α-Glutathione S-transferase (α-GST)
Clusterin	Mu gamma glutathione S-transferase (GSTYb1)
β2-microglobulin	N-acetyl-β-D-glucosaminidase (NAG)
Cystatin C	L-type fatty acid-binding protein (L-FABP)
Trefoil factor-3 (TFF3)	Interleukin-18 (IL-18)
Renal papillary antigen-1 (RPA-1)	Calbindin D28

released new guidance for qualifying biomarkers intended to support the drug development process (FDA, 2014a).

In 2014, the US FDA and EMA issued respective Letters of Support to encourage further nonclinical and exploratory clinical studies to evaluate the translatability for urinary NGAL and OPN (EMA, 2014; FDA, 2014b). Further nonclinical and clinical studies are warranted to demonstrate the translatability of kidney safety biomarkers. Translation for DIKI biomarkers is a rapidly evolving area; the scope of this chapter is to overview preclinical rodent (rat) and nonrodent (canine and nonhuman primates) kidney safety biomarkers and their applications in nonclinical studies for the management of DIKI liabilities in drug discovery and development.

To improve our understanding of the pathogenesis and early detection of DIKI, it is important to differentiate drug-induced kidney tissue damage from functional changes such as diuretics, angiotensin-converting enzyme inhibitors, and aquaretics that may be characterized as experimental agents for kidney dysfunction rather than toxicity. Currently, routine histopathology is the sole "gold standard" by which DIKI is established in preclinical settings (Lameire et al., 2008; Olson et al., 2009). The discovery (e.g., miRNA) and qualification of novel, noninvasive urinary protein biomarkers and early detection of compound-induced toxicities in renal safety assessment are hallmarks, for example, early increases in urinary KIM-1 concurrent to progressive proximal tubular epithelial cell degeneration/necrosis (Bonventre et al., 2010). It has become a regulatory authority-endorsed practice to consider the standard renal histopathological examination, sCr, and BUN along with real-time changes in novel DIKI biomarker in the urine as key indicators when characterizing reversible pathological processes in different compartments of the rat kidney. These

biomarkers represent three general classes: (i) kidney functional markers, (ii) kidney tissue injury leakage markers, and (iii) kidney tissue injury and repair response markers, as depicted in Figure 28.2. These following sections discuss the role and utility of traditional and novel urinary biomarkers as investigative tools for the early detection of DIKI in preclinical safety studies. Some of these new biomarkers do not simply act as an indicator or messenger of DIKI; they may also execute protective effects on kidney injury or other physiological/pathological functions that currently are not fully understood (Olson et al., 2009; Haschek et al., 2013; Gupta, 2014).

28.2 KIDNEY FUNCTIONAL BIOMARKERS (GLOMERULAR FILTRATION AND TUBULAR REABSORPTION)

28.2.1 Traditional Functional Biomarkers

The most established and widely used kidney safety biomarkers are indices of renal function (GFR) including sCr

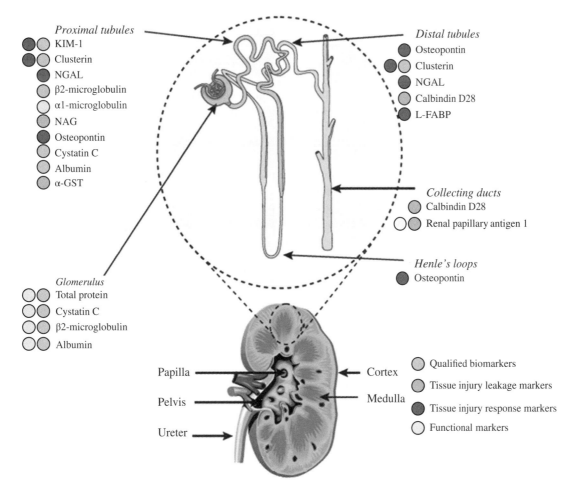

FIGURE 28.2 Select qualified and unqualified biomarkers used when monitoring for drug-induced kidney injury to specific nephron segments in rats. Adapted from Bonventre et al. (2010). (*See insert for color representation of the figure.*)

and BUN. Regulatory guidance specifically requests or anticipates routine inclusion of sCr and BUN in clinical pathology biochemistry panels for nonclinical compound safety studies and in clinical trials of pharmaceuticals (Anonymous, 2001).

28.2.1.1 Serum Creatinine (sCr)

The breakdown product of creatine phosphate is creatinine, which is primarily produced by muscle cells at a constant rate. The sCr is freely filtered by the glomerulus with minimal (~10%) tubular reabsorption, and it has been the standard assessment of renal function (GFR) in clinical and preclinical studies (Wyss and Kaddurah-Daouk, 2000). It is generally considered that a twofold increase of sCr is an indication of a ~50% decrease in GFR (Waikar et al., 2009; Bonventre et al., 2010). However, more recent studies have shown that up to 40% of creatinine may be actively secreted by the proximal tubule cells through OCT2/OATs and MATE1/MATE2K efflux transporters in rats (Tanihara et al., 2007; Vallon et al., 2012). Indeed, given the limitations, sCr may not be a sensitive or specific marker for an acute or subtle kidney injury.

28.2.1.2 Blood Urea Nitrogen

Urea is a purine nucleotide metabolism by-product and acts as an important water-soluble molecule to maintain medullary osmotic gradient that allows the formation of concentrated urine. The liver is the major site of urea formation, and similar to creatinine, GFR is a primary determinant of urea excretion; therefore decrements in GFR lead to increases in BUN levels. However, urinary urea or renal clearance of urea has traditionally been reported less than half of GFR, and increases in BUN only occur as a result of 75% damage of nephrons, indicating a substantial amount of filtered urea is reabsorbed by tubules (Haschek et al., 2013). Recent study has shown about 90% of the filtered urea is reabsorbed by the URAT1 and OAT4 transporters in the brush-border membrane and GLUT9 in the basolateral membrane (Hagos et al., 2007; Shima et al., 2010). Therefore, BUN has limited predictive value of minor AKI.

28.2.1.3 Urinalysis (Qualitative and Quantitative Panel)

Routine urinalysis is commonly performed in nonclinical toxicology studies. Urine samples are often taken at a single time point, during an overnight or timed collection period. Urinalysis parameters usually consist of the evaluation of physical parameters (urine volume, color, clarity, specific gravity, and osmolality), semiquantitative chemical examination by reagent strips (pH, glucose, protein, heme, ketones, bilirubin, urobilinogen, and nitrite), and sediment microscopic examination (different cell types, casts, abnormal crystals, and bacteria). Potentially, these qualitative and quantitative assays can provide general information on the hydration status, urogenital tract changes, acid–base balance,

and renal concentrating ability. Quantitative urinalysis is conducted to measure the absolute amount of substances excreted by kidneys over a specified time period. The parameters of interest usually include electrolytes (Na^+, Cl^-, K^+, Ca^+, and PO_4), total protein, albumin, and enzyme excretion. The data are generally normalized against urinary creatinine (relevant only if the test compound has no effect on urinary creatinine (uCr) excretion) or urine volume (normalize the effect of diluted or concentrated urine). Traditionally, uCr is used as the standard index for "normalization" of other urinary analytes or biomarkers to characterize changes in excreted urinary analytes or biomarkers (clearance) related to changes in GFR and/or changes in tubular reabsorption/secretion process. However, urinalysis is generally considered an imprecise tool in most toxicology studies due to its technical difficulties in collecting quality urine samples (contamination), timely analysis, large data variability, and data interpretation (Haschek et al., 2013).

28.2.2 Novel Functional Biomarkers

28.2.2.1 Serum and Urinary Cystatin C

Cystatin C, a nonglycosylated small molecular weight protein (13 kDa) that functions as a cysteine proteinases inhibitor, is continuously produced by all nucleated cells (Kaseda et al., 2007). Serum cystatin C is free filtered by the glomerulus and not secreted by the tubules, while it is reabsorbed and almost completely metabolized by the tubules. Therefore, increased serum cystatin C level implies decreased GFR (Vanmassenhove et al., 2013). Serum cystatin C has shown to be a sensitive and specific indicator of decreased GFR in rats, suggesting that serum cystatin C may be used as the principal surrogate measure for GFR (Bonventre et al., 2010). An increase in urinary cystatin C levels can indicate a possible impairment of proximal tubular reabsorption process (Togashi et al., 2012).

28.2.2.2 Urinary Total Protein and Albumin

Urinary total protein and albumin have been used for decades as glomerular injury biomarkers and, more recently, were qualified as measurements of glomerular filtration and tubular reabsorption function (Ferguson et al., 2008; Bonventre et al., 2010). Compared with blood concentrations of protein/albumin, a small amount of protein and albumin (microalbumin, which is below the albumin detection threshold by the conventional urinary dipstick: 30–300 mg/L) enters the filtrate by the glomerulus and is reabsorbed and subsequently catabolized in the normal kidney proximal tubule (Vaidya et al., 2008; Charlton et al., 2014). Therefore, increased urinary protein/albumin can reflect glomerular injury, tubular injury, or combined effects, though albuminuria can be observed in rats secondary to other effects such as dehydration or hypertensive conditions (Haschek et al., 2013).

28.2.2.3 *Urinary β2-Microglobulin (B2M)*

Urinary B2M is a component of the major histocompatibility complex (MHC) class I molecule (Sugita and Brenner, 1994). Urinary B2M is a single polypeptide chain (12 kDa), which is readily filtered by the glomerulus and reabsorbed by the proximal tubule and catabolized. Increased urinary B2M levels are associated with a variety of conditions that cause impaired tubular uptake function (Vaidya et al., 2008; Fuchs et al., 2012). However, it has been shown that B2M is unstable in acidic urine, and this may limit its clinical utility where α1-microglobulin (A1M) is preferred (Sugita and Brenner, 1994; Vaidya et al., 2008).

28.2.2.4 *Urinary α1-Microglobulin (A1M)*

The low molecular weight protein (27–33 kDa), A1M is produced by the liver, filtered by the glomerulus, and reabsorbed by the proximal tubule where it is catabolized;, and in the urine, increased concentrations are associated with a variety of conditions that cause reduced tubular function (Vaidya et al., 2008; Fuchs et al., 2012).

28.3 NOVEL KIDNEY TISSUE INJURY BIOMARKERS

As discussed, increases in the traditional kidney functional biomarkers (sCr, BUN) and serum cystatin-C or decreases in GFR do not necessarily associate with AKI. First, drug-induced changes in renal perfusion pressure (e.g., systemic arterial blood pressure), glomerular arteriolar resistance, and glomerular membrane permeability may result in decreases in GFR without DIKI (Haschek et al., 2013). Second, drug-induced changes in the proximal tubular OCT2 pathway (Ciarimboli et al., 2012) or urea transporters in the renal inner medulla (Sands, 1999) can introduce subsequent changes in sCr or BUN, respectively, that are not related to DIKI (false-positive signals). Third, in progressive or chronic renal injury, the remaining nephrons undergo anatomic and functional hypertrophy and offset the functional consequence of nephron loss, leading the functional biomarkers to underestimate the actual extent of kidney injury (false-negative signals). Furthermore, when drug-induced decrease in GFR is associated with DIKI, the change in kidney function is usually coincident with or following tissue injury (Haschek et al., 2013; Charlton et al., 2014). Although kidney functional biomarkers are not useful for prodromal prediction of DIKI, it may be helpful to estimate the impact of DIKI on a change of GFR.

More recently, kidney tissue injury biomarkers including urinary NAG, α-GST, calbindin D28, RPA-1, and other constituents of renal tubule epithelial cell brush border membranes and cytoplasm are used for the evaluation of drug-induced direct kidney tissue damage (de Geus et al., 2012; Peres et al., 2013; Charlton et al., 2014). Injured tubule epithelial cells release their membrane and cytoplasmic constituents into the tubule lumen where they accumulate and are detected in the urine. However, these biomarkers may be variably expressed in normal and diseased individual animals and humans. This wide biological variation and a tendency for instability in urine lower sensitivity and limit their usefulness to detect early and subtle tissue damage before substantial DIKI has occurred. There is also evidence that some of these biomarkers are filtered by the kidney and high values in the urine may reflect high plasma values rather than kidney injury (Vaidya et al., 2008; Bonventre et al., 2010; Haschek et al., 2013; Charlton et al., 2014).

28.3.1 Urinary *N*-Acetyl-β-D-Glucosaminidase (NAG)

The lysosomal enzyme, NAG is present in multiple tissues and not filtered by the kidney. Therefore, it is considered to be specific for renal tubular damage, although increases may be seen with glomerular, proximal tubular, or papillary injury (Bazzi et al., 2002; Vaidya et al., 2008). Increased urinary levels of NAG have typically been observed prior to increases in sCr or BUN in rat nephrotoxicity models. It usually increases rapidly following exposure to nephrotoxicants. However, there is noted variable expression in normal individual animals and humans, and increased urinary NAG levels have been reported without significant kidney damage (Erdener et al., 2005; Haschek et al., 2013; Charlton et al., 2014).

28.3.2 Urinary Glutathione *S*-Transferase α (α-GST)

The detoxification isoenzyme α-GST is mainly located in centrilobular hepatocytes and is a kidney proximal renal tubular cytosolic enzyme (Harrison et al., 1989; de Geus et al., 2012). When proximal renal tubular epithelial cells are damaged, their cytosolic contents enter the tubular fluid and appear in the urine for short periods around acute cellular injury. The α-GST is not stable in acid urine, and this may limit its utility in preclinical and clinical studies (Vaidya et al., 2008).

28.3.3 Urinary Renal Papillary Antigen 1 (RPA-1)

RPA-1 is a glycoprotein that has only been identified in the rat and generally localized to the collecting duct of kidney. The structure and function of RPA-1 are undetermined (Shaw, 2010). In rats, RPA-1 increased in urine and kidney tissue after treatment with nephrotoxic compounds (e.g., propyleneimine, indomethacin). Recently, RPA-1 was qualified for collecting duct injury in rats, but the human equivalent is unknown (Hildebrand et al., 1999; Price et al., 2010).

28.3.4 Urinary Calbindin D28

Calbindin D28 is multimeric membrane transporter protein involved in calcium transport in the kidney (Hemmingsen, 2000). It is expressed and located exclusively in the renal distal tubule and cortical collecting duct and has been used primarily as an immunohistochemical marker for distal renal tubules (Opperman et al., 1990; Kumar et al., 1994). Recent preclinical studies have suggested some utility as a toxicity biomarker for distal tubular and collecting duct injury (Thongboonkerd et al., 2005; Betton et al., 2012). In patients, increased urinary levels of calbindin D28 were found to correlate with distal tubule damage following cisplatin chemotherapy (Takashi et al., 1996). Overall, there is an interest for calbindin D28 as a specific regional marker for tubular injury, but detailed preclinical and clinical studies are required.

28.4 NOVEL BIOMARKERS OF KIDNEY TISSUE STRESS RESPONSE

Other promising kidney safety biomarkers appear to be involved with kidney tissue stress and repair responses occurring prior to overt cell injury or death. These molecules include KIM-1, clusterin, osteopontin, NGAL, L-FABP, and IL-18 (Vaidya et al., 2008; Bonventre et al., 2010; Urbschat et al., 2011; de Geus et al., 2012; Charlton et al., 2014). The upregulated small proteins appearance in urine may offer the promise of greater sensitivity over kidney functional and tissue injury/leakage biomarkers and greater utility to detect early stages of DIKI before moderate kidney tissue damage has occurred (prodromal utility) and utility to monitor DIKI reversibility (recovery utility) (Fuchs et al., 2012). Early transient increases of these urinary markers and return to baseline following acute kidney tubular injury were demonstrated in rats (de Geus et al., 2012). Qualification studies of these biomarkers suggest that they may produce fewer false-positive and false-negative DIKI signals than other candidate markers. Current experience with these urinary biomarkers is largely limited to well-controlled studies of drugs previously associated with DIKI (true positives) and untreated controls, and there is very limited experience or information available concerning changes associated with the presence of underlying kidney pathophysiology or disease (e.g., renal hypertension, papillary necrosis, age-related nephropathy).

Current regulatory opinion supports inclusion of these biomarkers in nonclinical safety studies supporting clinical trials, for purposes of monitoring DIKI in animal (e.g., rat) studies, and encourages exploratory inclusion in nonclinical studies (rodent and nonrodent species) and in clinical trials of pharmaceuticals where DIKI has been identified as a potential "cause for concern" (Dieterle et al., 2010a).

28.4.1 Urinary Kidney Injury Molecule-1 (KIM-1)

KIM-1 is a membrane glycoprotein (104 kDa) that is expressed at high levels on the apical surface of proximal tubule cells following injury, and the shedded renal KIM-1 fragments therefore can be detected in the urine (Ichimura et al., 1998; Vaidya et al., 2008). In rat kidneys subjected to toxicant or ischemia–reperfusion injury, KIM-1 upregulation was shown to be specific to proliferating cells in the proximal tubule S3 segment (Ichimura et al., 1998, 2004). A number of possible protective or repair functions for KIM-1 have been reported (Ichimura et al., 2008; Bonventre et al., 2010; Bonventre and Yang, 2011; Charlton et al., 2014; Yang et al., 2015). Many clinical and preclinical studies have demonstrated that KIM-1 is a sensitive and specific biomarker for renal tubular injury (Vaidya et al., 2008). However, whether KIM-1 may contribute to the development of renal injury or execute as a protective factor is unknown (Lim et al., 2013).

28.4.2 Urinary Clusterin

Clusterin is a glycoprotein that is synthesized by numerous tissues (Dvergsten et al., 1994). In the kidney, clusterin is highly expressed during renal development and is thought to play an antiapoptotic role (Hidaka et al., 2002; Ishii et al., 2007; Vaidya et al., 2008). It recently formed part of the PSTC preclinical submission for biomarker qualification, and a claim was made that it can be used to monitor drug-induced tubular alterations (Dieterle et al., 2010a, b). Although clusterin has been identified in a number of clinical and preclinical conditions, other biomarkers appear more sensitive.

28.4.3 Urinary Neutrophil Gelatinase-Associated Lipocalin (NGAL)

NGAL is a member of the lipocalin superfamily of proteins and is expressed by a wide range of tissues and cell types including neutrophils, epithelia, kidney, prostate, and respiratory and alimentary tracts (Kjeldsen et al., 1993; Vaidya et al., 2008). In rats, upregulated NGAL levels in proximal tubules, distal tubules, and the thick ascending limb of the loop of Henle and concurrent increases in urinary NGAL concentrations have been detected following acute DIKI (Singer et al., 2013). NGAL is a small molecule (25 kDa) that readily crosses the glomerulus and is reabsorbed by the proximal tubule. NGAL expression may be produced and released by the damaged kidney cells following nephrotoxicant-induced injury of different phenotypes. It has been suggested that NGAL may have a protective role or function against kidney injury and repair of damaged proximal tubule epithelia cells (Mishra et al., 2003; Schmidt-Ott, 2011; Charlton et al., 2014). Urinary levels generally correlate well with plasma/serum levels. However, given its widespread tissue distribution and ease of appearance in the urine, increased urinary NGAL may not uniquely aid the differentiation of compound-related multiorgan toxicities. The US FDA and EMA have issued Letters of Support to encourage sponsors to conduct nonclinical and exploratory clinical studies to further evaluate the translational relevance of changes in urinary NGAL (EMA, 2014).

28.4.4 Urinary Osteopontin (OPN)

OPN is known as a primary extracellular structural protein (44 kDa) constituent of the bone, and it is expressed in a wide variety of tissues with the highest levels in the bone and epithelial tissues. In rodents and human kidney, OPN expression is primarily restricted to the thick ascending limbs of the loop of Henle and the distal convoluted tubules (Hudkins et al., 1999). It is reported that OPN acts as an inhibitor of calcium oxalate formation and prevents mineral precipitation and stone formation and other functions (Asplin et al., 1998). OPN is also significantly upregulated in rodent models of kidney injury/disease such as ischemia–reperfusion injury and nephrotoxic drugs such as gentamicin, cisplatin, and cyclosporine (Xie et al., 2001). It has been also evaluated in a limited number of clinical and several preclinical studies to monitor for compound-induced acute damage to distal tubules and the thick ascending limb of the loop of Henle (Vaidya et al., 2008; Bonventre et al., 2010). More studies in rodents and humans are needed to determine whether OPN is an early diagnostic quantitative and sensitive indicator of AKI. Similar to OPN, the US FDA and EMA have issued Letters of Support to encourage further evaluations to determine the translational value for urinary OPN (EMA, 2014).

28.4.5 Urinary L-Type Fatty Acid-Binding Protein (L-FABP)

Fatty acid-binding proteins (FABPs) are low molecular weight (~15 kDa) cytoplasmic proteins expressed in all tissues with active fatty acid metabolism (Glatz and Van Der Vusse, 1996). L-FABP has been identified in the human kidney in the proximal tubule, and urinary L-FABP has been found to be a potential biomarker in preclinical models of renal interstitial injury associated with increased expression and urinary excretion of L-FABP (Kamijo et al., 2004, 2006) and clinical conditions (Maatman et al., 1992). Additional studies are needed to determine the utility of L-FABP when monitoring for DIKI (Vaidya et al., 2008).

28.4.6 Urinary Interleukin-18 (IL-18)

IL-18 is a proinflammatory cytokine involved in mediating inflammation in many organs (Jordan et al., 2001; Gracie et al., 2003). Recent studies have shown that renal IL-18 mRNA levels are significantly upregulated, cleaved in the proximal tubules, and detected in the urine following ischemia–reperfusion AKI and cisplatin-induced nephrotoxicity animal models (Melnikov et al., 2001; Leslie and

Meldrum, 2008). However, there have been few studies performed to validate the temporal expression patterns of urinary IL-18 for the early detection of AKI.

28.5 APPLICATION OF AN INTEGRATED RAT PLATFORM (AUTOMATED BLOOD SAMPLING AND TELEMETRY, ABST) FOR KIDNEY FUNCTION AND INJURY ASSESSMENT

To fully characterize potential nephrotoxicity of candidate drugs, it is important to understand the biomarkers for only one type of DIKI may not be useful for that induced by a different nephrotoxicant (tubulointerstitial toxicant damage vs. nontubular-specific damage). In the rat, remarkable changes in one kidney biomarker may be strengthened by the use of simultaneous changes in a panel of DIKI biomarkers associated with histopathology findings and/or other traditional DIKI biomarkers. It has been also reported that most drugs affect sCr by reducing GFR through damage to the kidney, altered renal hemodynamics, or prerenal dehydration (Haschek et al., 2013). However, limited information is available on comprehensive measurements of DIKI biomarkers while simultaneously measuring cardiac function (blood pressure and heart rate), renal hemodynamics (GFR; effective renal plasma flow, ERPF), excretory function (quantitative urinalysis), blood chemistry (including sCr and BUN), and renal histopathology in the same animal, for toxicodynamic comparison of the usefulness of these DIKI biomarkers for the early detection of renal dysfunction and/or notable morphological alterations caused by nephrotoxic agents (Chen et al., 2013). To achieve this, Chen and colleagues have developed an integrated rat platform consisting of an automated blood sampling and telemetry (ABST) system (Litwin et al., 2011), with added automated infusion and quantitative urine collection capabilities (Chen et al., 2011). It has been described for the first time as an integrated model of assessing drug-induced renal functional changes and kidney injury in the same animal (Chen et al., 2013).

The ABST system (Fig. 28.3) equipped with the capability for systemic infusion of two exogenous markers (inulin–FITC and para-aminohippuric acid, PAH) into ABST rats allows for direct measurements of GFR and ERPF by measuring plasma inulin–FITC and PAH concentrations in serially collected blood samples and qualification of values using Fick's equation as previously described (Sturgeon et al., 1998; Qi et al., 2004). In addition, the customized and optimized urine collection device minimizes feces contamination in order to enhance the accuracy of subsequent quantitative urinalysis and DIKI biomarkers assays, without interfering with automated blood sampling or telemetric data capture. Overall, this new apparatus enables simultaneous and comprehensive evaluations of drug-induced renal hemodynamic and excretory function changes with the

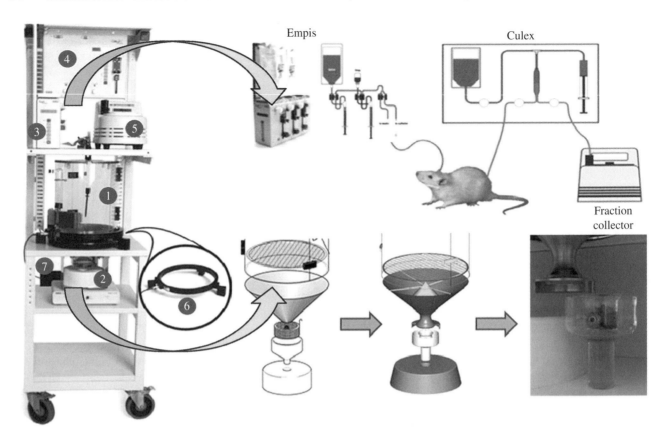

FIGURE 28.3   The automated blood sampler and telemetry (ABST) system. 1. Animal cage; 2. metabolic collection system; 3. Empis®
infusion pump; 4. blood sampling pump; 5. refrigerated blood storage unit; 6. circular telemetry antenna; 7. DSI telemetry receiver. BASi
Empis® automated infusion system for continuous infusion of inulin–FITC and PAH for GFR and ERPF measurements. The modified urine
collector minimizes feces contamination and provides clinical pathology quality urine samples. Chen et al. (2014). Reproduced with permission
from Elsevier.

potential to detect AKI as well as synchronized pharmacokinetic (PK) sampling and cardiovascular function by telemetry from the same conscious, free-moving rat. Importantly,
this multivariate assessment provides a tool for understanding drug-induced renal functional changes and potential mechanistic interactions between different organ systems
in rats.

The utility of the ABST system for DIKI detection has
been demonstrated in a recent cisplatin (CDDP) validation
study in rats (Chen et al., 2013). Briefly, conscious surgically prepared male Han Wistar rats, implanted with Data
Sciences International (DSI, St. Paul, MN) PhysioTel® mul-
tiplus radio transmitters (model TL11M2-C50-PXT) and
cannulated with jugular and femoral vein catheters by the
Charles River (Raleigh, NC), were given a single intraperitoneal dose of CDDP (15 mg/kg). GFR, ERPF, urinalysis,
DIKI biomarkers, CDDP pharmacokinetics, blood pressures, heart rate, and body temperature were measured in the
same vehicle- or CDDP-treated animals over 72 h. Plasma
chemistry (including Cr and BUN) and renal tissues were
examined at study termination. Cisplatin caused progressive
reductions of GFR, ERPF, heart rate, and body temperature

from day 1 (0–24 h). CDDP-treated rats showed significant
proteinuria and glucosuria from 48 h (2.8- and 19.5-fold vs.
control, respectively; $P<0.05$) with progressive and
maximum effects at 72 h (7.6- and 189.7-fold vs. control,
respectively; $P<0.05$), renal tubular changes in ABST rats
were characterized by mild to moderate acute tubular
necrosis (ATN), and minimal to mild dilatation of tubules
was observed in all CDDP-treated ABST rats.

The urine levels of eight DIKI biomarkers were analyzed
using multiplex immunoassays in 0–6, 6–24, 24–48, and
48–72 h postdose samples following a single dose of CDDP.
As shown in Table 28.1, urinary excretion for seven of the
eight biomarkers (uCr ratio) were elevated in a time-
dependent manner, with the maximum effects observed on
day 3 (fold change vs. control: α-GST, 352; albumin, 117;
GSTYb1, 20; clusterin, 41; lipocalin, 7; osteopontin, 26; and
KIM-1, 10; $P<0.05$). One marker, α-GST, showed an early
and significant increase at 6 h (α-GST: fivefold vs. control;
$P<0.05$), whereas α-GST and albumin increased at 24 h (α-
GST: sevenfold; albumin: fourfold vs. control; $P<0.05$)
after the administration of CDDP in ABST rats. No
significant change in RPA-1 excretion in the urine was seen

in this study. This study adds to the current understanding of CDDP action by demonstrating that early increases in urinary excretion of α-GST predict DIKI risk following acute exposure to CDDP in rats, before changes in traditional DIKI markers after or at 24 h are evident.

Scientific innovation has led to the development, qualification, and implementation of new technologies producing high-quality data, reductions in animal use, better derisking of potential clinical toxicities, and reductions in attrition in pharmaceutical development. Although CDDP validation results are quite promising in the present study, the renal capability of the ABST system should be fully qualified using nephrotoxic tool compounds (both positive and negative) from different drug classes in addition to chemotherapeutics (e.g., antimicrobials, analgesics, etc.). Ideally, these selected positive benchmark compounds have translational effects and are known to cause nephrotoxicity in both human and rat. In this way, the sensitivity, specificity, and predictivity of this ABST model could be calculated to demonstrate the usefulness of this new approach in preclinical drug safety assessment (Valentin et al., 2009). In addition to the direct evaluation of candidate drugs on renal hemodynamic function (GFR and ERPF), measuring excretion of a panel of qualified DIKI biomarkers in time-matched urine samples is particularly useful for earlier detection of concurrent minimal renal injury compared with traditional kidney injury markers or that that may occur prior to the appearance of histological lesions. Measuring a panel of markers is particular useful since the response of individual markers may vary depending not only on the extent and site of injury but also on the mode of action of the specific compound under investigation. Our data support the value of DIKI biomarkers for improved detection of kidney injury and highlight the importance of employing a biomarker panel as opposed to traditionally relying on a single indicator.

REFERENCES

Anonymous. (2001) Guidance for Industry: Safety Pharmacology Studies for Human Pharmaceuticals, U.S. Department of Health and Human Services Food and Drug Administration Center for Drug Evaluation and Research (CDER) Center for Biologics Evaluation and Research (CBER). Available at http://www.fda.gov/downloads/drugs/guidancecomplianceregulatoryinformation/guidances/ucm074959.pdf (accessed November 27, 2015).

Asplin JR, Arsenault D, Parks JH, Coe FL, Hoyer JR. (1998) Contribution of human uropontin to inhibition of calcium oxalate crystallization. *Kidney Int*. 53:194–199.

Bazzi C, Petrini C, Rizza V, Arrigo G, Napodano P, Paparella M, D'Amico G. (2002) Urinary N-acetyl-beta-glucosaminidase excretion is a marker of tubular cell dysfunction and a predictor of outcome in primary glomerulonephritis. *Nephrol Dial Transplant*. 17:1890–1896.

Betton GR, Ennulat D, Hoffman D, Gautier JC, Harpur E, Pettit S. (2012) Biomarkers of collecting duct injury in Han-Wistar and Sprague-Dawley rats treated with N-Phenylanthranilic acid. *Toxicol Pathol*. 40(4):682–694.

Bonventre JV, Yang L. (2011) Cellular pathophysiology of ischemic acute kidney injury. *J Clin Invest*. 121(11):4210–4221.

Bonventre JV, Vaidya VS, Schmouder R, Feig P, Dieterle F. (2010) Next-generation biomarkers for detecting kidney toxicity. *Nat Biotechnol*. 28:436–440.

Charlton JR, Portilla D, Okusa MD. (2014) A basic science view of acute kidney injury biomarkers. *Nephrol Dial Transplant*. 201:1–11.

Chen Y, Brott D, Bentley P, Thurman D, Kinter L, Bialecki RA. (2011) Urinary biomarkers for the early prediction of cisplatin-induced changes in kidney function and injury in rats. *Toxicologist*. 120(S2):742.

Chen Y, Brott D, Luo W, Gangl E, Kamendi H, Barthlow H, Lengel D, Fikes J, Kinter L, Valentin JP, Bialecki RA. (2013) Assessment of cisplatin-induced kidney injury using an integrated rodent platform. *Toxicol Appl Pharmacol*. 268(3):352–361.

Ciarimboli G, Lancaster CS, Schlatter E, Franke RM, Sprowl JA, Pavenstädt H, Massmann V, Guckel D, Mathijssen RHJ, Yang W, Pui CH, Relling MV, Herrmann E, Sparreboom A. (2012) Proximal tubular secretion of creatinine by organic cation transporter OCT2 in cancer patients. *Clin Cancer Res*. 18(4):1101–1108.

Dieterle F, Sistare F, Goodsaid F, Papaluca M, Ozer JS, Webb CP, Baer W, Senagore A, Schipper MJ, Vonderscher J, Sultana S, Gerhold DL, Phillips JA, Maurer G, Carl K, Laurie D, Harpur E, Sonee M, Ennulat D, Holder D, Andrews-Cleavenger D, Gu YZ, Thompson KL, Goering PL, Vidal JM, Abadie E, Maciulaitis R, Jacobson-Kram D, Defelice AF, Hausner EA, Blank M, Thompson A, Harlow P, Throckmorton D, Xiao S, Xu N, Taylor W, Vamvakas S, Flamion B, Lima BS, Kasper P, Pasanen M, Prasad K, Troth S, Bounous D, Robinson-Gravatt D, Betton G, Davis MA, Akunda J, McDuffie JE, Suter L, Obert L, Guffroy M, Pinches M, Jayadev S, Blomme EA, Beushausen SA, Barlow VG, Collins N, Waring J, Honor D, Snook S, Lee J, Rossi P, Walker E, Mattes W. (2010a) Renal biomarker qualification submission: a dialog between the FDA-EMEA and Predictive Safety Testing Consortium. *Nat Biotechnol*. 28(5):455–462.

Dieterle F, Perentes E, Cordier A, Roth DR, Verdes P, Grenet O, Pantano S, Moulin P, Wahl D, Mahl A, End P, Staedtler F, Legay F, Carl K, Laurie D, Chibout SD, Vonderscher J, Maurer G. (2010b) Urinary clusterin, cystatin C, beta2-microglobulin and total protein as markers to detect drug-induced kidney injury. *Nat Biotechnol*. 28(5):463–469.

Dvergsten J, Manivel JC, Correa-Rotter R, Rosenberg ME. (1994) Expression of clusterin in human renal diseases. *Kidney Int*. 45(3):828–835.

EMA. (2014) Letter of Support for PSTC Translational Drug-Induced Kidney Injury (DIKI) Biomarkers. Available at www.ema.europa.eu/docs/en_GB/document_library/Other/2014/11/WC500177133.pdf (accessed October 20, 2015).

Erdener D, Aksu K, Biçer I, Doğanavşargil E, Kutay FZ. (2005) Urinary N-acetyl-beta-D-glucosaminidase (NAG) in lupus nephritis and rheumatoid arthritis. *J Clin Lab Anal*. 19:172-176.

FDA. (2014a). Guidance for Industry and FDA Staff —Qualification Process for Drug Development Tools. Available at http://www.fda.gov/downloads/drugs/guidancecomplianceregulatoryinformation/guidances/ucm230597.pdf (accessed October 20, 2015).

FDA. (2014b). Biomarker Letter of Support. Available at. http://www.fda.gov/downloads/Drugs/DevelopmentApprovalProcess/DrugDevelopmentToolsQualificationProgram/UCM412843.pdf (accessed October 20, 2015).

Ferguson MA, Vaidya VS, Bonventre JV. (2008) Biomarkers of nephrotoxic acute kidney injury. *Toxicology*. 245:182–193.

Fuchs TC, Frick K, Emde B, Czasch S, von Landenberg F, Hewitt P. (2012) Evaluation of novel acute urinary rat kidney toxicity biomarker for subacute toxicity studies in preclinical trials. *Toxicol Pathol*. 40(7):1031–1048.

de Geus HR, Betjes MG, Bakker J. (2012) Biomarkers for the prediction of acute kidney injury: a narrative review on current status and future challenges. *Clin Kidney J*. 5(2):102-108.

Glatz JF, Van Der Vusse GJ. (1996) Cellular fatty acid-binding proteins: their function and physiological significance. *Prog Lipid Res*. 35:243–282.

Gracie JA, Robertson SE, McInnes IB. (2003) Interleukin-18. *J Leukoc Biol*. 73:213–224.

Gupta RC. (2014) *Biomarkers in Toxicology*. Chapter 56: Membrane Transporters and Transporter Substrates as Biomarkers for Drug Pharmacokinetics, Pharmacodynamics, and Toxicity/Adverse Events. Oxford: Academic Press. pp. 947–963.

Haase M, Bellomo R, Devarajan P, Schlattmann P, Haase-Fielitz A. (2009) NGAL meta-analysis investigator group accuracy of neutrophil gelatinase-associated lipocalin (NGAL) in diagnosis and prognosis in acute kidney injury: a systematic review and meta-analysis. *Am J Kidney Dis*. 54:1012–1024.

Hagos Y, Stein D, Ugele B, Burckhardt G, Bahn A. (2007) Human renal organic anion transporter 4 operates as an asymmetric urate transporter. *J Am Soc Nephrol*. 18:430–439.

Harpur E, Ennulat D, Hoffman D, Betton G, Gautier JC, Riefke B, Bounous D, Schuster K, Beushausen S, Guffroy M, Shaw M, Lock E, Pettit S; HESI Committee on Biomarkers of Nephrotoxicity. (2011) Biological qualification of biomarkers of chemical-induced renal toxicity in two strains of male rat. *Toxicol Sci*. 122(2):235–252.

Harrison DJ, Kharbanda R, Cunningham DS, McLellan LI, Hayes JD. (1989) Distribution of glutathione S-transferase isoenzymes in human kidney: basis for possible markers of renal injury. *J Clin Pathol*. 42:624–628.

Hemmingsen C. (2000) Regulation of renal calbindin-D28K. *Pharmacol Toxicol*. 87(Suppl. 3):5–30.

Hidaka S, Kranzlin B, Gretz N, Witzgall R. (2002) Urinary clusterin levels in the rat correlate with the severity of tubular damage and may help to differentiate between glomerular and tubular injuries. *Cell Tissue Res*. 310(3):289–296.

Hildebrand H, Rinke M, Schluter G, Bomhard E, Falkenberg FW. (1999) Urinary antigens as markers of papillary toxicity. II: application of monoclonal antibodies for the determination of papillary antigens in rat urine. *Arch Toxicol*. 73(4–5):233–245.

Hudkins KL, Giachelli CM, Cui Y, Couser WG, Johnson RJ, Alpers CE. (1999) Osteopontin expression in fetal and mature human kidney. *J Am Soc Nephrol*. 10:444–457.

Ichimura T, Bonventre JV, Bailly V, Wei H, Hession CA, Cate RL, Sanicola M. (1998) Kidney injury molecule-1 (KIM-1), a putative epithelial cell adhesion molecule containing a novel immunoglobulin domain, is up-regulated in renal cells after injury. *J Biol Chem*. 273(7):4135–4142.

Ichimura T, Hung CC, Yang SA, Stevens JL, Bonventre JV. (2004) Kidney injury molecule-1: a tissue and urinary biomarker for nephrotoxicant-induced renal injury. *Am J Physiol-Renal*. 286:F552–F563.

Ichimura T, Asseldonk EJ, Humphreys BD, Gunaratnam L, Duffield JS, Bonventre JV. (2008) Kidney injury molecule-1 is a phosphatidylserine receptor that confers a phagocytic phenotype on epithelial cells. *J Clin Invest*. 118(5):1657–1668.

Ishii A, Sakai Y, Nakamura A. (2007) Molecular pathological evaluation of clusterin in a rat model of unilateral ureteral obstruction as a possible biomarker of nephrotoxicity. *Toxicol Pathol*. 35(3):376–382.

Jordan JA, Guo RF, Yun EC, Sarma V, Warner RL, Crouch LD, Senaldi G, Ulich TR, Ward PA. (2001) Role of IL-18 in acute lung inflammation. *J Immunol*. 167(12):7060–7068.

Kamijo A, Kimura K, Sugaya T, Yamanouchi M, Hikawa A, Hirano N, Hirata Y, Goto A, Omata M. (2004) Urinary fatty acid-binding protein as a new clinical marker of the progression of chronic renal disease. *J Lab Clin Med*. 143:23–30.

Kamijo A, Sugaya T, Hikawa A, Yamanouchi M, Hirata Y, Ishimitsu T, Numabe A, Takagi M, Hayakawa H, Tabei F, Sugimoto T, Mise N, Omata M, Kimura K. (2006) Urinary liver-type fatty acid binding protein as a useful biomarker in chronic kidney disease. *Mol Cell Biochem*. 284:175–182.

Kaseda R, Iino N, Hosojima M, Takeda T, Hosaka K, Kobayashi A, Yamamoto K, Suzuki A, Kasai A, Suzuki Y, Gejyo F, Saito A. (2007) Megalin-mediated endocytosis of cystatin C in proximal tubule cells. *Biochem Biophys Res Commun*. 357:1130–1134.

Kellum JA, Levin N, Bouman C, Lameire N. (2002) Developing a consensus classification system for acute renal failure. *Curr Opin Crit Care*. 8:509–514.

Kjeldsen L, Johnsen AH, Sengelov H, Borregaard N. (1993) Isolation and primary structure of NGAL, a novel protein associated with human neutrophil gelatinase. *J Biol Chem*. 268:10425–10432.

Kumar R, Schaefer J, Grande JP, Roche PC. (1994) Immunolocalization of calcitriol receptor, 24-hydroxylase cytochrome P-450, and calbindin D28k in human kidney. *Am J Physiol Ren Physiol*. 266:477–485.

Lameire N, Van Biesen W, Hoste E, Vanholder R. (2008) The prevention of acute kidney injury: an in-depth narrative review Part 1: volume resuscitation and avoidance of drug- and nephrotoxin-induced AKI. *NDT Plus*. 1(6):392–402.

Leslie JA, Meldrum KK. (2008) The role of interleukin-18 in renal injury. *J Surg Res*. 145(1):170–175.

Lim AI, Tang SC, Lai KN, Leung JC. (2013) Kidney injury molecule-1: more than just an injury marker of tubular epithelial cells? *J Cell Physiol*. 228(5):917–924.

Lin X, Yuan J, Zhao Y, Zha Y. (2015) Urine interleukin-18 in prediction of acute kidney injury a systemic review and meta-analysis. *J Nephrol.* 28:7–16.

Litwin DC, Lengel DJ, Kamendi HW, Bialecki RA. (2011) An integrative pharmacological approach to radio telemetry and blood sampling in pharmaceutical drug discovery and safety assessment. *BioMed Eng Online.* 10:5.

Luyckx VA, Naicker S. (2008) Acute kidney injury associated with the use of traditional medicines. *Nat Clin Pract Nephrol.* 4:664–671.

Maatman RG, van de Westerlo EM, van Kuppevelt TH, Veerkamp JH. (1992) Molecular identification of the liver- and the heart-type fatty acid-binding proteins in human and rat kidney. Use of the reverse transcriptase polymerase chain reaction. *Biochem J.* 288(Pt 1):285–290.

Melnikov VY, Ecder T, Fantuzzi G, Siegmund B, Lucia MS, Dinarello CA, Schrier RW, Edelstein CL. (2001) Impaired IL-18 processing protects caspase-1-deficient mice from ischemic acute renal failure. *J Clin Invest.* 107(9):1145–1152.

Mishra J, Ma Q, Prada A, Mitsnefes M, Zahedi K, Yang J, Barasch J, Devarajan P. (2003) Identification of neutrophil gelatinase-associated lipocalin as a novel early urinary biomarker for ischemic renal injury. *J Am Soc Nephrol.* 14 (10):2534–2543.

Naughton CA. (2008) Drug-induced nephrotoxicity. *Am Fam Physician.* 78:743–750.

Olson S, Robinson S, Griffin R, Rapporteurs. (2009) *Accelerating the Development of Biomarkers for Drug Safety: Workshop Summary.* Chapter 4: Assessing and Predicting Kidney Safety. Editors: S Olson, S Robinson, and R Giffin, Rapporteurs. Washington, DC: National Academies Press. pp. 29–41.

Opperman LA, Pettifor JM, Ross FP. (1990) Immunohistochemical localization of calbindins (28K and 9K) in the tissues of the baboon Papio ursinus. *Anat Rec.* 228(4):425–430.

Peres LA, Cunha Júnior AD, Schäfer AJ, Silva AL, Gaspar AD, Scarpari DF, Alves JB, Girelli Neto R, Oliveira TF. (2013) Biomarkers of acute kidney injury. *J Bras Nephrol.* 35(3): 229–236.

Price SA, Davies D, Rowlinson R, Copley CG, Roche A, Falkenberg FW, Riccardi D, Betton GR. (2010) Characterization of renal papillary antigen 1 (RPA-1), a biomarker of renal papillary necrosis. *Toxicol Pathol.* 38(3):346–358.

Qi Z, Whitt I, Mehta A, Jin J, Zhao M, Harris RC, Fogo AB, Breyer MD. (2004) Serial determination of glomerular filtration rate in conscious mice using FITC-inulin clearance. *Am J Physiol Ren Physiol.* 286:F590–F596.

Redfern WS, Ewart L, Hammond TG, Bialecki R, Kinter L, Lindgren S, Pollard CE, Roberts R, Rolf MG, Valentin J-P. (2010) Impact and frequency of different toxicities throughout the pharmaceutical life cycle. *Toxicologist.* 114(S1):1081.

Rouse RL, Zhang J, Stewart SR, Rosenzweig BA, Espandiari P, Sadrieh NK. (2012) Comparative profile of commercially available urinary biomarkers in preclinical drug-induced kidney injury and recovery in rats. *Kidney Int.* 79(11):1186–1197.

Sands JM. (1999) Regulation of renal urea transporters. *JASN.* 10(3):635–646.

Schmidt-Ott KM. (2011) Neutrophil gelatinase-associated lipocalin as a biomarker of acute kidney injury—where do we stand today? *Nephrol Dial Transplant.* 26(3):762–764.

Shaw M. (2010) Cell-specific biomarkers in renal medicine and research. *Methods Mol Biol.* 641:271–302.

Shima JE, Komori T, Taylor TR, Stryke D, Kawamoto M, Johns SJ, Carlson EJ, Ferrin TE, Giacomini KM. (2010) Genetic variants of human organic anion transporter 4 demonstrate altered transport of endogenous substrates. *Am J Physiol Renal Physiol.* 299:F767–F775.

Singer E, Markó L, Paragas N, Barasch J, Dragun D, Müller DN, Budde K, Schmidt-Ott KM. (2013) Neutrophil gelatinase-associated lipocalin: pathophysiology and clinical applications. *Acta Physiol.* 207(4):663–672.

Sturgeon C, Sam AD, Law WR. (1998) Rapid determination of glomerular filtration rate by single-bolus inulin: a comparison of estimation analyses. *J Appl Physiol.* 84:2154–2162.

Sugita M, Brenner MB. (1994) An unstable beta 2-microglobulin: major histocompatibility complex class I heavy chain intermediate dissociates from calnexin and then is stabilized by binding peptide. *J Exp Med.* 180(6):2163–2171.

Takashi M, Zhu Y, Miyake K, Kato K. (1996) Urinary 28-kD calbindin-D as a new marker for damage to distal renal tubules caused by cisplatin-based chemotherapy. *Urol Int.* 56:174–179.

Tanihara Y, Masuda S, Sato T, Katsura T, Ogawa O, Inui K. (2007) Substrate specificity of MATE1 and MATE2-K, human multidrug and toxin extrusions/H(+)-organic cation antiporters. *Biochem Pharmacol.* 74(2):359–371.

Thongboonkerd V, Zheng S, McLeish KR, Epstein PN, Klein JB. (2005) Proteomic identification and immunolocalization of increased renal calbindin-D28k expression in OVE26 diabetic mice. *Rev Diabet Stud.* 2:19–26.

Togashi Y, Sakaguchi Y, Miyamoto M, Miyamoto Y. (2012) Urinary cystatin C as a biomarker for acute kidney injury and its immunohistochemical localization in kidney in the CDDP-treated rats. *Exp Toxicol Pathol.* 64:797–805.

Tonomura Y, Tsuchiya N, Torii M, Uehara T. (2010) Evaluation of the usefulness of urinary biomarkers for nephrotoxicity in rats. *Toxicology.* 273:53–59.

Urbschat A, Obermüller N, Haferkam A. (2011) Biomarkers of kidney injury. *Biomarkers.* 16(S1):S22–S30.

Vaidya VS, Ferguson MA, Bonventre JV. (2008) Biomarkers of acute kidney injury. *Annu Rev Pharmacol Toxicol.* 48:463–493.

Valentin JP, Bialecki RA, Ewart L, Hammond T, Leishmann D, Lindgren S, Martinez V, Pollard C, Redfern W, Wallis R. (2009) A framework to assess the translation of safety pharmacology data to humans. *J Pharmacol Toxicol Methods.* 60:152–158.

Vallon V, Eraly SA, Rao SR, Gerasimova M, Rose M, Nagle M, Anzai N, Smith T, Sharma K, Nigam SK, Rieg T. (2012) A role for the organic anion transporter OAT3 in renal creatinine secretion in mice. *Am J Physiol Renal Physiol.* 302(10):F1293–F1299.

Vanmassenhove J, Vanholder R, Nagler E, Van Biesen W. (2013) Urinary and serum biomarkers for the diagnosis of acute kidney injury: an in-depth review of the literature. *Nephrol Dial Transplant.* 28:254–273.

Waikar SS, Betensky RA, Bonventre JV. (2009) Creatinine as the gold standard for kidney injury biomarker studies? *Nephrol Dial Transplant.* 24(11):3263–3265.

Walter G, Smith G, Walker R. (2013) *Haschek and Rousseaux's Handbook of Toxicologic Pathology.* Chapter 29: Interpretation of Clinical Pathology Results in Non-Clinical Toxicology Testing. Third Edition. New York: Academic Press. pp. 853–882.

Ward PD, La D, McDuffie JE. (2013) *New Insights into Toxicity and Drug Testing.* Chapter 7: Renal Transporters and Biomarkers in Safety Assessment. Editor: S Gowder. Croatia: InTech. pp. 153–176.

Wyss M, Kaddurah-Daouk R. (2000) Creatine and creatinine metabolism. *Physiol Rev.* 80(3):1107–1213.

Xie Y, Sakatsume M, Nishi S, Narita I, Arakawa M, Gejyo F. (2001) Expression, roles, receptors, and regulation of osteopontin in the kidney. *Kidney Int.* 60:1645–1657.

Xin C, Yulong X, Yu C, Changchun C, Feng Z, Xinwei M. (2008) Urine neutrophil gelatinase-associated lipocalin and interleukin-18 predict acute kidney injury after cardiac surgery. *Ren Fail.* 30:904–913.

Yang L, Brooks CR, Xiao S, Sabbisetti V, Yeung MY, Hsiao LL, Ichimura T, Kuchroo V, Bonventre JV. (2015) KIM-1-mediated phagocytosis reduces acute injury to the kidney. *J Clin Invest.* 125(4):1620–1636.

29

CANINE KIDNEY SAFETY PROTEIN BIOMARKERS

MANISHA SONEE

Mechanistic & Investigative Toxicology, Discovery Sciences, Janssen Research & Development, L.L.C., Spring House, PA, USA

29.1 INTRODUCTION

The traditional renal biomarkers, BUN and sCr, are known to change after considerable kidney damage. Both these markers are not very sensitive and specific for early detection of kidney injury and impaired function as they are both affected by other renal and nonrenal factors. In the past few years, novel biomarkers have been explored and validated in rats by consortiums including the PSTC NWG and the HESI Committee for Nephrotoxicity Biomarkers. Urinary protein biomarkers including NAG, albumin, total protein, β2M, OPN, KIM-1, cystatin C, clusterin, TFF3, and RPA-1 have been approved as "qualified" safety biomarkers to detect drug-induced renal injury by FDA, EMA, and/or PMDA in preclinical toxicology studies (Dieterle et al., 2010; http://www.hesiglobal.org/i4a/pages/index.cfm?pageID=3541). More recently, the FDA and EMA issued "Biomarker Letter of Support" for two additional urinary biomarkers, osteopontin (OPN) and neutrophil gelatinase-associated lipocalin (NGAL), as markers of kidney proximal tubular injury based on evidence submitted by the PSTC NWG (http://www.fda.gov/downloads/Drugs/DevelopmentApprovalProcess/DrugDevelopmentToolsQualificationProgram/UCM412843.pdf; http://www.ema.europa.eu/docs/en_GB/document_library/Other/2014/11/WC500177133.pdf). Data from these submissions as well as from published literature suggested that the aforementioned biomarkers may be utilized when monitoring of drug-induced kidney injury in rodents during the drug development process. However, not much is known about the performance of these biomarkers in nonrodent species or their translation to humans. One of the most commonly used nonrodent species in regulatory toxicology studies is the canine (e.g., beagle dog). There has been very limited published data regarding the performance of the renal biomarkers in dogs as it is not a very commonly used animal model in academic research and represents a relatively expensive model for basic research purposes. If novel urinary biomarkers can detect renal injury and function in canines, it will not only help improve the drug development process but will also be beneficial to veterinary care for dogs. This chapter will focus on the current understanding of the performance of selected novel nephrotoxicity biomarkers in canines.

29.2 NOVEL CANINE RENAL PROTEIN BIOMARKERS

The canine urinary proteome has been reviewed previously (Nabity et al., 2011; Brandt et al., 2014). Limited data has been published on the performance of novel renal protein biomarkers in canine. Acute kidney injury induced in beagle dogs treated with gentamicin (80 mg/kg; intramuscular) for 9 days showed that NGAL and clusterin were significantly elevated as early as after one and three doses, respectively, and correlated well with the severity of the kidney lesions (Zhou et al., 2014). No significant changes in BUN and sCr were observed at these times, indicating the sensitivity of these markers in detecting AKI in dogs (Zhou et al., 2014). NGAL was also shown to be a sensitive tubular toxicity

Drug Discovery Toxicology: From Target Assessment to Translational Biomarkers, First Edition. Edited by Yvonne Will, J. Eric McDuffie, Andrew J. Olaharski, and Brandon D. Jeffy.

biomarker when dogs were treated with gentamicin (25 or 50 mg/kg; subcutaneous) for 9 days. Significant increases in urinary NGAL was observed after three and five doses in dogs treated with 50 mg/kg in the absence of any changes in BUN and sCr (Kai et al., 2013). Another recent study, where dogs were treated with subcutaneous injection of gentamicin (40 mg/kg) for 7 days, showed an increase with urinary cystatin C and decreased glomerular filtration rate (GFR) on day 4 before the other markers increased on the day 8 time point (Sasaki et al., 2014). In the previous studies, conducted by Zhou et al. (2014) and Kai et al. (2013), urinary cystatin C was not measured. Microarray analysis revealed that ~1000 genes were upregulated following intramuscular injection of gentamicin (40 mg/kg) for 10 days to dogs, of which 20 genes have been related to drug-induced kidney injury in rodents, canines, and/or humans (McDuffie et al., 2013). Clusterin and OPN were among the 18 genes that were increased in both kidney and urine of the gentamicin-treated dogs. Increased urinary KIM-1 mRNA was observed after six doses, preceding detectable levels of sCr and/or BUN, which correlated with the multifocal immunostaining of KIM-1 in the kidney corticomedullary tubular epithelial cells (McDuffie et al., 2013). The localization of canine aquaporin 1, aquaporin 2, calbindin D-28k, GST-α, and Tamm–Horsfall protein by immunohistochemistry staining of kidney histologic sections has been reported (Brandt et al., 2012).

Polymyxin B treatment to dogs also resulted in dose-related increases in urinary levels of NGAL along with urinary NAG, protein, and BUN (Burt et al., 2014). Inflammatory cytokines/chemokines in dog urine has positively identified acute kidney tubular injury in dogs earlier than sCr in cisplatin-treated dogs (McDuffie et al., 2010). Urinary NGAL was shown to be a sensitive and specific marker of naturally occurring (azotemic and nonazotemic forms) AKI in dogs and may be used to be a marker of mild and early stages of AKI (Segev et al., 2013). Urinary NGAL was highest in dogs with AKI as compared to other dogs with urinary tract infection or chronic kidney disease (CKD). However, this study also suggested that NGAL was increased in dogs with urinary tract infection or CKD as compared to healthy dogs. Additionally, the urinary NGAL levels was not only shown to be increased in dogs with AKI as compared to the non-AKI dogs (in a study conducted by Lee et al. (2012)) but also in dogs with CKD as compared to healthy dogs (Steinbach et al., 2014). Other studies have also implicated serum NGAL as a useful marker when evaluating dogs with CKD (Ahn and Hyun, 2013; Hsu et al., 2014). Increased levels of clusterin was observed in a population of dogs with leishmaniasis (a model of CKD) as compared to controls (Garcia-Martinez et al., 2012). Cystatin C were shown to be elevated in dogs with renal disease and correlated more strongly with GFR than serum creatinine (Wehner et al., 2008). However, another study found that serum cystatin C may not be superior as a marker of GFR (Almy et al., 2002). Further studies are required to determine the usefulness of these biomarkers in the detection of reversible drug-induced kidney injury and monitorable CKD in dogs.

29.3 EVALUATIONS OF NOVEL CANINE RENAL PROTEIN BIOMARKER PERFORMANCE

The PSTC NWG have also been evaluating the performance of selected novel kidney safety biomarkers by conducting studies in which dogs have been treated with various nephrotoxins (such as polymyxin B, amphotericin B). It is anticipated that these studies may enable better understanding of the translation of these biomarkers from nonrodent models to the clinical setting. Some of the preliminary findings from these studies were presented at scientific conferences (e.g., 2014 Society of Toxicology Annual Meeting, McDuffie et al., 2014). Results from these preliminary studies showed that the novel urinary protein biomarkers (e.g., NGAL, clusterin, etc.), which have been known to be altered in rats and humans, were also altered on kidney damage in dogs. Since the same set of biomarkers were not measured in all the studies, a direct comparison of the performance of each biomarker proved to be difficult within the studies presented. Both studies in the published literature and those being conducting with the PSTC indicate that the same set of biomarkers may not change under treatment with different nephrotoxicants. For example, urinary NGAL was increased in dogs treated with polymyxin B (Burt et al., 2014), but no changes in this marker was noted in amphotericin B-treated dogs (unpublished). The differences in performance of these biomarkers may be due to the differences in the type or the severity of the lesion. The lack of widely accepted canine biomarker assays may also complicate the comparison of biomarker performances among studies. As compared to rodent and human assays, there remains a lack of fully validated canine-specific assays. Therefore, considerable efforts have been devoted to the development and validation of biomarker assays that cross-react with canine proteins to support the measurement of analytes such as KIM-1 and clusterin.

29.4 CONCLUSION

In conclusion, the performance of a number of kidney toxicity biomarkers (mainly clusterin, NGAL, KIM-1, cystatin C, NAG, albumin, total protein, osteopontin, α-GST) are being currently evaluated in dogs treated with select nephrotoxins. Based on the current findings in the literature, few novel nephrotoxicity biomarkers (e.g., NGAL, clusterin) are already showing promise, with nephron specificity and sensitivity in detecting experimental kidney injury in dogs. With the development and validation of more canine-specific assays (e.g., alpha-GST, KIM-1), there will be an increased understanding of the performance of a panel of renal biomarkers in dogs. It is most likely that in the future a panel of translational biomarkers along with the traditional markers (sCr and BUN) will have the most clinical utility in assessing progressive and reversible nephrotoxicity in dogs and/or kidney diseases in nonclinical and clinical studies.

REFERENCES

Ahn HJ and Hyun C. Evaluation of serum neutrophil gelatinase-associated lipocalin (NGAL) activity in dogs with chronic kidney disease. *Vet Rec.* 173(18):452, Nov 9, 2013.

Almy FS, Christopher MM, King DP, and Brown SA. Evaluation of cystatin C as an endogenous marker of glomerular filtration rate in dogs. *J Vet Intern Med.* 16(1):45–51, Jan–Feb 2002.

Brandt LE, Bohn AA, Charles JB, and Ehrhart EJ. Localization of canine, feline, and mouse renal membrane proteins. *Vet Pathol.* 49(4):693–70, Jul 2012.

Brandt LE, Ehrhart EJ, Scherman H, Olver CS, Bohn AA, and Prenni JE. Characterization of the canine urinary proteome. *Vet Clin Pathol.* 43(2):193–205, June 2014.

Burt D, Crowell SJ, Ackley DC, Magee TV, and Aubrecht J. Application of emerging biomarkers of acute kidney injury in development of kidney-sparing polypeptide-based antibiotics. *Drug Chem Toxicol.* 37(2):204–12, Apr 2014.

Dieterle F, Sistare F, Goodsaid F, Papaluca M, Ozer JS, Webb CP, Baer W, Senagore A, Schipper MJ, Vonderscher J, Sultana S, Gerhold DL, Phillips JA, Maurer G, Carl K, Laurie D, Harpur E, Sonee M, Ennulat D, Holder D, Andrews-Cleavenger D, Gu YZ, Thompson KL, Goering PL, Vidal JM, Abadie E, Maciulaitis R, Jacobson-Kram D, Defelice AF, Hausner EA, Blank M, Thompson A, Harlow P, Throckmorton D, Xiao S, Xu N, Taylor W, Vamvakas S, Flamion B, Lima BS, Kasper P, Pasanen M, Prasad K, Troth S, Bounous D, Robinson-Gravatt D, Betton G, Davis MA, Akunda J, McDuffie JE, Suter L, Obert L, Guffroy M, Pinches M, Jayadev S, Blomme EA, Beushausen SA, Barlow VG, Collins N, Waring J, Honor D, Snook S, Lee J, Rossi P, Walker E, and Mattes W. Renal biomarker qualification submission: a dialog between the FDA-EMEA and Predictive Safety Testing Consortium. *Nat Biotechnol.* 28(5):455–62, May 10, 2010.

García-Martínez JD, Tvarijonaviciute A, Cerón JJ, Caldin M, and Martínez-Subiela S. Urinary clusterin as a renal marker in dogs. *J Vet Diagn Invest.* 24(2):301–6, March 2012.

Hsu WL, Lin YS, Hu YY, Wong ML, Lin FY, and Lee YJ. Neutrophil gelatinase-associated lipocalin in dogs with naturally occurring renal diseases. *J Vet Intern Med.* 28(2):437–42, Mar–Apr 2014.

Kai K, Yamaguchi T, Yoshimatsu Y, Kinoshita J, Teranishi M, and Takasaki W. Neutrophil gelatinase-associated lipocalin, a sensitive urinary biomarker of acute kidney injury in dogs receiving gentamicin. *J Toxicol Sci.* 38(2):269–77, 2013.

Lee YJ, Hu YY, Lin YS, Chang CT, Lin FY, Wong ML, Kuo-Hsuan H, and Hsu WL. Urine neutrophil gelatinase-associated lipocalin (NGAL) as a biomarker for acute canine kidney injury. *BMC Vet Res.* 8:248, Dec 2012.

McDuffie JE, Sablad M, Ma J, and Snook S. Urinary parameters predictive of cisplatin-induced acute renal injury in dogs. *Cytokine.* 52(3):156–62, Dec 2010.

McDuffie JE, Gao J, Ma J, La D, Bittner A, Sonee M, Wagoner M, and Snook S. Novel genomic biomarkers for acute gentamicin nephrotoxicity in dog. *OJMIP* 3:125–133, 2013.

McDuffie JE, Adler S, Phillips JA, Sonee M, Lynch K, Gautier J-C, and Burt D. Nonrodents Can Be Monitored, Too… Characterization of Novel Biomarkers of Drug-Induced Kidney Injury (DIKI) in Rats, Canines, Nonhuman Primates, and Humans. The Toxicologist. 80 Minute Workshop. Informational Session. 310(1):98, 2014.

Nabity MB, Lees GE, Dangott LJ, Cianciolo R, Suchodolski JS, and Steiner JM. Proteomic analysis of urine from male dogs during early stages of tubulointerstitial injury in a canine model of progressive glomerular disease. *Vet Clin Pathol.* 40(2):222–36, Jun 2011.

Sasaki A, Sasaki Y, Iwama R, Shimamura S, Yabe K, Takasuna K, Ichijo T, Furuhama K, and Satoh H. Comparison of renal biomarkers with glomerular filtration rate in susceptibility to the detection of gentamicin-induced acute kidney injury in dogs. *J Comp Pathol.* 151(2–3):264–70, 2014.

Segev G, Palm C, LeRoy B, Cowgill LD, and Westropp JL. Evaluation of neutrophil gelatinase-associated lipocalin as a marker of kidney injury in dogs. *J Vet Intern Med.* 27(6):1362–7, Nov–Dec 2013.

Steinbach S, Weis J, Schweighauser A, Francey T, and Neiger R. Plasma and urine neutrophil gelatinase-associated lipocalin (NGAL) in dogs with acute kidney injury or chronic kidney disease. *J Vet Intern Med.* 28(2):264–9, Mar–Apr 2014.

Wehner A, Hartmann K, and Hirschberger J. Utility of serum cystatin C as a clinical measure of renal function in dogs. *J Am Anim Hosp Assoc.* 44(3):131–8, May–Jun 2008.

Zhou X, Ma B, Lin Z, Qu Z, Huo Y, Wang J, and Li B. Evaluation of the usefulness of novel biomarkers for drug-induced acute kidney injury in beagle dogs. *Toxicol Appl Pharmacol.* 280(1):30–5, 2014.

30

TRADITIONAL KIDNEY SAFETY PROTEIN BIOMARKERS AND NEXT-GENERATION DRUG-INDUCED KIDNEY INJURY BIOMARKERS IN NONHUMAN PRIMATES

JEAN-CHARLES GAUTIER[1] AND XIAOBING ZHOU[2]

[1] Preclinical Safety, Sanofi, Vitry-sur-Seine, France

[2] National Center for Safety Evaluation of Drugs, Beijing, China

30.1 INTRODUCTION

Most preclinical studies to investigate biomarkers of nephrotoxicity have been conducted in rodents. However, nonhuman primate (NHP), mostly Cynomolgus monkey (*Macaca fascicularis*), is an appreciated nonrodent model for preclinical drug safety evaluation, especially in the case of biopharmaceuticals where it is often the only relevant species. Indeed the biological activity and the target tissue specificity of many biopharmaceuticals often preclude the use of standard toxicity testing species like rats and dogs. Another rationale to use NHPs is to help translation to human when studies in rat and dog fail to demonstrate tubular-specific toxicities (Infante et al., 2013). There is a need for kidney safety biomarkers more sensitive and specific than sCr and BUN in NHP model similar to the utilities described previously in other preclinical models (Gautier et al., 2014).

Very few studies have been published to date on kidney safety biomarker in NHPs. Limited species-specific biomarker assays are available in Cynomolgus monkeys, and human-specific reagents cross-reacting with monkeys are generally used for this purpose. A first study was published in Cynomolgus monkeys treated with a triple reuptake inhibitor producing specific lesions in distal tubules and collecting ducts (Guha et al., 2011). Kidney safety biomarkers were measured using a multiplex human assay cross-reacting with Cynomolgus in this study. Results showed a dose-proportional increase in the urinary excretion of calbindin D28 and clusterin without any statistically significant increases in sCr in compound-treated monkeys. Interestingly, the levels of biomarkers returned to baseline during the recovery period. The localization of calbindin in distal convoluted tubules, connecting ducts, and cortical and medullary collecting ducts of Cynomolgus monkey was confirmed by immunohistochemistry (Bauchet et al., 2011; Guha et al., 2011). These data demonstrated the utility of urinary calbindin D28 and clusterin to monitor the progression of drug-induced distal nephron injury and represent potential early biomarkers of drug-induced kidney injury (DIKI) in humans. As described in the case study by Burt et al. (2014), Cynomolgus monkeys were treated with the polypeptide antibiotic polymyxin B, which induces proximal tubular degeneration/regeneration and necrosis across species. Urinary KIM-1 was measured with a cross-reacting human singleplex immunoassay and was shown to be increased concomitantly with sCr and BUN in animals treated with polymyxin B. It was however not possible to correlate the biomarker data with kidney histopathology because the study was designed as nonterminal for NHPs to limit animal usage. Of note, KIM-1, clusterin, and osteopontin were found to be increased at the gene expression level in the kidney in a previous study conducted with Cynomolgus monkeys treated with the tubular nephrotoxicants gentamicin and everninomicin (Davis et al., 2004).

Drug Discovery Toxicology: From Target Assessment to Translational Biomarkers, First Edition. Edited by Yvonne Will, J. Eric McDuffie, Andrew J. Olaharski, and Brandon D. Jeffy.

30.2 EVALUATIONS OF NOVEL NHP RENAL PROTEIN BIOMARKER PERFORMANCE

The PSTC NWG (http://c-path.org/programs/pstc/#tools-) is currently testing systematically the cross-reactivity of human assays in NHPs and evaluating the diagnostic performance of new Cynomolgus-specific assays to measure kidney safety biomarkers in this species. Novel assay development (e.g., immunoprecipitation liquid chromatography–mass spectrometry (IP LC–MS)) as well as optimization of human singleplex immunoassays with <100% cross-reactivity to monkey urinary proteins is ongoing for some novel kidney safety biomarkers like urinary KIM-1 and urinary OPN in NHPs. Several exploratory studies have been conducted in NHPs treated with several proximal tubular nephrotoxicants (i.e., gentamicin and cisplatin) by the PSTC NWG to assess urinary changes in novel kidney safety biomarkers concurrent to remarkable kidney tubular lesions and immunohistochemistry staining. As an example, first unpublished results from a gentamicin study in NHPs indicate that the urinary protein biomarkers NAG, microalbumin, A1M, and clusterin may be useful sensitive biomarkers when monitoring for early, progressive drug-induced kidney tubular injury in Cynomolgus monkeys either in the absence of clinically relevant changes in BUN and/or sCr or in the presence of only slight increases in BUN and sCr levels following repeat dose administration of nephrotoxicant. Several dedicated definitive NHP studies are also ongoing in the context of the PSTC NWG to further support the translational utility of novel kidney safety biomarkers that show clinical potential, beside other kidney safety biomarker studies in rats and canine. Taken together, these studies will indicate whether kidney biomarker changes in NHPs are similar to those in humans.

30.3 NEW HORIZONS: URINARY MICRORNAs AND NEPHROTOXICITY IN NHPS

The evaluation of urinary microRNAs (miRNAs) as potential biomarkers of nephrotoxicity in NHPs is a recent advancement in discovery toxicology. Human TaqMan® Low Density Array (TLDA) cards contain miRNA probes which may be used when profiling miRNAs in Cynomolgus monkey, while next-generation sequencing (NGS) appears to be a more comprehensive approach for this purpose. Preliminary results obtained in the context of consortia like the HESI Biomarkers of Nephrotoxicity Committee and the PSTC NWG have shown that urinary miRNAs could potentially be used as sensitive biomarkers of nephrotoxicity in NHPs and support translation across species.

REFERENCES

Bauchet A.-L., Masson R., Guffroy M., and Slaoui M. Immunohistochemical identification of kidney nephron segments in the dog, rat, mouse, and cynomolgus monkey. *Toxicologic Pathology*; 39: 1115–1128 (2011).

Burt D., Crowell S. J., Ackley D. C., Magee T. V., and Aubrecht J. Application of emerging biomarkers of acute kidney injury in development of kidney-sparing polypeptide-based antibiotics. *Drug and Chemical Toxicology*; 37(2): 204–212 (2014).

Davis II J. W., Goodsaid F. M., Bral C. M., Obert L. A., Mandakas G., Garner II C. E., Collins N. D., Smith R. J., and Rosenblum I. Y. Quantitative gene expression analysis in a nonhuman primate model of antibiotic-induced nephrotoxicity. *Toxicology and Applied Pharmacology*; 200: 16–26 (2004).

Gautier J.-C., Gury T., Guffroy M., Khan-Malek R., Hoffman D., Pettit S., and Harpur E. Normal ranges and variability of novel urinary renal biomarkers in Sprague-Dawley Rats: comparison of constitutive values between males and females and across assay platforms. *Toxicologic Pathology*; 42(7): 1092–104 (2014).

Guha M., Heier A., Price S., Bielenstein M., Caccese R. G., Heathcote D. I., Simpson T. R., Stong D. B., and Bodes E. Assessment of biomarkers of drug-induced kidney injury in cynomolgus monkeys treated with a triple reuptake inhibitor. *Toxicological Sciences*; 120(2): 269–283 (2011).

Infante J. R., Rugg T., Gordon M., Rooney I., Rosen L., Zeh K., Liu, Burris H. A., and Ramanathan R. K. Unexpected renal toxicity associated with SGX523, a small molecule inhibitor of MET. *Investigational New Drugs*; 31(2): 363–369 (2013).

31

RAT KIDNEY MicroRNA ATLAS

AARON T. SMITH

Investigative Toxicology, Eli Lilly and Company, Indianapolis, IN, USA

31.1 INTRODUCTION

Recently microRNAs have garnered increasing interest as potential biofluid-based markers of organ injury which may have more advantageous characteristics than protein-based biomarkers. MicroRNA sequences that have variations in the base composition as compared to their reference sequence are termed isomiRs (Morin et al., 2008) and may have different tissue expression profiles with respect to their reference sequence. In order to begin to understand biofluid-based microRNA profiles resulting from toxic insult to organ(s), it is essential to understand the microRNA and isomiR composition of as many organs as possible. Further understanding of organ-specific and enriched microRNAs is one focus of the ILSI HESI Technical Committee's application of genomics to mechanism-based risk assessment (www.hesiglobal. org). Therefore, to address this need, the microRNA content of ~20 target organs of toxicologic interest was interrogated in five male and five female Sprague Dawley rats using Illumina microRNA sequencing, and the data were examined for tissue-specific and tissue-enriched microRNAs.

31.2 KEY FINDINGS

In this study the whole kidney as well as representative macrodissected medulla and cortex tissues of the kidney were examined. The medulla was found to contain only one tissue-specific microRNA, which mapped to miR-200b-3p and miR-429, while the whole kidney and cortex did not contain any tissue-specific microRNAs. The whole kidney, medulla,

and cortex shared many highly expressed microRNAs that are also expressed to a smaller degree in other organs/tissues examined (these were termed "tissue-enriched microRNAs"). These include, but are not limited to, miR-429, miR-6329, miR-30a5p, miR-196b-3p, miR-203b-3p, and miR-653-5p. It seems reasonable to conclude that a microRNA, which is uniquely expressed or is enriched in the kidneys of rats, would represent a candidate microRNA of interest for further characterization in rat. While this assumption seems reasonable, microRNAs that are not as tissue specific or enriched may also be considered for further characterization on a case-by-case basis. For example, if compound X is given to rats and only kidney toxicity is noted, then it may be practical to use a microRNA that is enriched in the kidney as a biomarker for kidney injury regardless of its expression in other tissues. In the presence of additional organ toxicities, this approach may not be rational. Consideration should be given with respect to the number of candidate microRNAs for characterization and the type of known and/or unknown kidney toxicants evaluated in discovery toxicology studies for biomarker characterization. Additionally, it is possible that different methods of analyzing the microRNA sequencing data generated in this study could lead to the discovery of previously unidentified kidney specific and enriched microRNAs. Moreover, microRNA sequencing is known to have biases due to the use of RNA ligases (Jayaprakash et al., 2011; Zhang and Lee, 2013), which does affect the quantitative capacity of microRNA sequencing and could potentially cause an inability to detect certain microRNAs. Therefore, additional kidney specific or enriched microRNAs may have yet to be discovered, and further

Drug Discovery Toxicology: From Target Assessment to Translational Biomarkers, First Edition. Edited by Yvonne Will, J. Eric McDuffie, Andrew J. Olaharski, and Brandon D. Jeffy.

characterization of the miRNA content of rat kidneys as well as different data analysis methods could yield more candidate biomarkers of kidney injury.

REFERENCES

Anitha D. Jayaprakash, Omar Jabadoz, Brian D. Brown, and Ravi Sachidanandam, Identification and remediation of biases in the activity of RNA ligases in small-RNA deep sequencing. *Nucleic Acids Research*, 2011. 39: p. 1–12.

Ryan D. Morin, Michael D. O'Connor, Malachi Griffith, Florian Kuchenbauer, Allen Delaney, Anna-Liisa Prabhu, Yongjun Zhao, Helen McDonald, Thomas Zeng, Martin Hirst, Connie J. Eaves, and Marco A. Marra, Application of massively parallel sequencing to microRNA profiling and discovery in human embryonic stem cells. *Genome Research*, 2008. 18: p. 610–621.

Zhaojie Zhang, Jerome E. Lee, Kent Riemondy, Emily M Anderson, and Rui Yi, High-efficiency RNA cloning enables accurate quantification of miRNA expression by deep sequencing. *Genome Biology*, 2013. 14(10): p. 13.

32

MicroRNAs AS NEXT-GENERATION KIDNEY TUBULAR INJURY BIOMARKERS IN RATS

HEIDRUN ELLINGER-ZIEGELBAUER[1] AND ROUNAK NASSIRPOUR[2]

[1] Investigational Toxicology, GDD-GED-Toxicology, Bayer Pharma AG, Wuppertal, Germany

[2] Biomarkers, Drug Safety Research and Development, Pfizer Inc., Andover, MA, USA

32.1 INTRODUCTION

Urinary protein biomarker (BM) research has achieved great advances in the detection of nephrotoxicity, most notably tubular toxicity in rats (EMA, 2009; Dieterle et al., 2010). As described elegantly (Harpur et al., 2011), although the newly identified protein BMs outperform traditional BMs in sensitivity and specificity, they are limited in terms of stability in biofluids or antibody availability to ensure translatability across preclinical species and humans (Antoine et al., 2013; Goodsaid, 2013). Ample evidence suggests that microRNAs (miRNAs) could complement the protein BMs by their unique characteristics such as stability in body fluids, relative ease of measurement with sensitive methods like quantitative polymerase chain reaction (qPCR), and conservation across species, which might to allow for translability (Dieterle and Sistare, 2010; Mall et al., 2013; Saikumar et al., 2014). Although qPCR has most commonly been used to profile the annotated miRNome, newer technologies that promise increased sensitivity and specificity are becoming available at astonishing rates. For example, NGS is gaining popularity and is successfully being used to characterize miRNA profiles in various tissues as well as biofluids. These advances in technology, such as the use of NGS in creating a rat miRNA atlas (Chapter 31), the have paved the way for identification of tissue-specific or at least tissue-enriched expression of subsets of miRNAs which adds to the degree of specificity required in characterizing body fluid miRNAs as kidney specific BMs released by induced cellular damage.

Therefore, renal-specific miRNAs released upon damage may be measured in urine and be used as BMs (Saikumar et al., 2012). Since the kidney is composed of several structural elements, including glomeruli, proximal and distal tubule regions connected via the Loop of Henle, and collecting ducts, it may be expected that at least some miRNAs show subregion-specific renal expression. Thus alterations in the levels of such miRNAs in urine may be associated with nephron-specific injury and provide further insights regarding certain kidney diseases or primary mechanisms of action for renal toxicants.

32.2 RAT TUBULAR miRNAs

Upon induction of renal tubular damage in rats with the reference toxin cisplatin, quite a few miRNAs were found to be increased in urine using a qPCR profiling technique (Kanki et al., 2014; Pavkovic et al., 2014). Since several of these miRNAs were detected in two independent studies intended to evaluate acute cisplatin nephrotoxicity among two different strains of rats, and could be confirmed with more sensitive qPCR assays run in parallel to miRNA standard curves, they may represent novel safety BMs for the renal tubules or at least for the kidney tissue in general. Candidate miRNA BMs suggested in this context include

Drug Discovery Toxicology: From Target Assessment to Translational Biomarkers, First Edition. Edited by Yvonne Will, J. Eric McDuffie, Andrew J. Olaharski, and Brandon D. Jeffy.

but are not limited to miR-15, -16, -20a, -192, -193, and -210 (Pavkovic et al., 2014). Additionally, global urinary miRNA expression profiling using both the more conventional qPCR TaqMan Low Density Array (TLDA-A, Life Technologies) method and NGS in control and tubular toxicant (gentamicin)-treated rats reported that although both platforms were able to identify drug-induced changes in miRNA expression, there is discordance in the level of sensitivity of detection. Nonetheless, the authors showed that differentially expressed urinary miRNAs can be detected by two distinct profiling platforms, qRT-PCR and miRNA-seq and that the three miRNAs detected by both technologies may be useful urinary BMs for tubular injury (Nassirpour et al., 2014). Ongoing and future comparisons of urinary miRNAs altered upon induction of subregion-specific injury may reveal whether these miRNAs can serve as renal tubular or general kidney BMs. *In situ* hybridization studies with probes detecting specific miRNAs will further aid to define miRNAs site-specific localization.

32.3 CONCLUSIONS

In any case, whether indicating tubular specific or general kidney injury, increased miRNA levels appear rather early upon injury induction and suggest more transient profile changes in concurrent urine samples when compared to relevant protein BMs (Kanki et al., 2014; Pavkovic et al., 2014), suggesting that they may increase sensitivity in detection of recovery. This could be a useful characteristic of renal miRNAs in preclinical recovery studies and when monitoring for drug-induced nephrotoxicity in clinical studies where there remains a lack of qualified soluble renal BMs indicative of recovery from kidney damage and/or dysfunction. Thus, relevant increases in renal-specific miRNAs in the urine may be taken as evidence for drug-induced renal damage, as has been observed for miR-122 profiles in blood from patients after acetaminophen-induced hepatotoxicity (Ward et al., 2014).

REFERENCES

Antoine, D.J., et al., Mechanistic biomarkers provide early and sensitive detection of acetaminophen-induced acute liver injury at first presentation to hospital. *Hepatology*, 2013. 58(2): p. 777–87.

Dieterle, F. and F. Sistare, Biomarkers of Acute Kidney Injury, in *Biomarkers: In Medicine, Drug Discovery, and Environmental Health*, V.S. Vaidya and J.V. Bonventre, Editors. 2010. Hoboken, NJ: John Wiley & Sons, Inc., p. 237–263.

Dieterle, F., et al., Renal biomarker qualification submission: a dialog between the FDA-EMEA and Predictive Safety Testing Consortium. *Nat Biotechnol*, 2010. 28(5): p. 455–62.

EMA, 2009. Final conclusions on the pilot Joint EMEA/FDA VXDS experience on Qualification of Nephrotoxicity biomarkers. http://www.c-path.org/pdf/EMAreport.pdf (accessed October 22, 2015).

Goodsaid F. *The Path from Biomarker Discovery to Regulatory Qualification*. 2013. San Diego, CA: Elsevier Inc.

Harpur, E., et al., Biological qualification of biomarkers of chemical-induced renal toxicity in two strains of male rat. *Toxicol Sci*, 2011. 122(2): p. 235–52.

Kanki, M., et al., Identification of urinary miRNA biomarkers for detecting cisplatin-induced proximal tubular injury in rats. *Toxicology*, 2014. 324: p. 158–68.

Mall, C., et al., Stability of miRNA in human urine supports its biomarker potential. *Biomark Med*, 2013. 7(4): p. 623–31.

Nassirpour, R., et al., Identification of tubular injury microRNA biomarkers in urine: comparison of next-generation sequencing and qPCR-based profiling platforms. *BMC Genomics*, 2014. 15: p. 485–500.

Pavkovic, M., B. Riefke, and H. Ellinger-Ziegelbauer, Urinary microRNA profiling for identification of biomarkers after cisplatin-induced kidney injury. *Toxicology*, 2014. 324: p. 147–57.

Saikumar, J., et al., Expression, circulation, and excretion profile of microRNA-21, -155, and -18a following acute kidney injury. *Toxicol Sci*, 2012. 129(2): p. 256–67.

Saikumar, J., K. Ramachandran, and V.S. Vaidya, Noninvasive micromarkers. *Clin Chem*, 2014. 60(9): p. 1158–73.

Ward, J., et al., Circulating microRNA profiles in human patients with acetaminophen hepatotoxicity or ischemic hepatitis. *Proc Natl Acad Sci U S A*, 2014. 111(33): p. 12169–74.

33

MicroRNAs AS NOVEL GLOMERULAR INJURY BIOMARKERS IN RATS

RACHEL CHURCH

University of North Carolina Institute for Drug Safety Sciences, Chapel Hill, NC, USA

33.1 INTRODUCTION

Nephrotoxicity subcommittees of biomarker consortia, PSTC and ILSI-HESI, have made great strides at advancing novel urinary protein biomarkers that surpass traditional renal biomarkers in their sensitivity and specificity for the detection of renal injury. To date, eight urinary proteins have been qualified for utilization as biomarkers of nephrotoxicity in preclinical rat studies by regulatory agencies in the United States, Europe, and Japan (Dieterle et al., 2010; Harpur et al., 2011). An additional two urinary protein biomarkers (osteopontin and neutrophil gelatinase-associated lipocalin) were recently acknowledged in a respective Letter of Support from the Food and Drug Administration (FDA) and European Medicine Agency (EMA) encouraging further exploration (EMA, 2014; FDA, 2014). Among qualified urinary biomarkers of renal injury, elevated levels of total protein, cystatin C, and β2-microglobulin may be used to predict drug-induced glomerular injury in rats; however these biomarkers can also become elevated in the urine following nonglomerular renal insults (Vlasakova et al., 2014). Similarly, although albuminuria is often considered a sensitive indicator of glomerular impairment, this biomarker is qualified only as an indicator of proximal tubule injury (Dieterle et al., 2010). The lack of biomarkers specific for glomerular injury represents a knowledge gap that has opened the door for exploration of novel biomarker candidates, including microRNA (miRNA) species.

33.2 RAT GLOMERULAR miRNAs

Research studying the role of miRNAs in the glomeruli has demonstrated the importance of these RNA species for proper renal function. In mice, podocyte-specific deletion of Dicer or Drosha, enzymes critical for miRNA processing, resulted in glomerular damage and progressive proteinuria (Shi et al., 2008; Zhdanova et al., 2011). Further, emerging *in vivo* data is also demonstrating the importance of miR-NAs in diseases that impair glomerular function, including glomerular nephritis and diabetic nephropathy, and some of these miRNAs are also altered in the urine. For instance, reduced levels of glomerular miR-26a and concurrently elevated urinary miR-26a levels were observed in mice with glomerular nephritis and patients with lupus nephritis (Ichii et al., 2014). Increased levels of miR-145 in the glomeruli and urine of diabetic mice and in the urine of diabetic patients with diabetic nephropathy have also been reported (Barutta et al., 2013). Collectively, this data suggests that glomerular injury results in the release of miR-NAs into the urine, and quantification of these species may be useful in the identification of biomarkers of drug-induced nephrotoxicity.

Two recent studies utilized TaqMan miRNA profiling to examine the release of specific miRNA species into the urine of rats following the administration of nephrotoxins that induce primary glomerular injury and secondary tubular damage (Church et al., 2014; Pavkovic et al., 2015). In one

Drug Discovery Toxicology: From Target Assessment to Translational Biomarkers, First Edition. Edited by Yvonne Will, J. Eric McDuffie, Andrew J. Olaharski, and Brandon D. Jeffy.

study, male Sprague Dawley rats given doxorubicin displayed significantly elevated levels of urinary miR-34c-3p prior to the appearance of protein biomarkers or histological lesions indicative of glomerular or tubular damage (Church et al., 2014). Urinary elevation of this miRNA persisted through the course of the study, and its performance as a biomarker was comparable to that of albuminuria. Importantly, quantification of miR-34c-3p in laser capture microdissected renal tissue demonstrated that expression of this miRNA was elevated in glomerular tissue of nontreated controls, compared to surrounding nonglomerular tissue, and was further elevated following administration of doxorubicin.

In the second study, male Wistar Kyoto and Sprague Dawley rats were administered nephrotoxic serum (NTS), which induced glomerular nephritis and tubular injury (Pavkovic et al., 2015). The authors identified five urinary miRNAs (miR-10a, -10b, -100, -211, and -486) whose elevation was suggested be to related to glomerular injury. Interestingly, localization of two of these species, miR-10b and miR-100, by *in situ* hybridization placed these species in distal segments of the nephron but not the glomeruli, although this was conducted in kidney tissue collected from untreated rats and does not rule out that expression of these miRNAs is induced in the glomeruli following treatment with NTS.

Why no overlapping candidate miRNAs were observed between these studies is unclear but may be attributed to differences in study design. For instance, the NTS study only evaluated miRNA alterations occurring on "A" TaqMan profiling cards, while candidate miR-34c-3p is located on "B" TaqMan cards. Therefore quantities of this miRNA were not assessed in the NTS study. Additionally, mechanistic differences in the production of glomerular injury occurring between doxorubicin and NTS may account for the study differences. In support of this hypothesis, Nassirpour et al. (2015) recently described changes in miRNA expression patterns in two different rodent models of glomerular injury (acute puromycin aminonucleoside nephropathy and passive Heymann nephritis). By employing two different modes of glomerular insult, oxidative stress and immune-mediated injury, miRNA changes in both isolated glomeruli and urine specimens allow for identification of urinary miRNA biomarkers that are specific to the glomerulus. Subsets of glomerular urinary miRNAs associated with these different modes of glomerular toxicity seem to be dependent on the mechanism of the induced injury, while nine miRNAs (miR-106a, 125a-5p, -17, -218, -223, -27b, -30c, -574-3p, and -196c) that changed early in both glomerular and urine specimens were common to both studies. The authors further showed that the miRNAs identified as mechanism-specific early glomerular injury biomarkers target key pathways and

transcripts relevant to the type of insult, while the insult-independent changes might serve as ideal glomerular injury biomarkers. Regardless, further research is necessary to qualify the utility of these miRNA species as candidate biomarkers of glomerular injury.

REFERENCES

Barutta, F, Tricarico, M, Corbelli, A, Annaratone, L, Pinach, S, Grimaldi, S, et al. Urinary exosomal microRNAs in incipient diabetic nephropathy. *PLoS ONE*. 8(11):e73798, 2013.

Church, RJ, McDuffie, JE, Sonee, M, Otieno, M, Ma, JY, Liu, X, et al. MicroRNA-34c-3p is an early predictive biomarker for doxorubicin-induced glomerular injury progression in male Sprague-Dawley rats. *Toxicology Research*. 3:384–394, 2014.

Dieterle, F, Sistare, F, Goodsaid, F, Papaluca, M, Ozer, JS, Webb, CP, et al. Renal biomarker qualification submission: a dialog between the FDA-EMEA and Predictive Safety Testing Consortium. *Nature Biotechnology*. 28(5):455–462, 2010.

EMA. 2014. Letter of Support for PSTC Translational Drug-Induced Kidney Injury (DIKI) Biomarkers. Available at: www.ema.europa.eu/docs/en_GB/document_library/Other/2014/11/WC500177133.pdf (accessed on October 20, 2015).

FDA. 2014. Biomarker Letter of Support. Available at: http://www.fda.gov/downloads/Drugs/DevelopmentApprovalProcess/DrugDevelopmentToolsQualificationProgram/UCM412843.pdf (accessed on October 20, 2015).

Harpur, E, Ennulat, D, Hoffman, D, Betton, G, Gautier, JC, Riefke, B, et al. Biological qualification of biomarkers of chemical-induced renal toxicity in two strains of male rat. *Toxicological Sciences*. 122(2):235–252, 2011.

Ichii, O, Otsuka-Kanazawa, S, Horino, T, Kimura, J, Nakamura, T, Matsumoto, M, et al. Decreased miR-26a expression correlates with the progression of podocyte injury in autoimmune glomerulonephritis. *PLoS ONE*. 9(10):e110383, 2014.

Nassirpour, R, Homer, BL, Mathur, S, Li, Y, Li, Z, Brown, T, et al. Identification of Promising Urinary MicroRNA Biomarkers in Two Rat Models of Glomerular Injury. *Toxicological Sciences*. 148(1):35–47, 2015.

Shi, S, Yu, L, Chiu, C, Sun, Y, Chen, J, Khitrov, G, et al. Podocyte-selective deletion of dicer induces proteinuria and glomerulosclerosis. *Journal of the American Society of Nephrology*. 19(11):2159–2169, 2008.

Vlasakova, K, Erdos, Z, Troth, SP, McNulty, K, Chapeau-Campredon, V, Mokrzycki, N, et al. Evaluation of the relative performance of 12 urinary biomarkers for renal safety across 22 rat sensitivity and specificity studies. *Toxicological Sciences*. 138(1):3–20, 2014.

Zhdanova, O, Srivastava, S, Di, L, Li, Z, Tchelebi, L, Dworkin, S, et al. The inducible deletion of Drosha and microRNAs in mature podocytes results in a collapsing glomerulopathy. *Kidney International*. 80(7):719–730, 2011.

34

INTEGRATING NOVEL IMAGING TECHNOLOGIES TO INVESTIGATE DRUG-INDUCED KIDNEY TOXICITY

BETTINA WILM[1] AND NEAL C. BURTON[2]

[1] *Department of Cellular and Molecular Physiology, The Institute of Translational Medicine, The University of Liverpool, Liverpool, UK*

[2] *iThera Medical, GmbH, Munich, Germany*

34.1 INTRODUCTION

The gold standard for measuring kidney function is to assess the glomerular filtration rate (GFR). This is often done by administering an agent, such as inulin, which is physiologically inert (i.e., does not affect kidney function) but is filtered by the glomerulus, is not bound to plasma protein, and is not subject to reabsorption, secretion, or metabolism in the kidney (Sandilands et al., 2013). Surrogate traditional renal safety biomarkers, such as blood urea nitrogen (BUN) and serum creatinine or the presence of protein in the urine, can also be used to inform the preclinical researcher about renal function without such labor-intensive procedures.

Blood and urine biomarker utility can be limited by the observation that they may reflect late or indirect changes in kidney function (i.e., BUN and serum creatinine). There is therefore an interest in having a more direct assessment of kidney function, which may provide an earlier indicator of disease (Sandilands et al., 2013). Kidney biopsy allows a clinician direct access to samples from the kidney. Histological analysis is a classical approach that can provide valuable information on the presence of renal pathologies at the cellular level (e.g., nephritis, nephropathy, and tubulointerstitial disease, to name just a few) (Weening and Jennette, 2012). More recently, *in situ* mass spectrometry has been demonstrated to allow the detection of a broad range of biomarkers in isolated tissue samples, including diagnostic information about disease severity or metabolic response to drug treatment (Nilsson et al., 2012). While biopsy-based diagnosis is very important in the diagnosis of renal cancer and can aid in the diagnosis of chronic renal disease (Dhaun et al., 2014), a less invasive procedure would be preferred for initial evaluation of the kidney. *In vivo* imaging has therefore gained the interest of clinicians. Ultrasound enables evaluation of structural abnormalities associated with renal disease, specifically the observation of vasculature and flow rates via Doppler and the visualization of microvasculature and perfusion through contrast enhancement with microbubbles (Cokkinos et al., 2013).

Other *in vivo* imaging techniques can be used to provide molecular information related to kidney function. Extensive reviews regarding *in vivo* imaging of renal physiology have been published (e.g., Beierwaltes et al., 2013). For this chapter, canonical molecular imaging modalities will be summarized briefly, and emerging modalities will be emphasized. Positron emission tomography (PET) has been used to assess renal function, benefitting from the synthesis of many different radioligands that display renal clearance. As a proof or principal, a radiotracer was shown to differentiate a healthy individual from a patient with drug-induced nephropathy (Phillips et al., 2014); however, in this case, the ideal time point for differentiation was 72-h postinjection. With optimization, more clinically relevant time frames could be established. Blood oxygen level-dependent (BOLD) magnetic resonance imaging (MRI),

Drug Discovery Toxicology: From Target Assessment to Translational Biomarkers, First Edition. Edited by Yvonne Will, J. Eric McDuffie, Andrew J. Olaharski, and Brandon D. Jeffy.

which provides information about deoxyhemoglobin concentration, can also provide functional kidney imaging and has been used in small clinical trials to demonstrate utility in characterizing renal disease (Inoue et al., 2011).

34.2 OVERVIEWS

Here we present and discuss a range of novel approaches combining emerging imaging technologies with detection of renal function in a range of animal models of induced kidney injury. It is anticipated that clinical translation is feasible in some cases with additional development.

It had previously been shown that renal function could be estimated through the measurement of dye clearance in conscious mice. These studies have been based on traditional approaches where inulin and the related derivative fructosan sinistrin, as well as diethylenetriamine pentaacetate (DTPA), have been used in various forms as renal clearance substances (Silkalns et al., 1973; Goates et al., 1990; Jung et al., 1992; McLachlan et al., 2001). Their detection involved enzymatic or colorimetric measurements of inulin and sinistrin from serum or urine or scintigraphy of radiolabeled DTPA.

Recent advances in detecting renal function include the development of fluorescently labeled novel compounds (hydrophilic pyrazine dyes, carbostyril 124-DTPA-Eu, fluorescein-labeled poly-D-lysine) and their detection via *in vivo* fluorescence imaging or through transcutaneous detection (Dorshow et al., 1998; Rabito et al., 2005; Rajagopalan et al., 2011; Poreddy et al., 2012). Both the hydrophilic pyrazine dyes (see Fig. 34.1) and the modified DTPA are cleared exclusively by glomerular filtration, and the elimination kinetics are not affected by tubular secretion or reabsorption. By contrast, the fluorescein-labeled poly-D-lysine is filtered through the glomeruli but also reabsorbed. By using intravital microscopy, the temporal ratio of freely filtered FITC-labeled inulin to not filtered Texas Red (TR)-labeled dextran signal was imaged, and subsequently quantified, with high spatial resolution in the renal vasculature and in the proximal and distal tubules (Yu et al., 2007). While this method allows visualization of the passage of the dyes through the vasculature and the tubular nephron components in the kidney, the surgery involved prevents a high-throughput approach. However, Yu and colleagues also measured the ratio of FITC–inulin to TR–dextran transcutaneously to quantify the elimination of the freely filtered inulin compound, which allowed them to determine the constant of the plasma clearance rate to calculate the GFR (Yu et al., 2007). They applied this approach to healthy control and rats with unilateral and bilateral ischemia, nephrectomy, or ureter ligation, respectively, to demonstrate the significant difference in GFR between the renal injury models. This technology was further advanced by the Gretz group, who developed a small carrier device containing a battery, LEDs and photodiodes, and a microcontroller that is attached to the shaved back of mice or rats for transcutaneous detection of elimination after FITC-sinistrin bolus injection as a measure of GFR (Pill et al., 2005, 2006; Schock-Kusch et al., 2009, 2011; Schreiber et al., 2012). Importantly, this advanced technology allows the FITC-sinistrin disappearance to be measured in animals that are conscious, which is of considerable advantage since anesthesia can negatively influence the GFR (Mazze et al., 1974; Fusellier et al., 2007). As the sampling rate is between 4 and 60 per minute, the collected data points generate the FITC-sinistrin elimination kinetics at a high resolution, using one- or two-compartment fitting (Schock-Kusch et al., 2011; Schreiber et al., 2012). The transcutaneous detection of GFR by proxy of FITC-sinistrin clearance has been used in a range of rodent kidney disease models to demonstrate that the change in the measured GFR is specific to the kidney damage. The animal models used include Dahl salt-sensitive hypertensive rats (Cowley et al., 2013) and mice with unilateral nephrectomy as well as a murine genetic model of nephronophthisis (Schreiber et al., 2012).

Recently, the measurement of FITC-sinistrin using the transcutaneous device (optical GFR) was directly compared with simultaneous GFR measurements obtained from gadolinium-based contrast agent filtration in the MRI (MRI-GFR; Zöllner et al., 2013). While discrepancies between the two detection methods were observed, the study demonstrated that MRI-GFR, similarly to optical GFR, allowed for visualization of the filtration function in healthy and unilateral nephrectomy rats (Zöllner et al., 2013).

While bulk detection of dyes is suggestive of filtration and elucidates renal function, imaging the distribution of dyes in the kidney might provide additional insight into renal

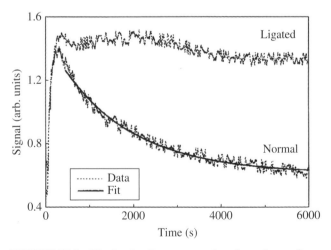

FIGURE 34.1 The *in vivo* fluorescence time dependence after a bolus injection of fluorescein-labeled poly-D-lysine in a single rat pre-kidney ligation (normal) and post-kidney ligation (ligated). The solid line is a single exponential fit to the measured data. Dorshow et al. (1998, figure 6). Reproduced with permission of SPIE.

function. Optical and optoacoustic imaging provide the ability to image renally filtered, near-infrared dyes noninvasively, at multiple time points during disease progression. In optical imaging, the near-infrared dyes are detected using a near-infrared small animal imaging system, which produces planar images of dye distribution. This is exemplified in a study where the near-infrared dye IRDye 800CW coupled to the RGD peptide (IRDye 800CW RGD) was administered to animals with antibody-induced glomerulonephritis (Du et al., 2012). The RGD peptide is recognized with high affinity by a group of integrin heterodimers among which are the αvβ integrins. Specifically the αvβ3 integrins have been shown to be upregulated in patients with glomerular nephropathies (Kanahara et al., 1994; Shikata et al., 1995; Jin et al., 1996; Patey et al., 1996; Roy-Chaudhury et al., 1997). Du and colleagues imaged animals with glomerular damage and controls daily after administration with IRDye 800CW RGD (on day 0) and detected a specific retention of the dye in the injured animals for 14 days, while in noninjured animals the dye had disappeared after 3 days (Du et al., 2012). This approach shows promise in the visualization of histological changes and damage in kidneys after nephrotoxic insult.

In a similar approach, IRDye 800CW imaging was utilized in imaging and detection of kidney function using optoacoustic imaging. Optoacoustic imaging allows detection of optically absorbing agents, but the ultrasound detection allows high spatial resolution (~150 μm) at depth (up to 2–3 cm) that is not possible in traditional optical imaging. Since IRDye 800CW had been shown to be excreted by rat kidneys without signs of toxicity (Marshall et al., 2010), Taruttis and colleagues imaged the distribution kinetics of IRDye 800CW carboxylate through mouse kidneys using multispectral optoacoustic tomography (MSOT; Taruttis et al., 2012). Specifically, IRDye 800CW carboxylate was found to enter quickly into the renal cortex, while appearance of the dye in the renal pelvis was delayed, indicating that the distribution of the dye over time in the kidney reflected the functional differences of the regions that were monitored (cortex vs. pelvis). Our group has recently for the first time used MSOT imaging to determine the clearance kinetics of IRDye 800CW carboxylate in mice with adriamycin-induced nephropathy (Scarfe et al., 2015). Specifically, 5 weeks after induction of glomerular damage by administration of adriamycin (doxorubicin), IRDye 800CW carboxylate was cleared at a slower rate from the renal cortex to the pelvis, when compared to saline-treated control mice.

34.3 SUMMARY

In summary, gold standard renal evaluation is labor-intensive, and surrogate markers can only provide indirect measures of renal function. Imaging therefore provides an attractive opportunity to study renal function noninvasively. Emerging technologies have the penetration and resolution to observe and measure renal filtration directly. With time, additional innovation may pave the way to translate these novel approaches into the clinic.

REFERENCES

Beierwaltes WH, Harrison-Bernard LM, Sullivan JC, and DL Mattson (2013). Assessment of renal function; clearance, the renal microcirculation, renal blood flow, and metabolic balance. *Comprehensive Physiology* 3(1):165–200.

Cokkinos DD, Antypa EG, Skilakaki M, Kriketou D, Tavernaraki E, and PN Piperopoulos (2013). Contrast enhanced ultrasound of the kidneys: what is it capable of? *International Journal of Biomedical Research* 2013:595873.

Cowley AW Jr, Ryan RP, Kurth T, Skelton MM, Schock-Kusch D, and N Gretz (2013). Progression of glomerular filtration rate reduction determined in conscious Dahl salt-sensitive hypertensive rats. *Hypertension* 62:85–90.

Dhaun N, Bellamy CO, Cattran DC, and DC Kluth (2014). Utility of renal biopsy in the clinical management of renal disease. *Kidney International* 85(5):1039–1048.

Dorshow RB, Bugaj JE, Burleigh BD, Duncan JR, Johnson MA, and WB Jones (1998). Noninvasive fluorescence detection of hepatic and renal function. *Journal of Biomedical Optics* 3(3):340–345.

Du Y, An S, Liu L, Li L, Zhou XJ, Mason RP, and C Mohan (2012). Serial non-invasive monitoring of renal disease following immune-mediated injury using near-infrared optical imaging. *PLoS ONE* 7(9): e43941.

Fusellier M, Desfontis J-C, Madec S, Gautier F, Debailleul M, and M Gogny (2007). Influence of three anesthetic protocols on glomerular filtration rate in dogs. *American Journal of Veterinary Research* 68:807–811.

Goates JJ, Morton KA, Whooten WW, Greenberg HE, Datz FL, Handy JE, Scuderi AJ, Haakenstad AO, and RE Lynch (1990). Comparison of methods for calculating glomerular filtration rate: technetium-99m-DTPA scintigraphic analysis, protein-free and whole-plasma clearance of technetium-99m-DTPA and iodine-125-iothalamate clearance. *Journal of Nuclear Medicine* 31:424–429.

Inoue T, Kozawa E, Okada H, Inukai K, Watanabe S, Kikuta T, Watanabe Y, Takenaka T, Katayama S, Tanaka J, and H Suzuki (2011). Noninvasive evaluation of kidney hypoxia and fibrosis using magnetic resonance imaging. *Journal of the American Society of Nephrology* 22(8):1429–1434.

Jin DK, Fish AJ, Wayner EA, Mauer M, Setty S, Tsilibary E, and Y Kim (1996). Distribution of integrin subunits in human diabetic kidneys. *Journal of the American Society of Nephrology* 7:2636–2645.

Jung K, Ilenke W, Schulze BD, Sydow K, Precht K, and S Klotzek (1992). Practical approach for determining glomerular filtration rate by single-injection inulin clearance. *Clinical Chemistry* 38:403–407.

Kanahara K, Yorioka N, Ogawa T, Taniguchi Y, Takemasa A, Hirabayashi A, and M Yamakido (1994). An immunohistochemical study of extracellular matrix components and integrins in human glomerular diseases. *Japanese Journal of Nephrology* 36:355–364.

Marshall MV, Draney D, Sevick-Muraca EM, and DM Olive (2010). Single-dose intravenous toxicity study of IRDye 800CW in Sprague-Dawley rats. *Molecular Imaging and Biology* 12:583–594.

Mazze RI, Cousins MJ, and GA Barr (1974). Renal effects and metabolism of isoflurane in man. *Anesthesiology* 40:536–542.

McLachlan AJ, Gross AS, Beal JL, Minns I, and SE Tett (2001). Analytical validation for a series of marker compounds used to assess renal drug elimination processes. *Therapeutic Drug Monitoring* 23:39–46.

Nilsson A, Forngren B, Bjurström S, Goodwin RJA, Basmaci E, Gustafsson I, Annas A, Hellgren D, Svanhagen A, Andrén PE, and J Lindberg (2012). In situ mass spectrometry imaging and ex vivo of renal crystalline deposits induced in multiple preclinical drug toxicology studies. *PLoS One* 7(10):e47353.

Patey N, Lesavrle P, Halbwachs-Mecarelli L, and LH Noel (1996). Adhesion molecules in human crescentic glomerulonephritis. *Journal of Pathology* 179:414–420.

Phillips E, Penate-Medina O, Zanzonico PB, Carvajal RD, Mohan P, Ye Y, Humm J, Gönen M, Kalaigian H, Schöder H, Strauss HW, Larson SM, Wiesner U, and Bradbury MS (2014). Clinical translation of an ultrasmall inorganic optical-PET imaging nanoparticle probe. *Science Translational Medicine.* 6(260):260ra149.

Pill J, Kraenzlin B, Jander J, Sattelkau T, Sadick M, Kloetzer H-M, Deus C, Kraemer U, and N Gretz (2005). Fluorescein-labeled sinistrin as marker of glomerular filtration rate. *European Journal of Medicinal Chemistry* 40:1056–1061.

Pill J, Issaeva O, Woderer S, Sadick M, Kränzlin B, Fiedler F, Klötzer H-M, Krämer U, and N Gretz (2006). Pharmacological profile and toxicity of fluorescein-labelled sinistrin, a novel marker for GFR measurements. *Naunyn-Schmiedeberg's Archives of Pharmacology* 373:204–211.

Poreddy AR, Neumann WL, Freskos JN, Rajagopalan R, Asmelash B, Gaston KR, Fitch RM, Galen KP, Shieh JJ, and RB Dorshow (2012). Exogenous fluorescent tracer agents based on pegylated pyrazine dyes for real-time point-of-care measurement of glomerular filtration rate. *Bioorganic & Medicinal Chemistry* 20(8):2490–2497.

Rabito CA, Chen Y, Schomacker KT, and MD Modell (2005). Optical, real-time monitoring of the glomerular filtration rate. *Applied Optics* 44:5956–5965.

Rajagopalan R, Neumann WL, Poreddy AR, Fitch RM, Freskos JN, Asmelash B, Gaston KR, Galen KP, Shieh J-J, and RB Dorshow (2011). Hydrophilic pyrazine dyes as exogenous fluorescent tracer agents for real-time point-of-care measurement of glomerular filtration rate. *Journal of Medicinal Chemistry* 54:5048–5058.

Roy-Chaudhury P, Hillis G, McDonald S, Simpson JG, and DA Power (1997). Importance of the tubulointerstitium in human glomerulonephritis. II. Distribution of integrin chains 1, a1 to 6 and av. *Kidney International* 52:103–110.

Sandilands EA, Dhaun N, Dear JW, and DJ Webb (2013). Measurement of renal function in patients with chronic kidney disease. *British Journal of Clinical Pharmacology* 76(4): 504–515.

Scarfe L, Rak-Raszewska A, Geraci S, Darssan D, Sharkey J, Huang J, Burton NC, Mason D, Ranjzad P, Kenny S, Gretz N, Lévy R, Park K, García-Fiñana M, Woolf AS, Murray P, and Wilm B (2015). Measures of kidney function by minimally invasive techniques correlate with histological glomerular damage in SCID mice with adriamycin-induced nephropathy. *Scientific Reports* 5:13601.

Schock-Kusch D, Sadick M, Henninger N, Kraenzlin B, Claus G, Kloetzer H-M, Weiß C, Pill J, and N Gretz (2009). Transcutaneous measurement of glomerular filtration rate using FITC-sinistrin in rats. *Nephrology, Dialysis, Transplantation* 24:2997–3001.

Schock-Kusch D, Xie Q, Shulhevich Y, Hesser Y, Stsepankou D, Sadick M, Koenig S, Hoecklin F, Pill J, and N Gretz (2011). Transcutaneous assessment of renal function in conscious rats with a device for measuring FITC-sinistrin disappearance curves. *Kidney International* 79:1254–1258.

Schreiber A, Shulhevich Y, Geraci S, Hesser J, Stsepankou D, Neudecker S, Koenig S, Heinrich R, Hoecklin F, Pill J, Friedemann J, Schweda F, Gretz N, and D Schock-Kusch (2012). Transcutaneous measurement of renal function in conscious mice. *American Journal of Physiology. Renal Physiology* 303:F783–F788.

Shikata K, Makino H, Morioka S, Kashitani T, Hirata K, Ota Z, Wada J, and YS Kanwar (1995). Distribution of extracellular matrix receptors in various forms of glomerulonephritis. *American Journal of Kidney Diseases* 25:680–688.

Silkalns GI, Jeck D, Earon J, Edelmann CM Jr, Chervu LR, Blaufox MD, and A Spitzer (1973). Simultaneous measurement of glomerular filtration rate and renal plasma flow using plasma disappearance curves. *Journal of Pediatrics* 83:749–757.

Taruttis A, Morscher S, Burton NC, Razansky D, and V Ntziachristos (2012). Fast multispectral optoacoustic tomography (MSOT) for dynamic imaging of pharmacokinetics and biodistribution in multiple organs. *PLoS ONE* 7(1):e30491.

Weening JJ and JC Jennette (2012). Historical milestones in renal pathology. *Virchows Archiv* 461(1):3–11.

Yu W, Sandoval RM, and BA Molitoris (2007). Rapid determination of renal filtration function using an optical ratiometric imaging approach. *American Journal of Physiology. Renal Physiology* 292:F1873–F1880.

Zöllner FG, Schock-Kusch D, Bäcker S, Neudecker S, Gretz N, and LR Schad (2013). Simultaneous measurement of kidney function by dynamic contrast enhanced MRI and FITC-sinistrin clearance in rats at 3 tesla: initial results. *PLoS ONE* 8(11): e79992.

35

IN VITRO TO *IN VIVO* RELATIONSHIPS WITH RESPECT TO KIDNEY SAFETY BIOMARKERS

Paul Jennings

Division of Physiology, Department of Physiology and Medical Physics, Medical University of Innsbruck, Innsbruck, Austria

35.1 RENAL CELL LINES AS TOOLS FOR TOXICOLOGICAL INVESTIGATIONS

Renal epithelial cells were originally used for viral replication and only later recognized for their potential in studying renal specific traits and chemical-induced toxicity. The porcine proximal tubule (PT)-like cell line, Lilly Laboratories cell porcine kidney 1 (LLC-PK1), and the canine collect duct-like cell line, Madin–Darby canine kidney (MDCK), are among the most widely studied renal cells (Gstraunthaler et al., 1985; Pfaller and Gstraunthaler, 1998). These lines belong to the so-called spontaneously immortalized cells, where immortalization was coincidental and the agent causing immortalization is unknown. The origin of such cell lines is assumed from the subsequent characterizations, for example, Na⁺-dependent apical transport of glucose, amino acids, and phosphate in LLC-PK1 cells (Handler et al., 1980; Gstraunthaler and Handler, 1987; Pfaller et al., 1990). A major disadvantage of these established cell lines is their animal origin and their somewhat dubious nephron segment pedigree. A number of human renal cell lines have been developed by introducing viral oncogenes either continuously or conditionally into primary cells of known origin. For example, the widely used HK-2 cell line was generated by transduction of human primary proximal tubular cells with human papillomavirus (HPV) early (E)6/E7 genes (Ryan et al., 1994). These cells retain many of the PT characteristics, such as brush-border enzyme activity, pH-dependent ammoniagenesis, and Na⁺-dependent/phlorizin-sensitive sugar transport

(Ryan et al., 1994; Wieser et al., 2008). However they do not become truly contact inhibited; they form a very low transepithelial electrical resistance (TEER), do not from domes, and after a few days at confluence begin to grow out from the monolayer (Aydin et al., 2008; Wieser et al., 2008). They also have an abnormal karyotype, are glycolytic, and have continuously changing transcriptome (Jennings et al., 2009). All of these undesired traits are most likely due to the inactivation of p53 (via the E6 protein), which creates a cancerous-type phenotype and an inability to become quiescence. In addition, the lack of p53-directed DNA repair mechanisms creates an unstable genetic background.

Conditional immortalization techniques have been used in an attempt to improve on permanent inactivation of the cells cell cycle arrest machinery; whereas tsA58 thermolabile SV40 TAg (SV40tsA58) is the most widely used. Expression of SV40tsA58 allows large T-antigen (TAg) expression under permissive conditions at 33°C but not at temperatures above 37°C (usually 39.5°C is recommended). TAg, like HPV E6, suppresses p53 activity, allowing continued proliferation. Cells cultured at temperatures above 37°C lose TAg expression allowing normal cell cycle regulation and differentiation. However, the inhibition of p53 during proliferation and amplification will unfortunately allow DNA errors to go unchecked and accumulate. Additionally, the culture of cells at 39°C is far from ideal for toxicological investigations due to the overexpression of heat shock proteins. A potentially better way to develop human cell lines is to overexpress the catalytic subunit of telomerase (Wieser et al., 2008).

Drug Discovery Toxicology: From Target Assessment to Translational Biomarkers, First Edition. Edited by Yvonne Will, J. Eric McDuffie, Andrew J. Olaharski, and Brandon D. Jeffy.

This enzyme increases telomere length so that telomeres do not reach the critical threshold length to initiate replicative senescence. Since human somatic cells express low levels of this enzyme, reintroduction effectively prevents replicative senescence without interfering with the normal activation of the p53 pathway. The human proximal tubular cell line, RPTEC/TERT1, was generated with human telomerase reverse transcriptase (hTERT) introduction (Wieser et al., 2008). It has been shown that this cell line maintains good proximal tubular characteristics, including parathyroid hormone cAMP stimulation, lack of arginine vasopressin cAMP stimulation, dome formation, formation of a stable TEER, and expression of PT tight junctions, for example, claudin 2 and claudin 10 (Wieser et al., 2008; Limonciel et al., 2012; Wilmes et al., 2014a). Also these cells switch from a high glycolytic metabolism to a low glycolytic metabolism with increased mitochondrial respiratory capacity and increased fatty acid oxidation after reaching confluence and entering p53-induced quiescence (Aschauer et al., 2013). The cells maintain a normal diploid karyotype over a wide range of population doublings (Wieser et al., 2008). However, hTERT expression alone is not guaranteed to produce a cell line, and a number of cell lines have been produced by combining SV40tsA58 and hTERT overexpression, including human podocytes (Delarue et al., 1991; Saleem et al., 2002; Sakairi et al., 2010) and PT cells (Wilmer et al., 2010). Interestingly some of these cell lines have been produced from viable cells that are shed into the urine (Wilmer et al., 2010). The use of exfoliated urinary cells potentially allows the investigations of donor-specific properties. However, as mentioned, one should be careful with karyotype abnormalities and genetic drifting when cells overexpress any viral oncogene such as the SV40 TAg.

An entirely new approach to generate renal epithelial cells is the use of inducible pluripotent stem cells (iPSC) (Takahashi and Yamanaka, 2006). iPSC offer several advantages over primary cell culture and cell lines in that the cells can be generated from specific donors where the genetic background is known the cells are mortal and thus have normal cell cycle regulation, target cells can be generated repeatedly from the same iPSC pool allowing biobanking of patient-derived cells, and multiple target cell lineages can be generated from the same donor, which will allow, for the first time, autologous co-culture experiments (Wilmes and Jennings, 2014). However, novel differentiation protocols are still being developed with some notable success(es). Araoka et al. have used a Wnt pathway activator, CHIR99021, together with a retinoic acid (RA) agonist (TTNPB) to generate intermediate mesoderm (IM) and subsequently different renal epithelial tubule-like cells (Araoka et al., 2014). Lam et al. used a similar strategy to produce cells with apically monociliated tubular structures that coexpressed Lotus tetragonolobus lectin, N-cadherin, and cadherin-16 (Lam et al., 2014). Song et al. demonstrated the possibility of

human iPSC differentiation into podocyte-like cells via stimulation with activin A, bone morphogenetic protein 7 (BMP7), and RA for 10 days (Song et al., 2012). iPSC-derived podocyte-like cells showed upregulation of the podocyte-specific markers podocin, nephrin, and Wilms tumor 1 (WT1) (Song et al., 2012). Additionally, the cells took up albumin and contracted in response to angiotensin II (Song et al., 2012). It is clear that more work will be required to optimize these approaches to increase yield, purity, stability, and differentiation status of the target cells; however iPSC holds great promise for investigating population dynamics and genetic basis for renal diseases.

35.2 MECHANISTIC APPROACHES AND *IN VITRO* TO *IN VIVO* TRANSLATION

Renal cell cultures have been used very successfully to investigate the mechanisms of clinical nephrotoxins (Jennings et al., 2008) and nephrotoxic compounds found in the environment and in food (Pfaller and Gstraunthaler, 1998; Limonciel and Jennings, 2014). In fact this is the major advantage of using *in vitro* cell cultures, as whole animal models are not very amenable to deep molecular investigation. Additionally, since we cannot conduct experiments on humans, the use of human-derived cell cultures is the only way to conduct explorative mechanistic research in cells and tissues of human origin. With the introduction of transcriptomics and other complementary -omic technologies, such as proteomics and metabolomics, it is now possible to conduct hypothesis-free investigations that produce high-content mechanistically rich-data. For example, the RPTEC/TERT1 model has been used with integrated -omics and biokinetics to further understand the mechanisms of prototype nephrotoxins, cyclosporine A (CsA), and cisplatin (Wilmes et al., 2013, 2014b). CsA was well tolerated up to 5 μM by the cells, which represents a high therapeutic concentration, but at 15 μM there was a severe CsA cellular accumulation over time leading to activation of the nuclear factor (erythroid-derived 2)-like 2 (NFE2L2 or Nrf2) oxidative stress pathway and the unfolded protein response pathway (Wilmes et al., 2013). At this high concentration, it was concluded that the P-glycoprotein-driven extrusion of CsA was overcome, leading to CsA accumulation in the mitochondria and endoplasmic reticulum (ER), causing both mitochondrial and ER disruptions. For cisplatin there was no clear threshold for accumulation, likely due to the fact that cisplatin uptake is primarily mediated by basolateral organic cation transporter 2 (OCT2) uptake, which is rate limiting (Wilmes et al., 2014b). Cisplatin exposure induced p53 and Nrf2 activation, while mammalian target of rapamycin (mTOR) signaling and eukaryotic initiation factor 2 (eIF2) signaling were suppressed (Wilmes et al., 2014b). There was also evidence for cisplatin-induced mitochondrial disturbances. These experiments demonstrate that integrated - omics together with biokinetics, when used together with appropriate stable differentiated cell culture models, can be extremely useful in understanding mechanisms of chemical-induced toxicity. Furthermore, combining *in vitro* responses with cellular compound concentration is necessary to build prediction models of responses in humans and move beyond hazard identification only. In addition the identification of mechanistic markers in *in vitro* experiments may also be applied to the clinical setting.

Lipocalin-2 (LCN2, aka NGAL) induction, was first demonstrated in kidney tissue in a mouse ischemia model (Kieran et al., 2003). Several *in vitro* studies have independently demonstrated that LCN2 mRNA and its protein product are induced upon exposure to nephrotoxins (Wu et al., 2013; Jennings et al., 2015). LCN2 is a 25-kDa protein and a member of the innate immune system; it functions to sequester bacterial iron-containing siderophores, thereby depleting bacteria of iron. Its inducibility in injured tissue is likely due to the fact that it contains one nuclear factor kappa-light-chain-enhancer of activated B cell (NF-κB) binding site and four signal transducer and activator of transcription 1 (STAT1) binding sites in its promoter region (Zhao and Stephens, 2013). It is known to be secreted in urine (primarily from PT cells), and increased urinary LCN2 correlates with decreased renal function under various settings (Ling et al., 2008; Hall et al., 2010; Wu et al., 2013). In a recent study with long-term RPTEC/TERT1 exposures to 9 nephrotoxins, it was shown that LCN2 is induced for the majority of these compounds (Aschauer et al., 2014). In the same inflammatory cluster, a less well-studied gene interleukin (IL)-19 was also identified (Aschauer et al., 2014; Jennings et al., 2015). It was demonstrated that IL-19 protein was also secreted by the cells and could be quantified in the cell culture medium. The potential of IL-19 to be detected in urine of patients with chronic kidney disease (CKD) was also investigated. Urinary IL-19 negatively correlated with glomerular filtration rate and positively correlated to both urinary LCN2 and NAG (Jennings et al., 2015). It was concluded that IL-19 is a novel biomarker of PT injury and a good example of a marker discovered *in vitro* with good translation as an *in vivo* biomarker.

35.3 CLOSING REMARKS

Human *in vitro* cell cultures from primary tissues, from cell lines, or iPSC derived are extremely useful tools for further understanding human physiology and how cells react to chemical-induced cell stress. However, we need to be a little more careful with cells that have cancerous phenotypes and severe chromosomal aberrations. In addition, the use of normal, non-proliferating cells best reflects the situation of the nephron *in vivo*. As discussed such cell systems can be used for mechanistic studies, hazard identification, risk assessment, and biomarker discovery. Finally, increasing the genetic background of *in vitro* models, for example, using iPSC-derived cells, will allow the assessment of population dynamics and individual susceptibilities, potentially ushering in a new era of personalized drug development.

REFERENCES

Araoka, T., Mae, S., Kurose, Y., Uesugi, M., Ohta, A., Yamanaka, S., and Osafune, K. (2014) Efficient and rapid induction of human iPSCs/ESCs into nephrogenic intermediate mesoderm using small molecule-based differentiation methods. *PloS ONE* 9, e84881.

Aschauer, L., Gruber, L. N., Pfaller, W., Limonciel, A., Athersuch, T. J., Cavill, R., Khan, A., Gstraunthaler, G., Grillari, J., Grillari, R., Hewitt, P., Leonard, M. O., Wilmes, A., and Jennings, P. (2013) Delineation of the key aspects in the regulation of epithelial monolayer formation. *Molecular and Cellular Biology* 33, 2535–2550.

Aschauer, L., Limonciel, A., Wilmes, A., Stanzel, S., Kopp-Schneider, A., Hewitt, P., Lukas, A., Leonard, M. O., Pfaller, W., and Jennings, P. (2014) Application of RPTEC/TERT1 cells for investigation of repeat dose nephrotoxicity: a transcriptomic study. *Toxicology In Vitro*. doi:10.1016/j.tiv.2014.10.005.

Aydin, S., Signorelli, S., Lechleitner, T., Joannidis, M., Pleban, C., Perco, P., Pfaller, W., and Jennings, P. (2008) Influence of microvascular endothelial cells on transcriptional regulation of proximal tubular epithelial cells. *American Journal of Physiology. Cell Physiology* 294, C543–C554.

Delarue, F., Virone, A., Hagege, J., Lacave, R., Peraldi, M. N., Adida, C., Rondeau, E., Feunteun, J., and Sraer, J. D. (1991) Stable cell line of T-SV40 immortalized human glomerular visceral epithelial cells. *Kidney International* 40, 906–912.

Gstraunthaler, G. and Handler, J. S. (1987) Isolation, growth, and characterization of a gluconeogenic strain of renal cells. *The American Journal of Physiology* 252, C232–C238.

Gstraunthaler, G., Pfaller, W., and Kotanko, P. (1985) Biochemical characterization of renal epithelial cell cultures (LLC-PK1 and MDCK). *The American Journal of Physiology* 248, F536–F544.

Hall, I. E., Yarlagadda, S. G., Coca, S. G., Wang, Z., Doshi, M., Devarajan, P., Han, W. K., Marcus, R. J., and Parikh, C. R. (2010) IL-18 and urinary NGAL predict dialysis and graft recovery after kidney transplantation. *Journal of the American Society of Nephrology* 21, 189–197.

Handler, J. S., Perkins, F. M., and Johnson, J. P. (1980) Studies of renal cell function using cell culture techniques. *The American Journal of Physiology* 238, F1–F9.

Jennings, P., Koppelstätter, C., Lechner, J., and Pfaller, W. (2008) Renal cell culture models: Contribution to the understanding of nephrotoxic mechanisms. In: *Clinical Nephrotoxins*, Springer, New York. pp 223–249.

Jennings, P., Aydin, S., Bennett, J., McBride, R., Weiland, C., Tuite, N., Gruber, L. N., Perco, P., Gaora, P. O., Ellinger-Ziegelbauer, H., Ahr, H. J., Kooten, C. V., Daha, M. R., Prieto, P.,

Ryan, M. P., Pfaller, W., and McMorrow, T. (2009) Inter-laboratory comparison of human renal proximal tubule (HK-2) transcriptome alterations due to Cyclosporine A exposure and medium exhaustion. *Toxicology In Vitro* 23, 486–499.

Jennings, P., Crean, D., Aschauer, L., Limonciel, A., Moenks, K., Kern, G., Hewitt, P., Lhotta, K., Lukas, A., Wilmes, A., and Leonard, M. O. (2015) Interleukin-19 as a translational indicator of renal injury. *Archives of Toxicology* 89, 101–106.

Kieran, N. E., Doran, P. P., Connolly, S. B., Greenan, M. C., Higgins, D. F., Leonard, M., Godson, C., Taylor, C. T., Henger, A., Kretzler, M., Burne, M. J., Rabb, H., and Brady, H. R. (2003) Modification of the transcriptomic response to renal ischemia/reperfusion injury by lipoxin analog. *Kidney International* 64, 480–492.

Lam, A. Q., Freedman, B. S., Morizane, R., Lerou, P. H., Valerius, M. T., and Bonventre, J. V. (2014) Rapid and efficient differentiation of human pluripotent stem cells into intermediate mesoderm that forms tubules expressing kidney proximal tubular markers. *Journal of the American Society of Nephrology* 25, 1211–1225.

Limonciel, A. and Jennings, P. (2014) A review of the evidence that ochratoxin A is an Nrf2 inhibitor: implications for nephrotoxicity and renal carcinogenicity. *Toxins* 6, 371–379.

Limonciel, A., Wilmes, A., Aschauer, L., Radford, R., Bloch, K. M., McMorrow, T., Pfaller, W., van Delft, J. H., Slattery, C., Ryan, M. P., Lock, E. A., and Jennings, P. (2012) Oxidative stress induced by potassium bromate exposure results in altered tight junction protein expression in renal proximal tubule cells. *Archives of Toxicology* 86, 1741–1751.

Ling, W., Zhaohui, N., Ben, H., Leyi, G., Jianping, L., Huili, D., and Jiaqi, Q. (2008) Urinary IL-18 and NGAL as early predictive biomarkers in contrast-induced nephropathy after coronary angiography. *Nephron Clinical Practice* 108, c176–c181.

Pfaller, W. and Gstraunthaler, G. (1998) Nephrotoxicity testing in vitro—what we know and what we need to know. *Environmental Health Perspectives* 106 Suppl 2, 559–569.

Pfaller, W., Gstraunthaler, G., and Loidl, P. (1990) Morphology of the differentiation and maturation of LLC-PK1 epithelia. *Journal of Cellular Physiology* 142, 247–254.

Ryan, M. J., Johnson, G., Kirk, J., Fuerstenberg, S. M., Zager, R. A., and Torok-Storb, B. (1994) HK-2: an immortalized proximal tubule epithelial cell line from normal adult human kidney. *Kidney International* 45, 48–57.

Sakairi, T., Abe, Y., Kajiyama, H., Bartlett, L. D., Howard, L. V., Jat, P. S., and Kopp, J. B. (2010) Conditionally immortalized human podocyte cell lines established from urine. *American Journal of Physiology. Renal Physiology* 298, F557–F567.

Saleem, M. A., O'Hare, M. J., Reiser, J., Coward, R. J., Inward, C. D., Farren, T., Xing, C. Y., Ni, L., Mathieson, P. W., and Mundel, P.
(2002) A conditionally immortalized human podocyte cell line demonstrating nephrin and podocin expression. *Journal of the American Society of Nephrology* 13, 630–638.

Song, B., Smink, A. M., Jones, C. V., Callaghan, J. M., Firth, S. D., Bernard, C. A., Laslett, A. L., Kerr, P. G., and Ricardo, S. D. (2012) The directed differentiation of human iPS cells into kidney podocytes. *PLoS ONE* 7, e46453.

Takahashi, K. and Yamanaka, S. (2006) Induction of pluripotent stem cells from mouse embryonic and adult fibroblast cultures by defined factors. *Cell* 126, 663–676.

Wieser, M., Stadler, G., Jennings, P., Streubel, B., Pfaller, W., Ambros, P., Riedl, C., Katinger, H., Grillari, J., and Grillari-Voglauer, R. (2008) hTERT alone immortalizes epithelial cells of renal proximal tubules without changing their functional characteristics. *American Journal of Physiology. Renal Physiology* 295, F1365–F1375.

Wilmer, M. J., Saleem, M. A., Masereeuw, R., Ni, L., van der Velden, T. J., Russel, F. G., Mathieson, P. W., Monnens, L. A., van den Heuvel, L. P., and Levtchenko, E. N. (2010) Novel conditionally immortalized human proximal tubule cell line expressing functional influx and efflux transporters. *Cell and Tissue Research* 339, 449–457.

Wilmes, A. and Jennings, P. (2014) The use of renal cell culture for nephrotoxicity investigations. In: *Predictive Toxicology*, Wiley-VCH Verlag GmbH & Co. KGaA, Weinheim. pp 195–216.

Wilmes, A., Limonciel, A., Aschauer, L., Moenks, K., Bielow, C., Leonard, M. O., Hamon, J., Carpi, D., Ruzek, S., Handler, A., Schmal, O., Herrgen, K., Bellwon, P., Burek, C., Truisi, G. L., Hewitt, P., Di Consiglio, E., Testai, E., Blaauboer, B. J., Guillou, C., Huber, C. G., Lukas, A., Pfaller, W., Mueller, S. O., Bois, F. Y., Dekant, W., and Jennings, P. (2013) Application of integrated transcriptomic, proteomic and metabolomic profiling for the delineation of mechanisms of drug induced cell stress. *Journal of Proteomics* 79, 180–194.

Wilmes, A., Aschauer, L., Limonciel, A., Pfaller, W., and Jennings, P. (2014) Evidence for a role of claudin 2 as a proximal tubular stress responsive paracellular water channel. *Toxicology and Applied Pharmacology* 279, 163–172.

Wilmes, A., Bielow, C., Ranninger, C., Bellwon, P., Aschauer, L., Limonciel, A., Chassaigne, H., Kristl, T., Aiche, S., Huber, C. G., Guillou, C., Hewitt, P., Leonard, M. O., Dekant, W., Bois, F., and Jennings, P. (2014) Mechanism of cisplatin proximal tubule toxicity revealed by integrating transcriptomics, proteomics, metabolomics and biokinetics. *Toxicology In Vitro*. doi:10.1016/j.tiv.2014.10.006.

Wu, K. D., Hsing, L. L., and Huang, Y. F. (2013) Concentration of plasma neutrophil gelatinase-associated lipocalin in patients with chronic kidney disease. *Clinical Laboratory* 59, 909–913.

Zhao, P. and Stephens, J. M. (2013) STAT1, NF-kappaB and ERKs play a role in the induction of lipocalin-2 expression in adipocytes. *Molecular Metabolism* 2, 161–170.

36

CASE STUDY: FULLY AUTOMATED IMAGE ANALYSIS OF PODOCYTE INJURY BIOMARKER EXPRESSION IN RATS

Jing Ying Ma

Molecular Pathology, Discovery Sciences, Janssen Research & Development, L.L.C., San Diego, CA, USA

36.1 INTRODUCTION

Increased desmin immunostaining in glomerular podocytes has been widely utilized as a sensitive podocyte injury marker and observed in various rat models of glomerular disease (Yaoita et al., 1990; Floege et al., 1992, 1997; Hoshi et al., 2002; Zou et al., 2006; Herrmann et al., 2012; Kakimoto, et al., 2013). Considering the heterogeneity of glomerular injuries in renal cortex (Sofue et al., 2012) and inherent bias of evaluating only a limited number of glomeruli when utilizing manual methods of image analysis, which could lead to arbitrary selection and improper results, new tools are warranted. Futhermore, semiautomated quantification of glomerular desmin-immunopositive area has been reported (Herrmann et al., 2012); yet, such methologies allow for random cell selections based on manual tracings to segregate glomeruli from other visible renal cell structures. Therefore, we aimed to develop a more accurate approach to consider when monitoring for invasive changes in glomerular-specific kidney safety biomarkers following nephrotoxicant exposure. In this case study, we describe a novel computational image analysis method to automatically detect glomeruli and to quantify areas and intensities of desmin immunolabeling in podocytes, particularly to detect enhanced desmin immunoreactivity in kidney sections from doxorubicin (DOX)-treated rats (Ma et al., 2015).

36.2 MATERIAL AND METHODS

Naive male Sprague-Dawley rats (5/group) received single intravenous injections of DOX (7.5 mg/kg) or vehicle (0.9% saline) once weekly. The scheduled termination was on study day 14. A complete necropsy was performed, and representative kidney tissues were collected in 10% neutral-buffered formalin (NBF). Kidney tissues were fixed in 10% NBF for 48h, processed, embedded in paraffin, sectioned at 4μm, and immunohistochemically stained with rabbit monoclonal antibody specific for desmin (Millipore, Cat# 04-585). In addition, the kidney sections were also stained with hematoxylin and eosin for routine histopathology evaluation. Quantitative image analysis was performed for desmin using Visiopharm VISIOmorph software (Visiopharm, Horsham, Denmark). IHC stained desmin slides were scanned at 40× magnification using NanoZoomer 2.0 HT (Hamamatsu Photonics, Middlesex, NJ, United States) whole-slide scanner, and virtual slides were imported into the Visiopharm® platform. Briefly, a linear threshold classifier, which was pretrained using HDAB-DAB feature combined with a mean filter and poly-local linear filter, detected glomeruli in whole-slide image during scanning with a 1024×1024-pixeled detection window. Next, boundaries of the detected glomeruli were automatically determined using postprocessing,

Drug Discovery Toxicology: From Target Assessment to Translational Biomarkers, First Edition. Edited by Yvonne Will, J. Eric McDuffie, Andrew J. Olaharski, and Brandon D. Jeffy.

including edge detection, smoothing, binarization, and morphology operation followed by the quantification of glomerular area, glomerular desmin-immunopositive area, and intensity, and then the percentage of desmin-immunopositive area was calculated. Two hundred glomeruli in average in the whole cortex including outer and juxtamedullary cortex were automatically recognized and analyzed for each rat. Results were expressed as mean±SD. Differences between groups were calculated by unpaired t test. The differences between groups were considered to be statistically significant when $p < 0.05$.

36.3 RESULTS

To quantitate glomerular desmin immunoreactivity, we developed a novel computational method to automatically detect glomeruli and to quantify the glomerular-immunopositive area and intensity. Here, a HDAB-DAB feature combined with a mean and poly-local linear filter trained the Visiopharm VISIOmorph software using a linear threshold classifier to recognize the glomeruli within the kidney tissue sections (Fig. 36.1). This feature sensitively detected moderately and minimally positively stained glomeruli.

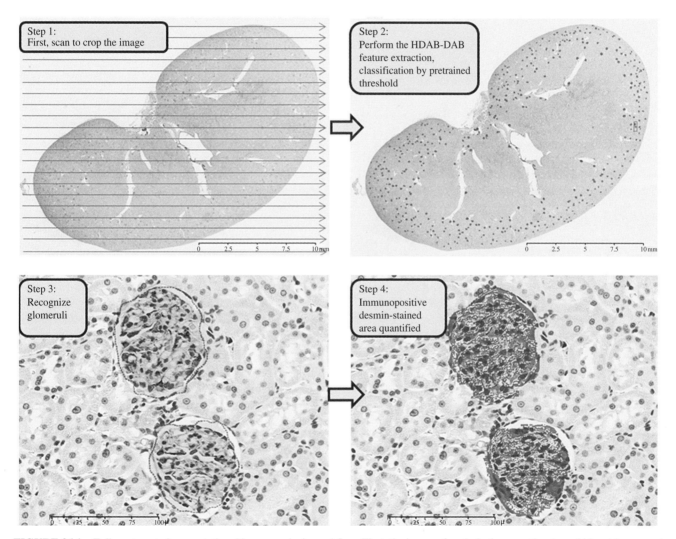

FIGURE 36.1 Fully automated computational image analysis workflow. First, the image of a whole tissue section (e.g., kidney) is scanned and subsequent subimages are processed (cropped). Second, the HDAB-DAB extraction feature combined with a mean and poly-local linear filter supports image classification using the pretrained linear threshold. This tool allows for candidate cropped subimage selection; the images reveal positively stained morphological structures (e.g., glomeruli and nonglomerular tissues—*see inset of recognized glomeruli*). Lastly, the algorithm automatically recognizes and quantifies the immunopositively stained area and intensity (such as the desmin immunostaining shown). (*See insert for color representation of the figure.*)

As a result, fully automated detection of the borders of the glomeruli was feasible using electronic postprocessing and subsequent quantification of glomerular desmin-immunopositive area and intensity. Minimal staining was detected in glomeruli of control rat kidneys while a significant increase in glomerular (whole cortex) desmin-immunopositive area ($p<0.001$) and intensity ($p<0.01$) in

the DOX-treated rats (Table 36.1). Marked differences in podocyte desmin immunoreactivity (area and intensity) between outer and juxtamedullary glomeruli in DOX-treated rats (Fig. 36.2 and Table 36.2) were also revealed. Statistically significant increases in desmin-positive area ($p<0.01$) and in desmin intensity ($p<0.05$) were observed in juxtamedullary glomeruli as compared with outer cortical glomeruli.

TABLE 36.1   Desmin Immunoreactivity in Whole Cortical Glomeruli

Group—Animal	Desmin (%)	Desmin (ODT)	Group—Animal	Desmin (%)	Desmin (ODT)
G1-1	1.36	136	G2-31	7.37	148
G1-2	2.49	138	G2-32	4.27	139
G1-3	1.29	139	G2-33	8.19	147
G1-4	0.73	135	G2-34	6.88	145
G1-5	1.52	138	G2-35	6.27	145
Mean ± SD	1.48 ± 0.64	137 ± 1.81	Mean ± SD	6.60 ± 1.48***	144.51 ± 3.49**

G1 (vehicle control group); G2 (DOX group); % (desmin-immunopositive area/whole cortical glomerular area); ODT (optical density thresholds for glomerular desmin-immunopositive staining intensity).
p Values (G1 vs. G2): ***$p<0.001$ or **$p<0.01$.

(a)   (b)

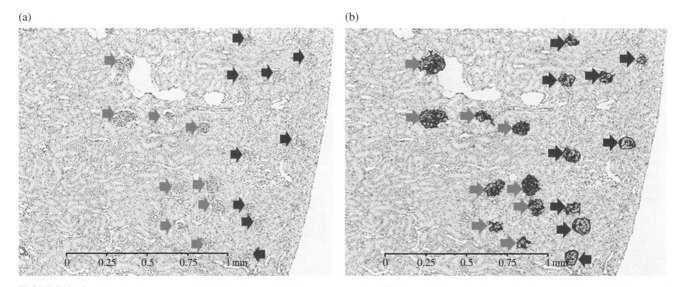

FIGURE 36.2   Desmin immunoreactivity in outer cortical (green arrows) and juxtamedullary (red arrows) glomeruli from a DOX-treated rat (a) and quantification step for assessing areas of enhanced desmin immunereactivity in damaged podocytes of juxtamedullary glomeruli (b). Glomeruli are recognized (green color); desmin-positive-labeled podocytes (red color). (*See insert for color representation of the figure.*)

TABLE 36.2   Desmin Immunoreactivity in Outer or Juxtamedullary Cortical Glomeruli

Treatment	Desmin (%)	Desmin (ODT)	Desmin (%)	Desmin (ODT)
DOX group—animal	Outer cortical glomeruli (A)		Juxtamedullary cortical glomeruli (B)	
G2-31	4.56	143	8.52	144
G2-32	3.57	137	6.08	141
G2-33	4.52	142	12.4	149
G2-34	5.63	143	8.07	146
G2-35	3.77	139	9.57	148
Mean ± SD	4.41 ± 0.81	140.80 ± 2.68	8.93 ± 2.32**	145.60 ± 3.21*

G2 (DOX group); % (desmin-immunopositive area/outer cortical glomerular area or juxtamedullary cortical glomerular area); ODT (optical density thresholds for desmin-immunopositive staining intensity).
p Values: **$p<0.01$ (A vs. B) or *$p<0.05$ (A vs. B).

36.4   CONCLUSIONS

We demonstrated the detection of DOX-related glomerular injury using desmin as a sensitive podocyte injury marker in the subacute nephrotoxicity in rats. The fully automated image analysis platform could feasibly support morphologic assessment of compound-induced glomerulopathy in bespoke discovery toxicology studies and/or retrospective analysis.

REFERENCES

Floege J, Alpers CE, Sage EH, Pritzl P, Gordon K, Johnson RJ, Couser WG. (1992) Markers of complement-dependent and complement-independent glomerular visceral epithelial cell injury in vivo. Expression of antiadhesive proteins and cytoskeletal changes. *Lab. Investig.* 67:486–497.

Floege J, Hackmann B, Kliem V, Kriz W, Alpers CE, Johnson RJ, Kühn KW, Koch K-M, Brunkhorst R. (1997) Age-related glomerulosclerosis and interstitial fibrosis in Milan normotensive rats: A podocyte disease. *Kidney Int.* 51, 230–243.

Herrmann A, Tozzo E, Funk J. (2012) Semi-automated quantitative image analysis of podocyte desmin immunoreactivity as a sensitive marker for acute glomerular damage in the rat puromycin aminonucleoside nephrosis (PAN) model. *Exp. Toxicol. Pathol.* 64:45–49.

Hoshi S, Shu Y, Yoshida F, Inagaki T, Sonoda J, Watanabe T, Nomoto K, Nagata M. (2002) Podocyte injury promotes progressive nephropathy in Zucker diabetic fatty rats. *Lab. Invest.* 82:25–35.

Kakimoto T, Kimata H, Iwasaki S, Fukunari A, Utsumi H. (2013) Automated recognition and quantification of pancreatic islets in Zucker diabetic fatty rats treated with exendin-4. *J. Endocrinol.* 216:13–20.

Ma JY, Church R, Kanerva J, Bittner A, McDuffie JE. (2015) Automated image analysis of podocyte desmin immunostaining in a rat model of sub-acute doxorubicin-induced glomerulopathy. *J J Biomark.* 1(2):008.

Sofue T, Kiyomoto H, Kobori H, Urushihara M, Nishijima Y, Kaifu K, Hara T, Matsumoto S, Ichimura A, Ohsaki H, Hitomi H, Kawachi H, Hayden MR, Whaley-Connell A, Sowers JR, Ito S, Kohno M, Nishiyama A. (2012) Early treatment with olmesartan prevents juxtamedullary glomerular podocyte injury and the onset of microalbuminuria in type 2 diabetic rats. *Am. J. Hypertens.* 25:604–611.

Yaoita E, Kawasaki K, Yamanoto T, Kihra I. (1990) Variable expression of desmin in rat glomerular epithelial cells. *Am. J. Pathol.* 136(4):899–908.

Zou J, Yaoita E, Watanabe Y, Yoshida Y, Nameta M, Li H, Qu Z, Yamamoto T. (2006) Upregulation of nestin, vimentin, and desmin in rat podocytes in response to injury. *Virchows Arch.* 448:485–492.

37

CASE STUDY: NOVEL RENAL BIOMARKERS TRANSLATION TO HUMANS

DEBORAH A. BURT

Biomarker Development and Translation, Drug Safety Research and Development, Pfizer Inc., Groton, CT, USA

37.1 INTRODUCTION

Depending on the type of injury, preclinical DIKI findings help to define safety margins for clinical studies. The margins are defined in part based on case-by-case evaluations of preclinical data including nephron-specific kidney histopathology findings and associated changes in traditional DIKI biomarkers (i.e., sCr, BUN) as well as reversibility. With the advancement of novel DIKI biomarkers, compound development programs may benefit from translational strategies that include supportive traditional DIKI biomarker data, which result in the inclusion of novel DIKI biomarker measurements, consultation with regulatory agencies on the monitoring and risk management plan, and consideration of safety margins with feasible stopping criteria based on clinical monitoring of fit-for-purpose biomarker panels. The following case study uniquely highlights options to consider when implementing DIKI assessment strategies.

37.2 IMPLEMENTATION OF TRANSLATIONAL RENAL BIOMARKERS IN DRUG DEVELOPMENT

Polypeptide antibiotics are essential for treatment of life-threatening gram-negative infections. AKI attributed to treatment severely limits their clinical application (Kubin et al., 2012). Traditional renal biomarkers including sCr, BUN, and total protein in urine lack the sensitivity and specificity needed to diagnose and monitor nephron-specific toxicities (Goodsaid et al., 2009). To overcome the limitations

of traditional markers, Burt et al. (2014) explored the utility of several emerging biomarkers of kidney injury (Ozer et al., 2010) for monitoring nephrotoxicity of polymyxin analogues in preclinical toxicity studies.

The biomarker strategy included identifying and evaluating several biomarkers of AKI for detection of nephrotoxicity caused by polymyxin B antibiotic in three preclinical models (Wistar rats, beagle dogs, and cynomolgus monkeys). Based on suitable species-specific reagent availabilities when the investigations were conducted, the biomarkers evaluated included urine concentration changes in creatinine, α-GST, total protein, KIM-1, NGAL, microalbumin, OPN, and/or glucose. Urinary KIM-1, urinary NGAL, and urinary NAG outperformed the other biomarkers and were selected for further evaluation.

The efficacious dose for polymyxin B in humans is 1.5 mg to 2.5 mg/kg/day (Falagas and Kasiakou, 2006). After polymyxin B administration to rats, dogs, and NHPs at 0.1–10 mg/kg/day, the results of standard biomarkers (sCr, BUN, and urinary total protein) were compared to the response of the emerging AKI biomarkers, KIM-1, NGAL, and NAG in urine. This resulted in biomarker profiles for each species that were correlated to histopathologic evidence of renal damage. In the rat, the standard length of a toxicity study for the polymyxin analogues was 2 weeks and included several sample collection time points for hematology and clinical chemistry as well as histopathology for each study. The earliest histopathology performed was at 1 week and showed tubular epithelium degeneration/regeneration in the low- and mid-dose groups and tubular necrosis in the high-dose group,

---

Drug Discovery Toxicology: From Target Assessment to Translational Biomarkers, First Edition. Edited by Yvonne Will, J. Eric McDuffie, Andrew J. Olaharski, and Brandon D. Jeffy.

indicating DIKI. Urinary total protein and BUN results showed a small (<2-fold) increase that was not dose dependent. Urinary total protein values increased at 48 h postdose but were transient, while BUN levels did not increase until 7 days post exposure. There were no sCr changes at any dose level throughout the study. Urinary NAG also had no response throughout but urinary NGAL, and urinary KIM-1 gave significant and dose-dependent increases at all dose levels and detected AKI as early as 48 h post exposure. Urinary NGAL was the most sensitive biomarker of AKI caused by polymyxin B in rats with levels peaking (13-fold increase) 48 h after the first dose. KIM-1 peaked 48 h post first dose with a significant (threefold) increase. The strong performance of urinary NGAL, which correlated with histopathology, led to the implementation of a 2-day study design in rats. Screening and rank ordering of polymyxin analogues with these abbreviated 2-day studies enabled fast and cost-effective drug development, reduced the numbers of animals and test compound needed, and eliminated the need for histopathologic examination of tissues. The selected lead compounds were also less nephrotoxic than polymyxin B.

Based on the rat toxicity studies, subsequent dog and nonhuman primate studies were performed with polymyxin B and polymyxin B analogues. The traditional biomarkers, sCr, BUN, and urinary total protein, were evaluated in addition to urinary NGAL, urinary KIM-1, and urinary NAG when species-specific assays were available. With these large animal studies, all of the classic and novel urinary and serum-based biomarkers showed dose-dependent increases, which correlated with histopathology in dogs (no histopathology was performed in the nonhuman primate studies), with no specific biomarker outperforming the rest. While the emerging AKI biomarkers evaluated did not provide added value in the dog and nonhuman primate after treatment with polymyxin B, this profiling did enable the development and implementation of short-term studies utilizing the emerging biomarkers as a screening tool for polymyxin analogues, most notably in rats.

The polymyxin program did not proceed into the clinic but translatability was still addressed using samples from patients treated with a variety of standard of care antibiotics (piperacillin/tazobactam, vancomycin, penicillin V, tobramycin, gentamicin, ampicillin, cefuroxime, clindomycin, clarithromycin, cefepime, daptomycin IV, bumetanide, vibramycin, and cyclophosphamide) compared to samples from healthy volunteers. This sample set was analyzed for sCr, BUN, urinary total protein, and urinary microalbumin as well as the emerging biomarkers urinary KIM-1, urinary NGAL, and urinary NAG. Much like the large animal studies with polymyxin B, the response for the antibiotic-treated cohort produced increases in all urinary- and serum-based biomarkers when compared to the healthy subjects. Although both traditional and novel biomarkers were able to detect AKI in humans, there is still the potential for the emerging biomarkers to add value as they may help to differentiate the area of injury within the kidneys (proximal tubules and/or distal tubules versus the glomeruli) and distinguish the development of treatment-related AKI onset in humans (Bonventre et al., 2010).

37.3   CONCLUSION

Biomarkers of AKI are essential for the detection and monitoring of nephrotoxicity in drug development. The rat is a commonly used species for toxicity screening; however, the classic serum-based biomarkers sCr and BUN were not able to effectively detect or monitor kidney injury after treatment with polymyxin analogues. The preclinical approach of evaluating a panel of AKI biomarkers and then selecting the most responsive marker (urinary NGAL) enabled fast and economical assessment of polymyxin analogues in rat. This enhanced compound screening capability, in turn, allowed faster progression of compounds to large animal studies. The species-specific biomarker responses to polymyxin B provided a solid platform for clinical translation and the similar biomarker profile of polymyxin B-treated dogs and nonhuman primates to antibiotic-treated human subjects showed translatability of the emerging biomarkers of AKI. Taken together, we demonstrated successful applications of emerging renal safety biomarkers for compound selection and managing risk of DIKI including the translational potentials for novel DIKI biomarkers to the clinic setting.

REFERENCES

Bonventre, J. V., et al. (2010). "Next-generation biomarkers for detecting kidney toxicity." *Nat Biotechnol* 28(5): 436–440.

Burt, D., et al. (2014). "Application of emerging biomarkers of acute kidney injury in development of kidney-sparing polypeptide-based antibiotics." *Drug Chem Toxicol* 37(2): 204–212.

Falagas, M. E. and S. K. Kasiakou (2006). "Toxicity of polymyxins: a systematic review of the evidence from old and recent studies." *Crit Care* 10(1): R27.

Goodsaid, F. M., et al. (2009). "Novel biomarkers of acute kidney toxicity." *Clin Pharmacol Ther* 86(5): 490–496.

Kubin, C. J., et al. (2012). "Incidence and predictors of acute kidney injury associated with intravenous polymyxin B therapy." *J Infect* 65(1): 80–87.

Ozer, J. S., et al. (2010). "A panel of urinary biomarkers to monitor reversibility of renal injury and a serum marker with improved potential to assess renal function." *Nat Biotechnol* 28(5): 486–494.

38

CASE STUDY: MICRORNAs AS NOVEL KIDNEY INJURY BIOMARKERS IN CANINES

CRAIG FISHER[1], ERIK KOENIG[2] AND PATRICK KIRBY[3]

[1] Drug Safety Evaluation, Takeda California Inc., San Diego, CA, USA

[2] Molecular Pathology, Takeda Boston, Takeda Pharmaceuticals International Co., Cambridge, MA, USA

[3] Drug Safety and Research Evaluation, Takeda Boston, Takeda Pharmaceuticals International Co., Cambridge, MA, USA

38.1 INTRODUCTION

Several publications have focused on miRNA tissue expression in rodent species and humans (Lagos-Quintana et al., 2002; Chen and Wang, 2013; Minami et al., 2014); however, far less data has been published on miRNA tissue expression in canines, an important nonclinical species used in human drug safety assessment. A recent study focused on generation of a comprehensive dog miRNA tissue atlas to identify candidate miRNA biomarkers of organ-specific injury and potentially improve toxicity prediction and monitoring for compounds intended for human use (Koenig et al., 2015, manuscript currently under review).

38.2 MATERIAL AND METHODS

A summary of the experimental study design is illustrated in Figure 38.1. Next-generation sequencing (miRNA-SEQ) analysis was conducted on 16 different tissues from naive male Marshall beagle dogs (n=5 males) using the Illumina Platform. Kidney tissues were not microdissected; therefore, findings represent an organ-wide view of miRNA expression.

38.3 RESULTS

Overall, miRNA-SEQ did not identified an kidney-specific miRNAs in dog kidney tissue; however, the kidney is unique as a target organ because urine can be a valuable biofluid for monitoring biomarkers. Therefore, the highest expressed renal miRNA may still be of value as urinary biomarkers of kidney injury. Renal miRNA demonstrating the highest levels of expression in the dog miRNA tissue atlas are listed in Table 38.1. The expression patterns of miR-10a, -10b, miR-22, miR-181a, miR-191, miR-192, and miR-378 observed in the dog kidney are consistent with previously reported literature on renal miRNA expression. In a study by Ichii et al., dog kidney was microdissected into renal cortex and medulla prior to miRNA-SEQ analysis and identified the highest expressed miRNAs in each section (Ichii et al., 2014). There was good correlation between the whole dog kidney (dog miRNA tissue atlas) and the highest expressed miRNA identified by Ichii et al. in the dog cortex (7/10 similar) and dog medulla (7/10 similar) with miR-10b-5p, miR-192-5p, miR-26a-5p, miR-22-3p, miR-181a, and miR-10a-5p being highly expressed in both renal cortex and medulla. Renal expression of miR-10a was identified as one of the most highly expressed miRNAs in the dog renal medulla (Ichii et al., 2014), while miR-192 was highly expressed in renal tubules of both rats and humans (Kito et al., 2015). In addition, kidney-enriched miR-10a and miR-192 were identified as potential circulating biomarkers of renal ischemia–reperfusion in rats (Wang et al., 2014), and miR-22 and miR-192 have been implicated as key contributors in the development of renal fibrosis (Long et al., 2013). miR-181a has been reported to be released into systemic circulation in patients with nephritic syndrome as well as kidney transplant rejection (Sui et al., 2014), miR-378 has been proposed

Drug Discovery Toxicology: From Target Assessment to Translational Biomarkers, First Edition. Edited by Yvonne Will, J. Eric McDuffie, Andrew J. Olaharski, and Brandon D. Jeffy.

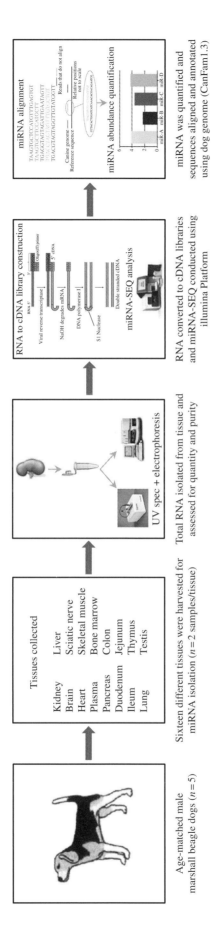

FIGURE 38.1 Dog miRNA tissue atlas methods. Five age-matched naive male Marshall beagle dogs were acquired via Bioreclamation LLC. Sixteen different tissues were harvested and snap frozen in RNAlater and stored at −80°C. Total RNA was isolated from tissue and quantity and purity were assessed by UV spectrophotometer and capillary electrophoresis, respectively. Following cDNA library construction, samples were analyzed for miRNA content using the Illumina-based miRNA-SEQ platform. miRNA were quantified and normalized to reads per million and sequences aligned and annotated using the dog genome (CanFam1.3).

TABLE 38.1   miRNA with the Highest Levels of Expression in the Dog Whole Kidney

Dog Renal miRNA
miR-10b
miR-10a
miR-30a
miR-192
miR-22
miR-181a
miR-26a
miR-143
miR-191
miR-378

as a biomarker for patients with renal cell carcinoma (Redova et al., 2012), and miR-191 and miR-378 are detected at higher levels in the urine of rats treated with cisplatin or gentamicin when compared to control animals (Kanki et al., 2014; Nassirpour et al., 2014). While miR-192 was highly expressed in the dog kidney, this miRNA was expressed at a high level in all 16 tissues investigated in the dog miRNA tissue atlas and therefore was not identified as a specific biomarker of kidney injury in the dog. Despite reports that miR-192 is highly expressed in rat kidney, miR-192 has been previously reported to be highly enriched in rat liver and was demonstrated to be released into systemic circulation following liver toxicant treatment (Laterza et al., 2009). Based on these results as well as results in the dog miRNA tissue atlas, miR-192 is not kidney specific in either dog or rat.

38.4   CONCLUSIONS

In conclusion, there were no significantly enriched kidney miRNAs identified in the dog tissue atlas; however, dog kidney tissue analyzed in the dog miRNA tissue atlas was not separated into renal cortex and medullar sections prior RNA isolation. This work is currently in progress and is anticipated to reveal unique miRNA expression patterns within these two nephron regions of dog kidney similar to those observed in other species (Tian et al., 2008; Ichii et al., 2014).

REFERENCES

Chen, D. B. and W. Wang (2013). "Human placental microRNAs and preeclampsia." *Biol Reprod* 88(5): 130.

Ichii, O., S. Otsuka, et al. (2014). "MicroRNA expression profiling of cat and dog kidneys." *Res Vet Sci* 96(2): 299–303.

Kanki, M., A. Moriguchi, et al. (2014). "Identification of urinary miRNA biomarkers for detecting cisplatin-induced proximal tubular injury in rats." *Toxicology* 324: 158–168.

Kito, N., Endo, K., Ikesue, M., et al. (2015). "miRNA profiles of tubular cells: diagnosis of kidney injury." *Biomed Res Int* 2015: 465479.

Koenig, E. M., Fisher, C., Bernard, H., Wolenski, F. S., Gerrein, J., Carsillo, M., Gallacher, M., Tse, A., Peters, R., Smith, A., Meehan, A., Tirrell, S., Kirby, P. *The Beagle Dog MicroRNA Tissue Atlas: Identifying Translatable Biomarkers of Organ Toxicity*. Cambridge, MA: Takeda Pharmaceuticals International Co.

Lagos-Quintana, M., R. Rauhut, et al. (2002). "Identification of tissue-specific microRNAs from mouse." *Curr Biol* 12(9): 735–739.

Laterza, O. F., L. Lim, et al. (2009). "Plasma MicroRNAs as sensitive and specific biomarkers of tissue injury." *Clin Chem* 55(11): 1977–1983.

Long, J., S. S. Badal, et al. (2013). "MicroRNA-22 is a master regulator of bone morphogenetic protein-7/6 homeostasis in the kidney." *J Biol Chem* 288(51): 36202–36214.

Minami, K., T. Uehara, et al. (2014). "miRNA expression atlas in male rat." *Sci Data* 1: 140005.

Nassirpour, R., S. Mathur, et al. (2014). "Identification of tubular injury microRNA biomarkers in urine: comparison of next-generation sequencing and qPCR-based profiling platforms." *BMC Genomics* 15: 485.

Redova, M., A. Poprach, et al. (2012). "Circulating miR-378 and miR-451 in serum are potential biomarkers for renal cell carcinoma." *J Transl Med* 10: 55.

Sui, W., H. Lin, et al. (2014). "Circulating microRNAs as potential biomarkers for nephrotic syndrome." *Iran J Kidney Dis* 8(5): 371–376.

Tian, Z., A. S. Greene, et al. (2008). "MicroRNA-target pairs in the rat kidney identified by microRNA microarray, proteomic, and bioinformatic analysis." *Genome Res* 18(3): 404–411.

Wang, J. F., Y. F. Zha, et al. (2014). "Screening plasma miRNAs as biomarkers for renal ischemia-reperfusion injury in rats." *Med Sci Monit* 20: 283–289.

39

NOVEL TESTICULAR INJURY BIOMARKERS

HANK LIN

Drug Safety Research and Development, Pfizer Inc., Cambridge, MA, USA

39.1 INTRODUCTION

Nonclinical safety assessment on candidate drug compounds is essential in the drug discovery and development process. Drug-induced testicular toxicity leads to the termination of the candidate drugs. Testicular toxicity can be manifested in multiple forms due to the differential sensitivity of various cell types comprising testicular tissue (i.e., germ, Leydig, and Sertoli cells) and how they respond to chemical insults. There are currently few translatable safety biomarkers for predicting testicular damage. Thus, testicular toxicities observed in preclinical toxicology studies are often evaluated using human sperm analysis methods; unique clinical studies may be necessary to determine if testicular toxicities identified in the preclinical animal models are also observed in the clinical setting. It has been hypothesized that potential testicular toxicity biomarkers may leak from damaged testicular tissue via seminiferous tubules into testicular interstitial fluid and subsequently into blood due to either loss of blood–testis barrier (BTB) integrity or germ cell-specific damage. Ideal testicular biomarkers must be testis specific, remarkably abundant in the testis, precise (easily measured during progressive and potentially reversible states of testicular toxicity), sensitive (normally absent or present at low levels in relevant biological fluids (i.e., sperm, blood, and/or urine)) under healthy conditions, and stable in biological samples (Campion et al., 2013).

39.2 THE TESTIS

The testis is involved in the production of sperm and hormones. It is comprised of two compartments: the interstitium and the seminiferous tubules. Leydig cells are interstitial cells that are responsible for the synthesis and secretion of testosterone. The production of testosterone is regulated by luteinizing hormone (LH) (Zirkin and Chen, 2000) and required for the maturation of spermatid (O'Donnell et al., 1994). In addition to Leydig cells, there are fibroblasts, macrophages, blood, and lymphatic vessels in the interstitium. The second compartment, seminiferous tubules, where spermatogenesis takes place, comprises of long convoluted tubules, and the seminiferous epithelium is made up of somatic Sertoli cells. The Sertoli cells provide support to surrounding germ cells via protein junction complexes (Pelletier, 2011). Coordination of "Sertoli cells–germ cells interactions" is essential for the movement of germ cells across the seminiferous epithelium during spermatogenesis. The BTB, known as Sertoli cell–Sertoli cell tight junctions, is formed via junction complexes between adjacent Sertoli cells that divide the seminiferous tubule into the basal and adluminal compartments (Pelletier and Byers, 1992). The BTB also acts as an immunological barrier to ensure the immune system doesn't act against the cell surface antigens of the haploid germ cells (Mital et al., 2011).

Drug Discovery Toxicology: From Target Assessment to Translational Biomarkers, First Edition. Edited by Yvonne Will, J. Eric McDuffie, Andrew J. Olaharski, and Brandon D. Jeffy.

39.3   POTENTIAL BIOMARKERS FOR TESTICULAR TOXICITY

Sperm analysis, hormone measurement in biological fluids (most often in blood), and testis histopathology are currently used for the evaluation of drug-induced testicular toxicity, but these methodologies have limitations (Sasaki et al., 2011). Sperm analysis and hormone measurement are commonly used as translatable indicators for assessment of testicular toxicity from preclinical to clinical testing. Sperm analyses, including counts, motility, and morphology, are highly variable and lack sensitivity (Cappon et al., 2013). Hormone measurements in serum/plasma, including testosterone, LH, and follicle-stimulating hormone (FSH), are often used to evaluate testicular toxicity; however, they are most reliable for only detecting severe toxicities (e.g., irreversible germinal epithelial damage). Testis histopathology remains the gold standard and most sensitive method to reliably identify and characterize drug-induced testicular toxicity, but this method is invasive, time consuming, and costly, which makes it impractical to longitudinally monitor for progressive testicular toxicity in animals, and it is similarly not feasible in the clinic setting. Thus, other more reliable and sensitive approaches are needed for the evaluation of testicular toxicity, including testis-specific biomarkers, temporal screening methods, and better assays for hormone measurements (Chapin, 2011). Development and validation of biomarkers would enable early detection of testicular toxicity and may possibly support longitudinal monitoring for treatment-related testicular toxicities (Stewart and Turner, 2005). A small number of testis-specific proteins have been previously investigated as potential biomarkers for evaluation of testicular toxicity, including inhibin B, androgen-binding protein, and SP22 (Klinefelter, 2008).

39.3.1   Inhibin B

Inhibin B, a dimeric glycoprotein, negatively regulates FSH and has been considered a biomarker of testicular toxicity for nearly two decades (Anderson and Sharpe, 2000). Circulating inhibin B concentrations have been demonstrated to be associated with the spermatogenic status and changes in Sertoli function (Sharpe et al., 1988), and increased inhibin B levels are generally reflective of germ cell proliferation (Petersen et al., 1999). It also has been used in clinics to separate male populations associated with infertility and their therapeutic responses, with a decrease in circulating inhibin B levels in men being indicative of severe spermatogenic abnormalities (Meeker et al., 2007). However, the utility of inhibin B as a testicular toxicity biomarker has been stalled by lack of sensitivity and technical problems with the assay (Chapin et al., 2013a). Analytical and inter-animal variability make it difficult to accurately and reliably detect testicular toxicant-associated reduction of inhibin B levels (Chapin et al., 2013b). The sensitivity of changes in circulating inhibin B is not as sensitive as histopathology, and more sensitive analysis methods are required to provide favorable performance in monitoring inhibin B changes in physiological and pathophysiological conditions. In clinics, serum inhibin B, when used in combination with FSH, is a more sensitive marker of spermatogenesis than FSH alone. However, the optimal level of inhibin B to assess male infertility has not been established (Kumanov et al., 2006).

39.3.2   Androgen-Binding Protein

Androgen-binding protein (ABP), a glycoprotein, is produced by the Sertoli cells and regulated under influence of FSH (Santiemma et al., 1992). Several studies have demonstrated that increased levels of ABP in serum, plasma, and interstitial fluid could be served as biomarkers for *in vivo* monitoring of germ cell and Sertoli cell toxicities (Rehnberg et al., 1989). However, the increased levels may not be due entirely to Sertoli cell damage but may rather be associated with secondary effects of spermatogenic disruption (Reader et al., 1991).

39.3.3   SP22

SP22, a sperm membrane protein, is expressed in postmeiotic germ cells and elongated spermatids. Studies have shown that sperm SP22 protein expression determined by ELISA is reduced after exposure to testicular toxicants (Kaydos et al., 2004). Antibodies to SP22 inhibit fertilization both *in vivo* and *in vitro* (Klinefelter et al., 2002), and SP22 is the only sperm protein that has been demonstrated in several studies to be a consistent marker of male infertility or impaired spermiogenesis (Klinefelter, 2008). However, a sensitive diagnostic methodology remains a challenge in measuring this low abundant sperm protein.

39.3.4   Emerging Novel Approaches

39.3.4.1   Circulating microRNAs   MicroRNAs (miRNAs) are short noncoding RNA species ~22 nucleotides, which transcriptionally regulate mRNA homeostasis, and have been implicated in the regulation of a wide range of biological processes (Ambros, 2004). Mammalian testicular miRNAs were discovered and have been demonstrated to be differentially expressed during testis development (Yan et al., 2007; Papaioannou and Nef, 2010). Testicular miRNAs are also required for the regulation of germ cell proliferation, while germ-specific deletion of Dicer, an endoribonuclease essential for miRNA formation, showed a decrease in the primordial germ cell numbers due to poor cell proliferation (Hayashi et al., 2008). Additionally, loss of Dicer in Sertoli cells results in a complete depletion of spermatozoa and ultimately results in testicular degeneration (Papaioannou et al., 2009). These studies indicate the miRNAs are crucial in the

normal development of spermatozoa. More recently, testicular miRNAs have been shown to report testicular toxicity (Dere et al., 2013). It is hypothesized that cell-type-specific miRNAs are secreted to the circulation following various types of damage (Salido-Guadarrama et al., 2014). Hence, it is possible that miRNAs, which are secreted, might be able to be used to detect testicular toxicity. A translational biomarker of seminiferous tubule damage could be used in nonclinical animal studies to define *in vivo* safety margins and then be further applied in clinical trials for monitoring testicular toxicity in patients. Two pharmaceutical companies, Pfizer and Eli Lilly, recently demonstrated that two miRNAs, miR-202 and miR-741, were elevated in the circulation following treatments of rats with known testicular toxicants, ethylene glycol monomethyl ether, 1,3-dinitrobenzene, and carbendazim; but circulating miR-202 and miR-741 were not detected following treatments with nonspecific ethylene glycol mono-*n*-butyl ether and 1,4-dinitrobenzene (Dere et al., 2013). These data suggest that circulating miRNAs can potentially be used as indicators of seminiferous tubular damage. Although additional evidence has demonstrated that serum or plasma miRNAs are promising biomarkers for a variety pathophysiological conditions; however, reliable and reproducible methodologies remain a great challenge, while circulating miRNA profiles from serum or plasma are still in need of significant technological advancement (Moldovan et al., 2014).

39.3.4.2 Sperm RNAs Sperm (spermatozoa, mature sperm cells derived from elongated spermatids) RNAs account for 40% RNA detected in the testis (Soumillon et al., 2013). An emerging research area is that of human sperm mRNAs as potential biomarkers of fertility. The genetic fingerprint from mRNA present in sperm of normal fertile men can be generated and that studying ejaculated sperm is a convenient method for investigating testis-specific infertility (Ostermeier et al., 2002). These sperm RNAs may be passively retained or actively involved in maintenance of chromatin structure, imprinting, gene silencing, and embryogenesis (Wang et al., 2004). Studies have identified altered sperm mRNA levels of transcripts including protamine and Bcl2 in oligozoospermic men (Steger et al., 2008). A 12-gene transcript reverse causal reasoning (RCR) array panel has been developed and validated for prediction of compounds, which may be Sertoli cell or germ cell toxicants. In particular, clusterin mRNA that was first identified in interstitial fluid was sensitive enough to detect testicular toxicity following exposure(s) to carbendazim prior to remarkable changes in retained spermatid heads (Pacheco et al., 2012). Utilization of RNA-based biomarkers in sperm represents sensitive and predictive candidate indicators of testicular damage. Therefore, the development of a sperm transcript panels may aid the diagnostic platform for evaluating fertility status and serve to a tool for longer-term monitoring strategies.

39.4 CONCLUSIONS

Without a doubt, development and validation of accurate and sensitive biomarkers for evaluation and monitoring the testicular toxicity would fill a current void in drug discovery toxicology. Novel promising technologies including circulating miRNAs and sperm RNAs are being evaluated for their magnitude of level changes in response to treatment-induced testis damage, with promising advantages including noninvasive, longitudinal, and translational monitoring. Taken together, once validated as useful platforms, next-generation testis-specific biomarkers would be useful for a wide range of purposes from monitoring in preclinical drug development studies to clinical trials and/or postmarketing risk assessment strategies.

REFERENCES

Ambros, V. (2004). "The functions of animal microRNAs." *Nature* 431(7006): 350–355.

Anderson, R. A. and R. M. Sharpe (2000). "Regulation of inhibin production in the human male and its clinical applications." *Int J Androl* 23(3): 136–144.

Campion, S., J. Aubrecht, K. Boekelheide, D. W. Brewster, V. S. Vaidya, L. Anderson, D. Burt, E. Dere, K. Hwang, S. Pacheco, J. Saikumar, S. Schomaker, M. Sigman, and F. Goodsaid (2013). "The current status of biomarkers for predicting toxicity." *Expert Opin Drug Metab Toxicol* 9(11): 1391–1408.

Cappon, G. D., D. Potter, M. E. Hurtt, G. F. Weinbauer, C. M. Luetjens, and C. J. Bowman (2013). "Sensitivity of male reproductive endpoints in nonhuman primate toxicity studies: a statistical power analysis." *Reprod Toxicol* 41: 67–72.

Chapin, R. E. (2011). "Whither the resolution of testicular toxicity?" *Birth Defects Res B Dev Reprod Toxicol* 92(6): 504–507.

Chapin, R. E., G. Weinbauer, M. S. Thibodeau, M. Sonee, L. P. Saldutti, W. J. Reagan, D. Potter, J. S. Moffit, S. Laffan, J. H. Kim, R. A. Goldstein, Z. Erdos, B. P. Enright, M. Coulson, and W. J. Breslin (2013a). "Summary of the HESI consortium studies exploring circulating inhibin B as a potential biomarker of testis damage in the rat." *Birth Defects Res B Dev Reprod Toxicol* 98(1): 110–118.

Chapin, R. E., J. D. Alvey, R. A. Goldstein, M. G. Dokmanovich, W. J. Reagan, K. Johnson, and F. J. Geoly (2013b). "The inhibin B response in male rats treated with two drug candidates." *Birth Defects Res B Dev Reprod Toxicol* 98(1): 54–62.

Dere, E., L. M. Anderson, M. Coulson, B. S. McIntyre, K. Boekelheide, and R. E. Chapin (2013). "SOT symposium highlight: translatable indicators of testicular toxicity: inhibin B, MicroRNAs, and sperm signatures." *Toxicol Sci* 136(2): 265–273.

Hayashi, K., S. M. Chuva de Sousa Lopes, M. Kaneda, F. Tang, P. Hajkova, K. Lao, D. O'Carroll, P. P. Das, A. Tarakhovsky, E. A. Miska, and M. A. Surani (2008). "MicroRNA biogenesis is required for mouse primordial germ cell development and spermatogenesis." *PLoS One* 3(3): e1738.

Kaydos, E. H., J. D. Suarez, N. L. Roberts, K. Bobseine, R. Zucker, J. Laskey, and G. R. Klinefelter (2004). "Haloacid induced alterations in fertility and the sperm biomarker SP22 in the rat are additive: validation of an ELISA." *Toxicol Sci* 81(2): 430–442.

Klinefelter, G. R. (2008). "Saga of a sperm fertility biomarker." *Anim Reprod Sci* 105(1–2): 90–103.

Klinefelter, G. R., J. E. Welch, S. D. Perreault, H. D. Moore, R. M. Zucker, J. D. Suarez, N. L. Roberts, K. Bobseine, and S. Jeffay (2002). "Localization of the sperm protein SP22 and inhibition of fertility in vivo and in vitro." *J Androl* 23(1): 48–63.

Kumanov, P., K. Nandipati, A. Tomova, and A. Agarwal (2006). "Inhibin B is a better marker of spermatogenesis than other hormones in the evaluation of male factor infertility." *Fertil Steril* 86(2): 332–338.

Meeker, J. D., L. Godfrey-Bailey, and R. Hauser (2007). "Relationships between serum hormone levels and semen quality among men from an infertility clinic." *J Androl* 28(3): 397–406.

Mital, P., B. T. Hinton, and J. M. Dufour (2011). "The blood-testis and blood-epididymis barriers are more than just their tight junctions." *Biol Reprod* 84(5): 851–858.

Moldovan, L., K. E. Batte, J. Trgovcich, J. Wisler, C. B. Marsh, and M. Piper (2014). "Methodological challenges in utilizing miRNAs as circulating biomarkers." *J Cell Mol Med* 18(3): 371–390.

O'Donnell, L., R. I. McLachlan, N. G. Wreford, and D. M. Robertson (1994). "Testosterone promotes the conversion of round spermatids between stages VII and VIII of the rat spermatogenic cycle." *Endocrinology* 135(6): 2608–2614.

Ostermeier, G. C., D. J. Dix, D. Miller, P. Khatri, and S. A. Krawetz (2002). "Spermatozoal RNA profiles of normal fertile men." *Lancet* 360(9335): 772–777.

Pacheco, S. E., L. M. Anderson, M. A. Sandrof, M. M. Vantangoli, S. J. Hall, and K. Boekelheide (2012). "Sperm mRNA transcripts are indicators of sub-chronic low dose testicular injury in the Fischer 344 rat." *PLoS One* 7(8): e44280.

Papaioannou, M. D. and S. Nef (2010). "microRNAs in the testis: building up male fertility." *J Androl* 31(1): 26–33.

Papaioannou, M. D., J. L. Pitetti, S. Ro, C. Park, F. Aubry, O. Schaad, C. E. Vejnar, F. Kuhne, P. Descombes, E. M. Zdobnov, M. T. McManus, F. Guillou, B. D. Harfe, W. Yan, B. Jegou, and S. Nef (2009). "Sertoli cell Dicer is essential for spermatogenesis in mice." *Dev Biol* 326(1): 250–259.

Pelletier, R. M. (2011). "The blood-testis barrier: the junctional permeability, the proteins and the lipids." *Prog Histochem Cytochem* 46(2): 49–127.

Pelletier, R. M. and S. W. Byers (1992). "The blood-testis barrier and Sertoli cell junctions: structural considerations." *Microsc Res Tech* 20(1): 3–33.

Petersen, P. M., A. M. Andersson, M. Rorth, G. Daugaard, and N. E. Skakkebaek (1999). "Undetectable inhibin B serum levels in men after testicular irradiation." *J Clin Endocrinol Metab* 84(1): 213–215.

Reader, S. C., C. Shingles, and M. D. Stonard (1991). "Acute testicular toxicity of 1,3-dinitrobenzene and ethylene glycol monomethyl ether in the rat: evaluation of biochemical effect markers and hormonal responses." *Fundam Appl Toxicol* 16(1): 61–70.

Rehnberg, G. L., R. L. Cooper, J. M. Goldman, L. E. Gray, J. F. Hein, and W. K. McElroy (1989). "Serum and testicular testosterone and androgen binding protein profiles following subchronic treatment with carbendazim." *Toxicol Appl Pharmacol* 101(1): 55–61.

Salido-Guadarrama, I., S. Romero-Cordoba, O. Peralta-Zaragoza, A. Hidalgo-Miranda, and M. Rodriguez-Dorantes (2014). "MicroRNAs transported by exosomes in body fluids as mediators of intercellular communication in cancer." *Onco Targets Ther* 7: 1327–1338.

Santiemma, V., P. Rosati, C. Guerzoni, S. Mariani, F. Beligotti, M. Magnanti, G. Garufi, T. Galoni, and A. Fabbrini (1992). "Human Sertoli cells in vitro: morphological features and androgen-binding protein secretion." *J Steroid Biochem Mol Biol* 43(5): 423–429.

Sasaki, J. C., R. E. Chapin, D. G. Hall, W. Breslin, J. Moffit, L. Saldutti, B. Enright, M. Seger, K. Jarvi, M. Hixon, T. Mitchard, and J. H. Kim (2011). "Incidence and nature of testicular toxicity findings in pharmaceutical development." *Birth Defects Res B Dev Reprod Toxicol* 92(6): 511–525.

Sharpe, R. M., I. A. Swanston, I. Cooper, C. G. Tsonis, and A. S. McNeilly (1988). "Factors affecting the secretion of immunoactive inhibin into testicular interstitial fluid in rats." *J Endocrinol* 119(2): 315–326.

Soumillon, M., A. Necsulea, M. Weier, D. Brawand, X. Zhang, H. Gu, P. Barthès, M. Kokkinaki, S. Nef, A. Gnirke, M. Dym, B. de Massy, T. S. Mikkelsen, and H. Kaessmann (2013). "Cellular source and mechanisms of high transcriptome complexity in the mammalian testis." *Cell Rep* 3(6): 2179–2190.

Steger, K., J. Wilhelm, L. Konrad, T. Stalf, R. Greb, T. Diemer, S. Kliesch, M. Bergmann, and W. Weidner (2008). "Both protamine-1 to protamine-2 mRNA ratio and Bcl2 mRNA content in testicular spermatids and ejaculated spermatozoa discriminate between fertile and infertile men." *Hum Reprod* 23(1): 11–16.

Stewart, J. and K. J. Turner (2005). "Inhibin B as a potential biomarker of testicular toxicity." *Cancer Biomark* 1(1): 75–91.

Wang, H., Z. Zhou, M. Xu, J. Li, J. Xiao, Z. Y. Xu, and J. Sha (2004). "A spermatogenesis-related gene expression profile in human spermatozoa and its potential clinical applications." *J Mol Med (Berl)* 82(5): 317–324.

Yan, N., Y. Lu, H. Sun, D. Tao, S. Zhang, W. Liu, and Y. Ma (2007). "A microarray for microRNA profiling in mouse testis tissues." *Reproduction* 134(1): 73–79.

Zirkin, B. R. and H. Chen (2000). "Regulation of Leydig cell steroidogenic function during aging." *Biol Reprod* 63(4): 977–981.

PART IX

BEST PRACTICES IN BIOMARKER EVALUATIONS

40

BEST PRACTICES IN PRECLINICAL BIOMARKER SAMPLE COLLECTIONS

JAQUELINE TARRANT

Development Sciences-Safety Assessment, Genentech Inc., South San Francisco, CA, USA

40.1 CONSIDERATIONS FOR REDUCING PREANALYTICAL VARIABILITY IN BIOMARKER TESTING

A critical consideration for applying biomarkers in toxicology studies is the recognition and control of variables in the preanalytical phase of testing. This is the phase that occurs before the sample is analyzed using biomarker testing method. It comprises biologic, procedure related, and *in vitro* variables that can alter the measured level of the analyte in the specimen. Failure to control the preanalytical phase introduces a potential source of error and variation in the data and can lead to misinterpretation of the biomarker response. Representative variables and possible effects on the measured biomarker level are summarized in Tables 40.1 and 40.2 for blood- and urine-based biomarkers, respectively. Biological variables, that is, the species, age, gender, and diurnal variation, are beyond the scope of this chapter and also make important contribution to data variability.

Careful understanding of the effects of each preanalytical variable on the biomarker data is complex and necessitates a staged approach conceptually similar to the fit-for-purpose analytical validation of a biomarker assay (see Chapter 41). In the early exploratory phase of biomarker investigation, standardization of procedures with a defined protocol for sample collection and handling will permit comparative interpretation and analysis of the data within study and/or between studies. Minimally, variables that should be experimentally evaluated to optimize sample collection for a specific biomarker will include matrix type, preservation (additives such as protease inhibitors), collection method, ambient condition (at room temperature or on wet ice), and sample stability. There are few published recommendations for sample collection and handling of toxicology biomarkers that would form the basis of standardized practice within drug development. However the consensus-based protocols recommended by public–private consortia to qualify biomarkers with government health authorities are an important source of best practice methods (FDA, 2015).

40.2 BIOLOGICAL SAMPLE MATRIX VARIABLES

Establishing the preferred matrix for analysis is a key first step prior to implementing a biomarker as measured concentrations in plasma and serum are not necessarily equivalent. Plasma is the noncellular component of blood prepared with the use of anticoagulants, for example, ethylenediaminetetraacetic acid (EDTA), sodium citrate, and heparin are the most common. Serum is the fluid product of coagulated blood that is devoid of clotting factors and cells. Serum could have artifactually higher values than plasma for biomarkers released from platelets during the clotting process (including several cytokines and growth factors). Peptide biomarkers may have enhanced stability in EDTA plasma from the chelation of metals to impairs protease-mediated biomarker degradation resulting in higher values than serum. On the other hand, chelation of metal cofactors could reduce the enzymatic detection signal in immunoassays or induce conformational changes in protein epitopes recognized by

Drug Discovery Toxicology: From Target Assessment to Translational Biomarkers, First Edition. Edited by Yvonne Will, J. Eric McDuffie, Andrew J. Olaharski, and Brandon D. Jeffy.

TABLE 40.1   Preanalytical Considerations for Blood Samples

Preanalytical Phase	Variable	Effects[a]
Sample		
Type	Matrix: serum versus plasma	Platelet proteins, exosomes, microparticles, RNA, and miRNA released upon clotting; biomarker stability during time taken for clotting
Treatment	Tube type and additives, for example, anticoagulants (EDTA, sodium citrate, heparin), clot activators, separator gels, RNase free, PAXgene tubes	Intended inhibition of coagulation; biomarker stability aided by EDTA inhibition of Ca- and Mg-dependent enzymes; absorption of biomarker to tube components; analytic interference by gel particles; improve stability of RNA
	Biomarker stabilizers, for example, acidification, protease inhibitors, nuclease inhibitors	Protection from biomarker degradation or modification *in vitro*
	Tube fill—correct ratio of blood to anticoagulant	Fibrin clots in plasma from overfilling; interference of tube additive in test method; hemolysis from underfilling
Quality	Hemolysis, lipemia, icterus	Assay interference; hemolysate contains biomarker or releases intracellular proteases that degrade the biomarker
	Fibrin clots	Assay interference
Exogenous substances	Drug, infectious agents	Assay interference, altered biological response of biomarker
Collection procedure		
Site	Arterial, central venous, peripheral venous, capillary	Different concentrations of some biomarkers in venous versus arterial blood and large versus small vessels
Method	Restraint	Stress and excitement (hypothalamus–pituitary–adrenal activation); muscle injury
	Anesthesia	Stress, blood pressure changes, blood gas disturbances, altered metabolism
	Needle and syringe, butterfly needle, vacuum tube, catheter	Stress, local trauma, hemolysis
	Needle gauge	Hemolysis with needles <21G
	Time taken to collect blood	Stress with longer collection times in conscious animals
	Sample mixing	Inadequate mixing can alter anticoagulant:blood ratio; vigorous mixing causes hemolysis
	Skill of phlebotomist	Collection time and animal stress; traumatic blood draw causing hemolysis and tissue contamination of sample
	Order of draw for multiple samples/tube types	Contamination of samples with tube additives due to carryover; differential concentration of biomarkers related to immediate effects of collection (i.e., platelet activation, vessel injury)
Volume	Blood volume/draw	Repeated and/or high-volume blood draws leading to reduced blood volume and compensatory cardiovascular effects and stress that alter biomarker levels
	Sequential sampling with short time interval	Procedure-related inflammation from previous blood draw; cumulative blood volume reduction (see previous text)
Sample processing		
	Processing time and temperature	Biomarker stability; fibrin clots from inadequate clotting time; hemolysis if centrifuged prior to complete clotting or exposed to temperature extremes; release of microparticles from blood cells
	Centrifugation: time, force, temperature, recentrifuging	Biomarker stability; hemolysis
Sample storage		
	Tube type	Evaporation of sample
	Temperature during short- and long-term storage	Biomarker stability
	Freeze–thaw stability	Biomarker stability

[a] Partial list of potential effects that can alter the measurable biomarker levels in blood.

TABLE 40.2 Preanalytical Considerations for Urine Samples[a]

Preanalytical Phase	Variable	Effects[a]
Sample		
Treatment	Biomarker stabilization, for example, refrigeration, acidification, RNase/DNase, protease inhibitors	Biomarker stability
Sample quality		
Contamination	Exogenous fecal, food, bacteria, environmental, for example, chemical environmental cleaning agents (e.g., Clidox)	Addition or dilution of biomarkers; biomarker stability; assay interference
	Blood (cystocentesis) and lower urinary tract cells (catheterization) introduced by collection method	
	Intrinsic sample differences: crystals, casts, epithelial, and blood cells; physicochemical properties (pH, albumin, hemoglobin)	
Collection procedure		
Site and method	Metabolism cage versus conventional cage, free catch, catheterization, cystocentesis	Contamination levels vary between methods
Timed collections	Urine: spot (random), short (≤4 h), and long/overnight (12–16 h) timed collections, relationship to light cycle	Variability in biomarker excretion
Ante- versus postmortem	Time after euthanasia	Biomarker stability and contamination
Sample processing	Whole urine or centrifuged (supernatant)	Potential for cellular contamination in whole urine (altering biomarker level; assay interference)

[a] Partial list of potential effects that can alter the measurable biomarker levels in urine.

immunoassay antibodies (Bowen and Remaley, 2014). As anticoagulants are distinctly different molecules, it is recommended to choose the optimal anticoagulant for both the biomarker and analytical method and keep the choice consistent. Selected blood collection tubes not only contribute a known anticoagulant to the matrix but also may contain additives such as separator gels, polymeric surfactants, and clot activators, factors that could interfere with the analytical method and/or postanalytic data processing (Bowen and Remaley, 2014). For example, heparin inhibits polymerase enzymes in a PCR assay, silica and surfactant additives can interfere with detection of low molecular weight molecules in mass spectrometry, and surfactants could interfere with surface binding of capture antibodies.

Protein and RNA biomarkers may have limited stability due to probable enzymatic degradation *in vitro*. Simple approaches such as rapid processing of samples, reducing the time at room temperature, and snap freezing in liquid nitrogen will assist in preserving the biomarker. Reduction in degradative enzymatic activity can be achieved by use of RNase-free tubes and nuclease inhibitors for RNA preservation, and protease inhibitor cocktails or specific protease inhibitors (e.g., aprotinin, a competitive serine protease inhibitor), chelation by anticoagulants, sample acidification for protein, and peptide preservation. Urine can be drained from cages into refrigerated collection units to reduce bacterial overgrowth in the sample, and acidification of urine could also assist in reducing enzymatic degradation

and maintaining the biomarker in solution. Limiting environmental, fecal, and food contamination of urine will also reduce introduction of infectious agents, exogenous proteins, and enzymes and alterations to the physicochemical properties of the matrix that may compromise biomarker stability.

Hemolysis is the membrane rupture and release of hemoglobin and other cytoplasmic constituents of the red blood cell into the extracellular space. It is the most important and common intrinsic sample quality attribute of serum or plasma in toxicology studies that can cause analytical interference (lipemia and icterus, less so). *In vitro* hemolysis is an artifact that can be introduced by numerous steps in sample collection and processing. *In vivo* hemolysis is far less common and can be related to drug formulation incompatibility, drug dependent or independent immune hemolytic anemia, or indirect red cell damage due to vascular toxicity or coagulopathy. Analytical interference caused by the hemolysate can affect multiple testing methods related to hemoglobin and/or the red cell constituents. Examples include interference due to the absorbance of hemoglobin in spectrophotometric assays, red cell enzymes that compete with enzymatic assays or degrade peptide and protein biomarkers, and nonspecific binding of the antibodies used in immunoassays to red cell components (Lippi et al., 2008). Artifact can also result from contribution of biomarkers contained in the red cell or dilution of the sample from the released fluid content of the cell. The effects of hemolysis can be vastly reduced by attention to adequate sample collection technique, recording of sample quality and association with sample data (including data censoring), and assessment of interference on the testing method.

40.3   COLLECTION VARIABLES

The collection method is an underappreciated source of pre-analytical variability and should be thoroughly defined in the study protocol. Short-term stress from common animal handling procedures in preclinical toxicity studies modify hematology and blood metabolite and clinical chemistry parameters and have been shown to alter the expression of immune function genes in rat liver (He et al., 2014). For blood-based biomarker testing, the levels measured can vary with the collection site, that is, from peripheral versus central and large versus small caliber vessels. This reflects the tissue source and half-life of the biomarker and the exchange of endogenous molecules (metabolites, proteins, and fluid) as blood circulate through different organs. The use of anesthesia may introduce untoward cardiovascular effects and some agents alter metabolism, for example, isoflurane suppresses circulating insulin levels. The choice of anesthetic should therefore be thoroughly researched and effects on the biomarker experimentally determined if information is lacking. In small animals, techniques such as trunk blood postdecapitation, cardiac puncture, and retro-orbital sinus collection may introduce contamination by tissue fluid, and muscle that may dilute the specimen, contribute extravascular (noncirculating) biomarkers, and activate the coagulation system. Poor phlebotomy technique can affect not only sample quality but also inexpert handling of the animal, and prolonged collection time will cause stress to the animal, activating the hypothalamus–pituitary–adrenal axis and potentially influencing biomarker levels.

40.4   SAMPLE PROCESSING AND STORAGE VARIABLES

Description of the sample processing and storage requirements for a biomarker is an opportunity to proactively avoid contamination and maintain specimen quality and biomarker stability. Limits should be set for suitable collection and processing times, ambient temperature, sample transport, separation into serum or plasma, aliquoting, and short- and long-term storage conditions. Centrifuge time, force, and temperature all need to be defined and kept constant. Recentrifuging specimens is not usually advisable as this can cause hemolysis. The use of appropriate personal protection equipment (e.g., disposable gloves) will also reduce or prevent sample contamination and deterioration by skin RNAses.

Acceptable short-term biomarker stability should be conducted in early fit-for-purpose biomarker assay validation (see Chapter 41). Freeze–thaw cycles cause ice crystal formation and reactivation of proteases that degrade molecules in the sample (Gillio-Meina et al., 2013). The effect on biomarker integrity may be method- and assay-dependent, particularly for immunoassays where degradation of the protein biomarker could either reduce, or via fragmentation, increase the number of epitopes detected by different antibody reagents. Short- and long-term storage is most consistent across many different biomarkers at temperatures of −80°C (Gillio-Meina et al., 2013).

REFERENCES

Bowen, R.A., Remaley, A.T. Interference from blood collection tube components on clinical chemistry assays. *Biochem Med* 2014; 24: 31–44.

FDA Biomarker Qualification Program. 2015 http://www.fda.gov/Drugs/DevelopmentApprovalProcess/DrugDevelopmentToolsQualificationProgram/ucm284076.htm (accessed October 23, 2015).

Gillio-Meina, C., Cepinskas, G., Cecchini, E.L., Fraser, D.D. Translational research in pediatrics II: blood collection, processing, shipping, and storage. *Pediatrics* 2013; 131: 754–766.

He, D.Y. Karbowski, C.M., Werner, J. Everds, N. et al. Common handling procedures conducted in preclinical safety studies result in minimal hepatic gene expression changes in Sprague-dawley rats. *PLoS One* 2014; 9: e88750.

Lippi, G., Blanckaert, N., Bonin, P., Green, S., et al. Hemolysis: an overview of the leading cause of unsuitable specimens in clinical laboratories. *Clin Chem Lab Med* 2008; 46: 764–772.

41

BEST PRACTICES IN NOVEL BIOMARKER ASSAY FIT-FOR-PURPOSE TESTING

KAREN M. LYNCH

Safety Assessment, GlaxoSmithKline, King of Prussia, PA, USA

41.1 INTRODUCTION

Recent advances in genomic, proteomic, and metabolomic research have resulted in an explosion of novel biomarkers that are increasingly being utilized across all stages of the drug development process to help guide decisions around the efficacy and safety of new therapeutic agents. Particularly in nonclinical toxicology studies, some of the newer exploratory biomarkers have the potential to be more sensitive or predictive indicators of target organ toxicities compared to current routine clinical pathology endpoints. These novel biomarkers are generally proteins and are quantitatively measured using immunoassay methodologies that span a variety of technological platforms. To fully evaluate the clinical or nonclinical utility of a biomarker, it is important that the quantitative method used to measure the biomarker is robust and of acceptable quality. This is achieved by conducting validation experiments that are designed to test the overall technical performance of a method.

41.2 WHY USE A FIT-FOR-PURPOSE ASSAY?

Validation is performed to demonstrate that an assay or method is suitable and reliable for its intended use. Defined validation requirements are specified in the International Conference on Harmonisation of Technical Requirements for Registration of Pharmaceuticals for Human Use

(ICH, 2005) and similar guidance from regulatory authorities (EMA, 2011; FDA, 2013) for bioanalytical methods that measure drug (pharmacokinetics and toxicokinetics) in biological samples as part of drug development programs. Food and Drug Administration (FDA)-approved diagnostic methods used in clinical laboratories are subject to strict regulations and require full validation as specified in Clinical and Laboratory Standards Institute (CLSI) method evaluation guidelines. The lack of available guidance for methods that measure protein biomarkers in both clinical and nonclinical (toxicology) biological samples as part of drug development has resulted in industry-wide "fit-for-purpose" validation concepts that incorporate regulatory guidelines and additional procedures applicable to antibody-based biomarker methods (Findlay et al., 2000; Miller et al., 2001; Lee et al., 2006; Lee, 2009; Lee and Hall, 2009; Valentin et al., 2011). The fit-for-purpose approach to validation allows flexibility for determining how much validation testing is required. Consequently, the FDA's Bioanalytical Method Validation Guidance for Industry was revised in 2013 and for the first time brings biomarker methods into scope and recommends a fit-for-purpose approach to validation. FDA expects rigorous and full validation of biomarkers that are used to support regulatory decisions and supports a limited or partial validation for exploratory or proof-of-concept biomarkers used in early drug development.

Collaborative efforts to establish standardized and harmonized validation practices across industry are needed to

Drug Discovery Toxicology: From Target Assessment to Translational Biomarkers, First Edition. Edited by Yvonne Will, J. Eric McDuffie, Andrew J. Olaharski, and Brandon D. Jeffy.

improve efficiency and set reasonable expectations of internal stakeholders and regulators (Khan et al., 2015). Although a fit-for-purpose approach allows for customizable validation practices, it has also led to a variety of interpretations of how to test the analytical performance of a method for nonclinical species. Pharmaceutical companies and contract research organizations establish internal validation protocols or plans based upon current recommendations (Stanislaus et al., 2012; Tomlinson et al., 2013). Public–private partnerships and consortiums that bring industry, academia, and regulators together to identify and qualify novel biomarkers also develop internal validation practices for the methods they use. Validation data are generally not reported with study data and are rarely included in publications that highlight the overall usefulness of safety or efficacy biomarkers.

41.3   OVERVIEW OF FIT-FOR-PURPOSE ASSAY METHOD VALIDATIONS

Method validation requirements are still evolving in the novel biomarker arena, especially with the introduction of multiplex immunoassay platforms (Ellington et al., 2010; Fu et al., 2010). Compared to pharmacokinetic assays where the drug is generally well characterized and thus successfully used as its own reference standard to measure "absolute" quantities of drug in biological samples, endogenous biomarker assays use unique reagents that cannot meet these standards. Biomarker reagents commonly employ competitive-binding or sandwich immunoassay formats with recombinant protein used to prepare standard curves against which the endogenous protein is quantified. Reactivity of antibodies often differs between recombinant protein and the endogenous protein because of posttranslational modifications such as glycosylation and endogenous protein binding. The production of antibodies and recombinant or purified protein standards yields reagents that are unique for each assay. Unlike FDA-approved diagnostic assays, reference or certified calibrators/standards for these emerging biomarker assays are not available and results are generally not translatable between methods that measure the same analyte. Commercial kits for these biomarkers are labeled as research grade or "for investigational use only" and, as such, can only measure "relative" quantities, not necessarily "accurate" quantities, of a protein in biological samples. For methods developed in-house or sourced commercially, however, relative accuracy can be inferred when validation testing shows good precision, reproducibility, and robustness. In nonclinical toxicology where studies are well designed and include concurrent control animals, sufficiently validated methods will provide evidence of treatment-related changes in biomarker values, even when it may be unclear how close the values are to the "true" values.

41.4   ASSAY METHOD SUITABILITY IN PRECLINICAL STUDIES

Demonstration that a method is suitable and reliable for use on preclinical toxicology studies requires careful evaluation of both the reagents and the intended biological sample or matrix. Acceptable reagents must adequately bind or react with the analyte of interest. The limited availability of species-specific reagents for nonclinical species often requires screening of clinical methods to find one with sufficient cross-reactivity to be useful. Once a method is identified, the validation process is progressive, beginning with the evaluation of accuracy and precision of the calibration curve and biological samples within and across multiple runs, followed by the assessment of matrix effects or dilution requirements, and long-term storage testing. The biological samples tested should have concentrations that span the range of the calibration curve. If normal levels in samples are too low to test at multiple concentrations, it may be necessary to spike samples with the method calibrator. As an alternative, preparation of homogenates from tissues rich in the analyte can also provide positive samples for validation testing.

41.5   BEST PRACTICES FOR ANALYTICAL METHODS VALIDATION

The sections in the succeeding text identify and recommend processes and procedures that are currently accepted as best practice and "fit for purpose" for ligand-binding or antibody-based quantitative methods that measure protein biomarkers in biological fluids for nonclinical (e.g., rodent, canine, nonhuman primate, minipig) species, particularly in the area of safety biomarkers. It should be recognized, however, that the diversity of methods and technologies/platforms (i.e., ELISA, Mesoscale Scale Discovery® (MSD), Luminex®-based, and automated clinical chemistry systems) used to measure biomarkers does not allow for "one-size-fits-all" guidance and may require alternative or additional validation practices. Validation practices for analytical methods and associated acceptance criteria described later and summarized in Table 41.1 are applicable for regulated (GLP) nonclinical safety studies.

41.5.1   Assay Precision

Precision is a measure of random error and is defined as the agreement among replicate measurements of a defined sample. It is generally expressed as the percentage coefficient of variation (%CV). %CV is calculated as

$$\%CV = \frac{\text{Standard deviation}}{\text{Mean}} \times 100$$

TABLE 41.1   Summary of Validation Procedures and Performance Goals for Quantitative Assays

Validation Parameter	Design/Plan	Target Performance Goals and Acceptance Criteria[a]

Calibration curve: accuracy and precision	• Perform duplicate analysis of standards across ≥3 analytical runs
	  • Calculate CV (%) for each nonzero standard concentration across all runs
	  • Calculate recovery (%) of observed (back-calculated) standard values compared to expected values
		CV ≤ 20%, CV ≤ 25% at LLOQ, and recovery of 80–120%

LLOQ	• Refer to calibration curve accuracy and precision data	The lowest standard concentration that demonstrates acceptable accuracy and precision CV ≤ 25% and recovery of 80–120%
ULOQ	• Refer to calibration curve accuracy and precision data	The highest standard concentration that demonstrates acceptable accuracy and precision CV ≤ 20% and recovery of 80–120%

Intra-assay precision	• Run samples (≥5 replicates) in a single run. Samples used for precision testing may include commercial QC samples, but precision testing should also include species-specific samples of the relevant matrix (e.g., serum, plasma, urine) for the intended study samples
	  • Test ≥3 samples at low, medium, and high concentrations across the calibration curve range
	  • Calculate mean, SD, and CV%
		CV ≤ 20%, CV ≤ 25% at LLOQ

Inter-assay precision	• Run samples (≥2 replicates) across ≥3 analytical runs. Samples used for precision testing may include commercial QC samples, but precision testing should also include species-specific samples of the relevant matrix (e.g., serum, plasma, urine) of the intended study samples
	  • Test ≥3 samples at low, medium, and high concentrations across the calibration curve range
	  • Calculate mean, SD, and CV%
		CV ≤ 20%, CV ≤ 25% at LLOQ

Dilutional linearity/ parallelism	• Prepare serial dilutions of multiple biological samples (≥3) containing high concentration of analyte (endogenous and/or spiked with commercial or native source) using assay-specific matrix
	  • Run dilutions in replicate (≥2) and evaluate linearity of dilution
	  • Calculate %recovery of expected versus observed concentrations for each dilution
		Observed mean values should recover within 80–120% of expected value

Normal ranges	Analyze naive species-specific and gender-specific samples for the estimation of range of values expected for control animals	$N = 10$–20/sex
Quality control	Run quality control samples (commercial kit supplied or spiked sample matrix) as part of precision testing during validation. This data can be used to confirm and/or establish QC ranges that will be used to monitor the assay when used on toxicology studies	Mean ± 2SD/3SD or ±20% nominal (expected) values
Stability	Samples (≥3) should be measured for analyte stability according to relevant sample storage conditions that study samples will be subjected to prior to analysis (i.e., matrix, gender, species-matched). Storage conditions may include but are not limited to: • Benchtop and/or refrigerated • Frozen (e.g., −20 or −70°C) over one or more durations (e.g., 1, 3, and 6 months, 2 years) • Multiple freeze/thaw cycles	Within 20% of original value

[a] Performance goals/acceptance criteria outlined earlier align with FDA/EMA Bioanalytical Method Validation Guidances and published best practices for method validation. However, technologies and/or platforms used may require additional or alternative validation plans and/or target criteria. All plans, procedures, and data should be documented in validation records.

- *Intra-assay precision*: to establish reproducibility/ repeatability of an assay by comparing replicates within a single run
- *Inter-assay precision*: to establish reproducibility of an assay by comparing replicates across multiple runs

41.5.2   Accuracy/Recovery

Accuracy refers to the closeness of agreement between a measured value and the reference or true value. Accuracy can be reported as %recovery, an estimation of the closeness of an observed (measured) result to its expected value. Recovery is expressed as a percent of the expected (theoretical or nominal) concentration:

$$\%\text{Recovery} = \frac{\text{Observed value}}{\text{Expected value}} \times 100.$$

41.5.3   Precision and Accuracy of the Calibration Curve

A calibration/standard curve is the relationship between instrument response and the expected concentration of analyte. To adequately define the relationship between concentration and response, a sufficient number of standards/ calibrators should be used, typically ≥4. The reproducibility and accuracy of the standard curve across multiple (≥3) runs should be confirmed by calculating interassay precision (%CV) and accuracy (%recovery of observed concentration compared to expected concentration) for each standard. Generally, a well-defined and optimized calibration curve will demonstrate acceptable precision (%CV ≤ 20%) and acceptable accuracy (%recovery of 80–120%).

41.5.4   Lower Limit of Quantification

Lower limit of quantification (LLOQ) is the lowest concentration limit of the assay that can be reported with acceptable accuracy and precision. LLOQ may be established after ≥3 assay runs using results from the standard curve. Generally, an LLOQ that demonstrates a %CV ≤ 20% and %recovery of 80–120% is considered acceptable. In cases when an analyte or biomarker normally measures within the reportable range of the standard curve, and if the biological response of the biomarker is expected to increase, the LLOQ may not impact results and an LLOQ value that demonstrates less precision (e.g., CV ≤ 25%) may be acceptable. Newer methods and technologies tend to include additional "anchor" standards at the high and low end of the standard curve, and, although they can help generate more robust concentration curves, they may not meet acceptable precision or accuracy limits.

41.5.5   Upper Limit of Quantification

Upper limit of quantification (ULOQ) is the highest concentration limit of the assay that can be reported with acceptable accuracy and precision. ULOQ is established after ≥3 assay runs using results from the standard curve. Generally, an ULOQ that demonstrates a %CV ≤ 20% and %recovery of 80–120% is considered acceptable. Samples with concentrations that fall at or above the ULOQ require sufficient dilution to bring concentrations into the reportable or dynamic range of the assay. Table 41.2 shows an example of an assessment of the precision and accuracy of a standard curve including the selection of LLOQ and ULOQ limits.

TABLE 41.2   NGAL (Rat, Meso Scale Discovery®) Standard Curve: Precision and Accuracy of Measured (Back-Calculated) Values versus Nominal Values

Nominal Value (ng/ml)	STD1 200	STD2 67	STD3 22	STD4 7	STD5 2	STD6 0.8	STD7 0.3
Plate 1	207	63.2	21.9	7.3	2.2	0.76	0.24
Plate 2	198	69.9	22.0	6.6	2.2	0.71	0.38
Plate 3	204	64.6	22.1	7.0	2.3	0.72	0.26
Plate 4	208	63.0	21.9	7.2	2.3	0.71	0.23
Plate 5	202	65.6	22.1	6.9	2.3	0.73	0.26
Plate 6	205	64.0	21.7	7.5	2.2	0.69	0.29
Plate 7	208	61.4	22.9	7.3	2.2	0.71	0.27
Plate 8	204	65.2	21.2	7.4	2.2	0.75	0.24
Mean	**204.5**	**64.6**	**22.0**	**7.1**	**2.2**	**0.72**	**0.27**
SD	**3.4**	**2.5**	**0.5**	**0.3**	**0.1**	**0.02**	**0.05**
%CV	**1.7**	**3.9**	**2.2**	**3.9**	**3.1**	**3.3**	**17.3**
%RE	**102**	**96**	**100**	**102**	**111**	**90**	**90**

↑
Meets criteria for ULOQ

↑
Meets criteria for LLOQ

41.5.6 Limit of Detection

Limit of detection (LOD) is the lowest concentration of an analyte that the bioanalytical procedure can reliably differentiate from background noise. There are several approaches for determining the LOD (ICH Harmonized Tripartite Guideline, 2005), but a common practice is to evaluate the variability of the analytical background response of blank samples. To estimate the LOD, run blank (e.g., assay buffer, zero calibrator) sample replicates (≥6) across one or more runs and calculate the mean background value ±2 SD or 3SD to define the LOD. Although commonly used to define the sensitivity of an assay, LOD should be used with caution because the value is defined in an inherently variable region of the curve and is based upon a user-defined calculation.

41.5.7 Precision Assessment for Biological Samples

Replicate measurement of the analyte in its natural or endogenous matrix is necessary to confirm the reliability of the method. The lack of available analyte-free natural matrices for preparing calibration curves has led to the common use of substituted matrices (e.g., PBS-BSA with 0.05% Tween® 20) that are optimized for the method reagents and commercial quality control (QC) materials, but these artificial matrices are not representative of the biological sample for which the method will be used. For this reason, precision testing using biological samples is an important component of validation.

Samples used for precision testing should include species-specific samples of the relevant matrix (e.g., serum, plasma, urine) for the intended study samples and should contain low, mid, and high concentrations across the calibration range of the method. Initial screening of freshly collected blood or urine samples from stock or normal animals is one way to select precision samples. In cases where an analyte concentration is low or below measurable limits under normal conditions, finding appropriate samples is challenging. Spiking samples with the method calibrator to make concentrations that span the range of the assay is an alternative approach during early validation testing. However, if the method calibrator is a recombinant protein that differs from the native protein due to posttranslational modifications and/or protein-binding characteristics, it is necessary to conduct precision studies using biological samples containing endogenous analyte when they become available. Preparation of homogenates from tissues rich in the biomarker of interest can be a useful source of native protein for spiking into biological samples.

Bioanalytical method validation guidelines recommend using a minimum of three samples at low, mid, and high concentrations across the calibration curve range for precision testing. Measurements should include the evaluation of precision or repeatability within a single analytical run (intra-assay) and across multiple runs (inter-assay). Reliable methods will show intra- and inter-assay precision of CV ≤ 20% at each concentration tested, except for concentrations near the LLOQ where CV ≤ 25% may be acceptable.

QC (see Section 41.5.9) samples that will be used to monitor the assay during routine use should also be subject to the intra- and inter-assay precision testing outlined earlier during the validation process.

41.5.8 Dilutional Linearity and Parallelism

Parallelism and dilutional linearity are defined in regulatory guidance documents (EMA, 2011; FDA, 2013) and publications of method validation practices (Findlay et al., 2000; Miller et al., 2001; Lee et al., 2006; Lee, 2009; Lee and Hall, 2009; Valentin et al., 2011) as distinct experiments, but they are similar in their shared goal to identify and mitigate potential matrix effects in the intended biological sample (e.g., serum, plasma, urine). Matrix effects are inherent nonspecific binding of biological proteins or other components with method reagents that can interfere with and suppress analytical signal causing a negative bias. Detecting potential matrix effects is an important component of validation to ensure valid results are reported for study samples. Commercial vendors do not always have access to samples from nonclinical species, and this aspect of validation generally falls on the end user of the method. If matrix effects are observed, dilution with a suitable diluent is usually enough to attenuate the interference.

Dilutional linearity generally refers to spiking a known amount of reference or purified analyte at concentrations within the calibration curve range to determine if the spiked amount is recovered when measured in the assay. This is a critical component in the development and validation of pharmacokinetic methods. For novel biomarkers in non-clinical use, the evaluation of dilutional linearity can be problematic because many of the common immunoassays lack readily available well-characterized, certified reference standards or purified proteins for spiking experiments. One alternative is to use the method calibrator for this evaluation. However, since many methods use recombinant proteins as a reference standard that is not identical to the native protein, caution should be used in assuming that the results are representative of endogenous analyte. Spiking samples with calibrator is a good first approach when the target biomarker concentration is low or unmeasurable under normal physiological conditions. Retrospective testing of study samples with induced or elevated concentrations should be considered when samples become available as described later.

Parallelism, or linearity of dilution in biological samples, is a practical approach in a fit-for-purpose validation for new biomarkers. Preparing and analyzing dilutions of individual samples containing measurable amounts of endogenous analyte serves two purposes. First, when matrix effects are observed, diluting out this interference will identify a minimal required dilution (MRD) for the study population (Table 41.3). Secondly, demonstration of a linear response, expressed as constant recovery of sample dilution results across the calibration curve range, infers precision and relative accuracy of the method for the intended study sample. Many commercial kits use surrogate (protein-based) matrices for the preparation of calibration curves and recommend this same matrix for diluting study samples. Linear dilutional response curves for biological samples in the surrogate matrix provide a useful means of confirming that the diluent used for the calibrator is also suitable for the samples. Recovery (%) of values corrected for dilution (observed) of $100 \pm 20\%$ of the nominal (expected) value is a commonly used acceptable range. The nominal value will be the highest value that falls within the calibration range and may be the neat value or, in cases where matrix effects were observed, the value at the MRD (Fig. 41.1).

41.5.9 Quality Control

QC samples are used to monitor assay performance during validation testing and after implementation of the new method. QC samples are included on each analytical run and used to determine if a run is acceptable or not. In some cases,

QC samples are supplied with commercial kits. In cases where QC samples are not provided, alternative commercial sources or internal preparation of QC samples is needed. One option is the preparation of species-specific sample pools containing the analyte of interest (spiked or endogenous), which are then divided into aliquots and stored (e.g., $-70°C$) for subsequent use as QC samples. Prior knowledge that the analyte is stable under frozen conditions is required when using these prepared QC samples. Homogenates prepared from tissues that are known to be a source of the protein biomarker can also be useful for preparing QC samples.

It is a common practice to establish QC ranges prior to routine use. Commercial QC vendors provide a target range of values but often recommend that each lab establish their own internal ranges to account for technology or platform differences. Internal preparation of QC samples also requires that ranges are established. To establish ranges for QC samples, analyze samples in replicates (≥ 2) over multiple days or runs (≥ 3). Mean $\pm 2SD/3SD$ or $\pm 20\%$ nominal (expected) values are commonly used to determine QC limits. Acceptance or rejection of an analytical run will be based upon QC sample results falling within these specified limits.

41.6 SPECIES- AND GENDER-SPECIFIC REFERENCE RANGES

Although toxicology studies incorporate baseline measurements and concurrent control animals as a standard part of study designs, early knowledge of normal concentrations for the target population may aid data interpretation and/or provide context for any aberrant or variable results observed on a study (Stanislaus et al., 2012; Tomlinson et al., 2013). The generation of statistical reference intervals (Friedrichs et al., 2012) is not practical for many toxicology or clinical pathology laboratories and is generally less informative than collection of data from a limited sample size (e.g., 10–20/sex), where a general estimation of ranges can provide useful information for interpretation of test article-related changes for new biomarkers in nonclinical toxicology studies. Prior to including novel biomarkers in nonclinical toxicology studies,

TABLE 41.3 Example of Endogenous Matrix Effect

| Dilution Factor | Concentration (ng/ml) × Dilution Factor (%Recovery) | | |
	Dog 1	Dog 2	Dog 3
5	19.9 (40%)	21.9 (46%)	—
10	29.5 (59%)	31.6 (66%)	39.2 (45%)
20	35.1 (70%)	38.6 (80%)	57.6 (67%)
40	42.1 (84%)	44.6 (93%)	71.6 (83%)
80	49.9 (nominal)	48.1 (nominal)	86.4 (nominal)
160	49.5 (99%)	55.9 (116%)	86.2 (100%)
320	44.9 (90%)	53.1 (110%)	82.9 (96%)

In this example, linearity of dilution identified matrix effects in dog urine samples. Minimum required dilution for dog urine was identified as 1:80, a dilution where recovery of the analyte shows reproducibility of results.

Rat serum	Expected	Observed	%recovery
Neat	146		
1:2	73.0	83.7	115
1:4	36.5	39.9	109
1:8	18.3	19.0	104
1:16	9.1	10.2	112
1:32	4.6	5.5	121

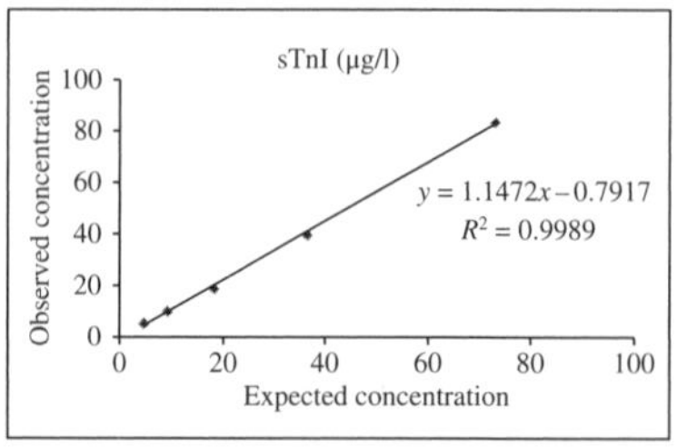

FIGURE 41.1 Rat skeletal troponin I (MSD®) assay: linearity of dilution.

it is advisable to determine a general range of normal values for the intended study population (e.g., strain, age, gender) under conditions similar to the study protocol (e.g., fasting vs. nonfasting, site of venipuncture, type of anesthesia), ideally with repeated measures from this control group at defined intervals (e.g., 3 weekly), to demonstrate intra-animal variability over time.

41.7 ANALYTE STABILITY

It is important to understand the stability of an analyte when samples will be stored at specific temperatures (e.g., room temperature, refrigerated, frozen) for variable lengths of time (hours, weeks, or months) prior to analysis. This storage evaluation should mimic the storage conditions that will be utilized for the nonclinical study samples. For circumstances when a request to measure an analyte is delayed (e.g., following completion of a study when pathology findings may warrant retrospective measurement of biomarkers), it is beneficial to determine longer-term frozen stability or the maximum amount of time that samples can be stored before analyte deterioration occurs (e.g., 6 months or 1 year and beyond). Testing samples subjected to multiple freeze/thaw cycles is also important during validation when it is known that limited study sample aliquots will be available and samples may need to be repeatedly thawed due to reanalysis or sharing of samples across labs or researchers.

41.8 ADDITIONAL METHOD PERFORMANCE EVALUATIONS

Some biomarker methods may require the evaluation of additional performance measures. The performance measures are generally conducted on a case-by-case basis and are evaluated based on laboratory experience with the technology/platform or with the species and sample type being studied. Some common performance measurements include but are not limited to:

- Comparison of the new method with a current or reference method.
- Lot-to-lot variability: Evaluation of multiple lot numbers (≥2) to demonstrate if consistent kit performance is maintained across reagent lot changes. Best practice, when possible, is to use the same lot of reagents for each study.
- Effect of sample type: If validation was originally performed for one sample type (e.g., serum) but another sample type (e.g., plasma) will be collected for the study, validation data (e.g., precision, linearity, and normal ranges) should be evaluated for the alternative sample type.

- Effect of interfering substances: Biological substances (e.g., hemoglobin, bilirubin, and lipids) or drug product that may potentially interfere with the assay should be spiked into matrix samples at expected concentrations to determine if their presence will impact accurate results. In some cases, assay interference information for endogenous substances may be provided by the vendor or in the package insert. Nonclinical toxicology study designs commonly incorporate fasting of animals overnight prior to blood collection, which avoids lipemia interferences in serum or plasma samples. Grossly hemolytic samples should not be analyzed or, if possible, a redraw sample should be requested. Careful attention to urine collection technique is important to prevent interferences from contamination (e.g., feces, food, bacteria). Other urine components anticipated or observed on a study (e.g., coloration from drug product or elevated levels of blood, protein, etc.) may require the evaluation of their interference potential.
- Biological variation: Assess repeated measurements in clinically healthy animals to understand intra-animal variation. Serial measurements at defined intervals can be used to derive individual variance of a parameter.

REFERENCES

Ellington A, Kullo IJ, Bailey KR, Klee G. Antibody-Based Protein Multiplex Platforms: Technical and Operational Challenges. *Clinical Chemistry* 2010; 56(2), 186–193.

European Medicines Agency (EMA). *Guidance for Industry Bioanalytical Method Validation.* EMEA, London, UK; 2011.

Findlay JWA, et al. Validation of Immunoassays for Bioanalysis: A Pharmaceutical Industry Perspective. *Journal of Pharmaceutical and Biomedical Analysis* 2000; 21, 1249–1273.

Food and Drug Administration (FDA). *Guidance for Industry: Bioanalytical Method Validation.* CDER, Rockville, MD; 2013.

Friedrichs KR, et al. ASVCP Reference Interval Guidelines: Determination of de Novo Reference Intervals in Veterinary Species and Other Related Topics. *Veterinary Clinical Pathology* 2012; 41(4), 441–453.

Fu Q, Zhu J, Van Eyk J. Comparison of Multiplex Immunoassay Platforms. *Clinical Chemistry* 2010; 56(2), 314–318.

ICH Harmonized Tripartite Guideline. Validation of Analytical Procedures: Text and Methodology, Q2(R1), November 2005. http://www.ich.org/fileadmin/Public_Web_Site/ICH_Products/Guidelines/Quality/Q2_R1/Step4/Q2_R1__Guideline.pdf (accessed November 28, 2015).

Khan MU, et al. Recommendations for Adaptation and Validation of Commercial Kits for Biomarker Quantification in Drug Development. *Bioanalysis* 2015; 7(2), 229–242.

Lee J. Method Validation and Application of Protein Biomarkers: Basic Similarities and Differences from Biotherapeutics. *Bioanalysis* 2009; 1(8), 1461–1474.

Lee J, Hall M. Method Validation of Protein Biomarkers in Support of Drug Development or Clinical Diagnosis/Prognosis. *Journal of Chromatography B* 2009; 877, 1259–1271.

Lee J, et al. Fit-for-Purpose Method Development and Validation for Successful Biomarker Measurement. *Pharmaceutical Research* 2006; 23(2), 312–328.

Miller KJ, et al. Workshop on Bioanalytical Methods Validation for Macromolecules: Summary Report. *Pharmaceutical Research* 2001; 18(9), 1373–1383.

Stanislaus D, et al. Society of Toxicologic Pathology Position Paper: Review Series: Assessment of Circulating Hormones in Nonclinical Toxicity Studies: General Concepts and Considerations. *Toxicologic Pathology* 2012; 40, 943–950.

Tomlinson L, et al. Best Practices for Veterinary Toxicologic Clinical Pathology, with Emphasis on the Pharmaceutical and Biotechnology Industries. *Veterinary Clinical Pathology* 2013; 42(3), 252–269.

Valentin MA, et al. Validation of Immunoassays for Protein Biomarkers: Bioanalytical Study Plan Implementation to Support Pre-clinical and Clinical Studies. *Journal of Pharmaceutical and Biomedical Analysis* 2011; 55, 869–877.

42

BEST PRACTICES IN EVALUATING NOVEL BIOMARKER FIT FOR PURPOSE AND TRANSLATABILITY

Amanda F. Baker

Arizona Health Sciences Center, University of Arizona, Tucson, AZ, USA

42.1 INTRODUCTION

Regulatory authorities often encourage the integration of biomarkers in drug development in order to accelerate the development of new investigational drugs and to personalize therapeutic regimens (Amur et al., 2008). The pitfalls and limitations in biomarker fit-for-purpose testing in the clinical setting have been reviewed previously (Drucker and Krapfenbauer, 2013). Here, we discuss various tactical aspects to consider when evaluating novel biomarker translatability. The evaluation procedures warrant careful attention in order to leverage the most appropriate traditional and novel biomarkers in the relevant biological matrices.

42.2 PROTOCOL DEVELOPMENT

Integration of biomarkers into clinical protocols requires incorporation of biomarker analysis into one of the study objectives. For novel biomarkers this is typically achieved by including an exploratory objective that examines the feasibility of measuring the biomarkers in the study population. Within the protocol a summary of preclinical or translational evidence for the biomarkers should be provided. When available, biomarker qualifications or letters of support from regulatory authorities should be referenced. A schedule of events describing when samples for biomarker analysis should be collected in context with all other clinical visits and tests should be included. When special collection, processing, or storage considerations are required, a separate

operations manual describing these procedures is useful and should be included as an appendix to the study protocol. Detailed information on the biomarker assay procedures is not included within the study protocol, but enough information should be provided for review bodies to determine the scientific merit of the proposed analysis method(s). A description of how the biomarker values will be interpreted and correlated with other measured studied outcomes should be included.

42.3 ASSEMBLING AN OPERATIONS TEAM

For translational biomarker research it is essential for the *sponsor* (e.g., pharmaceutical company) to form a *collaborative team of preclinical and clinical* scientists—the *operations team*—to take responsibility for *sample inventory management, assay validation, biobanking, data collection/ storage and analysis, and biomarker interpretation* (Fig. 42.1). This team should provide input on the schedule of study events related to sample collections and ensure informed consent forms reflect how samples from human subjects are intended to be used (i.e., downstream biomarker analysis). Prior to finalization of the *study protocol*, this team should develop standard operating procedures (SOP) or an operations manual to guide standardized sample collection, processing, shipping, and storage to ensure sample integrity and enable successful downstream biomarker analysis and interpretation.

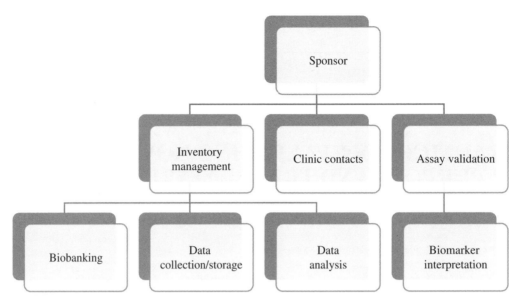

FIGURE 42.1 Operations team schematic. The study sponsor assembles a team to integrate sample collections at the point of clinical care with biomarker storage and analysis using validated assays. The samples may be submitted for immediate biomarker analysis and/or may be biobanked for future analysis. Data generated from biomarker analysis is collected and stored and forwarded to the data team for analysis using the statistical approaches described in the study sponsor protocol.

42.4 TRANSLATABLE BIOMARKER USE

The use of fit-for-purpose biomarkers in different phases of drug development has been described (Amur et al., 2008). The combination of traditional and novel preclinical safety (pharmacokinetic and/or pharmacodynamic) biomarkers is frequently used in phase I studies, using a translatable evidence-based approach. For example, novel rat nephron-specific injury biomarkers have been qualified to enable monitoring of candidate drug-induced acute tubular alterations in the rat with concurrent measurement of sCr or BUN concentrations (Dieterle et al., 2010). Recent successes in the identification of candidate drug-induced kidney injury (DIKI) biomarkers as well as the development and qualification of translational next-generation kidney safety biomarkers have been reviewed (Ennulat and Adler, 2015). Moreover, *sponsors* may demonstrate the benefits from fit-for-purpose biomarkers when rigorously evaluated in accordance to regulatory guidance (Table 42.1).

42.5 ASSAY SELECTION

As discussed earlier (see Chapter 41), assays should be analytically validated for all biomarkers investigated. When translating a biomarker from preclinical to clinical studies, the species specificity of the assay should be evaluated. Differences in amino acid sequence homology or posttranslational modifications between animal species and human proteins may result in the need for use of a species-specific assay. To appropriately interpret biomarker data, it is important to isolate cross-species assay differences from species differences in biomarker responsiveness.

42.6 BIOLOGICAL MATRIX SELECTION

Biomarkers are frequently measured in tissue biopsies, plasma, serum, and/or urine. Establishing the preferred matrix for biomarker analysis is paramount (see Chapter 40). Regardless of species, collection of whole blood for isolation of plasma and serum is semi-invasive and often associated with limited collection volumes (Diehl et al., 2001). Collection of urine is minimally invasive and allows for collection of large sample volumes. The kinetics of safety biomarkers has not been extensively studied; therefore, frequent sampling may be informative. However, sampling frequency in the clinical setting should be balanced with considerations for various factors including but not limited to patient comfort and ease of collection.

Understanding and controlling factors that ensure accurate and consistent biomarker results often supports prioritization of biomarkers to be included in a study and selection of study site. Gathering clinical research site-specific information regarding equipment available for sample processing, relevant certifications, and/or other regulatory issued documentations; education, training, and experience of personnel; and capabilities for sample storage and shipment are important factors to consider when determining whether site is capable of adhering to guidelines within the SOP/operations manual.

TABLE 42.1   Regulatory Guidance References

Regulatory Agency	Websites to Regulatory Guidances for Biomarkers in Drug Development
US FDA	http://www.fda.gov/downloads/drugs/guidancecomplianceregulatoryinformation/guidances/ucm267449.pdf
EMA	http://www.ema.europa.eu/docs/en_GB/document_library/Regulatory_and_procedural_guideline/2009/10/WC500004201.pdf
PMDA	https://www.pmda.go.jp/files/000153149.pdf

42.7   DOCUMENTATION OF PATIENT FACTORS

Multiple patient factors may impact biomarker expression. For example, hydration status, medication, diet, tobacco use, exercise, concomitant diseases, or comorbidities may confound interpretation of biomarker values. For tissue biopsies, anesthetic agents and tissue ischemia should be considered. Capturing this type of information at the time of sample collections may aid in identification of patient-related confounding factors that influence biomarker values and interpretation.

42.8   HUMAN SAMPLE COLLECTION PROCEDURES

42.8.1   Biomarkers in Human Tissue Biopsy and Biofluid Samples

For each priority biomarker, the amount of biofluid and/or tissue biopsy samples needed should be precalculated. The operations team should also decide whether extra biofluid sample volumes should be collected and banked for future retesting (as appropriate) and for bridging or cross-assay validation studies. If future exploratory studies are also in scope, the amount of tissue biopsy sample and number of biofluid aliquots per sample type should be predetermined for long-term biobanking.

42.8.1.1   Human Tissue Biopsy Collection(s)   Image-guided cutting needle biopsies are frequently used to assess disease-related organ damage. For drug safety biomarker assessment, every effort should be taken to minimize risk to the patient including identification of the least invasive method for tissue biopsies. The choice of needle size impacts risk of bleeding, discomfort, and pain for the patient, and the amount of tissue material obtained for biomarker assessment is organ dependent. In a study by Nicholson et al. (2000), three needle gauges (14, 16, and 18G) were compared for renal transplant biopsy. The 16G needle was found to offer the best balance between usefulness and patient comfort. For a review of issues related to liver biopsy techniques, the reader is referred to Strassburg and Manns (2006), Friedman (2004), and Copel et al. (2003). For skeletal muscle biopsies, the percutaneous biopsy technique has been reported to be safe and provide sufficient tissue for analysis (Shanely et al., 2014).

42.8.1.2   Human Biofluid Collection(s): Whole Blood   When collecting whole blood it is important to consider which collection methods should be used. For some patients, venipuncture may be difficult, especially if frequent collection time points are necessary. Catheters ports are an alternative collection option for some patients. However, guidance for flushing the catheter port should be provided as this may result in dilution of the sample or result in contamination with anticoagulant. Kontny et al. (2011) provide some guidance on development of standardized sampling procedures for pharmacokinetic analysis, which can also be used to guide collection procedures for biomarker analysis. A number of blood collection devices are available and may affect the biomarker analysis as mentioned previously (see Chapter 40) and reviewed by Bowen and Remaley (2014).

42.8.1.3   Human Biofluid Collection(s): Urine   For urine collections, it should be prespecified what type of collection is required. Random voids are easy to obtain but may be very dilute depending upon hydration status. First morning voids are more challenging to obtain in an outpatient setting. However, they are more concentrated. Midstream collections require instructing the patient on technique but minimize cellular contaminants. Timed urine collections should provide appropriate collection containers and instructions on how to store the container to minimize degradation of the sample or bacterial growth. For some patients, urine may need to be collected via a Foley catheter.

42.9   CHOICE OF COLLECTION DEVICE

Evidence-based protocols for sample collection should be developed and implemented for each biomarker analyzed. When multiple biomarkers are to be analyzed, multiple types of collection containers/tubes may be necessary.

42.9.1   Tissue Collection Device

Depending upon the downstream analysis technique, tissue should either be immediately frozen or placed in appropriate fixative agent. For RNA-based assays, preservation of tissue in a preservative such as RNA*later*® solution helps prevent degradation of the RNA during storage. When frozen samples will be used for downstream immunohistochemistry, optimization of tissue morphology is critical. Embedding biopsies in optimal cutting temperature (OCT) medium enables easy cryosectioning without remarkably compromising morphology. Numerous fixatives are available for long-term fixation of samples. Choice of fixative should be based on demonstrated performance for the antigens of interest. Time in fixatives should also be standardized. For example, formalin fixation time can impact antigen recognition for immunohistochemistry.

42.9.2   Plasma Collection Device

Plasma collection tubes with a variety of additives are commercially available (e.g., heparin, K_2 EDTA, sodium citrate). These additives have the potential to cause interference with some analytical assays. Thus, interference testing should be performed prior to implementing sample collection procedures. It should be noted that the collection tubes intended to generate plasma contain the optimal amount of the appropriate additive for the indicated sample volume. If tubes are not filled to this indicated volume, the concentration of additive will be higher than recommended and could pose problems during downstream analysis. After collection, sample tubes should be mixed by gentle inversion several times to ensure proper mixing of additive with the blood.

42.9.3   Serum Collection Device

Blood collected into sterile serum separator tubes should generally be allowed to sit at room temperature for 30–60 min to allow optimal clot formation. A longer time may be required for blood containing anticoagulants. The cellular components found in serum are highly influenced by the time it sits prior to centrifugation. Providing a guidance document on how to asses clot formation and a tracking sheet that documents time between blood collection and centrifugation may help assure quality sample processing.

42.9.4   Urine Collection Device

The choice of urine collection containers is based on the sample collection volume needed (spot or time collection) and need for sterility and preservative. Sterile containers are preferable to prevent microbial growth and contamination. Preservatives may be a source of assay interference but necessary for some biomarkers. A standardized set of instructions for clinical volunteers describing how urine should be collected may be useful to include in each sample collection kit.

42.10   SCHEDULE OF COLLECTIONS

Developing a sample collection schedule that is harmonized with the clinical schedule of events is helpful for ensuring samples are taken at appropriate intervals and collected in the correct collection containers and contain the specified amounts or volumes.

42.11   HUMAN SAMPLE QUALITY ASSURANCE

Having a knowledgeable representative from the operations team actively participate in a site initiation visit and other lines of communication (e.g., teleconferences) to answer key questions related to sample collections and processing methods is recommended. When necessary, touring the sample processing area during the visit may also provide valuable insights regarding site-specific workflow challenges. These steps are intended to assure the integrity of human-derived samples to further enable translational biomarker applications.

42.11.1   Monitoring Compliance to Sample Collection Procedures

If completion of sample collection and processing checklists are a requirement for the study, a plan should be developed for monitoring these source documents for compliance to SOP and guidance documents. Reviewing checklists after the first three to five samples are collected is highly recommended to identify errors in interpretation of the guidance documents early in the study and again at routine intervals.

42.11.2   Documenting Time and Temperature from Sample Collection to Processing

Workflow may impact consistency of time from sample collection to processing. The amount of time and the temperature and conditions (e.g., light exposure) may impact biomarker measurements due to degradation of analyte, lysis and contamination from cellular components within the matrix, bacterial growth in urine, etc. Determining acceptable standards for the time from collection to processing and optimal sample handling and methods for documenting adherence to these guidelines is recommended.

To document sample quality, assessment of sample parameters may be necessary. For example, hemolysis can interfere with some immunoassays. For urine, pH measurement, specific gravity, and microscopy analysis to assess white blood cell or bacterial contamination may be useful.

42.11.3   Optimal Handling and Preservation Methods

The stability of the biomarker over different temperatures and times may vary based on the biological matrix. For some biomarkers, it may be necessary to add preservatives, stabilizers, and/or buffers for biomarker stability. For studies where multiple biomarkers will be analyzed, it may be necessary to process the samples using separate, biomarker-specific protocols. Sample preservatives may need to be present when the sample is collected or may be added after further processing steps such as centrifugation. For plasma the potential for biomarker "contamination" from platelets or other blood components (e.g., peripheral blood leukocytes) should be determined. Centrifugation to remove these components may be required. For urine, removal of particulate matter and cellular contaminants may also need to be removed by centrifugation. Centrifugation speed and temperature may be biomarker specific.

42.11.4   Choice of Sample Storage Tubes

During assay validation the optimal storage preservation tubes should also be determined. Low binding tubes may be preferred for optimal recovery of some analytes. Sample size/aliquot volume should be sufficient for the downstream assay requirements, because freeze/thaw may compromise sample stability (see Chapter 41; page 483) and multiple amounts/aliquots may be needed. The number of samples has to be balanced with storage costs and availability.

42.11.5   Choice of Sample Labeling

For large studies, aliquot level barcodes are very helpful in tracking preservation method, aliquot volumes, and handling conditions. Ensuring that labels adhere to the tubes under all anticipated storage conditions is critical. Choosing barcodes that are broadly compatible with barcode readers at each analysis site should also be done in the planning phase of the project.

42.11.6   Optimal Sample Storage Conditions

Providing each collection site with standardized collection kits ensures consistency across sites. Kits may contain collection tubes, sample preservatives, sample preservation tubes, and sample labels.

After samples are processed they are frequently stored temporarily at the collection site until being shipped for analysis or long-term storage. Consideration should be given to what temperature is acceptable for temporary storage. Many sites only have −20°C freezers. The impact of storage at −20 versus −80°C should be assessed in advance to selecting clinical sites. The ability of the site to monitor the freezer temperatures where sample will be stored should also be evaluated.

42.12   LOGISTICS PLAN

Standardized sample inventory sheets can aid in tracking sample inventory at the clinical site and be used when tracking sample shipments to sponsor, analytical labs, or long-term storage facilities. The frequency of sample shipments should be determined based on storage space available at the site, stability of samples, and practical factors including personnel time required and cost. Temperature requirements and delivery times should be assessed. For overseas shipments, additional country-specific considerations are required. Sample shipping containers should be compliant with all federal regulations pertaining to shipment of biological materials. Sponsors should plan to provide appropriate sized shipping containers for each site and ensure that sites have access to dry ice or other materials required for shipments. A communication plan should be in place that notifies all relevant parties when shipments are made. To help document adherence to SOP and guidance procedures, a chain of custody form may be implemented. The chain of custody is a written record that traces possession of the sample from time of collection, processing, storage, shipment, analysis, and disposal.

42.13   DATABASE CONSIDERATIONS

Archiving the biomarker results in a database in association with clinical data is a requirement for many studies. Standards for data collection should be established prior to study initiation. Standard conventions for reporting time of day of sample collection, common analyte reporting units, and subject identification and sample identification conventions should be established. Capturing assay-related information may be valuable for some studies, including date of sample analysis, platform, and assay kit information.

42.14   CONCLUSIVE REMARKS

It is envisioned that the main advantage of the best practices discussed here would be to enable the delivery of predictive and translatable biomarkers using high-quality and ethically obtained human-derived samples.

REFERENCES

Amur S, Frueh FW, Lesko LJ, Huang SM. Integration and use of biomarkers in drug development, regulation and clinical practice: a US regulatory perspective. *Biomark Med*. 2008, 2(3):305–11.

Bowen RA, Remaley AT. Interferences from blood collection tube components on clinical chemistry assays. *Biochem Med (Zagreb)*. 2014, 24(1):31–44.

Copel L, Sosna J, Druskal JB, Kane RA. Ultrasound-guided percutaneous liver biopsy: indications, risks, and technique. *Surg Technol Int*. 2003, 11:154–60.

Diehl KH, Hull R, Morton D, Pfister R, Rabemampianina Y, Smith D, Vidal JM, van de Vorstenbosch C. A good practice guide to the administration of substances and removal of blood, including routes and volumes. *J Appl Toxicol*. 2001, 21(1):15–23.

Dieterle F, Sistare F, Goodsaid F, Papaluca M, Ozer JS, Webb CP, Baer W, Senagore A, Schipper MJ, Vonderscher J, Sultana S, Gerhold DL, Phillips JA, Maurer G, Carl K, Laurie D, Harpur E, Sonee M, Ennulat D, Holder D, Andrews-Cleavenger D, Gu YZ, Thompson KL, Goering PL, Vidal JM, Abadie E, Maciulaitis R, Jacobson-Kram D, Defelice AF, Hausner EA, Blank M, Thompson A, Harlow P, Throckmorton D, Xiao S, Xu N, Taylor W, Vamvakas S, Flamion B, Lima BS, Kasper P, Pasanen M, Prasad K, Troth S, Bounous D, Robinson-Gravatt D, Betton G, Davis MA, Akunda J, McDuffie JE, Suter L, Obert L, Guffroy M, Pinches M, Jayadev S, Blomme EA, Beushausen SA, Barlow VG, Collins N, Waring J, Honor D, Snook S, Lee J, Rossi P, Walker E, Mattes W. Renal biomarker qualification submission: a dialog between the FDA-EMEA and Predictive Safety Testing Consortium. *Nat Biotechnol*. 2010, 28(5):455–62.

Drucker E, Krapfenbauer K. Pitfalls and limitations in translation from biomarker discovery to clinical utility in predictive and personalized medicine. *EPMA J* 2013, 4(1):7.

Ennulat D, Adler S. Recent successes in the identification, development, and qualification of translational biomarkers: the next generation of kidney injury biomarkers. *Toxicol Pathol*. 2015, 43(1):62–9.

Friedman LS. Controversies in liver biopsy: who, where, when, how, why? *Curr Gastroenterol Rep*. 2004, 6(1):30–6.

Kontny NE, Hempet G, Boos J, Boddy AV, Krischke M. Minimization of the preanalytical error in plasma samples for pharmacokinetic analysis and therapeutic drug monitoring—using doxorubicin as an example. *Ther Drug Monit*. 2011, 33(6):766–71.

Nicholson ML, Wheatley TJ, Doughman TM, White SA, Morgan JD, Veitch PS, Furness PN. A prospective randomized trial of three different sizes of core-cutting needle for renal transplant biopsy. *Kidney Int*. 2000, 58(1):390–5.

Shanely RA, Zwetsloot KA, Triplett NT, Meaney MP, Farris GE, Nieman DC. Human skeletal muscle biopsy procedures using the modified Bergtrom technique. *J Vis Exp*. 2014, 10(91):51812.

Strassburg CP, Manns MP. Approaches to liver biopsy techniques-revisited. *Semin Liver Dis* 2006, 26(4):318–27.

43

BEST PRACTICES IN TRANSLATIONAL BIOMARKER DATA ANALYSIS

ROBIN MOGG[1] AND DANIEL HOLDER[2]

[1] *Early Clinical Development Statistics, Merck Research Laboratories, Upper Gwynedd, PA, USA*

[2] *Biometrics Research, Merck Research Laboratories, West Point, PA, USA*

43.1 INTRODUCTION

Biomarkers play a fundamental role throughout various phases of the drug development process. There are vast numbers of reports in the literature regarding potential biomarkers that may help identify efficacy and/or predictive drug-induced toxicity signals, aid in understanding the mechanism of action of a drug, or predict patients most likely to respond to treatment or those most at risk for disease occurrence or progression. Unfortunately, many of these studies lack sufficient rigor to fully translate into direct use during drug development studies. A general inconsistency of sampling, assay measurement, and statistical methods often lead to study results that are potentially biased, underpowered, or preferentially selected. There has been significant and increasing interest on the statistical perspective of biomarker qualification, including data analysis, and both preclinical and clinical trial designs (Buyse et al., 2011; Jenkins et al., 2011; Fosho et al., 2012), but no universal evidentiary standards exist that clearly define translational biomarker qualification. Generally speaking, qualification can be defined as the confirmation by rigorous statistical methods that a candidate biomarker fulfills a set of conditions that are necessary and sufficient for its use in the clinic (Buyse et al., 2010).

In 2014, the FDA issued a guidance document on the qualification of drug development tools (DDTs) (FDA, US Food and Drug Administration, 2014) that included biomarkers within its scope and detailed a process for qualifying DDTs for potential use in multiple clinical drug development programs within a specified context of use (COU). Specific evidentiary standards or statistical methods appropriate for biomarker qualification were not described, as acceptable standards and methods will vary depending on the category of biomarker considered for qualification (i.e., diagnostic, prognostic, predictive, or pharmacodynamic biomarkers), as well as the specific context in which the biomarker is intended to be used. Rather, the document provided a framework for regulatory interactions to guide the collection of data to support the qualification of a DDT. Within this framework, the DDT qualification and COU may be modified or expanded over time as additional data are collected. This principle is commonly referred to as a "rolling" qualification process (Editorial, 2010). There have been several recent initiatives to align data analysis approaches to support this progressive biomarker qualification process. Examples include a webinar series sponsored by the Critical Path (C-Path) Institute "Approaches to Statistical Analysis for Biomarker Qualification: Defining the Components to Drug Development Tool (DDT) Qualification" (Critical Path Institute webinar series, 2014) and a workshop focused on statistical methodology in biomarker research sponsored by the National Institute of Diabetes and Digestive and Kidney Diseases (NIDDK, 2014).

A generic operational strategy to qualify translational safety biomarkers using a two-stage approach of (Jenkins et al., 2011) initial exploratory studies to identify and characterize candidate biomarkers followed by (Fosho et al., 2012) confirmatory studies to build upon the evidence of the initial studies has been described (Matheis et al., 2011). We highlight key statistical considerations in these contexts.

Drug Discovery Toxicology: From Target Assessment to Translational Biomarkers, First Edition. Edited by Yvonne Will, J. Eric McDuffie, Andrew J. Olaharski, and Brandon D. Jeffy.

43.2 STATISTICAL CONSIDERATIONS FOR PRECLINICAL STUDIES OF SAFETY BIOMARKERS

As with any set of studies, statistical analysis starts with an understanding of the objectives. Preclinical biomarker studies often have the duel objectives of developing biomarkers both for use in animal studies as well as translation to the clinic. With regard to the latter use, analysis is usually regarded as exploratory, and thus there is less rigidity in terms of prespecification of statistical methods (Holder and Schipper, 2010). Nonetheless, animal studies can play a critical role in understanding biomarker mechanism and performance due to flexibility in study design and the ability to obtain a direct assessment of histopathology changes.

At a high level, the primary objective for statistical analysis of preclinical safety biomarker studies is to assess the association between biomarker levels and histopathological alterations. Although there are many useful and popular statistical methods to analyze this association (i.e., correlation, regression, receiver operating characteristic (ROC) curves), careful thought is needed to conduct a meaningful analysis and yield the most applicable outcomes.

Note that at the preclinical stage, biomarker assays are often not as well characterized as their clinical partners. There is no avoiding the fact that a poor assay almost always leads to confounded results. Where applicable, it is important to consider how to normalize biomarker values. For example, since urine excretion rates often differ greatly between animals (at baseline/at pre-dose and/or during disease progression/post-dose), it is usually prudent to normalize urinary biomarker concentrations to concurrent urinary creatinine concentrations. Also, since conditions can vary greatly between studies and days within a study, it can also be helpful to normalize biomarker values to concurrent control group animals. The changes in urine chemistry parameters may also be quite variable and much of the change likely reflects biological variation. Therefore, when urinary creatinine excretion is elevated after dosing, more weight on urine-based analyte excretion values rather than values normalized to urinary creatinine is warranted, to properly interpret the results. Pooling data across preclinical studies is common since individual studies often have limited sample sizes. However, comparability of measurements across studies should be considered carefully before doing so.

Another important consideration regarding assessment of association between biomarker and injury is whether the biomarker's intended use is diagnostic, prognostic, or as a biomonitoring tool. Diagnostic biomarkers need to show an association between the biomarker and histopathology findings at the time of biomarker collection. For prognostic biomarkers, the association is between the biomarker values and a future histopathological lesion severity. For biomonitoring purposes, it is useful to show that changes in the biomarker remarkably associate with relevant changes in histopathology changes.

Performance of a biomarker is best measured against a reference standard that is expected to reflect truth with a high degree of accuracy. For animal studies the standard is usually careful examination of morphology by a trained anatomic pathologist who is board certified by the American College of Veterinary Pathologists. Although highly accurate, this determination is not perfect, since pathologists cannot examine every section of an organ or tissue, some level of variability between the evaluations of pathologists is expected, and molecular signals may precede the ability to observe structural damage. Moreover, to use standard statistical methods, a consistent scoring system needs to be developed. For instance, in a rat nephrotoxicity study, a biomarker might have a strong association with kidney tubular degeneration and necrosis, but not other changes such as tubular regeneration (tubular basophilia). Before applying statistical techniques, careful choices must be made about which pathologies are counted as positive, which are not, and how to handle samples that have multiple pathologies. The data analyst must also choose whether injury severity should be incorporated into the analysis. Although the difference between a positive score and a nonpositive score may have the most relevance for translation to humans, modeling injury severity may increase the power to discern good markers.

When injury scores are dichotomized into positive and nonpositive, sensitivity and specificity are usually estimated and displayed in an ROC curve (Pepe, 2003). The area under the ROC curve (AUROC) is a common summary performance metric. Some advantages of this approach are that ROC curve estimation generally requires few assumptions, a plot of the ROC curve gives an easily interpreted visual display of results, and the AUROC is a global metric that does not require the specification of a threshold. Since both sensitivity and specificity can be greatly influenced by characteristics that are study specific such as severity or type of injury, it is advisable to compare candidate marker performance to that of standard markers. In rodent nephrotoxicity studies, candidate markers are commonly compared to sCr and BUN.

As described previously, AUROC can be useful to quantify a marker's ability to discriminate between samples from animals with and without injury. However, it does not make clear the extent to which the information provided by the candidate marker overlaps with information provided by the standard markers. One way to address this question is to compare the performance of a model that contains the standard markers together with the new candidate marker to a model that contains the standard markers without the new marker. The net reclassification index (NRI) and integrated discrimination index (IDI) (Pencina et al., 2008) are metrics that can be used to quantify and test the difference in performance between binary logistic regression models and thus assess the amount of information added by the new marker.

43.3 STATISTICAL CONSIDERATIONS FOR EXPLORATORY CLINICAL STUDIES OF TRANSLATIONAL SAFETY BIOMARKERS

Biomarker measurements cover a wide variety of data types. Relevant clinical data for exploratory studies minimally include baseline and longitudinal measurements in healthy volunteer and patient populations. Biomarker measurements in patients with and without known drug-induced toxicity are essential to characterize the predictive ability of biomarkers relative to current gold standards such as sCr and BUN for renal injury. The appropriate usage of a biomarker in the clinic, for example, the static measurement at a single point in time versus a dynamic change in the measurement over a specified time course, may not be obvious. Identifying and quantifying various sources of variability are essential objectives in exploratory studies and can aid in deciding on the appropriate biomarker usage. Understanding and minimizing analytical variability due to batch or laboratory effects, for example, are critical for consistency in results and clinical utility. Further, the ratio of the within-subject to the between-subject coefficient of variation (CV), also known as the index of individuality, can inform the appropriate biomarker usage (Fraser, 2004). A low index resulting from smaller within- versus between-subject variability favors utilizing a dynamic biomarker measure, while a high index suggests better utility of a static measurement (Lacher et al., 2005).

Estimation of reference intervals or limits of normality in healthy volunteer and patient populations are important elements of biomarker characterization. Several statistical methods can be used to establish reference intervals on either dynamic or static measurements within a specific population. Selected guidelines recommend a minimum of 120 measurements to establish reference intervals using nonparametric methods (Clinical and Laboratory Standards Institute, 2008). With fewer samples, alternative methods using distributional assumptions, bootstrap, or robust calculations can be used (Horn et al., 1998). Adequate consideration to identify potential confounding factors, such as gender, age, or ethnicity, on either the measurements themselves or variation in measurements is critical for thorough characterization. Examinations related to the choice of an appropriate data transformation or scale, or normalization of biomarker measurements, may provide benefits in reducing or eliminating potential confounding factors or decreasing variability.

There are numerous statistical methods to evaluate candidate biomarker performance related to drug-induced toxicity, including regression methods for quantitative drug-induced toxicity endpoints and classification methods for qualitative drug-induced toxicity endpoints. The most traditional data analysis approach to selecting markers involves regression models with single or multiple covariates. When there are several candidate biomarkers, selection methods, such as forward or backward stepwise selection or best subset selection, can be used to identify the most promising subsets of candidate biomarkers. With much larger candidate biomarker sets, penalized regression methods, such as ridge regression or least absolute shrinkage and selection operator (LASSO), offer advantages with respect to smaller prediction error and are analogous to Bayesian estimates under varying priors (Hastie et al., 2009). Alternatively, a principal component analysis can be used as a dimension reduction tool to transform a large multivariate, correlated matrix of observations into a set of linearly uncorrelated variables, where a small number of these linear combinations or composite biomarker measures can be used in a regression model to evaluate predictive performance.

As previously described, ROC curves are the most widely used classification tool to assess the clinical performance of biomarkers, including the trade-off between sensitivity and specificity and positive and negative predictive value. Regression methods, principal component analyses, or other robust procedures can be used to construct predictors or composite measures for classification that can be further evaluated within the context of ROC analyses (Greiner, 2000). Covariates can be accommodated by modeling the ROC curve as a function of covariates or combining biomarker and covariate information to improve discriminatory accuracy (Janes et al., 2009). Often, the current standard of an accepted drug-induced toxicity endpoint is an imperfect biomarker itself, in that it lacks appropriate sensitivity or specificity to be used as a true gold standard. An example of this is the use of serum creatinine to define renal injury. Assessing novel, potentially more sensitive biomarkers of renal toxicity using an ROC analysis against the imperfect gold standard of serum creatinine may prove detrimental in identifying useful, more sensitive biomarkers of renal toxicity. An alternative strategy to compare specificity and sensitivity may be to use exposure versus nonexposure to a known nephrotoxic drug that is expected to induce toxicity in some patients. Statistical methods to formally compare the sensitivities and specificities, or agreement, of two biomarkers, or a novel biomarker to a standard, on the same set of subjects are available (Hawass, 1997).

Less traditional but potentially highly informative statistical methods for combining large numbers of biomarkers include tree-based methods for regression and classification. There are potential advantages in that tree-based methods incorporate nonlinear relationships and interactions among large candidate biomarker sets. A thorough description of these and various other statistical learning methods has been reviewed previously (Hastie et al., 2009). These methods are often underutilized as predictive multivariate models due to their "black box" nature that often results in a lack of understanding of the variable relationships that drive the association to or prediction of the drug-induced toxicity endpoint. In addition, the performance of such methods with small to moderate sample sizes has not been well characterized.

Other statistical issues common in clinical trials, such as missing data and multiplicity considerations, remain important in exploratory biomarker studies. Missing data can arise for various reasons, and careful reflection of the underlying mechanism and assumptions adopted in analyses are required for a meaningful interpretation of results. With large sets of candidate biomarkers, methods to control the family-wise type I error rate such as Bonferroni or Hochberg will likely be highly conservative. Multiplicity adjustments to protect the false discovery rate (Benjamini and Hochberg, 1995) may offer a more powerful alternative while maintaining a statistically principled approach to bound error.

43.4   STATISTICAL CONSIDERATIONS FOR CONFIRMATORY CLINICAL STUDIES OF TRANSLATIONAL SAFETY BIOMARKERS

The performance of each candidate biomarker in exploratory clinical studies will guide selection of the most promising biomarkers for confirmation. Confirmation minimally requires repeated demonstration in an independent data set of the association of a candidate biomarker (or composite measure of biomarkers) with drug-induced toxicity. While internal cross-validation techniques (Hastie et al., 2009) are powerful approaches to minimize model overfitting and optimistic estimation of measures of performance such as specificity and sensitivity in the exploratory setting, they are generally inadequate for confirmation of biomarker performance. The confirmation stage must be clearly separated from the exploratory stage, with verification of results in an independent data source providing robust evidence for biomarker qualification.

Statistical methods for confirmation may be similar to, or a subset of, the methods described previously for exploratory purposes. In many contexts of use, qualification will also require demonstrating superior sensitivity in identifying drug-induced toxicity, while retaining acceptable or noninferior specificity, relative to the current standards. Demonstrating this may implicitly validate a threshold value for clinical use that was estimated during the exploratory phase based on performance criteria (Mandrekar and Sargent, 2009) such as the specificity, sensitivity, Youden's index, or cost of misclassification. Within confirmatory studies, clear hypotheses of interest, with individual analyses fully specified to address them, are required to ensure that study results carry the maximum confirmatory strength. Statistical methods for analysis, including plans to address missing data, maintain the family-wise type I error control with multiple hypothesis testing, and planned sensitivity analyses should be prospectively well defined and documented in a statistical analysis plan.

Study designs for predictive safety biomarker qualification may be retrospective or prospective in nature (Mandrekar and Sargent, 2009). Prospective studies are the gold standard for biomarker qualification, but they are often unfeasible due to expense and time. An alternative approach is to use a retrospective–prospective design, where a prospective analysis plan to evaluate prespecified biomarker hypotheses is carried out using retrospective data from a previous randomized clinical trial. The use of adaptive designs, including interim analyses for futility or additional learning, may prove useful in the "rolling" qualification process and should be carefully considered to improve the efficiency of the clinical qualification process, particularly when using prospective confirmatory studies.

43.5   SUMMARY

There are several key statistical considerations in the design and analysis of data from studies that assess translational safety biomarkers. There are no universal solutions, with evidentiary standards and appropriate statistical methods highly dependent on the specific context of biomarker use in the clinic. A staged approach that builds on evidence from exploratory to confirmatory studies is recognized as a useful operational strategy in a "rolling" biomarker qualification process. In the exploratory stage, it is important to characterize all aspects of biomarker usage, which may include combining two or more biomarkers, setting thresholds, identifying the relevant population, and an initial assessment of performance. Once a biomarker has been sufficiently characterized, it can be elevated to being evaluated in a confirmatory study. Hallmarks of good practices within a confirmatory setting include careful study design, with prospectively defined hypotheses and a detailed statistical analysis plan. Regulatory interactions to guide collection of data and statistical approaches to support qualification are paramount to efficient and robust clinical qualification.

REFERENCES

Benjamini Y, Hochberg Y (1995). Controlling the false discovery rate: a practical and powerful approach to multiple testing. *Journal of the Royal Statistical Society Series B*, 57(1): 289–300.

Buyse M, Sargent D, Grothey A, Matheson A, de Gramont A (2010). Biomarkers and surrogate endpoints—the challenge of validation. *Nature Reviews. Clinical Oncology*, 7: 309–317.

Buyse M, Michiels S, Sargent DJ, Grothey A, Matheson A, de Gramont A (2011). Integrating biomarkers in clinical trials. *Expert Review of Molecular Diagnostics*, 11(2): 171–182.

Clinical and Laboratory Standards Institute (2008). *"Defining, Establishing, and Verifying Reference Intervals in the Clinical Laboratory; Approved Guideline-Third Edition"*. CLSI document C28-A3 (ISBN1-56238-682-4). Wayne, PA: Clinical and Laboratory Standards Institute.

Critical Path Institute webinar series (2014). "Approaches to Statistical Analysis for Biomarker Qualification: Defining the Components to Drug Development Tool (DDT) Qualification". Seminars archived at http://c-path.org/approaches-to-statistical-analysis-for-biomarker-qualification-defining-the-components-to-drug-development-tool-ddt-qualification/ (accessed October 23, 2015).

Editorial (2010). Biomarkers on a roll. *Nature Biotechnology*, 28: 431.

FDA, US Food and Drug Administration (2014). Guidance for Industry and FDA Staff—Qualification Process for Drug Development Tools. Available at http://www.fda.gov/downloads/drugs/guidancecomplianceregulatoryinformation/guidances/ucm230597.pdf (accessed October 23, 2015).

Fosho M, Nagashima K, Sato Y (2012). Study designs and statistical analyses for biomarker research. *Sensors*, 12: 8966–8986.

Fraser CG (2004). Inherent biological variation and reference values. *Clinical Chemistry and Laboratory Medicine*, 42: 758–764.

Greiner M (2000). Principles and practical application of the receiver-operating characteristic analysis for diagnostic tests. *Preventive Veterinary Medicine*, 45: 23–41.

Hastie T, Tibshirani R, Friedman R (2009). *The Elements of Statistical Learning. Data Mining, Inference, and Prediction*. New York: Springer.

Hawass NE (1997). Comparing the sensitivities and specificities of two diagnostic procedures performed on the same group of patients. *The British Journal of Radiology*, 70: 360–366.

Holder D, Schipper M (2010). Statistical Issues in Biomarker Research. In Vaidya VS and Bonventre JV (Ed.), *Biomarkers in Medicine, Drug Discovery and Environmental Health*. Hoboken, NJ: John Wiley & Sons, Inc.

Horn PS, Pesce AJ, Copeland BE (1998). A robust approach to reference interval estimation and evaluation. *Clinical Chemistry*, 44(3): 622–631.

Janes H, Longton G, Pepe M (2009). Accommodating covariates in ROC analysis. *The Stata Journal*, 9(1): 17–39.

Jenkins M, Flynn A, Smart T, Harbron C, Sabin T, Ratnayake J, Delmar P, Herath A, Jarvis P, Matcham J (2011). A statistician's perspective on biomarkers in drug development. *Pharmaceutical Statistics*, 10: 494–507.

Lacher DA, Hughes JP, Carroll MD (2005). Estimate of biological variation in laboratory analytes based on the third national health and nutrition examination survey. *Clinical Chemistry*, 51(2): 450–452.

Mandrekar SJ, Sargent DJ (2009). Clinical trial designs for predictive biomarker validation: theoretical considerations and practical challenges. *Journal of Clinical Oncology*, 27(24): 4027–4034.

Matheis K, Laurie D, Andriamandroso C, Arber N, Badimon L, Benain X, Bendjama K, Clavier I, Colman P, Firat H, Goepfert J, Hall S, Joos T, Kraus S, Kretschmer A, Merz M, Padro T, Planatscher H, Rossi A, Schneiderhan-Marra N, Schuppe-Koistinen I, Thomann P, Vidal J, Molac B (2011). A generic operational strategy to qualify translational safety biomarkers. *Drug Discovery Today*, 16(13): 600–608.

National Institute of Diabetes and Digestive and Kidney (2014). Disease workshop "Towards Building Better Biomarkers". Workshop details archived at http://www.niddk.nih.gov/news/events-calendar/Pages/Toward-Building-Better-Biomarkers-Statistical-Methodology_12-2014.aspx (accessed October 23, 2015).

Pencina MS, D'Agostino RB Sr., D'Agostino RB Jr., Vasan RS (2008). Evaluating the added predictive ability of a new marker: from area under the ROC curve to reclassification and beyond. *Statistics in Medicine*, 27: 157–172.

Pepe MS (2003). *The Statistical Evaluation of Medical Tests for Classification and Prediction*. New York: Oxford University Press Inc.

44

TRANSLATABLE BIOMARKERS IN DRUG DEVELOPMENT: REGULATORY ACCEPTANCE AND QUALIFICATION

JOHN-MICHAEL SAUER, ELIZABETH G. WALKER AND AMY C. PORTER

Predictive Safety Testing Consortium (PSTC), Critical Path Institute (C-Path), Tucson, AZ, USA

44.1 SAFETY BIOMARKERS

Many academic and pharmaceutical industry scientists are currently involved in the discovery and biological validation of a multitude of novel biomarkers (Parekh et al., 2015). These biomarkers are intended to determine whether a patient is susceptible to a disease, already has a disease, or the extent to which a disease has progressed. In addition, biomarkers can be used to determine whether a patient is responding to a treatment, is experiencing adverse side effects related to the treatment, or whether a treatment has worked.

While much of the data generated for these novel biomarkers will be published in the open literature, the proof required to obtain regulatory biomarker qualification, that is, the demonstration of both the scientific utility and regulatory reliability of biomarkers in drug development, is significantly higher than what is required for peer-reviewed publication. The qualification of biomarkers is analogous to obtaining marketing authorization for a drug product or device in that there are high scientific and regulatory expectations. However, that is where the comparison ends. The same evidentiary standards applied to drug development cannot be applied to biomarker qualification, as the ultimate scientific goal is very different.

In addition, the relationship of the stakeholders in biomarker qualification is very different to the relationship of those in drug development. When a drug candidate receives formal regulatory approval and is marketed, the drug developer stands to reap the financial rewards associated with a successful product, while the health authority takes on the potential for additional risk to public health regardless of the financial success of the drug. However, drug developers, regulators, and the general public benefit when a biomarker or other drug development tool (DDT) is successfully qualified by virtue of the accelerated process and improved probability of developing efficacious and safe drugs through use of the biomarker. Thus, the relationship between DDT qualification submitters and health authorities has to be a collaborative relationship and differs significantly from a single company investing in a drug development program. Of course, during the scientific review of qualification data supporting a novel biomarker or DDT, the regulatory agencies retain their objective and independent assessment. But all stakeholders must invest in discussion around the appropriate study design, data analysis, and level of evidence necessary to support the proposed use of a novel biomarker. The precompetitive collaborative consortia models currently being applied in the biomarker space have created these unique, collaborative relationships across health authorities, pharmaceutical companies, academia, and patient groups. This concept has allowed for the sharing of costs, risks, and benefits necessary for the successful qualification of biomarkers.

Insufficient therapeutic index is a major cause of candidate attrition in drug development. The lack of appropriate prediction of safety liabilities results in unforeseen adverse events in clinical trials or the unwarranted abandonment of potentially safe and effective therapies. Throughout the drug discovery process, therapeutic and toxic exposures are determined, and clinical safety biomarkers are essential for

Drug Discovery Toxicology: From Target Assessment to Translational Biomarkers, First Edition. Edited by Yvonne Will, J. Eric McDuffie, Andrew J. Olaharski, and Brandon D. Jeffy.

maximizing therapeutic index/clinical safety in several ways. For example, safety biomarkers can be applied to address candidate selection and manage risk by monitoring the no observed adverse effect levels of exposure in preclinical and clinical studies. Safety biomarkers are also useful for assessing the human relevance of preclinical safety findings and enabling the development of safe or safer dosing paradigms. In nonclinical studies, target organ toxicity is assessed using histopathological analysis. However, in clinical trials, histopathological analysis is rarely available and biomarkers are critical to assess potential target tissue toxicity in humans. Thus, the most impactful safety biomarkers will be those used in clinical trials with direct translational ties to nonclinical safety studies.

Several consortia, primarily driven by pharmaceutical industry members and encouraged by health authorities, have been formed to evaluate and qualify safety biomarkers for use in early clinical drug development trials. In the remainder of this manuscript, we will focus on the safety biomarker qualification efforts of the Critical Path Institute's (C-Path) Predictive Safety Testing Consortium (PSTC), as well as the PSTC collaborations with the Foundation for the National Institutes of Health's Biomarkers Consortium's (FNIH BC) Kidney Safety Project (KSP) and the Innovative Medicines Initiative's (IMI) Safer and Faster Evidence-based Translation Consortium (SAFE-T). Both of these collaborations are driven by the common goal of modernizing safety science through the qualification of clinical safety biomarkers for use in drug development.

44.2 QUALIFICATION OF SAFETY BIOMARKERS

Qualification is a formal regulatory review and acceptance process at the US Food and Drug Administration (FDA) and the European Medicines Agency (EMA) whereby a conclusion is reached such that within the stated context of use (COU), a biomarker or other DDT or novel methodology can be used with regulatory certainty. Regulatory certainty is defined as the assurance to drug developers that these approaches will be accepted by regulatory authorities. A COU is analogous to a registered drug's label. According to the FDA, a COU is a comprehensive and clear statement that describes the manner of use, interpretation, and purpose of use of a biomarker in drug development (FDA, 2015a). A hypothetical COU is shown in the following. This COU has been proposed by the PSTC's Skeletal Myopathy Working Group (SKMWG) in response to discussions around how an ideal novel safety biomarker for drug-induced skeletal muscle injury could be most useful in drug development.

Hypothetical COU for novel safety biomarkers for pancreatic injury: The qualified biomarker(s) may be used to monitor for pancreatic safety in early clinical studies with new molecular entities (NMEs) that have been shown to cause pancreatic injury in animal toxicology studies. Ideally, the qualified biomarker(s) will also be translatable and will show a change in animal studies that can be monitored and then used to inform pancreatic safety in clinical studies. The qualified biomarker(s) will be used in conjunction with conventional markers of pancreatic injury (e.g., amylase and lipase) as a more sensitive and/or earlier biomarker of pancreatic injury. When a biomarker level greatly differs from a defined threshold, as seen in the absence (or presence) of an elevation of serum amylase activity, this would be considered an indicator of pancreatic injury. For NMEs with pancreatic pathology in animal toxicology studies, applying the biomarker(s) in the design of the initial single and multiple ascending dose studies would enable safer progression to clinical development. Use of the biomarker(s) could also enable or restrict the planned dose escalation to higher clinical exposures, depending on risk/benefit considerations. This is because the new biomarker(s) would detect pancreatic injury earlier and therefore increase confidence in escalating clinical exposures up to or exceeding the nonclinical no observed adverse effect level (NOAEL), provided that no change in concentration of biomarker(s) is seen in the clinical study. The absence of a significant change in the biomarker(s) in single and multiple ascending dose studies in healthy volunteers would signify no clinically relevant skeletal muscle injury at those exposures.

Regulatory qualification at FDA and EMA generally consists of a consultation and advice phase, followed by a review phase. In the case of the EMA, the process for scientific advice and opinion is utilized to give qualification advice and opinion. Qualification, on a fundamental level, involves a submitter articulating a COU for a novel DDT or methodology, which adds significant value to some aspect of drug development, and compiling the scientific data and evidence in support of that specific COU.

Following qualification, a guidance document on the uses and limitations of the biomarker, including the COU, is issued by the FDA and EMA. It is important to point out that the strategy for safety biomarker qualification utilizes a translational approach. Although the primary data for clinical qualification are the biomarker performance data from clinical studies, nonclinical data are used to underpin the clinical data and anchor the biomarker's response to a defined histopathological change. A positive qualification decision by regulatory authorities ensures a more efficient implementation of safety biomarkers and encourages researchers to utilize these biomarkers in the drug development process. Thus, qualification results in both increased scientific acceptance and regulatory certainty based on a weight of evidence argument. Qualified safety biomarkers should provide a clear and easily measurable indication of organ injury, giving all parties involved a standardized, reliable tool.

44.3 LETTER OF SUPPORT FOR SAFETY BIOMARKERS

The Letter of Support was established by the FDA and EMA in 2014 as a means to recognize the potential utility of exploratory biomarkers prior to qualification. C-Path's PSTC was the first biomarker submitter to receive a Letter of Support from the FDA and the EMA. The Letter of Support, as a regulatory outcome, resulted in part from the discussion between PSTC, EMA, and FDA and the realization that greater attention to promising biomarker programs would help facilitate the use of exploratory biomarkers.

The FDA has stated that the Letter of Support is an opportunity to recognize the potential utility of exploratory biomarkers prior to qualification (FDA, 2015b). A Letter of Support is issued from the FDA to a submitter who has assembled the necessary information about promising biomarkers. The letter briefly describes the views of the FDA's Center for Drug Evaluation and Research (CDER) on the potential value of a biomarker and encourages further evaluation of the biomarker. This letter does not connote qualification of a biomarker. It is meant to enhance the visibility of the biomarker, encourage data sharing, and stimulate additional studies on promising biomarkers which are not yet ready for qualification. The FDA's Letter of Support encourages the identification and qualification of new DDTs and has been recognized as an approach to overcome hurdles in drug development programs with the potential to enhance the availability of useful information about drug safety and efficacy.

The EMA has stated that based on qualification advice, the agency may propose a Letter of Support as an option, when the novel methodology under evaluation cannot yet be clinically qualified but is shown to be promising based on preliminary data (EMA, 2015a). A Letter of Support from the EMA aims to encourage data sharing and to facilitate studies supporting eventual qualification for the novel methodology under evaluation. These letters from the EMA include a high-level summary of the novel methodology, COU, available data, and ongoing and future investigations. Like the FDA, the EMA publishes Letters of Support on their website, in agreement with sponsors.

In each case where both the FDA and EMA have granted Letters of Support to PSTC, the intent of the letter, as well as the basic language, has been similar. However, although the goal of the Letter of Support mechanism is similar for both, the mechanism and regulatory infrastructure utilized to issue such letters differs. For the EMA, the Letter of Support is an integrated part of qualification and a result of qualification advice, while the FDA sees the Letter of Support as a product outside of the qualification process, although it may also be issued for a project pursuing qualification. While the final outcome may be the same, the process and program expectations to garner a Letter of Support are not identical between the FDA and EMA.

Regardless of its positioning, the Letter of Support is a significant step forward in helping to drive the qualification of exploratory biomarkers. This relatively straightforward approach has created numerous opportunities to share data from nonclinical and clinical studies utilizing exploratory biomarkers. For example, a broad data set from an exploratory clinical biomarker with a supporting nonclinical data set could result in the qualification of the biomarker, while dedicated prospective qualification studies or other approaches could be used to expand the COU for a given biomarker. To this end, it is essential that a centralized database be established where anonymized biomarker data from global academic and industry-sponsored trials can be collected, maintained, shared, and analyzed. This could result not only in the qualification of biomarkers but could also enable COU optimization and identify the impact of such interventions on drug safety.

In order to help drug development sponsors understand the value of including exploratory biomarkers in nonclinical studies and clinical trials, the PSTC has posted summary data packages on each of the biomarkers that have received a Letter of Support on the C-Path website (Critical Path Institute's Predictive Safety Testing Consortium, 2015a).

44.4 CRITICAL PATH INSTITUTE'S PREDICTIVE SAFETY TESTING CONSORTIUM

The PSTC is one of the twelve consortia comprising C-Path, a nonprofit organization launched in 2005 and dedicated to playing the role of a catalyst in the development of new approaches that advance medical innovation and regulatory science. This is achieved by leading teams that share data, knowledge, and expertise, resulting in sound, consensus-based science. Although C-Path has a number of funding models for its consortia, PSTC is funded by a grant from the FDA's Center for Federal Drug Administration and Industry Collaboration (CFIC) and members' contributions. Members' contributions consist of a membership fee and in-kind contributions that support the research required to drive the objectives of PSTC's working groups. PSTC is a unique, public–private partnership that brings pharmaceutical companies together to share and validate safety testing methods under the advisement of worldwide regulatory agencies, including the FDA, the EMA, and the Japanese Pharmaceuticals and Medical Devices Agency (PMDA) (Mattes and Walker, 2009; Dennis et al., 2013; Walker et al., 2013; Stephenson and Sauer, 2014; Critical Path Institute's Predictive Safety Testing Consortium, 2015b). All 18 corporate members of PSTC share a common goal: to find improved safety testing methods and approaches utilizing fluid-based safety biomarkers to accurately predict drug-induced tissue injury. Specifically, the primary goal of PSTC is the qualification of novel translational safety biomarkers for use in early clinical drug development trials in order to enable the safer investigation and development of new drug candidates.

As discussed previously, clinical safety biomarkers for use in early drug development trials are critically important because insufficient therapeutic index is a major cause of new drug failure. However, current biomarker standards for many drug-induced tissue injuries either do not exist or have significant limitations. Thus, there is a clear need for improved safety biomarkers for each of the target organs under investigation by the PSTC. PSTC's working groups are structured around target organs, including heart, liver, skeletal muscle, vasculature, kidney, and testis, with cross-functional teams working through common approaches such as microRNA analysis, assessment and categorization of pathological lesions, approaches to statistical analysis, and approaches to assay validation. Currently, there are six working groups in PSTC and a brief description of their objectives is provided in the following.

Nephrotoxicity Working Group (NWG): Conventional biomarkers of drug-induced kidney injury (DIKI) currently used in drug development lack sensitivity. The loss of kidney function that defines acute kidney injury (AKI) is most often detected by measurement of serum creatinine, which is slow to respond even in cases of severe kidney injury. Thus, there is a clear need for biomarkers that detect early DIKI to enable earlier intervention. The NWG, in collaboration with the FNIH BC KSP and the IMI SAFE-T DIKI work package, is working toward the clinical qualification of several urinary kidney safety biomarkers including osteopontin, clusterin, cystatin C, kidney injury molecule-1, N-acetyl-β-D-glucosaminidase, neutrophil gelatinase-associated lipocalin, total protein, and albumin. PSTC has already demonstrated the diagnostic utility of these biomarkers in rodents (EMA Final Conclusions, 2008; FDA Qualification, 2008; Bonventre et al., 2010; Dieterle et al., 2010a, b; Goodsaid and Papaluca, 2010; Mattes et al., 2010; Ozer et al., 2010; PMDA Record of Consultation, 2010; Sistare et al., 2010; Vaidya et al., 2010; Yu et al., 2010; Warnock and Peck, 2010; EMA Letter of Support, 2014; FDA Biomarker Letter of Support, 2014) and has an active research program in canines and nonhuman primates.

Skeletal Myopathy Working Group (SKMWG): Drug-induced skeletal muscle toxicity is becoming more prevalent as an issue in drug development likely due to the evaluation of novel pharmacological targets and the disease populations being investigated. Aspartate aminotransferase activity (AST) and creatine kinase (CK; serum CK activity), the traditional biomarkers of skeletal muscle toxicity, lack both specificity and sensitivity. Novel skeletal muscle biomarkers show promise as more sensitive and more specific biomarkers of drug-induced skeletal muscle injury. The SKMWG is working toward the clinical qualification of several skeletal muscle safety biomarkers including plasma/serum skeletal troponin I, myosin light chain 3, fatty acid-binding protein 3, and creatine kinase muscle type. It is hoped that these biomarkers will provide greater predictive accuracy in the diagnosis and monitoring of drug-induced skeletal muscle toxicity in drug development clinical trials (EMA Letter of Support, 2015a; FDA Biomarker Letter of Support, 2015c).

Hepatotoxicity Working Group (HWG): Standard biomarkers of drug-induced liver injury (DILI) utilized by the clinical community for many years include alanine aminotransferase (ALT) and AST. However, both the specificity and sensitivity of these markers are limited due to lack of correlation between changes in these liver enzymes and observable histopathological damage. Although these transaminases have proven to be excellent markers of hepatotoxicity, additional biomarkers that more fully inform prediction of DILI are desirable. For instance, new markers that help predict whether ALT increases will resolve or progress to more serious DILI, and markers that can better discriminate liver and skeletal muscle injury will help in the complex assessment of DILI in drug development. The HWG, in collaboration with the SAFE-T DILI work package, is working toward the clinical qualification of several liver safety biomarkers including plasma/serum miR-122, glutamate dehydrogenase, arginase, sorbitol dehydrogenase, and glutathione-S-transferase. Another objective of HWG's work has been to understand the potential hepatotoxic liability of drug candidates that are potent inhibitors of the bile salt efflux pump (BSEP) and devise strategies to mitigate the potential risk. The PSTC hopes to clarify several aspects associated with DILI, which is an important, complex issue in drug development.

Vascular Injury Working Group (VIWG): Currently, there are no biomarkers available to detect drug-induced vascular injury (DIVI) in humans. The VIWG in collaboration with the SAFE-T DIVI work package is characterizing several biomarkers that are diagnostic for inflammation, as well as endothelial cell and smooth muscle cell injury in nonclinical species and humans with the ultimate goal of qualifying these biomarkers for use in drug development. Although this group has successfully identified candidate biomarkers, differences in protein expression and function across humans and animals have limited the direct translation of these safety biomarkers. The qualification of DIVI biomarkers represents a challenge greater than that faced by other working groups due to the lack of a current clinical gold standard biomarker and the lack of direct translation of the clinical biomarkers being pursued (Mikaelian et al., 2014).

Testicular Toxicity Working Group (TWG): There are no biomarkers available for detecting drug-induced seminiferous tubule toxicity in the clinic, highlighting the value of work being done by the TWG. Currently, this group is focusing on the applicability of microRNA species as biomarkers of testicular injury. A focused research plan has been implemented with the ultimate goal of achieving clinical qualification of testicular safety biomarkers. The program is currently in the early discovery stage, working through much of the basic science associated with reliably quantifying microRNA in serum. Although this project's goal of providing biomarkers for drug-induced testicular injury is of significant value, the basic research around the quantification of microRNA-based biomarkers will also impact other biomarker discovery and qualification efforts.

Cardiac Hypertrophy Working Group (CHWG): Work is being completed by the CHWG to evaluate NT-proANP in rodents as a marker of drug-induced hemodynamic stress, which leads to changes in cardiac mass. The data collected indicates that NT-proANP can be used as a screening tool to identify clinical candidates with cardiac hypertrophy liabilities without resorting to ECG-gated magnetic resonance imaging (MRI) in nonclinical studies. Although NT-proANP may not be a candidate for biomarker qualification, the results of this work will have a fundamental impact on approaches used in investigational toxicology. This work points out one of the PSTC goals beyond regulatory endorsement: to impact the pharmaceutical industry's approach to toxicology (safety) in both drug discovery and development (Fielden et al., 2008, 2011).

Despite considerable advances in medicine and technology, many of the approaches and strategies used to evaluate drug safety have not changed in decades. The ultimate goal of the PSTC is to transform the current approach to drug safety testing and liaise with regulatory authorities to offer assurance to drug developers that these approaches will be accepted by regulatory authorities and thereby improve both the speed and precision of the drug development process. The PSTC has been successful in pursuit of this goal through the qualification of novel translational safety biomarkers.

In 2008, PSTC engaged in a joint process between the FDA and EMA to achieve the qualification of a biomarker. Utilizing this joint process, seven rodent kidney safety biomarkers were qualified by both agencies (EMA Final Conclusions, 2008; FDA Qualification, 2008). In 2010, these same kidney biomarkers were also qualified with Japan's PMDA (PMDA Record of Consultation, 2010). Following this series of qualifications, as additional biomarker qualification requests entered the consultation (i.e., qualification advice at EMA) phase, the regulatory expectations for evidentiary standards evolved. This resulted in a general bottleneck in the qualification process. Therefore, for the past several years, the PSTC has been working with the FDA and EMA to better define the requirements within the qualification process. An important initial step was the piloting of a mechanism by which regulatory authorities could recognize the potential utility of exploratory biomarkers prior to qualification, known as the Letter of Support. In 2014, the PSTC NWG received a Letter of Support from the FDA and EMA for two new kidney safety biomarkers (EMA Letter of Support, 2014; FDA Biomarker Letter of Support, 2014). And in 2015, the PSTC SKMWG received a Letter of Support for four new skeletal muscle injury biomarkers (EMA Letter of Support, 2015b; FDA Biomarker Letter of Support, 2015c). The Letter of Support represents a significant accomplishment in the regulatory authorities' armamentarium and has been compatible with PSTC's goal to assure all stakeholders greater clarity in the path to qualification. In addition, the Letter of Support opens the door to the potential for new opportunities via broader generation of use data to further impact the qualification process.

44.5 PREDICTIVE SAFETY TESTING CONSORTIUM AND ITS KEY COLLABORATIONS

No single company or research organization can independently change the approach to safety science. Thus, collaboration in precompetitive consortia like PSTC is an effective approach to impact the scientific and regulatory landscape that governs drug development. The large number of consortia actively involved in addressing gaps in the science and practice of drug development creates the opportunity to collaborate based on common objectives. However, it is interesting that although consortia are founded on the spirit of collaboration, cross-consortium collaborations are rare. A fundamental obstacle to collaboration appears to be "self-preservation" and the fear of losing relevance or advantage over rivals resulting in the demise of the consortium, or the desire to be the first consortium to succeed. Luckily, in some cases the benefit of collaboration outweighs the imagined liabilities, and strong leaders find common ground to achieve even larger objectives. Although it is likely that cross-consortium collaborations will continue to expand and become the accepted norm, the establishment of functional relationships between collaborative groups can be limited by legal, logistical, and cultural factors. Therefore, it is paramount that consortia organizers envision crucial collaborations at the project design stage and establish a collaborative framework at project inception.

By pooling resources and combining efforts, PSTC is working to improve the safety of newly created therapies, thereby expediting drug development and the regulatory approval process. This will have a positive, measurable impact on all stakeholders, including pharmaceutical companies, regulatory authorities, and patients. Cross-consortium collaboration provides the resources to radically impact safety science in the short term. For example, while focused on nonclinical and translational aspects of safety biomarker qualification, in some cases PSTC lacks the clinical expertise required for efficient clinical qualification of translational biomarkers. Therefore, in order to achieve their primary goal of qualifying safety biomarkers, PSTC has partnered with two important consortia, FNIHBC KSP and the IMI SAFE-T.

The PSTC collaborations with FNIH BC KSP and IMI SAFE-T are productive collaborations between consortia that share overlapping goals, as well as corporate members, in this case from the pharmaceutical industry. PSTC signed formal collaboration agreements with FNIH BC KSP on October 25, 2011, and with IMI SAFE-T on May 23, 2013. Although FNIH and SAFE-T have had a less formal relationship, PSTC's involvement with both consortia has helped enable the sharing of regulatory strategy and scientific approaches between these two groups. The following sections will discuss the collaborations that PSTC has established with the FNIH BC KSP and SAFE-T.

44.6 ADVANCING THE QUALIFICATION PROCESS AND DEFINING EVIDENTIARY STANDARDS

As previously stated, insufficient therapeutic index is a major cause of failure in drug development, and because many current safety biomarkers lack sufficient sensitivity and specificity to adequately evaluate the therapeutic index of new drugs, there is a critical need for improved safety biomarkers. However, the adoption of novel safety biomarkers through the qualification process has been slowed for two major reasons: (i) the regulatory and scientific expectations for qualification have been evolving as more experience is gained from this relatively new program and (ii) the inaccessibility of data from those using the biomarkers due to concerns over maintaining a competitive advantage and conservative legal positions around drug safety liability. Clearly defined evidentiary standards and access to appropriate data will dramatically accelerate the qualification of safety biomarkers (Amur et al., 2015).

The articulation of evidentiary standards will allow biomarker submitters to appropriately plan their qualification strategies and have more direct conversations with the FDA and EMA with the understanding that the level of evidence for qualification of a biomarker is directly related to factors such as the breadth of the stated COU, the implications for risk to patients if the biomarker "fails," and the predictability of the assay performance characteristics. At this point the obvious evidentiary gaps include:

1. Defined expectations around clinical data generation and prospective analysis
2. Statistical methodology expectations for confirmatory data analysis
3. Biomarker assay validation and performance expectations
4. Clinical and nonclinical data expectation for (translational) qualification of clinical safety biomarkers

The path to developing regulatory guidance on evidentiary standards for qualification of biomarkers will require involvement of all sectors of the bioscience research community including but not necessarily limited to industry, FDA, EMA, government research entities, academia, patient groups, and nonprofit organizations. As with the drug development and regulatory review processes, there will be a need for regulatory guidance documents focused on providing direction for critical elements of the overall biomarker qualification process. For example, there might be a guidance document specific to statistical methodology for biomarker qualification or one describing assay validation. This will be an iterative process that seeks to refine terminology, standards, language, etc., in parallel with incorporating lessons learned from ongoing biomarker qualification programs. This will require science-based discussions without attribution to enable an open dialogue with exchange of various expert perspectives. Those with specific expertise in current biomarker qualification, drug development, device development, clinical trial design, statistical methodology, analytic methodology, regulatory process and strategy, regulatory decision-making, data handling, data sharing, and database methodology must be included in order to ensure that information contributing to the framework for regulatory expectations and eventual guidance documents represents the application of scientific methods on the way to a regulatory outcome. Ultimately, with the attention of stakeholders, this process will provide the required underpinning to expedite the qualification of biomarkers and other DDTs and methodologies.

Several key aspects should be considered in creating scientific expectations specific to qualification of safety biomarkers. A brief list of considerations is presented in the following for safety biomarkers that are supported by both translational nonclinical data and clinical trial data:

1. Availability of sufficiently validated analytical assays to quantify biomarkers
2. Biological understanding of the biomarker including the specificity of the response to toxicological outcomes in the target tissue and other relevant tissues as well as the pharmacological effects of agents without toxicity in the target organ
3. Understanding of mechanism of the biomarker's biological response
4. Correlation of biomarker response to pathology and improved performance relative to other (standard) biomarkers
5. Consistent response across mechanistically different compounds with similar response; similar response across sex, strain, and species
6. Presence of dose–response and temporal relationship to the magnitude of response

The second issue slowing adoption of novel safety biomarkers is inaccessibility to data from those using the biomarkers. Therefore, multiple approaches should be considered to encourage drug development sponsors and academic centers to (i) capture data according to predetermined standards so that data sets across multiple contributors can be aggregated and (ii) share data through a protected mechanism in order to advance understanding of biomarker performance and contribute to robust decision-making about the regulatory acceptance of that biomarker. An approach being considered by C-Path is a proof-of-concept experiment whereby data from use of a prespecified set of biomarkers can be housed in a central data repository held by a neutral third party for the purpose of advancing the accumulation of needed evidence to enable regulatory decisions about the biomarkers. This experiment should demonstrate the value of a more collaborative approach that will expedite the timeline to achieving qualification of new biomarkers.

Finally, it is critical to identify the quickest path to qualification and the implementation of safety biomarkers in well-controlled clinical trials, because the only way to understand the advantages and disadvantages of a biomarker will be through its broad application.

REFERENCES

Amur S, LaVange L, Zineh I, Buckman-Garner S, Woodcock J. Biomarker qualification: Toward a multi-stakeholder framework for biomarker development, regulatory acceptance, and utilization. *Clin Pharmacol Ther*. 2015. 98:34–46.

Bonventre JV, Vaidya VS, Schmouder R, Feig P, Dieterle F. Next-generation biomarkers for detecting kidney toxicity. *Nat Biotechnol*. 2010. 28(5):436–40.

Critical Path Institute's Predictive Safety Testing Consortium. Regulatory successes. 2015a. Available at: http://c-path.org/programs/pstc/regulatory-successes/ Accessed May 20, 2015.

Critical Path Institute's Predictive Safety Testing Consortium. Translational safety strategies accelerating drug development: Collaborative structure. 2015b. Available at: http://c-path.org/programs/pstc/ Accessed May 20, 2015.

Dennis EH, Walker EG, Baker AF, Miller RT. Opportunities and challenges of safety biomarker qualification: Perspectives from the Predictive Safety Testing Consortium. *Drug Dev Res*. 2013. 74(2):112–26.

Dieterle F, Sistare F, Goodsaid F, Papaluca M, Ozer JS, Webb CP, Baer W, Senagore A, Schipper MJ, Vonderscher J, Sultana S, Gerhold DL, Phillips JA, Maurer G, Carl K, Laurie D, Harpur E, Sonee M, Ennulat D, Holder D, Andrews-Cleavenger D, Gu YZ, Thompson KL, Goering PL, Vidal JM, Abadie E, Maciulaitis R, Jacobson-Kram D, Defelice AF, Hausner EA, Blank M, Thompson A, Harlow P, Throckmorton D, Xiao S, Xu N, Taylor W, Vamvakas S, Flamion B, Lima BS, Kasper P, Pasanen M, Prasad K, Troth S, Bounous D, Robinson-Gravatt D, Betton G, Davis MA, Akunda J, McDuffie JE, Suter L, Obert L,

Guffroy M, Pinches M, Jayadev S, Blomme EA, Beushausen SA, Barlow VG, Collins N, Waring J, Honor D, Snook S, Lee J, Rossi P, Walker E, Mattes W. Renal biomarker qualification submission: A dialog between the FDA-EMEA and Predictive Safety Testing Consortium. *Nat Biotechnol*. 2010a. 28(5):455–62.

Dieterle F, Perentes E, Cordier A, Roth DR, Verdes P, Grenet O, Pantano S, Moulin P, Wahl D, Mahl A, End P, Staedtler F, Legay F, Carl K, Laurie D, Chibout SD, Vonderscher J, Maurer G. Urinary clusterin, cystatin C, [beta]2-microglobulin and total protein as markers to detect drug-induced kidney injury. *Nat Biotechnol*. 2010b. 28(5):463–9.

EMA. Final conclusions on the pilot joint EMEA/FDA VXDS experience on qualification of nephrotoxicity biomarkers. 2008. Available at: http://www.ema.europa.eu/docs/en_GB/document_library/Regulatory_and_procedural_guideline/2009/10/WC500004205.pdf Accessed May 20, 2015.

EMA. Letter of support for PSTC translational drug-induced kidney injury (DIKI) biomarkers osteopontin (OPN) and neutrophil gelatinase-associated lipocalin (NGAL). 2014. Available at: http://www.ema.europa.eu/docs/en_GB/document_library/Other/2014/11/WC500177133.pdf Accessed on May 20, 2015.

EMA. Qualification of novel methodologies for medicine development. 2015a. Available at: http://www.ema.europa.eu/ema/index.jsp?curl=pages/regulation/document_listing/document_listing_000319.jsp&mid=WC0b01ac0580022bb0 Accessed May 20, 2015.

EMA. Letter of support for skeletal muscle injury biomarkers. 2015b. Available at: http://www.ema.europa.eu/docs/en_GB/document_library/Other/2015/03/WC500184458.pdf Accessed on May 20, 2015.

FDA. Qualification of seven biomarkers of drug-induced nephrotoxicity in rats. 2008. Available at: http://www.fda.gov/downloads/Drugs/DevelopmentApprovalProcess/DrugDevelopmentToolsQualificationProgram/UCM285031.pdf Accessed on May 20, 2015.

FDA. Biomarker letter of support for urinary osteopontin (OPN) and neutrophil gelatinase-associated lipocalin (NGAL). 2014. Available at: http://www.fda.gov/downloads/Drugs/DevelopmentApprovalProcess/DrugDevelopmentToolsQualificationProgram/UCM412843.pdf Accessed on May 20, 2015.

FDA. Biomarker qualification context of use. 2015a. Available at: http://www.fda.gov/Drugs/DevelopmentApprovalProcess/DrugDevelopmentToolsQualificationProgram/ucm284620.htm Accessed on May 20, 2015.

FDA. Drug development tools (DDT) letters of support. 2015b. Available at: http://www.fda.gov/Drugs/DevelopmentApprovalProcess/ucm434382.htm Accessed on 20 May 2015.

FDA. Biomarker letter of support for plasma/serum myosin light chain 3 (Myl3), skeletal muscle troponin I (sTnI), fatty acid binding protein 3 (Fabp3), and creatine kinase, muscle type (CK-M, the homodimer CK-MM). 2015c. Available at: http://www.fda.gov/downloads/Drugs/DevelopmentApprovalProcess/DrugDevelopmentToolsQualificationProgram/UCM432653.pdf Accessed May 20, 2015.

Fielden MR, Nie A, McMillian M, Elangbam CS, Trela BA, Yang Y, Dunn RT II, Dragan Y, Fransson-Stehen R, Bogdanffy M, Adams SP, Foster WR, Chen S-J, Rossi P, Kasper P, Jacobson-Kram D,

Tatsuoka KS, Wier PJ, Gollub J, Halbert DN, Roter A, Young JK, Sina JF, Marlowe J, Martus H-J, Aubrecht J, Olaharski AJ, Roome N, Nioi P, Pardo I, Snyder R, Perry R, Lord P, Mattes W, Car BD. Interlaboratory evaluation of genomic signatures for predicting carcinogenicity in the rat. *Toxicol Sci.* 2008. 103(1):28–34.

Fielden MR, Adai A, Dunn RT II, Olaharski A, Searfoss G, Sina J, Aubrecht J, Boitier E, Nioi P, Auerbach S, Jacobson-Kram D, Raghavan N, Yang Y, Kincaid A, Sherlock J, Chen S-J, Car B. Development and evaluation of a genomic signature for the prediction and mechanistic assessment of nongenotoxic hepatocarcinogens in the rat. *Toxicol Sci.* 2011. 124(1):54–74.

Goodsaid F, Papaluca M. Evolution of biomarker qualification at the health authorities. *Nat Biotechnol.* 2010. 28(5):441–3.

Mattes WB, Walker EG. Translational toxicology and the work of the predictive safety testing consortium. *Clin Pharmacol Ther.* 2009. 85(3):327–30.

Mattes WB, Walker EG, Abadie E, Sistare FD, Vonderscher J, Woodcock J, Woosley RL. Research at the interface of industry, academia and regulatory science. *Nat Biotechnol.* 2010. 28(5):432–3.

Mikaelian I, Cameron M, Dalmas DA, Enerson BE, Gonzalez RJ, Guionaud S, Hoffmann PK, King NMP, Lawton MP, Scicchitano MS, Smith HW, Thomas RA, Weaver JL, Zabka TS; The Vascular Injury Working Group of the Predictive Safety Consortium. Nonclinical safety biomarkers of drug-induced vascular injury: Current status and blueprint for the future. *Toxicol Pathol.* 2014. 42(4):635–57.

Ozer JS, Dieterle F, Troth S, Perentes E, Cordier A, Verdes P, Staedtler F, Mahl A, Grenet O, Roth DR, Wahl D, Legay F, Holder D, Erdos Z, Vlasakova K, Jin H, Yu Y, Muniappa N, Forest T, Clouse HK, Reynolds S, Bailey WJ, Thudium DT, Topper MJ, Skopek TR, Sina JF, Glaab WE, Vonderscher J, Maurer G, Chibout S-D, Sistare FD, Gerhold DL. A panel of urinary biomarkers to monitor reversibility of renal injury and a serum marker with improved potential to assess renal function. *Nat Biotechnol.* 2010. 28(5):486–94.

Parekh A, Buckman-Garner S, McCune S, O'Neill R, Geanacopoulos M, Amur S, Clingman C, Barratt R, Rocca M, Hills I, Woodcock J. Catalyzing the critical path initiative: FDA's progress in drug development activities. *Clin Pharmacol Ther.* 2015. 97:221–33.

PMDA. Record of the consultation on pharmacogenomics/biomarkers. 2010. Available at: https://www.pmda.go.jp/files/000160006.pdf Accessed on May 20, 2015.

Sistare FD, Dieterle F, Troth S, Holder DJ, Gerhold D, Andrews-Cleavenger D, Baer W, Betton G, Bounous D, Carl K, Collins N, Goering P, Goodsaid F, Gu Y-Z, Guilpin V, Harpur E, Hassan A, Jacobson-Kram D, Kasper P, Laurie D, Lima BS, Maciulaitis R, Mattes W, Maurer G, Obert LA, Ozer J, Papaluca-Amati M, Phillips JA, Pinches M, Schipper MJ, Thompson KL, Vamvakas S, Vidal J-M, Vonderscher J, Walker E, Webb C, Yu Y. Towards consensus practices to qualify safety biomarkers for use in early drug development. *Nat Biotechnol.* 2010. 2(5):446–54.

Stephenson D, Sauer J-M. The Predictive Safety Testing Consortium and the coalition against major diseases. *Nat Rev Drug Discov.* 2014. 13(11):793–4.

Vaidya VS, Ozer JS, Dieterle F, Collings FB, Ramirez V, Troth S, Muniappa N, Thudium D, Gerhold D, Holder DJ, Bobadilla NA, Marrer E, Perentes E, Cordier A, Vonderscher J, Maurer G, Goering PL, Sistare FD, Bonventre JV. Kidney injury molecule-1 outperforms traditional biomarkers of kidney injury in preclinical biomarker qualification studies. *Nat Biotechnol.* 2010. 28(5):478–85.

Walker EG, Brumfield M, Compton C, Woosley R. Evolving global regulatory science through the voluntary submission of data a 2013 assessment. *Ther Innov Regul Sci.* 2013. 48(2):236–245.

Warnock DG, Peck CC. A roadmap for biomarker qualification. *Nat Biotechnol.* 2010. 28(5):444–5.

Yu Y, Jin H, Holder D, Ozer JS, Villarreal S, Shughrue P, Shi S, Figueroa DJ, Clouse H, Su M, Muniappa N, Troth SP, Bailey W, Seng J, Aslamkhan AG, Thudium D, Sistare FD, Gerhold DL. Urinary biomarkers trefoil factor 3 and albumin enable early detection of kidney tubular injury. *Nat Biotechnol.* 2010. 28(5):470–7.

PART X

CONCLUSIONS

45

TOXICOGENOMICS IN DRUG DISCOVERY TOXICOLOGY: HISTORY, METHODS, CASE STUDIES, AND FUTURE DIRECTIONS

Brandon D. Jeffy[1], Joseph Milano[2] and Richard J. Brennan[3]

[1] Exploratory Toxicology, Celgene Corporation, San Diego, CA, USA

[2] Milano Toxicology Consulting, L.L.C., Wilmington, DE, USA

[3] Preclinical Safety, Sanofi SA, Waltham, MA, USA

45.1 A BRIEF HISTORY OF TOXICOGENOMICS

The term "toxicogenomics" first appeared in the scientific literature in March 1999[1] when researchers from the National Institute of Environmental Health Sciences and the National Human Genome Research Institute described the development and application of a cDNA microarray, ToxChip v1.0, comprising 2090 human gene sequences chosen for their involvement in basic cellular processes and their known responses to different types of toxic insult (Nuwaysir et al., 1999). Although the development of microarray technology in the mid to late 1990s enabled the broad application of multiplexed gene expression analysis, the concept of using changes in gene expression as an early and sensitive surrogate measure of toxicological responses goes back much further.

It became recognized more than 50 years ago that certain toxicants, such as cytotoxic antibiotics, acted through the inhibition of messenger RNA (mRNA) synthesis from DNA (Gray et al., 1966; Clifford and Rees, 1967), and that, on the other hand, an increase in RNA synthesis was a characteristic of some hepatocarcinogens (Hawtrey and Nourse, 1964; Turner and Reid, 1964). Studies in experimental species,

demonstrating similarity across organs in RNA and protein biosynthesis, led to the conclusion that alterations in their synthesis by toxic agents could be evaluated concordantly (Witschi, 1972, 1973). It also had been recognized that toxic agents were able to upregulate the synthesis and activities of specific enzymes (Conney et al., 1957, 1960; Civen and Knox, 1959). In 1964, Lawrence Loeb and Harry Gelboin determined that the induction of microsomal enzyme activity by carcinogenic polycyclic hydrocarbons such as methylcholanthrene was mediated via altering the "gene action system" through increased mRNA synthesis (Loeb and Gelboin, 1964). They presciently postulated that "The relevance of these changes to carcinogenesis might be determined by a detailed analysis of the specific mRNA's synthesized in the presence of the carcinogen compared to the patterns of RNA formation elicited by the noncarcinogenic compound" (phenobarbital).

In the following years, the importance of drug-metabolizing enzymes, the regulation of their expression at the genetic level, and the association of gene regulatory loci such as the aryl hydrocarbon (Ah) locus with carcinogenesis and toxicity were established (Niwa et al., 1975; Thorgeirsson and Nebert, 1977; Kouri et al., 1978). The existence of distinct control mechanisms modulating differential expression of metabolizing enzymes for different inducers was also recognized (Oesch, 1976, 1980; Schmassmann and Oesch, 1978; Goujon et al., 1980). Further, the ability of gene

---

[1] As far as the authors can determine. Although a "concise review" by Farr and Dunn citing the term was submitted to *Toxicological Sciences* in September 1998, several months before the Nuwaysir et al. paper was submitted to *Molecular Carcinogenesis*, it was not published until July 1999.

Drug Discovery Toxicology: From Target Assessment to Translational Biomarkers, First Edition. Edited by Yvonne Will, J. Eric McDuffie, Andrew J. Olaharski, and Brandon D. Jeffy.

expression alterations by an inducer to alter sensitivity to the toxic effects of other toxicants through modulating their metabolic activation or detoxification was documented (Goldberger, 1974; Nebert and Felton, 1976; Nemoto and Takayama, 1980; Durnam and Palmiter, 1984).

With the invention of the Northern blot by James Alwine, David Kemp, and George Stark in 1977 (Alwine et al., 1977), quantitative analysis of the expression of specific genes became possible. This approach was soon applied to studying the differential expression of specific mRNA species in disease, toxicant response, and drug-induced lesions (Miles et al., 1981; Biggin et al., 1984; Raghow et al., 1984; Wong and Biswas, 1985; Ivy et al., 1988; Rataboul et al., 1989; Tsao et al., 1989).

In the 1990s, advances in polymerase chain reaction techniques led to the development of differential display (Liang and Pardee, 1992), serial analysis of gene expression (SAGE) (Velculescu et al., 1995), and other related multiplex expression analysis techniques. These expression profiling approaches enabled the simultaneous quantitative analysis of multiple mRNAs. Despite the promise of global gene expression approaches in revealing characteristic patterns of gene expression, and probing the mechanisms of stress and toxic response (e.g., Crawford and Davies, 1997; Gonzalez-Zulueta et al., 1998; Harris et al., 1998; Santiard-Baron et al., 1999; Caldwell et al., 2002), technical hurdles, including expertise dependency, large sample requirements, bias in transcript representation, and lack of robustness (Yamamoto et al., 2001; Cullen et al., 2002), limited the application of SAGE and its various derivatives in pharmaceutical R&D to screening for diagnostic biomarkers and novel targets (Bartlett, 2001).

In 1995 Todd et al. (1995) described the CAT-Tox (L) assay, taking a different, reductionist approach to leveraging gene expression analysis for mechanistic and predictive toxicology. The CAT-Tox (L) assay, marketed by Xenometrix Inc., employed a panel of 14 stress promoter– or response element–chloramphenicol acetyltransferase (CAT) fusion constructs stably integrated into a transformed human liver cell line (HepG2). Despite the somewhat limited range of responses monitored in the assay, its ability to classify chemicals by their modes of toxicity and in differentiating mechanisms was demonstrated for a variety of toxicants, including airborne pollutants (Vincent et al., 1997), DNA damagers (Beard et al., 1996), peroxisome proliferators (Lee et al., 1997), and heavy metals (Tully et al., 2000). Although commercial success for CAT-Tox (L) itself was limited, the proof of concept for utility of a sensitive and robust multiplexed gene expression profiling platform in predictive and mechanistic toxicology, and for classification of compounds by characteristic patterns of gene expression, was established. Furthermore, Xenometrix was granted a broad and controversial patent for the concept of "gene expression profiling" to characterize the biological effect of pharmacological

treatments, which, along with a similarly broad patent for expression array technologies held by Oxford Gene Technologies (founded by Edwin Southern in 1995), were a commercial hurdle in the early development of microarray technology (Stafford, 2006).

It was the evolution of high-density cDNA arrays, spotted onto nylon membranes (Zhao et al., 1995) and glass slides or other solid support media (Schena et al., 1995), which led to the broader adoption of gene expression profiling in drug discovery and development. Work in yeast by Patrick Brown at Stanford and others demonstrated that DNA microarrays could feasibly and reproducibly be employed to monitor gene expression patterns across an entire eukaryotic genome (DeRisi et al., 1997; Hauser et al., 1998). With the concurrent rapid expansion of published gene sequences for other species, including man, driven in large part by advances in sequencing technology in support of the Human Genome Project (Lander et al., 2001; Venter et al., 2001; Waterston et al., 2002; Gibbs et al., 2004), the potential applications in preclinical and clinical research, including safety assessment, were quickly recognized (Marton et al., 1998; Afshari et al., 1999, Pennie et al., 2000). By the end of 2000, scientists at AstraZeneca, Robert Wood Johnson Pharmaceutical Research Institute, and the US Environmental Protection Agency (EPA) National Health and Environmental Effects Research Laboratory (NHERL) had published studies using microarrays to investigate and discriminate mechanisms of toxicity across a variety of toxicants in human cell lines and rat *in vivo* studies (Burczynski et al., 2000; Holden et al., 2000; Nadadur et al., 2000). Investigative toxicology groups at Abbott (Waring et al., 2001), Pfizer (Bulera et al., 2001), and Eli Lilly (Baker et al., 2001) quickly followed suit with pilot studies in various systems demonstrating the promise of large-scale analysis of gene expression using microarray technology in hazard identification and mechanistic evaluation for drug safety.

The commercial potential of toxicogenomics also was recognized quickly. Companies such as Phase-1 Molecular Toxicology, Gene Logic, and Iconix Biosciences developed businesses around providing tools, services, and databases for toxicogenomic analyses. A particular focus of these companies was in the generation of compendia of gene expression profiles from drug and toxicant-treated model systems. These expression profiles were coupled to traditional clinical and histopathological measurements with the goal of mining the data to find specific gene expression patterns ("signatures") corresponding to the concurrent or future manifestation of toxic pathologies (Castle et al., 2002; Kier et al., 2004; Natsoulis et al., 2005). The Iconix DrugMatrix®[2] database, for example, contains expression profiles from multiple

---

[2] DrugMatrix was acquired by the National Toxicology Program in 2010 and has since been made freely available online along with the ToxFX® automated data analysis tool (https://ntp.niehs.nih.gov/drugmatrix/index.html).

tissues of rats treated *in vivo* with 638 different compounds at multiple doses and time points, along with histopathology, clinical pathology, and cheminformatics data for each experiment. One hundred forty-eight different gene expression signatures covering 98 distinct pathological or mode of action endpoints and covering a number of common target tissues of toxicity (including liver, kidney, heart, bone marrow, spleen, and skeletal muscle) are also stored in the database.

Classification approaches to derive gene signatures from multiple expression profiles have been widely applied in toxicogenomics to differentiate between different compound classes or modes of action (e.g., Hamadeh et al., 2002; Martin et al., 2007) to diagnose or predict pathological outcomes (e.g., Kier et al., 2004; Fielden et al., 2005; Natsoulis et al., 2008) and to provide early prediction of long-term effects such as carcinogenesis (Nie et al., 2006; Fielden et al., 2007). Despite enthusiasm in the scientific community for pharmacogenomic and toxicogenomic studies to advance personalized medicine, and drug discovery and development, a slew of concerns around the robustness, reproducibility, and accuracy of microarray technology in general, and around concordance in results between different platforms, raised serious doubts about the reliability of microarray expression data and the gene signatures derived from it (Ntzani and Ioannidis, 2003; Marshall, 2004; Shi et al., 2004).

These concerns were addressed, under FDA leadership, by a series of multicenter experiments by a consortium of academic, government, and commercial laboratories.

The Microarray Quality Control Consortium (MAQC) project demonstrated, in a series of publications, intraplatform consistency across various test sites and a high level of interplatform concordance in differentially expressed genes (Shi et al., 2006). A second phase of the project further investigated a broad range of approaches and algorithms for generating predictive signatures for disease and toxicity endpoints. It was demonstrated that predictability was more dependent on the endpoint, and the data available to describe it, than on the particular approach taken to derive the signature (Shi et al., 2010a). It was further shown that gene signatures derived using computational machine learning approaches nonetheless identified gene patterns relevant to the biology of the endpoint being predicted (Shi et al., 2010b). The MAQC projects provided tools and best practice guidance for microarray data generation, analysis, and reporting that alongside other forward-thinking efforts like the FDA Voluntary Genomic Data Submission program (Goodsaid et al., 2010) and the minimum information about a microarray experiment (MIAME) metadata standards (Brazma et al., 2001) provide a framework for ensuring data sharing, reproducibility, accuracy, and regulatory acceptance of microarray data in support of clinical trial proposals and drug registration applications.

Although toxicogenomics has not fulfilled some of the more wildly optimistic early predictions regarding its potential to replace traditional toxicity testing with a more sensitive, faster, accurate, and comprehensive readout of a compound's potential for causing harm, the utility of gene expression profiling to parse compound mechanism of toxicity, identify biomarkers of pharmacological and toxicological responses, and to give an early readout of the potential for certain long-term pathological responses has resulted in its widespread adoption in drug safety assessment.

45.2 TOOLS AND STRATEGIES FOR ANALYZING TOXICOGENOMICS DATA

The preface of *Microarray Biochip Technology* edited by Mark Schena (2000) compares the use of gene expression arrays to Galileo's first look at the cosmos in the seventeenth century (Schena, 2000). While this comparison harkened the promises of the Human Genome Project in the newly born field of genomics, the sobering complexity of biological systems led to our understanding that no single technology is sufficient to paint a complete picture of this complexity. This caveat holds true for the study of toxicological mechanisms. That said, a well-designed toxicogenomics study may provide insight into transcriptional perturbations that are the precursor of adaptive responses preceding tissue pathology. Here we will discuss tools and strategies for the analysis of toxicogenomics data.

The term "gene expression" is often used interchangeably to describe measurements of a gene transcript. Gene expression, however, is defined as the abundance of the functional product of a gene. This is most often assumed to be protein but may also be tRNA, rRNA, or microRNA (miRNA). This section will focus on the measurement and analysis of gene transcripts, the mRNA intermediate of the functional gene product. Many factors influence posttranscriptional control of gene expression, like regulation of mRNA transport out of the nucleus, mRNA stability, translational lag, protein degradation, and miRNA control (Larsson et al., 2013). Thus, the level of mRNA does not always correlate to the level of translated protein.

Understanding the data acquired from microarray analysis begins with the analyte. For the purpose of this discussion, we will describe the analyte of transcriptional assays focusing on the pervasive Affymetrix oligonucleotide microarray platform. These arrays measure the fluorescence signal of hybridized, labeled target using overlapping oligonucleotides complementary to a particular gene arranged on a glass substrate. As mentioned previously, the abundance of a gene's mRNA frequently does not positively correlate with the functional product; transcription may not equal translation. Oligonucleotide microarrays therefore do not directly measure gene expression. The output of a microarray experiment also represents (in most cases) the average mRNA hybridization signals of populations of cells in replicate samples. For the sake of simplicity and consistency, however, we will use the term "gene expression" for the measurements taken using this technology.

Reduction of candidate compound attrition has long been a goal of pharmaceutical R&D. Toxicity in both the preclinical and clinical settings has been the cause of great economic and opportunity cost and is the driver for investment in predictive tools to reduce this cost. Gene expression analyses in toxicology may be used to generate hypotheses regarding the molecular mechanism of a particular pathology, differentiate chemically related compounds, develop premonitory gene expression signatures, or to discover biomarkers of toxicity with improved sensitivity and specificity (Chen et al., 2012). Experimental design of toxicogenomics experiments should be determined by a specific set of questions. Mechanism-based studies seek to understand the underlying molecular events that lead to a particular tissue pathology (Milano et al., 2004); accordingly, the toxicity in question must be producible given the exposures selected. When using microarray analysis to discriminate between the toxic effects of compounds, dose selection must be based on an understanding of pharmacokinetics for *in vivo* experiments and cytotoxicity for *in vitro* experiments (McBurney et al., 2009, 2012).

Gene expression analysis begins with an experimental design in the appropriate test system to best address the experimental question. Acquisition of robust data and downstream analysis is dependent upon isolation of high-quality RNA to avoid systemic bias due to RNA degradation (Ohmomo et al., 2014). Total RNA quality may be monitored by agarose gel electrophoresis or analysis in a 2100 Bioanalyzer (Agilent Technologies). Assaying the starting material for RNA abundance may be performed on the level of a single transcript to an organism's entire transcriptome. Quantitative real-time PCR (qRT-PCR) is applied when assaying one to hundreds of gene transcripts. This technique is highly sensitive and specific in a well-controlled experiment.

Querying an organism's entire transcriptome may be accomplished using a variety of methods including SAGE, DNA microarrays, and the next-generation sequencing method, RNA-Seq (Anisimov, 2008; Wang et al., 2009). DNA microarrays are by far the most common tool in toxicogenomics and are available in many platforms that have been extensively studied for reproducibility within and between platforms (MAQC Consortium, 2006; Shi et al., 2006; Liu et al., 2012). The rest of this section will consider the use of the Affymetrix 3′ IVT oligonucleotide gene expression analysis platform.

Measurements of transcript abundance using the Affymetrix GeneChip® platform is based upon probe sets consisting of eleven overlapping 25-mer oligonucleotide pairs arrayed on a silicon substrate. Each probe is arranged in pairs of perfectly matched and corresponding single mismatched oligonucleotides against a target sequence. This arrangement allows for the subtraction of background hybridization to the mismatched probe and the measurement of multiple probes targeting a single gene, giving a better estimate of expression than a single probe. Total RNA or enriched mRNA is labeled by reverse transcription to cDNA then transcribed to cRNA in the presence of biotinylated ribonucleotides. After hybridization on the array, a streptavidin-bound fluorescent reporter is added to enable image capture and data acquisition. A single gene expression value based on fluorescence intensity is generated for each probe set by using a preprocessing algorithm for background correction, normalization, and summarization. A selection of algorithms is shown in Table 45.1. The strengths and weaknesses of various algorithms have been extensively reviewed elsewhere (Irizarry et al., 2006; Seo and Hoffman, 2006; Berger and Carlon, 2011).

Raw gene expression data preprocessing produces tens of thousands of intensity values across multiple experiments. The most common strategy for analysis of this quantity of data usually begins with an analysis of differentially expressed up- or downregulated genes. This may be accomplished with

TABLE 45.1 GeneChip Data Normalization and Summarization Algorithms

Algorithm	Method	Reference
RMA	Quantile normalization—raw intensity values are preprocessed to create equally distributed data between chips accounts for experimentally observed patterns for probe behavior. Only uses perfect match (PM) probes in signal calculation	Irizarry et al. (2003)
GC-RMA	Same as RMA but also accounts for GC content of individual probes	Wu et al. (2004)
PLIER	Same normalization as RMA with a multiplicative summarization model accounting for background from mismatch (MM) probes	Hubbell (2002)
MAS 5.0	Per chip normalization using spatial effect and MM subtraction for background subtraction and scale normalization	Hubbell et al. (2002)
dCHIP	Pools data from multiple arrays and applies a multiplicative model. MM probes are subtracted	Li and Wong (2001)
AffyILM	Relies on inputs from hybridization thermodynamics and uses an extended Langmuir isotherm model to compute transcript concentrations	Berger and Carlon (2011)

several well-established statistical tests like *t*-tests for comparison of two experimental groups, or, when comparing three or more experimental groups, one-way or two-way analysis of variance (ANOVA). Analyzing thousands of variables, in this case gene expression levels, poses a particular challenge when considering traditional *p*-value cutoffs in statistical testing (Manitoba Centre for Health Policy, 2008). For example, comparing gene expression values for a treated versus control response for a particular toxicity at a single time point using a *t*-test and a *p*-value cutoff of 0.05 would be expected to yield 500 differentially expressed genes in 10,000 purely by random chance. Reducing the impact of multiple comparisons may be accomplished with varying stringency using false discovery correction. Two methods commonly used for false discovery correction in gene expression analysis are the Bonferroni method and the false discovery rate correction described by Benjamini and Hochberg (1995; Abdi 2007).

Gene lists gathered from differential expression analysis may be further refined by using various clustering methods. Clustering is used to understand the differential expression, relatedness, and patterns within toxicogenomic data. The toxicologist may consult a biostatistician to assist with these analyses or use accessible tools available in gene expression analysis software. A basic understanding of these statistical methods is necessary, but a deep understanding of the mathematics involved in deriving relationships is not required. Commonly used methods include hierarchical clustering, principal component analysis (PCA), and self-organizing maps (SOM).

Hierarchical clustering may be used to show the relatedness of individual samples in an experiment and the relatedness of gene expression patterns (Eisen et al., 1998). Related gene expression patterns may be further explored to understand their biological significance to an observed toxicity.

PCA may be used to cluster samples or genes based on variables that best explain the difference in observations (Yeung and Ruzzo, 2001). PCA does not consider gene expression patterns, only the variance in gene expression to best explain the difference between observations.

SOM are also used to interpret gene expression patterns. SOM are useful for following patterns of expression that are closely related over time or dose (Tamayo et al., 1999; Cabrera et al., 2013).

There are several commercially available, free, and open-source bioanalytical tools to perform clustering analyses. If the user is comfortable with programming, open-source options include Bioconductor, which uses the R statistical programming language. Prepackaged freely available analysis software is available from the FDA in the form of ArrayTrack (Tong et al., 2003). There also are many commercially available platforms, including GeneSpring GX from Agilent Technologies and Array Studio from OmicSoft Corporation.

Further understanding of the toxicological relevance of gene expression data can be gained by functional analysis relating the gene lists gleaned from statistical filtering and clustering methods to biological functions through statistical enrichment and pathway and network analyses.

Enrichment analysis provides a biological context for differentially expressed genes by calculating the statistical likelihood that genes from predefined categories appear in a differential expression list by chance. Sets of related genes are curated from the scientific literature and biomedical knowledge bases. Categories may be defined by the gene's biological function, involvement in a particular biological process, or known association to a disease or pathological state for example.

Enrichment calculations are effectively a method of ranking gene sets within a category for those most overrepresented in a list of DEGs. This can be achieved using a number of statistical methods. The most commonly applied are the hypergeometric distribution (or Fisher's exact test) and Gene Set Enrichment Analysis (GSEA). A number of reviews are available describing and comparing methods for statistical enrichment (e.g., Abatangelo et al., 2009; Braun, 2014).

Categories used for functional analysis also may be based on a higher level of biological organization such as a network of gene interactions around a particular drug target, function, or process, or a signaling or metabolic pathway based on a well-characterized set of progressive, functional molecular interactions (Fig. 45.1). An additional benefit of data analysis using pathways or networks is the ability to visualize the interaction diagrams, which can be presented in an intuitively recognizable form and are often well annotated with a description of the process represented and its relationship to biological functions, diseases, or pathological states. The ability to overlay experimental data on the components of the network or pathway often is also available, providing a further visual demonstration of the effect of an experiment on the biological process illustrated (Fig. 45.1).

A number of publicly available resources exist with defined categorization of genes by biological process (e.g., Gene Ontology (GO) http://www.geneontology.org/), disease (e.g., Online Mendelian Inheritance in Man® (OMIM) http://www.omim.org/) or chemical interaction (e.g., Comparative Toxicogenomics Database (CTD) http://www.ctdbase.org/). Numerous public and commercially available options exist for functional and pathway analysis of gene lists. A nonexhaustive listing is given in Table 45.2. Commercially available analysis tools such as Ingenuity Pathway Analysis (http://www.ingenuity.com/), ToxWiz (www.camcellnet.com), or MetaCore (http://portal.genego.com) incorporate analysis across both public and proprietary gene.

A more investigative approach to functional analysis applies network reconstruction algorithms, which leverage comprehensive databases of molecular interactions (protein–protein, compound–protein, protein–gene, protein–enzyme, etc.) to

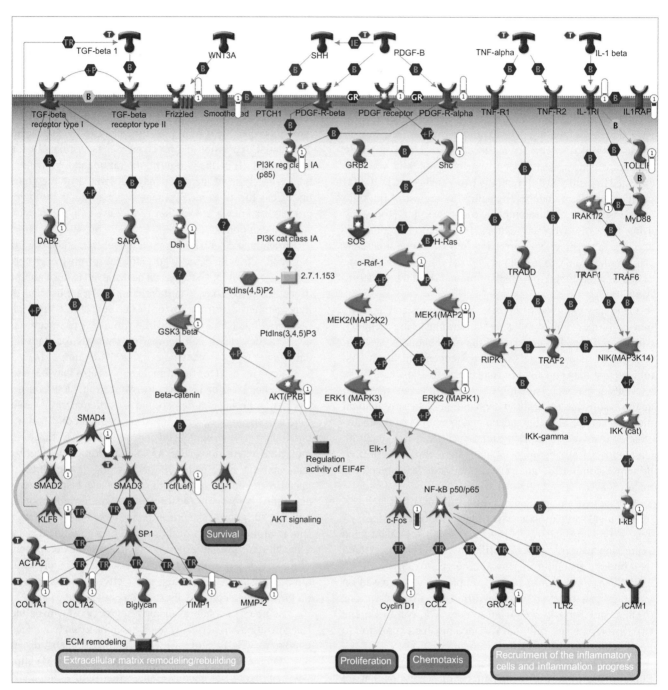

FIGURE 45.1 Canonical pathway map image representing processes involved in stellate cell activation and the development liver fibrosis. Map is from the Systems Toxicology Module of MetaCore (Thomson Reuters Inc.), full legend available at https://ftp.genego.com/files/MC_legend.pdf. Compounds are represented by hexagons, proteins by solid shapes representing different classes of protein, and enzymatic reactions by gray rectangles. Protein–protein, compound–protein, and compound–reaction interactions are shown as unidirectional arrows and a mechanism of interaction represented by letters in hexagonal boxes over the arrows. Red color of interactions and mechanisms represents a negative effect (inhibition, downregulation) and green a positive effect (activation, upregulation). "Thermometers" represent differential gene expression data for severe alcoholic hepatitis versus normal liver (data taken from Affò et al., 2013). Red color fill represents upregulation of the gene for the adjacent protein, blue—downregulation. Magnitude of the fill is semiquantitative (relative to the highest differential expression level on the map). Purple hexagons containing a white "T" represent map objects that correspond to known biomarkers of drug-induced liver fibrosis in the Systems Toxicology Module of MetaCore database. (*See insert for color representation of the figure.*)

TABLE 45.2 Public and Commercially Available Pathway Analysis Tools

Pathway Analysis Database	URL	Reference
Reactome (Public)	http://www.reactome.org/	Croft et al. (2014)
GSEA (Public)	http://www.broadinstitute.org/gsea/index.jsp	Subramanian et al. (2007)
Comparative Toxicogenomics Database (Public)	http://ctdbase.org/	Davis et al. (2013)
David (Public)	http://david.abcc.ncifcrf.gov/home.jsp	Huang da et al. (2009)
KEGG (Public)	http://www.genome.jp/kegg/	Kanehisa et al. (2012)
Pathway Commons	http://www.pathwaycommons.org/	Cerami et al. (2011)
ToxWiz (Commercial)	http://camcellnet.com/products/toxwiz/	
Ingenuity (Commercial)	http://www.ingenuity.com/	Krämer et al. (2014)
MetaCore (Commercial)	http://thomsonreuters.com/en/products-services/pharma-life-sciences/pharmaceutical-research/metacore.html	Ekins et al. (2006)

probe the physical and functional relationships between experimentally derived, or curated, gene lists. Such approaches can be used to identify novel regulatory networks, upstream driving factors, and downstream biomarkers of pathological mechanisms (Brennan et al., 2009; Gautier et al., 2013; Hoeng et al., 2014). Advanced network reconstruction techniques such as hidden nodes analysis and causal reasoning may be employed to identify mechanistically relevant network hubs and pathways that are not immediately revealed in the gene expression data alone. These approaches take into account the directionality of gene expression changes alongside curated information on protein–protein interactions, including their mechanism and effect, to "fill in the gaps" in the data to hypothesize specific causal mechanisms driving pathological effects (Dezso et al., 2009; Catlett et al., 2013; Enayetallah et al., 2013). Many pathway and network analysis tools also enable a systems biology approach, integrating transcriptomic, proteomic, metabolomic, and other data types to enhance mechanistic understanding of toxicity (Xu et al., 2008; Li et al., 2015a) (Fig. 45.2).

Gene expression signatures provide a method for identifying within a complex expression profile a specific pattern of changes that are characteristic of a particular phenotype. This approach has been employed to the greatest extent in oncology where signatures are used to aid differential diagnosis and to predict progression and treatment response (Chibon, 2013). In toxicology, gene expression signatures may be employed to monitor exposure, to diagnose or predict a compound-induced pathology, to identify a particular mode of action, or to differentiate between similar compounds with divergent toxicological effects or off-target pharmacology.

Identification of signatures in gene expression profiles can be achieved by a number of approaches that fall broadly into the two categories of unsupervised and supervised methods (Quackenbush, 2001). Generally, unsupervised methods cluster expression profiles together based on their level of similarity, whereas supervised methods employ machine learning algorithms to identify a pattern of gene expression that differentiates between experiments from predetermined classes. For the purpose of rapid deconvolution of gene expression patterns into readily interpretable predictions, supervised classification methods have the greatest utility.

Methods and algorithms for generating classification signatures have been widely discussed elsewhere (e.g., Natsoulis et al., 2005; Shi et al., 2010a); however a few key aspects of the process bear emphasizing here. Firstly, the suitability of gene expression profiling to differentiate between phenotypes of interest should be considered; a genomic signature may not provide useful discrimination of toxicants acting via acute systemic mechanisms, such as ion channel blockers or respiratory chain inhibitors, for example, whereas compounds interacting with nuclear receptors and transcription factors, or via mechanisms where affected cells respond with an adaptive transcriptional stress response, are likely to have robust and characteristic transcriptional patterns. The MAQC-II project demonstrated clearly that certain endpoints were distinguishable with good performance using a variety of different algorithms to derive signatures, whereas other endpoints did not generate well-performing signatures with any algorithm (Shi et al., 2010a).

Secondly, careful classification of the experiments used to train the signature is critical to its success. Ideally the experiments will be phenotypically anchored to the outcome in question using data generated in the same experiment, for example, a signature for renal tubule injury generated using expression data taken from experiments where compound-induced renal tubule damage was verified using standard histopathological evaluation (Fielden et al., 2005). Phenotypic anchoring is not always possible or feasible however, such as in the development of predictive signatures derived from short-term experiments for chronic outcomes, such as carcinogenesis (Nie et al., 2006; Fielden et al., 2007), or the development of signatures predicting a pharmacological property such as HMG-CoA reductase inhibition *in vivo* (Natsoulis et al., 2005). In these cases exhaustive curation of the training set compounds via careful literature evaluation

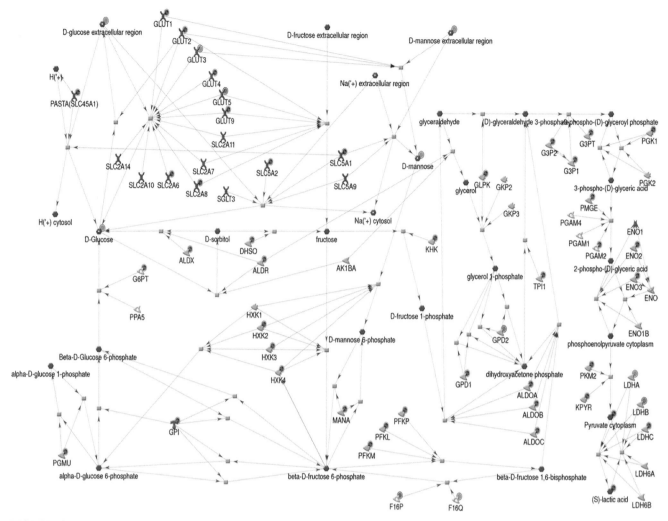

FIGURE 45.2 Dynamic metabolite and transcript changes in glycolysis and glucose transport after treatment of rats with cisplatin or gentamicin. Urine metabolites and kidney transcripts are indicated by a filled circle on the upper right-hand corner of each network object; a red dot in the circle indicates upregulation, whereas a blue dot indicates downregulation from the vehicle control group. Mixed colors indicate differential regulation patterns across dose and time groups. Small-molecule metabolites are represented by purple hexagons, and reactions are represented by gray rectangles. Gene/protein objects are represented by other various shapes and colors depending on their functional annotations. Green arrows represent activation, red arrows represent inhibition, and gray arrows represent other types of unspecified interactions (e.g., molecular transport or complex component binding). Xu et al. (2008). Reproduced with permission of American Chemical Society. (*See insert for color representation of the figure.*)

or anchoring to an orthogonal experiment such as a biochemical assay is needed.

Thirdly, the suitability of the organ or cell type used to monitor the toxicological response should be taken into account. For example, whereas an *in vitro* cell system may be useful to derive signatures for nuclear receptor activation (e.g., Yang et al., 2006) or genotoxicity (Rohrbeck et al., 2010; Li et al., 2015b), more complex organ toxicities may be less well reflected *in vitro*. Similarly, gene expression profiling in the liver is unlikely to provide robust classification of specific cardiotoxic outcomes. In some cases, however, cross-organ signatures of toxicity may be feasible, and indeed desirable, for example, in the prediction of

drug-induced liver injury in humans based on expression profiling in blood (Fannin et al., 2010). Such analyses should be undertaken with great care to the experimental design and careful validation of the resulting signatures.

Finally, as previously mentioned, some endpoints are more addressable by the signature approach than others, and a full understanding of the real-world performance characteristics of any given signature in terms of its ability to correctly identify both toxicants (sensitivity) and nontoxicants (specificity) is necessary in order to appropriately interpret the classification predictions generated on new data with novel compounds and treatments. Ideally this is accomplished with forward validation of the signature using a suitable independent data set with known classification, including both positive and negative treatments for the endpoint in question. The public availability of comprehensive, well-annotated databases of toxicogenomic profiles, DrugMatrix (https://ntp.niehs.nih.gov/drugmatrix/index.html), and TG-GATEs (http://toxico.nibiohn.go.jp/), may be of assistance in some cases; for other signatures of interest to a broader user group, a comprehensive validation effort through consortia such as the Predictive Safety Testing Consortium may be appropriate (Fielden et al., 2011).

When properly designed, executed, and validated, the use of genomic signatures can provide a rapid, sensitive, and powerful method to interpret toxicogenomic data and provide predictions on toxicological outcomes and mechanisms of action. Gene signatures can aid in the interpretation of complex gene expression data by deconvoluting them into discrete outcomes that are more readily comprehensible to pharmacologists and toxicologists nonexpert in gene expression analysis.

45.3 DRUG DISCOVERY TOXICOLOGY CASE STUDIES

Since the early 2000s, the use of toxicogenomic technologies in the field of drug discovery toxicology has skyrocketed. Early adopters of this technology were often overly optimistic about the potential applications of this approach, even though little was known about the applicability or even what types of toxicological questions could realistically be answered. Due to high cost, difficulty in conducting analysis of massive amounts of data, and inappropriate use of the technology, the field of toxicogenomics has often been accused of not living up to the original promise of what type of information could potentially be delivered (Chen et al., 2012). Although many obstacles to the routine use of toxicogenomics in drug discovery toxicology still exist, retrospective analyses of case studies show that even with these obstacles, use of the technology has made a positive impact on reducing attrition due to safety issues and contributing to mechanistic understanding of observed toxicities in exploratory toxicology studies (Foster et al., 2007). Many examples exist in the scientific literature, which conclusively demonstrate the value added to early drug discovery toxicology by toxicogenomics, including use of *in vitro* toxicogenomic approaches for guiding medicinal chemistry structure–activity relationship (SAR) and early characterization of on-target and off-target toxicities associated with a drug discovery program (Ryan et al., 2008).

A review of case studies in the use of toxicogenomics in drug discovery reveals three main categories of application: diagnostic toxicogenomics (biomarkers), predictive toxicogenomics, and mechanistic/investigative studies (Searfoss et al., 2005, Khor et al., 2006, Chen et al., 2012). In this

section, case study examples of each of these use categories will be discussed.

Diagnostic toxicogenomics can be defined as using gene expression patterns, signatures, and pathways analysis to identify a phenotypic toxic endpoint occurring at the time of tissue collection in the tissue or cells from which the RNA was extracted. This approach has the greatest utility when other primary biomarkers of toxicity (i.e., serum transaminases) were not measured or tissue samples for histopathology analysis are not available (small tissue sample, improper fixation, frozen tissue). While diagnostic toxicogenomic biomarkers/signatures have not demonstrated much utility in drug discovery due to availability of other less labor-intensive/cost-effective definitive methods for diagnosing toxicity (i.e., clinical chemistry, hematology, histopathology), the most valuable use of diagnostic toxicogenomics is to identify peripheral gene expression biomarkers of specific target organ injury in an easily accessible surrogate tissue such as circulating peripheral blood mononuclear cells (PBMCs). In using toxicogenomics in a diagnostic manner, multiple types of data analysis may be used, including induction/repression of individual genes of interest, analysis of specific toxicity pathways or de novo networks, or use of gene signatures.

Predictive toxicogenomics has greater potential value than diagnostic toxicogenomics. In the early days of toxicogenomics, however, predictive toxicogenomics was one of the most overhyped and least accurate uses of the technology. The premise underlying predictive toxicogenomics is that gene expression changes, which occur prior to the onset of a pathology endpoint, can be used to predict what endpoints will be modulated at some time after collection of the tissue (Foster et al., 2007). A second approach to predictive toxicogenomics is the use of gene expression in one model or species to predict what toxicities will occur in an alternative model—for example, monitoring gene expression in primary hepatocytes to predict what toxicities will occur in the liver during a repeat-dose *in vivo* toxicology study or the use of rat toxicogenomic data to predict whether or not a similar finding will occur in a human clinical trial. In predictive or prognostic toxicogenomics, the data analysis method most often employed is use of multigene, informatically derived gene signatures. An additional example of the use of predictive toxicogenomics is when data is generated in cells or tissue with one compound, and these data are used to predict the types of toxicity that will occur with a similar molecule or a compound from a different chemical series. Recent advances in computational technology with capabilities of combining chemical structures with toxicogenomic data to correlate SAR with a toxicological endpoint have proven to be useful (Ryan et al., 2008).

The third example of the use of toxicogenomic technologies in drug discovery is in investigative toxicology studies where mechanisms of toxicity are characterized. This type of mechanistic investigation is arguably the most useful application of toxicogenomics to drug discovery and is where the initial promise of the "toxicogenomics revolution" has borne out. With continual advances in pathway analysis and network generation *in silico*/computational tools for data analysis, understanding mechanisms of toxicity has become more routine and accessible to toxicologists with limited experience in bioinformatics. Moreover, mechanistic toxicogenomics offers the opportunity for drug discovery toxicologists to characterize mechanisms of toxicity associated with a particular molecule and then use the data to inform medicinal chemists which particular off targets to avoid, ensuring that the next generation of backup molecules in a program have a reduced risk of a particular toxic endpoint.

45.3.1 Case Studies: Diagnostic Toxicogenomics

As previously mentioned, diagnostic toxicogenomics has traditionally been used in primarily two distinct manners: (i) using global gene expression from RNA isolated from a tissue in order to diagnose a particular toxicity endpoint occurring at the time of collection in that tissue and (ii) using gene expression from an accessible surrogate tissue (usually whole blood or PBMCs) to diagnose injury in a different tissue occurring at the time of sample collection. The use of toxicogenomics as a diagnostic tool is often not warranted, as the gold standard clinical chemistry, hematology, and histopathology endpoints are generally sufficient for diagnosing a particular toxicity. There are, however, cases in which a diagnostic toxicogenomics approach has provided valuable data to supplement the common endpoints.

A diagnostic toxicogenomics liver injury study was designed to correlate gene expression changes with acetaminophen-induced hepatocellular necrosis across 36 strains of mouse (Harrill et al., 2009). In this study, 36 inbred strains of mouse were administered a single 300 mg/kg dose of acetaminophen, and livers were subsequently collected for histopathology and toxicogenomic analysis. Histopathology analysis revealed a liver necrosis severity gradient across strains, and PCA of liver global gene expression showed a correlation between gene expression changes and severity of liver necrosis. Analysis of the toxicogenomic data yielded a panel of 26 genes, which correlated well with liver necrosis in a strain-independent manner. Pathway analysis revealed that genes making up the liver necrosis panel were significantly enriched in pathways related to cell death and proliferation, including genes involved in cell-cycle regulation. This study demonstrated that a toxicogenomics approach was able to diagnose acute liver injury irrespective of confounding factors such as severity of histopathology findings or genetic background.

Another example of the use of toxicogenomic data for diagnostic purposes is identification of expression changes correlating with drug-induced renal tubular toxicity in rats (Jiang et al., 2007). In this study, nine proximal tubule toxicants and one glomerular toxicant were used as the positive class training/testing set, and microarray data was generated from RNA extracted from the kidneys. By using a binary support vector machines (SVM) classification algorithm, the authors were able to derive a proximal tubule necrosis diagnostic signature with 88% sensitivity and 91% specificity. One key observation was that, due to the nature of the classification algorithm, this diagnostic signature did not necessarily reveal any information regarding the affected pathways or mechanisms of toxicity; it is rather a statistical classifier in which the genes comprising the signature are accurate at providing a "yes" or "no" answer regarding renal tubular toxicity.

The most appropriate genes for classification may not be informative as regards the biological mechanism. In fact, when the authors randomly eliminated half of the genes on the array from inclusion in the diagnostic signature, signature performance was similar to the signature, which had been generated using all of the genes on the array, thus suggesting that a statistical approach to generating gene expression signatures does not depend on the toxicological mechanism underlying the pathology being diagnosed by the signature. This is a well-understood limitation of machine learning approaches. There are many nonoverlapping gene sets, probably due to coregulation, that may provide good classification performance for a given endpoint (Natsoulis et al., 2005, 2008). Due to the deliberately sparse nature of the signatures chosen by many algorithms, biological relevance is generally not extractable. When multiple nonoverlapping signatures for a given endpoint are analyzed, however, the combined gene lists do indicate that the signatures are derived from biologically relevant expression changes (Natsoulis et al., 2005; Shi et al., 2010b). Although histopathology and biochemical endpoints/serum biomarkers exist for diagnosing drug-induced renal toxicity, this example demonstrates the concept of generating and validating a diagnostic toxicogenomic with potential utility in numerous areas of drug discovery toxicology.

Many proponents of diagnostic toxicogenomics believe that the most valuable use of this approach in drug discovery toxicology is to be able to obtain a biological sample from an easily accessible surrogate tissue (i.e., blood) and use global gene expression data to diagnose a specific pathology in a less accessible tissue (i.e., liver). While much progress has been made in increasing both the number and accuracy of circulating nongenomic biomarkers of target organ toxicity, especially serum-based protein and miRNA biomarkers, genomic data obtained from PBMCs has multiple advantages over these other types of biomarkers—and some disadvantages as well. Advantages include the ability to obtain mechanistic gene expression data and to potentially identify specific types of target organ damage (i.e., centrilobular hepatocellular necrosis diagnosis by gene expression rather than a generic "liver injury" diagnosis by ALT/AST or miR-122 increases). Additionally, changes in genomic biomarkers may occur prior to the onset of irreversible damage and may also be more sensitive than traditional methods, such as blood urea nitrogen (BUN) and creatinine elevations in serum following massive renal injury. Disadvantages include the increased labor and cost of conducting a toxicogenomics experiment compared to traditional clinical chemistry.

In one example of the use of blood as a surrogate tissue for liver, publicly available gene expression data were mined in order to identify subsets of genes, which were commonly regulated in blood and liver following acetaminophen-induced hepatotoxicity. In this analysis, genomic data were obtained from a study in which male rats were treated with acetaminophen (150, 1500, or 2500 mg/kg) for either 6 or 24 h—at which time liver and blood were collected for RNA extraction and toxicogenomic analysis. The analysis revealed a panel of 760 genes at 6 h following acetaminophen administration, and 185 genes at the 24 h time point, which were regulated in a similar manner in both tissues. Moreover, genes in the 6 and 24 h panels were significantly enriched in ontologies associated with the known mechanism of acetaminophen-induced hepatotoxicity, including mitochondrial dysfunction and immune response (Zhang et al., 2012). In a follow-on human clinical study, patients were administered a 4 g bolus dose of acetaminophen, and blood was collected for transcriptome analysis (Fannin et al., 2010). While serum transaminase elevations were not observed, toxicogenomic analysis showed enrichment of genes involved in oxidative phosphorylation, mitochondrial function, and protein ubiquitination. Additionally, downregulation of mitochondrial functional genes was positively correlated with levels of the toxic metabolite of acetaminophen (NAPQI). Taken together, the data from these preclinical and clinical acetaminophen studies suggest that for some toxicants and organ systems, diagnostic use of toxicogenomics from blood may have utility in monitoring drug-induced toxicity in organs/systems, which are not readily accessible in preclinical safety studies or patients.

A final example of the use of diagnostic toxicogenomics in cross-tissue assessment of drug-induced injury is global gene expression changes in PBMCs in cynomolgus monkeys following exposure to astemizole (Park et al., 2008). Astemizole was used to induce cardiotoxicity, and blood was collected for RNA extraction and toxicogenomic analysis 24 h after drug administration. Using criteria of > or <1.5-fold change and statistical significance cutoff of $p < 0.05$ for genes perturbed by astemizole treatment, a total of 724 genes were identified. Using genes perturbed by astemizole in PBMCs, unsupervised hierarchical clustering analysis found that the astemizole-treated animals clustered by dose level and away from control animals. Pathway enrichment analysis found that genes transcriptionally regulated by astemizole in PBMCs were significantly enriched in cardiac injury categories, including cardiac hypertrophy, cardiac necrosis/cell death, cardiac dilation, cardiac fibrosis/transformation, and cardiac hypertrophy.

While diagnostic toxicogenomic biomarkers show great promise and utility in these types of proof-of-concept studies, translatability to routine use in the drug discovery testing paradigm or application to human health has not yet become mainstream.

45.3.2   Case Studies: Predictive Toxicogenomics

In the early 2000s, a few small private toxicogenomic companies such as Iconix Biosciences and Gene Logic emerged, partially with the intent of developing toxicogenomic technologies to a point where global gene expression data could routinely be used in a predictive manner. The business premise presented by these endeavors was that large databases, which integrate toxicogenomic data from hundreds of drugs/chemicals with traditional toxicology endpoints (clinical chemistry, hematology, histopathology, molecular pharmacology, etc.), could be used to generate patterns or signatures for very specific types of drug-induced injury or mechanisms of toxicity (Ganter et al., 2005; Natsoulis et al., 2005). If these patterns were generated by using toxicogenomic data, which was collected prior to the onset of a specific toxic endpoint in a particular tissue, a predictive signature could be generated. By comparing a toxicogenomic pattern generated from an unknown compound against the signature data patterns in the database, it would then be possible to match against these predictive signatures, and based on a positive predictive value score, estimate the likelihood of a particular toxicity occurring with the unknown compound. Although the toxicogenomics service business model was difficult to maintain, the efforts put forward by these companies resulted in driving the technology of predictive toxicogenomics forward to a point where this approach was considered valid and thus became integrated into the drug discovery toxicology testing paradigm. Case studies that demonstrate the accuracy of predictive toxicogenomic signatures are good proof-of-concept examples of how to best generate predictive signatures (i.e., types of algorithms to use, numbers of positive and negative class compounds needed to generate data, time points to use, etc.) but the published predictive signatures themselves are generally not widely used in drug discovery. One reason for the lack of use of predictive toxicogenomic signatures is that accuracy of the predictive signature may decrease if the test conditions used in the experiment are different from the test conditions used in generating the signature. For example, if data used to generate a predictive signature come from 4 to 14 day studies in male Sprague Dawley rats, the signature may not accurately predict outcome in a study in which an investigator obtains toxicogenomic data from a 7-day study in female Han Wistar rats (Fielden et al., 2011). Moreover, while many published predictive toxicogenomic signatures were derived using diverse positive and negative class compounds, there are many potential mechanisms driving a particular toxic endpoint, and all mechanisms may not be covered by the signature. Many examples exist however in which a robust set of data was used for derivation of predictive toxicogenomic signatures and cross- and forward validation studies have shown higher predictive sensitivity, specificity, and overall accuracy than other currently accepted methods.

Fielden et al. published a robust and well-validated predictive toxicogenomic signature for drug-induced renal tubular toxicity in rats (Fielden et al., 2005). Derivation of this signature was accomplished using a large data set consisting of 15 nephrotoxic and 49 nonnephrotoxic compounds. Rats were dosed daily for either 5 or 28 days, and dose levels were selected on the basis of the observation of nephrotoxicity determined by histopathological evaluation on day 28 that was not present on day 5. RNA was extracted from kidneys taken from animals dosed for 5 days, and these data were used to generate signatures to predict histopathology at 28 days.

Using a sparse linear programming algorithm, the authors generated a 38-gene signature. Split-sample cross-validation analysis revealed the signature to have predictive sensitivity and specificity of 83 and 94%, respectively. A separate forward validation experiment, using a set of 21 compounds not included in the training set used to derive the signature, demonstrated that the overall predictive accuracy of the signature was 76%.

Another example of use of a predictive toxicogenomics classifier is in the prediction of whether a compound is genotoxic or nongenotoxic with metabolic activation in human TK6 lymphoblastoid cells (Buick et al., 2015). A toxicogenomics signature of 65 genes (TGx-28.65) was derived using 28 reference compounds (Li et al., 2015b). The signature was used to classify compounds as either genotoxic or nongenotoxic in the presence of a metabolic activation system. For forward validation, known gentoxicants and nongenotoxicants not included in the training set were subjected to metabolic activation with rat S9; the gene signature was 100% accurate at 8 h. This predictive gene signature was further tested in an external validation study using a different cell line (HepaRG) and a different microarray platform (Affymetrix rather than the Agilent arrays used to generate the signature). Forward validation demonstrated 100% accuracy in classifying the genotoxicity of 15 compounds, suggesting that this classifier is highly robust and its use is not limited to one particular technology or cell line. While there are legitimate questions as to the utility of a gene expression signature to predict outcome in routinely run *in vitro* genotoxicity assays, this study is a demonstration of the

methodology used in validating predictive toxicogenomic signatures.

Predictive toxicogenomic signatures have also been investigated for prediction of nongenotoxic carcinogenicity (NGC). NGC is an issue for drug discovery and development due to the inability of traditional genotoxicity assays (i.e., Ames assay, *in vitro/in vivo* micronucleus or chromosome aberration assays) to positively identify this liability. While some mechanistic observations such as liver enzyme induction, hyperplasia, induction of cell proliferation, cytotoxicity/regeneration, or endocrine perturbations may indicate the potential for NGC, these endpoints have been found to have insufficient predictive value to replace the 2-year rodent carcinogenicity bioassay (Jacobs, 2005). Scientists at Iconix Biosciences derived a predictive toxicogenomic signature using 100 compounds classified as either nonhepatocarcinogens ($n = 75$) or nongenotoxic hepatocarcinogens ($n = 25$) (Fielden et al., 2007). Rats were treated for either 1, 3, 5, or 7 days with the compounds in the training set, and liver RNA was extracted 24 h after each time point. Expression values for 5443 probes on the Codelink UniSet Rat 1 Bioarray were mined for classifying gene signatures using an adjusted sparse linear programming algorithm (A-SPLP). The best signature consisted of 37 probes, and split-sample cross validation using a 60% test/40% training split showed a predictive sensitivity of 56% and specificity of 94%. To determine the actual performance of the predictive signature, 47 compounds not used in generating the signature (21 nongenotoxic hepatocarcinogens and 26 nonhepatocarcinogens) were used in forward validation. The signature had a predictive sensitivity of 86%, specificity of 81%, and overall accuracy of 83%. The accuracy demonstrated by this signature had better overall predictivity than seven nongenomic endpoints commonly used to predict nongenotoxic hepatocarcinogenicity. Follow-up studies were conducted by members of the Predictive Safety Testing Consortium (PSTC) Carcinogenicity Working Group in order to better understand interlaboratory variation in application (Fielden et al., 2008) and for transfer of the signature from a microarray format to a TaqMan RT-PCR-based format (Fielden et al., 2011). The extensive amount of data generated in this process clearly demonstrated the value of predictive toxicogenomics to early evaluation of carcinogenesis.

Despite numerous published examples of accurate premonitory gene expression changes prior to the onset of pathology, predictive toxicogenomic technologies have not gained routine use in the field of drug discovery toxicology. Although convincing data from published validation studies exist to suggest that gene expression may be able to predict toxicological endpoints, the relatively few outcomes for which accurate predictions have been made do not yet justify the cost and time needed to generate the toxicogenomic data.

45.3.3 Case Studies: Mechanistic/Investigative Toxicogenomics

The use of toxicogenomics as a tool in mechanistic or investigative toxicology studies in drug discovery and development is extensively documented. In general, mechanistic toxicogenomics can be defined as use of global gene expression data from cells or tissues exhibiting a certain pathology to understand the perturbation of biological pathways or networks underlying the toxicity. In pharmaceutical R&D, this approach has been successfully employed to determine whether a drug-induced pathology is pharmacologically or chemically mediated, is caused by on- or off-target activity, and to determine the clinical relevance or species specificity of safety signals. Most importantly, this approach enables a project toxicologist to formulate a mitigation strategy for a drug discovery project to avoid or more fully characterize a particular toxicity, particularly if the mechanistic data suggest the need to switch to a different chemical series or to optimize chemistry to increase specificity against a particular off target. The use of toxicogenomics in this scenario has proven powerful enough that it has rapidly become a standard approach in such scenarios.

Liguori et al. used toxicogenomics to characterize the mechanisms of idiosyncratic hepatotoxicity induced by the quinolone antibiotic trovafloxacin (Liguori et al., 2005). Human primary hepatocytes were treated with either trovafloxacin (known to induce idiosyncratic drug-induced liver injury (iDILI)) or five quinolone antibiotics, which do not cause iDILI. High-level analysis of microarray data from this study, including unsupervised hierarchical clustering and PCA, revealed that human hepatocytes treated with trovafloxacin had a global gene expression pattern, which differed from those induced by the other quinolone antibiotics. In-depth analysis of the significantly perturbed genes revealed enrichment in three primary functional categories believed to be involved in iDILI: mitochondrial function (mitochondrial ribosome proteins, mitofusion 1, oxidative stress genes, and bax), processing and regulation of transcription (from RNA polymerase II), and inflammation. The study provides additional molecular data in support of the hypothesis that mitochondrial toxicity and oxidative stress are contributing factors to iDILI. In a toxicogenomic study to examine mechanisms of immunotoxicity in a human T-cell line (Shao et al., 2013), 31 compounds comprising direct immunotoxicants, indirect immunotoxicants, and nonimmunotoxicants were evaluated. Pathway analysis identified canonical signaling pathways and GO category enrichments, which allowed grouping of compounds by structural similarity. Mechanisms underlying immunotoxicity mediated by individual compounds were identified. Pathway analysis revealed two primary mechanisms of direct immunotoxicity: cellular stress response and genes involved in antiapoptotic signaling. Other pathways and processes modulated by the direct-acting immunotoxicants included cell proliferation, gene transcription, protein translation, cholesterol/lipid metabolism, and immunomodulation. Such studies demonstrate clearly the utility of toxicogenomics in elucidating biological mechanisms underlying toxicities of concern to drug discovery.

At Merck Serono, a focused rat toxicogenomic study investigated mechanisms of teratogenicity induced by the calcium sensitizer/PDE III inhibitor, EMD82571 (Hewitt et al., 2014). Fetal malformations (exencephaly ± micrognathia and facial cleft) were observed following treatment of pregnant Wistar rats on gestational days 6–11. Liver findings including increased liver weight, congestion, inflammatory infiltrate, and centrilobular hypertrophy were also observed in adult female rats. Whole genome microarrays were run on four tissues from this study, including maternal liver, embryo liver, embryo bone, and whole embryo. Significantly perturbed genes were identified and functional analysis revealed that EMD82571 perturbed processes such as calcium metabolism, osteogenesis, differentiation and development, extracellular matrix and cytoskeleton, and ion channels/transporters.

iDILI is a concern for drug discovery toxicology due its unpredictability and the lack of mechanistic understanding of underlying processes. Since traditional toxicology endpoints in preclinical rodent and nonrodent studies often fail to pick up iDILI signals, novel methods—including mechanistic toxicogenomics—have been applied to understand iDILI. In one such study, a Reverse Causal Reasoning (RCR) approach was used to identify molecular and cellular mechanisms unique to drugs known to cause iDILI in humans (Laifenfeld et al., 2013). Nine compounds known to cause iDILI in humans, but not in preclinical species, and eight nonhepatotoxicants were administered to male Sprague Dawley rats as a single maximum tolerated dose. Animals were sacrificed 24 h after administration and liver RNA was extracted and hybridized to Affymetrix Rat Genome 230 2.0 whole genome microarrays. While none of the iDILI-inducing or nonhepatotoxic drugs caused liver injury in this study, gene expression data identified three primary pathways associated with the iDILI drugs. Enrichment in mitochondrial injury, inflammation, and endoplasmic reticulum stress were perturbed by the DILI-inducing drugs but not the nonhepatoxicants. Two-thirds of the iDILI compounds were enriched in all three mechanisms while individually, enrichment in mitochondrial toxicity, inflammatory response, and endoplasmic reticulum stress genes was seen with 66, 78, and 88%, respectively. As these three processes have been previously speculated to contribute to the etiology of iDILI, the authors ran a number of *in vitro* assays (mitochondrial toxicity assays using isolated rat liver mitochondria and ER stress CHOP mRNA assessment from HepG2 cells) to confirm the iDILI mechanisms identified by toxicogenomics. Disruption of mitochondrial oxygen consumption and increased CHOP mRNA expression were confirmed *in vitro*, suggesting that these mechanisms were likely contributing to the known hepatotoxicity observed in humans with these drugs.

The utility of toxicogenomics to understand transcriptional events that lead to unusual toxicity was demonstrated in the case of gamma-secretase inhibitors, which caused lesions in the small intestine in rats. Gamma secretase is a promiscuous aspartyl protease that is responsible for the cleavage of amyloid precursor protein to release amyloid beta, which deposits in regions of the brain and is the primary neuropathological hallmark of Alzheimer's disease. Among the many substrates for gamma secretase is the Notch receptor, which determines cellular differentiation in many tissues. Building on the work of Searfoss and coworkers, a group from AstraZeneca dosed Han Wistar rats for 5 days with gamma-secretase inhibitors from three chemical classes with varying potencies for the inhibition of Notch cleavage (Searfoss et al., 2003; Milano et al., 2004). The observed pathology in this study included goblet cell metaplasia in the small intestine suggesting marked changes in cellular differentiation. Microarray analysis of duodenum total RNA was performed on this 5-day time course experiment using the Affymetrix U34A rat oligonucleotide array. Gene expression profiling revealed upregulation of several transcripts in the Notch pathway on day 1 through day 5, including Delta1, furin, and NeuroD. Further scrutiny of gene expression in these samples using real-time PCR revealed the upregulation of a Notch pathway transcriptional activator of intestinal secretory cell differentiation, the rat homologue of the Drosophila atonal transcription factor, Rath1. Also, data for Rath1 mRNA abundance showed a correlation with *in vitro* Notch1 potency. This study provided both a possible biomarker for gamma-secretase inhibitor toxicity, Rath1, and mechanistic insight into the transcriptional events that led to the observed pathology.

45.3.4 Future Directions in Drug Discovery Toxicogenomics

The application of toxicogenomics in drug discovery has been continually evolving since the early days of transcriptome profiling. Toxicogenomic approaches were initially seen by many as a panacea for prediction and understanding of mechanisms of toxicity. As greater understanding of the uses and limitations of toxicogenomics was gained, a more focused approach developed in which very specific questions are now being asked with the technology and the appropriate and inappropriate uses of the technology have been better defined.

Continual improvements in data analysis tools have increased the accessibility of toxicogenomics to drug discovery toxicologists lacking a strong background in informatics. Affymetrix, for example, has developed a stepwise set of protocols, kits, algorithms, and data analysis tools enabling novice users to go from tissue to perturbed gene lists in a matter of days without any formal training. Commercially available functional analysis tools, such as MetaCore (Thomson Reuters), Ingenuity Pathways Analysis (Qiagen), ToxWiz (Cambridge Cell Networks), and Pathway Studio (Elsevier), have also made considerable improvement in user interfaces and workflows so that mapping toxicogenomic data to biological processes and canonical signaling pathways and generating novel networks is more intuitive. The content and quality of these and other toxicogenomic tools/databases (some publically available, Table 45.2) are rapidly evolving, making mechanistic interpretation of data accessible to any toxicologist with a computer and a few hours of time. Inclusion of miRNAs in pathways and networks in these tools also is allowing for a better understanding of how these small regulatory RNAs lead to changes in expression of specific genes and are impacting biological pathways and processes, providing a further improvement in mechanistic characterization of drug toxicity.

More specialized technologies applied to drug discovery are driving additional improvement in the toxicogenomics. For example, the use of laser capture microscopy allows the toxicologist to evaluate highly specific regions of tissue, as opposed to obtaining a heterogeneous tissue sample, which may contain large areas of normal tissue along with small areas of damage. As profiling technology continues to evolve, smaller and smaller amounts of biological sample are needed to generate whole genome transcriptome data. The development of RNA-seq, next-generation sequencing, small focused arrays, and novel whole genome microarrays (i.e., additional species, sense-transcript arrays) has made toxicogenomics more accessible, faster to generate actionable data, and considerably more affordable than even a few years ago. The same tools used for mRNA analysis are now being applied to miRNA analysis. The "miRNAome" can be interrogated in such a way that levels of hundreds of individual miRNAs can quickly be assessed. This additional layer of data can be integrated with mRNA expression profiles to more completely understand regulation of gene expression and to identify novel biomarkers of target organ toxicity.

An emerging subspecialty of toxicogenomics is toxicoepigenomics. Epigenetic modifications are now understood to be key regulators of gene transcription, and arrays probing DNA and histone methylation patterns are now used to better understand factors regulating mRNA expression following exposure to toxicants.

As new regulators of gene transcription (snoRNAs, piwi-interacting RNA, currently unknown regulators) are identified and characterized, there is no doubt that these will be integrated into toxicogenomic analyses and will contribute to better prediction and characterization of mechanisms of toxicity.

REFERENCES

Abatangelo, L., Maglietta, R., Distaso, A., et al. 2009. Comparative study of gene set enrichment methods. *BMC Bioinf.* 10:275.

Abdi H. 2007. Bonferroni and Šidák Corrections for Multiple Comparisons. In: *Encyclopedia of Measurement and Statistics* (ed Salkind, N.J.), Sage Publishing, Thousand Oaks. pp. 1–9.

Affò, S., Dominguez, M., Lozano, J.J., et al. 2013. Transcriptome analysis identifies TNF superfamily receptors as potential therapeutic targets in alcoholic hepatitis. *Gut* 62(3):452–460.

Afshari, C.A., Nuwaysir, E.F., Barrett, J.C. 1999. Application of complementary DNA microarray technology to carcinogen identification, toxicology, and drug safety evaluation. *Cancer Res.* 59(19):4759–4760.

Alwine, J.C., Kemp, D.J., Stark, G.R. 1977. Method for detection of specific RNAs in agarose gels by transfer to diazobenzyloxy-methyl-paper and hybridization with DNA probes. *Proc. Natl. Acad. Sci. U. S. A.* 74(12):5350–5354.

Anisimov SV. 2008. Serial analysis of gene expression (SAGE): 13 years of application in research. *Curr. Pharm. Biotechnol.* 9(5):338–350.

Baker, T.K., Carfagna, M.A., Gao, H., et al. 2001. Temporal gene expression analysis of monolayer cultured rat hepatocytes. *Chem. Res. Toxicol.* 214(9):1218–1231.

Bartlett, J. 2001. Technology evaluation: SAGE, Genzyme molecular oncology. *Curr. Opin. Mol. Ther.* 3(1):85–96.

Beard, S.E., Capaldi, S.R., Gee, P. 1996. Stress responses to DNA damaging agents in the human colon carcinoma cell line, RKO. *Mutat. Res.* 371(1–2):1–13.

Benjamini Y., Hochberg Y. 1995. Controlling the false discovery rate: a practical and powerful approach to multiple testing. *J. R. Stat. Soc. Ser. B Stat Methodol.* 57(1):289–300.

Berger F., Carlon E. 2011. From hybridization theory to microarray data analysis: performance evaluation. *BMC Bioinf.* 12:1–14.

Biggin, M., Farrell, P.J., Barrell, B.G. 1984. Transcription and DNA sequence of the BamHI L fragment of B95-8 Epstein-Barr virus. *EMBO J.* 3(5):1083–1090.

Braun, R. 2014. Systems analysis of high-throughput data. *Adv. Exp. Med. Biol.* 844:153–187.

Brazma, A., Hingamp, P., Quackenbush, J., et al. 2001. Minimum information about a microarray experiment (MIAME)-toward standards for microarray data. *Nat. Genet.* 29(4):365–371.

Brennan, R.J., Nikolskya, T., Bureeva, S. 2009. Network and pathway analysis of compound-protein interactions. *Methods Mol. Biol.* 575:225–247.

Buick, J., Moffat, I., Williams, A., et al. 2015. Integration of metabolic activation with a predictive toxicogenomics signature to classify genotoxic versus nongenotoxic chemicals in human TK6 cells. *Environ. Mol. Mutagen.* 56:520–534.

Bulera, S.J., Eddy, S.M., Ferguson, E., et al. 2001. RNA expression in the early characterization of hepatotoxicants in Wistar rats by high-density DNA microarrays. *Hepatology* 33(5):1239–1258.

Burczynski, M.E., McMillian, M., Ciervo, J., et al. 2000. Toxicogenomics-based discrimination of toxic mechanism in HepG2 human hepatoma cells. *Toxicol. Sci.* 58(2):399–415.

Cabrera S., Selman M., Lonzano-Bolaños A., et al. 2013. Gene expression profiles reveal molecular mechanisms involved in the progression and resolution of bleomycin-induced lung fibrosis. *Am. J. Physiol. Lung Cell. Mol. Physiol.* 304(9):L593–L601.

Caldwell, M.C., Hough, C., Fürer, S., et al. 2002. Serial analysis of gene expression in renal carcinoma cells reveals VHL-dependent sensitivity to TNFalpha cytotoxicity. *Oncogene* 21(6):929–936.

Castle, A.L., Carver, M.P., Mendrick, D.L. 2002. Toxicogenomics: a new revolution in drug safety. *Drug Discov. Today* 7(13):728–736.

Catlett, N.L., Bargnesi, A.J., Ungerer, S., et al. 2013. Reverse causal reasoning: applying qualitative causal knowledge to the interpretation of high-throughput data. *BMC Bioinf.* 14:340.

Cerami E.G., Gross B.E., Demir E., et al. 2011. Pathway commons, a web resource for biological pathway data. *Nucleic Acids Res.* 39(Database issue):D685–D690.

Chen, M., Zhang, M., Borlak, J., et al. 2012. Decade of toxicogenomic research and its contribution to toxicological science. *Toxicol. Sci.* 130(2):217–228.

Chibon, F. 2013. Cancer gene expression signatures—the rise and fall? *Eur. J. Cancer* 49(8):2000–2009.

Civen, M., Knox, W.E. 1959. The independence of hydrocortisone and tryptophan inductions of tryptophan pyrrolase. *J. Biol. Chem.* 234(7):1787–1790.

Clifford, J.I., Rees, K.R. 1967. The action of aflatoxin B1 on the rat liver. *Biochem. J.* 102(1):65–75.

Conney, A.H., Miller, E.C., Miller J.A. 1957. Substrate-induced synthesis and other properties of benzpyrene hydroxylase in rat liver. *J. Biol. Chem.* 228(2):753–766.

Conney, A.H., Davison, C., Gastel, R., et al. 1960. Adaptive increases in drug-metabolizing enzymes induced by phenobarbital and other drugs. *J. Pharmacol. Exp. Ther.* 130:1–8.

Crawford, D.R., Davies, K.J. 1997. Modulation of a cardiogenic shock inducible RNA by chemical stress: adapt73/PigHep3. *Surgery* 121(5):581–587.

Croft D., Mundo A.F., Haw R., et al. 2014. The reactome pathway knowledgebase. *Nucleic Acids Res.* 42(Database issue):D472–D477.

Cullen, P., Lorkowski, S., Bauer, O., et al. 2002. High-Throughput and Industrial Methods for mRNA Expression Analysis. In: *Analysing Gene Expression: A Handbook of Methods: Possibilities and Pitfalls* (eds Lorkowski, S. and Cullen, P.), Wiley-VCH Verlag GmbH & Co. KGaA, Weinheim, FRG, pp. 409–622.

Davis A.P., Murphy C.G., Johnson R., et al. 2013. The comparative toxicogenomics database: update 2013. *Nucleic Acids Res.* 41(Database issue):D1104–D1114.

DeRisi, J.L., Iyer, V.R., Brown, P.O. 1997. Exploring the metabolic and genetic control of gene expression on a genomic scale. *Science* 278(5338):680–686.

Dezso, Z., Nikolsky, Y., Nikolskaya, T., et al. 2009. Identifying disease-specific genes based on their topological significance in protein networks. *BMC Syst. Biol.* 3:36.

Durnam, D.M., Palmiter, R.D. 1984. Induction of metallothionein-I mRNA in cultured cells by heavy metals and iodoacetate: evidence for gratuitous inducers. *Mol. Cell. Biol.* 4(3):484–491.

Eisen M.B., Spellman P.T., Brown P.O., et al. 1998. Cluster analysis and display of genome-wide expression patterns. *Proc. Natl. Acad. Sci. U. S. A.* 95(25):14863–14868.

Ekins S, Bugrim A, Brovold L, et al. 2006. Algorithms for network analysis in systems-ADME/Tox using the MetaCore and MetaDrug platforms. *Xenobiotica* 36(10–11):877–901.

Enayetallah, A.E., Puppala, D., Ziemek, D., et al. 2013. Assessing the translatability of in vivo cardiotoxicity mechanisms to in vitro models using causal reasoning. *BMC Pharmacol. Toxicol.* 14:46.

Fannin, R.D., Russo, M., O'Connell, T.M., et al. 2010. Acetaminophen dosing of humans results in blood transcriptome and metabolome changes consistent with impaired oxidative phosphorylation. *Hepatology* 51(1):227–236.

Fielden, M., Eynon, B., Natsoulis, G., et al. 2005. A gene expression signature that predicts the future onset of drug-induced renal tubular toxicity. *Toxicol. Pathol.* 33:675–683.

Fielden, M., Brennan, R., Gollub, J. 2007. A gene expression biomarker provides early prediction and mechanistic assessment of hepatic tumor induction by nongenotoxic chemicals. *Toxicol. Sci.* 99(1):90–100.

Fielden, M., Nie, A., McMillon, M., et al. 2008. Interlaboratory evaluation of genomic signatures for predicting carcinogenicity in the rat. *Toxicol. Sci.* 103(1):28–34.

Fielden, M., Adai, A., Dunn, R., et al. 2011. Development and evaluation of a genomic signature for the prediction and mechanistic assessment of nongenotoxic hepatocarcinogens in the rat. *Toxicol. Sci.* 124(1):54–74.

Foster, W., Chen, S.-J., He, A., et al., 2007. A retrospective analysis of toxicogenomics in the safety assessment of drug candidates. *Toxicol. Pathol.* 35:621–635.

Ganter, B., Tugendreich, S., Pearson, C., et al. 2005. Development of a large-scale chemogenomics database to improve candidate selection and to understand mechanisms of chemical toxicity and action. *J. Biotechnol.* 119:219–244.

Gautier, L., Taboureau, O., Audouze, K. 2013. The effect of network biology on drug toxicology. *Expert Opin. Drug Metab. Toxicol.* 9(11):1409–1418.

Gibbs, R.A., Weinstock, G.M., Metzker, M.L., et al., 2004. Genome sequence of the Brown Norway rat yields insights into mammalian evolution. *Nature* 428(6982):493–521.

Goldberger, R.F. 1974. Autogenous regulation of gene expression. *Science* 183(4127):810–816.

Gonzalez-Zulueta, M., Ensz, L.M., Mukhina, G., et al. 1998. Manganese superoxide dismutase protects nNOS neurons from NMDA and nitric oxide-mediated neurotoxicity. *J. Neurosci.* 18(6):2040–2055.

Goodsaid, F.M., Amur, S., Aubrecht, J., et al. 2010. Voluntary exploratory data submissions to the US FDA and the EMA: experience and impact. *Nat. Rev. Drug Discov.* 9(6):435–445.

Goujon, F.M., Van Cantfort, J., Gielen, J.E. 1980. Comparison of aryl hydrocarbon hydroxylase and epoxide hydratase. Induction in primary fetal rat liver cell culture. *Chem. Biol. Interact.* 31(1):19–33.

Gray, G.D., Camiener, G.W., Bhuyan, B.K. 1966. Nogalamycin effects in rat liver: inhibition of tryptophan pyrrolase induction and nucleic acid biosynthesis. *Cancer Res.* 26(12):2419–2424.

Hamadeh, H.K., Bushel, P.R., Jayadev, S., et al. 2002. Prediction of compound signature using high density gene expression profiling. *Toxicol. Sci.* 67(2):232–240.

Harrill, A., Ross, P., Gatti, D., et al. 2009. Population-based discovery of toxicogenomics biomarkers for hepatotoxicity using a laboratory strain diversity panel. *Toxicol. Sci.* 110(1):235–243.

Harris, A.J., Shaddock, J.G., Manjanatha, M.G., et al. 1998. Identification of differentially expressed genes in aflatoxin B1-treated cultured primary rat hepatocytes and Fischer 344 rats. *Carcinogenesis* 19(8):1451–1458.

Hauser, N.C., Vingron, M., Scheideler, M., et al. 1998. Transcriptional profiling on all open reading frames of Saccharomyces cerevisiae. *Yeast* 14(13):1209–1221.

Hawtrey A.O., Nourse, L.D. 1964. The effect of 3'-methyl-4-dimethylaminoazobenzene on protein synthesis and DNA-dependent RNA-polymerase activity in rat-liver nuclei. *Biochim. Biophys. Acta* 80:530–532.

Hewitt, P., Singh, P., Kumar, A., et al. 2014. A rat toxicogenomics study with the calcium sensitizer EMD82571 reveals a pleiotropic cause of teratogenicity. *Reprod. Toxicol.* 47:89–101.

Hoeng, J., Talikka, M., Martin, F., et al. 2014. Case study: the role of mechanistic network models in systems toxicology. *Drug Discov. Today* 19(2):183–192.

Holden, P.R., James, N.H., Brooks, A.N., et al. 2000. Identification of a possible association between carbon tetrachloride-induced hepatotoxicity and interleukin-8 expression. *J. Biochem. Mol. Toxicol.* 14(5):283–290.

Huang da W., Sherman B.T., Lempicki R.A. 2009. Systematic and integrative analysis of large gene lists using DAVID bioinformatics resources. *Nat. Protoc.* 4(1):44–57.

Hubbell E. 2002. Guide to Probe Logarithmic Intensity Error (PLIER) Estimation. Affymetrix Technical Note. Available from: http://media.affymetrix.com/support/technical/technotes/plier_technote.pdf (accessed on October 23, 2015).

Hubbell E., Liu W.M., Mei R. 2002. Robust estimators for expression analysis. *Bioinformatics* 18(12):1585–1592.

Irizarry R.A., Hobbs B., Collin F., et al. 2003. Exploration, normalization, and summaries of high density oligonucleotide array probe level data. *Biostatistics* 4(2):249–264.

Irizarry R.A., Wu Z., Jaffee H.A. 2006. Comparison of Affymetrix GeneChip expression measures. *Bioinformatics* 22(7):789–794.

Ivy, S.P., Tulpule, A., Fairchild, C.R., et al. 1988. Altered regulation of P-450IA1 expression in a multidrug-resistant MCF-7 human breast cancer cell line. *J. Biol. Chem.* 263(35):19119–19125.

Jacobs, A. 2005. Prediction of 2-year carcinogenicity study results for pharmaceutical products: how are we doing? *Toxicol. Sci.* 88(1):18–23.

Jiang, Y., Gerhold, D., Holder, D., et al. 2007. Diagnosis of drug-induced renal tubular toxicity using global gene expression profiles. *J. Transl. Med.* 5:47.

Kanehisa M., Goto S., Sato Y., et al. 2012. KEGG for integration and interpretation of large-scale molecular data sets. *Nucleic Acids Res.* 40(Database issue):D109–D114.

Khor, T., Ibrahim, S., Kong, A.-N.. 2006. Toxicogenomics in drug discovery and drug development: potential applications and future challenges. *Pharm. Res.* 23(8):1659–1664.

Kier, L.D., Neft, R., Tang, L., et al. 2004. Applications of microarrays with toxicologically relevant genes (tox genes) for the evaluation of chemical toxicants in Sprague Dawley rats in vivo and human hepatocytes in vitro. *Mutat. Res.* 549(1–2):101–113.

Kouri, R.E., Rude, T.H., Joglekar, R., et al. 1978. 2,3,7,8-Tetrachlorodibenzo-p-dioxin as cocarcinogen causing 3-methyl-cholanthrene-initiated subcutaneous tumors in mice genetically "nonresponsive" at Ah locus. *Cancer Res.* 38(9):2777–2783.

Krämer A., Green J., Pollard J. Jr, et al. 2014. Causal analysis approaches in ingenuity pathway analysis. *Bioinformatics* 30(4):523–530.

Laifenfeld, D., Qiu, L., Swiss, R., et al. 2013. Utilization of causal reasoning of hepatic gene expression in rats to identify molecular pathways of idiosyncratic drug-induced liver injury. *Toxicol. Sci.* 137(1):234–248.

Lander, E.S., Linton, L.M., Birren, B., et al. 2001. Initial sequencing and analysis of the human genome. *Nature* 409(6822):860–921.

Larsson O., Tian B., Sonenberg N. 2013. Toward a genome-wide landscape of translational control. *Cold Spring Harb. Perspect. Biol.* 5(1):a012302.

Lee, M.J., Gee, P., Beard, S.E. 1997. Detection of peroxisome proliferators using a reporter construct derived from the rat acyl-CoA oxidase promoter in the rat liver cell line H-4-II-E. *Cancer Res.* 57(8):1575–1579.

Li, C., Wong, W.H. 2001. Model-based analysis of oligonucleotide arrays: expression index computation and outlier detection. *Proc. Natl. Acad. Sci. U. S. A.* 98(1):31–36.

Li H.H., Hyduke, D.R, Chen, R., et al. 2015a. Development of a toxicogenomics signature for genotoxicity using a dose-optimization and informatics strategy in human cells. *Environ. Mol. Mutagen.* 56(6):505–519.

Li, Z., Qin, T., Wang, K., et al. 2015b. Integrated microRNA, mRNA, and protein expression profiling reveals microRNA regulatory networks in rat kidney treated with a carcinogenic dose of aristolochic acid. *BMC Genomics* 16:365.

Liang, P., Pardee, A.B. 1992. Differential display of eukaryotic messenger RNA by means of the polymerase chain reaction. *Science* 257(5072):967–971.

Liguori, M., Anderson, M., Bukofzer, S., et al. 2005. Microarray analysis in human hepatocytes suggests a mechanism for hepatotoxicity induced by trovafloxacin. *Hepatology* 41:177–186.

Liu F., Kuo W.P., Jenssen T.K., et al. 2012. Performance comparison of multiple microarray platforms for gene expression profiling. *Methods Mol. Biol.* 802:141–155.

Loeb, L.A., Gelboin, H.V. 1964. Methylcholanthrene-induced changes in rat liver nuclear RNA. *Proc. Natl. Acad. Sci. U. S. A.* 52:1219–1226.

Manitoba Centre for Health Policy 2008. Concept: Multiple Comparisons. Available from: http://mchp-appserv.cpe.umanitoba.ca/viewConcept.php?conceptID=1049 (accessed on October 23, 2015).

The MAQC Consortium. 2006. The MicroArray Quality Control (MAQC) project shows inter- and intraplatform reproducibility of gene expression measurements. *Nat. Biotechnol.* 24(9):1151–1161.

Marshall, E. 2004. Getting the noise out of gene arrays. *Science* 306(5696):630–631.

Martin, M.T., Brennan, R.J., Hu, W., et al. 2007. Toxicogenomic study of triazole fungicides and perfluoroalkyl acids in rat livers predicts toxicity and categorizes chemicals based on mechanisms of toxicity. *Toxicol. Sci.* 97(2):595–613.

Marton, M.J., DeRisi, J.L., Bennett, H.A., et al. 1998. Drug target validation and identification of secondary drug target effects using DNA microarrays. *Nat. Med.* 4(11):1293–1301.

McBurney R.N., Hines W.M., Von Tungeln L.S., et al. 2009. The liver toxicity biomarker study: phase I design and preliminary results. *Toxicol. Pathol.* 37(1):52–64.

McBurney R.N., Hines W.M., VonTungeln L.S., et al. 2012. The liver toxicity biomarker study phase I: markers for the effects of tolcapone or entacapone. *Toxicol. Pathol.* 40(6):951–964.

Milano J., McKay J., Dagenais C., et al. 2004. Modulation of notch processing by gamma-secretase inhibitors causes intestinal goblet cell metaplasia and induction of genes known to specify gut secretory lineage differentiation. *Toxicol. Sci.* 82(1):341–358.

Miles, M.F., Hung, P., Jungmann, R.A. 1981. Cyclic AMP regulation of lactate dehydrogenase. Quantitation of lactate dehydrogenase M-subunit messenger RNA in isoproterenol-and N6,O2'-dibutyryl cyclic AMP-stimulated rat C6 glioma cells by hybridization analysis using a cloned cDNA probe. *J. Biol. Chem.* 256(23):12545–12552.

Nadadur, S.S., Schladweiler, M.C., Kodavanti, U.P. 2000. A pulmonary rat gene array for screening altered expression profiles in air pollutant-induced lung injury. *Inhal. Toxicol.* 12(12):1239–1254.

Natsoulis, G., El Ghaoui, L., Lanckriet, G., et al. 2005. Classification of a large microarray data set: algorithm comparison and analysis of drug signatures. *Genome Res.* 15:724–736.

Natsoulis, G., Pearson, C.I., Gollub, J.P., et al. 2008. The liver pharmacological and xenobiotic gene response repertoire. *Mol. Syst. Biol.* 4:175.

Nebert, D.W., Felton, J.S. 1976. Importance of genetic factors influencing the metabolism of foreign compounds. *Fed. Proc.* 35(5):1133–1141.

Nemoto, N., Takayama, S. 1980. Genetic differences between C57BL/6 and DBA/2 mice in the inductions of UDP-glucuronyl transferases for 3-hydroxybenzo(a)pyrene, p-nitrophenol, and bilirubin by 3-methylcholanthrene. *Toxicol. Lett.* 5(1):45–50.

Nie, A.Y., McMillian, M., Parker, J.B., et al. 2006. Predictive toxicogenomics approaches reveal underlying molecular mechanisms of nongenotoxic carcinogenicity. *Mol. Carcinog.* 45(12):914–933.

Niwa, A., Kumaki, K., Nebert, D.W., et al. 1975. Genetic expression of aryl hydrocarbon hydroxylase activity in the mouse. Distinction between the "responsive" homozygote and heterozygote at the Ah locus. *Arch. Biochem. Biophys.* 166(2):559–564.

Ntzani, E.E., Ioannidis, J.P. 2003. Predictive ability of DNA microarrays for cancer outcomes and correlates: an empirical assessment. *Lancet* 362(9394):1439–1444.

Nuwaysir, E.F., Bittner, M., Trent, J., et al. 1999. Microarrays and toxicology: the advent of toxicogenomics. *Mol. Carcinog.* 24(3):153–159.

Oesch, F. 1976. Differential control of rat microsomal "aryl hydrocarbon" monooxygenase and epoxide hydratase. *J. Biol. Chem.* 251(1):79–87.

Oesch, F. 1980. Influence of foreign compounds on formation and disposition of reactive metabolites. *Ciba Found. Symp.* 76:169–189.

Ohmomo H., Hachiya T., Shiwa Y., et al. 2014. Reduction of systematic bias in transcriptome data from human peripheral blood mononuclear cells for transportation and biobanking. *PLoS One* 9(8):e104283.

Park, H.-J., Seo, J.-W., Oh, J.-H., et al. 2008. Gene expression changes in peripheral blood mononuclear cells from cynomolgus monkeys following astemizole exposure. *Mol. Cell. Toxicol.* 4(4):323–330.

Pennie, W.D., Tugwood, J.D., Oliver, G.J., Kimber, I. 2000. The principles and practice of toxicogenomics: applications and opportunities. *Toxicol. Sci.* 54(2):277–283.

Quackenbush, J. 2001. Computational analysis of microarray data. *Nat. Rev. Genet.* 2(6):418–427.

Raghow, R., Gossage, D., Seyer, J.M., et al. 1984. Transcriptional regulation of type I collagen genes in cultured fibroblasts by a factor isolated from thioacetamide-induced fibrotic rat liver. *J. Biol. Chem.* 259(20):12718–12723.

Rataboul, P., Vernier, P., Biguet, N.F., et al. 1989. Modulation of GFAP mRNA levels following toxic lesions in the basal ganglia of the rat. *Brain Res. Bull.* 22(1):155–161.

Rohrbeck, A., Salinas, G., Maaser, K., et al. 2010. Toxicogenomics applied to in vitro carcinogenicity testing with Balb/c 3T3 cells revealed a gene signature predictive of chemical carcinogens. *Toxicol. Sci.* 118(1):31–41.

Ryan, T., Stevens, J., Thomas, C. 2008. Strategic applications of toxicogenomics in early drug discovery. *Curr. Opin. Pharmacol.* 8:654–660.

Santiard-Baron, D., Gosset, P., Nicole, A., et al. 1999. Identification of beta-amyloid-responsive genes by RNA differential display: early induction of a DNA damage-inducible gene, gadd45. *Exp. Neurol.* 158(1):206–213.

Schena M. 2000. A Highly Sensitive Microarray System for Differential Gene Expression Analysis. In: *Microarray Biochip Technology* (ed Schena, M.), Eaton Publishing, Natick. pp. ix–xi.

Schena, M., Shalon, D., Davis, R.W., et al. 1995. Quantitative monitoring of gene expression patterns with a complementary DNA microarray. *Science* 270(5235):467–470.

Schmassmann, H., Oesch, F. 1978. Trans-stilbene oxide: a selective inducer of rat liver epoxide hydratase. *Mol. Pharmacol.* 14(5):834–847.

Searfoss, G.H., Jordan, W.H., Calligaro, D.O., et al. 2003. Adipsin: a biomarker of gastrointestinal toxicity mediated by a functional g-secretase inhibitor. *J. Biol. Chem.* 278:46107–46116.

Searfoss, G., Ryan, T., Jolly, R. 2005. The role of transcriptome analysis in pre-clinical toxicology. *Curr. Mol. Med.* 5:53–64.

Seo J., Hoffman E.P.. 2006. Probe set algorithms: is there a rational best bet? *BMC Bioinf.* 7:395.

Shao, J., Katica, M., Schmeits, P., et al. 2013. Toxicogenomics-based identification of mechanisms for direct immunotoxicity. *Toxicol. Sci.* 135(2):328–346.

Shi, L., Tong, W., Goodsaid, F., et al. 2004. QA/QC: challenges and pitfalls facing the microarray community and regulatory agencies. *Expert Rev. Mol. Diagn.* 4(6):761–777.

Shi, L., Reid, L.H., Jones, W.D., et al., 2006. The MicroArray Quality Control (MAQC) project shows inter- and intraplatform reproducibility of gene expression measurements. *Nat. Biotechnol.* 24(9):1151–1161.

Shi, L., Campbell, G., Jones, W.D., et al., 2010a. The MicroArray Quality Control (MAQC)-II study of common practices for the development and validation of microarray-based predictive models. *Nat. Biotechnol.* 28(8):827–838.

Shi, W., Bessarabova, M., Dosymbekov, D., et al. 2010b. Functional analysis of multiple genomic signatures demonstrates that classification algorithms choose phenotype-related genes. *Pharmacogenomics J.* 10(4):310–323.

Stafford, P. 2006. Genomics, Transcriptomics, and Proteomics: Novel Detection Technologies and Drug Discovery. In: *Biochips as Pathways to Drug Discovery* (eds Carmen, A. and Hardiman, G.), CRC Press, Boca Raton. pp. 321–338.

Subramanian A., Kuehn H., Gould J., et al. 2007. GSEA-P: a desktop application for gene set enrichment analysis. *Bioinformatics* 23(23):3251–3253.

Tamayo P., Slonim D., Mesirov J., et al. 1999. Interpreting patterns of gene expression with self-organizing maps: methods and application to hematopoietic differentiation. *Proc. Natl. Acad. Sci. U. S. A.* 96(6):2907–2912.

Thorgeirsson, S.S., Nebert, D.W. 1977. The Ah locus and the metabolism of chemical carcinogens and other foreign compounds. *Adv. Cancer Res.* 25:149–193.

Todd, M.D., Lee, M.J., Williams, J.L., et al. 1995. The CAT-Tox (L) assay: a sensitive and specific measure of stress-induced transcription in transformed human liver cells. *Fundam. Appl. Toxicol.* 28(1):118–128.

Tong W., Cao X., Harris S., et al. 2003. ArrayTrack—supporting toxicogenomic research at the U.S. Food and Drug Administration National Center for Toxicological Research. *Environ. Health Perspect..* 111(15):1819–1826.

Tsao, M.S., Duong, M., Batist, G. 1989. Glutathione and glutathione S-transferases in clones of cultured rat liver epithelial cells that express varying activity of gamma-glutamyl transpeptidase. *Mol. Carcinog.* 2(3):144–149.

Tully, D.B., Collins, B.J., Overstreet, J.D., et al. 2000. Effects of arsenic, cadmium, chromium, and lead on gene expression regulated by a battery of 13 different promoters in recombinant HepG2 cells. *Toxicol. Appl. Pharmacol.* 168(2):79–90.

Turner, M.K., Reid, E. 1964. An anomalous effect of ethionine on ribonucleic acid synthesis. *Nature* 203:1174–1175.

Velculescu, V.E., Zhang, L., Vogelstein, B., et al. 1995. Serial analysis of gene expression. *Science* 270(5235):484–487.

Venter, J.C., Adams, M.D., Myers, E.W., et al. 2001. The sequence of the human genome. *Science* 291(5507):1304–1351.

Vincent, R., Goegan, P., Johnson, G., et al. 1997. Regulation of promoter-CAT stress genes in HepG2 cells by suspensions of particles from ambient air. *Fundam. Appl. Toxicol.* 39(1):18–32.

Wang Z., Gerstein M., Snyder M.. 2009. RNA-Seq: a revolutionary tool for transcriptomics. *Nat. Rev. Genet.* 10(1):57–63.

Waring, J.F., Ciurlionis, R., Jolly, R.A., et al. 2001. Microarray analysis of hepatotoxins in vitro reveals a correlation between gene expression profiles and mechanisms of toxicity. *Toxicol. Lett.* 120(1–3):359–368.

Waterston, R.H., Lindblad-Toh, K., Birney, E., et al. 2002. Initial sequencing and comparative analysis of the mouse genome. *Nature* 420(6915):520–562.

Witschi, H. 1972. A comparative study of in vivo RNA and protein synthesis in rat liver and lung. *Cancer Res.* 32(8):1686–1694.

Witschi, H. 1973. Qualitative and quantitative aspects of the biosynthesis of ribonucleic acid and of protein in the liver and the lung of the Syrian golden hamster. *Biochem. J.* 136(3):781–788.

Wong, D.T., Biswas, D.K. 1985. Mechanism of benzo(a)pyrene induction of alpha-human chorionic gonadotropin gene expression in human lung tumor cells. *J. Cell Biol.* 101(6):2245–2252.

Wu Z, Irizarry R.A., Gentleman R., et al. 2004. A Model Based Background Adjustment for Oligonucleotide Expression Arrays. Johns Hopkins University, Dept. of Biostatistics Working Papers 2004. Available from: http://biostats.bepress.com/jhubiostat/paper1/ (accessed on October 23, 2015).

Xu E.Y., Perlina A., Vu, H., et al. 2008. Integrated pathway analysis of rat urine metabolic profiles and kidney transcriptomic profiles to elucidate the systems toxicology of model nephrotoxicants. *Chem. Res. Toxicol.* 21(8):1548–1561.

Yamamoto, M., Wakatsuki, T., Hada, A., et al. 2001. Use of serial analysis of gene expression (SAGE) technology. *J. Immunol. Methods* 250(1–2):45–66.

Yang, Y., Abel, S.J., Ciurlionis, R., et al. 2006. Development of a toxicogenomics in vitro assay for the efficient characterization of compounds. *Pharmacogenomics* 7(2):177–186.

Yeung K.Y., Ruzzo W.L.. 2001. Principal component analysis for clustering gene expression data. *Bioinformatics* (9):763–774.

Zhang, L., Bushel, P., Chou, J., et al. 2012. Identification of identical transcript changes in liver and whole blood during acetaminophen toxicity. *Front Genet.* 3(162):1–10.

Zhao, N., Hashida, H., Takahashi, N., et al. 1995. High-density cDNA filter analysis: a novel approach for large-scale, quantitative analysis of gene expression. *Gene* 156(2):207–213.

46

ISSUE INVESTIGATION AND PRACTICES IN DISCOVERY TOXICOLOGY

Dolores Diaz[1], Dylan P. Hartley[2] and Raymond Kemper[3]

[1] Discovery Toxicology, Safety Assessment, Genentech, Inc., South San Francisco, CA, USA

[2] Drug Metabolism and Pharmacokinetics, Array BioPharma Inc., Boulder, CO, USA

[3] Discovery and Investigative Toxicology, Drug Safety Evaluation, Vertex Pharmaceuticals, Boston, MA, USA

46.1 INTRODUCTION

The objective of a discovery toxicology group is to select and progress molecules that have a safety profile that provides an optimal chance of successfully progressing through the IND process, into man, and through clinical trials. This is achieved by early involvement (target selection and target derisking), selecting the best possible chemical matter (lead optimization; screens and counterscreens), and safety characterization and issue investigation. The role of the discovery toxicologist in this setting is thus broad and diverse. Discovery toxicologists must understand the biology of the target/pathway and predict the ensuing toxicology that may occur when the system is modulated by a pharmaceutical agent (i.e., agonist, antagonist, inhibitor, or activator). In addition, the discovery toxicologist must possess a working knowledge of chemical motifs that can lead to off-target activities that could result in chemical-related toxicities. Throughout the drug discovery phase, the goal should be to enable rational decision-making, including molecule progression and molecule modification to mitigate toxicity and/or molecule/program termination for intractable targets as soon as possible.

Until relatively recently, toxicology was viewed as a drug development activity, and toxicologists were generally not involved in early discovery. The pharmaceutical industry has faced a high level of drug attrition due to toxicities that either manifested in early clinical trials (often resulting in an inability to achieve sufficient target coverage to interrogate efficacy), later in pivotal phase III trials, or, worse, in the postmarketing space, resulting in costly failures, black box warnings, and market withdrawals (Kola and Landis, 2004; Sasseville et al., 2004). These setbacks prompted pharmaceutical companies to explore and characterize the toxicity of pharmaceutical agents much earlier in a program. As such, toxicological limitations of a lead molecule are assessed in the discovery phase, where the expectation is that intractable targets might be identified early and discontinued and that rational lead optimization and investigative efforts would deliver candidate drugs with a higher probability of clinical success.

46.2 OVERVIEW OF ISSUE INVESTIGATION IN THE DISCOVERY SPACE

Assessment of potential safety risks by drug discovery toxicologists can and should be leveraged at the very early stages of drug discovery, including selection of the biological target and an analysis of the risk to benefit in a particular therapeutic area. Discovery toxicologists can review available information, either published or internal, to assess the safety of the chosen target, put this information into the context of the desired indication and patient population, and propose a derisking plan for the program. This information can be provided in the format of a safety target assessment and should be part of a complete and customized lead optimization strategy.

During the early stages of drug discovery, prior to the availability of the first chemical matter (hit-to-lead phase),

Drug Discovery Toxicology: From Target Assessment to Translational Biomarkers, First Edition. Edited by Yvonne Will, J. Eric McDuffie, Andrew J. Olaharski, and Brandon D. Jeffy.

issue investigation in discovery toxicology focuses on derisking the chosen target with regard to potential safety issues. One of the most useful safety deliverables at this early stage is a thorough characterization or phenotyping of relevant genetically engineered rodent models, typically mouse knockout (KO) models for inhibitor/antagonist-modulated targets. In addition, certain human-acquired diseases may also inform the biological phenotype. An important caveat with embryonic KO models is that the observed phenotype may reflect developmental effects that might not be seen in adult animals treated with a drug against the target (for more information see Chapter 10). When available, conditional KO animals may be a more relevant model for potential safety liabilities associated with engaging the target of interest. In addition, certain human genetic diseases can also aid in predicting biological phenotype. The intersection of genetic disease and pharmacological intervention on biological phenotype is well documented in the literature and in searchable databases (like the *Online Mendelian Inheritance in Man site*; www.OMIM.org). Overall, the assessment of the genotypic to phenotypic relationship of a target can help gauge the safety of the target, and it can provide a benchmark for interpretation of mechanisms of toxicity (on- vs. off-target) in future toxicology studies.

If there are specific safety concerns associated with the biology of the target, tool molecules can prove useful in interrogating these concerns. Tool molecules are by definition imperfect, sometimes lacking potency or selectivity and other times exhibiting poor oral bioavailability, precluding the use of *in vivo* models, although alternative dosing strategies, such as intraperitoneal dosing, can overcome high first-pass metabolism and/or low oral absorption of tool molecules. However, with a good understanding of their limitations, they can be used to rationally obtain useful safety information. An example of early safety exploration using a tool molecule is provided by Lee et al. (2014), where they used the tool bromo and extraterminal (BET) inhibitor JQ-1 to reveal multifocal effects on the hematopoietic and lymphoid organs in treated mice (Lee et al., 2014), therefore providing an early understanding of the potential safety profile of BET inhibitors. In general, early knowledge of potential target liabilities obtained using tool molecules can be very useful in interpreting toxicity findings and informing decision-making for a discovery program.

Early leads are also not ideal molecules to explore the safety of the target, but if the potential confounder of chemical-related toxicities is balanced with carefully chosen chemical diversity (e.g., with the use of two representative molecules from two different chemical series), one should be able to differentiate target-related findings from off-target effects. Alternatively, active and inactive enantiomers can be utilized to characterize pharmacologically versus nonpharmacologically mediated toxicities though these chemical tools are rarely available during early-stage hit-to-lead

campaigns. Ideally, these experiments can be done through testing specific hypothesis in *in vitro* mechanistic studies or in short-term toxicology studies (compound availability is typically limited at this early stage).

In addition to particular concerns about target safety that are usually derisked in the early stages of a program, and prior to safety characterization in dedicated pilot toxicology studies, toxicity issues can arise in early *in vivo* studies such as pharmacokinetic (PK) studies. Although this is uncommon in initial PK studies due to the low doses administered, when findings do arise they tend to be more serious due to the high toxicological potency. Safety signals are more likely to be observed in pharmacodynamic (PD) or PK/PD studies, which often employ dose escalation or in subsequent efficacy studies. In this case it is helpful to have a contingency plan in place to leverage the animals and samples in the study as much as possible to understand the safety issue. When unexpected toxicities are noticed in PK/PD or efficacy studies, the discovery toxicology representative should be notified to determine whether characterization of the toxicities is warranted. At termination of the study, gross necropsy observations or clinical pathology samples and tissue samples for microscopic examination may be useful in understanding the toxicology of the molecule. At the very least, it is helpful to obtain details of the findings such as a descriptive nature of the clinical signs, how many animals were affected, timing after dosing, and recovery and timing of recovery. These clinical observations, when coupled with plasma exposure, can prove useful to determine the nature of the toxicity, that is, whether the toxicity is related to the maximal achieved plasma concentration (C_{max}) or to the area under the curve (AUC). Although these observations are not typically used for decision-making, they can inform potential safety liabilities.

In addition, in some cases PK/PD and efficacy studies can be leveraged to obtain an early assessment of safety by adding specific safety endpoints to these studies or by conducting a broad toxicology exploration to inform potential safety issues early on (e.g., body weight, general blood clinical chemistry, hematological endpoints, or specific biomarkers of interest). This approach has the benefit of saving time and material, maximizing the use of animals, and also enabling the calculation of a "true preclinical therapeutic index (TI)" in mice or rats, since both toxicity and efficacy are obtained in the same species. Caveats to this approach include the relatively low doses typically used in these studies, which only allow for a limited exploration of the safety dose–response, and the fact that safety signals might be challenging to read in the background of a disease model. A couple of examples of an effective application of this approach are (i) the use of mouse xenograft models to determine that the retinal toxicity caused by nicotinamide phosphoribosyltransferase (NAMPT) inhibitors, which was first identified in rat safety studies, had no TI in mice (Sampath et al., 2015;

Zabka et al., 2015), and (ii) the use of similar models to determine the lack of TI for the intestinal toxicity caused by tankyrase inhibitors (Zhong et al., 2013).

At some point in the lead optimization phase, companies typically have a critical investment milestone (late-stage research transition, lead identification, or candidate nomination) after which safety failure is significantly more costly (i.e., post-initiation of *in vivo* IND-enabling studies). Enhanced safety derisking prior to this milestone transition is advantageous to inform a potential termination for an intractable or low probability of technical success (PTS) target, a serious liability for the chemical series, or simply an increased awareness of the safety issues prior to entering a more committed phase of drug discovery (e.g., IND-enabling studies). Safety information at this stage should provide solid derisking of the target, and a reasonable derisking of off-targets for the chemical series, particularly if only a single chemical series is being pursued. Although certain general guidelines can be established for drug discovery programs (i.e., KO phenotyping, rodent safety data with a tool

molecule, etc.), safety expectations should be program specific and balance a variety of factors including the clinical indication, medical need, desired safety profile (e.g., improved safety profile vs. the known competitors), competitive landscape, and first-in-class/best-in-class (FIC/BIC) strategy, among others.

46.3 STRATEGIES TO ADDRESS TOXICITIES IN THE DISCOVERY SPACE

Later in the candidate selection phase, short-term or pilot toxicology studies in rodents and nonrodents are typically conducted. When toxicities arise in animal studies, three primary questions come to mind: (i) what is the nature, severity, monitorability, reversibility and TI of the toxicity, (ii) is the toxicity likely to translate to humans, and (iii) what is the likely mechanism of toxicity (MOT), target related or off target? The answers to these critical questions will shape the safety strategy for the project (Fig. 46.1):

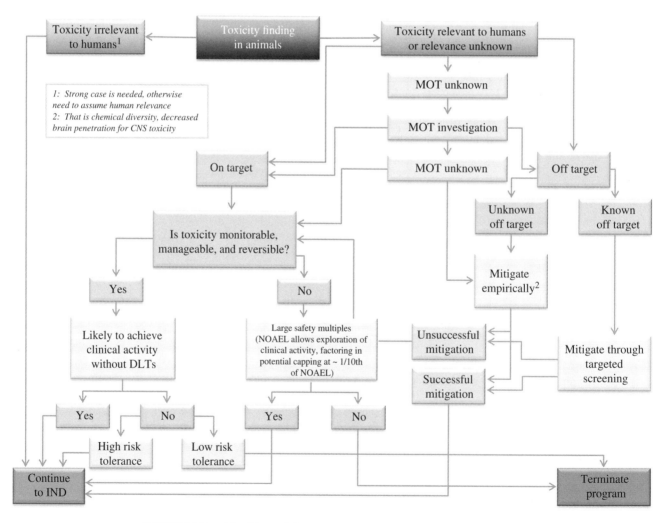

FIGURE 46.1 Algorithm to inform decision-making in compound progression.

1. **Human translation:** When adverse findings emerge in nonclinical animal studies, the relevance and implications for human safety are oftentimes unclear, but the default assumption is that a toxicity should be considered relevant to humans in the absence of strong evidence to the contrary. This paradigm is necessary to protect human safety. In some instances, a strong case can be made for lack of human relevance, for example, if the toxicity involves a mechanism that is not present in humans (a well-known example of this is the alpha 2μ-globulin nephropathy and carcinogenesis in male rats (Swenberg, 1993)) or if a toxicity is caused by a metabolite that is not formed in humans (Mutlib et al., 2000). In addition, if sufficient clinical information exists demonstrating lack of translation for a specific preclinical toxicity, this information can be convincingly leveraged to claim lack of human relevance; an example of this is provided by the mitogen-activated protein kinase (MEK) inhibitor-induced phosphorous dysregulation and soft-tissue mineralization in rats (Diaz et al., 2012). In some cases, *in vitro* systems that use human cells or tissues can provide a level of understanding about the potential human relevance of a particular toxicity. For example, if a molecule causes liver or bone marrow toxicity in preclinical models and if the toxicity can be recapitulated with a reasonable degree of confidence *in vitro* (i.e., using human bone marrow colony-forming unit (CFU) assays or human hepatocytes in culture), a parallel human *in vitro* system can be used to understand if the human cells or tissues, and by extension the human organism, might be less sensitive to the toxicity. Although the case for lower human sensitivity, or lack of human relevance, for a particular toxicity is hard to make in the absence of robust mechanistic information, these data can at least be used for internal decision-making.

2. **Mechanism of toxicity:** Off-target toxicities are often solvable through development of rational structure–activity relationships (SAR) for the finding of interest. In this regard, it is helpful to determine if the toxicity is related to the particular chemical series, since this can inform the potential to mitigate the toxicity through structural modifications. If a program has only one chemical series, this might require reinitiating high-throughput screening to look for new chemical matter and in some cases termination of the program if novel hits cannot be identified. However, there are numerous examples where off-target activity can be mitigated within the chemical series (i.e., through modification of functional groups to decrease human ether-à-go-go-related gene (hERG) inhibition (Carvalho et al., 2013)). If off-target toxicity is not related to chemical scaffold, it is typically easier to solve through structural modifications. In this regard,

it is important to have a balanced approach including efforts to solve the toxicity empirically, even with limited understanding of the molecular mechanism, and in parallel investigating the molecular mechanism, since this knowledge can help with screen development, understanding of human relevance, and monitoring strategies.

Knowing the particular molecular off-target can be very helpful in mitigating the toxicity (e.g., a certain off-target kinase or receptor) with empirical screening or through rational SAR exploration. In addition, sometimes it may be possible to model the off-target *in silico*, in which case SAR can be conducted prior to compound synthesis in conjunction with a computational chemist. A more indirect but practical approach to mitigating off-target toxicities is, for example, to limit brain penetration for off-target CNS hits (non-CNS programs) or to modulate PK to avoid high C_{max} concentrations through twice-a-day (BID) dosing or controlled-release formulations.

3. **Nature of the toxicity:** If a toxicity is monitorable and reversible, clinical entry is possible (at a cautiously low dose and with careful dose escalation and thorough monitoring for a toxicity with a low TI). This strategy would enable the exploration of target engagement and PK profile in humans in a manner that is safe for subjects and patients. On the other hand, for serious, irreversible, and/or nonmonitorable preclinical toxicities, caution must be exercised by the sponsor and the regulators to ensure the safety of the clinical trial subjects and patients. Typically this involves capping dose escalation and exposures at a dose that is sufficiently conservative (usually 1/10th of the no observed adverse effect level (NOAEL) in the most sensitive preclinical species). Oftentimes this does not allow for exploration of pharmacological activity, though, with appropriate biomarkers or PK/PD assays (e.g., *ex vivo* whole blood assays), an understanding of target engagement may still be gained.

46.4 CROSS-FUNCTIONAL COLLABORATIVE MODEL

Discovery toxicology project representatives are ideally positioned to be most effective when they are fully integrated members of research project teams from the inception of the project. In this role, they can design safety strategies that are tailored for each particular program and aligned with the particular indication, taking into consideration clinical input. In this effort, it is important to carefully consider a rational strategy for the specific program and avoid a "one-size-fits-all" approach. In the design and implementation of the early derisking strategies, discovery toxicologists should focus on how the experiments and data will inform decisions for the molecules and the program. In this regard, scenario planning for the different possible outcomes is helpful (i.e., if the decision is the same regardless of the outcome, the utility of the particular experiment or study should be questioned).

An enabling mindset on the part of the discovery toxicologist is very effective in research teams, with safety derisking strategies well aligned with the overall project goals. Ideally, the discovery toxicology plan should enable a path forward for the program through investigative efforts and lead optimization, or support a well-informed, science-based decision to terminate a chemical series or program as early as possible.

Interaction between discovery toxicology and **basic research** disciplines should occur at the earliest stages of a program's life cycle. It is essential for the discovery toxicologist to have an in-depth understanding of the biology of the target to be able to grasp the complexities of the potential safety findings and to design an effective and robust derisking strategy for each undesirable safety finding that may arise through the course of a program.

It is advisable for research teams to involve the discovery toxicologist even prior to official program entry into the portfolio, especially for targets that may be high risk or challenging from a safety perspective. Oftentimes the discovery toxicologist can provide useful context around the existing safety data (i.e., KO phenotypes) and suggestions as to what additional data should be generated at an early stage to derisk the target further. This guidance can take the form of a detailed target safety assessment and constitute the foundation of a detailed safety derisking plan as the program approaches portfolio entry. Continued collaboration with basic research is critical at all stages of the program to ensure a good awareness and understanding of the evolving biology of the target. This partnership can be especially effective during the investigation of toxicity findings.

Within research, collaboration with **translational groups** can also be greatly beneficial. PD and efficacy studies in relevant animal models are typically the first studies in which a new molecule is administered to animals at pharmacologically relevant doses in a multidose regimen (from several days to several weeks). These studies can be leveraged to identify early safety signals from which the discovery toxicologist can obtain valuable information. In addition, these studies can inform safety by adding dedicated toxicology endpoints.

Close collaboration between discovery toxicology and **medicinal chemistry** is critical. The implementation of rational and tailored *in vitro* lead optimization screening paradigms (including receptor, ion channels, and enzyme screens), in addition to genotoxicity, cytotoxicity, and issue-specific screens allows the selection of lead molecules with minimal off-target effects. Medicinal chemists can use the screening information to guide SAR into a more selective chemical space. When particular off-target issues are encountered, they can design tool molecules with the specific aim of understanding the structural drivers of toxicity (i.e., potent vs. nonpotent analogues, enantiomers, structurally diverse active molecules), which may enable the mitigation of these off-target effects. In addition, when *in vivo* toxicities are encountered, medicinal chemists can design useful tool molecules that will allow for investigation of the mechanism of toxicity (MOT), including less active or inactive analogues. It is also important that chemists and discovery toxicologists recognize potential toxicophores or structural alerts in molecules. Finally, computational chemists can model off-target activity, particularly when these off-target effects become significant obstacles for program progression, and provide rational structural input to guide the SAR.

Close collaboration with the **drug metabolism and pharmacokinetics** (DMPK) group is essential when entering animal testing. Achieving sufficient exposures in rodent and nonrodent species is critical to provide context for safety characterization and issue investigation. It is also critical to understand exposure relationships and safety margins, including the variables and assumptions that contribute to these calculations, especially when human efficacious dose projections are used. Oftentimes complexities arise in animal exposures that need to be investigated and understood, since they have implications for safety. These include altered exposure over time and the potential for CYP induction and inhibition (which can be investigated by analyzing liver CYP RNA expression and/or activity using *in vitro* assays), compound accumulation over time (which needs to be carefully considered when selecting doses for longer-term studies), sex differences, and individual animal variability in exposure and tissue distribution (e.g., CNS penetration for non-CNS targeted drugs).

Data from drug metabolism studies can also provide very valuable information for early compound derisking. These include *in vitro* metabolite profiling in human and preclinical species to determine toxicology species selection and inform the need to monitor metabolites with significant exposure and/or activity, characterizing metabolic interactions with safety implications such as CYP induction, time-dependent inhibition, covalent binding, metabolic stability, transporter activity, and target organ accumulation.

Close interaction with the **pharmaceutics** group is also crucial since formulations play an important role in enabling appropriate exposures to characterize safety. Exploration of a variety of formulations is often needed in order to achieve adequate exposures, especially with early tool compounds that lack refined pharmaceutical properties. Good formulations are also important to optimize stability and minimize variability in exposures. The selected formulations should also be well tolerated and lack associated histopathological findings that could confound data interpretation. In this regard, it is useful to consult the existing literature and also to keep a historical database of formulations used with their associated safety issues (Gad et al., 2006; Neervannan, 2006). In instances in which novel excipients are used, it may be prudent to include an additional study arm with a known vehicle control in order to properly assess the potential toxicities of the new excipients.

To ensure optimal *in vivo* pilot study design and smooth study conduct, it is helpful if the discovery toxicology, DMPK, and formulations representatives work closely together, especially in the case of pharmaceutically challenging molecules that may require exotic formulations to support *in vivo* studies. Although formulation can enable a compound with poor pharmaceutical properties, it is incumbent upon the team to understand the future development liabilities of challenging compounds. Preferably the team should select molecules with pharmaceutical properties that will allow the compound to be readily formulated and dosed to attain sufficient exposures for pharmacological and toxicological evaluation in preclinical species. In addition, specialized formulations (i.e., slow release) can potentially be leveraged to mitigate C_{max}-driven toxicities.

Clinical input is highly desirable in the establishment of the target safety profile for candidate molecules. Unfortunately, discovery project teams do not always seek this input until the point of candidate selection, when the acceptability of the candidate's safety profile is under close scrutiny. A good understanding of the acceptable safety characteristics for the intended indication and medical challenges of the target patient population, including likely polypharmacy, is critical in designing a tailored derisking strategy that incorporates informative go/no-go decision points. As the program progresses, it is important to maintain alignment with the clinical representative around the emerging toxicity findings (including those from competitors) and to discuss their acceptability for the intended indication and their ability to monitor and manage projected toxicities in humans.

Collaboration with **clinical drug safety** is also helpful in that it goes hand in hand with the toxicology–clinical interaction. After a molecule goes into development and reaches the clinic, it is important for the discovery toxicologist to remain aware of the adverse events and dose-limiting toxicities, such that the translation or lack thereof from animal findings can be better understood. This information feeds directly into the safety derisking of the backup program or of other programs in the same pathway and can be used to advance the knowledge around animal-to-human toxicity

translation, which is sorely needed in toxicology. In addition, if new safety findings arise in humans, the team can engage the discovery toxicologist to design and execute a collaborative investigative effort.

Throughout the discovery space, and when animal studies are performed, including phenotyping of genetically engineered rodent models, inclusion of clinical or histopathological endpoints in PK/PD or efficacy studies, investigative studies and early exploratory or pilot safety studies, the input and close collaboration of a **clinical and anatomical pathologist** is essential. The pathologists and discovery toxicologists must work closely together, and this close partnership can provide a very effective tag team with regard to team representation.

For issue investigation, discovery toxicologists can leverage external vendors and internal investigative toxicology laboratories, which can be particularly effective in developing specialized assays. In working with **internal investigative groups**, it is essential that these groups are fully engaged and aware of the big picture of the issue under investigation and that they are offered opportunities for team interaction and visibility. An expert working group model, including subject matter experts across different functions, can be very effective in investigating particular safety issues.

The discovery space typically ends after rodent and nonrodent pilot toxicology studies are completed, at which point the molecule formally enters development (i.e., IND-enabling GLP studies) and is handed over to a development team, which includes a **regulatory/development toxicologist**. Ideally, the discovery toxicologist, if different from the development toxicologist, will continue to participate in the development team at least until the IND is filed and will also contribute to the writing of the IND. The timing of the handoff of the molecule to the development toxicologist can vary greatly between companies but ideally should be smooth and deliberate to avoid the loss of important information. Close communication between the discovery and development toxicologist and carefully set expectations should ensure a seamless transition. Discovery toxicologists can also greatly benefit from gaining experience in the development space, ideally by directly taking a molecule into IND and beyond, and matrix organizations that allow and encourage this type of cross-training and adequately resource both spaces can greatly benefit by fostering the growth of well-rounded toxicologists that understand the continuum of drug discovery and development.

Next we present two case studies to illustrate issue investigation in the discovery toxicology space. The first case study covers the investigation of target-related safety findings for an oncology target; the second case study discusses the successful mitigation of off-target safety findings for a CNS program.

46.5   CASE-STUDIES OF ISSUE RESOLUTION IN THE DISCOVERY SPACE

Case Study 1

A case study that illustrates vigorous and elegant investigative work to address the mechanism (on- vs. off-target) and human relevance of several toxicities observed with NAMPT inhibitors is presented, which enabled critical internal decision-making for this program. NAMPT is an enzyme that catalyzes the rate-limiting step in the salvage pathway to generate nicotinamide adenine dinucleotide (NAD), which is a molecule critically involved in energy metabolism and many homeostatic functions. Inhibition of NAMPT leads to NAD depletion, followed by ATP depletion and loss of cellular viability. NAMPT has been pursued as an oncology target because cancer cells are highly dependent on the NAMPT salvage pathway for NAD generation, and in fact NAMPT is overexpressed in a number of cancer types (Sampath et al., 2015). Several small-molecule NAMPT inhibitors have been evaluated in clinical trials, but they failed to demonstrate sustained efficacy in cancer patients, likely due to their inability to achieve efficacious exposures due to dose-limiting thrombocytopenia (Hovstadius et al., 2002; Holen et al., 2008).

Genentech began pursuing NAMPT inhibitors with the acquisition of exclusive worldwide rights to Forma Therapeutics' early-stage small-molecule NAMPT inhibitors in mid-2011 (http://www.fiercebiotech.com/press-releases/forma-therapeutics-grants-genentech-exclusive-rights-acquire-pre-clinical-c-0). When the program was initiated, the intention was to administer nicotinic acid (NA) as an antidote to mitigate NAMPT inhibitor toxicity in normal cells and tissues (mainly thrombocytopenia) while preserving the antiproliferative activity in tumor cells. The basis for this strategy stems from the presence of a second NAMPT-independent NAD salvage pathway in normal cells (nicotinic acid phosphoribosyltransferase domain containing 1 (NAPRT1)), which can use NA to synthesize NAD. As this salvage pathway is absent in many tumors, it offered the potential of achieving a TI with NAMPT inhibitor and NA coadministration in NAPRT1-negative tumors. In the end, the emergence of two additional severe toxicities in rodents (cardiovascular and retinal toxicity), which were not sufficiently mitigated by NA coadministration and are likely translatable to humans, led to the conclusion that this target is not tractable from a safety perspective. An additional contributing factor was also the loss of activity in NAPRT1-negative tumors, likely due to the formation of NA metabolites by normal tissue (Sampath et al., 2015).

A critical question for the development of NAMPT inhibitors was whether NA supplementation would be able to mitigate the clinical dose-limiting toxicity of thrombocytopenia, such that higher doses would be tolerated allowing for higher and potentially efficacious exposures. In order to answer this question, an attempt was made to reproduce the clinically observed thrombocytopenia in rats, but without success. Administration of NAMPT inhibitors to rats was not associated with decreases in circulating platelets (as assessed by routine hematology and microscopic evaluation of hematopoietic tissue) in studies of up to 15 days of duration, up to maximal tolerated doses, and at systemic drug concentrations that exceeded that associated with thrombocytopenia in clinical studies, and despite clear lymphoid and erythroid depletion in these studies (Tarrant et al., 2015). This revealed that the rat (or the mouse) was not a suitable nonclinical model to recapitulate clinical thrombocytopenia associated with NAMPTi treatment. We further explored the utility of *in vitro* human megakaryocyte colony-forming assays (hMK-CFU), which is a validate model of clinically relevant drug-induced thrombocytopenia (Pessina et al., 2009). To ensure that the effect was target related, a set of chemical tools was designed consisting of three potent, selective, and structurally distinct NAMPT inhibitors (two competitor molecules, APO-866 and GMX-1778, and the internal lead molecule, GNE-617) and an inactive analogue of our lead molecule (GNE-643, target activity >100 times lower than GNE-617) (Oh et al., 2014; Zheng et al., 2014). All three NAMPT inhibitors caused potent inhibition of hMK colony formation ($IC_{50}s < 10$ nM), whereas the inactive analogue was >100 times less potent, demonstrating that this effect is target related. This approach illustrates the successful use of chemical tools like structurally distinct molecules and inactive analogues to explore whether a toxicity is on- or off-target.

Since the *in vitro* hMK-CFU assay recapitulated the human thrombocytopenia, this appeared to be a suitable model to test the hypothesis of toxicity mitigation by NA. For this, hMK-CFU cultures were cotreated with NAMPT inhibitor GEN-617 at increasing concentrations of NA, resulting in a clear cytoprotective effect of NA that manifested as a dose-dependent increase in IC_{50} values for NAMPT inhibitor cytotoxicity (Tarrant et al., 2015).

The effect of NAMPT inhibitors on platelets isolated from healthy human donors was also investigated. None of the three structurally distinct NAMPT inhibitors had an effect on platelet bioenergetics, aggregation, or activation up to a concentration of 50 μM, suggesting that the clinical thrombocytopenia was mainly due to effects on platelet production by megakaryocytes rather than effects on platelets (Tarrant et al., 2015). This approach illustrates the successful use of *in vitro* systems of primary human cells to recapitulate human toxicities when animals are not suitable models in order to evaluate MOT and also toxicity mitigation strategies.

In the course of exploring the rat as a potential model for thrombocytopenia, it was observed that NAMPT inhibitors caused lethality in rats after 3 days of dosing, which was associated with the presence of a cavitary transudative effusion that was consistent with congestive heart failure. Microscopic evaluation confirmed the presence of myocardial degeneration, although the severe nature of the toxicity compared to the mild nature of the histopathology finding suggested a primarily functional effect on the heart. This cardiac toxicity was also observed with the three structurally distinct NAMPT inhibitors (GNE-617, GMX-1778, and APO-866), and it was absent with the less potent NAMPT inhibitor GNE-643 at matching plasma exposures, indicating that the toxicity was target related.

To evaluate the human relevance of this toxicity, *in vitro* studies were designed using cardiac cells of rat and human origin. Primary rat cardiomyocytes recapitulated the rat toxicity, with NAMPT inhibitors causing potent functional toxicity associated with ATP depletion in these cells, whereas the iPSC-derived human cardiomyocytes were selected as the most appropriate model to assess human relevance, after exploring ESC-derived human cardiomyocytes and concluding that ESC-derived cells were not the most appropriate model because of decreased sensitivity to NAMPT inhibition likely due to a concurrent fibroblast population (unpublished observation). Structurally distinct NAMPT inhibitors were toxic to iPSC-derived cardiomyocytes at very low concentrations, whereas the inactive analogue was >100 times less potent. This finding suggested that the cardiac toxicity observed in rats was relevant to humans and that humans could potentially be even more sensitive given the higher sensitivity of human cardiomyocytes compared to rat cardiomyocytes. This approach illustrates the benefit of leveraging cellular models that maintain appropriate functionality and the importance of selecting the most appropriate *in vitro* model based on scientific rationale (i.e., iPSC vs. ESC) to evaluate toxicity findings.

Interestingly, one of the internal compounds (GNE-618) did not cause cardiac toxicity in rats. It was later discovered that this compound had very low potency in rat cardiomyocytes *in vitro*, but it was still very potent in human cardiomyocytes, suggesting that despite the lack of rat toxicity *in vivo*, this compound was likely to be cardiotoxic to humans (unpublished observation). This emphasizes the importance of assessing target potency in different species to understand and interpret toxicity findings, especially when questionable or inconsistent findings are encountered with the use of chemical tools.

Cytotoxicity to human cardiomyocytes was also associated with depletion of NAD+, confirming target inhibition at those dose levels. Cotreatment with NA also mitigated the cytotoxicity caused by NAMPT inhibitors to primary human cardiomyocytes, providing further evidence that this cytotoxicity was target related (unpublished observation).

Retinal toxicity was also observed with NAMPT inhibitors in both rat safety and mouse efficacy studies (studies with NAPMT inhibitors were not conducted in nonrodent species). The same toolkit was used to determine that the effect was target related. The use of *in vitro* systems consisted of rat and human retinal cell lines. The higher *in vitro* potency of NAMPT inhibitors in human retinal cell lines compared to a rat cell line also suggested that the retinal toxicity was relevant to humans (Zabka et al., 2015).

Mouse xenograft efficacy studies were leveraged to interrogate the existence of a TI for this retinal toxicity and the potential mitigation with NA coadministration. These *in vivo* studies revealed that no TI existed and that coadministration of NA provided no mitigation for this toxicity. The addition of specific safety endpoints to efficacy studies followed by sensible data interpretation can be a cost-effective and animal-use-conscious way of leveraging these studies to interrogate a particular toxicity.

Case Study 2

In the following case study, we describe an example of off-target toxicity mitigation for a CNS program. Several CNS-active beta-site amyloid precursor protein cleaving enzyme 1 (BACE-1) small-molecule inhibitors showed potent off-target binding in the mu (MOP) and kappa (KOP) opioid receptor assays (unpublished observation). Since these molecules are CNS penetrant, this raised concerns around potential opioid-like adverse effects in humans and potential implications for abuse liability. To understand the physiological relevance of the *in vitro* binding effects, functional cellular assays were conducted, revealing strong agonistic activity toward MOP and KOP. *In vivo* translation of the *in vitro* effects was corroborated by clinical signs observed in rats that were consistent with MOP effects (hypoactivity and decreased fecal production) and KOP effects (increased urine production). To further confirm that these *in vivo* effects were driven by the opioid receptor off-target hits, as opposed to other potential off-target effects, a charcoal transit study was conducted in mice, revealing that a representative BACE-1 inhibitor caused a significant decrease in charcoal transit, which was partially mitigated by the MOP antagonist naloxone.

Screening of follow-on molecules resulted in the identification of a structurally distinct and potent lead molecule with reduced opioid off-target effects. The mitigation of the toxicity was confirmed by the absence of clinical signs in rats at free brain drug levels that were comparable to the levels that resulted in opioid-like effects for the initial molecules. Furthermore, comparing the ratios of exposures with no observed adverse effects in rats to the exposures resulting in PD effects of similar magnitude for the different molecules (lowering of A-beta in the brain), a significant increase in safety margins for opioid off-target effects was confirmed for the lead molecule. This case study illustrates how off-target hits can be flagged early through *in vitro* binding screens, functional effects can be confirmed through follow-up functional cellular assays, *in vivo* translation can be assessed in animal studies through observation of clinical signs, mechanistic specificity can be investigative in dedicated functional *in vivo* studies, and finally, mitigation can be achieved by screening out the toxicity and ensuring the presence of an acceptable safety margin.

46.6 DATA INCLUSION IN REGULATORY FILINGS

When investigative work has been performed that contributes to the understanding and derisking of a particular toxicity, it is advisable to include these works in regulatory filings such as INDs/CTAs and regulatory updates. For this, it is important to clearly explain the finding or concern that led to the investigation and the objectives of the investigation. In addition, the experimental systems should be adequately described, and the results obtained and the interpretation of the data thoroughly explained. The summary section of regulatory filings (e.g., 2.6.2 and 2.6.6 of the IND) provides an opportunity to integrate findings across all of the studies and discuss the implications of the investigation. The discussion should contain the interpretation of the findings and their relevance in the context of human health and outline the need for specific monitoring and how these findings may impact the safe starting dose calculation. In addition, full reports from any *in vitro* or *in vivo* investigative studies should be included in the filing and referred to in the main section of the regulatory document.

In conclusion, this chapter illustrates how safety issue investigation can be effectively incorporated in the discovery space, both to address safety concerns around a target and observed safety findings in animal studies, with two illustrative examples. In addition, a collaborative framework that includes and leverages the diverse drug discovery and development functions has been discussed. In the end, these activities should contribute to the selection of candidate molecules with optimized and well-characterized safety profiles and ultimately drugs with an optimal chance of clinical success.

REFERENCES

Carvalho, J. F., Louvel, J., Doornbos, M. L., Klaasse, E., Yu, Z., Brussee, J., and AP, I. J. (2013). Strategies to reduce HERG K+ channel blockade. Exploring heteroaromaticity and rigidity in novel pyridine analogues of dofetilide. *Journal of Medicinal Chemistry* 56, 2828–2840.

Diaz, D., Allamneni, K., Tarrant, J. M., Lewin-Koh, S. C., Pai, R., Dhawan, P., Cain, G. R., Kozlowski, C., Hiraragi, H., La, N., Hartley, D. P., Ding, X., Dean, B. J., Bheddah, S., and Dambach, D. M. (2012). Phosphorous dysregulation induced by MEK small molecule inhibitors in the rat involves blockade of FGF-23 signaling in the kidney. *Toxicological Sciences* 125, 187–195.

Gad, S. C., Cassidy, C. D., Aubert, N., Spainhour, B., and Robbe, H. (2006). Nonclinical vehicle use in studies by multiple routes in multiple species. *International Journal of Toxicology* 25, 499–521.

Holen, K., Saltz, L. B., Hollywood, E., Burk, K., and Hanauske, A. R. (2008). The pharmacokinetics, toxicities, and biologic effects of FK866, a nicotinamide adenine dinucleotide biosynthesis inhibitor. *Investigational New Drugs* 26, 45–51.

Hovstadius, P., Larsson, R., Jonsson, E., Skov, T., Kissmeyer, A. M., Krasilnikoff, K., Bergh, J., Karlsson, M. O., Lonnebo, A., and Ahlgren, J. (2002). A Phase I study of CHS 828 in patients with solid tumor malignancy. *Clinical Cancer Research* 8, 2843–2850.

Kola, I. and Landis, J. (2004). Can the pharmaceutical industry reduce attrition rates? *Nature Reviews. Drug Discovery* 3, 711–715.

Lee, D. U., Katavolos, P., Sioson, C., Katewa, A., Pang, J., Choo, E., Ghilardi, N., Diaz, D., and Danilenko, D. M. (2014). Multifocal Defects in the Hematopoietic and Lymphoid Compartments in Mice Dosed with a Broad BET Inhibitor. In Society of Toxicology Proceedings; 954g, 23–27 March, Phoenix, AZ.

Mutlib, A. E., Gerson, R. J., Meunier, P. C., Haley, P. J., Chen, H., Gan, L. S., Davies, M. H., Gemzik, B., Christ, D. D., Krahn, D. F., Markwalder, J. A., Seitz, S. P., Robertson, R. T., and Miwa, G. T. (2000). The species-dependent metabolism of efavirenz produces a nephrotoxic glutathione conjugate in rats. *Toxicology and Applied Pharmacology* 169, 102–113.

Neervannan, S. (2006). Preclinical formulations for discovery and toxicology: physicochemical challenges. *Expert Opinion on Drug Metabolism & Toxicology* 2, 715–731.

Oh, A., Ho, Y. C., Zak, M., Liu, Y., Chen, X., Yuen, P. W., Zheng, X., Liu, Y., Dragovich, P. S., and Wang, W. (2014). Structural and biochemical analyses of the catalysis and potency impact of inhibitor phosphoribosylation by human nicotinamide phosphoribosyltransferase. *ChemBioChem* 15, 1121–1130.

Pessina, A., Parent-Massin, D., Albella, B., Van Den Heuvel, R., Casati, S., Croera, C., Malerba, I., Sibiril, Y., Gomez, S., de Smedt, A., and Gribaldo, L. (2009). Application of human CFU-Mk assay to predict potential thrombocytotoxicity of drugs. *Toxicology In Vitro* 23, 194–200.

Sampath, D., Zabka, T. S., Misner, D. L., O'Brien, T., and Dragovich, P. S. (2015). Inhibition of nicotinamide phosphoribosyltransferase (NAMPT) as a therapeutic strategy in cancer. *Pharmacology & Therapeutics* 151, 16–31.

Sasseville, V. G., Lane, J. H., Kadambi, V. J., Bouchard, P., Lee, F. W., Balani, S. K., Miwa, G. T., Smith, P. F., and Alden, C. L. (2004). Testing paradigm for prediction of development-limiting barriers and human drug toxicity. *Chemico-Biological Interactions* 150, 9–25.

Swenberg, J. A. (1993). Alpha 2u-globulin nephropathy: review of the cellular and molecular mechanisms involved and their implications for human risk assessment. *Environmental Health Perspectives* 101 Suppl 6, 39–44.

Tarrant, J. M., Dhawan, P., Singh, J., Zabka, T. S., Clarke, E., DosSantos, G., Dragovich, P. S., Sampath, D., Lin, T., McCray, B., La, N., Nguyen, T., Kauss, A., Dambach, D., Misner, D.L.,

Diaz, D., and Uppal, H. (2015). Preclinical models of nicotinamide phosphoribosyltransferase inhibitor-mediated hematotoxicity and mitigation by co-treatment with nicotinic acid. *Toxicol Mech Methods* 25(3), 201–211.

Zabka, T. S., Singh, J., Dhawan, P., Liederer, B. M., Oeh, J., Kauss, M. A., Xiao, Y., Zak, M., Lin, T., McCray, B., La, N., Nguyen, T., Beyer, J., Farman, C., Uppal, H., Dragovich, P. S., O'Brien, T., Sampath, D., and Misner, D. L. (2015). Retinal toxicity, in vivo and in vitro, associated with inhibition of nicotinamide phosphoribosyltransferase. *Toxicological Sciences* 144, 163–172.

Zheng, X., Baumeister, T., Buckmelter, A. J., Caligiuri, M., Clodfelter, K. H., Han, B., Ho, Y. C., Kley, N., Lin, J., Reynolds, D. J., Sharma, G., Smith, C. C., Wang, Z., Dragovich, P. S., Oh, A., Wang, W., Zak, M., Wang, Y., Yuen, P. W., and Bair, K. W. (2014). Discovery of potent and efficacious cyanoguanidine-containing nicotinamide phosphoribosyltransferase (Nampt) inhibitors. *Bioorganic & Medicinal Chemistry Letters* 24, 337–343.

Zhong, Y., Katavolos, P., Nguyen, T., Boggs, J., Sambrone, A., Kan, D., Merchant, M., Harstad, E., Diaz, D., Zak, M., Costa, M., and Schutten, M. (2013). Intestinal toxicity caused by a small molecule tankyrase inhibitor in mice is reversible. *Am Coll Toxicol Proc* 2013, P511.

ABBREVIATIONS

3D	three dimensional
3Rs	reduction, refinement and replacement of animal experimentation
Acetyl-CoA	acetyl coenzyme A
ADHD	attention deficit hyperactivity disorder
ADP	adenosine diphosphate
ADR	adverse drug reaction
AERS	Adverse Events Reporting System
ALF	acute liver failure
ALT	alanine aminotransferase
ATP	adenosine triphosphate
ATPase	ATP hydrolase
AUC	area under the curve
BBB	blood–brain barrier
BSEP	bile salt export pump
CAR	constitutive androstane receptor
CCCP	carbonyl cyanide m-cholorophenylhydrazone
CD_{50}	concentration causing 50% lethality
C_{eff}	Hu human efficacious dose (expressed as C_{max})
CL	clearance
clogP	logarithm of partition coefficient between n-octanol and water
C_{max}	maximal plasma concentration
C_{min}	minimal plasma concentration
CO_2	carbon dioxide
CoA	coenzyme A
C_{ss}	concentration at steady state
CNS	central nervous system
CPA	conditioned place aversion
CPP	conditioned place preference
DDL	drug discrimination learning
DfW	Derek for Windows
DILI	drug induced liver injury
EEG	electroencephalogram
EFPIA	European Federation of Pharmaceutical Industries and Associations
EMA	European Medicines Agency
EPA	US Environmental Protection Agency
ER	endoplasmic reticulum
F_0	proton channel of F-ATPase
F%	bioavailability
F_{abs}	fraction absorbed
FAD+	oxidized flavin adenine dinucleotide
FADH2	reduced flavin adenine dinucleotide hydroquinone
FAO	fatty acid oxidation
FDA	US Food and Drug Administration
FOB	functional observation battery
FR	fixed ratio of reinforcement
FXR	farnesoid X receptor
GABA	gamma-aminobutyric acid
GI	gastrointestinal
GLDH	glutamate dehydrogenase
GLP	good laboratory practices
HCA	high content imaging and analysis
HLA	human histocompatibility antigen
HLAED	Human Liver Adverse Effects Database
hpf	hours post fertilization
HTS	high-throughput screening
IC_{50}	concentration causing 50% inhibition

Drug Discovery Toxicology: From Target Assessment to Translational Biomarkers, First Edition. Edited by Yvonne Will, J. Eric McDuffie, Andrew J. Olaharski, and Brandon D. Jeffy.
© 2016 John Wiley & Sons, Inc. Published 2016 by John Wiley & Sons, Inc.

ICH	international conference on harmonisation	OATP	organic anion transporter
IP	intraperitoneal	OCHEM	Online Chemical Modeling Environment
IV	intravenous	OCT	ornithine carbamoyltransferase
IVSA	intravenous self-administration	PD	pharmacodynamics
IMI	Innovative Medicines Initiative	PhRMA	Pharmaceutical Research and Manufacturers of America
IV	intravenous		
KNN	k-nearest neighbor	PK	pharmacokinetics
LC_{50}	concentration causing 50% lethality	pK_a	ionization constant
LC-MS	liquid chromatography/mass spectrometry	PO	per os
lfabp 10a	liver-specific fatty acid binding protein 10a	PTZ	pentylenetetrazol
LFTs	liver function tests	ITC	International Transporter Consortium
LSD	lysergic acid diethylamide	PBPK	physiologically based pharmacokinetic
LTP	long-term potentiation	PPAR	peroxisome proliferator receptor
MEA	multielectrode array	QSAR	quantitative structure–activity relationship
MIE	molecular initiating event	ROS	reactive oxygen species
MRP2	multi-drug resistance protein type 2	RST	respiratory screening technology
MRT	mean residence time	SAR	structure–activity relationships
mtDNA	mitochondrially derived DNA	SC	subcutaneous
NAD+	oxidized nicotinamide adenine dinucleotide	SCH	sandwich cultured hepatocytes
NADH	reduced nicotinamide adenine dinucleotide	SLC	solute carrier
NAPQI	N-acetyl-p-benzoquinone imine	THC	tetrahydrocannabinol
NDMA	N-methyl-d-aspartate	TRAIL	tumor necrosis factor-related apoptosis-inducing ligand
nrf-2	transcription factor regulated by $NFE2L1$ gene		
		ULN	upper limit of normal reference range
NSAIDs	nonsteroidal anti-inflammatory drugs	V_{ss}	volume of distribution at steady state
NTCP	sodium-taurocholate cotransporting polypeptide	VTA	ventral tegmental area
		Zf	zebra fish

CONCLUDING REMARKS

I remember my first week at Pfizer in the investigative toxicology group based in La Jolla, CA, USA. I asked my supervisor, "What should I be working on?" He showed me a list of attrited compounds and simply said, "Fix that for me, please." Trained as a biochemist, familiar mostly with apoptosis and mitochondria, this task seemed far away from the training I had received, but who was I to protest? This was more than 10 years ago, and since then, I have become a well-rounded drug discovery toxicologist.

Over the years, I have spent time with the chemistry community to better understand that unfavorable physico-chemical properties can drive *in vitro* cytotoxicity and lower the threshold and severity for *in vivo* adverse events. Additionally, I have learned that even though we desire to predict early organ toxicities, we are far from accomplishing this aim. It is envisioned that we will soon determine whether complex 3D models and/or stem cell approaches will help us improve upon this goal.

Academic laboratories, biomedical research institutes, innovation centers, and consortia have evolved to provide inestimable support for the drug development process. Several companies remain engaged in the precompetitive space such as HESI, PSTC, and IMI Consortium, and these engagements have led to regulatory (i.e., FDA, EMA, and PMDA) acceptance of next-generation translational biomarkers intended to enable the drug discovery and development process.

When I started working in the emerging field of modern drug discovery toxicology in 2003, I wish I had known what I know today. We compiled over a decade of learnings into this book, comprehensively describing a road map of how to successfully apply safety assessment from target evaluation to translational biomarkers (*first-in-human*). Whereas several big pharma companies have fully embraced the drug discovery toxicology paradigm, many others will greatly benefit from the platforms, tools, and/or case studies described in this book. We thank our many colleagues around the world who provided chapters despite other pressing responsibilities. We also would like to thank Drs. Donna Dambach and Tomas Mo for helpful discussions on the content of the book. It is our hope that this book will foster a more rapid discovery of developable candidate drugs, bringing much needed novel medicines with improved safety profiles to patients in need.

On behalf of my coeditors, J. Eric McDuffie, Brandon D. Jeffy, and Andrew J. Olaharski, thank you.

Yvonne Will

Drug Discovery Toxicology: From Target Assessment to Translational Biomarkers, First Edition. Edited by Yvonne Will, J. Eric McDuffie, Andrew J. Olaharski, and Brandon D. Jeffy.
© 2016 John Wiley & Sons, Inc. Published 2016 by John Wiley & Sons, Inc.

INDEX

absorption, distribution, metabolism, and excretion (ADME)
  absorption properties, 84–85
  assessments, 20–22
  biodistribution, 83
  human polymorphisms, 87
  metabolism and drug-drug interaction, 85
  pharmacodynamic effectiveness, 83
  plasma protein binding, 84
  properties, 303, 308
  renal injury, 279
  target product profile, 84
  transporters, 88
accuracy, 484
acetylcysteine (AC), 418
acid sphingomyelinase knockout (ASMKO) mouse, 273
acute kidney injury (AKI), 365
  biomarkers, 466, 467
  candidate biomarkers, 432
  causes, 467
  cisplatin-induced, 281
  folic acid-induced, 281
  in humans, 280–281
  induce, 443
  I/R model, 280
  sepsis-induced, 281
acute liver failure (ALF), 96, 416
acute lung injury (ALI)
  in humans, 283
  models, 284
acute respiratory distress syndrome (ARDS), 283
acylcarnitines, DILI, 420
AD see Alzheimer's disease
adenomatous polyposis coli (APC), 277–278

adenosine triphosphate (ATP), 408
adrenomedullin (AM), 390
adverse drug reactions (ADRs), 195–196
adverse events, 148
afterload, 388
AKI see acute kidney injury
alanine aminotransferase (ALT), 96, 418
albuminuria, 452
aldehyde dehydrogenase 1 family, member A1 (ALDH1A1), 202
ALF see acute liver failure
ALI see acute lung injury
Alper's syndrome, 342
ALT see alanine aminotransferase
Alzheimer's disease (AD), 272
AM see adrenomedullin
α1-microglobulin (A1M), 435
aminoglycoside antibiotics, 366
amyloid precursor protein (APP), 272
analysis of covariance (ANCOVA) model, 320
androgen-binding protein (ABP), 472
angiotensin-converting enzyme (ACE), 150
angiotensin II (Ang II), 390–391
animal models
  cardiovascular disease, 268–270
  gastrointestinal injury, 273–278
  hepatic disease, 264–268
  human disease, 263
  nervous system disease, 270–273
  renal injury, 279–282
  respiratory disease, 282–285
ANP see atrial natriuretic peptide
antibiotic-associated diarrhea (AAD), 227
antibody-dependent cell-mediated cytotoxicity (ADCC) assays, 33

Drug Discovery Toxicology: From Target Assessment to Translational Biomarkers, First Edition. Edited by Yvonne Will, J. Eric McDuffie,
Andrew J. Olaharski, and Brandon D. Jeffy.
© 2016 John Wiley & Sons, Inc. Published 2016 by John Wiley & Sons, Inc.

antibody-dependent cell-mediated phagocytosis (ADCP) assays, 33–34
antibody drug conjugates, safety assessment
  developmental approach, 30
  new technologies evaluation, 33
  off-target toxicity, 32
  on-target toxicity, 30–32
antidrug antibodies (ADAs)
antitumor activity, 72
APC *see* adenomatous polyposis coli
APP *see* amyloid precursor protein
ARDS *see* acute respiratory distress syndrome
area under ROC curve (AUROC), 496
area under the curve (AUC), 67, 75
aristolochic acid, 365
asialoglycoprotein receptor (ASGPR), 39
aspartate aminotransferase (AST), 418
asthma, 283–285
ATP *see* adenosine triphosphate
atrial natriuretic peptide (ANP), 391–392
atypical acinar cell nodules (AACN), 246
AUROC *see* area under ROC curve
autoimmunity, 196
automated blood sampling and telemetry (ABST) system, 437–438

basicity, 60
Bcl-xL, 300, 301
benchmark dose (BMD), DO Mice, 320
benzodiazepines, 215
Bile duct loss, 266
bile salt export pump (BSEP), 63, 342
  inhibition of, 102–105
  and mitochondrial function, 105–108
biomarkers
  blood-based, 387–389
  cardiac hypertrophy, 390–392
  cell membrane leakage, 388
  definition, 387
  DILI, 416–423
  drug development, 387
  drug-induced vascular injury, 397–402
  EMPs, 401
  GI toxicology, 229–231
  of inflammation, 390
  miRNA, 252
  mouse population-based toxicology, 320–322
  pancreatic toxicity, 249–253
  peptide, 251–252
  secreted factors, 387
  skeletal muscle injury, drug-induced, 409–411
biopharmaceuticals, safety assessment
  ADC, 30, 33
  ADCC assays, 33
  ADCP assays, 33–34
  CDC assays, 33
  dose-ranging studies, 35
  immune system as target, 28–29
  immunogenicity, 33
  immunotoxicity testing, 34

  mAbs, 29–30
  modality-associated risks, 29
  off-target toxicity, 32
  on-target toxicity, 30–32
  species selection, 34–35
  target biology, 28
  unintended adverse consequences, 34
  warheads evaluation, 32–33
black box warning, 346, 392, 416, 497
blood-based biomarkers, 387–389
blood-brain barrier (BBB) function, 215
blood oxygen leveldependent (BOLD), 454
blood-testis barrier (BTB), 471
blood urea nitrogen (BUN), 431, 434
β2-microglobulin (B2M), 435
bone marrow
  animal and human clinical trials, 178
  detection of, stem and progenitor cell, 177
  hematopoiesis, 172–173
  hematotoxicity (*see* hematotoxicity)
  next generation of assays, 175
  proliferation or differentiation, 175–176
  toxicity test during drug development, 177–178
bone morphogenetic protein 7 (BMP7), 459
brain natriuretic peptide (BNP), 391–392
brain slice, electrophysiology, 217
brain uptake index (BUI), 89
BSEP *see* bile salt export pump
BTB *see* blood-testis barrier

CAD *see* cationic amphiphilic drug
calbindin D28
  kidney tissue injury biomarkers, 435–436
  NHP, 446
calcium transient measurements, 140–141
calprotectin, GI biomarkers, 231
CAMEO-96, 176
canine kidney safety protein biomarkers
  CKD, 444
  clusterin, 444
  cystatin C, 444
  inflammatory cytokines/chemokines, 444
  NGAL, 443, 444
  novel biomarkers, 443
  OPN, 443, 444
  performance evaluations, 444
  polymyxin B treatment, 444
  proteome, 443
  urinary protein biomarkers, 443
canonical pathway map, 516
cardiac action potential (AP)
  duration, 141
  effect on, 357
cardiac electrophysiologic effects
  AP/repolarization assays, 135–136
  ionic currents, 134–135
  proarrhythmia assays, 136
  stem cell-derived CM, 136
  subcellular techniques, 134

cardiac hypertrophy
  biomarker, 390–392
  diagnosis, 389
  echocardiography, 388, 389
  Frank-Starling mechanism, 388
  good laboratory practice, 389
  natriuretic peptides, 391–392
  NTproANP concentrations, 392
  physiologic, 388
  radiotelemetry, 389
  vasoactive peptides, 390–392
cardiac hypertrophy working group (CHWG), 504
cardiac ion channels
  high-throughput cardiac ion channel data, 137
  in silico approaches, 137–140
cardiac toxicity, 268–270
cardiovascular disease (CVD)
  advantages, 269–270
  animal models, 148–152
  from animal to human, 268–269
  drug attrition, 268
  drug-induced CV issues, 133
  hPSC-CMs, 350–352
  large *vs.* small molecules, 147–148
  limitations, 269–270
  myocardial contractility assessment, 144–147
  safety testing, 132
  stem cell-derived CMs, 140–141
  surrogate models, 269
  surrogates for humans, 149
  telemetry technology, 141–144
cationic amphiphilic drug (CAD), 318
CC mice, 318–319
  cost considerations, 325
  genome-wide association analysis, 324
  health status, 325–326
  model selection, 322–323
  sample size, 323
CD *see* Crohn's disease
CDC *see* complement dependent cytotoxicity
CD64, GI biomarkers, 231
CD44v6 antibody, 31
cell-based assays, for toxicity prediction, 7–8
cell culture systems, DIVI, 402
cell lines
  colon cancer, 278
  drug-induced liver injury, 336–337
  GI toxicology, 231
cell migration, GI tract, 236
cell models, of GI tract, 231–235
  cell lines, 231
  coculture systems, 232–233
  iPSCs, 232
  organs-on-chips, 235
  primary cells, 231
  3D organoids, 233–235
cell viability assays, GI tract, 236
central nervous system adverse events (CNS ADRs)
  drug abuse liability testing, 218–222

electrophysiological assays, 215
  neurotoxicity assays, 215
  pharmacological profiling, 215
  seizure liability testing, 216–218
  in vitro models, 215–216
  in vivo behavioral batteries, 214–215
cerivastatin, 62
CFC *see* colony-forming cell assay
CHB *see* cyanohydroxybutene
chemical inhibitors, 85
chemical structure-mediated toxicities, 17
chloroquine, 60
cholecystokinin (CCK)-1 receptor, 246, 266
choroid neovascularization (CNV), 210
chronic hypertension, 388
chronic kidney disease (CKD), 444
  diabetic nephropathy, 281
  hypertensive nephropathy, 281
  immune-mediated glomerular disease, 281–282
chronic obstructive pulmonary disease (COPD), 282, 283
chronic progressive nephropathy (CPN), 280
CHWG *see* cardiac hypertrophy working group
cisplatin, 438–439, 459
citrulline, GI biomarkers, 229–230
CK *see* creatine kinase
CKD *see* chronic kidney disease
CKM *see* creatine kinase M
clearance (CL) prediction, by PhRMA, 72, 75
Clinical and Laboratory Standards Institute (CLSI) method, 481
clusterin
  canine kidney safety protein biomarkers, 444
  kidney tissue stress response biomarkers, 436
  NHP, 446
CNV *see* choroid neovascularization
Coalition Against Major Diseases (CAMD), 6
coculture systems
  GI toxicology, 232–233
  inducible pluripotent stem cells, 340
colon cancer cell lines, 278
colony-forming cell (CFC) assay
  limitations, 175
  uses, 173–175
  in vitro/in vivo concordance, 175
colorectal cancer (CRC)
  animal models, 275, 277–278
  chemically/environmentally induced, 275, 277
  genetic manipulation, 277–278
  inflammation-induced, 277
  spontaneous, 275
complement activation, 197
complement dependent cytotoxicity (CDC), 33, 195
complex culture systems, 340
Comprehensive *In Vitro* Proarrhythmia Assay (CiPA), 132
conditioned place preference (CPP) test, 218, 222
constitutive androstane receptor (CAR), 85
COPD *see* chronic obstructive pulmonary disease
C-peptide, 252–253
CPN *see* chronic progressive nephropathy
CRC *see* colorectal cancer

C-reactive protein (CRP)
    cardiac hypertrophy biomarker, 390
    DIVI biomarkers, 401
creatine kinase (CK), 408
creatine kinase M (CKM), 409
Crohn's disease (CD), 275
cross-functional collaborative model
    clinical drug safety, 535
    clinical input, 535
    discovery toxicology, 533–534
    DMPK, 534
    internal investigative groups, 535
    medicinal chemistry, 534
    pharmaceutics group, 534–535
    translational groups, 534
cryopreservation, hepatocytes, 336
cultured cells, 217
cyanohydroxybutene (CHB)
    pancreas, 248
    peptide biomarkers, 251–252
cyclosporine A (CsA), 459
cynomolgus monkey *(Macaca fasciculata)*
    drug-induced kidney tubular injury, 447
    polymyxin B, 446
    treatment, 446
CYP1A2 *see* cytochrome P450 1A2
CYP3A4, 82
Cyp3a family, 305
CYP2D6, 87
cystatin C, 444
cytochrome P450 1A2 (CYP1A2), 303
cytochrome P450 liver enzymes, 21, 422
    induction, 85
    metabolism, 85–86
    transporter, 86
cytokine release, 197
cytotoxicity, 19–20, 537

DEN *see* diethylnitrosamine
diabetic nephropathy (DN), 281
diabetic neuropathy, 273
diagnostic toxicogenomics, 519–521
diamine oxidase (DAO), GI biomarkers, 230
DIAP *see* drug-induced AP
diethylenetriamine pentaacetate (DTPA), 455
diethylnitrosamine (DEN), 265
DILI *see* drug-induced liver injury
dilutional linearity, 485
dimethyl sulfoxide (DMSO) tolerant, 203
discovery toxicology
    application, 531–532
    case-studies, 536–538
    cross-functional collaborative model, 534–535
    early stages of, 530–531
    human translation, 533
    lead optimization phase, 532
    mechanism of toxicity, 533
    nature of toxicity, 533
    pharmaceutical industry, 530

    potential safety risks, 530
    regulatory filings, 538
    role of, 530
    safety strategy, 532
dofetilide, 150
DO mice
    cost considerations, 325
    genome-wide association analysis, 324
    health status, 326
    model selection, 322
    sample size, 322–324
doxorubicin (DOX), 462, 464
drug abuse liability testing
    abusing prescription drugs, 218
    CNS-active drug candidate, 218
    CPP test, 222
    DDL test, 219
    IVSA test, 219–222
    nonprecipitated withdrawal test, 219
drug attrition
    cardiac toxicity, 268
    GI toxicity, 273
    hepatic toxicity, 264
    nervous system toxicity, 270
    renal toxicity, 279
    respiratory toxicity, 282
drug development
    bone marrow toxicity test, 177–178
    hematotoxicity test, 173, 174
drug development tool (DDT), 500
drug discovery, toxicogenomics, 524
drug discrimination learning (DDL) test, 218, 219
drug-drug interactions (DDIs), 84, 168, 169
druggability, 83
drug-induced AP (DIAP), 243–244
drug-induced kidney injury (DIKI)
    BUN, 431
    Calbindin D28, 435–436
    candidate biomarkers, 432
    clusterin, 436
    detection, 431
    from functional changes, 433
    α-GST, 435
    histopathology, 433
    IL-18, 437
    integrated rat platform application, 437–439
    KIM-1, 436
    L-FABP, 437
    NAG, 435
    NGAL, 436–437
    novel functional biomarkers, 434–435
    OPL, 437
    preclinical investigation, 431
    qualified and unqualified biomarkers uses, 433
    qualify urinary biomarkers, 432
    renal injury biomarkers, 432
    routine urinalysis, 431
    RPA-1, 435
    sCr, 431

traditional functional biomarkers, 433–434
traditional kidney injury indices, 431, 432
translation for biomarkers, 432–433
drug-induced liver injury (DILI), 63
  acetylcysteine, 418
  acute liver failure, 416, 417
  acylcarnitines, 420
  alanine aminotransferase, 418
  aspartate aminotransferase, 418
  biomarkers, 416–423
  BSEP and mitochondrial function, 105–108
  BSEP inhibition, 102–105
  cell lines, 336–337
  cholestasis, 342
  complex cell models, 109–110
  CV toxicity, 264
  diagnosis, 416
  glutamate dehydrogenase, 419–420
  hepatocyte-like cells, 341–342
  high content image analysis, 108–109
  HMGB1, 420
  idiosyncratic, 334
  incidence, 416
  Keratin-18, 421
  mechanisms and susceptibility, 95–97
  microRNA-122, 421
  mitochondrial injury, 98–100
  murine models, 336
  piggyBac transposon system, 337–338
  predictable, 333–334
  primary human hepatocytes, 334–336
  reactive metabolite-mediated toxicity, 100–102
  relevance to human, 339
  safety assessment, 417
  self-replicating plasmids, 338
  in silico models, 114–117
  stem cells model, 337–338
  systems pharmacology, 117–119
  targeted analysis, 419
  types of, 333–334
  unbiased approaches, 419
  in vitro models, 422–423
  in Western world, 417
  zebrafish, 110–113
drug-induced nephrotoxicity (DIN), 365
drug-induced skeletal muscle injury
  creatine kinase, 408–409
  creatine kinase M, 409
  EMA, 411
  FABP3, 410
  Food and Drug Administration, 411
  future directions, 411–412
  miRNAs, 410–411
  Myl3, 409–410
  myoglobin, 410
  novel biomarkers, 409–411
  overview, 407–409
  parvalbumin, 410
  PDMA, 411

  regulatory endorsement, 411
  skeletal troponin I, 409
drug-induced vascular injury (DIVI)
  biomarkers, 397–402
  candidate biomarkers, 401
  case study, 403
  cell culture systems, 402
  histomorphologic outcomes, 398
  imaging modalities, 401
  Ingenuity Pathway Analysis, 401
  in preclinical drug development, 397–400, 403
  ultrasound, 401
  in vitro/in vivo systems, 402, 403
drug-likeness, 83
drug metabolism and pharmacokinetics (DMPK), 534
drug metabolism, genetically modified mouse, 303–306
drug testing, hPSC-CMs, 350–352
drug transporters, genetically modified mouse, 306–307

Ebola hemorrhagic fever (EHF), 319
echocardiography (ECG), cardiac hypertrophy, 388, 389
ECVAM *see* European Centre for the Validation of Alternative Methods
EEG recording, 217–218
Efficient Mixed-Model Association (EMMA), 324
electroretinography (ERG), 208
embryoid body (EB) methods, 347–349, 370–371
embryo lethal phenotype, 300, 301
embryonic kidney development
  hPSCs differentiation, 370
  IM-like cells, 370
  marker expression patterns, 369–370
  murine and human renal precursor structures, 371
  podocyte- or renal tubular-like cells, 371–372
  proximal tubular cells, 372–376
  self-organizing different cell types, 370–371
  UB-committed renal progenitor-like cells, 370
embryonic stem cells (ESCs), 299, 300
  drug-induced liver injury, 337
  3D culture, 340
enalaprilat, 150
end-diastolic volume (EDV), 388–389
endocrine pancreas, 242
  biomarkers, 252–253
  mechanisms of, 244–245
  in vitro toxicity, 255
endogenous polymerases, 47–48
endothelial microparticles (EMPs), 253, 256, 401
end-systolic volume (ESV), 388–389
enzymelinked immunosorbent assay (ELISA), 231, 251, 255
epithelial markers, 367
ERG *see* electroretinography
esophagus, 228
ET-1, 390–391
etomoxir, 119
European Centre for the Validation of Alternative Methods (ECVAM), 173, 175, 377
European Medicines Agency (EMA), 432, 452
  drug-induced skeletal muscle injury, 411
  Letter of Support, 502

excretion pumps inhibition, 63–64
exocrine pancreas, 242–244
    biomarkers, 250–251
    mechanisms of, 244
    serum biomarkers, 250–251
    in vitro toxicity, 254–255

FABP3 *see* fatty acid-binding protein 3
FACS *see* fluorescence-activated cell sorting
Fah system *see* fumarylacetoacetate hydrolase system
fatty acid-binding protein 3 (FABP3), 410
fatty acid-binding protein (FABP), 230
fatty acid oxidation (FAO), 98
fatty liver disease, 265–266
Fcγ receptors (FcγRs), 195
fibrotic lung disease, 285
FITC-sinistrin, 455
fit-for-purpose assay
    accuracy/recovery, 484
    analyte stability, 487
    biological samples precision, 485
    calibration/standard curve, 484
    dilutional linearity, 485
    exploratory biomarkers, 481
    FDA, 481
    LLOQ, 484
    LOD, 485
    nonclinical safety study, 482, 483
    parallelism, 485–486
    performance evaluations, 487
    precision, 482, 484
    preclinical study, suitable and reliable for, 482
    quality control, 486
    species- and gender-specific reference ranges,
      486–487
    ULOQ, 484
    uses, 484
    validation, 481, 482
flavin-containing monooxygenases (FMOs), 165, 169
fluorescein-labeled poly D-lysine, 455
fluorescence-activated cell sorting (FACS), 375
FOB *see* functional observational battery
Food and Drug Administration (FDA), 452
    DILI, 417
    drug development tools, 4, 6
    drug-induced skeletal muscle injury, 411
    fit-for-purpose approach, 481
    Letter of Support, 502
    regulatory science priorities, 4
    substrates, and inhibitors, 86
Frank-Starling mechanism, 388
fully automated computational image analysis
    desmin immunoreactivity, 464
    glomerular podocytes, 462
    material and methods, 462–463
    results, 463–464
    workflow, 463
fumarylacetoacetate hydrolase (Fah) system, 308
functional observational battery (FOB), 214–215

gastrointestinal (GI) toxicology, 227
    from animal to human, 273–274
    barrier integrity, 236
    biomarkers, 229–231
    calprotectin, 231
    CD64, 231
    cell-based *in vitro* assays, 235–236
    cell migration, 236
    cell models, 231–235
    cell viability assays, 236
    citrulline, 229–230
    coculture systems, 232–233
    colorectal cancer, 275, 277–278
    diamine oxidase, 230
    drug attrition, 273
    esophagus, 228
    FABP, 230
    hydroxyproline, 230
    idiopathic IBD, 275
    iPSCs, 232
    miRNAs, 230
    oral cavity, 228
    organs-on-chips, 235
    small and large intestine, 229
    stomach, 228–229
    TEER, 233, 235, 236
    3D organoids, 233–235
    trefoil factor, 230
    ulcer models, 274–275
    in vitro models, 234
    xenografts, 278
GeneChip data normalization, 514
gene expression signatures
    classification, 517
    identification of, 517
    liver, 518
gene targeting, genetically modified mouse, 299–300
genetically modified mouse
    applications, 302
    Bcl-xL, 300, 301
    B-Raf, 301
    CYP1A2, 303
    Cyp3a family, 305
    Cyp2e1 KO, 305
    DNA modification, 299
    drug metabolism, 303–306
    drug transporters, 306–307
    embryo lethal phenotype, 300, 301
    Fah system, 308
    gene targeting, 299–300
    GLP-1R agonists, 302
    humanized liver models, 308–309
    MEK inhibitor, 301
    midazolam, 304
    NTBC, 308
    nuclear receptors and coordinate induction, 307–308
    nuclease-based systems, 299
    PgP, 306
    phenotyping efforts, 298–300

PLK2 inhibitors, 302
retrospective evaluation, 298
RIPK1/3, 302–303
SKID, 308
tissue-specific, 298, 301
TK system, 308
UGTs, 306
uPA system, 308
use of, 300–303
Wilms' tumor, 304–305
genome-wide association studies (GWAS), 315, 324, 356
genotoxicity assessments, 22
gentamicin, NHP, 447
GLDH *see* glutamate dehydrogenase
glomerular filtration rate (GFR), 431
glucagon-like peptide receptor (GLP-1R) agonists, 302
glucose tolerance test (GTT), 252
glucose transporter-2 (GLUT2), 246
glutamate dehydrogenase (GLDH), 419–420
glutathione *S*-transferase α (α-GST), 435
G-protein-coupled receptor (GPCR), 19
granulocyte-macrophage (GM), 173
GTT *see* glucose tolerance test
gut microbiota, 227, 233
GWA analysis *see* genome-wide association analysis

HCC *see* hepatocellular carcinoma
health status, mouse population-based toxicology, 325–326
hematopoietic toxicity assessments, 20
hematotoxicity
    CFC assay, 173–175
    concordance, 175
    paradigm, 178–179
    prediction, 176–177
    test during drug development, 173, 174
hemolysis, 479–480
hepatic disease
    from animal to human, 264
    Bile duct loss, 266
    drug attrition, 264
    fatty liver disease, 265–266
    function effect, 264
    hepatotoxicity, 264–268
    liver cancer models, 265
    liver fibrosis, 266
    primary biliary cirrhosis, 266
    viral-induced inflammation, 266–267
hepatitis B virus (HBV), 266–267
hepatitis C virus (HCV), 266–267
hepatocellular carcinoma (HCC), 265
hepatocyte-like cells (HLCs)
    CRISPR/Cas9 knockout technology, 342
    HLA-associated toxicity, 342
    in toxicity assessment, 341–342
hepatocytes, 333
    cryopreservation, 336
    isolated, 336
hepatotoxicity, 264–268
hepatotoxicity working group (HWG), 503

HepG2 cell, 336–337, 342, 422
hESCs *see* human ESCs
high content image analysis, DILI, 108–109
high-content screening (HCS), 90
high lipoidal permeability drugs, 56–57
high-mobility group box-1 (HMGB1), 420
high-throughput screening (HTS), 99, 376
HLCs *see* hepatocyte-like cells
Huh7 cells, 336
human embryonic and induced pluripotent stem cell-derived
        cardiomyocytes, 140–141
human equivalent dose (HED), 6
human ESCs (hESCs), 367, 369, 374
human ether-a-go-go-related gene (hERG), 4, 346,
        387–388
humanized liver models, 308–309
human pharmacokinetic (PK) projection
    PhRMA initiatives, 72, 75–76
    plasma and tissue concentration-time profiles, 68–70
    plasma concentration-time profile prediction, 68, 69
    target tissue drug exposure, 70–74
human pluripotent stem cell (hPSC)
    advent of, 347–349
    cardiac differentiation, 347–349
    differentiation, 370
    reprogramming, 347
human pluripotent stem cell-derived cardiomyocytes
        (hPSC-CMs), 346
    cardiovascular disease modeling, 350–352
    drug testing, 350–352
    iPSC-based disease modeling and drug testing, 349–354
    iPSC-based drug discovery paradigm, 354–358
    limitations and challenges, 358–359
    new chemical entities, 83, 346, 347, 354, 357
    safety pharmacology and toxicological testing, 356–358
    traditional drug discovery paradigm, 354
human primary renal proximal tubular cells
        (HPTCs), 366, 377
human proximal tubular (hPT) cell line, 459
human proximal tubular (hPT) primary cultures
    advantages and limitations of, 165–166, 168–169
    drug design, screening and investigation of mechanism of
        action, 168, 169
    experimental models, 168
    genetic polymorphisms, 166
    implications of transport interactions, 168
    interindividual susceptibility, 166
    isolation methods, 164
    morphology, 165
    organic anion and organic cation transport pathways, 168
    phase I and phase II drug metabolism, 168
    toxicology studies in, 166–168
    validation, 165
    in vitro models (*see in vitro* models)
human telomerase reverse transcriptase (hTERT), 459
Huntington's disease (HD), 272
HWQ *see* hepatotoxicity working group
hydroxymethylglutaryl-coenzyme A (HMGCoA), 62, 300
hydroxyproline, GI biomarkers, 230

hypersensitivity
  assessment of, 195–197
  immune system, 195
hypertensive nephropathy, 281
hypertrophy
  cardiac (see cardiac hypertrophy)
  eccentric vs. concentric, 388

ICH M3(R2) guidelines, 147
ICH S7A, 131
ICH S7B, 131, 132
ICH S6(R1) guidelines, 147
ICH S6, regulatory guidance, 29
IDI see integrated discrimination index
idiopathic IBD, 275, 276
idiosyncratic drug-induced liver injury (IDILI), 267–268,
    334, 523
IHC see immunohistochemicals
IM-like cells, 370
immortalization techniques, 458
immune-mediated glomerular disease, 281–282
immune system
  compartments, 194
  effects of drugs, 195
  hypersensitivity, 195
  mechanisms, 194
  potential pathways, 194–195
  as target, 28–29
immunogenicity, 33, 197–198
immunohistochemical (IHC), pancreatic toxicity, 249
immunostimulation screening, siRNAs, 45–46
immunotoxicity assessment
  acquired immune responses, 194
  autoimmunity, 196
  drug allergy, 196
  FcγR, 195
  of hypersensitivity, 195–196
  immunogenicity, 197–198
  myelotoxicity, 198
  pseudoallergy, 196–197
  STS, 193, 194
  TDAR, 193, 194
  in vitro assays, 193, 194, 198
immunotoxicity testing, 34
indicator diffusion method (IDM), 89
inducible pluripotent stem cells (iPSCs), 459
  coculture models, 340
  complex culture systems, 340
  drug-induced liver injury, 337–338
  in drug screening, 339
  GI toxicology, 232
  integrative methods, 337–338
  nonintegrative methods, 338
  perfusion bioreactors, 340
  3D culture, 339–340
inducible pluripotent stem cells hepatocyte-like cells (iPSC HLC)
  differentiation methods, 335
  uses of, 338–339
inflammation-induced CRC, 277

inflammatory bowel disease (IBD), 319
  idiopathic, 275, 276
ingenuity pathway analysis, 401
inhibin B, 472
in silico models
  expert systems, 114
  intrinsic hepatotoxicity, 115
  knowledge-based expert system, 116
  pharmacology approach, 118
  for toxicity prediction, 6–7
  well-annotated hepatotoxicity, 116
insufficient therapeutic index, 500
integrated discrimination index (IDI), 496
integrated rat platform application, DIKI
  ABST system, 437–438
  CDDP, 438–439
  scientific innovation, 439
integrating novel imaging technology
  detecting renal function, 455
  IRDye 800CW imaging, 456
  optical and optoacoustic imaging, 456
  PET, 454
  in situ mass spectrometry, 454
  in vivo imaging techniques, 454
interleukin-18 (IL-18), 437
International Conference on Harmonization (ICH), 431
International Knockout Mouse Consortium (IKMC), 299–300
International Mouse Phenotyping Consortium (IMPC), 300
International Transporter Consortium (ITC), 105
interstitial pulmonary fibrosis (IPF), 285
Intestinal FABP (I-FABP), 230
intracranial self-administration test, 222
intraocular pressure (IOP) measurement, 207
intravenous self-administration (IVSA) test, 218–222
intrinsic cytotoxicity assessments, 19–20
intrinsic liver toxicity, 97
in vitro models
  advantages, 162
  biological processes and toxic responses, 163–164
  CNS ADRs, 215–216
  drug-induced liver injury, 422–423
  endocrine pancreatic toxicity, 255
  exocrine pancreatic toxicity, 254–255
  GI toxicology, 234
  kidneys, 162–163
  limitations of, 161–162
  for study mechanisms, 162–163
in vitro to in vivo translation
  CsA, 459
  lipocalin-2, 460
  renal cell cultures, 459
IPF see interstitial pulmonary fibrosis
iPSC-derived cardiomyocytes (iPSC-CMs), 347–349
  application, 348
  disease modeling and drug testing, 349–354
  drug discovery paradigm, 354–358
  long QT syndrome, 349
  model structural heart diseases, 349
  target identification and validation, 356

iPSCs *see* inducible pluripotent stem cells
IRDye 800CW imaging, 456
Irwin's test, 214–215
ischemic/reperfusion (I/R) model, 280
IVSA test *see* intravenous self-administration test

jaundice, 416
juxtamedullary cortical glomeruli, desmin
    immunoreactivity, 464

keratin-18 (K18), 421
ketamine hydrochloride, 207
kidney injury molecule-1 (KIM-1), 436, 446, 467
kidneys
    bioactivation pathway, 160
    cell populations, 161
    cortex, 161
    drugs and toxic chemicals, 160
    functional unit, 161
    glomerular filtration, 160
    inner medulla, 161
    nephrons, 161
    outer medulla, 161
    reabsorption and excretion, 160
    susceptibility factors, 161
    in vitro models, 162–163
kidney tissue injury biomarkers
    calbindin D28, 435–436
    α-GST, 435
    NAG, 435
    RPA-1, 435
kidney tissue stress response biomarkers
    clusterin, 436
    IL-18, 437
    KIM-1, 436
    L-FABP, 437
    NGAL, 436–437
    OPL, 437
knockout (KO) models *see* genetically modified mouse

LCD *see* liquid crystal displays
LD *see* linkage disequilibrium
Lilly Laboratories cell porcine kidney 1 (LLC-PK1), 458
limit of detection (LOD), 485
linkage disequilibrium (LD), 318
lipid permeability classification, 57
Lipinski's rule of 5, 83
lipocalin-2 (LCN2), 460
lipoidal/lipid permeability, 56
lipophilicity, 55–56, 60–61
liquid chromatography-mass spectrometry
    (LC-MS) approach
    miRNA biomarkers, 252
    peptide biomarkers, 251
liquid crystal displays (LCD), 205–206
liver
    cancer models, 265
    disease (*see* hepatic disease)
    fibrosis, 266

lobules, 333, 334
    toxicity, by chemicals, 97
LiverChip system, 341
LLNA *see* local lymph node
LLOQ *see* lower limit of quantification
local lymph node (LLNA), 196
locked nucleic acid (LNA), 47
LOD *see* limit of detection
long QT syndrome (LQTS), iPSC-CMs, 349
long-term potentiation (LTP), 215–216
lower limit of quantification (LLOQ), 484
L-type fatty acid-binding protein (L-FABP), 437
lysosomal storage diseases, 272–273

mAbs, 29–30
MA-EBOV *see* mouse-adapted strain of Ebola virus
magnetic resonance imaging (MRI), 454
MAQC *see* Microarray Quality Control Consortium
mast cell degranulation, 196–197
maximum tolerated concentration (MTC), 203
MDCK cell, 103–104
MDP *see* mouse diversity panel
MDZ *see* midazolam
MEFs *see* mouse embryonic fibroblasts
MEK inhibitor, 301
messenger RNA (mRNA), 230
metanephric mesenchyme (MM), 369–371
methylcellulose ATP assay, 176
microarray biochip technology, 513
Microarray Quality Control Consortium (MAQC), 513
micropatterned coculture (MPCC), 340
microphysiological systems, 216
microRNA-122 (miR-122), 421
microRNA (miRNA)
    circulation, 472–473
    dog tissue atlas methods, 468, 469
    drug-induced skeletal muscle injury, 410–411
    emerging biomarkers, 252
    GI biomarkers, 230
    glomerular injury, 452–453
    kidney expression, dog, 470
    ocular toxicity, 210
    protein biomarker, 450
    rat kidney atlas, 448–449
    rat tubular, 450–451
    tissue expression, in canines, 468
midazolam (MDZ), 304
miR-122 *see* microRNA-122 (miR-122)
miRNA *see* microRNA (miRNA)
mitochondrial injury, 98–100
molecular profiling, 90
mouse-adapted strain of Ebola virus (MA-EBOV), 319
mouse diversity panel (MDP), 317–318
    cost considerations, 325
    genome-wide association analysis, 324
    health status, 326
    model selection, 323
    sample size, 324
mouse embryonic fibroblasts (MEFs), 318, 323

mouse population-based toxicology, 314
  ANCOVA model, 320
  benchmark dose, 320
  biomarkers of sensitivity, 320–322
  candidate gene analysis, 324–325
  CC mice, 318–319
  cost considerations, 325
  DO mice, 319–320
  dose selection, 322
  EMMA, 324
  genome-wide association analysis, 324
  health status, 325–326
  linkage disequilibrium, 318
  model selection, 322–323
  mode of action, 322
  mouse diversity panel, 317–318
  personalized medicine, 320
  pharmaceutical safety science, 320–322
  pharmacogenetics, 314–316
  phenotyping, 324
  population variability, 314–316
  quantitative trait gene, 316, 317
  quantitative trait loci, 316, 317
  sample size, 323–324
  specific pathogen free, 325–326
multidrug resistance-associated protein 2 (MRP2), 103
multidrug resistance protein (MDR-1), 87
multielectrode array (MEA), 140, 216
murine models, 336
myelotoxicity *see* hematotoxicity
myocardial contractility assessment
  anesthetized animal models, 146
  conscious animal models, 146
  gold standard approaches, 144
  isolated tissue, 145
  translation to clinic, 146–147
  in vitro and *ex vivo* assays, 145
  in vivo assays, 145–146
myofibrils, 407
myoglobin, 410
myosin light chain 3 (Myl3), 409–410

N-Acetyl-*β*-ᴅ-Glucosaminidase (NAG), 435
N-acetylgalactosamine (GalNAc), 39
N-acetyl-*p*-benzoquinone imine (NAPQI), 101
NAFLD *see* nonalcoholic fatty liver disease (NAFLD)
National Cancer Institute Comparative Oncology Program, 275
natriuretic peptides (NPs), 391–392
NBF *see* neutral-buffered formalin
N-cadherin (N-CAD), 372
NCEs *see* new chemical entities
nephrotoxicity *see* drug-induced kidney injury (DIKI)
nephrotoxicity working group (NWG), 447, 503
nephrotoxic serum (NTS), 453
nervous system disease
  AD, 272–273
  from animal to human, 270
  diabetic neuropathy, 273
  drug attrition, 270
  HD, 272

lysosomal storage diseases, 272–273
  PD models, 270–272
net reclassification index (NRI), 496
neutral-buffered formalin (NBF), 462
neutrophil gelatinase-associated lipocalin (NGAL), 436–437
  canine kidney safety protein biomarkers, 443, 444
  performance, 467
  standard curve, 484
new chemical entities (NCEs), 83
  hPSC-CMs, 346–347
  safety pharmacology and toxicological testing, 357
  traditional drug discovery paradigm, 354, 355
next-generation sequencing (NGS), 447
Next-generation sequencing (miRNA-SEQ) analysis, 468
NGAL *see* neutrophil gelatinase-associated lipocalin
nicotinamide phosphoribosyltransferase
    (NAMPT) inhibitors, 536
Niemann-Pick disease (NPD), 272–273
2-(2-nitro-4-trifluoromethylbenzoyl)-1,3-cyclohexanedione
    (NTBC), 308
N-methyl-*N*-nitro-*N*-nitrosoguanidine (MNNG), 277
N-methyl-*N*-nitrosourea (MNU), 277
nonalcoholic fatty liver disease (NAFLD), 265–266
nonalcoholic steatohepatitis (NASH), 265
nongenotoxic carcinogenicity (NGC), 522
nonhuman primate (NHP)
  Cynomolgus monkey, 446, 447
  kidney safety biomarkers, 446
  KIM-1, 446
  novel renal protein biomarker performance evaluations, 447
  urinary microRNAs, 447
  uses, 446
nonparenchymal tissue, 242
nonsteroidal anti-inflammatory drugs (NSAIDs), 274
no observed adverse effect level (NOAEL), 5
NPD *see* Niemann-Pick disease (NPD)
NPs *see* natriuretic peptides (NPs)
NRI *see* net reclassification index
NTS *see* nephrotoxic serum
nuclear receptors, 307–308
nuclease-based systems, 299
nucleic acid-based microRNA, 210
NWG *see* nephrotoxicity working group

ocular toxicity
  emerging biomarkers of retinal toxicity, 210
  H&E staining, 209
  ocular reflexes and behaviors, 202–203
  rodents as model, 205–206
  routine eye examinations, 206–208
  in silico tools and strategies, 201–202, 208–210
  in vitro tools and strategies, 201–202, 208–210
  in vivo tools and strategies, 202–210
  zebrafish as model, 203–205
odd-skipped related (OSR)1, 369, 370
OKN *see* optokinetic nystagmus
OKR *see* optokinetic response
OMR *see* optomotor response
ophthalmoscopy, 207
optokinetic nystagmus (OKN), 202

optokinetic response (OKR), 202, 203
  rodents, assessing oculotoxicity, 205–206
  zebrafish, assessing oculotoxicity, 203–204
optomotor response (OMR), 202, 203
  rodents, oculotoxicity assessment, 205–206
  zebrafish, oculotoxicity assessment, 204
oral cavity, 228
oral GTT (OGTT), 252
organic anion transporter protein B1 (OATP1B1), 62
organic anion transporters (OATs), 165, 169
Organisation for Economic Co-operation and Development (OECD), 7
organs-on-chips
  CNS toxicity, 216
  GI toxicology, 235
OSR1-green fluorescence protein (GFP), 370
osteopontin (OPN), 437
  canine kidney safety protein biomarkers, 443, 444
  NHP, 446
osteoprotegerin (OPG), 390

pancreatic exocrine insufficiency (PEI), 243
pancreatic lipase immunoreactivity (PLI) assay, 251
pancreatic toxicity
  biomarkers, 249–253
  case study, 255–256
  CCK-1 receptor agonist, 246
  cyanohydroxybutene, 248
  definition, 242
  endocrine, 242, 244–245
  exocrine, 242–244
  immunohistochemical, 249
  interspecies/interstrain differences, 246–247
  microscopic appearance, 243
  microscopic findings in, 247, 248
  nonparenchymal, 242
  organotypic models, 256–257
  preclinical evaluation, 245–249, 253–257
  TEM, 249
  trypsinogen activation peptide, 249
  in vitro assessment, 253–257
  in vivo assessment, 247–249
  zymogen granulation depletion, 248, 249
parallelism, 485–486
Parkinson's disease (PD)
  animal models, 270–272
  PLK2 inhibitors, 302
parvalbumin, 410
patient-derived xenograft (PDX) models, 278
PBC *see* primary biliary cirrhosis
PD *see* Parkinson's disease
PDX models *see* patient-derived xenograft models
PEI *see* pancreatic exocrine insufficiency
pentylenetetrazole (PTZ), 217
peptide biomarkers, 251–252
perfusion bioreactors, 340, 422
personalized medicine, 320
PET *see* positron emission tomography
P-glycoprotein (Pgp), 63, 87, 306
Pharmaceutical Research and Manufacturers of America (PhRMA)

CL prediction, 72, 75
  concentration–time profile, 75–76
  lead commentaries, 76
  volume of distribution, 75
Pharmaceuticals and Medical Devices Agency (PDMA), 411, 502
PHHs *see* primary human hepatocytes
phosphatidylcholine, 59
phosphodiesterase type 5 (PDE5) inhibitors, 202
physicochemistry, drug molecules
  basicity and lipophilicity, 60
  compound concentration, by transporters, 61–63
  excretion pumps inhibition, 63–64
  high lipoidal permeability drugs, 56–57
  lipoidal/lipid permeability, 56
  lipophilicity, 55–56
  lipophilicity and PSA, 60–61
  metabolism and, 61
  polar surface area, 56
  toxicity prediction, 59–61
  volume of distribution and target access, 58–59
physiologically based pharmacokinetic (PBPK) model, 68–70
piggyBac transposon system, 337–338
plasma collection device, 492
PLI assay *see* pancreatic lipase immunoreactivity assay
PLK2 inhibitors *see* polo-like kinase 2 inhibitors
pluripotent stem cells (PSCs)
  into cells of renal lineage, 367–376
  embryonic kidney development, 369–372
podocytes, 365–366
polar surface area (PSA), 56, 60–61
polo-like kinase 2 (PLK2) inhibitors, 302
polyarteritis nodosa, 398
polymyxin B
  efficacious dose, 466
  NGAL, 444
  NHP, 446
polymyxin program, 467
positron emission tomography (PET), 454
preanalytical variability
  biological sample matrix variables, 477, 479–480
  in biomarker testing, 477
  for blood samples, 478
  collection method, 480
  hemolysis, 479–480
  protein and RNA biomarkers, 479
  sample processing and storage requirements, 480
  for urine samples, 479
predictable DILI, 333–334
Predictive Safety Testing Consortium (PSTC), 411
  CHWG, 504
  collaboration, 504–505
  critical path institute's, 502–503
  DILI, 419
  HWG, 503
  NWG, 447, 503
  pancreatic toxicity, 250
  qualification process, 505–506
  SKMWG, 503
  TWG, 503–504
  VIWG, 503

predictive toxicogenomics, 519, 521–522
pregnane X receptor (PXR), 85
preload, 388
primary biliary cirrhosis (PBC), 266
primary extracellular structural protein *see* osteopontin
primary human hepatocytes, 334–336
primary human hepatocytes (PHHs), 422
prolongation of QT interval, 387–388
prostate-specific membrane antigen (PSMA), 31
proximal tubular cells (PTCs), 366, 372
    embryonic development, 372
    hESC-based protocol, 374–375
    hiPSC-derived HPTC-like cells, 375–376
proximal tubule (PT), 366
PSCs *see* pluripotent stem cells
pseudoallergy, 196–197
PSTC *see* Predictive Safety Testing Consortium
PTCs *see* proximal tubular cells
PTZ *see* pentylenetetrazole

quantitative polymerase chain reaction (qPCR), 374–378, 450
quantitative trait gene (QTG)
    candidate gene analysis, 324–325
    mouse population-based toxicology, 316–318
quantitative trait loci (QTL)
    candidate gene analysis, 324–325
    CC mice, 318–319
    DO mice, 323, 324
    mouse population-based toxicology, 316, 317, 324

radioimmunoassay (RIA), 251
radiotelemetry, 389
rat glomerular miRNAs, 452–453
rat tubular microRNAs, 450–451
receiver operating characteristic (ROC) curves, 496
receptor-interacting protein kinase 1 and 3 (RIPK1 and RIPK3), 302–303
recombinant human acid sphingomyelinase (rhASM), 273
REGM *see* renal epithelial cell growth medium (REGM)
renal cell lines, 458–459
renal epithelial cell growth medium (REGM), 372
renal injury
    AKI, 280–281
    from animal to human, 279–280
    CKD, 281–282
    drug attrition, 279
renal *in vitro* models, 365
    for drug safety screening, 376–378
    hiPSC, 373
    predictive, 365, 367
renal papillary antigen 1 (RPA-1), 435
respiratory disease
    from animal to human, 282
    ARDS, 283
    asthma, 283–285
    COPD, 282, 283
    drug attrition, 282
    fibrotic lung disease, 285

respiratory screening technology (RST), 99
retinal toxicity, 201, 537
reverse causal reasoning (RCR), 523
rhASM *see* recombinant human acid sphingomyelinase (rhASM)
RIA *see* radioimmunoassay
ribose sugar modification, 43
RNAi, 39
rodents
    assessing ocular toxicity, 205–206
    chemically or environmentally induced CRC, 275, 277
    toxicology screening in, 46–47
routine eye examinations, 206–207
    ophthalmic equipment, 207
    ophthalmic examination procedure, 207
    trained examiner, 208

safety assessment
    cell-based assays for toxicity prediction, 7–8
    drug discovery, 4–5
    preclinical development, 5–6
    in silico models for toxicity prediction, 6–7
safety biomarkers
    academic and pharmaceutical industry, 500
    insufficient therapeutic index, 500
    Letter of Support, 502
    pharmaceutical industry, 501
    PSTC (*see* predictive safety testing consortium)
    qualification of, 500, 501
safety lead optimization, for small molecules
    ADME assessments, 20–22
    candidate lead optimization, 15
    chemical structure-mediated toxicities, 17
    discovery phase, 15, 16
    focused target organ assessments, 20
    genotoxicity assessments, 22
    intrinsic cytotoxicity assessments, 19–20
    prospective approaches, 17–18
    retrospective models, 22–23
    selectivity and secondary pharmacology assessments, 18–19
    target safety assessment, 16, 17
    undesired pharmacologically mediated toxicities, 17
    worst molecules, removal of, 15
safety pharmacology (SP), 130, 131
sarcomere
    eccentric *vs.* concentric hypertrophy, 388
    hPSC-CM, 359
sCr *see* serum creatinine
SEC *see* suspension expansion culture
seizure liability testing
    EEG recording, 217–218
    electrophysiological recording, 217
    precipitant models, 217
    in vivo zebra fish larvae, 216–217
sertoli cell, 471
serum and urinary cystatin C, 434
serum collection device, 492
serum creatinine (sCr), 431, 434
severe combined immunodeficiency (SKID) mice, 308

single nucleotide polymorphisms (SNPs), 315, 325
skeletal muscle injury, drug-induced
creatine kinase, 408–409
creatine kinase M, 409
EMA, 411
FABP3, 410
Food and Drug Administration, 411
future directions, 411–412
miRNAs, 410–411
Myl3, 409–410
myoglobin, 410
novel biomarkers, 409–411
overview, 407–409
parvalbumin, 410
PDMA, 411
regulatory endorsement, 411
skeletal troponin I, 409
skeletal myopathy working group (SKMWG), 503
skeletal troponin I (sTnI), 409
SKID mice *see* severe combined immunodeficiency mice
SKMWG *see* skeletal myopathy working group
slit lamp biomicroscope (SLBM), 207
small interfering RNAs (siRNAs)
candidate selection and development, 48–49
cell lines selection, 44
chemical modification, 42
conjugates for hepatic targets, 39
cross-species, 42
efficacy prediction, 41
endogenous polymerases, effects on, 47–48
immunostimulation screening, 45–46
lead optimization of, 44
monomeric metabolites, 47
off-target analysis, 44–45
off-target prediction, 41
screening, 42–45
sequence-based off targets, 41–42
target assessments, 40–41
target silencing evaluation, 44
toxicology screening in rodents, 46–47
transcription/translation, effects on, 48
in vitro safety evaluation, 48
SNPs *see* single nucleotide polymorphisms
SP22, 472
species-specific siRNAs, 40
specific pathogen free (SPF), 325–326
sperm RNAs, 473
spontaneously hypertensive rats (SHRs), 269, 270, 286
Sprague-Dawley rat, 256, 453, 462
standard toxicity studies (STS), 193
statistical methods
biomarker qualification, 495
for confirmatory clinical studies, 498
DDTs, 495
for exploratory clinical studies, 496–498
generic operational strategy, 495
for preclinical studies, 496
stem cell-derived CMs, 140–141

stem cell-derived renal cells, 365
aminoglycoside antibiotics, 366
embryonic kidney development, 369–372
hESC-based model, 377–378
PSCs into cells of renal lineage, 367–376
stem cells *see also* pluripotent stem cells
drug-induced liver injury, 337–338
inducible pluripotent , coculture models, 340
sTnI *see* skeletal troponin I
stomach ulcers, 275
streptozotocin (STZ), 281
STS *see* standard toxicity studies
surrogate models, 269
surrogate siRNAs, 40
suspension expansion culture (SEC), 177
symptomatic liver injury, 96

TAP *see* trypsinogen activation peptide
TaqMan® Low Density Array (TLDA), 447
TaqMan miRNA profiling, 452
target organ assessments, 20
target safety assessment
biological evaluation, of target, 16
biopharmaceuticals safety assessment, 28–29
components, 17
risk tolerance, 16
T-cell-dependent antibody response (TDAR), 193
TdP *see* torsades de pointes
TEER *see* transepithelial electrical resistance
telemetry technology, CV system
CNS evaluation, 143
CV SP evaluations, 142–143
electrooculogram, 143–144
exteriorized fluid-filled catheters, 142
gastrointestinal motility, 143
intraocular pressure, 144
respiratory function evaluation, 143
tail cuff methods, 142
TEM *see* transmission electron microscopy
testicular toxicity
ABP, 472
BTB, 471
circulating microRNAs, 472–473
inhibin B, 472
SP22, 472
sperm analysis methods, 471
sperm RNAs, 473
testis, 471
testicular toxicity working group (TWG), 503–504
3D culture, 339–340
3D organoids, 233–235
thymidine kinase (TK) system, 308
Timothy syndrome, 349
tissue collection device, 491–492
tissue-enriched microRNAs, 448
TLI *see* trypsin-like immunoreactivity (TLI)
TNF-α *see* tumor necrosis factor-α (TNF-α)
tonometry, 207

topological polar surface area (TPSA), 56
torsade de pointes (TdP), 4, 131, 346, 356–357
total body water (TBW), 58
toxicity prediction
  cell-based assays for, 7–8
  in silico models for, 6–7
toxic metabolites, 417
toxicogenomics
  application, 524
  case studies, 519–520
  diagnostic, 519–521
  history, 511–513
  mechanistic/investigative, 522–524
  predictive, 519, 521–522
  tools and strategy, 513–519
toxicokinetics (TK), 35
toxicology screening of siRNAs, in rodents, 46–47
traditional drug discovery paradigm, 354
traditional functional biomarkers, 433–434
  blood urea nitrogen, 434
  sCr, 434
  urinalysis, 434
transepithelial electrical resistance (TEER), 233, 235, 236, 458, 459
translatability
  assay selection, 490
  biofluid sample collection, 491
  biological matrix selection, 490
  choice of sample labeling, 493
  completion of sample collection and processing, 492
  database, 493
  documenting time and temperature, 492
  fit-for-purpose biomarkers, 490, 491
  logistics plan, 493
  operations team, 489–490
  optimal handling, 492–493
  optimal sample storage conditions, 493
  patient factors documentation, 491
  plasma collection device, 492
  preservation methods, 492–493
  protocol development, 489
  sample collection schedule, 492
  sample storage tubes, choice of, 493
  serum collection device, 492
  tissue biopsy collection, 491
  tissue collection device, 491–492
  urine collection device, 492
translational safety biomarkers
  for confirmatory clinical studies, 498
  for exploratory clinical studies, 496–498
  for preclinical studies, 496
transmission electron microscopy (TEM), 249
transporters, 61–63
trefoil factor (TF), GI biomarkers, 230
trifluoperazine, 61
trypsin-like immunoreactivity (TLI), 251
trypsinogen activation peptide (TAP), 249, 251

tsA58 thermolabile SV40 TAg (SV40tsA58), 458
tumor necrosis factor-α (TNF-α), 390
TWG *see* testicular toxicity working group
type 2 diabetes mellitus (T2DM)-induced DN, 281

UB-committed renal progenitor-like cells, 370
UDP-glucuronosyltransferases (UGTs), 306
ulcerative colitis (UC), 275–277
ulcer models, GI injury, 274–275
ULOQ *see* upper limit of quantification
ultrasound
  drug-induced vascular injury, 401
  integrating novel imaging technologies, 454
UNC Computational Genetics, 324
undesired pharmacologically mediated toxicities, 17
upper limit of quantification (ULOQ), 484
ureteric bud (UB), 369
urinary total protein and albumin, 434
urine collection device, 492
uroplasminogen activator (uPA) system, 308
US Food and Drug Administration (US FDA), 432

validation *see also* fit-for-purpose assay
  of CAMEO-96, 177
  hPT primary cultures, 165
vascular injury working group (VIWG), 503
vasoactive peptides, 390–392
ventral tegmental area (VTA), 222
vestibulo-ocular reflex (VOR), 203
video microscopy image-based analysis, 141
viral-induced inflammation, 266–267
Visiopharm VISIOmorph software, 463
visualmotor response (VMR), 204–205
VIWG *see* vascular injury working group
voltage-sensitive dyes, 141
VOR *see* vestibulo-ocular reflex (VOR)
VTA *see* ventral tegmental area (VTA)

Wajima allometric method, 68, 69
whole-cell patch clamp recordings, 140
whole cortical glomeruli, desmin immunoreactivity, 464
Wilms' tumor (WT), 304–305
Wistar rat, 246

xenobiotic metabolism, 333
  loss of, 336
xenograft models, 265
  gastrointestinal injury, 278
  mouse, 537

zebrafish, 110–113
  OKR, ocular toxicity assessment, 203–204
  OMR, ocular toxicity assessment, 204
  VMR, ocular toxicity assessment, 204–205
ZFIN, 111
zymogen granulation depletion, 248, 249